The Spectra and Dynamics of Diatomic Molecules

"The Spectra and Dynamics of Diatomic Molecules" is a revised and enlarged edition of the 1986 title "Perturbations in the Spectra of Diatomic Molecules".

The Spectra and Dynamics of Diatomic Molecules

Revised and Enlarged Edition

Hélène Lefebvre-Brion
Laboratoire de Photophysique Moléculaire
Université Paris-Sud
Orsay, France

Robert W. Field
Department of Chemistry
Massachusetts Institute of Technology
Cambridge, MA, USA

2004

ELSEVIER
ACADEMIC
PRESS
Amsterdam - Boston - Heidelberg - London - New York - Oxford
Paris - San Diego - San Francisco - Singapore - Sydney - Tokyo

ELSEVIER B.V.
Sara Burgerhartstraat 25
P.O. Box 211, 1000 AE Amsterdam
The Netherlands

ELSEVIER Ltd
The Boulevard, Langford Lane
Kidlington, Oxford OX5 1GB
UK

ELSEVIER Inc.
525 B Street, Suite 1900
San Diego, CA 92101-4495
USA

ELSEVIER Ltd
84 Theobalds Road
London WC1X 8RR
UK

2004: Revised and enlarged edition (of Perturbations in the Spectra of Diatomic Molecules)

ISBN: 0-12-441455-9 (hardbound)
ISBN: 0-12-441456-7 (paperback)

Transferred to Digital Printing 2008

Contents

1 Simple Spectra and Standard Experimental Techniques **1**

1.1 Rotation-Vibration-Electronic Spectra of Diatomic Molecules . . 2

1.1.1 Rotation-Vibration-Electronic Energy Levels and Standard Notation 3

1.1.2 Band Systems, Bands, and Branches 6

1.1.3 Rotational Structure: Fortrat Parabola 6

1.1.4 Some Strategies for Rotational Line Assignments 11

1.2 Experimental Techniques of Diatomic Molecule Spectroscopy . . 21

1.2.1 The Goals of a Spectroscopic Experiment 21

1.2.1.1 What and How Much? 22

1.2.1.2 Rotational Analysis of a Spectrum 24

1.2.1.3 Spectroscopic Strategies: Access and Selectivity 26

1.2.2 Classes of Spectroscopic Experiment 27

1.2.2.1 One Photon Resonant: Upward (Absorption Spectra) 27

1.2.2.2 One Photon Resonant: Downward (Emission Spectra) 33

1.2.2.3 Multiple Resonance Spectra 35

1.2.2.4 Photofragment and Coincidence Experiments . . 39

1.2.3 Molecule Sources 41

1.2.4 Frequency Calibration 43

1.2.5 Detection Techniques 45

1.2.6 Radiation Sources 47

1.3 References 55

2 Basic Models **61**

2.1 What Is a Perturbation? 62

2.2 Structural Models 69

2.3 Elementary Properties of Angular Momenta in Diatomic Molecules 72

2.3.1 Angular Momentum Components Defined by Normal and Anomalous Commutation Rules 73

2.3.2 Recipes for Evaluation of Molecule-Fixed Angular Momentum Matrix Elements 79

2.3.3 Euler Angles, $|JM\Omega\rangle$ Basis Functions, Direction Cosines, and Phases . . . 81
2.4 Estimation of Parameters in a Model Hamiltonian . . . 83
2.5 Data Compilations . . . 84
2.6 References . . . 85

3 Terms Neglected in the Born-Oppenheimer Approximation 87
3.1 The Born-Oppenheimer Approximation . . . 89
3.1.1 Potential Energy Curves . . . 90
3.1.2 Terms Neglected in the Born-Oppenheimer Approximation . . . 92
3.1.2.1 Electrostatic and Nonadiabatic Part of **H** . . . 92
3.1.2.1.1 Crossing or Diabatic Curves . . . 93
3.1.2.1.2 Noncrossing or Adiabatic Curves . . . 94
3.1.2.2 The Spin Part of **H** . . . 94
3.1.2.3 Rotational Part of **H** . . . 96
3.2 Basis Functions . . . 99
3.2.1 Hund's Cases . . . 101
3.2.1.1 Definition of Basis Sets . . . 103
3.2.1.2 Quantum Numbers, Level Patterns, and the Effects of Terms Excluded from $\mathbf{H}^{(0)}$ . . . 113
3.2.1.3 Intermediate Case Situations . . . 126
3.2.1.3.1 Introduction . . . 126
3.2.1.3.2 Examples . . . 127
3.2.1.4 Transformations Between Hund's Case Basis Sets 130
3.2.1.5 Spectroscopic vs. Dynamical Hund's Cases . . . 136
3.2.1.6 Relationship between Noncommuting Terms in **H** and the Most Appropriate Hund's Case . . . 137
3.2.2 Symmetry Properties . . . 138
3.2.2.1 Symmetry Properties of Hund's Case (a) Basis Functions . . . 138
3.2.2.2 Symmetry Properties of non-Hund's Case (a) Basis Functions . . . 145
3.2.3 Molecular Electronic Wavefunctions . . . 148
3.2.4 Matrix Elements between Electronic Wavefunctions . . . 156
3.3 Electrostatic Perturbations . . . 161
3.3.1 Diabatic Curves . . . 163
3.3.2 Approximate Representation of the Diabatic Electronic Wavefunction . . . 165
3.3.3 Adiabatic Curves . . . 168
3.3.4 Choice between the Diabatic and Adiabatic Models . . . 172
3.3.5 Electromagnetic Field-Dressed Diabatic and Adiabatic Potential Energy Curves . . . 177
3.4 Spin Part of the Hamiltonian . . . 180
3.4.1 The Spin-Orbit Operator . . . 181

3.4.2 Expression of Spin-Orbit Matrix Elements in Terms of One-Electron Molecular Spin-Orbit Parameters 183
3.4.2.1 Matrix Elements of the $\mathbf{l}_{zi} \cdot \mathbf{s}_{zi}$ Term 183
3.4.2.1.1 Diagonal Matrix Elements 184
3.4.2.1.2 Off-Diagonal Matrix Elements 187
3.4.2.2 Matrix Elements of the $\left(\mathbf{l}_i^+\mathbf{s}_i^- + \mathbf{l}_i^-\mathbf{s}_i^+\right)$ Part of $\mathbf{H}^{\mathrm{SO}}$ 190
3.4.3 The Spin-Rotation Operator 191
3.4.4 The Spin-Spin Operator 196
3.4.4.1 Diagonal Matrix Elements of $\mathbf{H}^{\mathrm{SS}}$: Calculation of the Direct Spin-Spin Parameter 196
3.4.4.2 Calculation of Second-Order Spin-Orbit Effects . 199
3.4.4.2.1 π^2 Configuration 201
3.4.4.2.2 $\pi^3\pi'$ (or $\pi^3\pi'^3$ and $\pi\pi'$) Configurations 201
3.4.4.3 Off-Diagonal Matrix Elements 202
3.4.5 Tensorial Operators 203
3.5 Rotational Perturbations 210
3.5.1 Spin-Electronic Homogeneous Perturbations 210
3.5.2 The **S**-Uncoupling Operator 212
3.5.3 The **L**-Uncoupling Operator 213
3.5.4 $^2\Pi \sim {}^2\Sigma^+$ Interaction 217
3.6 References 227

4 Methods of Deperturbation **233**
4.1 Variational Calculations 234
4.2 The Van Vleck Transformation and Effective Hamiltonians . . . 237
4.3 Approximate Solutions 243
4.3.1 Graphical Methods for Deperturbation 243
4.3.2 Direct Diagonalization Versus Algebraic Approaches . . . 247
4.4 Exact Solutions 248
4.4.1 Least-Squares Fitting 248
4.4.1.1 Linear Least-Squares Fitting 248
4.4.1.2 Nonlinear Least-Squares Fitting 251
4.4.1.3 Practical Considerations 255
4.4.1.4 Least Squares vs. Robust Estimator Fitting . . . 257
4.4.1.5 Types of Programs 258
4.4.2 Comparison Between Effective and True Parameters . . . 261
4.4.3 Coupled Equations 264
4.5 Typical Examples of Fitted Perturbations 267
4.5.1 An Indirect Heterogeneous Perturbation: NO $B^2\Pi \sim C^2\Pi \sim D^2\Sigma^+$ 267
4.5.2 A Strong Multistate Interaction in the NO Molecule . . . 269
4.6 References 272

5 Interpretation of the Perturbation Matrix Elements 275
5.1 Calculation of the Vibrational Factor . . . 278
5.1.1 Semiclassical Approximation . . . 279
5.1.2 Model Potentials . . . 285
5.1.3 Numerical Potentials and Vibrational Wavefunctions . . . 288
5.1.4 Some Remarks about "Borrowed" Computer Programs . . 291
5.1.5 Vibrational Assignment by the Matrix Element Method . 293
5.1.6 Homogeneous vs. Heterogeneous Perturbations . . . 305
5.2 Order of Magnitude of Electrostatic Perturbation Parameters: Interactions Between Valence and Rydberg States of the Same Symmetry . . . 307
5.2.1 Valence and Rydberg States . . . 308
5.2.2 Different Classes of Valence~Rydberg Mixing . . . 310
5.2.3 Electrostatic Perturbations Between Valence and Rydberg States . . . 312
5.2.4 Electrostatic Perturbations between Rydberg States Converging to Different States of the Ion . . . 314
5.3 Order of Magnitude of Spin Parameters . . . 315
5.3.1 Diagonal Spin-Orbit Parameters . . . 318
5.3.2 Off-Diagonal Spin-Orbit Parameters . . . 322
5.3.3 Spin-Spin Parameters . . . 323
5.4 Magnitudes of Rotational Perturbation Parameters . . . 325
5.5 Pure Precession Approximation . . . 327
5.6 R–Dependence of the Spin Interaction Parameters . . . 333
5.7 Beyond the Single-Configuration Approximation . . . 340
5.8 Identification and Location of Metastable States by Perturbation Effects . . . 341
5.9 References . . . 342

6 Transition Intensities and Special Effects 347
6.1 Intensity Factors . . . 348
6.1.1 Interrelationships between Intensity Factors . . . 348
6.1.2 General Formulas for One-Photon and Multi-Photon Transition Strengths . . . 355
6.1.2.1 One-Photon Transitions . . . 356
6.1.2.2 Two-Photon Transitions . . . 359
6.1.2.3 Three-Photon Transitions . . . 366
6.1.2.4 Comparisons between Excitation Schemes Involving Different Numbers of Photons . . . 367
6.2 Intensity Borrowing . . . 368
6.2.1 Perturbations by States with "Infinite" Radiative Lifetime; Simple Intensity Borrowing . . . 368
6.2.2 Multistate Deperturbation; The NO $^2\Pi$ States . . . 375
6.3 Interference Effects . . . 378
6.3.1 Perturbations between States of the Same Symmetry; Vibrational-Band Intensity Anomalies . . . 380

6.3.2 $\Delta\Lambda = \pm 1$ Perturbations; Rotational-Branch Intensity Anomalies . . . 386
6.3.3 Assignments Based on Pattern-Forming Rotational Quantum Numbers . . . 403
6.3.4 $\Sigma^+ \sim \Sigma^-$ Perturbations; Subband Intensity Anomalies in $\Sigma \leftrightarrow \Pi$ Transitions . . . 404
6.3.5 F_1 vs. F_2 Intensity Anomalies in $^2\Sigma$ States . . . 405
6.4 Forbidden Transitions; Intensity Borrowing by Mixing with a Remote Perturber . . . 406
6.5 Special Effects . . . 415
6.5.1 Differential Power Broadening . . . 416
6.5.2 Effects of Magnetic and Electric Fields on Perturbations . 418
6.5.3 Anticrossing, Quantum-Beat, and Double-Resonance Experiments . . . 427
6.5.4 Rydberg States and the Zeeman Effect . . . 439
6.5.5 Nonthermal Population Distributions; Chemical and Collisional Effects . . . 445
6.5.6 "Deperturbation" at High Pressure and in Matrices . . . 455
6.5.7 Matrix Effects . . . 458
6.6 References . . . 460

7 Photodissociation Dynamics 469
7.1 Photofragmentation . . . 470
7.2 Direct Dissociation . . . 471
7.2.1 Photodissociation: Wigner-Witmer Rules . . . 471
7.2.2 Photodissociation Cross Sections . . . 476
7.2.3 Photofragment Branching Ratios for Photodissociation . . 485
7.2.4 Photofragment Angular Distribution . . . 486
7.2.5 Alignment of the Photofragment . . . 491
7.3 Introduction to Predissociation . . . 493
7.4 Experimental Aspects of Predissociation . . . 495
7.4.1 Measurement of Lifetimes . . . 496
7.4.2 Measurement of Linewidths . . . 498
7.4.3 Energy Shifts . . . 503
7.4.4 Detection of Fragments . . . 505
7.5 Theoretical Expressions for Widths and Level Shifts . . . 505
7.6 The Vibrational Factor . . . 510
7.7 Mulliken's Classification of Predissociations . . . 514
7.8 The Electronic Interaction Strength . . . 518
7.8.1 Electrostatic Predissociation . . . 519
7.8.2 Spin-Orbit Predissociation . . . 520
7.8.3 Rotational or Gyroscopic Predissociation . . . 521
7.8.4 Hyperfine Predissociation . . . 521
7.9 Fano Lineshape . . . 522
7.10 Isotope Effects . . . 526
7.11 Examples of Predissociation . . . 528

7.11.1 Examples of Spin-Orbit Predissociation 528
7.11.2 Examples of Nonadiabatic Predissociation 531
7.12 Case of Intermediate Coupling Strength 535
7.13 Indirect (Accidental) Predissociation and Interference Effects . . 538
7.14 Some Recipes for Interpretation 543
7.15 References 545

8 Photoionization Dynamics 551
8.1 Direct Ionization 552
8.1.1 Photoelectron Spectroscopy 552
8.1.2 ZEKE Spectroscopy 557
8.1.3 Shape resonances 559
8.1.4 Cooper minima 561
8.2 Experimental Aspects of Autoionization 564
8.3 The Nature of Autoionized States 568
8.4 Autoionization Widths 569
8.5 Rotational Autoionization 572
8.6 Vibrational Autoionization 576
8.7 Spin-Orbit Autoionization 581
8.8 Electronic (or Electrostatic) Autoionization 586
8.9 Validity of the Approximations 588
8.10 Influence of Autoionization on ZEKE Peak Intensities 591
8.11 Photoelectron Angular Distribution, Photoion Alignment, and Spin Polarization 595
8.11.1 Photoelectron Angular Distribution 595
8.11.2 Photoion Alignment 600
8.11.3 Spin-Polarization 602
8.12 Competition between Autoionization and Predissociation 604
8.12.1 Superexcited State Decay Pathways 604
8.12.2 Theoretical Treatment 608
8.13 Coherent Control of Photofragmentation Product Branching Ratios 609
8.14 References 615

9 Dynamics 621
9.1 Dynamical Concepts, Tools, and Terminology 622
9.1.1 The Time-Dependent Picture: Terminology 623
9.1.2 Solution of the Time-Dependent Schrödinger Equation . . 624
9.1.3 Frequency Domain Spectra Treated as the Fourier Transform of the Autocorrelation Function 626
9.1.4 Dynamical Quantities 635
9.1.5 General Density Matrix Formulation of a Dynamical Experiment: Excitation, Evolution, and Detection Matrices . 639
9.1.6 Particle (Photon) vs. Wave Pictures of Spectroscopy . . . 643
9.1.7 Motion of the Center of the Wavepacket 644
9.1.8 Equations of Motion for Resonance Operators 646
9.1.9 The One-Color Pump-Probe Experiment 649

9.1.10 Crafted Pulses for Detailed Manipulation of Molecular Dynamics . . . 655
9.2 From Quantum Beats to Wavepackets . . . 656
9.2.1 Polarization Quantum Beats . . . 657
9.2.2 Population Quantum Beats . . . 658
9.2.3 Nuclear Wavepackets . . . 659
9.2.3.1 Vibrational Wavepackets . . . 661
9.2.3.2 Rotational Wavepackets . . . 667
9.2.4 Rydberg Wavepackets: Kepler and Precessional Periods . 668
9.3 Relaxation into a Quasi-Continuum: A Tool for Dimensionality Reduction . . . 671
9.3.1 The Complex-Energy Effective Hamiltonian . . . 672
9.3.1.1 Treatment of Two-State Interaction by Nondegenerate Perturbation Theory . . . 674
9.3.1.2 Treatment by Quasidegenerate Perturbation Theory: $\mathbf{2} \times \mathbf{2}$ Diagonalization . . . 675
9.3.2 Quantum Beats Between Two Decaying Quasi-Eigenstates 679
9.3.3 The Use of the Complex $\mathbf{H}^{\text{eff}}$ in Reduced-Dimension Models 681
9.4 *Beyond* the Spectra and Dynamics of Diatomic Molecules . . . 683
9.4.1 Basis States . . . 683
9.4.2 What is Deperturbation Anyway? . . . 684
9.4.3 Visualization of Dynamics . . . 685
9.4.4 Beyond Diatomic Molecules: Polyatomic Molecule Vibrational Dynamics . . . 687
9.4.5 Polyads . . . 689
9.4.6 Creation and Annihilation Operators . . . 690
9.4.7 Dynamics in State Space . . . 692
9.4.8 Number Operator . . . 693
9.4.9 Resonance Energy and Energy Transfer Rate Operators . 694
9.4.10 The Use of Expectation Values of Resonance Operators to Visualize Dynamic Processes . . . 697
9.4.11 $\mathbf{Q}$ and $\mathbf{P}$ . . . 700
9.4.12 Transformation Between Local and Normal Mode Limits . 702
9.4.12.1 Classical Mechanical Treatment . . . 703
9.4.12.2 The Morse Oscillator . . . 705
9.4.12.3 Quantum Mechanical Minimal Model for Two Anharmonically Coupled Local Stretch Morse Oscillators . . . 706
9.4.12.4 Transformation between 4-Parameter Forms of the Normal and Local Mode Basis Sets . . . 710
9.4.12.5 Transformation between 6-Parameter Forms of the Normal Mode and Local Mode $\mathbf{H}^{\text{eff}}$ . . . 714
9.4.13 From Quantum Mechanical $\mathbf{H}^{\text{eff}}$ to Classical Mechanical $\mathcal{H}(\mathbf{Q}, \mathbf{P})$ . . . 717
9.4.14 Polyatomic Molecule Dynamics . . . 733

9.4.15 Inter-System Crossing, Internal Conversion, and Intramolecular Vibrational Distribution 733
9.5 References 736

List of Tables

1.1 Typical Ground State Molecular Constants (Huber and Herzberg, 1979) . . . 5

2.1 Direction Cosine Matrix Elements for $\Delta J = 0^a$ . . . 77

3.1 Definitions and Approximations Associated with Different Types of Potential Energy Curves . . . 95
3.2 Off-Diagonal Matrix Elements of Total Hamiltonian between Unsymmetrized Basis Functions . . . 97
3.3 Hund's Cases . . . 112
3.4 Spin Eigenfunctions[a] of $\mathbf{S}^2$ for $\Sigma = S$ . . . 151
3.5 Comparison between Diabatic and Adiabatic Parameters . . . 171
3.6 Comparison between Two Types of Deperturbation of the G and I $^1\Pi$ States of SiO (in cm^{-1}) . . . 176
3.7 Comparison between Some Calculated and Observed Values for the First-Order Contribution to the Spin-Rotation Constant . . . 194
3.8 Semiempirical Expressions for Effective Spin-Spin Constants: Valence States of Homonuclear Molecules . . . 200
3.9 $^2\Pi(\pi) \sim^2 \Sigma^+(\sigma)$ Interaction Matrix[a] . . . 219

4.1 Mixing Coefficients for e–Parity Levels of the $^{14}N^{16}O$ $B^2\Pi$ $\sim C^2\Pi$ $\sim$ $L^2\Pi$ $\sim D^2\Sigma^+$ Perturbation[a] . . . 271

5.1 Calculated Electronic Factors for Trial Vibrational Numberings of the $e^3\Sigma^-$ Perturber of SiO $A^1\Pi$ (cm^{-1}) . . . 300
5.2 Perturbation Parameters for the BaS Molecule . . . 305
5.3 Comparison of Spin-Orbit Coupling Constants of Rydberg and Valence States (cm^{-1}) . . . 309
5.4 Some Values of the Electrostatic Parameter . . . 313
5.5 Dependence on n^* of the Electrostatic Perturbation Parameter . 314
5.6 Spin-Orbit Constants of Atoms and Ions (cm^{-1})[a] . . . 316
5.6 (Continued). . . . 317
5.7 Spin-Orbit Constants for Isovalent Molecules (cm^{-1})[a] . . . 321
5.8 Electronic Perturbation Parameters in the Single Configuration Approximation[a,b] . . . 324

5.9 Values of the Atomic Spin-Spin Parameter η (cm^{-1})^a 324
5.10 Effective Spin-Spin Constants for $^3\Sigma^-$ States of π^2 Configurations (cm^{-1}) 326
5.11 Some Spin-Orbit and Orbital Perturbation Matrix elements . . . 327

6.1 M–Independent $\langle\Omega' J'|\alpha|\Omega J\rangle$ Direction Cosine Matrix Elements . 391
6.2 Rotational Linestrengths for Case (c) $\frac{1}{2} - \frac{1}{2}$ Transitions 397
6.3 Rotational Linestrengths for $^3\Sigma^- - {}^1\Sigma^+$ Transitions 399
6.4 The NO $B^2\Pi_{1/2}$– $X^2\Pi_{1/2}$ Transition in the Gas Phase and in Inert Gas Matrices 456

7.1 Correlation between molecular and atomic wave functions for HX at R=∞ (X=Halogen)^c 475
7.2 Order of Magnitude of Predissociation Rates 499
7.3 Well-Characterized Weak Predissociations 500
7.4 Well-Characterized Strong Predissociations 501

8.1 Minimum n–values Required for Different Types of Autoionized States 573
8.2 Typical Examples of Autoionization Linewidths 578
8.3 The asymmetry parameter, β, for each angular momentum transfer, j_t 596
8.4 The geometrical factors for the β parameter for a Σ ion-core (from Table 1 of Thiel, 1983). 599

List of Figures

1.1 A diatomic molecule 4
1.2 A typical vibrational band of an electronic transition 7
1.3 A bandhead 9
1.4 The lines in the R and P branches 10
1.5 Lower and Upper State Rotational Combination Differences. . . . 12
1.6 Vibrational bands linked via a common vibrational level 14
1.7a A Loomis-Wood plot of an NbO $d^2\Delta_{3/2} - a^2\Delta_{3/2}$ sub-band . . . 16
1.7b A Loomis-Wood plot of an NbS $^2\Delta_{5/2} - {}^2\Delta_{5/2}$ sub-band 17
1.7c A one-sided Loomis-Wood plot 18
1.8 A vibrational progression of $R(J-1), P(J+1)$ doublets in an I_2 $B0_u^+ - X^1\Sigma_g^+(v' = 58, v'' = 93-98)$ dispersed fluorescence spectrum 20
1.9 Collision induced rotational relaxation 21
1.10 REMPI-TOF permits selective simultaneous recording of separate spectra for different cluster species or for different isotopomers 23
1.11 Lower state rotational combination differences for two consecutive vibrational levels. 25
1.12 Lower state rotational combination differences for two consecutive members of the same branch 25
1.13 Velocity Modulation Spectra (VMS) 30
1.14 Simultaneously recorded REMPI-TOF spectra of several isotopomers 32
1.15 Dispersed Fluorescence Spectrum of AgAu 33
1.16 v''–dependent $R(J-1) : P(J+1)$ intensity anomalies 35
1.17 Level diagrams for two-color double resonance experiments . . . 36
1.18 PHOFEX Spectra of Ketene 40

2.1 A comparison of the SiO $H^1\Sigma$ (0,0) and bands 62
2.2 Pictorial evidence that a perturbation affects the upper rather than lower electronic state. 63
2.3 Perturbations and predissociations affect absorption and emission line intensities in quite different ways 64
2.4 Relationships between molecule-fixed (xyz) and space-fixed (XYZ) axis systems. 83

3.1a Natural rotational quantum numbers for Hund's cases (a) and (b).118
3.1b (b) Plot of $T_{v,J} - BN(N+1)$ 119

3.2 e/f and total parity labeling of $^3\Sigma^+$ and $^1\Pi$ levels. 146
3.3 Branching diagram shows the number of states of each spin multiplicity 150
3.4 Electrostatic valence~Rydberg perturbations 162
3.5 Diabatic and adiabatic potential curves. 164
3.6 Variation of ΔG_v for the mutually interacting b′ and c′ $^1\Sigma_u^+$ states of N_2. 166
3.7 $^2\Delta$ and $^2\Pi$ potential energy curves of NO. 167
3.8 *Ab initio* adiabatic potential curves with double minima for the $^1\Sigma_g^+$ states of H_2 173
3.9 *Ab initio* $\partial/\partial R$ and $\partial^2/\partial R^2$ matrix elements between the E, F and G, K adiabatic states of H_2. 174
3.10 Perturbations between the $6p\pi$ and $11p\pi$ $^1\Pi_u$ Rydberg states of H_2. 175
3.11 Potential energy curves for the $1s\sigma_g$ and $2p\sigma_u$ states of H_2^+ . . . 179
3.12 Field-dressed potential energy curves for H_2^+ 180
3.13 Correlation diagram between case (a) and case (c). 189
3.14 Electronic coordinate system. 192
3.15 Σ states derived from π^2 and $\pi^3\pi'$ configurations. 199
3.16 Perturbations between $\Omega = \frac{3}{2}$ and $\frac{1}{2}$ sublevels of the $^{15}N^{80}Se$ $X^2\Pi_r$ state. 213
3.17 Anomalous Λ–doubling in the $X^2\Pi_r$ state of $^{15}N^{80}Se$ 214
3.18 Anomalous Λ–doubling in the CS^+ $A^2\Pi_i$ state 215
3.19 Potential energy curves of the excited states of CO. 217
3.20 Λ–doubling of $^2\Pi$ states and spin-doubling of $^2\Sigma^+$. 225

4.1 Plot of effective ΔB–value versus $J(J+1)$ for the $Si^{16}O$ $e^3\Sigma^-(v = 9) \sim A^1\Pi(v = 2)$ perturbation. 245
4.2 The ^{14}N ^{16}O $B^2\Pi$ $(v_B = 24) \sim C^2\Pi$ $(v_C = 6) \sim L^2\Pi$ $(v_L = 5) \sim D^2\Sigma$ $(v_D = 6)$ perturbation. 270

5.1 Isotope effect on a vibrational level 277
5.2 Bound level in a potential. 280
5.3 The regular Airy function $Ai(-Z)$. 281
5.4 Phase space mapping for the harmonic model. 282
5.5 Semiclassical calculation of vibrational overlap integrals 284
5.6 Two displaced harmonic oscillators. 287
5.7 Perturbations in the $^{28}Si^{18}O$ and $^{28}S^{16}O$ $A^1\Pi$ state. 293
5.8 Level shifts, reduced term values, and effective B–values for homogeneous versus heterogeneous perturbations. 297
5.9 Potential energy curves for the $X^1\Sigma^+$, $a^3\Pi_i$, $A'^1\Pi$, and $A^1\Sigma^+$ electronic states of the BaS molecule. 303
5.10 Interleaved $A'^1\Pi$ and $a^3\Pi$ levels of BaS. 304
5.11 Calculated $^2\Sigma^+$ $(3s\sigma)$ state of CH. 310
5.12 Rydbergized and complementary $^3\Pi_u$ states in O_2 311
5.13 CO $a^3\Pi_1 \sim a'^3\Sigma_1^+$ spin-orbit parameters 325

5.14 R–variation of the spin-orbit constant for the CO $a^3\Pi$ state. . . . 333
5.15 R–variation of the spin-orbit constant for the PO $X^2\Pi$ state. . . 335
5.16 Diagonal $A(R)$ and off-diagonal spin-orbit ($\langle AL^+\rangle$) and orbital ($\langle L^+\rangle$) perturbation parameters for the OH $X^2\Pi$ state 335
5.17 Correlation diagram showing the effect of an electric field on atomic $L - S$ terms 336
5.18a The diagonal and off-diagonal spin-orbit matrix elements as a function of the NO^+ internuclear separation 339
5.18b The dependence of the $\langle {}^3\Pi | L_+ | {}^3\Sigma^+ \rangle$ matrix element of NO^+ on the internuclear separation. 339

6.1 Parity and e/f labels for the rotational branches of a one-photon ${}^1\Pi - {}^1\Sigma^+$ transition. 360
6.2 Allowed two-photon pathways via the NO $A^2\Sigma^+$ virtual state for the two-photon O_{1e2e} and O_{1f2f} rotational branches. 365
6.3 Radiative decay rates ($\Gamma = 1/\tau$) for single e–parity rotational levels of CO $A^1\Pi$ ($v = 0$). 372
6.4 Mixing fractions in ${}^{30}Si^{32}S$ $A^1\Pi$ ($v = 5$). 373
6.5 Deperturbed potential curves for the mutually interacting ${}^2\Pi$ valence and Rydberg states of NO. 376
6.6 The vanishing of an entire band as a result of a homogeneous perturbation. 382
6.7 Schematic illustration of intensity interference effects. 385
6.8 Calculated perturbed and deperturbed intensities for N_2 ${}^1\Pi_u \leftarrow X^1\Sigma_g$ ($v = 0$) transitions. 387
6.9 Natural rotational quantum numbers for the NO $4f$ Rydberg complex. 401
6.10 Variation of the CO $a^3\Pi$ radiative lifetime versus J and Ω. . . . 410
6.11 The $a^3\Pi_0 - X^1\Sigma^+(0,0)$ sub-band of CS. 413
6.12 Level scheme for optical-optical double resonance on BaO $C^1\Sigma^+ \leftarrow A^1\Sigma^+ \leftarrow X^1\Sigma^+$. 416
6.13 OODR differential power broadening in BaO. 417
6.14 Effect of magnetic field and pressure on the CN $B^2\Sigma^+ - X^2\Sigma^+(0,0)$ R branch lines. 420
6.15 Calculated Zeeman patterns for the perturbed rotational levels of CN $A^2\Pi$ ($v = 10$) $\sim B^2\Sigma^+$ ($v = 0$). 422
6.16 g_J–values for ${}^{80}Se_2$ $B0_u^+$ 426
6.17 Difference in lifetime for σ^+ versus σ^- polarized fluorescence . . 426
6.18 H_2 $W^1\Sigma_g^+ \sim i^3\Pi_g$ anticrossing spectra. 428
6.19 Stark quantum beats in BaO $A^1\Sigma^+(v = 2, J = 1)$. 434
6.20 BaO $C^1\Sigma^+ - a^3\Sigma^+$ (2,12) fluorescence excited by OODR. 439
6.21 Angular Momentum Coupling Schemes for Rydberg States of a Diatomic Molecule with a closed-shell Molecular-Ion-Core. 441
6.22 Resolved Rotational (N^+) and Zeeman (m_l) Structures in the NO $15f - A^2\Sigma^+$ Electronic Transition 442

6.23 The Zeeman Patterns of $n = 6 - 30$ $nf(N^+ = 3)$ complexes of NO in a 1 Tesla Magnetic Field. 442
6.24 Calculated $J \rightarrow J+1$ rate constants for CaO $+N_2O$ collisions. . 452
6.25 Comparison of observed and deperturbed ΔG values for the NO $B^2\Pi_{1/2}$ state with ΔG values observed in a 5 K Ne matrix. . . . 457
6.26 Schematic of CN $A^2\Pi \leftrightarrow X^2\Sigma^+$ relaxation in a neon matrix. . . . 459
6.27 NO $a^4\Pi \rightarrow B^2\Pi$ intersystem crossing mechanism in Ar at 5 K. . 460

7.1 Transition from (Λ, S) to (J_1,J_2) coupling for the $^2P + {}^2S$ separated atom states. 474
7.2 Schematic description of the dissociation channels $Br^+(^3P_J) + H(^2S)$479
7.3 Reflection approximation 480
7.4 Potential energy curves of OH and Calculated photodissociation cross sections 481
7.5 Total photodissociation cross sections, $\sigma(E)$, for HCl 482
7.6 Schematic relativistic adiabatic potential curves of HI 483
7.7 $X*/(X*+X)$ photofragment branching ratios that result from photodissociation of HBr and HI molecules 484
7.8 Coordinate system defining the direction of the incident radiation $(+Z)$ and location of the photofragment detector 489
7.9 Intensity distributions for three important values of β: 2, 0, and -1. 489
7.10 Asymmetry parameters, β 491
7.11 M_J–sublevel populations for $J = 2$. (a) no polarization; (b) orientation; (c) alignment 492
7.12 Predissociation by rotation 494
7.13 Predissociation 494
7.14 Measured lifetimes of the CH $A^2\Delta$ state. 497
7.15 The diffuse 15-0 Band of the B 0_u^+–X 0_g^+ system of $^{78}Se_2$ 502
7.16 A Lorentzian lineshape 502
7.17 $\Delta^2G(v)$ differences in the O_2 $B^3\Sigma_u^-$ state. 504
7.18 Differential oscillator strengths 507
7.19 Pictorial descriptions of the phase difference between bound and continuum vibrational wavefunctions. 512
7.20 Example of an outer-limb curve crossing 514
7.21 Variation of linewidth versus N in the OD $A^2\Sigma^+$ state. 515
7.22 Schematic illustration of the variation of Γ versus v and J. . . . 516
7.23 Example of an inner-limb curve crossing 517
7.24 The potential energy curves of CH. 518
7.25 Example of $^2\Pi \sim {}^2\Pi$ Rydberg∼valence electrostatic interaction in NO. 520
7.26 Fano profiles for $q < 0, q = 0, q > 0$, and $q = \infty$ (and $-\infty$). . . . 522
7.27 The potential energy curves for some excited states of H_2. 524
7.28 Fano lineshape in H_2. 525
7.29 Reduced mass dependence of the reduced (dimensionless) linewidth.527

7.30 Jet absorption spectra of CO isotopomers in the region from 109100 to 109600 cm^{-1} 529
7.31 The potential energy curves of OH. 530
7.32 The potential energy curves of O_2^+. 531
7.33 Vibration dependence of the observed and calculated predissociation rates for the $C^2\Sigma_u^+$ state of $^{14}N_2^+$, $(^{14}N^{15}N)^+$, and $^{15}N_2^+$. . . 532
7.34 Calculated adiabatic potential energy curves for the $B^2\Sigma_u^+$ and $C^2\Sigma_u^+$ states of N_2^+. 532
7.35 Vibrational wavefunctions for the N_2^+ $C^2\Sigma_u^+ v = 4$ level and the isoenergetic continuum of $B^2\Sigma_u^+$. Isotopic dependence in the accumulation of $\langle \chi_E^B | W_{BC}^e(R) d/dR | \chi_4^C \rangle$ vibronic integral 534
7.36 Predissociation for a pair of states intermediate between adiabatic and diabatic coupling limits. 537
7.37 Variation of Γ versus $(H^e)^2$. 539
7.38 The potential energy curves of some states of N_2. 541
7.39 Schematic mechanism of indirect or accidental predissociation. . 541
7.40 Comparison between the Budó-Kovács and coupled equations models for indirect predissociation 542
7.41 Linewidth variation in the OH $A^2\Sigma^+(v = 8)$ level for the resolved F_1 and F_2 spin components. 544

8.1 A Photoelectron Spectrum (PES) of HCl 553
8.2 Differential cross sections for p→d and p→s transitions in both discrete and continuous spectra 554
8.3 Photoelectron spectrum of NO 556
8.4 Mechanism and timing of a ZEKE spectroscopy experiment . . . 558
8.5 A pair of Rydbergized and complementary states in O_2. 562
8.6 Vibrational state branching ratio for the photoionization of N_2 . 563
8.7 The 3d bound orbital and two examples of ϵf continuum wavefunctions 563
8.8 Relationship between full width at half maximum (FWHM) of individual lines and band contours. 566
8.9 Absorption spectrum showing the autoionization of the N_2 Hopfield Rydberg series. 567
8.10 Hydrogenic radial wavefunctions for successive 2*p* and 3*p* Rydberg orbitals 570
8.11 Schematic illustration of rotational autoionization in *para*–H_2 . . 575
8.12 Relative photoionization cross section of *para*–H_2, at 78 K, in the region of the ionization threshold. 577
8.13 Schematic illustration of vibrational autoionization of the H_2 $np\pi_u$ $^1\Pi_u$ state 578
8.14 Photoionization cross-section of N_2 580
8.15 Photoionization of HI showing spin-orbit autoionization 584
8.16 Energy level diagram for a $^2\Pi_i$ state 585
8.17 Calculated lifetimes of rotational levels of HCl $n = 93$ 585

8.18 Schematic illustration of the electrostatic autoionization of the $^1\Pi_u$ Rydberg states converging to the $A^2\Pi_u$ state of N_2^+ by the continuum of the $X^2\Sigma_g^+$ state. 587
8.19 The phase shift induced by the electric field of the core ion in a nonhydrogenic atom. 591
8.20 PES of HCl 593
8.21 ZEKE-PFI spectrum of HCl. 594
8.22 Photoion vibrational branching fractions and photoelectron asymmetry parameters for CO^+ $X^2\Sigma^+$ 599
8.23 Spectra illustrating various decay pathways for the H_2 $D^1\Pi_u$ $v =$ 8 level 605
8.24 Potential energy curves of HCl (solid curves) and HCl^+ (dashed curves). 606
8.25 Four spectra which sample different decay pathways for superexcited states of HCl 607
8.26 Calculated photoionization of HCl into the HCl^+ $X^2\Pi$ state . . . 608
8.27 Potential energy diagram of Na_2 displaying the control mechanism for direct ionization and dissociative ionization of Na_2. 611
8.28 Ratio of atomic to molecular products (Na^+/Na_2^+) as a function of the delay between the 340-nm pump and 540-nm probe pulses. 611
8.29 Detection of ionization (HI^+) and dissociation. 612
8.30 Potential energy diagram of HI showing the control mechanism for ionization vs. dissociation. 613
8.31 Calculated phase lag in the region of the 5d δ resonance of HI. . 614

9.1 Motion of a wavepacket in coordinate space 625
9.2 Relationships between features in the autocorrelation function, $\langle\phi|\phi(\tau)\rangle$, and lineshapes in the frequency domain spectrum, $I(\omega)$ 628
9.3 Photodissociation Spectra of ABC. 634
9.4 Dependence of the amplitude and phase of the survival and transfer probability on mixing angle in $\Psi_I(0)$ 638
9.5 The phase of a vibrational quantum beat 642
9.6 Wavepacket Interferometry in I_2 651
9.7 Vibrational wavepackets launched from a $v'' > 0$ initial state . . 662
9.8 Femtosecond Transition State (FTS) spectroscopy of I_2 664
9.9 The radial probability distribution of an $n \approx 30$ wavepacket in a hydrogen atom 668
9.10 Level anticrossings between states with complex energies in the strong and weak coupling limits 682
9.11 Comparison of survival probability to the resonance energy . . . 699
9.12 Contributions of DD1 and 3,245 to the transfer rate and accumulated population in mode 4 of HCCH 701
9.13 Phase space trajectories and polyad phase spheres for polyad $N =$ 3 of H_2O 721
9.14 Nested invariant tori and a surface of section 722

9.15 Evolution of the polyad phase sphere from the local mode to the normal mode limit 725
9.16 Polyads $N = 1$ through $N = 5$ for H_2O 726
9.17 The probability densities, $|\psi|^2$, in the top row are found to resemble the configuration space trajectories in the bottom row . . 726
9.18 Surfaces of section for the HCCH $[N_b, l] = [4, 0]$ and $[8, 0]$ polyads 730
9.19 Overview of the phase space and configuration space dynamics associated with the HCCH $[N_b = 22,\ l = 0]$ polyad 731
9.20 The *cis*-bend and *trans*-bend normal mode frequencies of HCCH tune into resonance at $N_b \approx 10$ 732

Preface

The title of this book is borrowed, slightly modified, from Gerhard Herzberg, whose scholarship, vision, and generosity pervade and enrich the field of Molecular Spectroscopy. We dedicate this book to his memory.

By adding the word "dynamics" to the title Herzberg chose for his book, and by putting it on an equal footing with "spectra", we express our belief that there is no fundamental difference between frequency and time domain spectra, no advantage in insight from an eigenstate vs. a wavepacket picture. There is profound complementarity. Herzberg taught us about the patterns that are likely to be found in a spectrum, how to use these patterns to make secure assignments, and how to build models that reproduce the measurements to the precision of the experiment. But patterns are made to be broken, and the breaking of standard patterns is the key to perceiving those dynamical features that demand explanation.

No spectrum, no matter how simple, is dynamics-free. No dynamical process, no matter how complex, fails to reveal its essential characteristics in one or a series of well designed spectroscopic experiments. The essential unity of spectrum and dynamics, of patterns of eigenstates and dynamical mechanisms, of simple few-level perturbations and multi-continua photofragmentation processes is the subject of this book.

The foundation for this book is "Perturbations in the Spectra of Diatomic Molecules." We hope that the spectroscopic community will find this book to be more than a corrected and 50% expanded version of the 1986 text. The focus is no longer on perturbations *per se*, but rather on all spectra of diatomic molecules, perturbed and regular, frequency domain and time domain. By putting the word "perturbations" in its title, we misleadingly narrowed the apparent scope of the previous book. In addition, we artificially restricted the treatment of intramolecular dynamics to *simple* decay processes such as predissociation, autoionization, and collision induced transitions. In modern experiments, femtosecond laser pulses create coherent superpositions of eigenstates, called *wavepackets*. The time evolution of a wavepacket is simultaneously more complicated than a single exponential decay yet more easily visualized, as ball-and-spring motions, than the equivalent information contained in a collection of level spacings and transition intensities in a frequency domain spectrum.

A frequency domain spectrum is a group of transitions between discrete energy eigenstates or between discrete states and continuum states. What is surprising and beautiful is the wealth of information about intramolecular motions contained in such a set of transitions. Eigenstates are *stationary* states, but they encode *dynamics* in subtle and often unexpected ways. Continua are dynamical entities, but in many ways they are modeled using the same terminology, concepts, and techniques as discrete eigenstates. Spectroscopic access to a continuum typically derives from an initially highly localized, perfectly know-able, and experiment-specific "pluck". This pluck and the mechanism of the ensuing dynamics are well described by the same basis states and coupling matrix elements as are used to represent the discrete and continuum eigenstates.

The spectra of diatomic molecules are neither as simple as one might naively expect nor as complex as one might believe is warranted for a many-body system. Small molecules may not be very smart, but they have one big advantage over experimentalists and theorists, they know their Hamiltonian!

This book is written for experimentalists by a theorist and for theorists by an experimentalist. We present the models and concepts needed by theorists to understand the spectroscopic language spoken by molecules as translated by experimentalists and the tools and terminology needed by experimentalists to communicate with both molecules and theorists.

Molecules are complex, many-body systems. But, especially for a short time after a specific, experimentalist-designed, localized excitation is created, the evolution of the excitation in the molecule is simple, mechanistic, visualizable, and describable, not stochastic, statistical, or chaotic. The vast majority of inter-particle and inter-state interactions negligibly affect the early time intramolecular dynamics.

The experimentally achievable localized excitations are typically described by one of the zero-order basis states (see Section 3.2), which are eigenstates of a part of the total molecular Hamiltonian. Localization can be in a part of the molecule or, more abstractly, in "state space". The localized excitations are often described by extremely bad quantum numbers. The evolution of initially localized excitations is often more complex and fascinating than an exponential decay into a nondescript bath or continuum in which all memory of the nature of the initial excitation is monotonically lost. The terms in the effective Hamiltonian that give birth to esoteric details of a spectrum, such as fine structure, lambda doubling, quantum interference effects (both lineshapes and transition intensity patterns), and spectroscopic perturbations, are the factors that control the evolution of an initially localized excitation. These factors convey causality and mechanism rather than mere spectral complexity.

An effective Hamiltonian contains all of the control parameters needed to describe a spectrum and any conceivable time domain experiment, not just the specific experiment by which these control parameters were determined. This is our unique message to the time-domain community. There exists a century of insights into relationships between observables, into ways to create, describe, and interrogate the post-preparation evolution of a localized excitation, and into the energy and intensity patterns on which unique assignments of a spectrum are based. Semi-classical theory provides a link between Quantum Mechanics, where information about motion and causality is non-transparently embedded in a time-independent Hamiltonian matrix, and Classical Mechanics, where motion is directly observed and observations of motion reveal the specific and spatially localized forces that influence that motion. It is always possible and often informative to convert a quantum mechanical effective Hamiltonian (in which all of the control parameters are determined by a fit to a spectrum) to a classical mechanical effective Hamiltonian (see Section 9.4.13).

Chapters 1 and 9 are entirely new. Chapter 1 is an abbreviated users guide to the techniques used to record and assign spectra. It is an extremely condensed and idiosyncratically selective overview of important spectroscopic instrumentation, tools, and techniques. Chapter 9 is intended as a bridge between the frequency and time domain communities and also as an opening of a window onto molecules larger than diatomic molecules, in which the ideas and tools developed to describe dynamics in small molecules can be fruitfully applied to the mechanistic understanding of dynamics in larger molecules. Section 9.4 is almost entirely devoted to vibrational dynamics in polyatomic molecules. We illustrate the interconnectedness of molecular structure parameters and coupling constants through examples of effective Hamiltonian based matrix element scaling rules, opposite limiting cases in which two parameters play complementary case-preserving and case-destroying roles, and approximate partitionings of the structure and dynamics of the entire molecule into those of weakly interacting subsystems.

Chapters 2-8 are significantly revised, corrected, and expanded (by 150 pages) versions of Chapters 1-7 of "Perturbations in the Spectra of Diatomic Molecules". The most substantial additions to these chapters are: the treatment of Hund's cases in Section 3.2.1; the expansion of Chapters 7 and 8, previously restricted to Predissociation and Autoionization, to include direct photofragmentation processes and photofragment angular distributions (Sections 7.1, 7.2, 8.1, and 8.10-13); and new Sections in Chapter 6 dealing with one-color multiphoton spectroscopy (Section 6.1.2), interference effects in two-photon transition amplitudes (with substantial input from Michael Hippler, Section 6.1.2.2), F_1 vs. F_2 interference effects in the predissociation of $^2\Sigma$ states, the effect of a strong magnetic field on Rydberg states (with substantial input from Dolorès Gauyacq, Section 6.5.4), and an updated and expanded section dealing with the relationship between perturbations ("doorway states") and collisional transfer between electronic states (with substantial input from Paul Dagdigian, Section 6.5.5). Pattern forming rotational quantum numbers and their use in assigning complex spectra, are discussed in Sections 3.2.1.2 and 6.3.3. Least squares and robust estimator fitting methods are compared in Section 4.4.1.4. The errors in the old Section 5.1.1 (now Section 6.1.1) on the interrelationships between factors related to transition intensities have been corrected. Section 3.3.5 is an introduction to electromagnetic field dressed potential curves. Schemes for coherent control of photofragment branching are introduced in Section 8.13.

Rydberg states have emerged as a major area of experimental and theoretical activity in the fifty years since Herzberg's book "The Spectra of Diatomic Molecules" was published. Therefore many of the examples in Chapters 3, 6, 7, and 8 deal with Rydberg states. Scattering theory (Multi-Channel Quantum Defect Theory, MQDT) is more appropriate than effective Hamiltonian theory to deal with interacting continuum states and the essential breakdown of the Born-Oppenheimer approximation when the Rydberg electron is far from the molecular ion-core. However, scattering theory is outside of the scope of this book, thus the treatment of MQDT in this book (Section 8.9) is only at an introductory level.

Hélène Lefebvre-Brion thanks Prof. Hajime Katô (Kobe, Japan), Prof. Alejandro Palma (Puebla, Mexico), Dr. Brenton Lewis (Canberra, Australia) and Prof. Nimrod Moiseyev (Haifa, Israel) for their hospitality during some weeks during which she has been able to work on this book. She thanks Dr. Brenton Lewis for an invitation to present a series of lectures to the Fifth Summer School on the Structure and Spectroscopy of Molecules in Canberra, Australia during January 1992. This was an opportunity for her to work on revising and extending the material in "Perturbations in the Spectra of Diatomic Molecules" for the present new book. She thanks for their answers to her questions on specific points, Drs. Alberto Beswick, Markus Drescher, Eric Charron, Claudina Cossart-Magos, Daniel Cossart, Gustav Gerber, Robert Gordon, Christian Jungen, Robert Le Roy, Frédéric Merkt, Georges Raşeev , François Rostas, Annick Suzor-Weiner, and Oleg Vasyutinskii. She realizes and regrets that many additional interesting papers concerning photodissociation and photoionization have not been referenced.

Robert Field thanks Richard F. Barrow, Thomas H. Bergeman, Padraic Burns, Daniel Cossart, Stephen L. Coy, F. Fleming Crim, Matthew P. Jacobson, Caroline C. Jarrold, David M. Jonas, Christian Jungen, Apostolos Kalemos, Michael E. Kellman, James L. Kinsey, William Klemperer, Roland Lefebvre, Robert H. Lipson, Anthony J. Merer, Frédéric Merkt, David C. Oertel, Stephen T. Pratt, Joëlle and François Rostas, Robert J. Silbey, Adam H. Steeves, Patrick H. Vaccaro, and James K. G. Watson for advice and criticism. Thomas H. Bergeman, Kyle L. Bittinger, Daniel Cossart, Michael E. Kellman, Olli Launila, and Michael D. Morse provided or generated figures. Daniel Cossart, Stephen L. Coy, Paul J. Dagdigian, Dolorès Gauyacq, and Michael Hippler generously provided substantial text. Robert Field is grateful for hospitality while writing sections of this book to Hai-Lung Dai (University of Pennsylvania), Hajime Katô (Kobe University), James L. Kinsey (Rice University), Stephen R. Leone (while at JILA in Boulder, Colorado), Li Li (Tsinghua University, Beijing), Frédéric Merkt (ETH Zürich), Joëlle Rostas (Laboratoire Photophysique Moléculaire), Timothy C. Steimle (Arizona State University), Guy Taïeb (Laboratoire Photophysique Moléculaire), Soji Tsuchiya (University of Tokyo), and Kaoru Yamanouchi (University of Tokyo).

This book owes its existence to Peter Giunta's determination, ingenuity, and artistry. Long before either H. Lefebvre-Brion or R. Field had the courage to imagine embarking on such an undertaking, Peter Giunta began preparing, in LaTeX, a corrected edition of "Perturbations in the Spectra of Diatomic Molecules".

Hélène Lefebvre-Brion
Orsay, France
helene.lefebvre-brion@ppm.u-psud.fr

Robert W. Field
Cambridge, Massachusetts
rwfield@mit.edu

From the Preface to "Perturbations in the Spectra of Diatomic Molecules"

(Chapter numbering revised to be consistent with the present numbering.)

Examined in sufficient detail, the spectrum of every diatomic molecule is full of surprises. These surprises or "perturbations" can be at least as interesting as the vast expanses of textbook spectra lying between the surprises. Perturbations are more than spectroscopic esoterica. We hope that our discussion of perturbations provides a useful and unified view of diatomic molecular structure. This is a book about the spectra of diatomic molecules, warts and all.

This book is for graduate students just beginning research, for theorists curious about what experimentalists actually measure, for experimentalists bewildered by theory, and for potential users of spectroscopic data in need of a user's guide. We have minimized recourse to abstract and elegant treatments (e.g., spherical tensors) wherever a simpler one (e.g., ladder operators) would suffice. We have worked through many examples rather than attempt to provide formulas and literature examples for all conceivable cases.

Chapters 2, 3, and 5 form the core of this book. Perturbations are defined and simple procedures for evaluating matrix elements of angular momentum operators are presented in Chapter 2. Chapter 3 deals with the troublesome terms in the molecular Hamiltonian that are responsible for perturbations. Particular attention is devoted to the reduction of matrix elements to separately evaluable rotational, vibrational, and electronic factors. Whenever possible the electronic factor is reduced to one– and two–electron orbital matrix elements. The magnitudes and physical interpretations of matrix elements are discussed in Chapter 5. In Chapter 4 the process of reducing spectra to molecular constants and the difficulty of relating empirical-parameters to terms in the exact molecular Hamiltonian are described. Transition intensities, especially quantum mechanical interference effects, are discussed in Chapter 6. Also included in Chapter 6 are examples of experiments that illustrate, sample, or utilize perturbation effects. The phenomena of predissociation and autoionization are forms of perturbation and are discussed in Chapters 7 and 8.

We want to thank the following experts who have helped us by reading and criticizing portions of the manuscript: M. Alexander, R.F. Barrow, N. Bessis, M. Broyer, L. Brus, M. Chergui, M. Child, D.L. Cooper, K. Dressler, C. Effantin, D. Gauyacq, A. Giusti-Suzor, R.A. Gottscho, G. Gouédard, M.

Graff, C. Green, J. T. Hougen, B. J. Howard, P. Houston, Ch. Jungen, A. Lagerqvist, C. Linton, S. McDonald, E. Miescher, J. Norman, S. Novick, I. Renhorn, J. Rostas, J. Schamps, H. Schweda, S. J. Silvers, J. Thoman, A. Tramer, P. Vaccaro, and J. F. Wyart. The errors and opacities that remain are ours, not theirs. We have not attempted to cite all relevant references; we apologize for our probably numerous failures to discuss the first and/or most important examples of phenomena, interpretations, and techniques. This book could never have been produced without the artistry of Mr. J. Lefèvre and the skillful and energetic word processing of Virginia Siggia, Kay Garcia, and Sarah M. Moore.

RWF is grateful to the Laboratoire de Photophysique Moléculaire of the Université de Paris–Sud for a visiting professorship in 1981, to S. Novick and B. Kohler of Wesleyan University and R.J. Saykally of the University of California, Berkeley, for their hospitality, and to project MAC of MIT for a grant of computer time. A NATO international travel grant greatly accelerated and enriched the process of preparing this book. RWF owes his fascination with molecules to W.A. Klemperer and the late H.P. Broida.

HLB has been greatly stimulated by many discussions with the experimentalists in the groups led by S. Leach (Orsay) and J. d'Incan (Lyon). Their numerous questions led her and Leach to organize the French meetings of the GESEM,* which provided an opportunity to develop some of the material included in this book. HLB is also grateful to the late A. Kastler, who invited her to lecture at the International Winter College in 1973 in Trieste. The lecture notes written for this purpose were the first draft for some sections of this book. Finally, her interest in perturbations goes back to a problem presented to her in 1963 by E. Miescher. This was the starting point for a fruitful and pleasant scientific exchange that is still underway. This is why this book is dedicated to E. Miescher.

*Groupe d'Etudes de Spectroscopie Electronique Moléculaire (1969–1979).

Chapter 1

Simple Spectra and Standard Experimental Techniques

Contents

1.1 Rotation-Vibration-Electronic Spectra of Diatomic Molecules . . . 2

1.1.1 Rotation-Vibration-Electronic Energy Levels and Standard Notation . . . 3

1.1.2 Band Systems, Bands, and Branches . . . 6

1.1.3 Rotational Structure: Fortrat Parabola . . . 6

1.1.4 Some Strategies for Rotational Line Assignments . . 11

1.2 Experimental Techniques of Diatomic Molecule Spectroscopy . . . 21

1.2.1 The Goals of a Spectroscopic Experiment . . . 21

1.2.1.1 What and How Much? . . . 22

1.2.1.2 Rotational Analysis of a Spectrum . . . 24

1.2.1.3 Spectroscopic Strategies: Access and Selectivity . . . 26

1.2.2 Classes of Spectroscopic Experiment . . . 27

1.2.2.1 One Photon Resonant: Upward (Absorption Spectra) . . . 27

1.2.2.2 One Photon Resonant: Downward (Emission Spectra) . . . 33

1.2.2.3 Multiple Resonance Spectra . . . 35

1.2.2.4 Photofragment and Coincidence Experiments 39

1.2.3 Molecule Sources . . . 41

1.2.4 Frequency Calibration . . . 43

1.2.5 Detection Techniques 45
1.2.6 Radiation Sources 47
1.3 References 55

1.1 Rotation-Vibration-Electronic Spectra of Diatomic Molecules

The field of diatomic molecule electronic spectroscopy is very old, broad, rich, and mature, yet astonishingly active and innovative (Herzberg, 1950; Huber and Herzberg, 1979; Bernath and McLeod, 2001). Few simple "textbook" spectra remain to be discovered and analyzed, though simple spectra are routinely rediscovered and reanalyzed as prerequisite to the launching of new experiments. However, perhaps the most important reason to learn how to extract the information encoded in these simplest possible molecular electronic spectra is to become able to recognize and focus on the features of the spectrum that are *not simple*. These not-so-simple features can yield fascinating insights into the mechanisms of intramolecular dynamical processes.

Even systems as seemingly simple as diatomic molecules often act as complex, many-body systems. Mechanistic understanding and insight, as opposed to mere empirical description, are based on the existence and discovery of patterns that owe their existence to approximate constants of motion. An approximate constant of motion is the eigenvalue of an operator that commutes with most, but not all, terms in the exact molecular Hamiltonian. Nonconservation of this quantity results in subtle rather than catastrophic corruption of the simple patterns on which spectroscopic assignments and mechanistic interpretations are based.

The simple spectra described in this chapter and the ideas presented in this book provide the language, concepts, and intuitive framework required for the design and interpretation of even the most elaborate and innovative modern molecular structure and dynamics experiments.

Although few *bona fide* examples of "textbook spectra" of gas phase diatomic molecules exist, we begin this book with an elementary description of a spectrum devoid of the esoteric details that encode the most interesting diatomic molecular "dynamics." The approach in this chapter is unashamedly phenomenological; the physical principles, derivations of formulas, and theoretical context will be presented in Chapters 2-9.

A *spectrum*, $I(\omega)$, is a plot of some sort of *signal strength* versus *frequency*. The signal can be in the form of either a decrease or an increase from a constant baseline level, can be due to either an absorption or emission of light, can be based on detection of photons, parent molecules, or photofragment species, and is recorded as an electrical current or voltage or the darkening of a photographic plate. The frequency is almost always expressed by spectroscopists in cm^{-1}

units,

$$E = h\nu = hc(1/\lambda) \tag{1.1.1}$$
$$1/\lambda = E/(hc) \tag{1.1.2}$$

where E is energy, h is Planck's constant, ν is frequency, $(1/\lambda)$ is waves per unit length or colloquially (but incorrectly) "wavenumber", and c is the speed of light. Spectroscopists seldom distinguish notationally between energy, frequency, and wavenumber. Almost every molecular constant derived from frequency domain rotation-vibration-electronic spectra is expressed in cm^{-1} units and referred to interchangeably as an energy, a frequency, or a wavenumber. We will follow that dubious but time-honored practice in this book.

1.1.1 Rotation-Vibration-Electronic Energy Levels and Standard Notation

Energy in diatomic molecules is distributed among three kinds of internal excitations: rotational, vibrational, and electronic. The *rotational energy* is described empirically by a truncated (denoted by ...) series in $J(J+1)$,

$$F(J) = B_v J(J+1) - D_v[J(J+1)]^2 + \ldots, \tag{1.1.3}$$

where F is the symbol traditionally used to denote rotational energy and J is the rotational quantum number (here, $J = 0, 1, 2, \ldots$). [We will show in later chapters that J can be either integer or half-integer and that the lowest allowed J–value might be larger than 0 or 1/2.] B_v is the *rotational constant* for the v–th vibrational level, represented empirically by a truncated power series in $v + 1/2$,

$$B_v = B_e - \alpha_e(v + 1/2) + \ldots \tag{1.1.4}$$
$$B_e(\text{cm}^{-1}) = \left(\frac{\hbar^2}{2}\right)\left(\frac{1}{hc}\right)\left(\frac{1}{\mu R_e^2}\right) = 16.85763\left[\mu(\text{amu})R_e(\text{Å})^2\right]^{-1}, \tag{1.1.5}$$

α_e is usually positive and on the order of $10^{-2}B_e$. D_v is the *centrifugal distortion constant* (typically on the order of $\approx 10^{-6}B_v$) for the v–th vibrational level. The reduced mass, μ, of the AB molecule is

$$\mu = \frac{M_A M_B}{M_A + M_B}. \tag{1.1.6}$$

R_e is the equilibrium internuclear distance. The B_v and D_v constants are weakly v–dependent, but strongly dependent on electronic state.

The *vibrational energy* is described empirically by a truncated power series in $v + 1/2$,

$$G(v) = \omega_e(v + 1/2) - \omega_e x_e(v + 1/2)^2 + \ldots, \tag{1.1.7}$$

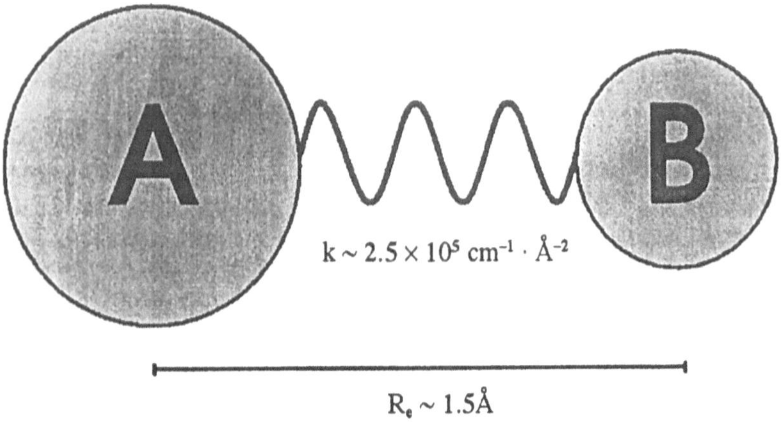

Figure 1.1: A diatomic molecule. Typical values of the harmonic force constant and equilibrium internuclear distance are $k/hc = 2.5 \times 10^5\ \text{cm}^{-1}/\text{Å}^2$ and $R_e = 1.5$ Å

where G is the symbol traditionally used to denote vibrational energy. v is the vibrational quantum number ($v = 0, 1, 2, \ldots$). ω_e is the harmonic vibrational frequency

$$\omega_e(\text{cm}^{-1}) = \left(\frac{h}{2\pi}\right)\left(\frac{1}{hc}\right)[k/\mu]^{1/2} = 4.1198 \times 10^8 \left[k(\text{erg} \cdot \text{Å}^{-2}) \big/ \mu(\text{amu})\right]^{1/2} \tag{1.1.8}$$

where k is the second derivative of the internuclear potential energy function, $V(R)$,

$$k = \left(\frac{d^2V(R)}{dR^2}\right)_{R=R_e} \tag{1.1.9}$$

evaluated at the equilibrium internuclear separation, R_e, which is the minimum of $V(R)$ [i.e., $dV/dR = 0$]. In Eq. (1.1.8) k is expressed in units of erg·Å^{-2} (typical value 5×10^{-11} erg·Å^{-2} = 5 mdyne/Å= 5×10^2 N/m, and $(k/hc) = 2.5 \times 10^5\ \text{cm}^{-1}/\text{Å}^2$) and μ is the reduced mass expressed in amu (see Fig. 1.1). The anharmonic vibrational constant, $\omega_e x_e$ (a single constant, not ω_e times x_e), expresses some of the effects of nonzero higher derivatives of $V(R)$, and is sometimes used to *estimate* the energy required to break the A–B bond

$$D^e_{\text{AB}} \approx \omega_e^2/4\omega_e x_e \tag{1.1.10}$$

(which is valid for a Morse potential, see Section 5.1.2). Typical values of $\omega_e x_e$ and D^e_{AB} are $10^{-2}\omega_e$ and $4 \times 10^4\ \text{cm}^{-1}$.

Vibrational-rotational molecular constants, and structural quantities (harmonic force constant, k, and equilibrium internuclear distance, R_e) derived from these constants are listed in Table 1.1 for the electronic ground states of three

Table 1.1: Typical Ground State Molecular Constants (Huber and Herzberg, 1979)

	H_2	I_2	HI	CsF
ω_e/cm^{-1}	4401.213	214.502	2309.014	352.56
$k/\text{cm}^{-1}\cdot\text{Å}^{-2(a)}$	$2.89 \times 10^{5(f)}$	$8.66 \times 10^{4(f)}$	$1.58 \times 10^{5(f)}$	6.13×10^4
$\omega_e x_e/\text{cm}^{-1}$	121.336	0.6147	39.6435	1.615
B_e/cm^{-1}	60.8530	0.037372	6.426365	0.184370
$R_e/\text{Å}^{(b)}$	0.7414	2.6663	1.6092	2.3454
α_e/cm^{-1}	3.0622	1.138×10^{-4}	0.1689	1.1756×10^{-3}
$\alpha_e(\text{calc.})^{(c)}$	2.08	1.19×10^{-4}	0.159	1.13×10^{-3}
D_e/cm^{-1}	4.71×10^{-2}	4.25×10^{-9}	2.07×10^{-4}	2.017×10^{-7}
$D_e\ (\text{calc.})^{(d)}$	4.65×10^{-2}	4.54×10^{-9}	1.99×10^{-4}	2.02×10^{-7}
D_e^0/cm^{-1}	36118	12443	24633	41540
$D_e^0\ (\text{calc.})^{(e)}$	39900	18700	33600	19200
μ/amu	0.503913	63.4522	0.999885	16.6223

[a] Eqs. (1.1.8) and (5.1.20) [b] Eqs. (1.1.5) and (5.1.24) [c] Pekeris equation for Morse oscillators, Eqs. (2.1.5) and (5.1.26) [d] Kratzer equation for Morse oscillators, Eqs. (2.1.4) and (5.1.25) [e] Eqs. (1.1.10) and (5.1.22) for Morse oscillators. [f] Note that $k_{AB} \approx [k_{AA}k_{BB}]^{1/2}$

covalently bound molecules (H_2, I_2, and HI) and one ionically bound molecule (CsF). Also included are molecular constants (α_e, D_e, D_e°) derived from other molecular constants via inter-relationships valid for a Morse oscillator (see Section 5.1.2). Note that the molecular constants for H_2 and HA hydrides are atypical, largely because of the unusually small reduced mass, μ.

The *electronic energy*, traditionally denoted by T_e or T_0, is not described by any standard equation, except in the case of Rydberg states, where a modified Rydberg equation

$$IP - T_0(n) = \frac{\Re_M}{(n-a)^2} \qquad (n \geq 2) \tag{1.1.11}$$

describes a series of electronic states of AB which converges to the $v^+ = 0$ level of an electronic state of the AB^+ ion (see Eq. (8.3.1) in which $IP = E_{\text{ion}}$ and $T_0(n) = E_n$). IP is the ionization potential of AB (see Sections 8.3 and 8.7). n is the principal quantum number, and a is the approximately n–independent "quantum defect," which is much more than a "fudge factor" that permits the use of Eq. (1.1.11) (see Section 8.3). T is the symbol used to denote electronic energy and the subscript e implies an energy difference between the potential energy minimum of the excited electronic state and that of the electronic ground state. The subscript 0 implies an energy difference between the respective $v = 0$ levels. $\Re_\infty = 109737.318\ \text{cm}^{-1}$ is the Rydberg constant. For an AB^+ ion of mass M, $\Re_M = \Re_\infty\left(\frac{M}{m_e+M}\right)$, where m_e is the mass of the electron.

1.1.2 Band Systems, Bands, and Branches

A rotation-vibration-electronic spectrum is a collection of transitions between an upper state, always denoted by a single prime, with energy

$$E'_{e',v',J'} = T'_0 + G'(v') + F'_{v'}(J'), \tag{1.1.12a}$$

and a lower state, always denoted by a double prime, with energy

$$E''_{e'',v'',J''} = T''_0 + G''(v'') + F''_{v''}(J''). \tag{1.1.12b}$$

The transition frequency is

$$E'_{e',v',J'} - E''_{e'',v'',J''} = \Delta T + \Delta G + \Delta F. \tag{1.1.13}$$

An electronic *band system* will consist of (v', v'') *vibrational bands*, and each vibrational band will contain many rotational lines, which may be organized into *rotational branches*, one branch for each value of $\Delta J = J' - J''$. Whenever a spectroscopist wants to indicate a change in anything (energy, a molecular constant, or a quantum number) using the symbol Δ, it is *always* implied that the lower state quantity ($''$) is subtracted from the upper state quantity ($'$), for example,

$$\Delta E = (T' - T'') + [G'(v') - G''(v'')] + [F'_{v'}(J') - F''_{v''}(J'')]. \tag{1.1.14}$$

It is conventional, in both spoken and written "spectroscopy-ese", always to say or write the quantum numbers of the upper state first (Mulliken, 1930; Jenkins, 1953; Mulliken, 1955; Sheppard, *et al.*, 1987; Schutte, *et al.*, 1997).

1.1.3 Rotational Structure: Fortrat Parabola

For all one-photon, electric dipole allowed electronic transitions, the rotational selection rule for transitions is $\Delta J = \pm 1, 0$. $\Delta J = J' - J'' = +1$ transitions belong to the R branch, and the symbol $R(J)$ specifies a transition (and the frequency of that transition) between $J'' = J$ (in the lower electronic state) and $J' = J + 1$ (in the upper state). $\Delta J = 0$ and -1 transitions are specified by $Q(J)$ and $P(J)$. The same symbol, e.g. $P(J)$, is used to denote an upward $J' = J - 1 \leftarrow J'' = J$ or a downward $J' = J - 1 \rightarrow J'' = J$ transition. The value of J that is specified by the quantity within parentheses in $R(J)$, $Q(J)$, and $P(J)$ is always J''. The frequencies of all transitions ("lines") in each rotational

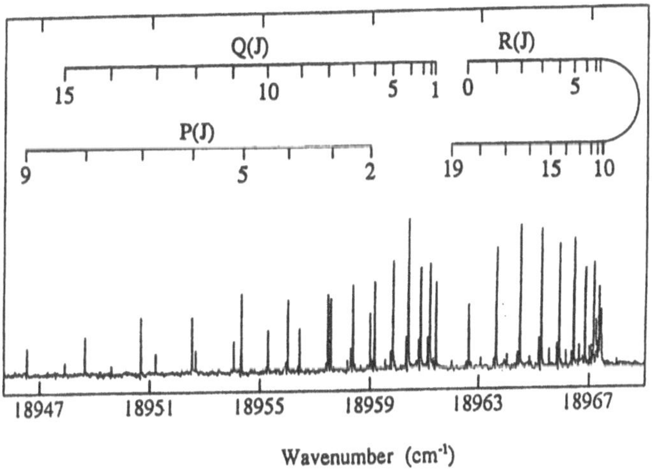

Figure 1.2: A typical vibrational band of an electronic transition, the (1,0) band of a $^{102}Ru^{12}C$ $^1\Pi \leftarrow X^1\Sigma^+$ transition. The assignments of rotational lines are distinguished by vertical tie lines. The lines in each R, Q, P rotational branch connect with a different horizontal tie line. The bandhead occurs in the R branch near J″ = 9 (from Langenberg, *et al.*, 1998).

branch of the (v', v'') band may be expressed by

$$R(J) = [T_0' - T_0''] + [G'(v') - G''(v'')] + [F_{v'}'(J+1) - F_{v''}''(J)]$$

$$\begin{aligned} R(J) = {} & [\Delta T_0 + \Delta G(v', v'')] + J^4(D'' - D') + J^3(2D'' - 6D') \\ & + J^2[(B' - B'') + (D'' - 13D')] \\ & + J^1(3B' - B'' - 12D') + J^0(2B' - 4D') \end{aligned} \quad (1.1.15a)$$

$$\begin{aligned} Q(J) = {} & [\Delta T_0 + \Delta G(v', v'')] + J^4(D'' - D') + J^3[2(D'' - D')] + \\ & J^2[(B' - B'') + (D'' - D')] + J^1(B' - B'') + J^0(0) \end{aligned} \quad (1.1.15b)$$

$$\begin{aligned} P(J) = {} & [\Delta T_0 + \Delta G(v', v'')] + J^4(D'' - D') + J^3[2(D'' + D')] \\ & + J^2[(B' - B'') + (D'' - D')] + J^1(-B'' - B') + J^0(0). \end{aligned} \quad (1.1.15c)$$

Three things should be noticed:

(1) Rotational branches form simple patterns that, in most cases, are easy to recognize (see Fig. 1.2).[†] These patterns consist of a series of lines with smoothly varying intensities and identical, approximately constant (neglecting

[†] It is tempting to say that spectroscopy begins with pattern recognition. When several simple patterns are overlaid, it is often not immediately apparent how to disentangle the overlapping patterns. However, it is almost always possible to recognize *fragments* of regularity. Some of these fragments will prove to be spurious while others, when represented by a simple polynomial equation, will be found to be real because they can be extrapolated to include a large number of spectral lines.

D', D'' contributions) *second differences*

$$\begin{aligned}\Delta^2 R(J) &\equiv [\Delta_1 R(J+1/2) - \Delta_1 R(J-1/2)] \equiv \\ &\quad [R(J+1) - R(J)] - [R(J) - R(J-1)] \\ &\approx 2(B' - B'') \qquad (1.1.16a)\\ \Delta^2 Q(J) &\approx 2(B' - B'') \qquad (1.1.16b)\\ \Delta^2 P(J) &\approx 2(B' - B''). \qquad (1.1.16c)\end{aligned}$$

The second differences may be measured directly from the spectrum without any knowledge whether J increases or decreases as one follows a branch toward higher frequency. Thus one immediately obtains a good estimate for $|B' - B''|$ neglecting contributions from D' and D''. The line intensities usually vary smoothly along a rotational branch because collisions cause rotational populations to approach Boltzmann equilibrium rapidly and rotational linestrength factors (see Section 6.1 and Whiting, *et al.*, 1980) are approximately linear in J.

(2) Depending on the sign of $B' - B''$, starting from the lowest possible J''–value the lines in *either* the R or P branch will get closer and closer together, pile up in an often unresolvable feature called the *bandhead*, and then continue toward higher J moving in the opposite direction in frequency (see Fig. 1.3). The bandhead forms when (neglecting centrifugal distortion),

$$0 = \frac{dR(J)}{dJ} \approx 2J(B' - B'') + 3B' - B'' \qquad (1.1.17a)$$

$$J^R_{\text{HEAD}} = \frac{3B' - B''}{2(B'' - B')} \geq 0 \text{ if } B'' > B' \qquad (1.1.18a)$$

$$R(J^R_{\text{HEAD}}) = \Delta T_0 + \Delta G(v', v'')] + \frac{1}{4}\left[\frac{(3B' - B'')^2}{B'' - B'}\right] + 2B' \qquad (1.1.19a)$$

or

$$0 = \frac{dP(J)}{dJ} = 2J(B' - B'') - (B' + B'') \qquad (1.1.17b)$$

$$J^P_{\text{HEAD}} = \frac{B' + B''}{2(B' - B'')} \geq 0 \text{ if } B'' < B' \qquad (1.1.18b)$$

$$\mathrm{P}(J^P_{\text{HEAD}}) = \Delta T_0 + \Delta G(v', v'')] - \frac{1}{4}\left[\frac{(B' + B'')^2}{B' - B''}\right]. \qquad (1.1.19b)$$

The bandhead conveys useful qualitative and quantitative information, but it may also conceal the low–J regions of the R, Q, and P branches. The qualitative information comes directly from the crude shape of a band. If a band has its head in the R branch, the band will appear to terminate abruptly on the high-frequency (blue) side and the rotational branches will extend, with increasing line spacings toward low-frequency (red). Such a band is said to be "red degraded". Red degradation implies $B' < B''$ and a bandhead in the R branch. Since most strong electronic transitions involve promotion of an electron from a bonding or nonbonding molecular orbital of the lower state into a nonbonding or antibonding molecular orbital of the upper state, $R'_e > R''_e$, thus $B' < B''$, therefore red degradation is more commonly observed than blue degradation. (Rydberg-Rydberg transitions typically do not form prominent bandheads, because $B' \approx B''$.) The quantitative information easily obtained from a bandhead is the energy of the bandhead. Provided that $\frac{1}{4}\left[\frac{(3B'-B'')^2}{B''-B'}\right]$ or $\frac{1}{4}\left[\frac{(B'+B'')^2}{B'-B''}\right]$ is sufficiently small and not strongly dependent on v' or v'', the bandhead provides a useful estimate of $[\Delta T_0 + \Delta G(v', v'')]$. Many vibrational analyses are initially performed, based solely on bandhead data.

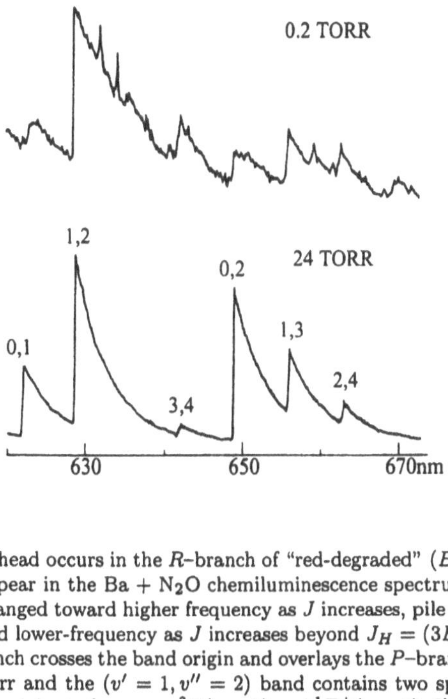

Figure 1.3: A bandhead occurs in the R–branch of "red-degraded" ($B' < B''$) BaO $A^1\Sigma^+ - X^1\Sigma^+$ bands that appear in the Ba + N_2O chemiluminescence spectrum. The low–J lines in the R–branch are arranged toward higher frequency as J increases, pile up near the bandhead, and then move toward lower-frequency as J increases beyond $J_H = (3B' - B'')/[2(B'' - B')]$. The returning R–branch crosses the band origin and overlays the P–branch. The top spectrum is recorded at 0.2 Torr and the ($v' = 1, v'' = 2$) band contains two spikes, which are $R(44)$ and $P(46)$ lines that originate from an $a^3\Pi(v = 0) \sim A^1\Sigma^+(v = 1)$ $J' = 45$ "doorway state" (see Section 6.5.5). In the bottom spectrum, recorded at 24 Torr, the spikes are absent due to collision induced rotational relaxation (from Field, 1976).

The formation of bandheads exacts a significant price in terms of the effort needed to establish the absolute J', J'' rotational assignments of all rotational lines in a vibrational band of an electronic transition. Valuable species identification, geometric structure, electronic structure, and dynamical information is locked up in the details of a band, and the *only* key to the lock is a rotational analysis.

(3) If one fits the lines in each possible branch fragment to the Fortrat

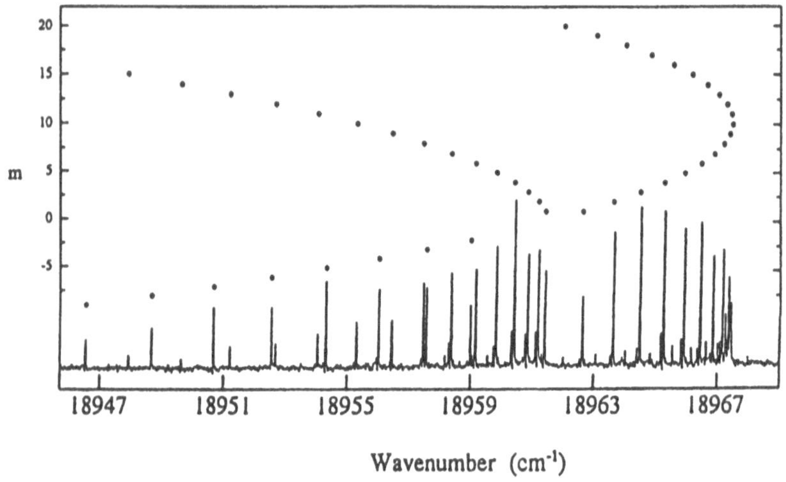

Figure 1.4: The lines in the R and P branches of the $^{102}Ru^{12}C$ $^1\Pi \leftarrow X^1\Sigma^+(1,0)$ band are unified by a single Fortrat parabola, which is described by the quadratic Fortrat equation, Eq. (1.1.20). The Q branch is representd by a different Fortrat equation (from Langenberg *et al.*, 1998).

equation (see pages 42-50, 168-192 of Herzberg, 1950), which has the form

$$\nu(m) = \mathrm{c} + \mathrm{d}m + \mathrm{e}m^2 + \ldots, \tag{1.1.20}$$

two desirable things happen. Segments of the same branch, isolated from each other by some interfering feature in the spectrum, may be joined together empirically (see Fig. 1.4). Surprisingly, the R and P branches of the same band are represented by the same polynomial expression,

$$\nu(m) = R(J) \text{ if } m = J'' + 1 \tag{1.1.21a}$$

$$\nu(m) = P(J) \text{ if } m = -J''. \tag{1.1.21b}$$

Thus, provided that a sufficient number of line frequencies in each branch fragment are measured accurately enough (in the vast majority of pattern recognition situations, accurate measurements are essential for distinguishing between genuine and spurious patterns) it is often possible to unify all lines in the R and P branches into a single *Fortrat parabola*, as shown in Fig. 1.4.

Unfortunately, the ***absolute rotational numbering*** cannot be established without additional information unless the low-J region is sufficiently well resolved to reveal a gap between the lowest-J first R and P lines, e.g. respectively $R(0)$ and $P(1)$ for a $^1\Sigma^+ - {}^1\Sigma^+$ transition. Note that at least one line predicted by the Fortrat equation, the $\nu(0)$ line, would be missing

$$\nu(2) - \nu(1) = R(1) - R(0) = 4B' - 2B'' \approx 2B \tag{1.1.22a}$$

$$\nu(1) - \nu(-1) = R(0) - P(1) = 2B' + 2B'' \approx 4B \tag{1.1.22b}$$

$$\nu(-1) - \nu(-2) = P(1) - P(2) = 4B' - 2B'' \approx 2B, \tag{1.1.22c}$$

but this *zero gap* at $\nu(0)$ is usually obscured by the bandhead, by the Q branch (for a $^1\Pi - {}^1\Sigma^+$ band), or by lines in the head-forming branch. In the absence of an unambiguously identifiable zero gap in the spectrum, the Fortrat equation *does not* imply an absolute value of m

$$\nu(m) = \mathrm{c} + \mathrm{d}m + \mathrm{e}m^2 + \ldots \tag{1.1.23}$$

Notice that, replacing m by $m+n$, it is again possible to describe the rotational lines by a new renumbered equation

$$\nu'(m+n) = \mathrm{c}' + \mathrm{d}'(m+n) + \mathrm{e}'(m+n)^2, \tag{1.1.24}$$

which can be expressed as a new polynomial in m,

$$\nu'(m+n) = (\mathrm{c}' + n\mathrm{d}' + n^2\mathrm{e}') + (\mathrm{d}' + 2\mathrm{e}'n)m + \mathrm{e}'m^2, \tag{1.1.25}$$

where $\nu'(m+n) = \nu(m)$ if one realizes that

$$\begin{aligned} \mathrm{e}' &= \mathrm{e} \\ \mathrm{d}' &= \mathrm{d} - 2n\mathrm{e} \\ \mathrm{c}' &= \mathrm{c} - n\mathrm{d} + n^2\mathrm{e}. \end{aligned} \tag{1.1.26}$$

The numbering of the lines described by a Fortrat equation is relative, not absolute. This means one needs some additional information beyond the Fortrat equation in order to establish the correct absolute numbering of the rotational lines.

1.1.4 Some Strategies for Rotational Line Assignments

The information required to eliminate the indeterminacy in absolute numbering of the R and P branch lines represented by the Fortrat equation can be obtained in a variety of ways:

1. Prior knowledge of B' or B''.
2. A reasonable guess of $B' + B''$.
3. Simultaneous analysis of two bands that share a common upper or lower electronic-vibrational level.
4. A Loomis-Wood rotational branch-picking pattern recognition scheme.
5. A spectral cross-correlation pattern recognition scheme.
6. A "pump-probe" experiment (see Section 1.2.1.2).

The way least likely to succeed is blind spectral simulation, which is the method most non-spectroscopists would normally try first.

(1) *Prior Knowledge of B' or B''*

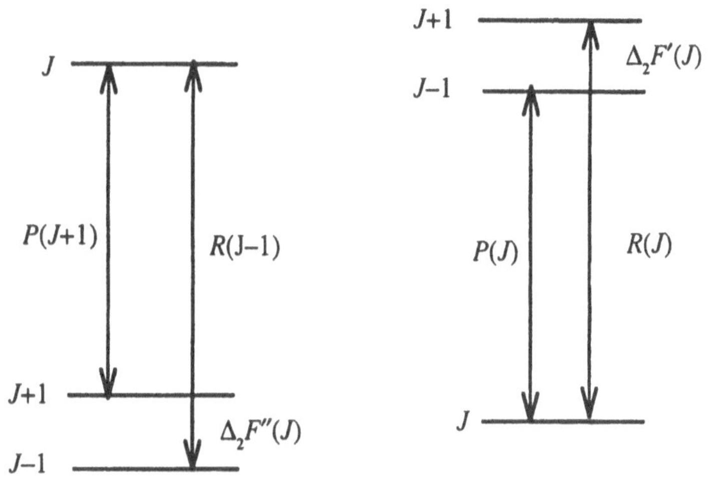

Figure 1.5: Lower and Upper State Rotational Combination Differences.

In the beginning, no B–values were known! Presently, it would be very unusual, if the *carrier* (the molecule responsible) of the spectrum is known, for neither B' nor B'' to be known. If B'' (and D'') is known, then one has a set of extremely accurate, *a priori* known rotational *combination differences* (see Fig. 1.5)

$$\begin{aligned}\Delta_2 F''(J'') &= F''(J''+1) - F''(J''-1) \\ &= B''[4J''+2] - D''[8J''^3 + 12J''^2 + 12J'' + 4] \\ &= R(J-1) - P(J+1). \end{aligned} \tag{1.1.27}$$

If instead B' is known, then

$$\begin{aligned}\Delta_2 F'(J') &= B'[4J'+2] - D''[8J'^3 + 12J'^2 + 12J' + 4] \\ &= R(J) - P(J). \end{aligned} \tag{1.1.28}$$

These combination differences can be used to find the correct absolute numbering of the Fortrat parabola. The Fortrat equation

$$\nu(m) = \mathrm{c} + \mathrm{d}m + \mathrm{e}m^2 + \ldots$$

specifies the pairs of R and P lines that correspond to known combination differences provided that the absolute numbering is correct, namely

$$m = -J'' \text{ for the } P \text{ branch}$$
$$m = J'' + 1 \text{ for the } R \text{ branch,}$$

thus,

$$\Delta_2 F''(J'') = \nu(m) - \nu(-m-1) \tag{1.1.29a}$$
$$\Delta_2 F'(J') = \nu(m-1) - \nu(-m). \tag{1.1.29b}$$

If the initial *absolute* numbering of the Fortrat parabola is incorrect (the *relative* numbering is assumed to be correct, but could easily be corrupted by inadequate precision in the wavenumber calibration of the measured spectrum), as it almost certainly will be, the set of *a priori* known $\Delta_2 F$ values will not match the set of trial combination differences computed from the Fortrat equation. It is necessary to examine a series of trial absolute numberings of the Fortrat equation, shifting m in integer steps, until the known combination differences match those derived from the Fortrat equation, either

$$\Delta_2 F''(J'') = \nu^{(n)}(m+n) - \nu^{(n)}(-m-1+n) \tag{1.1.30a}$$

or

$$\Delta_2 F'(J') = \nu^{(n)}(m-1+n) - \nu^{(n)}(-m+n) \tag{1.1.30b}$$

where

$$\nu^{(n)} = [\mathrm{c}^{(0)} - n\mathrm{d}^{(0)} + n^2\mathrm{e}^{(0)}] + [\mathrm{d}^{(0)} - 2n\mathrm{e}^{(0)}]m + \mathrm{e}^{(0)}m^2. \tag{1.1.31}$$

This procedure, especially when combined with a computer automated Loomis-Wood branch-picking procedure, (Launila, *et al.*, 1997; Winnewisser, *et al.*, 1989; McNaughton, *et al.*, 1991, freeware from Launila, 2003), is well suited for computerization. When the correct absolute rotational numbering is found, the entire set of $\Delta_2 F$ computed from *a priori* known (B', D') or (B'', D'') values will agree with the entire set (typically ~20 to 50 values) of $\Delta_2 F$ computed from the correctly numbered Fortrat equation (obtained by transforming the Fortrat equation that had been determined by a fit to rotational branch fragments). The quality of this agreement must be comparable to the measurement precision (typically $\sim 0.01\ \mathrm{cm}^{-1}$ for well-resolved lines in the visible and near ultraviolet regions). If proper attention is not paid to wavenumber calibration, this procedure will fail. When the measurements and fitting are done correctly, the result is an agreement between a large number of observed and predicted $\Delta_2 F$ values, the quality and redundancy of which is seldom encountered in other areas of physical science. One can be *certain* of a rotational analysis!

(2) *Reasonable Guess of B' or B''*

One reason for not knowing either B' or B'' could be that the identity of the carrier of the spectrum is uncertain. Even when the carrier is known, there might be several likely candidates for the upper or lower electronic-vibrational state. When several likely candidates exist for one of the electronic-vibrational states involved in the transition, an *a priori* known (or predicted by *ab initio* calculation) range of (B, D) values for the candidate state may be used, in

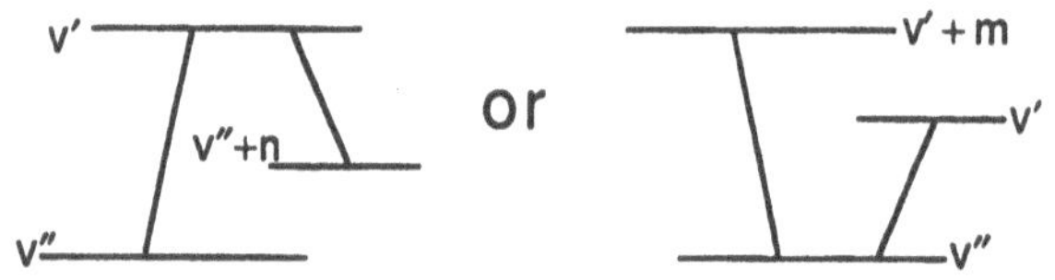

Figure 1.6: Vibrational bands linked via a common vibrational level in the upper or lower electronic state.

combination with the measured constant second differences

$$\Delta^2 R = \Delta^2 Q = \Delta^2 P = [\nu(m+1) - \nu(m)] - [\nu(m) - \nu(m-1)]$$
$$= 2e \approx 2(B' - B'') \quad (1.1.32)$$

and the readily observable red ($B' < B''$) vs. blue ($B' > B''$) degradation of the band, to yield a series of trial values for the unknown B value. For example, if B'' is approximately known and the band is red degraded, then

$$B_{\text{unknown}} \equiv B' = B''_{\text{known}} - |e|. \quad (1.1.33)$$

If, because of uncertain isotopic or vibrational assignments, there is a range of possible values of B_{known}, each trial value of B_{known} will be associated with a trial value of B_{unknown}

$$B_{\text{unknown}} = B_{\text{known}} + |e| \text{ or } - |e|.$$

Once B' and B'' are approximately known, it is possible to use the Fortrat equation to identify the low–J regions of the R and P branches, and to obtain sets of $\Delta_2 F'$ and $\Delta_2 F''$ rotational combination differences from which accurate values of the (B', D') and (B'', D'') constants can be obtained.

(3) *Simultaneous Analysis of Two Bands*

When neither B–value is known, a trial and error procedure is required. One such procedure involves the simultaneous rotational analysis of two bands which, on the basis of a vibrational analysis, are believed to involve a common upper or lower electronic-vibrational level, as illustrated by Fig. 1.6.

The key is the discovery of a set of $\Delta_2 F$ rotational combination differences common to the two bands. One begins by fitting branch fragments in both systems to Fortrat equations. Then trial sets of plausible absolute numberings of *both* Fortrat equations are used to compute sets of trial $\Delta_2 F$ combination differences. When a match is found between one trial set of combination differences from each band, the absolute rotational numbering of both bands is established. This procedure is readily computer-automated.

(4) *A Loomis-Wood Rotational Branch-Picking Pattern Recognition Scheme*

Lines that belong to the same rotational branch (or to a pair of branches related by a Fortrat equation) have smoothly varying relative intensities and nearly constant second differences. These properties are ideally suited for computer automated pattern recognition schemes, foremost among these is the

Loomis-Wood procedure (Loomis and Wood, 1928; Herzberg, 1950, p. 191; Kovács and Budó, 1953; Winnewiser, 1989; McNaughton, *et al.*, 1991; Launila, 2003), which maps a one-dimensional line list (line number, line frequency, line intensity)= (i, v_i, I_i) into a two-dimensional display in which rotational branches appear as more-or-less straight lines. The visibility of lines that belong to the same branch is enhanced by color coding the lines according to their relative intensity.

One begins by using human eye based pattern recognition to identify a series of lines that obviously belong to the same rotational branch. A polynominal function, $\nu(m)$, where m is an integer running number, is fitted to those lines. This polynomial serves as the 1D to 2D mapping function. One plots, for each integer value of m, the frequency differences $\nu_i - \nu(m)$ vs. m for all lines where $|\nu_i - \nu(m)| < \Delta$. Δ defines the width of the window about each of the predicted lines, $\nu(m)$, in which one is willing to search for additional lines in the selected branch or in a related branch. Successive lines in the mapping branch appear as a horizontal series of equally spaced points on the plot of $\nu_i - \nu(m)$ vs. m. Each line in the spectrum appears many times on the 2D plot. The mapping branch appears in several "orders", N, because $\nu(m+N) - \nu(m)$ vs. m will be a nearly straight line of slope $N\frac{d\nu}{dm}$ (see Fig. 1.7).

Bandheads are prominent features of a Loomis-Wood plot. A family of straight and curved lines emanates from the bandhead of the mapped branch on the Loomis-Wood plot. The straight lines actually intersect at the bandhead and correspond to various orders of the mapped branch. The curved lines are generated e.g. when the post-head R–branch overlaps with the pre-head segment of itself or with the P–branch described by the same Fortrat equation. Another prominent feature of Loomis-Wood plots is a horizontal strip free of points, which surrounds the mapped branch. The width of this strip is determined by the widths of lines in the mapped branch (instrumental resolution, Doppler width, nonradiative decay width, or unresolved hyperfine splittings). The Loomis-Wood plot can provide a survey map of J–dependent linewidth variations that encode valuable information about unimolecular dynamics.

When the mapping branch is extended through the band-origin region, it is often possible to uncover the "zero-gap" and thus deduce the absolute rotational assignments of the lines in the branch. Once this is accomplished, it is possible to derive the polynomial expressions that are expected to describe other, hitherto hidden or unrecognized rotational branches. Even extremely weak branches or the diagnostically important low–J regions of otherwise strong branches may be uncovered and freed of obscuration by overlapping unrelated lines, owing to the mapping from one into two dimensions.

(5) *Spectral Cross-Correlation Pattern Recognition*

When one considers both intensity and frequency information, the information content in a spectrum is very large: the number of resolution elements ($\sim 10^3$) times the instrument dynamic range (typically 8 bits = 256) divided by the noise (~2 bits). Spectra with an information content well in excess of the above estimated $\sim 10^5$ often require only a few minutes of data acquisition directly into a computer memory. Modern spectroscopy is ripe for new, model-

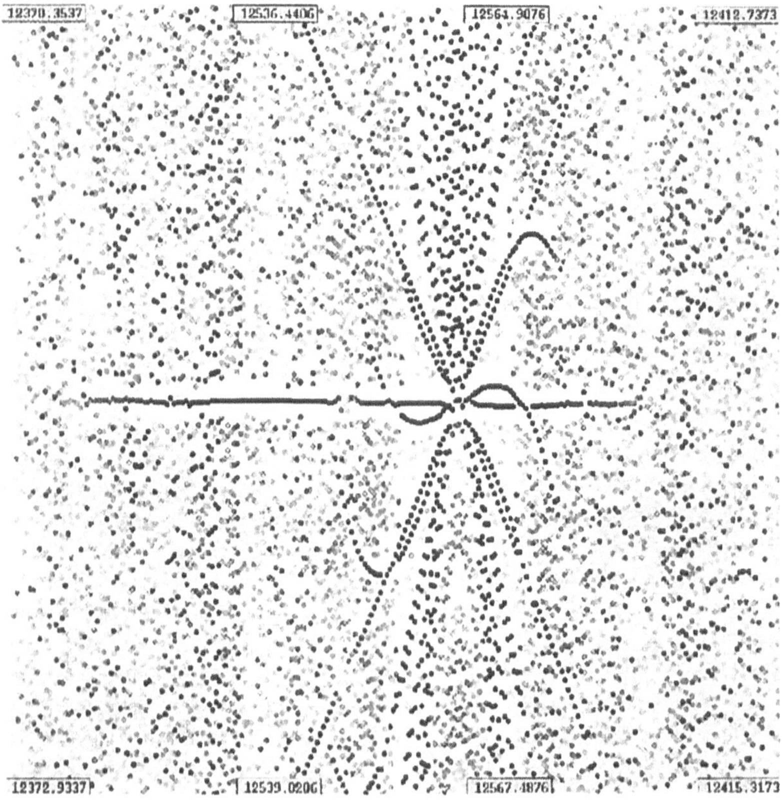

Figure 1.7a: A Loomis-Wood plot of an NbO $d^2\Delta_{3/2} - a^2\Delta_{3/2}$ sub-band in the 12370-12580 cm^{-1} spectral region. The plot, with a vertical span of 2.58 cm^{-1}, displays the P–branch and the head-forming R–branch. The boxes that label the horizontal axis along the top and bottom of the figure specify the wavenumber (cm^{-1}) extremes of the sampled spectral region surrounding the mapped branch. The horizontal axis is linear in the integer running number in the Fortrat equation and in fact, owing to the presence of the bandhead, is not even monotonic in cm^{-1}. The zero-gap (corresponding to missing lines belonging to four consecutive values of the Fortrat equation running number), located to the left of the R–head, is easily recognized. The bandhead is marked by the convergence of several orders (straight lines) of the R–branch (from Launila, *et al.*, 1997).

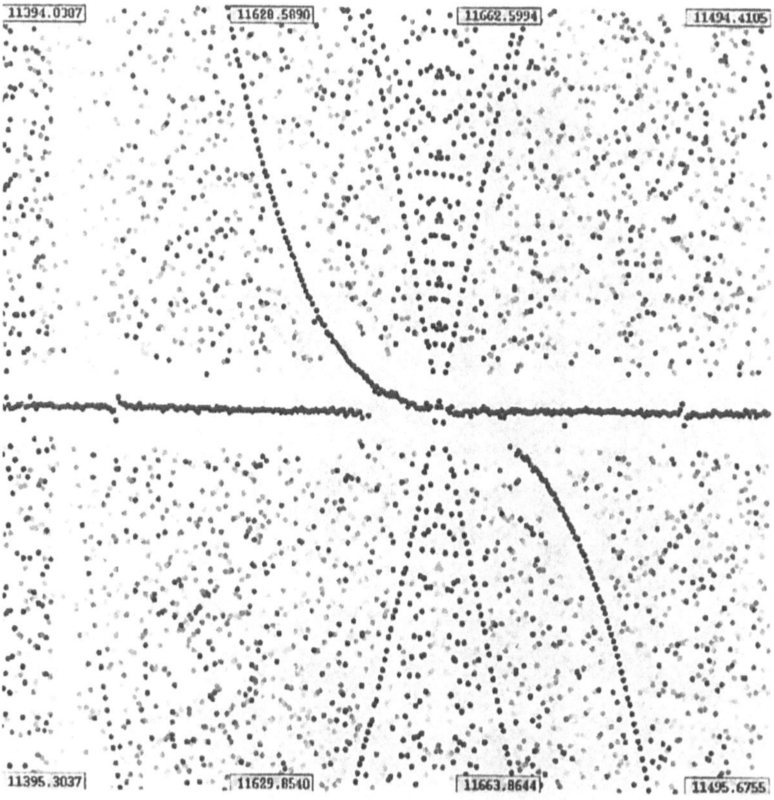

Figure 1.7b: A Loomis-Wood plot of an NbS $^2\Delta_{5/2} - {}^2\Delta_{5/2}$ sub-band in the 11400-11670 cm^{-1} region. The plot, with a vertical span of 1.26 cm^{-1}, displays the P– and head-forming R–branch. The zero gap (six missing Fortrat equation lines) lies to the left of the bandhead, but not as distant as in Fig. 1.7a. The bandhead generates a pair of crossing straight lines (different orders of the R–branch) and a pair of curved lines (due to overlap of the returning R–branch with itself and the P–branch). The region to the right of the bandhead shows the R–branch as a horizontal straight line and the P–branch as a downward going curved line. A very weak perturbation (matrix element of ~ 0.01 cm^{-1}) by a state with smaller B–value appears at $J' = 135.5$ in both the R– and P–branches. The horizontal white stripe (~ 0.1 cm^{-1} wide) shows that the linewidth is J–independent (from Launila and Taklif, unpublished).

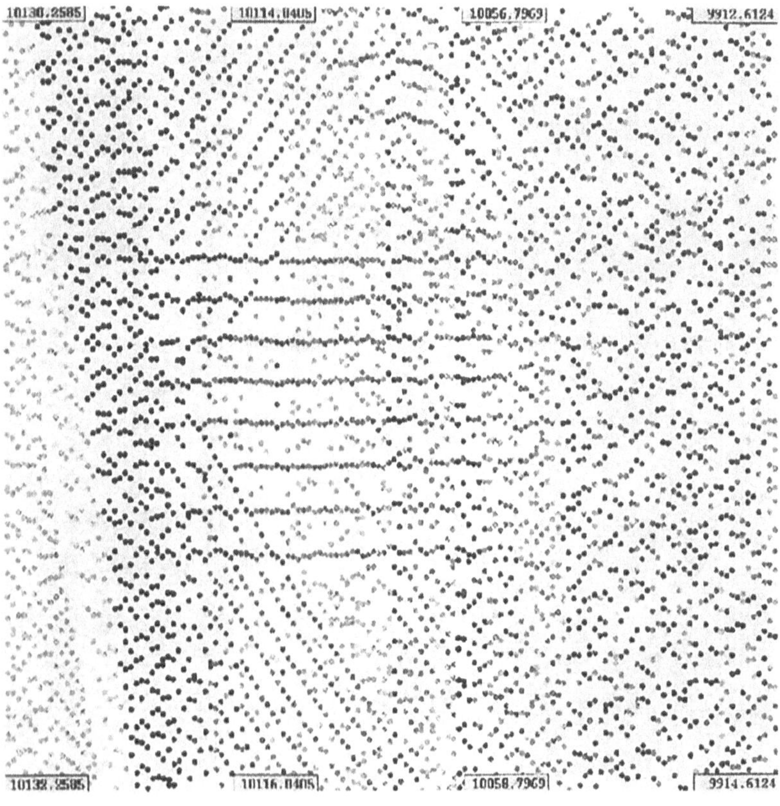

Figure 1.7c: A one-sided (R–branch region only) Loomis-Wood plot of a not yet analyzed band of ^{181}TaO in the 9900-10130 cm^{-1} region, with a vertical span of 2 cm^{-1}. The 8 horizontal lines correspond to the 8 expected $\Delta J = \Delta F$ hyperfine components for a nuclear spin of $I = 7/2$ $(2I + 1 = 8)$ (from Al-Khalili, *et al.*, 1999).

free pattern recognition schemes. It is essential that the scheme be model-free, because often the experimentalist has only limited, sometimes primitive, ideas about what kinds of information the spectrum contains. The idea of using one spectrum to define (without *a priori* knowledge of its nature or specifics) a pattern sought in another spectrum is not new. The above scheme for simultaneous analysis of two linked bands is a nice example, but in that scheme a specific type of pattern search (combination differences) is built into the procedure. Coy *et al.*, (1997) and Jacobson, *et al.*, (1997) have developed the "eXtended spectral Cross-Correlation" (XCC) method whereby n spectra are simultaneously processed to yield m *a priori* unknown patterns ($m > n$ is possible) which are exactly repeated (but with different overall relative intensities) in some of the spectra. Using this technique it has been possible to separate mixed isotopomer spectra into pure isotopomer spectra (by cross-correlating spectra recorded with different but not precisely known isotopic compositions) (Reid, *et al.*, 2000), to identify combination differences by taking advantage of precisely known intensity ratios for absorption transitions originating from a common J''–level or emission transitions from a common J'–level, and by changing the temperature to vary the relative intensities of different $[R(J), P(J)]$ or $[R(J-1), P(J+1)]$ pairs (Duan, *et al.*, 2003).

Another promising area for pattern recognition is when the pattern sought consists of many elements and the spectrum in which the pattern is to be located is very complex and congested, possibly because the spectrum consists of several spectrally shifted copies of the same many-line pattern. Neuhauser, *et al.*, (1998) have proposed the CRIES (Cross-Correlation Ionization Energy Spectroscopy) scheme for picking out many overlapped Rydberg series, each described by

$$\nu(n) = IP_{v^+J^+} - \frac{\Re}{n^2}, \qquad (1.1.34)$$

which are converging to different vibration (v^+), rotation (J^+) levels of the molecular ion-core.

(6) *Frequency Domain Pump-Probe Experiments*

It is possible to record a "dispersed fluorescence" spectrum where the only lines present are a vibrational progression of R, P doublets that correspond to a single set of lower electronic state rotational combination differences,

$$\Delta_2 F''(J') = [R(J-1) - P(J+1)], \qquad (1.1.35)$$

which originate from a single J'–value (see Fig. 1.8), or from collisionally populated neighboring J'–levels (see Fig. 1.9). Alternatively, a laser can be used to modulate the population in a single rotation-vibration-electronic level, thereby *labeling* it (Modulated Population Spectroscopy (MPS): Kaminsky, *et al.*, 1976). As a result, the only transitions in the spectrum that display labels will belong to the upper state combination difference that arises from the labeled J–value. These and other experiments will be discussed further in Section 1.2.

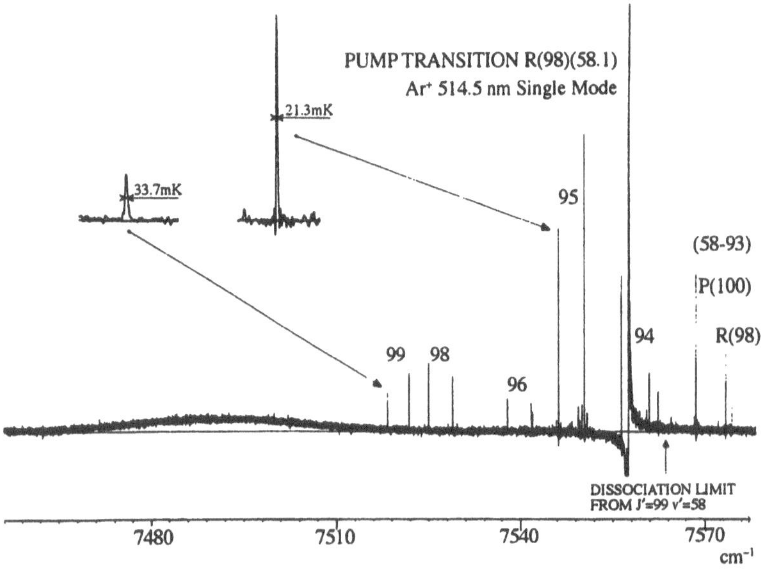

Figure 1.8: A vibrational progression of $R(J-1)$, $P(J+1)$ doublets in an I_2 $B0_u^+ - X^1\Sigma_g^+$ ($v' = 58, v'' = 93 - 98$) dispersed fluorescence spectrum, excited by a single mode Ar^+ laser and recorded by Fourier Transform Spectroscopy. The spectrum reveals lower state combination differences freed of overlapping, pattern-obscuring lines. Since the R, P spacings that originate from successive J'-levels change in steps of $4B''$, one need only measure the R, P splitting at an accuracy of $\sim 2B''$ in order to make secure J'-assignments. This spectrum demonstrates that, when the fluorescence is excited by a single mode laser and the fluorescence is collected in the forward direction, the spectrum exhibits sub-Doppler linewidths. This spectrum shows the highest vibrational levels of the I_2 $X^1\Sigma_g^+$ state, quasibound behind the $J'' = 100$ and 98 centrifugal barriers. The transition into $v'' = 99$, $J'' = 100$ is 50% broader than that into $v'' = 95$, $J'' = 100$. The broad feature centered at 7495 cm^{-1} is a transition into the $J'' = 100$ and 98 continua above the top of the centrifugal barrier. The vertical arrow marks the rotationless dissociation limit of the $X^1\Sigma_g^+$ state (from Bacis, *et al.*, 1980).

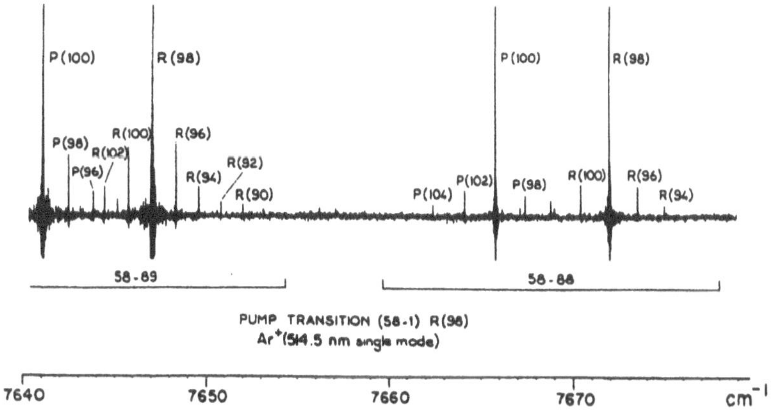

Figure 1.9: Collision induced rotational relaxation in the I_2 $B0_u^+ - X^1\Sigma_g^+$ ($v' = 58, v'' = 89$ and 88) bands. An Ar^+ laser excites the $v' = 58, J' = 99$ level. Collisions prior to fluorescence cause population to be transferred into odd J' $91 \leq J' \leq 103$ rotational levels. The $\Delta J'$ =even selection rule is a consequence of ortho,para nuclear spin conservation for a homonuclear molecule (from Bacis, *et al.*, 1980).

1.2 Experimental Techniques of Diatomic Molecule Spectroscopy

An extremely brief survey is given here of some of the experimental techniques and strategies of small molecule, gas phase, frequency domain, electronic spectroscopy. More complete discussions are found in Demtröder (1996).

There are many reasons that one might want to record, assign, and interpret the electronic spectrum of a diatomic molecule. These include: qualitative (which molecular species are present) and quantitative (what is the number density of a known quantum state of a known molecule) analysis, detection of trace constituents (wanted, as in analysis of ore samples for a precious metal, or unwanted, as in process diagnostics where specific impurities are known to corrupt an industrial process), detection of atmospheric pollutants, monitoring of transient species to optimize a combustion process by enhancing efficiency or minimizing unwanted byproducts, laboratory determinations of transition frequencies and linestrengths of interstellar molecules, and last but certainly not least, fundamental studies of molecular structure and dynamics.

1.2.1 The Goals of a Spectroscopic Experiment

Foremost is: record a spectrum! To accomplish this one needs a light source, a means of monochromatizing the light either before or after it interacts with the target molecules, a scheme for wavenumber calibration which is of absolute accuracy superior to the resolution of the experiment, a molecule source, and a signal detector. Often, a single device fulfills several of these requirements, such as, for example, a narrow spectral width tunable laser. In some highly selective experiments a molecule is required to jump through multiple hoops in order to produce a detected signal. The signal may be detected *indirectly*, through the detection of energized molecules, photofragmentation products (electron or ion or atom) singly or in temporal near coincidence, or *directly*, by monitoring the absorption or emission of light.

1.2.1.1 What and How Much?

Once a spectrum is recorded, it is sometimes necessary to figure out *what molecule was responsible for that spectrum*. This is a much less trivial and more frequently encountered problem than it might seem. One especially effective solution to the problem of species identification is the use of a mass spectrometer to select the mass of the ion injected into the resonance region or to detect only ions of a certain mass produced as a specific direct consequence of photoexcitation of the target molecule (Resonance Enhanced Multi-Photon Ionization spectroscopy with Time-Of-Flight detection: REMPI-TOF (Johnson, *et al.*, 1975; Johnson and Otis, 1981; Ashfold and Howe, 1994). This marriage of mass spectrometry and laser spectroscopy has made it possible to record selectively the spectrum of a single trace species in the presence of a complex mixture of transient species (van der Waals molecules, clusters (see Fig. 1.10)). Even without mass spectrometers there are tricks capable of revealing the identity of a mystery molecule. These include the known relative abundances of naturally occurring atomic isotopes, isotopic labeling, systematic variation of conditions (particularly with respect to ubiquitous contaminant gases containing H, C, N, and O), electronic fine structure (arising from the presence of one or more partly filled molecular orbitals) or hyperfine structure (non-zero nuclear spins), and odd-J/even-J intensity alternation (homonuclear diatomic molecules, permutation symmetry requirements for identical Fermion or Boson nuclei, see also Fig. 1.9) (see Table VI of Mulliken, 1931, and Table 15 of Herzberg, 1950).

Sometimes the goal is not to record a spectrum but to *monitor the concentration of a specific trace species*, such as CO emission in the exhaust from an automobile or unwanted OH radicals above the surface of a silicon substrate being plasma-etched. In such a case it would be unnecessary to scan over a continuous segment of a spectrum, nor would it be necessary to make rotational assignments. One would want to monitor absorption or induced fluorescence at a small number of well selected discrete frequencies: e.g., the peaks of two similar–J lines (peak–1, peak–2), selected to be free of overlap with absorption lines belonging to the most probable interfering species, plus one or two baseline frequencies (baseline (ν_1), baseline (ν_2)), at which no absorption from the target or an interfering species is expected. A set of signal ratios: i.e., peak–1/peak–2, peak–1/baseline, and peak–2/baseline, would enable the raw signal to be corrected for frequency independent attenuation of the probe beam (scattering by particulates, quasi-continuous absorption by a large molecule) as well as overlap of one of the monitored frequencies by a line from an unexpected trace species. Again, accurate frequency metrology is essential; the probe frequency must be

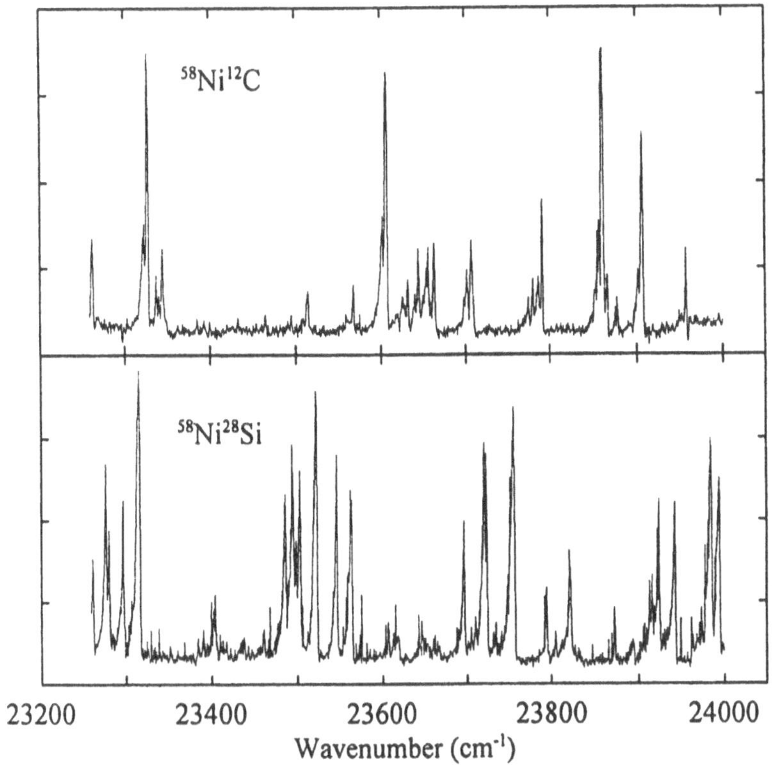

Figure 1.10: REMPI-TOF permits selective simultaneous recording of separate spectra for different cluster species or for different isotopomers. The spectra of NiC and NiSi were recorded (by resonant two-photon ionization using an ArF excimer laser to ionize) using laser ablation of a nickel target in a stream of carrier gas containing 3% CH_4. No intentional source of silicon was present (but the nickel sample had been roughened using SiC sandpaper) (from Brugh and Morse, 2002, and Lindholm, *et al.*, 2003).

accurately locked to the peak of the monitored transition.

1.2.1.2 Rotational Analysis of a Spectrum

When a spectrum is recorded, there are no molecule or quantum number labels attached to it. Sometimes the spectrum contains a previously unobserved band of a known species. More often it contains well known bands of well known species, but the experimentalist usually does not initially know to which of several plausible or implausible species the band belongs. The best way to answer this band identity question quickly and unambiguously is to perform a rotational analysis. It is often easier and *always* more conclusive to identify a band by the molecular constants (e.g., B', B'') derived from it than from the frequency location of the maximum intensity feature of the band. (In fact, incompletely resolved band intensity maxima are notoriously unreliable, because the position of the maximum can depend strongly on resolution, temperature, and pressure.)

One wants to be able to *assign a routine spectrum quickly* and to *assign a difficult spectrum reliably.* It is rare that either of these goals can be attained by starting with a spectrum simulation computer program. A simple spectrum is a band that is free of overlap by other bands and where most of the observed lines can be assembled into rotational branches that have smoothly varying intensities and constant second differences and may be accurately represented by one or more Fortrat equations (see Section 1.1.3). Often the combination of (i) $|\Delta B|$ determined from the coefficient of the quadratic term in the Fortrat equation, (ii) the sign of ΔB from the red vs. blue degradation of the band, and (iii) the wavenumber of the bandhead (either measured directly or obtained from the Fortrat equation) will be sufficient to identify the species and the specific upper and lower electronic-vibrational levels. The band identification can then be confirmed against $\Delta_2 F$ combination differences computed from a tabulated B–value. This procedure should be made sufficiently simple and automated to permit routine use by non-spectroscopists.

There are many reasons that a spectrum could be too complex for routine rotational analysis: overlap of the target spectrum by several other bands, extreme spectral congestion where frequent overlap of lines makes it difficult to recognize rotational branch fragments by eye, perturbations (level shifts and intensity anomalies, the main subject of Chapters 2-8), and electronic fine structure (the presence of several related sub-states, which occur whenever an electronic state arises from an electronic configuration with at least one partially filled molecular orbital). When a spectrum is complex, it is useful to have an experimental scheme that isolates certain key features of an assignment scheme in order to lead the experimenter unambiguously to correct assignments or to test whether a plausible assignment is in fact correct. Typically the question is one of *linkage*: (1) do two transitions share a common level? (2) what are some same–J $\Delta_2 F(J)$ rotational combination differences in different substates or in consecutive vibrational levels (see Fig. 1.11)? (3) What is the change in the $\Delta_2 F''(J)$ rotational combination differences observed via consecutive (rotationally unas-

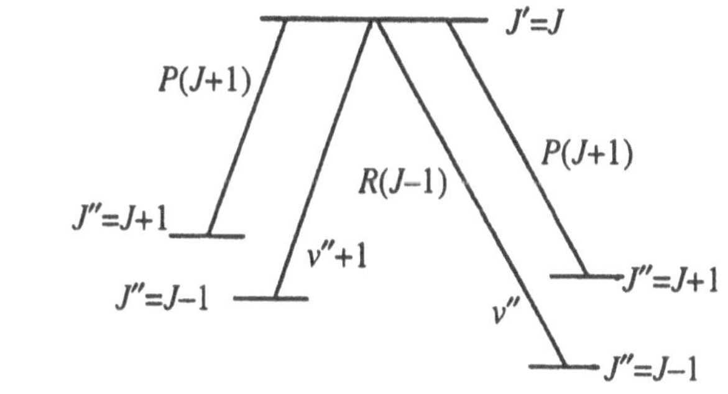

Figure 1.11: Lower state rotational combination differences for two consecutive vibrational levels.

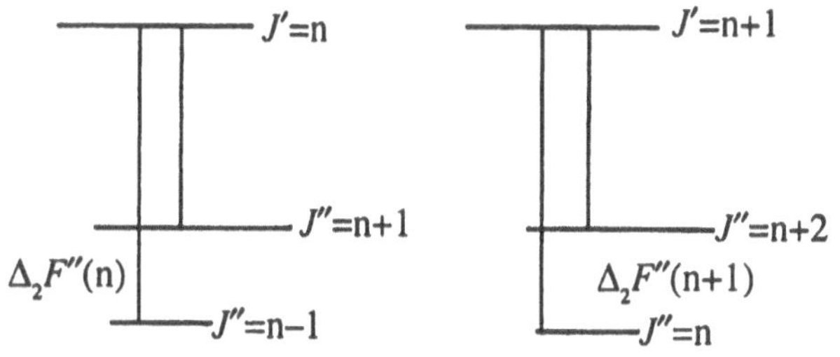

Figure 1.12: Lower state rotational combination differences for two consecutive members of the same branch. $\Delta_2F''(n+1) - \Delta_2F''(n) = 4B''$.

signed) members of a branch fragment (see Fig. 1.12)?

Questions of linkage are posed and answered by asking the molecule to satisfy successively two resonance conditions. Schemes which accomplish this include Dispersed Fluorescence Spectroscopy (DF, Section 1.2.2.2; a laser is tuned to excite a single line and the spectrum of the resulting molecular fluorescence is recorded), Modulated Population Spectroscopy (MPS, Section 1.2.2.3); an intense, fixed frequency, amplitude modulated PUMP laser is used to modulate the population in the upper and lower levels connected by the laser excited transition; the modulation is then detected by a frequency scanned PROBE laser), which is an example of Optical Optical Double Resonance (OODR, Section 1.2.2.3).

A dispersed fluorescence (DF) spectrum isolates individual pairs of $P, R,$ lines from which Δ_2F'' values may be directly measured. The change in Δ_2F'' values observed via consecutive (but unknown) J'-levels immediately gives a value for B'' because $\Delta_2F''(n+1) - \Delta_2F''(n) = 4B''$.

Other methods for dealing with a complex spectrum include *simplifying* it by cooling the molecule to ~5 K in a supersonic jet (Section 1.2.3) or selectively enhancing the relative intensities of certain low-J rotational lines (Magnetic Rotation Spectroscopy, Section 1.2.5).

1.2.1.3 Spectroscopic Strategies: Access and Selectivity

Often a spectroscopist must devise a strategy to gain access to a particular class of state or region of excitation energy to which access is prohibited by spectroscopic selection rules or by available excitation and detection technology. Sometimes *access* is not a problem but *selectivity* is. At high excitation energy, molecules dissociate or ionize, spectral lines broaden, and the resulting spectrum conceals the informative structures of overlapping structured continua. These problems of access and selectivity are solved by multi-color multiple resonance schemes (see Section 1.2.2.3), such as Stimulated Emission Pumping (SEP), Perturbation Facilitated Optical Optical Double Resonance (PFOODR), and Resonance Enhanced Multi-Photon Ionization (REMPI). SEP, which utilizes an upward PUMP transition followed by a downward DUMP transition provides access to high vibrational levels of the electronic ground state. PFOODR is an OODR scheme in which the intermediate level is a perturbed eigenstate (see Section 6.2) that contains a mixture of two different electronic states. Consequently states to which spectroscopic access would normally be prohibited by selection rules (e.g. $\Delta S = 0$, where S is the total electron spin) may be examined systematically. Two and three color, multiply resonant REMPI schemes, especially when one or more of the resonant steps involves two- or three-photon transitions, can give access to states at excitation energies well in excess of 50,000 cm^{-1}. The Vacuum Ultraviolet (VUV) $\lambda < 200$ nm and eXtreme Ultraviolet (XUV) $\lambda < 120$ nm [the "windowless" region starts beyond the absorption cut-off of LiF (105 nm) or MgF_2 (115 nm) windows] regions, previously dominated by large vacuum spectrographs and synchrotron facilities, are becoming increasingly accessible to "table-top" laser experiments (Hollenstein, *et al.*, 2000).

Insight into the detailed mechanisms of photofragmentation [photodissociation AB+$h\nu$ $\rightarrow$A + B (Chapter 7) and photoionization AB+$h\nu$ $\rightarrow$ $AB^+ + e^-$ (Chapter 8)] can be obtained when spectra of the energy regions above the lowest dissociation limit (D_0^0, which is the energy difference between the lowest existing vibrational-rotational level of the AB ground electronic state and the atoms A and B or atomic ions in their lowest energy state) or above the ionization potential (IP) are examined from a single, selectable, intermediate electronic-vibrational-rotational state. This eliminates the washing out of crucial information in the spectrum that can result when many structured continua originating from a thermal distribution of initial states are superimposed. Perfect intermediate state-selectivity also provides perfect knowledge of the internal energy and rigorously good quantum numbers (J and parity) of the photoactivated system. If some of the excitation steps are in the form of sufficiently short duration laser pulses, then initially well specified rotational, vibrational, and/or electronic wavepackets are formed (see Sections 9.2.3 and 9.2.4) and may be used to explore the fragmentation dynamics in the time domain (Schinke, 1993).

1.2.2 Classes of Spectroscopic Experiment

1.2.2.1 One Photon Resonant: Upward (Absorption Spectra)

There is a superficial similarity between *Direct Absorption* and *Laser Induced Fluorescence* (LIF) spectra. (See Section 1.2.5 for a discussion of direct vs. indirect detection schemes). In direct absorption, the molecular sample removes energy from the probe beam, either a "white light" continuum or a narrow spectral width tunable laser. The signal appears as a small decrease on a large "constant" background light intensity at the detector. In LIF, a tunable laser irradiates the molecular sample, and when the frequency of the laser coincides with an absorption line, the molecule converts some fraction of the incident coherent, unidirectional laser radiation into incoherent fluorescence, radiated into 4π steradians and containing many frequencies which correspond to the many possible $e', v', J' \rightarrow e'', v'', J''$ fluorescence transitions. The signal in LIF, usually collected in a direction perpendicular to the laser propagation direction, appears against a nearly perfectly dark background. This difference between absorption spectra (small decrease on a large background) and LIF spectra (signal on a near-zero background) is responsible for the enormous sensitivity advantage of LIF over the usual implementations of direct absorption spectroscopy.

Sensitivity, often confused with signal to noise ratio, may be defined as the ratio of signal *on*-resonance to the root mean square (rms) noise *off*-resonance. The sensitivity of LIF is primarily limited by extraneous background at the detector (dark current in a photomultiplier, cosmic rays, scattered laser light, background radiation from the molecule source, blackbody radiation, chemiluminescence, light emission from an electrical discharge). Unlike LIF, for which the noise off-resonance is typically much smaller than the noise on-resonance, the noise in a direct absorption experiment is approximately the same on- and off-resonance. The sensitivity of direct absorption is, at best, limited by the intrinsic quantum noise ("shot noise") of the large constant background

$$N_{\text{Quantum}} = (N_{\text{background}})^{1/2} \tag{1.2.1}$$

where N is the number of photons per detection period (time constant, reciprocal of electrical bandwidth of the detection system) and the signal (ΔI) to background ratio (SBR) would be, at best (i.e., for $\sim$ 100% absorption, or $\Delta I \approx N_{\text{background}}$)

$$\frac{\Delta I}{(N_{\text{background}})^{1/2}} \leq (N_{\text{background}})^{1/2}. \tag{1.2.2}$$

For 1mW of detected radiation at 500 nm, $(N_{\text{background}})^{1/2} \approx 2 \times 10^9$ in a 1 second detection period (1 Hz bandwidth), which would be an astonishingly large SBR (corresponding to a minimum detectable fractional absorption of

5×10^{-10}). However, for most light sources, especially lasers, the intensity fluctuations are on the order of 1%, resulting in a SBR $\sim 10^7$ times poorer than the quantum limit. Although clever tricks exist which can get rid of most of this excess noise, such as differential amplification, Frequency Modulation Spectroscopy [FM (Gehrtz, *et al.*, 1985; Bjorklund, 1980; Hall, *et al.*, 1981)], Cavity Ringdown Spectroscopy [CRDS (O'Keefe and Deacon, 1988; Scherer, *et al.*, 1997; Romanini and Lehmann, 1993)]; Magnetic Rotation Spectroscopy [MRS (Righi, 1898; Buckingham and Stephens, 1966; Litfin, *et al.*, 1980; Hinz, *et al.*, 1982)], no tricks are needed to remove the nonexistent excess noise in an LIF spectrum.

There is another important difference between direct absorption and LIF spectroscopies. This difference reflects a fundamental distinction between direct and indirect detection schemes. In a direct absorption spectrum, the signal integrated over the absorption line is determined by a fundamental molecular property, the integrated absorption cross section, σ_{ij}^0 (see Eqs. 6.1.11 and 6.1.13), times the number density (molecules/cm^3) of molecules in the lower level of the absorption transition. In an LIF spectrum, the detected signal is reduced by the product of the fluorescence quantum yield

$$\phi_{\text{fluor}} = \frac{k_{\text{fluor}}}{k_{\text{total}}}, \tag{1.2.3}$$

and the detector quantum efficiency, ϕ_{detector}. k_{fluor} is the rate of photon emission from the upper level of the probed transition and k_{total} is the total of all depopulation rates (dissociation, ionization, collisional quenching). ϕ_{detector} is more complicated than it might seem, because the spectrum of fluorescence from a single upper level often spans a wide wavelength region and the fraction of the total fluorescence detected is determined by an overlap integral of the product of the fluorescence branching ratios $(I(\nu_i)/I_{\text{total}})$ and with the λ–dependent response of the photon-detection system. A particularly troublesome (and often forgotten) problem is that the relative intensities of lines in an LIF spectrum can be significantly distorted by level specific variations of k_{total} and the λ–dependence of ϕ_{detector}. Thus the relative intensities of lines in an LIF spectrum can deviate from their intrinsic values (i.e., σ_{ij}°), as would always be available from a direct absorption spectrum. These effects can be particularly important since the relative line intensities often provide information crucial to the correct assignment of a spectrum or a correct mechanistic description of a dynamical process.

Direct absorption detection schemes are particularly ill suited for high resolution spectroscopic studies of molecular ions (Herzberg, 1971, Saykally and Woods, 1981). Owing to the extreme reactivity of ions, space-charge limitations on achievable molecular ion number density, and the inevitable background presence of vastly larger number densities of neutral species, the ability to record laboratory (as opposed to astronomical) spectra of ions has been dependent on the development of unique sources of transient species (e.g., flash discharge, applied to C_2^-, Herzberg and Lagerqvist, 1968) and tricks to disentangle relatively weak ion transitions from the vastly more intense neutral molecule transitions (owing to the greater abundance of neutral than ionic species), such as velocity modulation spectroscopy (Gudeman and Saykally, 1984; Gudeman, *et al.*, 1985). By applying an alternating longitudinal electric field to a dc discharge, the velocities of molecular ions are modulated whereas those of neutral species are, to first order, unmodulated (see Fig. 1.13). Viewed along the direction of the alternating electric field, the ion spectrum appears as two Doppler-split and amplitude modulated components. The magnitude of the Doppler shift provides unique measures of the mobility of molecular ions in a plasma and the magnitude and time-dependence of the axial electric field in a discharge (Radunsky and Saykally, 1988).

LIF (Ezekiel and Weiss, 1968; Cruse, *et al.*, 1973; Zare and Dagdigian, 1974; Kinsey, 1977) is an example of an *indirect* technique for the detection of a one-photon resonant upward transition. There are many other indirect detection techniques (optogalvanic, optothermal, photoacoustic, cavity ringdown), but Multi-Photon Ionization (MPI) is a special type of indirect technique uniquely well suited for combining absorption detection with other useful functionalities (see Section 1.2.1.1). In MPI, photo-ion detection replaces photon detection. The one-color, singly-resonant-enhanced $(n+m)$ REMPI [†] process consists of an n–photon resonant $e', v', J' \leftarrow e'', v'', J''$ excitation, followed by a further nonresonant m–photon excitation into the ionization continuum

$$\begin{aligned} AB + nh\nu &\rightarrow AB^* \\ AB^* + mh\nu &\rightarrow AB^+ + e^-. \end{aligned} \tag{1.2.4}$$

The key advantage of REMPI is that the resonant process creates two charged particles, either of which can be captured, manipulated, and detected with near unity quantum efficiency. The intrinsic disadvantage of REMPI is that the relative intensities of REMPI-detected spectral lines can be corrupted by the inherently nonlinear (and not proportional to the naively predicted I_{laser}^{n+m} integer power of I_{laser} either!) dependence on laser intensity, the competition between photoionization and photodissociation at the n and at the $n+m$ photon levels, and the sometimes complex and unpredictable dependence of the ion yield on the difference between the total excitation energy, $(n+m)h\nu$, and the ionization threshold (due to dynamically important structures in the continuum). Another disadvantage of $(n+m)$ REMPI schemes in which either n or $m \neq 1$ is that nonresonant excitation steps involving 2 or more photons require very high laser intensities. This can result in many unwanted effects (Ashfold and Howe, 1994; Signorell and Merkt, 1999) such as power broadening, Stark shifts, intensity perturbations due to accidental near-resonant enhancements of a few lines, suppression of bands by competition from multiphoton dissociation, and transition amplitude interference between nonresonant three-photon absorption and one-photon absorption of the 3ν radiation generated by four-wave mixing.

Nevertheless, the resonance enhanced production of AB^+ and e^- at the $t = 0$ arrival time of the excitation pulse, combined with the high efficiency

[†] For a two-color REMPI experiment, the notation $(n+m')$ would be used.

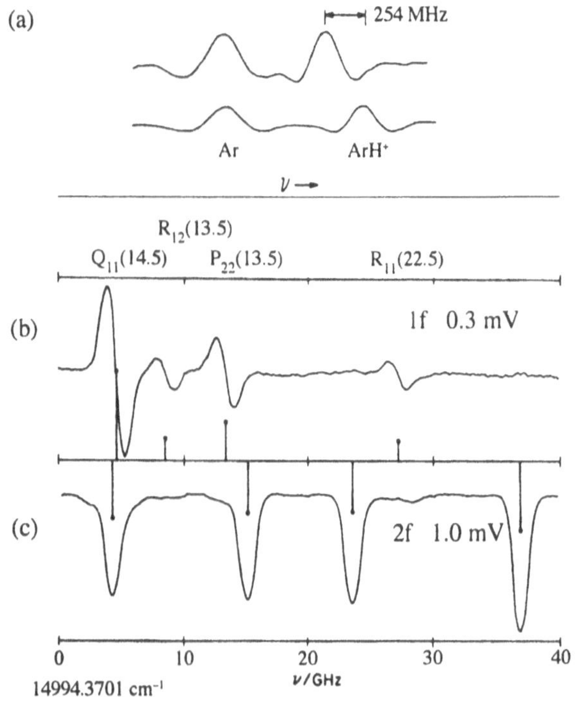

Figure 1.13: Velocity Modulation Spectra (VMS). An AC electric field applied along the laser propagation direction causes the velocity, hence the Doppler shift, of ionic but not neutral species to be modulated. Spectrum (a) shows ± 127 MHz $= 0.0042\ \text{cm}^{-1}$ Doppler shifts of a 2479.4113 cm^{-1} ArH^+ transition. Also shown is an Ar line that does not exhibit an electric field dependent Doppler shift. This spectrum was actually recorded at a fixed (DC) electric field using counter-propagating (frequency modulated) laser beams (from Haese, *et al.*, 1983). Spectrum (b) shows a portion of the VMS N_2^+ $A^2\Pi_u - X^2\Sigma_g^+$ (7,3) band. Since the AC electric field causes the N_2^+ line to be frequency modulated, phase sensitive detection at the modulation frequency (1f) results in a first derivative lineshape. Spectra (b) and (c) were recorded simultaneously. The N_2^+ lines are absent from spectrum (c), which was recorded in "population modulation" mode by phase sensitive detection at twice the modulation frequency (2f). Spectrum (c) contains a few lines belonging to a N_2 $B^3\Pi_g - A^3\Sigma_u^+$ First Positive band. The population of N_2 $A^3\Sigma_u^+$ is modulated at 2f because each half cycle of the AC field results in a full cycle modulation of the population of electronically excited N_2 molecules. (Spectra (b) and (c) are from Radunsky and Saykally, 1987.)

of collecting and detecting charged particles, opens the door to many ingenious tricks. These include multi-color, multi-resonant, energy and angular distribution analysis of the photoelectron, photoion-photoelectron coincidence schemes, and threshold ionization detection [Zero Electron Kinetic Energy: ZEKE (Müller-Dethlefs and Schlag, 1991; Merkt, 1997; Signorell and Merkt, 1999) and Mass Analyzed Threshold Ionization: MATI (Zhu and Johnson, 1991)], all discussed in Sections 1.2.2.3 and 1.2.2.4 below (see also Section 8.1.2). Foremost among these tricks is the opportunity to determine the mass of the resonantly produced ion: Resonance Enhanced Multi-Photon Ionization-Time-Of-Flight (REMPI-TOF) spectroscopy.

In a REMPI-TOF experiment, a pulsed supersonic beam of molecules is crossed by a pulsed laser beam. The ions produced when the laser is tuned to resonance with a molecular absorption transition are extracted from the resonance region by a weak pulsed electric field and injected into a Time-Of-Flight mass analyzer. All singly charged ions, on passing through a potential difference of ΔV volts, will have a kinetic energy $KE = \Delta V$ upon exiting the ion-acceleration region. Since $KE = \frac{1}{2}mv^2$, $v = (2\Delta V/m)^{1/2}$, thus each mass is accelerated to a different velocity and arrives at the ion-detector at mass-specific arrival times. One can selectively detect the spectrum of a specific AB molecule (discriminating against many other possibly more abundant species present in the supersonic jet) by collecting signal exclusively in the time window associated with the $(M_A + M_B)$ AB^+ ion (see Fig. 1.14). One can also record simultaneously the spectra of several species (e.g., isotopomers $^{40}Ca^{35}Cl$ and $^{40}Ca^{37}Cl$) by monitoring the ion signal in two (or more) time windows. The discovery of Buckminsterfullerene (Kroto, *et al.*, 1985), C_{60}, was a direct consequence of the combination of the sensitivity and species selectivity of MPI-TOF spectroscopy.

Degenerate four wave mixing (DFWM) spectroscopy is the final example to be discussed here of a one-photon resonant upward transition scheme (Abrams and Lind, 1978; Dreier and Rakestraw, 1990; Vaccaro, 1997). In a DFWM experiment, the sample is irradiated by three laser beams, all at exactly the same frequency. Two of the beams, which cross at a small angle, create a transient material grating. This grating, a periodic spatial modulation, can be of several types: molecular level population, index of refraction, density/temperature. The third beam (usually counter-propagating with one of the grating-generating beams) is diffracted by this grating, thereby creating a fourth laser-like beam of signal radiation (usually counter-propagating with the other of the grating-generating beams). The grating is only produced when the lasers are tuned into resonance with a molecular transition, thus the signal beam is only present on resonance. In the sense that the DFWM and LIF signals are produced against an essentially zero background, DFWM and LIF resemble each other. However, there are many important differences. DFWM is nonlinear, depending on $\left(\sigma_{ij}^{\circ}\right)^4$ and $(N_{\text{resonant}})^2$ (N_{resonant} is the number of molecules, within the beam overlap volume, which are in the initial state of the transition) (Dreier and Rakestraw, 1990; Zhang, *et al.*, 1992; Buntine, *et al.*, 1992), whereas LIF is linear in both σ_{ij}°

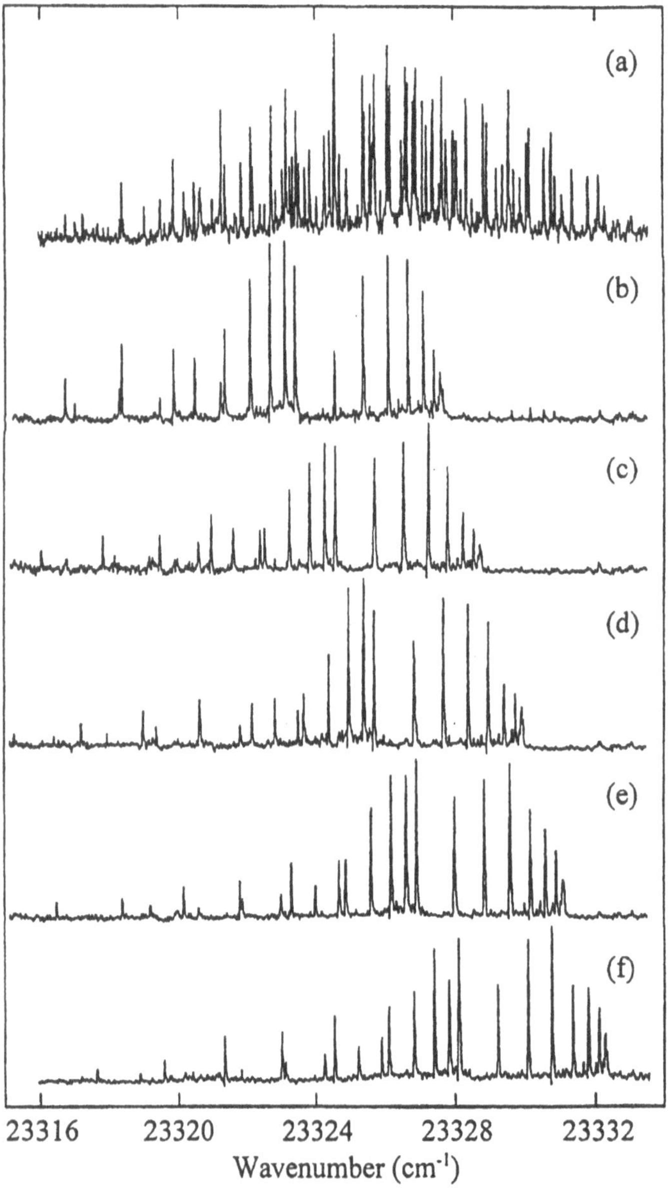

Figure 1.14: Simultaneously recorded REMPI-TOF spectra of several isotopomers of MoC. Spectra (b) - (f) correspond, respectively, to ^{98}MoC (24.1% natural abundance), ^{97}MoC (9.6%), ^{96}MoC (16.7%), ^{95}MoC (15.9%), and ^{94}MoC (9.3%). Spectrum (a) is the sum of spectra (b) - (f). It is *not* the spectrum that would have been obtained without TOF mass-selectivity because the summation does not include the spectra of ^{92}MoC (14.8%) and ^{100}MoC (9.6%) (from D. Brugh, Ph.D. thesis, 1997, courtesy of M. Morse; see also Brugh, *et al.*, 1998).

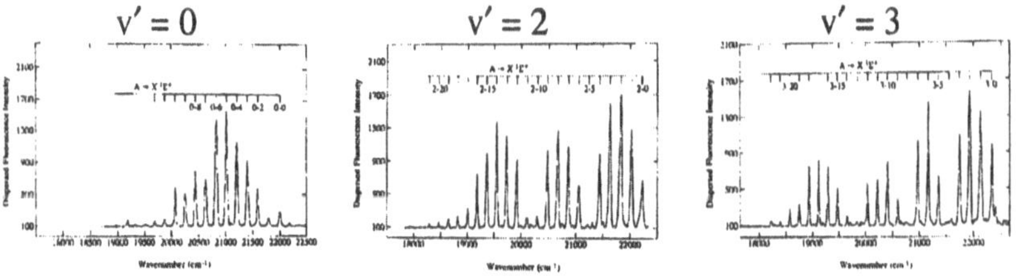

Figure 1.15: Dispersed Fluorescence Spectrum of AgAu. The AgAu A → $X^1\Sigma^+$ DF spectrum contains long $v' \rightarrow v''$ vibrational progressions because the bond length in the A–state is much longer than in the X–state. The nodal structures of the $v' = 0, 2$, and 3 vibrational states are displayed as intensity minima in the DF spectra. Since the vibrational quantum number is equal to the number of internal nodes in the wavefunction, a vibrational progression in an absorption or DF spectrum often reveals the absolute assignment of the initial vibrational state (from Fabbi, *et al.*, 2001).

and N_{resonant} (Cruse, *et al.*, 1973; Kinsey, 1977). Fluorescence from the upper state is irrelevant to DFWM but essential for LIF. Since the polarization states of four beams can be specified in a DFWM experiment, more information about rotational branch type and electronic symmetry is available from DFWM than LIF (Wasserman, *et al.*, 1998). The nonlinearity (both $(\sigma_{ij}^{\circ})^4$ and N_{resonant}^2) of DFWM can be a serious disadvantage, especially for weak transitions with small σ_{ij}° and for transient molecular species which cannot be produced at high number density.

1.2.2.2 One Photon Resonant: Downward (Emission Spectra)

Before the advent of lasers, the vast majority of molecular electronic spectroscopic data was obtained from analysis of emission spectra. Unlike absorption spectroscopy, recording an emission spectrum poses special problems that affect the decision whether to use a dispersive or Fourier Transform spectrometer. Dispersive spectrometers use ruled or holographic diffraction gratings. They can be used as either a scanning monochromator (single channel) or as a spectrograph with multiplex detection (photographic plate or multichannel diode array). The advantages of a Fourier Transform Spectrometer (FTS) over a dispersive instrument are high resolution, accurate wavenumber calibration, and meaningful relative intensities. However, in order to exploit the advantages of a FTS, the spectral emission source must satisfy stringent requirements: temporal intensity stability, spatial homogeneity and stability, stability of relative intensities for emission from different species that have spectral features within the selected wavelength region. The severe problem of weak molecular band emission occurring concomitant with strong atomic line fluorescence is discussed by Cossart and Elhanine, 1998.

Emission spectra are usually much more complicated than Direct Absorption spectra, because an absorption spectrum contains all possible upward transitions from a single initial vibration-electronic state, typically the ground state, whereas an emission spectrum contains all possible downward transitions from many initial upper states. Molecules can be excited to produce emission spectra by electric discharges, collisions with electrons, ions, or hyperthermal molecules, chemical reactions (chemiluminescence), heating (thermoluminescence), and by absorbing radiation (photoluminescence). There is minimal experimental control over the complexity of an emission spectrum, but many tricks to control the emitting species and to select classes of emitting states were once well known (Mitchell and Zemansky, 1961). The exquisite species and state specificity afforded by laser plus mass spectrometric techniques threatens the art of recording and analysis of non-laser-excited emission spectra with extinction born of unwarranted neglect. Novel discharge sources, especially for molecular ions, even doubly charged ions, N_2^{++}, (Cossart, *et al.*, 1985a), NO^{++} (Cossart, *et al.*, 1987), and CO^{++} (Cossart and Robbe, 1999) are definitely not a lost art (Cossart, *et al.*, 1985b, and Cossart and Cossart-Magos, 1990 and 1996, Cossart, 1994). However, the body of spectral data obtained from pre-laser emission spectra remains of comparable quality and vastly greater quantity than that obtained by all forms of modern laser spectroscopy.

Although Dispersed Fluorescence (DF) spectroscopy is probably better classified as a form of double resonance spectroscopy, DF is discussed here because it is a form of emission spectroscopy where all of the emission originates from a single, laser-populated, upper electronic-vibrational-rotational level, (e', v', J'). A DF spectrum typically contains two $[R(J'' = J' - 1), P(J'' = J' + 1)]$ or three $[R(J'-1), Q(J'), P(J'+1)]$ rotational transitions per electronic-vibrational e'', v'' level. Often there is a *progression* of vibrational bands, $[(v', v'' = n)$, $(v', v'' = n+1)$, $\ldots$ $(v', v'' = n+m)]$ where $v'' = n$ is the lowest vibrational level (band farthest to the blue) and $v'' = n+m$ is the highest vibrational level observable (limited either by the detector response or Franck-Condon factors) in the DF spectrum (see Fig. 1.8 and Fig. 1.15).

Sometimes the DF spectrum will contain transitions terminating on several different lower electronic states, each with its own progression of R, P doublets or R, Q, P triplets. DF spectra virtually assign themselves! If J' is initially known, then

$$R(J'-1) - P(J'+1) = \Delta_2 F''(J') = B''[4J'+2]. \tag{1.2.5}$$

The relative $R : Q : P$ intensities often reveal the identity of the lower electronic state. The spacings between members of the progression of R, P doublets are vibrational intervals

$$G''(n+1) - G''(n) = \omega_e'' - (2n+2)\omega_e'' x_e'', \tag{1.2.6}$$

and the relative intensities of vibrational bands are often sufficient to determine absolute vibrational assignments (see Section 5.1.5). Owing to the sparseness of most DF spectra, subtle effects [$R : P$ intensity anomalies Eq. (6.3.41) (see Fig. 1.16), level shifts, extra lines], indicative of spectroscopic perturbations by an otherwise unobserved state, are easily observed. These subtle effects often convey otherwise unobtainable structural and dynamical information.

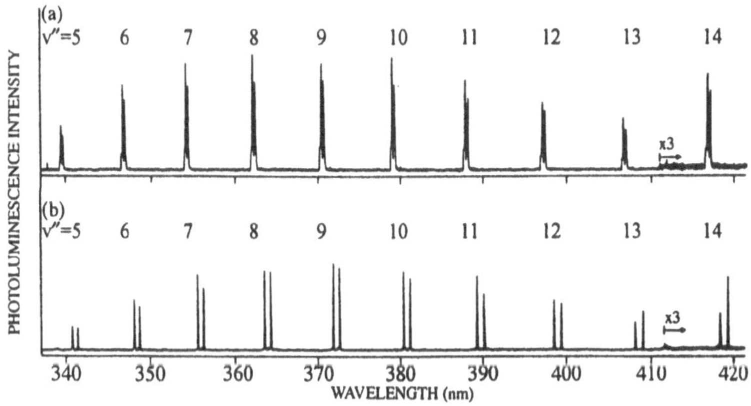

Figure 1.16: v''-dependent $R(J-1) : P(J+1)$ intensity anomalies in the BaO $C^1\Sigma^+$ $[v' = 0, J' = 18$ (top) and $J' = 48$ (bottom)] $-X^1\Sigma^+(v'' = 5-14)$ DF spectrum. In the top spectrum, the (higher frequency) R–line in each R, P doublet is $\sim 10\%$ more intense than the P–line. In the bottom spectrum, owing to a $B^1\Pi \sim C^1\Sigma^+$ upper state perturbation, an interference effect (see Section 6.3.2) causes the relative intensites of the R and P lines to be reversed in transitions to the $v'' = 13$ and 14 lower state vibational levels (from Gottscho, *et al.*, 1978).

1.2.2.3 Multiple Resonance Spectra

There are several reasons for accepting the extra experimental complexity of a pump/probe double resonance experiment. Foremost among these are: (i) a double resonance spectrum, Optical Optical Double Resonance (OODR): (Field, *et al.*, 1975); Modulated Population Spectroscopy (MPS): (Kaminsky, *et al.*, 1976), is always much simpler and easier to assign than a single resonance spectrum; (ii) states that would normally be unobservable or unresolvable in a single resonance experiment (because of electronic selection rules, vibrational Franck-Condon factors, limited laser tuning range or detector sensitivity range, superposition of transitions originating from different fine structure or lambda doublet components) may be systematically explored by double resonance (Perturbation-Facilitated-OODR (PFOODR): Li Li and Field, 1983; Gottscho, *et al.*, 1980); (iii) the resolution of double resonance spectra is typically limited by the laser spectral width rather than the Doppler linewidth (sub-Doppler-OODR: Gottscho, *et al.*, 1980); (iv) the absorption or emission of a microwave photon ($h\nu << kT$) can be detected via the emission of a visible or ultraviolet wavelength photon ($h\nu >> kT$) (Field, *et al.*, 1973); (v) double resonance often facilitates ultra-sensitive and ultra-selective detection schemes when, for example, the signal appears on a dark background, the signal is carried by ions, electrons, or easily photoionized Rydberg states that can be detected with considerably higher efficiency than photons (and with mass selectivity), or the pump/probe process can compete against or exploit the fast nonradiative decay of an intermediate or target state.

Most double resonance experiments exploit one of the four energy level schemes shown in Fig. 1.17. The variety of detection schemes is enormous, but most schemes may be divided into those which result in a signal on an essentially dark background vs. those which result in a dip in an essentially constant background level. By definition, the frequency of the PROBE (or DUMP) laser is scanned and that of the PUMP or DETECT laser is held fixed while a double resonance spectrum is being recorded.

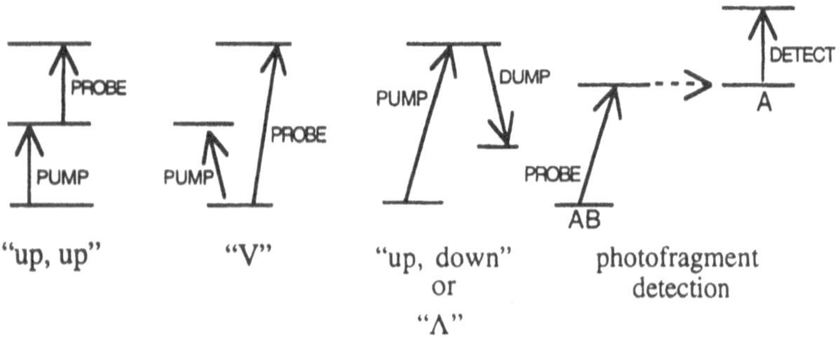

Figure 1.17: Level diagrams for two-color double resonance experiments.

The most common form of Optical Optical Double Resonance (OODR) (Field, *et al.*, 1975) uses the "up, up" level scheme and the OODR signal is usually carried by a UV photon emitted from the e^*, v^*, J^* upper level

$$\text{PUMP} \qquad e', v', J' \leftarrow e'', v'', J'' \tag{1.2.7a}$$

$$\text{PROBE} \qquad e^*, v^*, J^* \leftarrow e', v', J' \tag{1.2.7b}$$

$$\text{DETECT} \qquad e^*, v^*, J^* \rightarrow e, v, J. \tag{1.2.7c}$$

It is easy to detect this $e^*, v^*, J^* \rightarrow e, v, J$ signal-photon, free of interference from longer wavelength scattered PUMP or PROBE radiation and fluorescence from the e', v', J' intermediate level, because usually a significant fraction of the fluorescence from the e^*, v^*, J^* level is to the blue of both PUMP and PROBE. However, if the e^*, v^*, J^* level has a small fluorescence quantum yield owing to fast nonradiative decay processes, then a fluorescence-dip scheme can be used instead of UV fluorescence detection (Gottscho, *et al.*, 1980; Kaminsky, *et al.*, 1976; Kittrell, *et al.*, 1981). The fluorescence from the intermediate e', v', J' level is monitored (selected using an interference filter or a monochromator) and a small dip in the intensity of this fluorescence occurs whenever the PROBE laser causes the rate of the upward $e^*, v^*, J^* \leftarrow e', v', J'$ transition to be comparable to the spontaneous decay rate from the e', v', J' intermediate level. Fractional fluorescence dips of 1–10% are typical, but coherent versions of OODR , such as STImulated Rapid Adiabatic Passage (STIRAP) (Gaubatz, *et al.*, 1990; Schiemann, *et al.*, 1993), can produce 100% fluorescence dips.

Stimulated Emission Pumping (SEP) (Kittrell, *et al.*, 1981; Lawrance and Knight, 1982), a folded or "up, down" (Λ) version of OODR (see Fig. 1.17), is closely related to the fluorescence dip OODR detection scheme (Gottscho, *et al.*, 1980)). The DUMP (or, more properly, the PROBE) laser stimulates a downward transition

$$e', v', J' \longrightarrow e, v, J \tag{1.2.8}$$

from the PUMP-populated intermediate e', v', J' level. When the DUMP laser is tuned onto resonance with a downward transition, it stimulates emission in the forward direction. This stimulated emission in a direction to which the side-fluorescence detector is blind, competes with spontaneous fluorescence as the primary loss mechanism for the population in the e', v', J' level. As a result there is a dip in the detected side fluorescence.

Two-color, Resonance Enhanced Multi-Photon Ionization (REMPI) spectroscopy is similar to OODR, differing only in that the

$$e^*, v^*, J^* \leftarrow e', v', J' \tag{1.2.9a}$$

PROBE resonance is detected via the

$$\mathrm{AB}^+ + e^- \leftarrow \mathrm{AB}(e^*, v^*, J^*) \tag{1.2.9b}$$

photoionization process rather than the

$$e^*, v^*, J^* \rightarrow e, v, J + hv \tag{1.2.9c}$$

fluorescence process. A nice thing about ions and electrons is that they can be detected with nearly 100% efficiency, whereas photon detection is limited by both the solid angle of the light collection system and the quantum efficiency of the photodetector. In REMPI, either the PUMP, the PROBE, or a separate high intensity IONIZATION laser can be used to drive the ionization step. The advantage of using a separate ionization laser is that power broadening by either the PUMP or PROBE laser can be minimized and, by tuning the IONIZATION laser frequency close to the relevant ionization threshold, the ionization efficiency can be maximized.

There are many variants of REMPI. In most REMPI experiments one or more of the photoexcitation steps is a multiphoton transition. The rotation and electronic selection rules and relative intensity factors are quite different for a one-photon

$$e', v', J' \leftarrow e'', v'', J'' + h\nu \tag{1.2.10}$$

vs. a two-photon

$$e', v', J' \leftarrow\leftarrow e'', v'', J'' + 2h\nu \tag{1.2.11}$$

process. Signal strengths are also affected by quantum mechanical interference effects (see Section 5.3) between transition amplitudes into overlapping ionization or dissociation continua. Thus, intensities in REMPI spectra cannot be assumed to be trivially related to level populations and to tabulated rotational linestrength factors (Whiting, *et al.*, 1980; Herzberg, 1950).

Although not an essential part of REMPI spectroscopy, mass-selective detection (using either a time-of-flight or a quadrupole mass filter) of the REMPI signal confers enormous advantages at the cost of negligible loss of sensitivity. With mass-selection one can: (i) simultaneously record spectra of two (or more) isotopomers (see Fig. 1.14); (ii) exploit multiphoton ionization of a selected photofragment as a detection channel, for example

$$\begin{aligned} \mathrm{H} + \mathrm{F} &\leftarrow \mathrm{HF} + 2h\nu \\ \mathrm{F}^+ + e^- &\leftarrow \mathrm{F} + 3h\nu'; \end{aligned} \tag{1.2.12}$$

or (iii) selectively record the spectrum of a purposefully selected trace constituent in an ill-characterized mixture of other species that absorb strongly and continuously in the spectral region of interest.

Two especially important variants of REMPI (Johnson, *et al.*, 1975) spectroscopy are ionization-dip (Cooper, *et al.*, 1981) and Zero Electron Kinetic Energy (ZEKE) (Müller-Dethlefs and Schlag, 1991; Merkt, 1997; Signorell and Merkt, 1999) photoelectron spectroscopy. Ionization-dip REMPI spectroscopy is especially useful when one wants to record free←bound spectra from a single, selectable v, J level. Without such v, J selection, most of the oscillatory structure in a free←bound spectrum will be washed out. One potential problem with some ionization-dip schemes is that, if the ionization transition originates from the initial level of the free←bound transition being studied, there is a possibility that the observed linewidths will be distorted by power broadening (especially when the "free" final state is a weakly predissociated state with linewidth $\lesssim 1\mathrm{cm}^{-1}$).

ZEKE spectroscopy provides information about the electronic-vibrational-rotational structure of molecular ions. ZEKE is based on two properties of Rydberg states: (i) there will be a Rydberg series of the neutral molecule converging to every electronic-vibrational-rotational level of the molecular cation, and for all values of the principal quantum number $n > 200$, the Rydberg level lies within 2 cm^{-1} of the cation level to which the Rydberg series converges; (ii) the radiative and nonradiative decay rates of Rydberg states decrease very rapidly as n and l (l is the electronic orbital angular momentum) increase, typically as $\sim n^{-3}l^{-3}$ (Gallagher, 1994, pages 25 and 47), (see also Section 8.1.2).

It turns out that ZEKE is a bit of a misnomer (Signorell and Merkt, 1999) (except perhaps for anion ZEKE), because zero kinetic energy electrons (i.e. the electrons ejected *exactly* at each ionization threshold) are not actually detected. Delayed Pulsed Field Ionization (PFI) is an essential part of the usual neutral molecule detection process (Chupka, 1993; Vrakking and Lee, 1995; Held, *et al.*, 1998). At $t = 0$, many eigenstates of the neutral molecule that lie slightly below an e^+, v^+, J^+ ionization threshold are populated by a one-color or two-color excitation scheme. Even a minuscule DC electric field (either intentionally applied or present due to surface charges and nearby photoions) causes extreme l–mixing, mixing the optically accessible but relatively short-lived low–l states

($l \leq 3$) into a dense manifold of long-lived high-l ($3 < l \leq n - 1$) states. After the laser excitation pulse and the application of the weak l-mixing electric field, there is a waiting period during which all of the molecules in low-n, low-l states are allowed to decay radiatively, all of the very highest-n states are field-ionized by the weak electric field, and all of the ions and electrons produced accidentally and by field-ionization of the highest-n states are driven out of the detection region by a weak applied electric field. Now all that remains in the detection region are molecules in long-lived, high-l, intermediate-n ($\sim 200 < n < \sim 400$) states. These are detected (and mass analyzed by time-of-flight) when a pulse of higher electric field field-ionizes all of the Rydberg molecules with n-values *above* a selected threshold that remained within the detection region.

Thus, as the PROBE laser is scanned through the energy region of each ionization threshold, there will be a sharp peak located at a known energy below the e^+, v^+, J^+ ionization threshold. The peak is sharp because: (i) too far below the threshold only low-n states are populated and these either decay too rapidly or cannot be field ionized by the detection pulse, and, (ii) too close to the threshold all high-n states have been field-ionized by the weak field present during the waiting period.

The resolution of a ZEKE experiment can be selected by judicious choices of the electric field strengths applied during the waiting and detection periods. Careful minimization of stray electric fields, due either to charges on surfaces or a too high density of ions produced in the initial excitation step, permits the use of higher-n "ZEKE states", thereby enhancing the resolution of ZEKE spectra and the accuracy of derived ionization energies.

The sensitivity of ZEKE spectroscopy is high because, even though a small fraction of the molecules excited by the PROBE laser have n-values within the range capable of being ionized by the electric field detection pulse, the ions produced from these special "ZEKE states" are detected with 100% quantum efficiency against a perfectly dark background.

ZEKE spectroscopy can also be applied to anions (Zhao, *et al.*, 1994)

$$\mathrm{AB}^-(e'', v'', J'') + h\nu \rightarrow \mathrm{AB}(e', v', J') + e^- . \tag{1.2.13}$$

Unlike neutral molecule ZEKE, there are no Rydberg series of AB^- converging to photofragment thresholds that correspond to the ground electronic state of AB, thus true zero kinetic energy electrons are detected. Anion ZEKE can give valuable information about the relative energies of low-lying neutral molecule electronic states which, because of restrictive selection rules, cannot easily be obtained by spectroscopic study of the neutral molecule.

1.2.2.4 Photofragment and Coincidence Experiments

When a molecule in a well specified initial state (e'', v'', J'', M''_J, velocity $\vec{v}$) absorbs a photon and then falls apart

$$\mathrm{AB} + h\nu \rightarrow \mathrm{AB}^+ + e^- \qquad \text{photoionization} \tag{1.2.14}$$

$$\mathrm{AB} + h\nu \rightarrow \mathrm{A} + \mathrm{B} \qquad \text{photodissociation} \tag{1.2.15}$$

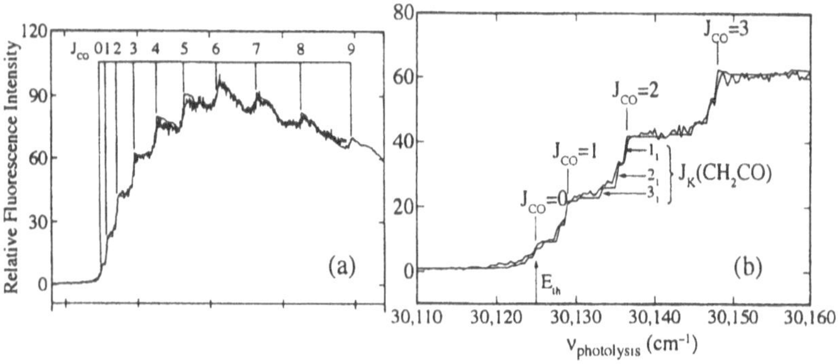

Figure 1.18: PHOFEX Spectra of Ketene $CH_2CO \rightarrow CH_2 + CO$. A photofragment fluorescence excitation (PHOFEX) spectrum is recorded by using laser induced fluorescence (LIF) to monitor the population of a specific rotation-vibration level of a photofragment while scanning the frequency of the photodissociation laser. The two spectra shown are obtained by monitoring the population in the CH_2 $\tilde{a}^1A_1$ $(v_1 = 0, v_2 = 0, v_3 = 0)$ $J_{K_aK_c} = 1_{01}$ rotational level via the $\tilde{b} - \tilde{a}$ $(0, 15, 0)$ $2_{11} \leftarrow (0, 0, 0)$ 1_{01} transition at 17757.76 cm^{-1} while scanning the photodissociation laser near the 30,125.0 cm^{-1} thermodynamic threshold for CH_2CO $\tilde{X}^1A_1(0, 0, \ldots 0)$ $0_{00} \rightarrow$ CO $X^1\Sigma^+(v = 0)$ $J = 0$ + CH_2 $\tilde{a}^1A_1(0, 0, 0)$ 0_{00}. Spectrum (a) shows that steps occur in the population of the monitored CH_2 rotational level every time a new photofragmentation channel (the next rotational level of CO) opens. The steps are spaced exactly by the known CO rotational spacing. Spectrum (b) shows the near threshold region where, in addition to the openings of successive CO rotational channels, a finer structure associated with the thermally populated $K_a = 1$ $J = 3, 2, 1$ excited rotational levels of ketene is observable. When the photofragmentation is initiated from a rotationally excited level of CH_2CO, the threshold for a particular pair of photofragment states (e.g., $J_{CO} = 2$, $J_{CH_2} = 1_{01}$) is lowered by the parent molecular rotational energy. The PHOFEX spectrum and a theoretical model are plotted respectively as heavy and light lines (from Chen, *et al.*, 1988).

information about the mechanism of the photofragmentation process (see Chapters 7 and 8) is obtained by measuring the angular distribution (Zare and Herschbach, 1963; Zare, 1972; Busch and Wilson, 1972) and/or the internal state distribution of either fragment or, in a *coincidence* experiment, various correlated properties of the two fragments produced by the same photofragmentation event.

A PHOtofragment Fluorescence EXcitation (PHOFEX) spectrum (Chen, *et al.*, 1988) is recorded by holding the MONITOR laser frequency fixed on a transition of a photofragment while scanning the frequency of the PROBE laser through a free←bound transition of the parent molecule (see Fig. 1.18). PHOFEX spectroscopy is particularly sensitive to the thresholds for producing various internal states of the photofragments.

An experimental setup identical to that for PHOFEX spectra can be used to measure correlations between two scalar (internal state population, kinetic energy) quantities or between a scalar and a vector quantity (angular distribution of fragment rotation $\vec{J}$ or velocity $\vec{v}$). By holding the PROBE laser frequency fixed and scanning the MONITOR laser through the Doppler profile of the fragment transition, one measures the fragment velocity distribution along the MONITOR laser propagation direction. The resulting transition profile, $\sigma(\nu)$, is typically far from Gaussian. The extreme red and blue edges of such a Doppler profile can be used to determine the total translational energy released in the AB+$h\nu \rightarrow$ A + B photofragmentation transition. The lineshape (maximum in the middle vs. two maxima near the edges) indicates whether the transition was of parallel ($\Delta\Omega = 0$) or perpendicular ($\Delta\Omega = \pm 1$) type, where parallel/perpendicular refers to the location of the electric dipole transition moment relative to the A $-$ B bond axis and Ω is the body-fixed projection of $\vec{J}$ onto the A $-$ B axis (see Sections 2.3 and 3.1.2.3) (Thoman, *et al.*, 1988; Buntine, *et al.*, 1991; Heck and Chandler, 1995).

Coincidence experiments provide the most detailed information about the photofragmentation process. Two particles that originate from the fragmentation of one parent molecule are detected, and the detection scheme sometimes records both temporal and spatial information about both photofragments. A segmented position-sensitive detector can be used to measure the arrival times and spatial locations of *two* particles on the detector. This information for one particle determines its vector velocity, $\vec{v}$, in the laboratory coordinate system. When such information is available for both particles that originate from the same event, (e.g., Photo-Ion Photo-electron COincidence, PIPECO (Tsal, *et al.*, 1975) or Velocity Imaging Photo-electron COincidence Spectroscopy (VIPCO) (Eland, *et al.*, 2000)) it is possible to compute the angular orientation and internuclear separation of the parent molecule in the laboratory frame at the instant of fragmentation. By collecting many such (A, B) or (AB^+, e^-) coincidence events, one determines the angular distribution of photofragments relative to the initial orientation of the parent (i.e., a partial wave analysis; s–wave is spherical, p–wave has one nodal plane, etc.). From this kind of information one learns the detailed nature of the forces acting between the two photofragments.

Coulomb Explosion methods are unique in spectroscopy because they, in principle, provide a direct image of the geometric structure of a molecule (Vager, 2001). This is accomplished without recourse to a quantum mechanical model by which geometric structure is inferred from the energy levels derived from spectroscopic data. The Coulomb explosion method is a coincidence experiment in which the times and positions of two (or more) atomic fragments arriving at a segmented position-sensitive detector are recorded. All of the (valence) electrons of the parent molecule are ejected at a precisely known time and location. This is accomplished using either a very intense picosecond pulse of UV radiation or by collision of a $>> 100keV$ molecular ion with a special foil. If the charge states of the fragments are known, then the ion arrival times and positions are sufficient to provide information about molecular geometry and orientation at the instant of the Coulomb explosion.

1.2.3 Molecule Sources

Most diatomic molecules are refractory, reactive, and heavy. Some are so fragile that room temperature collisions would destroy them. Molecules that dissociate into open shell atoms have many near degenerate electronic states at large internuclear separations. A variety of molecule sources exist which are ideally suited to obtaining high quality, readily analyzable spectra of most diatomic molecules.

Molecule sources may be divided into three types: thermal, supersonic jet, and cold collisions. Thermal sources (King furnace: King, 1926; Broida oven: West, *et al.*, 1975; electric or microwave discharge: Fehsenfeld, *et al.*, 1965; static gas cell) are simple, convenient, and versatile. However, when the rotational partition function, kT/B, approaches 10^3, each vibrational band will contain many hundreds of rotational lines, the spectrum will become congested, vibrational bands will overlap each other, and analysis will be difficult.

One of the most valuable characteristics of supersonic jet sources (Kantrowitz and Grey, 1950; Campargue, 1964; Anderson, *et al.*, 1966; Morse, 1996) is that they produce a beam of molecules that have been translationally, rotationally, and vibrationally cooled to a few Kelvin (Dyke, *et al.*, 1972; Smalley, *et al.*, 1975). Owing to the extremely short residence time ($\sim 1 - 10\mu s$) for molecules in the expansion region of the supersonic jet (Cross and Valentini, 1982; Gentry and Giese, 1978; Liu, *et al.*, 1996), where the molecules move from a high pressure region at $\gg 10^9$ collisions/sec to a collision free region, it is possible to combine jet cooling with a variety of extremely ungentle, molecule-synthesizing, external perturbations. These include photoablation, photolysis (Dietz, *et al.*, 1981; Morse, 1986; Monts, *et al.*, 1980), pyrolysis (Chen, *et al.*, 1986; Dunlop, *et al.*, 1988 and 1989), and discharge jet (Engelking, 1986; Ohshima and Endo, 1992; Anderson, *et al.*, 1996) sources.

In a photoablation jet (Dietz, *et al.*, 1981; Morse, 1986), a laser pulse is focussed onto a rotating solid rod or disc containing the atoms to be vaporized. The ablated material is entrained in an inert gas and swept through the expansion region. By controlling the residence time in the expansion region, the degree of clustering and/or the extent of reaction can be manipulated.

A photolysis jet (Monts, *et al.*, 1980) works by introducing a polyatomic parent molecule into the photolysis region just behind the jet orifice. After photolysis, the normally extremely reactive photolysis products are swept through the expansion region so quickly that their only collision partners are inert gas atoms.

In a pyrolysis jet, (Chen, *et al.*, 1986; Dunlop, *et al.*, 1988 and 1989) parent molecules are swept quickly through a very high temperature region just before exiting the jet. There is an art to the selection of a parent molecule optimal for production of the desired pyrolysis fragment (Chen, *et al.*, 1986). In a pyrolysis source, the weakest bond breaks first and subsequent reactive collisions are minimized.

In a discharge jet (Engelking, 1986; Ohshima and Endo, 1992; Anderson, *et al.*, 1996), a discharge is struck, usually between a pin electrode located just behind the jet exit orifice and the grounded body of the jet assembly, and the discharge causes a parent molecule to fragment, often into ions, in the expansion region. The correct choice of the polarity of a dc discharge relative to the molecule flow direction can be critical to the efficiency and pulse-to-pulse stability of the source (Anderson, *et al.*, 1996).

Molecules can be cooled to a few Kelvin by collisions with He gas at ~4 K (Ball and De Lucia, 1998), by being adsorbed onto large (superfluid) ^{4}He clusters ($\sim 10^6$ atoms) formed in a supersonic jet (Toennies and Vilesov, 1998; Stienkemeier, *et al.*, 1995; Goyal, *et al.*, 1992), or by being doped into crystalline para-H_2 (Oka, 1993). Although 4 K is well below the freezing point of all diatomic molecules, wall collisions are the primary loss mechanism from the gas phase for molecules injected into cold He gas. Molecules on ^{4}He clusters and in p-H_2 crystals typically exhibit sharp electronic-vibrational-rotational spectra because a large energy gap prevents exchange of energy between the molecule and the internal degrees of freedom of the bulk-quantum, condensed phase systems.

Atoms in an optical trap (Doppler cooling: Wineland, *et al.*, 1978; optical molasses: Chu, *et al.*, 1986; magneto-optic trap: Steane and Foot, 1991; Helmerson, *et al.*, 1992) are confined and cooled to translational temperatures on the order of $<< 1$ mK. "Ultracold collisions" between such trapped atoms permit the recording of bound←free spectra with resolution limited only by the translational temperature (1 mK, which corresponds to a frequency resolution of $\sim 7 \times 10^{-4}$ cm^{-1}) (Julienne and Mies, 1989; Lett, *et al.*, 1995; Burnett, *et al.*, 2002). This makes spectroscopically accessible the extremely long-range regions of potential energy curves ($R \gtrsim 10\text{Å} \approx 5R_e$) and otherwise only indirectly observable weakly bound or repulsive electronic states.

1.2.4 Frequency Calibration

Almost without exception, the usefulness of a spectrum is determined by the quality of its frequency calibration.

A spectrum contains an enormous quantity of information. This information can only be used when definitive feature-by-feature correspondences with a known spectrum can be established or, when the spectrum contains previously unobserved transitions, if the spectrum can be assigned. Some form of pattern recognition is essential. Every improvement in the accuracy of the frequency calibration will be rewarded by more facile identification and unambiguous assignment of the spectrum.

Different schemes are generally used to provide *relative* and *absolute* frequency calibration. Most relative frequency calibration schemes make use of some sort of étalon and most absolute calibration schemes are based on known frequencies of atomic (Harrison, 1969; Palmer, *et al.*, 1980; Palmer and Engleman, 1983) or molecular (I_2, Te_2, Br_2, small polyatomic molecules) transitions (Gerstenkorn and Luc, 1978, 1986; Cariou and Luc, 1980 and 1998; Gerstenkorn, *et al.*, 1985, 1993; Katô, 2000; Guelachvili and Rao, 1986; Guelachvili, *et al.*, 1996).

An étalon, reduced to its bare essentials, consists of two partially transmissive reflecting surfaces separated by a distance L. When a beam of monochromatic radiation is input to the étalon [see Demtröder (1996) pp. 140-152 for a discussion of Fabry-Perot étalons], and the frequency of this radiation is scanned, transmission through the étalon occurs whenever the input frequency, $\nu = c/\lambda$, is such that $nL = N(\lambda/2)$ where N is a large integer and n is the index of refraction of the material between the mirror surfaces (often air or quartz). The spacing between transmission maxima is the *free spectral range* of the étalon, $FSR = c/2nL$. Etalons used for relative frequency calibration range from a simple ~ 1 cm thick quartz substrate with $\sim$99% reflective surfaces to an elaborate, temperature and pressure stabilized assembly where the two reflective surfaces are on two separate substrates that can be positioned to achieve any desired *FSR*.

The comb of equally spaced (in cm^{-1}) transmission maxima from an étalon is useful as relative frequency markers to linearize the scan of a tunable laser. This makes it possible to interpolate between irregularly distributed absolute frequency standard lines (e.g., atomic lines, Fourier Transform Spectrometric Atlases of Molecular I_2, Br_2, Te_2, and small polyatomic molecules) and to measure frequency separations between features in the spectrum being recorded. Pattern recognition methods, such as Loomis-Wood (see Section 1.1.4) and searches for repeated combination differences, work best when applied to spectra in which the frequency separations between lines are determined at the highest possible precision (better than 1/10 the average line spacing).

Absolute frequency calibration is based on simultaneously recording the unknown spectrum and a spectrum of a suitable calibration species. The large number of calibration lines tabulated in atomic or molecular line atlases are in turn absolutely calibrated against a small number of even more accurately calibrated primary standard lines (Palmer, *et al.*, 1981; Diddams, *et al.*, 2000).

The accuracy of one primary standard (the frequency of a stabilized He-Ne laser) may be transferred to the unknown frequency of a cw tunable laser using a scanning wavemeter. The mirror spacing of an interferometer (étalon) is varied approximately linearly in time and the frequencies at which transmission peaks (fringes) of a He-Ne laser and an unknown laser occur are recorded simultaneously (Lee and Hall, 1976; Kowalski, *et al.*, 1976). The ratio of the reference and unknown fringe rates provides the ratio of the known and unknown frequencies. Other sorts of wavemeters exist which are capable of measuring the frequency of a pulsed laser. A Fizeau wavemeter (Morris, *et al.*, 1984) measures the spectral content of a single laser pulse by recording on a diode array the interference fringes produced by a parallel beam incident on a slightly wedged pair of quartz plates. The wavelength is determined to ~ 1 part in 10^7 from the combination of the average fringe spacing and the absolute fringe positions on the diode array.

A Fourier transform spectrometer is also a scanning wavemeter. It transfers the absolute frequency calibration of a reference laser to the frequencies of all lines in the FT spectrum. The I_2, Te_2, Br_2, and H_2O line atlases, with absolute accuracy on the order of 0.001cm^{-1}, are measured by FT spectroscopy relative to the more accurately known frequency of the He-Ne laser.

1.2.5 Detection Techniques

The days are long gone when all one needs to specify about a molecular spectrum is whether it is recorded in absorption or emission. Any list of important detection schemes will be incomplete. Any scheme for arranging detection schemes into categories will be rigid and arbitrary.

A spectrum may be detected *directly* by measuring the wavelength-dependent absorption or emission of electromagnetic radiation or *indirectly* by monitoring particles (photofragments, parent molecules, the energy content of parent molecules), transient changes in the bulk properties of a gas (conductivity, temperature, density, index of refraction), or changes in the wavevector or polarization state of probe beam(s) caused by transient spatial periodicities/anisotropies in the irradiated sample. Indirect schemes based on particle detection include Resonance Enhanced Multi-Photon Ionization (REMPI, detection of photo-ions (Johnson, *et al.*, 1975)), REMPI-Time-of-Flight (REMPI-TOF, detection of mass selected photo-ions), ZEKE (detection of "zero kinetic energy" photo-electrons (Müller-Dethlefs and Schlag 1991; Merkt 1997)), photoelectron spectroscopy (measurement of the kinetic energy spectrum of photoelectrons), Surface Electron Ejection by Laser Excited Metastable molecules (SEELEM): when molecules in a long-lived electronically excited state impact a metal surface, an electron is ejected provided that the vertical electronic excitation energy of the molecule exceeds the work function of the metal (Lichten, 1957 and 1960; Freund and Klemperer, 1965; Hemminger, *et al.*, 1976; Snè and Cheshnovsky, 1986); H atom translational spectroscopy (Krautwald, *et al.*, 1986; Segall, *et al.*, 1989), optothermal spectroscopy (the rate of energy transport by a molecular beam to a liquid helium cooled bolometer is altered (Gough, *et al.*, 1977)). Indirect schemes based on the bulk properties of the irradiated sample include photoacoustic spectroscopy: an acoustic wave is generated when some of the photoexcitation energy is converted by inelastic collisions to translational energy (Angus, *et al.*, 1975; Pao 1977); optogalvanic spectroscopy (transient change in the conductivity of a discharge caused by altering the number density of charge carriers through their equilibrium with high Rydberg species (Penning, 1928; Smyth, *et al.*, 1979; Pfaff, *et al.*, 1984)); and thermal lensing spectroscopy (Long, *et al.*, 1976). Indirect schemes based on transient spatial periodicities/anisotropies include polarization spectroscopy (Teets, *et al.*, 1976) and transient grating four-wave mixing schemes (Zhang, *et al.*, 1992; Buntine, *et al.*, 1992).

A direct absorption experiment is an example of a "flop-out" detection scheme (Ramsey, 1956). The usually small signal is detected on top of a usually much larger background. An LIF experiment is an example of a "flop-in" detection scheme. The signal is detected against an essentially dark background. The signal to background ratio is usually much larger for a flop-in than flop-out detection scheme. The main reason for this is that the noise associated with a "constant" background, N_b, is usually much larger than the shot noise limit, $N_b^{1/2}$, (1mW at 15,000 cm^{-1} is $\sim 3 \times 10^{15}$ photons/s, $N_b^{1/2} \sim 5 \times 10^7$, or $N_b^{1/2}/N_b \sim 2\times10^{-6}$ in 1 Hz bandwidth) (Moore, *et al.*, 1989; pp. 458-463). Typical amplitude noise for lasers is $\sim 1\%$ for a cw laser and $\sim 5\%$ pulse-to-pulse for nanosecond lasers. "Excess" or technical noise in most flop-out detected laser experiments is typically more than 10^4 times larger than the intrinsic or quantum limited noise level. A very effective way to approach the shot noise limit for the signal (on-resonance) to background (off-resonance) ratio is to arrange for a nearly perfect null of the detected signal when the tunable laser frequency is off-resonance. Magnetic Rotation Spectroscopy (MRS) (Righi, 1898; Buckingham and Stephens, 1966; Litfin, *et al.*, 1980; Hinz, *et al.*, 1982) and Frequency Modulation (FM) spectroscopy (Bjorklund, 1980; Hall, *et al.*, 1981; Gehrtz, *et al.*, 1985; Hall and North, 2000) are single-null schemes that permit absorption spectra to be recorded in flop-in rather than flop-out mode.

When a continuously tunable laser is used to record an LIF spectrum, the detection is "single channel." When a dispersed fluorescence spectrum is recorded using a CCD (Charge Coupled Device) camera (or a Fourier transform spectrometer), many resolution elements are recorded simultaneously and the detection is "multiplexed." This offers a considerable sensitivity advantage and also minimizes errors in measurements of relative intensities. Although spectroscopists have traditionally treated intensities as having importance secondary to that of frequencies, intensity information is valuable in characterizing dynamics and in the use of pattern-recognition schemes to reveal both spectroscopic assignments and dynamical mechanisms.

Sometimes a spectrum is recorded subject to temporal or spectral restrictions. As a laser is scanned through a complicated spectrum, features in the spectrum associated with different electronic states might be separated by time-gating (Pruett and Zare, 1975) and/or spectral-filtering schemes. A time-gated LIF spectrum of a long-lived upper state might be recorded free of interference from a short-lived upper state using a gated detector [Boxcar (Moore, *et al.*, 1989; p. 461)] to select only fluorescence that occurs in a late gate. LIF spectra associated with specific upper state vibrational and even rotational levels may be selectively detected using a monochromator to accept fluorescence only within a narrow spectral window (Nakagawa, *et al.*, 1983).

These temporally and/or spectrally selected spectra permit otherwise overlapping spectra to be disentangled. *Coincidence* schemes (see Section 1.2.2.4) carry this one step further. When two photofragments from the *same single molecule fragmentation event* are detected, information is obtained about the mechanism of the fragmentation or the geometric structure of the parent species. Coincidence measurements provide vector and vector-correlation information. The pre-fragmentation center-of-mass molecular frame orientation, the time-evolution of the forces between and relative orientations of the departing fragments (including fragments that are not detected) may be inferred from coincidence experiments. Formerly hypothesized cartoons of forces turning on suddenly as a curve crossing is traversed can be experimentally verified.

1.2.6 Radiation Sources

Perhaps the most important characteristic of a source of electromagnetic radiation for molecular spectroscopy is whether it is intended for use in a photons-as-bullets, molecules-as-targets *incoherent* experiment or a radiation-as-waves, molecules-as-nonlinear-wave-mixing-medium *coherent* experiment. As gas phase, small molecule, electronic spectroscopy was born in the pre-laser age when coherence-based experiments demanded exceptional ingenuity, it is not surprising that until very recently the bullets/targets incoherent picture has been dominant. This picture has had an important influence on the actual experiments performed and language used by spectroscopists to describe and design experimental schemes.

Aside from the possibility conferred by lasers of producing electromagnetic radiation of extremely narrow spectral width ($< 0.001\text{cm}^{-1}$) or extremely short temporal duration (~ 30fs), the coherence-relevant characteristics of longitudinal (frequency) and transverse (spatial) mode structure and the optical phase of laser radiation are often ignored. The coherence properties of laser radiation are not exploited in the vast majority of molecules/targets spectroscopic experiments. Lasers are treated as if they were sources of incoherent radiation, yet this radiation can possess uniquely useful properties of monochromaticity, intensity, and smooth (and often short) temporal envelope. Neglect of coherence properties can lead to some surprises, but effects not explainable by the bullets/targets picture are seldom evident in experiments not explicitly designed to exploit the coherence of the laser source.

Truly incoherent white light sources convenient for cw and transient absorption spectroscopy include cw, high pressure arc lamps ($\sim 2\ kW$, spectral irradiance $\approx 3\ mW/(\text{cm}^{-1} \cdot \text{cm}^2 \cdot \text{sr})$) and flashlamps ($\sim 1\ J, \sim 10\ \mu s$). Note that the spectral irradiance of a 1 mW He-Ne laser ($\Delta\nu \approx 1$ MHz, $d\Omega = 10^{-6}$sr, beam waist $w = 0.1$ cm) is $\sim 3 \times 10^9$ W/($\text{cm}^{-1} \cdot \text{cm}^2 \cdot$ sr), approximately 10^{12} times larger than that of the most robust incoherent cw radiation source.

Atomic discharge lamps are spectrally narrow ($\sim 0.1\ \text{cm}^{-1}$, Doppler broadened) radiation sources, sometimes accidentally coincident with the frequency of a molecular transition (Elander and Neuhaus, 1974; Dagnall and West, 1968; Budick, *et al.*, 1965). Atomic discharge lamps, in combination with filters (cold gas filter, colored glass filter, narrow band pass interference filter) selected to block unwanted atomic emission lines, are often used to excite selectively fluorescence via a single molecular absorption line. Excitation schemes based on accidental coincidences between atomic and molecular transitions were the basis for several historic pre-laser experiments in molecular spectroscopy. These include the recording of dispersed fluorescence spectra of I_2 (Steinfeld, *et al.*, 1965) (which in turn permitted the initial quantitative tests of Franck-Condon factors (Zare, 1964) computed from potential energy curves derived via the RKR method from vibrational, $G(v)$, and rotational, $B(v)$, constants), Hanle Effect (Hanle, 1924) and Level Crossing spectra (Zare, 1966 and 1971; McClintock, *et al.*, 1969) of Na_2, and optical-radio frequency double resonance spectra of OH (German and Zare, 1969) and CS (Silvers, *et al.*, 1970).

Atomic lamps are often very compact and rugged. This, combined with their monochromaticity, is a reason for their continued use in spectroscopy, in preference to more versatile and powerful tunable laser sources. Commercially available hollow cathode dc discharge lamps (Elander and Neuhaus, 1974; Dawson and Ellis, 1966) are convenient sources of emission lines from neutral and singly positively charged refractory metal atoms. Pen lamps (inert gases, Hg), ac discharges through a quartz capillary, are convenient for wavelength calibration of a grating spectrograph because the light from a pen lamp is easily imaged onto the spectrograph entrance slit. Electrodeless discharge sources (Dagnall and West, 1968; Davis and Braun, 1968), which consist of a source of microwave radiation (500W, 2450 MHz magnetron from a microwave oven), a tuned Evenson-Broida cavity (Fehsenfeld, *et al.*, 1965), and a sealed 1 cm diameter quartz tube containing some inert gas and a small quantity of the element from which the atomic emission is sought are convenient, inexpensive, and rugged.

The following classes of laser-based sources of electromagnetic radiation are briefly described here: (i) spectrally narrow, continuously tunable lasers; (ii) nonlinear wave-mixing processes such as Raman shifting, second harmonic generation (SHG), sum and difference frequency generation (SFG, DFG), optical parametric oscillation and amplification (OPO and OPA), and four-wave mixing (FWM); and (iii) soft X-ray and continuum generation by excitation with ultrashort pulse.

(i) *Spectrally narrow, continuously tunable lasers.* Examples include dye lasers, Titanium:Sapphire lasers, and diode lasers (see Section 5.7 of Demtröder, 1996). The gain media for dye and Ti:Sapphire lasers must be optically pumped, either by flashlamps or another cw or pulsed laser (e.g. Ar^+ ion laser, harmonics of a Nd:YAG laser, excimer laser). The gain medium is capable of supporting laser oscillation over a wide spectral region. Tuning is accomplished by inserting into the laser cavity one or more optical components (étalon, birefringent filter, grating, prism) that have high loss (typically by causing radiation to exit the cavity), except at the wavelength desired. Continuous tuning (avoiding discontinuous jumps between successive longitudinal modes of the laser cavity, spaced by $c/2\langle n \rangle L$, where c, $\langle n \rangle$, and L are respectively the speed of light, the average index of refraction seen by the radiation inside the laser cavity, and the length of the cavity) is achieved by continuously adjusting the optical length of the cavity, $\langle n \rangle L$, as the laser frequency is scanned. In order to tune the frequency of a tunable laser several nested tuning elements must be scanned in synchrony.

Energy is supplied to the gain medium of a diode laser by passing an electric current through the diode. The lasing frequency is tuned by controlling the temperature of the diode or by varying the current. Once again, continuous tuning requires continuous adjustment of the cavity length, thus at least one of the cavity end mirrors must be external to the diode.

Continuously tunable laser systems are complex, many-component systems. In a typical research laboratory, far more time is often spent aligning and trouble-shooting the laser system than in recording spectra.

(ii) *Nonlinear Wave-Mixing Processes.*

A variety of nonlinear wave-mixing schemes exist (see Section 5.8 of Demtröder, 1996, and Sections 7.1-7.3 of Levenson and Kano, 1988) by which laser-like beams of electromagnetic radiation may be generated at any desired wavelength, from the soft X-ray to the far infrared regions. "Laser-like" means that properties of the generated radiation (spatial mode, beam-waist and confocal parameter, spectral irradiance, coherence length, and optical phase) are calculable from corresponding properties of the generating laser sources.

Conservation of momentum, $\vec{p}$, or wave vector, $\vec{k}$,

$$\vec{p}(\omega) = \hbar\vec{k}(\omega) \tag{1.2.16}$$

is a central consideration in describing all wave-mixing processes. $\vec{k}(\omega)$ is a vector of magnitude

$$|k(\omega)| = n(\omega)\omega/c, \tag{1.2.17}$$

which points in the propagation direction of the beam with angular frequency ω, $n(\omega)$ is the index of refraction at ω of the medium, and c is the speed of light. The phase velocity , $v_\phi(\omega)$, of this beam is

$$v_\phi(\omega) = c/n(\omega) = \omega/|k(\omega)|. \tag{1.2.18}$$

If two beams of radiation with frequencies ω_1 and ω_2 are to be combined to generate radiation at ω_3, then momentum conservation requires

$$\vec{k}_1(\omega_1) \pm \vec{k}_2(\omega_2) = \vec{k}_3(\omega_3) \tag{1.2.19}$$

and energy conservation requires

$$\omega_1 \pm \omega_2 = \omega_3 \tag{1.2.20}$$

In some nonlinear wave-mixing schemes, all energy and momentum remains in the radiation field, but in others (e.g., Stokes Raman shifting, optically pumped lasers) some energy and momentum is exchanged between the radiation field and the material medium.

Certain crystals (those in which the unit cell lacks inversion symmetry) enable a variety of useful *three-wave mixing* processes: Second Harmonic Generation (SHG), $\omega_3 = \omega_1 + \omega_1$; Sum Frequency Generation (SFG), $\omega_3 = \omega_1 + \omega_2$; Difference Frequency Generation (DFG), $\omega_3 = \omega_1 - \omega_2$; and Optical Parametric Oscillation (OPO), $\omega_1 = \omega_2 + \omega_3$. In SHG, for example, energy and momentum (wavevector) conservation requires, for colinear propagation,

$$2n(\omega_1)\omega_1 = n(2\omega_1)2\omega_1$$

or

$$n(\omega_1) = n(2\omega_1). \tag{1.2.21}$$

Since the index of refraction of all materials is ω–dependent, the phase-matching requirement of Eq. (1.2.21) is not automatically satisfied. Failure to phase-match means that the amplitudes of the $2\omega_1$ wave generated at z and $z + dz$ along the $2\omega_1$ propagation direction will not add constructively.

Uniaxial (birefrigerent) crystals have an index of refraction that depends on the projection of the oscillating electric field of linearly polarized radiation on the unique axis (the optic axis) of the crystal. Consider the plane defined by the optic axis of the uniaxial crystal, $\vec{c}$, and the propagation direction of the ω_1 beam, which is described by the unit vector, $\hat{k} = \vec{k}(\omega_1)/|k(\omega_1)|$. Light polarized perpendicular to the $\vec{c}, \hat{k}$ plane is called the ordinary ray and propagates according to the ordinary index of refraction,

$$n_{\perp}(\omega) = n_0(\omega). \tag{1.2.22}$$

Light polarized in the $\vec{c}, \hat{k}$ plane propagates with an index of refraction that is dependent on the projection of the polarization direction on $\vec{c}$

$$n_{\|}(\omega, \theta) = \left[\frac{\cos^2\theta}{n_0^2(\omega)} + \frac{\sin^2\theta}{n_e^2(\omega)}\right]^{-1/2} \tag{1.2.23}$$

where θ is the angle between $\vec{c}$ and $\hat{k}$. When $\theta = \pi/2$, light polarized perpendicular to the $\vec{c}, \hat{k}$ plane is polarized parallel to the unique crystal axis and propagates according to n_e, the extraordinary index of refraction. "Type I" phase matching is achieved by selecting θ (tilting the crystal to adjust the projection of $\hat{k}$ on $\vec{c}$ while maintaining the polarization of ω_1 perpendicular to the $\hat{k}, \vec{c}$ plane) so that

$$n_{\perp}(\omega_1) = n_0(\omega_1) = n_{\|}(2\omega_1, \theta). \tag{1.2.24}$$

This results in a $2\omega_1$ beam that propagates in the same direction as the ω_1 beam but with linear polarization perpendicular to that of the ω_1 beam (Levenson and Kano, 1988, pp. 249-253).

If the crystal is close to, but not exactly at, the phase matching orientation relative to $\vec{k}(\omega_1)$, then a weak $2\omega_1$ beam will be generated propagating at a small angle relative to $\vec{k}(\omega_1)$. Minimization of this deviation angle provides a basis for servo-correction of the crystal orientation to maximize SHG. It also ensures that, as ω_1 is scanned, the $2\omega_1$ beam pointing direction is constant, which is good news for multiple resonance schemes that require careful superposition of several beams of laser radiation.

For sum and difference frequency generation, the ω_1 and ω_2 beams are superimposed using a dichroic mirror. The phase matching condition is

$$n(\omega_1 \pm \omega_2) = \frac{n(\omega_1) + n(\omega_2)}{2} + \frac{n(\omega_1) - n(\omega_2)}{2}\left(\frac{\omega_1 \mp \omega_2}{\omega_1 \pm \omega_2}\right) \tag{1.2.25}$$

where the second term is usually neglected. As for SHG, in Type I SFG and DFG, the $\omega_1 \pm \omega_2$ beam propagates in the same direction as the ω_1 and ω_2

beams, $\hat{k}$, the ω_1 and ω_2 beams are linearly polarized perpendicular to the $\hat{k}, \vec{c}$ plane, and the $\omega_1 \pm \omega_2$ beam is linearly polarized in the $\hat{k}, \vec{c}$ plane (polarized perpendicular to the ω_1, ω_2 beams).

The wavelength tuning range accessible via mixing in a particular crystal type is limited by absorption, and by the strength and wavelength dependence of the birefrigence $(n_e - n_O)$. The tuning range accessible using a particular crystal (i.e., the angle of the entrance face relative to the optic axis) is limited for SHG by clipping of the beam at the edge of the crystal and, for SFG and DFG, by walk-off of the ω_1, ω_2, and ω_3 beams owing to the n–dependent parallel displacements of the beams. These displacements increase as the angle of incidence on the entrance face of the crystal decreases from 90°.

Optical Parametric Oscillation is closely related to DFG and SFG. The input radiation, ω_1, is split into two frequencies

$$\omega_{\text{pump}} = \omega_{\text{signal}} + \omega_{\text{idler}} \tag{1.2.26}$$

where $\omega_{\text{signal}} > \omega_{\text{idler}}$. The names *signal* and *idler* are of traditional origin and have no special significance except that the properties (frequency, wavevector, mode structure) of the idler beam are completely determined by those of the pump and signal beams.

The ω_{signal} and ω_{idler} frequencies are tuned simultaneously by varying the phase matching condition (angle tuning or temperature tuning). By surrounding the crystal in a laser cavity, which is low-loss for either ω_{signal} or ω_{idler} (or both), the efficiency of the parametric splitting of ω_{pump} into ω_{signal} and ω_{idler} is maximized. Alternatively, the parametric process may be seeded (Optical Parametric Amplifier, OPA) by superimposing beams of radiation at ω_{pump} and ω_{signal} or ω_{idler}. However, even when the optical parametric process is seeded, the phase matching condition

$$\vec{k}(\omega_{\text{pump}}) = \vec{k}(\omega_{\text{signal}}) + \vec{k}(\omega_{\text{idler}}) \tag{1.2.27}$$

$$n(\omega_{\text{pump}}) \approx \frac{1}{2}\left[n(\omega_{\text{signal}}) + n(\omega_{\text{idler}})\right] \tag{1.2.28}$$

must be satisfied.

Nonlinear wave-mixing processes in condensed phase media are limited in spectral access by absorption and in efficiency by material damage thresholds. Nonlinear mixing processes in gases are not fundamentally restricted by absorption and material damage, but the isotropy of space (gases in field-free space are not birefringent) prohibits three-wave-mixing processes.

A wide variety of *four-wave-mixing* processes are important in the generation of laser-like beams of electromagnetic radiation and also in phase-coherent interrogation of the structural and dynamical properties of molecules. Since three beams of radiation, each with its own frequency, propagation direction, and polarization state, are combined to generate a fourth beam,

$$\vec{k}(\omega_1) + \vec{k}(\omega_2) + \vec{k}(\omega_3) = \vec{k}(\omega_4), \tag{1.2.29}$$

many possible mixing schemes exist (Wasserman, *et al.*, 1998). When a gas phase atom or molecule acts as the nonlinear medium in which the four-wave mixing occurs, strongly frequency dependent, anisotropic polarizabilities profoundly expand the range of possibilities for four-wave mixing processes.

Two four-wave mixing processes are briefly considered here: Vacuum Ultra-Violet (VUV) generation by frequency tripling in a supersonic jet of an inert gas (Hollenstein, *et al.*, 2000) and anti-Stokes Raman shifting in high pressure H_2 gas.

Since the index of refraction in an atomic inert gas at a frequency far from an absorption transition is very close to 1.00, the phase matching scheme for the frequency tripling process,

$$3\vec{k}(\omega_1) = \vec{k}(3\omega_1) \tag{1.2.30}$$

$$3n(\omega_1)\omega_1 = n(3\omega_1)3\omega_1$$

$$n(\omega_1) \approx n(3\omega_1), \tag{1.2.31}$$

is colinear. The $3\omega_1$ beam copropagates with the ω_1 beam. By exploiting a pair of supersonic jets in a single molecular beam apparatus, one jet of inert gas for generating the VUV radiation and another jet containing the molecule of interest, tunable VUV spectra may be recorded at wavelengths shorter than the transmission limit of the shortest wavelength window material (LiF, $\lambda_{\text{cutoff}} \approx$ 105 nm). The efficiency of the tunable VUV generation process may be increased by selecting one fixed frequency, ω_1, to be near a two-photon resonance [e.g., for an inert gas atom, $2\hbar\omega_1 \approx E(np^5(n+1)p\ {}^1D \leftarrow np^6\ {}^1S)$] in combination with one tunable frequency, ω_2, to generate $\omega_{\text{VUV}} = 2\omega_1 \pm \omega_2$ (Vidal, 1986).

Anti-Stokes Raman shifting in high pressure H_2 gas is a four-wave mixing process

$$2\vec{k}(\omega_1) = \vec{k}(\omega_{\text{Stokes}}) + \vec{k}(\omega_{\text{anti-Stokes}}) \tag{1.2.32}$$

$$\omega_{\text{Stokes}} = \omega_1 - \omega_v \tag{1.2.33}$$

$$\omega_{\text{anti-Stokes}} = \omega_1 + \omega_v, \tag{1.2.34}$$

where ω_v is the energy separation between the $v = 1$ and 0 levels of H_2,

$$\omega_v = 4155\ \text{cm}^{-1}.$$

The H_2 gas acts as a nonlinear medium that converts two ω_1 photons into one photon each at the Stokes and Anti-Stokes frequencies. Since the requirement for colinear phase matching

$$n(\omega_1) = \frac{1}{2}[n(\omega_1 + \omega_v) + n(\omega_1 - \omega_v)] \tag{1.2.35}$$

is not exactly satisfied (the difference of gas phase $n(\omega_1)$ values from 1.00 is proportional to pressure, and the pressure in a Raman-shifting H_2 cell is several atmospheres, much higher than the pressure used in supersonic jet VUV-generation schemes), Anti-Stokes radiation is generated on a diverging cone of small ($\sim 1^\circ$) apex angle. It turns out that the Stokes radiation is primarily generated colinear with the ω_1 radiation. The reason for this, discussed below, is that the primary mechanism for generating the Stokes radiation, Stimulated Raman gain, is not a four-wave mixing process.

One characteristic of nonlinear wave-mixing processes is that they are nonlinear! Once a lowest-order nonlinear process is brought above threshold, a cascade of higher-order processes often results. For example, the output from a Raman shifter typically contains several higher-order Stokes and Anti-Stokes Frequencies. Significantly, the teeth in the resultant comb of high order Stokes and Anti-Stokes frequencies differ by multiples of the $v = 1-0$ energy difference, not by the actual, anharmonically shifted vibrational overtone frequencies (e.g., the $2-0, 3-1, \ldots 3-0, 4-1$ frequencies are $2\omega_v - 4\omega_e x_e, 2\omega_v - 8\omega_e x_e, \ldots 3\omega_v - 10\omega_e x_e, 3\omega_v - 16\omega_e x_e$).

The H_2 molecule acts as a nonlinear medium for stepwise addition and subtraction of the $v = 1 - 0$ frequency to ω_1. Since the four-wave mixing process conserves both the energy and momentum of the radiation field, no net excitation of the nonlinear medium results.

$v = 0 \to v = 1$ excitations (as well as excitations to higher vibrational levels) result from processes that exchange energy and momentum between the material medium and the radiation field. Although laser-like beams of radiation are often generated by such processes, the spectral content and phase of the generated radiation is not as completely determined as for radiation generated by phase-matched four-wave mixing processes.

The first-order Stokes radiation is generated by a Stimulated Raman gain process. A perhaps over-simplified representation is the exchange of one-photon at ω_1 for one at $\omega_{\text{Stokes}} = \omega_1 - \omega_v$. If the ω_{Stokes} wave propagates in the same direction as the ω_1 wave, each exchange of one photon at ω_1 for one at ω_{Stokes} transfers $hc\omega_v$ energy and $+h\omega_v$ momentum to one molecule. However, even when the ω_1 radiation field is monochromatic, the ω_{Stokes} field is Doppler broadened. A molecule with velocity component v_z along the radiation propagation direction contributes a frequency

$$\omega_{\text{Stokes}} = \omega_{\text{Stokes}}(0) + \omega_v(v_z/c) \tag{1.2.36}$$

to the ω_{Stokes} wave. The Doppler width for the Stimulated Raman radiation is smaller than that for spontaneous Raman scattering by a factor of $\omega_v/\omega_{\text{Stokes}}$. If the Stimulated Raman gain medium is placed inside of a linear laser cavity, adjustment of the cavity length causes the frequency of the ω_{Stokes} radiation to tune continuously over an appreciable fraction of the Doppler width. However, in a phase-matched four-wave mixing scheme, there is no Doppler contribution to the spectral width of the generated wave.

The Stimulated Raman laser bears a superficial resemblance to an optically pumped gas phase molecular laser, but with one crucial difference. The Optically Pumped Laser (OPL) is a three level system (Koffend and Field, 1977). The PUMP transition is $e'v'J' \leftarrow e''v''J''$ and the lasing transition is $e'v'J' \to evJ$. If the pump laser is monochromatic, it selects a particular velocity in the $e'v'J'$ upper laser level. Thus there is a negligible Doppler width contribution to the gain bandwidth of the lasing transition (Koffend, *et al.*, 1978; Schweda, *et al.*, 1982). In order for an optically pumped laser to oscillate, the length of the laser cavity must be exactly an integer number of half wavelengths of the lasing transition. To tune an optically pumped laser over the normal Doppler lineshape of the $e'v'J' \rightarrow evJ$ lasing transition, both the pump frequency and the laser cavity length must be tuned in synchrony. The crucial differences between a three level laser system (an OPL) and a two level laser system (a stimulated Raman laser) are in the Doppler contribution to the gain bandwidth and how the frequency of a monochromatic pump laser (ω_1 or ω_{pump}) affects the frequency and efficiency of the output radiation (ω_{Stokes} or ω_{OPL}).

(iii) *Soft X-ray and continuum generation by excitation with ultrashort pulses.* Highly nonlinear processes demand extremely intense electromagnetic fields. Modern ultrashort ($\sim$ 30 femtoseconds) pulsed lasers offer the possibility of high peak field strength and pulse-to-pulse reproducibility (intensity, mode quality, spectral content) vastly superior to what had been available from nanosecond and picosecond lasers. Consequently, many high-order nonlinear processes are being put to use as standard, widely accessible as tools for spectroscopy (e.g., Nugent-Glandorf, *et al.*, 2002). Two examples are generation of soft X-rays by cascaded four-wave mixing (Zhou, *et al.*, 1996) and generation of ultrashort white light continuum pulses (Alfano, 1989).

Nugent-Glandorf, *et al.*, (2002) report an X-ray photoelectron spectrometer based on generation, in a pulsed jet of Ne, of up to the 65th harmonic of 800 nm (112 eV) from a Ti:sapphire laser (2 mJ, 80 fs). The high harmonic pulses generated in Ne contain typically 10^6 photons/pulse. The seemingly contradictory phase matching requirements to achieve a long cascade of four-wave mixing processes [$3\omega_1 = \omega_3, \omega_3 + 2\omega_1 = \omega_5, \omega_5 + 2\omega_1 = \omega_7$ and $2\omega_3 + \omega_1 = \omega_7$,] are met by focussing the ω_1 beam into the pulsed Ne (or Ar) jet. This provides a sufficiently wide range of input angles to satisfy simultaneously all of the phase matching requirements.

Ultrashort white light "supercontinuum" pulses (Alfano, 1989) are generated by focussing a $\sim 1\ mJ$, $\sim 100\ fs$ pulse into liquid H_2O. As is typical when a very high intensity pulse is tightly focussed into a condensed phase material, an out of control, usually destructive, sequence of nonlinear processes (self focussing, self phase-modulation, Coulomb explosion) ensues. The input pump pulse is efficiently converted into a short pulse of supercontinuum radiation emanating from the $\sim (10\ \mu m)^3$ focal volume. The term supercontinuum is used, because the spectral bandwidth far exceeds the Fourier transform of the pulse duration. This supercontinuum radiation may be used to probe a dynamical process at a precisely determined time after the process is initiated by an ultra short pump pulse. Even though the white light probe pulse is of short duration ($30\ fs \rightarrow\sim$ 500 cm^{-1}), the transient absorption spectral information written on it when it traverses the sample may be read, at considerably higher spectral resolution than the Fourier transform limit of the pulse duration, by a grating spectrograph equipped with a multichannel CCD detector.

1.3 References

Abrams, R. L., and Lind, R. C. (1978), *Opt. Lett.* **2**, 94.

Alfano, R. R. (editor), (1989), "The Supercontinuum Laser Source", Springer, New York.

Al-Khalili, A., Hällsten, U., and Launila, O. (1999), *J. Mol. Spectrosc.* **198**, 230.

Anderson, J. B., Andres, R. P., and Fenn, J. B. (1966), *Adv. Chem. Phys.* **10**, 275.

Anderson, D. T., Davis, S., Zwier, T. S., and Nesbitt, D. S. (1996), *Chem. Phys. Lett.* **258**, 207.

Angus, A. M., Marinero, E. E., and Colles, M. J. (1975), *Opt. Commun.* **14**, 233.

Ashfold, M. N. R., and Howe, J. D. (1994), *Annu. Rev. Phys. Chem.* **45**, 57.

Bacis, R., Churassy, S., Field, R. W., Koffend, J. B., and Vergès, J. (1980), *J. Chem. Phys.* **72**, 34.

Ball, C. D., and De Lucia, F. C. (1998), *Phys. Rev. Lett.* **81**, 305.

Bernath, P. F., and McLeod, S. (2001), *J. Mol. Spectrosc.* **207**, 287 (http://diref.uwaterloo.ca).

Bjorklund, G. C. (1980), *Opt. Lett.* **5**, 15.

Brugh, D. J., Ronningen T. J., and Morse, M. D. (1998), *J. Chem. Phys.* **109**, 7851.

Brugh, D. J., and Morse, M. D. (2002), *J. Chem. Phys.* **117**, 10703.

Buckingham, A. D., and Stephens, P. J. (1966), *Annu. Rev. Phys. Chem.* **17**, 399.

Budick, B., Novick, R., and Lurio, R. (1965), *Appl. Opt.* **4**, 229.

Buntine, M. A., Baldwin, D. P., Zare, R. N., and Chandler, D. W. (1991), *J. Chem. Phys.* **94**, 4672.

Buntine, M. A., Chandler, D. W., and Hayden, C. C. (1992), *J. Chem. Phys.* **97**, 707.

Burnett, K., Julienne, P. S., Lett, P. D., Tiesinga, E., and Williams, C. J. (2002), *Nature*, **416**, 225.

Busch, G. E., and Wilson, K. R. (1972), *J. Chem. Phys.* **56**, 3638.

Campargue, R. (1964), *Rev. Sci. Instrum.* **35**, 111.

Cariou, J., and Luc, P. (1980 and 1998), "Atlas du Spectre d'Absorption de la Molecule de Tellure: 18,500-21,200 cm^{-1} and 21,000-23,500 cm^{-1}," Laboratoire Aimé Cotton, CNRS II, 91405, Orsay.

Chen, I.-C., Green, W. H., and Moore, C. B. (1988), *J. Chem. Phys.* **89**, 314.

Chen, P., Colson, S. D., Chupka, W. A., and Berson, J. A. (1986), *J. Phys. Chem.* **90**, 2319.

Chu, S., Bjorkholm, J. E., Ashkin, A., and Cable, A. (1986), *Phys. Rev. Lett.* **57**, 314.

Chupka, W. (1993), *J. Chem. Phys.* **98**, 4520.

Cooper, D. E., Klincak, C. M., and Wessel, J. E. (1981), *Phys. Rev. Lett.* **46**, 324.

Cossart, D., Launay, F., Robbe, J.-M., and Gandara, G. (1985a), *J. Mol. Spectrosc.* **113**, 142.

Cossart, D., Cossart-Magos, C., Gandara, G., and Robbe, J.-M., (1985b), *J. Mol. Spectrosc.* **109**, 166.

Cossart, D., Bonneau, M., and Robbe, J.M., (1987) *J. Mol. Spectrosc.* **125**, 413.

Cossart, D., and Cossart-Magos, C. (1990), *J. Mol. Spectrosc.* **141**, 59.

Cossart, D. (1994), *J. Mol. Spectrosc.* **167**, 11.

Cossart, D., and Cossart-Magos, C. (1996), *Chem. Phys. Lett.* **250**, 128.

Cossart, D., and Elhanine, M., (1998) *Chem. Phys. Lett.* **285**, 83.

Cossart, D., and Robbe, J.-M. (1999), *Chem. Phys. Lett.* **311**, 248.

Coy, S. L., Jacobson, M. P., and Field, R. W. (1997), *J. Chem. Phys.* **107**, 8357.

Cross, J. B., and Valentini, J. J. (1982), *Rev. Sci. Instrum.* **53**, 38.

Cruse, H. W., Dagdigian, P. J., and Zare, R. N. (1973), *Farad. Disc. Chem. Soc.* **55**, 277.

Dagnall, R. M., and West, T. S. (1968), *Appl. Opt.* **7**, 1287.
Davis, D., and Braun, W. (1968), *Appl. Opt.* **7**, 2071.
Dawson, J. B., and Ellis, D. J. (1966), *Spectrochim. Acta* **23A**, 565.
Demtröder, W. (1996), "Laser Spectroscopy. Basic Concepts and Instrumentation", Springer, Berlin.
Diddams, S. A., Jones, D. J., Ye, J., Cundiff, S. T., Hall, J. L., Ranka, J. K., Windeler, R. S., Holzwarth, R., Udem, T., and Hänsch, T. W. (2000), *Phys. Rev. Lett.* **84**, 5102.
Dietz, T. G., Duncan, M. A., Powers, D. E., and Smalley, R. E. (1981), *J. Chem. Phys.* **74**, 6511.
Dreier, T., and Rakestraw, D. J. (1990), *Appl. Phys. B* **50**, 479.
Duan, Z., Field, R. W., Yamakita, N., and Tsuchiya, S. (2003), *Chem. Phys.*, submitted.
Dunlop, J. R., Karolczak, J., and Clouthier, D. J. (1988), *Chem. Phys. Lett.* **151**, 362.
Dunlop, J. R., Karolczak, J., and Clouthier, D. J. (1989), *Chem. Phys. Lett.* **154**, 613.
Dyke, T. R., Howard, B. J., and Klemperer, W. (1972), *J. Chem. Phys.* **56**, 2442.
Eland, J. H. D., Takahashi, M., and Hikosaka, Y. (2000), *Faraday Disc.* **115**, 119.
Elander, N., and Neuhaus, H. (1974), *Physica Scripta* **10**, 130.
Engelking, P. C. (1986), *Rev. Sci. Instrum.* **57**, 2274.
Ezekiel, S., and Weiss, R. (1968), *Phys. Rev. Lett.* **20**, 91.
Fabbi, J. C., Langenberg, J. D., Costello, Q. D., Morse, M. D., and Karlsson, L. (2001), *J. Chem. Phys.* **115**, 7543.
Fehsenfeld, F. C., Evenson, K. M., and Broida, H. P. (1965), *Rev. Sci. Instrum.* **36**, 294.
Field, R. W., English, A. D., Tanaka, T., Harris, D. O., and Jennings, D. A. (1973), *J. Chem. Phys.* **59**, 2191.
Field, R. W., Revelli, M., and Capelle, G. A. (1975), *J. Chem. Phys.* **63**, 3228.
Field, R. W. (1976), p. 261, in "Modern Spectroscopy: Modern Research", Vol. II. (K. N. Rao, ed.), Academic Press, NY.
Freund, R. S., and Klemperer, W. (1965), *J. Chem. Phys.* **43**, 2422.
Gallagher, T. F. (1994), "Rydberg Atoms," Cambridge University Press, Cambridge, UK.
Gaubatz, U., Rudecki, P., Schiemann, S., and Bergmann, K. (1990) *J. Chem. Phys.* **92**, 5363.
Gehrtz, M., Bjorklund, G. C., and Whittaker, E. A. (1985), *J. Opt. Soc. Am. B* **2**, 1510.
Gentry, W. R., and Giese, C. F. (1978), *Rev. Sci. Instrum.* **49**, 595.
German, K. R., and Zare, R. N. (1969), *Phys. Rev. Lett.* **23**, 1207.
Gerstenkorn, S., and Luc, P. (1978), "Atlas of the Absorption Spectra of the Iodine Molecule at 14,800-20,000 cm^{-1}," and "Complement: Assignments of the ($B(I_2)$–X) Iodine Lines," CNRS, Paris.
Gerstenkorn, S., Luc, P., and Raynal, A. (1985), "Atlas du Spectre d'Absorption de la Molecule de Brome ($^{79}Br_2$) 11,600-19,600 cm^{-1}," Laboratoire Aimé Cotton, CNRS II, 91405, Orsay.
Gerstenkorn, S. and Luc, P. (1986), "Identification des Transitions du System (B–X) de la Molecule d'Iode et Facteurs de Franck-Condon: 14,000-15,600 cm^{-1}", Laboratoire Aimé Cotton, CNRS II, 91405, Orsay.
Gerstenkorn, S., Luc, P., and Vergès, J. (1993), "Atlas du Spectre d'Absorption de la Molecule d'Iode: 7220 - 11,200 cm^{-1}," Laboratoire Aimé Cotton, CNRS II, 91405, Orsay.
Gottscho, R. A., Koffend, J. B., Field, R. W., and Lombardi, J. R. (1978), *J. Chem. Phys.* **68**, 4110.

Gottscho, R. A., Weiss, P. S., Field, R. W., and Pruett, J. G. (1980), *J. Mol. Spectrosc.* **82**, 283.
Gough, T. E., Miller, R. E., and Scoles, G. (1977), *Appl. Phys. Lett.* **30**, 338.
Goyal, S., Schutt, D. L., and Scoles, G. (1992), *Phys. Rev. Lett.* **69**, 933.
Gudeman, C. S., and Saykally, R. J. (1984), *Ann. Rev. Phys. Chem.* **35**, 387.
Gudeman, C. S., Martner, C., and Saykally, R. J. (1985), *Chem. Phys. Lett.* **122**, 108.
Guelachvili, G., and Rao, K. N. (1986), "Handbook of Infrared Standards with Spectral Maps and Transition Assignments Between 3 and 2600μm," Academic, London.
Guelachvili, G., Birk, M., Borde, Ch. J., Brault, J. W., Brown, L. R., Carli, B., Cole, A. R. H., Evenson, K. M., Fayt, A., *et al.*, (1996), *Spectrochem. Acta* **52A**, 717.
Haese, N. N., Pan, F.-S., and Oka, T. (1983), *Phys. Rev. Lett.* **50**, 1575.
Hall, G. E., and North, S. W. (2000) *Annu. Rev. Phys. Chem.* **51**, 243.
Hall, J. L., Hollberg, L., Baer, T. B., and Robinson, H. G. (1981), *Appl. Phys. Lett.* **39**, 680.
Hanle, W. (1924), *Z. Phys.* **30**, 93.
Harrison, G. R. (1969), "Wavelength Tables with Intensities in Arc, Spark, or Discharge Tube of more than 100,000 Spectrum Lines between 10,000 and 2,000 Å," MIT Press, Cambridge.
Heck, A. J. R., and Chandler, D. W. (1995), *Annu. Rev. Phys. Chem.* **46**, 335.
Held, A., Baranov, L. Y., Selzle, H. L., and Schlag, E. W. (1998), *Chem. Phys. Lett.* **291**, 318.
Helmerson, K., Martin, A., and Pritchard, D. E. (1992), *J. Opt. Soc. Am. B* **9**, 1988.
Hemminger, J. C., Wicke, B. G., and Klemperer, W. (1976), *J. Chem. Phys.* **65**, 2798.
Herzberg, G. H. (1950), "Molecular Spectra and Molecular Structure. I. Spectra of Diatomic Molecules," D. Van Nostrand, Princeton.
Herzberg, G. H., and Lagerquist, A. (1968), *Canad. J. Phys.* **48**, 2363.
Herzberg, G. H. (1971), *Rev. Chem. Soc.* **25**, 201.
Hinz, A., Pfeiffer, J., Bohle, W., and Urban, W. (1982), *Mol. Phys.* **45**, 1131.
Hollenstein, U., Palm, H., and Merkt, F. (2000), *Rev. Sci. Instrum.* **71**, 4023.
Huber, K. P., and Herzberg, G. (1979), "Molecular Spectra and Molecular Structure. IV. Constants of Diatomic Molecules," Van Nostrand-Reinhold, Princeton, New Jersey.
Jacobson, M. P., Coy, S. L., and Field, R. W. (1997), *J. Chem. Phys.* **107**, 8349.
Jenkins, F. A. (1953), *J. Opt. Soc. Am.* **43**, 425.
Johnson, P. M., Berman, M. R., Zakheim, D. (1975), *J. Chem. Phys.* **62**, 2500.
Johnson, P. M., and Otis, C. E. (1981), *Annu. Rev. Phys. Chem.* **32**, 139.
Julienne, P. S., and Mies, F. H. (1989), *J. Opt. Soc. Am. B* **6**, 2257.
Kaminsky, M. E., Hawkins, R. T., Kowalski, F. V., and Schawlow, A. L. (1976), *Phys. Rev. Lett.* **36**, 671.
Kantrowitz, A., and Grey, J. (1950), *Rev. Sci. Instrum.* **22**, 328.
Katô, H. (2000), "Doppler-free high resolution spectral atlas of iodine molecule 15,000-19,000 cm^{-1}," Japan Society for the Promotion of Science, Tokyo.
King, A. S. (1926), *J. Opt. Soc. Am.* **12**, 503.
Kinsey, J. L. (1977), *Annu. Rev. Phys. Chem.* **28**, 349.
Kittrell, C., Abramson, E., Kinsey, J. L., McDonald, S., Reisner, D. E., Katayama, D., and Field, R. W. (1981), *J. Chem. Phys.* **75**, 2056.
Koffend, J. B., and Field, R. W. (1977), *J. Appl. Phys.* **48**, 4468.
Koffend, J. B., Goldstein, S., Bacis, R., Field, R. W., and Ezekiel, S. (1978), *Phys. Rev. Lett.* **41**, 1040.
Kovács, I., and Budó, A. (1953), *Ann. Physik* **6**, 17.

Kowalski, F. V., Hawkins, R. T., and Schawlow, A. L. (1976), *J. Opt. Soc. Am.* **66**, 965.
Krautwald, H. J., Schneider, L., Welge, K. H., and Ashfold, M. N. R. (1986), *Farad. Disc. Chem. Soc.* **82**, 99.
Kroto, H. W., Heath, J. R., O'Brien, S.C., Curl, R. F., and Smalley, R. E. (1985), *Nature* **318**, 162.
Langenberg, J. D., DaBell, R. S., Shao, L., Dreesen, D., and Morse, M. D. (1998), *J. Chem. Phys.* **109**, 7863.
Launila, O., (1997), *Kosmos*; 135, Svenska Fysikersamfundet.
Launila, O., Schimmelpfennig, B., Fagerli, H., Gropen, O., Taklif, A. G., and Wahlgren, U. (1997), *J. Mol. Spectrosc.* **186**, 131.
Launila, O. (2003), http://kurslab.physics.kth.se/~olli/linuxdisp/.
Lawrance, W. D., and Knight, A. E. W. (1982), *J. Chem. Phys.* **76**, 5637.
Lee, S. A., and Hall, J. L. (1976), *Appl. Phys. Lett.* **29**, 367.
Lett, P. D., Julienne, P. S., and Phillips, W. D. (1995), *Annu. Rev. Phys. Chem.* **46**, 423.
Levenson, M. D., and Kano, S. S. (1988), "Introduction to Nonlinear Laser Spectroscopy", Academic Press, San Diego.
Lichten, W. (1957), *J. Chem. Phys.* **26**, 306.
Lichten, W. (1960), *Phys. Rev.* **120**, 848.
Li Li, and Field, R. W. (1983), *J. Phys. Chem.* **87**, 3020.
Lindholm, N. F., Brugh, D. J., Rothschopf, G. K., Sickafoose, S. M., and Morse, M. D. (2003), *J. Chem. Phys.* **118**, 2190.
Litfin, G., Pollock, C. R., Curl, R. F., and Tittel, F. K. (1980), *J. Chem. Phys.* **72**, 6602.
Liu, K., Fellers, R. S., Viant, M. R., McLaughlin, R. P., Brown, M. G., and Saykally, R. J. (1996), *Rev. Sci. Instrum.* **67**, 410.
Long, M. E., Swofford, R. L., and Albrecht, A. C. (1976), *Science* **191**, 183.
Loomis, F.W., and Wood, R.W. (1928), *Phys. Rev.* **32**, 223.
Ma, L.-S., Ye, J., Dubé, P. , and Hall, J. L. (1999), *J. Opt. Soc. Am. B* **16**, 2255.
McClintock, M., Demtröder, W., and Zare, R. N. (1969), *J. Chem. Phys.* **51**, 5509.
McNaughton, D., McGilvery, D., and Shanks, F. (1991), *J. Mol. Spectrosc.* **149**, 458.
Merkt, F. (1997), *Annu. Rev. Phys. Chem.* **48**, 675.
Mitchell, A. C. G., and Zemansky, M. W. (1961), "Resonance Radiation and Excited Atoms", Cambridge University Press, London.
Monts, D. L., Dietz, T. G., Duncan, M. A., and Smalley, R. E. (1980), *Chem. Phys.* **45**, 133.
Moore, J. H., David, C. C., Coplan, M. A., and Greer, S. C. (1989), "Building Scientific Apparatus. A Practical Guide to Design and Construction," Addison-Wesley, Redwood City, CA.
Morris, M. B., McIlrath, T. J., and Snyder, J. J. (1984), *Appl. Opt.* **23**, 3862.
Morse, M. D. (1986), *Chem. Rev.* **86**, 1049.
Morse, M. D. (1996), in Methods of Experimental Physics: Atomic, Molecular, and Optical Physics, Vol. II. Atoms and Molecules, (Dunning, F. B., and Hulet, R. eds.), Academic Press, Orlando, p. 26.
Müller-Dethlefs, K., and Schlag, E. W. (1991), *Annu. Rev. Phys. Chem.* **42**, 109.
Mulliken, R. S. (1930), *Phys. Rev.* **36**, 611.
Mulliken, R. S. (1931), *Rev. Mod. Phys.* **3**, 89.
Mulliken, R. S. (1955), *J. Chem. Phys.* **23**, 1997.
Nakagawa, J., Wormsbecher, R. F., and Harris, D. O. (1983), *J. Mol. Spectrosc.* **97**, 37.
Neuhauser, R., Siglow, K., and Neusser, H. J. (1998), *Phys. Rev. Lett.* **80**, 5089.
Nugent-Glandorf, L., Scheer, M., Samuels, D. A., Bierbaum, V., and Leone, S. R. (2002),

Rev. Sci. Instrum. **73**, 1875.
Ohshima, Y., and Endo, Y. (1992), *J. Mol. Spectrosc.* **153**, 627.
Oka, T. (1993), *Annu. Rev. Phys. Chem.* **44**, 299.
O'Keefe, A., and Deacon, D. A. G. (1988), *Rev. Sci. Instrum.* **59**, 2544.
Palmer, B. A., Keller, R. A., and Engleman, R., Jr. (1980), "An Atlas of Uranium Emission Intensities in a Hollow Cathode Discharge," Los Alamos Report LA-8251-MS, Los Alamos Scientific Laboratory, Los Alamos, New Mexico.
Palmer, B. A., Keller, R. A., Kowalski, F. V., and Hall, J. L. (1981), *J. Opt. Soc. Am.* **71**, 948.
Palmer, B. A., and Engleman, R., Jr. (1983), "Atlas of the Thorium Spectrum," Los Alamos Report LA-9615, Los Alamos Scientific Laboratory, Los Alamos, New Mexico.
Pao, Y.-H. (editor) (1977), "Optoacoustic Spectroscopy and Detection", Academic Press, New York.
Penning, F. M. (1928), *Physica* **8**, 137.
Pfaff, J., Begemann, M. H., and Saykally, R. J. (1984), *Mol. Phys.* **52**, 541.
Pruett, J. G., and Zare, R. N. (1975), *J. Chem. Phys.* **62**, 2050.
Radunsky, M. B., and Saykally, R. J. (1987), *J. Chem. Phys.* **87**, 898.
Radunsky, M. B., and Saykally, R. J. (1988), *Chem. Phys. Lett.* **152**, 419.
Ramsey, N. F. (1956), "Molecular Beams", pp. 6-7, Oxford University Press, Oxford.
Reid, J. P., Loomis, R. A., and Leone, S. R. (2000), *J. Phys. Chem.* **104**, 10139.
Righi, A. (1898), *Computes Rendus* **127**, 216, and **128**, 45.
Romanini, D., and Lehmann, K. K. (1993), *J. Chem. Phys.* **99**, 6287.
Saykally, R. J., and Woods, R. C. (1981), *Ann. Rev. Phys. Chem.* **32**, 403.
Scherer, J. J., Paul, J. B., O'Keefe, A., and Saykally, R. J. (1997), *Chem. Revs.* **97**, 25.
Schiemann, S., Kuhn, A., Steuerwald, S., and Bergmann, K. (1993), *Phys. Rev. Lett.* **71**, 3637.
Schinke, R. (1993), "Photodissociation Dynamics," Cambridge University Press, Cambridge, UK.
Schutte, C. J. H., Bertie, J. E., Bunker, P. R., Hougen, J. T., Mills, I. M., Watson, J. K. G., and Winnewisser, B. P. (1997), *Pure and Appl. Chem.* **69**, 1633.
Schweda, H. S., Chawla, G. K., and Field, R. W. (1982), *Opt. Commun.* **42**, 165.
Segall, J., Lavi, R., Wen, Y., and Wittig, C. (1989), *J. Phys. Chem.* **93**, 7287.
Sheppard, N., Willis, H. A., and Rigg, J. C. (1987), *Spectrochim. Acta* **43A**, 1.
Signorell, R., and Merkt, F. (1999), p. 479 in "The Role of Rydberg States in Spectroscopy and Photochemistry" (ed. Sandorfy, C.), Kluwer, Dordrecht.
Silvers, S., Bergeman, T., and Klemperer, W. (1970), *J. Chem. Phys.* **52**, 4385.
Smalley, R. E., Wharton, L., and Levy, D. H. (1975), *J. Chem. Phys.* **63**, 4977.
Smyth, K. C., Schenck, P. K., Mallard, W. G., and Travis, J. C. (1979), National Bureau of Standards Special Publication # 561, Proc. 10th Materials Res. Symp. on Characterization of High Temperature Vapors and Gases, NBS, Gaithersburg, MD, p. 865.
Snè, O., and Cheshnovsky, O. (1986), *Chem. Phys. Lett.* **130**, 53.
Steane, A. M., and Foot, C. J. (1991), *Europhys. Lett.* **14**, 231.
Steinfeld, J. I., Zare, R. N., Jones, L., Lesk, M., and Klemperer, W. (1965), *J. Chem. Phys.* **42**, 25.
Stienkemeier, F., Ernst, W. E., Higgins, J., and Scoles, G. (1995), *J. Chem. Phys.* **102**, 615.
Teets, R., Feinberg, R., Hänsch, T. W., and Schawlow, A. L. (1976), *Phys. Rev. Lett.* **37**, 683.

Thoman, J. W., Jr., Chandler, D. W., Parker, D. H., and Janssen, M. H. M. (1988), *Laser Chem.* **9**, 27.
Toennies, J. P., and Vilesov, A. F. (1998), *Annu. Rev. Phys. Chem.* **49**, 1.
Tolles, W. M., Nibler, J. W., McDonald, J. R., and Harvey, A. B. (1977), *Appl. Spectrosc.* **31**, 253.
Tsal, B. P., Baer, T., Werner, A. S., and Lin, S. F. (1975), *J. Phys. Chem.* **79**, 570.
Vaccaro, P. H. (1997), in "Nonlinear Spectroscopy for Molecular Structure Determination" (eds. Hirota, E., Field, R. W., Maier, J. P., and Tsuchiya, S.), Blackwell Scientific for IUPAC, London, p. 75.
Vager, Z. (2001), *Adv. At. Mol. Opt. Phys.* **45**, 203.
Vidal, C. R. (1986) in "Tunable Lasers" (Mollenauer, L. F., and White, S. C., editors), Springer Topics in Applied Physics **59**, Springer, p. 57.
Vrakking, M.-J. J., and Lee, Y. T. (1995), *J. Chem. Phys.* **102**, 8818 and 8833.
Wasserman, T. A., Vaccaro, P. H., and Johnson, B. R. (1998), *J. Chem. Phys.* **108**, 7713.
West, J. B., Bradford, R. S., Jr., Eversole, J. D., and Jones, C. R. (1975), *Rev. Sci. Instrum.* **46**, 164.
Whiting, E. E., Schadee, A., Tatum, J. B., Hougen, J. T., and Nicholls, R. W. (1980), *J. Mol. Spectrosc.* **80**, 249.
Wineland, D. J., Drullinger, R. E., and Walls, F. L. (1978), *Phys. Rev. Lett.* **40**, 1639.
Winnewisser, B. P., Reinstädtler, J., Yamada, K. M. T., and Behrend, J. (1989), *J. Mol. Spectrosc.* **136**, 12.
Zare, R. N., and Herschbach, D. R. (1963), *Proc. IEEE* **51**, 173.
Zare, R. N. (1964), *J. Chem. Phys.* **40**, 1934.
Zare, R. N. (1966), *J. Chem. Phys.* **45**, 4510.
Zare, R. N. (1971), *Acc. Chem. Res.* **4**, 361.
Zare, R. N. (1972), *Mol. Photochem.* **4**, 1.
Zare, R. N., and Dagdigian, P. J. (1974), *Science* **185**, 739.
Zhang, Q., Kandel, S. A., Wasserman, T. A. W., and Vaccaro, P. H. (1992), *J. Chem. Phys.* **96**, 1640.
Zhao, Y., Yourshaw, I., Reiser, G., Arnold, C. C., and Neumark, D. M. (1994), *J. Chem. Phys.* **101**, 6538.
Zhou, J., Peatross, J., Murnane, M., Kapteyn, H., and Christov, I. (1996), *Phys. Rev. Lett.* **76**, 752.
Zhu, L., and Johnson, P. (1991), *J. Chem. Phys.* **94**, 5769.

Chapter 2

Basic Models

Contents

2.1 What Is a Perturbation? 62

2.2 Structural Models 69

2.3 Elementary Properties of Angular Momenta in Diatomic Molecules 72

2.3.1 Angular Momentum Components Defined by Normal and Anomalous Commutation Rules 73

2.3.2 Recipes for Evaluation of Molecule-Fixed Angular Momentum Matrix Elements 79

2.3.3 Euler Angles, $|JM\Omega\rangle$ Basis Functions, Direction Cosines, and Phases 81

2.4 Estimation of Parameters in a Model Hamiltonian 83

2.5 Data Compilations 84

2.6 References 85

And now we enter the real world. The spectra of diatomic molecules can exhibit *fine structure* that arises from nonzero electron spin and electron orbital angular momenta. Some rotational-vibrational energy levels of one electronic state might be accidentally near degenerate with those of another electronic state (spectroscopic perturbations) or they might be embedded in the vibrational continuum of a repulsive electronic state (predissociation, Chapter 7) or the electronic ionization continuum of a Rydberg series (autoionization, Chapter 8). These effects demand the use of a molecular Hamiltonian that includes the effects of nonzero electron spin and orbital angular momentum. Although the unavoidable complexity of real molecular spectra is challenging to experimentalists and imposes major educational demands on graduate students, encoded in these non-textbook spectra we find many beautiful and enlightening dynamical effects.

2.1 What Is a Perturbation?

For many years the study of perturbations has been a main focus of the research of a small number of spectroscopists specializing in the electronic structure of diatomic molecules. The pioneering studies have been by Heurlinger (1917 - 1919), Kronig (1928), Van Vleck (1929, 1932, 1936), Mulliken (1930, 1931, 1932), Ittmann (1931), Dieke (1935), Schmid and Gerö (1935), Budó (1937), Budó and Kovács (1938), Kovács (1937, 1969), and Stepanov (1940, 1945). Lagerqvist, Barrow, Miescher, Herzberg, Ramsay, and Broida have been among the major workers in this field. The study of perturbations is an esoteric, usually avoided, but occasionally necessary pastime of a much larger group of spectroscopists, and an impenetrable mystery to most users of spectroscopic data. The label or footnote *perturbed*, when applied to a band in a data compilation (Herzberg, 1950; Huber and Herzberg, 1979; Rosen, 1970; Barrow, 1973, 1975, 1979, 1982; Pearse and Gaydon, 1976, Suarez, 1972; Kopp, *et al.*, 1974 and 1977; Bernath, 1997; Bernath and McLeod, 2001), has been like a quarantine poster warning against the use of that band because of a dangerous and perhaps contagious form of irregular behavior. In fact, irregular behavior is an excellent operational definition of perturbation phenomena.

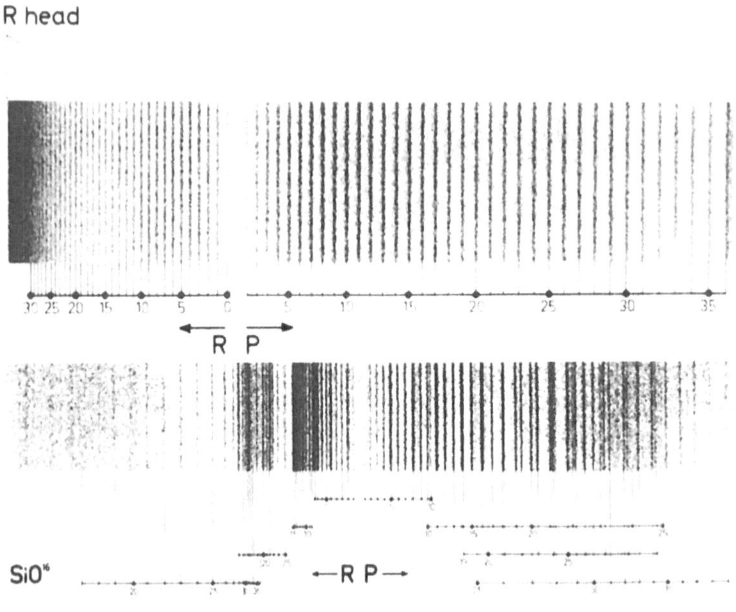

Figure 2.1: A comparison of the SiO $H^1\Sigma$ (0,0) (top) and (1,0) (bottom) bands. The (0,0) band at 1435 Å is perturbation-free, but perturbations in the $v = 1$ level of the $H^1\Sigma$ state cause the (1,0) band at 1413 Å to be shattered. (Courtesy I. Renhorn.)

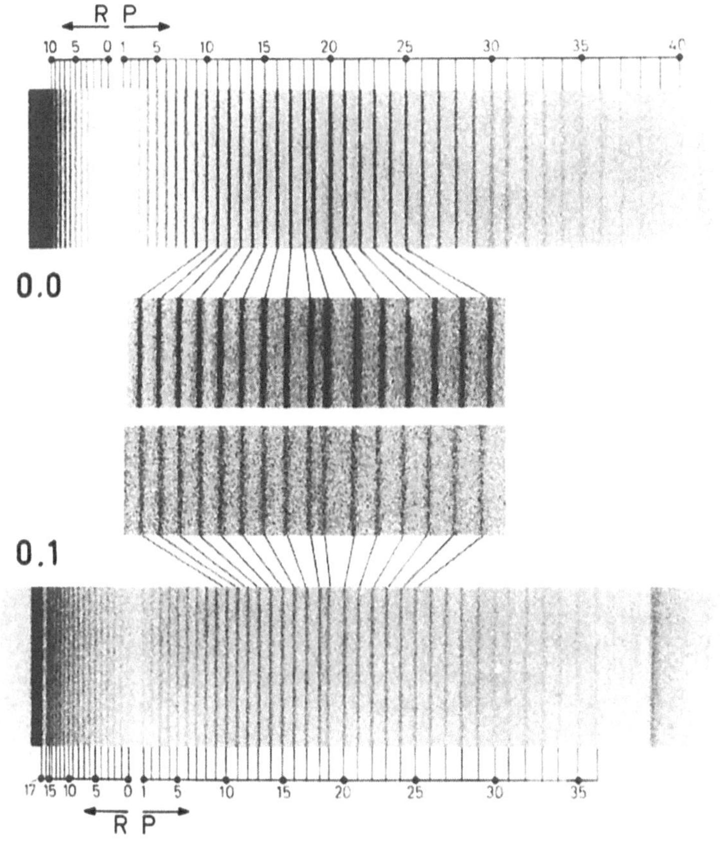

Figure 2.2: Pictorial evidence that a perturbation affects the upper rather than lower electronic state. The SiO $F^1\Sigma - X^1\Sigma$ (0,0) (1460 Å) and (0,1) (1487 Å) bands are weakly perturbed at $J' = 18$ and 19. The enlargements clearly show identical level shifts for the $P(19)$ line in the (0,0) and (0,1) bands (courtesy A. Lagerqvist).

One can define perturbations either pictorially, as deviations from a naive polynomial representation of the energy levels, or in terms of physical models of varying sophistication. Figure 2.1 shows spectra of two vibrational bands of perturbations in the SiO $H^1\Sigma - X^1\Sigma$ electronic system. The rotational structure of an isolated, well-resolved $^1\Sigma - {}^1\Sigma$ band is so simple that the difference between the perturbed and perturbation-free bands is obvious. Typically, the pattern of a band is not so easy to recognize, but when that pattern is disrupted by a few lines being shifted from their expected positions or deviating from a smoothly varying intensity distribution, it is obvious that something in the model has gone wrong and that the band is perturbed. Effects are sometimes restricted to two or three spectral lines associated with a single rotation-vibration level (Fig. 2.2).

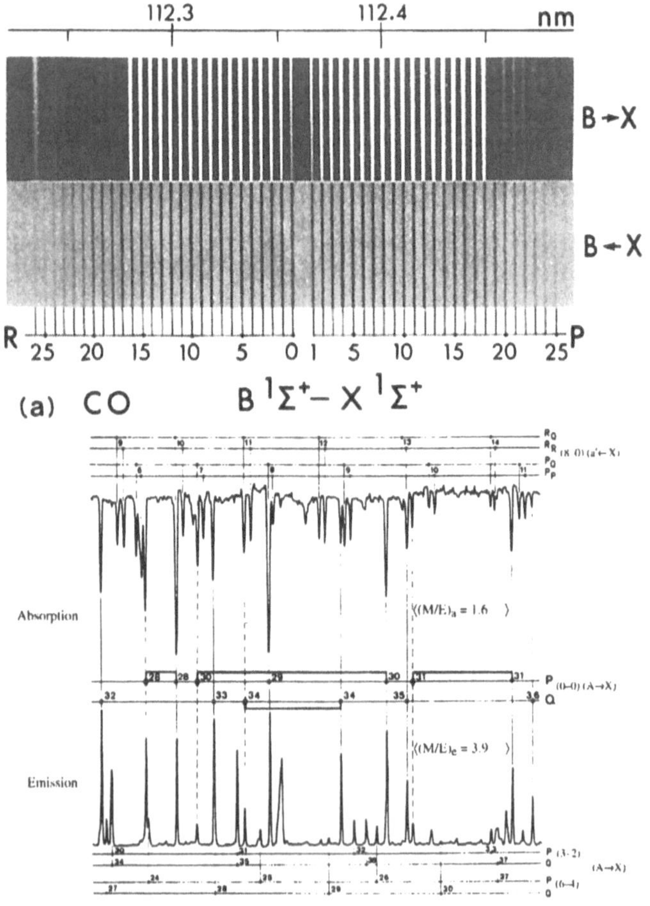

Figure 2.3: Perturbations and predissociations affect absorption and emission line intensities in quite different ways. Two pairs of absorption and emission spectra are shown. The first pair illustrates the disappearance of a weakly predissociated line in emission without any detectable intensity or lineshape alteration in absorption. The second pair shows that emission from upper levels with slow radiative decay rates can be selectively quenched by collision induced energy transfer. The opposite effect, selective collisional enhancement of emission from perturbed, longer-lived levels, is well known in CN $B^2\Sigma^+-X^2\Sigma^+(v' = 0, v'')$ emission spectra (see Fig. 6.14 and Section 6.5.5). (*a*) the CO $B^1\Sigma^+-X^1\Sigma^+(1, 0)$ band in emission (top) and absorption (bottom). The last strong lines in emission are $R(16)$ and $P(18)$. Emission from levels with $J' > 17$ is weak because the predissociation rate is larger than the spontaneous emission rate. (Courtesy F. Launay and J. Y. Roncin.) (*b*) The CO $A^1\Pi-X^1\Sigma^+(0, 0)$ band in emission (bottom) and absorption (top). The $a'^3\Sigma^+-X^1\Sigma^+(8, 0)$ band lines appear in absorption because the $A^1\Pi \sim a'^3\Sigma^+$ spin-orbit interaction causes a small amount of $A^1\Pi$ character to be admixed into the nominal $a'^3\Sigma^+$ levels. These $a'-X$ lines are absent from the emission spectrum because collisional quenching and radiative decay into $a^3\Pi$ compete more effectively with radiative decay into $X^1\Sigma^+$ from the long-lived $a'^3\Sigma^+$ state than from the short-lived $A^1\Pi$ state. In addition, collisions and radiative decay into $a^3\Pi$ cause the $P(31)$ extra line (E) (arising from a perturbation by $d^3\Delta$ $v = 4$) to be weakened in emission relative to the main line (M). (Courtesy F. Launay, A. Le Floch, and J. Rostas.)

Some perturbation effects can be dramatically dependent on the mode of observation of the spectrum, for example by a line being absent in emission versus apparently unperturbed in absorption or by being anomalously strong and weak under two different sets of excitation conditions (Fig. 2.3). It is even possible for all rotational lines in a given vibrational band to appear free of perturbation effects; yet, when an attempt is made to fit the band origin and rotational constants of that band into the regular pattern required for a vibrational progression, all lines are found to be strongly but smoothly affected by an unsuspected perturbation. Obviously, it is dangerous to rely on visual pattern recognition alone for detection or definition of perturbations.

The next step beyond pictorial patterns is a simple algebraic representation, which is based on the observed regularity of the pattern rather than any physical model. One expects that the energy levels E_{vJ} sampled in a band spectrum should be well represented by a simple, rapidly convergent, polynomial function of the rotational and vibrational quantum numbers, J and v. The Dunham expansion (1932),

$$E_{v,J} = \sum_{l,m=0}^{l_{max},m_{max}} Y_{lm} \left(v + \frac{1}{2}\right)^l [J(J+1)]^m \text{ †} \tag{2.1.1}$$

is an example of such a polynomial representation. Questions of the physical meaning of the Y_{lm} coefficients and the efficiency of the parametrization aside, one can regard a Dunham expansion as a convenient, empirical, model-free way of organizing a large quantity of spectral data. A polynomial expression is simply a translation of the expectation of smooth v, J–dependence into an algebraic form. Perturbations may be viewed, still in a model-independent sense, as a failure of a subset of observed v, J levels to be accommodated by the empirical energy level expression defined by the majority of sampled levels. Thus a band that appears unperturbed in isolation from other vibrational bands can be recognized as perturbed if its rotational constant or vibrational energy disagrees with an interpolation of

$$B(v) = \sum_l Y_{l1} \left(v + \frac{1}{2}\right)^l \tag{2.1.2}$$

or

$$G(v) = \sum_l Y_{l0} \left(v + \frac{1}{2}\right)^l \tag{2.1.3}$$

† Some of the relationships between traditional molecular constants and Dunham constants are: $Y_{10} = \omega_e, Y_{20} = -\omega_e x_e, Y_{30} = \omega_e y_e, Y_{01} = B_e, Y_{11} = -\alpha_e, Y_{21} = \gamma_e, Y_{02} = -D_e$, $Y_{12} = -\beta_e$. When the Y_{lm} constants of Eq. (2.1.1) are determined by a least squares fit to the observed energy levels, the usually assumed relationships between the Y_{lm} and properties of the potential energy curve [e.g., $Y_{10} \rightarrow \omega_e \rightarrow$ force constant (Eq. (5.1.20)), $Y_{01} \rightarrow B_e \rightarrow$ equilibrium internuclear distance (Eq. (5.1.24))] are only approximate [See Townes and Schawlow (1955), pp. 10-11].

However, a polynomial-based empirical definition of perturbations is limited in two important ways: there are no rules governing the permissible values of polynomial constants, and deviations from the polynomial expression have no meaning or predictive consequence.

Indeed, Dunham's energy-level formula [Eq. (2.1.1)] is based both on the concept of a potential energy curve, which rests on the separability of electronic and nuclear motions, and on the neglect of certain couplings between the angular momenta associated with nuclear rotation, electron spin, and electron orbital motion. The utility of the potential curve concept is related to the validity of the Born-Oppenheimer approximation, which is discussed in Section 3.1.

The potential energy curve concept provides many self-consistency checks between molecular constants, for example, the Kratzer relation,

$$Y_{02} = -4Y_{01}^3/Y_{10}^2, \tag{2.1.4}$$

which is valid for harmonic [see Eqs. (5.1.19)-(5.1.20)] or Morse [see Eqs. (5.1.21)-(5.1.24)] potentials, and the Pekeris relation

$$Y_{11} = 6Y_{01}^2\left[1-(-Y_{20}/Y_{01})^{1/2}\right]/Y_{10}, \tag{2.1.5}$$

which is valid for a Morse potential. When fitted constants fail such consistency checks, this can indicate the presence of a neglected perturbation. Conversely, if interrelationships between molecular constants derived from eigenvalues of a molecular potential are satisfied, then one has confidence in the deperturbed nature of the molecular constants from which the potential is derived.

Thus, the most useful definition of perturbed behavior is dependent on the potential curve concept. It is no longer sufficient that the energy levels be represented by an arbitrary polynomial in v or J; the coefficients in the polynomial expression must have values consistent with a simple mechanical oscillator in a well-behaved, isotopically invariant potential.

The study of perturbations involves going beyond the Born-Oppenheimer approximation. One starts with a zero-order model, $\mathbf{H}^0$, in which each electronic state is associated with a potential energy curve, $V(R)$, an electronic eigenfunction (which depends on internuclear distance), $\Phi(r,R)$, and a set of rotation-vibration eigenvalues and eigenfunctions, $E_{v,J}$ and $\chi_{v,J}(R)$. These zero-order $V(R)$, $\Phi(r,R)$, and $\chi_{v,J}(R)$ functions can only be defined by ignoring certain terms in the exact molecular Hamiltonian, $\mathbf{H}$. Foremost among these neglected terms, $\mathbf{H}'$, are those that couple the zero-order functions associated with different electronic states.

The explicit choice of which terms to neglect is not unique. For example, diabatic and adiabatic potentials are bases for two quite different but equally useful zero-order pictures. These two models and the reasons for selecting one over the other are discussed in Section 3.3. What appears as a large deviation from one zero-order model may be vanishingly small for another. Consequently, the definition of a perturbation depends critically on one's choice of zero-order model, which is often a matter of personal preference.

Once a zero-order model is selected, the effects of all terms ($\mathbf{H}'$) in the Hamiltonian ($\mathbf{H}$) neglected in the definition of the zero-order basis functions (ψ_i^0, which are eigenfunctions of $\mathbf{H}^0$) may be taken into account exactly. One constructs a deperturbation model in which perturbation parameters govern the strength of the perturbation or, in other words, the magnitudes of the differences between the (observable) eigenvalues of $\mathbf{H}$ (the perturbed levels) and the (unobservable but calculable) eigenvalues of $\mathbf{H}^0$ (the deperturbed levels). These perturbation parameters are matrix elements of $\mathbf{H}'$ between basis functions. By refining the molecular structural parameters that define both $\mathbf{H}^0$ and $\mathbf{H}'$, the deperturbation model is made to account for the properties of all observed levels. The deperturbation model should describe strongly perturbed and relatively perturbation free levels equally well.

The choice of zero-order model determines the names used to label the observed energy levels. These names are a conventionally agreed upon (Jenkins, 1953) set of shorthand labels for the basis functions ψ_j^0, not for the eigenfunctions ψ_i. Since the eigenfunctions of the exact $\mathbf{H}$ are linear combinations of basis functions

$$\psi_i = \sum_j c_{ij}\psi_j^0 \tag{2.1.6}$$

is not obvious how to label the levels compactly and unambiguously. It is customary to use the name of the basis function ψ_j^0, which has the largest coefficient $|c_{ij}|$ in the expansion of the ψ_i eigenfunction, as the name of the E_ith level. One must be careful not to be confused by the use of basis function names to specify both basis functions and the *nominal* (i.e., dominant) character of an eigenfunction.

It is important to emphasize that, although the label "perturbed" or "unperturbed" can depend on an arbitrary choice among several zero-order models (where each model in principle is equally capable of accounting for all molecular properties), there is no relativism in judging the quality of agreement between observed and calculated spectra. However, there are major differences in the ease of performing the calculations required by a particular deperturbation model or, more significantly, in the arduous process of refining a deperturbation model, starting from a fragmentary and incompletely assigned data base.

The primary purpose of a deperturbation should not be to obtain highly precise molecular constants. The most important objectives of a deperturbation are a perfect fit of the wavenumbers and (often) the relative intensities of the observed spectral lines, confirmation of line assignments, an assessment of the completeness of the deperturbation, proof of the mechanism of the perturbation, and a statistically rigorous determination of the values of the perturbation parameters (and their uncertainties). The quality of the fit establishes the credibility of the model and reduces the possibility of misassigned lines. Assignments of perturbed lines are invariably difficult because of the absence of the usual spectral pattern and the low intensity of the lines which contain maximum information. Often, a deperturbation model can guide new assignments of a few levels at a time, testing the consequences of a dubious assignment by extrapolation to the next level. In many cases reliable line assignments are more important than the molecular constants that reproduce the lines.

The values of the perturbation parameters are often useful for identifying the mechanism of interaction by showing that the interactions among one group of levels have the magnitude, relative to those among a different group of levels, required by the perturbation mechanism in question. The practical and theoretical value of these perturbation parameters is not generally appreciated. A major portion of this book is devoted to describing methods for determining unknown perturbation parameters, ranging from *ab initio* calculations to empirical interrelationships between seemingly unrelated observables.

In this book we regard the following phenomena as perturbations and attempt to display the similarities and relationships between the methods and parameters used to explain (deperturb) them:

1. Local perturbations between electronic states of different symmetry (Sections 3.4 and 3.5)

2. Perturbations between states of the same symmetry (treated using two different zero-order pictures, the diabatic picture, Section 3.3.1, and the adiabatic picture, Section 3.3.3).

3. Systematic interactions between basis states belonging to the same vibrational quantum number of one or several electronic states (evolution from one Hund's coupling case to another) (Section 3.2.1).

4. Local perturbations between substates belonging to different vibrational levels of the same electronic state (Section 3.5.2).

5. Interactions of a given electronic-vibrational level with all vibronic levels of a different symmetry (such as Λ-doubling, Sections 3.5.4 and 5.5).

6. Interaction between a bound rotation-vibration level of one electronic state and the vibrational continuum of another electronic state (predissociation, Chapter 7).

7. Interaction between a bound rotation-vibration level of the AB molecule and the electronic continuum associated with a particular rovibronic level of the AB^+ core ion (autoionization, Chapter 8).

The deperturbation models provide a uniform explanation of level patterns, transition intensities (Chapter 7), radiative lifetimes, linewidths, and systematic variations of coupling matrix elements with quantum numbers.

This book is intended as a user's guide for experimentalists and theorists. It is neither a formal, rigorous quantum mechanical treatment nor a descriptive review of perturbation phenomena. For experimentalists, we assume familiarity with Herzberg's *Spectra of Diatomic Molecules* (1950) and Hougen's monograph, *The Calculation of Rotational Energy Levels and Rotational Line Intensities in Diatomic Molecules* (1970). [See also Kovács, (1969); Katô, (1993); Bunker and Jensen, (1998); Bernath (1995); Brown (1998); Ogilvie (1998); and Brown and Carrington, (2003).] For theorists, we assume a curiosity about what can be measured readily and how measurable quantities can be related, often deviously, to calculable quantities. We explain how to get perturbation-related numbers from a spectrum, what the numbers tell us about molecular structure, and how to estimate perturbation parameters empirically, semiempirically, and semitheoretically.

In short, this book is an attempt to demystify perturbations by providing a minimally elegant but maximally unified view of the deperturbation process and by suggesting what can be learned from perturbations.

2.2 Structural Models

Perturbations are defined as deviations in the quantum-number variation of any observable from that predicted by a zero-order molecular structural model based on the Born-Oppenheimer approximation. This section is intended as an outline of the ingredients of molecular structural models.

The construction of a molecular structural model can be divided into the following steps:

1. Definition of the operator form of an exact microscopic Hamiltonian $\mathbf{H}$ (Section 3.1).

2. Definition of a complete set of basis functions (Section 3.2), including phase conventions (Section 3.2.2), and a scheme for naming the basis functions.

3. Construction of a matrix representation of the microscopic Hamiltonian appropriate to the specific structural problem under consideration. This matrix, constructed from a limited number of basis states, is of finite dimension. This is equivalent to treating the effect of all other energetically remote basis states by second- (or higher) order perturbation theory. The effect of these remote levels is introduced, either implicitly or explicitly, upon construction of an effective Hamiltonian $\mathbf{H}^{eff}$ by partitioning the true Hamiltonian $\mathbf{H}$ into two parts, $\mathbf{H}^{(0)}$ and $\mathbf{H}^{(1)}$ (Section 4.2).

4. Determination of the values of molecular constants and effective parameters by a least-squares fitting process, which involves comparison of properties of the eigenstates of $\mathbf{H}^{eff}$, computed for a given choice of constants, against the observed properties (Section 4.4).

5. An attempt to relate the effective parameters to the true molecular structural parameters calculated from the exact Hamiltonian (Chapter 5).

A clever choice of basis set causes the matrix representation of $\mathbf{H}$ to consist mostly of zeroes, the H_{ij} matrix elements of $\mathbf{H}$ that are exactly zero to be obvious by inspection, and enables many of the nonzero H_{ij} to be evaluated automatically using formulas or tables rather than numerical integration. It is necessary to refer to an important theorem that is the basis for the definition of a *a good quantum number* and for selection of a maximally useful set of *basis function labels. If two operators commute* for all functions Φ,

$$\mathbf{AB}\Phi - \mathbf{BA}\Phi = 0,$$

or, symbolically,

$$[\mathbf{A}, \mathbf{B}] = \mathbf{AB} - \mathbf{BA} = 0,$$

then it is possible, in principle, to find a complete set of functions that are simultaneously eigenfunctions of both operators. If two functions belong to different eigenvalues of $\mathbf{A}$,

$$\mathbf{A}\Phi_i = a_i\Phi_i, \qquad \mathbf{A}\Phi_j = a_j\Phi_j, \qquad a_i \neq a_j,$$

then

$$\langle i|\mathbf{B}|j\rangle = 0,$$

where $\mathbf{B}$ is any operator that commutes with $\mathbf{A}$. Thus, if an operator commutes with the exact Hamiltonian, its eigenvalues are rigorously good quantum numbers because there are only zero matrix elements of $\mathbf{H}$ between functions associated with two different values of the good quantum number.

There are only a few operators that commute with $\mathbf{H}$. Five important ones are

1. $\mathbf{J}^2$, where $\vec{\mathbf{J}}$ is the total angular momentum.
2. $\mathbf{J}_Z$, which is the projection of $\mathbf{J}$ on the laboratory Z–axis (eigenvalue M_J).
3. $\boldsymbol{\sigma}_v$, which is the symmetry operation of reflection, in the molecule-fixed coordinate system, through a plane containing the internuclear axis. $\boldsymbol{\sigma}_v$[†] is important in specifying the parity ($\pm$) of a state (see Section 3.2.2).
4. $\mathbf{i}$, the symmetry operation for homonuclear molecules corresponding to molecule-fixed inversion of the coordinates of all electrons and nuclei through the center of mass (g, u symmetry).
5. P_{ij}, the symmetry operation involving interchange of identical particles (nuclei or electrons). All particles are either Bosons or Fermions, and the total wavefunction must, respectively, be even or odd upon interchange of any pair of identical particles. The total wavefunction of a homonuclear molecule, exclusive of the nuclear spin part, is classified s or a according to whether it is even or odd with respect to nuclear exchange. Since electrons are Fermions, the total molecular wavefunction must be odd with respect to permutation of any two electrons. This requirement is satisfied by the determinantal form of the electronic wavefunction (see Section 3.2.4).

[†]Although parity is defined by the even/odd-ness of the complete molecular wavefunction under inversion of laboratory-frame coordinates of all electrons and nuclei (denoted either as I or E*), Hougen (1970) shows that $\boldsymbol{\sigma}_v$, applied to the complete electronic-vibration-rotation wavefunction, has the same effect as laboratory frame inversion.

It is easier to construct eigenfunctions of these operators than of **H**. This is the basis for factorization of **H** into separate blocks, labeled by J, M_J, and parity $\pm$, between which there are only zero matrix elements

$$H_{J,M_J,\pm;J',M'_J,\pm'} = \delta_{J,J'}\delta_{M_J,M'_J}\delta_{\pm,\pm'}E^{(0)}_{J,M_J,\pm}.$$

Although only a few operators commute with **H**, many operators (especially angular momentum operators) commute with parts of **H**.

The choice of a particular basis set is dictated by the nature of the desired partitioning of **H** into two parts,

$$\mathbf{H} = \mathbf{H}^{(0)}_{(a)} + \mathbf{H}^{(1)}_{(a)} \tag{2.2.1a}$$

or

$$\mathbf{H} = \mathbf{H}^{(0)}_{(b)} + \mathbf{H}^{(1)}_{(b)} \tag{2.2.1b}$$

The sets of basis functions [e.g., Hund's case (a) or (b)] corresponding to $\mathbf{H}^{(0)}_{(a)}$ and $\mathbf{H}^{(0)}_{(b)}$ can be quite different. This difference and the practical reason for a particular choice of $\mathbf{H}^{(0)}$ can be understood in terms of sets of mutually commuting operators. If several operators commute, then a set of simultaneous eigenfunctions of all commuting operators must exist. Hund's coupling cases (a), (b), (c), (d), and (e) (discussed in Section 3.2.1) correspond to five different mutually commuting sets of angular momentum operators [Mulliken, (1930 and 1931), Herzberg, (1950), Kovács, (1969)]. Each coupling case requires a different partitioning of **H** so that $\mathbf{H}^{(0)}$ commutes with all members of the appropriate set of angular momentum operators.

The terms *good quantum number* and *basis function label* are frequently used interchangeably, resulting in considerable confusion especially when observed levels are specified by their nominal basis function label. J is both a good quantum number and a label of the basis function [associated with eigenvalue $J(J+1)$] because the operator $\mathbf{J}^2$ commutes with $\mathbf{H}^{(0)}$ and $\mathbf{H}^{(1)}$ for all partitionings of **H**. The quantum number, Ω, associated with the projection of $\vec{\mathbf{J}}$ on the internuclear axis, $\mathbf{J}_z$, is not a good quantum number for **H** because $\mathbf{J}_z$ does not commute with $\mathbf{H}^{\dagger}$. However, Ω can be a basis function label because, for a specific partitioning of **H**, $\mathbf{J}_z$ commutes with $\mathbf{H}^{(0)}$.

Since many of the operators that appear in the exact Hamiltonian or in the effective Hamiltonian involve products of angular momenta, some elementary angular momentum properties are summarized in the next section. Matrix elements of angular momentum products are frequently difficult to calculate. A tremendous simplification is obtained by working with spherical tensor operator components and, in this way, making use of the Wigner-Eckart Theorem (Section 3.4.5). A more elementary but cumbersome treatment, based on Cartesian operator components, is presented in Section 2.3.

---

† The projection of **J** on the internuclear axis, $\mathbf{J}_z$, should not be confused with the projection of **J** on the laboratory Z-axis, $\mathbf{J}_Z$.

A major difficulty for molecular as opposed to atomic systems arises from the fact that two different reference axis systems are important, the molecule-fixed and the space-fixed system. Many perturbation related quantities require calculation of matrix elements of molecule-fixed components of angular momentum operators. Particular care is required with molecule-fixed matrix elements of operators that include an angular momentum operator associated with rotation of the molecule-fixed axis system relative to the space-fixed system. The molecule-fixed components of such operators have a physical meaning that is not intuitively obvious, as reflected by anomalous angular momentum commutation rules.

2.3 Elementary Properties of Angular Momenta in Diatomic Molecules

The angular momenta that can occur in a diatomic molecule, AB, include: spin of nucleus A or B, $\vec{\mathbf{I}}_A$, $\vec{\mathbf{I}}_B$; orbital angular momentum of the ith electron, $\vec{\mathbf{l}}_i$; spin of the ith electron, $\vec{\mathbf{s}}_i$; and rotation of the nuclear framework, $\vec{\mathbf{R}}$.

These fundamental angular momenta can couple to form:

- total nuclear spin — $\vec{\mathbf{I}} = \vec{\mathbf{I}}_A + \vec{\mathbf{I}}_B$ *
- total angular momentum of the ith electron — $\vec{\mathbf{j}}_i = \vec{\mathbf{l}}_i + \vec{\mathbf{s}}_i$
- total electronic orbital angular momentum — $\vec{\mathbf{L}} = \sum_i \vec{l}_i$
- total electron spin — $\vec{\mathbf{S}} = \sum_i \vec{\mathbf{s}}_i$
- total angular momentum — $\vec{\mathbf{F}} = \vec{\mathbf{R}} + \vec{\mathbf{L}} + \vec{\mathbf{S}} + \vec{\mathbf{I}}$
- total angular momentum exclusive of nuclear spin — $\vec{\mathbf{J}} = \vec{\mathbf{R}} + \vec{\mathbf{L}} + \vec{\mathbf{S}}$
- total angular momentum exclusive of nuclear and electron spin — $\vec{\mathbf{N}} = \vec{\mathbf{J}} - \vec{\mathbf{S}} = \vec{\mathbf{R}} + \vec{\mathbf{L}}$
- total electronic angular momentum — $\vec{\mathbf{J}}_a = \vec{\mathbf{L}} + \vec{\mathbf{S}}$ *
- total angular momentum exclusive of nuclear spin and electron orbital angular momentum. — $\vec{\mathbf{O}} = \vec{\mathbf{J}} - \vec{\mathbf{L}} = \vec{\mathbf{R}} + \vec{\mathbf{S}}$ *

With the exception of the angular momenta labeled by an asterisk, the notation employed is that specified by an international agreement among spectroscopists (Jenkins 1953, Mills *et al.*, 1988, Hirota, *et al.*, 1994). Various additional intermediate angular momenta, such as the electronic angular momenta of the ion core (Rydberg states)

$$\vec{\mathbf{J}}^+ = \vec{\mathbf{J}} - \vec{\mathbf{l}}$$
$$\vec{\mathbf{N}}^+ = \vec{\mathbf{N}} - \vec{\mathbf{l}}$$

or of electrons in orbitals well-localized on one atom (atoms with partially filled f orbitals), are sometimes specified in idiosyncratic notations.

Angular momentum basis functions, $|A\alpha M_A\rangle$, can be simultaneous eigenfunctions of three types of operators: the magnitude $\mathbf{A}^2$, the projection of $\vec{\mathbf{A}}$ onto the internuclear axis, $\mathbf{A}_z$, and the projection of $\vec{\mathbf{A}}$ onto the laboratory quantization axis, $\mathbf{A}_Z$. The basis function labels A, α, and M_A correspond to the eigenvalues of $\mathbf{A}^2$, $\mathbf{A}_z$, and $\mathbf{A}_Z$, respectively $\hbar^2 A(A+1), \hbar\alpha$, and $\hbar M_A$. The molecular fixed quantum number is generally represented by a Greek letter and, in the cases of $\mathbf{l}_z$ and $\mathbf{L}_z$, the *value of the quantum number* is also signified by a Greek letter (which becomes part of the name of the eigenfunction):

molecule-fixed projection of $\mathbf{I}_A$, $\mathbf{I}_B$, or $\mathbf{I}$	no standard notation
$\mathbf{l}_{iz}$	$\lambda_i = 0, 1, 2, \ldots (\sigma, \pi, \delta, \ldots)$
$\mathbf{s}_{iz}$	$\sigma_i = \pm\frac{1}{2}$
$\mathbf{R}_z$	zero
$\mathbf{j}_{iz}$	$\omega_i = \lambda_i + \sigma_i$
$\mathbf{L}_z$	$\Lambda = 0, 1, 2, \ldots (\Sigma, \Pi, \Delta, \ldots)$
$\mathbf{S}_z$	Σ
$\mathbf{F}_z$	no standard notation
$\mathbf{J}_z$	$\Omega = \Lambda + \Sigma$
$\mathbf{N}_z$	Λ.

2.3.1 Angular Momentum Components Defined by Normal and Anomalous Commutation Rules

An angular momentum operator may be defined by the commutation behavior of its space-fixed Cartesian components,

$$[\mathbf{A}_I, \mathbf{A}_J] = i\hbar \sum_K \epsilon_{IJK} \mathbf{A}_K, \tag{2.3.1}$$

where $\epsilon_{IJK} = 0$ if two coordinates are repeated, $+1$ if IJK is a cyclic permutation of XYZ, or -1 for an anticyclic permutation. Then, from these commutation rules, without ever specifying the differential operator form of $\vec{\mathbf{A}}$ or the functional form of $|A\alpha M_A\rangle$, all matrix elements of $\mathbf{A}^2$, $\mathbf{A}_X$, $\mathbf{A}_Y$, and $\mathbf{A}_Z$ in the $\mathbf{A}^2$, $\mathbf{A}_z$, $\mathbf{A}_Z$ eigenbasis, $|\mathrm{A}\alpha\mathrm{M_A}\rangle$, may be derived (Condon and Shortley, 1953, pp. 45-50):

$$\langle A'\alpha' M'_A|\mathbf{A}^2|A\alpha M_A\rangle = +\hbar^2 A(A+1)\delta_{A'A}\delta_{\alpha'\alpha}\delta_{M'_A M_A} \tag{2.3.2}$$

$$\langle A'\alpha' M'_A|\mathbf{A}_Z|A\alpha M_A\rangle = +\hbar M_A \delta_{A'A}\delta_{\alpha'\alpha}\delta_{M'_A M_A} \tag{2.3.3}$$

$$\begin{aligned}\langle A'\alpha' M'_A|\mathbf{A}_X \pm i\mathbf{A}_Y|A\alpha M_A\rangle &= \hbar \exp\left[i\left(\theta_{M_A} - \theta_{M_A\pm 1}\right)\right] \\ &\times \left[A(A+1) - M_A\left(M_A \pm 1\right)\right]^{1/2} \\ &\times \delta_{A'A}\delta_{\alpha'\alpha}\delta_{M'_A M_A\pm 1},\end{aligned} \tag{2.3.4}$$

where $\mathbf{A}_X + i\mathbf{A}_Y$ and $\mathbf{A}_X - i\mathbf{A}_Y$ are respectively raising and lowering operators, $\mathbf{A}_+$ and $\mathbf{A}_-$, and

$$\mathbf{A}_X = (1/2)\left(\mathbf{A}_+ + \mathbf{A}_-\right), \qquad \mathbf{A}_Y = -(i/2)\left(\mathbf{A}_+ - \mathbf{A}_-\right).$$

Further, θ_{M_A} and $\theta_{M_A\pm 1}$ are the arbitrary phase factors of the $|A\alpha M_A\rangle$ and $|A\alpha M_A \pm 1\rangle$ basis functions. If one requires that all matrix elements of $\mathbf{A}_X$ be real and positive, then each of the $2A+1$ $|A\alpha M_A\rangle$ basis functions for $M_A = -A, -A+1, \ldots, A$ must have the same phase factor. This is generally referred to as the Condon and Shortley phase convention (Condon and Shortley, 1953, p. 48).

It is possible to derive all matrix elements of molecule-fixed components of $\vec{\mathbf{A}}$ in the $|\alpha M_A\rangle$ basis from the commutation behavior of $\mathbf{A}_x, \mathbf{A}_y, \mathbf{A}_z$. Unfortunately, some angular momenta ($\mathbf{I}$, $\mathbf{L}$, $\mathbf{S}$, $\mathbf{J}_a$) obey normal commutation rules,

$$[\mathbf{A}_i, \mathbf{A}_j] = i\hbar \sum_k \epsilon_{ijk} \mathbf{A}_k, \tag{2.3.5}$$

while others ($\mathbf{R}$, $\mathbf{F}$, $\mathbf{J}$, $\mathbf{N}$, $\mathbf{O}$), *those describing rotations of the molecule-fixed coordinate system relative to the laboratory,* obey anomalous (i.e., reversed angular momentum) commutation rules,

$$[\mathbf{A}_i, \mathbf{A}_j] = -i\hbar \sum_k \epsilon_{ijk} \mathbf{A}_k. \tag{2.3.6}$$

For normal commutation, the matrix elements for $\mathbf{A}_i$ are analogous to those for $\mathbf{A}_I$,

$$\langle A'\alpha' M'_A | \mathbf{A}_z | A\alpha M_A \rangle = +\hbar\alpha \delta_{A'A} \delta_{\alpha'\alpha} \delta_{M'_A M_A}, \tag{2.3.7}$$

$$\langle A'\alpha' M'_A | \mathbf{A}^{\pm} | A\alpha M_A \rangle = +\hbar e^{i(\phi_\alpha - \phi_{\alpha\pm1})} [A(A+1) - \alpha(\alpha \pm 1)]^{1/2} \times \delta_{A'A} \delta_{\alpha'\alpha\pm1} \delta_{M'_A M_A}, \tag{2.3.8}$$

where the *molecule-fixed raising and lowering operators are denoted here* $\mathbf{A}^{\pm}$,

$$\mathbf{A}^{\pm} \equiv \mathbf{A}_x \pm i\mathbf{A}_y,$$

to distinguish them from space-fixed $\mathbf{A}_{\pm}$ operators. Similarly, for molecule-fixed components of $\vec{\mathbf{A}}$ obeying anomalous commutation rules, the $\mathbf{A}_z$ matrix elements are identical to Eq. (2.3.7), but

$$\begin{aligned}\langle A'\alpha' M'_A | \mathbf{A}_x \pm i\mathbf{A}_y | A\alpha M_A \rangle &= \langle A'\alpha' M'_A | \mathbf{A}^{\pm} | A\alpha M_A \rangle \\ &= \hbar e^{i(\phi_\alpha - \phi_{\alpha\mp1})} [A(A+1) - \alpha(\alpha \mp 1)]^{1/2} \\ &\times \delta_{A'A} \delta_{\alpha'\alpha\mp1} \delta_{M'_A M_A}.\end{aligned} \tag{2.3.9}$$

The raising and lowering roles of $\mathbf{A}^+$ and $\mathbf{A}^-$ are interchanged when $\mathbf{A}$ behaves anomalously. It would be reasonable to require that the $\mathbf{A}_x$ matrix elements for both normal and anomalous $\mathbf{A}$ be real and positive, thereby requiring the ϕ_α phase factors for all $2A+1$ basis functions to be equal. Such a phase choice is not in universal use. A standard phase convention has been proposed (Brown and Howard, 1976).

Brown and Howard (1976) have suggested that all molecule-fixed matrix elements be evaluated in terms of space-fixed operator components. The reasons for this are: the space-fixed components of all operators obey normal commutation rules; it is natural to adopt the Condon and Shortley phase convention for the space-fixed part of the basis functions [$\theta_{M_A} = \theta_{M_A \pm 1}$ for all M_A; $\mathbf{A}_{\pm}$ matrix elements are real and positive, see Eq. (2.3.4)]; and all remaining ambiguities about the relative phases of the molecule-fixed part of the basis functions [ϕ_α; see Eq. (2.3.9)] are settled by the choice of direction cosine matrix elements or rotation matrices used to transform between space- and molecule-fixed coordinate systems.

It is necessary to define precisely what is meant by a molecule-fixed component of any angular momentum that includes **R** (e.g., **R, N, O, J,** and **F**). Brown and Howard do this using spherical tensor methods; the following derivation is in terms of direction cosines. The direction cosine operator, $\boldsymbol{\alpha}$, is a 3×3 matrix, the components of which are defined by

$$\alpha_I^j = \hat{I} \cdot \hat{j}, \tag{2.3.10}$$

where $\hat{I}, \hat{J}, \widehat{K}$ and $\hat{i}, \hat{j}, \hat{k}$ are unit vectors defining right-handed Cartesian space- and molecule-fixed coordinate systems. Any operator which does not include **R** (e.g. **S, L, I**) does not cause relative rotation of the molecule- and space-fixed systems and therefore commutes with $\boldsymbol{\alpha}$. *For such operators only*, the transformation between space and molecule components is analogous to that for **S**,

$$\mathbf{S}_I = \sum_j \alpha_I^j \mathbf{S}_j \qquad \text{for} \quad I = X, Y, Z, +, - \tag{2.3.11a}$$

$$j = x, y, z, +, -$$

$$\mathbf{S}_X = \frac{1}{2}\left[\alpha_X^+ \mathbf{S}^- + \alpha_X^- \mathbf{S}^+\right] + \alpha_X^z \mathbf{S}_z \tag{2.3.11b}$$

$$\mathbf{S}_Y = \frac{1}{2}\left[\alpha_Y^+ \mathbf{S}^- + \alpha_Y^- \mathbf{S}^+\right] + \alpha_Y^z \mathbf{S}_z \tag{2.3.11c}$$

$$\mathbf{S}_Z = \frac{1}{2}\left[\alpha_Z^+ \mathbf{S}^- + \alpha_Z^- \mathbf{S}^+\right] + \alpha_Z^z \mathbf{S}_z \tag{2.3.11d}$$

where

$$\alpha_I^{\pm} \equiv \alpha_I^x \pm i\alpha_I^y. \tag{2.3.12}$$

As a specific example, consider the matrix elements of **J**·**S**, which is a product of two operators, one having anomalous, the other regular molecule-fixed commutation rules. The value of a specific matrix element, $\langle nJ\Omega M|\mathbf{J} \cdot \mathbf{S}|n'J'\Omega'M'\rangle$, cannot depend on whether **J**·**S** is written in terms of space-fixed components,

$$\mathbf{J} \cdot \mathbf{S} = \mathbf{J}_Z\mathbf{S}_Z + \mathbf{J}_X\mathbf{S}_X + \mathbf{J}_Y\mathbf{S}_Y = \mathbf{J}_Z\mathbf{S}_Z + \frac{1}{2}(\mathbf{J}_+\mathbf{S}_- + \mathbf{J}_-\mathbf{S}_+), \tag{2.3.13}$$

or molecule-fixed components,

$$\mathbf{J} \cdot \mathbf{S} = \mathbf{J}_z\mathbf{S}_z + \frac{1}{2}(\mathbf{J}^+\mathbf{S}^- + \mathbf{J}^-\mathbf{S}^+). \tag{2.3.14}$$

Inserting the Eqs. (2.3.11b-d) transformation from $\mathbf{S}_{SPACE}$ to $\mathbf{S}_{molecule}$ into Eq. (2.3.13) and comparing the coefficients of $\mathbf{S}_z, \mathbf{S}^-$, and $\mathbf{S}^+$ to those in Eq. (2.3.14), the following *definitions* of the molecule-fixed components of $\vec{J}$ are obtained

$$\mathbf{J}_z = \mathbf{J}_Z\alpha_Z^z + \frac{1}{2}(\mathbf{J}_+\alpha_-^z + \mathbf{J}_-\alpha_+^z), \tag{2.3.15a}$$

$$\mathbf{J}^+ = \mathbf{J}_Z\alpha_Z^+ + \frac{1}{2}(\mathbf{J}_+\alpha_-^+ + \mathbf{J}_-\alpha_+^+), \tag{2.3.15b}$$

$$\mathbf{J}^- = \mathbf{J}_Z\alpha_Z^- + \frac{1}{2}(\mathbf{J}_+\alpha_-^- + \mathbf{J}_-\alpha_+^-) \tag{2.3.15c}$$

where

$$\alpha_\pm^j \equiv \alpha_X^j \pm i\alpha_Y^j. \tag{2.3.16}$$

More concisely,

$$\mathbf{J}_i = \sum_I \mathbf{J}_I\alpha_I^i \qquad \text{for} \quad i = x, y, z, +, -, \tag{2.3.17}$$

where, for $i = +$ and $-$, the molecule-fixed $\mathbf{J}^\pm$ operators are implied.

The goal of this operator algebra is to obtain a simple and unambiguous prescription for evaluating the molecule-fixed matrix elements of $\mathbf{J}^\pm$. Consider the specific matrix element

$$\langle JM\Omega|\mathbf{J}^+|JM\Omega+1\rangle$$

Using Eq. (2.3.15b) and the completeness relationship,

$$\sum_{J',M',\Omega'} |J'M'\Omega'\rangle\langle J'M'\Omega'| = 1 \qquad \text{(the unit matrix)},$$

to expand the products of $\mathbf{J}_I$ and α_I^j operators, for example,

$$\langle JM\Omega|\mathbf{J}_+\alpha_-^+|JM\Omega+1\rangle = \sum_{J',M',\Omega'} \langle JM\Omega|\mathbf{J}_+|J'M'\Omega'\rangle \times \langle J'M'\Omega'|\alpha_-^+|JM\Omega+1\rangle, \tag{2.3.18}$$

noting that $\mathbf{J}_+$ selection rules require $J' = J$, $\Omega' = \Omega$, and $M' = M-1$, the summation reduces to a single term,

$$\langle JM\Omega|\mathbf{J}_+\alpha_-^+|JM\Omega+1\rangle = \langle JM\Omega|\mathbf{J}_+|JM-1\Omega\rangle \times \langle JM-1\Omega|\alpha_-^+|JM\Omega+1\rangle. \tag{2.3.19}$$

The space-fixed $\mathbf{J}_+$ matrix element is

$$\langle JM\Omega|\mathbf{J}_+|JM-1\Omega\rangle = \hbar[J(J+1) - M(M-1)]^{1/2},$$

Table 2.1: Direction Cosine Matrix Elements for $\Delta J = 0^{a}$

	$\langle JM'\Omega'\lvert\alpha\rvert JM\Omega\rangle$		
	$\Delta\Omega = \Omega' - \Omega$		
$\Delta M = M' - M$	+1	0	-1
+1	$\left\langle\alpha_{+}^{-}\right\rangle=\left\langle\alpha_{X}^{-}+i\alpha_{Y}^{-}\right\rangle=\left\langle\alpha_{X}^{x}-i\alpha_{X}^{y}+i\left(\alpha_{Y}^{x}-i\alpha_{Y}^{y}\right)\right\rangle$ $=\left\langle\alpha_{+}^{x}-i\alpha_{+}^{y}\right\rangle=\frac{[J(J+1)-M(M+1)]^{1/2}[J(J+1)-\Omega(\Omega+1)]^{1/2}}{J(J+1)}$	$\left\langle\alpha_{+}^{z}\right\rangle=\langle\alpha_{X}^{z}+i\alpha_{Y}^{z}\rangle$ $=\frac{\Omega[J(J+1)-M(M+1)]^{1/2}}{J(J+1)}$	$\left\langle\alpha_{+}^{+}\right\rangle=\left\langle\alpha_{X}^{+}+i\alpha_{Y}^{+}\right\rangle=\left\langle\alpha_{+}^{x}+i\alpha_{+}^{y}\right\rangle$ $=\frac{[J(J+1)-M(M+1)]^{1/2}[J(J+1)-\Omega(\Omega-1)]^{1/2}}{J(J+1)}$
0	$\left\langle\alpha_{Z}^{-}\right\rangle=\left\langle\alpha_{Z}^{x}-i\alpha_{Z}^{y}\right\rangle=\frac{M[J(J+1)-\Omega(\Omega+1)]^{1/2}}{J(J+1)}$	$\langle\alpha_{Z}^{z}\rangle=\frac{\Omega M}{J(J+1)}$	$\left\langle\alpha_{Z}^{+}\right\rangle=\left\langle\alpha_{Z}^{x}+i\alpha_{Z}^{y}\right\rangle=\frac{M[J(J+1)-\Omega(\Omega-1)]^{1/2}}{J(J+1)}$
-1	$\left\langle\alpha_{-}^{-}\right\rangle=\left\langle\alpha_{X}^{-}-i\alpha_{Y}^{-}\right\rangle=\left\langle\alpha_{-}^{x}-i\alpha_{-}^{y}\right\rangle$ $=\frac{[J(J+1)-M(M-1)]^{1/2}[J(J+1)-\Omega(\Omega+1)]^{1/2}}{J(J+1)}$	$\left\langle\alpha_{-}^{z}\right\rangle=\langle\alpha_{X}^{z}-i\alpha_{Y}^{z}\rangle$ $=\frac{\Omega[J(J+1)-M(M-1)]^{1/2}}{J(J+1)}$	$\left\langle\alpha_{-}^{+}\right\rangle=\left\langle\alpha_{X}^{+}-i\alpha_{Y}^{+}\right\rangle=\left\langle\alpha_{-}^{x}+i\alpha_{-}^{y}\right\rangle$ $=\frac{[J(J+1)-M(M-1)]^{1/2}[J(J+1)-\Omega(\Omega-1)]^{1/2}}{J(J+1)}$

[a] From Hougen (1970), Table 6.

and, using the $\Delta J = 0$ direction cosine matrix elements from Table 2.1 (see Hougen, 1970, Table 6) †

$$\langle JM-1\Omega|(\alpha_X^+ - i\alpha_Y^+)|JM\Omega+1\rangle = \frac{[J(J+1)-M(M-1)]^{1/2}[J(J+1)-\Omega(\Omega+1)]^{1/2}}{J(J+1)}.$$

Thus

$$\langle JM\Omega|\mathbf{J}_+\boldsymbol{\alpha}_-^+|JM\Omega+1\rangle = \frac{\hbar[J(J+1)-M(M-1)][J(J+1)-\Omega(\Omega+1)]^{1/2}}{J(J+1)}. \tag{2.3.20}$$

Similarly,

$$\langle JM\Omega|\mathbf{J}_-\boldsymbol{\alpha}_+^+|JM\Omega+1\rangle = \frac{\hbar[J(J+1)-M(M+1)][J(J+1)-\Omega(\Omega+1)]^{1/2}}{J(J+1)}, \tag{2.3.21}$$

and

$$\langle JM\Omega|\mathbf{J}_Z\boldsymbol{\alpha}_Z^+|JM\Omega+1\rangle = \frac{\hbar M^2[J(J+1)-\Omega(\Omega+1)]^{1/2}}{J(J+1)}, \tag{2.3.22}$$

thus

$$\begin{aligned}\langle JM\Omega|\mathbf{J}^+|JM\Omega+1\rangle &= \frac{\hbar J(J+1)[J(J+1)-\Omega(\Omega+1)]^{1/2}}{J(J+1)}\\ &= +\hbar[J(J+1)-\Omega(\Omega+1)]^{1/2}.\end{aligned} \tag{2.3.23}$$

The nonzero matrix elements for $\mathbf{J}_z$ and $\mathbf{J}^-$ are

$$\langle JM\Omega|\mathbf{J}_z|JM\Omega\rangle = +\hbar\Omega \tag{2.3.24}$$

$$\langle JM\Omega|\mathbf{J}^-|JM\Omega-1\rangle = +\hbar[J(J+1)-\Omega(\Omega-1)]^{1/2}. \tag{2.3.25}$$

The reversal of raising/lowering roles of $J^{\pm}$ is made obvious by

$$\langle JM\Omega\pm1|\mathbf{J}^{\mp}|JM\Omega\rangle = \hbar[J(J+1)-\Omega(\Omega\pm1)]^{1/2}.$$

This justifies the general statement of Eq. (2.3.9) that, for all operators $\mathbf{A}$ obeying anomalous molecule-fixed commutation rules, Eq. (2.3.6) (any operator that includes $\mathbf{R}$), the matrix elements of $\mathbf{A}^{\pm}$ are real and positive but that $\mathbf{A}^+$ and $\mathbf{A}^-$ act exactly as lowering and raising operators, respectively. This reversal of the lowering/raising roles of $\mathbf{A}^{\pm}$ may be viewed as arising from the direction cosine matrix elements [Eq. (2.3.17)] rather than any unusual property of the operator $\mathbf{A}$.

† The phase choices implicit in the definitions of $|JM\Omega\rangle$ basis functions and tabulated α_I^i direction cosine matrix elements are summarized in Section 2.3.3.

2.3.2 Recipes for Evaluation of Molecule-Fixed Angular Momentum Matrix Elements

Consider the basis set $|nJLS\Omega\Lambda\Sigma M_J\rangle$, where n is a shorthand for all remaining quantum numbers (e.g., electronic configuration and vibrational level) needed to specify the basis function uniquely. Although, for reasons discussed in Section 3.2, L cannot usually be specified, this is the Hund's case (a) basis. Inclusion of L here simplifies the following discussion and illustrates an important point, namely, that when a basis set is defined in terms of a sufficient number of magnitude and molecule-fixed projection quantum numbers, then all conceivable magnitude and molecule-fixed angular momentum matrix elements may be evaluated by elementary techniques. Matrix elements of $\mathbf{J}_a^2$ and $\mathbf{R}^2$ may be evaluated below as illustration of this point.

The matrix elements of the $\mathbf{J}_a$ operator are calculated here as a first step toward those of $\mathbf{R}$. Eigenvalues of the operator $\mathbf{J}_a^2$ do not appear in the list of quantum numbers which define the $|nJLS\Omega\Lambda\Sigma M_J\rangle$ basis. However,

$$\mathbf{J}_a = \mathbf{L} + \mathbf{S}, \tag{2.3.26}$$

$$\mathbf{J}_a^2 = \mathbf{L}^2 + \mathbf{S}^2 + \mathbf{L}\cdot\mathbf{S} + \mathbf{S}\cdot\mathbf{L}, \tag{2.3.27}$$

and, since $\mathbf{L}$ and $\mathbf{S}$ commute (because they act on different coordinates),

$$\mathbf{J}_a^2 = \mathbf{L}^2 + \mathbf{S}^2 + 2\mathbf{L}_z\mathbf{S}_z + (\mathbf{L}^+\mathbf{S}^- + \mathbf{L}^-\mathbf{S}^+). \tag{2.3.28}$$

Thus, $\mathbf{J}_a^2$ is expressed in terms of operators for which the matrix elements are easily evaluable. Further, $\mathbf{L}^2$, $\mathbf{S}^2$, and $\mathbf{L}_z\mathbf{S}_z$ have only diagonal matrix elements in the $|nJLS\Omega\Lambda\Sigma M_J\rangle$ basis, whereas the $\mathbf{L}^\pm\mathbf{S}^\mp$ terms follow $\Delta\Lambda = -\Delta\Sigma = \pm1, \Delta\Omega = 0, \Delta J = \Delta L = \Delta S = 0$ selection rules:

1. Diagonal matrix elements of $\mathbf{J}_a^2$:

$$\langle n'JLS\Omega\Lambda\Sigma M_J|\mathbf{J}_a^2|nJLS\Omega\Lambda\Sigma M_J\rangle = \hbar^2[L(L+1) + S(S+1) + 2\Lambda\Sigma]\delta_{n'n} \tag{2.3.29}$$

2. $\Delta\Lambda = -\Delta\Sigma = \pm1$ off-diagonal matrix elements of $\mathbf{J}_a^2$: the only term in $\mathbf{J}_a^2$ that gives rise to $\Delta\Lambda = -\Delta\Sigma = \pm1$ matrix elements is $\mathbf{L}^\pm\mathbf{S}^\mp$; thus the relevant matrix element is

$$\langle n'JLS\Omega, \Lambda\pm1, \Sigma\mp1, M_J|\mathbf{L}^\pm\mathbf{S}^\mp|nJLS\Omega\Lambda\Sigma M_J\rangle$$
$$= \hbar^2[L(L+1) - \Lambda(\Lambda\pm1)]^{1/2}[S(S+1) - \Sigma(\Sigma\mp1)]^{1/2}\delta_{n'n}. \tag{2.3.30}$$

If L cannot be specified in the basis set, then the selection rules remain unchanged and $L(L+1)$, wherever it appears, must be replaced by an unknown constant.

The operator $\mathbf{R}^2$ is important because it appears in the rotational Hamiltonian operator,

$$\mathbf{R}^2 = (\mathbf{J} - \mathbf{L} - \mathbf{S})^2, \tag{2.3.31}$$

or, taking advantage of the fact that matrix elements of $\mathbf{J}_a^2$ have just been worked out,

$$\mathbf{R}^2 = (\mathbf{J} - \mathbf{J}_a)^2 = \mathbf{J}^2 + \mathbf{J}_a^2 - \mathbf{J} \cdot \mathbf{J}_a - \mathbf{J}_a \cdot \mathbf{J}, \qquad (2.3.32)$$

and, since $[\mathbf{J}_i, \mathbf{J}_{ai}] = 0$, then

$$\mathbf{R}^2 = \mathbf{J}^2 + \mathbf{J}_a^2 - 2\mathbf{J}_z\mathbf{J}_{az} - (\mathbf{J}^+\mathbf{J}_a^- + \mathbf{J}^-\mathbf{J}_a^+). \qquad (2.3.33a)$$

Equation (2.3.33a) can be put into more symmetric form by recognizing that when either $\mathbf{J}_z$ or $\mathbf{J}_{az}$ operates on $|nJLS\Omega\Lambda\Sigma M_J\rangle$, the result is $+\Omega$, thus,

$$\mathbf{R}^2 = (\mathbf{J}^2 - \mathbf{J}_z^2) + (\mathbf{J}_a^2 - \mathbf{J}_{az}^2) - (\mathbf{J}^+\mathbf{J}_a^- + \mathbf{J}^-\mathbf{J}_a^+). \qquad (2.3.33b)$$

Since $\mathbf{J}$ (but not $\mathbf{J}_a$) is an operator that involves rotation of the molecule relative to the space-fixed coordinate system, $\mathbf{J}^+$ acts as a lowering operator on Ω, whereas $\mathbf{J}_a^+$ acts as a raising operator on Λ or Σ (and thus indirectly on $\Omega = \Lambda + \Sigma$). Arranged according to selection rules, the matrix elements of $\mathbf{R}^2$ are:

1. Diagonal matrix elements of $\mathbf{R}^2$:

$$\langle n'JLS\Omega\Lambda\Sigma M_J|\mathbf{R}^2|nJLS\Omega\Lambda\Sigma M_J\rangle$$
$$= \hbar^2[J(J+1) + L(L+1) + S(S+1) + 2\Lambda\Sigma - 2\Omega^2]\delta_{n'n} \qquad (2.3.34)$$

or, rearranged into a more symmetric form,

$$\hbar^2[J(J+1) - \Omega^2 + L(L+1) - \Lambda^2 + S(S+1) - \Sigma^2].$$

2. $\Delta\Omega = 0, \Delta\Lambda = -\Delta\Sigma = \pm 1$ off-diagonal matrix elements of $\mathbf{R}^2$:

$$\langle n'JLS\Omega\Lambda \pm 1\Sigma \mp 1 M_J|\mathbf{R}^2|nJLS\Omega\Lambda\Sigma M_J\rangle$$
$$= +\hbar^2[L(L+1) - \Lambda(\Lambda \pm 1)]^{1/2}[S(S+1) - \Sigma(\Sigma \mp 1)]^{1/2}\delta_{n'n}. \qquad (2.3.35)$$

3. $\Delta\Omega = \Delta\Lambda = \pm 1, \Delta\Sigma = 0$ off-diagonal matrix elements of $\mathbf{R}^2$:

$$\langle n'JLS\Omega \pm 1\Lambda \pm 1\Sigma M_J|\mathbf{R}^2|nJLS\Omega\Lambda\Sigma M_J\rangle$$
$$= -\hbar^2[J(J+1) - \Omega(\Omega \pm 1)]^{1/2}[L(L+1) - \Lambda(\Lambda \pm 1)]^{1/2}\delta_{n'n}. \qquad (2.3.36)$$

4. $\Delta\Omega = \Delta\Sigma = \pm 1, \Delta\Lambda = 0$ off-diagonal matrix elements of $\mathbf{R}^2$:

$$\langle n'JLS\Omega \pm 1\Lambda\Sigma \pm 1 M_J|\mathbf{R}^2|nJLS\Omega\Lambda\Sigma M_J\rangle$$
$$= -\hbar^2[J(J+1) - \Omega(\Omega \pm 1)]^{1/2}[S(S+1) - \Sigma(\Sigma \pm 1)]^{1/2}\delta_{n'n}. \qquad (2.3.37)$$

These results are summarized in Table 3.2.

In order to illustrate the unique convenience of the case (a) basis, consider matrix elements of $\mathbf{J}\cdot\mathbf{S}$ and $\mathbf{L}\cdot\mathbf{S}$ in the case (b) $|JLSN\Lambda M_J\rangle$ basis. The matrix elements of $\mathbf{J}\cdot\mathbf{S}$ (also $\mathbf{N}\cdot\mathbf{S}$ and $\mathbf{J}\cdot\mathbf{N}$) are evaluated easily by making the operator substitution:

$$\begin{aligned} \mathbf{N} &= \mathbf{J} - \mathbf{S} \\ \mathbf{N}^2 &= \mathbf{J}^2 + \mathbf{S}^2 - 2\mathbf{J}\cdot\mathbf{S}, \\ \mathbf{J}\cdot\mathbf{S} &= \frac{1}{2}(\mathbf{J}^2 + \mathbf{S}^2 - \mathbf{N}^2). \end{aligned} \tag{2.3.38}$$

Thus $\mathbf{J}\cdot\mathbf{S}$ has only diagonal matrix elements,

$$\begin{aligned} &\langle n'JLSN\Lambda M_J|\mathbf{J}\cdot\mathbf{S}|nJLSN\Lambda M_J\rangle = \\ &\qquad \frac{1}{2}\hbar^2[J(J+1)] + S(S+1) - N(N+1)]\delta_{n'n}. \end{aligned} \tag{2.3.39}$$

The story is quite different for $\mathbf{L}\cdot\mathbf{S}$ because no suitable operator replacement exists and, when attempting to apply standard raising and lowering operator techniques to matrix elements of

$$\mathbf{L}_z\mathbf{S}_z + \frac{1}{2}(\mathbf{L}^+\mathbf{S}^- + \mathbf{L}^-\mathbf{S}^+),$$

in the case (b) basis set, one finds that the matrix elements and selection rules for $\mathbf{S}_z$ and $\mathbf{S}^\pm$ are not obvious. The solution to this problem, discussed in Section 3.2, requires expression of the case (b) basis functions in terms of case (a) basis functions, for which evaluation of $\mathbf{L}\cdot\mathbf{S}$ matrix elements is easy.

2.3.3 Euler Angles, $|JM\Omega\rangle$ Basis Functions, Direction Cosines, and Phases

In dealing simultaneously with space- and molecule-fixed coordinate systems, one must explicitly define the transformations between coordinate systems and specify the relative phases of all basis functions. Larsson (1981) has reviewed the numerous phase conventions currently in use for rotating diatomic molecules. (See also Wolf, 1969.)

The normalized rotating molecule wavefunction, expressed in terms of the three Euler angles α, β, and γ (see Fig. 1.1 in Edmonds (1974), p.7) is

$$\begin{aligned} \langle\alpha\beta\gamma|JM\Omega\rangle &= \left(\frac{2J+1}{8\pi^2}\right)^{1/2} \mathcal{D}^J_{\Omega M}(\alpha,\beta,\gamma) \\ &= \left(\frac{2J+1}{8\pi^2}\right)^{1/2} \exp(iM\gamma)d^J_{\Omega M}(\beta)\exp(i\Omega\alpha) \end{aligned} \tag{2.3.40}$$

where $d^J_{\Omega M}$ is defined by Edmonds (1974, pp. 57-62).† The Euler angles relate the molecule-fixed axis system to the space-fixed system as follows (Edmonds,

†The notation $\langle\alpha\beta\gamma|JM\Omega\rangle$ is the translation, into Dirac notation, of the Schrödinger wavefunction

$$\psi_{JM\Omega}(\alpha,\beta,\gamma) = \langle\alpha\beta\gamma|JM\Omega\rangle.$$

One frequently sees $|JM\Omega\rangle$ defined as a function of α,β,γ, but strictly speaking $|JM\Omega\rangle$ is a vector, not a function of continuous variables.

1974, pp. 6-8): rotate the *molecule* first by γ about the space-fixed Z–axis, then by β about the space-fixed Y–axis, and finally by α about the space-fixed Z–axis. (Rotations are "active" and in the sense of advancing a right-handed screw along the specified axis.)

For a linear molecule, the position of the symmetry axis (the molecule-fixed z–axis) in space is specified by only two Euler angles, β and γ, which are respectively identical to the spherical polar coordinates θ and ϕ (see Fig. 2.4). The third Euler angle, α, which specifies the orientation of the molecule-fixed x– and y–axes, is unaffected by molecular rotation but appears explicitly as an Ω-dependent phase factor in the rotational basis functions [Eq. (2.3.41)]. Cartesian coordinates in space- and molecule-fixed systems are related by the geometrical transformation represented by the 3×3 direction cosine matrix (Wilson *et al.*, 1980, p. 286). The direction cosine matrix $\boldsymbol{\alpha}_I^i$ given by Hougen (1970, p. 18) is obtained by setting $\alpha = \pi/2$ (notation of Wilson *et al.*, 1980: $\theta = \beta, \phi = \gamma, \chi = \alpha = \pi/2$). The *direction cosine matrix* is expressed in terms of sines and cosines of θ and ϕ. Matrix elements $\langle J'M'\Omega'|\alpha_I^i|JM\Omega\rangle$, evaluated in the $|JM\Omega\rangle$ basis, of the direction cosines, are expressed in terms of the J, M, and Ω quantum numbers. The *direction cosine matrix elements* of Hougen (1970, p. 31), Townes and Schawlow (1955, p. 96), and Table 2.1 assume the basis set definition derived from Eq. (2.3.40) and the phase choice $\alpha = \pi/2$:

$$\left\langle \frac{\pi}{2}\theta\phi \middle| JM\Omega \right\rangle = \left(\frac{2J+1}{4\pi}\right)^{1/2} \mathcal{D}^J_{\Omega M}\left(\frac{\pi}{2}, \theta, \phi\right)$$

$$= \left(\frac{2J+1}{4\pi}\right)^{1/2} \exp(iM\phi) d^J_{\Omega M}(\theta) \exp\left(i\Omega\frac{\pi}{2}\right). \qquad (2.3.41)$$

The factor of $(2\pi)^{1/2}$ difference in normalization factor between Eqs. (2.3.40) and (2.3.41) occurs because Eq. (2.3.41) implies no integration over the angle α. The rotational basis function defined in Eq. (2.3.41) is identical to that used by Hougen (1970, p. 18) because

$$\mathcal{D}^J_{\Omega M}(\alpha\beta\gamma) = (-1)^{M-\Omega}\mathcal{D}^J_{M\Omega}(\alpha\beta\gamma).$$

Brown and Howard (1976) define $\langle(\pi/2)\theta\phi|JM\Omega\rangle$ in terms of Brink and Satchler's (1968) $\mathcal{D}^J_{M\Omega}{}^*(\alpha\beta\gamma)$. However, Brink and Satchler's $\mathcal{D}^J_{M\Omega}{}^*(\alpha\beta\gamma)$ corresponds to Edmonds' $\mathcal{D}^J_{M\Omega}{}^*(-\alpha-\beta-\gamma)$ (Brink and Satchler, 1968, p. 21) and, in Edmonds' notation,

$$\mathcal{D}^J_{M\Omega}{}^*(-\alpha-\beta-\gamma) = (-1)^{M-\Omega}\mathcal{D}^J_{M\Omega}(\alpha\beta\gamma) = \mathcal{D}^J_{\Omega M}(\alpha\beta\gamma);$$

thus the rotational basis functions used here are also identical to those of Brown and Howard (1976).

The result obtained in Section 2.3.1 that the matrix elements of the space-fixed $\mathbf{J}_\pm$ and molecule fixed $\mathbf{J}^\pm$ operator components are real and positive are consequences of the $|JM\Omega\rangle$ definition and α–phase choice.

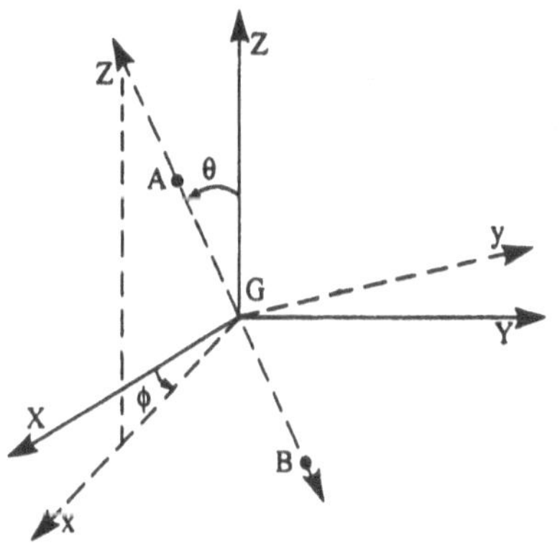

Figure 2.4: Relationships between molecule-fixed (xyz) and space-fixed (XYZ) axis systems. The origin of both coordinate systems is located at the AB molecule center of mass, G. Transformation from XYZ to xyz ("active rotation"; Larsson, 1981) requires rotation of the molecule by $\phi = \gamma$ about $+Z$, followed by clockwise rotation by $\theta = \beta$ about $+Y$, and then rotation by $\alpha = \pi/2$ (not shown) about Z. Since diatomic molecules have cylindrical symmetry about the internuclear axis, the third angle required to specify the orientation of xyz relative to XYZ is fixed and implicitly specifies a phase choice for certain molecule-fixed $\mathbf{A}^{\pm}$ matrix elements.

2.4 Estimation of Parameters in a Model Hamiltonian

Whenever a perturbation, predissociation, or autoionization is observed, its strength is governed by an off-diagonal matrix element. Although these matrix elements can frequently be calculated from *ab initio* wavefunctions or estimated semiempirically from other perturbation-related information, experimentalists typically treat the perturbation matrix element as a purely empirical parameter of no interest other than its capacity to account for the spectrum. This book is based on the premise that perturbation matrix elements have intrinsic molecular structural significance, that their magnitudes are predictable, and that their measured values often provide unexpected clues to a global description of the electronic structure of a molecule.

In Chapter 3, methods are described for using the electronic wavefunction to evaluate off-diagonal matrix elements of different terms that appear in the true microscopic Hamiltonian, **H**. Section 3.2.4 describes some simple procedures, based on the single-electronic-configuration approximation, for reducing many-electron matrix elements of **H** to simple molecular-orbital (Section 3.4.2) one- and two-electron matrix elements. Several examples where the single-configuration approximation fails but a two-configuration picture is adequate are discussed in Sections 5.7 and 7.11.1. Little insight is possible when it becomes necessary (as in large configuration-interaction *ab initio* computations) to go beyond two configurations.

In Sections 5.2, 5.3 and 5.4, the orders of magnitude of all types of perturbation parameters are related to known or easily estimable quantities in order to provide a basis for predicting their absolute magnitudes, relative sizes, and inter- or intramolecular trends.

The same electronic matrix elements that control the strength of perturbations also govern predissociation (Table 7.2) and, in "reduced" form, autoionization (Section 8.4).

The vibrational part of the **H** matrix elements is much easier to calculate than the electronic one. Methods for such calculations are described in Section 5.1 for bound~bound interactions and in Section 7.6 for bound~continuum vibrational interactions.

An understanding of observable properties is seldom trivial. Spectroscopic energy levels are, in principle, eigenvalues of an infinite matrix representation of **H**, which is expressed in terms of an infinite number of "true" deperturbed molecular constants. In practice, this matrix is truncated and the "observed" molecular constants are the "effective" parameters that appear in a finite-dimension "effective Hamiltonian." The Van Vleck transformation, so crucial for reducing **H** to a finite $\mathbf{H}^{\mathrm{eff}}$, is described in Section 4.2.

Experimentalists often treat the electronic part of **H** matrix elements, $\mathbf{H}^{\mathrm{el}}$, as a phenomenological "black box." They regard $\mathbf{H}^{\mathrm{el}}$ as an empirically determined variable parameter without structural significance and are overwhelmed by the difficulty of estimating $\mathbf{H}^{\mathrm{el}}$ or comparing it with calculated values (see Section 3.2.1 and Sections 3.3.1-3.3.3) . This cuts in the opposite direction as well: often theorists are unaware of or unable to utilize a class of observable electronic properties simply because the relationship between the observable quantity and the relevant matrix element of **H** is neither traditionally exploited nor explicitly defined.

The assumption implicit throughout this book is that the parameters used to fit or represent molecular transition frequencies and intensities contain insights into molecular structure. These insights can be more useful than the multi-digit fit parameters themselves, especially when simplifying assumptions are made and tested. Comparisons of observable or effective parameters to those obtained from an exact calculation (true parameters) or a simplified electronic structure model (one-electron orbital parameters) are seldom trivial or unique. The purpose of this book is to help experimentalists and theorists to go beyond molecular fit parameters to terms in the exact microscopic Hamiltonian on the one hand and to approximate electronic structure models on the other. Physical insight, not tables of spectral data and molecular constants, is the ultimate purpose of fundamental experimental and theoretical research.

2.5 Data Compilations

Barrow, R. F., ed. (1973, 1975, 1979, 1982), "Molécules Diatomiques. Bibliographie Critique de Données Spectroscopiques," Tables Internationales de Constantes, CNRS, Paris.

Bernath, P. F. (1997), "A Bibliographic Database for Diatomic Molecules," Northern Spectronics, Waterloo.
Bernath, P. F., and McLeod, S. (2001), *J. Mol. Spectrosc.* **207**, 287 (http://diref.uwaterloo.ca).
Herzberg, G. (1950), "Molecular Spectra and Molecular Structure. I. Spectra of Diatomic Molecules," Van Nostrand-Reinhold, Princeton, New Jersey, reprinted 1989 by Krieger, Malabar.
Huber, K. P., and Herzberg, G. (1979), "Molecular Spectra and Molecular Structure. IV. Constants of Diatomic Molecules," Van Nostrand-Reinhold, Princeton, New Jersey.
Kopp, I., Lindgren, R., and Rydh, B. (1974), "Table of Band Features of Diatomic Molecules in Wavelength Order," Institute of Physics, University of Stockholm, Tables Internationales de Constantes, Paris.
Kopp, I., Jansson, K., and Rydh, B. (1977), "Table of Band Features of Diatomic Molecules in Wavelength Order," Complement A1, Institute of Physics, University of Stockholm, Tables Internationales de Constantes, Paris.
Pearse, R. W. B., and Gaydon, A. G. (1976), "The Identification of Molecular Spectra," Chapman and Hall, London.
Rosen, B., ed. (1970), "Tables Internationales de Constantes Sélectionnées. 17. Données Spectroscopiques relatives aux Molécules Diatomiques," Pergamon, Oxford.
Suárez, C. B. (1972), "Bibliography of Spectra of Diatomic Molecules," 1960–1970, Universidad Nacional de La Plata.

2.6 References

Bernath, P. F. (1995), "Spectra of Atoms and Molecules," Oxford Univ. Press, Oxford.
Brink, D. M., and Satchler, G. R. (1968), "Angular Momentum," Oxford Univ. Press (Clarendon), London and New York.
Brown, J. M., and Howard, B. J. (1976), *Mol. Phys.* **31**, 1517; **32**, 1197.
Brown, J. M. (1998), "Molecular Spectroscopy," Oxford University Press, Oxford.
Brown, J. M., and Carrington, A. (2003), "Rotational Spectroscopy of Diatomic Molecules," Cambridge University Press, London and New York.
Budó, A. (1937), *Z. Phys.* **105**, 579.
Budó, A., and Kovács, I. (1938), *Z. Phys.* **109**, 393.
Bunker, P. R., and Jensen, P. (1998), "Molecular Symmetry and Spectroscopy", NRC Research Press, Ottawa.
Condon, E. U., and Shortley, G. H. (1953), "The Theory of Atomic Spectra," Cambridge Univ. Press, London and New York.
Dieke, G. H. (1935), *Phys. Rev.* **47**, 870.
Dunham, J. L. (1932), *Phys. Rev.* **41**, 713, 721.
Edmonds, A. R. (1974), "Angular Momentum in Quantum Mechanics," Princeton Univ. Press, Princeton, New Jersey.
Herzberg, G. (1950), "Molecular Spectra and Molecular Structure. I. Spectra of Diatomic Molecules", Van Nostrand-Reinhold, Princeton, New Jersey, reprinted 1989 by Krieger, Malabar.
Heurlinger, T. (1917), *Ark. Mat., Astron. Fys.* **12**, 1; (1918), *Phys. Z.* **19**, 316; (1919), *Phys. Z.* **20**, 188.
Hirota, E., Brown, J. M., Hougen, J. T., Shida, T., and Hirota, N. (1994), *Pure and Appl. Chem.* **66**, 571; reprinted (1994), *J. Mol. Spectrosc.* **168**, 628.

Hougen, J. T. (1970), "The Calculation of Rotational Energy Levels and Rotational Line Intensities in Diatomic Molecules," Nat. Bur. Stand. (U.S.), monograph 115.

Huber, K. P., and Herzberg, G. (1979), "Molecular Spectra and Molecular Structure. IV. Constants of Diatomic Molecules," Van Nostrand-Reinhold, Princeton, New Jersey.

Ittmann, G. P. (1931), *Z. Phys.* **71**, 616.

Jenkins, F. A. (1953), *J. Opt. Soc. Am.* **43**, 425.

Katô, H. (1993), "Energy Levels and Line Intensities of Diatomic Molecules," *Bull. Chem. Soc. Japan* **66**, 3203.

Kovács, I. (1937), *Z. Phys.* **106**, 431.

Kovács, I. (1969), "Rotational Structure in the Spectra of Diatomic Molecules," Amer. Elsevier, New York.

Kronig, R. L. (1928), *Z. Phys.* **50**, 347.

Larsson, M. (1981), *Phys. Scr.* **23**, 835.

Mills, I., Cvitas, T., Homann, K., Kallay, N., and Kuchitsu, K. (1988), "Quantities, Units, and Symbols in Physical Chemistry," Blackwell Scientific, Oxford.

Mulliken, R. S. (1930), *Rev. Mod. Phys.* **2**, 60, 506.

Mulliken, R. S. (1931), *Rev. Mod. Phys.* **3**, 89.

Mulliken, R. S. (1932), *Rev. Mod. Phys.* **4**, 1.

Ogilvie, J. P. (1998), "The Vibrational and Rotational Spectroscopy of Diatomic Molecules," Academic Press, San Diego.

Schmid, R., and Gerö, L. (1935), *Z. Phys.* **93**, 656.

Stepanov, B. I. (1940), *J. Phys. USSR* **2**, 81, 89, 197, 205, 381.

Stepanov, B. I. (1945), *J. Phys. USSR* **9**, 317.

Townes, C. H., and Schawlow, A. L. (1955), "Microwave Spectroscopy," McGraw-Hill, New York.

Wilson, E.B., Decius, J. C., and Cross, P. C. (1980), "Molecular Vibrations," Dover, New York.

Wolf, A. A. (1969), *Am. J. Phys.* **37**, 531.

Van Vleck, J. H. (1929), *Phys. Rev.* **33**, 467.

Van Vleck, J. H. (1932), *Phys. Rev.* **40**, 544.

Van Vleck, J. H. (1936), *J. Chem. Phys.* **4**, 327.

Chapter 3

Terms Neglected in the Born-Oppenheimer Approximation

Contents

3.1 The Born-Oppenheimer Approximation 89

3.1.1 Potential Energy Curves 90

3.1.2 Terms Neglected in the Born-Oppenheimer Approximation 92

3.1.2.1 Electrostatic and Nonadiabatic Part of **H** . 92

3.1.2.1.1 Crossing or Diabatic Curves . . . 93

3.1.2.1.2 Noncrossing or Adiabatic Curves . 94

3.1.2.2 The Spin Part of **H** 94

3.1.2.3 Rotational Part of **H** 96

3.2 Basis Functions 99

3.2.1 Hund's Cases 101

3.2.1.1 Definition of Basis Sets 103

3.2.1.2 Quantum Numbers, Level Patterns, and the Effects of Terms Excluded from $\mathbf{H}^{(0)}$. . 113

3.2.1.3 Intermediate Case Situations 126

3.2.1.3.1 Introduction 126

3.2.1.3.2 Examples 127

3.2.1.4 Transformations Between Hund's Case Basis Sets 130

3.2.1.5 Spectroscopic vs. Dynamical Hund's Cases 136

3.2.1.6 Relationship between Noncommuting Terms in **H** and the Most Appropriate Hund's Case 137

3.2.2 Symmetry Properties 138

3.2.2.1 Symmetry Properties of Hund's Case (a) Basis Functions . . . 138
3.2.2.2 Symmetry Properties of non-Hund's Case (a) Basis Functions . . . 145
3.2.3 Molecular Electronic Wavefunctions . . . 148
3.2.4 Matrix Elements between Electronic Wavefunctions 156
3.3 Electrostatic Perturbations . . . 161
3.3.1 Diabatic Curves . . . 163
3.3.2 Approximate Representation of the Diabatic Electronic Wavefunction . . . 165
3.3.3 Adiabatic Curves . . . 168
3.3.4 Choice between the Diabatic and Adiabatic Models . 172
3.3.5 Electromagnetic Field-Dressed Diabatic and Adiabatic Potential Energy Curves . . . 177
3.4 Spin Part of the Hamiltonian . . . 180
3.4.1 The Spin-Orbit Operator . . . 181
3.4.2 Expression of Spin-Orbit Matrix Elements in Terms of One-Electron Molecular Spin-Orbit Parameters . 183
3.4.2.1 Matrix Elements of the $\mathbf{l}_{zi} \cdot \mathbf{s}_{zi}$ Term . . . 183
3.4.2.1.1 Diagonal Matrix Elements . . . 184
3.4.2.1.2 Off-Diagonal Matrix Elements . . 187
3.4.2.2 Matrix Elements of the $\left(\mathbf{l}_i^+ \mathbf{s}_i^- + \mathbf{l}_i^- \mathbf{s}_i^+\right)$ Part of $\mathbf{H}^{\mathrm{SO}}$ . . . 190
3.4.3 The Spin-Rotation Operator . . . 191
3.4.4 The Spin-Spin Operator . . . 196
3.4.4.1 Diagonal Matrix Elements of $\mathbf{H}^{\mathrm{SS}}$: Calculation of the Direct Spin-Spin Parameter . 196
3.4.4.2 Calculation of Second-Order Spin-Orbit Effects . . . 199
3.4.4.2.1 π^2 Configuration . . . 201
3.4.4.2.2 $\pi^3\pi'$ (or $\pi^3\pi'^3$ and $\pi\pi'$) Configurations . . . 201
3.4.4.3 Off-Diagonal Matrix Elements . . . 202
3.4.5 Tensorial Operators . . . 203
3.5 Rotational Perturbations . . . 210
3.5.1 Spin-Electronic Homogeneous Perturbations . . . 210
3.5.2 The **S**-Uncoupling Operator . . . 212
3.5.3 The **L**-Uncoupling Operator . . . 213
3.5.4 $^2\Pi \sim {}^2\Sigma^+$ Interaction . . . 217
3.6 References . . . 227

3.1 The Born-Oppenheimer Approximation

The exact Hamiltonian **H** for a diatomic molecule, with the electronic coordinates expressed in the molecule-fixed axis system, is rather difficult to derive. Bunker (1968) provides a detailed derivation as well as a review of the coordinate conventions, implicit approximations, and errors in previous discussions of the exact diatomic molecule Hamiltonian.

Our goal is to find the exact solutions, ψ_i^T (T = Total), of the Schrödinger equation,

$$\mathbf{H}\psi_i^T = E_i^T \psi_i^T, \tag{3.1.1}$$

which correspond to the observed (exact) E_i^T energy levels. **H** is the nonrelativistic Hamiltonian, which may be approximated by a sum of three operators,

$$\mathbf{H} = \mathbf{T}^N(R,\theta,\phi) + \mathbf{T}^e(r) + V(r,R), \tag{3.1.2}$$

where $\mathbf{T}^N$ is the nuclear kinetic energy, $\mathbf{T}^e$ is the electron kinetic energy, V is the electrostatic potential energy for the nuclei and electrons (including $e^- - e^-$, $e^- - N$ and $N - N$ interactions), R is the internuclear distance, θ and ϕ specify the orientation of the internuclear axis (molecule-fixed coordinate system) relative to the laboratory coordinate system (see Section 2.3.3 and Fig. 2.4), and r represents all electron coordinates in the molecule-fixed system.

The nuclear kinetic energy operator is given by

$$\mathbf{T}^N(R,\theta,\phi) = \frac{-\hbar^2}{2\mu R^2}\left[\frac{\partial}{\partial R}\left(R^2\frac{\partial}{\partial R}\right) + \frac{1}{\sin\theta}\frac{\partial}{\partial\theta}\left(\sin\theta\frac{\partial}{\partial\theta}\right) + \frac{1}{\sin^2\theta}\frac{\partial^2}{\partial\phi^2}\right], \tag{3.1.3a}$$

where

$$\mu = \frac{M_\mathrm{A}M_\mathrm{B}}{M_\mathrm{A}+M_\mathrm{B}}$$

is the nuclear reduced mass, with M_A and M_B the masses of atoms A and B. $\mathbf{T}^N$ can be divided into vibrational and rotational terms,

$$\mathbf{T}^N(R,\theta,\phi) = \mathbf{T}^N(R) + \mathbf{H}^{ROT}(R,\theta,\phi). \tag{3.1.3b}$$

The electron kinetic energy operator is

$$\mathbf{T}^e(r) = \frac{-\hbar^2}{2m}\sum_i \nabla_i^2, \tag{3.1.3c}$$

where m is the electron mass and the summation is over all n electrons.[†]

[†] The part of $\mathbf{T}^e$,

$$\frac{-\hbar^2}{2(M_\mathrm{A}+M_\mathrm{B})}\sum_{i,j}\nabla_i\nabla_j,$$

the diagonal matrix elements of which contribute, for example, 38 cm^{-1} for the ground state of H_2 and 27 cm^{-1} for the B $^1\Sigma_u^+$ excited state of H_2 (Bunker, 1968), is neglected here.

To solve Eq. (3.1.1), it would be useful to write the total energy as a sum of contributions from interactions between different particles. In decreasing order of importance, there are: electronic energy, E^{el}, vibrational energy, $G(v)$, and rotational energy, $F(J)$. In fact, this separation is assumed whenever the expression

$$E^{T}_{app} = E^{el} + G(v) + F(J) \tag{3.1.4}$$

is used to represent observed energy levels. However, Eq. (3.1.4) is always an approximation. It is never possible to express E^T exactly as in Eq. (3.1.4). This means that it is not possible to separate $\mathbf{H}$ rigorously into terms corresponding to the different motions of the particles. The approximate wavefunction suggested by the desired but approximate energy expression, (Eq. (3.1.4)), is a product of two functions,

$$\psi^{BO}_{i,v} = \Phi_{i,\Lambda,S,\Sigma}(r;R)\chi_v(R,\theta,\phi), \tag{3.1.5}$$

where the first factor is the electronic wavefunction and the second is the vibration-rotation wavefunction. Λ is the projection of the electronic orbital angular momentum on the internuclear axis, S is the spin angular momentum, and Σ is its projection on the internuclear axis. The approximate solution [Eq. (3.1.5)] is called a Born-Oppenheimer (BO) product function. It corresponds to a solution where all of the couplings in $\mathbf{H}$ between nuclear and electronic motions are ignored.

3.1.1 Potential Energy Curves

Since $\mathbf{T}^N$ is smaller than $\mathbf{T}^e$ by the factor m/μ, it can be neglected initially. Φ is then the solution of the clamped nuclei electronic Schrödinger equation,

$$[\mathbf{T}^e(r) + V(r,R)]\Phi_i(r;R) = E^{el}_i(R)\Phi_i(r;R) \tag{3.1.6}$$

or

$$\mathbf{H}^{\mathrm{el}}\Phi_i = E^{el}_i\Phi_i.$$

If the approximate product function, $\Phi_i\chi$, is inserted into Eq. (3.1.1), after multiplying by Φ^*_i and integrating over the electronic coordinates r, one obtains

$$\langle\Phi_i|\mathbf{T}^N + \mathbf{T}^e + V|\Phi_i\rangle_r\,\chi = E^T\,\langle\Phi_i|\Phi_i\rangle_r\,\chi.$$

The electronic wavefunction is normalized to unity because it is a probability distribution function, $\langle\Phi_i|\Phi_i\rangle_r = 1$. If the effect of $\partial/\partial R$ [contained in $\mathbf{T}^N(R)$] on the electronic wavefunction, which certainly depends on R, is neglected, then

$$\langle\Phi_i|\mathbf{T}^N|\Phi_i\rangle_r\,\chi \simeq \mathbf{T}^N\,\langle\Phi_i|\Phi_i\rangle_r\,\chi = \mathbf{T}^N\chi.$$

Thus, Eq. (3.1.6) may be simplified to

$$\langle\Phi_i|\mathbf{T}_e + V|\Phi_i\rangle_r = E^{el}_i(R),$$

and the result used to obtain the nuclear Schrödinger equation,

$$[\mathbf{T}^N(R,\theta,\phi) + E_i^{el}(R)]\chi(R,\theta,\phi) = E^T\chi(R,\theta,\phi).$$

Since $\mathbf{T}^N(R,\theta,\phi) = \mathbf{T}^N(R) + \mathbf{H}^{\mathrm{ROT}}$, the radial and angular variables can be separated, as in the case of the hydrogen atom (Pauling and Wilson, 1935), as

$$\chi(R,\theta,\phi) = \chi_{v,J}(R)\mathcal{D}^J_{\Omega M}(\alpha = \pi/2, \beta = \theta, \gamma = \phi) = \chi_{v,J}(R)\,\langle\alpha\beta\gamma|JM\Omega\rangle\,,$$

where $\langle\alpha\beta\gamma|JM\Omega\rangle$ is the symmetric rotor function defined by Eq. (2.3.41) and $\chi_{v,J}$ is a vibrational eigenfunction of

$$\left[\mathbf{T}^N(R) + (\hbar^2/2\mu R^2)[J(J+1) - \Omega^2] + E_i^{el}(R)\right]\chi_{v,J}(R) = E^T\chi_{v,J}(R), \tag{3.1.7}$$

where $\Omega = \Lambda + \Sigma$.

E_i^{el} may be viewed as the potential energy curve in which the nuclei move, but it must be emphasized that potential energy curves do not correspond to any physical observable. They are a concept, derived from a specified set of assumptions for defining a particular type of approximate wavefunctions [Eq. (3.1.5)]. The observed levels are not exact energy eigenvalues of a given potential curve. In general, the separation between the electronic and nuclear motions, which constitutes the BO approximation, is convenient. But when the observed levels do not fit formulas such as Eq. (3.1.4), it is simply because the function [Eq. (3.1.5)] is a bad approximation in that particular case.

An exact solution, ψ_i^T, of the total Hamiltonian $\mathbf{H}$ must satisfy two conditions:

$$\begin{array}{lll} (i) & \left\langle\psi_i^T|\mathbf{H}|\psi_i^T\right\rangle = E_i^T & \\ (ii) & \left\langle\psi_j^T|\mathbf{H}|\psi_i^T\right\rangle = 0, & \text{for any } j \neq i. \end{array} \tag{3.1.8}$$

One then says that the Hamiltonian is diagonalized in the basis set $\{\psi^T\}$.

In principle, it is possible to express any exact solution as an infinite expansion over the BO product functions,

$$\psi_i^T = \sum_{j,v_j}^{\infty} c_{i,v_j}\Phi_j^{BO}\chi_{v_j}. \tag{3.1.9}$$

The coefficients of this expansion are determined by diagonalizing a matrix representation of the total Hamiltonian, constructed by evaluating matrix elements between BO basis functions.

It is most useful to define a basis set of the type of Eq. (3.1.5) for which the off-diagonal matrix elements of the total Hamiltonian are as small as possible. If *one term* of the expansion in Eq. (3.1.9) is sufficient, this means that the BO approximation is valid. Fortunately, when the BO approximation fails, often *only two terms* of the BO expansion are sufficient.

If large off-diagonal matrix elements of the Hamiltonian exist that couple *many* vibrational wavefunctions belonging to two different electronic states, then it is much more convenient to write and solve the coupled differential equations that describe these two states (see Section 4.4.3). The BO approximation fails whenever off-diagonal elements of $\mathbf{H}$, H_{ij}, connecting different eigenstates of the Eq. (3.1.5) form, are large compared to the difference between the diagonal elements, $H_{ii} - H_{jj}$.

3.1.2 Terms Neglected in the Born-Oppenheimer Approximation

This section deals with various types of nonzero off-diagonal matrix elements of $\mathbf{H}$ between approximate BO product basis functions. *In order to go beyond the BO approximation*, to try to obtain an exact solution, *it is necessary to use a BO representation*. In other words, the exact eigenvalues and eigenfunctions, which can be compared to observed energy levels, are expressed in terms of the BO representation, specifically as a linear combination of BO product functions.[†] Presently, $\{\psi^{BO}\}$ is the only available type of complete, rigorously definable basis set.

In the following, the off-diagonal matrix elements of

$$\mathbf{H} = \mathbf{H}^{\mathrm{el}} + \mathbf{T}^N(R) + \mathbf{H}^{ROT}$$

of the form

$$\left\langle \Phi_{i,\Lambda,S,\Sigma}\ \chi_{v_i,J} | \mathbf{H} | \Phi_{j,\Lambda',S',\Sigma'}\ \chi_{v_j',J} \right\rangle$$

will be discussed. The off-diagonal matrix elements of $\mathbf{H}^{\mathrm{el}}$ give rise to *electrostatic perturbations*. The off-diagonal matrix elements of $\mathbf{T}^N(R)$ give rise to *nonadiabatic interactions*. The off-diagonal matrix elements of $\mathbf{H}^{ROT}$ give rise to *rotational perturbations*. The total Hamiltonian discussed above does not include the relativistic part of the Hamiltonian. That contribution to $\mathbf{H}$ will be introduced as a phenomenological perturbation operator, $\mathbf{H}^{\mathrm{SO}}$, and will give rise to *spin-orbit perturbations*.

3.1.2.1 Electrostatic and Nonadiabatic Part of H

In Section 3.3 it will be shown that, to describe perturbations which result from neglected terms in the $\mathbf{H}^{\mathrm{el}} + T^N(R)$ part of the Hamiltonian, two different types of BO representations are useful. If a crossing (diabatic) potential curve representation is used, off-diagonal matrix elements of $\mathbf{H}^{\mathrm{el}}$ appear between the states of this representation. If a noncrossing (adiabatic) potential curve representation is the starting point, the $\mathbf{T}^N$ operator becomes responsible for perturbations.

[†] It is *never* appropriate to take linear combinations of vibrational basis states which belong to different potential curves unless the associated electronic and rotational factors are included.

3.1.2.1.1 Crossing or Diabatic Curves

If Φ_i and Φ_j are two different exact solutions of Eq. (3.1.6), then

$$\langle \Phi_i | \mathbf{H}^{\mathrm{el}} | \Phi_j \rangle = 0$$

It will be shown later that *exact* solutions of the electronic Schrödinger equation can give rise to double minimum potential curves. Such potentials can be inconvenient for treating some perturbation situations. It is often more convenient to start from *approximate* solutions of $\mathbf{H}^{\mathrm{el}}$ where potential curves, which would have avoided crossing for the exact $\mathbf{H}^{\mathrm{el}}$, actually cross. In such a case,

$$\langle \Phi_i^{app} | \mathbf{H}^{\mathrm{el}} | \Phi_j^{app} \rangle_r = H_{i,j}^e(R) \neq 0. \tag{3.1.10}$$

The expression for $\mathbf{H}^{\mathrm{el}}$ includes $\mathbf{T}^e(r)$ [Eq. (3.1.3c)] and $V(r,R)$, where

$$V(r,R) = V^{eN}(r,R) + V^{ee}(r) + V^{NN}(R);$$

$V^{eN}(r,R)$ is the Coulomb electron-nuclear attraction energy operator,

$$V^{eN}(r,R) = -\sum_{i=1}^{n} \left(\frac{Z_{\mathrm{A}}e^2}{r_{\mathrm{A}i}} + \frac{Z_{\mathrm{B}}e^2}{r_{\mathrm{B}i}} \right);$$

$V^{ee}(r)$ is the Coulomb interelectronic repulsion energy operator,

$$V^{ee}(r) = \sum_{i=1}^{n} \sum_{j>i} \frac{e^2}{r_{ij}},$$

where $j > i$ ensures that each repulsion between the ith and jth electrons is considered only once; and $V^{NN}(R)$ is the Coulomb internuclear repulsion energy operator,

$$V^{NN}(R) = Z_{\mathrm{A}} Z_{\mathrm{B}} e^2 / R.$$

Note that the negative sign of V^{eN} implies that it contributes to energy stabilization. Crossing curves are obtained by excluding parts of the spin-orbit term, $\mathbf{H}^{\mathrm{SO}}$, and of the interelectronic term, V^{ee}, from the $\mathbf{H}^{\mathrm{el}}$ operator.[†] The effect of V^{ee}, discussed in Section 3.3.2, is extremely important as it compromises the validity of the *single electronic configuration* picture which is often taken as synonymous with the diabatic potential curve picture.

In both crossing and noncrossing curve approaches, perturbations between levels of the same symmetry can occur. In the diabatic picture, these are usually called "electrostatic perturbations" because they arise from V^{ee}. In the adiabatic picture, they arise from $\mathbf{T}^N$, the nuclear kinetic energy operator, but are often misleadingly called electrostatic perturbations.

[†] It is not possible to give a unique definition of a diabatic potential curve without identifying the specific term in $\mathbf{H}^{\mathrm{el}}$ that is excluded. The impossibility of identifying such a term and the consequent nonuniqueness of the *a priori* definition of diabatic curves is discussed by Lewis and Hougen (1968), Smith (1969), and Mead and Truhlar (1982) (See also Section 3.3.2). Diabatic curves may be defined empirically (Section 3.3) by assuming a deperturbation model [e.g., that $H_{i,j}^e(R)$ is independent of R or, at most, varies linearly with R].

3.1.2.1.2 Noncrossing or Adiabatic Curves

Equation (3.1.7) was obtained by assuming that $\mathbf{T}^N(R)$ does not act on the electronic wavefunction. Actually, the Φ_i are functions of the nuclear coordinate, R. Adding to Eq. (3.1.7) the neglected R–dependent term

$$\langle \Phi_i | \mathbf{T}^N | \Phi_i \rangle_r ,$$

where integration over all electronic coordinates and the θ, ϕ nuclear coordinates is implied, one obtains

$$\begin{aligned}\left[\mathbf{T}^N(R) + (\hbar^2/2\mu R^2)[J(J+1) - \Omega^2] + \langle \Phi_i | \mathbf{T}^N | \Phi_i \rangle_r + E_i^{el}(R)\right] \chi_{v,J}(R) \\ = E^T \chi_{v,J}(R). \qquad (3.1.11)\end{aligned}$$

The potential curves defined by

$$E_i^{ad}(R) = E_i^{el}(R) + \langle \Phi_i | \mathbf{T}^N | \Phi_i \rangle_r$$

are called adiabatic potential curves, but the second term makes a much smaller contribution to the energy than $E^{el}(R)$ (Bunker, 1968).

The off-diagonal matrix element $\langle \Phi_i | \mathbf{T}^N | \Phi_j \rangle_r$, the nonadiabatic coupling term, is examined in Section 3.3.3. This type of matrix element appears between states of identical symmetry and gives rise to homogeneous perturbations.

3.1.2.2 The Spin Part of H

Equation (3.1.2) is the nonrelativistic Hamiltonian. This means that the spin-dependent part of the Hamiltonian ($\mathbf{H}^{\mathrm{SO}}$ spin-orbit and $\mathbf{H}^{\mathrm{SS}}$ spin-spin) has been neglected. The electronic angular momentum quantum numbers, which are well-defined for eigenfunctions of nonrelativistic adiabatic and diabatic potential curves, are Λ, Σ, and S (and redundantly, $\Omega = \Lambda + \Sigma$).

Off-diagonal matrix elements,

$$\langle \Phi_{i,\Lambda,\Sigma,\Omega} | \mathbf{H}^{\mathrm{SO}} + \mathbf{H}^{\mathrm{SS}} | \Phi_{j,\Lambda',\Sigma',\Omega} \rangle ,$$

can be nonzero between states of different Λ and S (but identical Ω) quantum numbers, corresponding to different solutions [Eq. (3.1.5)] of the nonrelativistic Hamiltonian [Eq. (3.1.2)]. Matrix elements of this type are discussed in Section 3.4. The spin-spin part, $\mathbf{H}^{\mathrm{SS}}$, of the spin Hamiltonian usually gives rise to matrix elements much smaller than those of $\mathbf{H}^{\mathrm{SO}}$ and $\mathbf{H}^{el}$.

The relativistic Hamiltonian may be defined by adding $\mathbf{H}^{\mathrm{SO}}$ to $\mathbf{H}^{\mathrm{el}}$. The eigenfunctions of this new Hamiltonian are the relativistic wavefunctions, $\Phi_{i,\Omega}$, which define the relativistic potential curves

$$\langle \Phi_{i,\Omega} | \{\mathbf{H}^{\mathrm{el}}(R) + \mathbf{H}^{\mathrm{SO}}(R)\} | \Phi_{i,\Omega} \rangle_r = E_i^r(R), \qquad (3.1.12)$$

where now the only good electronic angular momentum quantum number is $\Omega = \Lambda + \Sigma$.

Table 3.1 summarizes the different types of potential energy curves and the specific terms in $\mathbf{H}$ that are neglected in order to define the diabatic, adiabatic, relativistic-adiabatic, and relativistic-diabatic basis functions.

Table 3.1: Definitions and Approximations Associated with Different Types of Potential Energy Curves

Function type	Definition of Φ	Definition of $E_i(R)$	Curve type
Adiabatic functions	$\langle \Phi_i^{ad} \vert \mathbf{H}^{el} \vert \Phi_j^{ad} \rangle_r = 0$ for all $j \neq i$ $\langle \Phi_i^{ad} \vert \frac{\partial}{\partial R} \vert \Phi_j^{ad} \rangle_r \neq 0$	$E_i^{BO} = \langle \Phi_i^{ad} \vert \mathbf{H}^{el} \vert \Phi_i^{ad} \rangle_r$ $E_i^{ad} = E_i^{BO} + \langle \Phi_i^{ad} \vert T^N \vert \Phi_i^{ad} \rangle_r$	Born-Oppenheimer potential curve Adiabatic potential curve
Diabatic Functions	$\langle \Phi_i^{d} \vert \mathbf{H}^{el} \vert \Phi_j^{d} \rangle_r \neq 0$ for *specific* values of j $\langle \Phi_i^{d} \vert \frac{\partial}{\partial R} \vert \Phi_j^{d} \rangle_r = 0$ for all $i \neq j$	$E_i^{d} = \langle \Phi_i^{d} \vert \mathbf{H}^{el} \vert \Phi_i^{d} \rangle_r$	Diabatic potential curve
Relativistic adiabatic functions	$\langle \Phi_i^{ad\cdot r} \vert \mathbf{H}^{el} + \mathbf{H}^{SO} \vert \Phi_j^{ad\cdot r} \rangle_r = 0$ for all $j \neq i$ $\langle \Phi_i^{ad\cdot r} \vert \frac{\partial}{\partial R} \vert \Phi_j^{ad\cdot r} \rangle_r \neq 0$	$=$ $\mathrm{E}_i^{ad\cdot r} = \langle \Phi_i^{ad\cdot r} \vert \mathbf{H}^{el} + \mathbf{H}^{SO} + T^N \vert \Phi_i^{ad\cdot r} \rangle_r$	Relativistic adiabatic potential curve
Relativistic diabatic functions	$\langle \Phi_i^{d\cdot r} \vert \mathbf{H}^{SO} \vert \Phi_j^{d\cdot r} \rangle_r \neq 0$ for *one* value of j $\langle \Phi_i^{d\cdot r} \vert \frac{\partial}{\partial R} \vert \Phi_j^{d\cdot r} \rangle_r = 0$ for all $j \neq i$	$\mathrm{E}_i^{d\cdot r} = \langle \Phi_i^{d\cdot r} \vert \mathbf{H}^{el} + \mathbf{H}^{SO} \vert \Phi_i^{d\cdot r} \rangle_r$	Relativistic diabatic potential curve

3.1.2.3 Rotational Part of H

An expression for $\mathbf{H}^{\mathrm{ROT}}$ is given by Hougen (1970):

$$\mathbf{H}^{\mathrm{ROT}} = (1/2\mu R^2)\mathbf{R}^2 = (1/2\mu R^2)(\mathbf{R}_x^2 + \mathbf{R}_y^2),$$

where $\mathbf{R}$ is the nuclear rotation angular momentum operator. The nuclear motion is necessarily in a plane that contains the internuclear axis: thus $\vec{\mathbf{R}}$ is perpendicular to the z direction and $\mathbf{R}_z = 0$.

The total angular momentum, $\mathbf{J}$, is defined by

$$\vec{\mathbf{J}} \equiv \vec{\mathbf{R}} + \vec{\mathbf{L}} + \vec{\mathbf{S}}$$

and this definition can be used to reexpress $\mathbf{H}^{\mathrm{ROT}}$ in a convenient form,

$$\begin{aligned}\mathbf{H}^{\mathrm{ROT}} &= (1/2\mu R^2)[(\mathbf{J}_x - \mathbf{L}_x - \mathbf{S}_x)^2 + (\mathbf{J}_y - \mathbf{L}_y - \mathbf{S}_y)^2] \\ &= (1/2\mu R^2)[(\mathbf{J}^2 - \mathbf{J}_z^2) + (\mathbf{L}^2 - \mathbf{L}_z^2) + (\mathbf{S}^2 - \mathbf{S}_z^2) \\ &+ (\mathbf{L}^+\mathbf{S}^- + \mathbf{L}^-\mathbf{S}^+) - (\mathbf{J}^+\mathbf{L}^- + \mathbf{J}^-\mathbf{L}^+) - (\mathbf{J}^+\mathbf{S}^- + \mathbf{J}^-\mathbf{S}^+)],\end{aligned} \tag{3.1.13}$$

where

$$\mathbf{J}^\pm = \mathbf{J}_x \pm i\mathbf{J}_y \qquad \text{[cf. Eq.(2.3.9)]},$$
$$\mathbf{L}^\pm = \mathbf{L}_x \pm i\mathbf{L}_y, \qquad \mathbf{S}^\pm = \mathbf{S}_x \pm i\mathbf{S}_y \quad \text{[cf. Eq.(2.3.8)]}.$$

The first three terms of $\mathbf{H}^{\mathrm{ROT}}$ have diagonal matrix elements exclusively. This diagonal part of $\mathbf{H}^{\mathrm{ROT}}$ is the rotational energy of the $|JM\Omega\Lambda S\Sigma\rangle$ basis function. The eigenfunctions [defined by Eq. (2.3.40)] of the rotational eigenvalue equation,

$$(1/2\mu R^2)[(\mathbf{J}^2 - \mathbf{J}_z^2) + (\mathbf{L}^2 - \mathbf{L}_z^2) + (\mathbf{S}^2 - \mathbf{S}_z^2)]|JM\Omega\rangle = E^{\mathrm{ROT}}(R)\,|JM\Omega\rangle\,, \tag{3.1.14}$$

correspond to the R–dependent eigenvalues

$$E^{\mathrm{ROT}}(R) = (\hbar^2/2\mu R^2)[J(J+1) - \Omega^2 + S(S+1) - \Sigma^2 + L(L+1) - \Lambda^2]. \tag{3.1.15}$$

In general, electronic wavefunctions of *molecules* are not eigenfunctions of $\mathbf{L}^2$, hence $L(L+1)$ is not quantized, but the R–dependent quantity,

$$(\hbar^2/2\mu R^2)[L(L+1) - \Lambda^2],$$

may be incorporated into the electronic energy. Due to its μ–dependence, this term is responsible for part of the electronic isotope effect, the remainder arising from small terms neglected in the Hamiltonian given here (Bunker, 1968).

After integration of Eq. (3.1.14) over the vibrational coordinate, R,

$$E^{\mathrm{ROT}}(v,J) = B_v[J(J+1) - \Omega^2 + S(S+1) - \Sigma^2] \tag{3.1.16}$$

Table 3.2: Off-Diagonal Matrix Elements of Total Hamiltonian between Unsymmetrized Basis Functions

Neglected terms in the nonrelativistic Born-Oppenheimer approximation	Nature of the perturbation	Selection rules $\Delta J = 0$ and $g \not\leftrightarrow u$ always				Number of different spin-orbitals[d]
		$\Delta\Lambda$	$\Delta\Sigma$	ΔS	$\Delta\Omega$	
$\langle \Lambda, S, \Sigma, \Omega, v \vert \mathbf{H}^{el} \vert \Lambda, S, \Sigma, \Omega, v' \rangle \simeq \mathbf{H}^e \langle v \vert v' \rangle$	Electrostatic homogeneous	0	0	0	0	1 or 2
$\langle \Lambda, S, \Sigma, \Omega, v \vert \mathbf{T}^N \vert \Lambda, S, \Sigma, \Omega, v' \rangle$ $= \left\langle \Lambda, S, \Sigma, \Omega, v \left\vert -\frac{\hbar^2}{2\mu}\left(\frac{d^2}{dR^2} + \frac{2}{R}\frac{d}{dR}\right)\right\vert \Lambda, S, \Sigma, \Omega, v' \right\rangle$	Vibrational (or nonadiabatic) homogeneous	0	0	0	0	1
$\langle \Lambda, S, \Sigma, \Omega, v \vert \mathbf{H}^{SO} \vert \Lambda', S', \Sigma', \Omega, v' \rangle$ $= \langle v \vert v' \rangle \langle \Lambda, S, \Sigma, \Omega \vert \sum_i \hat{a}_i \mathbf{l}_i \cdot \mathbf{s}_i \vert \Lambda', S', \Sigma', \Omega \rangle$	Spin-Orbit homogeneous	0 or ±1 $\Sigma^\pm \sim \Sigma^\mp$	0 or ∓1	0^a or 1 0 or 1	0	0 or 1 1
$\langle \Lambda, S, \Sigma, \Omega, v \vert \mathbf{H}^{SS} \vert \Lambda', S', \Sigma', \Omega', v' \rangle$ $= \langle v \vert v' \rangle \langle \Lambda, S, \Sigma, \Omega \vert \mathbf{H}^{SS} \vert \Lambda', S', \Sigma', \Omega \rangle$	Spin-spin homogeneous	0 ±1 ±2 $\Sigma^\pm \sim \Sigma^\pm$	0 ∓1 ∓2	0^c, 1^b, or 2 0, 1, or 2 0, 1, or 2	0	0 1 2
$\left\langle \Lambda, S, \Sigma, \Omega, v \left\vert \frac{1}{2\mu R^2}\mathbf{L}^\pm \mathbf{S}^\mp \right\vert \Lambda \mp 1, S, \Sigma \pm 1, \Omega, v' \right\rangle$ $\simeq B_{vv'} \left\langle \Lambda, S, \Sigma, \Omega \left\vert \sum_i \mathbf{l}_i^\pm \sum_j \mathbf{s}_j^\mp \right\vert \Lambda \mp 1, S, \Sigma \pm 1, \Omega \right\rangle$	Spin-Electronic homogeneous	±1	∓1	0	0	1 or 2
$\left\langle \Lambda, S, \Sigma, \Omega, v \left\vert -\frac{1}{2\mu R^2}\mathbf{J}^\pm \mathbf{L}^\mp \right\vert \Lambda \pm 1, S, \Sigma, \Omega \pm 1, v' \right\rangle$ $\simeq -B_{vv'} \langle \Lambda, S, \Sigma, \Omega \vert \sum_i \mathbf{l}_i^\mp \vert \Lambda \pm 1, S, \Sigma, \Omega \pm 1 \rangle$ $\times [J(J+1) - \Omega(\Omega \pm 1)]^{1/2\ e}$	Heterogeneous (**L**-uncoupling)	±1	0	0	±1	1
$\left\langle \Lambda, S, \Sigma, \Omega, v \left\vert -\frac{1}{2\mu R^2}\mathbf{J}^\pm \mathbf{S}^\mp \right\vert \Lambda, S, \Sigma \pm 1, \Omega \pm 1, v' \right\rangle$ $= -B_{vv'}[S(S+1) - \Sigma(\Sigma \pm 1)]^{1/2}$ $\times [J(J+1) - \Omega(\Omega \pm 1)]^{1/2\ e}$	Heterogeneous (**S**-uncoupling)	0	±1	0	±1	1

[a]Zero if $\Omega = 0$ [b]Zero if $\Sigma = \Sigma' = 0$ [c]Zero if $\Sigma = \Sigma' = 0$ and $S = 0$ [d]Each state must be well represented by a single configuration.
[e]$B_{vv'} \equiv \langle v \vert \hbar^2/(2\mu R^2) \vert v' \rangle$.

where, following spectroscopic custom, B_v is expressed in cm^{-1} units as

$$B_v(cm^{-1}) = \langle v|(\hbar^2/2\mu R^2)|v\rangle\,(hc)^{-1}. \tag{3.1.17}$$

or

$$B_v(\mathrm{cm}^{-1}) = \frac{16.85763}{\mu(\mathrm{amu})}\langle v|R^{-2}|v\rangle\,(\text{Å}^{-2}).$$

The final three terms of the rotational operator in Eq. (3.1.13), which couple the orbital, spin, and total angular momenta, are responsible for perturbations between different electronic states:

1. $+(1/2\mu R^2)\mathbf{L}^{\pm}\mathbf{S}^{\mp}$ gives rise to *homogeneous* ($\Delta\Omega = 0$) spin-electronic perturbations between basis functions of the same Ω and S, but different Λ and Σ.

2. $-(1/2\mu R^2)\mathbf{J}^{\mp}\mathbf{S}^{\pm}$ is responsible for *heterogeneous* ($\Delta\Omega = \pm 1$) electronic-rotational† perturbations between basis states of different Ω having identical values of S and Λ, but different values of Σ. This operator is called the **S**-*uncoupling operator*.

3. $-(1/2\mu R^2)\mathbf{J}^{\mp}\mathbf{L}^{\pm}$ causes *heterogeneous* ($\Delta\Omega = \pm 1$) electronic-rotational† perturbations between states of different Ω having identical values of $\mathbf{S}$ and Σ, but different values of Λ. This operator is called the **L**-*uncoupling operator*.

These perturbation operators are examined in Section 3.5.

Following the nomenclature suggested by Hougen (1970), perturbations are classified with respect to the *signed* Ω quantum number. The selection rule, $\Delta\Omega = 0$, implies a *homogeneous perturbation*. The important feature of such perturbations is that the *interaction matrix element does not depend on the rotational quantum number, J*. The selection rule, $\Delta\Omega = \pm 1$, implies a *heterogeneous* perturbation. The important feature of such perturbations is that the *interaction matrix element depends on J* (usually approximately proportional to J). Table 3.2 summarizes the different perturbation selection rules. The matrix elements are written for the microscopic Hamiltonian, taking into account that L is not defined in a molecule, and in this sense differing slightly from those given in Chapter 2.

In all cases, the vibrational quantum number is no longer strictly defined because the perturbed level is a mixture of at least two vibrational levels, which belong to different electronic states [see Eq. (3.1.9)]. The only quantum number that remains rigorously well defined is J, because only J is a conserved observable ($\mathbf{J}^2$ commutes with $\mathbf{H}$). The selection rule common to all perturbations is

† Perturbations resulting from the $\mathbf{J}\cdot\mathbf{L}$ and $\mathbf{J}\cdot\mathbf{S}$ terms are frequently called "Coriolis" perturbations. This use of the name Coriolis is misleading and should be reserved for vibration-rotation interactions in polyatomic molecules.

$\Delta J = 0$; this means that, in the absence of nuclear spin and external perturbations (electric or magnetic fields, electromagnetic radiation fields, or collisions with other molecules), the total angular momentum of the molecule remains well defined. Even if the perturbation operator includes $\mathbf{J}^+$ or $\mathbf{J}^-$, this operator cannot change the value of J. Even in case (b), J (as well as N) remains well defined. Perturbations (denoted by $\sim$) correspond to an interaction between two levels, *as opposed to an electric dipole transition (denoted by $-$) between two levels*.

Mathematically, the conservation of the J quantum number on perturbation can be clearly shown, since no perturbation operator contains the θ and ϕ angular coordinates of the nuclei and consequently none can act on the rotational basis functions. As the rotational functions $|JM\Omega\rangle$ are orthogonal for different values of J, a matrix element of any perturbation operator, $\mathbf{H}'$, between two rotational functions is given by

$$\langle JM\Omega|\mathbf{H}'|J'M'\Omega\rangle = f(J)\,\langle JM\Omega|J'M'\Omega\rangle = f(J)\delta_{JJ'}\delta_{MM'}. \qquad (3.1.18)$$

For homonuclear molecules, the g or u symmetry is almost always conserved. Only external electric fields, hyperfine effects (Pique, *et al.*, 1984), and collisions can induce perturbations between g and u states. See Reinhold, *et al.*, (1998) who discuss how several terms that are neglected in the Born-Oppenheimer approximation can give rise to interactions between g and u states in hetero-isotopomers, as in the HD molecule. An additional symmetry will be discussed in Section 3.2.2: parity or, more usefully, the e and f symmetry character of the rotational levels remains well defined for both hetero- and homonuclear diatomic molecules. The matrix elements of Table 3.2 describe direct interactions between basis states. *Indirect* interactions can also occur and are discussed in Sections 4.2, 4.4.2 and 4.5.1. Even for indirect interactions the $\Delta J = 0$ and $e \not\leftrightarrow f$ perturbation selection rules remain valid (see Section 3.2.2).

3.2 Basis Functions

The delineation of Hund's angular momentum coupling cases in diatomic molecules has a long history. Herzberg (1950) defined cases (a)–(e) entirely in terms of vector precession pictures and without making any explicit distinction between molecular-ion-core and Rydberg-electron angular momenta. A vector precession picture is a cartoon that provides useful insights into the conserved quantities and dynamics that describe a particular scheme of coupled nuclear rotation, electron spin, and electron orbital angular momenta (for atoms, see Townes and Schawlow, pp. 120-124, 1955). Transformations between various angular momentum coupling arrangements exist, based on 3-j, 6-j, and 9-j angular momentum coupling coefficients (see Sections 3.2.1.1 and 3.2.1.5 and Watson, 1999). Nikitin and Zare (1994) base their Hund's cases on a hierarchical ordering of the $\mathbf{H}^{\mathrm{el}}$, $\mathbf{H}^{\mathrm{SO}}$, and $\mathbf{H}^{\mathrm{ROT}}$ terms in the total molecular $\mathbf{H}$. The nonspherical part of $\mathbf{H}^{\mathrm{el}}$ destroys the L quantum number that is associated with the orbital angular momentum operator, $\mathbf{L}^2$, and causes energy splittings between states belonging to different eigenvalues of $\mathbf{L}_z$ (Λ). $\mathbf{H}^{\mathrm{SO}}$ destroys the L and S quantum numbers and causes large energy splittings between components of a Λ-S multiplet state ($S > 0$) that belong to different eigenvalues of $\mathbf{L}_z + \mathbf{S}_z = \mathbf{J}_z$ ($\Lambda + \Sigma = \Omega$). $\mathbf{H}^{\mathrm{ROT}}$ destroys the Λ and Σ quantum numbers (eigenvalues of $\mathbf{L}_z$ and $\mathbf{S}_z$). Cases (a)–(e′) of Nikitin and Zare (1994) correspond to the six possible relative importance orderings of $\mathbf{H}^{\mathrm{el}}$, $\mathbf{H}^{\mathrm{SO}}$, and $\mathbf{H}^{\mathrm{ROT}}$:

	$\mathbf{H}^{\mathrm{el}}$	$\mathbf{H}^{\mathrm{SO}}$	$\mathbf{H}^{\mathrm{ROT}}$
(a)	strong	intermediate	weak
(b)	strong	weak	intermediate
(c)	intermediate	strong	weak
(d)	intermediate	weak	strong
(e)	weak	intermediate	strong
(e′)	weak	strong	intermediate.

This proposed relationship between the spectroscopically and dynamically most appropriate Hund's case and a ranking of terms in the molecular Hamiltonian is appealing, but is of insufficient numerical specificity, especially for Rydberg states. For Rydberg states it becomes necessary to partition the $\mathbf{H}^{\mathrm{el}}$ and $\mathbf{H}^{\mathrm{SO}}$ terms into ion-core and Rydberg electron contributions. Many more rankings of terms in the molecular Hamiltonian become possible. For example, Lefebvre-Brion (1990) has used a case (e) basis set to treat the Rydberg states of hydrogen halide, HX, molecules where the HX^+ ion-core is in a case (a) $^2\Pi$ state with a spin-orbit splitting that is large relative to off-diagonal matrix elements of $\mathbf{H}^{\mathrm{el}}$ for the Rydberg electron. This ranking of terms in the molecular Hamiltonian,

$$\mathbf{H}^{\mathrm{SO}}(\text{ion} - \text{core}) > \mathbf{H}^{\mathrm{ROT}} > \mathbf{H}^{\mathrm{el}}(\text{Rydberg } e^-),$$

corresponds to the case (e′) of Nikitin and Zare (1994). Brown and Carrington (2003) define case (e) similiarly to case (e′) of Nikitin and Zare (1994), and cite weakly bound levels of HeKr^+ as probably the first example of case (e) limit behavior in a non-Rydberg state (Carrington, *et al.*, 1996).

Nikitin and Zare (1994) also define two different sets of molecule frame projection quantum numbers according to whether the molecule frame projection of each component angular momentum is onto the internuclear axis, z $\left(\mathbf{L}_z, \mathbf{S}_z, \mathbf{j}_z, \mathbf{R}_z, \mathbf{J}_z, \rightarrow \tilde{\Lambda}, \tilde{\Sigma}, \tilde{\Omega}, 0, \tilde{\Omega},\right)$, or onto the total angular momentum axis, $\vec{\mathbf{J}}$ $\left(\hat{\mathbf{L}}_z, \hat{\mathbf{S}}_z, \hat{\mathbf{j}}_z, \hat{\mathbf{R}}_z, \hat{\mathbf{J}}_z, \rightarrow L_J, S_J, j_J, R, J\right)$. Eigenfunctions of the two classes of molecule frame angular momentum projection operators are expressed in different angular coordinates, thus one must be careful to evaluate correctly the effects of z-axis operators on rotation-axis basis states and *vice versa*. Fortunately, in the limit of very high–J,

$$L_J = N - R \tag{3.2.1a}$$

$$S_J = J - N \tag{3.2.1b}$$

$$j_J = J - R, \tag{3.2.1c}$$

the values of all of the rotation-axis projection quantum numbers are given by simple differences between angular momentum magnitude quantum numbers.

Watson (1999), unlike Nikitin and Zare (1994), makes an explicit distinction between the ion-core and the Rydberg electron. He proposes ion-core (a^+), (b^+), and (c^+), coupling sub-cases in combination with Rydberg electron (a), (b), (c), (d), and (e) coupling sub-cases. There would then be at least $3 \times 5 = 15$ possible (ion-core, Rydberg e^-) molecular coupling cases, e.g. (a^+, a), (a^+, b), ... (c^+, e), but many of these possibilities are unlikely to correspond to any physically realizable situation. Most of the 15 (ion-core, Rydberg e^-) molecular coupling schemes can occur in several different sub-cases according to whether the ion-core (A^+, A_z^+) and Rydberg-electron (b, b_z) angular momenta are combined in a coupled $\vec{\mathbf{A}} + \vec{\mathbf{b}} - \vec{\mathbf{C}}$ $|A^+ bCC_z\rangle$ or uncoupled $|A^+ A_z^+ bb_z\rangle$ form. For example, case (a^+, a) can exist in a fully uncoupled form, $|\Lambda^+ S^+ \Sigma^+ l \lambda s \sigma; \Omega J M_J\rangle$, and in a partially coupled form where S is a good quantum number, $|\Lambda^+ S^+ l \lambda s S \Sigma; \Omega J M_J\rangle$ (see Eq. (53) of Watson (1999)). Watson (1999), similarly to Nikitin and Zare (1999), also distinguishes between projection quantum numbers onto the molecular z-axis $(\mathbf{L}_z^+, \mathbf{S}_z^+, \mathbf{J}_z^+, \mathbf{l}_z, \mathbf{s}_z, \mathbf{j}_z \rightarrow \Lambda^+, \Sigma^+, \Omega^+, \lambda, \sigma, \omega)$ and onto the nuclear rotation axis, $\vec{\mathbf{R}}$ $(0, S_R, J_R, l_R, s_R, j_R)$. However, Watson's rotation-axis projections are onto $\vec{\mathbf{R}}$, whereas Nikitin's and Zare's rotation-axis projections are onto $\vec{\mathbf{J}}$.

3.2.1 Hund's Cases

Hund's cases are important because they tell the experimentalist what kind of patterns might be found in a spectrum, how to look for these patterns, and what inferences about quantum number assignments can be drawn from the patterns once they are detected. Hund's cases tell the dynamicist how to construct a reduced dimension picture of intramolecular processes. The reduction in dimensionality is based on the existence of approximate constants of motion, eigenvalues of operators that commute with *most* of the molecular Hamiltonian [see Sections 9.4.9 and 9.4.10]. Hund's cases, embodied in models of vectors precessing about other vectors, explain how information about molecule frame properties (e.g., a permanent magnetic dipole or an electric dipole transition moment) survives rotational averaging and becomes observable in the laboratory frame (and *vice versa*).

From a computational point of view, one is free to approach a spectrum from either a physically appropriate or a grossly inappropriate Hund's case. Each Hund's case comes equipped with a complete set of basis states. The accuracy with which the details in a spectrum are reproduced is not affected by the choice of Hund's case. However, the amount of labor, both by the experimentalist and the experimentalist's computer, is often profoundly affected by this choice.

The appropriateness of a particular Hund's case for a specific spectrum is reflected by the fraction of off-diagonal elements that fail the

$$\left|\frac{H_{ij}^{(1)}}{E_i^{(0)} - E_j^{(0)}}\right| < 1 \tag{3.2.2}$$

test for convergence of nondegenerate perturbation theory, where

$$\mathbf{H}^{(1)} \equiv \mathbf{H} - \mathbf{H}^{(0)}$$

$$\left\{\mathbf{E}_i^{(0)}\right\} \quad \text{are eigenvalues of } \mathbf{H}^{(0)}$$

(see Sections 4.2 and 4.3.2). An appropriate basis set is one in which most of the observed energy levels are nearly equal to a diagonal matrix element of $\mathbf{H}^{(0)}$. As a result, the *zero-order* pattern of energy levels and transition frequencies is easily recognizable in the spectrum.

Molecules are complex, many-body systems. However, patterns observable in the spectrum imply that the couplings between different parts of the molecule are sufficiently weak to justify building a basis set in which complete sets of quantum numbers are specified that describe the separate, weakly interacting parts of the molecule, as well as the quantum number dependent intramolecular interactions. Examples of such weak coupling situations in a diatomic molecule include: one Rydberg electron plus a (possibly open-shell) molecular-ion-core (Herzberg and Jungen, 1982; Jungen, *et al.*, 1989); an atom with incompletely filled core-like subshell (described by many-electron atom-core angular momenta, with subscript $_c$, e.g. $|f^N; J_c\Omega_c L_c S_c\rangle$, where the subscript c is a shorthand for atomic core) interacting with the cylindrically symmetric intramolecular electric field and the molecular orbitals built from the valence orbitals on both atoms (Field, 1982; Linton, *et al.*, 1983; Dulick and Field, 1985; Kaledin, *et al.*, 1994; Schamps, *et al.*, 1995; Kaledin, *et al.*, 1996); ultra-cold collisions between two open-shell atoms at very large internuclear separation, well described by separated atom angular momenta (Weiner, *et al.*, 1999; Bergeman, *et al.*, 2002).

When a molecule is excited by a sufficiently short duration pulse of radiation, the ultrafast pluck creates a highly localized, often quite simple and *a priori* knowable initial excitation describable by what one would normally consider to be extremely bad quantum numbers (see Sections 9.1.1, 9.1.4, 9.4.1 and 9.4.2). Rapid delocalization of a system prepared in an eigenstate, ψ_a, of an operator, $\mathbf{A}$,

$$\mathbf{A}\psi_a = a\psi_a \tag{3.2.3a}$$

$$\Psi(t=0) = \psi_a, \tag{3.2.3b}$$

that does not even approximately commute with the complete molecular Hamiltonian, $\mathbf{H}$, implies that the time evolution of such an initial preparation will not generate easily recognizable patterns in a high resolution frequency domain spectrum. Such a pattern would be subtly encoded in many eigenstates distributed over an energy region $\Delta E \sim \hbar/\tau_{\text{relax}}$, where τ_{relax} is the decay rate of the initial preparation (i.e., the initial decay rate of the survival probability, $P(t) = |\langle\Psi_a(t)|\Psi_a(0)\rangle|^2$, see Section 9.1.4).

There are a large number of possible nontraditional partitionings of molecules into weakly interacting subsystems or initially localized but rapidly evolving excitations. Major insights into intramolecular dynamics result when an experimental scheme capable of creating a new class of $t = 0$ localization is devised (e.g., Stimulated Emission Pumping, Kittrell, *et al.*, 1981).

3.2.1.1 Definition of Basis Sets

A choice of basis set implies a partitioning of the Hamiltonian, $\mathbf{H} = \mathbf{H}^{\mathrm{el}} + \mathbf{H}^{\mathrm{SO}} + \mathbf{T}^N(R) + \mathbf{H}^{\mathrm{ROT}}$, into two parts: a part, $\mathbf{H}^{(0)}$, which is fully diagonal in the selected basis set, and a residual part, $\mathbf{H}^{(1)}$. The basis sets associated with the various Hund's cases reflect different choices of the parts of $\mathbf{H}$ that are included in $\mathbf{H}^{(0)}$. Although in principle the eigenvalues of $\mathbf{H}$ are unaffected by the choice of basis, as long as this basis set forms a complete set of functions, one basis set is usually more convenient to use or better suited than the others for a particular problem. Convenience is a function of both the nature of the computational method and the relative sizes of electronic, spin-orbit, vibrational, and rotational energies. The angular momentum basis sets, from which Hund's cases (a)-(e) bases derive, are

$$(a)\ |nJS\Omega\Lambda\Sigma\rangle \tag{3.2.4a}$$

$$(b)\ |nJSN\Lambda(S_R)\rangle \tag{3.2.4b}$$

$$(c)\ |nJ[J_a]\Omega\rangle \tag{3.2.4c}$$

$$(d)\ \left|nJSN(S_R)J^+N^+S^+\Lambda^+l(l_R,s_R)\right\rangle^{\dagger} \tag{3.2.4d}$$

$$(e)\ \left|nJJ^+[J_a^+ \text{or } S^+ \text{and } \Lambda^+]\Omega^+lj(j_R)\right\rangle^{\dagger\dagger} \tag{3.2.4e}$$

where the label n stands for all remaining state labels (e.g. vibration, electronic configuration). Quantum numbers enclosed in parentheses are good only in the high rotation limit and are otherwise defined by the difference between two rotation-related quantum numbers that get large in the high rotation limit. Since problems involving laboratory-fixed electromagnetic fields are (mostly) beyond the scope of this book, the laboratory projection quantum number M_J is excluded from the basis set labels in Eqs. (3.2.4). The quantum numbers enclosed in brackets [ ] are often not well defined. However, inclusion of these

† Herzberg and Jungen (1982) introduce the quantum number N^+, which corresponds to the magnitude of the angular momentum $\mathbf{N}^+ = \mathbf{N} - \mathbf{l}$, where $\mathbf{l}$ is the orbital angular momentum of the electron in a Rydberg orbital and $\mathbf{N}^+$ is the total angular momentum, exclusive of spin, of the molecular ion-core. The N^+ rather than R quantum number is a more appropriate label for the rotational levels of Rydberg states, provided that off-diagonal matrix elements of the neglected $(\mathbf{L}^+)^{\pm}$ operator may be ignored.

†† $\mathbf{J}^+$ is the total angular momentum of the molecular ion-core (including electron spin). J^+ and Ω^+ are, respectively, the quantum numbers associated with the magnitude ($[J^+(J^+ + 1)]^{1/2}$) and projection of $\mathbf{J}^+$ (Ω^+) onto the molecular axis.

"extra" angular momenta can provide insights into the magnitudes and selection rules for coupling terms which are normally neglected. The quantum numbers describing the projection of an angular momentum onto the molecule frame rotation axis, $\vec{\mathbf{R}}$, are enclosed in parentheses because a basis set transformation (rotation by $\pi/2$ about an axis perpendicular to $\vec{\mathbf{R}}$ and z) must be performed in order to be able to make explicit use of these quantum numbers (see Eq. (3.2.12a)).

Some of the molecular frame $\vec{\mathbf{R}}$–axis quantum numbers (S_R, l_R, s_R and j_R) correspond to quantum numbers $\mathcal{L} = l_R$ and $\mathcal{S} = S_R$ that have been in widespread use in the literature. There is, however, an ambiguity concerning onto which molecule frame rotation axis ($\vec{\mathbf{J}}^+, \vec{\mathbf{N}}^+$, or $\vec{\mathbf{R}}$), the projection quantum number is defined. The subscript–R notation eliminates this ambiguity and we therefore recommend its use. Also, the lower-case forms of l_R and s_R conform to the convention that single-electron quantum numbers appear as lower-case letters.

The Hund's case (d) and (e) basis states listed above are appropriate for a specific class of problem: *one* Rydberg electron weakly coupled to a many-electron, possibly open-shell, molecular ion-core. It is rare that more than one electron can occupy an orbital that sufficiently escapes the influence of the nonspherical part of $\mathbf{H}^{\mathrm{el}}$ that the electron orbital angular momentum quantum number remains even approximately defined. Therefore, except for special situations, the only molecular situations that satisfy the prerequisites for Hund's cases (d) and (e), in which electron orbital and/or total angular momenta are specified, are Rydberg states. Two examples of special situations are (i) pure precession (see Section 5.5), where an atom-centered single-electron (Van Vleck, 1929; Mulliken and Christy, 1931) or single-hole (Gray, *et al.*, 1991) electronic structure *acts* as a Stark-split pure-l cluster of states, or, (ii) a molecule containing an open-core atom (the $4f$ orbitals in the rare earth atoms are core-like because their orbital radius is ~ 0.3Å, which is much smaller than the internuclear distance, R_e) in which the atom-centered (i.e., spherical) interelectronic interactions are vastly stronger than the effect of the nonspherical part of $\mathbf{H}^{\mathrm{el}}$, which in turn lifts the degeneracy of eigenstates of many-electron atomic-core angular momenta (L_c, S_c, J_c) causing them to be split into Ω_c components by the intramolecular Stark field, (Field, 1982; Kaledin, *et al.*, 1996).

The quantum numbers that are listed in any basis set must be eigenvalues of operators that form a set of mutually commuting operators. Watson (1999) analyzes the commutation rules among the magnitude, $\mathbf{A}^2$, and molecule frame component, $\mathbf{A}_\alpha$, angular momentum operators, where $\mathbf{A} = \mathbf{N}$, $\mathbf{N}^+$, $\mathbf{l}$, and explains why

$$[\mathbf{N}^2, (\mathbf{N}^+)^2] = [\mathbf{N}^2, \mathbf{l}^2] = [(\mathbf{N}^+)^2, \mathbf{l}^2] = 0 \tag{3.2.5a}$$

$$[\mathbf{l}^2, \mathbf{l}_z] = [\mathbf{l}^2, \mathbf{N}_z] = [\mathbf{l}^2, \mathbf{N}_z^+] = 0 \tag{3.2.5b}$$

$$[\mathbf{N}^2, \mathbf{N}_z] = [\mathbf{N}^2, \mathbf{l}_z] = [\mathbf{N}^2, \mathbf{N}_z^+] = 0 \tag{3.2.5c}$$

$$[(\mathbf{N}^+)^2, \mathbf{N}_z^+] = 0 \tag{3.2.5d}$$

but

$$[(\mathbf{N}^+)^2, \mathbf{l}_z] \neq 0 \tag{3.2.6a}$$

$$[(\mathbf{N}^+)^2, \mathbf{N}_z] \neq 0. \tag{3.2.6b}$$

This means that for the set of angular momentum operators ($\mathbf{N}^2$, $(\mathbf{N}^+)^2$, $\mathbf{l}^2$, $\mathbf{N}_z$, $\mathbf{N}_z^+$, $\mathbf{l}_z$), it is only possible to define two basis sets, a *coupled* basis, the members of which are simultaneously eigenstates of $\mathbf{N}^2, (\mathbf{N}^+)^2, \mathbf{l}^2, \mathbf{N}_z^+$ (and, in a restricted sense, $\mathbf{l}_R$), $|NN^+l\Lambda^+(l_R)\rangle$, and an *uncoupled* basis, the members of which are simultaneously eigenstates of $\mathbf{N}^2$, $\mathbf{l}^2, \mathbf{N}_z, \mathbf{l}_z$, and $\mathbf{N}_z^+ \equiv \mathbf{N}_z - \mathbf{l}_z$, $|Nl\Lambda\lambda, \Lambda^+ = \Lambda - \lambda\rangle$. It is never possible to define a basis set in which both the magnitude quantum number, N^+, and a projection quantum number, λ or Λ (or l_R), are good quantum numbers. (By a similar argument $(\mathbf{J}^+)^2$ does not commute with $\mathbf{l}_z$, $\mathbf{s}_z$, $\mathbf{j}_z$, and $\mathbf{J}_z$, hence the quantum number J^+ is incompatible with λ, σ, ω, and Ω.) Note that the non-commutation of $(\mathbf{N}^+)^2$ with $\mathbf{l}_z$ and $\mathbf{N}_z$ implies that, in a basis set in which any molecule-fixed projection of $\mathbf{l}$ (i.e., either λ or l_R) or $\mathbf{N}$ (i.e. either Λ or N_R) is a good quantum number, N^+ cannot also be a good quantum number. In such a case, N^+ should either be omitted from, or be enclosed in parentheses in the basis vector.†

The l_R quantum number is a label that conveys information about the projection of $\vec{\mathbf{l}}$ on the axis of nuclear rotation, $\vec{\mathbf{R}}$. In the limit of high R (or N), when $\vec{\mathbf{R}}$, $\vec{\mathbf{N}}$, and $\vec{\mathbf{N}}^+$ all point in approximately the same direction, l_R does in fact specify the integer valued projection of $\vec{\mathbf{l}}$ on $\vec{\mathbf{R}}$. However, l_R can always be regarded as an index that exactly specifies a difference in the magnitude quantum numbers,

$$l_R = N - N^+. \tag{3.2.7}$$

However, the expectation values of the operators derived by vector analysis to

† Basis states are labeled by a complete set of mutually compatible quantum numbers. These quantum numbers are eigenvalues of the members of a set of mutually commuting angular momentum magnitude and projection operators. Subtle difficulties arise when z–axis projection quantum numbers are specified for both ion-core, A_z^+, and Rydberg electron, α (hence for the total projection quantum number, $A_z = A_z^+ + \alpha$). Specification of a projection quantum number for both ion-core and Rydberg electron makes certain magnitude quantum numbers untenable. For example, if Σ^+ ($\mathbf{S}_z^+$) and σ ($\mathbf{s}_z$) are defined, then $\mathbf{J}^+$ ($\mathbf{J}^{+2}$) cannot be defined but $\mathbf{N}^+$ ($\mathbf{N}^{+2}$) can be defined. If Λ^+ ($\mathbf{L}_z^+ = \mathbf{N}_z^+$) and λ ($\mathbf{l}_z$) are defined, then neither $\mathbf{J}^+$ ($\mathbf{J}^{+2}$) nor $\mathbf{N}^+$ ($\mathbf{N}^{+2}$) can be defined, but a nonstandard magnitude quantum number, the eigenvalue of $(\mathbf{J}^+ - \mathbf{L}^+)^2$, can be defined.

give the projection of $\vec{\mathbf{l}}$ onto either $\vec{\mathbf{N}}$ or $\vec{\mathbf{N}}^+$,

$$\mathbf{l}\cdot\mathbf{N} = |\mathbf{l}||\mathbf{N}|\cos(\vec{\mathbf{l}},\vec{\mathbf{N}}) \tag{3.2.8a}$$

$$l_N = |\mathbf{l}|\cos(\vec{\mathbf{l}},\vec{\mathbf{N}}) = \frac{\mathbf{l}\cdot\mathbf{N}}{|\mathbf{N}|} \tag{3.2.8b}$$

$$|\mathbf{N}| = [N(N+1)]^{1/2} \approx N + 1/2 \tag{3.2.9}$$

$$\mathbf{N} - \mathbf{l} = \mathbf{N}^+ \tag{3.2.10a}$$

$$\mathbf{N}\cdot\mathbf{l} = \frac{\mathbf{N}^2 + \mathbf{l}^2 - (\mathbf{N}^+)^2}{2} \tag{3.2.10b}$$

$$l_N = \frac{N(N+1) + l(l+1) - N^+(N^++1)}{2(N+1/2)} \tag{3.2.11a}$$

$$l_{N^+} = \frac{N(N+1) - l(l+1) - N^+(N^++1)}{2(N^++1/2)} \tag{3.2.11b}$$

are equal to the required integer values of $l_N \approx l_{N^+} \approx l_R$ only in the high-N limit. For example, when $N = 5, l = 3$, and $N^+ = 3$ and 7, then $l_N = 2.73$ and -1.27 rather than the expected 2 and -2. This means that the l_R quantum number is a *useful label* in the $|NN^+l\Lambda^+\rangle$ basis set but is a *good quantum number in* the basis set defined by rotation of the quantization axis from along the internuclear axis to along $\vec{\mathbf{R}}$. This is accomplished by a coupled (molecule frame z-axis) $\to$ uncoupled transformation followed by an uncoupled$\to$coupled (molecule frame $\vec{\mathbf{R}}$-axis) unitary ($\mathbf{U}^{-1} = \mathbf{U}^+$) transformation

$$|ll_RN(N^+ = N - l_R)\Lambda^+\rangle = \sum_{N^+=N-l}^{N+l} U_{l;N^+}^{(lN\Lambda^+)}|l(l'_R = N - N^+)NN^+\Lambda^+\rangle \tag{3.2.12a}$$

$$|l(l'_R = N - N^+)N, N^+, \Lambda^+\rangle = \sum_{\lambda=-l}^{l} U_{N^+;\lambda}^{(lN\Lambda^+)}|l\lambda N\Lambda = \Lambda^+ + \lambda\rangle \tag{3.2.12b}$$

$$|l\lambda N\Lambda = \Lambda^+ + \lambda\rangle = \sum_{\lambda=-l}^{l} \Delta_{\lambda;l_R}^{l}|l(l'_R = N - N^+)NN^+\Lambda^+ = \Lambda - \lambda\rangle \tag{3.2.12c}$$

$$|ll_RN(N^+ = N - l_R)\Lambda^+\rangle = \sum_{N^+=N-l}^{N+l}\sum_{\lambda=-l}^{l} U_{l;N^+}^{(lN\Lambda^+)}U_{N^+;\lambda}^{(lN\Lambda^+)}\Delta_{\lambda;l_R}^{l} \times |l(l'_R = N - N^+)NN^+\Lambda^+\rangle \tag{3.2.12d}$$

$$U_{l;N^+}^{(lN\Lambda^+)} = \langle ll_RN(N^+ = N - l_R)\Lambda^+|l(l'_R = N - N^+)NN^+\Lambda^+\rangle \tag{3.2.12e}$$

$$U_{N^+;\lambda}^{(lN\Lambda^+)} = \langle l(l'_R = N - N^+)NN^+\Lambda^+|l\lambda N\Lambda = \Lambda^+ + \lambda\rangle \tag{3.2.12f}$$

$$\Delta^{l}_{\lambda;l_R} = d^{l}_{\lambda;l_R}(\pi/2) = \langle l\lambda N\Lambda = \Lambda^+ + \lambda | l l_R N \Lambda^+ \rangle \tag{3.2.13}$$

where

$$d^{l}_{\lambda;l_R}(\pi/2)$$

is the Wigner rotation matrix element (tabulated by Behkami, 1971) that corresponds to a rotation by $\pi/2$ about the body-fixed y–axis (perpendicular to both $\vec{\mathbf{R}}$ and the internuclear axis, z) [see Eqs. (114-116, 133) of Watson, 1999] and the $\langle|\rangle$ factors are vector coupling coefficients. Note that, in the $\vec{\mathbf{R}}$–axis basis set, N^+ appears in parentheses because the basis states are not eigenstates of $(\mathbf{N}^+)^2$ and the N^+ (bad) quantum number is only defined in the sense of a difference between two good quantum numbers, $N^+ = N - l_R$.

The transformation coefficients, $U^{(lN\Lambda^+)}_{l;N^+}$ and $U^{(lN\Lambda^+)}_{N^+;\lambda}$ in Eqs. (3.2.12a) - (3.2.12f), are expressed in the form suggested by Jungen and Raşeev (1998)

$$|i, \alpha\beta\rangle = \sum_j U^{(\alpha\beta)}_{i;j} |j, \alpha\beta\rangle.$$

The right subscripts specify the sets of quantum numbers (separated by a semicolon) that are *not common* to the $|i\rangle$ and $|j\rangle$ basis sets. The contents of the right superscript parentheses (optional) are those quantum numbers common to both basis sets that are needed to fully specify the 3-j or 6-j coefficients and/or the matrix elements expressed in the $|j\rangle$ basis.

The rotational Hamiltonian,

$$\mathbf{H}^{\text{ROT}} = B(R)\mathbf{R}^2, \tag{3.2.14}$$

where $B(R)$ is the rotational "constant" operator (Eq. (3.1.17)) and $\mathbf{R}$ is the nuclear rotational angular momentum, may be expressed in a form appropriate for each of the Hund's case (a)-(e) basis sets.

For **case (a)**:

$$\begin{aligned}\mathbf{H}^{\text{ROT}} &= B(R)(\mathbf{J} - \mathbf{L} - \mathbf{S})^2 \\ &= B(R)[\mathbf{J}^2 - \mathbf{J}_z^2 + \mathbf{S}^2 - \mathbf{S}_z^2 + (\mathbf{L}^2 - \mathbf{L}_z^2) \\ &\quad - \{(\mathbf{J}^+\mathbf{L}^- + \mathbf{J}^-\mathbf{L}^+) + (\mathbf{J}^+\mathbf{S}^- + \mathbf{J}^-\mathbf{S}^+) - (\mathbf{L}^+\mathbf{S}^- + \mathbf{L}^-\mathbf{S}^+)\}]\end{aligned} \tag{3.2.15}$$

where $(\mathbf{L}^2 - \mathbf{L}_z^2)$ is replaced by $\mathbf{L}_\perp^2$ and the unevaluated expectation value of $\text{B}(R)\mathbf{L}_\perp^2$ is implicitly included in the electronic energy. The zero-order rotational energy is $E^{\text{ROT}}_{(a)} = B(R)[J(J+1) - \Omega^2 + S(S+1) - \Sigma^2]$. The three pairs of terms in $\{\ \}$ give rise to off-diagonal matrix elements (respectively $\Delta\Omega = \Delta\Lambda = \pm1$, $\Delta\Omega = \Delta\Sigma = \pm1$, and $\Delta\Lambda = -\Delta\Sigma = \pm1$). The $(\mathbf{J}^+\mathbf{S}^- + \mathbf{J}^-\mathbf{S}^+)$ term operates within a case (a) ${}^{2S+1}\Lambda$ multiplet state, whereas the other two terms in $\{\ \}$

are responsible for off-diagonal matrix elements between case (a) ${}^{2S+1}\Lambda_\Omega$ and ${}^{2S+1}\Lambda \pm 1$ states.

For **case (b)**:

$$\begin{aligned}\mathbf{H}^{\mathrm{ROT}} &= B(R)(\mathbf{N}-\mathbf{L})^2 \\ &= B(R)[\mathbf{N}^2 - \mathbf{N}_z^2 + \mathbf{L}_\perp^2 - \{\mathbf{N}^+\mathbf{L}^- + \mathbf{N}^-\mathbf{L}^+\}] \end{aligned} \tag{3.2.16}$$

where the unevaluated expectation value of the $B(R)\mathbf{L}_\perp^2$ term is included in the electronic energy. Since $\mathbf{N} = \mathbf{R} + \mathbf{L}$ and $\vec{\mathbf{R}}$ is perpendicular to the internuclear axis,

$$\mathbf{N}_z \left| nJSN\Lambda(S_R) \right\rangle = \Lambda \left| nJSN\Lambda(S_R) \right\rangle, \tag{3.2.17}$$

and the zero-order rotational energy is

$$E_{(b)}^{\mathrm{ROT}} = B(R)[N(N+1) - \Lambda^2]. \tag{3.2.18}$$

The pair of terms in { } give rise to off-diagonal matrix elements ($\Delta J = \Delta N = 0, \Delta\Lambda = \pm 1$) between states belonging to different case (b) ${}^{2S+1}\Lambda$ states. The greater simplicity of $\mathbf{H}^{\mathrm{ROT}}$ in the case (b) than case (a) basis set comes with a price of requiring greater effort in dealing with matrix elements of some other terms in the total molecular $\mathbf{H}$. In particular, matrix elements of all operators that involve $\mathbf{S}_z$ are neither diagonal in the case (b) basis nor evaluable without resorting to tables of vector coupling coefficients. At low values of N, $S_R = J - N$ is a bad quantum number and therefore is listed enclosed in parentheses in the basis state.

For **case (c)**:

$$\begin{aligned}\mathbf{H}^{\mathrm{ROT}} &= B(R)(\mathbf{J}-\mathbf{J}_a)^2 \\ &= B(R)[\mathbf{J}^2 - \mathbf{J}_z^2 + \mathbf{J}_a^2 - \mathbf{J}_{az}^2 - \{\mathbf{J}^+\mathbf{J}_a^- + \mathbf{J}^-\mathbf{J}_a^+\}]. \end{aligned} \tag{3.2.19}$$

Since $\mathbf{J}_a = \mathbf{J} - \mathbf{R}$,

$$\mathbf{J}_{az} \left| nJ(J_a)\Omega \right\rangle = \Omega \left| nJ(J_a)\Omega \right\rangle \tag{3.2.20}$$

and the zero-order rotational energy is

$$E_{(c)}^{\mathrm{ROT}} = B(R)[J(J+1) - \Omega^2 + J_a(J_a+1) - \Omega^2]. \tag{3.2.21}$$

If the J_a quantum number is not well defined, then $\mathbf{J}_a^2 - \mathbf{J}_{az}^2 = \mathbf{J}_{a\perp}^2$ and the unevaluated expectation value of $B(R)\mathbf{J}_{a\perp}^2$ is implicitly included in the electronic energy. The pair of terms in { } in Eq. (3.2.19) give rise to off-diagonal matrix elements ($\Delta J = 0, \Delta J_a = 0, \Delta\Omega = \pm 1$) between different case (c) states, denoted $\Omega(J_a)$ or $\Omega({}^{2S+1}\Lambda)$ (or $J_{a\Omega}$, as for the open-core rare earth oxides and halides). The simplicity of $\mathbf{H}^{\mathrm{ROT}}$ in the case (c) basis set exacts a price in the difficulty of evaluating matrix elements of most operators that include $\mathbf{L}_z$, $\mathbf{S}^2$, or $\mathbf{S}_z$.

For **case (d)**:

$$\mathbf{H}^{\mathrm{ROT}} = B^+(R)(\mathbf{N}^+ - \mathbf{L}^+)^2$$
$$= B^+(R)[(\mathbf{N}^+)^2 - (\mathbf{N}_z^+)^2 + (\mathbf{L}_\perp^+)^2$$
$$- \{(\mathbf{N}^+)^+(\mathbf{L}^+)^- + (\mathbf{N}^+)^-(\mathbf{L}^+)^+\}] \tag{3.2.22}$$

where the usually unevaluable expectation value of the $B^+(R)(\mathbf{L}_\perp^+)^2$ term is implicitly included in the electronic energy, and the zero-order rotational energy is

$$E_{(\mathrm{d})}^{\mathrm{ROT}} = B^+(R)[N^+(N^+ + 1) - \Lambda^{+2}]. \tag{3.2.23}$$

If $l_R = N - N^+$, then

$$E_{(\mathrm{d})}^{\mathrm{ROT}} = B^+(R)[N(N+1) - 2Nl_R - l_R + l_R^2 - \Lambda^{+2}]. \tag{3.2.24}$$

The Eq. (3.2.24) form of $\mathbf{E}_{(\mathrm{d})}^{\mathrm{ROT}}$ is useful because rotational combination differences (see Section 1.1) often permit unambiguous determination of N (but not N^+) and the $-2l_R$ slope of a "reduced term value" plot of $E_{(\mathrm{d})}^{\mathrm{ROT}}(N) - BN(N+1)$ vs. N determines l_R, which in turn determines N^+ (see Section 6.3.3).

The $\mathbf{L}^+$ operator is suppressed in the usual form of the case (d) rotational Hamiltonian (because the ground state of most ion-cores is an isolated $^1\Sigma^+$ or $^2\Sigma^+$ state),

$$\mathbf{H}^{\mathrm{ROT}} = B^+(\mathbf{N}^+)^2 \tag{3.2.25}$$
$$E_{(d)}^{\mathrm{ROT}} = B^+N^+(N^+ + 1). \tag{3.2.26}$$

The rotational constant operator, $B^+(R)$, contains small contributions from rotation-electronic interactions with remote states (see Watson, 1999, and Section 4.2). The pair of terms in { } in Eq. (3.2.22) give rise to off-diagonal matrix elements ($\Delta J = \Delta N^+ = \Delta l = 0$, $\Delta\Lambda^+ = \pm 1$) between members of different Rydberg series built on different $^{2S^+ + 1}\Lambda^+$ ion-core states. The case (d) l–complex states are denoted $n^2l(^{2S^+ + 1}\Lambda^+)$. When $\mathbf{H}^{\mathrm{el}}$ is evaluated in a case (d) basis set, $\mathbf{H}^{\mathrm{el}}$ generates both energy differences and nonzero off-diagonal matrix elements between components of an l–complex (Jungen and Miescher, 1969; Watson, 1999) or supercomplex (Miescher, 1976; Jakubek and Field, 1994).

For **case (e)**:

$$\mathbf{H}^{\mathrm{ROT}} = B^+(R)[\mathbf{J} - \mathbf{j} - \mathbf{J}_a^+]^2 = B^+(R)[\mathbf{J}^+ - \mathbf{J}_a^+]^2$$
$$= B^+(R)[(\mathbf{J}^+)^2 - (\mathbf{J}_z^+)^2 + (\mathbf{J}_a^+)^2 - (\mathbf{J}_{az}^+)^2$$
$$- \{(\mathbf{J}^+)^+(\mathbf{J}_a^+)^- + (\mathbf{J}^+)^-(\mathbf{J}_a^+)^+\}] \tag{3.2.27}$$

where the usually unevaluable expectation value of the

$$B^+(R)[(\mathbf{J}_a^+)^2 - (\mathbf{J}_{az}^+)^2] = B^+(R)(\mathbf{J}_{a\perp}^+)^2 \tag{3.2.28}$$

term is implicitly included in the electronic energy, and the zero-order rotational energy is

$$E^{\mathrm{ROT}}_{(e)} = B^{+}(R)[J^{+}(J^{+}+1) - \Omega^{+2}]. \tag{3.2.29}$$

If the $\mathbf{J}_a^{+}$ operator is suppressed in the case (e) rotational Hamiltonian,

$$\mathbf{H}^{\mathrm{ROT}} = B^{+}(\mathbf{J}^{+})^2 \tag{3.2.30}$$

$$E^{\mathrm{ROT}}_{(e)} = B^{+}J^{+}(J^{+}+1) \tag{3.2.31}$$

The pair of terms in { } in Eq. (3.2.27) give rise to off-diagonal matrix elements ($\Delta J^{+} = \Delta j = \Delta J_a^{+} = 0$, $\Delta\Omega^{+} = \pm 1$) between members of different Rydberg series built on different $\Omega^{+}(J_a^{+})$ ion-core states. $\mathbf{H}^{\mathrm{el}}$ is not diagonal in the case (e) basis set.

Most treatments of molecular Rydberg states in the case (d) and (e) limits rest on a crucial simplification of the possible coupling schemes among the ion-core, Rydberg electron, and core+Rydberg electron angular momenta. When $S^{+} = 0$ or when $S^{+} = 1/2$ and $\Lambda^{+} = 0$, the case (d) and (e) basis functions become very similar. For example, if $S^{+} = 0$, then $N^{+} = J^{+}$ and the case (d) and (e) basis functions take on Watson's (b^{+}, d) and (b^{+}, e) forms, respectively, and the two basis sets are related by a simple case $(e) \leftarrow (d)$ transformation for the Rydberg electron, expressed in terms of { } 6-j coefficients,

$$|(b^{+},e)\rangle = |lsjJ^{+} = N^{+}\Lambda^{+}J\rangle = \sum_{N=N^{+}-l-s}^{N^{+}+l+s} U_{j;N}^{(lsjN^{+}\Lambda^{+}J)}|lsN^{+}\Lambda^{+}JN\rangle =$$

$$\sum_{N=N^{+}-l}^{N^{+}+l} (-1)^{J+N^{+}+l+s}[(2j+1)(2N+1)]^{1/2} \begin{Bmatrix} l & s & j \\ J & N^{+} & N \end{Bmatrix} |lsN^{+}\Lambda^{+}JN\rangle \tag{3.2.32a}$$

[Watson, 1999, Eq. (60)]. The reason that the transformation between two Hund's cases can be accomplished using a single (6-j) transformation coefficient rather than the usual product of three (3-j) transformation coefficients as are required for the $(a^{+}, e) \leftarrow (a^{+}, a)$ transformation in the general case when $S^{+} \neq 0$ and $\Lambda^{+} \neq 0$ (Jungen and Raşeev, 1998), is the abnormally small number of independent, nonzero ion-core quantum numbers ($S^{+} = 0$, therefore $N^{+} = J^{+}, \Omega^{+} = \Lambda^{+}$). In the not so quite so simple special case as $S^{+} = 0$, $\Lambda^{+} \neq 0$, the case $(b^{+}, d) \rightarrow$ case (b^{+}, e) transformation requires the replacement of two case (b^{+}, d) quantum numbers (N, S) by two case (b^{+}, e) quantum numbers (J^{+}, j),

$$|(b^{+},e)\rangle = |N^{+}S^{+}J^{+}lsjJ\rangle = \sum_{S=0}^{1}\sum_{N=J-S}^{J+S} U_{jJ^{+};NS}^{(N^{+}S^{+}J)}|N^{+}S^{+}lsSNJ\rangle.$$

In general, when both $S^{+} \neq 0$ and $\Lambda^{+} \neq 0$, the transformations between (ion-core, Rydberg e^{-}) coupling cases involve the replacement of as many as three quantum numbers by three quantum numbers (Jungen and Raşeev, 1998).

6-j coefficients are typically used in transformations between basis sets where one angular momentum magnitude quantum number is being replaced by another, as in Eq. (3.2.32a) where N in case (d) is replaced by j in case (e). When 3 or more angular momentum are being coupled to form a total angular momentum, several choices for intermediate angular momenta exist. In this $S^+ = 0$ example, for case (d) $\mathbf{N}^+ + \mathbf{l} = \mathbf{N}$, $\mathbf{N} + \mathbf{s} = \mathbf{J}$ thus $\mathbf{N}$ is the intermediate angular momentum, but for case (e) $\mathbf{l} + \mathbf{s} = \mathbf{j}$, $\mathbf{j} + \mathbf{N}^+ = \mathbf{J}$ thus $\mathbf{j}$ is the intermediate angular momentum. In general, if **a**, **b** and **c** are the three angular momenta being coupled to form **f**, and **d**=**a**+**b** and **e**=**a**+**c** are two of the three possible intermediate angular momenta, then

$$|abcd(a+b)f\rangle = \sum_{e=c-a}^{c+a} (-1)^{a+b+c+f}[(2d+1)(2e+1)]^{1/2} \begin{Bmatrix} a & b & d \\ f & c & e \end{Bmatrix} |abce(a+c)f\rangle. \tag{3.2.32b}$$

In contrast, 3-j coeficients (e.g. Section 3.2.1.4) are useful in transforming between coupled $|abcc_z\rangle$ and uncoupled $|aa_zbb_z\rangle$ basis sets.

Table 3.3 summarizes the five traditional Hund's cases and the mutually commuting terms included in $\mathbf{H}^{(0)}$. This set of mutually commuting terms is the part of the total $\mathbf{H}$ which is actually diagonalized exactly or assumed to be diagonal (as in model Hamiltonians used to fit spectroscopically observed energy levels) in the specified basis set. All of the basis function labels except n are eigenvalues of operators that commute with each other and with $\mathbf{H}^{(0)}$. This mutual commutation is a necessary condition for the existence of a set of basis functions that are simultaneously eigenfunctions of $\mathbf{H}^{(0)}$ and of each of the defining angular momenta. The basis function labels are good quantum numbers *with respect to* $\mathbf{H}^{(0)}$, but not necessarily with respect to $\mathbf{H}^{(1)}$. The only rigorously conserved quantities are E (the energy itself), and the quantum numbers J and parity (see Section 3.2.2).[†]

Even if a case (a) basis set provides an extremely poor zero-order representation of a low-n Rydberg complex, it is a complete set of basis states in terms of which the molecular electronic-vibrational-rotational $\mathbf{H}$ for the entire complex may be expressed. This case (a) $\mathbf{H}^{(a)}$ will contain both rotation-independent and rotation-dependent off-diagonal matrix elements. A unitary transformation diagonalizes $\mathbf{H}^{(a)}$ and provides eigen-energies and eigenvectors. Alternatively, a unitary transformation may be used to transform $\mathbf{H}^{(a)}$ into case (b), (d), or (e) form. (Trace invariance of $\mathbf{H}$ under a unitary transformation provides a basis for defining a case (d) or (e) potential curve as a simple average over the contributing case (a) potential curves.) There is no fundamental difference between such a procedure for describing a case (d) or (e) Rydberg complex or a case (a) or (b) $^{2S+1}\Lambda$ multiplet state.

[†] If the molecule has nonzero nuclear spin, then the magnitude of $\mathbf{F} = \mathbf{J} + \mathbf{I}$ but not of $\mathbf{J}$ is a rigorously good quantum number. Parity and J are spoiled in an electric field; only J is spoiled in a magnetic field.

Table 3.3: Hund's Cases

Coupling Case	Diagonalized Part of the Total Hamiltonian = $\mathbf{H}^{(0)}$	Good Quantum Numbers[a,e]	Good Approximation When[b]
Case (a)	$\mathbf{H}^{\text{el}} + B(\mathbf{J}^2 - \mathbf{J}_z^2 + \mathbf{S}^2 - \mathbf{S}_z^2) + B\mathbf{L}_\perp^2$	$n, J, \Omega, \Lambda, S, \Sigma$	$A\Lambda \gg BJ$
Case (b)	$\mathbf{H}^{\text{el}} + B(\mathbf{N}^2 - \mathbf{N}_z^2) + B\mathbf{L}_\perp^2$	n, J, N, Λ, S $(S_R = J - N)$	$A\Lambda \ll BJ$
Case (c)	$\mathbf{H}^{\text{el}} + B(\mathbf{J}^2 - \mathbf{J}_z^2 + \mathbf{J}_a^2 - \mathbf{J}_{az}^2)$	$n, J, \Omega, [J_a]$	$A\Lambda > [E^0(n\ ^{2S+1}\Lambda_\Omega) - E^0(n\ ^{2S-1}\Lambda_\Omega)]$
Case (d)[c]	$\mathbf{H}^{\text{el}(0)^f} + B^+((\mathbf{N}^+)^2 - (\mathbf{N}_z^+)^2) + B^+(\mathbf{L}_\perp^+)^2$	$n, J, S, N, J^+,$ N^+, S^+, Λ^+, l $(S_R = J - N)$ $(l_R = N - N^+)$ $(s_R = J^+ - N^+)$	$B^+J > [E^0(n\ ^{2S+1}\Lambda) - E^0(n\ ^{2S+1}\Lambda \pm 1)]$ and $A\Lambda \ll BJ$
Case (e)[c]	$\mathbf{H}^{\text{el}(0)^f} + B^+[(\mathbf{J}^+)^2 - (\mathbf{J}_z^+)^2 + (\mathbf{J}_a^+)^2 - (\mathbf{J}_{az}^+)^2]$	$n, J, J^+, [J_a^+],$ Ω^+, l, j [d] $(j_R = J - J^+)$	$B^+J^+ > [E^0(n^{2S+1}\Lambda_\Omega) - E^0(n^{2S+1}\Lambda_{\Omega\pm1})],$ $B^+J^+ > [E^0(n^{2S+1}\Lambda_\Omega) - E^0(n^{2S+1}(\Lambda\pm1)_{\Omega\pm1})]$ and $A\Lambda > [E^0(n^{2S+1}\Lambda_\Omega) - E^0(n^{2S-1}\Lambda_\Omega)]$

[a] Good quantum numbers are associated with operators that commute with $\mathbf{H}^{(0)}$ and with each other.

[b] A and B are the spin-orbit and rotational constants, defined respectively by Eqs. (3.4.6) and (3.1.17). B^+ is the rotational constant of the ion-core.

[c] Angular momentum operators associated with the ion-core will be denoted as $\mathbf{A}^+$ $[(\mathbf{A}^+)^2, \mathbf{A}_z^+, (\mathbf{A}^+)^\pm$, respectively magnitude squared, projection on the bond axis, and molecule frame raising/lowering operator]. There is potential for confusion with the molecule fixed raising operator for the total molecular angular momentum, $\mathbf{A}^+$. However, whenever ion-core angular momenta $\mathbf{A}^+$ appear in combination with other operators, e.g. $[(\mathbf{A}^+)^+\mathbf{B}^- - (\mathbf{A}^+)^-\mathbf{B}^+]$, ambiguity will be minimized by enclosing the ion-core operator in ( ).

[d] j is the quantum number associated with the total angular momentum of the Rydberg electron.

[e] The quantum numbers of the form A_R specify the projection of $\vec{\mathbf{A}}$ on $\vec{\mathbf{R}}$, see Watson (1999) and Nikitin and Zare (1994). The A_R quantum numbers are "good" in the high–J limit and, always, according to a ranking in order of $\langle \mathbf{A_R} \rangle$ values (see Eq. 3.2.11). We encourage the use of l_R and S_R rather than the previously used $\mathcal{L}$ and $\mathcal{S}$ quantum numbers because the molecule frame axis into which the angular momenta are projected is specified, and this makes explicit the transformation relationships between angular momentum basis sets (Watson, 1999).

[f] $\mathbf{H}^{\text{el}}$ is diagonal in case (a), (b), and (c) basis sets where the quantum numbers Λ^+ (or Ω^+) and λ (or ω) are defined. $\mathbf{H}^{\text{el}}$ is not diagonal in case (d) and (e) basis sets. Due to Coulomb and exchange interactions of the Rydberg electron with the ion-core, at both short and long range, isoconfigurational states are split by $\mathbf{H}^{\text{el}}$. A spherical-ized and therefore diagonal and degenerate $\mathbf{H}^{\text{el}(0)}$ is defined as the average (R–dependent) energy of the $2(2l+1)(2S^+ + 1)(2 - \delta_{\Lambda^+,0})$ case (a) basis states that result when a core-nonpenetrating Rydberg electron in an nl orbital is combined with a $^{2S^+ +1}\Lambda^+$ ion-core state (see Eq. (3.2.52a)). When, as for core-penetrating Rydberg orbitals, l is ill-defined, there is not a unique way to define $\mathbf{H}^{\text{el}(0)}$.

Experimental $G(v)$ and $B(v)$ data may be readily inverted by the RKR procedure (see Section 5.1.3) to obtain the potential energy curves [and, by applying a case (a)→(d) or (e) transformation to $\mathbf{H}^{(a)}$ for cases (d) and (e), the off-diagonal molecular field anisotropy (i.e., differences between the clamped nuclei electronic energies for $(^{2S^{+}+1}\Lambda^{+})nl\lambda$ and $(^{2S^{+}+1}\Lambda^{+})nl\lambda \pm 1$ case (a) states) coupling terms (see Eq. (3.2.47), Eqs. (3.2.48a) - (3.2.48c), and Eqs. (3.2.70) - (3.2.71))] that are appropriate to express **H** in a case (b), (d), or (e) basis set.

Disagreements about the ability to define a potential curve arise from the different meanings of the words "Born-Oppenheimer approximation" in *ab initio* quantum chemical calculations and a scattering theory (Multichannel Quantum Defect Theory) treatment of the electronic structure and dynamics of Rydberg states. The "clamped nuclei" version of the Born-Oppenheimer approximation (see Section 3.1.1) provides an ambiguous recipe for computing the energies of electronic states on a grid of internuclear distances, and these electronic energies, parametrically dependent on R, are known as Born-Oppenheimer potential energy curves. The "clamped nuclei" picture provides another crucial simplfication: there is no difference between the laboratory and molecule-fixed coordinate systems because clamped nuclei neither rotate or vibrate. In scattering theory, the artifice of clamped nuclei is not needed. Space is divided into two regions, a "Born-Oppenheimer" region near the nuclei where the electrons move so fast relative to the nuclei that the nuclei are *de facto* clamped, and a "non-Born-Oppenheimer" region far from the nuclei where electron motion becomes comparable to or slower than nuclear motion. Scattering theory provides a far more complete and unified representation of Rydberg states (electron-ion scattering at "negative energy") than clamped nuclei potential curves. However, scattering theory is mostly outside of the intended scope of this book. The clamped nuclei picture is especially useful to experimentalists and quantum chemists in providing a computational framework for studies of low-lying electronic states, especially valence states and low-n Rydberg complexes.

3.2.1.2 Quantum Numbers, Level Patterns, and the Effects of Terms Excluded from $\mathbf{H}^{(0)}$.

Case (a). For case (a), the good quantum numbers are J, S, Λ, and Σ. Ω is also good, but since $\Omega = \Lambda + \Sigma$, it is redundant. The $|nJS\Lambda\Sigma\rangle$ basis states are eigenfunctions of the electronic Hamiltonian and the part of $\mathbf{H}^{\mathrm{ROT}}$ given in Eq. (3.1.14). $\mathbf{H}^{(0)}$ excludes $\mathbf{H}^{\mathrm{SO}}$ and part of $\mathbf{H}^{\mathrm{ROT}}$. Case (a) electronic wave functions correspond exactly to those obtained by most *ab initio* calculations, which rest on the Born-Oppenheimer assumption of the separability of nuclear and electronic motions. An advantage of the case (a) basis is that the absence of intermediate angular momenta (N, J_a) and the presence of the maximum number of molecule-fixed angular momentum z–components $(\Lambda, \Sigma, \Omega)$ enable $\mathbf{H}^{\mathrm{SO}}$ and $\mathbf{H}^{\mathrm{ROT}}$ matrix elements to be evaluated using elementary raising and lowering operator techniques. For these reasons, the case (a) basis is the preferred basis throughout this book.

The energy-level pattern associated with the case (a) $\mathbf{H}^{(0)}$ consists of groups of $2S+1$ Ω–substates ($|\Lambda| - S \leq \Omega \leq |\Lambda| + S$), all multiplet components having the same rotational constant, rotational energy $E_J = BJ(J+1)$,[†] and potential energy curve. In order for the case (a) energy-level pattern to be a good approximation to that of a real molecule, it is necessary that

$$|A\Lambda| \gg BJ(2S)^{1/2} \tag{3.2.33a}$$

where

$$\begin{aligned} |A\Lambda| = |\,\langle n, |\Lambda|, S, \Sigma = S, \Omega = S + |\Lambda||\mathbf{H}^{\text{SO}}|n, |\Lambda|, S, \Sigma, \Omega\rangle \\ - \langle n, |\Lambda|, S, \Sigma' = S-1, \Omega' = S + |\Lambda| - 1|\mathbf{H}^{\text{SO}}|n, |\Lambda|, S, \Sigma', \Omega'\rangle| \end{aligned} \tag{3.2.33b}$$

$$\begin{aligned} BJ(2S)^{1/2} \cong |\langle n, J, |\Lambda|, S, \Sigma = S, \Omega = S + |\Lambda||BJ^{-}\mathbf{S}^{+} \\ |n, J, |\Lambda|, S, \Sigma' = S-1, \Omega' = S + |\Lambda| - 1\rangle|. \end{aligned} \tag{3.2.33c}$$

That is, values of the off-diagonal matrix elements of the $B\mathbf{J}^{\pm}\mathbf{S}^{\mp}$, **S**-uncoupling operator, within a ${}^{2S+1}\Lambda_{\Omega}$ multiplet state, must be much smaller than the energy separation between multiplet components that result from differences between diagonal matrix elements of $\mathbf{H}^{\text{SO}}$. Since matrix elements of the S-uncoupling operator are proportional to J, at sufficiently high–J case (a) ceases to be as good an approximation as case (b).

Case (b). The good case (b) quantum numbers are J, S, N, and Λ. In replacing Σ of case (a) by N of case (b), the dimension of the basis is unchanged because for each J, Λ, S there are $2S+1$ possible N or Σ values. An example of an expansion of a case (b) basis function, $|nJSN\Lambda\rangle$, in terms of the $2S+1$ case (a) functions, $|nJS\Sigma\Lambda\Omega\rangle$, of different Σ values, is given in Section 3.2.1.4 (see Eqs. (3.2.61a) and (3.2.61b)). Σ is not defined in case (b). This means that $\vec{\mathbf{S}}$ does not precess around the internuclear axis. This decoupling of $\vec{\mathbf{S}}$ from the z–axis is caused by the "S-uncoupling" $B\mathbf{J}^{\pm}\mathbf{S}^{\mp}$ operator. The coordinate system in which the spin variables are quantized becomes the molecule frame system in which the quantization axis is the direction of the nuclear rotation angular momentum, $\vec{\mathbf{R}}$, rather than the molecular axis. However, this coordinate system is inconvenient for *ab initio* calculations. Although $\vec{\mathbf{S}}$ is decoupled from the molecule fixed z–axis, it is still quantized in the molecule fixed and *not* the laboratory fixed coordinate system (M_S is not a good quantum number). A situation where $\vec{\mathbf{S}}$ is fully decoupled from the molecule frame is discussed in Section 6.5.4.

The case (b) basis functions are defined as eigenfunctions of $\mathbf{H}^{\text{el}}$ and the $B(\mathbf{N}^2 - \mathbf{N}_z^2)$ part of $\mathbf{H}^{\text{ROT}}$. The part of $\mathbf{H}^{\text{ROT}}$ that is included in $\mathbf{H}^{(0)}$ has only diagonal case (b) matrix elements, and the rotational energy is $E_N = B[N(N+1) - \Lambda^2]$ rather than $B[J(J+1) - \Omega^2 + S(S+1) - \Sigma^2]$. Each N–value is $2S+1$–fold degenerate, with J–values in the range $N - S \leq J \leq N + S$. This degeneracy is lifted by the spin-rotation term (Section 3.4.3),

$$\mathbf{H}^{\text{SR}} = \gamma\mathbf{N}\cdot\mathbf{S}, \tag{3.2.34}$$

[†]More correctly, the rotational energy is $E^{\text{ROT}}_{J\Omega S\Sigma} = B[J(J+1) - \Omega^2 + S(S+1) - \Sigma^2]$.

which has only diagonal matrix elements in case (b) because

$$\mathbf{N}\cdot\mathbf{S} = \frac{1}{2}\left(\mathbf{J}^2 - \mathbf{N}^2 - \mathbf{S}^2\right). \tag{3.2.35}$$

Neglecting $\mathbf{H}^{\mathrm{SR}}$, from which the maximum resultant splitting of the J–components of a given N–value is

$$E^{(0)}(N, J = N + S) - E^{(0)}(N, J = N - S) = \gamma S(2N + 1), \tag{3.2.36}$$

and provided that the levels are labeled by N rather than J, the $J = N - S$ stack of J–levels is identical to the $J = N - S + 1$, $J = N - S + 2, \ldots$, and $J = N + S$ stacks, except when $N - \Lambda < S + \frac{1}{2}$, for which the J–levels with $N - \Lambda < 0$ are absent.

Each Hund's case is associated with a pattern-forming rotational quantum number, a problematic term in $\mathbf{H}^{\mathrm{ROT}}$ that generates off-diagonal matrix elements within a systematically near degerate group of basis states, and terms in $\mathbf{H}^{(0)}$ that lift the degeneracy of the basis states coupled by this problematic term in $\mathbf{H}^{\mathrm{ROT}}$. For case (a) the pattern-forming rotational quantum number is J,

$$E^{\mathrm{ROT}}_{(a)} = B[J(J + 1) - \Omega^2 + S(S + 1) - \Sigma^2],$$

the spin-uncoupling term

$$-B[\mathbf{J}^+\mathbf{S}^- + \mathbf{J}^-\mathbf{S}^+]$$

is most problematic because it has nonzero off-diagonal matrix elements within a ${}^{2S+1}\Lambda$ multiplet state, the degeneracies of which are lifted by

$$-B[\mathbf{J}_z^2 + \mathbf{S}_z^2] + A\mathbf{L}_z \cdot \mathbf{S}_z.$$

The case (a) form of $\mathbf{H}^{\mathrm{ROT}}$ may be rearranged into a form that better illustrates the effect of the transformation to case (b),

$$\mathbf{H}^{\mathrm{ROT}} = B[\mathbf{J}^2 + \mathbf{S}^2 + \mathbf{L}_\perp^2 - \mathbf{L}_z^2 - 2\mathbf{J}\cdot\mathbf{S} - (\mathbf{N}^+\mathbf{L}^- + \mathbf{N}^-\mathbf{L}^+)] \tag{3.2.37}$$

In the case (b) basis set, the $-2B\mathbf{J}\cdot\mathbf{S}$ term is diagonal because

$$\mathbf{J}\cdot\mathbf{S} = \frac{1}{2}[\mathbf{N}^2 - \mathbf{J}^2 - \mathbf{S}^2].$$

All terms in the modified case (a) form of $\mathbf{H}^{\mathrm{ROT}}$ can be evaluated in the case (b) basis set,

$$\begin{aligned} E^{\mathrm{ROT}}_{(b)} &= B[J(J+1) + S(S+1) - \Lambda^2] - B[J(J+1) + S(S+1) - N(N+1)] \\ &= B[N(N+1) - \Lambda^2]. \end{aligned} \tag{3.2.38}$$

The result is that all of the "wrong-form" case (a)-like terms in $E^{\mathrm{ROT}}_{(b)}$ are cancelled by terms from $-B\mathbf{J}\cdot\mathbf{S}$ and the expected $B[N(N+1) - \Lambda^2]$ form for $E^{\mathrm{ROT}}_{(b)}$ is obtained.

Deviations from the case (b) level pattern are caused by $\mathbf{H}^{SO}$, $\mathbf{H}^{SS}$, and by the excluded portion of $\mathbf{H}^{ROT}$,

$$\begin{aligned}\mathbf{H}^{ROT} - B(\mathbf{N}^2 - \mathbf{N}_z^2) &= B[-2\mathbf{N}\cdot\mathbf{L} + 2\mathbf{N}_z\mathbf{L}_z] \\ &= -B[\mathbf{N}^+\mathbf{L}^- + \mathbf{N}^-\mathbf{L}^+]. \end{aligned} \qquad (3.2.39)$$

As for case (a), the unevaluated expectation value of the quantity

$$B\left[\mathbf{L}^2 - \mathbf{N}_z\cdot\mathbf{L}_z\right] = B\mathbf{L}_\perp^2$$

is usually incorporated in the electronic energy (Section 3.1.2.3). In the case (b) basis, selection rules for $\mathbf{H}^{SO}$ and $B\mathbf{N}\cdot\mathbf{L}$ are

	$\mathbf{H}^{SO}$	$-B[\mathbf{N}^+\mathbf{L}^- + \mathbf{N}^-\mathbf{L}^+]$
ΔJ	0	0
ΔS	$0,\pm1$	0
ΔN	$0,\pm1$	0
$\Delta\Lambda$	$0,\pm1$	±1

(3.2.40)

The $B(\mathbf{N}^+\mathbf{L}^-+\mathbf{N}^-\mathbf{L}^+)$ $\mathbf{L}$-uncoupling term has no effect *within* a case (b) multiplet state, but has nonzero matrix elements between electronic states differing in Λ by ±1. The $\Delta\Lambda = \pm1$ interactions have matrix elements roughly proportional to BN and, when

$$BN \gg |E^0(n\Lambda) - E^0(n'\Lambda\pm1)|, \qquad (3.2.41)$$

these $\Delta\Lambda = \pm1$ matrix elements are responsible for the transition from case (b) toward case (d).

In order to evaluate matrix elements of the $A\mathbf{L}\cdot\mathbf{S}$ form of $\mathbf{H}^{SO}$ and the $-\frac{2}{3}\lambda[\mathbf{S}^2-3\mathbf{S}_z^2]$ form for $\mathbf{H}^{SS}$ (see Sections 3.4.1 and 3.4.4) within a case (b) $^{2S+1}\Lambda$ multiplet state, it is necessary to evaluate $\Delta J = \Delta S = \Delta\Lambda = 0$, $\Delta N = 0,\pm1$ matrix elements of $\mathbf{S}_z$,

$$\langle nJSN\Lambda|\mathbf{S}_z|nJSN\Lambda\rangle$$

and

$$\langle nJSN\Lambda|\mathbf{S}_z|nJSN\pm1\Lambda\rangle$$

in a basis set in which the $\mathbf{S}_z$ operator is not diagonal (see Section 3.2.1.4). This is accomplished by expressing the case (b) basis functions in terms of case (a) basis functions,

$$|nJSN\Lambda\rangle = \sum_{\Sigma=-S}^{S} U_{N;\Sigma}^{(JS)}|nJS\Sigma\Lambda\rangle \qquad (3.2.42)$$

$$U_{N;\Sigma}^{(JS)} = \langle JS\Sigma|JSN\rangle$$

where $\langle JS\Sigma|JSN\rangle$ is a vector coupling coefficient. The matrix elements of $\mathbf{H}^{\mathrm{SO}}$ in the case (b) basis are given by Kovács (p. 55, 1969) as

$$\langle nJSN\Lambda|\mathbf{H}^{\mathrm{SO}}|nJSN\Lambda\rangle = A\Lambda^2\left[\frac{J(J+1)-N(N+1)-S(S+1)}{2N(N+1)}\right] \quad (3.2.43)$$

$$\begin{aligned}&\langle nJSN\Lambda|\mathbf{H}^{\mathrm{SO}}|nJSN+1\Lambda\rangle\\ &\quad = A\Lambda\{[(N+1)^2-\Lambda^2][(J+N+1)(J+N+2)-S(S+1)]\}^{1/2}\\ &\quad\times\frac{[S(S+1)-(J-N)(J-N-1)]^{1/2}}{2(N+1)[(2N+1)(2N+3)]^{1/2}}.\end{aligned} \quad (3.2.44)$$

The effect of the Eq. (3.2.43) diagonal matrix element of $\mathbf{H}^{\mathrm{SO}}$ is important only at low J in $\Lambda \neq 0$ states and causes the $J = N + S$ and $J = N - S$ levels of the same N–value to be shifted by $A\Lambda^2[S/(N+1)]$ and $-A\Lambda^2(S/N)$, respectively. The $\Delta N = \pm 1$ off-diagonal matrix elements cause the low–N levels of a $\Lambda \neq 0$ multiplet state to approach the case (a) energy-level pattern. $\mathbf{H}^{\mathrm{SO}}$ is also responsible for $(\Delta J = 0, \Delta S = 0, \pm 1, \Delta\Lambda = 0, \pm 1, \Delta N = 0, \pm 1)$ interactions between two case (b) electronic states.

The difference between the energy-level patterns associated with cases (a) and (b) is best illustrated by plotting the energy levels of $^2\Sigma$ and $^2\Pi$ states versus $J(J+1)$, the natural case (a) quantum number, and versus $N(N+1)$, the natural case (b) quantum number. The traditional method of plotting rotational energy levels is the "reduced term value" plot, $E_Q - BQ(Q+1)$ versus $Q(Q+1)$ where Q is the natural or "pattern forming" rotational quantum number. On such a plot the energy levels will appear as straight, nearly horizontal lines, provided that the energy levels have reached the Hund's case limiting behavior that corresponds to the selected natural quantum number. The $^2\Sigma$ state, which is necessarily at the case (b) limit for all J and N, appears as a single horizontal line in Fig. 3.1b but as a pair of upward $(J = N - \frac{1}{2})$ and downward $(J = N + \frac{1}{2})$ curving lines in Fig. 3.1a. The $^2\Pi$ state, which evolves from case (a) limiting behavior at low J to case (b) at high J, appears in Fig. 3.1a as a pair of upward $(\Omega = \frac{3}{2})$ and downward $(\Omega = \frac{1}{2})$ curving lines that are nearly straight at low J. In Fig. 3.1b, the $^2\Pi$ levels appear as a pair of curves, widely separated at low N and approaching a single horizontal straight line at high N. The existence of a natural, pattern forming, rotational quantum number for each Hund's case and the simplicity of the limiting behavior displayed by the reduced term value plot illustrate a major reason that spectroscopists find Hund's coupling cases useful.

Note that there is no difference between cases (a) and (b) for singlet states. A difference between cases (a) and (b) can occur only for electronic states that can have several values of Σ (or, equivalently, of $N - J$).

Case (d). The traditional (simplified) case (d) limit is approached for Rydberg complexes of relatively high values of l (nonpenetrating states) where the centrifugal barrier keeps the Rydberg electron far from the core, and the core has $(S^+ = 0$, any $\Lambda^+)$, or $(S^+ = 1/2, \Lambda^+ = 0)$ symmetry. As shown by Jungen and Raşeev, (1998) and Watson, (1999), when *both* S^+ and Λ^+ are nonzero,

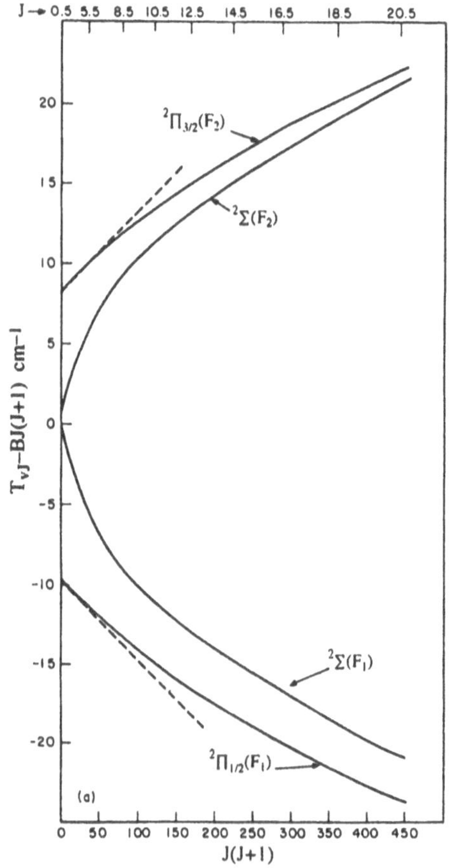

Figure 3.1a: Natural rotational quantum numbers for Hund's cases (a) and (b). Reduced term value plots for $^2\Sigma$ ($B = 1.0$ cm^{-1}) and $^2\Pi_r$ ($B = 1.0$ cm^{-1}, $A = 20.0$ cm^{-1}). (a) Plot of $T_{v,J} - BJ(J+1)$ versus $J(J+1)$ displays case (a) limiting behavior for the $^2\Pi$ state at very low J. The dotted lines illustrate the $\pm B^2/A$ corrections to the near case (a) effective B–values (See Section 3.5.4). The $^2\Sigma$ state does not exhibit case (a) behavior even at low J (at $J = 0$ the limiting slopes of the $^2\Sigma$ F_1 and F_2 curves are $-\infty$ and $+\infty$).

the angular momentum coupling is far more complicated and can be described by a multiplicity of combinations of ion-core (a^+, b^+, c^+) and Rydberg electron (a, b, c, d, e) subcases. When there are five non-zero angular momenta being coupled together $\left(\vec{\mathbf{S}}^+, \vec{\boldsymbol{\Lambda}}^+, \vec{\mathbf{R}}, \vec{\mathbf{l}}, \text{ and } \vec{\mathbf{s}}\right)$ there are many possible choices of intermediate angular momenta (and projection quantum numbers), and these choices are reflected by the numerous (ion-core, Rydberg electron) coupling cases and sub-cases.

At low-N^+ and low-n, Rydberg states typically exhibit a case (b) level pat-

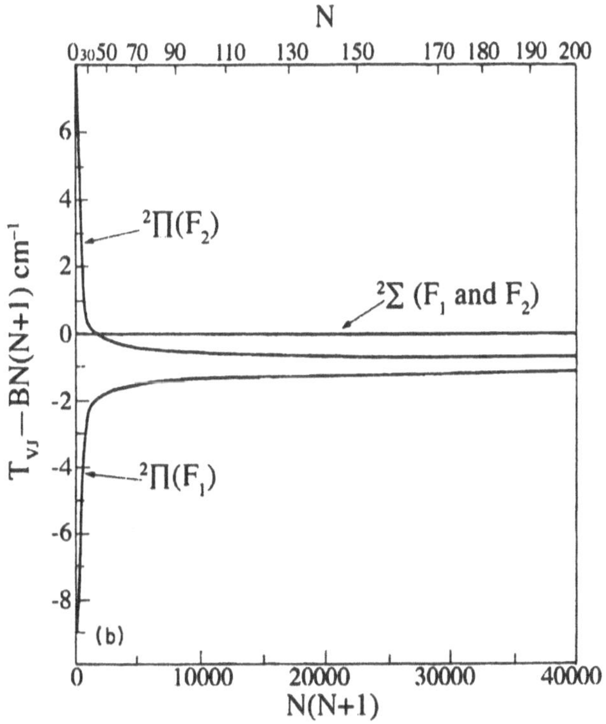

Figure 3.1b: (b) Plot of $T_{v,J} - BN(N+1)$ versus $N(N+1)$ displays case (b) limiting behavior for $^2\Sigma$ at all N and for $^2\Pi$ at very high N.

tern, which evolves toward case (d) as N^+ increases. This transition from case (b) to (d) level patterns occurs at lower N^+ as n increases. Low-l Rydberg states tend to form profoundly mixed-l supercomplexes (Miescher, 1976; Fredin, *et al.*, 1987), which evolve to the case (d) energy level pattern only at very high N^+. Higher-l Rydberg states tend to form nearly pure-l complexes, even at low rotation.

Case (d) is an atom-like coupling case in the sense that the l-quantum number is well defined (eigenstates of $\mathbf{l}^2$) despite the fact that any distortion from spherical symmetry (as is always present in $\mathbf{H}^{\text{el}}$ for a diatomic molecule) causes the magnitude of $\mathbf{l}$, *but not its projection onto the z–axis*, to become poorly defined. Case (d) is appropriate for high–n and $l \gtrsim 2$ Rydberg states because the Rydberg electron is in an orbital that is so large that the non-atom-like character (nonsphericity) of the molecular core is so weakly sampled by the Rydberg electron that the atom-like l becomes an approximately good quantum number and the molecular energy level pattern approaches that of case (d).

In the traditional (simplified) case (d) limit, the good quantum numbers are drawn from the partially redundant set, $nJSN(S_R)J^+N^+S^+\Lambda^+sl(s_R, l_R)$, redundant set, $nJSN(S_R)J^+N^+S^+\Lambda^+sl(s_R,l_R)$, where the quantum numbers with subscript R specify the projection of an angular momentum on the direction of the nuclear rotation angular momentum, $\vec{\mathbf{R}}$,

$$S_R = J - N \tag{3.2.45a}$$

$$s_R = (J - N - J^+ + N^+) = \mathcal{S}^\dagger \tag{3.2.45b}$$

$$l_R = N - N^+ = \mathcal{L}^\dagger. \tag{3.2.45c}$$

Watson (1999) analyzes two subcases of case (d), which he calls (a^+,d) and (b^+,d). The good quantum numbers in the former are $\Lambda^+S^+\Sigma^+J^+\Omega^+ls(J - s_R)J$ ($2S^+ + 1$ values of Ω^+) and in the latter subcase $\Lambda^+S^+N^+J^+ls(J - s_R)J$ ($2S^+ + 1$ values of N^+). Consider an example in which the ion-core has ${}^1\Lambda$ symmetry, then the nonzero quantum numbers in both (a^+,d) and (b^+,d) sets of basis functions are reduced to Λ^+N^+lsNJ, which is the set of quantum numbers traditionally used to describe a Hund's case (d) Rydberg state. The larger set of quantum numbers, expressed in the form specified by Watson (1999) rather than the more familiar quantum numbers used here, is needed to uniquely specify the transformations between (core, Rydberg) composite cases when $S^+ \neq 0$.

For an ion-core in a ${}^1\Lambda$ state, the case (b) basis state is specified by $|nl, \Lambda = \Lambda^+ + \lambda, S = s = 1/2, NJ\rangle$ and there are $2l+1$ possible signed values of Λ (there are $2l + 1$ unique $|\Lambda|$ states; for each nonzero value of $|\Lambda|$ there are + and − parity components) and 2 values of N for each set of values of l, Λ^+, and J. The case (d) basis set is specified by $|nl\Lambda^+N^+, S = s = 1/2, NJ\rangle$. For a specific set of l, Λ^+, and J values there are $2l+1$ N^+ values for each value of N (provided that $N \geq l$) and two values of N for each value of N^+, a total of $2(2l+1)$ states, exactly the same as for the case (b) situation. When l is defined (and $S^+ = 0$), Clebsch-Gordan coefficients allow transformation from case (b)→(d), just as for case (a)→(b) (see Sections 3.2.4 and 8.4).

In case (b) the pattern-forming rotational quantum number is N and the l–uncoupling term,

$$-B[\mathbf{N}^+\mathbf{l}^- + \mathbf{N}^-\mathbf{l}^+]$$

is most problematic because it has nonzero off-diagonal matrix elements between components of the same ${}^2l({}^{2S^++1}\Lambda^+)$ Rydberg complex, the degeneracies of which are lifted by $\mathbf{H}^{\mathrm{el}}$(Rydberg). The case (b) form of $\mathbf{H}^{\mathrm{ROT}}$ may be rearranged into a form that better illustrates the effect of a transformation to case (d) (or, more specifically, Watson's (1999) case (b^+,d))

$$\mathbf{H}^{\mathrm{ROT}} = B[\mathbf{N}^2 + \mathbf{l}^2 + (\mathbf{L}_\perp^+)^2 - (\mathbf{L}_z^+)^2 - 2\mathbf{N}\cdot\mathbf{l} - [(\mathbf{N}^+)^+(\mathbf{L}^+)^- + (\mathbf{N}^+)^-(\mathbf{N}^+)^+)].$$

In the case (d) basis, the $-2B\mathbf{N}\cdot\mathbf{l}$ term is diagonal because

$$\mathbf{N}\cdot\mathbf{l} = \frac{1}{2}[\mathbf{N}^2 + \mathbf{l}^2 - \mathbf{N}^{+2}].$$

† The symbol $\mathcal{L}$ was introduced by Weizel (1931) and $\mathcal{L}$ and $\mathcal{S}$ were used by Jungen and Miescher (1969). Other notations for the quantity denoted by $\mathcal{L}$, l_J and l_R, have been introduced by Nikitin and Zare (1994) and Watson (1999), respectively.

All terms in the modified case (b) form of $\mathbf{H}^{\text{ROT}}$ can be evaluated in the case (d) basis set,

$$\begin{aligned} E_{(d)}^{\text{ROT}} &= B[N(N+1) + l(l+1) - \Lambda^{+2}] - B[N(N+1) + l(l+1) - N^{+}(N^{+}+1)] \\ &= B[N^{+}(N^{+}+1) - \Lambda^{+2}]. \end{aligned}$$

The result is that all of the wrong-form case (b)-like terms in $E_{(d)}^{\text{ROT}}$ are cancelled by terms from $-2B\mathbf{N}\cdot\mathbf{l}$ and the expected $B[N^{+}(N^{+}+1) - \Lambda^{+2}]$ form for $E_{(d)}^{\text{ROT}}$ is obtained. However, the transformation from case (b) to (d) converts the diagonal and degeneracy-lifting case (b) $\mathbf{H}^{\text{el}}$ into non-diagonal form in case (d). This is illustrated by the unitary (orthogonal) transformation of a two-state $\mathbf{H}^{\text{el}}$ matrix

$$\mathbf{H}^{\text{el}} = \begin{pmatrix} +\Delta/2 & 0 \\ 0 & -\Delta/2 \end{pmatrix} \tag{3.2.46a}$$

$$\widetilde{\mathbf{H}}^{\text{el}} = \mathbf{U}^{\dagger}\mathbf{H}^{\text{el}}\mathbf{U} \tag{3.2.46b}$$

$$\mathbf{U} = \begin{pmatrix} \cos\theta & \sin\theta \\ -\sin\theta & \cos\theta \end{pmatrix} \tag{3.2.46c}$$

$$\widetilde{\mathbf{H}}^{\text{el}} = \begin{pmatrix} (\cos^2\theta - \sin^2\theta)\Delta/2 & -(\cos\theta\sin\theta)\Delta \\ -(\cos\theta\sin\theta)\Delta & -(\cos^2\theta - \sin^2\theta)\Delta/2 \end{pmatrix} \tag{3.2.46d}$$

if $\theta = \pi/4$

$$\widetilde{\mathbf{H}}^{\text{el}}(\pi/4) = \begin{pmatrix} 0 & -\Delta/2 \\ -\Delta/2 & 0 \end{pmatrix}. \tag{3.2.46e}$$

Thus a diagonal $\mathbf{H}^{\text{el}}$ with diagonal elements differing by Δ is converted into a non-diagonal $\widetilde{\mathbf{H}}^{\text{el}}$. The size of the case (d)-destroying off-diagonal matrix element of $\widetilde{\mathbf{H}}^{\text{el}}$, $\Delta/2$, is proportional to the original case (b)-preserving difference in diagonal energies, Δ. When the basis set for a ${}^2l({}^{2S^{+}+1}\Lambda^{+})$ Rydberg complex contains more than two members, the case (b)→(d) angular momentum unitary ($\mathbf{U}^{-1} = \mathbf{U}^{\dagger}$) transformation, $\mathbf{U}$(case b→case d), uniquely specifies the case (d) basis states in terms of the case (b) basis states, e.g., for the j–th case (d) state, schematically

$$|d, j\rangle = \sum_{k} U_{jk}(b \to d)|b, k\rangle \tag{3.2.47}$$

where

$$U_{jk}(b \to d) = \langle b, k|d, j\rangle$$

is a vector coupling coefficient. Thus the matrix elements of any operator, in particular the $\mathbf{H}^{\text{el}}$ term in the total $\mathbf{H}$, may be evaluated in the case (d) basis

$$\mathbf{H}^{(d)} = \mathbf{U}(b \to d)\mathbf{H}^{\text{el(b)}}\mathbf{U}^{\dagger}(b \to d) \tag{3.2.48a}$$

$$\mathbf{H}_{jk}^{\text{el(d)}} = \sum_{n,m} U_{jn}\mathbf{H}_{nm}^{\text{el(b)}}U_{mk}^{\dagger} \tag{3.2.48b}$$

but $\mathbf{H}^{\mathrm{el(b)}}$ is completely diagonal

$$\mathbf{H}^{\mathrm{el(b)}}_{nm} = E^{\mathrm{el}}_n \delta_{nm}$$

thus

$$\mathbf{H}^{\mathrm{el(d)}}_{jk} = \sum_n U_{jn} E^{\mathrm{el}}_n U^{\dagger}_{nk} = \overline{E}\delta_{jk} + \sum_n U_{jn} U^{\dagger}_{nk} \left(E^{\mathrm{el}}_n - \overline{E}^{\mathrm{el}}\right) \tag{3.2.48c}$$

where the average energy, $\overline{E}^{\mathrm{el}}$, is the degeneracy-weighted average electronic energy within the ${}^2l({}^{2S^+ +1}\Lambda^+)$ Rydberg complex. This form of $\mathbf{H}^{\mathrm{el(d)}}$ makes clear that the magnitudes of the off-diagonal elements of $\mathbf{H}^{\mathrm{el(d)}}$ are proportional to the difference of each individual case (b) E^{el}_n energy from the average energy. This is the reason why $\mathbf{H}^{\mathrm{el}}$ is split into an atom-like (spherical) part for the Rydberg electron, $\mathbf{H}^{\mathrm{el(0)}} (\mathbf{H}^{\mathrm{el(0)}}_{nm} = \overline{E}\delta_{nm})$, and a non-atom-like term, $\mathbf{H}^{\mathrm{el(1)}} \equiv \mathbf{H}^{\mathrm{el}} - \mathbf{H}^{\mathrm{el(0)}}$, that expresses the strengths of all remaining electrostatic and exchange interactions of the Rydberg electron with the ion-core. No approximation is being made here because $\mathbf{H}^{\mathrm{el(1)}}$ is defined as the exact $\mathbf{H}^{\mathrm{el}}$ minus $\mathbf{H}^{\mathrm{el(0)}}$.

Each N^+–level of a case (d) ${}^{2S+1}[{}^2l({}^{2S^+ +1}\Lambda^+)]$ Rydberg complex is a $(2l+1)(2S+1)$–fold degenerate multiplet, consisting of $2l+1$ N–sublevels ($N^+ - l \leq N \leq N^+ + l$), each of which is $(2S+1)$–fold degenerate, corresponding to J–values $N - S \leq J \leq N + S$. The $(2l+1)$–fold spatial degeneracy is lifted by $\mathbf{H}^{\mathrm{el(1)}}$. The $(2S+1)$–fold spin degeneracy is lifted by $\mathbf{H}^{\mathrm{SR}} = \gamma \mathbf{N} \cdot \mathbf{S}$, which is fully diagonal in the case (d) basis set,

$$\mathbf{H}^{\mathrm{SR}} = (\gamma/2)[\mathbf{J}^2 - \mathbf{N}^2 - \mathbf{S}^2]. \tag{3.2.49}$$

This form of $\mathbf{H}^{\mathrm{SR}}$ for case (d) is identical to $\mathbf{H}^{\mathrm{SR}}$ expressed in the case (b) basis set (Eq. (3.2.37)).

The spin-orbit operator (see Section 3.4.1) for the Rydberg electron

$$\mathbf{H}^{\mathrm{SO}}(\text{Rydberg}) = a\mathbf{l}\cdot\mathbf{s} = a\left[\mathbf{l}_z\mathbf{s}_z + \frac{1}{2}(\mathbf{l}^+\mathbf{s}^- + \mathbf{l}^-\mathbf{s}^+)\right] \tag{3.2.50}$$

$$= (a/2)\left[\mathbf{j}^2 - \mathbf{l}^2 - \mathbf{s}^2\right] \tag{3.2.51a}$$

has nonzero $\Delta N = 0$ and $\Delta N = \pm 1$ (and $\Delta S = \pm 1$) matrix elements in case (d) because neither $\mathbf{j}^2$ nor $\mathbf{l}_z$ and $\mathbf{s}_z$ are defined in the case (d) basis set. Thus $a\mathbf{l}\cdot\mathbf{s}$ lifts both electron-orbital and -spin degeneracies, thus, at low N^+, forces a transition toward case (a). However, in the molecule fixed $\vec{\mathbf{R}}$–axis case (d) basis set, $\mathbf{H}^{\mathrm{SO}}$(Rydberg) takes a form that is extremely convenient for evaluating matrix elements

$$\mathbf{H}^{\mathrm{SO}}(\text{Rydberg}) = a l_R s_R + (a/2)[\mathbf{l}^+\mathbf{s}^- + \mathbf{l}^-\mathbf{s}^+]_R, \tag{3.2.51b}$$

where the selection rules for the first term are $\Delta l_R = \Delta s_R = 0$ and for the second term are $\Delta l_R = -\Delta s_R = \pm 1$. Thus, the high-$N$ limit for the molecule fixed z–axis form of the case (d) basis set is $\langle \mathbf{H}^{\mathrm{SO}}(\mathrm{Rydberg})\rangle = a l_R s_R$, where $a \overset{\propto}{\sim} n^{-3} l^{-3}$. The $\Delta l_R = -\Delta s_R = \pm 1$ matrix elements of $\mathbf{H}^{\mathrm{SO}}(\mathrm{Rydberg})$ force the Rydberg electron toward case (e), but this force is usually strongly opposed by the much larger energy differences between the l_R components of a case (d) Rydberg complex, due to the non-atom-like part of $\mathbf{H}^{\mathrm{el}}$.

The z–axis and $\vec{\mathbf{R}}$–axis forms of $\mathbf{H}^{\mathrm{ROT}}$ in the case (d) basis set are related by a term, $-2B^+\mathbf{N}\cdot\mathbf{l}$, that has off-diagonal matrix elements in the $\vec{\mathbf{R}}$–axis form but is diagonal in the z–axis form.

In the $\vec{\mathbf{R}}$–axis form of case (d) (Watson's (b^+, d)), l_R and N_R (and N_R^+) are good but N^+ is not a good quantum number because $(\mathbf{N}^+)^2$ does not commute with any component of $\vec{\mathbf{l}}$ or $\vec{\mathbf{N}}$. Thus it is necessary to replace $(\mathbf{N}^+)^2$ by $(\mathbf{N}-\mathbf{l})^2$ in the case (d) $\vec{\mathbf{R}}$–axis form of $\mathbf{H}^{\mathrm{ROT}}$,

$$\mathbf{H}^{\mathrm{ROT}} = B^+(\mathbf{N}-\mathbf{l})^2 = B^+[\mathbf{N}^2 + \mathbf{l}^2 - 2\mathbf{N}\cdot\mathbf{l}].$$

In this form of $\mathbf{H}^{\mathrm{ROT}}$, the pattern-forming quantum number is N rather than N^+ and $\mathbf{N}\cdot\mathbf{l}$ is problematic because it gives rise to nonzero off-diagonal matrix elements within a $^2l(^{2S^++1}\Lambda^+)$ Rydberg complex,

$$\langle n\Lambda^+ l l_R N N_R N_R^+(N^+)|\mathbf{N}\cdot\mathbf{l}|n\Lambda^+ l l_R N N_R N_R^+(N^+)\rangle = N_R l_R \qquad (3.2.52a)$$

$$\left\langle n\Lambda^+ l l_R N N_R = N_R^+ + l_R, N_R^+(N^+)|\frac{1}{2}\mathbf{N}^+\mathbf{l}^-\right.$$
$$\left.|n\Lambda^+ l l_R - 1, N, N_R = N_R^+ + l_R - 1, N_R^+(N^+)\right\rangle$$
$$= \frac{1}{2}[N(N+1) - N_R(N_R-1)]^{1/2}[l(l+1) - l_R(l_R-1)]^{1/2}. \qquad (3.2.52b)$$

The diagonal, linear-in-N term, $\sim N l_R$, preserves the case (d) $\vec{\mathbf{R}}$–form basis in opposition to the destructive effect of the off-diagonal term, of magnitude $2^{-1/2}N^{1/2}[l(l+1) - l_R(l_R-1)]^{1/2}$. But the $N^{1/2}$ dependence of the off-diagonal term is eventually overwhelmed, in the high–N limit, by the linear-in-N diagonal term. Thus the $\vec{\mathbf{R}}$–form basis set becomes approximately good in the high–N limit. However, the off-diagonal $-2B^+\mathbf{N}\cdot\mathbf{l}$ term in the $\vec{\mathbf{R}}$-form basis set has only diagonal matrix elements in the z–form case (d) basis, and its matrix elements exactly cancel the terms in $E_{(d)}^{\mathrm{ROT}}$ that depend on N and l rather than N^+,

$$E_{(d)}^{\mathrm{ROT}} = B^+[N(N+1) + l(l+1) - N(N+1) - l(l+1) + N^+(N^++1)].$$

Case (c) is unusual in that its basis functions are specified by the fewest quantum numbers. In case (c), both $\mathbf{H}^{\mathrm{el}}$ and $\mathbf{H}^{\mathrm{SO}}$ are included in $\mathbf{H}^{(0)}$. $\mathbf{H}^{\mathrm{el}} + \mathbf{H}^{\mathrm{SO}}$ is the relativistic electronic Hamiltonian. In case (c), separate Ω–components of a single case (a) multiplet state become separate electronic states, *each with its own potential curve*. In addition, if two case (a) states of different symmetry, say $^1\Sigma^+$ and $^3\Pi$, interact via $\mathbf{H}^{\mathrm{SO}}$, then the two corresponding case (c) 0^+ potential curves could exhibit an avoided crossing. The only remaining good quantum number in case (c) is Ω, and the basis functions are (often unspecifiable) linear combinations of case (a) functions of the same Ω but different Λ, Σ, and S values. The rotational energies, as for case (a), are given by $B[J(J+1) - \Omega^2]$. The **L**-uncoupling and **S**-uncoupling operators that mix basis states of different Ω are combined into a single $\mathbf{J}_a$–uncoupling operator (Section 2.3.2) since neither Λ nor Σ is defined (see Veseth, 1973a). The sum of the **L**- and **S**-uncoupling perturbation operators can be written as

$$-2B(\mathbf{J} \cdot \mathbf{J}_a - \mathbf{J}_z \mathbf{J}_{az})$$

with

$$\mathbf{J}_a \equiv \mathbf{L} + \mathbf{S}$$

and has nonzero $\Delta\Omega = \pm 1$ matrix elements.

Case (c) arises when off-diagonal spin-orbit matrix elements are large relative to the energy separations of case (a) electronic basis states that belong to the same Ω value. Case (c) is often important for molecules that include one atom from at least the fourth row (K through Kr) of the periodic table. A good example is the I_2 molecule (Mulliken, 1971), for which perturbations among the I_2 $X^1\Sigma_g^+$, $a1_g$, and $a'0_g^+$ states are due to $B[\mathbf{J}^+\mathbf{J}_a^- + \mathbf{J}^-\mathbf{J}_a^+]$ matrix elements (Martin *et al.*, 1983). Weakly bound molecules composed of light atoms with a single hole in a *p*-orbital, such as Ne_2^+, are well described by the case (c) limit when the interatomic binding energy is smaller than the spin-orbit splitting between the $^2P_{1/2}$ and $^2P_{3/2}$ separated atom states. Case (c) can be a good approximation even for light molecules with an $S^+ \neq 0, \Lambda^+ \neq 0$ ion-core in high–n Rydberg states (See Fig. 3.13 and Section 8.7).

Often, strong spin-orbit coupling in a diatomic molecule arises from one (heavy) open-shell atom. When an isolated open-shell atom is in the strong spin-orbit limit, $\vec{\mathbf{J}}_a \equiv \vec{\mathbf{L}} + \vec{\mathbf{S}}$, J_a remains a good quantum number despite the destruction of the atomic L_a and S_a orbital and spin angular momentum quantum numbers by $\mathbf{H}^{\mathrm{SO}}$ owing to large $\Delta L_a = \pm 1$, $\Delta S_a = \pm 1$ off-diagonal matrix elements. When the atomic J_a quantum number is approximately good in a case (c) molecule, (and in the absence of J_a–uncoupling perturbations)

$$\mathbf{H}^{\mathrm{ROT}} = B(\mathbf{J}^2 - \mathbf{J}_z^2 + \mathbf{J}_a^2 - \mathbf{J}_{az}^2) \tag{3.2.53}$$

$$E^{\mathrm{ROT}}(J, J_a, \Omega) = B[J(J+1) - \Omega^2 + J_a(J_a+1) - \Omega^2] \tag{3.2.54}$$

but when J_a is not a good quantum number, the unevaluated expectation value of the

$$B[\mathbf{J}_a^2 - \mathbf{J}_{az}^2] = B\mathbf{J}_{a\perp}^2 \tag{3.2.55}$$

term is implicitly included in the electronic energy.

Different groups of states of the same molecule can separately exhibit case (a), (b), or (c) behavior. Furthermore, within any given electronic state, it is possible for the coupling limit to change from case (a) to case (c) as R, the internuclear distance, increases, because, near a dissociation asymptote, the interval between electronic states can become small compared to the spin-orbit splittings of the separated atoms (see Section 5.6).

True relativistic *ab initio* calculations have been done for hydride molecules such as TlH (Pyykkö and Desclaux, 1976) and have been extended to heavy nonhydride molecules, such as I_2 (de Jong, *et al.*, 1997). Other types of calculations involve treating the spin-orbit operator as a perturbation and adding its effects to the nonrelativistic potential curves (for example, Ne_2, Cohen and Schneider, 1974; LiHg, Gleichmann and Hess, 1994).

The *case (e)* limit is approached for Rydberg states of high n values converging to a molecular ion state with $\Lambda^+ \neq 0$ and $S^+ \neq 0$[†] The Rydberg states with low values of the n quantum number are well described in Hund's case (a). Then, as n increases and the energy separation between singlet and triplet Rydberg states becomes small compared to the spin-orbit splitting of the ion-core, the Rydberg states must be described in Hund's case (c) (see Fig. 3.13 for the example of a $^2\Pi$ ion core). Eventually, when n is sufficiently large, the energy separations between case (c) Rydberg states that belong to the same value of n and J but to different values of Ω become very small compared with the ion-core spin-orbit splitting. As a result, profound mixing among these Rydberg states occurs, caused by both the s–uncoupling, $-B^+[\mathbf{J}^+\mathbf{s}^- + \mathbf{J}^-\mathbf{s}^+]$, and the l–uncoupling, $-B^+[\mathbf{J}^+\mathbf{l}^- + \mathbf{J}^-\mathbf{l}^+]$, operators. This is an extension beyond the situation described by case (d), for which there is only extreme l–uncoupling for the Rydberg electron. For *case (e)*, the combination of l– and s–uncoupling effects is treated as j–uncoupling (Lefebvre-Brion, 1990 and 1995), $-B^+[\mathbf{J}^+\mathbf{j}^- + \mathbf{J}^-\mathbf{j}^+]$ (with $\mathbf{j}=\mathbf{l}+\mathbf{s}$), for the Rydberg electron. $\mathbf{H}^{(0)}$ contains $\mathbf{H}^{el(0)}$ (atom-like part only), $\mathbf{H}^{SO}$, and $\mathbf{H}^{ROT}$, thus the basis set is specified by the quantum numbers $n, l, j, J^+, \Omega^+, \Lambda^+, S^+, \Sigma^+, J$. This corresponds to Watson's (1999) case (a^+, e) in which $j_R = J - J^+$ is also defined. The three angular momentum operators, $\mathbf{J}$, $\mathbf{J}^+$, and $\mathbf{j}$, are related by $\mathbf{J} = \mathbf{J}^+ + \mathbf{j}$. The total number of states when n, l, and J are specified is the same in Hund's case (a), where the states are labeled by Λ, Σ and S, as in Hund's case (e), where the states are labeled by Ω^+, J^+ and j. For example, for a $^2\Pi$ ion-core and $l = 2$, there will be respectively, for each parity, four, eleven, sixteen, and nineteen substates for $J = 0, 1, 2$ and 3 [descended from case (a) states: $d\pi\ ^1\Sigma^\pm, d\pi\ ^3\Sigma^\pm, d\sigma\ ^{1,3}\Pi, d\pi\ ^{1,3}\Delta, d\delta\ ^{1,3}\Pi$ and $d\delta\ ^{1,3}\Phi$; see Lefebvre-Brion, 1990 and 1995, and Section 8.7 for more details].

The only examples of Rydberg states treated in a Hund's case (e) basis set include molecules which have a $^2\Pi_i$ ion-core in the case (a) limit, for example, the hydrogen halides: HCl (Zhu, *et al.*, 1993; Drescher, *et al.*, 1993), HBr (Wales, *et al.*, 1996), HI (Mank, *et al.*, 1991), CO_2 (Cossart-Magos, *et al.*, 1995) and CS_2 (Cossart-Magos, *et al.*, 1997). This combination of a case (a) ion-core with a case (e) Rydberg electron is called (a^+, e) by Watson (1999). It would be interesting to treat an $(a^+ \rightarrow b^+, e)$ example where the ion-core evolves from case (a) toward case (b) at high-J. Then each ion-core state would be described by a superposition of Ω^+–values rather than a single Ω^+–value (Jungen and Raşeev,

[†] Core-non-penetrating Rydberg states of molecules that contain heavy atoms and have a $^1\Sigma^+$ ion-core ground state, such as HgF or BaF, are likely to exhibit a level pattern closer to case (b^+, e) than (b^+, d). The $^2F_{7/2} - {}^2F_{5/2}$ spin-orbit splitting in Ba^+ is 225 cm^{-1} for $4f$ and 16 cm^{-1} for $11f$, which is in reasonable agreement with the n^{-3}–scaling prediction of $225(4/11)^3 = 11$ cm^{-1}.

1998). Rydberg states built on the N_2^+ $A^2\Pi_u$ state, the NO^+ $b^3\Pi$ state, or the O_2^+ $a^4\Pi_u$ state would be interesting examples.

Brown and Carrington (2003) cite $HeKr^+$ as an example of case (e) coupling. In their Fig. 10.77, the rotational energy levels built on the Kr^+ $^2P_{3/2}$ atomic-ion state resemble the case (e) $BR(R+1)$ $(R = J \pm 3/2, J \pm 1/2)$ level pattern more closely than the case (c) $B[J(J+1)+j_a(j_a+1)-2\Omega^2]$ $(j_a = 3/2, \Omega = \pm 1/2, \pm 3/2)$ level pattern.

3.2.1.3 Intermediate Case Situations

3.2.1.3.1 Introduction

No Rydberg state ever exactly corresponds to a pure Hund's case, except at $n = \infty$ or 3, but rather to a situation in an intermediate Hund's case, and it is necessary to choose, to describe it, a basis set corresponding to any one of the pure Hund's cases. The energy terms to be added in each Hund's case are different depending on this choice of basis set. To clarify these choices, consider first the simple case of atoms.

In atoms, let us consider the two possible coupling cases, $L-S$ and $j-j$ coupling. If one uses an $L-S$ coupling basis set, it is necessary for treating the intermediate case to add the spin-orbit interaction evaluated in the $L-S$ basis set. If one starts from the $j-j$ coupling basis set, some part of the electronic energy is missing corresponding to the energy difference between the states of different multiplicity.

In molecules, starting from a case (d) basis set, the part of the electronic energy that must be added corresponds to the energy difference between the states of different Λ due to the different Coulomb and exchange interactions with the core [see Section 3.3.2], a difference which can be expressed, like the example shown by Herzberg and Jungen (1972) for a p–complex, as a function of the difference between the quantum defects of σ and π.

For case (c), it is the same as in atoms. Figure 3.13 illustrates the two possibilities. Starting from a case (a) basis set, going from left to right on the figure (n increasing), one must add the effect of the spin-orbit, which increases with n. If one starts from a case (c) basis set, going from right to left (n decreasing), one must introduce the effect of the energy difference between singlet and triplet states, an effect proportional to n^{-3}, which depends on the exchange integral, K, and is expressed, for example, as a function of the difference between the quantum defects of singlet and triplet states (see Eq. (8.7.5)).

In case (e), one has the sum of the effects of cases (d) and (c) and this is why, starting from a case (e) basis set, the neglected part of the electronic energy corresponds to the energy differences between all of the case (a) components (see Eq. 17 of Lefebvre-Brion, 1995).

3.2.1.3.2 Examples

Consider an example intermediate between cases (a) and (d). The case (d) limit corresponds to a pure-l Rydberg orbital (an eigenfunction of $\mathbf{l}^2$) bound to an atom-like ion-core. Molecular-ion-cores are not spherical, thus the case (d) limit energy level pattern $[E_{N^+,l_R} = B^+N^+(N^++1)]$ is approached only at high-N^+. The case (d) pattern is approached at high-N^+ even when the low-N^+ states have strongly l-mixed character. Low-l orbitals are strongly core-penetrating, thus they are most strongly affected by the non-sphericity of the ion-core. The result is mixed-l supercomplexes [NO s~d: Miescher, 1976, and Fredin, *et al.*, 1987; BaF s~p~d~f: Jakubek, *et al.*, 1994]. However, at high-N^+, the effect of $\Delta l = 0$, $\Delta\lambda = \pm 1$ l-uncoupling matrix elements overpowers the $\Delta l \neq 0$, $\Delta\lambda = 0$ l-mixing matrix elements that are due to ion-core non-sphericity. This is illustrated by the simplified model,

$$\mathbf{H}^{\text{eff}} = \begin{matrix} E_{ns\sigma} + B\mathbf{N}^2 & H_{s\sigma,d\sigma} & 0 \\ \text{sym} & E_{nd\sigma} + B\mathbf{N}^2 & -B6^{1/2}[N(N+1)]^{1/2} \\ \text{sym} & \text{sym} & E_{nd\pi} + B\mathbf{N}^2 \end{matrix}$$

which shows that at high-N, the s~d mixing is undone. The repulsion between the $nd\sigma$ and $nd\pi$ basis states, due to the l-uncoupling interaction ($\propto N$) can shift the $d\pi \sim d\sigma$ mixed states away from the $s\sigma$ state, thereby reducing the $s\sigma \sim d\sigma$ mixing caused by $\mathbf{H}_{s\sigma,d\sigma}$.

The spherical-$|nl\rangle$, spherical ion-core limit resembles the electronic structure of an alkali atom. It corresponds to an electronic Hamiltonian, $\mathbf{H}^{\text{el}(0)}$, in which the $2l+1$ λ-components $(-l \leq \lambda \leq l)$ (eigenfunctions of both $\mathbf{l}^2$ and $\mathbf{l}_z$) in a case (a) or (b) basis set are exactly degenerate or, in a case (d) basis set, the $2l+1$, $-l \leq l_R \leq l$, $l_R \equiv N - N^+$, l_R–components (eigenstates of $\mathbf{H}^{\text{el}(0)} + \mathbf{H}^{\text{ROT}}$) that belong to the same value of N^+ are exactly degenerate.

The effects of non-sphericity ($\mathbf{H}^{\text{el}(1)} = \mathbf{H}^{\text{el}} - \mathbf{H}^{\text{el}(0)}$, i.e., all parts of the exact $\mathbf{H}^{\text{el}}$ that are not included in $\mathbf{H}^{\text{el}(0)}$) may be treated by perturbation theory (Field, unpublished). These effects are manifest in a case (a) or (b) basis set as splittings, due to diagonal elements of $\mathbf{H}^{\text{el}(1)}$, of the $2l+1$ λ-components into $l+1$ non-degenerate $|\lambda|$-substates. In contrast, the non-spherical $\mathbf{H}^{\text{el}(1)}$ term affects a case (d) basis set in two ways: *(i) diagonal* matrix elements of $\mathbf{H}^{\text{el}(1)}$ lift the $2l+1$-fold l_R–degeneracy and create a high–N^+ limit level pattern that appears as a splitting of each N^+ rotational level into $l+1$ $|l_R|$–sublevels;[†] *(ii)* matrix elements of $\mathbf{H}^{\text{el}(1)}$ *off-diagonal* in N^+ (but diagonal in N) cause l_R–mixing (equivalent to N^+–mixing), with the result that a transition toward case (a) or (b) occurs at low–N. In the case (a) or (b) picture, the lifting of degeneracy is due to a λ–dependence of the diagonal elements of $\mathbf{H}^{\text{el}(1)}$. Owing to the cylindrical symmetry of $\mathbf{H}^{\text{el}(1)}$, matrix elements of $\mathbf{H}^{\text{el}(1)}$ off-diagonal in λ vanish. In the case (d) picture, the lifting of degeneracy is due to both diagonal and off-diagonal matrix elements of $\mathbf{H}^{\text{el}(1)}$. $\mathbf{H}^{\text{el}(1)}$ contains higher-order terms

[†] In contrast to λ, where the basis states with well defined parity (see Section 3.2.2) have the form $2^{1/2}[|\lambda\rangle \pm |-\lambda\rangle]$, for l_R, each of the signed-l_R $|l, l_R\rangle$ states has parity $(-1)^{l-l_R}$.

(especially the quadrupole moment) in the multipole expansion[†] of the ion-core intramolecular electric field (and polarizability) that are neglected when the case (d) basis functions are defined as eigenfunctions of $\mathbf{l}^2$ and $\mathbf{l}_z$ [see Jungen and Miescher, 1969, and minor corrections in the Appendix of Cheung, *et al.*, 1983, for an example of an isolated $4f$ Rydberg complex.]

The potential energy curve for a pure case (d) l–complex (with a $^1\Sigma^+$ ion-core) could, in principle (Field, unpublished), be computed *ab initio* as the degeneracy-weighted average of the l+1 different case (a) λ–component potential curves,

$$V_l(R) = \frac{1}{2l+1}\left[V_\sigma(R) + 2V_\pi(R) + 2V_\delta(R) + \ldots\right]. \tag{3.2.56}$$

If l is not a good quantum number for the clamped nuclei (non-rotating) electronic wavefunction, then the *ab initio* $V_\lambda(R)$ curves for all of the components of a supercomplex (within a unity range of the effective principal quantum number) could be averaged to define a single potential curve for the entire supercomplex, e.g., for an $s \sim p \sim d \sim f$ supercomplex, there would be four σ potentials ($V_{1\sigma}$, $V_{2\sigma}$, $V_{3\sigma}$, $V_{4\sigma}$), three π potentials, two δ potentials, and one ϕ potential

$$V_n(R) = \frac{1}{1+3+5+7}[V_{1\sigma}(R) + V_{2\sigma}(R) + \ldots 2V_\phi(R)].$$

Each of the λ–dependent $V_\lambda(R)$ potential curves could have a different R–dependence. Owing to the n^{-3}–scaling of all hydrogenic orbital expectation values of the form

$$\langle nl|r^{-\sigma}|nl\rangle \underset{\sim}{\propto} n^{-3} \tag{3.2.57}$$

for integer values of $\sigma \geq 2$ (see page 117 of Condon and Shortley, 1953), and the n^{-3}–scaling of the energy difference between members of two Rydberg series (Herzberg and Jungen, 1972; see also Section 8.4)

$$\begin{aligned} E_{n\lambda} - E_{n'\lambda'} &= -\frac{\Re}{(n-a_\lambda)^2} + \frac{\Re}{(n'-a_{\lambda'})^2} \\ &\approx \frac{2\Re(a_{\lambda'} - a_\lambda)}{\overline{n}^3}, \end{aligned} \tag{3.2.58}$$

where a_λ and $a_{\lambda'}$ are (R–dependent) quantum defects and $\overline{n} = (n+n')/2$, the R–dependent energy difference between potential curves (assuming that $|n-n'| << \overline{n}$),

$$V_{nl\lambda}(R) - V_{nl}(R) = \frac{2\Re}{n^3}[a_l(R) - a_{l\lambda}(R)], \tag{3.2.59}$$

[†]The coefficients of each multipole term in this multipole expansion are necessarily dependent on r, the distance of the Rydberg electron from the molecular-ion center of mass.

can be used to derive the R–dependent difference between R–dependent quantum defect functions,

$$a_l(R) - a_{l\lambda}(R) = \frac{n^3}{2\Re}[V_{nl\lambda}(R) - V_{nl}(R)]. \tag{3.2.60a}$$

Ross and Jungen (1999) have used differences between *ab initio* ion-core and Rydberg state potential curves to determine quantum defect functions. Thus the R–dependent quantum defect functions associated with the atom-like $\mathbf{H}^{\text{el}(0)}$ and the molecular $\mathbf{H}^{\text{el}(1)}$ are, respectively, $a_l(R)$ and $a_{l\lambda}(R)$.

In Section 8.5 a specific example of the evolution of a singlet ($S = 0$) p–complex from case (a) toward case (d) is discussed. In a case (a) basis set $\mathbf{H}^{\text{el}}$ is diagonal but, due to the non-sphericity of the ion-core experienced by the Rydberg electron, $\mathbf{H}^{\text{el}(1)}$, the zero-order potential energy curves of the $^1\Sigma$ and $^1\Pi$ states are split by

$$V_{np\pi}(R) - V_{np\sigma}(R) = \frac{2\Re}{n^3}[a_\sigma(R) - a_\pi(R)]. \tag{3.2.60b}$$

The same-J e–symmetry levels of the case (a) $^1\Sigma^+$ and $^1\Pi$ states are coupled by off-diagonal matrix elements of the $-B^+[\mathbf{J}^+\mathbf{l}^- + \mathbf{J}^-\mathbf{l}^+]$ term in $\mathbf{H}^{\text{ROT}}$. The l–uncoupling drives the molecule toward case (d), but this is opposed by the difference in the diagonal matrix elements of $\mathbf{H}^{\text{el}(1)}$ for the $^1\Pi$ and $^1\Sigma^+$ states, as expressed by the R–dependent difference between the R–dependent a_π and a_σ quantum defect functions. Eventually, at high $\mathbf{J}^+$, the l-uncoupling term overwhelms the resistance of the $\mathbf{H}^{\text{el}(1)}$ term, and the level pattern approaches case (d). When the same situation is treated in a case (d) basis set, $\mathbf{H}^{\text{ROT}}$ is diagonal but $\mathbf{H}^{\text{el}(1)}$ is no longer diagonal. As illustrated by Eqs. (3.2.46a) - (3.2.46e) and Eq. (3.2.48c), the off-diagonal matrix elements of $\mathbf{H}^{\text{el}(1)}$ are proportional to $[a_\sigma(R) - a_\pi(R)]n^{-3}$. Thus, in case (d), $\mathbf{H}^{\text{el}(1)}$ drives the molecule toward case (a), but this is opposed by the $B^+(4J+2)$ difference in zero-order energies (diagonal matrix elements of $\mathbf{H}^{\text{ROT}}$) for the two same–J, e–symmetry, $N^+ = J \pm 1$ rotational levels. The opposing roles of $\mathbf{H}^{\text{el}(1)}$ and $\mathbf{H}^{\text{ROT}}$ are reversed when the same problem is approached in a case (a) rather than a case (d) basis set.

Similarly, if two same–Ω case (a) basis states that belong to different multiplicities of the same (nonrelativistic) electronic configuration are exactly degenerate, $\mathbf{H}^{\text{SO}}$ will cause these two states to be completely mixed. The resultant eigenfunctions of the relativistic $\mathbf{H}^{\text{el}}$ are then perfect case (c) basis functions. However, when the case (a) basis functions are not exactly degenerate, the term in $\mathbf{H}^{\text{el}}$ responsible for this splitting, (the exchange term from the $1/r_{ij}$ interelectronic repulsion, see Section 3.2.4) although diagonal in the case (a) basis, must have both diagonal and nondiagonal matrix elements in the case (c) basis. In Section 8.5 a specific example of an interacting $^1\Pi \sim {}^3\Pi_1$ pair of $\Omega = 1$ basis functions is examined. An electrostatic exchange splitting in the case (a) picture is shown to be related in the case (c) picture to a difference in the quantum defects for two $\Omega = 1$ Rydberg series converging to the $\Omega = \frac{1}{2}$ and $\frac{3}{2}$ components of a $^2\Pi$ ion-core state.

3.2.1.4 Transformations Between Hund's Case Basis Sets

Throughout this book, matrix elements of various perturbation operators are evaluated in the case (a) basis set. To illustrate the irrelevance of the choice of basis, it is instructive to see that it is always possible to express the basis functions of one Hund's case in terms of those of another. Several methods exist whereby the basis functions of any Hund's case may be expanded in terms of those of case (a).

One especially simple procedure for defining transformations between cases (b), (c), (d), or (e) and case (a) is to express the effective Hamiltonian in the case (a) basis, adjust the values of crucial parameters in order to create the degeneracies appropriate for the desired limiting Hund's case [$A = 0$ in $A\Lambda\Sigma$ for case (b), $A = 0$ *and* $E^{\rm el}_{\lambda=l} = E^{\rm el}_{\lambda=l-1} = \cdots = E^{\rm el}_{\lambda=0}$ for case (d), $E^{\rm el}(^{2S+1}\Lambda_\Omega) = E^{\rm el}(^{2S-1}\Lambda_\Omega)$ for case (c)†, and $E^{\rm el}(^{2S+1}\Lambda_\Omega) = E^{\rm el}(^{2S+1}\Lambda_{\Omega-1}) = E^{\rm el}(^{2S+1}(\Lambda-1)_{\Omega-1}) = E\,(^{2S-1}\Lambda_\Omega)$ for case (e)] and then diagonalize (see Sections 8.4 and 8.6). The transformations thus obtained numerically may be used to transform any matrix expressed in a case (a) basis to the desired non-case (a) basis, regardless of whether the molecular parameters that define the actual matrix result in a level structure that is near any limiting case.

Another systematic procedure makes use of Clebsch-Gordan (CG) coefficients. This use of CG coefficients is very familiar in atomic physics, for example, when the $|LSJM_J\rangle$ basis functions are expressed in terms of $|LM_LSM_S\rangle$ or $|j_1j_2m_1m_2\rangle$ basis functions (Condon and Shortley, 1953, pp. 73 and 284), two angular momenta, $\mathbf{L}$ and $\mathbf{S}$ or $\mathbf{j}_1$ and $\mathbf{j}_2$, are coupled to form the resultant total angular momentum, $\mathbf{J}$. Similarly, for a molecule in case (b), $\mathbf{N}$ and $\mathbf{S}$ are coupled to give $\mathbf{J}$, but because of the anomalous commutation rules for the molecule-fixed components of $\mathbf{J}$, the coupling $\mathbf{N} + \mathbf{S} = \mathbf{J}$ must be replaced by $\mathbf{N} = \mathbf{J} - \mathbf{S}$ (Brown and Howard, 1976). Case (b) functions, $|N, \Lambda, S, J\rangle$, may be expressed as linear combinations of case (a) functions, $|\Lambda, S, \Sigma, J, \Omega\rangle$, using CG coefficients to replace Σ in case (a) by N in case (b), as follows:

$$|J, S, N, \Lambda\rangle = \sum_{\Sigma=-S}^{+S} U^{(JS\Lambda)}_{N;\Sigma}\, |J, \Omega, S, \Sigma, (\Lambda = \Omega - \Sigma)\rangle\,. \tag{3.2.61a}$$

$$U^{(JS\Lambda)}_{N;\Sigma} = (J, \Omega, S, -\Sigma = \Lambda - \Omega | J, S, N, \Lambda)\,. \tag{3.2.61b}$$

Instead of CG coefficients it is often more convenient to use 3-j coefficients. These are related to the CG coefficients as follows:

$$(J, \Omega, S, -\Sigma = \Lambda - \Omega | J, S, N, \Lambda) = (-1)^{J-S+\Lambda}(2N+1)^{1/2} \begin{pmatrix} J & S & N \\ \Omega & -\Sigma & -\Lambda \end{pmatrix} \tag{3.2.62}$$

---

†Strictly speaking, there is no unique way of specifying the case (c) limit. However, case (c) corresponds to a situation where $\mathbf{H}^{\rm SO}$ is more important than differences between electrostatic terms. Thus, if one sets all isoconfigurational exchange splittings to zero, one has a computationally useful definition of the case (c) limit.

For example, consider the two $\Omega = 1/2$ and $3/2$ case (a) $^2\Pi$ basis functions. If $A = 0$, the two case (a) basis functions $|^2\Pi_{1/2}\rangle$ and $|^2\Pi_{3/2}\rangle$ of the same J become almost completely mixed. Two case (b) $^2\Pi$ spin-components result: the F_1 component corresponds *by convention* to $J = N + \frac{1}{2}$, and the F_2 component to $J = N - \frac{1}{2}$. Except at the lowest J values, the F_i indices label the energy order of the eigenvalues for a given J, with $i = 1$ the lowest and $i = 2S + 1$ the highest. For a $\Lambda > 0$ state, there are $2S+1$ e–symmetry and $2S+1$ f–symmetry components; the F_{ie}, F_{if} indices are determined by the energy ordering within the e- or f-symmetry same–J group of levels. The general expression for both the F_1 and F_2 case (b) $^2\Pi$ basis functions is

$$\begin{aligned}|N, 1, 1/2, J\rangle &= |1, 1/2, -1/2, J, 1/2\rangle\,(J, 1/2, 1/2, 1/2|J, 1/2, N, 1)\\ &\quad + |1, 1/2, 1/2, J, 3/2\rangle\,(J, 3/2, 1/2, -1/2|J, 1/2, N, 1)\,. \qquad (3.2.63)\end{aligned}$$

An extremely useful listing of 3-j coefficients is found in Edmonds (1974, p. 125). Some of the properties of 3-j coefficients are discussed in Section 3.4.5. For the special case of $|J - N| = \frac{1}{2}$, there is a convenient closed-form expression for the 3-j coefficients,

$$\begin{pmatrix} J+\frac{1}{2} & J & \frac{1}{2} \\ M & -M-\frac{1}{2} & \frac{1}{2} \end{pmatrix} = (-1)^{J-M-\frac{1}{2}} \left[\frac{J-M+\frac{1}{2}}{(2J+2)(2J+1)}\right]^{1/2}. \qquad (3.2.64)$$

For $N = J + \frac{1}{2}$ (the F_2 level),

$$\begin{aligned}\left|J+\frac{1}{2}, 1, \frac{1}{2}, J\right\rangle &= (-1)^{J+\frac{1}{2}}(2J+2)^{1/2}\left[\begin{pmatrix} J & \frac{1}{2} & J+\frac{1}{2} \\ \frac{1}{2} & \frac{1}{2} & -1 \end{pmatrix}|^2\Pi_{1/2}\rangle\right.\\ &\quad \left. + \begin{pmatrix} J & \frac{1}{2} & J+\frac{1}{2} \\ \frac{3}{2} & -\frac{1}{2} & -1 \end{pmatrix}|^2\Pi_{3/2}\rangle\right], \qquad (3.2.65)\end{aligned}$$

where the first 3-j coefficient is [Eq. (3.2.64) with $M = -1$]

$$\begin{pmatrix} J & \frac{1}{2} & J+\frac{1}{2} \\ \frac{1}{2} & \frac{1}{2} & -1 \end{pmatrix} = (-1)^{J+1/2}\left[\frac{(J+\frac{3}{2})}{(2J+2)(2J+1)}\right]^{1/2} \qquad (3.2.66)$$

and the second 3-j coefficient is [Eq. (3.2.64) with $M = 1$]

$$\begin{aligned}\begin{pmatrix} J & \frac{1}{2} & J+\frac{1}{2} \\ \frac{3}{2} & -\frac{1}{2} & -1 \end{pmatrix} &= \begin{pmatrix} J+\frac{1}{2} & J & \frac{1}{2} \\ -1 & \frac{3}{2} & -\frac{1}{2} \end{pmatrix}\\ &= (-1)^{J-3/2}\left[\frac{(J-\frac{1}{2})}{(2J+2)(2J+1)}\right]^{1/2}, \qquad (3.2.67)\end{aligned}$$

thus

$$|^2\Pi, F_2\rangle = \left[\frac{J+\frac{3}{2}}{2J+1}\right]^{1/2} |^2\Pi_{1/2}\rangle - \left[\frac{J-\frac{1}{2}}{2J+1}\right]^{1/2} |^2\Pi_{3/2}\rangle . \tag{3.2.68a}$$

For $J = \frac{1}{2}$, only the F_2 component exists. The F_1 component for $J > \frac{1}{2}$ is derived by constructing a function which is orthogonal to the above F_2 function. The F_1 function $\left(N = J - \frac{1}{2}\right)$ is

$$|^2\Pi, F_1\rangle = \left[\frac{J-\frac{1}{2}}{2J+1}\right]^{1/2} |^2\Pi_{1/2}\rangle + \left[\frac{J+\frac{3}{2}}{2J+1}\right]^{1/2} |^2\Pi_{3/2}\rangle . \tag{3.2.68b}$$

In Section 3.5.4, these expressions will be derived by diagonalizing the matrix that includes the spin-uncoupling interaction between the $^2\Pi$ case (a) basis functions. That the $J = 1/2$ F_1 level does not exist is made clear by inserting $J = 1/2$ into Eq. (3.2.68b) and noting that $|^2\Pi, F_1, J = 1/2\rangle = |^2\Pi_{3/2}, J = 1/2\rangle$ cannot exist because it would violate the $J \geq \Omega$ requirement.

The transformation between cases (a^+, a) and (a^+, e) is of special importance for Rydberg series built on a case (a^+) $^{2S^++1}\Lambda^+$ ion-core with $S^+ \neq 0$ and $\Lambda^+ \neq 0$. The most frequently encountered example of case $(a^+,\ e)$ behavior is for an ion-core in a $^2\Pi$ state. Lefebvre-Brion (1990, 1995) derived the transformation between cases $(a^+,\ a)$ and $(a^+,\ e)$

$$|(a^+, a)\rangle = |S^+(\Sigma^+ = \Omega^+ - \Lambda^+)\Lambda^+\Omega^+ l\lambda s(\sigma = \omega - \lambda)\omega J(\Omega = \Omega^+ + \omega)\rangle \tag{3.2.69a}$$

and

$$|(a^+, e)\rangle = |J^+S^+(\Sigma^+ = \Omega^+ - \Lambda^+)\Lambda^+\Omega^+ lsjJ\rangle \tag{3.2.69b}$$

for the specific case of a $^2\Pi$ ion-core. The transformation consists of two steps. First, case $(a^+,\ c)$ basis states are constructed as linear combinations of case $(a^+,\ a)$ basis states that belong to different values of S ($S = S^+ + 1/2$ and $S = S^+ - 1/2$) and to the same value of $\Lambda = \Lambda^+ + \lambda$ (see Section 3.4.2 for evaluation of spin-orbit matrix elements and Fig. 5.3 of Lefebvre-Brion, 1995). Then the case $(a^+,\ c) \leftrightarrow (a^+,\ e)$ transformation is accomplished by a sum over terms involving products of two 3–j coefficients (replace the case $(a^+,\ c)$ Λ, Ω quantum numbers by the case $(a^+,\ a)$ Ω^+, j quantum numbers). Particular care must be taken to construct basis states that are properly symmetrized with respect to the $\sigma_v(xz)$ operator (see Section 3.2.2) whenever $\Lambda^+ + \lambda = 0$ (Σ^+ and Σ^- states) and $\Omega^+ + \omega = 0$ (0^+ and 0^- states).

Jungen and Raşeev (1998) derived a general case $(a^+,\ a) \leftrightarrow (a^+,\ e)$ transformation, expressed as a properly symmetrized sum over terms involving products of three 3–j coefficients,

$$\begin{aligned} |(a^+,\ e)\rangle &= |J^+S^+\Omega^+\Lambda^+(\Sigma^+ = \Omega^+ - \Lambda^+)lsjJ\rangle \\ &= \sum_{S,\Lambda,\Omega} U^{(S^+\Lambda^+ lsJ)}_{J^+\Omega^+ j;S\Lambda\Omega} |\Lambda^+S^+ lsS\Lambda(\Sigma = \Omega - \Lambda)\Omega J\rangle . \end{aligned} \tag{3.2.69c}$$

where the transformation coefficients, $U^{(S^+\Lambda^+lsj)}_{J^+\Omega^+j;S\Lambda\Omega}$ (replace the case (a^+, a) quantum numbers S, Λ, Ω by the case (a^+, e) quantum numbers J^+, Ω^+, j) are expressed in the notation summarized below Eq. (3.2.13).

This is a sequence of three coupled↔uncoupled transformations, described by vector coupling or 3–j coefficients

$$\begin{aligned}&\langle j_1 m_1 j_2 m_2 | j_1 j_2 j_3 - m_3 = m_1 + m_2\rangle \\ &\qquad = (-1)^{j_1 - j_2 - m_3}(2j_3+1)^{1/2}\begin{pmatrix} j_1 & j_2 & j_3 \\ m_1 & m_2 & m_3 = -m_1 - m_2\end{pmatrix}.\end{aligned}$$

Two important properties of 3–j coefficients are that they are unchanged by cyclic permutations of the three columns but are multiplied by a phase factor $(-1)^{j_1+j_2+j_3}$ for either an anti-cyclic exchange of columns or a reversal of sign of all projection quantum numbers in the bottom row.

The three required transformations to go from case (a^+, a) to (a^+, e) are

(i) coupled (S) → uncoupled (Σ^+)

$$\begin{aligned}|s\sigma S^+\Sigma^+\rangle &= \sum_{S=|S^+-s|}^{S^++s} |sS^+S\Sigma\rangle\langle sS^+S\Sigma|s\sigma S^+\Sigma^+\rangle \\ &= \sum_{S=|S^+-s|}^{S^++s} |sS^+S\Sigma\rangle[2S+1]^{1/2}(-1)^{s-S^++\sigma+\Sigma^+}\begin{pmatrix} s & S^+ & S \\ \sigma & \Sigma^+ & -\sigma-\Sigma^+\end{pmatrix}\end{aligned} \tag{3.2.69d}$$

(ii) uncoupled (λ) → coupled (j)

$$\begin{aligned}|lsj\omega\rangle &= \sum_{\omega=-j}^{j} |l\lambda s\sigma = \omega-\lambda\rangle\langle l\lambda s\sigma = \omega - \lambda|lsj\omega\rangle \\ &= \sum_{\omega=-j}^{\omega=-j} |l\lambda s\sigma = \omega-\lambda\rangle[2j+1]^{1/2}(-1)^{l-s+\omega}\begin{pmatrix} l & s & j \\ \lambda & \sigma & -\omega\end{pmatrix}.\end{aligned} \tag{3.2.69e}$$

This transformation must be expressed in a form suited for the case (a^+, a) → (a^+, e) transformation in which Ω^+ is common to both basis sets and the sum on ω is replaced by a sum on $\Omega = \Omega^+ + \omega$. The range of this sum is limited by the value of S specified in the previous sum on S. Specifically, if $S = S^+ - 1/2$, then $\Omega^+ - j + 1 \le \Omega \le \Omega^+ + j - 1$ but if $S = S^+ + 1/2$, then $\Omega^+ - j \le \Omega \le \Omega^+ + j$. Thus the sum on Ω is over the range $\Omega^+ - j + [S^+ - S + 1/2] \le \Omega \le \Omega^+ + j -$

$[S^+ - S + 1/2]$,

$$|lsj\omega = \Omega - \Omega^+, \Omega^+\rangle = \sum_{\Omega=\Omega^+-j+[S^+-S+1/2]}^{\Omega^++j-[S^+-S+1/2]} |l\lambda s\sigma = \Omega - \Omega^+ - \lambda, \Omega^+\rangle \times [2j+1]^{1/2} \times (-1)^{l-s+\Omega-\Omega^+} \begin{pmatrix} l & s & j \\ \lambda & \Omega - \Omega^+ - \lambda & \Omega^+ - \Omega \end{pmatrix} \quad (3.2.69f)$$

(iii) uncoupled (Ω) (reversed j) $\rightarrow$ coupled (J^+)

$$|J\bar{j}J^+\Omega^+\rangle = \sum_{\Omega=\Omega^+-j+[S^+-S+1/2]}^{\Omega^++j-[S^+-S+1/2]} (-1)^{j+\omega}|J\Omega j - \omega\rangle[2J^+ + 1]^{1/2} \times (-1)^{J-j-\omega+\Omega} \begin{pmatrix} J & j & J^+ \\ \Omega & -\omega & \omega - \Omega \end{pmatrix}. \quad (3.2.69g)$$

The overbar on the j indicates angular momentum subtraction, because $\vec{\mathbf{J}}^+ = \vec{\mathbf{J}} - \vec{\mathbf{j}}$, hence the factor of $(-1)^{j+\omega}$ and the reversal of the sign of ω.

The above sequence of three transformations must be recast in a form appropriate for the case $(a^+,\ a) \rightarrow (a^+,\ e)$ transformation, which requires sums on S, Ω, and Λ. The sums on S and Ω have already been implemented, and all that remains is a sum on Λ. The range of Λ is constrained by the values of S and Ω specified by the sums on S and Ω in which the sum on Λ is nested. The range of Λ constrained by S and Ω is $\Omega^+ - l \leq \Omega - S \leq \Lambda \leq \Omega + S \leq \Lambda^+ + l$.

Putting everything together

$$|(a^+,\ e)\rangle = \sum_{S=|S^+-1/2|}^{S^++1/2} \sum_{\Omega=\Omega^+-j+[S^+-S+1/2]}^{\Omega^++j-[S^+-S+1/2]} \sum_{\Lambda=\Omega-S\leq\Lambda^+-l}^{\Omega+S\leq\Lambda^++l} U^{(S^+\Lambda^+lsJ)}_{J^+\Omega^+j;S\Lambda\Omega}|S^+\Lambda^+\Omega^+l\lambda = \Lambda - \Lambda^+ s\omega = \Omega - \Omega^+ J\rangle \quad (3.2.70)$$

where

$$U^{(S^+\Lambda^+lsJ)}_{J^+\Omega^+j;S\Lambda\Omega} = [(2S+1)(2j+1)(2J^++1)]^{1/2}(-1)^{S^+-\Omega^++l+\Lambda+J+\Omega+1} \times \begin{pmatrix} s & S^+ & S \\ \Sigma - \Sigma^+ & \Sigma^+ & -\Sigma \end{pmatrix} \begin{pmatrix} l & s & j \\ \Lambda - \Lambda^+ & \Sigma - \Sigma^+ & \Omega^+ - \Omega \end{pmatrix} \times \begin{pmatrix} J^+ & j & J \\ \Omega^+ & \Omega - \Omega^+ & -\Omega \end{pmatrix} \quad (3.2.71)$$

and the phase factor has been simplified from

$$(-1)^{s-S^++\sigma+\Sigma^++l-s+\sigma+\lambda+j+\omega+J-j-\omega+\Omega}$$

recognizing that $2j$, $2s$, 2ω, 2σ, and $2(J - S^+)$ are odd integers. This result, except for an inconsequential overall factor of -1 agrees with the first term in the case $(a) \leftrightarrow (e)$ transformation that appears in Eq. (12) of Jungen and Raşeev (1998).

Equations (3.2.70 and 3.2.71) give the explicit linear combination of signed-Ω^+, Λ^+, Ω, Λ, ω, λ case (a^+, a) basis states that corresponds to each signed-Ω^+, Λ^+ case (a^+, e) basis state. However (see Section 3.2.2) it is necessary to construct basis states that have well defined symmetry with respect to the $\boldsymbol{\sigma}_v$ operator, in particular overall parity,

$$\boldsymbol{\sigma}_v \psi = \pm\psi,$$

and, *uniquely for* $\Lambda = 0$ *and* $\Lambda^+ = 0$ basis states, Σ^+, Σ^- symmetry,

$$\boldsymbol{\sigma}_v |\Lambda = 0^\pm\rangle = \pm|\Lambda = 0^\pm\rangle.$$

For example, a $\Lambda = 0$ state with definite Σ^+, Σ^- symmetry is constructed by taking the linear combination

$$2^{-1/2}\left[|\Lambda^+ = |\lambda|, \lambda = -|\lambda|\rangle \pm |\Lambda^+ = -|\lambda|, \lambda = |\lambda|\rangle\right]$$

and the resultant Σ^+ and Σ^- states are distinct electronic-states with different electronic energies. It is necessary to defer discussion of the construction of properly symmetrized basis states to Section 3.2.2. However, Eq. (12) of Raşeev and Jungen (1998) is a general expression that incorporates total parity Σ^+, Σ^- symmetry for $\Lambda = 0$ and/or $\Lambda^+ = 0$ basis states, and 0^+, 0^- symmetry for $\Omega = 0$ and/or $\Omega^+ = 0$ basis states. As a result, the case $(a^+, a) \leftrightarrow (a^+, e)$ transformation is expressed as a three term sum of products of three $3\text{–}j$ coefficients, where the second and/or third terms (and the modified normalization factors) are only present when $\Lambda = 0$ and/or $\Lambda^+ = 0$.

The non-symmetrized transformation matrix elements derived here as Eqs. (3.2.70) and (3.2.71), $U^{(S^+\Lambda^+lsJ)}_{(J^+\Omega^+j;S\Lambda\Omega)}$ must be replaced by elements of Raşeev's and Jungen's (1998) symmmetrized transformation matrix, denoted by $\widetilde{\mathbf{U}}$, when it is necessary to transform the $\mathbf{H}$ matrix from the case (a^+, a) to the (a^+, e) basis set (see Eqs. 3.2.46–3.2.48 for a similar example) (Lefebvre-Brion, 1995):

$$\begin{aligned}&\langle nJ^+\Omega^+lsjJ|\mathbf{H}|nJ^{+\prime}\Omega^{+\prime}lsjJ\rangle \\ &= [E_\infty + E(J^+,\Omega^+)]\delta_{J^+,J^{+\prime}}\delta_{\Omega^+,\Omega^{+\prime}} \\ &+ \sum_{S=|S^+-1/2|}^{S^++1/2} \sum_{\Omega}\sum_{\Lambda} \widetilde{U}^{(S^+\Lambda^+lsJ)}_{J^+\Omega^+j;S\Lambda\Omega}\epsilon_{n,l,\lambda,\Lambda,S,\Omega}\widetilde{U}^{(S^+\Lambda^+lsJ)}_{J^{+\prime}\Omega^{+\prime};S\Lambda\Omega},\end{aligned} \tag{3.2.72}$$

where the limits of the sums on Ω and Λ are as specified in Eq. (3.2.70) and $\widetilde{U}$ incorporates Σ^+, Σ^- symmetry for $\Lambda^+ = 0$ and/or $\Lambda = 0$. $E_\infty + E(J^+,\Omega^+)$ is the energy of the J^+,Ω^+ rotational level of the ion core $(n \to \infty)$ and $\epsilon_{n,l,\lambda,\Lambda,S,\Omega}$

is the electronic binding energy of the Rydberg electron in the case (a^+, a) limit (diagonal matrix element of the case (a^+, a) $\mathbf{H}$),

$$\epsilon_{n,l,\lambda,\Lambda,S,\Omega} = -\frac{\Re}{[n - a_{l,\lambda,\Lambda,S,\Omega}(R)]^2} \tag{3.2.73}$$

and $a(R)$ is the internuclear distance dependent quantum defect defined in Section 8.3.

3.2.1.5 Spectroscopic vs. Dynamical Hund's Cases

Fragmentation processes

$$\mathrm{AB} \rightarrow \mathrm{AB}^+ + e^- \text{(see Chapter 8)}$$
$$\mathrm{AB} \rightarrow \mathrm{A}^+ + \mathrm{B} \quad \text{(see Chapter 7)}$$

and their inverse are best described by a linear combination of molecular wavefunctions in which all of the quantum numbers of the photofragments are specified (including, when appropriate, projections on a molecule-frame or laboratory axis). Scattering theory is often more appropriate and convenient than effective Hamiltonian Theory (the primary approach developed in this book) to describe fragmentation, inelastic scattering,

$$\mathrm{A}^* + \mathrm{B} \rightarrow \mathrm{A} + \mathrm{B}^*,$$

dissociative recombination,

$$\mathrm{AB}^* + e^- \rightarrow \mathrm{A} + \mathrm{B},$$

and associative ionization,

$$\mathrm{A} + \mathrm{B} \rightarrow \mathrm{AB}^+ + e^-$$

processes. The ion-core + Rydberg electron basis sets (e.g. (a^+, e), (b^+, d), etc.), discussed in Sections 3.2.1-3.2.4 and by Watson (1999) and Jungen and Raşeev (1998), are well suited for ion $+e^-$ scattering. Separated atom basis sets (Nikitin, 1965; Nikitin and Umanskii, 1984) are designed to deal with molecular dissociation into open-shell atoms or inelastic scattering in collisions between open-shell atoms. Other examples in which the spin and orbital angular momentum quantum numbers of atoms provide insight into molecular electronic structure and dynamics include pure precession (Section 5.5) in diatomic hydrides and case (c) models of molecules with a single p-orbital hole [e.g. HCl^+ (see Fig. 7.1), HeKr^+ (Carrington, *et al.*, 1996)] or incompletely filled core-like $4f$ orbitals (rare earth monoxides and monofluorides, Field, 1982; Kaledin, *et al.*, 1996).

The Hund's coupling cases discussed in most detail in this book express hierarchies of coupling among the molecular angular momenta **R**, **S**, and **L** and, when appropriate, the Rydberg electron angular momenta, **l** and **s**, that provide insights into the energy level patterns of bound diatomic molecules.

These *spectroscopic* Hund's cases should not be confused with a generalization to *dynamical* Hund's cases (Nikitin, 1965) to deal with molecular dissociation into *open-shell* atoms or atomic ions (Nikitin and Umanskii, 1984). In particular, the Hund's cases (d) and (e) discussed here [Watson's (1999) cases ($(a^+, b^+$, or $c^+)$, d or e)] are specialized to express the coupling of a single Rydberg electron with molecular ion-core rovibronic states which correspond to the $n \to \infty$ limit of a Rydberg series. Nikitin's dynamical Hund's coupling cases are expressed in terms of the couplings between the electron spin and orbital angular momenta of the two separated atoms produced in the molecular dissociation process (see end of Section 7.2.1).

3.2.1.6 Relationship between Noncommuting Terms in H and the Most Appropriate Hund's Case

The molecular Hamitonian

$$\mathbf{H} = \mathbf{H}^{\rm el} + \mathbf{H}^{\rm ROT} + \mathbf{H}^{\rm SO} + \mathbf{H}^{\rm SS} + \mathbf{H}^{\rm SR} \tag{3.2.74}$$

can be equally well expressed in any complete basis set. Effective Hamitonian matrices, even matrices of enormous dimension, can be generated automatically and diagonalized effortlessly. So why do so many Hund's cases remain in use? The simple answer is that spectra are not born with line assignments attached. Assignments are based on the ability to recognize patterns. These patterns are of two fundamental types: (i) patterns guaranteed by a *fundamental* symmetry based on the commutation of an operator with the exact **H** (e.g. $\mathbf{J}^2$, which is why rotational combination differences form the rigorous foundation upon which all rotational analyses are built); (ii) patterns generated by an *approximate* symmetry that is based on the commutation of operators with a part of **H** [e.g. $(\mathbf{N}^+)^2$, which is a pattern-forming but non-rigorous quantum number that tells the spectroscopist how to arrange the observed spectrum into a pattern (see reduced term value plots, Fig. 3.1a and Fig. 3.1b) that aids the analysis].

The reason there are so many Hund's cases is that each Hund's case corresponds to an arrangement of terms in **H** in order of relative importance. For each arrangement there is a different Hund's case. Since each Hund's case is associated with a complete set of commuting angular momentum operators, explicitly defined transformations between any two Hund's case basis sets may be specified independent of the details of a particular molecular example.

It is instructive to list the terms in **H**, the specific orderings of which lead to various familiar and unfamiliar Hund's cases (Nikitin and Zare, 1994; Watson, 1999). It is important to note that any two incompatible (i.e., non-commuting) terms in **H** always express their antagonism in the same way. Suppose $\mathbf{H}^{\rm i}$ and $\mathbf{H}^{\rm j}$ are two such non-commuting terms. (More specifically, $\mathbf{H}^j$ comprises all terms in **H** that do not commute with $\mathbf{H}^i$, and *vice versa.*) Then, if $\mathbf{H}^{\rm i}$ is included in $\mathbf{H}^{(0)i}$, $\mathbf{H}^{(0)i}$ generates differences in the energies, $E_1^{(0)i}$ and $E_2^{(0)i}$, that are connected by non-zero off-diagonal matrix elements of $\mathbf{H}^j, H_{12}^j$, and *vice versa*: $\mathbf{H}^{(0)j} \to |E_1^{(0)j} - E_2^{(0)j}|$ vs. H_{12}^i. Whether $\mathbf{H}^i$ or $\mathbf{H}^j$ "wins" and defines the more appropriate basis set depends on whether

$$\left|E_1^{(0)i} - E_2^{(0)i}\right| \gg H_{12}^j \qquad (\mathbf{H}^i \text{“wins”}) \tag{3.2.75}$$

or

$$\left|E_1^{(0)j} - E_2^{(0)j}\right| \gg H_{12}^i \qquad (\mathbf{H}^j \text{“wins”}). \tag{3.2.76}$$

This operator antagonism is also manifest in dynamics through the equation of motion for expectation values (see Section 9.1.7)

$$i\hbar \frac{d\langle \mathbf{H}^i \rangle}{dt} = \langle | [\mathbf{H}^i, \mathbf{H}] | \rangle = \langle | [\mathbf{H}^i, \mathbf{H}^j] | \rangle. \tag{3.2.77}$$

There is no unique way of specifying the potentially antagonistic terms in **H**.

Various orderings of antagonistic terms in **H** have been discussed by Watson (1999), Nikitin and Zare (1994), and Jungen (Wüest, 1999). Recent Optical-Optical-Millimeterwave spectra of H_2 (Osterwalder, *et al.*, 2000) illustrate a succession of couplings and recouplings into angular momentum coupling cases that result as all electronic energy splittings scale as n^{*-3}, where n^* is the effective principal quantum number. Even the hyperfine structure of the H_2^+ $X^2\Sigma_g^+$ ion-core eventually makes an n^*-independent contribution to the fine/hyperfine structure.

3.2.2 Symmetry Properties

A detailed discussion of the symmetry properties of Hund's case (a) basis functions, $|nJ\Omega\Lambda S\Sigma\rangle$, is presented in Section 3.2.2.1 followed in Section 3.2.2.2 by a brief summary of the symmetry properties of other classes of basis functions, such as those that are expressed in |ion-core; Rydberg electron⟩ form and those that specify molecule-fixed $\vec{\mathbf{R}}$–axis ($\mathbf{S}_R$, $\mathbf{l}_R$, $\mathbf{s}_R$, $\mathbf{j}_R$) rather than z–axis projection quantum numbers.

3.2.2.1 Symmetry Properties of Hund's Case (a) Basis Functions

Up to this point, basis set wavefunctions with signed values of molecule-fixed z–axis projection quantum numbers Λ, Σ, and Ω (or, for a single orbital, λ, σ, and ω) have been discussed. However, the electronic basis functions associated with positive and negative signs of $(\Lambda, \Sigma, \Omega)$ are degenerate. In order to construct functions with a well-defined value of $|\Omega|$, it is necessary to combine functions with specific signed values of Λ, Σ and Ω. For example, a $^1\Pi_1$ state is doubly degenerate, the $\Lambda = +1$ and -1 components leading, respectively, to $\Omega = +1$ and -1 basis functions. These two Ω–components, degenerate in energy for the nonrotating molecule, are not properly symmetrized eigenfunctions of the total Hamiltonian, **H**. Only linear combinations of these $\pm\Omega$ functions have well-defined symmetry. This symmetry is called *parity*, and $\boldsymbol{\sigma}_v$ is the operator used to classify basis functions as belonging to either even or odd parity. The total Hamiltonian commutes with the σ_v symmetry operator (see Hougen, 1970), which means that the only nonzero off-diagonal matrix elements of $\mathbf{H}$ are between basis functions that belong to the same eigenvalue of σ_v. The eigenvalues of σ_v are +1 and −1 and correspond, respectively, to *even* and *odd total parity*.

It is sufficient for our discussion of diatomic molecules to use the σ_v operator to construct electronic-vibration-rotation wavefunctions of definite parity and to determine the parity of a properly symmetrized wavefunction. However, the standard definition of parity is based on the behavior of the total molecular wavefunction upon replacement of the coordinates of all particles by their negatives in a *laboratory-fixed* coordinate system with origin at the molecular center of mass. The operator that accomplishes this laboratory frame inversion has been denoted as $\mathbf{I}$ or $\mathbf{E}^*$. The existence of parity as a rigorous symmetry for molecules is based on the isotropy of space, or more precisely, the invariance of the eigenvalues of the molecular $\mathbf{H}$ when $\mathbf{H}$ is expressed in either a right- or left-handed coordinate system. Hougen (1970) illustrates the effects of geometric symmetry operations ($\sigma_v(xz), \mathrm{C}_2(y), \mathrm{C}_\infty^\epsilon(z), \mathrm{S}_\infty^\epsilon(z)$, and $\mathbf{i}$) on the *molecule-fixed* coordinates of electrons, nuclei, and nuclear displacements. In order to determine the effects of these geometric symmetry operations on the *laboratory*-fixed coordinates of all particles, it is also necessary to consider the effects of these symmetry operations on the Euler angles, which express the relative orientations of the molecule-fixed and laboratory-fixed coordinate systems. The Euler angles appear as the arguments of the rotational wavefunction. Since the σ_v, $\mathbf{i}$, and $\mathrm{S}_\infty^\epsilon$ operators reverse the handedness of a coordinate system and such an operation cannot be expressed in terms of Euler angles, it is conventional to define the effect of a handedness-reversing operation on the Euler angles by considering only the pure rotation obtained when such an operator is multipled by $\mathbf{i}$; thus $\sigma_v(xz) \rightarrow \mathrm{C}_2(y)$, $\mathbf{i} \rightarrow \mathbf{E}$, ($\mathbf{E}$ is the identity operator), $\mathrm{S}_\infty^\epsilon(z) \rightarrow \mathrm{C}_\infty^{\epsilon+\pi}(z)$. When the effects of $\sigma_v(xz)$ on all electrons, nuclei, nuclear displacements, *and* the rotational wavefunctions are considered, the laboratory coordinates of all particles and displacement vectors are replaced by their negatives, which is identical to the effect of the space inversion operator ($\mathbf{I}$ or $\mathbf{E}^*$, hence the equivalence of $\sigma_v(xz)$ and $\mathbf{I}$).

Before using the σ_v operator to classify symmetrized basis functions as even versus odd, it is necessary to insert a note of caution about four distinct types of symmetry classifications, each frequently and misleadingly called "parity," and each associated with the effect of the σ_v operator on *different parts* of the basis function.

These four distinct, σ_v–related symmetry classifications are: *(i)* total parity (effect of σ_v on the complete electronic-vibration-rotation wavefunction); *(ii)* e/f–symmetry (total parity of the complete wavefunction exclusive of a $(-1)^J$ or $(-1)^{J-1/2}$ rotational factor, respectively for molecules with an even or odd number of electrons); *(iii)* the intrinsic even/odd symmetry under $\sigma_v(xz)$ of the electronic wavefunction, exclusively for $\Lambda = 0$ states, Σ^+ or Σ^-; and *(iv)* the high–J (or case (b)) limiting behavior of the spatial coordinates of the electronic wavefunction under $\sigma_v(xz)$ reflection in a plane that contains the internuclear axis and is perpendicular to $\vec{\mathbf{J}}$ (approximately along the y–axis in the high–J limit), denoted $\Lambda(A')$ for even and $\Lambda(A'')$ for odd (Alexander, *et al.*, 1988).

The A', A'' electronic reflection symmetry provides a basis for interpreting differences in the photodissociative production or collisional depopulation rates of the Λ–doublet component in which the singly occupied $\lambda \neq 0$ orbital is aligned in a plane that contains the axis of rotation (A'') vs. that where the singly occupied $\lambda \neq 0$ orbital lies in the plane of rotation (A'). In the case (b) limit, the parity of $\Pi(A')$ levels is $(-1)^N$ and of $\Pi(A'')$ levels is $(-1)^{N+1}$. Intermediate between cases (a) and (b), when N is not a good quantum number, the above (N, parity)$\leftrightarrow$ A', A'' relationship, based on a correlation diagram definition of N, hence of A', A'', may be obtained from the energy ranks within a specified value of J and $+,-$ parity. For an isolated $^{2S+1}\Lambda$–state intermediate between cases (a) and (b), the lowest energy fine structure component of specified $J, \pm$ is $N = J - S$ (F_1) and the highest energy component is $N = J + S$ (F_{2S+1}).

The $\boldsymbol{\sigma}_v$ operator corresponds to a reflection through the molecule-fixed xz plane. If it operates on the spatial and spin coordinates of all electrons, the nuclear displacement vectors, and the rotational wavefunction (expressed in terms of Euler angles, which specify the orientation of the molecule-fixed coordinate system relative to a laboratory-fixed coordinate system), then the eigenvalues of $\boldsymbol{\sigma}_v$ label the *total parity*, $\pm$, of a rotating molecule basis function,

$$|n, J, |\Omega|, |\Lambda|, S, |\Sigma|\rangle = (2)^{-1/2} \left[|n, J, \Omega, \Lambda, S, \Sigma\rangle \pm |n, J, -\Omega, -\Lambda, S, -\Sigma\rangle\right]. \tag{3.2.78}$$

When $\boldsymbol{\sigma}_v$ operates on a symmetrized basis function, $|n, J, |\Omega|, \Lambda, S, \Sigma, \rangle$, the result is $+1$ or -1 times the original symmetrized basis function that is expressed as one of the two symmetrized functions shown on the right hand side of Eq. (3.2.78). Which of the $+$ or $-$ linear combinations is associated with the $+1$ or -1 eigenvalue of $\boldsymbol{\sigma}_v$ is not obvious and must be established by careful analysis, as follows.

The total parity of a given class of levels (F_i fine structure component for Σ–states, upper versus lower Λ–doublet component for Π–states) is found to alternate with J. The second type of label, often loosely called the e/f symmetry, factors out this $(-1)^J$ or $(-1)^{J-1/2}$ J–dependence (Brown *et al.*, 1975) and becomes a *rotation-independent label.* (Note that e/f is not the parity of the symmetrized nonrotating molecule $|\Lambda S\Sigma\rangle$ basis function. In fact, for half-integer S, it is not possible to construct eigenfunctions of $\boldsymbol{\sigma}_v$ in the form $[|\Lambda, S, \Sigma\rangle \pm |-\Lambda, S, -\Sigma\rangle]$, because, for half-integer S, $\boldsymbol{\sigma}_v$ converts $[|\Lambda S\Sigma\rangle + |-\Lambda S - \Sigma\rangle]$ into $\pm\,[|\Lambda S\Sigma\rangle \; - |-\Lambda S - \Sigma\rangle]$ and *vice versa.*) The third type of parity label arises when $\boldsymbol{\sigma}_v$ is allowed to operate only on the spatial coordinates of all electrons, resulting in a classification of $\Lambda = 0$ states according to their intrinsic Σ^+ or Σ^- symmetry. Only $|\Lambda = 0\rangle$ basis functions have an intrinsic parity of this last type because, unlike $|\Lambda| > 0\rangle$ functions, they cannot be put into $[|\Lambda\rangle \pm | - \Lambda\rangle]$ symmetrized form. The peculiarity of this $\Sigma^{\pm}$ symmetry is underlined by the fact that the selection rule for spin-orbit perturbations (see Section 3.4.1) is $\Sigma^+ \leftrightarrow \Sigma^-$, whereas for all types of electronic states and all perturbation mechanisms (except hyperfine perturbations) the total parity and e/f selection rules are $+ \leftrightarrow +, - \leftrightarrow -, e \leftrightarrow e, f \leftrightarrow f$.

The effect of $\boldsymbol{\sigma}_v(xz)$ on electron spatial (molecule-fixed) coordinates is

$$\boldsymbol{\sigma}_v(xz)[x, y, z] = [x, -y, z]. \tag{3.2.79}$$

For a right-handed spherical polar coordinate system (see Fig. 3.14), which is defined so that the $+z, +y$, and $+x$-axes correspond respectively to polar and azimuthal angles $[\theta, \phi]$ of $[0, \phi], [\pi/2, \pi/2]$, and $[\pi/2, 0]$, the effect of $\boldsymbol{\sigma}_v(xz)$ becomes

$$\boldsymbol{\sigma}_v(xz)[r, \theta, \phi] = [r, \theta, -\phi]. \tag{3.2.80}$$

Note that here θ, ϕ are coordinates of the electron in the molecule-fixed coordinate system; θ, ϕ do not specify, nor are they affected by, the orientation of the molecule-fixed coordinate system relative to the laboratory-fixed system. Since the ϕ–dependence of an orbital angular momentum basis function of the one-electron ($|\lambda\rangle$) or many-electron ($|\Lambda\rangle$) type can be expressed in terms of a factor $e^{i\lambda(\phi+\phi_0)}$ or $e^{i\Lambda(\phi+\phi_0)}$, where ϕ_0 is an arbitrary phase factor, the effect of $\boldsymbol{\sigma}_v(xz)$ on a molecular orbital becomes

$$\begin{aligned}\boldsymbol{\sigma}(xz)|\lambda\rangle &= \boldsymbol{\sigma}_v\left[f(r,\theta)e^{i\lambda(\phi+\phi_0)}\right]\\ &= f(r,\theta)e^{-i\lambda\phi}e^{i\lambda\phi_0}\\ &= e^{2i\lambda\phi_0}\left|-\lambda\right\rangle,\end{aligned}$$

which, for the specific phase choice $\phi_0 = n\pi$, becomes

$$\boldsymbol{\sigma}_v(xz)|\lambda\rangle = +|-\lambda\rangle. \tag{3.2.81}$$

This is the phase convention commonly used in *ab initio* molecular electronic structure calculations.

Unfortunately, diatomic molecule electronic spectroscopists most commonly use the phase convention

$$\boldsymbol{\sigma}_v|\lambda\rangle = (-1)^{\lambda}|-\lambda\rangle, \tag{3.2.82}$$

which corresponds either to the use of $\boldsymbol{\sigma}(yz)$ rather than $\boldsymbol{\sigma}(xz)$ or to the use of $\boldsymbol{\sigma}(xz)$ with $\phi_0 = \left(n+\frac{1}{2}\right)\pi$. Then

$$\boldsymbol{\sigma}_v(yz)[r, \theta, \phi] = [r, \theta, \pi - \phi] \tag{3.2.83}$$

and thus

$$\boldsymbol{\sigma}_v e^{i\lambda\phi} = e^{i\lambda(\pi-\phi)} = (-1)^{\lambda}e^{-i\lambda\phi}. \tag{3.2.84}$$

This phase convention is similar to what is often called the *Condon and Shortley phase convention*, which specifies that:

1. $\boldsymbol{\sigma}_v(xz)\left|A, \alpha, M_A\right\rangle = (-1)^{A-\alpha}\left|A, -\alpha, M_A\right\rangle$

$$\tag{3.2.85a}$$

where $\mathbf{A}$ is any defined molecular angular momentum with magnitude $[A(A+1)]^{1/2}$ and α and M_A are its molecule–(z) and space–(Z) fixed projections.

2. Matrix elements

$$\langle A, \alpha, M_A \pm 1 \,|\mathbf{A}_\pm|\, A, \alpha, M_A\rangle = +\hbar\,[A(A+1) - M_A(M_A \pm 1)]^{1/2} \quad (3.2.85b)$$

$$\langle A, \alpha \pm 1, M_A \,|\mathbf{A}^\pm|\, A, \alpha, M_A\rangle = +\hbar\,[A(A+1) - \alpha(\alpha \pm 1)]^{1/2} \quad (3.2.85c)$$

are real and positive where

$$\mathbf{A}_\pm = \mathbf{A}_X \pm i\mathbf{A}_Y, \qquad \mathbf{A}^\pm = \mathbf{A}_x \pm i\mathbf{A}_y,$$

are space-fixed (for any angular momentum) and molecule-fixed (only intrinsically molecule-fixed angular momenta, such as **l**, **s**, **L**, **S**, **I**) raising and lowering operators. See Section 2.3.1 and Brown and Howard (1976) for evaluation of matrix elements involving molecule-fixed components of the **R**, **N**, **J**, and **F** angular momenta.

A good reason for the widespread use of this phase convention is that, whenever the magnitude and molecule-fixed z–axis projection of an angular momentum are both specified, its transformation properties are automatically and uniformly specified. For example, if σ_v acts on spin or rotational functions,

$$\sigma_v\,|S\Sigma\rangle = (-1)^{S-\Sigma}\,|S - \Sigma\rangle \quad (3.2.86)$$

$$\sigma_v\,|J\Omega M\rangle = (-1)^{J-\Omega}\,|J - \Omega M\rangle.^{\dagger} \quad (3.2.87)$$

Since the quantum numbers l or L are seldom well-defined in a one-electron valence orbital or a many-electron case (a) basis set, and since, for a linear

† Equation (3.2.87) is a direct consequence of the definition [Eq. (2.3.41)],

$$\begin{aligned}\left\langle \frac{\pi}{2}\theta\phi|JM\Omega\right\rangle &= [(2J+1)/4\pi]^{1/2}\mathcal{D}^J_{\Omega M}\left(\frac{\pi}{2},\theta,\phi\right)\\ &= [(2J+1)/4\pi]^{1/2}\exp(iM\phi)d^J_{\Omega M}(\theta)\exp(i\Omega\pi/2);\end{aligned}$$

and the equivalence, for diatomic molecules, of σ_v and **I** (inversion of all space-fixed coordinates). If θ and ϕ are the space-fixed spherical polar coordinates of the intermolecular axis (as distinct from the previous discussion of $|l\lambda\rangle$ where θ, ϕ were the molecule–fixed coordinates of the electron), then

$$\sigma_v[\theta,\phi] = \mathbf{I}[\theta,\phi] = [\pi-\theta,\phi+\pi].$$

Now, to derive Eq. (3.2.87),

$$\begin{aligned}\sigma_v\left\langle \frac{\pi}{2}\theta\phi|JM\Omega\right\rangle &= \left[\frac{2J+1}{4\pi}\right]^{1/2}\exp[iM(\phi+\pi)]d^J_{\Omega M}(\pi-\theta)\exp\left(i\Omega\frac{\pi}{2}\right)\\ &= \left[\frac{2J+1}{4\pi}\right]^{1/2}(-1)^M\exp(iM\phi)d^J_{\Omega M}(\pi-\theta)(-1)^\Omega\exp\left(-i\Omega\frac{\pi}{2}\right).\end{aligned}$$

From Edmonds [1974, p. 60, 4.2.4],

$$d^J_{\Omega M}(\pi-\theta) = (-1)^{J-\Omega}d^J_{M,-\Omega}(\theta),$$

and from Edmonds [1974, p. 60, 4.2.6],

$$d^J_{M,-\Omega}(\theta) = (-1)^{M+\Omega}d^J_{-\Omega,M}(\theta),$$

molecule, there is no intrinsic distinction between the molecule-fixed x– and y–axes (σ_{xz} versus σ_{yz}), nor reason to choose the phase of $|\lambda\rangle$ so that it is real and positive at any particular value of ϕ, the confusion over the transformation properties of the orbital angular momentum basis functions $|\lambda\rangle$ and $|\Lambda\rangle$ is understandable.

The choice of a phase convention is a matter of taste. However, the convention adopted must be internally consistent. See Brown and Howard (1976) for a discussion of the Condon and Shortley (1953) phase convention and molecule- versus space-fixed angular momentum components; see Larsson (1981) for a brief but comprehensive summary of all of the most frequently encountered phase conventions. Throughout this book, an attempt has been made to use the phase conventions of Eq. (3.2.82), Eq. (3.2.85a), Eq. (3.2.86), and Eq. (3.2.87), and

$$\sigma_v(xz)\left|\lambda, s=\frac{1}{2},\sigma\right\rangle = (-1)^{\lambda+1/2-\sigma}\left|-\lambda,\frac{1}{2},-\sigma\right\rangle^{\dagger} \tag{3.2.88}$$

$$\sigma_v(xz)\left|\Lambda^{\pm}S\Sigma\right\rangle = (-1)^{\Lambda+S-\Sigma+s}\left|-\Lambda^{\pm}S-\Sigma\right\rangle \tag{3.2.89}$$

$$\sigma_v(xz)\left|J\Omega M\Lambda^{\pm}S\Sigma\right\rangle = (-1)^{J-\Omega+\Lambda+S-\Sigma+s}\left|J-\Omega M-\Lambda^{\pm}S-\Sigma\right\rangle \tag{3.2.90}$$

where $s=1$ for a Σ^- state and $s=0$ for all other states. Note that Eq. (3.2.90) is consistent with

$$\sigma_v(xz)\left|J\Omega M\Lambda^{\pm}S\Sigma\right\rangle = (-1)^{J-S+s}\left|J-\Omega M-\Lambda^{\pm}S-\Sigma\right\rangle \tag{3.2.91}$$

of Larsson (1981), because $(-1)^{S-\Sigma} = (-1)^{\Sigma-S+2(S-\Sigma)} = (-1)^{\Sigma-S}$, since $2(S-\Sigma)$ is an even integer. Replacing $S-\Sigma+\Lambda-\Omega$ in Eq. (3.2.90) by $\Sigma-S+\Lambda-\Lambda-\Sigma=-S$, one obtains Larsson's $(-1)^{J-S+s}$ factor.

One example of how one ensures that the one-electron phase convention of Eq. (3.2.88) is compatible with the many-electron phase convention of Eq. (3.2.89) will be discussed in Section 3.2.4.

Having specified the σ_v transformation properties of all case (a) basis functions, it becomes possible to return to the problem of constructing basis functions with well-defined parity. Two types of parity eigenfunctions are important, those labeled according to their total parity, $\pm$, and those labeled according to

---

one obtains

$$d^J_{\Omega M}(\pi-\theta) = (-1)^{J+M}d^J_{-\Omega,M}(\theta),$$

hence

$$\begin{aligned}\sigma_v\left\langle\frac{\pi}{2}\theta\phi|JM\Omega\right\rangle &= \left[\frac{2J+1}{4\pi}\right]^{1/2}\exp(iM\phi)d^J_{-\Omega,M}(\theta)\exp\left(-i\Omega\frac{\pi}{2}\right)(-1)^{J-\Omega}(-1)^{2(M+\Omega)}\\ &= (-1)^{J-\Omega}\left\langle\frac{\pi}{2}\theta\phi|J,M,-\Omega\right\rangle.\end{aligned}$$

† It is confusing but customary to use the same symbol, σ, for the one-electron spin quantum number [Eq. (3.2.88)] and the symmetry with respect to $\sigma_v(xz)$ of the spatial part of the many-electron wavefunction for Σ–states.

their e/f symmetry. Brown *et al.* (1975) define $\binom{e}{f}$ basis functions as those with total parity $(\pm)(-1)^J$ for molecules with an even number of electrons and $(\pm)(-1)^{J-1/2}$ for molecules with an odd number of electrons.

Two types of basis functions must be considered:

1. $\Lambda = 0, \Omega = 0$ (even number of electrons)

$$\sigma_v \left|J, \Omega = 0, \Lambda = 0^{\pm}, S, \Sigma = 0\right\rangle = (-1)^{J-S+s} \left|J, 0, 0^{\pm}, S, 0\right\rangle .$$

Odd-multiplicity Σ–states always have exactly one $\Omega = 0$ basis function. This function transforms into itself under σ_v; that is, it has a well-defined parity. The relationships between $\pm$ and $\Sigma^{\pm}$ and between e/f and $\Sigma^{\pm}$ are as follows:

$$+ \text{ total parity: } \left|{}^{2S+1}\Sigma_0^s, +\right\rangle \quad \begin{Bmatrix} \underrightarrow{\text{even } J} & {}^1\Sigma^+, {}^3\Sigma_0^-, {}^5\Sigma_0^+, \text{etc.} \\ \underrightarrow{\text{odd } J} & {}^1\Sigma^-, {}^3\Sigma_0^+, {}^5\Sigma_0^-, \text{etc.} \end{Bmatrix} \quad (3.2.92a)$$

$-$ total parity: $\left|{}^{2S+1}\Sigma_0^s, -\right\rangle$ {Roles of even and odd J reversed in Eq. (3.2.92a)}

e levels: $\left|{}^{2S+1}\Sigma_0^s, J, e\right\rangle$ ${}^1\Sigma^+, {}^3\Sigma_0^-, {}^5\Sigma_0^+$, etc. all J (3.2.92b)

f levels: $\left|{}^{2S+1}\Sigma_0^s, J, f\right\rangle$ ${}^1\Sigma^-, {}^3\Sigma_0^+, {}^5\Sigma_0^-$, etc. all J. (3.2.92c)

2. All other states

$$\sigma_v \left[\left|J\Omega\Lambda^{\pm}S\Sigma\right\rangle \pm \left|J - \Omega - \Lambda^{\pm}S - \Sigma\right\rangle\right] = \pm(-1)^{J-S+s} \left[\left|J\Omega\Lambda^{\pm}S\Sigma\right\rangle \pm \left|J - \Omega - \Lambda^{\pm}S - \Sigma\right\rangle\right] \quad (3.2.93a)$$

$$\pm \text{ total parity: } \left|{}^{2S+1}\Lambda_\Omega, J, \pm\right\rangle = (2)^{-1/2} \left[\left|J\Omega\Lambda^{\pm}S\Sigma\right\rangle \pm (-1)^{J-S+s} \left|J - \Omega - \Lambda^{\pm}S - \Sigma\right\rangle\right] \quad (3.2.93b)$$

$$e/f \text{ symmetry (odd number of electrons): } \left|{}^{2S+1}\Lambda_\Omega, J, {}^e_f\right\rangle = (2)^{-1/2} \left[\left|J\Omega\Lambda^{\pm}S\Sigma\right\rangle \pm (-1)^{-S+s+\frac{1}{2}} \left|J - \Omega - \Lambda^{\pm}S - \Sigma\right\rangle\right] \quad (3.2.94)$$

$$e/f \text{ symmetry (even number of electrons): } \left|{}^{2S+1}\Lambda_\Omega, J, {}^e_f\right\rangle = (2)^{-1/2} \left[\left|J\Omega\Lambda^{\pm}S\Sigma\right\rangle \pm (-1)^{-S+s} \left|J - \Omega - \Lambda^{\pm}S - \Sigma\right\rangle\right] . \quad (3.2.95)$$

The $\pm$ and e/f labels are really two different bookkeeping devices for the same physical property, but the e/f labels are more convenient, mainly for optical transition and perturbation selection rules.

1. *Selection rules for electric dipole transitions*: in terms of e/f levels, the rules are

$$\left.\begin{array}{ll} Q \text{ branch:} & e \leftrightarrow f \\ P \text{ or } R \text{ branch:} & e \leftrightarrow e \quad \text{and} \quad f \leftrightarrow f \end{array}\right\} u \leftrightarrow g$$

2. *Selection rules for perturbations*;

$$\Delta J = 0: \qquad e \leftrightarrow e \quad \text{and} \quad f \leftrightarrow f\} \quad u \leftrightarrow u \text{ and } \quad g \leftrightarrow g.$$

These selection rules are valid for all types of perturbations treated in this book. For perturbations due to the hyperfine Hamiltonian, which are beyond the scope of this book, the above selection rules are no longer rigorously valid (see Section 7.8.4). Figure 3.2 illustrates the utility of e/f compared to $\pm$ labels.

The value of e/f parity labels extends even to atom-molecule collisional processes. There is a strong propensity for conservation of e/f–symmetry in J–changing collisions (Davis and Alexander, 1983, and Yang and Dagdigian, 1998) (See Section 6.5.5).

3.2.2.2 Symmetry Properties of non-Hund's Case (a) Basis Functions

The preceding discussion is focussed on the parity and e/f symmetry of case (a) basis states (Eq. 3.2.4a). Since it is often useful to construct basis sets of the form $|(\text{ion-core}); (\text{Rydberg electron})\rangle$ where the ion-core can be in cases a^+, b^+, or c^+ and the Rydberg electron can be in cases a, b, c, d, or e (Watson 1999), the effects of $\sigma_v(xz)$ on each of these basis states are summarized here (Field, unpublished):

ion-core :

case (a^+) $$\sigma_v(xz)\,|J^+\Omega^+\Lambda^+S^+\Sigma^+s^+\rangle = (-1)^{J^+-\Omega^++\Lambda^++S^+-\Sigma^++s^+} \times |J^+-\Omega^+-\Lambda^+S^+-\Sigma^+s^+\rangle \tag{3.2.96a}$$

case (b^+) $$\sigma_v(xz)\,|J^+N^+\Lambda^+S^+s^+(S_R^+)\rangle = (-1)^{N^+-\Lambda^++s^+} \times |J^+N^+-\Lambda^+S^+s^+(S_R^+)\rangle \tag{3.2.96b}$$

case (c^+) $$\sigma(xz)\,|J^+J_a^+\Omega^+s^+\rangle = (-1)^{J^+-\Omega^++s^+}\,|J^+J_a^+-\Omega^+s^+\rangle \tag{3.2.96c}$$

All of the ion-core quantum numbers appear with a right superscript $^+$. The orbital reflection symmetry index is $s^+ = 1$ for $\Lambda^+ = 0\ \Sigma^-$ or $\Omega^+ = 0\ 0^-$ states

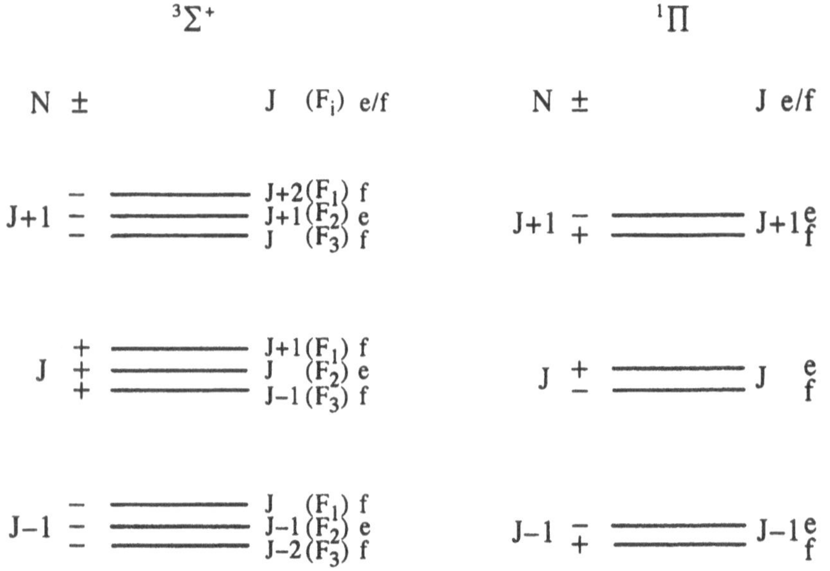

Figure 3.2: e/f and total parity labeling of $^3\Sigma^+$ and $^1\Pi$ levels. Levels are labeled assuming that J is an even integer. The $^3\Sigma^+$ levels illustrate the case (b) pattern in which $2S+1$ levels of different J–values and the same total parity form an N–multiplet. The total parity alternates with N for Σ states, whereas the e/f pattern is invariant. The total parities for the Λ–doublet components of the $^1\Pi$ state alternate with J, first + then − at higher energy, but the e/f labels do not alternate. The lack of alternation of the e/f symmetries with N or J makes e/f a more convenient label than total parity. Note that the ordering of the F_i levels within a given N for $^3\Sigma^+$ and e above f for $^1\Pi$ are arbitrarily chosen for this figure. The usual pattern of same–N J–levels for a $^3\Sigma$ state is that, at small N, the F_1 and F_3 components are close together $[F_1 - F_3 \simeq \gamma(2N+1)]$ and removed by 2λ from the F_2 component [case (a)] but, at large N, the F_1 and F_3 levels move apart and become separated from F_2 by an asymptotically N–independent splitting $[F_2 - (F_1+F_3)/2 \simeq \lambda$, case (b)].

and $s^+ = 0$ otherwise.

Rydberg-electron :

case (a) $$\sigma_v(xz)\,|l\lambda s\sigma\rangle = (-1)^{l-\lambda+s-\sigma}\,|l-\lambda s-\sigma\rangle^{\dagger} \tag{3.2.97a}$$

case (b) $$\sigma_v(xz)\,|l\lambda s(s_R)\rangle = (-1)^{l-\lambda}\,|l-\lambda s(s_R)\rangle^{\dagger} \tag{3.2.97b}$$

case (c) $$\sigma_v(xz)\,|j\omega ls\rangle = (-1)^{j-\omega}\,|j-\omega ls\rangle \tag{3.2.97c}$$

case (d) $$\sigma_v(xz)\,|l(l_R)s(s_R)\rangle = (-1)^{l}\,|l(l_R)s(s_R)\rangle \tag{3.2.97d}$$

case (e) $$\sigma_v(xz)\,|j(j_R)ls\rangle = (-1)^{l}\,|j(j_R)ls\rangle \tag{3.2.97e}$$

† There is an appearance of incompatibility between the $(-1)^{l-\lambda}$ and $(-1)^{\lambda}$ factors in the

The spin of a single electron is always $s = 1/2$. No orbital reflection symmetry index is needed for a Rydberg orbital because all $\lambda = 0$ states have Σ^+ symmetry and the reflection symmetry of $\omega = 0$ states is $(-1)^j$.

When $\Lambda^+ \neq 0$ and $\Lambda^+ + \lambda = 0$ in cases $|(\mathrm{a}^+ \text{or } \mathrm{b}^+); (\mathrm{a} \text{ or } \mathrm{b})\rangle$, or $\Omega^+ \neq 0$ and $\Omega + \omega = 0$ in case $|\mathrm{c}^+; \mathrm{c}\rangle$, an additional overall symmetry index, s_t, is needed to specify the energetically distinct $\Lambda = 0$ Σ^+ and Σ^- or $\Omega = 0$ 0^+ and 0^- states. In order to obtain a basis state of definite total $\Lambda = 0$ $\Sigma^\pm$ or $\Omega = 0$ $0^\pm$ reflection symmetry, it is necessary to construct orbitally symmetrized basis states of the form,

$$2^{-1/2}\left[|\Lambda^+; \lambda = -\Lambda^+\rangle \pm |-\Lambda^+; \lambda = \Lambda^+\rangle\right]$$

or

$$2^{-1/2}\left[|\Omega^+; \omega = -\Omega^+\rangle \pm |-\Omega^+; \omega = \Omega^+\rangle\right].$$

The correspondences between the $\pm$ sign in the symmetrized basis states and the $\Sigma^\pm$ or $0^\pm$ symmetry of the $\Lambda = \Lambda^+ + \lambda = 0$ or $\Omega = \Omega^+ + \omega = 0$ states are determined by the product of the effects of $\boldsymbol{\sigma}_v(xz)$ on the ion-core and Rydberg orbital factors of the overall wavefunction. For example, in case $(\mathrm{a}^+;\ \mathrm{b})$ the + linear combination gives Σ states of $(-1)^{S^+ + l + |\Lambda^+|}$ $\Sigma^\pm$–symmetry (and the overall parity is given by an additional factor of $(-1)^{J^+}$), in case $(\mathrm{b}^+;\ \mathrm{b})$ the + combination gives Σ states of $(-1)^l$ $\Sigma^\pm$–symmetry (and the overall parity is given by an additional factor of $(-1)^{N^+}$), and in case $(\mathrm{c}^+;\ \mathrm{c})$ the + combination gives $\Omega = 0$ states of $(-1)^j$ $0^\pm$ symmetry (and the overall parity is given by an additional factor of $(-1)^{J^+}$).

Note that the parity of the Rydberg electron case (d) or (e) part of the basis state is determined by l alone, that the values of the rotation axis $_R$ projection quantum numbers S_R, l_R, s_R, and j_R do not enter into the specification of parity, and the signs of the rotation axis projection quantum numbers are not reversed upon application of $\boldsymbol{\sigma}_v(xz)$.

A case $(\mathrm{b}^+; \mathrm{d})$ basis state with definite overall parity, for example, may be constructed based on the effect of $\boldsymbol{\sigma}_v(xz)$ on the overall basis state,

$$\begin{aligned}\boldsymbol{\sigma}_v(xz)\,&|J^+N^+\Lambda^+S^+s^+(S_R^+); l(l_R)s(s_R)\rangle \\ &= (-1)^{N^+-\Lambda^++l}\,|J^+N^+ - \Lambda^+S^+s^+(S_R^+); l(l_R)s(s_R)\rangle, \qquad (3.2.98)\end{aligned}$$

however, since $N = N^+ + l_R$,

$$(-1)^{N^+-\Lambda^++l} = (-1)^{N+l-\Lambda^+-l_R},$$

and, since $J^+ = N^+ + S_R^+$ and $J = N^+ + S_R^+ + s_R$

$$(-1)^{N^+-\Lambda^++l} = (-1)^{J-S_R^+-s_R-\Lambda^++l}.$$

$\boldsymbol{\sigma}_v(xz)$ reflection symmetry for single orbitals when l is respectively defined or not defined (Eq. 3.2.88) (see Hougen, 1970, p. 19). Basis states of well defined parity may be constructed following either prescription.

The alternative forms of the $\sigma_v(xz)|\mathrm{b}^+;\mathrm{d}\rangle$ parity requirement are useful because, in the absence of $\mathbf{H}^{\mathrm{SO}}$ effects, N is a rigorously good quantum number whereas N^+ is not, and, in the presence of spin-orbit effects, J is a rigorously good quantum number whereas N^+ and N are not. Case (b^+;b) $\sim$ case (b^+;d) perturbations between core-penetrating and nonpenetrating Rydberg states are diagonal in N but not in N^+.

3.2.3 Molecular Electronic Wavefunctions

Hund's case (a) basis functions will be used here for estimating values of matrix elements, because these wavefunctions are conveniently calculated *ab initio*. In the examples given here, single-configuration representations of the electronic wavefunctions will be used frequently. Except for Rydberg states, the single-configuration approximation is often invalid, but it is usually a good starting point for estimates of perturbation matrix elements. In Section 5.7 one example is discussed where the single-configuration approximation is not sufficient, even for the ground state, to explain observed perturbations, but where a two-configuration model is adequate. Note also that the single-configuration representation, even if it is valid at small internuclear distances, can be invalid at large distances. As R increases the effect of contributions from other configurations becomes more and more important, and this can explain, for example, the variation with R of electronic quantities such as the spin-orbit coupling constant, A, or the spin-spin parameter, λ (see Section 5.6).

Several different notations for molecular orbitals are commonly used. The numbering $1\sigma, 2\sigma, \ldots$ adopted by most theoreticians corresponds to the energy order of the orbitals for each value of λ and not to any principal quantum number. Note that this notation is distinct from the λnl (e.g., $\sigma 1s, \sigma 2p$) notation, which specifies the composition of each molecular orbital in terms of the dominant atomic orbital $(1s, 2p)$ contribution. The $nl\lambda$ notation (e.g., $3p\sigma$) is reserved for Rydberg molecular orbitals. The designations σ and σ^* or π and π^* are often used to distinguish between bonding (unstarred, excess electron density between atoms) and antibonding (starred, excess electron probability density behind rather than between atoms) orbitals.[†] The term spin-orbital will be used when it is advantageous to specify both the spatial and spin parts of the one-electron wave function. Each orbital also has a well-defined value and sign for λ, the projection of the one-electron orbital angular momentum on the internuclear axis. In the notation used here, the right $\pm$ superscript on the Greek letter gives the sign of λ. For example, π^+ denotes $\lambda = +1$, π^- denotes $\lambda = -1$. Another notation has been used by Field *et al.* (1975) in which the main numeral specifies the signed value of λ and the superscript is the sign of σ($+$ for α, $-$ for β). The correspondence between these two notations (on the left, Field's notation; on the right, the notation used in this book) is summarized

[†] For homonuclear molecules, the distinction between bonding $(\sigma_g, \pi_u, \delta_g)$ and antibonding $(\sigma_u, \pi_g, \delta_u)$ orbitals is made (indirectly) by the g, u labels, and the $*$ symbol becomes superfluous.

as follows:

$$1^+ = \pi^+\alpha \qquad 1^- = \pi^+\beta$$
$$-1^+ = \pi^-\alpha \qquad -1^- = \pi^-\beta.$$

Still another notation designates β-spin by a bar over the wavefunction, α functions being unbarred; for example, $\bar{\pi}^+ = \pi^+\beta$, or $-\bar{1} = \pi^-\beta$ (Field *et al.*, 1972).

The electronic wavefunction, constructed from the $\phi_i (i = a, b, \ldots, n)$ spin-orbitals, is written in the form of a determinant (or a sum of determinants),

$$\Phi(r_1, \ldots, r_n) = (n!)^{-1/2} \begin{vmatrix} \phi_a(1) & \phi_a(2) & \ldots & \phi_a(n) \\ \phi_b(1) & \phi_b(2) & \ldots & \phi_b(n) \\ \vdots & \vdots & \vdots & \vdots \\ \phi_n(1) & \phi_n(2) & \ldots & \phi_n(n) \end{vmatrix}. \tag{3.2.99}$$

This determinantal form expresses the antisymmetry of the wavefunction with respect to interchange of two identical particles. The determinantal wavefunction will always be specified by an abbreviated notation that lists only the diagonal of the determinant,

$$\Phi = |\phi_a(1)\phi_b(2) \ldots \phi_n(n)| = |\phi_a\phi_b \ldots \phi_n|. \tag{3.2.100}$$

A molecular *electronic configuration* specifies only the total occupancies of molecular orbitals. It does not specify how the spins and angular momentum components of the individual electrons in these orbitals are coupled to form total molecular angular momenta, such as S, Σ, and Λ. When a given orbital has its maximum permissible occupancy, it is said to be *full* and to form a *closed shell*. It takes two electrons to fill a $\sigma(\lambda = 0)$ orbital and four to fill any $\lambda > 0$ orbital, two each in $\lambda > 0$ and $\lambda < 0$ *subshells*. In order to specify $\Lambda = \Sigma_i\lambda_i$, it is necessary to know the occupancy of each subshell. The Λ-values that can arise for a given electronic configuration may be determined by inspecting the possible subshell occupancies. Once the possible λ-values and the associated subshell occupancies are recognized, the eigenfunctions of $\mathbf{S}^2$ compatible with each Λ-value are obtained simply by counting the number of unpaired electrons, N_u (which is the same as the number of half-filled subshells). The possible values of S range from $S = N_u/2$ down to $S = 0$ (N_u even) or $S = \frac{1}{2}$ (N_u odd) in steps of 1.

When a configuration consists exclusively of closed shells, values of $\Lambda = 0$ and $S = 0$ are the only ones possible, giving rise to a single $^1\Sigma^+$ electronic state. For example,

$$(1\sigma_g)^2(1\sigma_u)^2(2\sigma_g)^2(2\sigma_u)^2(3\sigma_g)^2(1\pi_u)^4$$

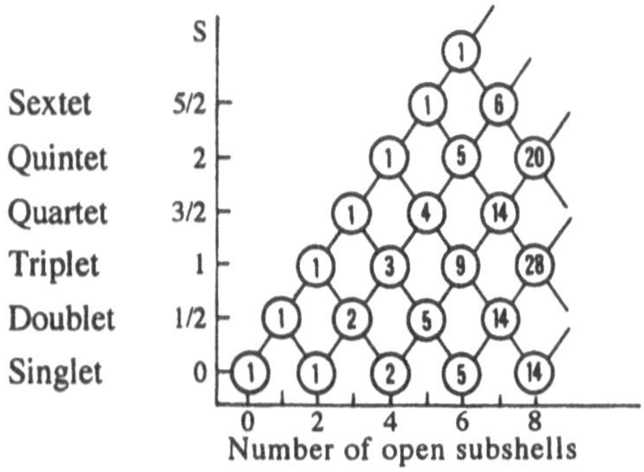

Figure 3.3: Branching diagram shows the number of states of each spin multiplicity versus the number of open (i.e., partly filled) subshells. It is frequently convenient to construct n-electron determinantal wavefunctions by adding a spin-orbital to an already known $(n-1)$-electron wavefunction. The explicit spin wavefunctions are summarized in Table 3.4.

is the configuration of the N_2 $X^1\Sigma_g^+$ electronic ground state. If a configuration containing an odd total number of electrons has only one open (half-filled) subshell, an $S = \frac{1}{2}$ (doublet) state is the sole result. For example, removing one electron from the N_2 $X^1\Sigma_g^+$ $3\sigma_g$ or $1\pi_u$ orbital gives rise to the N_2^+ $X^2\Sigma_g^+$ or $A^2\Pi_u$ state, respectively. A configuration with more than one open subshell gives rise to several electronic states.

The "branching diagram" (Fig. 3.3) shows how many eigenfunctions of $\mathbf{S}^2$ can be constructed from a given number of open subshells. The numbers in circles specify the number of states of a given multiplicity. For example, with two open subshells, one triplet and one singlet function result. Then, if an additional electron is added to an empty orbital, yielding three open subshells, the branching diagram shows that one doublet state is derived from the singlet parent and one quartet state and one doublet state are derived from the triplet parent.

Consider first the trivial case of the π^2 configuration. The following table gives all possible distributions of these two electrons among the two subshells: π^+ and π^- (the right $\pm$ superscript specifies whether $\lambda > 0$ or $\lambda < 0$; the spin projection is not yet specified).

π^+	π^-	$\Lambda = \sum_i \lambda_i$		Number of open subshells
2	0	2	Δ state	0
0	2	-2		
1	1	0	Σ state	2

The Δ state, which has no unpaired electron, can only be a singlet state. The Σ states, which are associated with two unpaired electrons, include one singlet

Table 3.4: Spin Eigenfunctions[a] of $\mathbf{S}^2$ for $\Sigma = S$

Number of Open subshells	Eigenfunctions
One	
1 doublet: $S = \frac{1}{2}, \Sigma = \frac{1}{2}$	$[\alpha(1)]$
Two	
1 singlet: $S = 0, \Sigma = 0$	$(2)^{-1/2}[\alpha(1)\beta(2) - \beta(1)\alpha(2)]$
1 triplet: $S = 1, \Sigma = 1$	$[\alpha(1)\alpha(2)]$
Three	
2 doublets: $S = \frac{1}{2}, \Sigma = \frac{1}{2}$	$(2)^{-1/2}[\alpha(1)\alpha(2)\beta(3) - \alpha\beta\alpha]$
	$(6)^{-1/2}[2\beta(1)\alpha(2)\alpha(3) - \alpha\alpha\beta - \alpha\beta\alpha]$
1 quartet: $S = \frac{3}{2}, \Sigma = \frac{3}{2}$	$[\alpha(1)\alpha(2)\alpha(3)]$
Four	
2 singlets: $S = 0, \Sigma = 0$	$\frac{1}{2}[\alpha(1)\beta(2)\alpha(3)\beta(4) + \beta\alpha\beta\alpha - \beta\alpha\alpha\beta - \alpha\beta\beta\alpha]$
	$(12)^{-1/2}[2\alpha(1)\alpha(2)\beta(3)\alpha(4) + 2\beta\beta\alpha\alpha - \beta\alpha\alpha\beta$
	$-\alpha\beta\alpha\beta - \beta\alpha\beta\alpha - \alpha\beta\beta\alpha]$
3 triplets: $S = 1, \Sigma = 1$	$(2)^{-1/2}[\alpha(1)\alpha(2)\alpha(3)\beta(4) - \alpha\alpha\beta\alpha]$
	$(6)^{-1/2}[-2\alpha(1)\beta(2)\alpha(3)\alpha(4) + \alpha\alpha\alpha\beta + \alpha\alpha\beta\alpha]$
	$(12)^{-1/2}[-3\beta(1)\alpha(2)\alpha(3)\alpha(4) + \alpha\alpha\beta\alpha + \alpha\alpha\alpha\beta$
	$+\alpha\beta\alpha\alpha]$
1 quintet: $S = 2, \Sigma = 2$	$[\alpha(1)\alpha(2)\alpha(3)\alpha(4)]$

[a] Eigenfunctions for up to six open subshells can be found in Yamazaki (1963a). Analytic formulas for constructing electronic interaction matrices between basis functions that belong to the same value of S (similar to the example for $^2\Pi$ states) are given in Yamazaki (1963b).

and one triplet state.

The linear combination of Slater determinantal basis states that correspond to any $S \leq 2$ $|\Omega, \Lambda, S, \Sigma\rangle$ basis state may be constructed from the information in Table 3.4 by the following procedure:

1. List, in standard order, the orbital labels (e.g., σ, π^+, π^{*+}, etc.) for all open shell orbitals specified by the electron configuration (e.g. $\sigma\pi^3\pi^*$) of the desired electronic state. For each orbital with $\lambda_i \neq 0$, consider both possibilities: λ_i^+ and λ_i^-. Be sure that $\sum_i \lambda_i = \Lambda$. If $\Lambda = 0$, there may be more than one way to specify the occupied orbitals (e.g., $\pi^+\pi^{*-}$ or $\pi^-\pi^{*+}$ for the $\pi\pi^*$ configuration); linear combinations of these different starting points will give rise to Σ^+ or Σ^- states. If $\Lambda \neq 0$, these different starting points (e.g., $\pi^+\pi^-\pi^{*+}$ and $\pi^+\pi^+\pi^{*-}$ for $\pi^2\pi^*$) will give rise to different electronic states.

2. For the $\Sigma = S$ substate of a $\Lambda > 0$ state, construct Slater determinants by combining the spin functions specified in Table 3.4 with the open-shell orbital products from step 1 above, always in the same order. *Each* spin

function must be combined with *each* orbital product. The specific linear combination of Slater determinants to which each electronic state belongs must be determined by solution of an electronic secular determinant (see Eqs. (3.2.103-3.2.103b), (3.2.104), (3.2.106) and Eqs. (3.2.107)-(3.2.110c) for the $\pi_g^2\pi_u$ $^2\Pi$ example).

3. For a $\Sigma < S$ basis state, apply $\mathbf{S}^- = \sum_i \mathbf{s}_i^-$ to the $\Sigma = S$ basis state.

Table 3.4 lists the spin functions, each of which appears along the diagonal of one determinant constructed from unpaired molecular spatial wavefunctions. For example, if there are two half-filled subshells, $\phi_a(1)$ and $\phi_b(2)$, using Table 3.4, the singlet wavefunction is written as a sum of two determinants:

$$(2)^{-1/2}\{|\phi_a(1)\alpha(1)\phi_b(2)\beta(2)| - |\phi_a(1)\beta(1)\phi_b(2)\alpha(2)|\}.$$

For a filled subshell, the spin functions are evidently $\alpha(1)\beta(2)$. Using the simplified notation of Eq. (3.2.100);

$$^1\Delta \qquad \begin{array}{ll} |\pi^+\alpha\pi^+\beta| & \Lambda = +2 \\ |\pi^-\alpha\pi^-\beta| & \Lambda = -2 \end{array} \qquad \Sigma = 0 \tag{3.2.101a}$$

$$^1\Sigma^+ \qquad (2)^{-1/2}\{|\pi^+\alpha\pi^-\beta| - |\pi^+\beta\pi^-\alpha|\} \qquad \Sigma = 0^\dagger \tag{3.2.101b}$$

$$^3\Sigma^- \qquad \begin{array}{ll} |\pi^+\alpha\pi^-\alpha| & \Sigma = +1 \\ (2)^{-1/2}\{|\pi^+\alpha\pi^-\beta| + |\pi^+\beta\pi^-\alpha|\} & \Sigma = 0^\dagger \\ |\pi^+\beta\pi^-\beta| & \Sigma = -1 \end{array} \tag{3.2.101c}$$

Table 3.4 only specifies the spin function for the maximum value of $\Sigma = S$, $|S = 1, \Sigma = 1\rangle = ||\alpha(1)\alpha(2)||\rangle$, but the wavefunctions for $-S \le \Sigma < +S$ can be derived by application of the $\mathbf{S}^-$ operator:

$$\mathbf{S}^-|S,\Sigma\rangle = +\hbar\,[S(S+1) - \Sigma(\Sigma-1)]\,|^{1/2}\,|S,\Sigma-1\rangle\,.$$

The + sign is a consequence of the sign convention of Eq. (3.2.85c). The result of $\mathbf{S}^-$ operating on a determinantal wavefunction is

$$\mathbf{S}^-|\mathrm{det}| = \sum_i \mathbf{s}^-(i)\left|\cdots\phi_i(i)^{\alpha(i)}_{\beta(i)}\cdots\right| = \sum_i (1-c_i)\,|\cdots\phi_i(i)\beta(i)\cdots|\,, \tag{3.2.102}$$

†The two $\Sigma = 0$ Slater determinants appear with the same sign for $^3\Sigma^-$ and opposite signs for $^1\Sigma^+$. This may be demonstrated by applying $\mathbf{S}^-$ (Eq. (3.2.102)) to the $\Sigma = +1$ $^3\Sigma^-$ determinant, $|\pi^+\alpha\pi^-\alpha|$.

where c_i is equal to 1 or 0 if the original spin function of $\phi_i(i)$ is $\beta(i)$ or $\alpha(i)$, respectively. The wavefunction must be renormalized after each application of $\mathbf{S}^-$. The orthogonality between wavefunctions of identical Σ but different S is often useful (but sometimes insufficient) to find the proper eigenfunctions of $\mathbf{S}^2$.

If another open shell is added to the π_g^2 configuration – for example, a singly occupied π_u orbital – the $\pi_g^2\pi_u$ configuration results. The possible states are shown in table form.

π_g^2 Parent State	π_g^+	π_g^-	π_u^+	π_u^-	$\Lambda = \Sigma\lambda_i$	Number of Open Subshells	$\pi_g^2\pi_u$ State
$^1\Delta_g$	2	0	1	0	3	1	$^2\Phi_u$
	0	2	0	1	-3		
$^1\Delta_g$	2	0	0	1	1	1	$^2\Pi_u$
	0	2	1	0	-1		
$^1\Sigma_g^+$, $^3\Sigma_g^-$	1	1	1	0	1	3	one $^4\Pi_u$,
	1	1	0	1	-1		two $^2\Pi_u$

Evidently, the $\pi_g^2\pi_u$ configuration gives rise to three independent $^2\Pi_u$ states. One is formed from the $^1\Delta_g$ parent. The other two originate from the $(\pi_g^+)(\pi_g^-)$ subshell occupancy, one associated with each of the $^3\Sigma_g^-$ and $^1\Sigma_g^+$ parent states. This is consistent with the branching diagram in Fig. 3.3. The determinantal wavefunctions can be derived easily for these three $^2\Pi_u$ states, using this scheme. When the singlet $^1\Sigma_g^+$ parent state ($\Lambda_1 = 0, \Sigma_1 = 0, S_1 = 0$) is coupled with a π_u electron $(\Lambda_2 = +1, \Sigma_2 = +\frac{1}{2})$, a doublet state results (because $S = \Sigma_{\max} = \Sigma_1 + \Sigma_2 = \frac{1}{2}$), which must be $^2\Pi_u$ (because $\Lambda = \Lambda_1 + \Lambda_2 = 1$). The wavefunction is

$$\left|^2\Pi(1), \pi_g^2\pi_u, \Omega = \frac{3}{2}\right\rangle$$
$$= (2)^{-1/2}\left\{\left|\cdots\pi_u^+\alpha\pi_g^+\alpha\pi_g^-\beta\right| - \left|\cdots\pi_u^+\alpha\pi_g^+\beta\pi_g^-\alpha\right|\right\}. \quad (3.2.103)$$

Another doublet state arises from the $^3\Sigma_g^-$ parent. There are two ways to construct a $\Sigma = \frac{1}{2}, \Lambda = 1$ state from $^3\Sigma_g^-$, one each from the $^3\Sigma_0^-$ and $^3\Sigma_1^-$ parent substates. Choosing the $^3\Sigma_0^-$ parent, $\Sigma_1 = 0, \Lambda_1 = 0$, $\left\{\left|\pi_g^+\alpha\pi_g^-\beta\right| + \left|\pi_g^+\beta\pi_g^-\alpha\right|\right\}$, and combining it with an electron in the $\pi_u^+\alpha$ $(\Sigma_2 = \frac{1}{2}, \Lambda_2 = 1)$ spin-orbital, the result is

$$\Phi\left(^3\Sigma_0^- \to \Pi_{3/2}\right) = \left\{\left|\pi_u^+\alpha\pi_g^+\alpha\pi_g^-\beta\right| + \left|\pi_u^+\alpha\pi_g^+\beta\pi_g^-\alpha\right|\right\}$$
$$\Sigma = \Sigma_1 + \Sigma_2 = \frac{1}{2}, \qquad \Lambda = \Lambda_1 + \Lambda_2 = 1, \qquad \Omega = \Sigma + \Lambda = \frac{3}{2}. \quad (3.2.103a)$$

When $^3\Sigma_1^-, \Sigma_1 = 1, \Lambda_1 = 0\left(\left|\pi_g^+\alpha\pi_g^-\alpha\right|\right)$ is combined with the $\pi_u^+\beta$

$\left(\Sigma_2 = -\frac{1}{2}, \Lambda_2 = 1\right)$ spin-orbital, the result is

$$\begin{aligned}\Phi\left({}^3\Sigma_1^- \to \Pi_{3/2}\right) &= \left|\pi_u^+\beta\pi_g^+\alpha\pi_g^-\alpha\right| \\ \Sigma = \Sigma_1 + \Sigma_2 = \frac{1}{2}, \qquad \Lambda &= \Lambda_1 + \Lambda_2 = 1, \qquad \Omega = \Sigma + \Lambda = \frac{3}{2},\end{aligned} \tag{3.2.103b}$$

but neither of the states thus generated is an eigenstate of $\mathbf{S}^2$, because they are mixtures of ${}^2\Pi_{3/2}$ and ${}^4\Pi_{3/2}$. From these two $\Sigma = \frac{1}{2}, \Lambda = 1$ functions, two eigenfunctions of $\mathbf{S}^2$ can be constructed, a doublet,

$$\begin{aligned}&\left|{}^2\Pi_u(2), \pi_g^2\pi_u, \Omega = \frac{3}{2}\right\rangle \\ &= (6)^{-1/2}\left[2\left|\pi_u^+\beta\pi_g^+\alpha\pi_g^-\alpha\right| - \left|\pi_u^+\alpha\pi_g^+\alpha\pi_g^-\beta\right| - \left|\pi_u^+\alpha\pi_g^+\beta\pi_g^-\alpha\right|\right],\end{aligned} \tag{3.2.104}$$

and a quartet,

$$\begin{aligned}&\left|{}^4\Pi_u, \pi_g^2\pi_u, \Omega = \frac{3}{2}\right\rangle \\ &= (3)^{-1/2}\left[\left|\pi_u^+\alpha\pi_g^+\alpha\pi_g^-\beta\right| + \left|\pi_u^+\alpha\pi_g^+\beta\pi_g^-\alpha\right| + \left|\pi_u^+\beta\pi_g^+\alpha\pi_g^-\alpha\right|\right].\end{aligned} \tag{3.2.105}$$

The spin parts of the ${}^2\Pi_u$ $\left(\Omega = \frac{3}{2}\right)$ functions are taken from Table 3.4. Here the spatial part has been chosen as $\pi_u^+(1)\pi_g^+(2)\pi_g^-(3)$. Any permutation of the order of the orbitals defines the ${}^2\Pi(1)$ and ${}^2\Pi(2)$ basis states differently and with different energies. The spin part of ${}^4\Pi_u$ $\Omega = \frac{3}{2}$ is derived either by applying $\mathbf{S}^-$ to the spin function for $\left|{}^4\Pi_u, \Omega = \frac{5}{2}\right\rangle$ from Table 3.4 or by requiring that $\left|{}^4\Pi_u, \Omega = \frac{3}{2}\right\rangle$ be orthogonal to $\left|{}^2\Pi_u, \Omega = \frac{3}{2}\right\rangle$. A third ${}^2\Pi_u$ state is derived when the ${}^1\Delta_g$ parent, $S_1 = 0, \Lambda_1 = 2\left(\left|\pi_g^+\alpha\pi_g^+\beta\right|\right)$, is combined with an electron in the $\pi_u^-\alpha$ $\left(\Sigma_2 = \frac{1}{2}, \Lambda_2 = -1\right)$ spin-orbital,

$$\left|{}^2\Pi_u(3), \Omega = \frac{3}{2}\right\rangle = \left|\pi_u^-\alpha\pi_g^+\alpha\pi_g^+\beta\right|. \tag{3.2.106}$$

One might expect that the relative energies of the three ${}^2\Pi$ states would mirror the relative energies of the parent states, namely, that the ${}^2\Pi_u(2)$ and ${}^2\Pi_u(1)$ basis functions derived from the lowest $\left({}^3\Sigma_g^-\right)$ and highest $\left({}^1\Sigma_g^+\right)$ energy parent states would become, respectively, the lowest- and highest-lying ${}^2\Pi$ states. In fact, the observed ${}^2\Pi_u$ states are found to be mixtures of the three ${}^2\Pi_u$ functions:

$$\begin{aligned}\left|{}^2\Pi_{ui}(\mathrm{obs})\right\rangle &= c_{1i}\left|{}^2\Pi_u(1)\right\rangle + c_{2i}\left|{}^2\Pi_u(2)\right\rangle + c_{3i}\left|{}^2\Pi_u(3)\right\rangle \\ i &= 1, 2, 3.\end{aligned} \tag{3.2.107}$$

The c_{ji} coefficients are obtained by setting up and diagonalizing a 3×3 secular energy matrix for the three isoconfigurational ${}^2\Pi_u$ states. The diagonal matrix elements have the expected energy order, but the off-diagonal matrix elements that originate from the electrostatic part (e^2/r_{ij}) of the electronic Hamiltonian couple the Slater determinants of basis functions with identical $|\Lambda, S, \Sigma\rangle$ but that differ by two spin-orbitals. Using the rules given in the next section (Section 3.2.4), the matrix elements are

$$\begin{aligned}
\langle {}^2\Pi(2)|\, \mathbf{H}^{\mathrm{el}}\, |{}^2\Pi(2)\rangle &= E\left({}^2\Pi(2)\right) = E\left({}^3\Sigma_g^-, \pi_g^2\right) + a - \frac{1}{2}\left[K^{(0)}_{\pi_u\pi_g} + K^{(2)}_{\pi_u\pi_g}\right] \\
\langle {}^2\Pi(3)|\, \mathbf{H}^{\mathrm{el}}\, |{}^2\Pi(3)\rangle &= E\left({}^2\Pi(3)\right) = E\left({}^1\Delta_g, \pi_g^2\right) + a - K^{(2)}_{\pi_u\pi_g} \\
\langle {}^2\Pi(1)|\, \mathbf{H}^{\mathrm{el}}\, |{}^2\Pi(1)\rangle &= E\left({}^2\Pi(1)\right) = E\left({}^1\Sigma_g^+, \pi_g^2\right) + a + \frac{1}{2}\left[K^{(0)}_{\pi_u\pi_g} + K^{(2)}_{\pi_u\pi_g}\right] \\
\langle {}^2\Pi(1)|\, \mathbf{H}^{\mathrm{el}}\, |{}^2\Pi(2)\rangle &= \left(3^{1/2}/2\right)\left(K^{(0)}_{\pi_u\pi_g} - K^{(2)}_{\pi_u\pi_g}\right) \\
\langle {}^2\Pi(1)|\, \mathbf{H}^{\mathrm{el}}\, |{}^2\Pi(3)\rangle &= (2)^{1/2} J^{(2)}_{\pi_u\pi_g} \\
\langle {}^2\Pi(2)|\, \mathbf{H}^{\mathrm{el}}\, |{}^2\Pi(3)\rangle &= -\left(\frac{3}{2}\right)^{1/2} K^{(0)}_{\pi_u\pi_g}
\end{aligned} \tag{3.2.108}$$

where the quantity a represents the Coulombic interaction energy of the π_u electron with the field of the other electrons. The $K^{(0)}$ and $K^{(2)}$ exchange and $J^{(2)}$ Coulomb integrals are defined, using the abbreviated notation,

$$\langle \phi_a\phi_b|\, \phi_c\phi_d\rangle \equiv \int \phi_a^*(1)\phi_b(1)\frac{e^2}{r_{12}}\phi_c^*(2)\phi_d(2)d\tau_1 d\tau_2 \tag{3.2.109}$$

$$K^{(0)}_{\pi_u\pi_g} = \left\langle \phi_{\pi_g^+}\phi_{\pi_u^+} \middle| \phi_{\pi_u^+}\phi_{\pi_g^+}\right\rangle = \left\langle \phi_{\pi_g^+}\phi_{\pi_u^+} \middle| \phi_{\pi_u^-}\phi_{\pi_g^-}\right\rangle \tag{3.2.110a}$$

$$K^{(2)}_{\pi_u\pi_g} = \left\langle \phi_{\pi_g^+}\phi_{\pi_u^-} \middle| \phi_{\pi_u^-}\phi_{\pi_g^+}\right\rangle = \left\langle \phi_{\pi_g^-}\phi_{\pi_u^+} \middle| \phi_{\pi_u^+}\phi_{\pi_g^-}\right\rangle \tag{3.2.110b}$$

$$J^{(2)}_{\pi_u\pi_g} = \left\langle \phi_{\pi_g^+}\phi_{\pi_g^-} \middle| \phi_{\pi_u^-}\phi_{\pi_u^+}\right\rangle = \left\langle \phi_{\pi_g^-}\phi_{\pi_g^+} \middle| \phi_{\pi_u^+}\phi_{\pi_u^-}\right\rangle \tag{3.2.110c}$$

The off-diagonal element between the ${}^2\Pi(2)$ and ${}^2\Pi(3)$ basis states associated with the ${}^3\Sigma_g^-$ and ${}^1\Delta_g$ parents is so large that the energy order of the roots of the secular equation is not that expected from the order of the diagonal elements. Consequently, the highest and lowest energy ${}^2\Pi$ states are predominantly mixtures of ${}^3\Sigma_g^-$ and ${}^1\Delta_g$ parents, even though the highest energy basis state is that associated with the ${}^1\Sigma_g^+$ parent.

An example has been discussed by Dixon and Hull (1969), who apply these ideas to the three states arising from the O_2^+ $\pi_u^3\pi_g^2$ configuration. The eigenfunctions obtained are, starting with the lowest state, the $A^2\Pi_u$ state,

$$\begin{aligned}
|A^2\Pi_u\rangle &= -0.20\,|{}^2\Pi(1)\rangle + 0.66\,|{}^2\Pi(2)\rangle + 0.72\,|{}^2\Pi(3)\rangle, \\
|2^2\Pi_u\rangle &= 0.91\,|{}^2\Pi(1)\rangle + 0.39\,|{}^2\Pi(2)\rangle - 0.15\,|{}^2\Pi(3)\rangle, \\
|3^2\Pi_u\rangle &= 0.37\,|{}^2\Pi(1)\rangle + 0.74\,|{}^2\Pi(2)\rangle - 0.57\,|{}^2\Pi(3)\rangle.
\end{aligned} \tag{3.2.111}$$

The values of the c_{ij} coefficients are very similar to those for the excited $^2\Pi$ states of NO (Field, *et al.*, 1975) and of PO (Roche and Lefebvre-Brion, 1973), but the signs of the c_{ij} depend on the choice of phases for the electronic functions.

Since Rydberg states can be viewed as an electron in a Rydberg orbital coupled to the many-electron electronic state of the molecular-ion core, it is possible to construct the Rydberg states of O_2 which belong to series converging to the three $^2\Pi$ electronic states of O_2^+ described in Eq. (3.2.111) and the $^4\Pi$ state described in Eq. (3.2.105). For example, one $^3\Pi_u$ state belongs to the configuration $\pi_u^3\pi_g^2 3s\sigma_g$, which has four open subshells ($\pi_u^-, \pi_g^+, \pi_g^-$, and $3s\sigma_g$; π_u^+ is full because it is doubly occupied). Table 3.4 specifies that four open shells give three distinct triplet spin functions. Two of these triplet spin functions arise from doublet ion-core states, and one arises from the $^4\Pi$ state. Let us examine this last state. Taking the first open shell as $3s\sigma_g$, it has the spins in the other three subshells arranged as a quartet with $\Sigma_1 = 3/2$ (a single Slater determinant) and another quartet with $\Sigma_1 = 1/2$ (a linear combination of three determinants). Letting the spin in the $3s\sigma_g$ orbital define Σ_2, $\Sigma_2 = -1/2$ in combination with the $\pi_u^3\pi_g^2$ quartet $\Sigma_1 = 3/2$ determinant and $\Sigma_2 = +1/2$ in combination with the three quartet $\Sigma_1 = 1/2$ determinants, one obtains a triplet state with $\Sigma = 1$, which is the $^3\Pi_u$ state descended from a $^4\Pi_u$ ion-core. This state has the wave function (Lefebvre-Brion, 1992)

$$\begin{aligned}(12)^{-\frac{1}{2}}\{-3&\left|3s\sigma_g\beta\pi_u^+\alpha\pi_u^+\beta\pi_u^-\alpha\pi_g^+\alpha\pi_g^-\alpha\right| \\ &+\left|3s\sigma_g\alpha\pi_u^+\alpha\pi_u^+\beta\pi_u^-\beta\pi_g^+\alpha\pi_g^-\beta\right| + \left|3s\sigma_g\alpha\pi_u^+\alpha\pi_u^+\beta\pi_u^-\alpha\pi_g^+\beta\pi_g^-\alpha\right| \\ &+\left|3s\sigma_g\alpha\pi_u^+\alpha\pi_u^+\beta\pi_u^-\alpha\pi_g^+\alpha\pi_g^-\beta\right|\}\end{aligned} \quad (3.2.111a)$$

This wave function is a first approximation for the $H^3\Pi_u$ Rydberg state of O_2 (Nishinati, *et al.*, 1984).

The two $^3\Pi_u$ states, which arise from the two $^2\Pi_u$ ion core states, have the spin functions given in Table 3.4 (first and second lines of the triplet spin functions). The first triplet state is derived from the first doublet spin function by adding a $\Sigma_2 = +1/2$, $3s\sigma_g$ orbital to the $\Sigma_1 = 1/2$ spin doublet function. The second triplet state arises similarly from the second doublet spin function (there is a change of sign for the total wavefunction, but as is well known, the absolute phase of a wave function is not defined). The wave function for the $^3\Pi_u$ state, which arises from the $^2\Pi_u$ (3) ion-core state, is trivial to write, because the ion-core has only one open shell, and the $^3\Pi_u$ (3) state has only two open shells (see Table 3.4).

3.2.4 Matrix Elements between Electronic Wavefunctions

The rules given by Slater must be used when evaluating matrix elements of any operator in a Slater determinantal representation (Slater, 1960, p. 291). These rules reduce the matrix elements of one- and two-electron operators between many-electron Slater determinantal functions to simple sums over spatial orbital integrals.[†]

Determinants have the property that they change sign when any two adjacent rows or columns are interchanged. Since electrons are Fermions, the total wavefunction must change sign upon all possible permutations of electrons. One finds that Slater determinantal basis functions are properly antisymmetrized with respect to electron interchange and that one needs only to keep track of the order of spin-orbitals along the diagonal in order to ascertain whether two determinants, constructed from the same spin-orbitals, are identical or differ by -1. Two determinants differing by an odd (even) number of adjacent spin-orbital permutations are identical except for a factor of $-1(+1)$.

The procedure for evaluating determinantal matrix elements requires that all spin-orbitals common to both determinantal functions be permuted so that both the common spin-orbitals and the orbitals that differ in the spatial part but are identical in the spin part appear in the same locations. It is necessary to specify all common orbitals. For example, when a matrix element between states belonging to the two configurations $\sigma\pi^3$ and π^2 is to be calculated, it is important not to be confused by the abbreviated notation π^2, which suppresses the usually irrelevant closed σ shell, instead of the more complete designation, $\sigma^2\pi^2$.

If two determinantal functions, Φ_i and Φ_j, differ by only one spin-orbital, ϕ_i in Φ_i and ϕ_j in Φ_j, then a one-electron operator, $\mathbf{Op}_1$, can have a nonzero matrix element

$$\mathbf{Op}_1 \equiv \sum_{i=1}^{n} \mathbf{Op}_1(i)$$

$$\begin{aligned} \langle \Phi_i | \mathbf{Op}_1 | \Phi_j \rangle &= (-1)^P \int \phi_i^*(1) \mathbf{Op}_1(1) \phi_j(1) d\tau_1 \\ &= (-1)^P \langle \phi_i | \mathbf{Op}_1 | \phi_j \rangle , \end{aligned} \tag{3.2.112}$$

where P is the number of permutations required to match the order of appearance of the two sets of spin-orbitals. Diagonal matrix elements of $\mathbf{Op}_1$ are given by

$$\langle \Phi_i | \mathbf{Op}_1 | \Phi_i \rangle = \sum_{k=1}^{n} \langle \phi_k | \mathbf{Op}_1 | \phi_k \rangle \tag{3.2.113}$$

If two determinantal functions differ by two spin-orbitals, then their matrix elements for all one-electron operators will be zero. However, a two-electron operator,

$$\mathbf{Op}_{12} = \sum_{j>i=1}^{n} \mathbf{Op}_{12}(i,j),$$

[†] It is necessary to assume that the spin-orbitals are mutually orthogonal.

can have nonzero matrix elements. Denoting the unique spin-orbitals as ϕ_i and ϕ_j in Φ_{ij} and ϕ_k and ϕ_l in Φ_{kl}, then

$$\langle\Phi_{ij}|\mathbf{Op}_{12}|\Phi_{kl}\rangle = (-1)^P \left[\langle\phi_i\phi_k|\mathbf{Op}_{12}|\phi_j\phi_l\rangle - \langle\phi_i\phi_l|\mathbf{Op}_{12}|\phi_j\phi_k\rangle\right], \quad (3.2.114)$$

where, by convention [Eq. (3.2.109)],

$$\langle\phi_i\phi_k|\mathbf{Op}_{12}|\phi_j\phi_l\rangle \equiv \int \phi_i^*(1)\phi_k(1)\mathbf{Op}_{12}(1,2)\phi_j^*(2)\phi_l(2)d\tau_1 d\tau_2. \quad (3.2.115)$$

In the frequently encountered situation when $\mathbf{Op}_{12}$ does not operate on electron spin wavefunctions, then one of the two integrals in Eq. (3.2.114) vanishes whenever the spin parts of ϕ_i and ϕ_j or ϕ_k and ϕ_l are not identical. When $\phi_i = \phi_k, \phi_j = \phi_l$, and $\mathbf{Op}_{12}(1,2) = e^2/r_{12}$, then

$$\langle\phi_i\phi_i|\mathbf{Op}_{12}|\phi_j\phi_j\rangle \equiv J_{ij}, \qquad \langle\phi_i\phi_j|\mathbf{Op}_{12}|\phi_j\phi_i\rangle \equiv K_{ij},$$

where J_{ij} and K_{ij} are direct (or Coulomb) and exchange integrals. J_{ij} is the classical electrostatic repulsion between the charge densities associated with orbitals ϕ_i and ϕ_j. There is no classical analog for K_{ij}, but it is always true that $J_{ij} > K_{ij}$.

Diagonal matrix elements of $\mathbf{Op}_{12}$ are given by

$$\langle\Phi_i|\mathbf{Op}_{12}|\Phi_i\rangle = \sum_{j>i=1}^{n} \left[\langle\phi_i\phi_i|\mathbf{Op}_{12}|\phi_j\phi_j\rangle - \langle\phi_i\phi_j|\mathbf{Op}_{12}|\phi_j\phi_i\rangle\right], \quad (3.2.116)$$

and matrix elements of $\mathbf{Op}_{12}$ between functions differing by one spin-orbital are

$$\langle\Phi_i|\mathbf{Op}_{12}|\Phi_j\rangle = (-1)^P \sum_{k=1}^{n} \left[\langle\phi_i\phi_j|\mathbf{Op}_{12}|\phi_k\phi_k\rangle - \langle\phi_i\phi_k|\mathbf{Op}_{12}|\phi_k\phi_j\rangle\right], \quad (3.2.117)$$

where P is the number of permutations required to match the orbital orders in Φ_i and Φ_j. Matrix elements of $\mathbf{Op}_{12}$ between functions differing by two spin-orbitals are given by Eq. (3.2.114), and those between functions differing by more than two orbitals are necessarily zero.

The *absolute* sign of an off-diagonal matrix element cannot be determined, since it depends arbitrarily on the chosen phase for the determinantal wave function, namely, on the order of the spin-orbitals. However, the *relative* sign of *two* off-diagonal matrix elements can often be determined experimentally. Thus, care must be taken to define the phases of the wavefunctions consistently.

In Section 5.3 it will become clear why it is useful to reduce matrix elements between Slater determinantal functions to matrix elements between individual molecular orbitals.

Some operators that will be used here contain the orbital and spin angular momenta of the electrons. By definition,

$$\mathbf{l}_z(1)\,|p\lambda(1)\rangle = \hbar\lambda\,|p\lambda(1)\rangle$$

or

$$\mathbf{l}_z \left|\pi^+\right\rangle = +\hbar \left|\pi^+\right\rangle \qquad \mathbf{l}_z \left|\pi^-\right\rangle = -\hbar \left|\pi^-\right\rangle . \tag{3.2.118}$$

For the spin,

$$\begin{gathered} \mathbf{s}_z(1)\left|s\sigma(1)\right\rangle = \hbar\sigma\left|s\sigma\right\rangle \\ \mathbf{s}_z\left|\alpha\right\rangle = +\hbar/2\left|\alpha\right\rangle \qquad \mathbf{s}_z\left|\beta\right\rangle = -\hbar/2\left|\beta\right\rangle . \end{gathered} \tag{3.2.119}$$

Raising and lowering operators have the effect

$$\mathbf{s}^+\left|\beta\right\rangle = +\hbar\left|\alpha\right\rangle, \qquad \mathbf{s}^+\left|\alpha\right\rangle = 0,$$

and

$$\mathbf{s}^-\left|\alpha\right\rangle = +\hbar\left|\beta\right\rangle, \qquad \mathbf{s}^-\left|\beta\right\rangle = 0. \tag{3.2.120}$$

The effect of $\mathbf{S}^-$ on a determinantal function was illustrated by Eq. (3.2.102).

In general, the l value of a molecular orbital is not well defined. However, in some cases, such as for Rydberg orbitals, l has nearly an integer value. In such a case,

$$\mathbf{l}^\pm\left|l\lambda\right\rangle = \hbar\left[l(l+1) - \lambda(\lambda\pm1)\right]^{1/2}\left|l,\lambda\pm1\right\rangle, \tag{3.2.121}$$

and, for a p orbital,

$$\begin{aligned} \mathbf{l}^+\left|p\sigma\right\rangle &= (2)^{1/2}\hbar\left|p\pi^+\right\rangle \quad \text{and} \quad \mathbf{l}^-\left|p\pi^+\right\rangle = (2)^{1/2}\hbar\left|p\sigma\right\rangle \\ \mathbf{l}^-\left|p\sigma\right\rangle &= (2)^{1/2}\hbar\left|p\pi^-\right\rangle \quad \text{and} \quad \mathbf{l}^+\left|p\pi^-\right\rangle = (2)^{1/2}\hbar\left|p\sigma\right\rangle \\ \mathbf{l}^+\left|p\pi^+\right\rangle &= 0 \qquad\qquad\qquad\;\; \text{and} \quad \mathbf{l}^-\left|p\pi^-\right\rangle = 0. \end{aligned} \tag{3.2.122}$$

Until now it has been assumed that the spatial part of all many-electron Σ-state basis functions is an eigenfunction of the $\boldsymbol{\sigma}_v$ operator. However, using the phase convention of Eq. (3.2.82),

$$\boldsymbol{\sigma}_v(xz)\left|\lambda\right\rangle = (-1)^\lambda\left|-\lambda\right\rangle,$$

it appears that there can be no one-electron wavefunction (i.e., orbital) of Σ^- symmetry. Evidently, $\Sigma^\pm$ symmetry is a property of many-electron wavefunctions. In order to understand the origin of states of Σ^- symmetry and to demonstrate the compatibility of Eq. (3.2.82) with

$$\boldsymbol{\sigma}_v(xz)\left|\Lambda^\pm\right\rangle = (-1)^{\Lambda+s}\left|-\Lambda^\pm\right\rangle$$

(where $s = 1$ for Σ^- and $s = 0$ for Σ^+ and all $\Lambda \neq 0$), it is necessary to examine the many-electron form of the $\boldsymbol{\sigma}_v$ operator. Let

$$\boldsymbol{\sigma}_v = \prod_{i=1}^{n} \boldsymbol{\sigma}_v^i,$$

where $\boldsymbol{\sigma}_v^i$ operates on the ith electron and, for the spatial part of the wavefunction only,

$$\begin{aligned}\boldsymbol{\sigma}_v(xz)\,|\lambda_1\cdots\lambda_n| &= (-1)^{\Sigma_i^n \lambda_i}\,|-\lambda_1\cdots-\lambda_n| \\ &= (-1)^{\Lambda}\,|-\lambda_1\cdots-\lambda_n| \end{aligned} \tag{3.2.123}$$

or, for the complete wavefunction (see Eq. (3.2.81)),

$$\begin{aligned}\boldsymbol{\sigma}_v(xz)\,|\lambda_1\sigma_1,\cdots,\lambda_n\sigma_n| &= (-1)^{\Sigma_i^n(\lambda_i+1/2-\sigma_i)}\,|-\lambda_1-\sigma_1,\cdots,-\lambda_n-\sigma_n| \\ &= (-1)^{\Lambda-\Sigma+n/2}\,|-\lambda_1-\sigma_1,\cdots,-\lambda_n-\sigma_n| \end{aligned} \tag{3.2.124}$$

The correct $\Sigma^{\pm}$ symmetry appears when, for $\Lambda = 0$, the orbitals on the right side of Eq. (3.2.123) are rearranged to match the order of those on the left side, keeping track of the number of adjacent orbital permutations required.

Consider, as an example, the simplest configuration known to give rise to a Σ^- state, π^2. The determinantal wavefunctions corresponding to $^1\Delta, {}^1\Sigma^+$, and $^3\Sigma^-$ were given as (Eqs. (3.2.101a) - (3.2.101c)). Adopting a standard order for the spin-orbitals, $(\pi^+\alpha, \pi^+\beta, \pi^-\alpha, \pi^-\beta)$, and always rearranging the determinantal functions so that the spin-orbitals appear in this order, then

$$\begin{aligned}\boldsymbol{\sigma}_v\,|^3\Sigma_1^-\rangle &= \boldsymbol{\sigma}_v\,|\pi^+\alpha\pi^-\alpha| = (-1)^0\,|\pi^-\beta\pi^+\beta| \\ &= -\,|\pi^+\beta\pi^-\beta| = -\,|^3\Sigma_{-1}^-\rangle,\end{aligned}$$

which is consistent with Eq. (3.2.89), namely,

$$\begin{aligned}\boldsymbol{\sigma}_v\,|^3\Sigma_1^-\rangle &= \boldsymbol{\sigma}_v\,|\Lambda = 0^-, S = 1, \Sigma = 1\rangle \\ &= (-1)^{0+1-1+1}\,|0^-, 1, -1\rangle = (-1)\,|^3\Sigma_{-1}^-\rangle.\end{aligned}$$

Similarly,

$$\begin{aligned}\boldsymbol{\sigma}_v\,|^1\Sigma^+\rangle &= (2)^{-1/2}\left[\boldsymbol{\sigma}_v\,|\pi^+\alpha\pi^-\beta| - \boldsymbol{\sigma}_v\,|\pi^+\beta\pi^-\alpha|\right] \\ &= (2)^{-1/2}\left[-\,|\pi^-\beta\pi^+\alpha| + |\pi^-\alpha\pi^+\beta|\right] \\ &= (2)^{-1/2}\left[|\pi^+\alpha\pi^-\beta| - |\pi^+\beta\pi^-\alpha|\right] = +\,|^1\Sigma^+\rangle,\end{aligned}$$

which is consistent with

$$\boldsymbol{\sigma}_v\,|^1\Sigma^+\rangle = (-1)^{0+0+0+0}\,|^1\Sigma^+\rangle.$$

Finally,

$$\begin{aligned}\boldsymbol{\sigma}_v\,|^1\Delta_2\rangle &= -\,|\pi^-\beta\pi^-\alpha| = |\pi^-\alpha\pi^-\beta| = +\,|^1\Delta_{-2}\rangle \\ &= (-1)^{2+0+0+0}\,|^1\Delta_2\rangle = +\,|^1\Delta_{-2}\rangle.\end{aligned}$$

Thus, at least for the π^2 configuration, Eq. (3.2.82) and Eq. (3.2.89) are self-consistent.

The Σ^-–states generally arise from configurations containing partially filled $\lambda > 0$ orbitals, for example

$$\left|{}^4\Sigma^-_{3/2}\right\rangle = |\lambda^+\alpha\lambda^-\alpha\sigma\alpha|$$

and

$$\left|{}^2\Sigma^{\pm}_{1/2}\right\rangle = \left[|\lambda^+\alpha\lambda^+\beta\,(2\lambda)^-\alpha| \pm |\lambda^-\alpha\lambda^-\beta\,(2\lambda)^+\alpha|\right].$$

It is not possible to construct a Σ^- state from $\lambda = 0$ orbitals alone or from a combination of $\lambda = 0$ orbitals with only *one* $\lambda \neq 0$ orbital. When dealing with perturbations involving Σ^- states, it is useful to remember this configurational property.

3.3 Electrostatic Perturbations

Electrostatic perturbations occur between states of identical electronic symmetry (Fig. 3.4), i.e., those with identical values of Λ, Σ and S. In Section 3.1.2, two types of off-diagonal matrix elements were shown to connect these states: those of the $\mathbf{H}^{\rm el}$ and $\mathbf{T}^N$ operators,

$$\langle 1, \Lambda, \Sigma, S|\mathbf{H}|2, \Lambda, \Sigma, S\rangle = \left\langle\Phi_1|\mathbf{H}^{\rm el}|\Phi_2\right\rangle + \left\langle\Phi_1|T^N|\Phi_2\right\rangle. \tag{3.3.1}$$

Here the electronic wavefunctions can be taken as nonsymmetrized or symmetrized. The matrix elements are identical in either case. The ideal starting point would be a basis set, Φ_1 and Φ_2, that minimizes the values of the Eq. (3.3.1) off-diagonal matrix elements. Unfortunately, it is not possible to find solutions of the electronic Hamiltonian for which *both* terms of Eq. (3.3.1) are zero. Two possible types of deperturbed or zeroth-order electronic functions may be defined (see also Table 3.1):

1. *Diabatic* functions for which, by definition,

$$\left\langle\Phi_1^d|\mathbf{T}^N|\Phi_2^d\right\rangle_r = 0,$$

but then

$$\left\langle\Phi_1^d|\mathbf{H}^{\rm el}|\Phi_2^d\right\rangle_r = H_{12}^e(R) \neq 0. \tag{3.3.2}$$

This latter relation means that the diabatic potential curves associated with these Φ_i^d functions can cross. The noncrossing rule between states of identical symmetry applies only for exact solutions of the electronic Hamiltonian.

2. *Adiabatic* functions are precisely those that diagonalize the electronic Hamiltonian. By definition,

$$\left\langle\Phi_1^{ad}|\mathbf{H}^{\rm el}|\Phi_2^{ad}\right\rangle_r = 0,$$

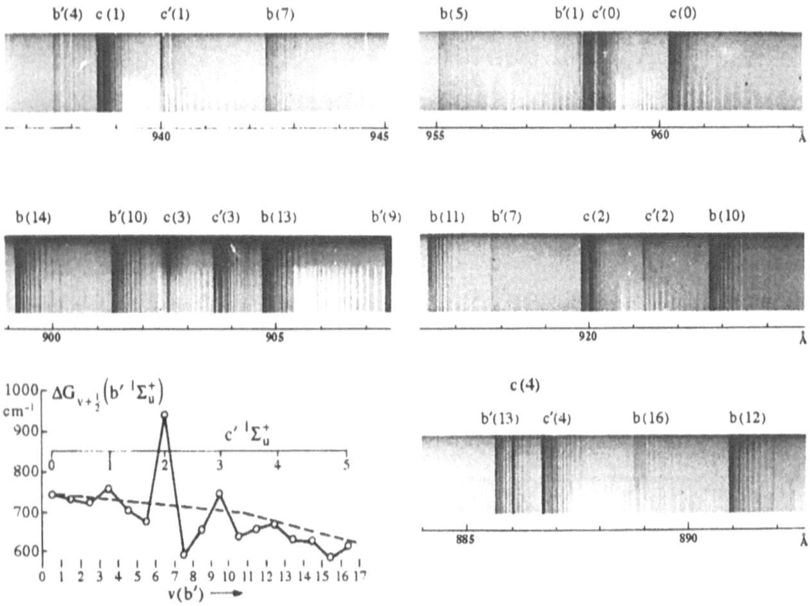

Figure 3.4: Electrostatic valence∼Rydberg N_2 $b'^1\Sigma_u^+(v_{b'}) \sim c'^1\Sigma_u^+(v_{c'})$ and $b^1\Pi_u(v_b) \sim c^1\Pi_u(v_c)$ perturbations. Each segment of the absorption spectrum (from Yoshino, *et al.*, 1979) shows several perturbing states near a nominal $v_{c'} = 0 - 4$ $c'^1\Sigma_u^+$ level. The b and b′ valence states are perturbed by the c and c′ Rydberg states (nominally the $^1\Pi_u$ and $^1\Sigma_u^+$ components of a $3p$ complex) as well as by higher Rydberg states. The $\Delta G_{v+1/2}$ plot for the $b'^1\Sigma_u^+$ state (from Dressler, 1969; see also Fig. 3.6) in the lower left corner shows the massive level shifts that had made it difficult to recognize the electronic state parentage of the observed singlet vibronic levels of N_2. The largest positive deviations of the observed ΔG values from the smoothly varying deperturbed value (dashed line) occur when the $v_{c'} = 1, 2$ and 3 levels are sandwiched between the $v_{b'} = 3$ and 4, 6 and 7, and 9 and 10 levels, respectively. These b′ ∼c′ perturbations are discussed in Section 6.3.1 and further illustrated in Fig. 3.6.

but then

$$\langle \Phi_1^{ad} | \mathbf{T}^N | \Phi_2^{ad} \rangle_r \neq 0. \tag{3.3.3}$$

In principle, whatever the initial model, after introducing the vibronic coupling terms corresponding to the chosen type of deperturbed potential curves, the experimental energy levels are obtained by diagonalizing one or the other type of interaction matrix. One example will be discussed later (Section 3.3.4).

If the deperturbed curves intersect and are characterized by very different molecular constants, then they are diabatic curves (see Fig. 3.5). If the crossing is avoided, adiabatic curves are involved, and one of these curves can have a double minimum. One frequently finds that, in the region of an avoided crossing, the adiabatic wavefunction changes electronic character abruptly and the derivative of the electronic wavefunction with respect to R can be large. In fact, it is this derivative that controls the size of the matrix elements of the $\mathbf{T}^N$ operator [Eq. (3.3.10)].

In the older literature (Dieke, 1935; Kovács, 1969), there is some confusion. Only the $\mathbf{T}^N$ operator has been assumed to connect states of the same symmetry, and the potential curves of these interacting states have been assumed to cross. Neither assumption is correct.

3.3.1 Diabatic Curves

The vibronic interaction between the level v_1 of the diabatic potential curve $V_1^d(R)$ and the level v_2 of another diabatic curve $V_2^d(R)$ is reduced to

$$H_{1,v_1;2,v_2} = \langle \Phi_1^d \chi_{v_1}^d | \mathbf{H}^{\text{el}} | \Phi_2^d \chi_{v_2}^d \rangle \tag{3.3.4}$$

since, by definition,

$$\langle \Phi_1^d | \mathbf{T}^N | \Phi_2^d \rangle = 0.$$

In the diabatic model, the electronic part of the matrix element $H_{1,v_1;2,v_2}$ is often assumed to be independent of R. Then the nuclear and electronic coordinates can be separated in the integration of Eq. (3.3.4). By integration over the electronic coordinates, one obtains

$$H_{1,v_1;2,v_2} = H^e \langle v_1^d | v_2^d \rangle , \tag{3.3.5}$$

where

$$H^e = \langle \Phi_1^d | \mathbf{H}^{\text{el}} | \Phi_2^d \rangle$$

and

$$\langle v_1^d | v_2^d \rangle = \int \chi_{v_1}^{d*}(R) \chi_{v_2}^d(R) dR.$$

Indeed, as for any electronic quantity, the value of H^e actually depends on R. However, this dependence is usually weak. Equation (3.3.5) holds even if the R–dependence of H^e is a linear function of the R–centroid (Halevi, 1965) defined by

$$\bar{R}_{ij} = \frac{\langle v_i | R | v_j \rangle}{\langle v_i | v_j \rangle}. \tag{3.3.6a}$$

The significance of the R–centroid is illustrated as follows. The electronic matrix element, $H^e(R)$, may be expanded in a power series about an arbitrarily chosen internuclear distance, R' (most usefully, $R' = R_C$, the internuclear distance at which the two potential curves cross),

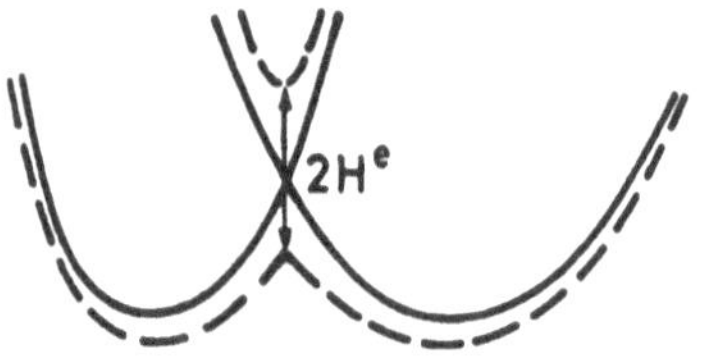

Figure 3.5: Diabatic and adiabatic potential curves. The diabatic curves (solid lines) cross at R_c and are defined by neglecting the part of $\mathbf{H}^{el}$ that causes the adiabatic curves (dotted lines) to avoid crossing by $2H^e$ at R_C.

$$H^e(R) = H^e(R') + \left(\frac{dH^e}{dR}\right)_{R=R'}(R-R') + \frac{1}{2}\left(\frac{d^2H^e}{dR^2}\right)_{R=R'}(R-R')^2. \tag{3.3.7a}$$

Then the vibrational matrix elements of $H^e(R)$ are expressed in terms of R^n–centroids,

$$\overline{R^n_{ij}} = \frac{\langle v_i|R^n|v_j\rangle}{\langle v_i|v_j\rangle} \tag{3.3.6b}$$

$$\begin{aligned}\langle v_i|H^e(R)|v_j\rangle &= H^e(R')\langle v_i|v_j\rangle + \left(\frac{dH^e}{dR}\right)_{R=R'}\left[\langle v_i|R|v_j\rangle - R'\langle v_i|v_j\rangle\right]\\ &+ \frac{1}{2}\left(\frac{d^2H^e}{dR^2}\right)_{R=R'}\left[\langle v_i|R^2|v_j\rangle - 2R'\langle v_i|R|v_j\rangle + R'^2\langle v_i|v_j\rangle\right]\end{aligned}$$

$$\begin{aligned}\frac{\langle v_i|H^e(R)|v_j\rangle}{\langle v_i|v_j\rangle} &= H^e(R') + \left(\frac{dH^e}{dR}\right)_{R=R'}\left[\overline{R_{ij}} - R'\right]\\ &+ \frac{1}{2}\left(\frac{d^2H^e}{dR^2}\right)_{R=R'}\left[\overline{R^2_{ij}} - 2R'\overline{R_{ij}} + R'^2\right].\end{aligned} \tag{3.3.7b}$$

Now, making the R–centroid *approximation*, which is quite distinct from the R^n–centroid expansion, namely,

$$\bar{R} \simeq \frac{\overline{R^n_{ij}}}{\overline{R^{n-1}_{ij}}} = \frac{\langle v_i|R^n|v_j\rangle}{\langle v_i|R^{n-1}|v_j\rangle}$$

[or, in other words, $\overline{R^n} = \left(\overline{R}\right)^n$], then Eq. (3.3.7b) becomes

$$\begin{aligned}\frac{\langle v_i|H^e(R)|v_j\rangle}{\langle v_i|v_j\rangle} &= H^e(R') + \left(\frac{dH^e}{dR}\right)_{R=R'}(\overline{R_{ij}} - R')\\ &+ \frac{1}{2}\left(\frac{d^2H^e}{dR^2}\right)_{R=R'}(\overline{R_{ij}} - R')^2,\end{aligned}$$

which is identical to Eq. (3.3.7a) where R is set equal to $\overline{R_{ij}}$.

For near-degenerate vibrational levels of any two crossing potential curves, the R–centroid has the convenient property of being nearly equal to R_C, the R–value where the two curves cross (Schamps, 1977). Thus,

$$\overline{R} = \overline{R^k} \Big/ \overline{R^{k-1}} = R_C,$$

and, setting $R' = R_C$,

$$\frac{\langle v_i|H^e|v_j\rangle}{\langle v_i|v_j\rangle} = H^e(R_C).$$

The R–centroid approximation has repeatedly been tested numerically. For perturbations between levels of potentials that intersect exactly once, the R–centroid approximation can be regarded as more accurate than experimentally measurable matrix elements.

The validity of the R–centroid approximation is based on a stationary phase argument (see Section 5.1.1) (Tellinghuisen 1984). For two vibrational states, $|v_i\rangle$ and $|v_j\rangle$, at nearly identical total energy, the vibrational overlap integral

$$I(R') = \int_0^{R'} \langle v_i|R\rangle \langle R|v_j\rangle \, dR$$

accumulates only near $R' = R_C$, where the two vibrational wavefunctions oscillate at the same spatial frequency. This stationary phase argument works equally well at low and high v and for R_C near or far from vibrational turning points, provided that there is only one intersection between the potential curves within the range of R defined by the vibrational turning points at the total energy of the perturbation.

Approximate deperturbed curves can be derived from unperturbed vibrational levels far from the energy of the curve crossing region. The overlap factor between vibrational wavefunctions is calculable numerically. (Note that a Franck-Condon factor is the absolute magnitude squared of the overlap factor.) From Eq. (3.3.5) and the experimental value of $H_{1,v_1;2,v_2}$, an initial trial value for H^e can be deduced. If the value of H^e is as large as the value of ω_e (see Table 4.4), then electrostatic interactions strongly perturb the entire set of vibrational levels. Figure 3.6 shows the irregular pattern of ΔG values for the perturbed $^1\Sigma_u^+$ states of N_2. Final deperturbed curves are obtained by diagonalization of the matrix, including all vibrational levels of both states, as described in Section 4.4.

3.3.2 Approximate Representation of the Diabatic Electronic Wavefunction

The single-configuration approximation is an approximate representation for the diabatic electronic function. Such an approximation is expected to result in

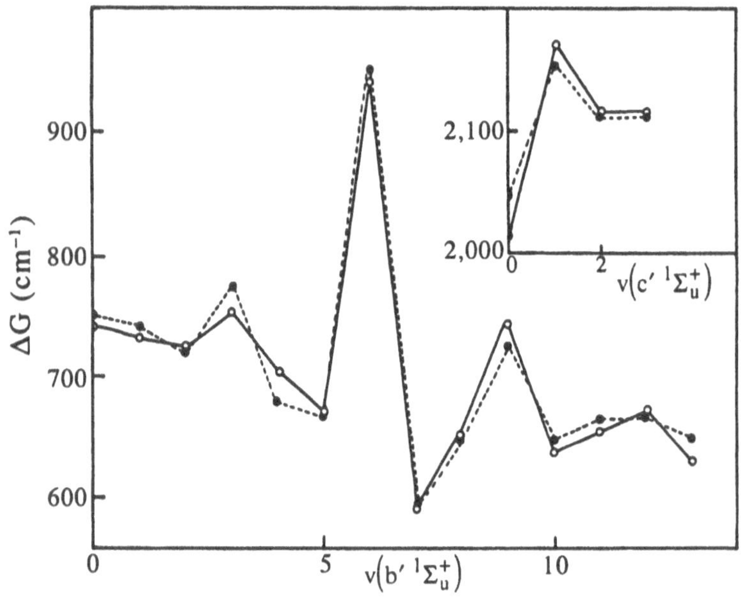

Figure 3.6: Variation of ΔG_v for the mutually interacting b′ and $c'^1\Sigma_u^+$ states of N_2. The solid and dashed lines correspond, respectively, to the observed and calculated (Lefebvre-Brion, 1969) values. The deperturbed c′ ($v = 2$) and b′ ($v = 7$) levels are nearly degenerate and interact strongly (see Table 5.4). This accounts for the largest ΔG anomalies.

smooth R–variation of the electronic wavefunction. In the case where the configurations of the two interacting states differ by two spin-orbitals, the electronic matrix element of the $\mathbf{T}^N$ operator between these two approximate functions is exactly zero, since it can be demonstrated that the $\mathbf{T}^N$ operator acts as a one-electron operator (Section 3.2.4) (Sidis and Lefebvre-Brion, 1971). However, as the electronic Hamiltonian, $\mathbf{H}^{\text{el}}$, contains a two-electron operator,

$$\sum_{i<j} \frac{e^2}{r_{ij}},$$

$\mathbf{H}^{\text{el}}$ can have nonzero matrix elements. If the unique orbitals are ϕ_a, ϕ_b and ϕ_c, ϕ_d for the two electronic configurations, then Eq. (3.2.54) and Eq. (3.2.55) give

$$H^e \propto \langle \phi_a\phi_c |\, 1/r_{12} | \phi_b\phi_d \rangle - \langle \phi_a\phi_d |\, 1/r_{12} | \phi_b\phi_c \rangle . \qquad (3.3.8)$$

If the spin parts of ϕ_a and ϕ_b are identical, both terms in Eq. (3.3.8) are nonzero. Otherwise, the single nonzero term is the one in which the spins of electron 1 and those of electron 2 are identical (i.e., $\langle \phi_a\phi_c |\, 1/r_{12} | \phi_b\phi_d \rangle = 0$

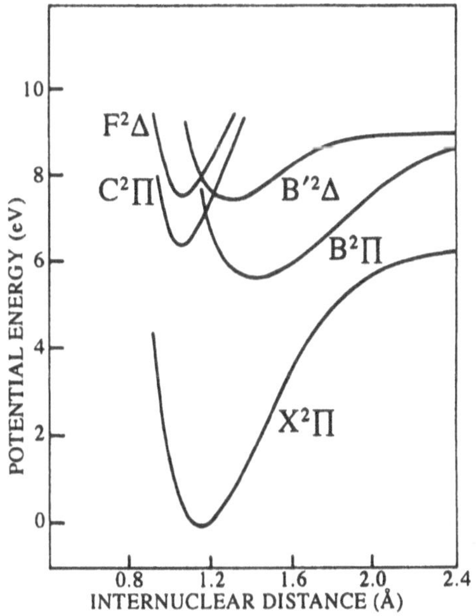

Figure 3.7: $^2\Delta$ and $^2\Pi$ potential energy curves of NO. The $C^2\Pi$ Rydberg state is homogeneously perturbed by the $B^2\Pi$ valence state. Similarly, the $F^2\Delta$ Rydberg state is perturbed by the $B'^2\Delta$ valence state.

unless the spin parts of ϕ_a and ϕ_b, respectively, match those of ϕ_c and ϕ_d). In the case where the two configurations differ by only one *orbital*, but still by two *spin-orbitals*, more complicated formulas apply[†]. This situation would occur for two Rydberg states of the same symmetry that belong to series converging to ion-core states of different multiplicities (see Section 5.2.4 for an example of $^3\Sigma_u^-$ Rydberg states belonging to series converging to $^2\Sigma_g^-$ and $^4\Sigma_g^-$ ion cores).

It will be shown in Section 5.2 that electrostatic perturbations occur frequently between states whose configurations differ by two orbitals, especially between Rydberg and valence states. An example from the NO spectrum is discussed here (Fig. 3.7). The $B'^2\Delta$ valence state of the NO molecule can be represented by the configuration $\sigma 2p\ \pi 2p^4\ \pi^* 2p^2$. The $F^2\Delta$ state belongs to a Rydberg series that converges to the ground state of the NO^+ ion and is represented by the $\sigma 2p^2\ \pi 2p^4\ 3d\delta$ configuration. The configurations of the $B'^2\Delta$ and $F^2\Delta$ states differ by two orbitals that have different spin functions. The electrostatic interaction is given by

$$H^e = \langle \sigma 2p\ \pi^* 2p |\ 1/r_{12} | 3d\delta\ \pi^* 2p \rangle .$$

[†] The added complication arises from the necessity to express the wavefunctions for states with multiple open subshells as properly symmetrized sums of Slater determinantal functions. H^e will then include off-diagonal matrix elements between several pairs of Slater determinants.

This integral has been evaluated *ab initio* and found equal to 300 cm^{-1} (Felenbok and Lefebvre-Brion, 1966). (The one-center part of this integral is approximately the nonzero atomic integral $\langle sp|\ 1/r_{12}|dp\rangle$.) This calculated value is in fair agreement with the "semiexperimental" value of 450 cm^{-1} found by a deperturbation procedure (Jungen, 1966). Note that this electrostatic interaction is responsible not only for perturbations between states of identical symmetry but also for predissociation (Section 7.8.1) and auto-ionization (Section 8.8).

Owing to numerical difficulties associated with minimizing the radial ($\partial/\partial R$ and $\partial^2/\partial R^2$) couplings, other ways to theoretically calculate diabatic states have been proposed (Yarkony, 2000; Köppel, *et al.*, 1984). Since, in the region of the avoided crossing, some characteristics of the wavefunctions are interchanged, several molecular properties will reflect this change. Consequently, diabatic states can be defined by requiring the smoothness of the R–dependence of a molecular property such as the dipole moment or electronic transition moment. The transformation matrix between the diabatic and adiabatic states can be obtained by diagonalizing the adiabatic dipole (Weiner and Meyer, 1981) or quadrupole (Li, *et al.*, 1997) moment matrix.

3.3.3 Adiabatic Curves

In the adiabatic model, the matrix element between the v_1 level of the first adiabatic curve $V_1^{ad}(R)$ and the v_2 level of the second adiabatic curve $V_2^{ad}(R)$ (see Fig. 3.5) is reduced to

$$H_{1,v_1;2,v_2} = \langle \Phi_1^{ad}\chi_{v_1}^{ad}|\mathbf{T}^N|\Phi_2^{ad}\chi_{v_2}^{ad}\rangle, \tag{3.3.9}$$

and

$$\langle \Phi_1^{ad}|\mathbf{H}^{\mathrm{el}}|\Phi_2^{ad}\rangle = 0.$$

$\mathbf{T}^N$ is the nuclear kinetic energy operator, which is expressed in the molecular frame as

$$\mathbf{T}^N = -\frac{\hbar^2}{2\mu R^2}\left\{\frac{\partial}{\partial R}\left[R^2\frac{\partial}{\partial R}\right] - \mathbf{R}^2\right\} = -\frac{\hbar^2}{2\mu}\left[\frac{\partial^2}{\partial R^2} + \frac{2}{R}\frac{\partial}{\partial R}\right] + \frac{\hbar^2}{2\mu R^2}\mathbf{R}^2.^{\dagger}$$

Let us ignore the $\mathbf{R}^2$ rotational part (**R**=**J**-**L**-**S**) of this operator, which leads to off-diagonal matrix elements that are proportional to $J(J+1)$ but still very small compared to the matrix elements of the remaining radial term (Leoni, 1972). The effect of the derivatives with respect to R on the electronic and vibrational wavefunctions, both of which depend on R, is given by

$$\frac{\partial^2(\Phi\chi)}{\partial R^2} = \chi\frac{\partial^2\Phi}{\partial R^2} + \Phi\frac{d^2\chi}{dR^2} + 2\frac{\partial\Phi}{\partial R}\frac{d\chi}{dR}$$

†Recall that R is the internuclear distance and **R** is the nuclear rotation angular momentum.

and

$$\frac{2}{R}\frac{\partial}{\partial R}\Phi\chi = \frac{2}{R}\chi\frac{\partial \Phi}{\partial R} + \frac{2}{R}\Phi\frac{d\chi}{dR}.$$

Combining this result with Eq. (3.3.9) yields

$$\begin{aligned} H_{1,v_1;2,v_2} = &-\frac{\hbar^2}{2\mu}\left\langle \chi_{v_1}^{ad}\left|\left\langle \Phi_1^{ad}\left|\frac{\partial^2}{\partial R^2} + \frac{2}{R}\frac{\partial}{\partial R}\right|\Phi_2^{ad}\right\rangle_r\right|\chi_{v_2}^{ad}\right\rangle_R \\ &-\frac{\hbar^2}{2\mu}\langle \Phi_1^{ad}|\Phi_2^{ad}\rangle_r \left\langle \chi_{v_1}^{ad}\left|\frac{d^2}{dR^2} + \frac{2}{R}\frac{d}{dR}\right|\chi_{v_2}^{ad}\right\rangle_R \\ &-\frac{\hbar^2}{\mu}\left\langle \chi_{v_1}^{ad}\left|\left\langle \Phi_1^{ad}\left|\frac{\partial}{\partial R}\right|\Phi_2^{ad}\right\rangle_r\right|\frac{d}{dR}\chi_{v_2}^{ad}\right\rangle_R. \end{aligned} \tag{3.3.10}$$

In Eq. (3.3.10), the second term is zero after integration over the electronic coordinates r, since Φ_1^{ad} and Φ_2^{ad} are two different solutions of the same equation and must therefore be orthogonal. The off-diagonal matrix elements in Eq. (3.3.10) are often called nonadiabatic corrections to the energies.

The vibrational wavefunction is often written as $\chi = \xi/R$, where ξ is normalized with respect to dR (as opposed to R^2dR as for χ). Then the derivative of the vibrational function with respect to R results in two terms. One of these terms exactly cancels the term in $(2/R)(\partial\Phi/\partial R)$ of Eq. (3.3.10), and the matrix element simplifies to

$$\begin{aligned} H_{1,v_1;2,v_2}(\text{cm}^{-1}) = &\frac{-16.8576}{\mu(\text{amu})}\left\langle \xi_{v_1}^{ad}\left|\left\langle \Phi_1^{ad}\left|\frac{\partial^2}{\partial R^2}\right|\Phi_2^{ad}\right\rangle_r (\text{Å}^{-2})\right|\xi_{v_2}^{ad}\right\rangle_R \\ &-\frac{33.7152}{\mu(\text{amu})}\left\langle \xi_{v_1}^{ad}\left|\left\langle \Phi_1^{ad}\left|\frac{\partial}{\partial R}\right|\Phi_2^{ad}\right\rangle_r (\text{Å}^{-1})\right|\frac{d}{dR}\xi_{v_2}^{ad}\right\rangle_R (\text{Å}^{-1}). \end{aligned} \tag{3.3.11}$$

In the case where an avoided crossing is being represented by adiabatic curves, a relation between electronic matrix elements for basis functions belonging to adiabatic versus diabatic curves can be derived easily (Bandrauk and Child, 1970; Oppenheimer, 1972), as shown below.

Adiabatic potential curves can be obtained by diagonalizing, at a grid of R–values, the configuration-interaction matrix. This matrix is constructed in a particular diabatic (single-configuration) basis. The off-diagonal configuration-interaction matrix element is the familiar diabatic coupling term, $H_{12}^e(R)$, which involves integration over electronic coordinates at a fixed value of internuclear distance [Eq. (3.1.10)]. The configuration-interaction secular equation is

$$\begin{vmatrix} E_1^d(R) - E & H_{12}^e(R) \\ H_{12}^e(R) & E_2^d(R) - E \end{vmatrix} = 0 \tag{3.3.12}$$

where $E_i^d(R)$ is defined by the fixed-R integral over electronic coordinates,

$$E_i^d(R) = \langle \Phi_i^d|\mathbf{H}^{\text{el}}|\Phi_i^d\rangle_r.$$

The resultant eigenstates are found to be linear combinations of diabatic electronic functions, for which the configuration interaction mixing coefficients are explicitly dependent on internuclear distance,

$$\begin{aligned} \Phi_1^{ad} &= \cos\theta(R)\Phi_1^d - \sin\theta(R)\Phi_2^d, \\ \Phi_2^{ad} &= \sin\theta(R)\Phi_1^d + \cos\theta(R)\Phi_2^d. \end{aligned} \tag{3.3.13}$$

These expressions imply that the functions Φ_i^{ad} are orthogonal. At the crossing point $R = R_C, \theta = \pi/4$.

The vertical energy separation between two interacting diabatic potentials can be assumed to vary linearly with R in the crossing region,

$$E_1^d(R) - E_2^d(R) = a(R - R_C).$$

Now, as the Φ_i^d functions are diabatic, the derivative with respect to R acts only on the coefficients of the linear combinations in Eq. (3.3.13), thus

$$\left\langle \Phi_1^{ad} \left| \frac{\partial}{\partial R} \right| \Phi_2^{ad} \right\rangle_r = \left[\sin^2\theta + \cos^2\theta\right] \frac{\partial\theta}{\partial R} = \frac{\partial\theta}{\partial R}.$$

By definition, the adiabatic functions Φ_1^{ad} and Φ_2^{ad} diagonalize the electronic Hamiltonian. Using Eq. (3.3.13), one finds ($H^e \equiv H_{12}^e$)

$$\begin{aligned} &\langle \Phi_1^{ad} | \mathbf{H}^{\text{el}} | \Phi_2^{ad} \rangle_r \\ &\qquad = +E_1^d \sin\theta\cos\theta - H^e \sin^2\theta + H^e \cos^2\theta - E_2^d \sin\theta\cos\theta = 0, \end{aligned}$$

from which the R–dependence of θ near R_C may be determined,

$$\frac{\sin\theta\cos\theta}{\cos^2\theta - \sin^2\theta} = \frac{1}{2}\tan 2\theta = \frac{H^e}{E_2^d - E_1^d} = -\frac{H^e}{a(R - R_C)}.$$

Thus,

$$\theta = \frac{1}{2}\tan^{-1}\left[\frac{-2H^e}{a(R - R_C)}\right] = \frac{1}{2}\cot^{-1}\left[\frac{a(R - R_C)}{-2H^e}\right],$$

and since

$$\frac{d}{dx}\cot^{-1}\left(\frac{x}{c}\right) = \frac{-c}{c^2 + x^2}$$

then

$$\left\langle \Phi_1^{ad} \left| \frac{\partial}{\partial R} \right| \Phi_2^{ad} \right\rangle_r = \frac{\partial\theta}{\partial R} = \frac{aH^e}{4(H^e)^2 + a^2(R - R_C)^2}.$$

Table 3.5: Comparison between Diabatic and Adiabatic Parameters

Interacting States	Molecule	H^e (cm^{-1})	Maximum value of $W^e(R)$ (Å^{-1})	Width (Å) FWHM of $W^e(R)$
$B^2\Sigma_u^+ \sim C^2\Sigma_u^+$	N_2^+[a]	10,000	3.0	0.34
$E, F^1\Sigma_g^+ \sim G, K^1\Sigma_g^+$	H_2[b]	$\sim$ 3,000	3.4	0.26
$G^1\Pi \sim I^1\Pi$	SiO[c]	$\sim$ 400	6.4	0.16
$C^3\Pi_u \sim C'^3\Pi_u$	N_2[d]	1,000	$\sim$ 20.4	0.05
$B^3\Sigma_u^- \sim B'^3\Sigma_u^-$	O_2[e]	4,000	21.6	0.06

[a]Roche and Lefebvre-Brion, 1975.
[b]Dressler, *et al.*, 1979.
[c]Lagerqvist and Renhorn, 1979 (semiexperimental value).
[d]Robbe, 1978.
[e]Yoshimine and Tanaka, 1978.

Defining $b = H^e/a$, then

$$\left\langle \Phi_1^{ad} \left| \frac{\partial}{\partial R} \right| \Phi_2^{ad} \right\rangle_r = \frac{b}{4b^2 + (R - R_C)^2} = W^e(R). \tag{3.3.14}$$

If the diabatic coupling matrix element, H^e, is R–independent, this $\partial/\partial R$ matrix element between two adiabatic states must have a Lorentzian R–dependence with a full width at half maximum (FWHM) of $4b$. Evidently, the adiabatic electronic matrix element $W^e(R)$ is not R–independent but is strongly peaked near R_C. Its maximum value occurs at $R = R_C$ and is equal to $1/4b = a/4H^e$. Thus, if the diabatic matrix element H^e is large, the maximum value of the electronic matrix element between adiabatic curves is small. This is the situation where it is convenient to work with deperturbed adiabatic curves. On the contrary, if H^e is small, it becomes more convenient to start from diabatic curves. Table 3.5 compares the values of diabatic and adiabatic parameters. The deviation from the relation, $W^e(R)_{\max} \times \text{FWHM} = 1$, is due to a slight dependence of H^e on R and a nonlinear variation of the energy difference between diabatic potentials. When $W^e(R)$ is a relatively broad curve without a prominent maximum, the adiabatic approach is more convenient. When $W^e(R)$ is sharply peaked, the diabatic picture is preferable. The first two cases in Table 3.5 would be more convenient to treat from an adiabatic point of view. The description of the last two cases would be simplest in terms of diabatic curves. The third case is intermediate between the two extreme cases and will be examined later (see Table 3.6).

To obtain the second adiabatic electronic matrix element of Eq. (3.3.11), the ket, $(\partial/\partial R)\left|\Phi_2^{ad}\right\rangle$, is expanded using the complete set, $\left\{\Phi_j^{ad}\right\}$, of adiabatic functions

$$\begin{aligned}
\frac{\partial}{\partial R}\left|\Phi_2^{ad}\right\rangle &= \sum_i \left|\Phi_i^{ad}\right\rangle \left\langle \Phi_i^{ad}\left|\frac{\partial}{\partial R}\right|\Phi_2^{ad}\right\rangle \\
\frac{\partial^2}{\partial R^2}\left|\Phi_2^{ad}\right\rangle &= \frac{\partial}{\partial R}\frac{\partial}{\partial R}\left|\Phi_2^{ad}\right\rangle = \sum_i \left\{\frac{\partial}{\partial R}\left|\Phi_i^{ad}\right\rangle\right\} \left\langle \Phi_i^{ad}\left|\frac{\partial}{\partial R}\right|\Phi_2^{ad}\right\rangle \\
&\quad + \sum_i \left|\Phi_i^{ad}\right\rangle \frac{\partial}{\partial R}\left\langle \Phi_i^{ad}\left|\frac{\partial}{\partial R}\right|\Phi_2^{ad}\right\rangle \\
\left\langle \Phi_1^{ad}\left|\frac{\partial^2}{\partial R^2}\right|\Phi_2^{ad}\right\rangle &= \sum_{i\neq 1,2}\left[\left\langle \Phi_1^{ad}\left|\frac{\partial}{\partial R}\right|\Phi_i^{ad}\right\rangle \left\langle \Phi_i^{ad}\left|\frac{\partial}{\partial R}\right|\Phi_2^{ad}\right\rangle\right. \\
&\quad \left. + \frac{\partial}{\partial R}\left\langle \Phi_1^{ad}\left|\frac{\partial}{\partial R}\right|\Phi_2^{ad}\right\rangle\right].
\end{aligned}$$

The second summation reduces to a single term because the adiabatic functions are orthonormal (Hobey and McLachlan, 1960). In the simple case where only two electronic states interact (Eq. (3.3.13)), one can assume that the matrix elements of $\partial/\partial R$ connecting either of these two states with other states are negligible and, from Eq. (3.3.14),

$$\left\langle \Phi_1^{ad}\left|\frac{\partial^2}{\partial R^2}\right|\Phi_2^{ad}\right\rangle_r = \frac{\partial}{\partial R}\left\langle \Phi_1^{ad}\left|\frac{\partial}{\partial R}\right|\Phi_2^{ad}\right\rangle_r = \frac{\partial}{\partial R}W^e(R).$$

This calculated matrix element of $\partial^2/\partial R^2$, acting on the electronic wavefunctions for the E, F and G, K states of H_2 (Fig. 3.8), is displayed in Fig. 3.9 and is seen not to deviate appreciably from the derivative of a Lorentzian curve. Its contribution to the $H_{1,v_m;2,v_n}$ vibronic matrix element [Eq. (3.3.11)] is generally smaller than the contribution due to the $\partial/\partial R$ operator acting on the electronic functions, but it is in no case negligible.

3.3.4 Choice between the Diabatic and Adiabatic Models

If the approximate deperturbed potential curves cross, they are diabatic curves. One can assume an interaction matrix element given by Eq. (3.3.5) and carry out a complete deperturbation.

The choice of an adiabatic picture leads to difficulties when one of the potentials has a double minimum (see Fig. 3.5). The vibrational level separations of such a curve do not vary smoothly with vibrational quantum number, as do the levels of a single minimum potential. In the separate potential wells (below the barrier), the levels approximately follow two different smooth curves. However, above the potential barrier the separation between consecutive energy levels oscillates. The same pattern of behavior is found for the rotational constants below and above the potential barrier. In addition, the rotational levels above the barrier do not vary as $B_vJ(J+1)$. An adiabatic deperturbation of the (E,F+G,K) $^1\Sigma_g^+$ states of H_2 has been possible (Dressler *et al.*, 1979) only because the adiabatic curves were known from very precise *ab initio* calculations.

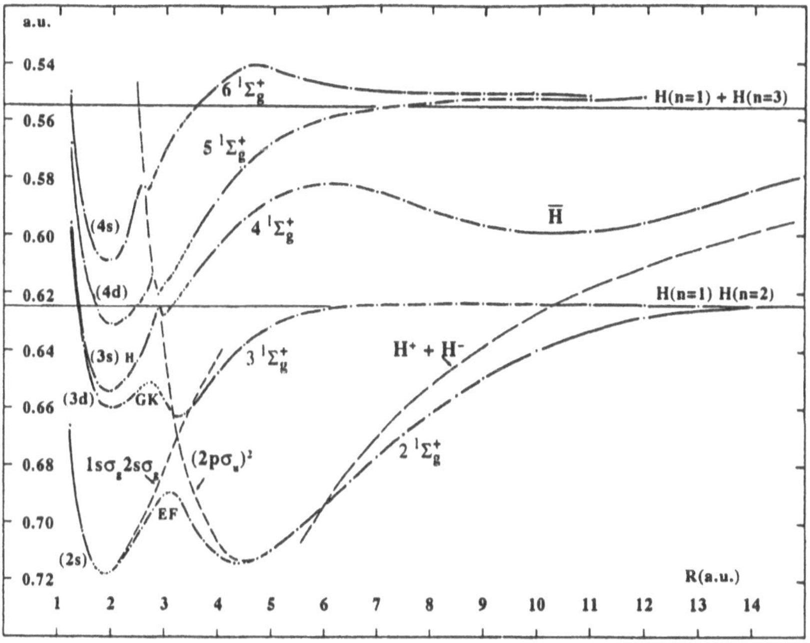

Figure 3.8: *Ab initio* adiabatic potential curves with double minima for the $^1\Sigma_g^+$ states of H_2 (dash-dot line). The diabatic $1s\sigma_g 2s\sigma_g$ and $(2p\sigma_u)^2$ potentials are also plotted (dashed line). The (nl) labels on the small-R potential mimima denote the dominant $1s\sigma_g\ nl\sigma_g$ configuration (Wolniewicz and Dressler, 1977). Note that more accurate results on the 4 $^1\Sigma_g^+$ H, $\overline{\text{H}}$ state have been presented by Wolniewicz and Dressler (1979) and by Dressler and Wolniewicz (1981).

If the approximate deperturbed curves do not cross or have similar spectroscopic constants, the most convenient starting point is an adiabatic approach. Two situations must be considered:

1. *The adiabatic curves result from an apparently avoided crossing.* This means that the diabatic curves belong to very different electronic configurations. The coupling matrix element, $W^e(R)$, can be assumed to have a Lorentzian shape [Eq. (3.3.14)]. This is the situation for the G and I states of SiO.

2. *The adiabatic curves correspond to configurations that differ by only one orbital.* Rydberg states belonging to series which converge to the same state of the ion fall into this category. Then the matrix element of $\partial/\partial R$ is generally R–independent (or linear in R).[†] One example is the perturbations observed between identical-symmetry Rydberg states of H_2 that converge to the same state of H_2^+ (the ground state) but to different vibrational levels of this state (Herzberg and Jungen, 1972). As the density of electronic levels increases near

[†] If the unique orbital changes its character with R (e.g., "Rydbergization", Section 5.2.2), the coupling matrix element behaves similarly to case (1).

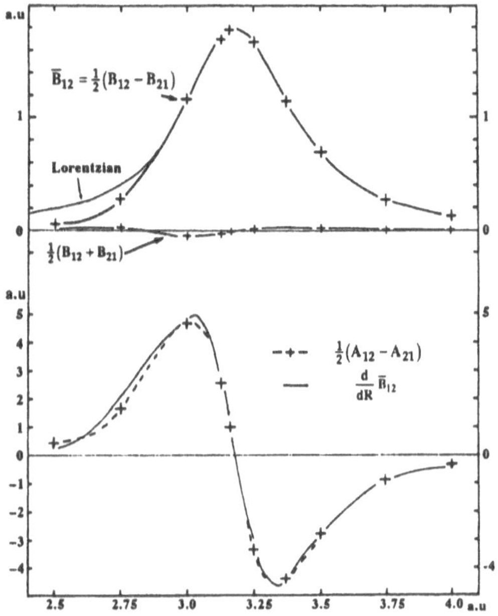

Figure 3.9: *Ab initio* $\partial/\partial R$ and $\partial^2/\partial R^2$ matrix elements between the E, F and G, K adiabatic states of H_2: $B_{12} \equiv \left\langle \Phi_{1ad} \middle| \frac{\partial}{\partial R} \middle| \Phi_{2ad} \right\rangle_r$, $A_{12} \equiv \left\langle \Phi_{1ad} \middle| \frac{\partial^2}{\partial R^2} \middle| \Phi_{2ad} \right\rangle_r$ where Φ_1 and Φ_2 are the adiabatic electronic wavefunctions for the E, F and G, K double-minimum states, respectively. Except for the smallest R values, $\bar{B}_{12}(R)$ is Lorentzian. The relationships $B_{12} = -B_{21}$ and $\frac{1}{2}(A_{12} - A_{21}) = \frac{d}{dR}\bar{B}_{12}$ are not satisfied exactly because Φ_{1ad} and Φ_{2ad} are, in these calculations, not exactly orthogonal (Dressler, *et al.*, 1979).

the ionization limit, the $v = 1$ level of the ${}^1\Pi_u$ $n = 11$ Rydberg state is nearly degenerate with $v = 2$ of the ${}^1\Pi_u$ $n = 6$ Rydberg state (Fig. 3.10). The electronic factor is found to be nearly independent of R, $\langle \Phi_1^{ad} | \partial/\partial R | \Phi_2^{ad} \rangle_r$ = constant. The vibrational wavefunctions of the two interacting states belong to virtually identical potentials; thus the vibrational factor is zero except for $\Delta v = 1$, which, in the harmonic approximation, is proportional to $(v+1)^{1/2}$,

$$\left\langle \chi_{v_1} \middle| \frac{\partial}{\partial R} \middle| \chi_{v_2} \right\rangle (\text{Å}^{-1}) = \left(\frac{\mu(\text{amu})\omega\,(\text{cm}^{-1})}{67.4304} \right)^{1/2} (v_1 + 1)^{1/2}, \qquad (3.3.15)$$

with $v_2 = v_1 + 1$ Another approach to the treatment of these Rydberg~Rydberg interactions is given in Section 8.6, using, as zero-order electronic wavefunctions, those of a single electron in the field of the ion core (Jungen and Atabek, 1977).

The vibrational eigenstates of neither the diabatic nor the adiabatic potential curve exactly represent the observed levels. Interaction matrix elements between these zero-order levels (eigenstates of either diabatic or adiabatic potentials)

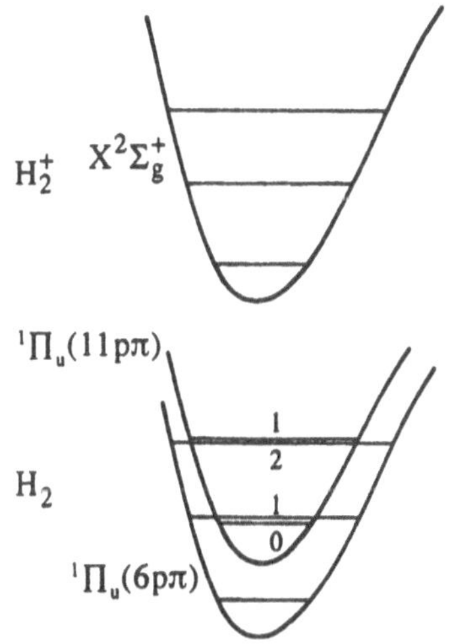

Figure 3.10: Perturbations between the $6p\pi$ and $11p\pi$ $^1\Pi_u$ Rydberg states of H_2. Two Rydberg series converging to different vibrational levels of the H_2^+ $X^2\Sigma_g^+$ state interact via nonzero $\Delta v = \pm 1$ vibrational matrix elements of the $\partial/\partial R$ operator.

must be added in order to reproduce the observed levels. Table 3.6 summarizes the results of two types of deperturbation approaches, diabatic and adiabatic, to the same problem, a pair of interacting $^1\Pi$ states of SiO.

Recall that the purpose of a deperturbation calculation is to obtain a model, consisting of a pair of deperturbed potentials and an interaction matrix element (either R–dependent or R–independent), that exactly reproduces the observed rotation-vibration energy levels. Whether this model is diabatic or adiabatic has no effect on the quality of the agreement between observed and calculated levels. Where the two approaches differ is in the complexity of the calculation (size of matrix to be diagonalized, number of fitting iterations required, etc.) and in the magnitudes of the differences between observed and zero-order deperturbed levels (the eigenvalues of the deperturbed potentials). Table 3.6 displays the differences (obs - dep) between the observed levels (far right column), and the *zero-order* diabatic (second column) and *zero-order* adiabatic (fourth column) levels. The energies of the zero-order adiabatic levels are closer to those of the observed levels; thus the adiabatic picture is a better starting point for an iterative deperturbation calculation.

The observed levels in Table 3.6 may be obtained from the diabatic potentials represented by the $T_e, \omega_e, \omega_e x_e$, and R_e constants, which generate the deperturbed levels via $T_v = T_e + \omega_e\left(v+\frac{1}{2}\right) - \omega_e x_e\left(v+\frac{1}{2}\right)^2$; an R–independent electronic matrix element, $H^e = 365$ cm^{-1}; and vibrational overlap factors calculated using the vibrational eigenfunctions of the deperturbed diabatic potentials. Similarly, the observed levels may be computed from the adiabatic potentials, a Lorentzian interaction term [Eq. (3.3.14)] $W^e(R)$ with $b = 0.1014$

Table 3.6: Comparison between Two Types of Deperturbation of the G and I $^1\Pi$ States of SiO (in cm^{-1})

	Diabatic deperturbation		Adiabatic deperturbation		
State	Diabatic levels	obs - dep	Adiabatic levels	obs - dep	Observed levels[a]
G 0	70180	− 78	70099	3	70102
G 1	71056	− 91	70966	− 1	70965
G 2	71919	− 87	71819	13	71832
G 3	72769	−150	72658	−39	72619
G 4	73606	−163	73482	−39	73443
G 5	74431	−177	74292	−38	74254
G 6	75243	−192	75087	−36	75051
G 7	76042	−194	75866	−18	75848
I 0	71689	89	71788	−10	71778
I 1	72564	151	72675	40	72715
I 2	73423	170	73546	47	73593
I 3	74264	200	74401	63	74464
I 4	75089	236	75241	84	75325
I 5	75896	276	76066	106	76172

Spectroscopic constants for deperturbed curves (cm^{-1})[b]

Diabatic curves	T_e	ω_e	$\omega_e x_e$	R_e(Å)
G	69,734.7	890.024	6.57	1.6139
I	71,245.0	892.014	8.66	1.6548

Interaction matrix element: $H^e = 365\ cm^{-1}$

Adiabatic curves	T_e	ω_e[c]	$\omega_e x_e$[c]	R_e (Å)
G	69,633.0	881.07	7.02	1.6155
I	71,334.0	902.84	7.97	1.6526

Interaction matrix element:[d] $b = 0.1014$ Å
$R_C = 1.915$ Å

[a]Lagerqvist and Renhorn, (1979)

[b]Lefebvre-Brion (unpublished calculations, 1981).

[c]Approximate values, since the adiabatic levels are obtained by direct numerical integration of the adiabatic curves constructed from the diabatic curves and H^e.

[d]The adiabatic interaction parameter, $W^e(R)$, is defined by Eq. (3.3.14) in terms of b and R_C.

Å, $R_C = 1.915$ Å, and vibrational factors calculated using eigenfunctions of the deperturbed adiabatic potentials.

Finally, the eigenfunctions for the observed levels are obtained, in either representation, by diagonalizing the complete interaction matrix consisting of all vibrational levels of the two potentials in the diabatic picture,

$$\Psi = \sum_{v_1} a_{v_1} \Phi_1^d \chi_{v_1}^d + \sum_{v_2} b_{v_2} \Phi_2^d \chi_{v_2}^d, \tag{3.3.16}$$

or, in the adiabatic picture,

$$\Psi = \sum_{v_1} c_{v_1} \Phi_1^{ad} \chi_{v_1}^{ad} + \sum_{v_2} d_{v_2} \Phi_2^{ad} \chi_{v_2}^{ad}. \tag{3.3.17}$$

Equations (3.3.16 and 3.3.17) suggest one final indicator of whether the diabatic or adiabatic approach is preferable. The better approach is the one for which the sum over deperturbed functions involves fewer terms, especially if one term is dominant (for example, $a_i \geq 2^{-1/2}$). An adiabaticity parameter

$$\gamma \equiv H^e / \Delta G^{ad}$$

has been introduced by Dressler (1983 and 1985), where ΔG^{ad} is the vibrational frequency of the higher-energy member of the pair of adiabatic electronic states. Near adiabatic behavior occurs for $\gamma \gg 1$; near diabatic behavior occurs for $\gamma \ll 1$. For $\gamma \simeq 1.0$, which corresponds to the case for the SiO G and I $^1\Pi$ states, the mixing of the vibrational basis functions is large in both diabatic and adiabatic descriptions.

3.3.5 Electromagnetic Field-Dressed Diabatic and Adiabatic Potential Energy Curves

Most of the experiments discussed in this book are performed using continuous wave or nanosecond-pulsed lasers at $I < 10^9$ W/cm^2 intensities, which correspond to electromagnetic radiation electric field strengths of $\mathcal{E} < 9 \times 10^5$ V/cm. Such electric fields are weak relative to typical intramolecular field strengths, $\mathcal{E}^0 = 1 a.u. = 5.14 \times 10^9$ V/cm and are appropriately treated perturbatively (see Sections 6.1.2 and 6.5.3).

When experiments are performed in the intense $I > 10^{11}$ W/cm^2 electromagnetic fields typically provided by picosecond or femtosecond pulsed lasers, it is necessary to solve the time-dependent Schrödinger equation in a non-perturbative approach (see Section 9.1.2). The total Hamiltonian is

$$\mathbf{H} = \mathbf{H}^\circ + \vec{\mathcal{E}}(t) \cdot \left(\sum_i e\vec{r}_i \right) \tag{3.3.18}$$

where $\mathbf{H}^\circ$ is the field-free Hamiltonian of Eq. (3.1.1) and

$$\vec{\mathcal{E}}(t) = \hat{\mathbf{k}} \mathcal{E}^0 \cos \omega t \tag{3.3.19}$$

is the electric field strength of an electromagnetic radiation field that is linearly polarized along the laboratory Z–axis (unit vector $\vec{\mathbf{k}}$). If $\mathcal{E}^0 = 1$ a.u., then $I = 3.51 \times 10^{16}$ W/cm^2 (a 10mJ, 10fs pulse focused to 100μ diameter gives $I = 1.3 \times 10^{16}$ W/cm^2).

It is convenient to treat intense electromagnetic field problems in the *dressed molecular states picture* (see review by Giusti-Suzor, *et al.*, (1995)). This picture allows one to think about intense field problems in a framework that closely resembles the weak field, diabatic or adiabatic states picture that is the primary focus of this book. In the dressed states picture the photon energy is added to, or subtracted from, the field-free potential energy curves. One obtains field-dressed potential curves.

The electromagnetic radiation field is taken into account by adding the energy of the photons to the various molecular potential curves, $V_i(R)$. If the photon number is initially N, when n photons are absorbed, the remaining number of photons is $N - n$. The resultant field-dressed diabatic state has potential energy

$$V_{i,n} = V_i(R) + (N - n)\hbar\omega. \tag{3.3.20}$$

Each field free potential curve, $V_i(R)$, generates a family of field-dressed potentials, $V_{i,n}(R)$, with the result that many intersections between potential curves occur that are not present in the field-free case. The spectroscopic and dynamical consequences of these field-dressed curve crossings are understood using exactly the same methods presented in this book for field-free intersecting potentials.

Figures 3.11 and 3.12 illustrate the weak-field and field-dressed pictures of photodissociation of the H_2^+ molecular ion. In Fig. 3.11, one-, two-, and three-photon transitions from $v = 2$ of the $X^2\Sigma_g^+(1s\sigma_g)$ bound state to the $^2\Sigma_u^+(2p\sigma_u)$ continuum are shown. The 1- and 3-photon transitions are electronically $u \leftarrow g$ allowed, but, in a perturbative picture, the 1-photon transition has a vastly smaller Franck-Condon factor than the 3-photon transition. The 2-photon transition to the $^2\Sigma_u^+$ state is ($g \nleftrightarrow u$)forbidden, but the $^2\Sigma_u^+$ state acts as a "virtual intermediate state" that permits excitation of the continuum of the $X^2\Sigma_g^+$ state.

Figure 3.12 shows that the $X^2\Sigma_g^+$ $n = 0$ curve is predissociated radiatively by the dressed $^2\Sigma_u^+$ $n = 1$ and $n = 3$ curves. The coupling between the crossing, field-dressed, diabatic states (solid curves in Fig. 3.12) is the radiative interaction,

$$(\mathcal{E}^0/2)\hat{\mathbf{k}} \cdot \mu(R_c),$$

where $\mu(R_c)$ is the electronic transition moment evaluated at the crossing point, R_c. If this 2-state radiative interaction is diagonalized, one obtains non-crossing, field-dressed adiabatic potential curves (the dotted and the dashed curves of Fig. 3.12). In the adiabatic picture, the coupling between the vibrational states of the two curves is the non-adiabatic (nuclear kinetic energy) coupling due to the R–dependence of the electronic wavefunctions near R_c.

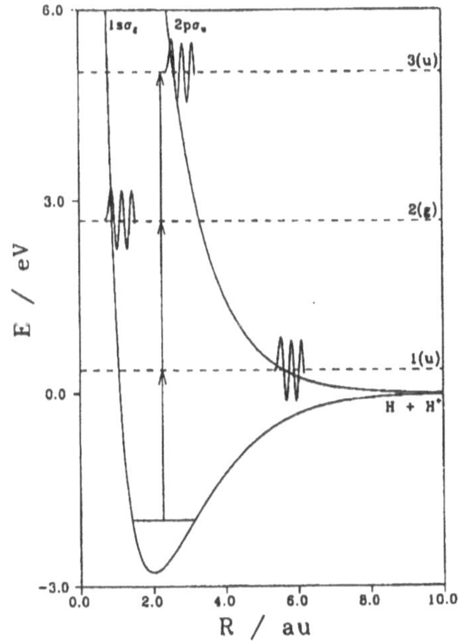

Figure 3.11: Potential energy curves for the $1s\sigma_g$ and $2p\sigma_u$ states of H_2^+ and the continuum nuclear wavefunctions (wavepackets) initially produced after absorption of 1, 2, or 3 532nm (2.33eV) wavelength photons from the $v_0 = 2$ level of the $X^2\Sigma_g^+$ electronic ground state (from Giusti-Suzor, *et al.*, 1995).

As the intensity of the radiation field increases, the radiative interaction deforms the field-dressed adiabatic potentials. For example, if the laser intensity increases from 1×10^{13} W/cm^2 to 4×10^{13} W/cm^2, the $v = 3$ level of the $X^2\Sigma_g^+$, $n = 0$ curve becomes unbound because the adiabatic barrier to dissociation is lowered. This photodissociation mechanism is called *bond-softening* (Bucksbaum, *et al.*, (1990)). As the laser intensity increases, the field-dressed adiabatic curves become progressively decoupled (the electronic wavefunction become more weakly R-dependent) and the molecule can become trapped in a minimum on the upper adiabatic potential. In Fig. 3.12, the $v = 12$ level of the $X^2\Sigma_g^+$, $n = 0$ state can get trapped in the minimum of the upper adiabatic $g, n = 0 \sim u,\ n = 1$ potential. This is called *population trapping* (Giusti-Suzor and Mies, 1992; Zavriyev, *et al.*, 1993; Yao and Chu, 1993; Aubanel, *et al.*, 1993). Finally, consider the $v = 2$ level of the $g, n = 0$ diabatic potential. The energy of this level lies above the $u, n = 1$ separated atom asymptote and this state can predissociate by tunneling through the $g, n = 0 \sim u,\ n = 1$ adiabatic barrier. As the radiation field strength increases, this barrier gets lower and narrower. An alternative tunneling path for $v = 2$ photofragmentation is via the $u, n = 3$ curve. This corresponds to 3-photon absorption. Thus, as for bound-bound multiphoton excitation transitions, multiphoton transitions in the continuum are also possible. This strong-field phenomenon is known as *Above Threshold Dissociation* (ATD) (Giusti-Suzor, *et al.*, 1990; Bucksbaum, *et al.*, 1990). For a review, see Posthumus and McCann, (2001).

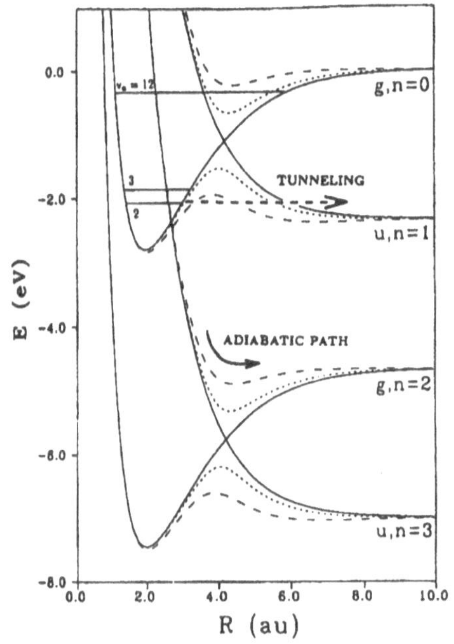

Figure 3.12: Field-dressed potential energy curves for H_2^+ interacting with a 532nm laser field. The field-dressed diabatic curves are shown as full lines. The field-dressed adiabatic curves, shown as dotted and dashed curves, correspond respectively to laser intensities of 1×10^{13} W/cm^2 and 4×10^{13} W/cm^2 (from Giusti-Suzor, *et al.*, 1995).

3.4 Spin Part of the Hamiltonian

As in atoms, relativistic terms due to the interaction between the spin and orbital angular momenta of nuclei and electrons of the molecule must be added to the electronic Hamiltonian. There is also a magnetic interaction energy created by the orbital motion of the electrons and the rotational motion of the electrically charged nuclei. The relativistic effects consist mainly of three parts:

1. Interaction between the spin and orbital angular momenta of the electrons, $\mathbf{H}^{SO}$ = spin-orbit operator.

2. Interaction between the electron spin and the rotational angular momenta of the nuclei, $\mathbf{H}^{SR}$ = spin-rotation operator.

3. Interaction between the spins of different electrons, $\mathbf{H}^{\mathrm{SS}}$ = spin-spin operator.

These interactions remove the spin degeneracy of the levels of an electronic state and give rise to the multiplet splitting or the so-called zero-field splitting (i.e., the splitting that appears even in the absence of any external magnetic field). They are also responsible for mixing states of the same or different multiplicities, thus providing a mechanism for observation of nominally forbidden transitions (see Section 6.4), homogeneous ($\Delta\Omega = 0, \Delta S = 0, \pm 1$) perturbations (see Section 3.4.2), predissociations (see Sections 7.8.2 and 7.11.1), and autoionizations (see Section 8.7).

3.4.1 The Spin-Orbit Operator

The form of the spin-orbit Hamiltonian has been given by Van Vleck (1951) and is an extension to diatomic molecules of the solution for the relativistic equation originally derived for a two-electron atom:

$$\begin{aligned}\mathbf{H}^{\mathrm{SO}} = \frac{g_S \mu_{\mathrm{B}}}{c} \sum_i \bigg\{ & \frac{Z_{\mathrm{A}} e}{r_{i\mathrm{A}}^3} [(\mathbf{r}_i - \mathbf{r}_{\mathrm{A}}) \times \mathbf{v}_i/2] \cdot \mathbf{s}_i \\ & + \frac{Z_{\mathrm{B}} e}{r_{i\mathrm{B}}^3} [(\mathbf{r}_i - \mathbf{r}_{\mathrm{B}}) \times \mathbf{v}_i/2] \cdot \mathbf{s}_i \bigg\} \quad \text{direct spin-orbit interaction} \\ & - \frac{g_S \mu_{\mathrm{B}}}{c} \sum_{i \neq j} \frac{e}{r_{ij}^3} \left[(\mathbf{r}_i - \mathbf{r}_j) \times \left(\frac{1}{2}\mathbf{v}_i - \mathbf{v}_j \right) \right] \cdot \mathbf{s}_i \quad \text{spin-other-orbit interaction}\end{aligned} \tag{3.4.1}$$

where $\mathbf{v}_i$ is the velocity of the ith electron measured in the molecule fixed coordinate frame, g_S is the Landé electronic factor, which is approximately 2, and μ_B is the Bohr magneton, $e\hbar/2mc$. If $\mathbf{v}_i$ is replaced by $\mathbf{p}_i/m$ where $\mathbf{p}_i$ is the momentum of the electron measured in the molecule fixed coordinate frame and if the angular momentum operator $\mathbf{l}_i$ is introduced in the form

$$\mathbf{l}_{i\mathrm{A}} = (1/\hbar)(\mathbf{r}_{i\mathrm{A}} \times \mathbf{p}_i), \qquad \mathbf{l}_{i\mathrm{B}} = (1/\hbar)(\mathbf{r}_{i\mathrm{B}} \times \mathbf{p}_i)$$

then the expression for the spin-orbit Hamiltonian becomes

$$\mathbf{H}^{\mathrm{SO}} = \frac{\alpha^2}{2} \sum_i \left\{ \frac{Z_{\mathrm{A}}}{r_{i\mathrm{A}}^3} \mathbf{l}_{i\mathrm{A}} \cdot \mathbf{s}_i + \frac{Z_{\mathrm{B}}}{r_{i\mathrm{B}}^3} \mathbf{l}_{i\mathrm{B}} \cdot \mathbf{s}_i \right\} - \frac{\alpha^2}{2} \sum_{i \neq j} \frac{1}{r_{ij}^3} (\mathbf{r}_{ij} \times \mathbf{p}_i)(\mathbf{s}_i + 2\mathbf{s}_j) \tag{3.4.2}$$

where α is the fine-structure constant, $\alpha = e^2/\hbar c = 1/137.036$. If the matrix elements of the operators in the Eq. (3.4.2) summations are evaluated in atomic units, the final result must be multiplied by $109737.3\alpha^2 = 5.8436$ in order to obtain a value in cm^{-1}.

The first part of the Eq. (3.4.2) form of $\mathbf{H}^{\mathrm{SO}}$ represents the spin-orbit coupling of each electron in the field of the two bare nuclei with charges Z_{A} and Z_{B}. It is a single-electron operator.

The second part, the spin-other-orbit interaction, is due to interelectronic interactions and has the effect of partially counterbalancing the field of the bare nuclei. Its sign is opposite to that of the first part. This operator is a two-electron operator because of the r_{ij}^{-3} term.

It can be shown (Veseth, 1970) that all electron-nuclear distances, $r_{i\mathrm{K}}$, can be referred to a common origin, and, neglecting only the contribution of spin-other-orbit interactions between unpaired electrons, the two-electron part of the spin-orbit Hamiltonian can be incorporated into the first one-electron part as a screening effect. The spin-orbit Hamiltonian of Eq. (3.4.2) can then be written as

$$\mathbf{H}^{\mathrm{SO}} = \sum_i \hat{a}_i \mathbf{l}_i \cdot \mathbf{s}_i \qquad \text{with} \quad \hat{a}_i \mathbf{l}_i = \sum_{\mathrm{K}} \frac{\alpha^2}{2} \frac{Z_{eff,\mathrm{K}}}{r_{i\mathrm{K}}^3} \mathbf{l}_{i\mathrm{K}} \tag{3.4.3}$$

where $\mathbf{l}_{i\mathrm{K}}$ is the orbital angular momentum of electron i about nucleus K and $Z_{eff,\mathrm{K}}$, the effective charge of the K'th nucleus, is less than $\mathrm{Z_K}$ because the spin-other-orbit part has the effect of screening the nuclear charge by typically 20-50%. [See the atomic calculations of Blume and Watson (1962, 1963) and Blume *et al.*, (1964).] Note that $\hat{a}$ is an operator that acts only on the radial part of the wavefunction.

In the form of Eq. (3.4.3), $\mathbf{H}^{\mathrm{SO}}$ is a single-electron operator. It is prudent to avoid discussion of further simplified effective forms of this operator, such as $A\mathbf{L}\cdot\mathbf{S}$, because these forms of $\mathbf{H}^{\mathrm{SO}}$ have led to many errors in the literature. The selection rules for the microscopic $\mathbf{H}^{\mathrm{SO}}$ operator are based on its symmetry with respect to the operations of the point group of the diatomic molecule, $C_{\infty v}$(heteronuclear) or $D_{\infty h}$(homonuclear). The $\hat{a}$ factor of $\mathbf{H}^{\mathrm{SO}}$ is invariant under all symmetry operations. The $\mathbf{l}_i \cdot \mathbf{s}_i$ term may be expanded,

$$\mathbf{l}_i \cdot \mathbf{s}_i = \mathbf{l}_{iz} \cdot \mathbf{s}_{iz} + \frac{1}{2}\left(\mathbf{l}_i^+ \mathbf{s}_i^- + \mathbf{l}_i^- \mathbf{s}_i^+\right), \qquad \text{where } \mathbf{l}_{iz} = \frac{\hbar}{i}\left(x\frac{\partial}{\partial y} - y\frac{\partial}{\partial x}\right) = \frac{\hbar}{i}\frac{\partial}{\partial \phi} \tag{3.4.4}$$

and, as has been shown in Section 3.2.2, the $\partial/\partial\phi$ operator changes its sign (as does ϕ) upon reflection through *any* plane containing the two nuclei. Consequently, as expected, for $\Sigma^{\pm}$ states the spin-orbit operator must have zero diagonal matrix elements but, less obviously, nonzero off-diagonal matrix elements between $\Sigma^{\pm}$ and $\Sigma^{\mp}$ states. Ignorance of this $\Sigma^+ \sim \Sigma^-$ spin-orbit matrix element selection rule has led to many errors in the literature.

The selection rules for matrix elements of $\mathbf{H}^{\mathrm{SO}}$ are summarized as follows (Kayama and Baird, 1967), see also Eq. (3.4.46):

$$\begin{gathered} \Delta J = 0 \quad \Delta\Omega = 0 \quad g \leftarrow\backslash\rightarrow u \quad e \leftarrow\backslash\rightarrow f \quad \Sigma^+ \leftrightarrow \Sigma^- \\ \Delta S = 0 \;\text{ or }\; \Delta S = \pm 1 \quad \Delta\Lambda = \Delta\Sigma = 0 \;\text{ or }\; \Delta\Lambda = -\Delta\Sigma = \pm 1. \end{gathered} \tag{3.4.5}$$

In the single-configuration limit, if the two interacting states belong to the same configuration, then $\Delta\Lambda = \Delta\Sigma = 0$ or, if the two states differ by at most one spin-orbital, then $\Delta\Lambda = -\Delta\Sigma = \pm 1$. One example where the single-configuration approximation is not sufficient will be analyzed later (Section 5.7). If the more complex form of $\mathbf{H}^{SO}$ [Eq. (3.4.2)] is used, the selection rules are identical but, because the spin-other-orbit interaction is a two-electron operator, additional nonzero matrix elements between configurations that differ by two spin-orbitals must be considered. Also, the spin-other-orbit interaction will contribute significantly to diagonal spin-orbit constants in cases where the direct spin-orbit interaction is zero [e.g., $^2\Delta$ states arising from a $\sigma\pi^2$ configuration (Lefebvre-Brion and Bessis, 1969), $^3\Delta_g$ derived from a $\pi_u^2\pi_g^2$ configuration (Veseth, 1973b)].

For diagonal matrix elements (diagonal in all quantum numbers including S), we can define A, the spin-orbit coupling constant as

$$\langle \Lambda, \Sigma, S, \Omega, v \left| \mathbf{H}^{SO} \right| \Lambda, \Sigma, S, \Omega, v \rangle = A_{\Lambda,v} \Lambda \Sigma. \tag{3.4.6}$$

This means that fine-structure levels are expected to be equally spaced (by $A\Lambda$); however, second-order spin-orbit effects can distort the equidistant fine-structure pattern.

For $\Lambda > 0$ states, second-order spin-orbit effects cause two types of J-independent level shifts: proportional to $\Lambda\Sigma$ (the form of the diagonal spin-orbit matrix element) or to $3\Sigma^2 - S(S+1)$ (the form of the diagonal spin-spin matrix element). Second-order spin-orbit effects are unobservable for $^1\Sigma$ and $^2\Sigma$ states because they result in no new level splittings or changes in existing separations. For Σ states with $S > \frac{1}{2}$, second-order spin-orbit effects are observable only as shifts similar to the form of the spin-spin interaction. See Levy (1973) for a demonstration of this fact.

3.4.2 Expression of Spin-Orbit Matrix Elements in Terms of One-Electron Molecular Spin-Orbit Parameters

Using Eq. (3.4.3) for the spin-orbit operator and a single-configuration representation for the electronic states, and Section 3.2.4 for the expressions of the relevant matrix elements, it is possible to relate observable spin-orbit matrix elements to one-electron orbital integrals, which are called molecular spin-orbit parameters.

3.4.2.1 Matrix Elements of the $\mathbf{l}_{zi} \cdot \mathbf{s}_{zi}$ Term

When two states belong to the same values of the Λ and Σ quantum numbers, the only part of $\mathbf{H}^{SO}$ [Eq. (3.4.4)] by which they can interact is $\mathbf{l}_{zi} \cdot \mathbf{s}_{zi}$. The selection rules are $\Delta\Lambda = \Delta\Sigma = 0$ and $\Delta S = 0$ or ±1.

If the two interacting states have the same value of S and $\Lambda \neq 0$, then the matrix element has the form of the diagonal matrix element [Eq. (3.4.6)]. From the effect of the $\mathbf{s}_z$ operator, which results in matrix elements with opposite signs for α and β spins, the effect of $\Sigma_i \mathbf{l}_{zi} \cdot \mathbf{s}_{zi}$ on closed shells is identically zero by inspection. Only open shells give a nonzero contribution; therefore the summation over i in Eq. (3.4.3) need only include open shells.

3.4.2.1.1 Diagonal Matrix Elements

Four examples are given for Π states: $\pi^1\ {}^2\Pi, \pi^3\ {}^2\Pi, \sigma\pi\ {}^3\Pi$, and $\delta\sigma\pi\ {}^4\Pi$.

$^2\Pi$ States

Case of π^1 *Configuration*

For the substate $^2\Pi_{3/2}$, the wavefunction is

$$\left|\Lambda = 1, \Sigma = \frac{1}{2}, \Omega = \frac{3}{2}\right\rangle = |\pi^+\alpha\rangle .$$

For the substate $^2\Pi_{1/2}$, the wavefunction is

$$\left|\Lambda = 1, \Sigma = -\frac{1}{2}, \Omega = \frac{1}{2}\right\rangle = |\pi^-\beta\rangle .$$

From Eq. (3.4.6),

$$\langle {}^2\Pi_{3/2} |\mathbf{H}^{\mathrm{SO}}| {}^2\Pi_{3/2}\rangle = A/2 \quad \langle {}^2\Pi_{1/2} |\mathbf{H}^{\mathrm{SO}}| {}^2\Pi_{1/2}\rangle = -A/2 \tag{3.4.7}$$

and

$$\langle {}^2\Pi_{3/2} |\mathbf{H}^{\mathrm{SO}}| {}^2\Pi_{3/2}\rangle = \langle \pi^+\alpha |\hat{a}\mathbf{l}_z\mathbf{s}_z| \pi^+\alpha\rangle = \frac{1}{2}\langle \pi |\hat{a}| \pi\rangle = \frac{1}{2}a_\pi \tag{3.4.8}$$

$$\langle {}^2\Pi_{1/2} |\mathbf{H}^{\mathrm{SO}}| {}^2\Pi_{1/2}\rangle = \langle \pi^+\beta |\hat{a}\mathbf{l}_z\mathbf{s}_z| \pi^+\beta\rangle = -\frac{1}{2}\langle \pi |\hat{a}| \pi\rangle = -\frac{1}{2}a_\pi . \tag{3.4.9}$$

From Eq. (3.4.7)and Eq. (3.4.8), $A({}^2\Pi, \pi) = a_\pi$; a_π is a molecular spin-orbit parameter, analogous to the atomic $\zeta_{\mathrm{K}}(nl)$ parameter defined by

$$\zeta_{\mathrm{K}}(nl) = \left\langle nl \left| \frac{\alpha^2}{2}\frac{Z_{eff,\mathrm{K}}}{r_{\mathrm{K}}^3} \right| nl \right\rangle .$$

As a_π is *always a positive quantity* (the orbital π appears twice in the matrix element), $A > 0$: the $^2\Pi$ state is *regular* for the π configuration, as is well known.

Case of π^3 *Configuration*

For the substate $^2\Pi_{3/2}$, the wavefunction is $|\pi^+\alpha\pi^+\beta\pi^-\alpha\rangle$.
For the substate $^2\Pi_{1/2}$, the wavefunction is $|\pi^+\alpha\pi^+\beta\pi^-\beta\rangle$.
The contribution of the $\pi^+\alpha\pi^+\beta$ subshell to the spin-orbit matrix element is zero; thus

$$\langle {}^2\Pi_{3/2} |\mathbf{H}^{\mathrm{SO}}| {}^2\Pi_{3/2}\rangle = \langle \pi^-\alpha |\hat{a}\mathbf{l}_z\mathbf{s}_z| \pi^-\alpha\rangle = -\frac{1}{2}a_\pi . \tag{3.4.10a}$$

From Eq. (3.4.7),

$$A\left({}^2\Pi, \pi^3\right) = -a_\pi ; \tag{3.4.10b}$$

thus the state is *inverted.*

$^3\Pi$ States

Case of $\sigma\pi$ Configuration

For $^3\Pi_2$ the wavefunction is $|\Lambda = 1, \Sigma = 1, \Omega = 2\rangle = |\sigma\alpha\pi^+\alpha\rangle$.

For $^3\Pi_1$ the wavefunction is $|\Lambda = 1, \Sigma = 0, \Omega = 1\rangle = 2^{-1/2}\left[|\sigma\alpha\pi^+\beta\rangle + |\sigma\beta\pi^+\alpha\rangle\right]$.

For $^3\Pi_0$ the wavefunction is $|\Lambda = 1, \Sigma = -1, \Omega = 0\rangle = |\sigma\beta\pi^+\beta\rangle$.

Only the π electron results in a nonzero contribution from the $\mathbf{l}_z$ operator. A and a_π are defined by Eqs. (3.4.6) and (3.4.8),

$$\langle ^3\Pi_2 | \mathbf{H}^{\mathrm{SO}} | ^3\Pi_2 \rangle = A(^3\Pi)$$

and

$$\langle ^3\Pi_2 | \mathbf{H}^{\mathrm{SO}} | ^3\Pi_2 \rangle = \langle \pi^+\alpha | \hat{\mathbf{a}}\mathbf{l}_z\mathbf{s}_z | \pi^+\alpha \rangle = \frac{1}{2}a_\pi \tag{3.4.11}$$

thence

$$A\left(^3\Pi, \sigma\pi\right) = \frac{1}{2}a_\pi. \tag{3.4.12}$$

In the approximation where the orbital shapes are assumed to be identical for the molecule and positive ion, the following experimental values and Eqs. (3.4.10b) and (3.4.12),

$$A\left(\mathrm{B}^3\Pi_g, \sigma_g\pi_g, \mathrm{N}_2\right) = 42.2\ \mathrm{cm}^{-1} = a_{\pi_g}/2$$
$$A\left(\mathrm{A}^2\Pi_u, \pi_u^3, \mathrm{N}_2^+\right) = -80\ \mathrm{cm}^{-1} = -a_{\pi_u},$$

show that the ratio of the two observed magnitudes of A is not far from the predicted value of -2. In this homonuclear example, the equality of $a_{\pi_g} \approx a_{\pi_u}$ is expected to be approximately valid (see Section 5.3.1).

In general it is easiest to derive a relationship between the many-electron A_Λ parameter and the one-electron a_λ parameter from the diagonal matrix elements of the $|\Lambda, S, \Sigma = S\rangle$ basis function.[†] Consider, for example, the $^4\Pi$ state belonging to a $\delta\sigma\pi$ configuration (e.g., the VO molecule; Cheung *et al.*, 1982). For three open subshells (see Table 3.4);

$$|^4\Pi_{5/2}\rangle = \left|\Lambda = 1, S = \frac{3}{2}, \Sigma = \frac{3}{2}, \Omega = \frac{5}{2}\right\rangle$$
$$= |\delta^+\alpha\sigma\alpha\pi^-\alpha\rangle,$$
$$\langle ^4\Pi_{5/2} | \mathbf{H}^{\mathrm{SO}} | ^4\Pi_{5/2} \rangle = A\Lambda\Sigma = A(1)\left(\frac{3}{2}\right) = \frac{3}{2}A(^4\Pi),$$

[†] The choice of $\Sigma = S$ is advantageous because the required linear combination of Slater determinants is simplest for this case (often a single Slater determinant).

and

$$\langle{}^4\Pi_{5/2}\left|\mathbf{H}^{\mathrm{SO}}\right|{}^4\Pi_{5/2}\rangle = \left\langle \delta^+\alpha\sigma\alpha\pi^-\alpha \left| \sum_i \hat{a}_i \mathbf{l}_{zi}\mathbf{s}_{zi} \right| \delta^+\alpha\sigma\alpha\pi^-\alpha \right\rangle$$
$$= (2)\left(\frac{1}{2}\right)\langle\delta|\hat{a}|\delta\rangle - (1)\left(\frac{1}{2}\right)\langle\pi|\hat{a}|\pi\rangle;$$

thus

$$A({}^4\Pi, \delta\sigma\pi) = \frac{1}{3}(2a_\delta - a_\pi).$$

Another example concerns the $a^4\Pi_u$ state of O_2^+ which belongs to the $\pi_u^3\pi_g^2$ configuration. For $\Sigma = 3/2$,

$$\langle{}^4\Pi_{5/2}\left|\mathbf{H}^{\mathrm{SO}}\right|{}^4\Pi_{5/2}\rangle = A\Lambda\Sigma = A \times 1 \times 3/2$$
$$= \langle\pi_u^+\alpha\pi_u^+\beta\pi_u^-\alpha\pi_g^+\alpha\pi_g^-\alpha\left|\mathbf{H}^{\mathrm{SO}}\right|\pi_u^+\alpha\pi_u^+\beta\pi_u^-\alpha\pi_g^+\alpha\pi_g^-\alpha\rangle = -\frac{1}{2}a_{\pi_u}$$

thus

$$A\left({}^4\Pi\right) = -\frac{1}{3}a_{\pi_u}.$$

The O_2 $^3\Pi_u$ Rydberg state with a $^4\Pi_u$ core and an outer $3s\sigma_g$ electron (the $H^3\Pi_u$ state, see the end of Section 3.2.3), for which the wavefunction is specified in Eq. (3.2.111a), has a spin-orbit constant:

$$A\left({}^3\Pi_u\right) = \langle{}^3\Pi_2\left|\mathbf{H}^{\mathrm{SO}}\right|{}^3\Pi_2\rangle = \frac{1}{12}\left[-9 \times \frac{1}{2} - \frac{1}{2}\right]a_{\pi_u}$$
$$= -\frac{5}{4} \times \frac{1}{3}a_{\pi_u} = +\frac{5}{4}A\left({}^4\Pi\right)^\dagger$$

In Section 3.2.3 it has been shown that the three other $^3\Pi_u$ wavefunctions with $^2\Pi_u$ cores are similar to the doublet wavefunctions except for the supplementary $3s\sigma_g$ orbital. With this simple argument

$$\langle{}^3\Pi_2(1)\left|\mathbf{H}^{\mathrm{SO}}\right|{}^3\Pi_2(1)\rangle = \langle{}^2\Pi_{3/2}(1)\left|\mathbf{H}^{\mathrm{SO}}\right|{}^2\Pi_{3/2}(1)\rangle$$

and

$$\langle{}^3\Pi_2(2)\left|\mathbf{H}^{\mathrm{SO}}\right|{}^3\Pi_2(2)\rangle = \langle{}^2\Pi_{3/2}(2)\left|\mathbf{H}^{\mathrm{SO}}\right|{}^2\Pi_{3/2}(2)\rangle$$

and

$$\langle{}^3\Pi_2(3)\left|\mathbf{H}^{\mathrm{SO}}\right|{}^3\Pi_2(3)\rangle = \langle{}^2\Pi_{3/2}(3)\left|\mathbf{H}^{\mathrm{SO}}\right|{}^2\Pi_{3/2}(3)\rangle$$

† Exp $A({}^4\Pi_u, O_2^+) = -47.8\ \mathrm{cm}^{-1}$, from which $A(H^3\Pi_u, O_2) = -59.8\ \mathrm{cm}^{-1}$, exp $-60\ \mathrm{cm}^{-1}$ (Tonkyn, *et al.*, 1989).

where $^3\Pi(1)$ is the wavefunction with a $^2\Pi(1)$ core (descended from $\pi_g^2\ ^1\Sigma_g^+$) given by Eqs. (3.2.103) - (3.2.103b) and $^3\Pi(2)$ with a $^2\Pi(2)$ core (descended from $\pi_g^2\ ^3\Sigma_g^-$) given by Eq. (3.2.104), and $^3\Pi(3)$ with a $^2\Pi(3)$ core (descended from $\pi_g^2\ ^1\Delta_g$) given by Eq. (3.2.106), since $3s\sigma_g$ does not contribute to the spin-orbit constant. Because

$$\left\langle ^2\Pi_{3/2} \left| \mathbf{H}^{\mathrm{SO}} \right| ^2\Pi_{3/2} \right\rangle = A\Lambda\Sigma = A(^2\Pi) \times 1 \times 1/2$$

and

$$\left\langle ^3\Pi_2 \left| \mathbf{H}^{\mathrm{SO}} \right| ^3\Pi_2 \right\rangle = A(^3\Pi) \times 1 \times 1,$$

thus

$$A(^3\Pi(1)) = (1/2)A(^2\Pi(1))$$

and

$$A(^3\Pi(2)) = (1/2)A(^2\Pi(2))$$

and

$$A(^3\Pi(3)) = (1/2)A(^2\Pi(3)).$$

Actually the wave functions for three of the four observable $^3\Pi_u$ Rydberg states are linear combinations of three primitive $^2\Pi$ wavefunctions as given by Eq. (3.2.107). One of them corresponds to the experimentally observed O_2 $J^3\Pi_u$ state (Nishinati, *et al.*, 1984).

Expressions for the more complicated cases appear in the literature. See, for the $\pi^3\pi^{*2}$ configuration of the NO molecule, the paper by Field *et al.*, (1975).

3.4.2.1.2 Off-Diagonal Matrix Elements

Nonzero spin-orbit matrix elements for $\Delta\Lambda = \Delta\Sigma = 0$ but $\Delta S = 1$ occur mainly between states belonging to the same configuration.† When such states are far from each other, the effect causes only second-order effects called "isoconfigurational" spin-orbit effects. These isoconfigurational interactions are of greatest importance in understanding the evolution at strong spin-orbit toward the case (c) limit. The first example is for the $\sigma\pi$ configuration, which gives rise to one $^3\Pi$ and one $^1\Pi$ state. As the $\Delta\Omega$ selection rule for the spin-orbit operator is *always* $\Delta\Omega = 0$, only the states having $\Omega = 1$ $(\Lambda = 1, \Sigma = 0)$ can

† In Section 5.1.5 the notation $A_{12} = \left\langle \psi_1 \left| \mathbf{H}^{\mathrm{SO}} \right| \psi_2 \right\rangle$ is introduced for off-diagonal matrix elements of $\mathbf{H}^{\mathrm{SO}}$ (including the vibrational part of the total wavefunction).

interact:

$$\begin{aligned}
&\langle {}^3\Pi_1, v|\mathbf{H}^{\mathrm{SO}}|{}^1\Pi_1, v'\rangle \\
&= \left\langle \pi^+\sigma 2^{-1/2}(\alpha\beta + \beta\alpha), v \left| \sum_i \hat{a}_i \mathbf{l}_{zi}\mathbf{s}_{zi} \right| \pi^+\sigma 2^{-1/2}(\alpha\beta - \beta\alpha), v' \right\rangle \\
&= \frac{1}{2}\left[\frac{1}{2}\langle\pi|\hat{a}|\pi\rangle + \frac{1}{2}\langle\pi|\hat{a}|\pi\rangle\right]\langle v|v'\rangle \\
&= \frac{1}{2}a_\pi \langle v|v'\rangle . \qquad (3.4.13)
\end{aligned}$$

Since the two states belong to the same configuration, the following hypothesis can be adopted: the electronic wavefunctions for the ${}^3\Pi$ and ${}^1\Pi$ states are constructed from identical orbitals, and the two states have identical potential curves; consequently,

$$\langle {}^3\Pi_1, v|\mathbf{H}^{\mathrm{SO}}|{}^1\Pi_1, v'\rangle \approx A({}^3\Pi)\delta_{vv'}. \qquad (3.4.14)$$

When the two interacting states are relatively far apart, this spin-orbit mixing will result in second order energy shifts which destroy the equidistant splittings of the ${}^3\Pi$ substates (Fig. 3.13).

When the interacting states are close to each other relative to $A({}^3\Pi)$, then the mixing is very strong and it is an example of what happens as the case (c) limit is approached. Even at the case (c) limit it is important to note that certain substates (in this case, ${}^3\Pi_0$ and ${}^3\Pi_2$) might remain well-described by a single case (a) basis function while the neighboring substates are at the case (c) limit where the quantum number S is undefined.

The series of N_2 $ns\sigma_g$ Rydberg states converging to the N_2^+ $A^2\Pi_u$ state (Ogawa and Tanaka, 1962) is an excellent example of the evolution from case (a) to case (c). The interval between the triplet and singlet Π_u states of the configuration $\pi_u p\ ns\sigma_g$ is given by $2K$, where K is the exchange integral:

$$K = \langle \pi_u p(1) ns\sigma_g(1)\,|1/r_{12}|\, ns\sigma_g(2)\pi_u p(2)\rangle . \qquad (3.4.15)$$

At small n–values, only the ${}^1\Pi_u - X^1\Sigma_g$ series can be seen in absorption. The ${}^3\Pi_1$ and ${}^1\Pi_1$ states are mixed by the constant interaction $A = \frac{1}{2}a_{\pi_u p}$ [Eq. (3.4.14)]. As n increases, the interval between these two states, $2K$, which is proportional to n^{-3}, decreases while $a_{\pi_u p}$ remains constant. When n becomes infinite, the singlet and triplet basis functions become degenerate and the resulting eigenstates are completely singlet∼triplet mixed and separated by $2A$, as shown in Fig. 3.13.

Consequently, in addition to the ${}^1\Pi_u - X^1\Sigma_g^+$ series (Worley's third Rydberg series), a ${}^3\Pi_{1_u} - X^1\Sigma_g^+$ Rydberg series appears in absorption at high n values. The observed separation between the two series reaches a constant value of 80 cm^{-1}, which is just the value of the spin-orbit coupling constant for the N_2^+ ion (which is inverted and thus opposite to the situation illustrated in Fig. 3.13).

A second example is given by the interaction between the ${}^3\Sigma^-$ and ${}^1\Sigma^+$ states that belong to a π^2 configuration. Note again that Eq. (3.4.6) is only

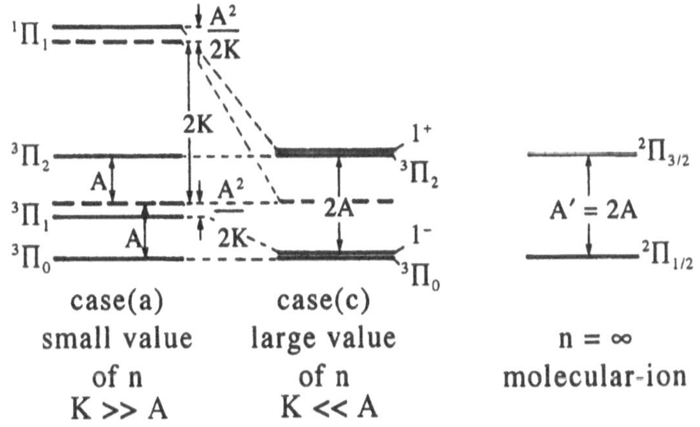

Figure 3.13: Correlation diagram between case (a) and case (c). Convergence of $^3\Pi$ and $^1\Pi$ Rydberg series to a $^2\Pi$ ion core. At small n, the exchange splitting, $K \propto n^{-3}$, is large compared to the spin-orbit splitting of $^3\Pi$ (A) and the $^1\Pi_1 \sim {}^3\Pi_1$ ($-A$) interaction. The symmetric case (a) $^3\Pi$ spin splitting is slightly distorted. At large n, K (proportional to n^{-3}) has become much smaller than A (which arises entirely from the core π orbital) and the case (c) level pattern results. A' is the spin-orbit constant of the molecular ion-core $^2\Pi$ state. The level labels 1^+ and 1^- denote $|1^\pm\rangle = 2^{-1/2}\left[|^3\Pi_1\rangle \pm |^1\Pi_1\rangle\right]$.

valid for $\Delta S = 0$ diagonal matrix elements. For Σ states, a nonzero matrix element occurs between only those states of different multiplicity that behave differently under the σ_v operation (see selection rules given in Section 3.4.1). This has often been forgotten in the literature (see Tinkham and Strandberg, 1955; Veseth, 1970). The matrix element between the $^3\Sigma_0^- e$ and $^1\Sigma_0^+ e$ basis functions from the $\sigma^2\pi^2$ configuration is given by

$$\begin{aligned}
&\langle {}^3\Sigma_0^- e, \Lambda = 0, \Sigma = 0, |\mathbf{H}^{\mathrm{SO}}| \, {}^1\Sigma_0^+ e, \Lambda = 0, \Sigma = 0, \rangle \\
&= \left\langle \pi^+\pi^-(2)^{-1/2}(\alpha\beta + \beta\alpha) \middle| \sum_i \hat{a}_i \mathbf{l}_{iz}\mathbf{s}_{iz} \middle| \pi^+\pi^-(2)^{-1/2}(\alpha\beta - \beta\alpha) \right\rangle \\
&= \frac{1}{2}\left\{2\frac{a_\pi}{2} + 2\frac{a_\pi}{2}\right\} = a_\pi
\end{aligned} \tag{3.4.16}$$

If one assumes that the orbitals in common for the $\sigma^2\pi^2$ $^3\Sigma^-$ and $^1\Sigma^+$ and the $\sigma\pi^3$ $^3\Pi$ states are identical, then

$$\langle {}^3\Sigma_0^- e |\mathbf{H}^{\mathrm{SO}}| \, {}^1\Sigma_0^+ e \rangle = -2A\left({}^3\Pi, \sigma\pi^3\right).$$

Thus, an approximate value for the unknown $^3\Sigma^- \sim {}^1\Sigma^+$ interaction can be obtained from an observable diagonal spin-orbit constant. As will be discussed later (Sections 3.4.4 and 5.3.3), second-order spin-orbit effects of this type contribute significantly to the effective spin-spin interaction in $^3\Sigma^-$ states.

All matrix elements of the $\mathbf{l}_{zi}\mathbf{s}_{zi}$ part of the spin-orbit Hamiltonian can be calculated by using unsymmetrized wavefunctions. The $\mathbf{l}_{zi}\mathbf{s}_{zi}$ matrix elements between properly symmetrized states (eigenfunctions of the σ_v operator) are identical to those between unsymmetrized wavefunctions.

To summarize, matrix elements of the $\mathbf{l}_{zi}\mathbf{s}_{zi}$ part of $\mathbf{H}^{\mathrm{SO}}$ are important mainly between states belonging to the same configuration. If these states lie far apart in energy, only minor perturbation effects (second-order shifts) result. If the interacting basis states are near degenerate, the result is evolution from case (a) toward case (c). Local $\Delta v \neq 0$ perturbations between electronic states belonging to the same configuration are rare and usually small, because their appearance requires differences between the potential curves of isoconfigurational states or an unusually rapid R–variation of the spin-orbit interaction. However, isoconfigurational $^1\Pi \sim {}^3\Pi$ perturbations with $\Delta v \neq 0$ have been observed for BeO (Lavendy *et al.*, 1984), BeS (Pouilly *et al.*, 1982; Cheetham *et al.*, 1965), AsN (Perdigon and Femelat, 1982), and DCl (Huber and Alberti, 1983).

3.4.2.2 Matrix Elements of the $\left(\mathbf{l}_i^+\mathbf{s}_i^- + \mathbf{l}_i^-\mathbf{s}_i^+\right)$ Part of $\mathbf{H}^{\mathrm{SO}}$

Matrix elements of the total spin-orbit Hamiltonian between basis states differing by $\Delta\Lambda = \pm 1$, $\Delta\Sigma = \mp 1$, for a *given signed* value of Ω, may be calculated using only the $\frac{1}{2}\mathbf{l}_i^+\mathbf{s}_i^-$ or $\frac{1}{2}\mathbf{l}_i^-\mathbf{s}_i^+$ part of $\mathbf{H}^{\mathrm{SO}}$. Again, it is unnecessary to use symmetrized basis functions (except when one of the states involved is a $\Sigma_0^\pm$ state):

$$\begin{aligned}
&\left\langle 2^{-1/2}[(\Lambda,\Sigma) \pm (-\Lambda,-\Sigma)] \left| \frac{1}{2}\sum_i \hat{a}_i \left(\mathbf{l}_i^+\mathbf{s}_i^- + \mathbf{l}_i^-\mathbf{s}_i^+\right) \right| 2^{-1/2}[(\Lambda+1,\Sigma-1)\right.\\
&\qquad \left. \pm(-\Lambda-1,-\Sigma+1)]\right\rangle\\
&\qquad = \langle \Lambda,\Sigma,+\Omega|\mathbf{H}^{\mathrm{SO}}|\Lambda+1,\Sigma-1,+\Omega\rangle = \left\langle +\Omega \left| \frac{1}{2}\sum_i \hat{a}_i \mathbf{l}_i^-\mathbf{s}_i^+ \right| +\Omega \right\rangle\\
&\qquad = \langle -\Lambda,-\Sigma,-\Omega|\mathbf{H}^{\mathrm{SO}}|-\Lambda-1,-\Sigma+1,-\Omega\rangle\\
&\qquad = \left\langle -\Omega \left| \frac{1}{2}\sum_i \hat{a}_i \mathbf{l}_i^+\mathbf{s}_i^- \right| -\Omega \right\rangle\\
&\qquad = \left\langle \Omega_f^e \left| \frac{1}{2}\sum_i \hat{a}_i \left(\mathbf{l}_i^+\mathbf{s}_i^- + \mathbf{l}_i^-\mathbf{s}_i^+\right) \right| \Omega_f^e \right\rangle. \qquad (3.4.17)
\end{aligned}$$

When $2S+1$ is odd, all $\Sigma^\pm$ states have only one $\Omega = 0$ substate. Consequently, all $\mathbf{H}^{\mathrm{SO}}$ matrix elements between a $\Sigma_0^\pm$ basis function and a *symmetrized* Π_0 basis function must be obtained by multiplying the result from the third or fourth line of Eq. (3.4.17) by $2^{1/2}$. An example will be treated in Section 3.5.3.

The following example is for states where $\Omega \neq 0$, for the interaction between $^2\Pi_{1/2}$ and $^2\Sigma_{1/2}^+$ states.

The wavefunction for the $^2\Pi_{1/2}$ state from a π^1 configuration is $|\Lambda = 1, \Sigma = -\frac{1}{2}, \Omega = +\frac{1}{2}\rangle = |\pi^+\beta\rangle$; the wavefunction for the $^2\Sigma_{1/2}^+$ state from a

σ^1 configuration is $|\Lambda = 0, \Sigma = +\frac{1}{2}, \Omega = +\frac{1}{2}\rangle = |\sigma\alpha\rangle$. Thus,

$$\left\langle {}^2\Pi_{1/2}, v \left| \mathbf{H}^{\mathrm{SO}} \right| {}^2\Sigma^+_{1/2}, v' \right\rangle = \left\langle \pi^+\beta \left| \frac{1}{2}\hat{a}\mathbf{l}^+\mathbf{s}^- \right| \sigma\alpha \right\rangle \langle v|v'\rangle$$
$$= \frac{1}{2}\left\langle \pi^+ \left| \hat{a}\mathbf{l}^+ \right| \sigma \right\rangle \langle v|v'\rangle = \frac{1}{2}a_+ \langle v|v'\rangle . \qquad (3.4.18)$$

The notation a_+ is used to avoid confusion with the a magnetic hyperfine constant. If one assumes that the σ and π molecular orbitals have the same well-defined value for l and that the potential curves of the two states are identical, $\langle v|v'\rangle = \delta_{vv'}$, then (see Eq. (3.2.121))

$$a_+ = \langle \pi^+|\hat{a}\mathbf{l}^+|\sigma\rangle = [l(l+1) - 0(0+1)]^{1/2}\langle \pi|\hat{a}|\pi\rangle , \qquad (3.4.19)$$

or, from $A\left({}^2\Pi, \pi\right) = a_\pi \equiv \langle \pi|\hat{a}|\pi\rangle$,

$$a_+ = [l(l+1)]^{1/2} A({}^2\Pi, \pi). \qquad (3.4.20)$$

If the value of l is 1, $a_+ = 2^{1/2}A$. This is known as part of the pure precession hypothesis, which is approximately true only for hydrides, Rydberg states of an $l > 0$ complex, and certain highly ionic molecules (see Section 5.5). Although the pure precession hypothesis is frequently applied to homonuclear molecules, there is no justification for this.

These interactions between states of different symmetries, which arise from the $\frac{1}{2}\left(\mathbf{l}^+\mathbf{s}^- + \mathbf{l}^-\mathbf{s}^+\right)$ part of the spin-orbit Hamiltonian, are very common. They often occur between states having $\Delta S = 1$. One example, dealing with the interactions between the excited states of the CO molecule, is given in Section 5.3.2. Another example shows that such an interaction explains the predissociation of the $A^2\Sigma^+$ state of OH (Section 7.11.1).

3.4.3 The Spin-Rotation Operator

This operator accounts for the interaction between the electron spins and the magnetic field created by nuclear motion. As the nuclei are heavy, their angular velocity is approximately m/M times smaller than the angular velocity of the electrons. Consequently, except for light molecules, the spin-rotation interaction is very small compared to the spin-orbit interaction. The microscopic Hamiltonian from Kayama and Baird (1967) and Green and Zare (1977) has the case (a) form,

$$\mathbf{H}^{\mathrm{SR}} = -2g_S\mu_B\frac{\mu_n M}{I}\left\{\sum_i\sum_{\mathrm{K}} Z_{\mathrm{K}}\left(\frac{R_{G\mathrm{K}}\cos\theta_{i\mathrm{K}}}{r_{i\mathrm{K}}^2}s_i\right)\right\}\cdot\mathbf{R}. \qquad (3.4.21)$$

The nuclear magneton, μ_n, is equal to $(m/M)\mu_B$, where m/M is the ratio of the mass of the electron to the proton, 1/1836, I is the molecular moment of inertia, Z_{K} is the nuclear charge, $\theta_{i\mathrm{K}}$ is the angle between the electron coordinate $r_{i\mathrm{K}}$ and $R_{G\mathrm{K}}$ as defined in Fig. 3.14, and $R_{G\mathrm{K}}$ is the distance of nucleus K from the

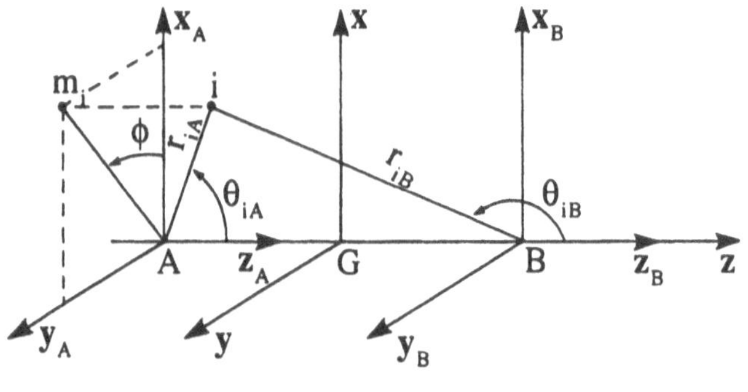

Figure 3.14: Electronic coordinate system. The electron, at i, is located relative to three coordinate systems located respectively at atom A, at the center of mass G, and at atom B; m_i is the projection of i onto the $x_{A,B}, y_{A,B}$ plane; ϕ locates the electron relative to the xz plane ($0 \leq \phi \leq 2\pi$); when $\phi = 0$, electron is in the xz plane; when $\phi = \pi/2$, electron is in the yz plane. The angles θ_{iA} and $\theta_{iB}(0 \leq \theta \leq \pi)$ locate the electron relative to the $+z_A$ and $+z_B$ axes.

center of mass. If the moment of inertia, I, is related to the rotational constant, B_v, the result for a diatomic molecule is

$$\mathbf{H}^{\mathrm{SR}} = -1.05 \times 10^{-4} B_v \left\{ \sum_i \left(\frac{Z_A R_{GA} \cos\theta_{iA}}{r_{iA}^2} + \frac{Z_B R_{GB} \cos\theta_{iB}}{r_{iB}^2} \right) \mathbf{s}_i \right\} \cdot \mathbf{R} \tag{3.4.22}$$

where B_v is in cm^{-1}, all distances are in atomic units, and the matrix elements of $\mathbf{H}^{\mathrm{SR}}$ are in cm^{-1}. The angles are defined in Fig. 3.14.

The effective form for the spin-rotation Hamiltonian is given by Brown and Watson (1977) as

$$\mathbf{H}^{SR} = \gamma \mathbf{R} \cdot \mathbf{S} = \gamma(\mathbf{J} - \mathbf{L} - \mathbf{S}) \cdot \mathbf{S} = \gamma(\mathbf{N} - \mathbf{L}) \cdot \mathbf{S}, \tag{3.4.23}$$

which is convenient for the case (a) basis.

An alternative, commonly used form of $\mathbf{H}^{\mathrm{SR}}$ [convenient for case (b)],

$$\mathbf{H}^{\mathrm{SR}} = \gamma \mathbf{N} \cdot \mathbf{S}, \tag{3.4.24}$$

differs from Eq. (3.4.23) by $\gamma \mathbf{L} \cdot \mathbf{S}$. Since the matrix elements of $\mathbf{H}^{\mathrm{SO}}$ within any multiplet state (i.e., $\Delta S = 0$) have the form $A\mathbf{L} \cdot \mathbf{S}$, the choice of whether to adopt Eq. (3.4.23) or Eq. (3.4.24) for $\mathbf{H}^{\mathrm{SR}}$ affects the value of A by $\pm\gamma$, so that

$$A[Eq.(3.4.23)] = A[Eq.(3.4.24)] + \gamma.$$

However, since $\gamma \ll A$, this choice is immaterial for $\Lambda \neq 0$ states of all molecules except perhaps H_2. For a $^3\Pi_u$ state of H_2, Jette and Miller (1974) used the fact that the isotopic dependence of γ (proportional to μ^{-1}) is different from that of A (independent of μ) to separate $\gamma = -24$ MHz from $A = -3741$ MHz.

The diagonal matrix elements of $\mathbf{H}^{\mathrm{SR}}$ in the case (a) basis are

$$\langle \Lambda S \Sigma \Omega | \mathbf{H}^{\mathrm{SR}} | \Lambda S \Sigma \Omega \rangle = \gamma \left[\Sigma^2 - S(S+1) \right] \qquad (3.4.25)$$

if Eq. (3.4.23) is used, or

$$\langle \Lambda S \Sigma \Omega | \mathbf{H}^{\mathrm{SR}} | \Lambda S \Sigma \Omega \rangle = \gamma \left[\Omega \Sigma - S(S+1) \right] \qquad (3.4.26a)$$

if Eq. (3.4.24) is used. Equation (3.4.24) is particularly useful for evaluating matrix elements of $\mathbf{H}^{\mathrm{SR}}$ in the case (b) basis,

$$\langle NJS\Lambda | \mathbf{H}^{\mathrm{SR}} | NJS\Lambda \rangle = (\gamma/2)[J(J+1) - N(N+1) - S(S+1)], \qquad (3.4.26b)$$

because $\mathbf{N}\cdot\mathbf{S}$ has only diagonal matrix elements in the case (b) basis.

The spin-rotation Hamiltonian has $\Delta\Omega = \Delta\Sigma = \pm 1, \Delta\Lambda = 0, \Delta S = 0$ off-diagonal matrix elements,

$$\begin{aligned} &\langle \Lambda, S, \Sigma, \Omega, v | \mathbf{H}^{\mathrm{SR}} | \Lambda, S, \Sigma, \pm 1, \Omega \pm 1, v \rangle \\ &\quad = \left\langle \Sigma, \Omega, v \left| \frac{\gamma}{2} \mathbf{J}^{\pm} \mathbf{S}^{\mp} \right| \Sigma \pm 1, \Omega \pm 1, v \right\rangle \\ &\quad = (\gamma_v / 2) \left[J(J+1) - \Omega(\Omega \pm 1) \right]^{1/2} [S(S+1) - \Sigma(\Sigma \pm 1)]^{1/2}. \qquad (3.4.27) \end{aligned}$$

This off-diagonal matrix element connects the same basis states as a term in $\mathbf{H}^{\mathrm{ROT}}$ (S-uncoupling operator), but its sign is opposite to that of the B_v contribution. The off-diagonal B_v term appears with a negative sign while γ appears with a positive sign because of the phase convention [Eq. (3.2.85c) and Eq. (3.2.86)],

$$\mathbf{S}^{\pm} | S, \Sigma \rangle = +\hbar [S(S+1) - \Sigma(\Sigma \pm 1)]^{1/2} | S, \Sigma \pm 1 \rangle. \qquad (3.4.28)$$

Zare *et al.* (1973) adopt a phase convention that requires a negative sign in Eq. (3.4.28). However, regardless of phase convention, the $\Delta\Sigma = \Delta\Omega = \pm 1, \Delta\lambda = 0$ matrix elements always have the form

$$+ \text{ or } - (B - \gamma/2)[J(J+1) - \Omega(\Omega \pm 1)]^{1/2} [S(S+1) - \Sigma(\Sigma \pm 1)]^{1/2}.$$

Example. Consider the calculation of γ_{true} for $^3\Sigma^-$ of O_2. From Eq. (3.4.27) one obtains

$$\langle {}^3\Sigma^-_{+1} | \mathbf{H}^{\mathrm{SR}} | {}^3\Sigma^-_0 \rangle = (\gamma/2)[2J(J+1)]^{1/2}.$$

Following Eq. (3.4.22) and noting that for homonuclear molecules $R_{GA} = -R/2 = -R_{GB}$,

$$\begin{aligned} \langle {}^3\Sigma^-_{+1} | \mathbf{H}^{\mathrm{SR}} | {}^3\Sigma^-_0 \rangle = &+ \ 1.05 \times 10^{-4} (B_{\mathrm{O}_2}) (Z_{\mathrm{O}}) (R_{\mathrm{O}_2}) \\ &\times \sum_i \left\langle {}^3\Sigma^-_{+1} \left| \frac{\cos\theta_{i\mathrm{A}}}{r_{i\mathrm{A}}^2} \mathbf{s}_i^+ \right| {}^3\Sigma^-_0 \right\rangle \left(\frac{1}{2} \right) [J(J+1)]^{1/2}. \end{aligned}$$

Table 3.7: Comparison between Some Calculated and Observed Values for the First-Order Contribution to the Spin-Rotation Constant

	Molecule and State						
Parameter	H_2 $c^3\Pi_u$	OD $X^2\Pi$	HF^+ $X^2\Pi$	HCl^+ $X^2\Pi$	NO $X^2\Pi$	SiO $a^3\Pi$	CS $a^3\Pi$ $(v=0)$
$\gamma(calc) \times 10^4 \text{cm}^{-1}$	8.67[a]	3.9[b]	7.6[b]	2.9[b]	1.12[e]	0.09[e]	-
$\gamma(depert) \times 10^4 \text{cm}^{-1}$	8.0[a]	16.4[c]	-	80.0[d]	-175.0[f]	-	-51[g]

[a] Jette and Miller (1974); [b] Green and Zare (1977); [c] Coxon (1975); [d] Brown, *et al.*, (1979); [e] Lefebvre-Brion (unpublished results); [f] Amiot, *et al.*, (1978); [g] Cossart, *et al.*, (1977).

The wavefunction for ${}^3\Sigma^-_{+1}$ is $|\Lambda = 0, \Sigma = 1, \Omega = 1\rangle = |\pi_g^+ \alpha \pi_g^- \alpha\rangle$, and the wavefunction for ${}^3\Sigma^-_0$ is $|\Lambda = 0, \Sigma = 0, \Omega = 0\rangle = 2^{-1/2}\left|\left[|\pi_g^+ \alpha \pi_g^- \beta| + |\pi_g^+ \beta \pi_g^- \alpha|\right]\right\rangle$; thus,

$$\sum_i \left\langle {}^3\Sigma^-_{+1} \left| \frac{\cos\theta_{i\text{A}} \mathbf{s}_i^+}{r_{i\text{A}}^2} \right| {}^3\Sigma^-_0 \right\rangle = 2^{-1/2}\left[\left\langle \pi_g^- \alpha \left| \frac{\cos\theta_{1\text{A}}}{r_{1\text{A}}^2} \mathbf{s}_1^+ \right| \pi_g^- \beta \right\rangle + \left\langle \pi_g^+ \alpha \left| \frac{\cos\theta_{1\text{A}}}{r_{1\text{A}}^2} \mathbf{s}_1^+ \right| \pi_g^+ \beta \right\rangle\right] = 2^{1/2}\left\langle \pi_g \left| \frac{\cos\theta_{1\text{A}}}{r_{1\text{A}}^2} \right| \pi_g \right\rangle,$$

where the π_g^- and π_g^+ matrix elements are identical because the $\cos\theta_{1\text{A}}/r_{1\text{A}}^2$ operator does not operate on the ϕ coordinate. The integral over the π_g molecular orbital may be simplified by expressing π_g in terms of the atomic orbitals, π_A and π_B,

$$\pi_g = [2(1-S)]^{-1/2}[\pi_\text{A} - \pi_\text{B}],$$

where S is the atomic orbital overlap integral. The $\langle \pi_\text{A} | \cos\theta_{1\text{A}}/r_{1\text{A}}^2 | \pi_\text{A}\rangle$ one-center integral is zero by symmetry (odd integrand). The two-center integral $\langle \pi_\text{A} | \cos\theta_{1\text{A}}/r_{1\text{A}}^2 | \pi_\text{B}\rangle$ is negligible. The integral $\langle \pi_\text{B} | \cos\theta_{1\text{A}}/r_{1\text{A}}^2 | \pi_\text{B}\rangle$ may be estimated by a point charge approximation to have the value $1/R^2$ (Kayama and Baird, 1967). With $R_{\text{O}_2} = 2.282$ a.u., $B_{\text{O}_2} = 1.44566 \text{ cm}^{-1}$, and neglecting S (overlap);

$$\gamma = 2.66 \times 10^{-4} \text{ cm}^{-1} = 8 \text{ MHz}.$$

A more exact calculation, which does not make the point charge approximation, gives 3.74 MHz. The experimental value of $\gamma_\text{eff} = -8.4 \times 10^{-3} \text{ cm}^{-1}$ is much larger due to second-order effects.

In view of the approximate relationship

$$\frac{\langle \mathbf{H}^\text{SR}\rangle}{\langle \mathbf{H}^\text{SO}\rangle} \simeq -\frac{m}{M} \propto \frac{\gamma}{A}, \tag{3.4.29}$$

Van Vleck (1929) has suggested a scheme for determining γ from A. In general, Van Vleck's approximation is not justified. As pointed out by Green and Zare (1977), the relevant radial integrals, respectively

$$\langle \phi_A | 1/r_{1A}^3 | \phi_A \rangle$$

for the spin-orbit operator and

$$\langle \phi_A | \cos\theta_{1B}/r_{1B}^2 | \phi_A \rangle$$

for the spin-rotation operator, act in different spatial regions. Furthermore, the integration often includes different orbitals for the two operators. For example, in the case of a $\sigma\pi$ $^3\Pi$ state, the value of A is determined by the π orbital and the value of γ depends on both orbitals. This is particularly striking for the $1s\sigma_g\ np\pi_u$ $^3\Pi_u$ state of H_2, where γ is determined nearly exclusively by the $1s\sigma_g$ orbital. The observed value of γ cannot be compared directly to the calculated value, because γ_{obs} is an effective constant that includes direct (or first-order) contributions from $\mathbf{H}^{SR}$ and second-order from $\mathbf{H}^{SO}$ and $\mathbf{H}^{ROT}$.

Second-order effects arising from the product of matrix elements involving $\mathbf{J}^+\ \mathbf{L}^-$ and $\mathbf{L}^+\ \mathbf{S}^-$ operators have the same form as $\gamma\mathbf{J}^+\mathbf{S}^-$. In the case of H_2, the second-order effect seems to be smaller than the first-order effect, but in other molecules this second-order effect will be more important than the first-order contribution to the spin-rotation constant. These second-order contributions can be shown to increase in proportion with spin-orbit effects, namely roughly as Z^2, but the direct spin-rotation interaction is proportional to the rotational constant. For $^2\Pi$ states, γ is strongly correlated with A_D, the spin-orbit centrifugal distortion constant [see definition, Eq. (5.6.6)], and direct evaluation from experimental data is difficult. On the other hand, the main second-order contribution to γ is often due to a neighboring $^2\Sigma^+$ state. Table 3.7 compares calculated with deperturbed values of $\gamma \cdot \gamma^{\text{eff}}$ of a $^2\Pi$ state may be deperturbed with respect to $^2\Sigma^+$ by

$$\gamma^{\Pi}(\text{deperturbed}) = \gamma^{\text{eff}} + p^{\Pi}(^2\Sigma^+)/2,$$

where $p^{\Pi}(^2\Sigma^+)$ is defined by Eq. (3.5.32) (see also Coxon, 1975 and Brown *et al.*, 1979). γ(deperturbed) includes not only the first-order $\mathbf{H}^{SR}$ contribution but second-order ($\mathbf{H}^{SO} \times \mathbf{H}^{ROT}$) contributions from all other states *except* the nearby $^2\Sigma^+$ (see Section 5.5).

For the $a^3\Pi$ state of CS, the second-order contribution to γ has been determined to be about 8×10^{-4} cm^{-1} and subtracted from the effective γ value. The resulting γ value is certainly not due to the first-order contribution (cf. the calculated value, $\gamma = 9 \times 10^{-6}$ cm^{-1}, for SiO $b^3\Pi$, Table 3.7) but to other high-order effects.

Perturbations between electronic states resulting from $\mathbf{H}^{SR}$ are usually so weak (comparable in magnitude to hyperfine perturbations for light molecules) as to be virtually undetectable.

3.4.4 The Spin-Spin Operator

The spin-spin operator represents the interaction energy between the magnetic dipoles associated with the spins of two different electrons. It is a rather complicated two-electron operator,

$$\mathbf{H}^{\mathrm{SS}} = -\alpha^2 \sum_{i<j} \frac{1}{r_{ij}^5} \left[3\left(\mathbf{r}_{ij}\cdot\mathbf{s}_i\right)\left(\mathbf{r}_{ij}\cdot\mathbf{s}_j\right) - \left(\mathbf{s}_i\cdot\mathbf{s}_j\right) r_{ij}^2\right]. \tag{3.4.30}$$

The "contact" term, which gives an identical contribution for all components of a given multiplet, is neglected here. The selection rules for this operator are

$$\Delta S = 0, \pm 1, \pm 2$$
$$\Delta\Sigma = -\Delta\Lambda = 0 \text{ or } \pm 1 \text{ or } \pm 2$$

and always,

$$\Delta\Omega = 0.$$

The spin-spin interaction is zero for Σ states with $S \leq \frac{1}{2}$. The other selection rules for the $\mathbf{H}^{\mathrm{SS}}$ operator are $g \sim g$ or $u \sim u$, but the selection rule $\Sigma^{\pm} \sim \Sigma^{\pm}$ is opposite to that for the spin-orbit operator, which is $\Sigma^{\pm} \sim \Sigma^{\mp}$. Note, however, that the spin-spin interaction is zero between triplet and singlet states if both of them are Σ states (for example, a $^3\Sigma_0^+$ state has only f levels and the universal selection rule for perturbations is $e \not\leftrightarrow f$; thus $^1\Sigma^{\pm} \sim {}^3\Sigma^{\pm}$ $\mathbf{H}^{\mathrm{SS}}$ perturbations are $e \not\leftrightarrow f$ forbidden; see the end of Section 3.4.5).

Second-order spin-orbit effects result in matrix elements which have the same form as the spin-spin operator. Symbolically,

$$[(\mathbf{S}\cdot\mathbf{L})][(\mathbf{L}\cdot\mathbf{S})] \equiv (\mathbf{S})(\mathbf{S}).$$

Experimentally, this second-order spin-orbit effect is indistinguishable from the direct spin-spin interaction.

3.4.4.1 Diagonal Matrix Elements of H^{SS}: Calculation of the Direct Spin-Spin Parameter

The usual form of the effective spin-spin Hamiltonian, derived by application of the Wigner-Eckart theorem with $\Delta S = \Delta\Sigma = 0$ (see Eq. (3.4.49) of Section 3.4.5), is

$$\mathbf{H}^{\mathrm{SS}} = \frac{2}{3}\lambda\left(3\mathbf{S}_z^2 - \mathbf{S}^2\right), \tag{3.4.31a}$$

which has nonzero case (a) matrix elements,

$$\langle S, \Sigma|\mathbf{H}^{\mathrm{SS}}|S, \Sigma\rangle = \frac{2}{3}\lambda[3\Sigma^2 - S(S+1)]. \tag{3.4.31b}$$

Note that in the literature, two other symbols are also used:

$$\frac{2}{3}\lambda = -C_\Lambda = \epsilon.$$

The splitting of Σ states with $S > \frac{1}{2}$ is due to the spin-spin effect. For example, for $^3\Sigma$ states,

$$E\left(^3\Sigma_{\pm1}\right) - E\left(^3\Sigma_0\right) = 2\lambda. \tag{3.4.31c}$$

For states other than Σ states, the spin-spin interaction contributes only to the asymmetry of the spin-orbit splitting.

Expressed in the form of Eq. (3.4.30), the calculation of matrix elements of the spin-spin operator is not trivial. For $^3\Sigma$ states, only the following terms in $\mathbf{H}^{\mathrm{SS}}$ give rise to nonzero matrix elements:

$$\mathbf{H}^{\mathrm{SS}}(^3\Sigma) = -\alpha^2 \sum_{i<j} \left(3z_{ij}^2 - r_{ij}^2\right)/r_{ij}^5 \left[\mathbf{s}_{iz}\mathbf{s}_{jz} - \frac{1}{4}\left(\mathbf{s}_i^+\mathbf{s}_j^- + \mathbf{s}_i^-\mathbf{s}_j^+\right)\right]. \tag{3.4.32}$$

For example, the direct spin-spin interaction of a $\pi^2\ {}^3\Sigma^-$ state may be derived as follows (see Kayama, 1965). From Eq. (3.4.31b),

$$\langle ^3\Sigma_0^- | \mathbf{H}^{\mathrm{SS}} | ^3\Sigma_0^- \rangle = -\frac{4}{3}\lambda.$$

Now, again evaluating in the one-electron spin-orbital basis for the π^2 configuration,

$$\begin{aligned}
&\langle (2)^{-1/2}\left\{|\pi^+\alpha\pi^-\beta| + |\pi^+\beta\pi^-\alpha|\right\} |\mathbf{H}^{\mathrm{SS}}| (2)^{-1/2}\left\{|\pi^+\alpha\pi^-\beta| + |\pi^+\beta\pi^-\alpha|\right\}\rangle \\
&= -\alpha^2\left\{\left\langle \pi^+\alpha(1)\pi^+\alpha(1) \left| \frac{3z_{12}^2 - r_{12}^2}{r_{12}^5}(\mathbf{s}_{1z}\mathbf{s}_{2z})\right| \pi^-\beta(2)\pi^-\beta(2)\right\rangle\right. \\
&\quad \left. - \left\langle \pi^+\alpha(1)\pi^+\beta(1) \left| \frac{3z_{12}^2 - r_{12}^2}{r_{12}^5}\frac{1}{4}\left(\mathbf{s}_1^+\mathbf{s}_2^- + \mathbf{s}_1^-\mathbf{s}_2^+\right)\right| \pi^-\beta(2)\pi^-\alpha(2)\right\rangle\right\}.
\end{aligned} \tag{3.4.33}$$

After evaluating the effects of the spin operators on simple spin-orbitals, Eq. (3.4.33) reduces to

$$\frac{\alpha^2}{2}\langle \pi^+(1)\pi^+(1) | \left(3z_{12}^2 - r_{12}^2\right)/r_{12}^5 | \pi^-(2)\pi^-(2)\rangle.$$

The atomic spin-spin parameter, η_{A}, is defined in terms of the p_{A} orbitals by

$$\begin{aligned}
\eta_{\mathrm{A}} &= -\frac{\alpha^2}{16}\langle p_{\mathrm{A}}^+ p_{\mathrm{A}}^+ | \left(3z_{12}^2 - r_{12}^2\right)/r_{12}^5 | p_{\mathrm{A}}^- p_{\mathrm{A}}^-\rangle \\
&= (\alpha^2/20)\int_0^\infty p_{\mathrm{A}}^2(r_1)dr_1 \int_{r_1}^\infty \frac{p_{\mathrm{A}}^2(r_2)}{r_2^3}dr_2 \geq 0
\end{aligned} \tag{3.4.34}$$

where, if the integral is evaluated in atomic units and the factor $\alpha^2/20 = 0.58436$, the final result for η_A is in cm^{-1}. For hydride molecules, AH, it is a good approximation to reduce the π molecular orbital to the p_A atomic orbital. Then

$$\langle {}^3\Sigma_0^- | \mathbf{H}^{SS} | {}^3\Sigma_0^- \rangle = -8\eta$$

or

$$\lambda\left({}^3\Sigma^-, \text{AH}\right) = 6\eta.$$

In the case of the homonuclear molecule, A_2, if two-center integrals are neglected, the semiempirical expression

$$\lambda\left({}^3\Sigma^-, \text{A}_2\right) = 3\eta.$$

is obtained.

For a configuration $\sigma\pi^3$, there can be an off-diagonal contribution from the spin-spin interaction, with $\Delta\Lambda = -\Delta\Sigma = \pm 2$ matrix elements between the two components of a ${}^3\Pi_0$ sublevel, which results in a nonzero Λ doubling at $J = 0$,

$$\begin{aligned} E({}^3\Pi_{0f}) - E({}^3\Pi_{0e}) &= -2\alpha^{\dagger} = +2C^{\delta} \\ &= 2[\langle \sigma\alpha\sigma\beta | \mathbf{H}^{SS} | \pi^- \alpha\pi^+ \beta \rangle - \langle \sigma\alpha\pi^+ \beta | \mathbf{H}^{SS} | \pi^- \alpha\sigma\beta \rangle]. \end{aligned} \tag{3.4.35}$$

The second integral, which is an exchange-type integral (see Section 3.2.4), is much smaller than the first, and consequently the Λ–doubling difference of Eq. (3.4.35) is always positive (Horani *et al.*, 1967).

Table 3.8 gives semiempirical expressions for the direct spin-spin part of several effective spin-spin constants and some numerical examples for valence states of homonuclear molecules where the π^* orbital is assumed to be an antibonding orbital. For Rydberg states, the contribution of the Rydberg π' orbital can be neglected.

In Table 5.9 some numerical values for the atomic spin-spin parameter are reported, calculated, or deduced from fitting of atomic fine-structure spectra. These values are only a few hundredths of reciprocal centimeters for atoms heavier than first-row atoms.

† This spin-spin constant, α, should not be confused with the fine-structure constant $\alpha = e^2/\hbar c$. The direct spin-spin contribution to the Λ doubling, α, cannot be separated from a second-order spin-orbit term. All contributions are combined in the o_v parameter, which is the coefficient of the term in the effective Hamiltonian (Brown *et al.* 1979; Brown and Merer, 1979) with a $\Delta\Sigma = -\Delta\Lambda = \pm 2$ selection rule,

$$o_v\left(\mathbf{S}_x^2 - \mathbf{S}_y^2\right) = (o_v/2)\left(\mathbf{S}_+^2 + \mathbf{S}_-^2\right),$$

where

$$o_v = -\alpha + \text{second-order spin-orbit}.$$

3.4.4.2 Calculation of Second-Order Spin-Orbit Effects

As has already been mentioned, second-order spin-orbit effects result in matrix elements that have the same form as the spin-spin operator. Consequently, $\lambda_{eff} = \lambda^{SS} + \lambda^{SO}$, where λ^{SS} is the direct spin-spin parameter and λ^{SO} is the second-order spin-orbit contribution. The main second-order contributions are due to the nearest states; often these are the states that belong to the same configuration as the state under consideration. Moreover, as these nearby states are in general spectroscopically well characterized, it is often relatively easy to estimate semiempirically the contribution of these nearby states to observed spin-spin constants. These contributions are called *isoconfigurational* second-order spin-orbit effects. Some selected examples are given in the following (see Fig. 3.15).

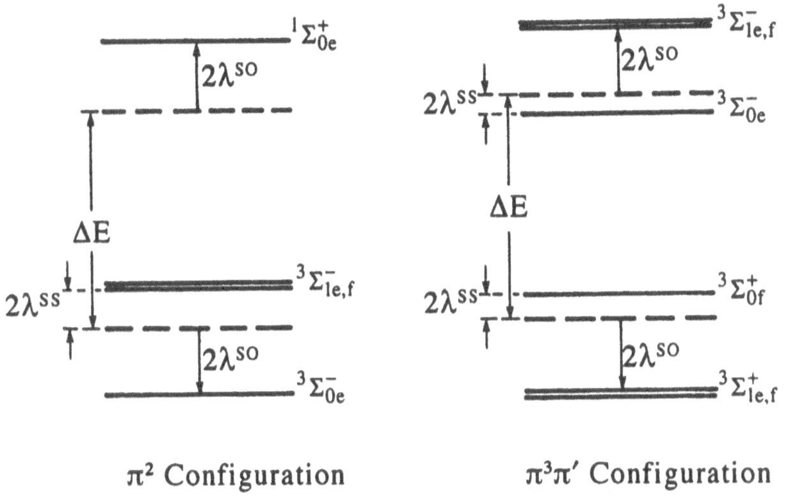

Figure 3.15: Σ states derived from π^2 and $\pi^3\pi'$ configurations. For the π^2 states, the location of the $\Omega = 0$ levels, deperturbed with respect to the isoconfigurational spin-orbit interaction, is indicated by the dashed line. Since the deperturbed location of $^3\Sigma^-_0$ is below $^3\Sigma^-_1, \lambda^{ss} > 0$. For the π^2 configuration, the only Σ^+ state that can interact with $^3\Sigma^-$ always lies at higher energy, and the isoconfigurational spin-orbit contribution to λ is $\lambda^{SO} > 0$. For $\pi^3\pi'$, the dashed lines denote the deperturbed $\Omega = 1$ levels. On the figure, $\lambda^{SS}(^3\Sigma^+) = -\lambda^{SS}(^3\Sigma^-) < 0$. For $\pi^3\pi'$ there are two Σ states contributing, with opposite signs, to each λ^{SO}. The figure shows only the $^3\Sigma^- \sim {}^3\Sigma^+$ mutual contribution, which is negative for $\lambda(^3\Sigma^+)$ [and of equal magnitude and positive for $\lambda(^3\Sigma^-)$] if, as usual, $^3\Sigma^-$ lies above $^3\Sigma^+$.

Table 3.8: Semiempirical Expressions for Effective Spin-Spin Constants: Valence States of Homonuclear Molecules

Configuration	Order of states[a]	State	Direct $\mathbf{H}^{SS}$	Isoconfigurational second-order $\mathbf{H}^{SO}$	Numerical examples (cm^{-1})			
						$\lambda^{SS}_{calc.}$	$\lambda^{SS}_{calc.}$ (iso)	$\lambda^{eff}_{exp.}$
π^2	$^1\Sigma^+$, $^1\Delta$, $^3\Sigma^-$	$^3\Sigma^-$	3η	$\frac{2[A(^3\Pi,\sigma\pi^3)]^2}{E(^1\Sigma^+)-E(^3\Sigma^-)}$	$X^3\Sigma_g^-$ O_2[b]	0.75	1.34	1.98
$\pi^3\pi^*$	$^1\Sigma^+$, $^1\Delta$, $^1\Sigma^-$, $^3\Sigma^-$, $^3\Delta$, $^3\Sigma^+$	$^3\Sigma^+$	-3η	$\frac{1}{2}\frac{[2A(^3\Delta,\pi^3\pi^*)-2A(^3\Pi,\sigma\pi^*)]^2}{E(^3\Sigma^+)-E(^3\Sigma^-)}$	$A^3\Sigma_u^+$ N_2[c]	-0.42	-0.22	-1.33
		$^3\Sigma^-$	3η	$\frac{1}{2}\frac{[2A(^3\Delta)-2A(^3\Pi)]^2}{E(^3\Sigma^-)-E(^3\Sigma^+)}$	$B^3\Sigma_u^-$ N_2[c]	0.42	0.22	0.66
		$^3\Delta$	3η	$-\frac{1}{2}\frac{[2A(^3\Delta)-2A(^3\Pi)]^2}{E(^3\Delta)-E(^1\Delta)}$	$W^3\Delta_u$ N_2	0.42	0.29	0.67[d]
$\pi\pi^*$ or $\pi^3\pi^{*3}$	$^1\Sigma^+$, $^1\Delta$, $^3\Sigma^-$, $^3\Sigma^+$, $^3\Delta$, $^1\Sigma^-$	$^3\Sigma^+$	0	$\frac{2[A(^3\Delta,\pi\pi^*)]^2}{E(^1\Sigma^-)-E(^3\Sigma^+)}$	$A^3\Sigma_u^+$ O_2[e]	0	-5.5	-5.1
		$^3\Sigma^-$	6η	$\frac{2[A(^3\Delta)]^2}{E(^1\Sigma^+)-E(^3\Sigma^-)}$	$B^3\Sigma_u^-$ O_2[e]	1.38	0	1.63
		$^3\Delta$	0	0				
$\sigma\pi^2$	$^2\Sigma^+$, $^2\Sigma^-$, $^4\Sigma^-$	$^4\Sigma^-$	0	$\frac{1}{6}\frac{[A(^2\Pi,\sigma^2\pi)]^2}{E(^2\Sigma^+)-E(^4\Sigma^-)}$	$b^4\Sigma_g^-$ O_2^+	0	0.34[f]	0.21[g]
$\sigma\pi$ or $\sigma\pi^3$	$^1\Pi$, $^3\Pi$	$^3\Pi$	$-\frac{3}{2}\eta$	$+\frac{1}{2}\frac{[A(^3\Pi)]^2}{E(^1\Pi)-E(^3\Pi)}$	$B^3\Pi_g$ N_2	-0.20[h]	0.09	0.21[i]
$\pi^3\pi^{*2}$ [j]	$^2\Pi$, $^2\Pi$, $^4\Pi$	$^4\Pi$	2η	$\frac{1}{4}\frac{[\langle ^2\Pi_{3/2}\lvert\mathbf{H}^{SO}\rvert ^4\Pi_{3/2}\rangle]^2}{E(^2\Pi)-E(^4\Pi)}$	$a^4\Pi_u$ O_2^+[g]	0.62	0.17	0.95

[a]The ordering of the states within each configuration corresponds to the energy order in the single-configuration approximation (see Recknagel, 1934).

[b]See Table 5.10. [c]Sink *et al.* (1975). [d]Effantin, *et al.*(1979a). [e]Field and Lefebvre-Brion (1974). [f]From calculated term values of Beebe, *et al.*, (1976). [g] Brown, *et al.*, (1981).

[h]Value obtained using the pure precession approximation. (Sections 3.5.4 and 5.5). Calculated value is -0.08 cm^{-1} (Lefebvre-Brion unpublished calculation).

[i]Effantin, *et al.*, (1979b).

[j]In addition to the states listed, there is a third $^2\Pi$ state and a $^2\Phi$ state. The value of λ^{SO}_{calc}(iso) for O_2^+ $a^4\Pi_u$ calculated by Brown, *et al.*, (1981) takes into account the spin-orbit interaction with only the $A^2\Pi_u$ state.

3.4.4.2.1 π^2 Configuration

The π^2 configuration gives rise to three states, $^3\Sigma^-$ (the lowest state), $^1\Delta$, and $^1\Sigma^+$ states (see Section 3.2.3). Spin-orbit interaction is possible only between the $^3\Sigma_0^-$ and $^1\Sigma_0^+$ basis functions that have the same value of Ω and the same e/f symmetry (both are e). The other selection rules, $\Delta S = 1$ and $\Sigma^+ \sim \Sigma^-$, are also satisfied. Since $\Delta\Lambda = \Delta\Sigma = 0$, only the $\mathbf{l}_{zi}\mathbf{s}_{zi}$ part of the spin-orbit operator is relevant and, as has been shown in Eq. (3.4.16),

$$\langle {}^3\Sigma_0^- | \mathbf{H}^{\mathrm{SO}} | {}^1\Sigma_0^+ \rangle = a_\pi \approx 2A({}^3\Pi, \sigma\pi^3).$$

Thus $^1\Sigma_0^+$ pushes $^3\Sigma_0^-$ down by

$$\frac{\left[\langle {}^3\Sigma_0^- | \mathbf{H}^{\mathrm{SO}} | {}^1\Sigma_0^+ \rangle\right]^2}{E({}^1\Sigma^+) - E({}^3\Sigma^-)} = \frac{4A^2}{E({}^1\Sigma^+) - E({}^3\Sigma^-)} = 2\lambda^{SO}, \tag{3.4.36}$$

since, by convention, $E({}^3\Sigma_1) - E({}^3\Sigma_0) = 2\lambda$. In Table 5.10, values of effective spin-spin constants for $^3\Sigma^-$ states of π^2 configurations are collected. For molecules with one atom from beyond the second row of the periodic table (Li-Ne), the second-order spin-orbit contribution always outweighs the direct spin-spin contribution. See, for example NBr (Sakamaki, *et al.*, 1998).

3.4.4.2.2 $\pi^3\pi'$ (or $\pi^3\pi'^3$ and $\pi\pi'$) Configurations

The $\pi^3\pi'$, $\pi^3\pi'^3$, and $\pi\pi'$ configurations give rise to six states: $^3\Sigma^+$, $^3\Delta$, $^3\Sigma^-$, $^1\Sigma^-$, $^1\Delta$ and $^1\Sigma^+$. The ordering of these states is given in Table 3.8 in the single-configuration approximation (Recknagel, 1934). Let us consider the spin-orbit interaction between the $^3\Sigma^+$, which is the lowest state (of $\pi^3\pi'$), and the $^3\Sigma^-$ state, which lies above the $^3\Sigma^+$ state ($\Delta S = 0$ with $\Sigma^+ \sim \Sigma^-$). The component $^3\Sigma_{0f}^-$ has a different parity than the $^3\Sigma_{0e}^-$ component and consequently cannot interact with it, but the $^3\Sigma_{1e}^+$ component interacts with the $^3\Sigma_{1e}^-$ component and these two components repel each other (see Fig. 3.15). There is an identical interaction between the $^3\Sigma_{1f}^+$ and $^3\Sigma_{1f}^-$ components. Consequently, the λ^{SO} value of the $^3\Sigma^+$ state is negative, and the λ^{SO} value of the $^3\Sigma^-$ state is equal in magnitude but positive if only this $^3\Sigma^+ \sim {}^3\Sigma^-$ effect is considered. However, one must also take into account the effect of the higher-energy isoconfigurational $^1\Sigma_{0f}^-$ state, which repels the $^3\Sigma_{0f}^+$ component and thereby causes the λ_{eff} value of $^3\Sigma^+$ to become less negative. Similarly, the higher-energy isoconfigurational $^1\Sigma_{0e}^+$ state repels the $^3\Sigma_{0e}^-$ component and thereby increases the λ_{eff} value of the $^3\Sigma^-$ state. Putting the $\mathbf{H}^{\mathrm{SO}}$ interactions in terms of the a_π orbital parameter, and using the phase choice of Field and Lefebvre-Brion (1974):

$$\begin{aligned} \langle {}^3\Sigma_1^+ | \mathbf{H}^{\mathrm{SO}} | {}^3\Sigma_1^- \rangle &= -\frac{1}{2}a_\pi - \frac{1}{2}a_{\pi'}, \\ \langle {}^1\Sigma_0^- | \mathbf{H}^{\mathrm{SO}} | {}^3\Sigma_0^+ \rangle = \langle {}^1\Sigma_0^+ | \mathbf{H}^{\mathrm{SO}} | {}^3\Sigma_0^- \rangle &= +\frac{1}{2}a_\pi - \frac{1}{2}a_{\pi'} \end{aligned} \tag{3.4.37a}$$

where this result is valid for the $\pi^3\pi'$ configuration only. As the spin-orbit coupling constant of the isoconfigurational $\pi^3\pi'\ ^3\Delta$ state is given by

$$2A(^3\Delta, \pi^3\pi') = -\frac{1}{2}a_\pi + \frac{1}{2}a_{\pi'}$$

and that of the $^3\Pi$ state of the $\sigma\pi'$ configuration is given by

$$A(^3\Pi, \sigma\pi') = +\frac{1}{2}a_{\pi'},$$

the semiempirical expressions,

$$\langle ^3\Sigma_1^+ | \mathbf{H}^{\mathrm{SO}} | ^3\Sigma_1^- \rangle = 2A(^3\Delta) - 2A(^3\Pi) = -\frac{1}{2}(a_\pi + a_{\pi'})$$
$$\langle ^1\Sigma_0^\mp | \mathbf{H}^{\mathrm{SO}} | ^3\Sigma_0^\pm \rangle = -2A(^3\Delta) = \frac{1}{2}(a_\pi - a_{\pi'})$$

may be used to estimate the isoconfigurational $\mathbf{H}^{\mathrm{SO}}$ contributions to λ_{eff} for the $^3\Sigma^+$ and $^3\Sigma^-$ states.

For valence states of homonuclear molecules, the π' orbital is the antibonding counterpart, π_g, of the bonding π_u orbital. One then finds that $a_{\pi_u} \simeq a_{\pi_g}$ ($a_\pi \approx a_{\pi^*}$). Thus the interaction between the $^3\Sigma^+$ and $^3\Sigma^-$ states will dominate the λ^{SO} values for the $\pi^3\pi^*$ configuration and is given in Table 3.8. For example, in the P_2 molecule, it has been observed that $\lambda\left(a^3\Sigma_u^+\right) = -3.26\ \mathrm{cm}^{-1}$ and $\lambda\left(b^3\Sigma_u^-\right) = +3.20\ \mathrm{cm}^{-1}$ (Brion, *et al.*, 1974; Brion and Malicet, 1976). For this heavy (third-row) molecule, the direct spin-spin parameter is negligible.

For the $\pi\pi'$ (or $\pi^3\pi'^3$) configuration, with the phase choice of Field and Lefebvre-Brion (1974):

$$\langle ^3\Sigma_1^+ | \mathbf{H}^{\mathrm{SO}} | ^3\Sigma_1^- \rangle = -\frac{1}{2}a_\pi + \frac{1}{2}a_{\pi'}$$
$$\langle ^1\Sigma_0^\pm | \mathbf{H}^{\mathrm{SO}} | ^3\Sigma_0^\mp \rangle = -\frac{1}{2}a_\pi - \frac{1}{2}a_{\pi'}.^\dagger \qquad (3.4.37b)$$

Therefore, in valence states of homonuclear molecules belonging to the $\pi\pi^*$ (or $\pi^3\pi^{*3}$) configuration, the λ^{SO} effect from the $^1\Sigma^-$ state is dominant. The $^1\Sigma^+$ state of this configuration is generally very high-lying, and its effect has been neglected in Table 3.8. For more details, see Field and Lefebvre-Brion (1974). For Rydberg states, where π' is a Rydberg orbital, $a_{\pi'}$ is in general negligible relative to a_π, where π is a valence orbital.

3.4.4.3 Off-Diagonal Matrix Elements

Off-diagonal matrix elements, especially between states belonging to two different configurations, can give rise to perturbations. Such an example is the

† Note that the signs of the off-diagonal matrix elements in Eq. (3.4.37b) depend on the phase choice (i.e., the selected "standard order" of the spin-orbitals). However, the overall signs of the matrix elements in Eq. (3.4.37b) do not affect the sign or magnitude of the λ^{SO} values which are derived from the square of these matrix elements.

perturbation of the $B^2\Sigma^+$ state of CN by a $^4\Sigma^+$ state, which causes a barely detectable shift of about 0.02 cm^{-1} (Coxon *et al.*, 1975; Miller *et al.*, 1976). This may be due to a direct $\mathbf{H}^{SS}$ effect or, more plausibly, to a second-order spin-orbit effect. For example, a second-order interaction via a $^2\Sigma^-$ state would have the form

$$\langle ^2\Sigma^+|\mathbf{H}^{SO}|^4\Sigma^+\rangle = \frac{\langle ^2\Sigma^+|\mathbf{H}^{SO}|^2\Sigma^-\rangle\langle ^2\Sigma^-|\mathbf{H}^{SO}|^4\Sigma^+\rangle}{E - E(^2\Sigma^-)}$$

where E is the average energy of the interacting $^2\Sigma^+$ and $^4\Sigma^+$ levels. Recall that for this type of indirect or second-order interaction, the selection rule is $\Delta S = 0, \pm 1, \pm 2$. Another example of a very weak perturbation due to $\mathbf{H}^{SS}$ has been detected between the $a'^3\Sigma^+$ and $a^3\Pi$ states of CO (Effantin *et al.*, 1982). This interaction must involve a combination of a direct $\mathbf{H}^{SO}$ perturbation *and* a direct $\mathbf{H}^{SS}$ perturbation. Both types of perturbation follow a $\Delta\Omega = 0$ selection rule, but the signs of the $\Omega = 0$ and $\Omega = 1$ perturbation matrix elements are the same for $\mathbf{H}^{SO}$ (see Eq. (3.4.46)) and opposite for $\mathbf{H}^{SS}$ (see Eq. (3.4.49)). This is an example where the relative signs of off-diagonal matrix elements have observable consequences. Owing to partial cancellation between $\mathbf{H}^{SO}$ and $\mathbf{H}^{SS}$ interaction terms in one Ω and addition of interaction terms in the other Ω, the absolute value of the $\Omega = 0 : \Omega = 1$ ratio of interaction strengths is different from the ratio of 3-j coefficients in either Eq. (3.4.46) ($\mathbf{H}^{SO}$) or Eq. (3.4.49) ($\mathbf{H}^{SS}$). Thus values for *both* spin-orbit and spin-spin perturbation matrix elements could be determined from the spectroscopic data.

Similarly to the $^2\Sigma^+ \sim {}^4\Sigma^+$ interaction, a spin-spin interaction is possible between $^5\Sigma$ and $^1\Pi$ states as assumed, for example, in NH by Smith, *et al.*, (1976). For the spin-orbit operator, recall again the $\Delta S \leq 1$ selection rule; however, second-order spin-orbit effects can mix $^5\Sigma$ and $^1\Pi$ states as follows:

$$\langle ^5\Sigma_1|\mathbf{H}^{SO}|^1\Pi_1\rangle = \frac{\langle ^5\Sigma_1|\mathbf{H}^{SO}|^3\Pi_1\rangle\langle ^3\Pi_1|\mathbf{H}^{SO}|^1\Pi_1\rangle}{E - E(^3\Pi)},$$

where E is the average energy of the interacting $^5\Sigma$ and $^1\Pi$ levels.

3.4.5 Tensorial Operators

Energy, which is the observable quantity associated with the Hamiltonian operator, is a pure number or, more precisely, a scalar quantity. Therefore, the Hamiltonian must be a scalar operator. In this section, the prescriptions for constructing a scalar operator from combinations of more complicated operators, such as vector angular momenta, and for evaluating matrix elements of these composite scalar operators are reviewed briefly.

Some operators, such as the interelectronic electrostatic interaction e^2/r_{ij}, are obviously scalar quantities. Others are scalar products of two tensorial operators. A tensor of rank zero is a scalar. A tensor of rank one is a vector. There are several ways of combining two vector operators: the scalar product

$$\vec{\mathbf{U}} \cdot \vec{\mathbf{V}} = \mathbf{U}_x\mathbf{V}_x + \mathbf{U}_y\mathbf{V}_y + \mathbf{U}_z\mathbf{V}_z \tag{3.4.38}$$

yields a scalar quantity, whereas the vector product

$$\vec{\mathbf{U}} \times \vec{\mathbf{V}} = (\mathbf{U}_y\mathbf{V}_z - \mathbf{U}_z\mathbf{V}_y)\,\hat{x} + (\mathbf{U}_z\mathbf{V}_x - \mathbf{U}_x\mathbf{V}_z)\,\hat{y} + (\mathbf{U}_x\mathbf{V}_y - \mathbf{U}_y\mathbf{V}_x)\,\hat{z} \tag{3.4.39}$$

yields a vector quantity. These schemes can be generalized to combinations of tensors of any rank yielding composite tensors of any rank. The reason for constructing operators of well-defined tensorial character is that the evaluation of matrix elements of such operators is tremendously simplified.

The general expression for the *scalar product* of two tensorial operators of rank k is

$$\mathbf{T}_0^{(0)}\left(\vec{\mathbf{U}}, \vec{\mathbf{V}}\right) = \sum_{q=-k}^{+k} (-1)^q \mathbf{U}_q^{(k)} \mathbf{V}_{-q}^{(k)}, \tag{3.4.40}$$

where q is the *spherical* tensor (as opposed to Cartesian) component of the operator. For a vector operator, the relationship between spherical and Cartesian components is, as exemplified for $\vec{\mathbf{r}}$, the position vector of the electron (e.g., the transition moment operator, $e\vec{\mathbf{r}}$),

$$\mathbf{T}_0^{(1)}(\vec{\mathbf{r}}) = z = r\cos\theta \tag{3.4.41a}$$

$$\mathbf{T}_{+1}^{(1)}(\vec{\mathbf{r}}) = -(2)^{-1/2}(x + iy) = -r(\sin\theta)e^{i\phi} \tag{3.4.41b}$$

$$\mathbf{T}_{-1}^{(1)}(\vec{\mathbf{r}}) = (2)^{-1/2}(x - iy) = r(\sin\theta)e^{-i\phi}, \tag{3.4.41c}$$

or, for $\vec{\mathbf{S}}$

$$\mathbf{T}_0^{(1)}(\vec{\mathbf{S}}) = \mathbf{S}_z \tag{3.4.42a}$$

$$\mathbf{T}_{+1}^{(1)}(\vec{\mathbf{S}}) = -(2)^{-1/2}\mathbf{S}^+ \tag{3.4.42b}$$

$$\mathbf{T}_{-1}^{(1)}(\vec{\mathbf{S}}) = +(2)^{-1/2}\mathbf{S}^- \tag{3.4.42c}$$

The power of the Wigner-Eckart theorem (Messiah, 1960, p. 489; Edmonds, 1974, p. 75) is that it relates one nonzero matrix element to another, thereby vastly reducing the number of integrals that must either be explicitly evaluated or treated as a variable parameter in a least-squares fit to spectral data. For example, consider $\mathbf{S}^{(k)}$, a tensor operator of rank k that acts exclusively on spin variables. The Wigner-Eckart theorem requires

$$\left\langle S, \Sigma, \left|\mathbf{S}_q^{(k)}\right| S', \Sigma' \right\rangle = (-1)^{S-\Sigma} \begin{pmatrix} S & k & S' \\ -\Sigma & q & \Sigma' \end{pmatrix} \left\langle S \left\|\mathbf{S}^{(k)}\right\| S' \right\rangle, \tag{3.4.43}$$

where $\begin{pmatrix} S & k & S' \\ -\Sigma & q & \Sigma' \end{pmatrix}$ is a 3-j coefficient (introduced in Section 3.2.1.4) and $\langle S \|\mathbf{S}^{(k)}\| S' \rangle$ is a *reduced matrix element* that depends on neither Σ', Σ, nor q.

If a matrix element is evaluated for a particular value of Σ and Σ' by another method (see Section 3.4.1), for example for the maximum value of $\Sigma = S$ and

$\Sigma' = S'$, the $\left\langle S\Sigma'' \left\| \mathbf{S}_q^{(k)} \right\| S'\Sigma''' \right\rangle$ matrix elements for all values of Σ'' and Σ''' are related to the first matrix element by the ratio of two 3-j coefficients,

$$\frac{\left\langle S, S \left| \mathbf{S}_q^{(k)} \right| S', S' \right\rangle}{\left\langle S, \Sigma'' \left| \mathbf{S}_q^{(k)} \right| S', \Sigma''' \right\rangle} = (-1)^{(S-S)-(S-\Sigma'')} \begin{pmatrix} S & k & S' \\ -S & q & S' \end{pmatrix} \Big/ \begin{pmatrix} S & k & S' \\ -\Sigma'' & q & \Sigma''' \end{pmatrix}. \tag{3.4.44}$$

Equation (3.4.44) illustrates one use of the Wigner-Eckart theorem, the expression of matrix element ratios as simple ratios of 3-j coefficients. Another important application is that selection rules for nonzero matrix elements of any operator are immediately evident. The 3-j coefficient $\begin{pmatrix} J_1 & J & J_2 \\ M_1 & M & M_2 \end{pmatrix}$ can be non-zero when

1. $M_1 = M = M_2 = 0$, provided that $J_1 + J + J_2$ is even.

2. The triangle rule is satisfied,

$$|J_1 - J_2| \leq J \leq J_1 + J_2,$$

   which means that it must be possible to construct a triangle with sides of length J_1, J, J_2.

3. $M_1 + M + M_2 = 0$. This rule can be understood by making the correspondences

$$\begin{aligned} J \to l, \qquad & M \to m \\ |J, M\rangle \to |l, m\rangle &= \mathbf{Y}_{lm} \\ \langle lm| = \mathbf{Y}_{lm}^* &= (-1)^{-m}\mathbf{Y}_{l-m} \\ \mathbf{Y}_{lm}(\theta,\phi) &= (-1)^m \left[\frac{(2l+1)(l-m)!}{4\pi(l+m)!}\right]^{1/2} \mathbf{P}_l^m(\cos\theta)e^{im\phi} \\ \mathbf{T}_m^{(l)}(r,\theta,\phi) &= f(r)\mathbf{Y}_{lm}(\theta,\phi). \end{aligned}$$

Then the matrix element, $\left\langle l'm' \left| \mathbf{T}_m^{(l)} \right| l''m'' \right\rangle$, may be expressed as the $r^2 \sin\theta$ $dr\ d\theta\ d\phi$ volume integral,

$$\left\langle l'm' \left| \mathbf{T}_m^{(l)} \right| l''m'' \right\rangle = K \int e^{i(m-m'+m'')\phi} d\phi \times \int r^2 f(r) dr \int \mathbf{P}_{l'}^{-m'} \mathbf{P}_l^m \mathbf{P}_{l''}^{m''} \sin\theta d\theta,$$

where K is a constant. This form of the integral optimally displays the Δm selection rules, namely, that the $d\phi$ integral will vanish unless

$$m - m' + m'' = 0$$

which is identical to the nonvanishing condition for the 3-j coefficients of Eq. (3.4.43). Now, using Edmonds' result (1974, page 63) for the $\sin\theta d\theta d\phi$ integral over the product of three spherical harmonics, one obtains

$$\left\langle l'm' \left| \mathbf{T}_m^{(l)} \right| l''m'' \right\rangle = (-1)^{l'-m'} \begin{pmatrix} l' & l & l'' \\ -m' & m & m'' \end{pmatrix} \left\langle l' \left\| \mathbf{T}^{(l)} \right\| l'' \right\rangle, \qquad (3.4.45a)$$

where the reduced matrix element

$$\left\langle l' \left\| \mathbf{T}^{(l)} \right\| l'' \right\rangle$$
$$= (-1)^{-l'} \left[\frac{(2l'+1)(2l+1)(2l''+1)}{4\pi} \right]^{1/2} \begin{pmatrix} l' & l & l'' \\ 0 & 0 & 0 \end{pmatrix} \int r^2 f(r) dr \qquad (3.4.45b)$$

is a radial integral times some purely l, l', l''–dependent factors. Note that the electric dipole transition moment operator for atoms transforms as $\mathbf{T}_m^{(1)}$; thus Eq. (3.4.45b) requires that $\Delta l = 0, \pm 1$ and $\Delta m = 0, \pm 1$, but Eq. (3.4.45a) forbids $\Delta l = 0$. Similarly, for electric quadrupole transitions, $\mathbf{T}_m^{(2)}$ permits $\Delta l = 0, \pm 2$ and $\Delta m = 0, \pm 1, \pm 2$.

The 3-j coefficients also identify groups of matrix elements that have identical magnitudes. Even or cyclic permutations,

$$\begin{pmatrix} J_1 & J & J_2 \\ M_1 & M & M_2 \end{pmatrix} \longrightarrow \begin{pmatrix} J_2 & J_1 & J \\ M_2 & M_1 & M \end{pmatrix},$$

of the columns of the 3-j symbol leave its value unchanged. Odd permutations of the columns or a sign reversal of all lower row entries are accounted for by the phase factor $(-1)^{J_1+J+J_2}$.

Consider the spin-orbit operator,

$$\mathbf{H}^{\mathrm{SO}} = \sum_i \hat{a}_i \mathbf{l}_i \cdot \mathbf{s}_i,$$

which is a sum of scalar operators, each of which is a scalar product of two vector operators. Thus,

$$\mathbf{H}^{\mathrm{SO}} = \sum_q (-1)^q \sum_{i=1}^{n} \hat{a}(i) \mathbf{l}_{-q}(i) \mathbf{s}_q(i).$$

Matrix elements between two n–electron wavefunctions can be factored into spin and spatial (including all information about orbital angular momenta) parts. It has been shown frequently that the Wigner-Eckart theorem can be applied as follows (Langhoff and Kern, 1977; McWeeny, 1965; Cooper and Musher, 1972):

$$\langle S, \Sigma, \Lambda, \Omega | \mathbf{H}^{\mathrm{SO}} | S', \Sigma', \Lambda', \Omega \rangle$$
$$= (-1)^{S-\Sigma} \begin{pmatrix} S & 1 & S' \\ -\Sigma & q & \Sigma' \end{pmatrix} \langle S, \Lambda \| \mathbf{H}^{\mathrm{SO}} \| S', \Lambda' \rangle, \qquad (3.4.46)$$

where the reduced matrix element is reduced with respect to Σ but not with respect to Λ and Ω (which involve spatial rather than spin coordinates). From the 3-j coefficient, the selection rules

$$q = \Sigma - \Sigma'$$
$$|S-1| \leq S' \leq S+1 \qquad \text{or} \quad \Delta S = 0, \pm 1,$$

stated without proof in Section 3.4.1, are verified. For $q = 0$ (diagonal and off-diagonal elements of the $\mathbf{l}_z\mathbf{s}_z$ part of $\mathbf{H}^{\mathrm{SO}}$),

$$\Sigma' = \Sigma \qquad \text{because} \qquad \Sigma' - q - \Sigma = 0,$$

and for $q = \pm 1$,

$$\Sigma' = \Sigma \pm 1.$$

Note also that, for $S' = S$ and $\Sigma' = \Sigma$,

$$\langle \Lambda S \Sigma | \mathbf{H}^{\mathrm{SO}} | \Lambda S \Sigma \rangle = (-1)^{S-\Sigma} \begin{pmatrix} S & 1 & S \\ -\Sigma & 0 & \Sigma \end{pmatrix} \langle \Lambda S \| \mathbf{H}^{\mathrm{SO}} \| \Lambda S \rangle$$

where $\begin{pmatrix} S & 1 & S \\ -\Sigma & 0 & \Sigma \end{pmatrix}$ is proportional to Σ, which is consistent with Eq. (3.4.6).

It is now possible to examine a specific example, the spin-orbit part of the $^4\Sigma^- \sim {}^4\Pi$ interaction. The $\mathbf{H}^{\mathrm{SO}}$ operator has three types of $\Delta\Omega = 0$ matrix elements between $|\Lambda = 1\ {}^4\Pi\rangle$ and $|{}^4\Sigma^-\rangle$ basis functions:

$$\begin{aligned}
|{}^4\Pi_{3/2}\rangle &= \left|\Lambda = 1, \Sigma = \frac{1}{2}\right\rangle \sim \left|{}^4\Sigma^-_{3/2}\right\rangle = \left|\Lambda = 0, \Sigma = \frac{3}{2}\right\rangle \\
|{}^4\Pi_{1/2}\rangle &= \left|\Lambda = 1, \Sigma = -\frac{1}{2}\right\rangle \sim \left|{}^4\Sigma^-_{1/2}\right\rangle = \left|\Lambda = 0, \Sigma = \frac{1}{2}\right\rangle \\
|{}^4\Pi_{-1/2}\rangle &= \left|\Lambda = 1, \Sigma = -\frac{3}{2}\right\rangle \sim \left|{}^4\Sigma^-_{-1/2}\right\rangle = \left|\Lambda = 0, \Sigma = -\frac{1}{2}\right\rangle .
\end{aligned}$$

(There are also the same three types of $|\Lambda = -1\ {}^4\Pi\rangle \sim |{}^4\Sigma^-\rangle$ matrix elements.) In parameterizing the $^4\Pi \sim {}^4\Sigma^-$ matrix, it would be desirable to represent the three types of spin-orbit interactions by a single perturbation parameter.

Applying Eq. (3.4.46) with $S = S' = \frac{3}{2}$,

$$\left\langle {}^4\Sigma^-_{3/2} \left| \mathbf{H}^{\mathrm{SO}} \right| {}^4\Pi_{3/2} \right\rangle = \left\langle S = \frac{3}{2}, \Sigma = \frac{3}{2} \left| \mathbf{H}^{\mathrm{SO}} \right| S = \frac{3}{2}, \Sigma = \frac{1}{2} \right\rangle$$
$$= (-1)^{(3/2)-(3/2)} \begin{pmatrix} \frac{3}{2} & 1 & \frac{3}{2} \\ -\frac{3}{2} & 1 & \frac{1}{2} \end{pmatrix} \langle {}^4\Sigma^- \| \mathbf{H}^{\mathrm{SO}} \| {}^4\Pi \rangle$$
$$= (-1)^0 \times \left[-(10)^{-1/2} \right] \langle {}^4\Sigma^- \| \mathbf{H}^{\mathrm{SO}} \| {}^4\Pi \rangle$$

$$\left\langle {}^4\Sigma^-_{1/2} \left| \mathbf{H}^{\mathrm{SO}} \right| {}^4\Pi_{1/2} \right\rangle = \left\langle \frac{3}{2}, \frac{1}{2} \left| \mathbf{H}^{\mathrm{SO}} \right| \frac{3}{2}, -\frac{1}{2} \right\rangle$$
$$= (-1)^{(3/2)-(1/2)} \begin{pmatrix} \frac{3}{2} & 1 & \frac{3}{2} \\ -\frac{1}{2} & 1 & -\frac{1}{2} \end{pmatrix} \langle {}^4\Sigma^- \| \mathbf{H}^{\mathrm{SO}} \| {}^4\Pi \rangle$$
$$= (-1) \times \left(\frac{2}{15} \right)^{1/2} \langle {}^4\Sigma^- \| \mathbf{H}^{\mathrm{SO}} \| {}^4\Pi \rangle$$

$$\left\langle {}^4\Sigma^-_{-1/2} \left| \mathbf{H}^{\mathrm{SO}} \right| {}^4\Pi_{-1/2} \right\rangle = \left\langle \frac{3}{2}, -\frac{1}{2} \left| \mathbf{H}^{\mathrm{SO}} \right| \frac{3}{2}, -\frac{3}{2} \right\rangle$$
$$= (-1)^{(3/2)+(1/2)} \begin{pmatrix} \frac{3}{2} & 1 & \frac{3}{2} \\ +\frac{1}{2} & 1 & -\frac{3}{2} \end{pmatrix} \langle {}^4\Sigma^- \| \mathbf{H}^{\mathrm{SO}} \| {}^4\Pi \rangle$$
$$= (-1)^2 \left[-(10)^{-1/2} \right] \langle {}^4\Sigma^- \| \mathbf{H}^{\mathrm{SO}} \| {}^4\Pi \rangle$$

$$\frac{\left\langle {}^4\Sigma^-_{3/2} \left| \mathbf{H}^{\mathrm{SO}} \right| {}^4\Pi_{3/2} \right\rangle}{\left\langle {}^4\Sigma^-_{1/2} \left| \mathbf{H}^{\mathrm{SO}} \right| {}^4\Pi_{1/2} \right\rangle} = (3)^{1/2} \Big/ 2$$

$$\frac{\left\langle {}^4\Sigma^-_{3/2} \left| \mathbf{H}^{\mathrm{SO}} \right| {}^4\Pi_{3/2} \right\rangle}{\left\langle {}^4\Sigma^-_{-1/2} \left| \mathbf{H}^{\mathrm{SO}} \right| {}^4\Pi_{-1/2} \right\rangle} = 1.$$

The ${}^4\Pi \sim {}^4\Sigma^-$ example involved matrix elements evaluated using individual $|\Omega\Lambda S\Sigma\rangle$ basis functions, in which the sign of $\Omega = \Lambda + \Sigma$ is specified, rather than the symmetrized,

$$(2)^{-1/2}[|\Omega\Lambda\Sigma\rangle \pm | -\Omega - \Lambda - \Sigma\rangle]$$

e/f linear combinations, in which the range of Ω is $S - |\Lambda| \leq \Omega \leq S + |\Lambda|$. Off-diagonal case (a) matrix elements in these two basis sets are identical, but with one important exception for odd-multiplicity systems. For odd-multiplicity systems (even number of electrons, integer S), each Ω–component is always doubly degenerate, except for $\Lambda = \Omega = 0$ which is nondegenerate. Matrix elements between nondegenerate $\Omega = 0$ and degenerate Ω basis functions in the signed-Ω basis, for example,

$$\begin{aligned} |^3\Pi_0\rangle &= |\Lambda = 1, \Sigma = -1\rangle \\ |^3\Sigma_0^+\rangle &= |\Lambda = 0, \Sigma = 0\rangle \\ \langle ^3\Pi_0 |\mathbf{H}^{\mathrm{SO}}|\, ^3\Sigma_0^+\rangle &= (2)^{1/2}\alpha \end{aligned} \tag{3.4.47}$$

(α is a reduced matrix element), are always a factor of $(2)^{1/2}$ smaller than the corresponding quantity in the symmetrized basis,

$$\left\langle ^3\Pi_{0f} |\mathbf{H}^{\mathrm{SO}}|\, ^3\Sigma_{0f}^+ \right\rangle = 2\alpha \tag{3.4.48}$$

The spin-spin operator is a scalar product of two tensorial operators of rank 2, one of which operates on spin coordinates only, the other (spherical harmonics of rank 2) of which affects only spatial coordinates. $\mathbf{H}^{\mathrm{SS}}$ can be factored into its spin and spatial parts, and the Wigner-Eckart theorem can be used to evaluate matrix elements of the spin part. Considering only diagonal matrix elements,

$$\langle \Lambda, S, \Sigma, \Omega |\mathbf{H}^{\mathrm{SS}}|\, \Lambda, S, \Sigma, \Omega\rangle = (-1)^{S-\Sigma} \begin{pmatrix} S & 2 & S \\ -\Sigma & 0 & \Sigma \end{pmatrix} \langle \Lambda, S \,\|\mathbf{H}^{\mathrm{SS}}\|\, \Lambda, S\rangle, \tag{3.4.49}$$

the $\begin{pmatrix} S & 2 & S \\ -\Sigma & 0 & \Sigma \end{pmatrix}$ 3-j coefficient is proportional to $3\Sigma^2 - S(S+1)$, which is consistent with Eq. (3.4.31b). ΔS and $\Delta\Sigma$ selection rules for nonzero off-diagonal matrix elements of $\mathbf{H}^{\mathrm{SS}}$ are obtained from the triangle rule,

$$\begin{gathered} S - 2 \le S' \le S + 2 \\ \Delta S = 0, \pm 1, \pm 2, \end{gathered}$$

and

$$\langle \Lambda, S, \Sigma, \Omega |\mathbf{H}^{\mathrm{SS}}|\, \Lambda', S', \Sigma', \Omega\rangle = (-1)^{S-\Sigma} \begin{pmatrix} S & 2 & S' \\ -\Sigma & q & \Sigma' \end{pmatrix} \langle \Lambda, S \,\|\mathbf{H}^{\mathrm{SS}}\|\, \Lambda', S'\rangle \tag{3.4.50}$$

where, for

$q = 0$	$\Sigma' = \Sigma$	$\Delta\Sigma = 0$ (and $\Delta\Lambda = 0$ because $\Delta\Omega = 0$)
$q = \pm 1$	$\Sigma' = \Sigma \pm 1$	$\Delta\Sigma = \pm 1 = -\Delta\Lambda$
$q = \pm 2$	$\Sigma' = \Sigma \pm 2$	$\Delta\Sigma = \pm 2 = -\Delta\Lambda$.

For the special case where $S' = S + 1$, the sum $S + 2 + S'$ is odd; consequently, the $\Sigma = \Sigma' = 0$ matrix element is zero. For example,

$$\langle ^1\Sigma_0^+ |\mathbf{H}^{\mathrm{SS}}|\, ^3\Sigma_0^+\rangle = 0.$$

3.5 Rotational Perturbations

There are three terms in the rotational part of the Hamiltonian, $\mathbf{H}^{\text{ROT}}$, that are neglected in the Born-Oppenheimer approximation:

1. $(1/2\mu R^2)(\mathbf{L}^+\mathbf{S}^- + \mathbf{L}^-\mathbf{S}^+)$, which causes *spin-electronic homogeneous* $(\Delta\Omega = 0)$ perturbations.
2. $-(1/2\mu R^2)(\mathbf{J}^+\mathbf{S}^- + \mathbf{J}^-\mathbf{S}^+)$, which is the **S**-*uncoupling* operator.
3. $-(1/2\mu R^2)(\mathbf{J}^+\mathbf{L}^- + \mathbf{J}^-\mathbf{L}^+)$, which is the **L**-*uncoupling* operator.

The last two terms give rise to *heterogeneous* $(\Delta\Omega = \pm 1)$ perturbations.

3.5.1 Spin-Electronic Homogeneous Perturbations

For the off-diagonal elements of the $(1/2\mu R^2)(\mathbf{L}^+\mathbf{S}^- + \mathbf{L}^-\mathbf{S}^+)$ operator, $\mathbf{H}^{\text{SE}}$, the selection rules are the same $(\Delta\Omega = 0, \Delta\Lambda = -\Delta\Sigma = \pm 1)$ as for the part of the spin-orbit operator,

$$\frac{1}{2}\sum_i \hat{a}_i(\mathbf{l}_i^+\mathbf{s}_i^- + \mathbf{l}_i^-\mathbf{s}_i^+),$$

except that, for $\mathbf{H}^{\text{SE}}$, $\Delta S = 0$ *only*. An important but subtle difference between these two operators is that

$$\mathbf{L}^+\mathbf{S}^- = \sum_i \mathbf{l}_i^+ \sum_j \mathbf{s}_j^- = \sum_i \mathbf{l}_i^+\mathbf{s}_i^- + \sum_{i<j} \mathbf{l}_i^+\mathbf{s}_j^-. \tag{3.5.1}$$

In contrast, the spin-orbit operator only includes a summation $\sum_i \mathbf{l}_i^+\mathbf{s}_i^-$ involving the same electron for both $\mathbf{l}^+$ and $\mathbf{s}^-$. The terms in $\mathbf{H}^{\text{SE}}$, $\sum_{i<j} \mathbf{l}_i^+\mathbf{s}_j^-$, are two-electron operators and can cause nonzero matrix elements between wavefunctions that differ by two spin-orbitals.

Moreover, the operator $B(R) = \hbar^2/2\mu R^2$ acts on the vibrational part of the wavefunction,

$$\langle v|B(R)|v'\rangle = \langle v|\hbar^2/2\mu R^2|v'\rangle \tag{3.5.2}$$

$\langle v|\hbar^2/2\mu R^2|v'\rangle$ can be approximated by the value $(\hbar^2/2\mu R_C^2)\langle v|v'\rangle$, where R_C is the crossing point of the two electronic potential curves. If the two interacting electronic states have identical potential curves (a prerequisite for pure precession), then

$$B_{vv'} \equiv \langle v|B(R)|v'\rangle \approx B_v\delta_{vv'}.$$

When the two potential curves are well-characterized, the $B_{vv'}$ matrix elements may be calculated numerically. For *nearly* identical potentials, the $B_{vv'}$ matrix elements behave much less like $\delta_{vv'}$ functions than do $\langle v|v'\rangle$ vibrational overlaps.

As an example, consider the interaction between the $^3\Pi_1$ state ($\sigma\pi^4\pi'$ configuration) and the $^3\Sigma_1^+$ state ($\sigma^2\pi^3\pi'$ configuration). For $B(R)\mathbf{L}\cdot\mathbf{S}$, as for the $\mathbf{H}^{\mathrm{SO}}$ operator, the interacting basis wavefunctions must have the same value of Ω. For this example, the $\Omega = \pm 1$ matrix elements of the $\mathbf{L}^+\mathbf{S}^-$ part of $\mathbf{H}^{\mathrm{SE}}$ will be evaluated.

Let ($\pi^+\alpha\ \pi^+\beta\ \pi^-\alpha\ \pi^-\beta\ \sigma\alpha\ \sigma\beta\ \pi'^+\alpha\ \pi'^+\beta\ \pi'^-\alpha\ \pi'^-\beta$) be the standard reference order of spin-orbitals,

$$\begin{aligned}
|^3\Pi_1\rangle &= 2^{-1/2}\{|\pi^+\alpha\pi^+\beta\pi^-\alpha\pi^-\beta\sigma\alpha\pi'^+\beta| - |\pi^+\alpha\underline{\pi^-\beta}\pi^-\alpha\underline{\pi^+\beta}\sigma\beta\pi'^+\alpha|\} \\
&= 2^{-1/2}[\det 1 - \det 2] \\
|^3\Sigma_1^+\rangle &= 2^{-1/2}\{-|\pi^+\alpha\pi^+\beta\pi^-\alpha\underline{\sigma\beta}\underline{\sigma\alpha}\pi'^-\alpha| - |\pi^+\alpha\underline{\pi^-\beta}\underline{\pi^-\alpha}\sigma\alpha\sigma\beta\pi'^+\alpha|\} \\
&= 2^{-1/2}[-\det 3 - \det 4]
\end{aligned}$$

where the underlined spin-orbitals have been permuted out of standard order. Note that determinants 1 and 3 differ by orbitals at the fourth ($\pi^-\beta \to \sigma\beta$) and sixth ($\pi'^+\beta \to \pi'^-\alpha$) positions, 2 and 4 differ only at the fourth ($\pi^+\beta \to \sigma\alpha$) position, 1 and 4 differ at the sixth position ($\pi'^+\beta \to \pi'^+\alpha$) and by ($\pi^+\beta \to \sigma\beta$), which can be brought into registration by an even number of permutations, and 2 and 3 differ at the sixth position $\left(\pi'^+\alpha \to \pi'^-\alpha\right)$ and by $\left(\pi'^-\beta \to \sigma\alpha\right)$. The matrix elements between determinants 1 and 3 and determinants 2 and 3 are zero because $\mathbf{l}^\pm$ cannot change π'^+ to π'^-. The remaining combinations are

$$\begin{aligned}
&\left\langle -\det 2 \left| \sum_{i,j} \mathbf{l}_i^+\mathbf{s}_j^- \right| -\det 4 \right\rangle \\
&= \langle \pi^+\beta|\mathbf{l}_1^+\mathbf{s}_1^-|\sigma\alpha\rangle - \langle \pi^+\alpha\pi^+\beta|\mathbf{l}_1^+\mathbf{s}_2^-|\sigma\alpha\pi^+\alpha\rangle - \langle \sigma\beta\pi^+\beta|\mathbf{l}_2^+\mathbf{s}_1^-|\sigma\alpha\sigma\beta\rangle \\
&= (1-1-1)\langle \pi^+|\mathbf{l}^+|\sigma\rangle \equiv -b
\end{aligned}$$

$$\begin{aligned}
\left\langle \det 1 \left| \sum_{i,j} \mathbf{l}_i^+\mathbf{s}_j^- \right| -\det 4 \right\rangle &= -\left\langle \pi'^+\beta|\mathbf{s}^-|\pi'^+\alpha\right\rangle \langle \pi^+\beta|\mathbf{l}^+|\sigma\beta\rangle \\
&= -\langle \pi^+|\mathbf{l}^+|\sigma\rangle \equiv -b
\end{aligned}$$

thus

$$\begin{aligned}
\left\langle {}^3\Pi_{1_f^e,v} \left|\mathbf{H}^{\mathrm{SE}}\right| {}^3\Sigma^+_{1_f^e,v'} \right\rangle &= \langle {}^3\Pi_1, \Omega = +1, v|B(R)\mathbf{L}^+\mathbf{S}^-|{}^3\Sigma_1^+, \Omega = +1, v'\rangle \\
&= (2)^{-1/2}(2)^{-1/2}(-b-b)\langle v|B(R)|v\rangle = -bB_{vv'}.
\end{aligned} \tag{3.5.3}$$

The $^3\Pi_1 \sim {}^3\Sigma_1^+$ matrix element of $\mathbf{H}^{\mathrm{SO}}$ is much easier to evaluate because $\mathbf{H}^{\mathrm{SO}}$ is a one-electron operator,

$$\left\langle {}^3\Pi_{1_f^e,v} \left|\mathbf{H}^{\mathrm{SO}}\right| {}^3\Sigma^+_{1_f^e,v} \right\rangle = \frac{1}{4}a_+ \langle v|v'\rangle. \tag{3.5.4}$$

For light molecules, the $\mathbf{H}^{SE}$ contribution from Eq. (3.5.3) cannot be separated from the much larger $\mathbf{H}^{SO}$ contribution [Eq. (3.5.4)] except by isotope effects and, much more importantly, by the fact that other fine-structure components of $^3\Sigma^+$ and $^3\Pi$ interact with different J–dependences for the b parameter (see Table 5.8). For example, in the CO molecule, the interaction matrix elements between $\mathrm{a}^3\Pi_1$ $v = 4$ and $\mathrm{a}'^3\Sigma_1^+ v' = 0$ can be separated into

$$\langle {}^3\Pi_1, v \left| \mathbf{H}^{SO} \right| {}^3\Sigma_1^+, v' \rangle = 10.57 \text{ cm}^{-1}$$

and

$$\langle {}^3\Pi_1, v \left| \mathbf{H}^{SE} \right| {}^3\Sigma_1^+ v' \rangle = -0.1158 \text{ cm}^{-1}$$

(Field, *et al.*, 1972).

3.5.2 The S-Uncoupling Operator

The **S**–uncoupling operator, $-(1/2\mu R^2)(\mathbf{J}^+\mathbf{S}^- + \mathbf{J}^-\mathbf{S}^+)$, has the selection rules $\Delta S = 0, \Delta\Omega = \Delta\Sigma = \pm 1$, and $\Delta\Lambda = 0$. Generally, this operator mixes different components of the same multiplet electronic state. It is responsible for the evolution, as J increases, from Hund's case (a) to case (b).

When the **S**-uncoupling operator acts between two components of a multiplet state that belong to the same vibrational quantum number, then the vibrational part of the $B(R)\mathbf{L}\cdot\mathbf{S}$ matrix element is

$$\langle v|B(R)|v\rangle \equiv B_v.$$

However, if by chance there is a near degeneracy between the Ω spin-component of the vth level and the $\Omega' = \Omega \pm 1$ components of the $(v+1)$th level of the same electronic state, then the **S**-uncoupling operator can cause a perturbation between these levels. In the harmonic approximation and using the phase choice that all vibrational wavefunctions are positive at the inner turning point,

$$B_{vv'} = \langle v \left| \hbar^2/2\mu R^2 \right| v+1 \rangle = 2\left(\frac{B_e^3}{\omega_e}\right)^{1/2} (v+1)^{1/2}. \qquad (3.5.5)$$

Note that nonzero $\Delta v \neq 0$ matrix elements within an electronic state require an operator that is R–dependent. The total interaction from the $-2B(R)\mathbf{J}\cdot\mathbf{S}$ term is then given by

$$\begin{aligned} &\langle {}^{2S+1}\Lambda_{\Omega,v}| - B(R)(\mathbf{J}^+\mathbf{S}^- + \mathbf{J}^-\mathbf{S}^+)|{}^{2S+1}\Lambda_{\Omega+1,v+1}\rangle \\ &= -2\left(\frac{B_e^3}{\omega_e}\right)^{1/2} (v+1)^{1/2}[J(J+1) - (\Omega+1)\Omega]^{1/2} \\ &\quad \times [S(S+1) - (\Omega-\Lambda+1)(\Omega-\Lambda)]^{1/2} \qquad (3.5.6) \end{aligned}$$

with the same phase convention as given in Eq. (3.4.27) for the $\mathbf{S}^{\pm}$ operator. The rotational levels will cross if the spin splitting between the spin substates of one vibrational level is slightly smaller than the vibrational interval. This can occur for heavy molecules.

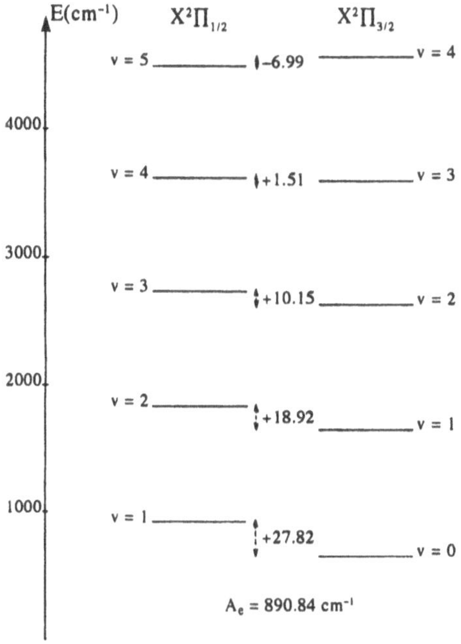

Figure 3.16: Perturbations between $\Omega = \frac{3}{2}$ and $\frac{1}{2}$ sublevels of the $^{15}N^{80}Se$ $X^2\Pi_r$ state. The $J = \frac{1}{2}$ energy offset of each $\Omega = \frac{1}{2}, v + 1$ level above the $\Omega = \frac{3}{2}, v$ level is indicated.

An example of such a perturbation due to this **S**-uncoupling operator has been observed by Jenouvrier, *et al.*, (1973) in the ground electronic state of the NSe molecule, between the $v = 3$ level of the $X^2\Pi_{1/2}$ substate and the $v = 2$ level of the $X^2\Pi_{3/2}$ substate. The position of the levels is shown in Fig. 3.16. Figure 3.17 displays an anomaly in the $X^2\Pi_{3/2}$ substate Λ–doubling, which has borrowed its large value from the $X^2\Pi_{1/2}$ substate. Expression Eq. (3.5.6) accounts very well for all perturbation effects in this case. However, the actual potential curves are anharmonic, and matrix elements of $B(R)$ are small but nonzero for $|\Delta v| > 1$. A $\Delta v = 2$ perturbation of this type has been reported in the HgH molecule by Veseth (1972), in Cl_2 by Ishiwata, *et al.*, (1991), and in ClF by Alekseev, *et al.*, (1999).

3.5.3 The L-Uncoupling Operator

The **L**-uncoupling operator, $-(1/2\mu R^2)(\mathbf{J}^+\mathbf{L}^- + \mathbf{J}^-\mathbf{L}^+)$, which is responsible for the evolution as J increases from Hund's case (a) to case (d), causes numerous perturbations between states that differ by $\Delta\Omega = \Delta\Lambda = \pm 1$ and with $\Delta S = 0$. This specific type of rotational perturbation is often called a *gyroscopic perturbation*.

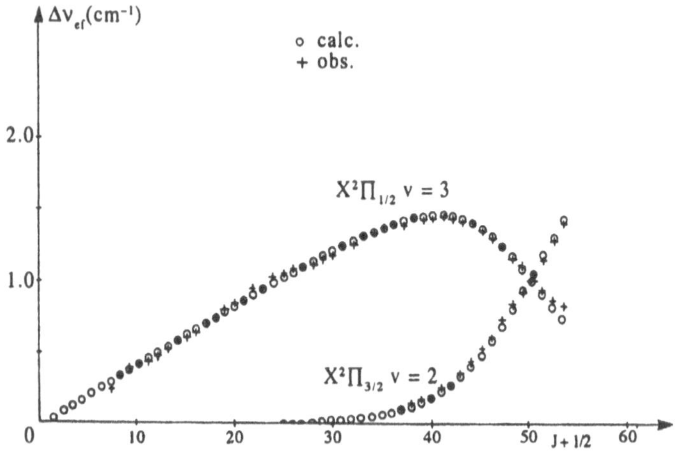

Figure 3.17: Anomalous Λ–doubling in the $X^2\Pi_r$ state of $^{15}N^{80}Se$, due to S-uncoupling perturbations (From Jenouvrier, *et al.*, 1973.)

Since $\mathbf{J}^+\mathbf{L}^- = \mathbf{J}^+\sum_i \mathbf{l}_i^-$, the **L**-uncoupling operator is a one-electron operator, and consequently, in the single-configuration approximation, the configurations describing the two interacting states can differ by no more than one spin-orbital. The electronic part of the perturbation matrix element is then proportional to the same $\langle\pi^+|\mathbf{l}^+|\sigma\rangle$ or b parameter that appeared in the spin-electronic perturbation. However, owing to the presence of the $\mathbf{J}^+$ operator, the total matrix element of the $B(R)\mathbf{J}^+\mathbf{L}^-$ operator between $\langle\Omega-1|$ and $|\Omega\rangle$ is proportional to $[J(J+1)-\Omega(\Omega-1)]^{1/2}$.

An example of this type of interaction between $^2\Pi_{3/2}$ and $^2\Sigma$ states of identical v is given in Section 3.5.4. If the two interacting states are sufficiently far part in energy, this type of interaction gives rise to the splitting between the e and f components of a degenerate Λ state. This splitting is called Λ–doubling. The Λ–doubling of a $^1\Pi_1$ state, due to interaction with a $^1\Sigma_0$ state via the L-uncoupling operator, is treated by Hougen (1970). The Λ–doubling of Δ states, which is a fourth order effect in perturbation theory, has been discussed by Brown, *et al.*, (1987).

Rydberg states that belong to the same l–complex are mixed by the L-uncoupling operator (see the end of Section 6.3.2 and Section 8.5). This most frequently involves states of the same value of v. At high n, very weak perturbations between states belonging to two Rydberg complexes of different nominal l values have been observed [in NO, $\Delta v = 0, 8f \sim 9s$ and $6f \sim 6d$ perturbations were analyzed by Miescher (1976)]. These different l–complexes of identical v values converging to the same ion-core limit form a supercomplex, and consequently an l–mixing results, due to the anisotropy of the molecular-ion field. The l–mixing is caused by the non-sphericity of the ion core, i.e. by the fact that the molecular ion has an electric dipole moment (in a heteronuclear molecule), which mixes Rydberg orbitals differing by $\Delta l = \pm 1$ (Watson, 1994), and a quadrupole moment, which mixes orbitals differing by $\Delta l = \pm 2$ (Jungen, 1970).† This is a purely electrostatic effect due to $\mathbf{H}^{\text{el}}$ which preserves the value of Λ ($\Delta S = 0, \Delta\Sigma = 0, \Delta\Lambda = 0$). The l–mixing effect is distinct from the l uncoupling effect, which preserves the value of l but destroys the value of Λ.

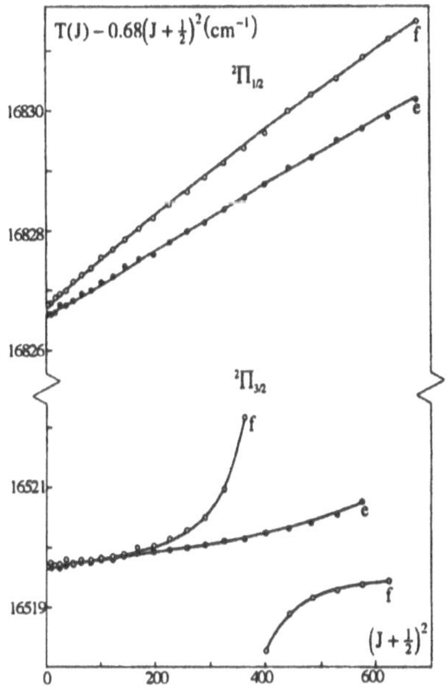

Figure 3.18: Anomalous Λ–doubling in the CS^+ $A^2\Pi_i$ state due to L-uncoupling interaction with the $X^2\Sigma^+$ state. This is a reduced term-value plot for the CS^+ $A^2\Pi_i$ $v = 5$ level, which is perturbed by $X^2\Sigma^+$ $v = 13$. In order to expand the vertical scale, $0.68(J + \frac{1}{2})^2$ is subtracted from the observed term value (zero of energy at $v = 0, N = 0$ of $X^2\Sigma^+$). A plot of $T - B_{\text{eff}}(J + \frac{1}{2})^2$ versus $(J + \frac{1}{2})^2$ should show four nearly horizontal straight lines for a case (a) $^2\Pi$ state, with Λ-doubling in $^2\Pi_{1/2}$ and $^2\Pi_{3/2}$ proportional to J and J^3, respectively. The f–component of $X^2\Sigma^+$ ($F_2, N = J + \frac{1}{2}$) crosses $^2\Pi_{3/2f}$ near $J \simeq 19$ and disrupts the expected J^3 Λ–doubling pattern. The $^2\Pi_{3/2e}$ component is crossed at a J–value beyond the range of observed lines. The solid lines are calculated from the constants given by Gauyacq and Horani (1978, Table 8) and the circles represent observed spectral lines. (From Gauyacq and Horani, 1978.)

Therefore, owing to the effects of the l–mixing electrostatic perturbation, the L-uncoupling matrix element is slightly different from zero between states of different nominal l, since for a nonspherical system l is no longer a good

† The partitioning of ion-core non-sphericity into multipole moments depends on the choice of origin of coordinates, but for rotating molecules the choice of origin at the center of mass seems obligatory. In prolate spheroidal (elliptical) coordinates, the electronic Schrödinger equation separates (in a region outside the ion-core) into radial and angular equations, but the good angular momentum quantum number, $\tilde{l}$, has no simple relationship to the spherical quantum number, l (see Appendix B of Arif, *et al.*, 1996).

quantum number (see Jungen and Atabek, 1977, p. 5607). However, note that the goodness of l is destroyed by an electrostatic term, not by the $\mathbf{l}^+$ or $\mathbf{L}^+$ operator.

If the energy interval between the Λ–components of a Rydberg l–complex is similar to the magnitude of the vibrational interval, degeneracy can occur, for example, between a $\Pi(np\pi)$ $v = 0$ level and a $\Sigma(np\sigma)$ $v = 1$ level. Equation (3.5.5) shows that a nonzero interaction can occur between these two levels.

L-uncoupling perturbations also occur between non-Rydberg states of two configurations that differ by a single orbital. Many examples of perturbations between nearly degenerate $^2\Pi_{3/2}$ and $^2\Sigma_{1/2}$ levels of different v are known in the literature. These perturbations result in a large and strongly J–dependent Λ–doubling, which changes sign at the perturbation culmination. A striking example is shown in Fig. 3.18, where strong perturbations of the CS^+ $A^2\Pi_{3/2}(v' = 5)$ component are shown. In this case, extra lines due to the perturbing states are not observed. However, it has been shown (Gauyacq and Horani, 1978), that the state responsible for the perturbation is a high vibrational level ($v'' = 13$) of the $X^2\Sigma^+$ state.

Consider now a perturbation between a $\sigma\pi^4\pi'$ $^1\Pi_1$ state and a $\sigma^2\pi^3\pi'$ $^1\Sigma_0^-$ state. Such an example appears in the spectrum of CO (see Fig. 3.19; Field, *et al.*, 1972). It is easy to see from the method presented in Section 3.2.3 that the $\pi^3\pi'$ configuration gives rise to two $^1\Sigma$ states from $\pi^+(\pi^-)^2\pi'^+$ and $(\pi^+)^2\pi^-\pi'^-$ subshell occupancies. The $^1\Sigma^-$ and $^1\Sigma^+$ wavefunctions, properly symmetrized with respect to the σ_v operator acting only on the spatial part of the wavefunction, must be constructed as linear combinations of the four possible $\Lambda = 0, \Sigma = 0$ Slater determinants:

$$
\begin{aligned}
|^1\Pi_1\rangle &= (2)^{-1/2}\{|\pi^+\alpha\pi^+\beta\pi^-\alpha\pi^-\beta\sigma\alpha\pi'^+\beta| - |\pi^+\alpha\pi^+\beta\pi^-\alpha\pi^-\beta\sigma\beta\pi'^+\alpha|\} \\
&= (2)^{-1/2}\{\text{det } 1 - \text{ det } 2\} \\
|^1\Sigma_0^-\rangle &= \frac{1}{2}\{|\pi^+\alpha\pi^+\beta\pi^-\alpha\sigma\alpha\sigma\beta\pi'^-\beta| - |\pi^+\alpha\pi^+\beta\pi^-\beta\sigma\alpha\sigma\beta\pi'^-\alpha| \\
&\quad - |\pi^+\alpha\pi^-\alpha\pi^-\beta\sigma\alpha\sigma\beta\pi'^+\beta| + |\pi^+\beta\pi^-\alpha\pi^-\beta\sigma\alpha\sigma\beta\pi'^+\alpha|\} \\
&= \frac{1}{2}\{\text{det } 3 - \text{ det } 4 - \text{ det } 5 + \text{ det } 6\}.
\end{aligned}
$$

The only determinants of $|^1\Pi\rangle$ that differ from one of $|^1\Sigma^-\rangle$ by a single spin-orbital are det 1 $\sim$ det 5 ($\pi^+\beta \to \sigma\beta$) and det 2 $\sim$ det 6 ($\pi^+\alpha \to \sigma\alpha$). Thus, the nonzero matrix elements of $\mathbf{L}^+ = \sum_i \mathbf{l}_i^+$ are

$$
\begin{aligned}
\langle +\text{det } 1|\mathbf{l}^+| - \text{det } 5\rangle &= -(-1)^P \langle \pi^+\beta|\mathbf{l}^+|\sigma\beta\rangle = +b \qquad (P \text{ is odd}) \\
\langle -\text{det } 2|\mathbf{l}^+| + \text{det } 6\rangle &= -(-1)^{P'} \langle \pi^+\alpha|\mathbf{l}^+|\sigma\alpha\rangle = +b \qquad (P' \text{ is odd});
\end{aligned}
$$

thus

$$
\langle ^1\Pi_1^+|\mathbf{L}^+|^1\Sigma_0^-\rangle = \frac{1}{2}(2)^{-1/2}(b + b) = (2)^{-1/2}b.
$$

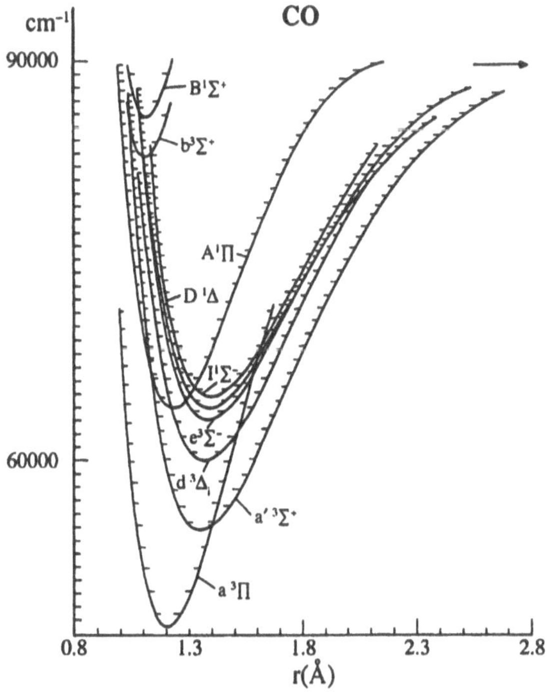

Figure 3.19: Potential energy curves of the excited states of CO. (From Tilford and Simmons, 1972.)

Transforming to the e/f parity basis, the above matrix element must be multiplied by $2^{1/2}$ since a Σ_0 function is involved. The final result is

$$\left\langle {}^1\Pi_{1f}, v|\mathbf{H}^{\text{ROT}}|{}^1\Sigma^-_{0f}, v'\right\rangle = -b\,\langle v|B(R)|v'\rangle\,[J(J+1)]^{1/2} \tag{3.5.7}$$

3.5.4 $^2\Pi \sim {}^2\Sigma^+$ Interaction

Matrix element expressions for spin-orbit and rotational interactions will be developed here in order to :

1. Treat the evolution from case (a) to case (b) for $^2\Pi$ states and thereby provide an alternative derivation of the expressions for case (b) functions in terms of case (a) functions previously derived in Section 3.2.1.4 using Clebsch-Gordan coefficients.

2. Construct the interaction matrix between $^2\Pi$ and $^2\Sigma^+$ states.

Using Eq. (3.1.16) and Eq. (3.4.6) to calculate the rotational and spin-orbit energies of the case (a) $^2\Pi_{3/2}$ state (without e/f symmetrization),

$$
\begin{aligned}
|^2\Pi_{3/2}\rangle &= \left|\Lambda = 1, S = \frac{1}{2}, \Sigma = \frac{1}{2}, \Omega = +\frac{3}{2}\right\rangle, \\
E\left(^2\Pi_{3/2}, v, J\right) &= T_\Pi^v + A_v/2 + B_v\left[J(J+1) - \frac{9}{4} + \frac{3}{4} - \frac{1}{4}\right] \\
&= T_\Pi^v + A_v/2 + B_v\left[\left(J + \frac{1}{2}\right)^2 - 2\right]. \qquad (3.5.8)
\end{aligned}
$$

For the $^2\Pi_{1/2}$ state, $|\Lambda = 1, S = \frac{1}{2}, \Sigma = -\frac{1}{2}, \Omega = +\frac{1}{2}\rangle$,

$$
\begin{aligned}
E\left(^2\Pi_{1/2}, v, J\right) &= T_\Pi^v - A_v/2 + B_v\left[J(J+1) - \frac{1}{4} + \frac{3}{4} - \frac{1}{4}\right] \\
&= T_\Pi^v - A_v/2 + B_v\left(J + \frac{1}{2}\right)^2 \qquad (3.5.9)
\end{aligned}
$$

The only nonzero off-diagonal matrix element between these two substates, which differ by $\Delta\Omega = \Delta\Sigma = 1$, is given by the **S**-uncoupling part of the rotational Hamiltonian. Using Table 3.2, one obtains

$$
\begin{aligned}
&\langle ^2\Pi_{3/2}, v|\mathbf{H}^{\mathrm{ROT}}|^2\Pi_{1/2}, v\rangle \\
&\quad = \langle J, \Lambda, S, \Sigma + 1, \Omega + 1 |-(1/2\mu R^2)\mathbf{J}^-\mathbf{S}^+| J, \Lambda, S, \Sigma, \Omega\rangle \\
&\quad = \left\langle v, J, 1, \frac{1}{2}, \frac{1}{2}, \frac{3}{2}, \right| - (1/2\mu R^2)\mathbf{J}^-\mathbf{S}^+ \left| v, J, 1, \frac{1}{2}, -\frac{1}{2}, \frac{1}{2}\right\rangle \\
&\quad = -(\hbar^2/2\mu)\langle v|R^{-2}|v\rangle\left[\frac{3}{4} - \left(-\frac{1}{2}\right)\left(\frac{1}{2}\right)\right]^{1/2}\left[J(J+1) - \left(\frac{1}{2}\right)\left(\frac{3}{2}\right)\right]^{1/2} \\
&\quad = -B_v\left[J(J+1) - \frac{3}{4}\right]^{1/2} = -B_v\left[\left(J + \frac{1}{2}\right)^2 - 1\right]^{1/2}. \qquad (3.5.10)
\end{aligned}
$$

It is necessary to assume that the two substates, $^2\Pi_{1/2}$ and $^2\Pi_{3/2}$, have identical potential curves, and thus the same vibrational wavefunctions. The other part of the **S**-uncoupling operator, $\mathbf{J}^+\mathbf{S}^-$, acts between $^2\Pi_{-3/2}\left(\Omega = -\frac{3}{2}\right)$ and $^2\Pi_{-1/2}\left(\Omega = -\frac{1}{2}\right)$, and gives the same matrix element. There are no off-diagonal matrix elements of the **S**-uncoupling operator between $\Omega > 0$ and $\Omega < 0$ $^2\Pi$ basis functions. Thus the e and f $^2\Pi$ basis functions, which are linear combinations of Ω and $-\Omega$ basis functions, have the same matrix elements of the rotational operator as the separate signed-Ω functions.

When the eigenvalues of the 2×2 $^2\Pi_{1/2}$, $^2\Pi_{3/2}$ matrix are found (see the matrix excluding $^2\Sigma$ in Table 3.9), the well-known result of Hill and Van Vleck (1928) is obtained. The energy levels, expressed as a function of $Y = A_v/B_v$,

Table 3.9: $^2\Pi(\pi) \sim ^2\Sigma^+(\sigma)$ Interaction Matrix[a]

e/f[b]	$^2\Pi_{1/2}$	$^2\Pi_{3/2}$	$^2\Sigma^+_{1/2}$
$^2\Pi_{1/2}$	$T_v^\Pi - \frac{A_v}{2} + B_v\left(J+\frac{1}{2}\right)^2$	$-B_v\left[\left(J-\frac{1}{2}\right)\left(J+\frac{3}{2}\right)\right]^{1/2}$	$\frac{a}{2} + \left[1 \mp \left(J+\frac{1}{2}\right)\right] bB_v$
$^2\Pi_{3/2}$		$T_v^\Pi + \frac{A_v}{2} + B_v\left[\left(J+\frac{1}{2}\right)^2 - 2\right]$	$-bB_v\left[\left(J-\frac{1}{2}\right)\left(J+\frac{3}{2}\right)\right]^{1/2}$
$^2\Sigma^+_{1/2}$			$T_v^\Sigma + B_v\left[\left(J+\frac{1}{2}\right)^2 \mp \left(J+\frac{1}{2}\right)\right]$

[a] Assuming identical potential curves for the $^2\Pi$ and $^2\Sigma^+$ states and that $v_\Pi = v_\Sigma$.
[b] The differences between the matrices for e and f levels are in the $^2\Pi_{1/2} \sim ^2\Sigma^+_{1/2}$ element and in the $^2\Sigma^+_{1/2} \sim ^2\Sigma^+_{1/2}$ element. The top sign is for e, the bottom for f.

are

$$E = T_\Pi^v + B_v\left[\left(J+\frac{1}{2}\right)^2 - 1\right] \pm (B_v/2)\left[Y(Y-4) + 4\left(J+\frac{1}{2}\right)^2\right]^{1/2}, \tag{3.5.11}$$

which can be rewritten as

$$E = T_\Pi^v + B_v\left[\left(J+\frac{1}{2}\right)^2 - 1\right] \pm \left[-\frac{1}{2}(A_v - 2B_v)\right]\left(1 + \frac{B_v^2\left[\left(J+\frac{1}{2}\right)^2 - 1\right]}{\left[-\frac{1}{2}(A_v - 2B_v)\right]^2}\right)^{1/2}. \tag{3.5.12}$$

In the case where B/A is very small, the radical can be expanded according to

$$(1+x)^{1/2} \simeq 1 + x/2,$$

resulting in an approximate expression for $^2\Pi$ levels intermediate between cases (a) and (b), but nearer the case (a) limit,

$$\begin{aligned} E\left(^2\Pi_{1/2}\right) &= T_\Pi^v - (1/2)(A_v - 2B_v) \\ &\quad + B_v[1 - B_v/(A_v - 2B_v)]\left[\left(J+\frac{1}{2}\right)^2 - 1\right] + \ldots \\ E\left(^2\Pi_{3/2}\right) &= T_\Pi^v + (1/2)(A_v - 2B_v) \\ &\quad + B_v\left[1 + B_v/(A_v - 2B_v)\right]\left[\left(J+\frac{1}{2}\right)^2 - 1\right] + \ldots \end{aligned}$$

These energy-level formulas may be expressed in a form similar to that in pure case (a) (i.e., without $^2\Pi_{1/2} \sim {}^3\Pi_{3/2}$ mixing) by introducing a B_{eff} constant for each Ω substate:

$$E\left(^2\Pi_{1/2}\right) = T_\Pi^v - \frac{1}{2}(A_v - 2B_v) + B_{1eff}\left[\left(J + \frac{1}{2}\right)^2 - 1\right] \quad (3.5.13)$$

$$E\left(^2\Pi_{3/2}\right) = T_\Pi^v + \frac{1}{2}(A_v - 2B_v) + B_{2eff}\left[\left(J + \frac{1}{2}\right)^2 - 1\right] \quad (3.5.14)$$

The difference between B_{1eff} and B_{2eff}, $2B_v^2/A_v$, is related to the value of the A constant. Each substate is doubly degenerate; the e and f levels of each substate have the same energy in the absence of interaction with $^2\Sigma^+$ or $^2\Sigma^-$ states.

Nearer the case (b) limit, if A is very small compared to BJ, the $^2\Pi$ eigenstates can be expressed in terms of case (b) functions, which can be obtained simply by setting $A = 0$ in the 2×2 $^2\Pi$ matrix. The energy eigenvalues for $BJ \gg A$ are, from Eq. (3.5.11),

$$E = T_\Pi^v + B_v\left[\left(J + \frac{1}{2}\right)^2 - 1\right] \pm B_v\left(J + \frac{1}{2}\right)\left[1 + \frac{Y(Y-4)}{8\left(J + \frac{1}{2}\right)^2}\right] + \ldots, \quad (3.5.15)$$

Equation (3.5.15) may be rearranged into case (b) form simply by replacing J with $N \mp \frac{1}{2}$ [$J = N - \frac{1}{2}$ for top sign in Eq. (3.5.15), $J = N + \frac{1}{2}$ for bottom sign]. The resulting energy level expressions are

$$F_2\left(N = J + \frac{1}{2}\right) \qquad E_2 = T_\Pi^v + B_v(N^2 - 1) + B_vN + B_vY(Y-4)/8N$$
$$= T_\Pi^v + B_vN(N+1) - B_v + B_vY(Y-4)/8N \quad (3.5.16)$$

which, except for the last term (negligible when $BN \gg A$), is the F_2 energy level expression of Eq. (3.2.68a), and

$$F_1\left(N = J - \frac{1}{2}\right) \qquad E_2 = T_\Pi^v + B_v[(N+1)^2 - 1]$$
$$- B_v(N+1) - B_vY(Y-4)/8(N+1)$$
$$= T_\Pi^v + B_vN(N+1) - B_v - B_vY(Y-4)/8(N+1), \quad (3.5.17)$$

which, except for the last term, is the F_1 energy level expression of Eq. (3.2.68b).

Starting from a $^2\Pi$ matrix expressed in terms of case (a) basis functions, case (b) energy level expressions have been derived for $BJ \gg A$. This means that the eigenfunctions are almost exactly the case (b) basis functions. Alternatively, in the case (b) basis the **S**-uncoupling operator, $-2B(R)\mathbf{J}\cdot\mathbf{S}$, can be replaced by $-B\left(\mathbf{N}^2 - \mathbf{J}^2 - \mathbf{S}^2\right)$ because

$$(\mathbf{N})^2 = (\mathbf{J} - \mathbf{S})^2$$
$$2\mathbf{J}\cdot\mathbf{S} = \mathbf{N}^2 - \mathbf{J}^2 - \mathbf{S}^2.$$

Thus, it is evident that the only nonzero case (b) matrix elements of the **S**-uncoupling operator are $\Delta N = \Delta J = \Delta S = 0$ and that all off-diagonal matrix elements of this operator between F_1 and F_2 basis functions vanish. The departure from case (b) at low J is caused by $\Delta N = \pm 1$ matrix elements of the $\mathbf{H}^{\mathrm{SO}}$ and $\mathbf{H}^{\mathrm{SS}}$ operators.

Consider the levels of an isolated $^2\Sigma^+$ state. The case (a) basis function that corresponds to the $\Omega = +\frac{1}{2}$ component of this state is

$$\left|\Lambda = 0, \Sigma = \frac{1}{2}, \Omega = \frac{1}{2}\right\rangle$$

while the $\Omega = -\frac{1}{2}$ component is

$$\left|\Lambda = 0, \Sigma = -\frac{1}{2}, \Omega = -\frac{1}{2}\right\rangle.$$

The energies of both components are identical, given by

$$E_\Sigma = T_\Sigma^v + B_v\left[J(J+1) - \frac{1}{4} + \frac{3}{4} - \frac{1}{4}\right] = T_\Sigma^v + B_v\left(J + \frac{1}{2}\right)^2. \qquad (3.5.18)$$

These two components differ by $\Delta\Omega = +1$ and interact via a nonzero matrix element of the **S**-uncoupling operator. This interaction can *never* be neglected, since the two interacting components have the same energy. The matrix element,

$$\left\langle \Lambda = 0, \Sigma = \frac{1}{2}\right| - \frac{1}{2\mu R^2}\mathbf{J}^-\mathbf{S}^+ \left|\Lambda = 0, \Sigma = -\frac{1}{2}\right\rangle = -B_v\left[J(J+1) + \frac{1}{4}\right]^{1/2}$$
$$= -B_v\left(J + \frac{1}{2}\right), \qquad (3.5.19)$$

causes the two case (a) components of $^2\Sigma^+$ belonging to the same J–value to become completely mixed. This gives rise to states that are e and f symmetry components,

$$\left|^2\Sigma_e^+\right\rangle = (2)^{-1/2}[|J, \Omega, \Lambda, \Sigma\rangle + |J, -\Omega, -\Lambda, -\Sigma\rangle]$$
$$= (2)^{-1/2}\left[\left|J, \frac{1}{2}, 0, \frac{1}{2}\right\rangle + \left|J, -\frac{1}{2}, 0, -\frac{1}{2}\right\rangle\right] \qquad (3.5.20a)$$

$$\left|^2\Sigma_f^+\right\rangle = (2)^{-1/2}[|J, \Omega, \Lambda, \Sigma\rangle - |J, -\Omega, -\Lambda, -\Sigma\rangle]$$
$$= (2)^{-1/2}\left[\left|J, \frac{1}{2}, 0, \frac{1}{2}\right\rangle - \left|J, -\frac{1}{2}, 0, -\frac{1}{2}\right\rangle\right] \qquad (3.5.20b)$$

Taking the matrix element of Eq. (3.5.19) into account, the energy eigenvalues are

$$E(^2\Sigma_e^+) = T_\Sigma^v + B_v\left[J(J+1) + \frac{1}{4} - \left(J + \frac{1}{2}\right)\right],$$

or, if $N = J - \frac{1}{2}$ (F_1 levels),

$$E\left({}^2\Sigma_e^+\right) = T_\Sigma^v + B_v[N(N+1)] \qquad (3.5.21)$$

and

$$E\left({}^2\Sigma_f^+\right) = T_\Sigma^v + B_v\left[J(J+1) + \frac{1}{4} + \left(J + \frac{1}{2}\right)\right],$$

or, if $N = J + \frac{1}{2}$ (F_2 levels),

$$E\left({}^2\Sigma_f^+\right) = T_\Sigma^v + B_v[N(N+1)]. \qquad (3.5.22)$$

This is the usual case (b) energy level expression for ${}^2\Sigma^+$ states. The $e(F_1)$ and $f(F_2)$ levels of the same N are degenerate, but the $e(F_1)$ and $f(F_2)$ levels of the same J are separated by $2B\left(J+\frac{1}{2}\right)$.

This e/f degeneracy between the two same–N components will be lifted by interaction with a ${}^2\Pi$ state. If the potential curves of the ${}^2\Sigma^+$ and ${}^2\Pi$ states are identical and the configurations of the ${}^2\Sigma^+$ and ${}^2\Pi$ states are σ^1 and π^1, respectively, then this interaction takes a particularly simple form. The discussion through the end of this chapter deals exclusively with these σ^1 and π^1 configurations. Perturbation selection rules for unsymmetrized basis functions require that the following interactions be considered. The ${}^2\Pi_{1/2}$ state experiences two types of $\Delta\Omega = 0$ interactions with ${}^2\Sigma_{1/2}^+$: *spin-orbit*,

$$\left\langle \Lambda = 1, \Sigma = -\frac{1}{2}, \Omega = \frac{1}{2}, v_\Pi \left|\mathbf{H}^{\mathrm{SO}}\right| \Lambda = 0, \Sigma = \frac{1}{2}, \Omega = \frac{1}{2}, v_\Sigma \right\rangle = \frac{a_+}{2}\langle v_\Pi | v_\Sigma\rangle \qquad (3.5.23)$$

where $a_+ = \langle \pi | \hat{a}\mathbf{l}^+ | \sigma\rangle$ and $\langle v_\Pi | v_\Sigma\rangle = \delta_{v_\Pi v_\Sigma}$; and *spin-electronic*,

$$\left\langle 1, -\frac{1}{2}, \frac{1}{2}, v_\Pi \left|(1/2\mu R^2)\mathbf{L}^+\mathbf{S}^-\right| 0, \frac{1}{2}, \frac{1}{2}, v_\Sigma \right\rangle = B_{v_\Pi v_\Sigma} b \qquad (3.5.24)$$

where $b = \langle \pi | \mathbf{l}^+ | \sigma\rangle$ and $B_{v_\Pi v_\Sigma} = B_v \delta_{v_\Pi v_\Sigma}$. The ${}^2\Pi_{1/2}$ state also experiences a $\Delta\Omega = \pm 1$ interaction via the **L**-uncoupling operator with the $\Omega = -\frac{1}{2}$ component of the ${}^2\Sigma^+$ state,

$$\left\langle 1, -\frac{1}{2}, \frac{1}{2}, v_\Pi \left| -\frac{1}{2\mu R^2}\mathbf{J}^-\mathbf{L}^+ \right| 0, -\frac{1}{2}, -\frac{1}{2}, v_\Sigma \right\rangle = -B_{v_\Pi v_\Sigma} b\left(J + \frac{1}{2}\right). \qquad (3.5.25)$$

In the following, it is assumed that $v_\Pi = v_\Sigma$. It is more convenient to construct the ${}^2\Pi \sim {}^2\Sigma^+$ interaction matrix between e/f–symmetrized basis functions. The e and f ${}^2\Sigma^+$ basis functions are given by Eqs. (3.5.20).

The corresponding symmetrized ${}^2\Pi$ functions are:

$$\left|{}^2\Pi^e_{1/2}\right\rangle = (2)^{-1/2}\left[\left|J,\frac{1}{2},1,-\frac{1}{2}\right\rangle + \left|J,-\frac{1}{2},-1,\frac{1}{2}\right\rangle\right] \quad (3.5.26a)$$

$$\left|{}^2\Pi^f_{1/2}\right\rangle = (2)^{-1/2}\left[\left|J,\frac{1}{2},1,-\frac{1}{2}\right\rangle - \left|J,-\frac{1}{2},-1,\frac{1}{2}\right\rangle\right] \quad (3.5.26b)$$

$$\left|{}^2\Pi^e_{3/2}\right\rangle = (2)^{-1/2}\left[\left|J,\frac{3}{2},1,\frac{1}{2}\right\rangle + \left|J,-\frac{3}{2},-1,-\frac{1}{2}\right\rangle\right] \quad (3.5.27a)$$

$$\left|{}^2\Pi^f_{3/2}\right\rangle = (2)^{-1/2}\left[\left|J,\frac{3}{2},1,\frac{1}{2}\right\rangle - \left|J,-\frac{3}{2},-1,-\frac{1}{2}\right\rangle\right]. \quad (3.5.27b)$$

The eigenvalues of the ${}^2\Pi_e$ and ${}^2\Pi_f$ matrices are identical with each other and with those of the $({}^2\Pi_{1/2}, {}^2\Pi_{3/2})$ and $({}^2\Pi_{-1/2}, {}^2\Pi_{-3/2})$ matrices [Eq. (3.5.11)]. The off-diagonal ${}^2\Pi \sim {}^2\Sigma^+$ elements are

$$\begin{aligned}
\left\langle {}^2\Pi^e_{1/2}\left|\mathbf{H}^{SO} + \mathbf{H}^{ROT}\right|{}^2\Sigma^{+e}\right\rangle &= a_+/2 + bB_v - bB_v(J+\frac{1}{2}) \\
&= a_+/2 - bB_v(J-\frac{1}{2}) \\
\left\langle {}^2\Pi^f_{1/2}\left|\mathbf{H}^{SO} + \mathbf{H}^{ROT}\right|{}^2\Sigma^{+f}\right\rangle &= a_+/2 + bB_v + bB_v(J+\frac{1}{2}) \\
&= a_+/2 + bB_v(J+\frac{3}{2}).
\end{aligned} \quad (3.5.28)$$

Note that the crucial difference between the $e \sim e$ and $f \sim f$ ${}^2\Pi \sim {}^2\Sigma^+$ matrix elements arises from an interference effect between $\Delta\Omega = +1$ and $\Delta\Omega = -1$ interactions that is not evident from Eqs. (3.5.24) and (3.5.25).[†] The reader

[†]This interference effect, which occurs only for even multiplicity $\Pi \sim \Sigma$ and $\Sigma \sim \Sigma$ matrix elements of $\mathbf{H}^{ROT}$, is the second of two cases where extreme care must be taken in going from the unsymmetrized, signed-Ω basis to the e/f symmetrized basis.

For odd-multiplicity systems, all off-diagonal matrix elements involving a Σ_0 basis function increase by a factor of $2^{1/2}$ in going from the unsymmetrized to the symmetrized basis.

For even-multiplicity systems, matrix elements between basis states of $\Omega > 0$ and $\Omega < 0$ cause problems. The $\mathbf{H}^{ROT}$ operator is responsible for both $\Delta\Omega = \Delta\Lambda = \pm 1$ and $\Delta\Omega = \Delta\Lambda = 0$ matrix elements. These two types of matrix elements are quite distinct in the signed basis, yet couple the same two basis states in the symmetrized basis. For example,

$$\left\langle {}^2\Sigma^+_{+1/2}|\mathbf{H}^{ROT}|{}^2\Pi_{-1/2}\right\rangle = -(J+\frac{1}{2})B_{v_\Sigma v_\Pi} b$$

yet

$$\left\langle {}^2\Sigma^+_{|1/2|}{}^e_f|\mathbf{H}^{ROT}|{}^2\Pi_{|1/2|}{}^e_f\right\rangle = \left[1 \mp \left(J+\frac{1}{2}\right)\right] B_{v_\Sigma v_\Pi} b.$$

This sort of interference effect (resulting in an e/f–dependent matrix element in the symmetrized basis) can occur via $\mathbf{H}^{ROT}$ only between two $\Lambda \leq 1$ $|\Omega| = \frac{1}{2}$ levels.

It is interesting and significant that both of these differences between the signed-Ω and e/f basis sets are crucial to the existence of Λ–doubling. For odd-multiplicity, Λ–doubling is *caused* by the odd number of Σ_0 basis functions. For even multiplicity, Λ–doubling is *caused* by an interference effect unique to $\mathbf{H}^{ROT}$ matrix elements involving at least one $\Sigma_{1/2}$ basis function.

should verify that the $e \sim f$ matrix elements for the $^2\Pi, {}^2\Sigma^+$ basis functions are identically zero, consistent with the $e \not\leftrightarrow f$ perturbation selection rule.

For $^2\Pi_{3/2}$, only the **L**-uncoupling interaction with $^2\Sigma$ is nonzero,

$$\left\langle 1, \frac{1}{2}, \frac{3}{2}, v_\Pi \right| - (1/2\mu R^2)\,\mathbf{J}^-\mathbf{L}^+ \left| 0, \frac{1}{2}, \frac{1}{2}, v_\Sigma \right\rangle = -bB_v \left[\left(J-\frac{1}{2}\right)\left(J+\frac{3}{2}\right)\right]^{1/2}$$

and

$$\left\langle ^2\Pi^e_{3/2} \right| \mathbf{H}^{\mathrm{ROT}} \left| ^2\Sigma^{+e} \right\rangle = \left\langle ^2\Pi^f_{3/2} \right| \mathbf{H}^{\mathrm{ROT}} \left| ^2\Sigma^{+f} \right\rangle = -bB_v \left[\left(J-\frac{1}{2}\right)\left(J+\frac{3}{2}\right)\right]^{1/2} \tag{3.5.29}$$

These $^2\Pi \sim {}^2\Sigma^+$ matrix elements are collected in Table 3.9.

The $^2\Pi \sim {}^2\Sigma^+$ interaction is different for the e and f levels. This difference gives rise to Λ–doubling in $^2\Pi$ and a spin-rotation splitting of the $\Delta N = 0$ degeneracy of the $^2\Sigma^+$ state. If the $^2\Sigma^+$ state is sufficiently far from the $^2\Pi$ state, it is possible to use second-order perturbation theory to evaluate the Λ–doubling. The $^2\Pi^e_{1/2}$ level is repelled by the $^2\Sigma^+$ state with an energy shift of

$$\Delta T_{e1/2} = \left[a_+/2 - bB_v\left(J-\frac{1}{2}\right)\right]^2 \Big/ (E_\Pi - E_\Sigma). \tag{3.5.30a}$$

The $^2\Pi^e_{1/2}$ component is lowered in energy if the $^2\Sigma^+$ state lies above $^2\Pi$, as shown in Fig. 3.20. The $^2\Pi^f_{1/2}$ level is also repelled, but with a different energy shift:

$$\Delta T_{f1/2} = \left[(a_+/2) + bB_v\left(J+\frac{3}{2}\right)\right]^2 \Big/ (E_\Pi - E_\Sigma). \tag{3.5.30b}$$

If a_+ and bB_v have the same signs, then $|\Delta T_f| > |\Delta T_e|$. The Λ–doubling contribution to $^2\Pi_{1/2}$ (more precisely, to the $\langle ^2\Pi_{1/2}|\mathbf{H}|^2\Pi_{1/2}\rangle$ matrix element) is given by

$$\begin{aligned}\Delta\nu_{fe}\left(^2\Pi_{1/2}\right) = \Delta T_{f1/2} - \Delta T_{e1/2} &= \frac{(a_+bB_v + 2b^2B_v^2)\left[(J+\frac{3}{2}) + (J-\frac{1}{2})\right]}{E_\Pi - E_\Sigma} \\ &= \frac{(2J+1)(bB_v)(a_+ + 2bB_v)}{E_\Pi - E_\Sigma}.\end{aligned} \tag{3.5.31}$$

If, as is typical, $2bB_v \ll a_+$, then

$$\Delta\nu_{fe}\left(^2\Pi_{1/2}\right) = \frac{2a_+bB_v}{E_\Pi - E_\Sigma}\left(J+\frac{1}{2}\right) = p\left(J+\frac{1}{2}\right) \tag{3.5.32}$$

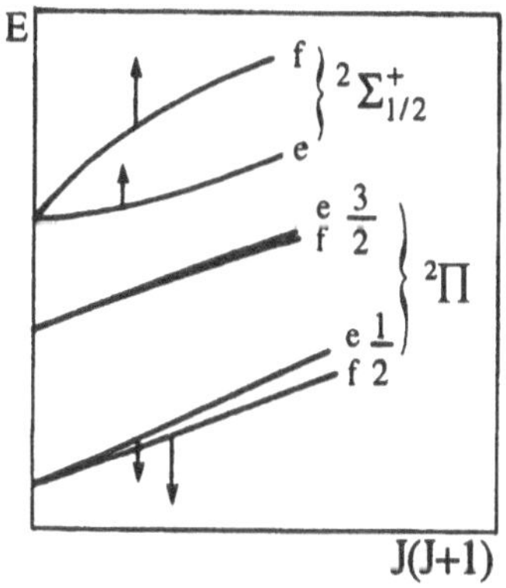

Figure 3.20: Λ–doubling of $^2\Pi$ states and spin-doubling of $^2\Sigma^+$. The e/f–dependent interaction between $^2\Pi$ and $^2\Sigma$ states is shown schematically. The $^2\Sigma^+$ state, if plotted versus $N(N+1)$ rather than $J(J+1)$, would show spin-rotation splitting identical in magnitude but opposite in sign to the $^2\Pi_{1/2}$ Λ–doubling. A $^2\Pi \sim {}^2\Sigma^-$ interaction would result in a similar pattern of Λ–doubling and spin-splitting, except that the shifts of e vs. f levels are reversed.

The $^2\Sigma$ e and f levels will be shifted by an amount equal and opposite to that of the $^2\Pi$ e and f levels (Fig. 3.20), and

$$\begin{aligned}\Delta\nu_{fe}\left({}^2\Sigma^+, J\right) &= E\left({}^2\Sigma^+_f, N = J + \frac{1}{2}\right) - E\left({}^2\Sigma^+_e, N = J - \frac{1}{2}\right)\\ &= 2\left(J + \frac{1}{2}\right)B_v + \left[2a_+ bB_v/\left(E_\Sigma - E_\Pi\right)\right]\left(J + \frac{1}{2}\right)\end{aligned}$$

or

$$\Delta\nu_{fe}\left({}^2\Sigma^+, N\right) = \left[2a_+ bB_v/\left(E_\Sigma - E_\Pi\right)\right]\left(N + \frac{1}{2}\right) \equiv -\gamma\left(N + \frac{1}{2}\right). \quad (3.5.33)$$

If the $^2\Pi$ and $^2\Sigma^+$ states (with identical potential curves) in question interact appreciably only with each other and not with other $^2\Sigma^\pm$ or $^2\Pi$ states, then $p = \gamma$. When the pair of interacting $^2\Pi$ and $^2\Sigma$ states are found to have $p = \gamma$, it is said that they are in a "unique perturber" relationship, which has *nothing* to do with the pure precession hypothesis, discussed below and in Section 5.5.

There is one other mechanism by which the $^2\Pi \sim {}^2\Sigma^+$ interaction results in Λ–doubling in the $^2\Pi$ state. The $\langle {}^2\Pi_{3/2}|\mathbf{H}|{}^2\Pi_{3/2}\rangle$ matrix element is affected in a *parity-independent* sense by the $\Pi \sim \Sigma$ interaction,

$$\Delta T_{f3/2} = \Delta T_{e3/2} = \frac{b^2B_v^2\left[\left(J - \frac{1}{2}\right)\left(J + \frac{3}{2}\right)\right]}{\left(E_\Pi - E_\Sigma\right)}, \quad (3.5.34)$$

but this makes no direct contribution to the $^2\Pi_{3/2}$ Λ–doubling. However (see Section 4.2 for a discussion of the Van Vleck transformation), the

$\langle {}^2\Pi_{3/2}|\mathbf{H}|{}^2\Pi_{1/2}\rangle$ matrix element is affected in a parity-dependent manner,

$$\langle {}^2\Pi_{e3/2}|\mathbf{H}|{}^2\Pi_{e1/2}\rangle = -\left[\left(J-\frac{1}{2}\right)\left(J+\frac{3}{2}\right)\right]^{1/2}\left[B_v + p/4 - q\left(J-\frac{1}{2}\right)\Big/2\right] \tag{3.5.35a}$$

$$\langle {}^2\Pi_{f3/2}|\mathbf{H}|{}^2\Pi_{f1/2}\rangle = -\left[\left(J-\frac{1}{2}\right)\left(J+\frac{3}{2}\right)\right]^{1/2}\left[B_v + p/4 + q\left(J+\frac{3}{2}\right)\Big/2\right] \tag{3.5.35b}$$

where

$$q = 2b^2B_v^2/\left(E_\Pi - E_\Sigma\right) \tag{3.5.36}$$

and p is defined by Eq. (3.5.32). Now, using second-order perturbation theory to estimate the parity-dependent effect of Eqs. (3.5.35a) - (3.5.35b) on the Λ–doubling in ${}^2\Pi_{3/2}$,

$$\Delta\nu_{fe}\left({}^2\Pi_{3/2}\right) \approx 2\left(J-\frac{1}{2}\right)\left(J+\frac{3}{2}\right)\left(J+\frac{1}{2}\right)B_vq/A_v, \tag{3.5.37}$$

then it is evident that Λ–doubling in ${}^2\Pi_{3/2}$ is proportional to $\sim qJ^3/Y$. (It can also be shown by second-order perturbation theory that the ${}^2\Pi_{1/2}$ Λ–doubling transferred to ${}^2\Pi_{3/2}$ by S-uncoupling is $\sim pJ^3/Y^2$.)

In Van Vleck's pure precession hypothesis (Van Vleck, 1929; Mulliken and Christy, 1931; see Section 5.5),

$$a_+ = [l(l+1)]^{1/2}A_v \tag{3.5.38}$$

and

$$b = [l(l+1)]^{1/2}. \tag{3.5.39}$$

Then

$$p = \gamma = \frac{2ABl(l+1)}{E_\Pi - E_\Sigma} \tag{3.5.40}$$

and

$$q = \frac{2B^2l(l+1)}{E_\Pi - E_\Sigma}. \tag{3.5.41}$$

For example, the spin-rotation splitting (called "ρ–doubling" by Van Vleck) of the OH $A^2\Sigma^+$ state is given mainly by interaction with the inverted $X^2\Pi$ state. Consequently, for A< 0 and $E_\Pi - E_\Sigma < 0$, one predicts $\gamma > 0$. The e–levels of the $A^2\Sigma^+$–state are well above the f–levels, in qualitative agreement with the pure precession prediction. Consequently p and q have the same sign for $A > 0$ and p and q have opposite signs for $A < 0$: this is the case for the $X^2\Pi$

states of OH, SH (see de Vivie, *et al.*, 1988), SeH and TeH (see Fink, *et al.*, 1989) hydrides. In the case of the CH molecule, the configuration of the $X^2\Pi$ ground state is $\sigma^2\pi$ and the sign and magnitude of the Λ–doubling of this state is determined by the difference between the contributions from the two states $B^2\Sigma^-$ and $C^2\Sigma^+$ which belong to the configuration $\sigma\pi^2$ ($\sigma\pi^+\pi^-$, three open shells, see Table 3.4) (see Hinkley, *et al.*, 1972).

Equations (3.5.40 and 3.5.41) are also valid for a π^3 $^2\Pi$ state. In the case of a $X^2\Sigma^+$ state interacting with an inverted $^2\Pi_i$ state, γ is negative. (An incorrect conclusion has been given by Childs, *et al.*, 1987, but corrected by Simard, *et al.*, 1999.)

3.6 References

Alekseev, V. A., Setser, D. W., and Tellinghuisen, J. (1999), *J. Mol. Spectrosc.* **195**, 162.

Alexander, M. H., Andresen, P., Bacis, R., Bersohn, R., Comes, F. J., Dagdigian, P. J., Dixon, R. N., Field, R. W., Flynn, G. W., Gericke, K.-H., Howard, B. J., Huber, J. R., King, D. S., Kinsey, J. L., Kleinermanns, K., Luntz, A. C., MacCaffery, A. J., Pouilly, B., Reisler, H., Rosenwaks, S., Rothe, E., Shapiro, M., Simons, J. P., Vasudev, R., Wiesenfeld, J. R., and Zare, R. N. (1988), *J. Chem. Phys.* **89**, 1749.

Amiot, C., Bacis, R., and Guelachvili, G. (1978), *Can. J. Phys.* **56**, 251.

Arif, M., Jungen, C., and Roche, A. L. (1996), *J. Chem. Phys.* **106**, 4102.

Aubanel, E. E., Gauthier, J. M., and Bandrauk, A. D. (1993), *Phys. Rev. A* **48**, 2145.

Bandrauk, A. D., and Child, M. S. (1970), *Mol. Phys.* **19**, 95.

Beebe, N. H. F., Thulstrup, E. W., and Andersen, A. (1976), *J. Chem. Phys.* **64**, 2080.

Behkami, A. N. (1971), *Nucl. Data Tables* **10**, 1.

Bergeman, T., Julienne, P. S., Williams, C. J., Tiesinga, E., Manaa, M. R., Wang, H., Gould, P. L., and Stwalley, W. C. (2002), *J. Chem. Phys.* **117**, 7491.

Blume, M., and Watson, R. E. (1962), *Proc. R. Soc. London Ser. A* **270**, 127.

Blume, M., and Watson, R. E. (1963), *Proc. R. Soc. London Ser. A* **271**, 565.

Blume, M., Freeman, A. J., and Watson, R. E. (1964), *Phys. Rev. A* **134**, 320.

Brion, J., and Malicet, J. (1976), *J. Phys. B* **9**, 2037.

Brion, J., Malicet, J., and Guenebaut, H. (1974), *Can. J. Phys.* **52**, 2143.

Brown, J. M., Hougen, J. T., Huber, K. P., Johns, J. W. C., Kopp, I., Lefebvre-Brion, H., Merer, A. J., Ramsay, D. A., Rostas, J., and Zare, R. N. (1975), *J. Mol. Spectrosc.* **55**, 500.

Brown, J. M., and Howard, B. J. (1976), *Mol. Phys.* **31**, 1517.

Brown, J. M., and Watson, J. K. G. (1977), *J. Mol. Spectrosc.* **65**, 65.

Brown, J. M., and Merer, A. J. (1979), *J. Mol. Spectrosc.* **74**, 488.

Brown, J. M., Colbourn, E. A., Watson, J. K. G., and Wayne, F. D. (1979), *J. Mol. Spectrosc.* **74**, 294.

Brown, J. M., Milton, D. J., Watson, J. K. G., Zare, R. N., Albritton, D. L., Horani, M., and Rostas, J. (1981), *J. Mol. Spectrosc.* **90**, 139.

Brown, J. M., Cheung, A. S.-C., and Merer, A. J. (1987), *J. Mol. Spectrosc.* **124**, 464.

Brown, J. M., and Carrington, A. (2003), "Rotational Spectroscopy of Diatomic Molecules," Cambridge University Press, Cambridge.

Bucksbaum, P. H., Zavriyev, A., Muller, H. G., and Schumacher, D. W. (1990), *Phys.*

Rev. Lett. **64**, 1883.
Bunker, P. R. (1968), *J. Mol. Spectrosc.* **28**, 422.
Carrington, A., Pyne, C. H., Shaw, A. M., Taylor, S. M., Hutson, J. M., and Law, M. M. (1996), *J. Chem. Phys.* **105**, 8602.
Cheetham, C. J., Gissane, W. J. M., and Barrow, R. F. (1965), *Trans. Faraday Soc.* **61**, 1308.
Cheung, A. S. C., Taylor, A. W., and Merer, A. J. (1982), *J. Mol. Spectrosc.* **92**, 381.
Cheung, W. Y., Chupka, W. A., Colson, S. D., Gauyacq, D., Avouris, P., and Wynne, J. J. (1983), *J. Chem. Phys.* **78**, 3625.
Childs, W. J., Poulsen, O., and Steimle, T. C. (1987), *J. Chem. Phys.* **88**, 598.
Cohen, J. S., and Schneider, B. (1974), *J. Chem. Phys.* **61**, 3230.
Condon, E. U., and Shortley, G. H. (1953), "The Theory of Atomic Spectra," Cambridge Univ. Press, London and New York.
Cooper, I. L., and Musher, J. I. (1972), *J. Chem. Phys.* **57**, 1333.
Cossart, D., Horani, M., and Rostas, J. (1977), *J. Mol. Spectrosc.* **67**, 283.
Cossart-Magos, C., Lefebvre-Brion, H., and Jungen, M. (1995), *Mol. Phys.* **85**, 821.
Cossart-Magos, C., Lefebvre-Brion, H., Jungen, M., and Launay, F. (1997), *J. Chem. Phys.* **107**, 1308.
Coxon, J. A. (1975), *J. Mol. Spectrosc.* **58**, 1.
Coxon, J. A., Ramsay, D. A., and Setser, D. W. (1975), *Can. J. Phys.* **53**, 1587.
Davis, S. L., and Alexander, M. H. (1983), *J. Chem. Phys.* **78**, 800.
Dieke, G. H. (1935), *Phys. Rev.* **47**, 870.
Dixon, R. N., and Hull, S. E. (1969), *Chem. Phys. Lett.* **3**, 367.
Drescher, M., Brockhinke, A., Böwering, N., Heinzmann, U., and Lefebvre-Brion, H. (1993), *J. Chem. Phys.* **99**, 2300.
Dressler, K. (1969), *Can. J. Phys.* **47**, 547.
Dressler, K., Gallusser, R., Quadrelli, P., and Wolniewicz, L. (1979), *J. Mol. Spectrosc.* **75**, 205.
Dressler, K., and Wolniewicz, L. (1981), *J. Mol. Spectrosc.* **86**, 534.
Dressler, K. (1983), *Ann. Isr. Phys. Soc.* **6**, 141.
Dressler, K. (1985), in "Photophysics and Photochemistry above 6eV" (F. Lahmani, ed.), p. 327, Elsevier Science Publishers, Amsterdam.
Dulick, M., and Field, R. W. (1985), *J. Mol. Spectrosc.* **113**, 105.
Edmonds, A. R. (1974), "Angular Momentum in Quantum Mechanics," Princeton Univ. Press, Princeton, New Jersey.
Effantin, C., d'Incan, J., Bacis, R., and Vergès, J. (1979a), *J. Mol. Spectrosc.* **76**, 204.
Effantin, C., Amiot, C., and Vergès, J. (1979b), *J. Mol. Spectrosc.* **76**, 221.
Effantin, C., Michaud, F., Roux, F., d'Incan, J., and Vergès, J. (1982), *J. Mol. Spectrosc.* **92**, 349.
Felenbok, P., and Lefebvre-Brion, H. (1966), *Can. J. Phys.* **44**, 1677.
Field, R. W., Wicke, B. G., Simmons, J. D., and Tilford, S. G. (1972), *J. Mol. Spectrosc.* **44**, 383.
Field, R. W., and Lefebvre-Brion, H. (1974), *Acta Phys. Hung.* **35**, 51.
Field, R. W., Gottscho, R. A., and Miescher, E. (1975), *J. Mol. Spectrosc.* **58**, 394.
Field, R. W. (1982), *Berichte Bunsenges. Phys. Chem.* **86**, 771.
Fink, E. M., Setzer, K. D., Ramsay, D. A., and Vervloet, M. (1989), *J. Mol. Spectrosc.* **138**, 19.
Fredin, S., Gauyacq, D., Horani, M., Jungen, C., Lefebvre, G., and Masaou-Seeuws, F. M. (1987), *Mol. Phys.* **60**, 825.
Gauyacq, D., and Horani, M. (1978), *Can. J. Phys.* **56**, 587.

Giusti-Suzor, A., He, X., Atabek, O., and Mies, F. H. (1990), *Phys. Rev. Lett.* **64**, 515.
Giusti-Suzor, A., and Mies, F. H. (1992), *Phys. Rev. Lett.* **68**, 3869.
Giusti-Suzor, A., Mies, F. H., Di Mauro, L. F., Charron, E., and Yang, B. (1995), *J. Phys. B* **28**, 309.
Gleichmann, M. M., and Hess, B. A. (1994), *J. Chem. Phys.* **101**, 9691.
Gray, J. A., Li, M., Nelis, T., and Field, R. W. (1991), *J. Chem. Phys.* **95**, 7164.
Green, S., and Zare, R. N. (1977), *J. Mol. Spectrosc.* **64**, 217.
Halevi, P. (1965), *Proc. Phys. Soc.* **86**, 1051.
Herzberg, G. (1950), "Spectra of Diatomic Molecules," Van Nostrand-Reinhold, Princeton, New Jersey, reprinted in 1989 by Krieger, Malabar.
Herzberg, G., and Jungen, C. (1972), *J. Mol. Spectrosc.* **41**, 425.
Herzberg, G., and Jungen, C. (1982), *J. Chem. Phys.* **77**, 5876.
Hill, E. L., and Van Vleck, J. H. (1928), *Phys. Rev.* **32**, 250.
Hinkley, R. K., Hall, J. A., Walker, T. E. H., and Richards, W. G. (1972), *J. Phys. B* **5**, 204.
Hobey, W. D., and McLachlan, A. D. (1960), *J. Chem. Phys.* **33**, 1694.
Horani, M., Rostas, J., and Lefebvre-Brion, H. (1967), *Can. J. Phys.* **45**, 3319.
Hougen, J. T. (1970), *Monogr.* 115, Nat. Bur. Stand. (US) Washington D.C.
Huber, K. P., and Alberti, F. (1983), *J. Mol. Spectrosc.* **97**, 387.
Ishiwata, T., Ishiguro, A., and Obi, K. (1991), *J. Mol. Spectrosc.* **147**, 300.
Jakubek, Z. J., and Field, R. W. (1994), *Phys. Rev. Lett.* **72**, 2167.
Jenouvrier, A., Pascat, B., and Lefebvre-Brion, H. (1973), *J. Mol. Spectrosc.* **45**, 46.
Jette, A. N., and Miller, T. A. (1974), *Chem. Phys. Lett.* **29**, 547.
de Jong, W. A., Visscher, D., and Nieuwport, W. C. (1997), *J. Chem. Phys.* **107**, 9045.
Jungen, C. (1966), *Can. J. Phys.* **44**, 3197.
Jungen, C., and Miescher, E. (1969), *Can. J. Phys.* **47**, 1769.
Jungen, C. (1970), *J. Chem. Phys.* **53**, 4168.
Jungen, C., and Atabek, O. (1977), *J. Chem. Phys.* **66**, 5584.
Jungen, C., Dabrowski, I., Herzberg, G., and Kendall, D. J. W. (1989), *J. Chem. Phys.* **91**, 3926.
Jungen, C., and Raşeev, G. (1998), *Phys. Rev. A* **57**, 2407.
Kaledin, L. A., Bloch, J. C., McCarthy, M. C., Gurvich, L. V., and Field, R. W. (1994), *Mol. Phys.* **83**, 881.
Kaledin, A. L., Heaven, M. C., Field, R. W., and Kaledin, L. A. (1996), *J. Mol. Spectrosc.* **179**, 310.
Kayama, K. (1965), *J. Chem. Phys.* **42**, 622.
Kayama, K., and Baird, J. C. (1967), *J. Chem. Phys.* **46**, 2604.
Kittrell, C., Abramson, E., Kinsey, J. L., McDonald, S., Reisner, D. E., Katayama, D., and Field, R. W. (1981), *J. Chem. Phys.* **75**, 2056.
Köppel, H., Domke, W., and Cederbaum, L. S. (1984), *Adv. Chem. Phys.* **57**, 59.
Kovács, I. (1969), "Rotational Structure in the Spectra of Diatomic Molecules," Am. Elsevier, New York.
Lagerqvist, A., and Renhorn, J. (1979), *Phys. Scr.* **19**, 289.
Langhoff, S. R., and Kern, C. W. (1977), in "Methods of Electronic Structure Theory" (H. F. Schaefer, III, ed.), Chap. 10, p. 381. Plenum, New York.
Larsson, M. (1981), *Phys. Scr.* **23**, 835.
Lavendy, H., Pouilly, B., and Robbe, J. M. (1984), *J. Mol Spectrosc.* **103**, 379.
Lefebvre-Brion, H. (1969), *Can. J. Phys.* **47**, 541.
Lefebvre-Brion, H., and Bessis, N. (1969), *Can. J. Phys.* **47**, 2727.
Lefebvre-Brion, H. (1990), *J. Chem. Phys.* **93**, 5898.

Lefebvre-Brion, H. (1992), in "Atomic and Molecular Physics and Quantum Physics" (H. A. Bachor, K. Kumar and B. A. Robson, eds.) p.389, World Scientific, Singapore.

Lefebvre-Brion, H. (1995), in "High Resolution Laser Photoionization and Photoelectron Studies" edited by I. Powis, T. Baer and C. Y. Ng, John Wiley & Sons, p. 171.

Leoni, M. (1972), thesis, ETH, Zürich (unpublished).

Levy, D. H. (1973), *Adv. Magn. Reson.* **6**, 1.

Lewis, J. K., and Hougen, J. T. (1968), *J. Chem. Phys.* **48**, 5329.

Li, Y., Petsalakis, D., Liebermann, H.-P., Hirsh, G., and Buenker, R. J. (1997), *J. Chem. Phys.* **106**, 1123.

Linton, C., Dulick, M., Field, R. W., Carette, P., Leyland, P. C., and Barrow, R. F. (1983), *J. Mol. Spectrosc.* **102**, 441.

Mank, A., Drescher, M., Huth-Fehre, T., Böwering, N., Heinzmann, U., and Lefebvre-Brion, H. (1991), *J. Chem. Phys.* **95**, 1676.

Martin, F., Churassy, S., Bacis, R., Field, R. W., and Vergès, J. (1983), *J. Chem. Phys.* **79**, 3725.

McWeeny, R. (1965), *J. Chem. Phys.* **42**, 1717.

Mead, C. A., and Truhlar, D. G. (1982), *J. Chem. Phys.* **77**, 6090.

Messiah, A. (1960), "Mécanique Quantique," Dunod, Paris; (1961), "Quantum Mechanics," North-Holland Publ., Amsterdam.

Miescher, E. (1976), *Can. J. Phys.* **54**, 2074.

Miller, T. A., Freund, R. S., and Field, R. W. (1976), *J. Chem. Phys.* **65**, 3790.

Mulliken, R. S. (1971), *J. Chem. Phys.* **55**, 288.

Mulliken, R. S., and Christy, A. (1931), *Phys. Rev.* **38**, 87.

Nikitin, E. E. (1965), *J. Chem. Phys.* **43** 744.

Nikitin, E. E., and Umanskii, S. Y. (1984), "Theory of the Slow Atomic Collisions," Springer Verlag, Berlin, p. 82.

Nikitin, E. E., and Zare, R. N. (1994), *Mol. Phys.* **82**, 85.

Nishinati, E., Tanaka, I., Tanaka, K., Kato, T., and Koyano, I. (1984), *J. Chem. Phys.* **81**, 3429.

Ogawa, M., and Tanaka, Y. (1962), *Can. J. Phys.* **40**, 1593.

Oppenheimer, M. (1972), *J. Chem. Phys.* **57**, 3899.

Osterwalder, A., Seiler, R., and Merkt, F. (2000), *J. Chem. Phys.* **113**, 7939.

Pauling, L., and Wilson, E. B. (1935), "Introduction to Quantum Mechanics," McGraw-Hill, New York.

Perdigon, P., and Femelat, B. (1982), *J. Phys. B* **15**, 2165.

Pique, J. P., Hartman, F., Bacis, R., Churassy, S., and Koffend, J. B. (1984), *Phys. Rev. Lett.* **52**, 267.

Posthumus, J. H., and McCann, J. F. (2001), in "Molecules and Clusters in Intense Laser Fields" (J. Posthumus, ed.) p. 27, Cambridge University Press.

Pouilly, B., Robbe, J. M., Schamps, J., Field, R. W., and Young, L. (1982), *J. Mol. Spectrosc.* **96**, 1.

Pyykkö, P., and Desclaux, J. P. (1976), *Chem. Phys. Lett.* **42**, 545.

Recknagel, A. (1934), *Z. Phys.* **87**, 375.

Reinhold, E., Hogervorst, W., and Ubachs, W. (1998), *Chem. Phys. Lett.* **296**, 411.

Rice, S. F., Martin, H., and Field, R. W. (1985), *J. Chem. Phys.* **82**, 5023.

Roche, A. L., and Lefebvre-Brion, H. (1973), *J. Chem. Phys.* **59**, 1914.

Roche, A. L., and Lefebvre-Brion, H. (1975), *Chem. Phys. Lett.* **32**, 155.

Robbe, J. M. (1978), thesis, Université des Sciences et Techniques de Lille, Lille, France (unpublished).

Ross, S. C., and Jungen, C. (1994), *Phys. Rev. A* **49**, 4353.
Sakamaki, T., Okabayashi, T., and Tanimoto, M. (1998), *J. Chem. Phys.* **109**, 7169.
Schamps, J. (1977), *J. Quant. Spectrosc. Radiat. Transfer* **17**, 685.
Schamps, J., Bencheikh, M., Barthelat, C., and Field, R. W. (1995), *J. Chem. Phys.* **103**, 8004.
Sidis, V., and Lefebvre-Brion, H. (1971), *J. Phys. B* **4**, 1040.
Simard, B., Jakubek, Z., Niki, H., and Balfour, W. J. (1999), *J. Chem. Phys.* **111**, 1483.
Sink, M. L., Lefebvre-Brion, H., and Hall, J. A. (1975), *J. Chem. Phys.* **62**, 1802.
Slater, J. C. (1960), "Quantum Theory of Atomic Structure," Vol. 1, McGraw-Hill, New York.
Smith, F. T. (1969), *Phys. Rev.* **179**, 111.
Smith, W. H., Brzozowski, J., and Erman, P. (1976), *J. Chem. Phys.* **64**, 4628.
Tellinghuisen, J. (1984), *J. Mol. Spectrosc.* **103**, 455.
Tilford, S. G., and Simmons, J. D. (1972), *J. Phys. Chem. Ref. Data* **1**, 147.
Tinkham, M., and Strandberg, M. W. P. (1955), *Phys. Rev.* **97**, 937.
Tonkyn, R. G., Winniczek, J. W., and White, M. G. (1989), *J. Chem. Phys.* **91**, 6632.
Townes, C. H., and Schawlow, A. L. (1955), "Microwave Spectroscopy", McGraw-Hill, New York.
Van Vleck, J. H. (1929), *Phys. Rev.* **33**, 467.
Van Vleck, J. H. (1951), *Rev. Mod. Phys.* **23**, 213.
Veseth, L. (1970), *Theor. Chim. Acta* **18**, 368.
Veseth, L. (1972), *J. Mol. Spectrosc.* **44**, 251.
Veseth, L. (1973a), *J. Phys. B* **6**, 1473.
Veseth, L. (1973b), *Mol. Phys.* **26**, 101.
de Vivie, R., Marian, C. M., and Peyerimhoff, S. D. (1988), *Mol. Phys.* **63**, 3.
Wales, N. P. L., Buma, W. J., de Lange, C. A., and Lefebvre-Brion, H. (1996), *J. Chem. Phys.* **105**, 2978.
Watson, J. K. G. (1994), *Mol. Phys.* **81**, 277.
Watson, J. K. G. (1999), pp. 293-327, in "The Role of Rydberg States in Spectroscopy and Photochemistry", (C. Sandorfy, ed.), Kluwer, Netherlands.
Weiner, J., Bagnato, V. S., Zilio, S., and Julienne, P. S. (1999), *Rev. Mod. Phys.* **71**, 1.
Weizel, W. (1931), "Bandenspecktren", Academische Verlagsgesellscheft, Leipzig.
Werner, H. J., and Meyer, W. (1981), *J. Chem. Phys.* **74**, 5802.
Wolniewicz, L., and Dressler, K. (1977), *J. Mol. Spectrosc.* **67**, 416.
Wolniewicz, L., and Dressler, K. (1979), *J. Mol. Spectrosc.* **77**, 286.
Wüest, A. (1999), Diplomarbeit, Laboratorium für Physikalische Chemie, ETH Zürich and Laboratoire Aimé Cotton, Université de Paris-Sud.
Yamazaki, M. (1963a), *Sci. Rep. Kanezawa Univ.* **8**, 371.
Yamazaki, M. (1963b), *Sci. Rep. Kanezawa Univ.* **8**, 397.
Yang, X., and Dagdigian, P. J. (1998), *J. Chem. Phys.* **109**, 8920.
Yao, G., and Chu, S. I. (1993), *Phys. Rev. A* **48**, 485.
Yarkony, D. R. (2000), p. 459, in "Computational Molecular Spectroscopy", (editors Jensen, P., and Bunker, P. R.), John Wiley and Sons, Chichester.
Yoshimine, M., and Tanaka, Y., private comm.; (1976), *J. Chem. Phys.* **64**, 2254.
Yoshino, K., Freeman, D. E., and Tanaka, Y. (1979), *J. Mol. Spectrosc.* **76**, 153.
Zare, R. N., Schmeltekopf, A. L., Harrop, W. J., and Albritton, D. L. (1973), *J. Mol. Spectrosc.* **46**, 37.
Zavriyev, A., Bucksbaum, P. H., Squier, J., and Saline, F. (1993), *Phys. Rev. Lett.* **70**, 1077.
Zhu, Y.-F., Grant, E. R., and Lefebvre-Brion, H. (1996), *J. Chem. Phys.* **99**, 2287.

Chapter 4

Methods of Deperturbation

Contents

4.1 **Variational Calculations** **234**
4.2 **The Van Vleck Transformation and Effective Hamiltonians** **237**
4.3 **Approximate Solutions** **243**
4.3.1 Graphical Methods for Deperturbation 243
4.3.2 Direct Diagonalization Versus Algebraic Approaches 247
4.4 **Exact Solutions** **248**
4.4.1 Least-Squares Fitting 248
4.4.1.1 Linear Least-Squares Fitting 248
4.4.1.2 Nonlinear Least-Squares Fitting 251
4.4.1.3 Practical Considerations 255
4.4.1.4 Least Squares vs. Robust Estimator Fitting 257
4.4.1.5 Types of Programs 258
4.4.2 Comparison Between Effective and True Parameters 261
4.4.3 Coupled Equations 264
4.5 **Typical Examples of Fitted Perturbations** **267**
4.5.1 An Indirect Heterogeneous Perturbation: NO $B^2\Pi \sim C^2\Pi \sim D^2\Sigma^+$ 267
4.5.2 A Strong Multistate Interaction in the NO Molecule 269
4.6 **References** **272**

This chapter deals with the process of obtaining molecular parameters from perturbed spectra. The terms in the Hamiltonian responsible for perturbations were discussed in Chapter 3, and the magnitudes of the off-diagonal matrix elements of these terms are interpreted in Chapter 5. The goal of a deperturbation calculation is not simply a phenomenological model that reproduces a specified set of spectral line frequencies and/or intensities; rather, the goal is a physical model, expressed in terms of potential energy curves and R–dependent interaction parameters, which is not merely capable of reproducing all spectral data from which the model parameters were derived, but should also be capable of predicting additional properties of the observed levels and the characteristics of not yet observed levels.

There are two sides to the deperturbation process: the Hamiltonian model and the partially assigned and analyzed spectrum. The molecular Hamiltonian operator must be organized in a way that, of the infinite number of molecular bound states and continua, emphasis is placed on the finite subset of levels sampled in the spectrum under analysis. A computational model is constructed in which the unknown information about the relevant states is represented by a set of adjustable parameters that may be systematically varied, until an acceptable match between calculated and observed properties is obtained.

The choice of a computational model depends in part on personal taste (matrix diagonalization versus numerical integration of partial differential equations, adiabatic versus diabatic potentials, Hund's coupling case), but the nature, quality, and completeness of the spectral data dictate the specific content of the model. This means that a great deal of work must be done on the raw spectra before the complete deperturbation model is set up and the final fitting process begun. Fragments of spectra must be partially assigned and the derived data reduced to phenomenological *effective molecular constants*. Traditional trial-and-error and graphical procedures are extremely useful at this initial stage.

Once lines are organized into branches and branches into subbands, computational models start to become useful by suggesting the identities of unassigned features and interrelationships between partially assigned features. The assembly of subbands into multiplet systems, vibrational levels into electronic states, and electronic states into configurations or Rydberg series often requires a succession of progressively more elaborate and extensive deperturbation models. At each stage, the parameters from one partial deperturbation model serve as initial parameter estimates for a more global and complete deperturbation.

The deperturbation methods discussed in this chapter range from non-adiabatic solutions of the complete Schrödinger equation (Section 4.1) to straight-line plots of quantities derived from observed energy levels (Section 4.3.1). Three computational techniques will be discussed: (1) variational calculations in which either the absolute energy (*ab initio* calculations, Section 4.1) or the sum of the squared deviations between observed and calculated energies (least-squares fitting, Section 4.4.1) is minimized in a finite basis set computational procedure; (2) perturbation theory in which the exact Hamiltonian is replaced by an approximate one, but where the Van Vleck (or "contact") transformation (Section 4.2) is used to replace an infinite basis set representation of $\mathbf{H}^{\text{approx}}$ by a finite-dimension effective Hamiltonian; (3) numerical solution of a system of coupled differential equations (Section 4.4.3).

4.1 Variational Calculations

The exact wavefunction ψ_i, associated with the energy level E_i, can, in principle, be expressed as a linear combination of the members of a complete set of basis functions. The mixing coefficients of each basis function are obtained by applying the variational principle [Eq. (3.1.9)]. As complete basis sets are infinite, the expansion of ψ_i must in practice be truncated and, consequently, the finite basis variational wavefunction obtained is only an approximation to ψ_i. The finite basis set can be the Born-Oppenheimer or, preferably, the adiabatic (i.e., solutions of the electronic Hamiltonian including the diagonal nuclear kinetic energy corrections, see Table 3.1) basis. The nonadiabatic vibronic wavefunctions (excluding rotation) have the form

$$\psi_i(r,R) = \sum_{n=1}^{q} \sum_{v_n=0}^{p_n} \Phi_n(r,R)\chi_{v_n}(R)c_{i,v_n}, \tag{4.1.1}$$

where the n–summation is over electronic states and the v_n–summation is over vibrational levels (for the basis states associated with electronic or vibrational continua, the summation is replaced by integration).

Application of the variational principle to the energy results in a system of linear equations of order $q + \sum_{n=1}^{q} p_n$:

$$0 = \sum_{n=1}^{q} \sum_{v_n=0}^{p_n} \left(H_{m,v_m;n,v_n} - E\delta_{m,v_m;n,v_n}\right) c_{v_n} \tag{4.1.2}$$

for $m = 1, \ldots, q$ and $v_m = 0, \ldots, p_m$ where

$$H_{m,v_m;n,v_n} = \langle \Phi_m \chi_{v_m} | \mathbf{H} | \Phi_n \chi_{v_n} \rangle \,. \tag{4.1.3}$$

Equation (4.1.2) is written assuming that the members of the $\{\Phi\}$ and $\{\chi\}$ basis sets are orthogonal and normalized. The condition for the existence of a nontrivial set of c_{v_n} is that E has a value that satisfies the secular equation

$$0 = \left| \sum_{n=1}^{q} \sum_{v_n=0}^{p_n} \left(H_{m,v_m;n,v_n} - E\delta_{m,v_m;n,v_n}\right) \right| . \tag{4.1.4}$$

The number of E values satisfying Eq. (4.1.4) is equal to $q + \sum_{n=1}^{q} p_n$, although some eigenvalues may be degenerate. A different set of c_{v_n} is associated with each E_i value; these sets of mixing coefficients are distinguished by adding an index to specify the eigenvalue; hence c_{i,v_n} is the mixing coefficient for the $\Phi_n \chi_{v_n}$ basis function in the ψ_i eigenfunction.

When one does a variational calculation to find the eigenfunctions of the exact molecular $\mathbf{H}$, the only quantum numbers available to label these eigenstates are *symmetry labels* (eigenvalues of an operator that commutes with $\mathbf{H}$) and the *energy order* within a given set o f symmetry labels. Only the index i appears on ψ in Eq. (4.1.1), not two indices to reflect the electronic *and* vibrational parentage of ψ. Although it is possible to label each ψ_i by its *nominal character* that corresponds to the largest $|c_{i,v_n}|^2$ in the Eq. (4.1.1) double summation, the only unambiguous (i.e., basis-set-independent) label is the energy order.

Very few nonadiabatic *ab initio* calculations have been performed. They require very accurate adiabatic calculations as a starting point. Variational nonadiabatic calculations have been completed for H_2^+ and H_2. In the $H_2 X^1\Sigma_g^+$ state, for which nonadiabatic effects are very small, the computed energy levels approach spectroscopic accuracy ($|\text{obs} - \text{calc}| \ll 1\text{cm}^{-1}$). For example, the nonadiabatic correction obtained by Bishop and Cheung (1978) with a 963 term wave function is only 0.42 cm^{-1} for H_2 $X^1\Sigma_g^+ v = 0$. Better results have been obtained by Wolniewicz (1995).

The electronic-vibration levels of H_2^+ and the vibrational levels of the H_2 $X^1\Sigma_g^+$ state are of little relevance to the perturbation problems discussed in this book. However, the E,F and G,K double-minimum $^1\Sigma_g^+$ states of H_2, already discussed in Section 3.3.3, are typical examples of homogeneous perturbations. Dressler, *et al.*, (1979) have calculated, in a very accurate adiabatic basis set, the off-diagonal matrix elements necessary for a nonadiabatic variational calculation. The eigenvalues of the 42 × 42 secular determinant were obtained and found to differ from experimentally observed energy levels. This difference was only a few reciprocal centimeters for the lowest levels, but the error increased significantly for the highest levels. This is a consequence of the finite dimension of the variational calculation, in particular the exclusion of the higher Rydberg states.

An alternative method to obtain the nonadiabatic wavefunctions [Eq. (4.1.1)], the coupled equation approach, will be discussed in Section 4.4.3. It has been used for an excited $^1\Sigma_g^+$ state of H_2 and the error is now smaller than 1 cm^{-1} for the lowest vibrational levels (Yu and Dressler, 1994). Multichannel Quantum Defect Theory (MQDT), discussed in Chapter 8, has also been used with success for the same problem by Ross and Jungen (1994). Finally, a variational numerical approach (Wolniewicz, 1996), gives very good results for H_2.

Although, for molecules heavier than H_2, *ab initio* adiabatic or Born-Oppenheimer electronic wavefunctions are now sufficiently accurate to be used as a starting point for nonadiabatic calculations, semiempirical variational procedures continue to be helpful. The diagonal and off-diagonal matrix elements of Eq. (4.1.4) are taken as unknown parameters. The values of these parameters are then adjusted so as to minimize the squared deviation between the calculated eigenvalues and the observed energy levels. However, since there are $N(N+1)/2$ independent elements in a symmetric N by N matrix and only N eigenvalues, it is necessary to build in some constraints whereby the large number of matrix elements are expressed in terms of a small number of unknown molecular constants. For example, the diagonal matrix elements could be expressed as a Dunham expansion,

$$E_{vJ} = T_e + \sum_{lm} Y_{lm} \left(v + \frac{1}{2}\right)^l [J(J+1)]^m, \tag{4.1.5}$$

many off-diagonal elements would be zero by symmetry, and the remaining off-diagonal elements would be interrelated by expressing them as a product of an explicitly calculable factor times an unknown variational parameter. In this way, a small number of variable parameters serves to define a model Hamiltonian.

It is not assured, just because the eigenvalues of a model Hamiltonian reproduce a group of spectroscopic energy levels, that the eigenfunctions of this Hamiltonian are identical to the eigenfunctions of the exact **H** or that they would reproduce other properties of the observed levels. If a model Hamiltonian, least-squares fitting procedure had been used for the $^1\Sigma_g^+$ states of H_2, the obtained fitting parameters would not yield matrix elements identical to those accurately calculable from *ab initio* wavefunctions. The reason for this is that the fitting model is a finite-order Hamiltonian matrix, whereas an infinite-order matrix is required to represent the exact molecular **H**.

The errors associated with the replacement of an infinite dimension **H** matrix by a finite dimension *effective* $\mathbf{H}^{\text{eff}}$ must be, in principle, small and of estimable magnitude. The replacement of **H** by $\mathbf{H}^{\text{eff}}$ is equivalent to a partitioning of an infinite basis set into two classes: *class 1* includes those basis functions that belong to the proper symmetry species and have energy expectation values in the energy region sampled by the spectrum under analysis; *class 2* includes the infinite number of remaining basis functions that are energetically remote from the class 1 functions, and their effect on the class 1 functions may therefore be treated by nondegenerate perturbation theory. The procedure for truncating the infinite **H** into a finite $\mathbf{H}^{\text{eff}}$ is known as the *Van Vleck* or the *contact transformation.*

4.2 The Van Vleck Transformation and Effective Hamiltonians

The Van Vleck transformation is an approximate block diagonalization procedure. It allows one, in effect, to throw away an infinite number of "unimportant" (class 2) basis functions after taking into account, through second-order nondegenerate perturbation theory, the effect of these ignorable functions on the finite number of "important" (class 1) functions. The Van Vleck transformation, **T**, is defined by

$$\mathbf{T}^\dagger \mathbf{H}\mathbf{T} = \widetilde{\mathbf{H}} = \left(\begin{array}{c|c} \widetilde{\mathbf{H}}_2 & \widetilde{\mathbf{H}}_{1\sim 2} \\ \hline \widetilde{\mathbf{H}}_{2\sim 1} & \widetilde{\mathbf{H}}_1 \end{array}\right). \qquad (4.2.1)$$

This transformation has the effect of reducing the size of the class 1$\sim$class 2 matrix element, $\widetilde{H}_{1\sim 2}$, relative to the corresponding $H_{1\sim 2}$ elements of the exact **H** so that the error made in computing the energy of a class 1 level resulting from neglect of a nonzero $\widetilde{H}_{1\sim 2}$ matrix element is significantly smaller than the error made by neglecting $H_{1\sim 2}$ itself. Thus it is a good approximation to define an effective Hamiltonian matrix as

$$\mathbf{H}^{\text{eff}} \equiv \widetilde{\mathbf{H}}_1, \qquad (4.2.2)$$

where only the class 1 portion of $\widetilde{\mathbf{H}}$ is retained.

The approximate block diagonalization produced by the Van Vleck transformation may be understood by first illustrating how nondegenerate perturbation theory approximately transforms the two-level problem into two almost decoupled one-level blocks. Consider

$$\mathbf{H} = \mathbf{H}^{(0)} + \mathbf{H}^{(1)}, \tag{4.2.3}$$

where the $\psi_i^{(0)}$ are eigenfunctions of $\mathbf{H}^{(0)}$ with eigenvalues $E_i^{(0)}$. The normalized eigenfunctions, correct through first-order perturbation theory, are

$$\begin{aligned} \psi_1 &\simeq c_1\psi_1^{(0)} + \psi_2^{(0)} H_{12}^{(1)}/(E_1^{(0)} - E_2^{(0)}) \\ \psi_2 &\simeq c_2\psi_2^{(0)} + \psi_1^{(0)} H_{21}^{(1)}/(E_2^{(0)} - E_1^{(0)}) \end{aligned} \tag{4.2.4}$$

where c_1 and c_2 are normalization factors,

$$c_1 = c_2 = \left[1 - \left|H_{12}^{(1)}\Big/\left(E_1^{(0)} - E_2^{(0)}\right)\right|^2\right]^{1/2} \simeq 1 - \frac{1}{2}\left|H_{12}^{(1)}\Big/\left(E_1^{(0)} - E_2^{(0)}\right)\right|^2.$$

If ψ_1, ψ_2 define the transformation

$$\mathbf{T}^\dagger \mathbf{H} \mathbf{T} = \widetilde{\mathbf{H}},$$

then the matrix elements of $\widetilde{\mathbf{H}}$ are

$$\begin{aligned} \widetilde{H}_{11} = \left\langle \psi_1 | \widetilde{\mathbf{H}} | \psi_1 \right\rangle &= E_1^{(0)} + H_{11}^{(1)} + \left|H_{12}^{(1)}\right|^2 \Big/ \left(E_1^{(0)} - E_2^{(0)}\right) \\ &+ \left[H_{12}^{(1)}/\left(E_1^{(0)} - E_2^{(0)}\right)\right]^2 \left[H_{11}^{(1)} - H_{22}^{(1)} - 2H_{12}^{(1)}\right] \end{aligned} \tag{4.2.5a}$$

$$\begin{aligned} \widetilde{H}_{22} = \left\langle \psi_2 | \widetilde{\mathbf{H}} | \psi_2 \right\rangle &= E_2^{(0)} + H_{22}^{(1)} - \left|H_{12}^{(1)}\right|^2 \Big/ \left(E_1^{(0)} - E_2^{(0)}\right) \\ &+ \left[H_{12}^{(1)}/\left(E_1^{(0)} - E_2^{(0)}\right)\right]^2 \left[H_{11}^{(1)} - H_{22}^{(1)} + 2H_{12}^{(1)}\right] \end{aligned} \tag{4.2.5b}$$

$$\begin{aligned} \widetilde{H}_{12} = \left\langle \psi_1 | \widetilde{\mathbf{H}} | \psi_2 \right\rangle &= \left[H_{12}^{(1)}/\left(E_1^{(0)} - E_2^{(0)}\right)\right] \left[\left(H_{22}^{(1)} - H_{11}^{(1)}\right)\right. \\ &\left. - \left|H_{12}^{(1)}\right|^2 \Big/ \left(E_1^{(0)} - E_2^{(0)}\right)\right]. \end{aligned} \tag{4.2.6}$$

The first three terms on the right-hand side of Eqs. (4.2.5) are the zeroth-, first-, and second-order terms from nondegenerate perturbation theory. The lowest-order term in the off-diagonal matrix element, Eq. (4.2.6), is second-order in the perturbation term, $\mathbf{H}^{(1)}$, and would therefore give rise to a fourth-order correction to the energy. Thus neglect of $\widetilde{H}_{12}$ would result in a less serious energy error than neglect of H_{12}.

An original derivation of the Van Vleck transformation was given by Löwdin (1951). The following derivation is adapted from Herschbach (1956) and Wollrab

(1967). It is useful to express the $\mathbf{H}$ and $\widetilde{\mathbf{H}}$ matrix representations of the Hamiltonian as

$$\mathbf{H} = \mathbf{H}^{(0)} + \lambda \mathbf{H}^{(1)} \tag{4.2.7}$$

$$\widetilde{\mathbf{H}} = \widetilde{\mathbf{H}}^{(0)} + \lambda \widetilde{\mathbf{H}}^{(1)} + \lambda^2 \widetilde{\mathbf{H}}^{(2)} + \ldots \tag{4.2.8}$$

where λ is an order-sorting parameter. Any unitary matrix can be expressed as

$$\mathbf{T} = e^{i\lambda \mathbf{S}} = 1 + i\lambda \mathbf{S} - (\lambda^2/2)\mathbf{S}^2 + \ldots \tag{4.2.9}$$

where $\mathbf{S}$ is a Hermitian matrix. A specific unitary transformation, $\widetilde{\mathbf{H}} = \mathbf{T}^\dagger \mathbf{H} \mathbf{T}$, called the *contact transformation*, is sought whereby all class 1 $\sim$ class 2 matrix elements in $\widetilde{\mathbf{H}}$ are smaller, by a factor of $\left[H_{12}^{(1)} \Big/ \left(E_1^{(0)} - E_2^{(0)}\right)\right]^2$, than the corresponding matrix element in $\mathbf{H}$. Writing out the effect of the general unitary transformation,

$$\begin{aligned}\widetilde{\mathbf{H}} &= \mathbf{H}^{(0)} + \lambda \widetilde{\mathbf{H}}^{(1)} + \lambda^2 \widetilde{\mathbf{H}}^{(2)} \\ &= \left[1 - i\lambda \mathbf{S} - (\lambda^2/2)\mathbf{S}^2\right] \left(\mathbf{H}^{(0)} + \lambda \mathbf{H}^{(1)}\right) \left[1 + i\lambda \mathbf{S} - (\lambda^2/2)\mathbf{S}^2\right],\end{aligned} \tag{4.2.10}$$

the terms with identical powers of λ can be collected,

$$\lambda^0: \quad \widetilde{\mathbf{H}}^{(0)} = \mathbf{H}^{(0)} \tag{4.2.11}$$

$$\lambda^1: \quad \widetilde{\mathbf{H}}^{(1)} = \mathbf{H}^{(1)} + i(\mathbf{H}^{(0)}\mathbf{S} - \mathbf{S}\mathbf{H}^{(0)}) \tag{4.2.12}$$

$$\lambda^2: \quad \widetilde{\mathbf{H}}^{(2)} = i(\mathbf{H}^{(1)}\mathbf{S} - \mathbf{S}\mathbf{H}^{(1)}) + \mathbf{S}\mathbf{H}^{(0)}\mathbf{S} - \frac{1}{2}\left(\mathbf{H}^{(0)}\mathbf{S}^2 - \mathbf{S}^2\mathbf{H}^{(0)}\right). \tag{4.2.13}$$

Equations (4.2.11 and 4.2.13) allow specific requirements to be built into $\mathbf{S}$ so that $\mathbf{T}$ transforms $\mathbf{H}$ into the approximately block diagonal form desired for $\widetilde{\mathbf{H}}$. For clarity, class 1 and 2 basis functions will be denoted by Roman (a, b) and Greek (α, β) letters, respectively. The goal is to make all first-order class 1$\sim$class 2 matrix elements of $\widetilde{\mathbf{H}}$ vanish,

$$\widetilde{H}_{a\alpha}^{(1)} = 0,$$

without causing any mixing among the class 1 (or class 2) basis functions. This is accomplished by requiring

$$S_{ab} = 0 \quad \text{(no mixing among class 1 functions)} \tag{4.2.14}$$

$$S_{\alpha\beta} = 0 \quad \text{(no mixing among class 2 functions)} \tag{4.2.15}$$

$$S_{a\alpha} = \frac{iH_{a\alpha}^{(1)}}{E_a^{(0)} - E_\alpha^{(0)}} \tag{4.2.16}$$

and inserting **Eqs. (4.2.14)-(4.2.16)** into Eq. (4.2.12). When Eq. (4.2.16) is inserted into Eq. (4.2.12),

$$\begin{aligned}\widetilde{H}'_{a\alpha} &= H_{a\alpha}^{(1)} + i\left[\left(\mathbf{H}^{(0)}\mathbf{S}\right)_{a\alpha} - \left(\mathbf{S}\mathbf{H}^{(0)}\right)_{a\alpha}\right] \\ &= H_{a\alpha}^{(1)} - \left[\frac{E_a^{(0)} H_{a\alpha}^{(1)}}{E_a^{(0)} - E_\alpha^{(0)}} - \frac{H_{a\alpha}^{(1)} E_\alpha^{(0)}}{E_a^{(0)} - E_\alpha^{(0)}}\right] = 0,\end{aligned}$$

because

$$\mathbf{H}^{(0)}|a\rangle = E_a^{(0)}|a\rangle$$
$$\mathbf{H}^{(0)}|\alpha\rangle = E_\alpha^{(0)}|\alpha\rangle.$$

Equations (4.2.14 and 4.2.15) cause the class 1 blocks of **H** and **S** (hence **T**) to commute (also the class 2 blocks), hence the only effect of **T** on the class 1 block of **H** is to fold into it some information from the off-diagonal $\mathbf{H}_{1\sim 2}$ block. Equation (4.2.16) forces the only nonzero elements of the second term in Eq. (4.2.12), $i(\mathbf{H}^{(0)}\mathbf{S}-\mathbf{S}\mathbf{H}^{(0)})$, to cancel exactly all interblock elements of $\mathbf{H}^{(1)}$. Thus the lowest-order interblock matrix elements of $\widetilde{\mathbf{H}}$ occur in the second-order term, $\widetilde{\mathbf{H}}^{(2)}$, and may be neglected.

Equations (4.2.11 and 4.2.16) completely define the Van Vleck transformation. The matrix elements in the class 1 block of $\widetilde{\mathbf{H}}$ are

$$\widetilde{H}_{ab} = E_a^{(0)}\delta_{ab} + \lambda H_{ab}^{(1)} + \frac{\lambda^2}{2}\sum_\alpha \left[\frac{H_{a\alpha}^{(1)}H_{\alpha b}^{(1)}}{E_a^{(0)} - E_\alpha^{(0)}} + \frac{H_{a\alpha}^{(1)}H_{\alpha b}^{(1)}}{E_b^{(0)} - E_\alpha^{(0)}}\right] + \ldots \quad (4.2.17)$$

The first two terms in Eq. (4.2.17) are identical to the corresponding terms in the exact **H**. The correction terms introduced by the Van Vleck transformation appear both on and off diagonal in $\widetilde{\mathbf{H}}$, in contrast to the strictly on-diagonal location of energy correction terms from ordinary, nondegenerate second-order perturbation theory. Often, the off-diagonal correction term is replaced by a single summation over an averaged energy denominator,

$$\widetilde{H}_{ab}^{(2)} = \sum_\alpha \frac{H_{a\alpha}^{(1)}H_{\alpha b}^{(1)}}{\left[\left(E_a^{(0)} + E_b^{(0)}\right)/2 - E_\alpha^{(0)}\right]} \quad (4.2.18)$$

which is negligibly different from the form of Eq. (4.2.17) if

$$\left|E_a^{(0)} - E_b^{(0)}\right| \ll \left|\left(E_a^{(0)} + E_b^{(0)}\right)/2 - E_\alpha^{(0)}\right|.$$

The third-order terms, $\widetilde{\mathbf{H}}^{(3)}$, may be obtained by retaining additional powers of λ in Eq. (4.2.10),

$$\begin{aligned}\widetilde{H}_{ab}^{(3)} &= \sum_{\alpha,\beta}\left[\frac{H_{a\alpha}^{(1)}H_{\alpha\beta}^{(1)}H_{\beta b}^{(1)}}{\left(E_a^{(0)} - E_\alpha^{(0)}\right)\left(E_\beta^{(0)} - E_b^{(0)}\right)}\right] \\ &- \frac{1}{2}\sum_{\alpha,c}\left[\frac{H_{ac}^{(1)}H_{c\alpha}^{(1)}H_{\alpha b}^{(1)}}{\left(E_c^{(0)} - E_\alpha^{(0)}\right)\left(E_b^{(0)} - E_\alpha^{(0)}\right)} + \frac{H_{a\alpha}^{(1)}H_{\alpha c}^{(1)}H_{cb}^{(1)}}{\left(E_a^{(0)} - E_\alpha^{(0)}\right)\left(E_c^{(0)} - E_\alpha^{(0)}\right)}\right].\end{aligned} \quad (4.2.19)$$

It is important to note that the summations implied in the Van Vleck transformation are over all electronic and all vibrational levels of class 2. The perturbation summations are restricted neither to only the nearest electronic state of a particular symmetry species nor to the bound vibrational levels. It is necessary, in principle, to integrate over the electronic (ionization) and nuclear (dissociation) continua.

The Van Vleck transformation specifies the functional form (Zare, *et al.*, 1973; Hill and Van Vleck, 1928; Mulliken and Christy, 1930; Horani, *et al.*, 1967; Freed, 1966; Miller, 1969; Brown, *et al.*, 1979; Brown and Merer, 1979) of the effective Hamiltonian in terms of a small number of parameters from $\mathbf{H}^{(0)}$ and $\mathbf{H}^{(1)}$ having explicit mechanical (rotation-vibration constants) or electromagnetic $(T_e, A, \lambda, \gamma)$ significance and a much larger group of second-order parameters from $\widetilde{\mathbf{H}}^{(2)}$ $(o, p, q, D, H, L, A_D, \text{etc.})$ of significance ranging from explicit to phenomenological fitting parameters.

Values for many of the parameters in $\mathbf{H}^{\text{eff}}$ cannot be determined from a spectrum, regardless of the quality or quantity of the spectroscopic data, because of correlation effects. When two parameters enter into the effective Hamiltonian with identical functional forms, only their sum can be determined empirically. Sometimes it is possible to calculate, either *ab initio* or semiempirically, the value of one second-order parameter, thereby permitting the other correlated parameter to be evaluated from the spectrum. Often, although the parameter definition specifies a summation over an infinite number of states, the largest part or the explicitly vibration-dependent part of the parameter may be evaluated from an empirically determined electronic matrix element times a sum over calculable vibrational matrix elements and energy denominators (Wicke, *et al.*, 1972).

Brown, *et al.*, (1979) and Brown and Merer (1979) have proposed that $\mathbf{H}^{\text{eff}}$ be expressed in terms of a minimal set of fitting parameters. In this way, the same $\mathbf{H}^{\text{eff}}$ would be used by most spectroscopists and the task of reproducing a spectrum from a set of published constants would be simplified. This proposal of a universal set of fitting parameters will not necessarily lead to the reduction of spectra to phenomenological rather than physical constants, because each of the fitting parameters is defined explicitly in terms of the parameters from $\mathbf{H}^{(0)}, \mathbf{H}^{(1)}$, and $\mathbf{H}^{(2)}$. There is, however, a danger that users of universal $\mathbf{H}^{\text{eff}}$ matrices will neither study the assumptions made in the derivation of $\mathbf{H}^{\text{eff}}$ nor be prepared to exploit the significance of or the implicit relationships among the fitting parameters.

Before concluding this section it will be useful to discuss two examples of the Van Vleck transformation: centrifugal distortion in a $^3\Pi$ state and Λ–doubling in a $^2\Pi$ state.

Although centrifugal distortion is not a perturbation effect, a derivation of the form of the centrifugal distortion terms in $\mathbf{H}^{\text{eff}}$ provides an excellent illustration of the Van Vleck transformation. If the vibrational eigenfunctions of the *nonrotating* molecular potential, $V(R)$ rather than $[V(R) + J(J+1)\hbar^2/2\mu R^2]$, are chosen as the vibrational basis set, then the rotational "constant" becomes an operator,

$$\langle v|B(R)|v'\rangle = (\hbar^2/2\mu)\,\langle v|R^{-2}|v'\rangle = B_{vv'} \neq 0,$$

which is not diagonal in the vibrational basis set.

The rotational Hamiltonian for a $^3\Pi$ state has matrix elements

$$\langle ^3\Pi_0, v|\mathbf{H}^{\mathrm{ROT}}|^3\Pi_0, v'\rangle = B_{vv'}[J(J+1)+1] \tag{4.2.20a}$$
$$\langle ^3\Pi_1, v|\mathbf{H}^{\mathrm{ROT}}|^3\Pi_1, v'\rangle = B_{vv'}[J(J+1)+1] \tag{4.2.20b}$$
$$\langle ^3\Pi_2, v|\mathbf{H}^{\mathrm{ROT}}|^3\Pi_2, v'\rangle = B_{vv'}[J(J+1)-3] \tag{4.2.20c}$$
$$\langle ^3\Pi_0, v|\mathbf{H}^{\mathrm{ROT}}|^3\Pi_1, v'\rangle = -B_{vv'}[2J(J+1)]^{1/2} \tag{4.2.20d}$$
$$\langle ^3\Pi_0, v|\mathbf{H}^{\mathrm{ROT}}|^3\Pi_2, v'\rangle = 0 \tag{4.2.20e}$$
$$\langle ^3\Pi_1, v|\mathbf{H}^{\mathrm{ROT}}|^3\Pi_2, v'\rangle = -B_{vv'}[2J(J+1)-4]^{1/2} \tag{4.2.20f}$$

The Van Vleck transformation incorporates the effect of all $\Delta v \neq 0$ matrix elements into the $\Delta v = 0$ block of the $^3\Pi$ Hamiltonian. The perturbation summation

$$\sum_{v'} \frac{\langle v|B(R)|v'\rangle \langle v'|B(R)|v\rangle}{E_v^0 - E_{v'}^0} \equiv -D_v \tag{4.2.21}$$

defines D_v (Zare, *et al.*, 1973; Albritton, *et al.*, 1973), but this summation must be carried out by summing over the bound levels and integrating over the continuum. The matrix elements of the transformed Hamiltonian, $\tilde{\mathbf{H}}$, require summation over the $\Omega = 0, 1$, and 2 levels of the v' level:

$$\left\langle ^3\Pi_0, v|\tilde{\mathbf{H}}|^3\Pi_0, v\right\rangle = -D_v[J(J+1)+1]^2 - D_v[2J(J+1)],$$

where the first term comes from the $\Omega = 0 \rightarrow 0 \rightarrow 0$ path and the second from $\Omega = 0 \rightarrow 1 \rightarrow 0$. The centrifugal distortion terms appear in $\mathbf{H}^{\mathrm{eff}}$ as follows:

$$\left\langle ^3\Pi_0, v|\tilde{\mathbf{H}}|^3\Pi_0, v\right\rangle = -D_v[x^2+4x+1] \tag{4.2.22a}$$
$$\left\langle ^3\Pi_1, v|\tilde{\mathbf{H}}|^3\Pi_1, v\right\rangle = -D_v[x^2+6x-3] \tag{4.2.22b}$$
$$\left\langle ^3\Pi_2, v|\tilde{\mathbf{H}}|^3\Pi_2, v\right\rangle = -D_v[x^2-4x+5] \tag{4.2.22c}$$
$$\left\langle ^3\Pi_0, v|\tilde{\mathbf{H}}|^3\Pi_1, v\right\rangle = +D_v[2(x+1)(2x)^{1/2}]^{\dagger} \tag{4.2.22d}$$
$$\left\langle ^3\Pi_0, v|\tilde{\mathbf{H}}|^3\Pi_2, v\right\rangle = -D_v[4x(x-2)]^{1/2} \tag{4.2.22e}$$
$$\left\langle ^3\Pi_1, v|\tilde{\mathbf{H}}|^3\Pi_2, v\right\rangle = +D_v[2(x-1)(2x-4)^{1/2}]^{\dagger} \tag{4.2.22f}$$
$$x \equiv J(J+1).$$

† Note that the centrifugal distortion matrix elements for $\Delta\Omega =$ odd appear with + signs. The $\Delta\Omega =$ odd centrifugal distortion matrix elements result, via the Van Vleck transformation, from a product of one $\Delta\Omega = \pm 1$ and one $\Delta\Omega = 0$ matrix element of $\mathbf{H}^{\mathrm{ROT}}$, whereas the $\Delta\Omega =$ even centrifugal distortion matrix element result from the square of a $\Delta\Omega = 0$ and/or a $\Delta\Omega =$

It is evident from Eq. (4.2.22) that the correct form of the centrifugal distortion terms is *not* given by the intuitive prescription: replace B_v everywhere it appears in Eq. (4.2.20) by $B_v - D_v J(J+1)$. Alternative methods for evaluating the Eq. (4.2.21) perturbation summation have been proposed by Tellinghuisen (1973) and Kirschner and Watson (1973).

As discussed in Sections 3.5.4 and 5.5 , the Λ–doubling in a $^2\Pi$ state can result from interactions with remote $^2\Sigma^+$ and $^2\Sigma^-$ states via $\mathbf{H}^{\mathrm{ROT}}$ and $\mathbf{H}^{\mathrm{SO}}$. The Van Vleck transformation defines [Eqs. (5.5.1a) - (5.5.3a)] three second-order parameters (o, p, and q) that appear in the $^2\Pi$ block of $\widetilde{\mathbf{H}}^{(2)}$. These second-order $^2\Pi \sim {}^2\Sigma$ interaction parameters cause both e/f independent level shifts as well as Λ–doubling. The e/f–dependent terms all arise from the e/f–dependence of the

$$\left\langle {}^2\Pi_{1/2}{}^{e}_{f}, v_\Pi \left| \mathbf{H}^{\mathrm{ROT}} \right| {}^2\Sigma^{s}{}^{e}_{f}, v_\Sigma \right\rangle = \left[1 \mp (-1)^s \left(J + \frac{1}{2}\right)\right] B_{v_\Pi, v_\Sigma} \langle \Pi | L^+ | \Sigma \rangle$$

matrix element ($s = 0$ for Σ^+ and $s = 1$ for Σ^-). This means that e/f–dependent terms can arise in $\widetilde{\mathbf{H}}^{(2)}$ if: (1), the $\mathbf{H}^{\mathrm{ROT}}$ operator appears once (p) or twice (q) in $H^{(1)}_{a\alpha} H^{(1)}_{\alpha b}$; and (2), either a or $b = \left|{}^2\Pi_{1/2}\right\rangle$. This explains why the o parameter cannot contribute to Λ–doubling in $^2\Pi$ states and why e/f–dependence appears in the $\left\langle {}^2\Pi_{1/2} \left| \widetilde{\mathbf{H}}^{(2)} \right| {}^2\Pi_{1/2} \right\rangle$ and $\left\langle {}^2\Pi_{1/2} \left| \widetilde{\mathbf{H}}^{(2)} \right| {}^2\Pi_{3/2} \right\rangle$ matrix elements.

4.3 Approximate Solutions

At an early stage in the analysis of a perturbed spectrum, graphical methods based on approximate algebraic solutions to the secular determinant [Eq. (4.1.4)] can be very useful, both as an aid in pattern recognition and in estimation of parameter values. At some point before completion of the analysis it is usually advantageous to switch from a graphical/algebraic approach to one based on exact diagonalization of $\mathbf{H}^{\mathrm{eff}}$ matrices. The labor in setting up an algebraic model for anything larger than a two-level interaction problem, even when the equations are presented in a convenient format (Kovács, 1969), is often greater than required to write a suitable $\mathbf{H}^{\mathrm{eff}}$ subroutine. In addition, a statistically rigorous least-squares fitting procedure is often better able to make use of fragmentary data than a graphical or algebraic procedure.

4.3.1 Graphical Methods for Deperturbation

There are many extremely useful graphical presentations of spectral data. Most are based on several of the following ideas: (1) straight-line plots, (2) scale

odd matrix element of $\mathbf{H}^{\mathrm{ROT}}$. The sign for the $\Delta\Omega =$ odd matrix elements is a consequence of the phase choice used in this book that results in positive, real values for nonzero matrix elements of $\mathbf{J}^{\pm}$ operators. The opposite phase choice made by Freed (1966) and used by Wicke, *et al.*, (1972) results in positive signs for the $\mathbf{H}^{\mathrm{ROT}}$ matrix elements in Eqs. (4.2.20d) and (4.2.20f), hence negative signs in Eqs. (4.2.22d) and (4.2.22f).

expansion by subtracting away the expected, unperturbed behavior, (3) display of large-scale regularities, (4) the center-of-gravity rule.

Prior to the days when spectral line assignments could be checked by double-resonance or resolved-fluorescence techniques, assignments were based on both pattern recognition and redundancy. By constructing suitable combination differences, the spectroscopist is able to go directly from the spectrum to level separations in either the upper or lower electronic state without assuming any model (except that dipole selection rules restrict ΔJ to $\pm 1, 0$). This means that combination difference graphical presentations can serve the dual purposes of displaying a structural feature of one electronic state and verifying assignments by plotting the corresponding values of measured combination differences obtained from two bands thought to have an upper or lower level in common.

Figure 3.18 is a reduced term-value plot for the CS^+ $A^2\Pi_i(v = 5)$ state perturbed by $X^2\Sigma^+$ $(v = 13)$. The form of the plot $[T - B(J+\frac{1}{2})^2$ versus $(J+\frac{1}{2})^2]$ is appropriate for a near case (a) $^2\Pi$ state. The scale expansion achieved by subtracting 0.68 $(J + \frac{1}{2})^2$ from all of the energy levels (this amounts to 460 cm^{-1} by $J = 25.5$) permits the ~ 2 cm^{-1} perturbation level shifts in $^2\Pi_{3/2}$ at $J = 18.5$ and 19.5 to remain visible and allows the values of the effective B_v and p_v $^2\Pi$ constants to be determined from the slopes and splitting of the nearly straight $^2\Pi_{1/2}$ e– and f–level plots. The pattern of the perturbation is unmistakably that of a Σ state (large splitting between e and f levels of the same J value) with $B_\Sigma > B_\Pi$ (level shift first positive, then negative). The maximum level shift provides a good estimate (lower bound) for the $B\mathbf{L}^+\mathbf{J}^-$ $^2\Pi_{3/2} \sim {}^2\Sigma^+$ matrix element $[2.5 \text{ cm}^{-1}/(J + \frac{1}{2}) \simeq 0.13$ at $J = 19]$.

Figure 5.7 is a crossing diagram [E versus $J(J + 1)$] that shows the J–values at which the vibrational levels of the $Si^{16}O$ and $Si^{18}O$ $A^1\Pi$ state are crossed by various perturbers (Field, *et al.*, 1976). The different classes of perturber ($^3\Sigma^-$ versus $^1\Sigma^-$ versus $^{1,3}\Delta$) were immediately evident from the patterns of level crossings in the e and f components of the $A^1\Pi$ state. The overall pattern of perturbations, including the expected locations of undetected weak perturbations or multiple crossings as shown by the nearly straight tie lines connecting the crossing points for each class of perturber, is very clear. This sort of diagram is very useful for arranging the perturbing levels into electronic states and establishing their relative vibrational numberings.

Several other kinds of information are available from the information in Fig. 5.7. Two consecutive vibrational levels of $A^1\Pi$ are crossed by the same vibrational level of $e^3\Sigma^-$. The Q branch (F_2) crossing occurs in $v_A = 1$ and 0 at $J = 48.9$ and 61.5. Since the deperturbed $A^1\Pi$ term energies at these two J–values are known by interpolation, accurate values for $B(e^3\Sigma^-)$ and $E(e^3\Sigma^-)$ can respectively be determined from the slope and intercept of the straight line drawn through these $v_A = 1$, $J = 48.9$ and $v_A = 0$, $J = 61.5$ term values. Alternatively, the J–values of all three $^1\Pi \sim {}^3\Sigma^-(F_1, F_2, F_3)$ crossings are determined accurately at many perturbations, for example, (56.5, 61.5, 66.0) in $Si^{16}O$ $A^1\Pi$ $v = 0$. Each $^3\Sigma^-$ N–level consists of three near-degenerate J–components. The perturbation selection rule is $\Delta J = 0$, thus the F_1 and F_3

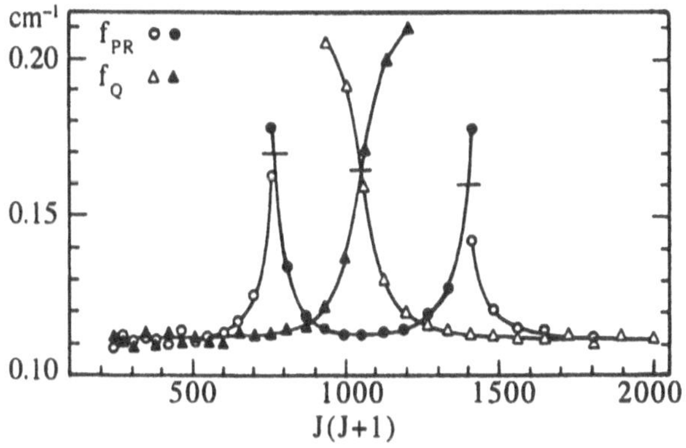

Figure 4.1: Plot of effective ΔB–value versus $J(J+1)$ for the $Si^{16}O$ $e^3\Sigma^-(v=9) \sim A^1\Pi(v=2)$ perturbation. The quantities plotted [defined by Eqs. (Eq. (4.3.1)) and (Eq. (4.3.2))] are related to the difference between the rotational constants of the lower and upper states, $B'' - B$. When a perturbation occurs, the effective rotational constant of the perturbed state approaches the average of the B–values of the perturbed and perturbing states, $(B' + B_p)/2$. The horizontal marks locate the exact level crossing point, from which B_p may be determined. (From Field, *et al.*, 1976.)

crossings involve $N = J - 1$ and $N = J + 1$ levels, and the energy separation between these levels is

$$B(^3\Sigma^-)\,[(J_3+1)(J_3+2) - (J_1-1)J_1\,] = B(^1\Pi)\,[J_3(J_3+1) - J_1(J_1+1)\,]\,. \tag{4.3.1}$$

Thus,

$$\begin{aligned} B(^3\Sigma^-) &= B(^1\Pi)\frac{J_3(J_3+1) - J_1(J_1+1)}{(J_3+1)(J_3+2) - (J_1-1)J_1} \\ &= B(^1\Pi)\;0.826. \end{aligned} \tag{4.3.2}$$

Finally, the absence of a detectable perturbation at a crossing point predicted by drawing nearly horizontal tie lines between corresponding perturbations on Fig. 5.7 (no unfilled Δ on $v = 1$ of $Si^{16}O$ near $J = 70$) implies that either the expected perturbing state would have "$v = -1$" or that the matrix element is less than

$$|H_{\text{min}}| < [\delta E_{\text{min}} \Delta B(J+1)]^{1/2}\,, \tag{4.3.3}$$

where δE_{min} is the minimum detectable level shift, $\Delta B = \left|B(^1\Pi) - B(\Delta)\right|$, and J is the J–value just below the predicted noninteger crossing point.

Figure 4.1 is based on Gerö's (1935) extra-line method for determining the B–value of the perturbing state, B_p. It is based on the observation that a perturbation disrupts the regular series of lines in a branch. One series of lines includes the "main" lines below the crossing point and a few "extra lines" (see Section 6.2.1) above the crossing. This series of lines appears to break off abruptly at or above the crossing and is replaced by another, seemingly unrelated, series of lines starting just below or at the crossing. The discontinuity between these two series of lines is usually represented in tables of spectral lines by a column shift. Often there are two lines, one in each column, for each J-value at or near the crossing. Since the members of a given series appear to be related to each other, it is reasonable to use combination difference relationships to determine the effective B-values in the local J-region. Gerö (1935) derived

$$\frac{R(J-2) - R(J-1) + P(J) - P(J+1)}{4J} = B'' - B' - 2J^2(D'' - D') \tag{4.3.4a}$$

$$\frac{Q(J-1) - Q(J)}{2J} = B'' - B' - 2J^2(D'' - D') \tag{4.3.4b}$$

from the usual expressions for the frequencies of R, Q, and P branch lines, for example (neglecting centrifugal distortion),

$$R(J-2) = \nu_0 + B'(J-1)(J) - B''(J-2)(J-1) \tag{4.3.5}$$

$$\Delta R = R(J-2) - R(J-1) = -2JB' + 2(J-1)B'' \tag{4.3.6}$$

$$\Delta P = P(J) - P(J+1) = -2JB' + 2(J+1)B'' \tag{4.3.7}$$

$$\Delta R + \Delta P = -4JB' + 4JB''. \tag{4.3.8}$$

If $(\Delta R + \Delta P)/4J$ or $\Delta Q/2J$ is plotted versus J^2 for an unperturbed band, a straight line is obtained with intercept and slope, respectively, of $B'' - B'$ and $D'' - D'$.

Figure 4.1 shows how the $B'' - B'$ versus $J(J+1)$ plot is affected by a $^1\Pi \sim {}^3\Sigma^-$ perturbation in the upper state. Far above and below the crossings (at $J' = 26.6$ and 36.1 in the P and R branches for the e-levels, at $J' = 31.3$ in the Q branch for the f-levels), the $B'' - B'$ value is about 0.111 cm^{-1}. The limiting $B'' - B'$ value for the $\Delta Q/2J$ plot (filled triangles) from the $J' > 31$ continuation of the low-J series of lines is ~ 0.22 cm^{-1}, implying that $B(^3\Sigma^-)$ is ~ 0.11 cm^{-1} smaller than $B(^1\Pi)$. A more accurate measure of $B(^3\Sigma^-)$ is obtained from the intersection between the low-J and high-J $\Delta Q/2J$ curves at $B'' - B' = 0.163$ cm^{-1}. At the exact crossing between $^1\Pi_f$ and $^3\Sigma^-(F_2)$ levels, the effective B'-value is $[B(^1\Pi) + B(^3\Sigma^-)]/2$, thus $B(^1\Pi) - B(^3\Sigma^-) = 0.104$ cm^{-1}. The $B(^3\Sigma^-)$ values obtained from the plot at the $^1\Pi_e \sim {}^3\Sigma^-(F_1$ and $F_3)$ crossings are slightly smaller and larger than at the F_2 crossing, indicating that $^1\Pi_e$ simultaneously feels the effects of both F_1 and F_3 $^3\Sigma^-$ e-components. Even though the F_1 and F_3 perturbations are not isolated from each other, the average of the two $B(^3\Sigma^-)$-values obtained graphically at these crossings is identical to the value from the isolated F_2 crossing.

The center of gravity rule can be useful for obtaining parameter estimates, locating extra lines, and detecting the presence of an unsuspected second perturber. When main and extra lines are observed for more than two J-values at a level crossing, then a plot of the average energy,

$$\overline{E(J)} = \left[E_{\mathrm{MAIN}}(J) + E_{\mathrm{EXTRA}}(J)\right]/2, \tag{4.3.9}$$

versus $J(J+1)$ has slope and intercept

$$\overline{B} = \left(B_{\mathrm{MAIN}} + B_{\mathrm{EXTRA}}\right)/2 \tag{4.3.10}$$

$$\overline{E} = \left[E_{\mathrm{MAIN}}(0) + E_{\mathrm{EXTRA}}(0)\right]/2. \tag{4.3.11}$$

This is true regardless of the perturbation mechanism (J–dependent or J–independent matrix elements) because the trace of a matrix is representation invariant; the sum of the basis function (i.e., deperturbed) energies is equal to the sum of the eigenvalues. Since the approximately unperturbed B_{MAIN} and $E_{\mathrm{MAIN}}(0)$ values are usually known from a relatively perturbation-free portion of the band, the constants for the perturbing state can be inferred from Eqs. (4.3.10) and (4.3.11). If the average energy plot shows any deviation from linearity, this implies either an incorrect line identification or an additional perturber.

Kovács (1969) gives many examples of approximate graphical solutions to single- and multiple-state perturbation problems.

4.3.2 Direct Diagonalization Versus Algebraic Approaches

Perturbation theory is an extremely useful analytic tool. It is almost always possible to treat a narrow range of J–values in a multistate interaction problem by exactly diagonalizing a two-level problem after correcting, by nondegenerate perturbation theory or a Van Vleck transformation, for the effects of other nearby perturbers. Such a procedure can enable one to test for the sensitivity of the data set to the value of a specific unknown parameter.

There is a long history in molecular spectroscopy of elegant algebraic solutions for the eigenvalues and eigenfunctions of complicated secular equations. Whenever the secular determinant is larger than 3×3, algebraic solution requires the use of nondegenerate perturbation theory, which can only be valid when a critical off-diagonal matrix element H_{ab} is smaller than a zero-order energy difference, $E_a^0 - E_b^0$. Most algebraic solutions describe limiting behavior very well, but seldom describe all observable levels, from low to high J. It is not very satisfying when one must use different models to account for several portions of a single band. Perturbations make the situation even more difficult, necessitating frequent switches from one algebraic model to another.

There is a good historical reason for the algebraic tradition in spectroscopy. Without computers, exact matrix diagonalizations were impossibly time consuming. There was no choice but to resort to algebraic formulas. Now there is a choice.

4.4 Exact Solutions

4.4.1 Least-Squares Fitting

One starts with a model, either a set of equations or a matrix to be diagonalized. The least-squares fitting process involves comparison of the model predictions to a set of experimental data, followed by adjustment of the model parameters to minimize the squared deviations between the experimental data and model predictions. The experimental data may be of several types, either separately or in combination: spectral line frequencies, term values, combination differences, relative intensities, magnetic g–values, etc. Each input datum can be weighted according to its measurement precision and/or the probability that it was a result of a correct assignment; however, one often finds all input data given equal weights, regardless of quality, type, or reliability.

The goal of a least-squares fitting process is to obtain a statistically rigorous and unbiased set of best possible model-defining parameters, with their uncertainties and correlations, for the specific combination of model and input data. If the model is inappropriate or the data set contains assignment or typographical errors, the least-squares procedure is at best only capable of providing a warning (systematic residuals, equality of fit poorer than measurement precision). The least-squares fitting process is often highly interactive; inflexible, few-parameter, diagnostic fits in the early stages followed by maximally flexible, multiparameter, model-refining fits in the final stages.

The treatment of least-squares fitting given here is superficial and nonrigorous. Albritton, *et al.*, (1976); Marquardt (1963), Wentworth (1965); and Press, *et al.*, (1996) discuss the least-squares process more completely.

4.4.1.1 Linear Least-Squares Fitting

Consider a model

$$F(A_{1i}, \ldots, A_{ni}, X_1, \ldots, X_p) = y_i \tag{4.4.1}$$

defined by p initially unknown parameters X_k and n known coefficients that constitute the name (e.g., quantum numbers) of the data point whose measured value, y_i^{obs}, is to be compared to the computed value, y_i. For example, the rotation-vibration levels of a $^1\Sigma^+$ state are described by a model

$$E(v, J) = T_e + \sum_{l=0}^{l_{\max}} \sum_{m=0}^{m_{\max}} Y_{lm} \left(v + \frac{1}{2}\right)^l [J(J+1)]^m, \tag{4.4.2}$$

consisting of two A–values per level (J_i and v_i) and $(l_{\max} + 1)(m_{\max} + 1) + 1$ unknown parameters. The correspondences between the quantities in the general model [Eq. (4.4.1)] and the specific example [Eq. (4.4.2)] are

$$y_i \leftrightarrow E(v_i, J_i)$$
$$A_{1i} \quad \text{and} \quad A_{2i} \leftrightarrow v_i \quad \text{and} \quad J_i$$
$$X_k \leftrightarrow Y_{lm}.$$

It is conceivable that the model would be asked to account for properties of the v_i, J_i level in addition to $E(v_i, J_i)$ (such as level population, zero-field g–value, etc.): in such case the data point would be specified by an additional index indicating which property is to be computed, but this is beyond the scope of the present discussion.

A model is said to be linear if

$$\frac{\partial y_i}{\partial X_k} = B_{ik}$$

for all i and k. This means that all higher derivatives, $\partial^n y_i/\partial X_k^n$, are zero. The model defined by Eq. (4.4.2) is linear where

$$\partial y_i/\partial X_k \equiv \partial E(v_i, J_i)/\partial Y_{lm} = (v + \frac{1}{2})^l [J(J+1)]^m.$$

The presence of $l, m > 1$ powers of v and J is irrelevant to whether the model is linear. Least-squares fits to a linear model have the convenient property of converging, in a single step, to a unique best set of parameter values.

The equations for a linear model with one dependent (y) and two independent (v, J) variables [for example, Eq. (4.4.2)] have the form

$$y_i = \sum_{m=1}^{p} B_{im} X_m, \tag{4.4.3}$$

where p is the number of parameters and each B_{im} is an explicit function of the two independent variables (v_i and J_i). A matrix formulation of Eq. (4.4.3) is useful,

$$\mathbf{BX} = \mathbf{y}, \tag{4.4.4}$$

where $\mathbf{y}$ and $\mathbf{X}$ are column matrices having respectively N (the number of input data points) and p rows and $\mathbf{B}$ is a matrix with N rows and p columns.

The sum of the squared residuals, $\mathbf{\Phi}$, is

$$\mathbf{\Phi} \equiv \sum_{i=1}^{N} \left(y_i^{\text{obs}} - y_i^{\text{calc}}\right)^2 = \sum_{i=1}^{N} \left(y_i^{\text{obs}} - \sum_{m=1}^{p} B_{im} X_m\right)^2. \tag{4.4.5}$$

The set of parameter values sought is that for which $\mathbf{\Phi}$ is minimized,

$$\frac{\partial \mathbf{\Phi}}{\partial X_m} = 0 \qquad \text{for} \quad m = 1, \ldots, p \tag{4.4.6}$$

or, taking the derivative of Eq. (4.4.5),

$$0 = \sum_{i=1}^{N} \frac{\partial}{\partial X_m} \left(y_i^{\text{obs}} - \sum_{m=1}^{p} B_{im} X_m\right)^2, \tag{4.4.7}$$

$$0 = \sum_{i=1}^{N} \sum_{l=1}^{p} \left(B_{il} B_{im} X_l - y_i^{obs} B_{im}\right), \tag{4.4.8}$$

or

$$\sum_{i=1}^{N}\sum_{l=1}^{p} B_{il}B_{im}X_l = \sum_{i=1}^{N} y_i^{obs}B_{im} \qquad \text{for} \qquad m = 1,\ldots,p. \tag{4.4.9}$$

This system of p equations in p unknowns can be rewritten in matrix notation as

$$\mathbf{B}^t\mathbf{B}\mathbf{X} = \mathbf{B}^t\mathbf{y}^{\text{obs}}, \tag{4.4.10}$$

where the right superscript t means transpose. Then

$$\mathbf{W} \equiv \mathbf{B}^t\mathbf{B} \tag{4.4.11}$$

defines the square $p \times p$ matrix $\mathbf{W}$, called the *normal matrix*, which has the property

$$\mathbf{X} = \mathbf{W}^{-1}\mathbf{B}^t\mathbf{y}^{\text{obs}}. \tag{4.4.12}$$

The left-hand side of Eq. (4.4.12) contains the desired optimized parameter values and the right-hand side consists entirely of measured ($\mathbf{y}^{\text{obs}}$) and calculable ($\mathbf{W}^{-1}$ and $\mathbf{B}$) quantities, provided that the columns of $\mathbf{B}$ are linearly independent (determinant of $\mathbf{W} \neq 0$) so that $\mathbf{W}^{-1}$ exists.

The quality of a nonweighted linear least-squares fit is given by the *standard deviation*,

$$\sigma \equiv \left(\frac{\Phi}{N-p}\right)^{1/2} \tag{4.4.13}$$

where Φ is the sum of the squared residuals and $N-p$ is the number of degrees of freedom (number of data points minus number of varied parameters). σ has the same units as the input data (cm^{-1}) and, if the fit is acceptable, should be comparable to the measurement uncertainty, δy. The standard error for each fitted parameter is given, in the same units as the parameter itself, by

$$\delta X_i = \sigma\left[\left(\mathbf{W}^{-1}\right)_{ii}\right]^{1/2} \tag{4.4.14}$$

where $\sigma\mathbf{W}^{-1}$ is known as the *variance-covariance matrix*. In a fit with more than 10 degrees of freedom ($N-p > 10$), the true value of the parameter is expected to fall within the interval $X_i^{\text{fit}} - 2\delta X_i \leq X_i^{\text{true}} \leq X_i^{\text{fit}} + 2\delta X_i$ at the 95% confidence level.

The off-diagonal elements of $\sigma\mathbf{W}^{-1}$ reflect the correlations between fitted values of pairs of parameters. The *correlation matrix*, a matrix with ones on diagonal and numbers between $+1$ and -1 off-diagonal,

$$C_{ij} \equiv W_{ij}^{-1} / \left(W_{ii}^{-1}W_{jj}^{-1}\right)^{1/2}, \tag{4.4.15}$$

reflects the sign and magnitude of the error in the jth parameter resulting from an error in the ith parameter. The true values of two parameters that have

$C_{ij} \simeq 0$ are likely to fall anywhere in the rectangle defined by $X_i \pm 2\delta X_i$, $X_j \pm 2\delta X_j$, whereas the true values of two parameters with $C_{ij} \simeq 1$ are likely to fall along the $\delta X_i \simeq \delta X_j$ diagonal. Thus the *joint confidence region* for two strongly correlated constants is much smaller than the naive estimate from the uncorrelated standard errors.

In most real experiments, some measurements are made with greater precision than others (strong unblended lines versus weak or blended lines; a combined fit to microwave and optical data). A *weighted* least-squares procedure is appropriate. Typically, each measurement is weighted by the square of the reciprocal of its estimated uncertainty.

$$w_{ii} = (\delta y_i)^{-2}. \tag{4.4.16}$$

A weight matrix, w, is defined by

$$w_{ij} = \delta_{ij}\sigma^2 \delta y_i^{-2}, \tag{4.4.17}$$

where w is a diagonal matrix and σ^2 is the *a priori* undetermined [see Eq. (4.4.19)] variance of a measurement with unit weight. The weighted least-squares equations are very similar to the unweighted ones:

$$\mathbf{X} = (\mathbf{B}^t\ \mathbf{wB})^{-1}\mathbf{B}^t\mathbf{wy}^{\text{obs}} \tag{4.4.18}$$

$$\sigma^2 = \frac{1}{N-p}\sum_{i=1}^{N}\left(y_i^{\text{obs}} - y_i^{\text{calc}}\right)^2 / \delta y_i^2, \tag{4.4.19}$$

where σ^2, the *variance* of the fit, is a dimensionless number with value near 1 if the estimates of δy_i are reasonable and the fitting model is appropriate. The uncertainties in the parameters,

$$\delta X_i = \sigma\left[\left(\mathbf{B}^t\mathbf{wB}\right)_{ii}^{-1}\right]^{1/2}, \tag{4.4.20}$$

do not depend on the initial estimate of σ^2 because the σ in Eq. (4.4.20) is cancelled by a σ^{-1} from $\left[(\mathbf{B}^t\mathbf{wB})_{ii}^{-1}\right]^{1/2}$.

4.4.1.2 Nonlinear Least-Squares Fitting

When the model [Eq. (4.4.1)] is nonlinear,

$$\frac{\partial^n y_i}{\partial X_k^n} \neq 0 \quad \text{for} \quad n \geq 2,$$

the least-squares procedure is considerably more complicated because it is no longer possible to write a matrix equation analogous to Eq. (4.4.12) or Eq. (4.4.18). It is still possible to find a set of parameter values, **X**, that minimizes the square of the residuals, $\mathbf{\Phi}$, but the process is iterative, and it is possible that a false convergence will occur to a local minimum of $\mathbf{\Phi}$.

A typical nonlinear model encountered in spectroscopy is one where the spectral line frequencies, y_i, are computed from differences between eigenvalues (specified by two groups of quantum numbers) of two effective Hamiltonian matrices, defined, in turn, by the molecular constants, X_j. For example, consider the 2×2 matrix of a $^2\Pi$ state:

$$\begin{aligned}\mathbf{H} &= \begin{vmatrix} T_0 + A/2 + B(x^2-2) & -B(x^2-1)^{1/2} \\ -B(x^2-1)^{1/2} & T_0 - A/2 + Bx^2 \end{vmatrix} \\ &= T_0 \begin{vmatrix} 1 & 0 \\ 0 & 1 \end{vmatrix} + A \begin{vmatrix} \frac{1}{2} & 0 \\ 0 & -\frac{1}{2} \end{vmatrix} + B \begin{vmatrix} x^2-2 & -(x^2-1)^{1/2} \\ -(x^2-1)^{1/2} & x^2 \end{vmatrix} \\ &= \sum_{m=1}^{p} X_m \mathbf{H}_m, \end{aligned} \tag{4.4.21}$$

where $x = J + \frac{1}{2}$. This way of writing $\mathbf{H}$ as a pseudolinear form (Chedin and Cihla, 1967) takes advantage of the Hellmann-Feynman theorem (Hellmann, 1937; Feynman, 1939),

$$\left(\mathbf{U}^t \frac{\partial \mathbf{H}}{\partial X_k} \mathbf{U}\right)_{ii} = \frac{\partial E_i}{\partial X_k} \tag{4.4.22}$$

where

$$\mathbf{H}|i\rangle = E_i|i\rangle$$

and $\mathbf{U}$ is the unitary matrix that diagonalizes $\mathbf{H}$. The $\mathbf{H}_m$ matrices in Eq. (4.4.21) are identical to the $\partial\mathbf{H}/\partial X_k$ in Eq. (4.4.22). The convenient point about the Hellmann-Feynman theorem is that it provides all of the $\partial E_i/\partial X_k$ derivatives after a single diagonalization of $\mathbf{H}$, rather than the $p+1$ diagonalizations that would otherwise be needed to compute the derivatives by finite differences from

$$\frac{\partial E_i}{\partial X_k} \approx \frac{E_i\left(X_k^0\right) - E_i\left(X_k\right)}{X_k^0 - X_k}, \tag{4.4.23}$$

where the X_k^0 are the initial estimates of the unknown parameter values. Matrix diagonalizations are usually the most time consuming portion of a nonlinear least-squares fit.

The utility of Eq. (4.4.22) may be illustrated very simply. At very high J-values, a $^2\Pi$ state will approach the case (b) limit, at which point there is essentially no information in the spectrum from which the spin-orbit constant, A, may be determined. The $\mathbf{U}$ matrix for the case (a)→(b) transformation at the high–J limit is

$$\mathbf{U} \approx \begin{vmatrix} 2^{-1/2} & 2^{-1/2} \\ -2^{-1/2} & 2^{-1/2} \end{vmatrix}.$$

Applying this transformation to the matrix $\mathbf{H}_m$ for $m = T_0$,

$$\mathbf{U}^\dagger \mathbf{H}_{T_0} \mathbf{U} = \mathbf{H}_{T_0} = \begin{vmatrix} 1 & 0 \\ 0 & 1 \end{vmatrix},$$

and to the $\mathbf{H}_A$ coefficient matrix,

$$\mathbf{U}^\dagger \mathbf{H}_A \mathbf{U} = \begin{vmatrix} 0 & \frac{1}{2} \\ \frac{1}{2} & 0 \end{vmatrix},$$

it is evident that $\partial E_i/\partial T_0 = (\mathbf{U}^\dagger \mathbf{H}_{T_0} \mathbf{U})_{ii} = 1$ at both the case (a) and (b) limits but $\partial E_i/\partial A = (\mathbf{U}^\dagger \mathbf{H}_A \mathbf{U})_{ii} = 0$ in the case (b) limit. The sensitivity of the eigenvalues to the value of the T_0 parameter is approximately J–independent, but the eigenvalues become insensitive to A as the case (b) limit is approached at high J.

It is now possible to derive a nonlinear equation analogous to Eq. (4.4.12):

$$y_i^{\text{calc}} = (\mathbf{U}^\dagger \mathbf{H} \mathbf{U})_{ii} = \left[\mathbf{U}^\dagger \left(\sum_{m=1}^{p} X_m \mathbf{H}_m\right) \mathbf{U}\right]_{ii} = \sum_{m=1}^{p} (\mathbf{U}^\dagger \mathbf{H}_m \mathbf{U})_{ii} X_m$$
$$\equiv \sum_{m=1}^{p} B_{im} X_m. \quad (4.4.24a)$$

where

$$B_{im} = (\mathbf{U}^\dagger \mathbf{H}_m \mathbf{U})_{ii}. \quad (4.4.24b)$$

Equation (4.4.24a) looks exactly like Eq. (4.4.3), but the hidden difference is that the B_{im} coefficients depend implicitly, through the $\mathbf{U}$ matrix, on the unknown parameters, X_m. It will be necessary to resort to an iterative procedure whereby the approximate $B_{im}^{(n)}$ matrix elements are recalculated after each adjustment of the parameter values,

$$\mathbf{X}^{(n)} = \mathbf{X}^{(n-1)} + \delta^{(n-1)}, \quad (4.4.25)$$

and the least-squares procedure is recast as a series of iterative calculations of δ so that $\mathbf{X}^{(n)}$ converges to $\mathbf{X}$. Numerical convergence of the procedure is achieved when the $\mathbf{X}^{(n)}$ and $\mathbf{X}^{(n-1)}$ are found to differ by less than a preset convergence criterion.

The matrix formulation of the iterative process is derived by a series of steps identical to Eqs. (4.4.5) - (4.4.12),

$$\mathbf{X}^{(0)} + \delta^{(0)} = \mathbf{X}^{(1)} = [\mathbf{W}^{(0)}]^{-1} \mathbf{B}^{(0)t} \mathbf{y}^{\text{obs}}, \qquad \text{where} \quad \mathbf{W}^{(0)} \equiv \mathbf{B}^{(0)t} \mathbf{B}^{(0)}. \quad (4.4.26)$$

The right-hand side of Eq. (4.4.26) is expressed entirely in terms of known ($\mathbf{y}^{\text{obs}}$) and calculable ($\mathbf{W}^{(0)}, \mathbf{B}(0)$) quantities, but since the model is nonlinear, the next iteration requires that the $\mathbf{B}^{(1)}$ matrix and y_i^{calc} be recalculated by Eqs.

(4.4.24) using the new parameter values, $\mathbf{X}^{(1)}$, and the new matrix $\mathbf{U}^{(1)}$, which diagonalizes the new $\mathbf{H}^{(1)} \equiv \sum_{m=1}^{p} \mathbf{H}_m X_m^{(1)}$; moreover, the true minimum of Φ may not have been reached after one iteration. The iterative process described by Eqs. (4.4.24a) - (4.4.26) is called the *Gauss-Newton method* (Marquardt, *et al.*, 1961) and is *implicitly* based on a truncated Taylor series expansion,

$$\mathbf{y}^{\mathrm{calc}(1)} \simeq \mathbf{y}^{\mathrm{calc}(0)} + \sum_{m=1}^{p} \left(\frac{\partial \mathbf{y}^{\mathrm{calc}(0)}}{\partial X_m} \right)_{X_m = X_m^{(0)}} \delta_m^{(0)}. \tag{4.4.27}$$

Its convergence properties are excellent near the minimum of the $\Phi(\mathbf{X})$ hypersurface, but poor far from the minimum where the nonlinear dependence of Φ on the X_i values is most important.

An alternative approach is based on a *method of steepest descents*. The Eq. (4.4.26) definition of $\boldsymbol{\delta}$ is replaced by

$$\boldsymbol{\delta}_g = - \left(\frac{\partial \Phi}{\partial X_1}, \ldots, \frac{\partial \Phi}{\partial X_p} \right)^t \tag{4.4.28}$$

where the subscript g signifies that $\boldsymbol{\delta}_g$ points in the direction of the negative *gradient* of the $\Phi(\mathbf{X})$ hypersurface and the superscript t means transpose ($\boldsymbol{\delta}$ is a $1 \times p$ column matrix). The $\partial \Phi / \partial X_m$ derivatives are computed from

$$\frac{\partial \Phi}{\partial X_m} = \left[\mathbf{W}^{(0)} \mathbf{X}^{(0)} - \mathbf{B}^{(0)t} \mathbf{y}^{\mathrm{obs}} \right]_m . \tag{4.4.29}$$

The steepest descent method is very effective far from the minimum of Φ, but is always much less efficient than the Gauss-Newton method near the minimum of Φ. Marquardt (1963) has proposed a hybrid method that combines the advantages of both Gauss-Newton and steepest descent methods. Marquardt's method, combined with the Hellmann-Feynman pseudolinearization of the Hamiltonian energy level model, is the method of choice for most nonlinear molecular spectroscopic problems.

Birss (1983) has proposed a modification of the nonlinear fitting process that takes advantage of an extension of the Hellmann-Feynman theorem to computation of second derivatives,

$$\begin{aligned} \frac{\partial^2 E_i}{\partial X_k \partial X_l} &= 2 \sum_{j \neq i} \frac{\langle i | \mathbf{U}^\dagger \mathbf{H}_k \mathbf{U} | j \rangle \langle j | \mathbf{U}^\dagger \mathbf{H}_l \mathbf{U} | i \rangle}{E_i - E_j} \\ &= 2 \sum_{j \neq i} \frac{\left(\mathbf{U}^\dagger \mathbf{H}_k \mathbf{U} \right)_{ij} \left(\mathbf{U}^\dagger \mathbf{H}_l \mathbf{U} \right)_{ji}}{E_i - E_j}, \end{aligned} \tag{4.4.30}$$

where $\mathbf{U}$ defines the transformation that diagonalizes $\mathbf{H}$, and $\mathbf{H}_m$ is defined by Eq. (4.4.21). The availability of higher derivatives enables the most important nonlinearities in $\Phi(\mathbf{X})$ to be dealt with explicitly by the fitting algorithm, yielding faster convergence.

Curl (1970) has proposed a *diagnostic least-squares* procedure that allows prior knowledge about the physically reasonable values (and their estimated 90% confidence intervals) for several X_i to be incorporated as constraints on the usual least-squares fitting procedure. When spectroscopic data are inadequate to determine all of the parameters in $\mathbf{H}^{\text{eff}}$, the normal matrix, $\mathbf{W} = \mathbf{B}^t\mathbf{B}$, will be nearly singular. Since it is impossible to compute the inverse, $\mathbf{W}^{-1}$, of a singular $\mathbf{W}$ matrix, an unmodified least-squares procedure must fail. Curl's procedure is sufficiently flexible that, if it turns out that the input data are capable of determining the value of a parameter more accurately than the prior estimate of its uncertainty, then that parameter is allowed to vary freely; otherwise the normal equation, $\mathbf{W}$, and parameter correction, $\boldsymbol{\delta}$, matrices are modified to remove the singularity by constraining the parameter. The variation ranges of more than one parameter can be constrained, and the standard errors and correlations between the unconstrained parameters are computed in a statistically rigorous manner.

Curl's method for recognizing an ill-determined parameter involves examination of the eigenvalues and eigenvectors of the normal matrix, $\mathbf{W}$. If $\mathbf{W}$ is nearly singular, then it must have at least one very small eigenvalue, λ_i. The eigenvector of $\mathbf{W}$ corresponding to this smallest λ_i is, χ_i, the specific linear combination of parameters X,

$$\chi_i = \sum_j a_{ji} X_j,$$

which is so poorly determined that roundoff error may prohibit or slow down convergence of the fit to the optimum value of χ_i. Lees (1970) has discussed the diagnostic significance of the eigenvalues and eigenvectors of $\mathbf{W}$ and has shown that $P + \log \lambda_i$ is the number of significant figures to which χ_i can be determined, where P is the number of significant figures stored by the computer.

4.4.1.3 Practical Considerations

The end results of a fit of the parameters in $\mathbf{H}^{\text{eff}}$ to a set of spectroscopic data are: (1) a set of molecular constants, standard errors, and correlation coefficients for the constants; (2) a quality of fit indicator (σ); (3) a numerical model capable of reproducing the fitted data set without systematic error larger than the measurement precision; and (4) a model capable of both extrapolating to unobserved levels and computing properties other than eigenenergies from the eigenvectors of $\mathbf{H}^{\text{eff}}$.

The fitted parameters may or may not be of intrinsic interest. It is important to remember that the parameter values obtained depend on the specific $\mathbf{H}^{\text{eff}}$ model, the parameters allowed to vary in the fit, the fixed values chosen for parameters not allowed to vary, and the specific subset of spectral data fitted. Generation of multidigit molecular constants should not normally be the primary objective nor the stopping point of a spectroscopic investigation.

Even though molecular constants are model-dependent (see Section 4.4.2) and often strongly correlated, it is important to report these constants with one or two more digits than are statistically determined (Watson, 1977; Tellinghuisen, 1995; LeRoy, 1998). These extra digits are required to reproduce the fitted data. The ability to reproduce a spectrum (typically ~1000 lines reproduced by ~20 constants), predict unobserved lines, and calculate seemingly unrelated properties is one of the most valuable and least model-dependent products of the spectrum-fitting process.

Before starting a multiparameter fitting process, it is important to perform diagnostic tests on the model Hamiltonian subroutine and on the input data set. A strongly recommended computer code testing strategy is to show that the $\mathbf{H}^{\text{eff}}$, molecular constants, and spectral lines in a published paper are self-consistent. The input data set should be examined for typographical and assignment errors by running the fitting program for at most one iteration, allowing it to vary only one obviously well determined parameter. The quality of the fit will be very poor, but the systematics of the residuals [discontinuities in (obs - calc) versus J] will identify most of the questionable lines.

It is of diagnostic value to perform a series of fits in which progressively larger numbers of parameters are simultaneously optimized. At each stage, the systematics of the residuals can suggest which additional parameter should be allowed to vary. If any fitted parameter is within twice its fitted uncertainty of zero or some prior estimate of its value, its value should be set to zero or to the prior estimate and that parameter should be temporarily removed from the fit. Sometimes, near convergence, a parameter that was poorly determined at an earlier stage (fewer parameters varied, larger σ) can be fitted.

When an additional parameter is allowed to vary, the quality of the fit should improve. If $\sigma(p)$ and $\sigma(p+1)$ are the standard deviations of the fit when p and $p+1$ parameters are varied, it is reasonable to vary the $(p+1)$th parameter when

$$\sigma(p+1) < \sigma(p)\left(\frac{N-p}{N-p-1}\right)^{1/2}. \tag{4.4.31}$$

Note that, if one choice of a $(p+1)$th parameter fails the Eq. (4.4.31) test, there may be a different $(p+1)$th parameter that will pass it. It is time to stop looking for new parameters when the variance of the fit is near one (the lines are being fit within their estimated measurement precision) and when there are no systematic residuals.

One problem unique to perturbed spectra concerns the correspondence between the spectroscopist's and a computer's names for a given energy level. An eigenstate of $\mathbf{H}^{\text{eff}}$ may be unambiguously labeled by specifying three quantities: its good quantum numbers, J and e/f, and its energy rank within the group of eigenstates belonging to the same values of the good quantum numbers. For example, an $\mathbf{H}^{\text{eff}}$ containing $^3\Delta, {}^1\Delta, {}^3\Sigma^+, {}^3\Sigma^-, {}^1\Sigma^+, {}^1\Sigma^-$, and $^1\Pi$ basis functions has 18 eigenstates for each value of $J \geq 3$, nine each of e and f symmetry. It is possible to compute, from the eigenvectors of $\mathbf{H}^{\text{eff}}$, the fractional characters of each basis function present in each eigenfunction. If

$$\left(\mathbf{U}^\dagger\mathbf{H}\mathbf{U}\right)_{ij} = E_i\delta_{ij}, \tag{4.4.32}$$

then the *fractional* character of the kth basis function in the ith eigenfunction is $|U_{ki}|^2$. The *nominal* character of an eigenstate, the spectroscopist's name for that state, is the *leading character* determined by the largest of the $|U_{ki}|^2$ among all basis functions k contributing to the ith eigenstate. All J–levels of the same energy rank do not belong to the same nominal character. This is certainly true for $J < 3$ because not all basis functions exist at low–J ($J \geq \Omega$ is required). Rank changes also occur at higher J–values when level crossings occur. When the energy of the $^1\Pi$ basis function, $H_{^1\Pi,^1\Pi}$, overtakes the $^1\Sigma^-$ basis function energy, $H_{^1\Sigma^-,^1\Sigma^-}$, from below (because $B_{^1\Pi} > B_{^1\Sigma^-}$), the energy rank of the nominal $^1\Pi$ level must change. The parameter variation process during a fit can be improperly trapped by an incorrect energy ranking. What started as a correct ranking for the initial parameter estimates becomes incorrect after an iteration. The computer then thinks it has a level of nominal perturber $\left(^1\Sigma^-\right)$ rather than main line $\left(^1\Pi\right)$ character and adjusts the otherwise weakly determined constants of the perturber to accommodate the incorrectly ranked level. The possibility of computer-generated ranking errors can be eliminated if levels are labeled by their nominal character, which is almost always evident from a rotational analysis supplemented by level-shift and relative-intensity information. Le Floch, *et al.*, (1987) have devised a character/rank dual-labeling scheme.

Fitted parameters are often strongly correlated. This can be a consequence of the structure of the data set or it can be an intrinsic property of the $\mathbf{H}^{\text{eff}}$ model (for example, γ and A_D in a $^2\Pi$ state, Brown, *et al.*, 1979). The former effect can be minimized by supplementing the data set. A combined fit to optical and microwave data or to two electronic transitions sharing a common state (e.g., $^2\Pi - {}^2\Sigma^+$ and $^2\Sigma^+ - {}^2\Sigma^+$ systems) can be very effective. Another approach is to replace the two correlated parameters by two new parameters that are the sum and difference of the original parameters (e.g., $B' + B''$ and $B' - B''$). The correlation matrix [Eq. (4.4.15)] and the magnitudes of the eigenvalues of the normal matrix [Eq. (4.4.11)] provide useful insights (Albritton, *et al.*, 1976; Curl, 1970). Isotope relationships, computed D_v–values, and other semiempirical constraints are frequently used to minimize both data set and intrinsic correlation effects.

4.4.1.4 Least Squares vs. Robust Estimator Fitting

The satisfactory convergence of a least-squares fit is often frustrated by a small number of ill-fit data points or "outliers". Exclusion of the outliers from the fit is often undesirable because the outliers contain information about the fit model to which the remainder of the data set is insensitive. This is particularly true for the J–values in the neighborhood of a level crossing. Sometimes, when the dark perturbing state consists of several fine structure components, an otherwise unsampled fine structure component of the dark state will cross through the rotational term curve of the bright state without any tell-tale signs of a level crossing (a pair of consecutive–J residuals of opposite signs). Level crossings which are incorrectly assigned (in terms of either energy rank or nominal character) or unsuspected can trap a fit model in a false local minimum of the variance. The majority of such traps are set by the inflexible treatment of outliers by an essential feature of the least squares fitting procedure. As the fit approaches convergence and the intersections of bright and dark term curves are iteratively shifted (usually by trial and error) to the correct locations, a series of term values appear to be outliers, often with residuals of opposite signs on successive model attempts.

Least squares fitting is sensitive to outliers because it is based on minimization of a *figure of de-merit*, the sum of reciprocal-squared uncertainty-weighted squared deviations. An alternative approach, known to statisticians as "robust estimators", (Section 15.7 of Press, *et al.*, (1996); Dabrowski, *et al.*, (1998)) maximizes a user-specified *figure of merit* (Ruckstuhl, *et al.*, 1993; Ruckstuhl and Dressler, 1994 and 1996). For each data point, information about empirically-based estimates of measurement uncertainty and intensity-based estimates of the importance of each data point to the fit model may be embedded in the figure of merit (Jacobson, *et al.*, 1997). In this way, the small number of data points near a level crossing, which are essential to the fit model, contribute increasingly importantly to the figure of merit as the fit approaches convergence. Dabrowski, *et al.*, (1998) and Watson (2003) convert a standard, "reciprocal squared uncertainty" weighted ($1/\sigma_i^2, \sigma_i$ is the user-specified uncertainty of the i–th datum), nonlinear least squares fitting computer program into robust estimator form merely by modifying the weights assigned to each data point after each iteration of the fit. The weighting function,

$$w_i = \frac{1}{\sigma_i^2 + \alpha r_i^2}, \tag{4.4.33}$$

where r_i is the (obs-calc) for the i–th data point after the just completed iteration of the fit, provides a new set of weights $\{w_i\}$ for the next iteration of the fit. This weighting function has the effect of temporarily reducing the weight of an ill-fit measurement and permanently de-weighting erroneous measurements. Dabrowski, *et al.*, (1998) and Watson (2003) report that the choice $\alpha = 1/3$ is a good compromise between robustness and an acceptable rate of convergence of the fit. It becomes unnecessary to devise special treatment strategies for sensitive data points to prevent the fit model from being trapped in a false local minimum.

Experience with the use of the robust estimator approach in the fitting of spectroscopic data is limited. The robust estimator procedures presently in use appear to require more computer time than least square procedures, but the reduced incidence of trapping makes robust estimators (and hybrid least squares/robust schemes, e.g., Ruckstuhl, *et al.*, 2001) very attractive.

4.4.1.5 Types of Programs

At least three classes of spectrum-fitting programs exist that are suitable for reducing line spectra to molecular constants: the term-value method (Åslund, 1974), the direct approach (Zare, *et al.*, 1973), and merging of separate band-by-band fits (Albritton, *et al.*, 1977; Coxon, 1978; Tellinghuisen, 1996).

The *term-value method* (Åslund, 1974) has the effect of focusing on a single electronic state. Transitions from many bands are converted, in a model-independent manner, to a set of term values. Many of the term values are multiply determined from several branches of many bands. The term-value approach reduces an enormous number of observed transitions to a much smaller number of preaveraged energy levels. In many cases, especially when one state is perturbed, the term-value approach can aid the assignment process and vastly simplify part of the least-squares parameter optimization process. The term value approach has been criticized by Zare, *et al.*, (1973) and Albritton, *et al.*, (1977).

The *direct approach* had been used by spectroscopists (Birss, *et al.*, 1970; Stern, *et al.*, 1970; Merer and Allegretti, 1971; Field and Bergeman, 1971; Brand, *et al.*, 1971; Meakin and Harris, 1972) for several years before Zare, *et al.*, (1973) presented a detailed description of both model Hamiltonians and numerical procedures. The direct approach is capable of accepting transitions (electronic, vibrational, or pure rotational), term values, combination differences, radio-frequency transitions between Λ–doublets, in short any sort of energy-level information. All information is treated in a statistically rigorous, properly weighted, uncorrelated, and unbiased manner. As many $\mathbf{H}^{\text{eff}}$ matrices are set up as are needed to calculate all observed energy levels and transition frequencies. A direct approach fit to many bands can require a lot of computer memory and time because of the need to set up, store, invert, and diagonalize many huge matrices.

The method of *merging* separate band-by-band fits can simplify the fitting process with no sacrifice in accuracy of parameter determination. One of the reasons for considering a simultaneous multiband fit is that parameter correlations and standard errors can be significantly reduced. The merge process takes the band-by-band fitted parameters and variance-covariance matrices and combines them to obtain global constants that are identical to those that would have been obtained from a global fit.

Several spectrum-fitting programs are described in the literature (Lefebvre-Brion, 1969; Zare, *et al.*, 1973; Johns and Lepard, 1975; Le Floch, *et al.*, 1987). Some of these programs are designed to handle transitions between virtually any pair of electronic states but are not set up for perturbation situations where actual level crossings occur within the sampled range of J–values. Other programs are designed for a specific problem and would require major modifications by a user wishing to study another class of state or to fit a data set of different structure. Specific fitting models, such as the "unique perturber" approximation (Zare, *et al.*, 1973), are often valuable in that they reduce the number of free parameters and, in some cases, prevent the mechanical significance of the major constants (E_v, B_v, D_v) from being contaminated by intrinsically correlated second-order constants (o, p, q). However, by attributing all perturbation effects to a single remote perturber and treating these effects by a Van Vleck transformation (Section 4.2) with constraints, the parameters obtained from the fit may not have the meaning implied by the microscopic definitions of the parameters (Brown, *et al.*, 1979). It is important that every spectroscopic laboratory develop its own spectrum-fitting programs, usually by extensively modifying one of the widely distributed classes of program.

Most of the previous discussion has focused on programs that fit rotational lines. Often, in the case of homogeneous perturbations, there are so many mutually interacting vibrational levels that a fit to individual rotational levels (rather than to G_v and B_v) could be contemplated only after developing a model which accounts for the nonrotating molecule vibrational structure. In such a model, the concept of potential-energy curves plays a crucial role. These curves may be either Morse (Lefebvre-Brion, 1969) or RKR curves (Stahel, *et al.*, 1983). One assumes a perturbation interaction of the form

$$H_{1,v_1;2,v_2} = H^e_{12} \langle v_1|v_2\rangle , \tag{4.4.34}$$

and the potential curves and vibrational overlap integrals are recalculated after each iteration of the vibrational constants. The rotational "constants" of the perturbed vibrational levels are also calculated after each iteration from

$$B_i^{\text{calc}} = \sum_n \sum_{v_n} \left|c_{i,v_n}\right|^2 B_{v_n}, \tag{4.4.35}$$

where the summations are over electronic states (n) and vibrational levels of the nth state (v_n), B_{v_n} is the deperturbed rotational constant of the v_n level,

$$B_{v_n} = (\hbar^2/2\mu) \left\langle v_n|R^{-2}|v_n\right\rangle , \tag{4.4.36}$$

and $\left|c_{i,v_n}\right|^2$ is the mixing fraction of the v_n basis function in the ith eigenfunction as obtained from the unitary $\left(\mathbf{U}^{-1} = \mathbf{U}^\dagger\right)$ transformation, $\mathbf{U}^\dagger\mathbf{H}\mathbf{U}$, that diagonalizes $\mathbf{H}$,

$$\left|c_{i,v_n}\right|^2 = \left|U_{v_n,i}\right|^2 . \tag{4.4.37}$$

The B_i^{calc} are compared to the observed values, B^{obs}, obtained by extrapolating the observed levels to $J = 0$. The $B^{\text{obs}} - B^{\text{calc}}$ residuals allow the R_e–values of all interacting electronic states to be refined, whereas the vibrational data alone only determine differences in R_e–values.

This multivibrational-level homogeneous deperturbation approach, in which the potential curves and vibrational integrals ($\langle v_i|v_j\rangle$, B_{v_n}) are recalculated after each iteration of a least-squares fitting process, resembles the way by which constants obtained from band-by-band local perturbation analyses are globally deperturbed. An initial series of fits is performed to obtain partially deperturbed G_v and B_v values. An initial set of RKR potentials are generated, and values for all quantities that are calculable from the potential curves are obtained (D_v, $\langle v_i|v_j\rangle$, $\langle v_i|R^{-2}|v_j\rangle$, A_D, etc.). These calculated values are then input into the fitting program as constraints on many weakly determined or highly correlated constants. The fit is repeated, new RKR curves and derivable constants are calculated, and the process is iterated to self-consistency. The multilevel and band-by-band deperturbation approaches both use potential-energy curves to place constraints on a myriad of seemingly unrelated molecular constants and perturbation parameters, thereby vastly reducing the number of degrees of freedom in the spectrum-fitting model. The use of such a tight model enables model inadequacies and unsuspected perturbations to be detected, and, most importantly, it provides a set of deperturbed molecular constants with clearly separate mechanical and magnetic meanings (Zare, *et al.*, 1973).

4.4.2 Comparison Between Effective and True Parameters

The parameters obtained by fitting the observed energy levels to the eigenvalues of an effective Hamiltonian are not unique. Their values depend on: (1) the precision and completeness of the input data; (2) the size of the matrix that is actually diagonalized; (3) the choice of model Hamiltonian; and (4) choices such as which parameters to vary, which parameters will be held fixed at values different from zero, and the imposition of constraints on varied parameters. Spectroscopists seldom agree on the meaning of the term "deperturbed" or on what information belongs in $\mathbf{H}^{\text{eff}}$.

This book was written to help spectroscopists understand the relationship between the exact molecular Hamiltonian, effective Hamiltonians used in fitting spectral data, and the molecular parameters obtained from both spectra and *ab initio* calculations. Although the general ideas for constructing effective Hamiltonians (Section 4.2) and several examples appropriate to special cases (for example the $^2\Sigma^+ \sim {}^2\Pi$ interaction in Sections 3.5.4 and 5.5) are discussed, no attempt is made here to present a complete and universal effective Hamiltonian for diatomic molecules. Brown, *et al.*, (1979) derive an effective Hamiltonian that should be the starting point for the fitting of most non–$^1\Sigma$, perturbation-free, diatomic molecular spectra. Other less general, effective Hamiltonians have been proposed, by De Santis, *et al.*, (1973) for $^3\Sigma$ states, by Brown and Milton (1976) for $S \geq \frac{3}{2}$ Σ–states, and by Brown and Merer (1979) for $S \geq 1$ Π–states.

The idea of an effective Hamiltonian for diatomic molecules was first articulated by Tinkham and Strandberg (1955) and later developed by Miller (1969) and Brown, *et al.*, (1979). The crucial idea is that a spectrum-fitting model (for example Eq. 18 of Brown, *et al.*, 1979) be defined in terms of the minimum number of *linearly independent* fit parameters. These fit parameters have no physical significance. However, if they are defined in terms of sums of matrix elements of the exact Hamiltonian (see Tables I and II of Brown, *et al.*, 1979) or sums of parameters appropriate to a special limiting case (such as the "unique perturber approximation," see Table III of Brown, *et al.*, 1979, or pure precession, Section 5.5), then physically significant parameters suitable for comparison with the results of *ab initio* calculations are usually derivable from fit parameters.

The following example illustrates the critical dependence of the values of the fit parameters on the dimension of the effective Hamiltonian matrix.

Consider the secular equation for the 2 × 2 problem,

$$\begin{vmatrix} E_{v_1} - E & H_{1,v_1;2,v_2'} \\ H_{1,v_1;2,v_2'} & E_{v_2'} - E \end{vmatrix} \tag{4.4.38}$$

where $H_{1,v_1;2,v_2'}$ is independent of J, as in homogeneous electrostatic perturbations, and E_{v_1} and $E_{v_2'}$ are the zero-order energies (i.e., deperturbed) of the v_1 and v_2' vibrational levels belonging to electronic states 1 and 2. Of course, in the exact, infinite-dimension Hamiltonian matrix, the $1, v_1$ and $2, v_2'$ basis functions will have nonzero off-diagonal matrix elements with other, energetically remote, basis functions. If all but the $H_{1,v_1;2,v_2'}$ off-diagonal element are ignored, the E_{v_1}, $E_{v_2'}$, and $H_{1,v_1;2,v_2'}$ parameters in Eq. (4.4.38) become *effective parameters*, $E_{v_1}^{\text{eff}}, E_{v_2'}^{\text{eff}}, H_{v_1,v_2'}^{\text{eff}}$. The effective parameters differ from the *true parameters* by perturbation summations defined by the Van Vleck transformation [Eqs. (4.2.17) and (4.2.19)],

$$E_{v_1}^{\text{eff}} = E_{v_1}^{\text{true}} + \sum_{n,v_n \neq 1,v_1} \frac{(H_{1,v_1;n,v_n})^2}{E_{v_1} - E_{v_n}} \tag{4.4.39a}$$

$$E_{v_2'}^{\text{eff}} = E_{v_2'}^{\text{true}} + \sum_{n,v_n \neq 2,v_2'} \frac{(H_{2,v_2';n,v_n})^2}{E_{v_2'} - E_{v_n}} \tag{4.4.39b}$$

$$H_{v_1,v_2'}^{\text{eff}} = H_{v_1,v_2'}^{\text{true}} + \sum_{n,v_n \neq 1,v_1 \text{ or } 2,v_2'} \frac{H_{1,v_1;n,v_n} H_{n,v_n;2,v_2'}}{\left(E_{v_1} + E_{v_2'}\right)/2 - E_{v_n}}, \tag{4.4.40}$$

where corrections through second order are included. Thus, after a 2×2 *local deperturbation* is performed by fitting the observed levels to the $E_{v_1}, E_{v_2}, H_{1,v_1;2,v_2'}$ *effective parameters* in Eq. (4.4.38), the perturbation summations in Eqs. (4.4.39a) - (4.4.40) can be evaluated [Eq. (4.4.34)] to yield the *more completely deperturbed* $E_{v_1}^{\text{true}}, E_{v_2'}^{\text{true}}, H_{v_1,v_2'}^{\text{true}}$ *parameters*. Lagerqvist and Miescher (1958) were the first to employ such a two-step deperturbation process to the G_v and B_v constants of the NO $B^2\Pi \sim C^2\Pi$ valence~Rydberg interacting states.

However, Eqs. (4.4.39a) - (4.4.40)) can give erroneous results. Gallusser and Dressler (1982) have performed a simultaneous, multistate deperturbation on the NO $B^2\Pi$ ($v = 0$–37) and $L^2\Pi$ ($v = 0$–11) valence states and $C^2\Pi$ ($v = 0$–9), $K^2\Pi$ ($v = 0-4$), and $Q^2\Pi$ ($v = 0-3$) Rydberg states (see Section 6.2.2). They diagonalized 69 × 69 matrices rather than Van Vleck transform-corrected 2 × 2 matrices and found that in some cases the third-order correction to $H_{1,v_1;2,v_2'}^{\text{eff}}$,

$$H_{1,v_1;2,v_2'}^{\text{eff}} = H_{1,v_1;2,v_2'}^{\text{true}} + \sum_{n',v_n' \neq 1,v_1} \sum_{n,v_n \neq 2,v_2'} \frac{H_{1,v_1;n',v_n'} H_{n',v_n';n,v_n} H_{n,v_n;2,v_2'}}{\left(E_{v_1} - E_{v_n'}\right)\left(E_{v_2'} - E_{v_n}\right)}; \tag{4.4.41}$$

is larger than the first- and second-order contributions. In Eq. (4.4.41) the unprimed and primed state labels correspond to valence and Rydberg states, respectively. The second-order term is omitted from Eq. (4.4.41) because the deperturbed basis set is defined to consist of a group of prediagonalized $^2\Pi$ valence states (all $H_{1,v_1;n,v_n} = 0$) and a separately prediagonalized group of Rydberg states (all $H_{2,v_2';n',v_n'} = 0$). The only nonzero off-diagonal matrix elements of $\mathbf{H}^{\text{true}}$ are valence~Rydberg ($H_{v,v'} \neq 0$).

Gallusser and Dressler (1982) found that the first-order interaction between the $L^2\Pi$ ($v = 2$) and $K^2\Pi$ ($v = 0$) *basis states* calculated using Eq. (4.4.34) was $H^{\text{true}}_{L,2;K,0} = 0.07\ \text{cm}^{-1}$, although a matrix element of $H^{\text{true}}_{L,2;K,0} = 5\ \text{cm}^{-1}$ was obtained from a 2×2 diagonalization (Dressler and Miescher, 1981) of the interaction between the *nominal* L,2 and K,0 levels. A single term was found to dominate the Eq. (4.4.41) summation,

$$
\begin{aligned}
H_{L,2;C,5} &= 40\ \text{cm}^{-1} \\
H_{C,5;B,21} &= 180\ \text{cm}^{-1} \\
H_{B,21;K,0} &= 39\ \text{cm}^{-1} \\
E_{L,2} - E_{C,5} &= 443\ \text{cm}^{-1} \quad (\Omega = 3/2) \\
E_{K,0} - E_{B,21} &= 197\ \text{cm}^{-1} \quad (\Omega = 3/2) \\
\frac{H_{L,2;C,5} H_{C,5;B,21} H_{B,21;K,0}}{(E_{L,2} - E_{C,5})(E_{K,0} - E_{B,21})} &= 3.2\ \text{cm}^{-1}
\end{aligned}
$$

It is a rather disturbing, yet not uncommon, situation where a third-order correction term is 70 times larger than the first-order term.

Another surprising result is found by Ngo, *et al.*, (1974) for some apparent homogeneous Rydberg~Rydberg perturbations in the PO molecule. Since Rydberg states converging to the same limit have nearly identical potential energy curves and differ by a single electron orbital, both the H^e (See Section 3.3.4) and $\langle v|v' \neq v\rangle$ factors in the perturbation matrix element should be very small. The observation of $P^{16}O$ $A^2\Sigma^+(v = 12) \sim H^2\Sigma^+(v = 0)$ ($\mathrm{H}_{\mathrm{A},12;\mathrm{H},0} = 11.5\ \text{cm}^{-1}$) and $A^2\Sigma^+(v = 9) \sim G^2\Sigma^+(v = 0)$ ($\mathrm{H}_{\mathrm{A},9;\mathrm{G},0} = 30.0\ \text{cm}^{-1}$) perturbations corresponds to a perturbation between two Rydberg states with $\Delta v \gg 1$ and Rydberg orbitals with different $l\lambda$ values. (The configurations of the PO Rydberg states are: $[\mathrm{PO}^+\ X^1\Sigma^+]nl\lambda$ with $A^2\Sigma^+$ $4s\sigma$, $G^2\Sigma^+$ $4p\sigma$, $H^2\Sigma^+$ $3d\sigma$.) A nonzero $H^{\text{eff}}_{1,v_1;2,v_2'}$ matrix element could arise via second-order Van Vleck corrections

$$
H^{\text{eff}}_{1,v_1;2,v_2'} = 0 + \sum_{i,v_i \neq 1,v_1 \text{ or } 2,v_2'} \frac{H_{1,v_1;i,v_i} H_{i,v_i;2,v_2'}}{\left(E_{v_1} + E_{v_2'}\right)/2 - E_{v_i}}. \tag{4.4.42}
$$

For the PO examples, the leading terms in the perturbation summations are

$$
H^{\text{eff}}_{\mathrm{A},12;\mathrm{H},0} \simeq \frac{H_{\mathrm{A},12;\mathrm{F},8} H_{\mathrm{F},8;\mathrm{H},0}}{\left(E_{\mathrm{A},12} + E_{\mathrm{H},0}\right)/2 - E_{\mathrm{F},8}} = \frac{(36.4)(100.9)}{25,415 - 25,276} = 26.4\ \text{cm}^{-1}
$$

$$
H^{\text{eff}}_{\mathrm{A},9;\mathrm{H},0} \simeq \frac{H_{\mathrm{A},9;\mathrm{F},3} H_{\mathrm{F},3;\mathrm{G},0}}{\left(E_{\mathrm{A},9} + E_{\mathrm{G},0}\right)/2 - E_{\mathrm{F},3}} = \frac{(21.7)(172.0)}{21,669 - 21,509} = 23.3\ \text{cm}^{-1}
$$

where $F^2\Sigma^+$ is a valence state (see Table 5.4). These results for $\mathbf{H}^{\text{eff}}$ do not exactly reproduce the experimental values because only the largest term in the perturbation sum is considered here.

The next subsection shows how the solution of a system of coupled differential equations can yield molecular parameters that approach closer to the true parameter values than the usual matrix diagonalization approach.

4.4.3 Coupled Equations

Instead of searching for an eigenfunction of $\mathbf{H}$ in the form of Eq. (4.1.1), where the unknown *coefficients* c_{i,v_n} describe a linear combination of members of a complete set of known vibrational wavefunctions, it is possible to obtain a solution of the type

$$\psi_i(r, R) = \sum_{n=1}^{q} \mathbf{\Phi}_n(r, R)\chi'_{n,i}(R) \tag{4.4.43}$$

where the $\chi'_{n,i}$ are unknown *numerical* functions and the summation is over electronic states only. The standard approach Eq. (4.1.1) attempts to represent $\chi'_{n,i}$ as a superposition of the approximate χ_{v_n} vibrational wavefunctions associated with the nth potential curve,

$$\chi'_{n,i}(R) = \sum_{v_n=0}^{p_n} \chi_{v_n}(R)\, c_{i,v_n}. \tag{4.4.44}$$

The vibrational wavefunctions χ_{v_n} are analytic or numerical solutions of q *uncoupled* (independent) vibrational Schrödinger equations (see Section 5.1.3),

$$\left[\frac{-\hbar^2}{2\mu}\frac{d^2}{dR^2} + V_n(R) + \frac{\hbar^2}{2\mu R^2}J(J+1) - E\right]\chi_{v_n}(R) = 0. \tag{4.4.45}$$

The two cases where the $\mathbf{\Phi}_n$ are the diabatic or the adiabatic electronic wavefunctions will be discussed here. In the case where the $\mathbf{\Phi}_n$ are the diabatic basis functions, the *coupled* equations are written as

$$\left[\frac{-\hbar^2}{2\mu}\frac{d^2}{dR^2} + V_n^d(R) + \frac{\hbar^2}{2\mu R^2}J(J+1) - E\right]\chi_n'^d(R) = \sum_{m\neq n} H^e_{m,n}(R)\chi_m'^d(R) \tag{4.4.46}$$

where

$$H^e_{m,n}(R) = \int \mathbf{\Phi}_m^{d*}(r, R)\mathbf{H}^{el}\mathbf{\Phi}_n^d(r, R)dr. \tag{4.4.47}$$

The number of coupled equations is equal to the number, q, of different electronic state (or substate) potential curves.

Given a set of diabatic potential curves (V_n^d) and electronic interaction functions $(H_{m,n}^e)$, the coupled equations may be integrated numerically to find the energies of the bound levels. The numerical integration may be done by Fox-Goldwin's procedure (see references in Atabek and Lefebvre, 1980) or by the renormalized Numerov method of Johnson (1978) as employed by Stahel, *et al.*, (1983). If the V_n^d and $H_{m,n}^e$ were derived from a least-squares fit of the eigenvalues of a matrix $\mathbf{H}^{\mathrm{eff}}$ to the observed levels, then the energies obtained from the coupled equations method will all be lower than those obtained by the diagonalization method. This is easily explained because the coupled equations take into account *all* bound vibrational levels and a portion of the vibrational continuum associated with each coupled electronic state. The matrix method excludes some of the higher vibrational levels and all of the continuum. The net result is that, in the coupled-equations approach, the vibrational functions implicitly added to the basis set (all at energies above the observed and fitted levels) push all of the eigenvalues to lower energy. The only limitation on the coupled equations approach is in the limits of integration over R.

An attempt to combine the coupled-equations approach with a least-squares, matrix-diagonalization, parameter-iteration scheme was made by Stahel, *et al.*, (1983) for the multistate $N_2{}^1\Sigma_u^+ \sim {}^1\Sigma_u^+$ and ${}^1\Pi_u \sim {}^1\Pi_u$ valence~Rydberg perturbations. The energy differences between the coupled equation solutions and experimentally observed energy levels were added to the experimental values. These fictitious observed levels were then refitted by the usual matrix diagonalization method. The new V_n^d and H_{mn}^e functions obtained from the matrix approach, when introduced into the coupled equations, were found to give energy levels in good agreement with experiment. Interestingly, the electronic interaction parameters were found to be slightly smaller than previously. Since the basis set used in the coupled equations is more nearly complete than in the matrix diagonalization approach, the hybrid matrix/coupled-equations $\mathrm{V}_n^d(R)$ and $\mathrm{H}_{m,n}^e(R)$ functions are more completely deperturbed (closer to the true functions) than the ordinary $\mathbf{H}^{\mathrm{eff}}$ matrix solutions. This is true even though $\mathbf{H}^{\mathrm{eff}}$ included 30 levels for the ${}^1\Pi_u$ states and 41 levels for the ${}^1\Sigma_u^+$ states. The only feature lacking in the coupled equation solution by Stahel, *et al.*, (1983) is the influence of the higher electronic Rydberg states. This influence could be introduced using multichannel quantum defect theory (MQDT) (Raoult, 1987) (see Chapter 8).

A rigorous least-squares fitting procedure for the coupled-equations approach has been proposed by Dunker and Gordon (1976) for the case of a potential of a van der Waals complex.

The coupled equations have also been solved using the adiabatic electronic basis set for the H_2 ${}^1\Sigma_g^+$ states (Glass-Maujean, *et al.*, 1983). In this basis set, the d/dR derivative, acting on the unknown $\chi_m^{\prime ad}$ vibrational functions, appears in the second term on the right-hand side of the coupled equations, namely,

$$\left[\frac{-\hbar^2}{2\mu}\frac{d^2}{dR^2} + V_n^{ad}(R) + \frac{\hbar^2}{2\mu R^2}J(J+1) - E\right]\chi_n^{'ad}(R)$$
$$= \sum_{m\neq n} + \left[H_{m,n}^{(1)}(R)\chi_m^{'ad}(R) + 2H_{m,n}^{(2)}(R)\frac{d}{dR}\chi_m^{'ad}(R)\right] \quad (4.4.48)$$

where

$$H_{m,n}^{(1)} = \frac{\hbar^2}{2\mu}\int \Phi_m^{ad*}(r,R)\left(\frac{\partial^2}{\partial R^2} + \frac{2}{R}\frac{\partial}{\partial R}\right)\Phi_n^{ad}(r,R)dr$$

and

$$H_{m,n}^{(2)} = \frac{\hbar^2}{2\mu}\int \Phi_m^{ad*}(r,R)\frac{\partial}{\partial R}\Phi_n^{ad}(r,R)dr.$$

Therefore, before solving the coupled equations, it was necessary to transform to the diabatic electronic basis in order to eliminate this $\partial/\partial R$ derivative. To date there has been only one example for bound states, the H_2^+ three-particle system (Hunter and Pritchard, 1967), where the coupled equations were solved directly without such a transformation.

The coupled-equation method has also been used to calculate the widths and energy shifts arising from the interaction between a discrete and a repulsive state (predissociation, see Section 7.12) (Child and Lefebvre, 1978; Shapiro, 1982; Torop, *et al.*, 1987). An example of indirect predissociation, where two discrete states interact with the same repulsive state giving three coupled equations, has also been treated in the diabatic basis (see Section 7.13) (Lefebvre-Brion and Colin, 1977). The adiabatic basis coupled equations have been solved numerically without transformation to the diabatic basis. This has been done for the homogeneous photopredissociation of the OH $^2\Pi$ states by van Dishoeck, *et al.*, (1984), and more recently, using the complex scaling technique, for the predissociation of the CO $B^1\Sigma^+$ and $C^1\Sigma^+$ Rydberg states by Li, *et al.*, (1998). In an alternative approach, using a complex absorbing potential and the discrete variable representative (DVR) method, the predissociation widths of the CO $B^1\Sigma^+$ state are calculated in both diabatic and adiabatic representations (Monnerville and Robbe, 1999).

The coupled-equation numerical vibrational wavefunctions are difficult to visualize. A major difficulty is that their number of nodes is not simply related to a vibrational quantum number; hence prior expectations about the node count cannot be used to establish correspondences between observed and calculated levels. Johnson (1978) has presented a method for counting the number of nodes of the calculated wave functions to avoid inadvertently skipping over an eigenstate. The $\chi_n^{'ad}(R)$ basis functions for the H_2 $^1\Sigma_g^+$ states have been expressed in the form of Eq. (4.4.44) as a linear combination of the known $\chi_{v_n}^{ad}$ adiabatic vibrational functions. These $\chi_n'(R)$ functions were used to compute the nonadiabatic radiative lifetimes for all bound $J = 0$ vibrational levels of the EF, GK, and $H\overline{H}$ $^1\Sigma_g^+$ electronic states (Glass-Maujean, *et al.*, 1983).

The power of the coupled equation method is that, in principle, if the electronic interaction parameters are either calculated *ab initio* or taken as adjustable parameters, the microscopic vibrational Hamiltonian can be used without introducing a finite-basis effective Hamiltonian matrix. Scattering concepts have penetrated into the field of molecular spectroscopy (Mies, 1980). The coupled equations and MQDT methods are both borrowed from scattering theory and are extensively used to treat perturbations and problems of dissociation and ionization (see Chapters 7 and 8).

4.5 Typical Examples of Fitted Perturbations

4.5.1 An Indirect Heterogeneous Perturbation: NO $B^2\Pi \sim C^2\Pi \sim D^2\Sigma^+$

The NO $C^2\Pi$ and $D^2\Sigma^+$ states are the π and σ components of a $3p$ Rydberg complex belonging to a series which converges to the NO^+ $X^1\Sigma^+$ ion core,

$$5\sigma^2\ 1\pi^4\ 3p \begin{cases} \nearrow \pi\ C^2\Pi \\ \searrow \sigma\ D^2\Sigma^+. \end{cases}$$

The $B^2\Pi$ state is the lowest of three $^2\Pi$ valence states arising from the $5\sigma^2\ 1\pi^3\ 2\pi^2$ configuration (Section 3.2.3). The configurations of the $B^2\Pi$ and $D^2\Sigma^+$ states differ by two orbitals, which means that the B$\sim$D heterogeneous interaction is forbidden in the single configuration approximation. However, the electrostatic $B^2\Pi \sim C^2\Pi$ interaction is configurationally allowed (Section 3.3.2), and it is the admixture of $C^2\Pi$ character into both $B^2\Pi$ and $D^2\Sigma^+$ states that is responsible for the nominal B$\sim$D perturbation. This is an indirect effect of the form

$$\langle \mathrm{B}|\mathbf{H}^{\mathrm{el}}|\mathrm{C}\rangle\,\langle \mathrm{C}|\mathbf{H}^{\mathrm{ROT}}|\mathrm{D}\rangle$$

but, owing to the near-degeneracy of the three interacting vibronic levels, this problem cannot be treated by a second-order Van Vleck correction to $H^{\mathrm{eff}}_{\mathrm{B,D}}$ [Eqs. (4.2.17), (4.4.40), and (4.4.41)]. A direct diagonalization of the two 5×5 matrices ($B^2\Pi_{1/2}$, $B^2\Pi_{3/2}$, $C^2\Pi_{1/2}$, $C^2\Pi_{3/2}$, $D^2\Sigma^+$; one matrix each for e and f levels) is required. One simplifying feature is that the pure precession approximation (Section 5.5) is expected to be valid for the C$\sim$D interaction because these states differ by a single $3p$ Rydberg orbital. This means that (for $v_{\mathrm{C}} = v_{\mathrm{D}}$ interactions) reliable prior estimates are possible for the spin-orbit $[\langle v_{\mathrm{C}}|v_{\mathrm{D}}\rangle\, a_+ \simeq 2^{1/2} A\,(C^2\Pi)]$ and l-uncoupling

$$(\langle v_{\mathrm{C}}|B|v_{\mathrm{D}}\rangle\, b \simeq 2^{1/2}\left[B\,(C^2\Pi) + B\,(D^2\Sigma^+)\right]/2)$$

[Eqs. (3.4.18), (3.4.19), (3.5.24), (3.5.28), (3.5.29)] perturbation parameters. The 5×5 $\mathbf{H}^{\mathrm{eff}}$ matrices were constructed from a 3×3 C$\sim$D interaction matrix

(Table 3.9) augmented by a 2×2 $\mathrm{B}^2\Pi$ matrix and a single homogeneous $H_{\mathrm{B}\sim\mathrm{C}}$ interaction parameter ($H_{\mathrm{B}\sim\mathrm{D}}$ is held fixed at zero because B~D heterogeneous interactions are configurationally forbidden).

The B~C~D perturbations were first successfully treated by Jungen and Miescher (1968) using a simple and pedagogical model. They used a fixed value of the B~C perturbation parameter, obtained from previous analyses of isolated B~C interactions (Lagerqvist and Miescher, 1968) combined with calculated vibrational overlap factors (Felenbok and Lefebvre-Brion, 1966). Rather than diagonalize the 5×5 matrices in a single step, a three-step diagonalization procedure was adopted. First, a transformation from case (a) to case (b) was performed in order to eliminate off-diagonal $^2\Pi_{1/2} \sim {}^2\Pi_{3/2}$ elements [Eqs. (3.5.16) and (3.5.17)]. The C~D matrix elements in the transformed representation are, for the e–levels,

$$H_{\mathrm{C}(F_2)\sim\mathrm{D}} = \left[\left(J+\frac{3}{2}\right)\Big/(2J+1)\right]^{1/2} a_+/2 \tag{4.5.1a}$$

$$H_{\mathrm{C}(F_1)\sim\mathrm{D}} = -\left[\left(J-\frac{1}{2}\right)\Big/(2J+1)\right]^{1/2}\left[a_+/2 - 2bB_v\left(J+\frac{1}{2}\right)\right],$$

and for the f–levels,

$$H_{\mathrm{C}(F_2)\sim\mathrm{D}} = \left[\left(J+\frac{3}{2}\right)\Big/(2J+1)\right]^{1/2}\left[a_+/2 + 2bB_v\left(J+\frac{1}{2}\right)\right] \tag{4.5.1b}$$

$$H_{\mathrm{C}(F_1)\sim\mathrm{D}} = -\left[\left(J-\frac{1}{2}\right)\Big/(2J+1)\right]^{1/2} a_+/2.$$

Next, the B~C interaction is diagonalized. The result is eight groups of B~C mixed $^2\Pi$ levels (four e, four f, the e and f levels having identical energies and mixing coefficients), which can be denoted $|^2\Pi,\ J,\ e/f, k\rangle$ where k ($k = 1,2,3,4$) is the energy rank. The transformed basis functions are then

$$\begin{aligned}|^2\Pi, J, e/f, k\rangle = {} & c_{k,\mathrm{C}(F_1)}(J)\,|\mathrm{C}^2\Pi, F_1\rangle + c_{k,\mathrm{C}(F_2)}(J)\,|\mathrm{C}^2\Pi, F_2\rangle \\ & + c_{k,\mathrm{B}(F_1)}(J)\,|\mathrm{B}^2\Pi, F_1\rangle + c_{k,\mathrm{B}(F_2)}(J)\,|\mathrm{B}^2\Pi, F_2\rangle\end{aligned} \tag{4.5.2}$$

where the mixing coefficients, c, are determined by the two transformations. The interaction matrix elements between the $|\mathrm{D}^2\Sigma^+\rangle$ and the two-step transformed $|^2\Pi, J, e/f, k\rangle$ functions depend only on the $c_{k,\mathrm{C}(F_1)}$ and $c_{k,\mathrm{C}(F_2)}$ mixing coefficients and the Eqs. (4.5.1a) - (4.5.1b) C~D matrix elements (the B~D interactions are assumed to be zero). The remaining 2×2 interactions between $\mathrm{D}^2\Sigma^+$ and the eight $|^2\Pi, J, e/f, k\rangle$ functions are diagonalized in the final step of the deperturbation process. The matrix elements obtained from the separate 2×2 diagonalizations were found to have the relative magnitudes predicted by Eqs. (4.5.1a) and (4.5.2).

The a_+ and b parameters obtained by Jungen and Miescher (1968) account very well for the observed level shifts and intensities. These parameters agree satisfactorily with those obtained by Amiot and Vergès (1982) for a larger and more precisely measured set of spectral lines and using a direct diagonalization fit to the full 5×5 $\mathbf{H}^{\text{eff}}$ matrices in which all parameters were simultaneously varied. The b-value obtained by Amiot and Vergès (1982) indicates that, as expected, the pure precession approximation, $b = 2^{1/2}$, is obeyed by members of a Rydberg complex

$$b^{\text{obs}} = 1.36.$$

Another example of a third-order effect is reported by Cheung, *et al.*, (1986). An OODR-MPI experiment has revealed a weak, high-J perturbation between the NO $K^2\Pi$ ($v_K = 2$) and the $F^2\Delta$ ($v_F = 3$) level. This interaction between two Rydberg states of different symmetry can be explained by an indirect mechanism involving the valence $B^2\Pi$ and $B'^2\Delta$ states. This nominal $4p\pi$ $K^2\Pi$ ($v_K = 2$) level contains a small admixture of the $B^2\Pi$ ($v_B = 29$) level (see Section 6.2.2 and Fig. 6.5). A homogeneous interaction occurs between the $3d\delta$ $F^2\Delta$ ($v_F = 3$) level and $B'^2\Delta$ ($v_{B'} = 9$) (see Section 3.3.2 and Fig. 3.7). Since the $B^2\Pi_{3/2}$ and $B'^2\Delta_{3/2}$ valence substates interact via $\mathbf{H}^{\text{SO}}$ (Field, *et al.*, 1975), this interaction between the $F^2\Delta(3d\delta)$ and $K^2\Pi(4p\pi)$ nominal Rydberg levels is a third-order effect where the effective interaction matrix element is given by

$$H_{\mathrm{F,3;K,2}} = \frac{H^{el}_{\mathrm{F,3;B',9}} H^{SO}_{\mathrm{B',9;B,29}} H^{el}_{\mathrm{B,29;K,2}}}{\left(E^0_{\mathrm{F,3}} - E^0_{\mathrm{B',9}}\right)\left(E^0_{\mathrm{K,2}} - E^0_{\mathrm{B,29}}\right)}. \tag{4.5.3}$$

4.5.2 A Strong Multistate Interaction in the NO Molecule

The example treated in Section 4.5.1 dealt with the interaction between the $B^2\Pi$, $C^2\Pi$, and $D^2\Sigma^+$ states of NO. A two-step treatment yielded deperturbed molecular constants and a physical delineation of all processes affecting this three-state interaction. Here, a four-state interaction, also in NO, will be discussed. Strong interactions among three $^2\Pi$ states, the two valence states $B^2\Pi$ ($v = 24$) and $L^2\Pi$ ($v = 5$), and the $C^2\Pi$ ($v = 6$) Rydberg state, combined with the crossing of all three $^2\Pi$ states by the $D^2\Sigma^+$ ($v = 6$) Rydberg state result in an unusually complicated and confusing pattern of perturbed rotational levels. Figure 4.2 shows the rotational term values deduced from transitions observed in absorption (Lagerqvist and Miescher, 1966; Dressler and Miescher, 1981).

A numerical deperturbation of the NO $B^2\Pi$, $L^2\Pi$, and $C^2\Pi$ states has been performed by Gallusser and Dressler (1982) in which many vibrational levels were simultaneously deperturbed (see Section 6.2.2). Another deperturbation, which included the $D^2\Sigma^+$ state, has been performed (Lefebvre-Brion, unpublished calculation). In this four-state deperturbation, only one vibrational level from each state was treated. The deperturbed levels are represented by the straight lines on Fig. 4.2 (solid lines for $^2\Pi$ states, dashed lines for $D^2\Sigma$). The *effective* interaction matrix elements obtained from the four-state deperturba-

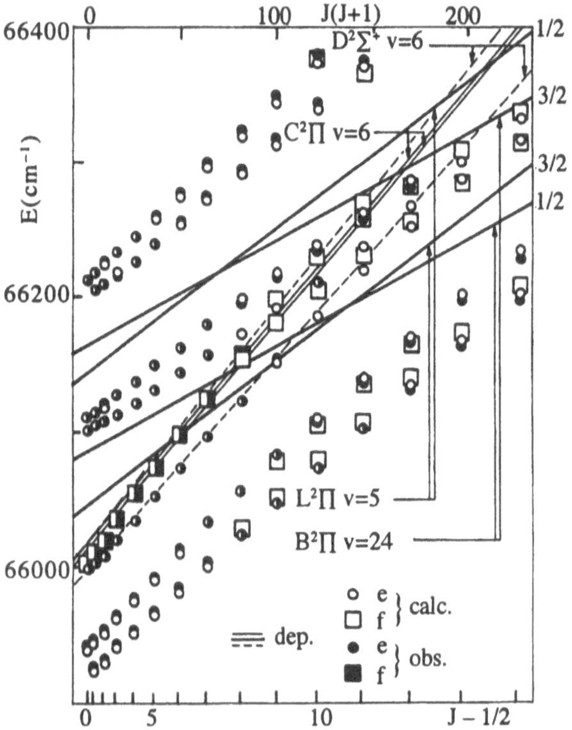

Figure 4.2: The ^{14}N ^{16}O B$^2\Pi$ ($v_{\rm B} = 24$) $\sim$ C$^2\Pi$ ($v_{\rm C} = 6$) $\sim$ L$^2\Pi$ ($v_{\rm L} = 5$) $\sim$ D$^2\Sigma$ ($v_{\rm D} = 6$) perturbation. The deperturbed $^2\Pi_{1/2}$ and $^2\Pi_{3/2}$ levels are plotted as solid lines (heavy lines for the valence B$^2\Pi$ and L$^2\Pi$ states, light lines for the Rydberg C$^2\Pi$ state). The D$^2\Sigma^+$ e and f levels are plotted as dashed lines. The observed and calculated e–levels are indicated by solid and open circles, respectively. The observed (solid squares) and calculated (open squares) f–levels are shown only when the Λ–doubling is larger than 2.5 cm^{-1}.

tion are

$$H^{\rm BC}_{24,6} = 66.7 \text{ cm}^{-1}$$
$$H^{\rm LC}_{5,6} = 102.6 \text{ cm}^{-1}$$

which are very close to the *true* values of 66 cm^{-1} and 114 cm^{-1} obtained by Gallusser and Dressler (1982).

Table 4.1: Mixing Coefficients for e–Parity Levels of the $^{14}N^{16}O$ $B^2\Pi \sim C^2\Pi \sim L^2\Pi \sim D^2\Sigma^+$ Perturbation[a]

Eigenstate index[b]	$B^2\Pi_{3/2}$	$L^2\Pi_{1/2}$	$B^2\Pi_{1/2}$	$C^2\Pi_{3/2}$	$C^2\Pi_{1/2}$	$L^2\Pi_{3/2}$	$D^2\Sigma^+$
$J = 0.5$							
1				does not exist			
3	0.0000	0.4074	0.3728	0.0000	-0.8335	0.0000	-0.0186
5	0.0000	0.0092	0.0101	0.0000	-0.0133	0.0000	0.9998
7	0.0000	-0.4211	0.8867	0.0000	0.1908	0.0000	-0.0026
9				does not exist			
11				does not exist			
13	0.0000	0.8103	0.2733	0.0000	0.5184	0.0000	-0.0033
J=8.5							
1	-0.2078	-0.1929	-0.1905	0.6213	0.3414	-0.5933	0.1697
3	-0.0929	0.3627	0.4624	0.2899	-0.6622	-0.3384	-0.0944
5	0.0160	0.0727	0.1665	-0.0074	-0.0642	0.1499	0.9696
7	0.2045	-0.4858	0.7851	-0.0898	0.2828	-0.1159	-0.0658
9	0.6858	0.0741	-0.2203	-0.3511	-0.0818	-0.5790	0.1023
11	-0.6302	-0.2862	-0.0585	-0.5843	-0.1899	-0.3653	0.0813
13	0.1967	-0.7089	-0.2319	0.2392	-0.5640	0.1701	0.0279
J=12.5							
1	0.2334	0.2298	0.3058	-0.5249	-0.3524	0.6087	-0.1703
3	0.0977	-0.2748	-0.6360	-0.2859	0.4865	0.4352	0.0510
5	-0.1799	0.4164	-0.5924	0.0305	-0.3254	-0.1119	-0.5692
7	-0.4325	0.3920	-0.1810	0.1779	-0.1943	0.3361	0.6664
9	-0.7095	-0.2687	0.2704	0.1668	0.1620	0.3611	-0.4084
11	0.4257	0.3189	0.0777	0.6789	0.2792	0.3743	-0.1789
13	-0.1786	0.6088	0.1978	-0.3484	0.6256	-0.2100	-0.0390

[a]From a calculation by Lefebvre-Brion, part of which has been published by Kimman, *et al.*(1985).

[b]Odd-eigenstate indices correspond to the e–parity levels. The eigenvalues for each value of J are ranked in order of energy (index 1 for the lowest eigenstate, index 13 for the highest).

Each observed rotational eigenstate may be expressed as an explicit linear combination of deperturbed basis substates. The mixing coefficients are obtained from the eigenvectors of the four-state effective Hamiltonian (which are generated when the Hamiltonian is numerically diagonalized). For each value of J (except $J = \frac{1}{2}$) there are seven e–parity and seven f–parity basis substates (three $\Omega = \frac{1}{2}$ and three $\Omega = \frac{3}{2}$ $^2\Pi$ functions and one $^2\Sigma^+$ function). The eigenstates of the deperturbation Hamiltonian are specified by J and an integer between 1 and 14 (even = f, odd = e, 1 and 2 lowest energy, 13 and 14 highest energy). The mixing coefficients for the case (a) basis functions in the $J = 0.5, 8.5$, and 12.5 e–parity eigenfunctions are given in Table 4.1. For $J = 0.5$, only $\Omega = \frac{1}{2}$ basis states exist and the observed levels are exclusively of $\Omega = \frac{1}{2}$ character. At higher J–values, **S**-uncoupling within the $^2\Pi$ states causes Ω to become poorly defined. For example, the $J = 12.5e$ level number 7 has

$$\begin{aligned}\overline{\Omega} \equiv \langle J_z \rangle &= 3/2\left[(0.4325)^2 + (0.1779)^2 + (0.3361)^2\right] \\ &\quad + 1/2\left[(0.3920)^2 + (0.1810)^2 + (0.1943)^2 + (0.6664)^2\right] \\ &= 0.83.\end{aligned}$$

The electronic and vibrational character of this level is also poorly defined

22.0%	$B^2\Pi$	v_B=24
26.7%	$L^2\Pi$	v_L=5
6.9%	$C^2\Pi$	v_C=6
44.4%	$D^2\Sigma^+$	v_D=6

and no basis state accounts for more than 50% of its character. Thus even the *nominal* vibronic character of this level is poorly defined. The only meaningful labels that can be attached to the observed levels are their good quantum numbers: J, parity, and energy. Nevertheless, the observable properties of each level may be calculated from those of the basis states properly weighted by the mixing coefficients obtained from the eigenvectors of the deperturbation Hamiltonian.

4.6 References

Albritton, D. L., Harrop, W. J., Schmeltekopf, A. L., and Zare, R. N. (1973), *J. Mol. Spectrosc.* **46**, 25.

Albritton, D. L., Schmeltekopf, A. L., and Zare, R. N. (1976), *in* "Molecular Spectroscopy: Modern Research" Vol. 2, p. 1, (K. N. Rao, ed.), Academic Press, New York.

Albritton, D. L., Schmeltekopf, A. L., and Zare, R. N. (1977), *J. Mol. Spectrosc.* **67**, 132.

Amiot, C., and Vergès, J. (1982), *Phys. Scr.* **25**, 302 (Note a sign error in the off-diagonal matrix element between $^2\Pi_{1/2}$ and $^2\Pi_{3/2}$).

Åslund, N. (1974), *J. Mol. Spectrosc.* **50**, 424.

Atabek, O., and Lefebvre, R. (1980), *Chem. Phys.* **52**, 199.
Birss, F. W., Brown, J. M., Cole, A. R. H., Lofthus, A., Krishnamachari, S. L. N. G., Osborne, G. A., Paldus, J., Ramsay, D. A., and Watmann, L. (1970), *Can. J. Phys.* **48**, 1230.
Birss, F. W. (1983), *J. Mol. Spectrosc.* **99**, 133.
Bishop, D. M., and Cheung, L. M. (1978), *Phys. Rev. A* **18**, 1846.
Brand, J. C. D., Jones, V. T., and DiLauro, C. (1971), *J. Mol. Spectrosc.* **40**, 616.
Brown, J. M., and Milton, D. J. (1976), *Mol. Phys.* **31**, 409.
Brown, J. M., and Merer, A. J. (1979), *J. Mol. Spectrosc.* **74**, 488.
Brown, J. M., Colbourn, E. A., Watson, J. K. G., and Wayne, R. D. (1979), *J. Mol. Spectrosc.* **74**, 294.
Chedin, A., and Cihla, Z. (1967), *Cah. Phys.* **219**, 129.
Cheung, W. Y., Chupka, W. A., Colson, S. D., Gauyacq, D., Avouris, P. , and Wynne, J. J. (1986), *J. Phys. Chem.* **90**, 1086.
Child, M. S., and Lefebvre, R. (1978), *Chem. Phys. Lett* **55**, 213.
Coxon, J. A. (1978), *J. Mol. Spectrosc.* **72**, 252.
Curl, R. F. (1970), *J. Comput. Phys.* **6**, 367.
Dabrowski, I., Tokaryk, D. W., and Watson, J. K. G. (1998), *J. Mol. Spectrosc.* **189**, 95.
De Santis, D., Lurio, A., Miller, T. A., and Freund, R. S. (1973), *J. Chem. Phys.* **58**, 4625.
van Dishoeck, E. F., van Hemert, M. C., Allison, A. C., and Dalgarno, A. (1984), *J. Chem. Phys.* **81**, 5709.
Dressler, K., and Miescher, E. (1981), *J. Chem. Phys.* **75**, 4310.
Dressler, K., Gallusser, R., Quadrelli, P., and Wolniewicz, L. (1979), *J. Mol. Spectrosc.*. **75**, 205.
Dunker, A. M., and Gordon, R. G. (1976), *J. Chem. Phys.* **64**, 4984.
Felenbok, P., and Lefebvre-Brion, H. (1966), *Can. J. Phys.* **44**, 1677.
Feynman, R. P. (1939), *Phys. Rev.* **56**, 340.
Field, R. W., and Bergeman, T. H. (1971), *J. Chem. Phys.* **54**, 2936.
Field, R. W., Gottscho, R., and Miescher, E. (1975), *J. Mol. Spectrosc.* **58**, 394.
Field, R. W., Lagerqvist, A., and Renhorn, I. (1976), *Phys. Scr.* **14**, 298.
Freed, K. F. (1966), *J. Chem. Phys.* **45**, 4214.
Gallusser, R., and Dressler, K. (1982), *J. Chem. Phys.* **76**, 4311.
Gerö, L. (1935), *Z. Phys.* **93**, 669.
Glass-Maujean, M., Quadrelli, P., Dressler, K., and Wolniewicz, L. (1983), *Phys. Rev. A* **28**, 2868.
Hellmann, H. (1937), "Einführung in die Quantenchemie," Franz Deuticke, Vienna.
Herschbach, D. (1956), unpublished notes.
Hill, E. L., and Van Vleck, J. H. (1928), *Phys. Rev.* **32**, 250.
Horani, M., Rostas, J., and Lefebvre-Brion, H. (1967), *Can. J. Phys.* **45**, 3319.
Hunter, G., and Pritchard, H. O. (1967), *J. Chem. Phys.* **46**, 2153.
Jacobson, M. P., Coy, S. L., and Field, R. W. (1997), *J. Chem. Phys.* **107**, 8349.
Johns, J. W. C., and Lepard, D. W. (1975), *J. Mol. Spectrosc.* **55**, 374.
Johnson, B. R. (1978), *J. Chem. Phys.* **69**, 4678.
Jungen, C., and Miescher, E. (1968), *Can. J. Phys.* **46**, 987.
Kimman, J., Lavollée, M., and Van der Wiel, M. J. (1985), *Chem. Phys.* **97**, 137.
Kirschner, S. M., and Watson, J. K. G. (1973), *J. Mol. Spectrosc.* **47**, 234.
Kovács, I. (1969), "Rotational Structure in the Spectra of Diatomic Molecules," Am. Elsevier, New York.
Lagerqvist, A., and Miescher, E. (1958), *Helv. Phys. Acta* **31**, 221.

Lagerqvist, A., and Miescher, E. (1966), *Can. J. Phys.* **44**, 1525.
Lees, R. M. (1970), *J. Mol. Spectrosc.* **33**, 124.
Lefebvre-Brion, H. (1969), *Can. J. Phys.* **47**, 541.
Lefebvre-Brion, H., and Colin, R. (1977), *J. Mol. Spectrosc.* **65**, 33.
Le Floch, A. C., Launay, F., Rostas, J., Field, R. W., Brown, C. M., and Yoshino, K. (1987), *J. Mol. Spectrosc.* **121**, 337.
Le Roy, R. J. (1998), *J. Mol. Spectrosc.* **191**, 223.
Li, Y., Buenker, R. J., and Hirsch, G. (1998), *Theor. Chem. Acc.* **100**, 112.
Löwdin, P. O. (1951), *J. Chem. Phys.* **19**, 1396.
Marquardt, D. W. (1963), *J. Soc. Indust. Appl. Math.* **11**, 431.
Marquardt, D. W., Bennett, R. G., and Burrell, E. J. (1961), *J. Mol. Spectrosc.* **7**, 269.
Meakin, P., and Harris, D. O. (1972), *J. Mol. Spectrosc.* **44**, 219.
Merer, A. J., and Allegretti, J. M. (1971), *Can. J. Phys.* **49**, 2859.
Mies, F. M. (1980), *Mol. Phys.* **41**, 953, 973.
Miller, T. A. (1969), *Mol. Phys.* **16**, 105.
Monnerville, M., and Robbe, J. M. (1999), *Eur. Phys. J. D* **5**, 381.
Mulliken, R. S., and Christy, A. (1930), *Phys. Rev.* **38**, 87.
Ngo, T. A., DaPaz, M., Coquart, B., and Couet, C. (1974), *Can. J. Phys.* **52**, 154.
Press, W.H., Teukolsky, S. A., Vetterling, W. T., and Flannery, B. P. (1996), "Numerical Recipes: The Art of Scientific Computing," Cambridge University Press, Cambridge.
Raoult, M. (1987), *J. Chem. Phys.* **87**, 4736.
Ross, S. C., and Jungen, C. (1994), *Phys. Rev. A* **50**, 4618.
Ruckstuhl, A. F., Stahel, W. A., and Dressler, K. (1993), *J. Mol. Spectrosc.* **160**, 434.
Ruckstuhl, A. F. and Dressler, K. (1994), *J. Mol. Spectrosc.* **168**, 185.
Ruckstuhl, A. F., and Dressler, K. (1996), *J. Mol. Spectrosc.* **176**, 185.
Ruckstuhl, A. F., Jacobson, M. P., Field, R. W., and Dodd, J. A. (2001), *J. Quant. Spectrosc. Radiat. Transf.* **68**, 179.
Shapiro, M. (1982), *Chem. Phys. Lett.* **91**, 12.
Stahel, D., Leoni, M., and Dressler, K. (1983), *J. Chem. Phys.* **79**, 2541.
Stern, R. C., Gammon, R. H., Lesk, M. E., Freund, R. S., and Klemperer, W. A. (1970), *J. Chem. Phys.* **52**, 3467.
Tellinghuisen, J. (1973), *Chem. Phys. Lett.* **18**, 544.
Tellinghuisen, J. (1995), *J. Mol. Spectrosc.* **173**, 308.
Tellinghuisen, J. (1996), *J. Mol. Spectrosc.* **179**, 299.
Tinkham, M., and Strandberg, M. W. P. (1955), *Phys. Rev.* **97**, 937.
Torop, L., McCoy, D. G., Blake, A. J., Wang, J., and Scholz, T. (1987), *J. Quant. Spectrosc. Radiat. Tranfer* **38**, 9.
Watson, J. K. G. (1977), *J. Mol. Spectrosc.* **66**, 500.
Watson, J. K. G. (2003), *J. Mol. Spectrosc.* **219**, 326.
Wentworth, W. E. (1965), *J. Chem. Educ.* **42**, 96.
Wicke, B. G., Field, R. W., and Klemperer, W. (1972), *J. Chem. Phys.* **56**, 5758.
Wollrab, J. E. (1967), "Rotational Spectra and Molecular Structure," Appendix 7, Academic Press, New York.
Wolniewicz, L. (1995), *J. Chem. Phys.* **103**, 1792.
Wolniewicz, L. (1996), *J. Chem. Phys.* **105**, 10691.
Yu, S., and Dressler, K. (1994), *J. Chem. Phys.* **101**, 7692.
Zare, R. N., Schmeltekopf, A. L., Harrop, W. J., and Albritton, D. L. (1973), *J. Mol. Spectrosc.* **46**, 37.

Chapter 5

Interpretation of the Perturbation Matrix Elements

Contents

5.1 Calculation of the Vibrational Factor 278

5.1.1 Semiclassical Approximation 279

5.1.2 Model Potentials 285

5.1.3 Numerical Potentials and Vibrational Wavefunctions 288

5.1.4 Some Remarks about "Borrowed" Computer Programs 291

5.1.5 Vibrational Assignment by the Matrix Element Method 293

5.1.6 Homogeneous vs. Heterogeneous Perturbations . . . 305

5.2 Order of Magnitude of Electrostatic Perturbation Parameters: Interactions Between Valence and Rydberg States of the Same Symmetry 307

5.2.1 Valence and Rydberg States 308

5.2.2 Different Classes of Valence~Rydberg Mixing 310

5.2.3 Electrostatic Perturbations Between Valence and Rydberg States 312

5.2.4 Electrostatic Perturbations between Rydberg States Converging to Different States of the Ion 314

5.3 Order of Magnitude of Spin Parameters 315

5.3.1 Diagonal Spin-Orbit Parameters 318

5.3.2 Off-Diagonal Spin-Orbit Parameters 322

5.3.3 Spin-Spin Parameters 323

5.4 Magnitudes of Rotational Perturbation Parameters 325

5.5 Pure Precession Approximation 327

5.6 R–Dependence of the Spin Interaction Parameters 333

5.7 **Beyond the Single-Configuration Approximation** 340
5.8 **Identification and Location of Metastable States by Perturbation Effects** 341
5.9 **References** 342

The magnitudes of perturbation matrix elements are seldom tabulated in compilations of molecular constants. If deperturbed diagonal constants are listed, then the off-diagonal perturbation parameters should be listed as well, even though they cannot, without specialized narrative footnotes, be accommodated into the standard tabular format of such compilations. Without specification of at least the electronic part of the interaction parameters, it is impossible to reconstruct spectral line frequencies or intensities; thus the deperturbed diagonal constants by themselves have no meaning.

The purpose of this chapter is to show how to extract useful information from perturbation matrix elements. The uses of this information range from tactical (vibrational and electronic assignments) to insight into the global electronic structure of a molecule or family of molecules. It is tempting to suggest that perturbation matrix elements can contain at least as much structural information and vastly more dynamical information than the usual molecular constants.

The magnitude of the perturbation interaction, $H_{1,v_i,J;2,v_j,J}$, is determined by the product of two factors, one electronic and the other vibrational, both of which can depend on J. Recall (Section 3.3.1) that if the electronic factor varies linearly with internuclear distance in the neighborhood of the R–value of the crossing between the potential curves of states 1 and 2, R_C, then

$$H_{1,v_i,J;2,v_j,J} = H_{12}^e\left(\overline{R}_{v_i,v_j,J}\right)\langle v_i,J|v_j,J\rangle\,,$$

where the R–centroid for each pair of near-degenerate perturbing levels has the value

$$\overline{R}_{v_i,v_j} = \frac{\langle v_i|R|v_j\rangle}{\langle v_i|v_j\rangle} \simeq R_C$$

with a precision better than ±5%[†] Thus perturbations between states 1 and 2 sample $H_{12}^e(R,J)$ at $R = R_C$ and the magnitude of the electronic factor is nearly independent of v_i, v_j, provided that the curves cross exactly once and that the interacting vibrational levels are near-degenerate.

The observability of a perturbation between levels $1, v_i, J \sim 2, v_j, J$ requires that *both* the electronic and vibrational factors be nonzero. For example,

---

[†] The vibrational part of H_{12} is not always simply a vibrational overlap integral. For example, matrix elements of $\mathbf{H}^{\mathrm{ROT}}$ include the factor

$$\begin{aligned}\langle v_i,J|\mathbf{B}(R)|v_j,J\rangle &= (\hbar/4\pi\mu c)\,\langle v_i,J|R^{-2}|v_j,J\rangle\\ &\simeq \langle v_i,J|v_j,J\rangle\,R_C^{-2}\hbar\big/4\pi\mu c\\ &= 16.85763\frac{\langle v_i,J|v_j,J\rangle}{R_c^2[\text{Å}^2]\mu[\text{amu}]}\ [\text{cm}^{-1}].\end{aligned}$$

although the $J = 12.5$ levels of the $B^2\Pi_{1/2}$ ($v = 18$) and $B'^2\Delta_{5/2}$ ($v = 1$) states† of $^{15}N^{18}O$ are separated by 0.3 cm^{-1} and $H^e_{B_{3/2}\sim B'_{3/2}}$ is estimated to be > 18 cm^{-1}, no perturbation was detectable within the 0.1 cm^{-1} measurement precision (Field, *et al.*, 1975). This nonperturbation is explained by the computed $v_B = 18 \sim v_{B'} = 1$ vibrational overlap factor of 8×10^{-3}. Alternatively, if a perturbation is observed when a computed vibrational factor is prohibitively small, then the electronic or vibrational identification of either the perturbed or perturbing state is incorrect. For example, the $^{31}P^{16}O$ $D^2\Pi$ ($v = 0$) level is perturbed by what is now known (Coquart, *et al.*, 1974; Ghosh, *et al.*, 1975) to be a highly excited vibrational level of $B'^2\Pi$ (Verma and Dixit, 1968; Verma, 1971). Upon isotopic substitution, the $^{31}P^{18}O$ $D^2\Pi$ ($v = 0$) level is found to be perturbed by a level of $B'^2\Pi$ with T_{ve} nearly identical to that found for the $^{31}P^{16}O$ perturber. Verma initially assumed the same vibrational quantum number of the perturber for both isotopic species and concluded from the absence of a large isotope shift that $v_{\text{pert}} = 0$ (Verma and Dixit, 1968). However, if a potential curve for the perturber is derived from the incorrect vibrational numbering and molecular constants, the $v_D = 0 \sim v_{\text{pert}} = 0$ vibrational overlap is computed to be too small for any perturbation to have been detectable. It turns out that the $^{31}P^{16}O$ perturber of $D^2\Pi$ ($v = 0$) is $B'^2\Pi$ ($v = 24$) and, by chance, the isotope shift for the B' state is nearly equal to the $\Delta G(24.5)$ value so that the perturber for $^{31}P^{18}O$ is actually $B'^2\Pi$ ($v = 25$) (Verma, 1971; Coquart, *et al.*, 1974; Ghosh, *et al.*, 1975). This situation is illustrated by Fig. 5.1.

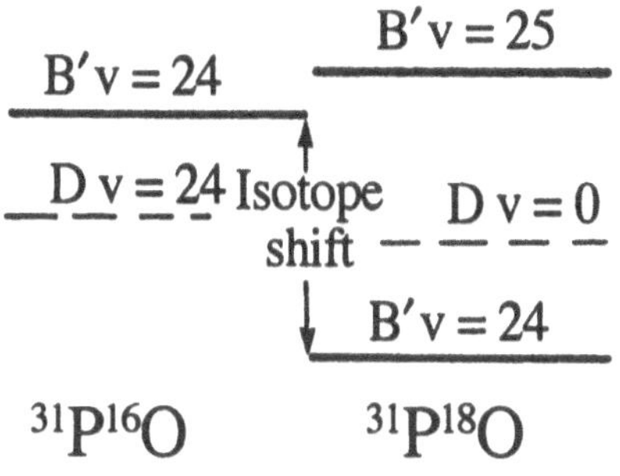

Figure 5.1: Isotope effect on a vibrational level of the $B'^2\Pi$ state of the PO molecule.

The electronic factor can also provide an explanation for a nonperturbation and a test of a vibrational assignment (Robbe, *et al.*, 1981). However, this factor can only be computed *ab initio* or, sometimes, estimated semiempirically. This dependence on theoreticians for the electronic factor is in contrast to the situation for the vibrational factor, which is routinely calculable by the

† Direct perturbations between *basis functions* differing in Ω by ± 2 are rigorously forbidden. However, at the $J = 12.5$ crossing between *nominal* $B_{1/2}$ and $B'_{5/2}$ substates, spin-uncoupling has introduced significant $\Omega = \frac{3}{2}$ basis function character into both nominal levels.

experimentalist using widely available computer programs. Although *ab initio* computations of $H^e_{12}(R_C)$ are not always accessible to the spectroscopist, the fact that the electronic factor, for an important class of perturbations considered here, is vibration-independent is the basis of a powerful method for determining vibrational and electronic assignments of perturbers. This "matrix element method" is described in Section 5.1.5.

The ultimate test of a deperturbation consists of: (1) a demonstration that all observed perturbation matrix elements have the v, J–dependence required by the factorization into electronic and vibrational parts; (2) agreement between the observed and *ab initio* values of $H^e_{12}(R_C)$; (3) verification that the molecular constants for both electronic states are internally consistent [isotope shifts, D_v values calculable from $G(v)$ and B_v functions] (Gottscho, *et al.*, 1979; Kotlar, *et al.*, 1980).

Section 5.1 and Sections 5.2-5.4 deal respectively with the vibrational and electronic factors of the perturbation matrix element. The vibrational wavefunctions and matrix elements may be computed semiclassically (Section 5.1.1), derived analytically for special potentials, such as the Morse potential (Section 5.1.2), or obtained numerically via the RKR inversion followed by numerical integration (Section 5.1.3). In discussing the electronic factor, it is useful to distinguish between electrostatic perturbations (Section 5.2), which appear mainly between valence and Rydberg states, and spin or rotational perturbations, which appear mainly between different valence states (Sections 5.3 and 5.4). The distinction between valence and Rydberg states is discussed in Section 5.2.1. Electrostatic perturbations are divided into valence~Rydberg interactions (Sections 5.2.2 and 5.2.3) and Rydberg~Rydberg interactions (Section 5.2.4). Section 5.3 shows for the spin parameters, and Section 5.4 for the rotational parameters how comparisons of corresponding perturbations for isovalent molecules provide global insights into molecular electronic structure. The pure precession approximation is discussed in Section 5.5. After brief discussion of the origin of the R–variation of the electronic factor (Section 5.6) and the validity of the single configuration approximation (Section 5.7), it is shown how the study of perturbations can help to identify and locate metastable states (Section 5.8).

5.1 Calculation of the Vibrational Factor

Most spectroscopists are comfortable using RKR and Franck-Condon programs as black boxes. Since these programs are widely available, inexpensive to run, and generate more accurate vibrational factors than semiclassical or model potential approaches, the application-minded reader is advised to skip Sections 5.1.1 and 5.1.2.

5.1.1 Semiclassical Approximation

The power of this semiclassical approach is that it enables derivation of analytic expressions for the overlap integrals between vibrational eigenfunctions of two potentials. Since, for perturbations, one often knows one potential very well and the other potential poorly or not at all, these analytic relationships can provide a direct link between observed vibrational overlap integrals and some features of the poorly known potential, namely, the location of the crossing between the two potentials, E_C and R_C, and the slope of the unknown potential at R_C.

The physical idea behind this method is that, at R_C, if (E_{1,v_i}, χ_{v_i}) and (E_{2,v_j}, χ_{v_j}) are associated, respectively, with potentials 1 and 2, and $E_{1,v_i} \approx E_{2,v_j}$, then the vibrational wavefunctions χ_{v_i} and χ_{v_j} oscillate at the same frequency near R_C. The magnitude of the overlap integral is determined largely in the region where the two rapidly oscillatory functions oscillate at the same frequency (Tellinghuisen, 1984). The length of this region (how long it takes for the phase of χ_{v_i} to change by π relative to that of χ_{v_j}) is long or short depending on whether V_1 and V_2 have similar or different slopes at R_C. Three factors contribute to the magnitude of $\langle v_i|v_j\rangle$: the difference in slopes at R_C, the product of the maximum amplitudes of χ_{v_i} and χ_{v_j} near R_C, and the phase of χ_{v_i} relative to χ_{v_j} at R_C. The amplitude of χ at R_C is related to the classical probability density, which is inversely proportional to the velocity (v) times the oscillation period, $1/\overline{\omega}$; thus the semiclassical amplitude will be proportional to

$$(\overline{\omega}/\mathrm{v})^{1/2} = \overline{\omega}^{1/2}(\mu/2)^{1/4}[E - V(R_C)]^{-1/4}$$

or

$$(\overline{\omega}/\mathrm{v})^{1/2} = |\overline{\omega}\mu/\hbar k(R_C)|^{1/2}$$

where μ is the reduced mass and k is the wavenumber, defined below.

Given a potential energy curve, it is possible to locate (iteratively) the vibrational energy levels using the semiclassical quantization condition

$$\int_a^b k(R)dR = (v + \frac{1}{2})\pi \tag{5.1.1}$$

$$k(R) = p(R)/\hbar = \left\{[2\mu]^{1/2}/\hbar\right\}[E - V(R)]^{1/2} \tag{5.1.2}$$

where p is the classical momentum associated with the kinetic energy, $E - V(R)$, and where a and b, respectively, are the left and right classical turning points [at which the total energy, E, is equal to the potential energy, $V(R)$] (Fig. 5.2). The eigenfunctions may be approximated by the semiclassical JWKB wavefunction (J=Jeffreys, W=Wentzel, K=Kramers, B=Brillouin),

$$\chi^{\mathrm{JWKB}}(R) = N[k(R)]^{-1/2}\sin[\phi(R) + \pi/4] \tag{5.1.3}$$

$$\phi(R) = \int_a^R k(R')dR', \tag{5.1.4}$$

where N is a normalization factor. This χ^{JWKB} is a good approximation to the exact wavefunction between a and b but *not too close* to either turning point (see Merzbacher, 1998, p. 115).

The JWKB wavefunction defined by Eq. (5.1.3) is appropriate for either bound (two turning points) or unbound states (one turning point), provided that R is restricted to the region where $E > V(R)$. The normalization factor for the unbound χ^{JWKB} at energy E is

$$N_E = |2\mu/\pi\hbar^2|^{1/2}, \tag{5.1.5}$$

which corresponds to unity probability of finding the system with energy between E and $E + dE$. For a bound level, the normalization factor is

$$N_v = |2\mu\hbar\overline{\omega}/\pi\hbar^2|^{1/2}, \tag{5.1.6}$$

which corresponds to unity probability of finding the system in level v. N_E and N_v are related (discussed in Section 7.5) by

$$\hbar\overline{\omega} = \left(\frac{N_v}{N_E}\right)^2, \tag{5.1.7}$$

where $\hbar\overline{\omega}$ is the local energy-level spacing, which can be approximated by

$$\hbar\overline{\omega} \simeq \frac{E_{v+1} - E_{v-1}}{2}$$

or computed from the semiclassical quantization condition [Eq. (5.1.1)]

$$\hbar\overline{\omega} = \frac{\partial E}{\partial v} = \left(\frac{\partial v}{\partial E}\right)^{-1} = \left\{\frac{\partial}{\partial E}\left[\frac{1}{\pi}\int_a^b k(R)dR - \frac{1}{2}\right]\right\}^{-1}$$

$$\hbar\overline{\omega} = \left(\frac{1}{\pi}\int_a^b \frac{dk}{dE}dR\right)^{-1} = \frac{\pi\hbar^2}{\mu}\left[\int_a^b k^{-1}(R)dR\right]^{-1} \tag{5.1.8}$$

since

$$\frac{dk}{dE} = \frac{\mu}{\hbar^2}\frac{1}{k}. \tag{5.1.9}$$

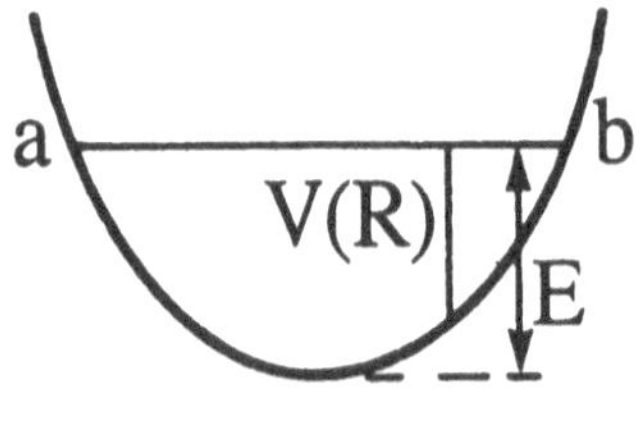

Figure 5.2: Bound level in a potential.

In order to take care of the problem that χ^{JWKB} is not a good approximation near the turning points, it is necessary to define $\chi^{\mathrm{USC}}(R)$, USC indicating uniform semiclassical wavefunction (Miller and Good, 1953; Langer, 1937; Miller, 1968), which is uniformly valid close to or far from the turning points

$$\chi^{\mathrm{USC}}(R) = \pi^{1/2} N \left[\frac{Z(R)}{k(R)^2}\right]^{1/4} \mathrm{Ai}[-Z(R)] \tag{5.1.10}$$

$$Z(R) = \left[\frac{3}{2}\int_a^R k(R')dR'\right]^{2/3} = \left[\frac{3}{2}\phi(R)\right]^{2/3}. \tag{5.1.11}$$

$Ai[-Z(R)]$ is an Airy function, which is a decreasing exponential at large positive argument ($Z < 0, k$ imaginary, nonclassical region) and oscillatory at large negative argument ($Z > 0, k$ real, classical region). For $Z > 1.5$,

$$\mathrm{Ai}[-Z(R)] \longrightarrow \pi^{-1/2} Z(R)^{-1/4} \sin[\phi(R) + \pi/4].$$

Thus χ^{USC} is constructed so that when R is in the classical region but far from a turning point, $\chi^{\mathrm{USC}}(R) = \chi^{\mathrm{JWKB}}(R)$. Note also that for a linear potential,

$$V(R) = -FR$$

(F is an R-independent force between nuclei), the function [Eq. (5.1.10)] is the exact eigenfunction.

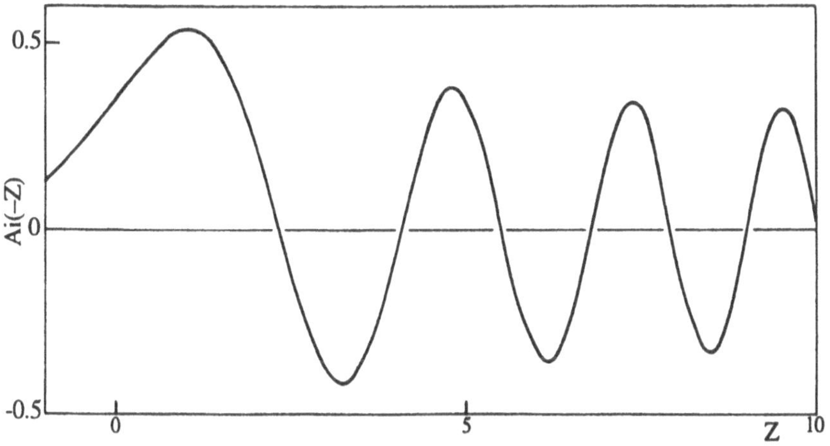

Figure 5.3: The regular Airy function $\mathrm{Ai}(-Z)$.

A typical Airy function, $\mathrm{Ai}(-Z)$, is plotted in Fig. 5.3; note, however, that the argument of the Airy function that appears in χ^{USC} is a function of R; thus a plot of $\mathrm{Ai}[-Z(R)]$ vs. R does not behave exactly like Fig. 5.3.

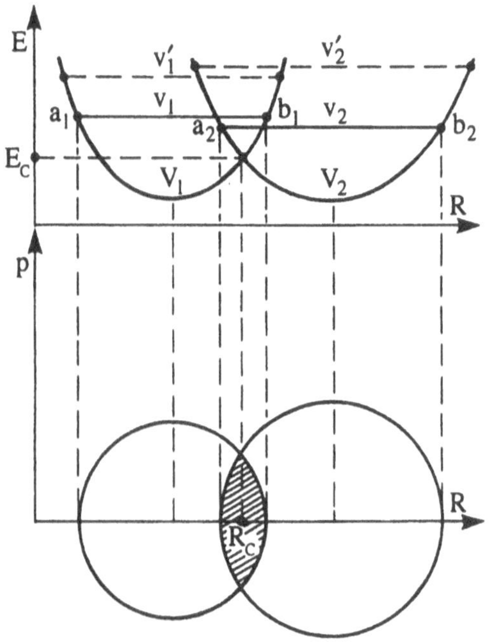

Figure 5.4: Phase space mapping for the harmonic model. Two pairs of exactly degenerate interacting levels are shown. Since V_2 is less steep than V_1, $\omega_2 < \omega_1$ and $(v_1' - v_1) < (v_2' - v_2)$.

The immediate objective is to obtain an analytic expression for vibrational overlap integrals, $\langle v_i | v_j \rangle$. Conveniently, the integral over the product of the Airy functions can be shown (Connor, 1981) to be approximately another Airy function (in fact, rigorously true for linear potentials). If

$$E = E_{1,v_i} = E_{2,v_j},$$

then

$$\langle v_i | v_j \rangle = \left| \frac{2(\hbar\overline{\omega}_1)(\hbar\overline{\omega}_2)\mu}{\hbar^2 k(R_C)\,|(dV_1/dR) - (dV_2/dR)|_{R=R_C}} \right|^{1/2} \xi(E)^{1/4} \mathrm{Ai}[-\xi(E)] \tag{5.1.12}$$

$$\hbar k(R_C) = (2\mu)^{1/2}(E - E_C)^{1/2} \tag{5.1.13}$$

$$\xi(E) = \left[\left(\frac{3}{2}\right) \phi(E) \right]^{2/3} \tag{5.1.14}$$

$$\phi(E) = \int_{a_2}^{R_C} k_2(R)dR + \int_{R_C}^{b_1} k_1(R)dR, \tag{5.1.15}$$

where E_C and R_C are the energy and R value of the curve crossing. Equation (5.1.15) is valid for the specific case where the crossing point lies between the minima of the two potential curves (Fig. 5.4). The Airy function that appears in Eq. (5.1.12) is a function of E rather than R. Note that if $\xi(E) < 0$ (ϕ is imaginary), $\langle v_i|v_j\rangle$ is an exponentially decreasing function of $E_C - E$, whereas if $\xi(E) > 0$ (ϕ is real), the overlap is an oscillatory function of $E - E_C$. $\phi(E)$ is real if $E > E_C$; thus, as E approaches E_C from below, $|\langle v_i|v_j\rangle|$ increases monotonically to an absolute maximum just above $E = E_C$ and then passes through a series of increasingly smaller relative maxima. For $\xi(E) > 1.5$ ($E > E_C, \phi$ real),

$$\langle v_i|v_j\rangle \to \left|\frac{2(\hbar\overline{\omega}_1)(\hbar\overline{\omega}_2)\mu}{\hbar(2\mu)^{1/2}\,|(dV_1/dR)-(dV_2/dR)|_{R=R_C}}\right|^{1/2} \times \pi^{-1/2}(E-E_C)^{-1/4}\sin[\phi(E)+\pi/4] \tag{5.1.16}$$

and for $\xi(E) < -1.5$ ($E_C > E, i\phi = \Phi < 0, \Phi$ is real),

$$\langle v_i|v_j\rangle \to \left|\frac{2(\hbar\overline{\omega}_1)(\hbar\overline{\omega}_2)\mu}{\hbar(2\mu)^{1/2}\,|(dV_1/dR)-(dV_2/dR)|_{R=R_C}}\right|^{1/2} \times \frac{\pi^{-1/2}}{2}(E_C-E)^{-1/4}e^{-\Phi(E)}. \tag{5.1.17}$$

Figure 5.4 is a pictorial description of $\phi(E)$ for $E \geq E_C$. In phase space, the classical motion of a system bound by a harmonic potential appears as a circle[†] with radius

$$p = (2\mu)^{1/2}[E - V(R_{min})]^{1/2}$$

centered at $p = 0, R = R_{min}$. The two circles in Fig. 5.4 intersect at $R = R_C$ and their overlap, shown shaded, is $2\phi(E)$. Obviously, a single experimentally measured perturbation matrix element will be insufficient to determine the unknown potential. This semiclassical method is based on the observed pattern of $\langle v_i|v_j\rangle$ versus v_i and upon isotopic substitution. The same sort of information forms the basis of a fully quantum mechanical, but trial-and-error, method for characterizing the unknown potential (Section 5.1.5).

Note that the three factors mentioned at the beginning of this section as determining the size of $\langle v_i|v_j\rangle$ appear explicitly in Eq. (5.1.12). The overlap integral contains

$$\left|\frac{dV_1}{dR} - \frac{dV_2}{dR}\right|_{R=R_C}^{-1/2}, \qquad \left[\frac{\overline{\omega}\mu}{\hbar k(R_C)}\right]^{1/2},$$

and a factor, $\phi(E)$, expressing the phase relationship at R_C.

[†]The motion in phase space is a circle rather than an ellipse only when $\mu\omega = 1$.

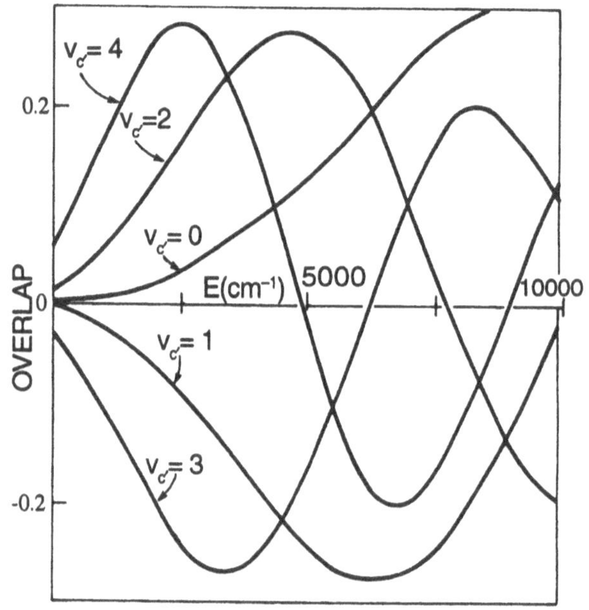

Figure 5.5: Semiclassical calculation of vibrational overlap integrals between successive vibrational levels of the N_2 $c'^1\Sigma_u^+$ Rydberg state and the bound region of the $b'^1\Sigma_u^+$ valence state where the vibrational energy in the b′–state is treated as a continuous variable. Both states are represented by Morse potentials. The zero of energy is taken as $v = 0, J = 0$ of $b'^1\Sigma_u^+$. For the E values corresponding to the bound vibrational levels of the $b'^1\Sigma_u^+$ state, the values calculated using Eq. (5.1.12) reproduce the exact values (Table V of Lefebvre-Brion, 1969) to two decimal places. (Courtesy R. Lefebvre.)

The $(E - E_C)^{-1/4}\xi(E)^{1/4}Ai[-\xi(E)]$ terms in Eq. (5.1.12) contain all information about the locations of the maxima and zeroes in the oscillatory variation of $\langle v_i|v_j\rangle$ with E, but the constant term,

$$\left|\frac{2(\hbar\overline{\omega}_1)(\hbar\overline{\omega}_2)\mu}{\hbar(2\mu)^{1/2}\left|(dV_1/dR) - (dV_2/dR)\right|_{R=R_C}}\right|^{1/2}$$

determines the magnitudes of $\langle v_i|v_j\rangle$ at its relative maxima. Absolute magnitudes of the vibrational overlap cannot be measured directly, but they can be inferred from the measured $H_{1,v_i;2,v_j}$ if an *ab initio* value of H_{12}^e is available. Alternatively, if sets of $H_{1,v_i;2,v_j}$ are available for two isotopic species, then the isotopic ratios of relative maxima in $\langle v_i|v_j\rangle$ versus E are obtainable from Eq. (5.1.12). Since the respective reduced mass-dependence of $k, \overline{\omega}, \phi(E)$, and $\xi(E)$ are $\mu^{+1/2}, \mu^{-1/2}, \mu^{+1/2}$, and $\mu^{+1/3}$, the first maximum value of $\langle v_i|v_j\rangle$ is proportional to $\mu^{-1/6}$,

$$\frac{\langle v_i|v_j\rangle_{E_{max,l}}}{\langle v_i|v_j\rangle_{E_{max,h}}} = \left(\frac{\mu_l}{\mu_h}\right)^{-1/6} \tag{5.1.18}$$

The accuracy of the semiclassical approach to $\langle v_i|v_j\rangle$ (typically better than $\pm 10\%$) is not limited to the case $E_{1,v_i} = E_{2,v_j}$.[†] This approach is particularly valuable for displaying the general qualitative form of $\langle v_i|v_j\rangle$ versus E (see Fig. 5.5 for an example) and for relating certain crucial features of the experimentally well-known oscillatory behavior of $H_{1,v_i;2,v_j}$ (Lagerqvist and Miescher, 1958, Fig. 16) to $E_C, R_C, (dV_2/dR)|_{R=R_C}$, and $a_2 - R_C$ or $b_2 - R_C$, where V_2 is the unknown potential. However, the semiclassical approach has not been widely applied to bound~bound interactions, its major use having been bound~free interactions (Section 7.6).

5.1.2 Model Potentials

When an electronic state is known only through its perturbations of a better known state, frequently only the energy and rotational constant of one or two vibrational levels of unknown absolute vibrational numbering can be determined. If the information available is insufficient to generate a realistic potential energy curve, then one has no choice but to adopt a model potential and exploit relationships between Dunham (Y_{lm}) and other derived constants (vibrational overlaps), which are rigorously valid for the model potential and approximately valid for general potentials.

The two most useful primitive model potentials are the harmonic and Morse oscillators,

$$V^{\text{harmonic}}(R)/hc = \frac{1}{2}(k/hc)(R - R_e)^2 \text{ cm}^{-1} \tag{5.1.19}$$

$$k/hc = (4\pi^2 c/h)\omega_e^2\mu = 2.966016 \times 10^{-2}\omega_e^2\mu \text{ cm}^{-1}\text{Å}^{-2} \tag{5.1.20}$$

where ω_e is in reciprocal centimeters, μ is in atomic mass units (Aston units, $^{12}C = 12$), R is in Ångstroms, and V/hc is in reciprocal centimeters, and

$$V^{\text{Morse}}(R)/hc = D^e\{1 - \exp[-\beta(R - R_e)]\}^2 \tag{5.1.21}$$

where D^e is the dissociation energy

$$D^e = \omega_e^2/4\omega_e x_e \text{ cm}^{-1} \tag{5.1.22}$$

$$\beta = (8\pi^2 c\mu\omega_e x_e/h)^{1/2} = 2.435576 \times 10^{-1}(\mu\omega_e x_e)^{1/2} \text{ Å}^{-1} \tag{5.1.23a}$$

or

$$\beta = (2\pi^2 c\mu\omega_e^2/D^e h)^{1/2} = 1.2177881 \times 10^{-1}(\mu\omega_e^2/D^e)^{1/2} \text{ Å}^{-1} \tag{5.1.23b}$$

[†] When $E_1, v_i \neq E_2, v_j$, it is necessary before applying Eq. (5.1.12) to vertically shift one of the potentials to ensure strict energy degeneracy. The new curve crossing point obtained in this way is called a pseudo-crossing point (Miller, 1968).

where D^e, ω_e, and $\omega_e x_e$ are in reciprocal centimeters, and μ is in atomic mass units. For both the harmonic and Morse oscillators (and general potentials),

$$R_e\ (\text{Å}) = (8\pi^2 c\mu B_e/h)^{-1/2} = 4.105805\ (\mu B_e)^{-1/2}\ \text{Å} \tag{5.1.24}$$

where B_e is in reciprocal centimeters.

In order to construct a harmonic or Morse potential from spectroscopic data, respectively, two (R_e and k) or three (D^e, R_e, and β) independent constants must be determined. The usual routes to these constants, through B_e, ω_e, and $\omega_e x_e$, are often impassable and it is necessary to exploit special relationships, two of the most useful of which are due to Kratzer (Kratzer, 1920),

$$D_e = 4B_e^3/\omega_e^2 \tag{5.1.25}$$

(D_e is the centrifugal distortion constant), and to Pekeris (Pekeris, 1934),

$$\omega_e x_e = \alpha_e^2\omega_e^2/36B_e^3 + \alpha_e\omega_e/3B_e + B_e, \tag{5.1.26}$$

where

$$\alpha_e \approx B_v - B_{v+1}.$$

Equation (5.1.25) is valid for both harmonic and Morse potentials; Equation (5.1.26) is valid for the Morse potential only.

Isotope effects are another source of information. For a Morse oscillator,

$$\begin{aligned} T^l_{ve} - T^h_{(v+n)e} &= G_l(v) - G_h(v+n) \\ &\simeq (\Delta\mu/\mu_h)\{n\omega_{el}/2 + \\ &\quad (v+1/2)[\omega_{el}/2 - \omega_e x_{el}(v+1/2-2n)]\} - n\omega_{el}{}^{\dagger} \end{aligned} \tag{5.1.27}$$

$$B_{v,l} - B_{v+n,h} \simeq (\Delta\mu/\mu_h)[B_{el} - (3/2)(v+1/2+n)\alpha_{el}] + n\alpha_{el}{}^{\dagger} \tag{5.1.28}$$

$$\Delta\mu = \mu_h - \mu_l \tag{5.1.29}$$

where h and l refer to the heavy and light isotopic molecule. When the vibrational quantum number of the perturber is the same for both isotopic species ($n = 0$), Eqs. (5.1.27) and (5.1.28) reduce to simpler and more useful forms. Recall that, for the PO $B'^2\Pi \sim D^2\Pi$ perturbation, an apparent small isotope shift for the perturber of $D^2\Pi$ ($v = 0$) suggested that $n = v = 0$, but the correct assignment was $v = 24, n = 1$.

† These isotope shift equations are derived by expanding the ratio

$$Y^l_{mn}/Y^h_{mn} = (\mu_l/\mu_h)^{-(m/2+n)} = (1 - \Delta\mu/\mu_h)^{-(m/2+n)} \simeq 1 + (m/2+n)\Delta\mu/\mu_h.$$

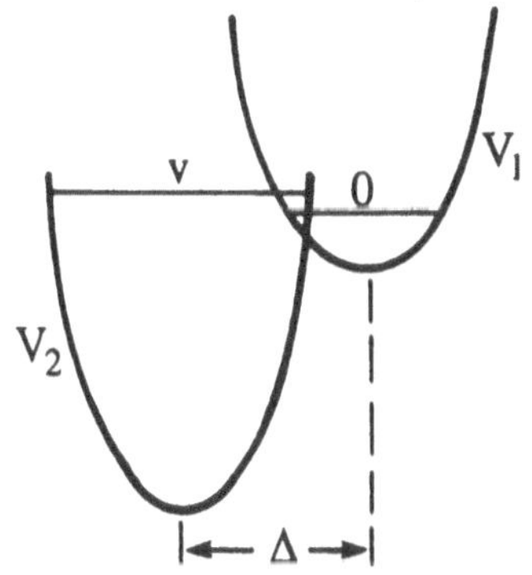

Figure 5.6: Two displaced harmonic oscillators.

Once a model potential is derived, it is possible to verify and refine this potential, making use of observed perturbation matrix elements and calculated overlap integrals between the vibrational levels of the two interacting electronic states. Although it is usually more convenient to input the analytic form of $V(R)$ into a numerical integration program to calculate overlap integrals (Section 5.1.3), analytic expressions exist for harmonic and Morse $\langle v_i|v_j\rangle$ factors.

Even for two harmonic potentials, when $\omega_{e1} \neq \omega_{e2}$ and $R_{e1} \neq R_{e2}$, the analytic form for $\langle v_{1i}|v_{2j}\rangle$ is very complicated. For $\omega_{e1} = \omega_{e2}$ and $\Delta = R_{e1} - R_{e2}$ (Fig. 5.6),

$$\langle v_1 = 0|v_2 = v\rangle = \exp(-d^2/2)d^v(v!)^{-1/2} \tag{5.1.30}$$

$$d = \left(\frac{c\mu\omega_e}{2\hbar}\right)^{1/2} \Delta \tag{5.1.31}$$

where ω_e is in reciprocal centimeters. If $v_1 \neq 0$, formulas are given by Katriel, (1970), but the sign of d is incorrect (Schamps, 1977). The sign of the vibrational overlap integral, although dependent on arbitrary phase choices for the two vibrational wavefunctions, is best fixed by what appears to be a universal phase choice: all vibrational wavefunctions are positive at the inner turning point, $\chi_v(R_{\rm min}) > 0$ for all v (see Section 5.1.4).

The vibrational wavefunctions for a Morse oscillator can be expressed analytically in terms of Whittaker functions (Felenbok, 1963) or in terms of polygamma functions (Matsumoto and Iwamoto, 1993).

The harmonic potential is a model of last resort for diatomic molecules. Its behavior at $R = 0$ and $R = \infty$ is unphysical, as is the sign of α_e. Exact diatomic molecule vibrational wavefunctions for levels above $v = 0$, except for their number of nodes, differ from harmonic oscillator eigenfunctions (Hermite polynomials with an exponential factor) in that they are not symmetric about R_e and, increasingly so at high v, are skewed toward the outer turning point.

The Morse potential displays the qualitative features of realistic molecular potentials and serves adequately in vibrational overlap computations for levels up to about 10% of D^e (the Morse D^e, not the true D^e). The Morse curve rises sharply at small R, but not steeply enough. This deficiency is often minimized by splicing a short-R AR^{-12} segment onto V^{Morse}. The Morse potential goes asymptotically to a dissociation limit at large R, but with an R–dependence that is at variance with the required power law interaction of the separated atoms (Le Roy, 1973). One can splice a long range tail of the form $D^e - C_n R^{-n}$ onto V^{Morse}, where D^e is the measured dissociation energy (which can be quite different from the Morse D^e, determined from $\omega_e^2/4\omega_e x_e$). The value of v_{Morse}, where the long range tail begins, is a parameter to be determined.

One problem with splicing long–R and short–R extensions onto V^{Morse} is that dV/dR and higher derivatives become discontinuous at R_{splice}. Such discontinuities cause difficulty with numerical integration of the vibrational Schrödinger equation. Lagrange polynomial interpolation or, even better, spline techniques exist by which an intermediate splicing function is used to ensure continuity of $V(R), dV/dR$, and as many higher derivatives as the degrees of freedom of the splicing function permit.

In addition to the Morse potential there exist other model potentials defined by three adjustable parameters: for example, the Varshni potential (Varshni, 1957)

$$V^{\mathrm{Varshni}}(R)/hc = D^e \left\{1 - \frac{R_e}{R} exp[-\alpha(R - R_e)]\right\}^2 \qquad (5.1.32)$$

which, in certain cases, can be a better approximation to the RKR curve than the Morse potential.

The value of model potentials is not the availability of analytic formulas for vibrational overlap integrals; rather, it is the ability to pull together fragmentary information of diverse types into a coherent picture in order to test the observable consequences (location and strength of other perturbations) of alternative assignment schemes (Section 5.1.5). RKR and numerical integration comprise the most convenient way to obtain $\langle v_i|v_j\rangle$ factors and other quantities derivable from $V(R)$ [e.g., centrifugal distortion constants (Albritton, *et al.*, 1973)].†

5.1.3 Numerical Potentials and Vibrational Wavefunctions

The Rydberg-Klein-Rees (RKR) procedure is the most widely used method for deriving $V(R)$ from the $G(v)$ and $B(v)$ functions of diatomic molecules. Countless RKR computer programs of independent origin exist, with differences primarily in the way a singularity at the upper limit of integration of Eqs. (5.1.39a or b) and (5.1.40a or c) is handled, all of which give essentially identical results (see Mantz, *et al.*, 1971 for the CO $X^1\Sigma^+$ potential), except

† A program to calculate the energies and wavefunctions of bound states of any potential and the Franck-Condon factors between the states of two potentials has been written by Le Roy (see the WWW site: http://theochem.uwaterloo.ca/~leroy/).

for those based on the more accurate second-order JWKB quantization condition (Kirschner and Watson, 1974). The RKR potential is generated as pairs of turning points, $R_-(E)$ and $R_+(E)$, at specified energies [usually $E_n = G(n/4)$ for $-2 \le n \le 4v_{\max}$, where $v_{\max}$ is the highest observed v]. The RKR potential is not an analytic function, but can be compactly represented as such by least-squares fitting a suitable $V(R)$ function to the turning points.

RKR is based on Eq. (5.1.1), the semiclassical quantization condition; thus the $V^{\text{RKR}}(R)$ it generates is not identical to the "true" Born-Oppenheimer potential, $V^{\text{BO}}(R)$. The deperturbed experimental energies, $E_{v,J}^{\text{depert}}$, are, by definition, the eigenvalues of $V^{\text{BO}}(R)$, $E_{v,J}^{\text{BO}}$. Thus, the small difference between V^{BO} and V^{RKR} is manifest in small (typically 0.1 cm^{-1}) and systematic discrepancies between $E_{v,J}^{\text{depert}}$ and $E_{v,J}^{\text{RKR}}$. These discrepancies are typically 10-100 times smaller in magnitude than the unknown difference between partially and fully deperturbed experimental energies. The inverse perturbation approach (IPA) systematically distorts V^{RKR} to minimize the deviations between the partially deperturbed observed $E_{v,J}$ and the exact eigenvalues of V^{IPA}, $E_{v,J}^{\text{IPA}}$ (Kosman and Hinze, 1975; Vidal and Scheingraber, 1977). However, as long as the $E_{v,J}$ are not fully deperturbed, V^{IPA} may be no better an approximation to V^{BO} than V^{RKR}.

Many detailed derivations of the RKR equations have been published (Zare, 1964; Miller, 1971; Elander, *et al.*, 1979). The unknown potential energy for a rotating diatomic molecule,

$$V_J(R) = V_0(R) + J(J+1)\hbar^2/2\mu R^2, \tag{5.1.33}$$

is a function of both R and J. The function

$$A(E,J) = \int_{R_-(E)}^{R_+(E)} [E - V_J(R)]dR \tag{5.1.34}$$

is easily shown to have the convenient properties that

$$\frac{\partial A}{\partial E} = \int_{R_-(E)}^{R_+(E)} 1dR = R_+(E) - R_-(E) \tag{5.1.35}$$

and

$$\begin{aligned}\frac{\partial A}{\partial J} &= \frac{(2J+1)\hbar^2}{2\mu} \int_{R_-(E)}^{R_+(E)} \frac{1}{R^2}dR \\ &= (2J+1)\frac{\hbar^2}{2\mu}\left[\frac{1}{R_-(E)} - \frac{1}{R_+(E)}\right]\end{aligned} \tag{5.1.36a}$$

and, for $J = 0$,

$$\frac{\partial A}{\partial J} = \frac{\hbar^2}{2\mu}\left[\frac{1}{R_-(E)} - \frac{1}{R_+(E)}\right]. \tag{5.1.36b}$$

Thus, if it is independently possible to define $A(E, J)$ in terms of $G(v)$ and $B(v)$ functions, Eqs. (5.1.35) and (5.1.36b) could be used to determine $V_0(R)$. The key to the procedure is that Eq. (5.1.34) has a form very similar to the semiclassical quantization condition [Eq. (5.1.1)],

$$\left[(2\mu)^{1/2}/\hbar\right]\int_{R_-(E)}^{R_+(E)}[E - V_J(R)]^{1/2}dR = \left(v + \frac{1}{2}\right)\pi. \tag{5.1.37}$$

Nontrivial manipulation of Eqs. (5.1.34) and (5.1.35) yields

$$A(E, J) = 2\pi\int_{v(0)=v_{\min}}^{v(E)}[E - E(v, J)]^{1/2}dv. \tag{5.1.38a}$$

$$A(E, J) = 2\pi\int_{v_{\min}}^{v(E)}[E - G(v) - B(v)J(J+1)]^{1/2}dv. \tag{5.1.38b}$$

Taking $\partial/\partial E$ and $\partial/\partial J$,

$$\frac{\partial A}{\partial E} = \pi\int_{v_{\min}}^{v(E)}[E - G(v) - B(v)J(J+1)]^{-1/2}dv \tag{5.1.39a}$$

$$\frac{\partial A}{\partial J} = \pi\int_{v_{\min}}^{v(E)}[E - G(v) - B(v)J(J+1)]^{-1/2}\frac{\partial E(v, J)}{\partial J}dv \tag{5.1.40a}$$

At $J = 0$

$$\frac{\partial A}{\partial E} = \pi\int_{v_{\min}}^{v(E)}[E - G(v)]^{-1/2}dv \tag{5.1.39b}$$

and

$$\frac{\partial E(v, J)}{\partial J} = (2J+1)B(v) - (4J^3 + 6J^2 + 2J)D(v) \tag{5.1.40b}$$

so, for $J = 0$

$$\frac{\partial A}{\partial J} = \pi\int_{v_{\min}}^{v(E)}B(v)[E - G(v)]^{-1/2}dv. \tag{5.1.40c}$$

Note that the integrals in Eqs. (5.1.39b) and (5.1.40c) become infinite at the upper limit of integration, but this singularity is integrable by various techniques (Mantz, *et al.*, 1971). Note also that, although $G(-1/2) = 0$,

$$E_{v,0} = Y_{00} + G(v), \tag{5.1.41}$$

(Kaiser, 1970) where (Dunham, 1932)

$$Y_{00} = \frac{B_e - \omega_e x_e}{4} + \frac{\alpha_e\omega_e}{12B_e} + \left(\frac{\alpha_e\omega_e}{12B_e}\right)^2\frac{1}{B_e} \tag{5.1.42}$$

(Y_{00} is zero for harmonic and Morse oscillators) and

$$v_{\text{min}} \cong -1/2 - Y_{00}/\omega_e. \tag{5.1.43}$$

Thus, knowledge of $G(v)$ is sufficient to determine $R_+(E) - R_-(E)$ from Eqs. (5.1.35) and (5.1.39b). Equations (5.1.36b and 5.1.40c), based on *both* $G(v)$ and $B(v)$, then determine $R_+(E)$ and $R_-(E)$. There is nothing about the RKR equations that restricts the energies at which turning points are computed to those corresponding to integer values of v. As noted by Friedman-Hill and Field (1991), the RKR algorithm does not presuppose a Dunham polynomial expression for the energy levels. Any functional form which represents the input data can be used.

The RKR potential may be tested against the input $G(v)$ and $B(v)$ values by exact solution of the nuclear Schrödinger equation [see Wicke and Harris, 1976, review and compare various procedures, e.g., Numerov-Cooley numerical integration (Cooley, 1961), finite difference boundary value matrix diagonalization (Shore, 1973), and the discrete variable representation (DVR) (Harris, *et al.*, 1965)]. $G(v) + Y_{00}$ typically deviates from $E_{v,J=0}$ by < 1 cm^{-1} except near dissociation. B_v may be computed from $\chi_{v,J=0}(R)$ by

$$B(v) = (\hbar^2/2\mu)\left\langle v, J = 0 \left| R^{-2} \right| v, 0 \right\rangle \tag{5.1.44}$$

and is typically found to agree with experimental values to 1 part in 10^5 except near dissociation. Alternatively, the completeness of the deperturbation of the experimental data may be tested by comparison of observed and computed (Albritton, *et al.*, 1973; Brown, *et al.*, 1973; Tellinghuisen, 1973; Kirschner and Watson, 1973) centrifugal distortion constants or $G(v), B(v)$ values for other isotopic species.

5.1.4 Some Remarks about "Borrowed" Computer Programs

The relative signs of off-diagonal matrix elements are often important (see Section 6.3), even though the phase of any given basis function is arbitrary. The sign of a computed vibrational overlap integral (or other vibrational matrix element) depends on the phase choices for $\chi_{v_i}(R)$ and $\chi_{v_j}(R)$ used in the computer program. A frequently used phase convention (not recommended!) is that the vibrational wave functions with even or odd quantum numbers have opposite signs at the inner turning point. An alternative convention, (the "universal" phase choice: highly recommended!) for which all $\chi(R_{\text{min}}) > 0$, can be more convenient when dealing with the vibrational continuum. The important point here is that the choice of one or the other convention has no physical consequence, but it is necessary to know which convention one has chosen and to use it self-consistently in all computations. It could be disastrous to calculate $\langle v_i | v_j \rangle$ integrals with a program following the $\chi(R_{\text{min}}) > 0$ convention and then calculate $\left\langle v_i \left| R^{-2} \right| v_j \right\rangle$ integrals with another program that requires alternating signs.

Remember that for every observable quantity, the sign of each wavefunction (electronic and vibrational) always appears *twice*. This is the reason why the phase convention adopted is irrelevant provided that *the same* phase convention is used both times each wavefunction appears. It is dangerous to derive an observable quantity, using one matrix element taken from the literature without knowing the phase convention that has been used (see one example in Section 6.2.1).

The following example illustrates the necessity for self-consistent phases (as manifest in the sign of off-diagonal matrix elements) in order to predict correctly which of two transitions is more intense. Let basis functions ϕ_2 and ϕ_3 interact with each other (see Sections 6.2.1 and 6.3):

$$\begin{aligned} E_2 &= \frac{H_{22}+H_{33}}{2} + \frac{H_{23}^2}{H_{22}-H_{33}} \qquad \Psi_2 = \phi_2 + \frac{H_{23}}{H_{22}-H_{33}}\phi_3 \\ E_3 &= \frac{H_{22}+H_{33}}{2} - \frac{H_{23}^2}{H_{22}-H_{33}} \qquad \Psi_3 = \phi_3 - \frac{H_{23}}{H_{22}-H_{33}}\phi_2. \end{aligned} \tag{5.1.45}$$

Assume that ϕ_2 and ϕ_3 have identical transition probabilities to basis function $\phi_1, P_{12}^0 = P_{13}^0$,

$$P_{12}^0 \propto |\langle\phi_1|\mu|\phi_2\rangle|^2 \qquad P_{13}^0 \propto |\langle\phi_1|\mu|\phi_3\rangle|^2, \tag{5.1.46}$$

then, even though $P_{12}^0 = P_{13}^0$, the perturbed intensities, P_{12} and P_{13}, will not be equal:

$$P_{12} = P_{12}^0 + \left(\frac{H_{23}}{H_{22}-H_{33}}\right)^2 P_{13}^0 + \frac{2H_{23}}{H_{22}-H_{33}}\langle\phi_1|\mu|\phi_2\rangle\langle\phi_1|\mu|\phi_3\rangle \tag{5.1.47a}$$

$$P_{13} = P_{13}^0 + \left(\frac{-H_{23}}{H_{22}-H_{33}}\right)^2 P_{12}^0 - \frac{2H_{23}}{H_{22}-H_{33}}\langle\phi_1|\mu|\phi_2\rangle\langle\phi_1|\mu|\phi_3\rangle. \tag{5.1.47b}$$

Note that whether the higher energy perturbed level has greater ϕ_2 or ϕ_3 character has nothing to do with any arbitrary choice of phase factors. It depends only on the sign of $(H_{22} - H_{33})$. However, the relative intensities of the transitions into the higher- versus lower-energy perturbed level depend on the product of six signed quantities [three of which, $H^e, \mu_{12}(R)$, and $\mu_{13}(R)$, can be determined *ab initio*]:

$$H_{23} = \langle 2|3\rangle H^e \tag{5.1.48a}$$

$$\langle\phi_1|\mu|\phi_2\rangle = \langle 1|2\rangle \mu_{12}(\overline{R}_{12}) \tag{5.1.48b}$$

$$\langle\phi_1|\mu|\phi_3\rangle = \langle 1|3\rangle \mu_{13}(\overline{R}_{13}) \tag{5.1.48c}$$

Each vibrational wavefunction appears twice; thus the phase of the vibrational wave function is irrelevant to whether the upper or the lower eigenstate has the larger transition intensity to level 1. However, it is crucial to use internally consistent vibrational phases when computing the three sets of overlap integrals in order to correctly predict whether the transition into the higher or lower energy perturbed level is more intense (Section 6.3.1).

It is important to realize that most computer programs work with

$$\xi(R) = R\chi(R),$$

where $\chi(R)$ is the vibrational eigenfunction, which means that integration must be over dR, not R^2dR. This point is illustrated by noting the difference between Eqs. (3.3.10) and (3.3.11) (Section 3.3.3). It is common practice to combine computer subroutines from various sources, for example, eigenvalue search, numerical generation of radial eigenfunctions, matrix element evaluation, etc. No matter how widely a subroutine is used, it is essential to check for compatibility between subroutines.

5.1.5 Vibrational Assignment by the Matrix Element Method

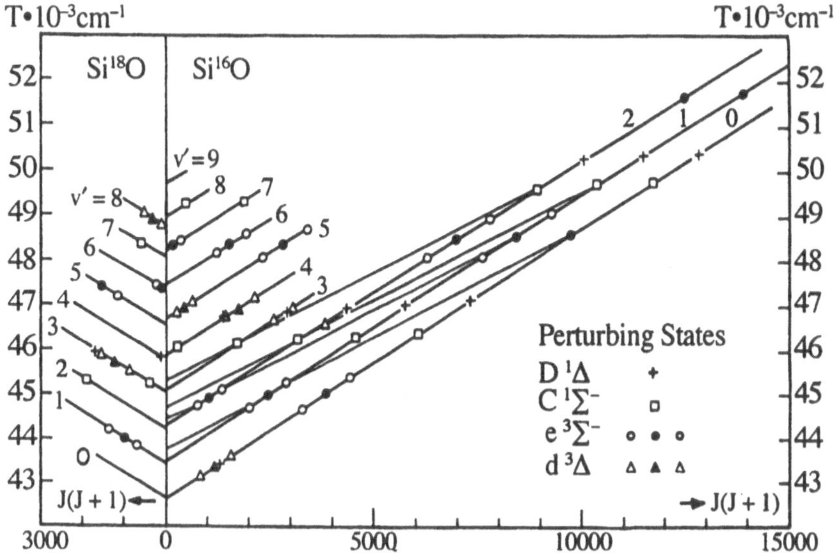

Figure 5.7: Perturbations in the $^{28}Si^{18}O$ and $^{28}S^{16}O$ $A^1\Pi$ state. The J–values where each $A^1\Pi$ vibrational level crosses a perturber are marked with a different symbol for each type of perturbing state. The signs of the level shifts near each level crossing indicate that $A^1\Pi$ overtakes each perturber from below as J increases. This means that the same v–level of a perturbing state will cross the v_A and $v_A - 1$ levels of $A^1\Pi$ at low J and high J, respectively. Two consecutive vibrational levels of the $Si^{16}O$ $e^3\Sigma^-$ and $C^1\Sigma^-$ states are shown, each of which crosses two vibrational levels of $A^1\Pi$. [From Field, *et al.*, (1976).]

Often, the energies and rotational constants of several vibrational levels of an otherwise unknown electronic state are obtained from analysis of perturbations of a better known electronic state. Figure 5.7 illustrates the ease with which the perturbing levels may usually be grouped into classes corresponding to separate electronic states and how, *within each class*, the *relative* vibrational numbering of the observed levels is established. There are two sources of information that enable determination of the *absolute* vibrational numbering of the perturbers: isotope shifts and the magnitudes of observed perturbation matrix elements. The use of isotope shifts is considerably better known (Herzberg, 1950, pp. 162-166) than that of matrix elements.

The matrix element method is based on the fact that if two potential curves intersect exactly once (at $E = E_C, R = R_C$), then the matrix element connecting *near-degenerate* perturbing levels can be factored (see Section 3.3.1),

$$H_{1,v_1;2,v_2} = H_{12}^e(R_C)\,\langle v_1|v_2\rangle\,,$$

and the electronic matrix element is independent of vibrational level

$$\frac{H_{1,v_1;2,v_2}}{\langle v_1|v_2\rangle} \simeq \frac{H_{1,v_1+n;2,v_2+m}}{\langle v_1+n|v_2+m\rangle} \simeq H_{12}^e(R_C). \tag{5.1.49}$$

The correctness of a trial vibrational numbering of electronic state 2 is then indicated by the constancy of the ratio of the observed matrix element to the calculated vibrational overlap. This method is similar to the Franck-Condon method for establishing the absolute vibrational numbering. There, one compares observed transition intensities to sets of Franck-Condon factors calculated for a series of trial numberings. The key to the success of both methods is the existence of at least one relative minimum or maximum among the observed $H_{1,v;2,v}$ matrix elements or transition intensities.

Consider a well-characterized electronic state whose *absolutely* numbered vibrational levels are perturbed by a set of levels of another electronic state for which energies, rotational constants, and a *relative* vibrational numbering are known. A set of trial absolute numberings for the perturber is obtained, starting with the numbering that assigns the quantum number $v = 0$ to the lowest energy observed perturbing level. This first trial numbering corresponds to trial T_v and B_v functions,

$$T_v^0 = T_e^0 + \omega_e^0\left(v+\frac{1}{2}\right) - \omega_e x_e\left(v+\frac{1}{2}\right)^2 \tag{5.1.50}$$

$$B_v^0 = B_e^0 - \alpha_e\left(v+\frac{1}{2}\right). \tag{5.1.51}$$

New T_v and B_v functions are obtained by adding n ($n = 1, 2, 3, \ldots$, some larger integer) to each trial vibrational quantum number of the perturber

$$v + n = v_n \tag{5.1.52a}$$

$$T_{v_n}^n = T_e^n + \omega_e^n\left(v+n+\frac{1}{2}\right) - \omega_e x_e\left(v+n+\frac{1}{2}\right)^2 \tag{5.1.52b}$$

$$B_{v_n}^n = B_e^n - \alpha_e\left(v+n+\frac{1}{2}\right) \tag{5.1.52c}$$

and, adjusting the T_e^n, ω_e^n, and B_e^n constants,

$$T_e^n = T_e - n\omega_e - n^2\omega_e x_e \tag{5.1.53a}$$
$$\omega_e^n = \omega_e + 2n\omega_e x_e \tag{5.1.53b}$$
$$B_e^n = B_e + n\alpha_e \tag{5.1.53c}$$

so that the transformed functions

$$T_{v_n}^n = T_{v+n}^n = T_v^0$$
$$B_{v_n}^n = B_{v+n}^n = B_v^0$$

change only the name of a level, not its observed T_v or B_v value. Each set of constants $(T_e^n, \omega_e^n, \omega_e x_e, B_e^n, \alpha_e)$ defines an RKR potential energy curve and a set of vibrational wavefunctions. It is a routine and inexpensive matter to generate tens of trial RKR curves and, for each trial curve, calculate a set of vibrational overlap factors with the single, well-defined RKR curve of the better known, perturbed electronic state.

Each time the trial numbering is increased by 1, the trial potential is shifted to $\sim \omega_e$ lower energy and to slightly smaller internuclear distance,

$$\delta R = -\alpha_e (2B)^{-3/2} \hbar \mu^{-1/2}. \tag{5.1.54}$$

Consequently, the point of intersection between the potentials of the two interacting states is shifted. It is not possible to derive a general expression for the direction and magnitude of the shift in E_C or R_C. However, in Section 5.1.1 it was shown that the vibrational overlap increases monotonically as the energy of the interacting pair of levels approaches E_C from below, reaches a maximum just above E_C, and then undergoes weakly damped oscillations [proportional to $(E - E_C)^{-1/4}$], increasingly closely spaced. Thus, if E_C is above the highest observed perturbation pair, the observed matrix elements will vary monotonically with v and the matrix element method will almost certainly fail. Similarly, if the lowest observed perturbation pair is far above E_C, where the vibrational overlap oscillates so rapidly that successive maxima are separated by less than ω_e, the matrix element method will probably fail. Even when E_C is bracketed by observed perturbation pairs, if the absolute numbering suggested is such that the lowest observed perturber has $v > 15$, this assignment must be regarded as uncertain to about ± 1 quantum. The RKR curve is very sensitive to extrapolated ω_e and B_e values, and the overlaps calculated for observed high-v levels can be significantly affected by extrapolation errors.

When the matrix element method fails, two possibilities for establishing the vibrational numbering remain, *ab initio* $H^e(R)$ functions and isotope shifts. When E_C appears to lie above the highest observed perturbing level, isotope shifts are the method of choice. However, if $H^e(R)$ is available, then a modified matrix element method may prove successful. Each trial numbering determines R_C^{trial}, hence $H^e\left(R_C^{\text{trial}}\right)$. The calculated vibrational overlap should be equal to the observed perturbation matrix element divided by $H^e\left(R_C^{\text{trial}}\right)$. However, if the R–dependence of H^e matches the effect of a shift in E_C^{trial}, R_C^{trial} on the calculated $\langle v_1|v_2\rangle$ elements, the vibrational numbering will be ambiguous by ± 1 quantum. When the lowest sampled vibrational level of the perturber is > 10, isotope shifts are no more reliable than matrix elements.

The first rigorous test of the matrix element method was with the $A^1\Pi \sim (C^1\Sigma^-, D^1\Delta, d^3\Delta,$ and $e^3\Sigma^-)$ perturbations of the $^{28}Si^{16}O$ and $^{28}Si^{18}O$ molecules (Field, *et al.*, 1976). Vibrational assignments derived from isotope shifts were found to be identical to those deduced from the vibrational variation of observed matrix elements. Previously, the near-perfect vibration-independence of the electronic part of perturbation matrix elements observed between levels of well-characterized potentials had been experimentally demonstrated for $(a'^3\Sigma^+, d^3\Delta, e^3\Sigma^-,$ and $I^1\Sigma^-) \sim (A^1\Pi$ and $a^3\Pi)$ perturbations of CO (Field, *et al.*, 1972). The first use (Field, 1974) of the matrix element method was in identifying the electronic $(^3\Pi$ and $^1\Pi)$ and vibrational perturbers of the $A^1\Sigma^+$ state of CaO (Hultin and Lagerqvist, 1951), SrO (Almkvist and Lagerqvist, 1950), and BaO (Lagerqvist, *et al.*, 1950).

The perturbations of the SiO $A^1\Pi$ state are summarized in Fig. 5.7. The J–values where the $A^1\Pi$ and perturber basis function energies become equal are marked using a different symbol for each type of electronic perturber. This not necessarily integer J–value is called the culmination of the perturbation and is easily recognized because:

1. The energy shift of the main line [by definition, the main line corresponds to the eigenfunction that has the largest perturb*ed* basis function fractional character and is usually more intense than extra lines (see Section 6.2.1)] switches sign as one passes through this J–value.

2. The main and extra lines reach their minimum energy separation,

$$|E_{\text{main}} - E_{\text{extra}}| = 2\,|H_{12}|\,. \tag{5.1.55}$$

3. The intensities of the main and extra lines become equal.

Grouping the perturbers into classes is not as trivial as it may seem from the figure.

The qualitative pattern of the perturbers of a given vibrational level provides clues to the electronic identify of the perturber, particularly when the perturber is a Σ–state of any multiplicity. These clues consist of:

1. As J increases, does the perturber approach from lower energy (larger B–value) or higher energy (smaller B–value)? This is indicated by the sign of the level shifts above and below the culmination.

2. Is the perturbation matrix element J–dependent (heterogeneous perturbation or **S**-uncoupling in either the perturbed or perturbing state)? Figure 5.8 contrasts the level shifts resulting from J–independent perturbation matrix elements with those from matrix elements proportional to J.

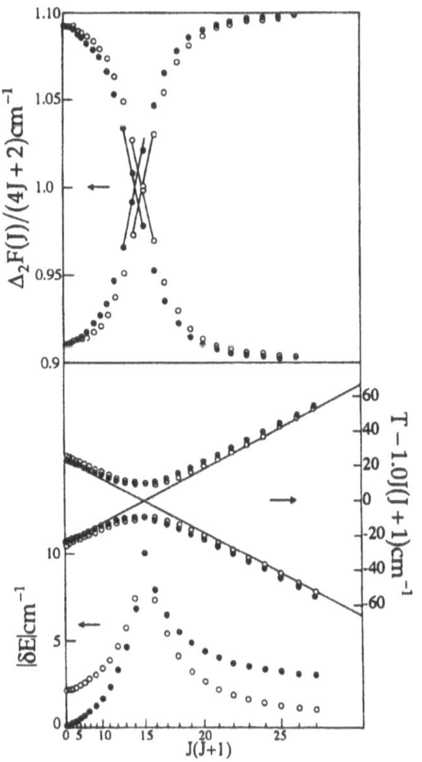

Figure 5.8: Level shifts, reduced term values, and effective B-values [$\Delta_2F(J)$] for homogeneous versus heterogeneous perturbations. Two states, with deperturbed B-values of $B_1^0 = 1.1$ and $B_2^0 = 0.9\,\text{cm}^{-1}$, cross at exactly $J = 15$. The open circles correspond to a homogeneous interaction where the interaction matrix element is $H_{12} = 10.0\,\text{cm}^{-1}$ for all J-values. The filled circles correspond to a heterogeneous interaction with $H_{12}(J) = 0.6455[J(J+1)]^{1/2}\ \text{cm}^{-1}$ which is chosen so that, at the crossing point ($J = 15$), the perturbation matrix elements for both the homogeneous and heterogeneous interactions are identical. The level-shift plot for the homogeneous interaction is symmetric about $J = 15$ and asymptotically approaches $|\delta E| = 0$ at high J. The level shifts for the heterogeneous interaction are not symmetric about $J = 15$ and will asymptotically approach $|\delta E| = (0.6455)^2/\Delta B = 2.08\ \text{cm}^{-1}$. The reduced term values (see Fig. 3.18 and Section 4.3.1), $T_{i,J} - J(J+1)(B_1^0 + B_2^0)/2$, where $T_{1,J}$ and $T_{2,J}$ are the perturbed term values, are plotted versus $J(J+1)$ and form high- and low-energy series of levels. The high-energy series is nominally state 2 at $J < 15$ and goes smoothly over into nominally state 1 at $J > 15$ (vice versa for the low-energy series). The deperturbed term values are shown as a pair of straight lines that cross at $J = 15$. The effective B-value plot for the high-energy series of levels goes from $\sim 0.91\ \text{cm}^{-1}$ at low J to $\sim 1.10\,\text{cm}^{-1}$ at high J (vice versa for the low-energy series). The $\Delta_2F(J)$ values asymptotically approach B_1^0 and B_2^0 at high J for the homogeneous mchanism but approach $B_1^0 + \delta$ and $B_2^0 - \delta$ for the heterogeneous mechanism, where $\delta \approx (0.6455)^2/\left(T_2^0 - T_1^0\right)$. The B-value plots cross at $(B_1^0 + B_2^0)/2$ (see Fig. 4.1), but the J-value of the B-value crossing for the heterogeneous case is ~ 0.5 J-unit lower than that of the actual level crossing. See Section 5.1.6 for further discussion.

It is important to note that when perturbations by two or more states are not well separated, it is difficult to distinguish between the J–independent and J–dependent cases. In addition, the actual matrix element J–dependence is often more complicated than either of these two simple cases, particularly for weak perturbations ($H_{12} \lesssim 0.1\text{cm}^{-1}$) and at high J ($J \approx A/B$).

3. When the perturbed state has two near-degenerate parity components for each J ($\Lambda \geq 1$), then it is possible to distinguish between three classes of perturbers: single-parity perturbers ($^1\Sigma^{\pm}$ has e or f parity only, $^{2S+1}\Sigma$ has widely separated e and f components belonging to the same J and only one total parity for all of the near degenerate J components of a given N, etc.), states with significant Λ–doubling [usually only Π states or $\Omega < 2$ states in case (c)], and states without Λ–doubling. Kovács (1969, pp. 227-284) illustrates the qualitative perturbation patterns for 42 combinations of states with $S \leq \frac{3}{2}, \Lambda \leq 2$.

4. On the $E_{v,J}$ versus $J(J+1)$ plot, all crossings for a particular perturbation sequence (constant $\Delta v = v_{\text{main}} - v_{\text{perturber}}$) fall onto a family of approximately straight lines with slope

$$\left(\frac{dE_{\text{cross}}}{d[J(J+1)]}\right)_{\Delta v} = B_{\text{main}}(v) - \Delta G_{\text{main}}\left(v - \frac{1}{2}\right)\frac{\delta B_{\Delta v}}{\delta G_{\Delta v}} \tag{5.1.56}$$

where

$$\delta B_{\Delta v} = B_{\text{main}}(v) - B_{\text{perturber}}(v - \Delta v)$$
$$\delta G_{\Delta v} = \Delta G_{\text{main}}\left(v - \frac{1}{2}\right) - \Delta G_{\text{perturber}}\left(v - \frac{1}{2} - \Delta v\right).$$

Note that since $\delta B_{\Delta v}$ and $\delta G_{\Delta v}$ usually have the same sign, the slopes of these constant Δv perturbation tie lines will be smaller than both B_{main} and $B_{\text{perturber}}$. Thus, once several families of perturber are identified using clues 1-3, the remaining perturbers can be tentatively grouped by inspecting the $E_{v,J}$ versus $J(J+1)$ plot for approximately horizontal tie lines.

Even after the perturbers are grouped into classes and the relative vibrational numbering within a class established, it may still be premature to apply the matrix element method. Knowledge of the perturber's electronic symmetry is necessary, because this determines the J–dependence of the perturbation matrix element. The matrix element at the perturbation culmination that one determines from a local graphical treatment of the J–levels near the crossing can be quite different from the one obtained from a least-squares fit of energy levels ($T_{v,J}$) for all J–values to a model Hamiltonian. It is the *value* of the matrix element in the deperturbation model, not the *local magnitude* of the matrix element, to which the matrix element method applies. In order to illustrate this point, three types of perturbations of the SiO $A^1\Pi$ state will be discussed: $A^1\Pi \sim e^3\Sigma^-$, a J–independent interaction; $A^1\Pi \sim D^1\Delta$, an explicitly J–dependent interaction; $A^1\Pi \sim d^3\Delta$, an implicitly J–dependent interaction.

On Fig. 5.7 the $A^1\Pi \sim e^3\Sigma^-$ crossings appear in groups of three. The f–parity levels of $A^1\Pi$ (Q–branch lines of the $A^1\Pi$–$X^1\Sigma^+$ system) are perturbed by a single class of $e^3\Sigma^-$ levels; these are the f–parity, $\Omega = \Sigma = 1$ F_2 levels. The local matrix element at the $A^1\Pi \sim e^3\Sigma^-(F_2)$ crossing, determined from one-half the interpolated closest approach of main and extra lines, is exactly equal to the matrix element in the least-squares model Hamiltonian,

$$A_{10^-} = \langle {}^3\Sigma_1^- f\,|\mathbf{H}^{\mathrm{SO}}|\,{}^1\Pi f\rangle .$$

The e–parity levels of $A^1\Pi$ (R and P branch lines of A-X) are perturbed by two classes of $e^3\Sigma^-$ levels; these are the e–parity, about 50-50 mixed† $|\Omega| = |\Sigma| = 1$ and $\Omega = \Sigma = 0$ F_1 and F_3 levels,

$$\begin{aligned} \left|{}^3\Sigma^-, J, F_1\right\rangle &\simeq (2)^{-1/2}\left[\left|{}^3\Sigma_0^-, J\right\rangle + \left|{}^3\Sigma_{|\Omega|=1}^-, J\right\rangle\right] \\ \left|{}^3\Sigma^-, J, F_3\right\rangle &\simeq (2)^{-1/2}\left[\left|{}^3\Sigma_0^-, J\right\rangle - \left|{}^3\Sigma_{|\Omega|=1}^-, J\right\rangle\right]. \end{aligned} \quad (5.1.57)$$

Since only the ${}^3\Sigma_1^-$ character of the F_1 and F_3 $e^3\Sigma^-$ components can interact via $\mathbf{H}^{\mathrm{SO}}$ with $A^1\Pi$, the local matrix element at the e–parity crossings is a factor of $(2)^{-1/2}$ smaller than the matrix element in the model Hamiltonian,

$$H\left({}^1\Pi_e \sim {}^3\Sigma_e^-, F_1 \quad \text{or} \quad F_3\right) = (2)^{-1/2} A_{10^-} \quad (5.1.58)$$

A_{10^-} may be obtained from a local graphical treatment of any one of the ${}^1\Pi \sim {}^3\Sigma^-$ (F_1, F_2, F_3) crossings, provided it is certain that the perturber is a component of a ${}^3\Sigma^-$ state and proper account is taken of the $(2)^{-1/2}$ factor.

Table 5.1 shows that the vibrational numbering of the SiO $e^3\Sigma^-$ state is unambiguously determined by the matrix element method. The weighted rms deviations from the supposedly constant average electronic factor for the various trial numberings are 55% $(v_{\mathrm{pert}} + 3)$, 26%$(v_{\mathrm{pert}} + 1)$, 5%(v_{pert}), 37%$(v_{\mathrm{pert}} - 1)$, and 31% $(v_{\mathrm{pert}} - 3)$. One expects the matrix element method to work because the $A^1\Pi$ and $e^3\Sigma^-$ potential curves cross near $v_{\mathrm{A}} = 2.5 \sim v_{\mathrm{e}} = 9$ ($E_C \simeq 44{,}600\ \mathrm{cm}^{-1}$, $R_C = 1.50$ Å), which is within the range of vibrational levels

†The 50-50 mixing of $\Omega = 0$ and $|\Omega| = 1$ basis functions shown in Eq. (5.1.57) is a limit reached when $BJ \gg \lambda$. First-order perturbation theory for the correction to the case (a) wavefunctions implies that the mixing coefficient of $\left|{}^3\Sigma_{|\Omega|=1}\right\rangle$ in the nominal $\left|{}^3\Sigma_0\right\rangle$ function is

$$\begin{aligned} C_{01} &= \frac{\langle {}^3\Sigma_0|H'|{}^3\Sigma_{|\Omega|=1}\rangle}{\langle {}^3\Sigma_1|H^0|{}^3\Sigma_1\rangle - \langle {}^3\Sigma_0|H^0|{}^3\Sigma_0\rangle} \\ &= \frac{-B_\Sigma 2[J(J+1)]^{1/2}}{(0B_\Sigma - \frac{4}{3}\lambda) - (2B_\Sigma + \frac{2}{3}\lambda)} \\ &= 2[J(J+1)]^{1/2}\frac{B_\Sigma}{2\lambda + 2B_\Sigma} \end{aligned}$$

which implies that the mixing is essentially complete when $J > |\lambda/B| + 1$.

Table 5.1: Calculated Electronic Factors for Trial Vibrational Numberings of the $e^3\Sigma^-$ Perturber of SiO $A^1\Pi$ (cm^{-1})

			$\langle {}^1\Pi, v_\Pi \lvert \mathbf{H}^{SO} \rvert {}^3\Sigma_1^-, v_\Sigma \rangle / \langle v_\Pi \vert v_\Sigma \rangle$				
Isotopic species	v_Σ[a]	v_Π	$v_\Sigma + 3$	$v_\Sigma + 1$	v_Σ	$v_\Sigma - 1$	$v_\Sigma - 3$
$Si^{16}O$	9	2	225.(2)[b]	44.6(4)[b]	25.1(2)[b]	17.1(2)[b]	20.0(2)[b]
	14	6	87.(4)	68.8(28)	23.7(10)	29.4(12)	33.6(13)
	15	7	63.(1)	24.6(2)	23.3(2)	40.4(4)	24.2(2)
$Si^{18}O$	8	1	297.(10)	51.5(15)	26.2(8)	15.5(5)	9.8(3)
	13	5	109.(5)	32.1(14)	23.7(11)	23.9(11)	67.5(30)
	14	6	81.(1)	28.0(3)	23.6(2)	29.1(3)	32.5(3)

[a]Vibrational quantum number of $e^3\Sigma^-$ state as determined from isotope shifts.
[b]Numbers in parentheses are five-standard-deviation uncertainties in the last digits of the fitted perturbation matrix element.

sampled (Field, *et al.*, 1976). As expected, A_{10^-} goes through a maximum at an energy, $v_A = 5 \sim v_e = 13$, slightly above the A–e curve crossing.

The $A^1\Pi \sim D^1\Delta$ interaction is explicitly J–dependent,

$$\langle {}^1\Delta, J \vert \mathbf{H}^{ROT} \vert {}^1\Pi, J \rangle = -\beta_{12}[J(J+1)-2]^{1/2}. \tag{5.1.59}$$

There are two crossings for each $v_A \sim v_D$ pair, one for each parity. Since neither the $A^1\Pi$ nor the $D^1\Delta$ state has appreciable Λ-doubling, the e– and f–parity perturbations culminate at approximately the same J–value. The local matrix elements for the e– and f–parity crossings should be identical and yield β_{12} upon division by $[J(J+1)-2]^{1/2} \simeq J$. The crucial problem here is the ability to distinguish a perturbation by ${}^1\Delta$ from one by another double-parity state, such as ${}^1\Pi, {}^3\Pi_2, {}^3\Delta_1, {}^3\Delta_2$, or ${}^3\Delta_3$, for which the J–dependence is not $\approx J^1$. For SiO, A~D perturbations occur over a range of J–values, 11.4-112.9, that is larger than the range of derived β_{12} values, 0.039-0.084 cm^{-1}, from which the J–dependence of the local matrix element has been removed; success of the matrix element method may therefore be construed as proof of both the absolute vibrational numbering and the electronic assignment.

The $A^1\Pi \sim d^3\Delta$ interaction is explicitly J–independent,

$$\langle {}^3\Delta_1, J \lvert \mathbf{H}^{SO} \rvert {}^1\Pi \rangle = A_{12}, \tag{5.1.60}$$

but implicitly J–dependent because the strengths of the local interactions with the three spin-components of ${}^3\Delta$ depend on the $\lvert {}^3\Delta_1 \rangle$ basis function character, C_{1i}, present in the $\lvert {}^3\Delta, J, F_i \rangle$ eigenfunctions,

$$\lvert {}^3\Delta, J, F_i \rangle = C_{1i} \lvert {}^3\Delta_1 \rangle + C_{2i} \lvert {}^3\Delta_2 \rangle + C_{3i} \lvert {}^3\Delta_3 \rangle \tag{5.1.61}$$

$$\langle {}^1\Pi, J \lvert \mathbf{H}^{SO} \rvert {}^3\Delta, J, F_i \rangle = C_{1i} A_{12} \tag{5.1.62}$$

(for $^3\Delta, N = J - 2 + i$, nominal $\Omega = i$ for $A_\Delta > 0$ and nominal $\Omega = 4 - i$ for $A_\Delta < 0$). The $|^3\Delta_1\rangle$ mixing coefficients depend on J, A_Δ, and B_Δ, for example, using first-order perturbation theory for the correction to the wavefunction

$$|C_{12}| = \left| \frac{\langle ^3\Delta_1|\mathbf{H}|^3\Delta_2\rangle}{\langle ^3\Delta_1|\mathbf{H}^0|^3\Delta_1\rangle - \langle ^3\Delta_2|\mathbf{H}^0|^3\Delta_2\rangle} \right|$$

$$= \left| \frac{\langle ^3\Delta_1 |(-B\mathbf{J}^+\mathbf{S}^-)|^3\Delta_2\rangle}{\{-2A_\Delta + B_\Delta[J(J+1) - 0]\} - \{0A_\Delta + B_\Delta[J(J+1) - 2]\}} \right|,$$

where

$$\mathbf{H}^{\circ} = E_v(n, \Lambda, S) + B_v[J(J+1) - \Omega^2 + S(S+1) - \Sigma^2] + \mathbf{H}^{SO}$$

and, neglecting the $2B_\Delta$ in the denominator,

$$|C_{12}| \approx [2J(J+1) - 4]^{1/2} \frac{B_\Delta}{2A_\Delta}. \tag{5.1.63}$$

Analytic expressions for the mixing coefficients are given by Kovács (1969, pp. 68–70). In order to apply the matrix element method to a $^3\Delta$ perturber, it is necessary to observe perturbations by at least two of the three $^3\Delta$ spin components and to infer the relative vibrational numbering of these spin components. There are generally four possible relative numberings, corresponding to whether A_Δ is positive or negative, and whether $|2A_\Delta|$ is larger or smaller than ω_e ($^3\Delta$). If the electronic configuration of a plausible $^3\Delta$ perturber is known, the sign and magnitude of A_Δ can be estimated with sufficient accuracy to eliminate all but one of these relative numberings. Success of the matrix element method for $^1\Pi \sim {}^3\Delta$ perturbations implies determination of both the absolute vibrational numbering and the diagonal spin-orbit constant of the $^3\Delta$ state. For SiO, the matrix element method suggests a vibrational numbering in agreement with that deduced from isotope shifts even though the $d^3\Delta \sim A^1\Pi$ curve crossing occurs just at the highest vibrational levels sampled ($v_A = 8 \sim v_d = 19, E_C \simeq 49{,}000\ \text{cm}^{-1}, R_C = 1.42$ Å).

The $A^1\Pi \sim$ ($C^1\Sigma^-$, $D^1\Delta$, $d^3\Delta$, and $e^3\Sigma^-$) perturbations of the $^{28}Si^{32}S$ and $^{30}Si^{32}S$ molecules (Harris, *et al.*, 1982) are analogous to those observed for SiO except that the vibrational quantum numbers of the SiS perturbing levels are considerably higher than those for SiO, $v = 14 - 31$ versus $v = 6 - 19$. The quality of perturbation data is generally insufficient to determine more than the minimum number of equilibrium constants, $T_e, \omega_e, \omega_e x_e, B_e$, and α_e. This is true even for SiS $A^1\Pi$, where the number of perturbations (83) and range of vibrational levels sampled is far greater than for SiO $A^1\Pi$. Neglect of higher-order constants (especially γ_e) and the length of the extrapolation to $v = -\frac{1}{2}$ lead to a larger uncertainty in R_e. This is particularly serious when the perturbation matrix elements sampled do not include the first maximum, which occurs slightly above the curve crossing (E_C, R_C), as is the case for SiS $A^1\Pi(v = 0 - 11)$. [Ferber, *et al.*, (2000) find that the matrix element method fails for the NaK

$B^1\Pi \sim c^3\Sigma^+$ perturbations, also because all of the observed perturbed levels lie below the energy of the inner-wall intersection of the B-state and c-state potential curves. The monotonic energy dependence of the measured spin-orbit matrix elements prohibits a unique vibrational numbering of the $c^3\Sigma^+$ state.] For an inner-wall curve crossing, an error in R_C can cause a large change in the apparent location of (E_C, R_C). Consequently, the matrix element method cannot distinguish between the $v-1, v$, and $v+1$ absolute vibrational numberings. In addition, *ab initio* values of the electronic perturbation parameters (Robbe, *et al.*, 1981) cannot be used to resolve this numbering ambiguity because, even with the correct vibrational assignment, extrapolation uncertainty could cause the calculated vibrational overlap to be erroneous by a factor of two.

It must be emphasized that the reason the matrix element method begins to fail for SiS $A^1\Pi$ has nothing to do with the strength of the perturbation interactions or the degree to which the perturbers remain isolated from each other. In general, as the interactions between basis states become stronger, one must increase the dimension of the effective Hamiltonian used to account for the observed details. This does not necessarily imply a vast increase in the number of variable perturbation parameters, provided that the potential energy curves of all but one of the interacting states are known well enough so that a deperturbation procedure, based on matrix element factorization, can be initiated. One example will be discussed briefly, BaS $A^1\Sigma^+ \sim A'^1\Pi \sim a^3\Pi$ (Cummins, *et al.*, 1981)]. A similar case, the NO $B^2\Pi \sim C^2\Pi \sim D^2\Sigma^+$ interaction, was discussed in Section 4.5.

The lowest electronic states of the BaS molecule are summarized in Fig. 5.9. Figure 5.10 shows the $A'^1\Pi$ and $a^3\Pi_i$ levels responsible for perturbations of BaS $A^1\Sigma^+(v = 0-2)$ [Cummins, *et al.*, 1981]. The spin-orbit constant $A(a^3\Pi) = -186$ cm^{-1} is almost as large as $\omega_e(a^3\Pi) = 260$ cm^{-1}, and the $A'^1\Pi \sim a^3\Pi_1$ spin-orbit interaction, $A_{11} = 178$ cm^{-1}, is only slightly smaller than the deperturbed separation of these two states, $\Delta T_e = 260$ cm^{-1}. Figure 5.10 shows that the $a^3\Pi(v = 12) \sim A'^1\Pi$ $(v = 12)$ perturbation complex spans more than 500 cm^{-1} and is interleaved with components of the $v = 10-14$ $A' \sim a$ complexes. Each nominal $A'^1\Pi, v_{A'}$ and $a^3\Pi_\Omega, v_a$ level is a complicated mixture of $A'^1\Pi$, $a^3\Pi_{0,1,2}$ basis functions belonging to several vibrational quantum numbers.

A partial prediagonalization procedure was employed in the fitting of the BaS $A^1\Sigma^+$ perturbations. Since a $|^1\Sigma^+\rangle$ basis function interacts only with $|^1\Pi\rangle$ and $|^3\Pi_0\rangle$ basis functions, the interaction of each $A^1\Sigma^+, v_\Sigma$ level with a nominal $^{2S+1}\Pi_{\Omega,v_\Pi}$ level was represented by a sum of two contributions: interactions with $|a^3\Pi_0, v_a = v_\Pi\rangle$ and $\left|A'^1\Pi, v_{A'} = v_\Pi\right\rangle$. Since the $A'^1\Pi$ and $a^3\Pi$ potentials are nearly identical, it is a good approximation to neglect all $v_a \neq v_{A'}$ interactions and consider only the dominant vibrational component of the nominal $^{2S+1}\Pi_{\Omega,v_\Pi}$ function. The success of this approach is indicated by the constancy of the a and b parameters derived from the observed A~a and A~A′ perturbations (Table 5.2). It is important to note that this deperturbation was aided considerably by knowledge of the lower vibrational levels of the $A'^1\Pi$ and $a^3\Pi$ states derived from $B^1\Sigma^+ - A'^1\Pi$ and $B^1\Sigma^+ - a^3\Pi_1$ fluorescence spectra.

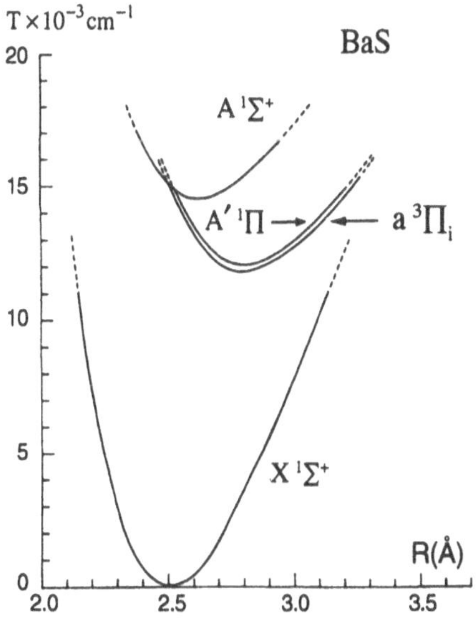

Figure 5.9: Potential energy curves for the $X^1\Sigma^+$, $a^3\Pi_i$, $A'^1\Pi$, and $A^1\Sigma^+$ electronic states of the BaS molecule. Solid lines indicate the energy range over which the potentials are determined by Cummins, *et al.*(1981).

Ferber, *et al.*, (2000) made a systematic study of the matrix element method for the NaK $B^1\Pi \sim c^3\Sigma^+$ and $b^3\Pi \sim c^3\Sigma^+$ perturbations. It is a difficult test for the matrix element method because the lowest sampled vibrational level of the $c^3\Sigma^+$ state is $v_0 >> 0$ (the final derived value is $v_0 = 20$), the perturbing levels of the $b^3\Pi$ state are $v_b = 60$, 62, and 63, and the initial vibrational numbering for the bound part of the $a^3\Sigma^+$ state (the lower state in all of the observed fluorescence from the $b^1\Pi \sim c^3\Sigma^+ \sim b^3\Pi$ perturbation complex) was incorrect. Ferber, *et al.* found that the matrix element method will surely fail if the sampled perturbation matrix elements vary nearly monotonically with energy because all mutually perturbed levels lie below the energy of the potential curve intersection. However, if an approximate value of the electronic part of the matrix element is known from semi-empirical or *ab initio* calculations, then the unknown absolute vibrational numbering is reliably established by comparing the observed and predicted vibrational matrix elements. The near-monotonic overlap integrals are exponentially dependent on the correct value of v_0. Finally, whenever there is at least one strong local minimum in the experimental matrix elements (as there is for the $b^3\Pi \sim c^3\Sigma^+$ perturbations), the absolute vibrational numbering of the unknown state may be securely established. Of course, if any aspect of the assignment scheme upon which the matrix element method is founded (incorrect assignment of a "known" state, inadequate separation of multiple perturbations), the vibrational matrix element method will fail.

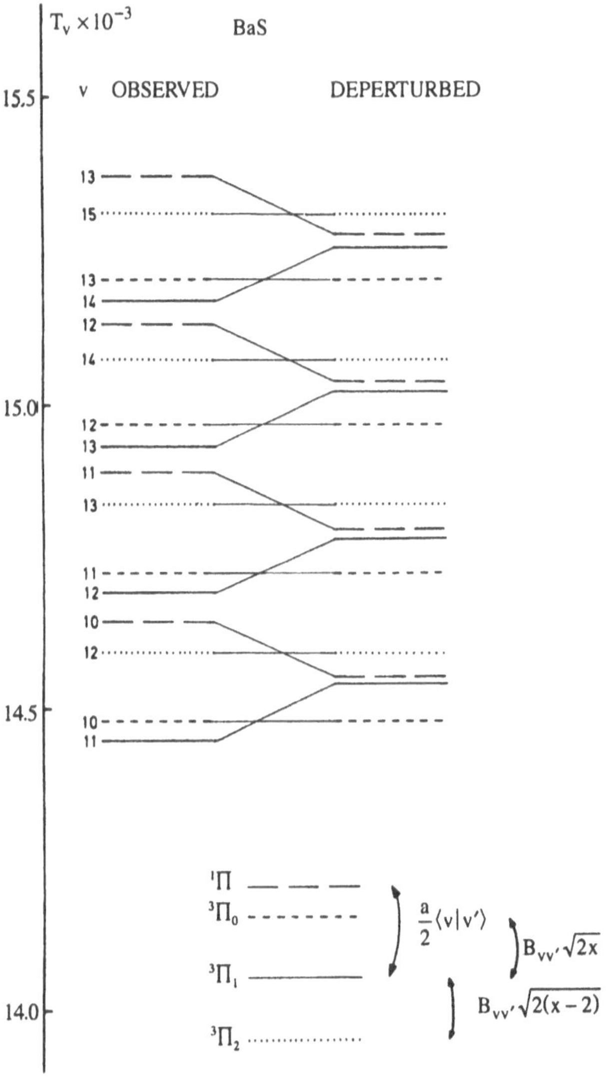

Figure 5.10: Interleaved $A'^1\Pi$ and $a^3\Pi$ levels of BaS. Energy-level diagram shows the observed and deperturbed energies of the rotationless vibrational levels of the BaS $A'^1\Pi$ and $a^3\Pi_i$ states that interact with $v = 0-2$ of $A^1\Sigma^+$. The grouping together of $^3\Pi_2$ $(v+1)$, $^1\Pi$ $(v-1)$, $^3\Pi_1$ $(v+1)$ and $^3\Pi_0$ (v) levels is suggestive of Hund's case (c) coupling. The interactions between case (a) basis functions are shown schematically at the bottom of the figure. Levels are indicated as follows: $^1\Pi$, long dashes; $^3\Pi_0$, short dashes; $^3\Pi_1$, solid lines; $^3\Pi_2$, dotted lines.

Table 5.2: Perturbation Parameters for the BaS Molecule

(a) $A^1\Sigma^+ \sim A'^1\Pi$ orbital perturbation parameters

$A^1\Sigma^+(v)$	$A'^1\Pi(v')$	$B_{vv'}(cm^{-1})$	b^a	$\overline{R}_{vv'}(Å)$
0	10	0.01131	1.08(2)	2.530
1	11	0.01963	1.11(1)	2.533
2	12	0.02470	1.07(2)	2.536
			Average: 1.09(1)[a,b]	

(b) $A^1\Sigma^+ \sim a^3\Pi$ spin-orbit perturbation parameters

$A^1\Sigma^+(v)$	$a^3\Pi_0(v')$	$\langle v\|v'\rangle$	$a_+(cm^{-1})^a$	$\overline{R}_{vv'}(Å)$
0	10	0.11177	189.6(1)	2.530
0	11	-0.07977	176.4(6)	2.513
1	12	-0.15080	183.(4)	2.516
2	13	-0.20760	182.(4)	2.519
			Average: 183.(5)	

[a] 1σ uncertainty in parentheses.

[b] This value of b does not allow for J–independent admixture of $|^3\Pi_1\rangle$ character into the nominal $^1\Pi$ eigenfunction,

$$|^1\Pi\rangle_{\text{nominal}} \simeq 0.89\,|^1\Pi_1\rangle + 0.46\,|^3\Pi_1\rangle\,.$$

Thus, the fully deperturbed value of b is

$$b = \frac{1.09}{0.89} = 1.21.$$

5.1.6 Homogeneous vs. Heterogeneous Perturbations

Bender, *et al.*, (1986) report that it is not possible, *from spectroscopic wavenumber data alone*, to distinguish between homogeneous (J–independent perturbation matrix element) and heterogeneous (matrix element $\propto J$) perturbation mechanisms, even when the perturbation is isolated and main and extra lines are observed for many J–values in the region of the intersection of the rotational term curves. Least squares fits to the observed rotational term values are equally satisfactory for homogeneous-only and heterogeneous-only fit models, despite the impression to the contrary conveyed by Fig. 5.8. However, the homogeneous vs. heterogeneous question can be decided when additional information is available, such as relative intensity measurements for main and extra lines in both R and P branches, or an accurate interpolated perturbation-free B°–value for either state.

One of the keys to the characterization of an isolated perturbation is a determination of the non-integer J–value, J_x, at which the unperturbed rotational term curves intersect. For a homogeneous perturbation, J_x coincides with the noninteger J–value where the energy separation between the perturbed rotational term curves reaches its minimum value, J_{min}. Ignoring centrifugal distortion,

$$\begin{aligned}
\Delta T^\circ &= T_2^\circ(0) - T_1^\circ(0) \\
\Delta B^\circ &= B_2^\circ - B_1^\circ \\
J_{\text{min}} &= [-\Delta T^\circ/\Delta B^\circ]^{1/2} - 1/2.
\end{aligned}$$

Since $J_{\text{min}} = J_x$ and J_x is the J–value where

$$T_1^\circ(0) + B_1^\circ J_x(J_x+1) = T_2^\circ(0) + B_2^\circ J_x(J_x+1)$$

it is possible to accurately locate J_{min} from a term value plot and, from it, determine $T_1^\circ(J_x)$ and $T_2^\circ(J_x)$. If $T_1^\circ(0)$ or $T_2^\circ(0)$ can be estimated from the relatively perturbation-free low–J region of the main state term curve (or by a second-order perturbation correction to the observed $J = 0$ term value), then an accurate value for either B_1° or B_2° is obtained. Since the J-by-J average of main and extra lines is described by

$$\frac{E_+(J) + E_-(J)}{2} = \frac{T_1^\circ(0) + T_2^\circ(0)}{2} + \frac{B_1^\circ + B_2^\circ}{2} J(J+1),$$

the other deperturbed B°–value is obtained.

However, for a heterogeneous perturbation, where

$$H_{12} = \beta[J(J+1)]^{1/2}$$

the empirically observable J_{min} value is

$$J_{\text{min}} = \left[-(2\beta^2 + \Delta T^\circ \Delta B^\circ)/\Delta B^{\circ 2}\right]^{1/2} - 1/2$$

and J_{min} is slightly smaller than the empirically useful J_x value

$$J_{\text{min}} = J_x - \beta^2 |\Delta B^\circ|^{-3/2} |\Delta T^\circ|^{-1/2}.$$

If the difference between J_x and J_{min} is ignored and J_{min} is used to extract a value for B_1° or B_2°, that value will be in error by $\beta^2/\Delta T^\circ$. The difference in B° values derived from this incorrect J_x value will be larger than the true $|\Delta B^\circ|$ by $2\beta^2/|\Delta T^\circ|$. A least squares fit of a heterogeneous perturbation using the inappropriate homogeneous interaction model will yield deperturbed B°–values shifted from the true values by $\pm\beta^2/|\Delta T^\circ|$. This means that, if an interpolated value of B_1° or B_2° is known to an accuracy better than $\beta^2/|\Delta T^\circ|$ prior to any fit to the perturbed data, then the quality of fit (with one B°–value constrained) would discriminate between the homogeneous and inhomogeneous models. However, typical values are $\beta \approx 0.1$ cm^{-1} and $\Delta T \approx 50$ cm^{-1}, thus B° needs to be predicted to an accuracy of ~ 0.0002 cm^{-1}.

An alternative method (Barnes, *et al.*, 1997) is to use the relative intensities of main and extra lines. If one of the interacting states is "bright" and the other "dark", then the J–value at which main and extra lines have equal intensity is J_x for *both* homogeneous and heterogeneous perturbations. The equal intensity J–value, J_{eq}, will coincide with the J–value of minimum separation between main and extra lines for the homogeneous case, but *not* for the heterogeneous case. When both of the interacting states are appreciably bright, the test for coincidence between J_{min} and J_{eq} cannot be used. However, intensity interference effects (see Section 6.3.2) provide unambiguous discrimination between the homogenous and heterogeneous interaction models. If the interaction is homogeneous, both interacting states are accessed by either a parallel or a perpendicular type transition. As a result the main and extra line intensity ratios will be the same in R and P branches

$$\frac{R_{\text{extra}}(J-1)}{R_{\text{main}}(J-1)} = \frac{P_{\text{extra}}(J+1)}{P_{\text{main}}(J+1)}$$

(for a perturbation in the upper state). However, if the interaction is heterogeneous and transitions into both interacting states have intrinsic intensity, then one transition will be of parallel type and the other of perpendicular type. The $R, P \parallel, \perp$ intensity interference mechanism discussed in Section 6.3.2 requires that

$$\frac{R_{\text{extra}}(J-1)}{R_{\text{main}}(J-1)} \neq \frac{P_{\text{extra}}(J+1)}{P_{\text{main}}(J+1)}.$$

In modern laser experiments, relative intensity measurements have accuracy sufficient to distinguish homogeneous from heterogeneous interactions, even when both main and extra states are bright.

5.2 Order of Magnitude of Electrostatic Perturbation Parameters: Interactions Between Valence and Rydberg States of the Same Symmetry

In light molecules, homogeneous perturbations due to electrostatic interaction (the e^2/r_{ij} operator) are often very strong. Some values of the electrostatic interaction have been listed in Table 3.5 for the case where H^e values, representing the interaction between crossing diabatic curves, are particularly large (400 cm^{-1} $< H^e <$ 10,000 cm^{-1}). The large sizes of these tabulated H^e values indicate that an adiabatic approach to these interactions would be more convenient, but the diabatic picture is equally valid and often more insightful.

5.2.1 Valence and Rydberg States

Most frequently interactions between states of the same electronic symmetry are between a Rydberg state and a valence state and are often referred to in the literature as valence~Rydberg mixing. In the following, a distinction is drawn between these two classes of electronic states.

First, the valence-shell states or valence states are those associated with electronic configurations in which all occupied molecular orbitals are comparable to or smaller in size than the typical 2 Å molecular internuclear distance. These valence molecular orbitals are constructed from atomic orbitals with principal quantum numbers, n, less than or equal to the maximum n involved in the ground state configurations of the constituent atoms. Transitions from the molecular ground state to an excited valence state frequently involve promotion of an electron from a bonding to an antibonding valence orbital. In such a case, R_e of this excited valence state will be larger and ω_e smaller than for the ground state.

Second, molecular Rydberg states are excited states similar to atomic Rydberg states. The excited electron is in an orbital that can be constructed from atomic orbitals with principal quantum numbers larger than those of the ground state of the constituent atoms (see Section 8.3). The first Rydberg state of H_2 has $n = 2$, whereas $n = 3$ for the lowest Rydberg states of molecules composed of atoms from the second row (Li through F) of the periodic table. The mean radius and other properties of a Rydberg orbital may be estimated using formulas derived for hydrogenic orbitals (Condon and Shortley, 1953, p. 117) by simply replacing the principal quantum number, n, by $n^* \equiv n - a$, where a is the quantum defect. The mean radius in atomic units (1 a.u. =0.52917 Å) is

$$\langle r \rangle_{n^*,l} = \frac{n^{*2}}{Z}\left\{1 + \frac{1}{2}\left[1 - \frac{l(l+1)}{n^{*2}}\right]\right\}, \tag{5.2.1}$$

where Z is the charge of the ion core ($Z = 1$ for neutral molecules). $\langle r \rangle$ rapidly becomes very large as n^* increases. For example,

$$\begin{aligned} \langle r \rangle_{3p} &= 6.9 \text{ a.u.} = 3.7 \text{ Å} \quad \text{with} \quad a_p = 0.7 \\ \langle r \rangle_{4f} &= 18 \text{ a.u.} = 9.5 \text{ Å} \quad \text{with} \quad a_f = 0.0. \end{aligned}$$

Thus, the Rydberg electron neither strongly perturbs the core molecular orbitals nor contributes to bonding; consequently, Rydberg potential curves are very similar to the potential curve of the ion to which the Rydberg series converges. The energy levels follow the Rydberg formula:

$$E_R = E_{\text{ion}} - \frac{\Re}{n^{*2}} = E_{\text{ion}} - \frac{\Re}{(n-a)^2}. \tag{5.2.2}$$

where $\Re$ is the Rydberg constant (=109,737.318 cm^{-1}).

The spin-orbit multiplet splitting of Rydberg states may originate either from the outer Rydberg electron or the ion core ($\Lambda_{\text{core}} \geq 1, S_{\text{core}} \geq \frac{1}{2}$). The

Table 5.3: Comparison of Spin-Orbit Coupling Constants of Rydberg and Valence States (cm^{-1})

Molecule	Rydberg state	Rydberg orbital	A(atom)[a]	A_R(obs)	Valence state	A_V(obs)
NO	$C^2\Pi$	$3p\pi$	$2.17(3p_O)$	3.2^b	$B^2\Pi$	$28.^c$
NO	$H'^2\Pi$	$3d\pi$	$0.02(3d_O)$	0.96^d	$L^2\Pi$	$-104.^c$
PO	$D^2\Pi$	$4p\pi$	$6.13(4p_P)$	25.1^e	$B'^2\Pi$ ($v = 24$)	32.3^e

[a]The difference between the spin-orbit constants of the main constituent *atomic* orbital, $A(\text{atom}, nl)$, in the Rydberg molecular orbital and that of the molecular Rydberg state, A_R, is due to "penetration" of the Rydberg MO into the molecular core, i.e., to the contribution of the $(n - l)$ atomic orbital responsible for the orthogonality between the Rydberg MO and the molecular core orbitals.

[b]Ackermann and Miescher (1968).

[c]Dressler and Miescher (1965)

[d]Suter (1969).

[e]Couet (1968).

contribution from the Rydberg orbital ($Z = 1$) is proportional to $\langle r^{-3}\rangle$

$$\langle r^{-3}\rangle = \frac{Z^3}{n^{*3}(l+1)\left(l+\frac{1}{2}\right)l} \propto (n^*)^{-3}. \tag{5.2.3}$$

Consequently, the spin-orbit constants for $\lambda \neq 0$ Rydberg states will be smaller than those of the valence states constructed from orbitals of smaller n^* values. In Table 5.3, spin-orbit constants of valence and Rydberg states of the same molecule are compared.

Another readily estimable quantity for Rydberg states is the energy interval between same–$nl\lambda$ states of different multiplicities. This splitting is governed by an exchange integral, K, between a compact inner orbital and a diffuse Rydberg orbital. To a good approximation, $K \propto (n^*)^{-3}$. This can be used to distinguish a Rydberg state from a valence state. For example, in N_2 the singlet-triplet interval is 10^4 cm^{-1} for the $\sigma_g\pi_g$ $a^1\Pi_g$ and $B^3\Pi_g$ valence states and 870 cm^{-1} for the $\sigma_g 3p\sigma_u$ $c_4'^1\Sigma_u^+$ and $D^3\Sigma_u^+$ Rydberg states. (However, K can also be very small for valence states of highly ionic molecules where the orbitals involved are highly localized on separate atoms.) The $(n^*)^{-3}$ dependence of A and K has been used in Section 3.4.2 to illustrate the convergence of $^1\Pi$ and $^3\Pi$ states to the $^2\Pi_{3/2}$ and $^2\Pi_{1/2}$ ion-core substates.

Note that a Rydberg state is generally well described by a single electronic configuration of the form

$$(AB^+)(nl\lambda)$$

where (AB^+) is the configuration of the core-ion, $nl\lambda$ is the special notation reserved for Rydberg orbitals, and n has the value that appears in Eq. (5.2.2).

l is only an approximate quantum number because the molecular potential in which the Rydberg electron moves is distorted from a spherical potential.

5.2.2 Different Classes of Valence~Rydberg Mixing

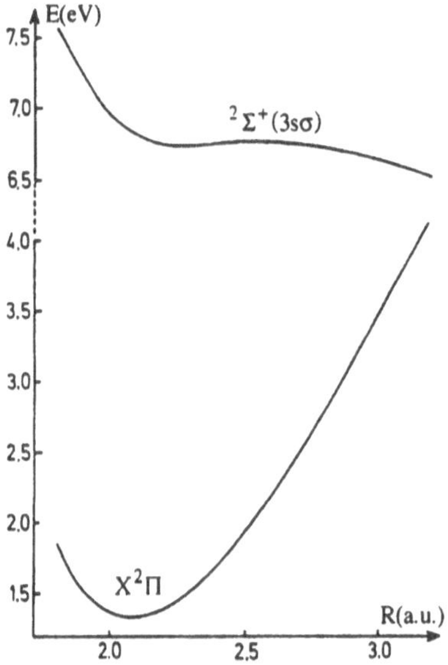

Figure 5.11: Calculated $^2\Sigma^+$ $(3s\sigma)$ state of CH. (Lefebvre-Brion, unpublished calculation).

Two types of mixing between Rydberg and valence states can be distinguished:

1. *The configurations of the Rydberg and valence states differ by two orbitals.* In the diabatic approximation, an electrostatic interaction takes place and mixes the vibrational levels of these two states. In the adiabatic representation, the coefficients of the two configurations vary with R, and, through the d/dR operator, this R-dependence of the configuration mixing coefficients is the origin of nonzero matrix elements between adiabatic states (see Section 3.3). For example, at short internuclear distance the O_2 valence $B^3\Sigma_u^-$ and Rydberg B′ (or E) $^3\Sigma_u^-$ states arise from a mixture of the $\pi_u^3\pi_g^3$ and $\pi_u^4\pi_g 3p\pi_u$ configurations (Katayama, *et al.*, 1977).

2. *The configurations of the Rydberg and valence states differ by only one orbital.* The adiabatic representation is appropriate. In this picture, only the orbital, not the configuration mixing coefficients, changes with R. For the lowest Rydberg state, this phenomenon has been called "Rydbergization" because the molecular orbital that, at dissociation, is an atomic orbital becomes a Rydberg molecular orbital at small internuclear distances (Mulliken, 1976). The character of the orbital changes smoothly with R. Rydbergization occurs in the first Rydberg state of many AH hydrides. The orbital, which is $3s_A$ at

short R, becomes at intermediate R an antibonding orbital ($2p\sigma_A - 1s_H$), and finally becomes localized on the $1s_H$ orbital at infinite R. The resulting state is, in general, unbound. Examples include CH $\pi^0 3s\sigma$ $^2\Sigma^+$ (see Fig. 5.11), OH $\pi^2 3s\sigma$ $^2\Sigma^-$ (Lefebvre-Brion, 1973) and NH $\pi^1 3s\sigma$ $^1\Pi$ (Mulliken, 1972) states. This explains the fact that the first (lowest-n) states of the Rydberg $s\sigma$ series have not been observed in CH, NH, and OH.

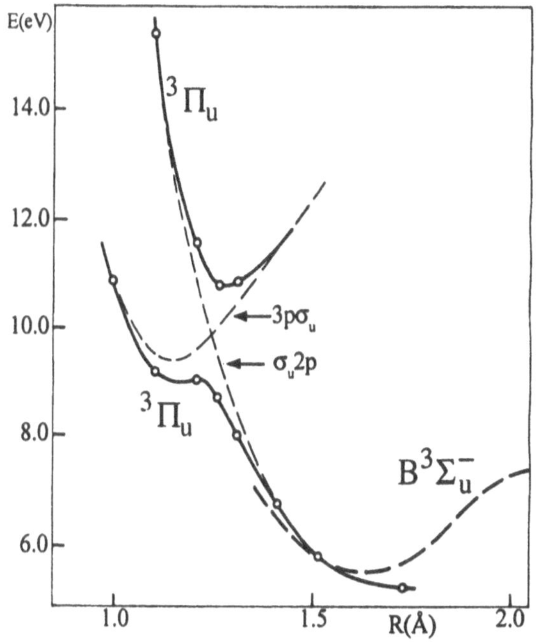

Figure 5.12: Rydbergized and complementary $^3\Pi_u$ states in O_2 arising from configurations with a $\pi_u^4\pi_g$ core and a σ_u external orbital. The solid lines depict the adiabatic potential curves calculated by Buenker and Peyerimhoff (1975). The lower curve is the Rydbergized state, which has an external orbital that evolves from a $3p\sigma_u$ Rydberg orbital at small R to a $\sigma_u 2p$ antibonding orbital at large R. The upper curve is the complementary state, associated with an external orbital that is $\sigma_u 2p$ at small R and $3p\sigma_u$ at large R. The dashed lines represent the diabatic states that would be obtained if the external orbital were frozen as R varies. More complete potential curves are shown in Figure 3 of England, *et al.*, (1996).

The existence of states complementary to those for which Rydbergization occurs as R decreases, in which the molecular orbital becomes a Rydberg orbital at large R, has been suggested by *ab initio* calculations. For example, in O_2 the configuration $\pi_u^4\pi_g 3p\sigma_u$ gives rise to a $^3\Pi_u$ state that is a Rydberg state at small R, but since the $3p\sigma_u$ orbital becomes a $\sigma_u 2p$ antibonding orbital at large R (see correlation diagram of Herzberg, 1950, p. 329), the state is repulsive and responsible, in part, for predissociation of the $B^3\Sigma_u^-$ state. The complementary $^3\Pi_u$ state is the $F^3\Pi_u$ state observed by Chang and Ogawa (1972). The isoconfigurational g $^1\Pi_u$ state has also been observed (Katayama, *et al.*, 1977). The energy separation between the Rydbergized and the complementary state is generally large: 2 × 8550 cm^{-1} for the $^3\Pi_u$ states of O_2 (obs: 2 × 7110 ±50 cm^{-1}, England *et al.*, 1996) (see Fig. 5.12), 2 × 5000 cm^{-1} for the $^2\Sigma^+$ states of NO [the $\sigma^2 3p\sigma$ configuration of the $D^2\Sigma^+$ state becomes the $\sigma^2\sigma^*$ configuration to which the repulsive $A'^2\Sigma^+$ state has been attributed (Gallusser, 1976)].[†] In principle, a weak coupling can occur between the pair of complementary states, which in the adiabatic picture results from the d/dR operator acting on the *orbital* and not on the configuration-mixing coefficients.

This second class of valence~Rydberg interaction (configurations differing by one orbital) is often confused with the first class. Some theoreticians consider that the diabatic picture is valid for states belonging to configurations which differ by a single orbital. To compute such an *ab initio* diabatic potential curve in this case, it is necessary to freeze the external orbital as R varies (see dashed curves of Fig. 5.12). Even in a self-consistent field approximation, i.e. a single-configuration approximation, the calculated curves do not cross (see full curves of Fig. 5.12) and can be considered to be adiabatic curves. H^e, the half-distance at closest approach between these two adiabatic curves, is often very large as seen above. Consequently the adiabatic curves are a better approximation. Another example of Rydbergization is the interaction between the $V^1\Sigma^+$ state of HCl (configuration $\sigma\pi^4\sigma^*$) and the $4s\sigma$ Rydberg state with the $A^2\Sigma^+$ excited ionic core (configuration $\sigma\pi^4 4s\sigma$) which has been calculated to have an interaction matrix element of 7000 cm^{-1} (Yencha, *et al.*, 1993). In the following subsection, only the first class of valence~Rydberg mixing, for which the configurations differ by two electrons, is considered.

5.2.3 Electrostatic Perturbations Between Valence and Rydberg States

Some of the highest energy valence states lie at similar energies to those of Rydberg states and numerous valence~Rydberg perturbations result. Table 5.4 reviews several cases treated in the literature. The values of H^e are obtained either by deperturbation or by *ab initio* calculations. Prediction of the strength of H^e is, in general, difficult. A detailed numerical example was discussed in Section 3.3.2 for the case of interacting NO $^2\Delta$ states. The interaction strength between pairs of states belonging to the same pair of configurations but to different symmetries can vary by an order of magnitude. In the case of the O_2 $\pi_u^3\pi_g^3$ and $\pi_u^4\pi_g\ 3p\pi_u$ configurations, the interaction between the $^3\Sigma_u^-$ states is 10 times larger than that between the $^3\Delta_u$ states (see Table 5.4). This results from partial cancellation between two relatively large integrals for $^3\Delta_u$ states that add in the $^3\Sigma^-$ case [cf. Eq. (3.3.8)]. For molecules composed of atoms from the second row of the periodic table, the diabatic potential curves of valence and Rydberg states generally have very different spectroscopic constants. This is particularly true when the valence state is an ionic state (examples include

[†]The complementary state may be located above the ionization limit. It may then appear as a "shape resonance" in the photoionization cross-section (Morin, *et al.*, 1982). (See Section 8.1.3.)

Table 5.4: Some Values of the Electrostatic Parameter

Mol.	Symm.	Valence States Name	Valence States Config.	Rydberg states Name	Rydberg states Config.	H^e [a] (cm^{-1})	Ref. [b]
N_2	$^1\Pi_u$	b	$\sigma_u\pi_u^4\sigma_g^2\pi_g$	c	$\sigma_u^2\pi_u^4\sigma_g 3p\pi_u$	1100	1
		b	$\sigma_u\pi_u^4\sigma_g^2\pi_g$	o	$\sigma_u^2\pi_u^3\sigma_g^2 3s\sigma_g$	1000	
	$^1\Sigma_u^+$	b′	$\sigma_u^2\sigma_g^2\pi_u^3\pi_g$	c′	$\sigma_u^2\sigma_g\pi_u^4 3p\sigma_u$	900	
CO	$^3\Sigma^+$	a′	$\sigma^2\pi^3\pi^*$	b	$\sigma\pi^4 3s\sigma_g$	700	2
	$^1\Sigma^+$	D′	$\sigma^2\pi^3\pi^*$	B	$\sigma\pi^4 3s\sigma_g$	2900	3
NO	$^2\Sigma^+$	I	$\sigma\pi^{*2}$	H	$\sigma^2 3d\sigma$	700	4
		I	$\sigma\pi^{*2}$	E	$\sigma^2 4s\sigma$	50	
		I	$\sigma\pi^{*2}$	M	$\sigma^2 4p\sigma$	100	
O_2	$^3\Sigma_u^-$	B	$\pi_u^3\pi_g^3$	E	$\pi_u^4\pi_g 3p\pi_u$	4030	5
	$^3\Delta_u$	A′	$\pi_u^3\pi_g^3$	e′	$\pi_u^4\pi_g 3p\pi_u$	770	
	$^1\Sigma_u^+$	f′	$\pi_u^3\pi_g^3$	f	$\pi_u^4\pi_g 3p\pi_u$	1600	
	$^1\Delta_u$		$\pi_u^3\pi_g^3$		$\pi_u^4\pi_g 3p\pi_u$	2640	
	$^3\Sigma_u^+$	A	$\pi_u^3\pi_g^3$	D	$\pi_u^4\pi_g 3p\pi_u$	420	
	$^3\Pi_g$		$\sigma_g\pi_g^3$	C	$\sigma_g^2\pi_g 3s\sigma_g$	620	
	$^1\Pi_g$		$\sigma_g\pi_g^3$	d	$\sigma_g^2\pi_g 3s\sigma_g$	630	
BeH	$^2\Sigma^+$	C	$\sigma\sigma^{*2}$		$\sigma^2 3s\sigma$	(660)	6
PO	$^2\Sigma^+$	F	$\sigma\pi^{*2}$	A	$\sigma^2 4s\sigma$	410	7
		F	$\sigma\pi^{*2}$	G	$\sigma^2 4p\sigma$	540	

[a]Values in parentheses are calculated values.

[b]References: (1) Stahel, *et al.*(1983); (2) from Stepanov (1940) and RKR curve calculations; (3) Tchang-Brillet, *et al.*(1992); (4) Gallusser (1976); (5) Lewis, *et al.*(2000); (6) Lefebvre-Brion and Colin (1977) (see also Petsalakis, *et al.*(1992); (7) Ngo, *et al.*(1974).

perturbations in H_2, HF, F_2, Cl_2). For heavier molecules, the potential curves of valence and Rydberg states become very similar and the deperturbation process becomes more difficult (e.g., SiO, PO).

The interaction between a valence state and successive Rydberg states of the same series (same values of l and λ, but different n) decreases proportionally to $(n^*)^{-3/2}$. Consequently, the product of $H_n^e \times (n^*)^{3/2}$ should be constant (cf. Section 8.4). Table 5.5 shows how well this law is obeyed for the NO molecule. This is one of the basic concepts of the multichannel quantum defect theory (Dill and Jungen, 1980). Thus, because of the typically large diabatic interaction matrix elements, the adiabatic picture is often the best starting point for treating the interaction between a valence state and the *first* Rydberg state of a given series (for example, the E,F; G,K interaction in H_2). This adiabatic picture becomes increasingly difficult to apply for the interaction of higher vibrational levels of the valence state with the higher-n members of the Rydberg series because the strength of the valence~Rydberg interactions becomes weaker (and the diabatic picture becomes easier to apply).

Table 5.5: Dependence on n^* of the Electrostatic Perturbation Parameter

Molecule	Symmetry	Configuration of Valence State	Configuration of Rydberg State	$H_n^e(\text{cm}^{-1})$	$n^{*3/2}H_n^e(\text{a.u.})$
NO	$^2\Pi^a$	B $\pi^3\pi^{*2}$	C $\pi^4 3p\pi$	1383	0.0208
			K $\pi^4 4p\pi$	804	0.0214
			Q $\pi^4 5p\pi$	595	0.0237
		L $\pi^3\pi^{*2}$	C $\pi^4 3p\pi$	550	0.0083
			K $\pi^4 4p\pi$	250	0.0066
			Q $\pi^4 5p\pi$	200	0.0080
	$^2\Delta^b$	B′ $\sigma\pi^{*2}$	F $\sigma^2 3d\delta$	450	0.0103
			N $\sigma^2 4d\delta$	400	0.0142

[a]Gallusser and Dressler, (1982).
[b]Lefebvre-Brion, (1969).

5.2.4 Electrostatic Perturbations between Rydberg States Converging to Different States of the Ion

Electrostatic perturbations can occur between two Rydberg states that belong to series converging to different states of the ion. One example, the interaction between $^1\Pi_u$ states of the N_2 molecule (Stahel, *et al.*, 1983), has been studied. The $c^1\Pi_u$ state has the configuration $\sigma_u^2\sigma_g\pi_u^4 3p\pi_u$ and belongs to a series converging to the $X^2\Sigma_g^+$ ground state of N_2^+. The $o^1\Pi_u$ state has the configuration $\sigma_u^2\sigma_g^2\pi_u^3 3s\sigma_g$ and belongs to a series converging to the $A^2\Pi_u$ state of N_2^+. These two Rydberg states are strongly perturbed by the $b^1\Pi_u$ valence state, but they also interact with each other. The H^e effective interaction parameter between the two Rydberg states has been found to be about 100 cm^{-1}. The same type of interaction is responsible for the autoionization of the upper members of Worley's third series (of which $o^1\Pi_u$ is the first member) by the continuum of the $X^2\Sigma_g^+$ state of N_2^+ (see Section 8.8).

It is possible for Rydberg states of the same symmetry, converging to different multiplet states of the same ionic configuration, to interact strongly. In such a case, the configurations of the interacting Rydberg states appear to differ by only the Rydberg orbital, but the multiplicities of the ion cores are different. The interacting states differ by two *spin*-orbitals: the Rydberg spin-orbital and one of the core spin-orbitals. The electrostatic interaction is expressed in terms of two-electron integrals, as in Eq. (3.3.8), where the ϕ_a and ϕ_b orbitals are identical. This type of interaction can give rise to autoionization: for example, the autoionization of the Rydberg states of O_2 converging to the O_2^+ $\sigma_g\pi_g^2$ $B^2\Sigma_g^-$ state by the ionization continuum of the $\sigma_g\pi_g^2$ configuration of the $b^4\Sigma_g^-$ state of O_2^+ (Dehmer and Chupka, 1975). The interaction between two $^3\Sigma_u^-$ Rydberg states belonging to series that converge to these $^2\Sigma_g^-$ and $^4\Sigma_g^-$ ion core states is given by

$$\langle (\mathrm{b}^4\Sigma_g^-)\, np\sigma_u \left|\mathbf{H}^{el}\right| (\mathrm{B}^2\Sigma_g^-)\, n'p\sigma_u \rangle = \frac{2\times 2^{1/2}}{3}\left[\langle \pi_g np\sigma_u|\pi_g n'p\sigma_u\rangle - \langle \sigma_g np\sigma_u|\sigma_g n'p\sigma_u\rangle\right],$$

where the wavefunctions are

$$\left|(\mathrm{b}^4\Sigma_g^-)\, np\sigma_u\right\rangle = 12^{-1/2}\left|np\sigma_u\sigma_g\pi_g^+\pi_g^-\right|[-3\beta\alpha\alpha\alpha + \alpha\alpha\beta\alpha + \alpha\alpha\alpha\beta + \alpha\beta\alpha\alpha]$$
$$\left|(\mathrm{B}^2\Sigma_g^-)\, n'p\sigma_u\right\rangle = 6^{-1/2}\left|n'p\sigma_u\sigma_g\pi_g^+\pi_g^-\right|[-2\alpha\beta\alpha\alpha + \alpha\alpha\alpha\beta + \alpha\alpha\beta\alpha].$$

(See Table 3.4 for the case of four open subshells.)

It was suggested in Section 5.2.2 that a valence antibonding orbital can become, at small R, a Rydberg orbital of the same symmetry (Rydbergization). Thus, the valence state can be considered as the first member of a Rydberg series converging to an excited state of the ion. The π_g antibonding orbital of N_2 becomes similar to a $3d\pi_g$ Rydberg orbital near R_e (Mulliken, 1976). The N_2 b′ $^1\Sigma^+$ valence state, arising from the configuration $\sigma_g^2\pi_u^3\pi_g$, could thus be considered as the first member of the series "$d\pi$" converging to the N_2^+ $A^2\Pi_u$ state. (Unfortunately, this valence state is not satisfactorily described by this single configuration.) With this point of view, the $(n^*)^{-3/2}$ matrix element scaling rule, which is valid for Rydberg~Rydberg interactions, could also be applied to valence states to predict approximate interaction parameters for perturbations and/or autoionizations in the upper members of the series of which the first member is a valence state (see Section 8.8).

Values are collected in Table 5.4 for the electronic part of the electrostatic interaction. It is clear that this interaction will give rise either to perturbations or to predissociations depending on whether the interaction occurs between discrete states or discrete and continuum states. In the above example, concerning the N_2 $^1\Pi_u$ states of Worley's third series, it was shown that the electrostatic interaction can also give rise to autoionizations (Chapter 8).

5.3 Order of Magnitude of Spin Parameters

In Section 3.4.2, spin-orbit matrix elements are expressed, in the single-configuration approximation, in terms of molecular spin-orbit parameters. These molecular parameters can also be related to atomic spin-orbit parameters. In Table 5.6, some values are given for atomic spin-orbit constants, $\zeta(nl)$. Sections 5.3.1 and 5.3.3 describe semiempirical methods for estimating molecular spin-spin parameters and both diagonal and off-diagonal spin-orbit parameters.

Table 5.6: Spin-Orbit Constants of Atoms and Ions (cm^{-1})[a]

(a) $\zeta(np)$

				B^+	15.2	C^+	42.7	N^+	87.5	O^+	*169	F^+	327.1
				$2s2p$	$2p$		$2p^2$		$2p^3$		$2p^4$		
Li	0.2	Be	2.0	B	10.7	C	29.0	N	*73.3	O	151	F	269.3
$2p$		$2s2p$		$2s^2 2p$		$2p^2$		$2p^3$		$2p^4$		$2p^5$	
		Be^-	0.7	B^-	5.3	C^-	-	N^-	55	O^-	121	F^-	-
		$2s^2 2p$		$2s^2 2p^2$				$2p^4$		$2p^5$			
		Mg^+	61	Al^+	124.9	Si^+	191.3	P^+	313.5	S^+	*482.8	Cl^+	664
		$3p$		$3s3p$		$3p$		$3p^2$		$3p^3$		$3p^4$	
Na	11.5	Mg	40.5	Al	74.7	Si	148.9	P	*275.2	S	382.4	Cl	587.3
$3p$		$3s3p$		$3s^2 3p$		$3p^2$		$3p^3$		$3p^4$		$3p^5$	
		Mg^-	13.0	Al^-	48	Si^-	-	P^-	188	S^-	326	Cl^-	-
		$3s^2 3p$		$3s^2 3p^2$				$3p^4$		$3p^5$			
		Ca^+	149	Ga^+	920	Ge^+	1178	As^+	1693	Se^+	[2310]	Br^+	2560
		$4p$		$4s4p$		$4s^2 4p$		$4p^2$		$4p^3$		$4p^4$	
K	38	Ca	87	Ga	(464)	Ge	940	As	*1500	Se	1690	Br	2457
$4p$		$4s4p$		$4s^2 4p$	551	$4p^2$	*880	$4p^3$	(1201)	$4p^4$	(1658)	$4p^5$	(2215)
							(800)						
				Ga^-	387	Ge^-	-	As^-	1000	Se^-	1519	Br^-	-
				$4s^2 4p^2$				$4p^4$		$4p^5$			
		Sr^+	534	In^+	2368	Sn^+	2834	Sb^+	[3700]				
		$5p$		$5s5p$		$5s^2 5p$		$5p^2$					
Rb	158	Sr	387	In	1475	Sn	*2097.3	Sb	*3400	Te	(3383)	I	(4303)
$5p$		$5s5p$		$5p$	(1183)	$5p^2$	(1855)	$5p^3$	(2593)	$5p^4$		$5p^5$	5068
								Sb^-	2000	Te^-	3338		
								$5p^4$		$5p^5$			
		Ba^+	1127					Bi^+	[11,540]				
		$6p$						$6p^2$					
Cs	369	Ba	832	Tl	5195	Pb	*7294	Bi	*10,100	Po	(8608)	At	(10,607)
$6p$		$6s6p$		$6p$	(3410)	$6p^2$		$6p^3$		$6p^4$		$6p^5$	

[a]Values in parentheses are calculated values from Froese-Fischer (1972, 1977). Other calculated Hartree-Fock values can be found in Fraga, *et al.*, (1976), Table VII (2). For $\zeta(np)$, the experimental values are deduced from the observed splitting, $\Delta E = E_{J_{max}} - E_{J_{min}}$, of the lowest-energy $L > 0$ multiplet listed in Moore's (1971) tables of atomic energy levels. For both $p^N\ ^3P(N = 2 \text{ or } 4)$ and $^2P(N = 1 \text{ or } 5)$, $\zeta(np) = \frac{2}{3}|\Delta E|$. Values with asterisks are, for $2p$ and $3p$, interpolated values from Ishiguro and Kobori (1967) and, for np $n > 3$, are deduced from a matrix diagonalization fit to observed levels which exhibit coupling intermediate between $L - S$ and $j - j$ (Condon and Shortley, 1953). Values in brackets are from Sobel'man (1972). For the $\zeta(nd)$ table, the first value listed for the $3d$ positive ion corresponds to the $3d^N 4s$ configuration where N is the number of $3d$ electrons in the neutral atom (Johansson, *et al.*, 1980). The second value is obtained by a fit of the observed levels from the $3d^{N+1}$, $3d^N 4s$, and $3d^{N-1}4s^2$ configurations (Shadmi, *et al.*, 1968) and corresponds to the $3d^N 4s$ value. The values for doubly positive ions are for the $3d^N$ configurations (Shadmi, *et al.*, 1969). Values in brackets are calculated for $d^9\ ^2D$ states using $\zeta(nd) = \frac{2}{5}\left|E_{J_{max}} - E_{J_{min}}\right|$ and energy levels from Moore (1971).

Table 5.6: (Continued).

(b) $\zeta(nd)$

$N=1$		$N=2$		$N=3$		$N=4$		$N=5$		$N=6$		$N=7$		$N=8$		$N=9$		$N=10^*$	
		Ti^{2+}	119	V^{2+}	167	Cr^{2+}	233	Mn^{2+}	317	Fe^{2+}	509	Co^{2+}	539	Ni^{2+}	677	Cu^{2+}	833	$Zn^{2+}(3d^9 4s)$	1102
Se^+	56	Ti^+	96	V^+	152	Cr^+	224	Mn^+	392	Fe^+	416	Co^+	536	Ni^+	672	Cu^+	824	$Zn^+(3d^{10} 4p)$	583
	79		117								409				670		827	$Zn(3d^{10} 4s4p)$	386
Sc	(77)	Ti	(123)	V	(177)	Cr	(243)	Mn	(325)	Fe	(417)	Co	(530)	Ni	(663)	Cu	(318)	$Cd^{2+}(4d^9 5s)$	2306
$3d$		$3d^2$		$3d^3$		$3d^4$		$3d^5$		$3d^6$		$3d^7$		$3d^8$		$3d^9$		$Cd^+(4d^{10} 5p)$	1655
																		$Cd(4d^{10} 5s5p)$	1142
		Zr^+	363	Nb^+	502	Mo^+	663			Ru^+	1052	Rh^+	1278	Pd^+	1527	Ag^+	[1330]	$Hg^{2+}(5d^9 6s)$	6222
Y	(261)	Zr	(387)	Nb	(524)	Mo	(677)	Tc	(850)	Ru	(1038)	Rh	(1253)	Pd	(1495)	Ag	(1767)	$Hg^+(5d^{10} 6p)$	6082
$4d$		$4d^2$		$4d^3$		$4d^4$		$4d^5$		$4d^6$		$4d^7$		$4d^8$		$4d^9$		$Hg(5d^{10} 6s6p)$	4265
Lu	(1158)	Hf	1219	Ta	1699	W	2085	Re	2545	Os	3045	Ir	3617	Pt	4221	Au	(5097)		
$5d$		$5d^2$	(1578)	$5d^3$	(1995)	$5d^4$	(2432)	$5d^5$	(2901)	$5d^6$	(3378)	$5d^7$	(3905)	$5d^8$	(4475)	$5d^9$			

(c) $\zeta(nf)$

$N=1$		$N=2$		$N=3$		$N=4$		$N=5$		$N=6$		$N=7$	
		Ce^+	526	Pr^+	649	Nd^+	772	Pm^+	911	Sm^+	1039		
La	(556)	Ce	(687)	Pr	(820)	Nd	(957)	Pm	(1103)	Sm	(1263)	Eu	(1438)
$4f$		$4f^2$		$4f^3$		$4f^4$		$4f^5$		$4f^6$		$4f^7$	
$N=8$		$N=9$		$N=10$		$N=11$		$N=12$		$N=13$		$N=14^*$	
Gd^+	1347	Tb^+	1578	Dy^+	1791	Ho^+	2044	Er^+	2216	Tm^+	2505	$Yb^{2+}(4f^{13} 6s)$	2914
Gd	(1611)	Tb	(1802)	Dy	(2010)	Ho	(2233)	Er	(2471)	Tm	(2728)	$Yb^+(4f^{14} 5d)$	549
$4f^8$		$4f^9$		$4f^{10}$		$4f^{11}$		$4f^{12}$		$4f^{13}$		$Yb(4f^{14} 5d6s)$	313

For $\zeta(4d)$, the three configurations $4d^{N+1}, 4d^N 5s$, and $4d^{N-1}5s^2$ of the positive ions are fitted (Shadmi, 1961). The values listed correspond to the $4d^N 5s$ configuration.

For $\zeta(5d)$, values are obtained by fitting the $5d^N 6s^2, 5d^{N+1}6s$, and $5d^{N+2}$ configurations of the neutral atoms. The values for $\zeta(5d)$ depend strongly on the number of d–electrons and the values given are for the $5d^N 6s^2$ configuration (Wyart, 1978).

For $\zeta(4f)$, the values tabulated for the positive ions are deduced from a fit to experimental energy levels belonging to $4f^N 5d$ and $4f^N 6s$ configurations and should not differ significantly from the values for the $4f^N 6s^2$ configurations of the neutral atoms (Wyart and Bauche-Arnoult, 1981).

Values of $\zeta(5f)$ are not tabulated here. See Blaise, *et al.*(1980).

Values for negative ions are taken from Sambe and Felton (1976) and Hotop and Lineberger (1975).

*The $N = 10$ group samples the spin-orbit coupling constants for nd (obtained from $d^9 s\ ^3D$ using $\zeta(nd) = \frac{2}{5}\left|E_{J_{MAX}} - E_{J_{MIN}}\right|$) and $(n+1)p$ (obtained from $d^{10}sp\ ^3P$ and $d^{10}p\ ^2P$ using $\zeta(np) = \frac{2}{3}\left|E_{J_{MAX}} - E_{J_{MIN}}\right|$) orbitals. (Moore, 1971).

*The $N = 14$ group samples the nf (obtained from $f^{13}s\ ^3F$ using $\zeta(f) = \frac{2}{7}\left|E_{J_{MAX}} - E_{J_{MIN}}\right|$) and $(n+1)d$ orbitals (Martin, *et al.*, 1978).

5.3.1 Diagonal Spin-Orbit Parameters

A semiempirical calculation of molecular spin-orbit constants can be made, using the method of Ishiguro and Kobori (1967).

In the simplest case where the bonding and antibonding π molecular orbitals can be expanded in terms of one atomic p orbital on each atom, we have

$$|\pi\rangle = C_\mathrm{A}|p\pi_\mathrm{A}\rangle + C_\mathrm{B}|p\pi_\mathrm{B}\rangle \qquad \text{for the } \pi \text{ bonding orbital,}$$
$$|\pi^*\rangle = C_{\mathrm{A}^*}|p\pi_\mathrm{A}\rangle + C_{\mathrm{B}^*}|p\pi_\mathrm{B}\rangle \qquad \text{for the } \pi^* \text{ bonding orbital,}$$

and

$$a_\pi = \left\langle \pi \left| \mathbf{H}^\mathrm{SO} \right| \pi \right\rangle = C_\mathrm{A}^2 \zeta_\mathrm{A}(p) + C_\mathrm{B}^2 \zeta_\mathrm{B}(p)$$
$$a_{\pi^*} = \left\langle \pi^* \left| \mathbf{H}^\mathrm{SO} \right| \pi^* \right\rangle = C_{\mathrm{A}^*}^2 \zeta_\mathrm{A}(p) + C_{\mathrm{B}^*}^2 \zeta_\mathrm{B}(p), \qquad (5.3.1)$$

where $\zeta_\mathrm{A}(p)$ and $\zeta_\mathrm{B}(p)$ are the atomic spin-orbit parameters given in Table 5.6,

$$\zeta_\mathrm{K}(p) = \left\langle p_\mathrm{K} \left| \frac{\alpha^2}{2} \frac{Z_{\mathrm{eff,K}}}{r_\mathrm{K}^3} \right| p_\mathrm{K} \right\rangle .$$

This semiempirical approximation neglects two-center integrals such as

$$\left\langle p_\mathrm{A} \left| \frac{\alpha^2}{2} \frac{Z_{\mathrm{eff,K}}}{r_\mathrm{K}^3} \right| p_\mathrm{B} \right\rangle, \quad \mathrm{K} = \mathrm{A} \text{ or } \mathrm{B},$$

which, due to the r_K^{-3} dependence of the spin-orbit operator, are very small. Their contribution is always less than 10% of the one-center contribution. Even in *ab initio* computations of spin-orbit constants using Eq. (3.4.2), the two-center integrals are frequently neglected because the contributions of the one- and the two-electron two-center integrals tend to cancel each other (see, for example, Richards, *et al.*, 1981). Note that C_K^2 is slightly different from the fraction of the electron located on the K atom, Mulliken's "atomic population" (Mulliken, 1955). The Mulliken populations are given in this case by

$$C_\mathrm{A}^2 + C_\mathrm{A} C_\mathrm{B} S_\mathrm{AB} \qquad \text{the atom A population}$$
$$C_\mathrm{B}^2 + C_\mathrm{A} C_\mathrm{B} S_\mathrm{AB} \qquad \text{the atom B population}$$

where

$$S_\mathrm{AB} = \langle p\pi_\mathrm{A} | p\pi_\mathrm{B} \rangle , \qquad \text{the atomic overlap integral.}$$

Normalization of the molecular orbital requires that

$$C_\mathrm{A}^2 + C_\mathrm{B}^2 + 2C_\mathrm{A} C_\mathrm{B} S_\mathrm{AB} = 1.$$

For homonuclear molecules,

$$|\pi_u\rangle = C_u \left[|p\pi_\mathrm{A}\rangle + |p\pi_\mathrm{B}\rangle \right] \qquad \text{with} \quad C_u = \left[2\left(1 + S_\mathrm{AB}\right) \right]^{-1/2} \le 2^{-1/2}$$

and

$$|\pi_g\rangle = C_g\left[|p\pi_A\rangle - |p\pi_B\rangle\right] \qquad \text{with} \quad C_g = \left[2\left(1 - S_{AB}\right)\right]^{-1/2} \geq 2^{-1/2}.$$

If the overlap integral is neglected, $C_u = C_g = 2^{-1/2}$ and

$$a_{\pi_u} = a_{\pi_g} = \zeta_A(p). \tag{5.3.2}$$

This gives [see Eq. (3.4.12)]

$$A\left(B^3\Pi_g, \sigma_g\pi_g, N_2\right) = \frac{a_{\pi_g}}{2} \simeq \frac{\zeta_N(p)}{2} = 36.6\ \text{cm}^{-1}(\text{exp}: 42.2\ \text{cm}^{-1}).^{\dagger}$$

Similarly,

$$A\left(A^2\Pi_u, \pi_u^3, N_2^+\right) = -a_{\pi_u} = -[\zeta_N(2p) + \zeta_{N^+}(2p)]/2$$
$$= -80\ \text{cm}^{-1} \quad (\text{exp}: -80\ \text{cm}^{-1}).$$

Also,

$$A\left(b^3\Pi_g, \sigma_g\pi_g, P_2\right) = +138\ \text{cm}^{-1}(\text{exp}: +131\text{cm}^{-1})$$

(Brion, *et al.*, 1977) and

$$A\left(X^2\Pi_u, \pi_u^3, P_2^+\right) = -294\ \text{cm}^{-1}(\text{exp}: -260 \pm 20\ \text{cm}^{-1})$$

(Malicet, *et al.*, 1976). This last example illustrates that the spin-orbit constants increase as the weight of the molecule increases. For hydrides, the semi-empirical method works very well, because the hydrogen atom does not contribute to

---

† Note that neglect of S_{AB} leads to a slight underestimate for a_{π_g} and overestiamte for a_{π_u}. Using

$$A(W^3\Delta_u, \pi_u^3\pi_g, N_2) = [a_{\pi_g} - a_{\pi_u}]/4 = 5.7\ \text{cm}^{-1}$$

and the semiempirical value of $a_{\pi_g} = 84.5\ \text{cm}^{-1}$ from $A(B^3\Pi_g, \sigma_g\pi_g, N_2)$, one obtains

$$a_{\pi_u} = 61.7\ \text{cm}^{-1} \qquad (\text{from } \zeta_N(2p) = 73.3\ \text{cm}^{-1}).$$

Similarly, for the O_2 molecule, the value of $a_{\pi_g} = 184\ \text{cm}^{-1}$ deduced from

$$A(F^3\Pi_u, \pi_u^4\pi_g 3p\sigma_u) = 92\ \text{cm}^{-1}$$

is slightly larger than the value of $\zeta_O(2p) = 151\ \text{cm}^{-1}$. Also, the value of $a_{\pi_u} = 108\ \text{cm}^{-1}$ obtained from the equality

$$A(A'\ ^3\Delta_u, \pi_g^3\pi_u^3)_{\text{exp}} = -73\ \text{cm}^{-1} = -\frac{1}{4}[a_{\pi_g} + a_{\pi_u}]$$

is smaller than $\zeta_O(2p)$. However, this semiempirical method should not be expected to give more than a best guess for the value of the molecular spin-orbit constant. This method assumes that the open-shell atomic orbitals are not significantly altered (in the core region) by the molecular field and that the single-configuration approximation is valid.

the spin-orbit interaction. For example, $A(X^2\Pi, TeH) = -3383\ cm^{-1}$ (exp : $-3816.4\ cm^{-1}$ ($v = 0$) Fink, *et al.*, (1989)).

For heteronuclear molecules, *ab initio* calculations of the Self-Consistent-Field (SCF) molecular orbitals are usually necessary to obtain the values of the C_k coefficients. The values of the spin-orbit constants depend on the effective charge on each atom, a positive charge causing a contraction of the atomic orbitals and an increase in ζ(atomic); a negative charge causes an opposite effect. In the simple case of an expansion of the molecular orbitals in terms of a large number of atomic orbitals associated with a single l–value on each atom (Lefebvre-Brion and Moser, 1966),

$$|\pi\rangle = \sum_{i=1}^{r} C_{i\mathrm{A}} \left| il\pi_{\mathrm{A}} \right\rangle + \sum_{j=1}^{s} C_{j\mathrm{B}} \left| jl\pi_{\mathrm{B}} \right\rangle$$

$$a_\pi = \left[\sum_i^r \left(C_{i\mathrm{A}}^2 + \sum_{i \neq j}^{r} C_{i\mathrm{A}} C_{jA} S_{i\mathrm{A},j\mathrm{A}} \right) \right] \zeta_{\mathrm{A}}(l) + \left[\sum_j^s \left(C_{j\mathrm{B}}^2 + \sum_{k \neq j}^{s} C_{j\mathrm{B}} C_{k\mathrm{B}} S_{j\mathrm{B},k\mathrm{B}} \right) \right] \zeta_{\mathrm{B}}(l) \tag{5.3.3}$$

where the S terms are the overlap between Slater-type atomic orbitals located on the same atom. More sophisticated expressions for a_π are required when a molecular orbital has mixed–l or mixed–n character.

It is interesting to relate the spin-orbit constants for corresponding states of molecules having the same number of valence electrons (isovalent series). An example is given for molecules with 10 valence electrons for the $^3\Pi$ state of the $\sigma\pi^*$ configuration for which

$$A(^3\Pi, \sigma\pi^*) = \frac{1}{2} a_{\pi^*} = \frac{1}{2} \left(C_{\mathrm{A}}^{*2} \zeta_{\mathrm{A}} + C_{\mathrm{B}}^{*2} \zeta_{\mathrm{B}} \right).$$

Since, by definition, the π^* antibonding orbital has a higher energy than that of the π bonding orbital, it is reasonable to expect that the π^* orbital will be concentrated on the less electronegative atom, but this localization is by no means complete. Table 5.7 shows that the dominant atomic character of the molecular π^* orbital is in qualitative agreement with this picture. In contrast to the case for the carbon and silicon compounds, the less electronegative atom in the germanium compounds has the larger atomic spin-orbit constant. This explains the large value predicted for the $A(^3\Pi)$ spin-orbit constant of GeO and GeS. The calculated values given in Table 5.7 have been obtained by more sophisticated calculations than the semiempirical estimates. The importance of configuration interaction must be stressed, especially for heavy molecules. A small mixing with a configuration in which the π orbital is present will give a large contribution to the spin-orbit constant if this π orbital is concentrated on the heavier atom. Furthermore, it must be kept in mind that second-order effects, neglected in the results of simple calculations listed in Table 5.7, can

Table 5.7: Spin-Orbit Constants for Isovalent Molecules (cm^{-1})[a]

	$^3\Pi(\sigma\pi^*$ configuration)			$^3\Delta(\pi^3\pi^*$ configuration)			
	Calculated[a]					$\zeta_A(np)$ (less electronegative atom)	$\zeta_B(np)$ (more electronegative atom)
	SCF	CI	Exp.	Semiempirical calculated	Exp.		
CO	35.0	39.5[b]	41.51[c]	-16	-16[c]	$\zeta_C(2p) = 24$	$\zeta_O(2p) = 150$
CS	64.5	92.5[d]	92.76[e]	-45	-61[e]		$\zeta_S(3p) = 382$
CSe	(200)	(330)[d]	326.0[f]	(-208)	-		$\zeta_{Se}(4p) = 1690$
CTe	(420)	(690)[d]	-	(-420)	-		$\zeta_{Te}(5p) = 3380$
SiO	58.5	61.5[g]	73.0[h]	0	8[l]	$\zeta_{Si}(3p) = 150$	
SiS	78	85[i]	< 96[j]	-43	-		
SiSe	(170)	(203)[i]	-	(-285)	-		
SiTe	(270)	(340)[i]	-	(-600)	-		
GeO	(420)	(423)[i]	-	(146)	-	$\zeta_{Ge}(4p) = 940$	
GeS	(440)	(447)[i]	-	(103)	-		
PN	88[k]	-	-	31[k]	-	$\zeta_P(3p) = 275$	$\zeta_N(2p) = 73$
AsN	(670)[k]	-	-	(190)[k]	-	$\zeta_{As}(4p) = 1500$	

[a]Values in parentheses are values estimated from calculations on lighter molecules.
[b]Hall, *et al.*(1973); [c]Field, *et al.*(1972); [d]Robbe and Schamps (1976); [e]Cossart and Bergeman (1976); [f]Lebreton, *et al.*(1973); [g]Robbe, *et al.*(1979); [h]Bredohl, *et al.*(1974); [i]Robbe, *et al.*(1981); [j]Linton (1980); [k]Gottscho *et al.*(1978); [l]Field, *et al.*(1976).

be important for heavy molecules, leading to an asymmetric splitting of the $^3\Pi$ states.

Another example is illustrated by the $^3\Delta$ states of the same type of molecule. The $^3\Delta$ state has the $\sigma^2\pi^3\pi^*$ configuration, and the spin-orbit constant is given by

$$A(^3\Delta) = \frac{1}{2}\left\langle ^3\Delta_3 \left| \mathbf{H}^{SO} \right| ^3\Delta_3 \right\rangle = \frac{1}{4}\left(a_{\pi^*} - a_\pi\right). \tag{5.3.4}$$

Using the expressions for

$$a_\pi = C_A^2\zeta_A + C_B^2\zeta_B$$
$$a_{\pi^*} = C_A^{*2}\zeta_A + C_B^{*2}\zeta_B \simeq C_B^2\zeta_A + C_A^2\zeta_B,$$

one expects, as previously, that if A is the more electropositive atom, then $C_B^2 > C_A^2$, since the antibonding π^* orbital tends to be more localized on this atom. Thus

$$\left|A(^3\Delta)\right| = -\frac{(C_A^2 - C_B^2)}{4}\left|\zeta_A - \zeta_B\right|. \tag{5.3.5}$$

If the more electropositive atom has the smaller $\zeta_A(np)$ value, $A\left(^3\Delta\right)$ will be negative, and *vice versa*. For homonuclear molecules, Eq. (5.3.5) predicts a very small value for $A(^3\Delta)$; for example, A(W $^3\Delta$, N_2) = 5.82 cm^{-1} (Effantin, *et al.*, 1979). Note that, because

$$C_A^2 - C_B^2 = \left(C_A^2 + C_AC_BS_{AB}\right) - \left(C_B^2 + C_AC_BS_{AB}\right),$$

$C_A^2 - C_B^2$ is simply the difference between the Mulliken populations of the two atoms in either the π or π^* orbital. Using Eq. (5.3.5) and the spin-orbit parameters listed in Table 5.7, for carbon compounds the atomic population of π^* on carbon is semiempirically calculated to be 0.75 for CO and 0.84 for CS, leaving 0.25 or 0.16 for the more electronegative atom. The atomic population difference is thus approximately 0.5, and this justifies the formula for $A(^3\Delta)$ recommended by Field, *et al.*, (1976). For silicon and germanium compounds, the atomic population is 0.87 on the less electronegative atom and 0.13 on the other atom. Thus the orbital localization is slightly larger $\left(C_\mathrm{A}^2 - C_\mathrm{B}^2 = 0.74\right)$. Using Eq. (5.3.5), semiempirical estimates of $\mathrm{A}(^3\Delta)$ for $\pi^3\pi^*$ configurations are given in Table 5.7 and shown to be in good agreement with experiment. Another example where the mixing coefficients $\mathrm{C_A}$ and $\mathrm{C_B}$ can be deduced from experimental values of both diagonal and off-diagonal spin-orbit parameters is given for the $\mathrm{A}^4\Pi$ and $\mathrm{d}^2\Pi$ states (configuration $\delta^2\pi$) of CrN (Zhou, *et al.*, 1997). A similar analysis of the spin-orbit structure $\delta^3\pi$ configuration $^3\Pi$ and $^3\Phi$ states is reported by Langenberg, *et al.*., (1998).

5.3.2 Off-Diagonal Spin-Orbit Parameters

A similar semiempirical method can be used to estimate off-diagonal spin-orbit parameters. Consider the a_+ parameter, which occurs for the example in Eq. (3.4.18):

$$a_+ = \langle \pi^+ | \hat{a}\mathbf{l}^+ | \sigma \rangle .$$

Assuming simple expressions for molecular orbitals,

$$|\pi^+\rangle = C_\mathrm{A}|p\pi_\mathrm{A}^+\rangle + C_\mathrm{B}|p\pi_\mathrm{B}^+\rangle$$

and

$$|\sigma\rangle = C'_\mathrm{A}|p\sigma_\mathrm{A}\rangle + C'_\mathrm{B}|p\sigma_\mathrm{B}\rangle + d_\mathrm{A}|s\sigma_\mathrm{A}\rangle + d_\mathrm{B}|s\sigma_\mathrm{B}\rangle.$$

By definition,

$$\mathrm{l}^+|p\sigma_\mathrm{K}\rangle = 2^{1/2}|p\pi_\mathrm{K}\rangle \qquad \text{and} \quad \mathrm{l}^+|s\sigma_\mathrm{K}\rangle = 0.$$

Neglecting the two-center integrals,

$$\begin{aligned} \langle \pi^+ | \hat{a}\mathbf{l}^+ | \sigma \rangle &= 2^{1/2}\left\{C_\mathrm{A}C'_\mathrm{A}\langle p_\mathrm{A}|\hat{a}|p_\mathrm{A}\rangle + C_\mathrm{B}C'_\mathrm{B}\langle p_\mathrm{B}|\hat{a}|p_\mathrm{B}\rangle\right\} \\ &= 2^{1/2}\left\{C_\mathrm{A}C'_\mathrm{A}\zeta_\mathrm{A}(p) + C_\mathrm{B}C'_\mathrm{B}\zeta_\mathrm{B}(p)\right\} \end{aligned} \tag{5.3.6}$$

Note that $\langle\sigma|\hat{a}\mathrm{l}^-|\pi^+\rangle = \langle\pi^+|\hat{a}\mathrm{l}^+|\sigma\rangle = a_+$, since the spin-orbit operator is Hermitian. In the homonuclear case, if the $s\sigma$ contribution to the σ orbital and the S_AB overlap are neglected, then $C_\mathrm{A} = C'_\mathrm{A} = 2^{-1/2}, C_\mathrm{B} = C'_\mathrm{B} = \pm 2^{-1/2}$ (where the + sign is for u orbitals and the − sign for g orbitals), and

$$\langle \pi_g^+ | \hat{a}\mathbf{l}^+ | \sigma_g \rangle = \langle \pi_u^+ | \hat{a}\mathbf{l}^+ | \sigma_u \rangle = 2^{1/2}\zeta_\mathrm{A}(p). \tag{5.3.7}$$

This result may be extended to the case of a molecular orbital expanded in terms of many atomic orbitals similarly to that for diagonal spin-orbit constants [see Eq. (5.3.3)]. The spin interactions in the CO molecule are an example of such a calculation (Hall, *et al.*, 1973). If the different excited states of CO are represented by the same set of molecular orbitals, many off-diagonal spin-orbit interactions can be expressed in terms of the same molecular spin-orbit parameters. Perturbations known for many years (Schmid and Gerö, 1937) occur between the vibrational levels of the $a^3\Pi$ and $A^1\Pi$ states, which belong to the $\pi^4\sigma\pi^*$ configuration, and those of the $a'^3\Sigma^+$, $e^3\Sigma^-$, $d^3\Delta$, $I^1\Sigma^-$, and $D^1\Delta$ states of the $\pi^3\sigma^2\pi^*$ configuration (Fig. 3.19). These interactions are expressed in terms of a_+ and b parameters in Table 5.8. Figure 5.13 illustrates the evaluation of $a_+ \left(a^3\Pi \sim a'^3\Sigma^+\right)$ both semiempirically, using a minimal SCF basis set, and fully *ab initio* , using configuration-interaction wave functions and different molecular orbital basis sets for the two states. This figure shows surprising agreement between the semiempirical method and the experimental value, when both calculations are performed at the R-centroid of the intersection between the potential curves of the two interacting states. Note the large R-variation of the a_+ parameter, which implies that the value of a_+ will depend on the location of the curve crossings for different pairs of electronic states. This R-dependence of a_+ is related to the R-variation of the atomic orbital composition of the σ and π molecular orbitals (see Section 5.6).

5.3.3 Spin-Spin Parameters

The direct spin-spin parameter has been expressed in Section 3.4.4.1 in terms of the atomic η parameter. Its value is given in Table 5.9 for some atoms. It decreases rapidly with atomic weight, and for second-row atoms and beyond, the second-order spin-orbit contribution to the effective spin-spin constant is dominant. This contribution can be estimated using the semiempirical formulas given in Section 3.4.4. Recall that Eq. (3.4.31c), $2\lambda = E^\circ(^3\Sigma_{\pm 1}) - E^\circ(^3\Sigma_0)$. For the lowest $^3\Sigma^-$ state of the π^2 configuration one obtains [Eq. (3.4.36)], for example,

$$\lambda^{\mathrm{SO}} = \frac{\left[\langle ^3\Sigma_0^- |\mathbf{H}^{\mathrm{SO}}| ^1\Sigma_0^+\rangle\right]^2/2}{E^0(^1\Sigma^+) - E^0(^3\Sigma_0^-)} \simeq \frac{\left[A\left(^3\Pi, \sigma\pi^3\right)\right]^2}{E^0(^1\Sigma^+) - E^0(^3\Sigma_1^-)}. \tag{5.3.8}$$

However, Eq. (5.3.8) is not applicable if $A(^3\Pi, \sigma\pi^3)$ is unknown or if the $^3\Sigma_0^- \sim {}^1\Sigma^+$ spin-orbit interaction is comparable to $E^0(^1\Sigma^+) - E^0(^3\Sigma_0^-)$, the $^1\Sigma^+/^3\Sigma_0^-$ energy separation excluding the second-order spin-orbit contribution. When experimental $A(^3\Pi)$ values are not available, they may be estimated semiempirically as described in Section 5.3.2 using the following formulas. For homonuclear molecules,

$$A\left(\sigma_g\pi_g{}^3\Pi\right) = \frac{1}{2}\langle\pi_g|\mathbf{H}^{\mathrm{SO}}|\pi_g\rangle = C_g^2\zeta_{\mathrm{A}}(p).$$

For heteronuclear AB molecules,

$$A(\sigma\pi^*\ {}^3\Pi) = \frac{1}{2}\langle\pi^*|\mathbf{H}^{\mathrm{SO}}|\pi^*\rangle = \left[C_{\mathrm{A}}^2\zeta_{\mathrm{A}}(p) + C_{\mathrm{B}}^2\zeta_{\mathrm{B}}(p)\right]/2.$$

Table 5.8: Electronic Perturbation Parameters in the Single Configuration Approximation[a,b]

$$\left\langle {}^3\Pi_{1f}^{e}, v \left| \left[\sum_{i=1}^{6} \hat{a}_i \vec{\mathbf{l}}_i \cdot \vec{\mathbf{s}}_i + B\left(\mathbf{L}^+\mathbf{S}^- + \mathbf{L}^-\mathbf{S}^+\right) \right] \right| {}^3\Sigma_{1f}^{+e}, v' \right\rangle = \tfrac{1}{4} a_+ \langle v|v'\rangle - B_{vv'} b$$

$$\left\langle {}^3\Pi_{1f}^{e}, v \left| \left[\sum_{i=1}^{6} \hat{a}_i \vec{\mathbf{l}}_i \cdot \vec{\mathbf{s}}_i + B\left(\mathbf{L}^+\mathbf{S}^- + \mathbf{L}^-\mathbf{S}^+\right) \right] \right| {}^3\Sigma_{1f}^{-e}, v' \right\rangle = -\tfrac{1}{4} a_+ \langle v|v'\rangle + B_{vv'} b$$

$$\left\langle {}^3\Pi_{1f}^{e}, v \left| \left[\sum_{i=1}^{6} \hat{a}_i \vec{\mathbf{l}}_i \cdot \vec{\mathbf{s}}_i + B\left(\mathbf{L}^+\mathbf{S}^- + \mathbf{L}^-\mathbf{S}^+\right) \right] \right| {}^3\Delta_{1f}^{e}, v' \right\rangle = \tfrac{1}{4} 2^{1/2} a_+ \langle v|v'\rangle - 2^{1/2} B_{vv'} b$$

$$\left\langle {}^3\Pi_0, f, v \left| \sum_{i=1}^{6} \hat{a}_i \vec{\mathbf{l}}_i \cdot \vec{\mathbf{s}}_i \right| {}^1\Sigma_0^-, f, v' \right\rangle = \tfrac{1}{4} 2^{1/2} a_+ \langle v|v'\rangle$$

$$\left\langle {}^3\Pi_{2f}^{e}, v \left| \sum_{i=1}^{6} \hat{a}_i \vec{\mathbf{l}}_i \cdot \vec{\mathbf{s}}_i \right| {}^1\Delta_{2f}^{e}, v' \right\rangle = \tfrac{1}{4} 2^{1/2} a_+ \langle v|v'\rangle$$

$$\left\langle {}^1\Pi_{1f}^{e}, v \left| \sum_{i=1}^{6} \hat{a}_i \vec{\mathbf{l}}_i \cdot \vec{\mathbf{s}}_i \right| {}^3\Sigma_{1f}^{+e}, v' \right\rangle = \tfrac{1}{4} a_+ \langle v|v'\rangle$$

$$\left\langle {}^1\Pi_{1f}^{e}, v \left| \sum_{i=1}^{6} \hat{a}_i \vec{\mathbf{l}}_i \cdot \vec{\mathbf{s}}_i \right| {}^3\Sigma_{1f}^{-e}, v' \right\rangle = -\tfrac{1}{4} a_+ \langle v|v'\rangle$$

$$\left\langle {}^1\Pi_{1f}^{e}, v \left| \sum_{i=1}^{6} \hat{a}_i \vec{\mathbf{l}}_i \cdot \vec{\mathbf{s}}_i \right| {}^3\Delta_{1f}^{e}, v' \right\rangle = -\tfrac{1}{4} 2^{1/2} a_+ \langle v|v'\rangle$$

$$\left\langle {}^1\Pi_1, f, v \right| - B(\mathbf{J}^+\mathbf{L}^- + \mathbf{J}^-\mathbf{L}^+) \left| {}^1\Sigma_0^- f, v' \right\rangle = -B_{vv'} b X^{1/2}$$

$$\left\langle {}^1\Pi_{1f}^{e}, v \right| - B(\mathbf{J}^+\mathbf{L}^- + \mathbf{J}^-\mathbf{L}^+) \left| {}^1\Delta_{2f}^{e}, v' \right\rangle = B_{vv'} b (X-2)^{1/2}$$

[a]The $^3\Pi$ and $^1\Pi$ states belong to the $1\pi^4 5\sigma 2\pi$ configuration. The $^3\Sigma^+, {}^3\Sigma^-, {}^3\Delta, {}^1\Sigma^-$, and $^1\Delta$ states belong to $1\pi^3 5\sigma^2 2\pi$. The one-electron orbital integrals, a_+ and b, and the vibrational integral, $B_{vv'}$, are defined as follows:

$$a_+ = \langle 1\pi|\hat{a}\mathbf{l}^+|5\sigma\rangle \quad b = \langle 1\pi|\mathbf{l}^+|5\sigma\rangle \quad B_{vv'} = \left\langle v \left| \frac{\hbar^2}{2\mu R^2} \right| v' \right\rangle \quad X = J(J+1)$$

[b]Not all possible matrix elements for $^3\Delta \sim {}^3\Pi$ and $^3\Sigma^\pm \sim {}^3\Pi$ perturbations are given. The others may be obtained by applying the Wigner-Eckart theorem and using the above results to define the reduced matrix elements.

Table 5.9: Values of the Atomic Spin-Spin Parameter η (cm^{-1})[a]

	C	N	O	F
2p	0.08	(0.14)	0.27	(0.35)
	Si	P	S	Cl
3p	(0.02)	(0.03)	(0.04)	(0.06)
	Ge	As	Se	Br
4p	(0.01)	(0.02)	(0.03)	(0.04)

[a]Values in parentheses are calculated values from Froese-Fischer (1972). Experimental values for C and O are taken from Kayama and Baird (1967).

When the semiempirical A value is comparable to the observed $^1\Sigma^+, {}^3\Sigma_0^-$ energy separation, an exact 2×2 diagonalization is required to obtain a calculated value of λ^{SO} (or to compute a semiempirical value of A from the observed $\lambda(\exp)$ value). The deperturbed energy separation is obtained from

$$\left[E^0({}^1\Sigma^+) - E^0({}^3\Sigma_0^-)\right]_{\mathrm{depert}} = \left[E({}^1\Sigma^+) - E({}^3\Sigma_0^-)\right]_{\mathrm{obs}} - 2\Delta$$

where

$$\Delta = \left[E(^3\Sigma_1^-) - E(^3\Sigma_0^-)\right]_{\text{obs}}.$$

This assumes that the $^3\Sigma_1^-$ level is not shifted from its deperturbed position by spin-orbit interactions with $^3\Sigma^+$ or $^3\Pi$ states from other configurations. The λ^{SO} (calculated) values in Table 5.10, obtained from

$$\lambda^{\text{SO}}(\text{calc}) = \left\{\left(\left\langle {}^3\Sigma_0^- \left|\mathbf{H}^{\text{SO}}\right|^1\Sigma^+\right\rangle\right)^2 + \left[E^0\left(^1\Sigma^+\right) - E^0\left(^3\Sigma_0^-\right)\right]^2_{\text{depert}}/4\right\}^{1/2}/2,$$

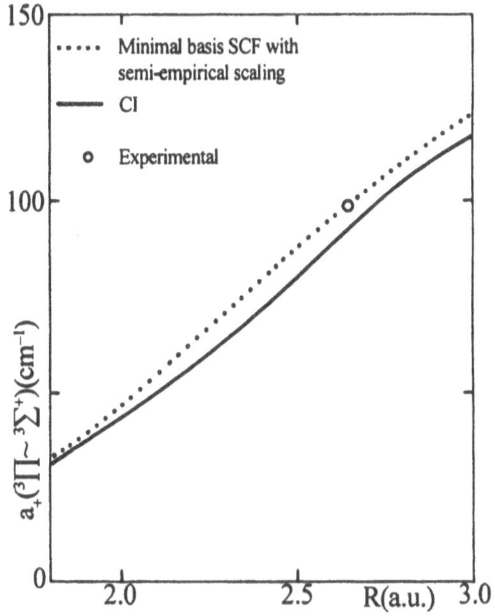

Figure 5.13: CO $a^3\Pi_1 \sim a'^3\Sigma_1^+$ spin-orbit parameters calculated at various levels of approximation. [From Schamps (1973).]

5.4 Magnitudes of Rotational Perturbation Parameters

The operator $\mathbf{L}^+ = \sum_i \mathbf{l}_i^+$, where $\mathbf{l}$ is the electronic angular momentum, $\mathbf{l}_\text{G}$, is defined with respect to the center of mass G of the molecule,

$$\vec{\mathbf{l}}_\text{G} = \vec{\mathbf{r}}_\text{G} \times \vec{\mathbf{p}},$$

where $\mathbf{p}$ is the momentum of the electron. Because, in general, atomic orbitals are centered at the nucleus of the respective atom, one can write

$$\vec{\mathbf{l}}_G = \vec{\mathbf{l}}_A - \frac{m_B}{m_A + m_B}\left(\vec{\mathbf{R}} \times \vec{\mathbf{p}}\right) = \vec{\mathbf{l}}_B + \frac{m_A}{m_A + m_B}\left(\vec{\mathbf{R}} \times \vec{\mathbf{p}}\right) \tag{5.4.1}$$

where m_A and m_B are the atomic masses and $\vec{\mathbf{R}}$ is the internuclear distance vector pointing from A to B.

Table 5.10: Effective Spin-Spin Constants for $^3\Sigma^-$ States of π^2 Configurations (cm^{-1})

	$\langle ^1\Sigma^+ \vert \mathbf{H}^{SO} \vert ^3\Sigma_0^- \rangle^a$	$[E_{^1\Sigma^+} - E_{^3\Sigma_0^-}]_{obs}$	$[E^0_{^1\Sigma^+} - E^0_{^3\Sigma_0^-}]_{depert}$	$\lambda^{SO}(\text{calc})^b$	$\lambda(\text{exp})$
O_2	94[c] (95)	13,120.9[c]	13113	1.35 (1.38)	1.98[c]
SO	159[c] (172)	10,469.3[c]	10448	4.84 (5.66)	5.26[c]
S_2	231[c] (240)	7961±15[d]	7914	13.49 (14.6)	11.82[c]
SeO	(630)	9685.4[c]	9354	(84.1)[k]	82.95[c]
SeS	((650))	7794±15[e]	7400	((112))[k]	98.4[f]
Se_2	(1060)	7957.2[g]	6933	(310)[k]	256.0[g]
TeO	(1223)	9966±10[h]	8390	(343)[k]	394 ± 5[h]
TeS	((1181))	8457±10[h]	6801	((388))[k]	414±3[h]
TeSe	((1591))	8785±5[i]	6313	((720))[k]	618±2[i]
Te_2	(2122)	9600.0[j]	5650	(1296)[k]	987.4[c]

[a]Values of $\left\langle ^1\Sigma^+ \vert \mathbf{H}^{SO} \vert ^3\Sigma_0^- \right\rangle$ without parentheses are derived from the spin-orbit constant of the $\pi_u^4\pi_g\sigma_u$ $^3\Pi_u$ state. Values in single parentheses are calculated using atomic orbital mixing coefficients $C_g = 0.792$ for the π_g^* orbital of homonuclear molecules or $C_O = 0.71$ and $C_X = 0.837$ for the π' orbital of monoxide molecules, XO. Values in double parentheses are calculated using $C = 0.792$ for both nonoxygen atoms.

[b]Direct spin-spin contributions have been calculated to be 0.75 cm^{-1} (O_2), 0.28 cm^{-1} (SO), and 0.13 cm^{-1} (S_2) (Wayne and Colbourn, 1977).

[c]Huber and Herzberg (1979). A value of 100 cm^{-1} has been deduced from the spin-orbit perturbation between $X^3\Sigma_g^-(v = 28)$ and $b^1\Sigma_g^+(v = 19)$ of O_2 (Jongma, *et al.*, 1999).

[d]Barnes, *et al.*(1979). [e]Winter, *et al.*(1980). [f]MacDonald, *et al.*(1982). [g]Prosser, *et al.*(1982).

[h]Winter, *et al.*(1982). [i]Fink, private communication. [j]Vergès, *et al.*(1982).

[k]Calculated from level shifts obtained by diagonalizing the 2 × 2 $^1\Sigma \sim ^3\Sigma_0^-$ matrix.

compare favorably with experimental values,

$$\lambda(\text{exp}) = \left[E\left(^3\Sigma_1^-\right) - E\left(^3\Sigma_0^-\right)\right]_{obs}/2.$$

Table 5.11: Some Spin-Orbit and Orbital Perturbation Matrix elements

	CO^a	CS^b	SiO^c	SiS^d	CS^{+e}
	$^1\Sigma^- \sim {}^1\Pi$	$^3\Sigma^- \sim {}^3\Pi$	$^1\Sigma^- \sim {}^1\Pi$	$^1\Sigma^- \sim {}^1\Pi$	$^2\Pi \sim {}^2\Sigma^+$
b (unitless)	0.2	0.14	0.71	0.357	-0.08
	$^3\Sigma^- \sim {}^1\Pi$	$^3\Sigma^- \sim {}^3\Pi$	$^3\Sigma^- \sim {}^1\Pi$	$^3\Sigma^- \sim {}^1\Pi$	$^2\Pi \sim {}^2\Sigma^+$
a_+ (cm^{-1})	100	312	100	362	290

[a]Field, *et al.*(1972).
[b]Cossart, *et al.*(1977).
[c]Field, *et al.*(1976).
[d]Harris, *et al.*(1982).
[e]Gauyacq and Horani (1978).

This expression shows clearly that $\vec{l}_G$ depends on internuclear distance and that one-electron two-center integrals *cannot be neglected* in calculating matrix elements of l_G^+ (Colbourn and Wayne, 1979). This situation is quite different from that for the off-diagonal spin-orbit parameters. The b parameters are given (Section 5.3.2) by:

$$b = \langle \pi^+ | l_G^+ | \sigma \rangle = C_A C_{A'} \langle p\pi_A | l_G^+ | p\sigma_A \rangle + C_B C_{B'} \langle p\pi_B | l_G^+ | p\sigma_B \rangle + C_A C_{B'} \langle p\pi_A | l_G^+ | p\sigma_B \rangle + C_{A'} C_B \langle p\pi_B | l_G^+ | p\sigma_A \rangle \tag{5.4.2}$$

where l_G^+ must be replaced by its form given in Eq. (5.4.1). Consequently, a semiempirical evaluation of b is never valid. As distinct from $\hat{a}l^+, l_G^+$ does not act only in the neighborhood of each nucleus. Only if the σ and π molecular orbitals are well-represented by *one p* atomic orbital centered at the center of mass, (i.e., if there is no mixing with orbitals with $l_G > 1$) it is possible to say that $|b| \leq 2^{1/2}$. Since, for AH hydride molecules, the center of mass is located very near the heavy atom nucleus, $\vec{l}_A \simeq \vec{l}_G$ and semiempirical relationships between b and l_A are approximately valid. This, in part, explains the success of the *pure precession* approximation (Section 5.5) for AH molecules.

It has been pointed out (Robbe and Schamps, 1976) that there is no general reason that the $\hat{a}l^+$ and l^+ matrix elements should be simply proportional to each other. For $\hat{a}l^+$, the contribution from the atom with the larger atomic spin-orbit parameter is dominant; for l^+, the p orbitals of both atoms play equal roles. Table 5.11 compares some values for matrix elements of these two operators. For a_+, the values of CO and SiO are similar, as are the values for CS and SiS. For b, the strong R–dependenceof this parameter prohibits any simple predictions. Even the sign changes when one passes from CS to CS^+.

5.5 Pure Precession Approximation

The hypothesis of pure precession (Van Vleck, 1929) is often used in the estimation of Λ–doubling constants. These constants, in the o, p, q notation suggested by Mulliken and Christy (1931), are introduced into the effective Hamiltonian matrix for Π–states in order to represent the J– and Ω–dependence of the splitting between the e– and f–symmetry Λ–components caused by the interaction between the Π–state and energetically remote Σ–states (Zare, *et al.*, 1973; Horani, *et al.*, 1967; Brown and Merer, 1979).

This discussion is intended to distinguish the levels of approximation often hidden behind the name "pure precession." For clarity, only $^2\Sigma \sim {}^2\Pi$ interactions will be discussed. The considerably more complicated $^3\Sigma \sim {}^3\Pi$ case is treated by Brown and Merer (1979).

The o, p, and q constants may be defined in a form that is semiempirically evaluable by second-order perturbation theory (see the discussion of the Van Vleck transformation in Section 4.2). Using the notation suggested by Rostas, *et al.*, (1974) and Merer, *et al.*, (1975),

$$o_v^{\Pi}\left({}^2\Sigma^s\right)^{\dagger} = \frac{1}{4}\sum_{{}^2\Sigma, v'} \frac{(-1)^s \left|\left\langle {}^2\Pi, v \left| \sum_i \hat{a}_i \mathbf{l}_i^+ \mathbf{s}_i^- \right| {}^2\Sigma^s, v' \right\rangle\right|^2}{E_{\Pi,v} - E_{\Sigma,v'}} \tag{5.5.1a}$$

$$p_v^{\Pi}\left({}^2\Sigma^s\right) = 2\sum_{{}^2\Sigma, v'} (-1)^s \left\langle {}^2\Pi, v \left| \sum_i a_i \mathbf{l}_i^+ \mathbf{s}_i^- \right| {}^2\Sigma^s, v' \right\rangle \times \frac{\left\langle {}^2\Sigma^s, v' \left| \frac{\hbar^2}{2\mu R^2} \sum_i \mathbf{l}_i^- \right| {}^2\Pi, v \right\rangle}{E_{\Pi,v} - E_{\Sigma,v'}} \tag{5.5.2a}$$

$$q_v^{\Pi}\left({}^2\Sigma^s\right) = 2\sum_{{}^2\Sigma, v'} \frac{(-1)^s \left|\left\langle {}^2\Pi, v \left| \left(\hbar^2/2\mu R^2\right) \sum_i \mathbf{l}_i^+ \right| {}^2\Sigma^s, v' \right\rangle\right|^2}{E_{\Pi,v} - E_{\Sigma,v'}} \tag{5.5.3a}$$

where s is zero for $^2\Sigma^+$ states and 1 for $^2\Sigma^-$ states and the operators are understood to act on unsymmetrized, signed-Ω basis functions. Since remote levels of both $^2\Sigma^+$ and $^2\Sigma^-$ states contribute to the Λ–doubling of a given $^2\Pi$ level, the o, p, and q parameters may be partitioned into parts arising from $^2\Sigma^+$ and $^2\Sigma^-$ states, as follows:

$$o_v^{\Pi} = o_v^{\Pi}\left({}^2\Sigma^+\right) + o_v^{\Pi}\left({}^2\Sigma^-\right) \tag{5.5.1b}$$

$$p_v^{\Pi} = p_v^{\Pi}\left({}^2\Sigma^+\right) + p_v^{\Pi}\left({}^2\Sigma^-\right) \tag{5.5.2b}$$

$$q_v^{\Pi} = q_v^{\Pi}\left({}^2\Sigma^+\right) + q_v^{\Pi}\left({}^2\Sigma^-\right). \tag{5.5.3b}$$

In the *unique perturber* approximation (Zare, *et al.*, 1973), only one $^2\Sigma$ electronic state is considered to be responsible for the Λ–doubling of the $^2\Pi$ state. If the interacting $^2\Pi$ and $^2\Sigma$ states belong to the π^1 and σ^1 configurations, then the matrix elements of the exact many-electron wavefunctions may be replaced by the one-electron matrix elements (see Section 3.5.4),

$$a_+ = \left\langle \pi^+ \left| \hat{a}\mathbf{l}^+ \right| \sigma \right\rangle$$

$$b = \left\langle \pi^+ \left| \mathbf{l}^+ \right| \sigma \right\rangle$$

† $o_v^{\Pi}\left({}^2\Sigma^s\right)$ is included even though it makes no contribution to the Λ–doubling in $^2\Pi$ states.

and

$$\left\langle {}^2\Pi, v \left| \sum_i \hat{a}l_i^+ s_i^- \right| {}^2\Sigma^+, v' \right\rangle = a_+ \left(\overline{R}_{vv'}\right) \langle v|v'\rangle \tag{5.5.4}$$

$$\left\langle {}^2\Pi, v \left| \left(\hbar^2/2\mu R^2\right) \sum_i l_i^+ \right| {}^2\Sigma^+, v' \right\rangle = b\left(\overline{R}_{vv'}\right) B_{vv'}, \tag{5.5.5}$$

where a_+ and b are dependent on the R–centroid which is v'–dependent, unlike the situation for interactions between all near-degenerate pairs of vibrational levels belonging to two potentials that cross at R_C. Inserting Eqs. (5.5.4) and (5.5.5) into Eqs. (5.5.1a), (5.5.2a), and (5.5.3a),

$$o_v^\Pi = \frac{1}{4}\sum_{v'} \frac{\left[a_+\left(\overline{R}_{vv'}\right)\langle v|v'\rangle\right]^2}{E_{\Pi,v} - E_{\Sigma,v'}} \tag{5.5.1c}$$

$$p_v^\Pi = 2\sum_{v'} \frac{a_+\left(\overline{R}_{vv'}\right) b\left(\overline{R}_{vv'}\right)\langle v|v'\rangle B_{vv'}}{E_{\Pi,v} - E_{\Sigma,v'}} \tag{5.5.2c}$$

$$q_v^\Pi = 2\sum_{v'} \frac{\left[b\left(\overline{R}_{vv'}\right) B_{vv'}\right]^2}{E_{\Pi,v} - E_{\Sigma,v'}}. \tag{5.5.3c}$$

If one makes the generally very bad approximation that a_+ and b are independent of $\overline{R}_{vv'}$ (thus defining $\overline{a}_+, \overline{b}$), then the perturbation summations

$$o_v^\Pi = \frac{1}{4}\overline{a}_+^2 \sum_{v'} \frac{[\langle v|v'\rangle]^2}{E_{\Pi,v} - E_{\Sigma,v'}} \tag{5.5.1d}$$

$$p_v^\Pi = 2\overline{a}_+\overline{b}\sum_{v'} \frac{\langle v|v'\rangle B_{vv'}}{E_{\Pi,v} - E_{\Sigma,v'}} \tag{5.5.2d}$$

$$q_v^\Pi = 2\overline{b}^2\sum_{v'} \frac{\left(B_{vv'}\right)^2}{E_{\Pi,v} - E_{\Sigma,v'}} \tag{5.5.3d}$$

may be evaluated numerically. This approach should account for the v–dependence (over a small range of v, especially at low v, see Wicke, *et al.*, 1972) but not the magnitudes of o_v^Π, p_v^Π, and q_v^Π.

In the special case when the unique interacting pair of ${}^2\Pi \sim {}^2\Sigma^+$ states have identical potential energy curves, the v' summations reduce to a single $v = v'$ term and

$$o_v^\Pi = \frac{1}{4}\frac{\left[a_+\left(\overline{R}_{vv}\right)\right]^2}{\Delta E_{\Pi\Sigma}} \tag{5.5.1e}$$

$$p_v^\Pi = 2\frac{a_+\left(\overline{R}_{vv}\right) b\left(\overline{R}_{vv}\right) B_v}{\Delta E_{\Pi\Sigma}} \tag{5.5.2e}$$

$$q_v^\Pi = 2B_v^2\frac{\left[b\left(\overline{R}_{vv}\right)\right]^2}{\Delta E_{\Pi\Sigma}} \tag{5.5.3e}$$

where

$$\Delta E_{\Pi\Sigma} = E_{\Pi,v} - E_{\Sigma,v}.$$

At this level of approximation, a convenient estimate of o_v^Π is obtained from

$$o \simeq p^2/4q.$$

Even in this special case, the $\overline{R}$–dependence of a_+ and b cannot be ignored.

An important and frequently encountered result, often mistakenly taken as evidence for "pure precession," is that, in the *unique perturber, identical potential curve* limit, the effective spin-rotation constant of the $^2\Sigma^+$ state, γ_v, is equal to p_v^Π. This result is a direct consequence of the second-order perturbation theoretical definition of the contribution of a $^2\Pi$ state to the spin-rotation splitting in a $^2\Sigma$ state (see Section 3.5.4):

$$\begin{aligned}\gamma_v^{\text{eff}} &= -p_v^{\Sigma^+}(^2\Pi) \\ &= 2\sum_{^2\Pi,v'} \frac{\langle ^2\Sigma^+, v\,|\Sigma_i \hat{a}_i \mathbf{l}_i^- \mathbf{s}_i^+|\,^2\Pi, v'\rangle \langle ^2\Pi, v'\,|(\hbar^2/2\mu R^2)\,\Sigma_i \mathbf{l}_i^+|\,^2\Sigma^+, v\rangle}{E_{\Pi,v} - E_{\Sigma,v'}}\end{aligned} \tag{5.5.6}$$

which, in the unique perturber, identical potential limit reduces to

$$\gamma_v^{\text{eff}} = -p_v^{\Sigma^+}(^2\Pi) = +p_v^\Pi(^2\Sigma). \tag{5.5.7}$$

Finally, the *pure precession approximation* requires, in addition to the unique perturber, identical potential assumptions, that the interacting $^2\Pi$ and $^2\Sigma$ states are each well described by a single configuration, that these configurations are identical except for a single spin-orbital, and that this spin-orbital is a pure atomic $|nl\lambda\rangle$ orbital so that

$$\mathbf{l}^\pm|nl\lambda\rangle = [l(l+1) - \lambda(\lambda \pm 1)]^{1/2}|nl\lambda \pm 1\rangle. \tag{5.5.8}$$

Then, for $l = 1$ (a p-complex),

$$a_+ = \langle \pi^+\,|\hat{a}\mathbf{l}^+|\,\sigma\rangle = \zeta(np)2^{1/2} = A_v^\Pi 2^{1/2} \tag{5.5.9}$$

$$b = \langle \pi^+\,|\mathbf{l}^+|\,\sigma\rangle = 2^{1/2} \tag{5.5.10}$$

and, for the case of "*simple* pure precession" where all orbitals except the unique π/σ pair are completely full,

$$o_v^\Pi(^2\Sigma^+) = \frac{1}{2}\,A_v^2/\,\Delta E_{\Pi\Sigma} \tag{5.5.1f}$$

$$p_v^\Pi(^2\Sigma^+) = 4A_vB_v/\,\Delta E_{\Pi\Sigma} \tag{5.5.2f}$$

$$q_v^\Pi(^2\Sigma^+) = 4B_v^2/\,\Delta E_{\Pi\Sigma} \tag{5.5.3f}$$

$$p_v^\Pi/\,q_v^\Pi = A_v/\,B_v. \tag{5.5.11}$$

Similar equations for o, p, and q may be derived for "*generalized* pure precession" situations involving two or more open shells. The validity of these pure precession equations depends on many assumptions, yet experimentalists are very quick to use them to "explain" the sign or even the magnitude of observed Λ–doubling and spin-rotation constants or to use these constants to "infer" l_{eff} values for valence-shell σ and π orbitals.

For the OH radical, the values of p and q for the $X^2\Pi\left(\sigma^2\pi^3\right)$ ground state can be attributed to a unique perturber interaction with the $A^2\Sigma^+\left(\sigma\pi^4\right)$ state. The pure precession approximation simply ignores the contribution of the $\sigma 1s_{\text{H}}$ atomic orbital to the pσ molecular orbital. For all hydrides, the $\sigma 1s_{\text{H}}$ orbital makes a negligible contribution to $\mathbf{H}^{\text{SO}}$ and $\mathbf{BL}^+$ matrix elements. For OH, the H–atom contributions to a_+ and b are $4 \times 10^{-4}\%$ and 1%, respectively (Hinkley, *et al.*, 1972).

For the CH radical, the simple Eqs. (5.5.4) and (5.5.5) are not valid for the interaction of $X^2\Pi\left(\sigma^2\pi\right)$ with the $B^2\Sigma^-\left(\sigma\pi^2\right)$ and $C^2\Sigma^+\left(\sigma\pi^2\right)$ states. Although the $B^2\Sigma^-$ and $C^2\Sigma^+$ states are well represented by the single $\sigma\pi^2$ configuration, both states must be written as sums of Slater determinants,

$$\left|{}^2\Sigma^-_{1/2}\right\rangle = 6^{-1/2}[2|\pi^+\alpha\pi^-\alpha\sigma\beta| - |\pi^+\alpha\pi^-\beta\sigma\alpha| - |\pi^+\beta\pi^-\alpha\sigma\alpha|] \tag{5.5.12}$$

$$\left|{}^2\Sigma^+_{1/2}\right\rangle = 2^{-1/2}[|\pi^+\alpha\pi^-\beta\sigma\alpha| - |\pi^+\beta\pi^-\alpha\sigma\alpha|], \tag{5.5.13}$$

and the matrix elements become (Hinkley, *et al.*, 1972)

$$\left\langle {}^2\Pi_{1/2}, v \left| \sum_i \hat{a}_i \mathbf{l}_i^+ \mathbf{s}_i^- \right| {}^2\Sigma^-, v' \right\rangle = +(6)^{-1/2} a_+ \langle v|v'\rangle \tag{5.5.14}$$

$$\left\langle {}^2\Pi_{1/2}, v \left| \sum_i \hat{a}_i \mathbf{l}_i^+ \mathbf{s}_i^- \right| {}^2\Sigma^+, v' \right\rangle = +(2)^{-1/2} a_+ \langle v|v'\rangle \tag{5.5.15}$$

$$\left\langle {}^2\Pi_{3/2}, v \left| \mathbf{B} \sum_i \mathbf{l}_i^+ \right| {}^2\Sigma^-_{1/2}, v' \right\rangle = +\left(\frac{3}{2}\right)^{1/2} bB_{vv'} \tag{5.5.16}$$

$$\left\langle {}^2\Pi_{3/2}, v \left| \mathbf{B} \sum_i \mathbf{l}_i^+ \right| {}^2\Sigma^+_{1/2}, v' \right\rangle = -(2)^{-1/2} bB_{vv'}. \tag{5.5.17}$$

This generalization to cases where the interacting Π and Σ states are derived from more complex configurations than π^1 and σ^1 is the basis for *generalized pure precession*. The p and q parameters for CH $X^2\Pi$ are

$$p_v^{\Pi}\left({}^2\Sigma^-\right) = -B_v b a_+ / \Delta E_{\Pi\Sigma} \tag{5.5.18a}$$

$$p_v^{\Pi}\left({}^2\Sigma^+\right) = -B_v b a_+ / \Delta E_{\Pi\Sigma} \tag{5.5.18b}$$

$$q_v^{\Pi}\left({}^2\Sigma^-\right) = -3B_v^2 b^2 / \Delta E_{\Pi\Sigma} \tag{5.5.18c}$$

$$q_v^{\Pi}\left({}^2\Sigma^+\right) = B_v^2 b^2 / \Delta E_{\Pi\Sigma}. \tag{5.5.18d}$$

Predictions based on generalized pure precession depend on the same assumptions as ordinary pure precession. The idea of generalized pure precession is especially useful for treating interactions between Rydberg states built upon non-$^1\Sigma^+$ ion cores.

For SiH, the repulsive $^2\Sigma^-$ state makes the dominant contribution to the Λ–doubling of the $X^2\Pi$ $(\sigma^2\pi)$ ground state (Cooper and Richards, 1981). In order to use the Eqs. (5.5.1a) - (5.5.3a) definitions of p^Π and q^Π and the Eqs. (5.5.14) - (5.5.17) matrix elements, it is necessary to make assumptions about the R–dependence of a_+ and b and to replace the summation over vibrational levels of $^2\Sigma^-$ by an integration over the vibrational continuum [see Eq. (7.5.9)].

In the case of highly polar, heteronuclear, nonhydride diatomic molecules, if the interacting states belong to configurations differing by a single orbital that is highly localized on the same atom, then the p^Π, q^Π, and γ^Σ constants are likely to have values close to the pure precession predictions. However, this must not be taken as evidence that the interacting states are components of an l–complex. For example, the $A^2\Pi$ (π^1) and $B^2\Sigma^+$ (σ^1) states of the alkaline earth monohalides certainly belong to the unique perturber, identical potential limit and satisfy the requirement that

$$p^\Pi = \gamma^\Sigma.$$

Furthermore, using

$$p_v^\Pi = 2A_v B_v l_{\text{eff}}\left(l_{\text{eff}} + 1\right)/\Delta E_{\Pi\Sigma} \tag{5.5.19}$$

to define l_{eff}, one finds $l_{\text{eff}} = 1.05, 1.12, 1.23$, and 1.38 for CaF, CaCl, CaBr, and CaI, respectively (Bernath, 1980). This suggests that the $A^2\Pi$ and $B^2\Sigma^+$ states of the CaX molecules are of predominantly Ca^+ $4p$ character, but with a small admixture of Ca^+ $3d$. This use of Eq. (5.5.19) to define l_{eff} is unsound because it assumes that the amount of $p \sim d$ mixing is the same for both σ and π orbitals. In fact, the $p \sim d$ mixing is much more extensive in σ than π orbitals. This $p \sim d$ mixing is the mechanism by which the metal-centered orbital polarizes to avoid the negatively charged X^- ion and is much more important for $p\sigma$ than $p\pi$. It is dangerous to use the satisfaction of a pure precession prediction as evidence for the validity of the numerous pure precession assumptions, especially when there is good reason to suspect that the assumptions are invalid (HCl^+; Brown and Watson, 1977). Likewise, it is risky to use the magnitudes or signs of second-order parameters, such as p, q, or γ, to infer configurational parentage, molecular orbital composition, or whether a multiplet $\Lambda > 0$ state is regular or inverted (BeF; Cooper, *et al.*, 1981).

The pure precession approximation is *not* guaranteed to be valid for homonuclear diatomic molecules, although it frequently yields predictions in excellent agreement with observations (see, for example the L–uncoupling interaction between $^1\Delta_g$ and $^1\Pi_g$ states of O_2 (Lewis, *et al.*, 1999)). Equation (5.3.7) explains the success of the pure precession approximation as far as spin-orbit matrix elements are concerned. However, a simple atom-centered evaluation of the $l^\pm$ orbital angular momentum matrix elements is unjustified, especially when the molecular center of mass is far from both nuclei (See Section 5.4 and Colbourn and Wayne, 1979 and Robbe and Schamps, 1976).

For valence states of nonhydride molecules, there is no reason to expect that the generalized pure precession approximation should be valid. In contrast, Rydberg orbitals, because of their large size and nonbonding, single-center, near-spherical, atomic-like character, are almost invariably well-described by the pure precession picture in terms of nl–complexes.

5.6 R–Dependence of the Spin Interaction Parameters

Spin-orbit, spin-spin, and spin-rotation constants are found experimentally to depend on v and J. This reflects an implicit R–dependence of the electronic matrix elements. Such effects may be calculated *ab initio* and then compared against the experimentally determined R–variation inferred from the v, J dependence of spin constants. The most sophisticated $f(v, J) \leftrightarrow F(R)$ inversion methods are those of Watson (1979), Coxon (1975), and Bessis, *et al.* (1984).

There is no reason to expect that diagonal and off-diagonal spin constants will vary with R in the same way, except in some special cases. Consider first the R–dependence of the diagonal spin-orbit constants. This variation originates from two sources:

1. variation of a specific open-shell $\lambda \neq 0$ molecular orbital with R, and

2. variation of the dominant configuration with R.

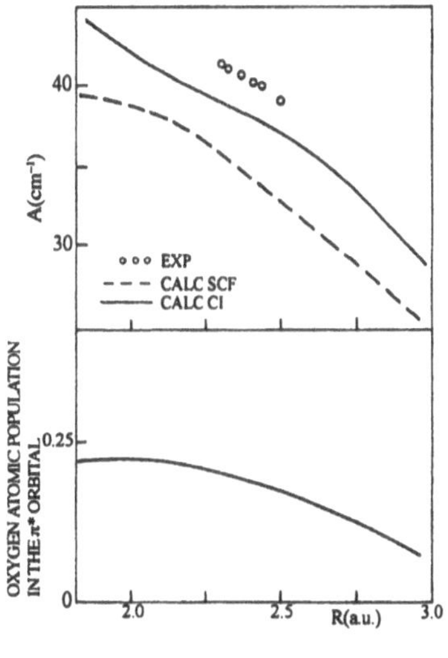

Figure 5.14: R–variation of the spin-orbit constant for the CO $a^3\Pi$ state. Experimental values from Field, *et al.*(1972), calculated values from Hall, *et al.*(1973).

In Fig. 5.14, the calculated R–dependence of the spin-orbit constant of the $a^3\Pi$ state of CO, $A(a^3\Pi)$ is shown. This constant, in the single-configuration approximation, is related to the one-electron matrix element, $\langle \pi^* |\hat{a}l_z| \pi^* \rangle = 2A(a^3\Pi)$. The main contribution to the spin-orbit constant comes from the oxygen atom, even though $C_C^2 > C_O^2$ [compare $\zeta_C(2p) = 30$ cm^{-1} to $\zeta_O(2p) = 150$ cm^{-1}] and the value of A varies with R roughly as the fractional oxygen $2p$ atomic population in the π^* orbital. At large internuclear distances, the π^* orbital becomes a carbon $2p\pi$ atomic orbital. For CO $a^3\Pi$ there is no significant difference between the values of $A(a^3\Pi)$ calculated using single-configuration (SCF) or many-configuration configuration interaction (CI) wavefunctions.

In Fig. 5.15, the calculated variation of the spin-orbit constant of the $X^2\Pi$ state of PO is shown to be in good agreement with experiment. In this case, the dominant configuration at equilibrium internuclear distance is π^* where the π^* molecular orbital has primarily P($3p$) character. This character does not change appreciably with R. In contrast, the percentage of the π^* configuration in the CI wavefunction diminishes very rapidly with R to allow PO $X^2\Pi$ to dissociate into the correct separated atom states, and its R–dependence is mirrored by that of the $A(X^2\Pi)$ constant.

Finally, in a third example, both types of A versus R variation occur, the first at small internuclear separation, the second at large separation. Figure 5.16 illustrates the variation of A for the OH $X^2\Pi$ state (Coxon and Foster, 1982). Near R_e, the dominant configuration is $\sigma^2\pi^3$. The magnitude of the spin-orbit constant associated with the π orbital, $A(X^2\Pi) = -a_\pi$, increases slightly with R from $a_\pi = \zeta_{O^-}(2p) = 120$ cm^{-1}, the value of the constant for the oxygen negative ion, to $a_\pi = \zeta_O(2p) = 150$ cm^{-1}, that of the neutral oxygen atom. Near the $O(^3P)$+$H(^2S)$ dissociation limit, the spin-orbit splitting of OH $X^2\Pi$ should correspond to the $^3P_1 - {}^3P_2$ splitting of the oxygen atom ^3P state. Consequently, $|A(X^2\Pi)|$ decreases at large R, going to the asymptotic value $\frac{2}{3}\zeta_O = 100$ cm^{-1} (Langhoff, *et al.*, 1982).

Except in the previous example, only the variation of A in the neighborhood of R_e has been considered. The correlation of molecular spin-orbit splittings to those of atoms at infinite internuclear distance is very difficult to establish. It is necessary to take into account not only the configuration mixing due to electronic correlation but also mixing due to off-diagonal matrix elements of the spin-orbit operator between near-degenerate states of identical Ω. This corresponds to working in a (J_1, J_2) coupling scheme (Herzberg, 1950, p. 319) where Ω is the sum $|M_{J_1} + M_{J_2}|$ for the two atoms (see Fig. 5.17 and Section 7.2.1).

Assuming that an electronic property varies linearly with both R (near R_e) and v (near $v = 0$), approximate formulas relating the R– and v–dependence can be deduced. For example, if one assumes

$$A(v) = A_e - \alpha_A \left(v + \frac{1}{2}\right) = \langle v|A(R)|v\rangle\,, \tag{5.6.1}$$

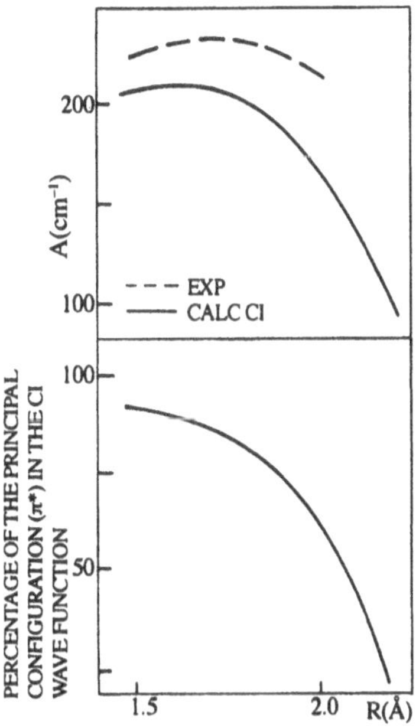

Figure 5.15: R-variation of the spin-orbit constant for the PO $X^2\Pi$ state. Experimental values form Verma and Singhal (1975), calculated values from Roche and Lefebvre-Brion (1973).

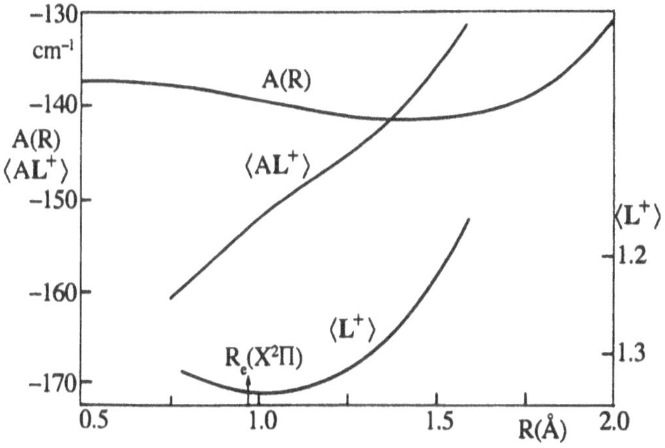

Figure 5.16: Diagonal $[A(R)]$ and off-diagonal spin-orbit ($\langle AL^+ \rangle$) and orbital ($\langle L^+ \rangle$) perturbation parameters for the OH $X^2\Pi$ state (from Coxon and Foster, 1982) and the OD $X^2\Pi \sim A^2\Sigma^+$ interaction. (From Coxon and Hammersley, 1975.)

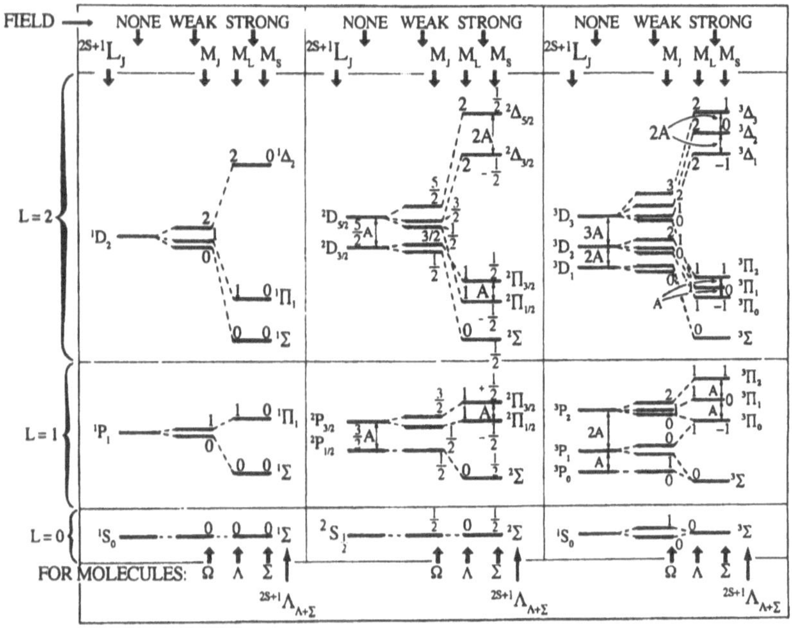

Figure 5.17: Correlation diagram showing the effect of an electric field on atomic $L-S$ terms (from Jevons, 1932). This electric field could be externally applied, splitting the atomic terms into J, M_J (weak field) or M_L, M_S (strong field) components, or it could be the cylindrically symmetric molecular field associated with a nearby atom, causing the atomic terms to be split into case (c) J_a, Ω (weak field) or case (a) Λ, S (strong field) terms. Two effects are illustrated: the correlation of atomic spin-orbit splittings [shown in terms of $A \equiv \zeta(L, S)$] with molecular spin-orbit intervals (ΛA), and the splitting of atomic $L-S-J$ terms into molecular $\Lambda-S-\Omega$ states. For example, Ar(^{1}S) + Na (^{2}P) and Kr (^{1}S) + Au (^{2}P) correlate with $^2\Pi$ and $^2\Sigma^+$ states of NaAr and AuKr (Walliman, *et al.*, to be published). This sort of diagram provides a framework for understanding the ligand field splitting of an M^+ (d^N or f^N) $L-S$ core into MX molecular states by X^- (^{1}S) (see Section 3.2.1). The Σ^+/Σ^- symmetry of molecular states depends on whether $L + \Sigma_{i=1}^N l_i$ is even or odd (Herzberg, 1950, p. 318). The correlations between the ^{3}D $|J, M_J\rangle$ and $^3\Pi$ $|\Omega\rangle$ components are as follows: $|3,0\rangle \leftrightarrow |0^{\mp}\rangle, |2,2\rangle \leftrightarrow |2\rangle, |2,1\rangle \leftrightarrow |1\rangle, |2,0\rangle \leftrightarrow |0^{\pm}\rangle$ (the +/− symmetry of $\Omega = 0$ states is determined by whether $J + \Sigma_i l_i$ is even/odd).

where

$$A(R) = A_e + \left(\frac{dA}{dR}\right)_{R=R_e} (R - R_e), \tag{5.6.2}$$

the $\Delta v = 0$ matrix element of $(R - R_e)^m$ between vibrational functions belonging to the same electronic state can be expressed in terms of Dunham coefficients from the expansion of $V(R)$ in powers through the cubic term of $(R - R_e)/R_e$

(for example, see Table I of James, 1964):

$$\langle v\,|(R-R_e)|\,v\rangle = -\frac{3a_1R_eB_e}{\omega_e}\left(v+\frac{1}{2}\right) \tag{5.6.3}$$

and

$$a_1 = -\frac{\alpha_e\omega_e}{6B_e^2} - 1. \tag{5.6.4}$$

Taking the matrix element of Eq. (5.6.2) and recognizing α_A as the coefficient of $-\left(v+\frac{1}{2}\right)$,

$$\alpha_A = -\frac{R_e}{2B_e}\left(\frac{dA}{dR}\right)_{R=R_e}\left(\alpha_e + \frac{6B_e^2}{\omega_e}\right). \tag{5.6.5}$$

If the linear approximation of $A(v)$ is inadequate, more complete formulas have been given by De Santis, *et al.*, (1973) and by Zaidi and Verma (1975).

The R–dependence of A also accounts for the slight J–dependence of $A(v)$ similar to the centrifugal distortion of $B(v)$,

$$A_D \equiv 2A_J = 4\left(\frac{dA}{dR}\right)_{R=R_e}\frac{B_e^2}{\omega_e^2}R_e. \tag{5.6.6}$$

Alternatively (Veseth, 1970; Merer, 1972),

$$A_D = -2D_e\alpha_A\left(\alpha_e + 6B_e^2/\omega_e\right)^{-1}. \tag{5.6.7}$$

The recipe (Zare, *et al.*, 1973) for adding the A_D term to the effective Hamiltonian matrix [expressed in the case (a) basis] is to replace the constant B_v where it appears along the diagonal ($\langle J\Omega\Lambda\Sigma|\mathbf{H}|J\Omega\Lambda\Sigma\rangle$) by

$$B_v + A_D\Lambda\Sigma$$

and by

$$B_v + A_D\Lambda\left(\Omega - \Lambda \pm \frac{1}{2}\right)$$

where B_v appears in the $\langle J\Omega \pm 1\Lambda\Sigma \pm 1|\mathbf{H}|J\Omega\Lambda\Sigma\rangle$ location. Recall that, for $^2\Pi$ states, the A_D and γ^{Π} (spin-rotation) parameters (see Section 3.4.3) are statistically correlated and cannot be simultaneously determined. The problem of separating A_D from γ is discussed by Brown and Watson (1977).

The R–dependence of matrix elements of the $\mathbf{l}_{zi}\mathbf{s}_{zi}$ part of $\mathbf{H}^{\mathrm{SO}}$ can be related to the R–variation of diagonal spin-orbit constants. For example, if Eq. (5.6.2) is used, and by analogy with Eq. (3.4.14),

$$\begin{aligned}&\langle{}^1\Pi_1, v=0|\mathbf{H}^{\mathrm{SO}}|{}^3\Pi_1, v=1\rangle \\ &\qquad\approx A_e\,\langle v=0|v=1\rangle + \left(\frac{dA}{dR}\right)_{R=R_e}\langle v=0\,|(R-R_e)|\,v=1\rangle. \end{aligned}\tag{5.6.8}$$

The first term is zero if the two potential curves of the isoconfigurational $^1\Pi_1$ and $^3\Pi_1$ states are assumed to be similar cf. Eq. (3.4.14). The second term can be evaluated from Eq. (5.6.5) if the α_A constant is known from the $A(v)$ variation. An example of such a nonzero interaction matrix element (Eq. (5.6.8)) is given for DCl (Huber and Alberti, 1983). Also see Section 3.4.2.

The variation of orbitally or configurationally off-diagonal spin-orbit perturbation parameters cannot be related to that of a diagonal spin-orbit constant because the off-diagonal matrix element is taken between two different orbitals which do not necessarily have the same R-dependence. This is illustrated by Fig. 5.16 where the variation with R of the $\langle A^2\Sigma^+ |\mathbf{H}^{SO}| X^2\Pi\rangle$ matrix element for OD is given. In the single configuration approximation, the $A^2\Sigma^+$ state belongs to $\sigma\pi^4$. The electronic part of the interaction matrix element between this state and the $X^2\Pi_i$ $(\sigma^2\pi^3)$ state is then equal to $\langle\pi^+ |\hat{a}l^+|\sigma\rangle$ and its absolute value decreases with R (and v) because, in order to dissociate correctly, the σ orbital becomes a $1s_H$ orbital at large R. This R-variation is in the opposite sense to that of $A(X^2\Pi)$.

Another example is from an *ab initio* study of the $b^3\Pi$ and $a^3\Sigma^+$ states of NO^+ (Hutter, *et al.*, 1994). The $b^3\Pi$ state, with the $5\sigma 1\pi^4 2\pi$ configuration, has a positive value of A. At larger internuclear distance, the $5\sigma 1\pi^3 2\pi^2$ configuration (see Section 3.2.3) contributes more and more to the $b^3\Pi$ state, which makes a negative contribution to the spin-orbit matrix element, and eventually even produces a change in sign. This is seen in Fig. 5.18a in which the diagonal spin-orbit matrix element of $^3\Pi_0$, by definition equal to $-A$ (Section 3.4.2.1.1), is plotted vs. internuclear distance.

The off-diagonal matrix elements of $\mathbf{H}^{SO}$ and $\mathbf{L}^+$, taken between the $b^3\Pi$ and the $a^3\Sigma^+$ $(5\sigma^2 1\pi^3 2\pi)$ states, are small at equilibrium internuclear distance, because the 5σ and 1π orbitals are localized on different atoms (5σ on the nitrogen and 1π on the oxygen). The $\mathbf{L}^+$ matrix element even becomes negative at short internuclear distance, but it increases with R due to the evolution of the 5σ orbital character (see Fig. 5.18b, where matrix elements of $\mathbf{L}_{+1} = -2^{-1/2}\mathbf{L}^+$ are plotted). At the dissociation limit ($R \rightarrow \infty$), the coupling between the $a^3\Sigma^+$ and $b^3\Pi$ states vanishes because the $a^3\Sigma^+$ state goes adiabatically to the ground state dissociation $N(^4S) + O^+(^4S)$ asymptote, while the $b^3\Pi$ state goes to an excited dissociation product asymptote, $N^+(^3P) + O(^3P)$.

In practice, the R-dependence of off-diagonal constants is less important than that of diagonal constants. In Section 3.3.1 it was shown that if the potential curves of the interacting states cross (at R_C), the value of the electronic part of the off-diagonal interaction between *near-degenerate* vibrational levels is equal to the value computed at R_C,

$$H_{1,v_i;2,v_j} = H_{12}^e(R_C)\,\langle v_i|v_j\rangle\,.$$

Thus, even though H_{12}^e may vary rapidly with R, perturbations will not be very sensitive to this variation.

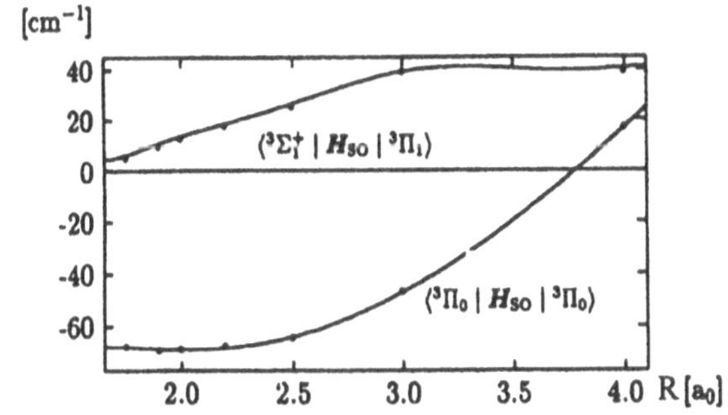

Figure 5.18a: The diagonal and off-diagonal spin-orbit matrix elements of the $b^3\Pi$ state with the $a^3\Sigma^+$ state as a function of the NO^+ internuclear separation (from Hutter, *et al.*, 1994).

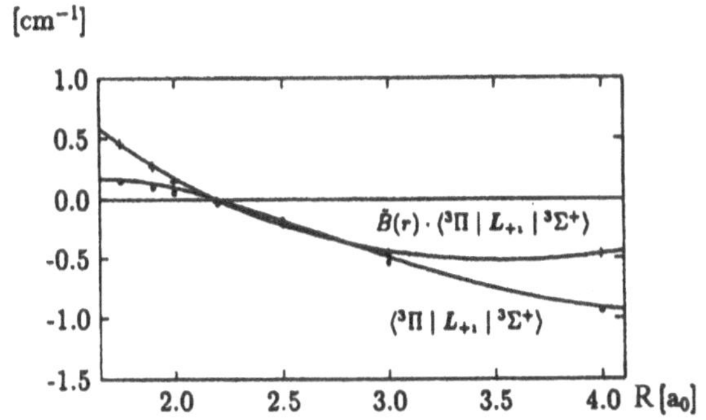

Figure 5.18b: The dependence of the $\langle^3\Pi|\mathbf{L}_+|^3\Sigma^+\rangle$ matrix element of NO^+ on the internuclear separation. The rotational coupling matrix element is shown (from Hutter, *et al.*, 1994). The matrix elements that are actually plotted are of the spherical tensor form of the operator, $\mathbf{L}_{+1} = -2^{-1/2}\mathbf{L}^+$, which has sign opposite to that for $\mathbf{L}^+$ discussed in the text.

The R– or v–dependence of second-order constants $(o, p, q, \gamma$, etc.) is more important than that of the off-diagonal perturbation parameters themselves (as determined from a direct fit to perturbed levels) because, in their definitions as sums over the interactions with *energetically remote levels* [Eqs. (5.5.1a) - (5.5.3a)], the off-diagonal matrix elements are taken between nondegenerate vibrational levels. Thus the R–centroid, $\overline{R}_{vv'}$, will vary with v and v' and will not be equal to R_C. Since a_+ and b are not independent of $\overline{R}_{vv'}$ and $\overline{R}_{vv'}$ is not independent of v' at fixed v, it is not permissible to remove the electronic factor from the summation over remote perturbers. Furthermore, the denominator of each term in the perturbation summation depends on an energy difference. This can cause a strong v, J–dependence of the second-order constants, especially when the vibrational levels of the state under consideration are near the energy of the curve crossing, E_C. It is frequently necessary to introduce centrifugal distortion correction terms for p and q, p_D and q_D, similar to the A_D term (for OH, see Coxon and Foster, 1982).

Effantin, *et al.*, (1982), Amiot and Islami (1986), and Wada and Kanamori (1999), in their analysis of CO $a^3\Pi \sim a'^3\Sigma^+$ perturbations, found it necessary to include in their $\mathbf{H}^{\text{eff}}$ fit model, in addition to the spin-orbit (α_0) and L–uncoupling (β_0) off-diagonal interaction parameters, a direct spin-spin interaction parameter (ϵ) as well as a second-order $\mathbf{H}^{\text{SO}} \times \mathbf{H}^{\text{ROT}}$ interaction term (p_3). The p_3 parameter is not a centrifugal correction to either α_0 or β_0. It is a cross term between the implicit R–dependence of the spin-orbit coupling constant of the $c^3\Pi$ state and the explicit R–dependence of the $B(R)\mathbf{J}\cdot\mathbf{L}$ perturbation operator.

5.7 Beyond the Single-Configuration Approximation

Until this point, the single-configuration approximation for each electronic state has been assumed.[†] The C_2 $b^3\Sigma_g^- \sim X^1\Sigma_g^+$ interaction is one example where it is necessary to represent one of the interacting states by a mixture of two configurations. Another example involving predissociation is discussed in Section 7.11.1.

For many years, the ground state of the C_2 molecule had been assumed to be a $^3\Pi_u$ state, in part because the relative energies of the manifolds of singlet and triplet states were unknown. The discovery of perturbations between the $X^1\Sigma_g^+$ and $b^3\Sigma_g^-$ states (Ballik and Ramsay, 1963) allowed the relative energies of singlet and triplet states to be determined.

The C_2 $b^3\Sigma_g^-$ state belongs predominantly to the $1\pi_u^2\ 3\sigma_g^2$ configuration, which differs by more than one orbital from the $1\pi_u^4\ 3\sigma_g^0$ dominant configuration of the $X^1\Sigma_g^+$ state. Nevertheless, the matrix element $\langle {}^1\Sigma_g^+ | \mathbf{H}^{\text{SO}} | {}^3\Sigma_g^- \rangle$ is nonzero if $X^1\Sigma_g^+$ contains a significant admixture of the $1\pi_u^2 3\sigma_g^2$ configuration. The ${}^1\Sigma_g^+ (\pi^2) \sim {}^3\Sigma_g^- (\pi^2)$ interaction matrix element is given by Eq. (3.4.16). As the mixing coefficient of the $1\pi_u^2 3\sigma_g^2$ configuration in the $X^1\Sigma_g^+$ wavefunction is about 0.16, the electronic matrix element is

$$0.16\ \zeta_{\text{C}}(2p) = 4.6\ \text{cm}^{-1},$$

which compares quite favorably with the experimental value, 4.8 cm^{-1} (Langhoff, *et al.*, 1977).

Configuration mixing is often necessary to account for spin-orbit interactions between states with configurations which differ by more than one orbital (for

[†] This assumption is central to the discussion of perturbation phenomena from a diabatic point of view. In contrast, the adiabatic approach is intrinsically a mixed-configuration picture.

example, the spin-orbit interaction between the $\pi_u^3\pi_g^3 B^3\Sigma_u^-$ and $\pi_u^4\pi_g\sigma_u$ $B''^3\Pi_u$ states of S_2, Green and Western, 1996).

5.8 Identification and Location of Metastable States by Perturbation Effects

The location of metastable states is often made possible by observation of intercombination systems. For example, the intensity of the Cameron system in CO ($a^3\Pi -$ $X^1\Sigma^+$) is borrowed from singlet transitions through spin-orbit mixing (see Section 6.4). When intercombination transitions are not observed, it is often possible to locate the states of different multiplicity from that of the ground state by observation of spin-orbit $\Delta S \neq 0$ perturbations.

The example of the C_2 molecule given in the previous section is a very important one. A similar result has been demonstrated in the case of the MgO molecule (Ikeda, *et al.*, 1977). Thanks to perturbations in the $v \geq 3$ levels of the $X^1\Sigma^+$ state by the $a^3\Pi$ state, the electronic ground state has been definitely proven to be $X^1\Sigma^+$ and the $a^3\Pi$ state has been found to lie above the $X^1\Sigma^+$ state at $T_e = 2623$ cm^{-1}.

It was possible to locate the lowest singlet states of OH^+ and OD^+, $a^1\Delta$ and $b^1\Sigma^+$, from their perturbations of the $A^3\Pi_i - X^3\Sigma^-$ system (Merer, *et al.*, 1975). The X, a, and b states all belong to the $\sigma^2\pi^2$ configuration, whereas $A^3\Pi_i$ belongs to $\sigma\pi^3$. $^3\Pi$ states are both sensitive detectors of perturbers as well as good discriminants of perturber symmetry. The vibrational numbering of the $b^1\Sigma^+$ state was proven by the absence of a perturbation in $A^3\Pi$ ($v = 0$) of OH^+ and OD^+.

The NO spectrum is one of the most completely known among all diatomic molecules. However, it was not until 1980 (Miescher, 1980) that it became possible to locate the quartet states precisely with respect to the well-known doublet states. Small perturbations between the $B^2\Pi$ state ($v = 1$) and the $b^4\Sigma^-$ state ($v = 0$) had been predicted by Miescher on the basis of *ab initio* calculations (reported by Field, *et al.*, 1975) to be characterized by interaction matrix elements on the order of 5 cm^{-1}. Such perturbations have been observed (Miescher, 1980) and, based on direct $b^4\Sigma^- \leftarrow X^2\Pi$ (Dyer, *et al.*, 1993) and $a^4\Pi \leftarrow X^2\Pi$ (Copeland, *et al.*, 1995; Drabbels, *et al.*, 1995) excitation spectra, $b^4\Sigma^- \rightarrow a^4\Pi$ emission spectra (Huber and Vervloet, 1988), and $a^4\Pi \leftarrow X^2\Pi$ excitation detected by $B^2\Pi \rightarrow X^2\Pi$ emission due to perturbation-facilitated $a^4\Pi \sim B^2\Pi$ collisional transfer (Cosby, *et al.*, 1997), a complete and accurate picture of the NO quartet states and the locations of all b$\sim$B and a$\sim$B perturbations have been assembled.

Heavy molecules are, in principle, more favorable for detecting spin-orbit perturbations and, in this way, locating metastable states. In the NS molecule, which is isovalent with NO, a perturbation matrix element between $b^4\Sigma^-$ and $B^2\Pi$ of 8 cm^{-1} has allowed the $^4\Sigma^-$ state to be located (Jenouvrier and Pascat, 1980). In the NSe molecule, the $^4\Pi$ state has been detected by its interaction with $^2\Pi$ states (Daumont, *et al.*, 1976).

The details of the location and perturbations of metastable states are important for understanding and possibly exploiting collision-induced energy transfer between metastable and short-lived, strongly radiating levels (see Section 6.5.5).

5.9 References

Ackermann, F., and Miescher, E. (1968), *Chem. Phys. Lett* **2**, 351.

Albritton, D. L., Harrop, W. J., Schmeltekopf, A. L., and Zare, R. N. (1973), *J. Mol. Spectrosc.* **46**, 25.

Almkvist, G., and Lagerqvist, A. (1950), *Ark. Fys.* **2**, 233.

Amiot, C., and Islami, K. (1986), *J. Mol. Spectrosc.* **118**, 363.

Ballik, E. A., and Ramsay, D. A. (1963), *Astrophys. J.* **137**, 61, 84.

Barnes, I., Becker, K. H., and Fink, E. H. (1979), *Chem. Phys. Lett.* **67**, 314.'

Barnes, M., Clouthier, D. J., Hajigeorgiou, P. G., Huang, G., Kingston, C. T., Merer, A. J., Metha, G. F., Peers, J. R. D., and Rixon, S. J. (1997), *J. Mol. Spectrosc.* **186**, 374.

Bender, D., Schaefer, S. H., and Tiemann, E. (1986), *J. Mol. Spectrosc.* **116**, 286.

Bernath, P. F. (1980), Ph.D. Thesis, MIT, Cambridge, Massachusetts.

Bessis, N., Hadinger, G., and Tergiman, Y. S. (1984), *J. Mol. Spectrosc.* **107**, 343.

Blaise, J., Wyart, J. F., Conway, J. G., and Worden, E. F. (1980), *Phys. Scr.* **22**, 224.

Bredohl, H., Cornet, R., Dubois, I., and Remy, F. (1974), *J. Phys. B*, **7**, L66.

Brion, J., Malicet, J., and Merienne-Lafore, M. F. (1977), *Can. J. Phys.* **55**, 68.

Brown, J. D., Burns, G., and Le Roy, R. J. (1973), *Can. J. Phys.* **51**, 1664.

Brown, J. M., and Merer, A. J. (1979), *J. Mol. Spectrosc.* **74**, 488.

Brown, J. M., and Watson, J. K. G. (1977), *J. Mol. Spectrosc.* **65**, 65.

Buenker, R. J., and Peyerimhoff, S. D. (1975), *Chem. Phys. Lett.* **34**, 225.

Chang, H. C., and Ogawa, M. (1972), *J. Mol. Spectrosc.* **44**, 405.

Colbourn, E. A., and Wayne, F. D. (1979), *Mol. Phys.* **37**, 1755.

Condon, E. U., and Shortley, G. H. (1953), "The Theory of Atomic Spectra," Cambridge Univ. Press, London and New York.

Connor, J. N. L. (1981), *J. Chem. Phys.* **74**, 1047.

Cooley, J. W. (1961), *Math Comput.* **XV**, 363.

Cooper, D. L., and Richards, W. G. (1981), *J. Chem. Phys.* **74**, 96.

Cooper, D. L., Prosser, S. J., and Richards, W. G. (1981), *J. Phys. B* **14**, 487.

Copeland, R. A., Dyer, M. J., Huestis, D. L., and Slanger, T. G. (1995), *Chem. Phys. Lett.* **236**, 350.

Coquart, B., DaPaz, M., and Prud'homme, J. C. (1974), *Can. J. Phys.* **52**, 177.

Cosby, P. C., Copeland, R. A., Williamson, D. G., Gaudin, G., Dyer, M. J., Huestis, D. L., and Slanger, T. G. (1997), *J. Chem. Phys.* **107**, 2249.

Cossart, D., and Bergeman, T. M. (1976), *J. Chem. Phys.* **65**, 5462.

Cossart, D., Horani, M., and Rostas, J. (1977), *J. Mol. Spectrosc.* **67**, 283.

Couet, C. (1968), *J. Chim. Phys. Phys. Chim. Biol.* **65**, 1241.

Coxon, J. A. (1975), *J. Mol. Spectrosc.* **58**, 1.

Coxon, J. A., and Hammersley, R. E. (1975), *J. Mol. Spectrosc.* **58**, 29.

Coxon, J. A., and Foster, S. C. (1982), *J. Mol. Spectrosc.* **91**, 243.

Cummins, P. G., Field, R. W., and Renhorn, I. (1981), *J. Mol. Spectrosc.* **90**, 327.

Daumont, D., Jenouvier, A., and Pascat, B. (1976), *Can. J. Phys.* **54**, 1292.

Dehmer, P. M., and Chupka, W. A. (1975), *J. Chem. Phys.* **62**, 4525.
De Santis, D., Lurio, A., Miller, T. A., and Freund, R. S. (1973) *J. Chem. Phys.* **58**, 4625.
Dill, D., and Jungen, C. (1980), *J. Phys. Chem.* **84**, 2116.
Drabbels, M., Morgan, C. G., and Wodtke, A. M. (1995), *J. Chem. Phys.* **103**, 7700.
Dressler, K., and Miescher, E. (1965), *Astrophys. J.* **141**, 1266.
Dunham, J. L. (1932), *Phys. Rev.* **41**, 721.
Dyer, M. J., Faris, G. W., Cosby, P. C., Huestis, D. L., and Slanger, T. G. (1993), *Chem. Phys.* **171**, 237.
Effantin, C., d'Incan, J., Bacis, R., and Vergès, J. (1979), *J. Mol. Spectrosc.* **76**, 204.
Effantin, C., Michaud, F., Roux, F., d'Incan, J., and Vergès, J. (1982), *J. Mol. Spectrosc.* **92**, 349.
Elander, N., Hehenberger, M., and Bunker, P. R. (1979), *Phys. Scr.* **20**, 631.
England, J. P., Lewis, B. R., Gibson, S. T., and Ginter, M. L. (1996), *J. Chem. Phys.* **104**, 2765.
Felenbok, P. (1963), *Ann. Astrophys.* **26**, 393.
Ferber, R., Pazyuk, E. A., Stolyarov, A. V., Zaitsevskii, A., Kowalczyk, P., Chen, H., Wang, H., and Stwalley, W. C. (2000), *J. Chem. Phys.* **112**, 5740.
Field, R. W., Wicke, B. G., Simmons, J. D., and Tilford, S. G. (1972), *J. Mol. Spectrosc.* **44**, 383.
Field, R. W. (1974), *J. Chem. Phys.* **60**, 2400.
Field, R. W., Gottscho, R. A., and Miescher, E. (1975), *J. Mol. Spectrosc.* **58**, 394.
Field, R. W., Lagerqvist, A., and Renhorn, I. (1976), *Phys. Scr.* **14**, 298.
Fink, E. H., Setzer, K. D., Ramsay, D. A., and Vervloet, M. (1989), *J. Mol. Spectrosc.* **138**, 19.
Fraga, S., Saxena, K. M. S., and Karwowski, J. (1976), "Handbook of Atomic Data," Physical Sciences Data 5, Elsevier, Amsterdam.
Friedman-Hill, E. J., and Field, R. W. (1991), *J. Chem. Phys.* **96**, 2444.
Froese-Fischer, C. (1972), *At. Data* **4**, 301.
Froese-Fischer, C. (1977), "The Hartree-Fock Method for Atoms," Wiley, New York.
Gallusser, R. (1976), Thesis, Physical Chemistry Laboratory, ETH Zurich, Switzerland.
Gallusser, R., and Dressler, K. (1982), *J. Chem. Phys.* **76**, 4311.
Gauyacq, D., and Horani, M. (1978), *Can. J. Phys.* **56**, 587.
Ghosh, S., Nagaraj, S., and Verma, R. D. (1975), *Can. J. Phys.* **54**, 695.
Gottscho, R. A., Field, R. W., and Lefebvre-Brion, H. (1978), *J. Mol. Spectrosc.* **70**, 420.
Gottscho, R. A., Field, R. W., Dick, K. A., and Benesch, W. (1979), *J. Mol. Spectrosc.* **74**, 435.
Green, M. E., and Western, C. M. (1996), *J. Chem. Phys.* **104**, 848.
Hall, J. A., Schamps, J., Robbe, J. M., and Lefebvre-Brion, H. (1973), *J. Chem. Phys.* **59**, 3271.
Harris, D. O., Engerholm, G. G., and Gwinn, W. D. (1965), *J. Chem. Phys.* **43**, 1515.
Harris, S. M., Gottscho, R. A., Field, R. W., and Barrow, R. F. (1982), *J. Mol. Spectrosc.* **91**, 35.
Herzberg, G. H. (1950), "Molecular Spectra and Molecular Structure. I. Spectra of Diatomic Molecules," D. Van Nostrand, Princeton.
Hinkley, R. K., Hall, J. A., Walker, T. E. H., and Richards, W. G. (1972), *J. Phys. B* **5**, 204.
Horani, M., Rostas, J., and Lefebvre-Brion, H. (1967), *Can. J. Phys.* **45**, 3319.
Hotop, H., and Lineberger, W. C. (1975), *J. Phys. Chem. Ref. Data* **4**, 539.

Huber, K. P., and Herzberg, G. (1979), "Constants of Diatomic Molecules," Van Nostrand-Reinhold, New York.
Huber, K. P., and Alberti, F. (1983), *J. Mol. Spectrosc.* **97**, 387.
Huber, K. P., and Vervloet, M. (1988), *J. Mol. Spectrosc.* **129**, 1.
Hultin, M., and Lagerqvist, A. (1951), *Ark. Fys.* **2**, 471.
Hutter, S. J., Hess, B. A., Marian, C. M., and Samzow, R. (1994), *J. Chem. Phys.* **100**, 5617.
Ikeda, T., Wong, N. B., Harris, D. O., and Field, R. W. (1977), *J. Mol. Spectrosc.* **68**, 452.
Ishiguro, E., and Kobori, M. (1967), *J. Phys. Soc. Jpn.* **22**, 263.
James, T. C. (1964), *J. Chem. Phys.* **41**, 631.
Jenouvrier, A., and Pascat, B. (1980), *Can. J. Phys.* **58**, 1275.
Jevons, W. (1932), "Band Spectra of Diatomic Molecules," Univ. Press, Cambridge, England.
Johansson, S., Litzén, U., Sinzelle, J., and Wyart, J. F. (1980), *Phys. Scr* **21**, 40.
Jongma, R. T., Shi, S., and Wodtke, A. (1999), *J. Chem. Phys.* **111**, 2588.
Kaiser, E. W. (1970), *J. Chem. Phys.* **53**, 1686.
Katayama, D. H., Ogawa, S., Ogawa, M., and Tanaka, Y. (1977), *J. Chem. Phys.* **67**, 2132.
Katriel, J. (1970), *J. Phys. B* **3**, 1315.
Kayama, K., and Baird, J. C. (1967), *J. Chem. Phys.* **46**, 2604.
Kirschner, S. M., and Watson, J. K. G. (1973), *J. Mol. Spectrosc.* **47**, 234.
Kirschner, S. M., and Watson, J. K. G. (1974), *J. Mol. Spectrosc.* **51**, 321.
Kosman, W. M., and Hinze, J. (1975), *J. Mol. Spectrosc.* **56**, 93.
Kotlar, A. J., Field, R. W., Steinfeld, J. I., and Coxon, J. A. (1980), *J. Mol. Spectrosc.* **80**, 86.
Kovács, I. (1969), "Rotational Structure in the Spectra of Diatomic Molecules," American Elsevier, Elsevier, New York.
Kratzer, A. (1920), *Z. Phys.* **3**, 289.
Lagerqvist, A., Lind, E., and Barrow, R. F. (1950), *Proc. Phys. Soc., London, Sect. A.* **63**, 1132.
Lagerqvist, A., and Miescher, E. (1958), *Helv. Phys. Acta* **31**, 221.
Langenberg, J. D., DaBell, R. S., Shao, L., Dreessen, D., and Morse, M. D. (1998), *J. Chem. Phys.* **109**, 7863.
Langer, R. E. (1937), *Phys. Rev.* **51**, 669.
Langhoff, S. R., Sink, M. L., Pritchard, R. H., Kern, C. W., Strickler, S. J., and Boyd, M. J. (1977), *J. Chem. Phys.* **67**, 1051.
Langhoff, S. R., Sink, M. L., Pritchard, R. H., and Kern, C. W. (1982), *J. Mol. Spectrosc.* **96**, 200.
Lebreton, J., Bosser, G., and Marsigny, L. (1973), *J. Phys. B* **6**, L226.
Lefebvre-Brion, H., and Moser, C. M. (1966), *J. Chem. Phys.* **44**, 2951.
Lefebvre-Brion, H. (1969), *Can. J. Phys.* **47**, 541.
Lefebvre-Brion, H. (1973), *J. Mol. Struct.* **19**, 103.
Lefebvre-Brion, H., and Colin, R. (1977), *J. Mol. Spectrosc.* **65**, 33.
Le Roy, R. J. (1973), in "Molecular Spectroscopy" (R. F. Barrow, D. A. Long, and D. J. Millen, eds.) Vol. 1, Chemical Society, London, Specialist Periodical Report, Chap. 3, p. 113.
Lewis, B. R., Gibson, S. T., Morrill, J. S., and Ginter, M. L. (1999), *J. Chem. Phys.* **111**, 186.
Lewis, B. R., Gibson, S. T., Banerjee, S. S., and Lefebvre-Brion, H. (2000), *J. Chem.*

Phys. **113**, 2214.
Linton, C. (1980), *J. Mol. Spectrosc.* **80**, 279.
MacDonald, C. A., Eland, J. H. D., and Barrow, R. F. (1982), *J. Phys. B* **15**, L93.
Malicet, J., Brion, J., and Guenebaut, M. (1976), *Can. J. Phys.* **54**, 907.
Mantz, A. W., Watson, J. K. G., Rao, K. N., Albritton, D. L., Schmeltekopf, A. L., and Zare, R. N. (1971), *J. Mol. Spectrosc.* **39**, 180.
Martin, W. C., Zalubas, R., and Hagan, L. (1978), "Atomic Energy Levels – The Rare Earth Elements," NSRDS-NBS-60, U. S. Govt. Printing Office, Washington.
Matsumoto, A., and Iwamoto, K. (1993), *J. Quant. Spectrosc. Radiat. Transf.* **50**, 103.
Merer, A. J. (1972), *Mol. Phys.* **23**, 309.
Merer, A. J., Malm, D. N., Martin, R. W., Horani, M., and Rostas, J. (1975), *Can. J. Phys.* **53**, 251.
Merzbacher, E. (1998), "Quantum Mechanics," John Wiley and Sons, New York.
Miescher, E. (1980), *J. Chem. Phys.* **73**, 3088.
Miller, S. C., and Good, R. H. (1953), *Phys. Rev. A* **91**, 174.
Miller, W. H. (1968), *J. Chem. Phys.* **48**, 464.
Miller, W. H. (1971), *J. Chem. Phys.* **54**, 4174.
Moore, C. E. (1971), "Atomic Energy Levels," Nat. Bur. Stand. NSRDS-NBS-35, Washington, D.C.
Morin, P., Nenner, I., Adam, M. Y., Hubin-Franskin, M. J., Delwiche, J., Lefebvre-Brion, H., and Giusti-Suzor, A. (1982), *Chem. Phys. Lett.* **92**, 609.
Mulliken, R. S. (1955), *J. Chem. Phys.* **23**, 1833.
Mulliken, R. S. (1972), *Chem. Phys. Lett.* **14**, 141.
Mulliken, R. S. (1976), *Acc. Chem. Res.* **9**, 7.
Mulliken, R. S., and Christy, A. (1931), *Phys. Rev.* **38**, 87.
Ngo, T. A., DaPaz, M., Coquart, B., and Couet, C. (1974), **52**, 154.
Pekeris, C. L. (1934), *Phys. Rev.* **45**, 98.
Petsalakis, I. D., Theodorakopoulos, G., and Nicolaides, C. A. (1992), *J. Chem. Phys.* **97**, 7623.
Prosser, S. J., Barrow, R. F., Effantin, C., d'Incan, J., and Vergès, J. (1982), *J. Phys. B* **15**, 4151.
Richards, W. G., Trivedi, H. P., and Cooper, D. L. (1981), "Spin-Orbit Coupling in Molecules," Oxford Univ. Press (Clarendon), London and New York.
Robbe, J. M., and Schamps, J. (1976), *J. Chem. Phys.* **65**, 5420.
Robbe, J. M., Schamps, J., Lefebvre-Brion, H., and Raşeev, G. (1979), *J. Mol. Spectrosc.* **74**, 375.
Robbe, J. M., Lefebvre-Brion, H., and Gottscho, R. A. (1981), *J. Mol. Spectrosc.* **85**, 215.
Roche, A. L., and Lefebvre-Brion, H. (1973), *J. Chem. Phys.* **59**, 1914.
Rostas, J., Cossart, D., and Bastien, J. R. (1974), *Can. J. Phys.* **52**, 1274.
Sambe, H., and Felton, R. H. (1976), *Chem. Phys.* **13**, 299.
Schamps, J. (1973), thesis, Université des Sciences et Techniques de Lille, Lille, France (unpublished).
Schamps, J. (1977), *J. Quant. Spect. Radiat. Transfer* **17**, 685.
Schmid, R., and Gerö, L. (1937), *Z. Phys.* **106**, 205.
Shadmi, Y. (1961), *Bull. Res. Counc. Isr. Sect. F* **9**, 141.
Shadmi, Y., Oreg., J., and Stein, J. (1968), *J. Opt. Soc. Am.* **58**, 909.
Shadmi, Y., Caspi, E., and Oreg, J. (1969), *J. Res. Nat. Bur. Stand, Sect. A* **73**, 173.
Shore, B. W. (1973), *J. Chem. Phys.* **59**, 6450.
Sobel'man, I. I. (1972), "Introduction to the Theory of Atomic Spectra," Pergamon,

Oxford.
Stahel, D., Leoni, M., and Dressler, K. (1983), *J. Chem. Phys.* **79**, 2541.
Stepanov, B. I. (1940), *J. Phys. (USSR)* **2**, 205.
Suter, R. (1969), *Can. J. Phys.* **47**, 881.
Tchang-Brillet, W.-Ü L., Julienne, P. S., Robbe, J.-M., Letzelter, C., and Rostas, F. (1992), *J. Chem. Phys.* **96**, 6735.
Tellinghuisen, J. (1973), *Chem. Phys. Lett.* **18**, 544.
Tellinghuisen, J. (1984), *J. Mol. Spectrosc.* **103**, 455.
Van Vleck, J. H. (1929), *Phys. Rev.* **33**, 467.
Varshni, Y. P. (1957), *Revs. Mod. Phys.* **29**, 664.
Vergès, J., Effantin, C., Babaky, O., d'Incan, J., Prosser, S. J., and Barrow, R. F. (1982), *Phys. Scr.* **25**, 338.
Verma, R. D., and Dixit, M. N. (1968), *Can. J. Phys.* **46**, 2079.
Verma, R. D. (1971), *Can. J. Phys.* **49**, 279.
Verma, R. D., and Singhal, R. S. (1975), *Can. J. Phys.* **53**, 411.
Veseth, L. (1970), *J. Phys. B* **3**, 1677.
Vidal, C. R., and Scheingraber, H. (1977), *J. Mol. Spectrosc.* **65**, 46.
Wada, A., and Kanamori, H. (1999), *J. Mol. Spectrosc.* **200**, 196.
Wallimann, F., Leutwyler, S., and Field, R. W., *to be published.*
Watson, J. K. G. (1979), *J. Mol. Spectrosc.* **74**, 319.
Wayne, F. D., and Colbourn, E. A. (1977), *Mol. Phys.* **34**, 1141.
Wicke, B. G., Field, R. W., and Klemperer, W. (1972), *J. Chem. Phys.* **56**, 5758.
Wicke, B. G., and Harris, D. O. (1976), *J. Chem. Phys.* **64**, 5236.
Winter, R., Barnes, I., Fink, E. H., Wildt, J., and Zabel, F. (1980), *Chem. Phys. Lett.* **73**, 297.
Winter, R., Barnes, I., Fink, E. H., Wildt, J., and Zabel, F. (1982), *J. Mol. Struct.* **80**, 75.
Wyart, J. F. (1978), *Phys. Scr.* **18**, 87.
Wyart, J. F., and Bauche-Arnoult, C. (1981), *Phys. Scr.* **22**, 583.
Yencha, A. J., Kaur, D., Donovan, R. J., Kvaran, A., Hopkirk, A., Lefebvre-Brion, H., and Keller, F. (1993), *J. Chem. Phys.* **99**, 4986.
Zaidi, H. R., and Verma, R. D. (1975), *Can. J. Phys.* **53**, 420.
Zare, R. N. (1964), *J. Chem. Phys.* **40**, 1934.
Zare, R. N., Schmeltekopf, A. L., Harrop, W. J., and Albritton, D. L. (1973), *J. Mol. Spectrosc.* **46**, 37.
Zhou, C., Balfour, W. J., and Qian, C. X. W. (1997), *J. Chem. Phys.* **107**, 4473.

Chapter 6

Transition Intensities and Special Effects

Contents

6.1 Intensity Factors 348

6.1.1 Interrelationships between Intensity Factors 348

6.1.2 General Formulas for One-Photon and Multi-Photon Transition Strengths 355

6.1.2.1 One-Photon Transitions 356

6.1.2.2 Two-Photon Transitions 359

6.1.2.3 Three-Photon Transitions 366

6.1.2.4 Comparisons between Excitation Schemes Involving Different Numbers of Photons . . 367

6.2 Intensity Borrowing 368

6.2.1 Perturbations by States with "Infinite" Radiative Lifetime; Simple Intensity Borrowing 368

6.2.2 Multistate Deperturbation; The NO $^2\Pi$ States . . . 375

6.3 Interference Effects 378

6.3.1 Perturbations between States of the Same Symmetry; Vibrational-Band Intensity Anomalies 380

6.3.2 $\Delta\Lambda = \pm 1$ Perturbations; Rotational-Branch Intensity Anomalies 386

6.3.3 Assignments Based on Pattern-Forming Rotational Quantum Numbers 403

6.3.4 $\Sigma^+ \sim \Sigma^-$ Perturbations; Subband Intensity Anomalies in $\Sigma \leftrightarrow \Pi$ Transitions 404

6.3.5 F_1 vs. F_2 Intensity Anomalies in $^2\Sigma$ States 405

6.4 Forbidden Transitions; Intensity Borrowing by Mixing with a Remote Perturber 406

6.5 Special Effects 415

6.5.1 Differential Power Broadening 416
6.5.2 Effects of Magnetic and Electric Fields on Perturbations 418
6.5.3 Anticrossing, Quantum-Beat, and Double-Resonance Experiments 427
6.5.4 Rydberg States and the Zeeman Effect 439
6.5.5 Nonthermal Population Distributions; Chemical and Collisional Effects 445
6.5.6 "Deperturbation" at High Pressure and in Matrices 455
6.5.7 Matrix Effects 458
6.6 References 460

6.1 Intensity Factors

6.1.1 Interrelationships between Intensity Factors

The relationships between measurable quantities related to absolute transition probability (e.g. absorption cross section, molar absorption or extinction coefficient, radiative lifetime) and the fundamental quantities used to describe and inter-relate the observable quantities are fraught with difficulties of unit conversions and internally consistent treatments of initial- and final-state degeneracies. Several excellent papers on this subject exist (Hilborn, 1982 and 2002, Larsson, 1983, Tatum, 1967, Schadee, 1978, and Whiting, *et al.*, 1980). Much of Section 6.1.1 is based on or checked against Hilborn (1982 and 2002), although slightly different notation and definitions are used.

It is a well established convention in spectroscopy to describe a transition by specifying the quantum numbers of the upper state first, regardless of whether the upper state is the initial or final state of the observed transition. It is not possible to follow this convention in defining most of the quantities that specify the absolute strength of a transition. The reason for this is the necessity to include initial and/or final state degeneracy factors (g_i and g_f) in the definitions of some of the measures of transition strength.

There are a variety of properties related to transition intensities that can be affected by perturbations. The most important of these are the following isolated-molecule quantities.

1. *Transition moment for electric dipole transitions* (in the dipole length approximation) is defined as,

$$\mu_{i,v_i;j,v_j} = \langle i, v_i|\boldsymbol{\mu}|j, v_j\rangle \tag{6.1.1}$$

$$\vec{\mu} \equiv e\sum_{k=1}^{n}\vec{r}_k, \tag{6.1.2}$$

where n is the number of electrons. $\vec{\mu}$ is a one-electron vector operator and its matrix elements may be simplified using the Wigner-Eckart Theorem and standard angular momentum techniques (see Section 6.1.2). The notation for the *electronic transition moment function*, recommended by Whiting, *et al.*(1980), is

$$R_e(\overline{R}_{v_i,v_j}) = \langle i, v_i|\mu|j, v_j\rangle / \langle v_i|v_j\rangle \tag{6.1.3}$$

if the R–centroid approximation is valid. This approximation is not always valid for transition intensities (Noda and Zare, 1982; Tellinghuisen, 1984; see also Section 6.4). The symbol μ, with or without subscripts, represents a matrix element of $\boldsymbol{\mu}$ between electronic-vibration functions, whereas R_e represents an electronic matrix element. Although the same notation is used for equilibrium internuclear distance, the electronic transition moment typically appears as $R_e(\overline{R}_{v_i,v_j})$ which means that the R–dependent transition moment function is evaluated at $R = \overline{R}_{v_i,v_j}$. Note that μ is a joint property of two (non-rotating) electronic-vibration eigenstates. Since μ_{i,v_i,j,v_j} describes an eigenstate-to-eigenstate ($|n_i, \Lambda_i, S_i, \Sigma_i\rangle\, |v_i\rangle \leftrightarrow |n_j, \Lambda_j, S_j, \Sigma_j\rangle\, |v_j\rangle$) transition, no degeneracy factors appear and the same transition moment describes the $j \leftarrow i$ and $j \rightarrow i$ processes.

2. *Radiative lifetime*, τ_i, is a property of a single level, whereas μ_{ij} describes a transition between two levels. $1/\tau_i$ is the rate at which the population of level i decays,

$$N_i(t) = N_i^0 \exp(-t/\tau_i), \tag{6.1.4}$$

where N_i^0 is the population of level i at $t = 0$. In the absence of external fields, all of the M_J components of the i–th electronic-vibration-rotation eigenstates have the same radiative lifetime. No degeneracy factors appear in the definition of τ_i.

3. *Transition probabilities*, such as the *Einstein spontaneous emission coefficient*, A_{ij}, are defined so that, in the absence of collisions, nonradiative decay processes (see Chapters 7 and 8), and stimulated emission the upper level, i, decays at a rate

$$\tau_i^{-1} = \sum_j A_{ij}. \tag{6.1.5}$$

Note that, in the absence of collisions, the population of a single level decays as a single exponential, despite the appearance of a summation over individual Einstein A–coefficients in Eq. (6.1.5). Even if one selectively monitors fluorescence at the frequency of the $i \rightarrow j$ transition or the appearance of level j, the measured decay constant will be τ_i^{-1} and not A_{ij}. A_{ij} is related to μ_{ij} by

$$A_{ij} = |\mu_{ij}|^2 \nu_{ij}^3 [8\pi^2/(3\hbar\epsilon_0)], \tag{6.1.6a}$$

where (for rotating molecule eigenstates) a sum over lower state M_{J_j}–components is implicit in $|\mu_{ij}|^2$

$$|\mu_{ij}|^2 = \sum_{M_j=M_i-1}^{M_i+1} |\langle iM_i|\,\boldsymbol{\mu}\,|jM_j\rangle|^2 \tag{6.1.6b}$$

$$A_{ij} = 3.137 \times 10^{-7} |\mu_{ij}|^2 \nu_{ij}^3 \quad \mathrm{s}^{-1} \tag{6.1.6c}$$

where μ is in Debyes (1 D = 0.3935 a.u. $= 1 \times 10^{-18}$esu $\cdot$ cm $= 3.336 \times 10^{-30}$ Coulomb $\cdot$ m), ϵ_0 is the permittivity of vacuum (8.8542×10^{-12}s$^4 \cdot A^2 \cdot$kg$^{-1} \cdot$m^{-3}), and the transition frequency, ν_{ij}, is in reciprocal centimeters.

Other intensity factors include:

4. *Einstein stimulated emission/absorption*, B_{ij}, is defined as

$$B_{ij} = |\mu_{ij}|^2/(6\hbar^2 \epsilon_0 c), \tag{6.1.7a}$$

where $B_{ij}\rho(\nu_{ij})$ is the $i \rightarrow j$ transition rate for a single molecule stimulated by a radiation field with energy density ρ (energy per unit volume per unit frequency, in cm^{-1} units) and frequency centered at the molecular resonance frequency, ν_{ij}.† The implicit assumption is made that the initial level, i, is treated as a single M_{J_i} component. The M_J–degeneracy of the initial level always appears in the population factor, N_i,

$$W_{ij} = N_i B_{ij} \rho(\nu_{ij}) \tag{6.1.7b}$$

where W_{ij} is the rate, molecules/s, or photons/s participating in the $i \rightarrow j$ stimulated process. The stimulated absorption and stimulated emission rate processes are related by

$$B_{ij} = \frac{g_j}{g_i} B_{ji}, \tag{6.1.7c}$$

where g_j and g_i are the total degeneracies of the j and i levels (including the $2J+1$ M_J–degeneracy). For μ in Debye units and ρ in J/(m$^3 \cdot$ cm^{-1}),

$$B_{ij} = 6.283 \times 10^8 |\mu_{ij}|^2 \quad \mathrm{J^{-1} \cdot m^3 \cdot cm^{-1} \cdot s^{-1}}. \tag{6.1.7b}$$

5. *Oscillator strength*, f_{ij}, is defined, for the $i \rightarrow j$ transition, as

$$f_{ij} = |\mu_{ij}|^2 \nu_{ij} [4\pi m_e c/(3\hbar e^2)], \tag{6.1.8a}$$

and for the $j \rightarrow i$ transition,

$$f_{ji} = -\frac{g_i}{g_j} f_{ij}, \tag{6.1.8b}$$

where m_e and e are the mass and charge of the electron and, for μ in Debye and ν_{ij} in cm^{-1},

$$f_{ij} = 4.703 \times 10^{-7} \nu_{ij} |\mu_{ij}|^2. \tag{6.1.8c}$$

There is an oscillator strength sum rule that is instructive but frequently misunderstood. The sum of the f–values for all transitions that involve orbital promotions of the same electron sum to 1. If, as in the Ca 4s^2 ^1S ground

† It is implicit in the definition of the Einstein B–coefficient that $\rho(\nu)$ is constant over the linewidth of the molecular transition.

state, there are two electrons in the same (4s) orbital, the oscillator strength sum will be 2. However, the oscillator strength sum for an initial state that is electronically excited includes terms that are negative for downward transitions and positive for upward transitions. Thus it is possible for the oscillator strength to be much larger than 1 for a single upward transition from a singly occupied orbital in an electronically excited state.

6. The *Cross-Section* for the $i \to j$ transition is expressed as one of three frequently confused quantitites, $\sigma_{ij}(\nu - \nu_0), \sigma_{ij}(0)$, and σ_{ij}°, where $\sigma_{ij}(\nu - \nu_0)$ is the lineshape function (units of area), $\sigma_{ij}(0)$ is the cross section at line center (units of area), and σ_{ij}° is the integrated cross section (units of area times frequency). One commonly encounters Lorentzian

$$L(\nu; \nu_0, \Delta\nu) = \frac{1}{\pi} \frac{(\Delta\nu/2)}{(\Delta\nu/2)^2 + (\nu - \nu_0)^2} \tag{6.1.9a}$$

and Gaussian

$$G(\nu; \nu_0, \Delta\nu) = \left[\frac{4\ln 2}{\pi}\right]^{1/2} \frac{1}{\Delta\nu} \exp\left[-(\nu - \nu_0)^2 (4 ln2/\Delta\nu^2)\right] \tag{6.1.10a}$$

lineshapes, where ν_0 is the line center and $\Delta\nu$ is the FWHM of the line. Both $L(\nu; \nu_0, \Delta\nu)$ and $G(\nu; \nu_0, \Delta\nu)$ are normalized to 1,

$$\int_{-\infty}^{\infty} L(\nu; 0, \Delta\nu) d\nu = 1 \tag{6.1.9b}$$

$$\int_{-\infty}^{\infty} G(\nu; 0, \Delta\nu) d\nu = 1 \tag{6.1.10b}$$

regardless of choice of frequency unit (Hz, rad/s, or cm^{-1}). For a Lorentzian lineshape

$$\sigma_{ij}(\nu) = \sigma_{ij}^{\circ} L(\nu; \nu_0, \Delta\nu) \tag{6.1.9c}$$

and at line-center

$$\sigma_{ij}(\nu_0) = \sigma_{ij}^{\circ} \left[\frac{2}{\pi\Delta\nu}\right] \tag{6.1.9d}$$

and for a Gaussian lineshape

$$\sigma_{ij}(\nu) = \sigma_{ij}^{\circ} G(\nu; \nu_0, \Delta\nu) \tag{6.1.10c}$$

$$\sigma_{ij}(\nu_0) = \sigma_{ij}^{\circ} \left[\frac{4\ln 2}{\pi}\right]^{1/2} \frac{1}{\Delta\nu}. \tag{6.1.10d}$$

The integrated cross section (in $\text{cm}^2 \cdot \text{cm}^{-1}$) is related to the transition moment (in Debyes)

$$\sigma_{ij}^{\circ} = 4.166 \times 10^{-19} \nu_{ij} |\mu_{ij}|^2 \text{cm}^2 \cdot \text{cm}^{-1} \tag{6.1.11a}$$

$$\sigma_{ji}^{\circ} = \frac{g_i}{g_j} \sigma_{ij}^{\circ} \tag{6.1.11b}$$

and, for a Lorentzian lineshape

$$\sigma_{ij}(\nu_0) = 2.652 \times 10^{-19} \frac{\nu_{ij}}{\Delta\nu} |\mu_{ij}|^2 \text{ cm}^2 \tag{6.1.12a}$$

or a Gaussian lineshape

$$\sigma_{ij}(\nu_0) = 3.914 \times 10^{-19} \frac{\nu_{ij}}{\Delta\nu} |\mu_{ij}|^2 \text{ cm}^2. \tag{6.1.12b}$$

There is an alternative path to the integrated cross section

$$\sigma_{ij}^{\circ} = \frac{1}{4}\lambda^2 \frac{A_{ij}}{2\pi c} \tag{6.1.13a}$$

$$\sigma_{ij}^{\circ} = 1.327 \times 10^{-12} \frac{A_{ij}}{\nu_{ij}^2} \text{ cm}^2 \cdot \text{cm}^{-1} \tag{6.1.13b}$$

for ν_{ij} in cm^{-1}. If the lineshape is Lorentzian, all of the linewidth is associated with the upper level, and the sum over all lower states is dominated by a single term (the j–th lower state), $\sum_k A_{ik} \approx A_{ij}$, then

$$\Delta\nu = \frac{1}{2\pi c \tau_i} = \frac{A_{ij}}{2\pi c} \tag{6.1.14}$$

$$\sigma_{ij}(\nu_0) = \frac{1}{2\pi}\lambda^2. \tag{6.1.15}$$

This is a remarkable result because it implies that the linecenter cross section for an absorption transition from the electronic ground state to several of the lowest lying electronically excited states, including forbidden transitions, will be approximately equal to the square of the wavelength. However, in order for this to be true, other sources of line broadening, such as collision broadening

$$\Delta\nu \approx [10\text{MHz/Torr}][1/30000 \text{ MHz/cm}^{-1}]p, \tag{6.1.16}$$

where p is the pressure in Torr, or Doppler broadening

$$\Delta\nu \approx 7.2 \times 10^{-7} \nu_0 [T/M]^{1/2} \tag{6.1.17}$$

where T is the temperature in Kelvin and M is the mass of the molecule in amu, must be negligible relative to the Eq. (6.1.14) linewidth.

As a point of reference, the following quantities correspond to a fully allowed transition, $f = 1$, at $\nu_{ij} = 20{,}000 \text{ cm}^{-1}$ $(g_i = g_j = 1)$:

$$\begin{aligned}
\mu_{ij} &= 10.3 \text{ D} \\
A_{ij} &= 2.67 \times 10^8 \text{ s}^{-1} \\
\tau_i &= 3.75 \times 10^{-9} \text{ s} \quad (\text{if } \Sigma_k A_{ik} \approx A_{ij}, \text{i.e. all } A_{ik} \ll A_{ij} \text{ for } k \neq j) \\
B_{ij} &= 6.67 \times 10^{10} \text{ J}^{-1} \cdot \text{m}^3 \cdot \text{cm}^{-1} \cdot \text{s}^{-1} \\
\sigma_{ij}^0 &= 8.86 \times 10^{-13} \text{ cm}^2 \cdot \text{cm}^{-1} \\
\Delta\nu &= 1.42 \times 10^{-3} \text{ cm}^{-1} \\
\sigma_{ij}(\nu_0) &= 3.98 \times 10^{-10} \text{ cm}^2.
\end{aligned}$$

(for a Lorentzian lineshape) [1 Megabarn $= 10^{-18}\text{cm}^2$.]

In order to measure an absorption cross section, it is necessary to have a source of tunable radiation (or spectrometer) that has a spectral resolution narrower than the width of the molecular line (natural and/or Doppler) being measured. Inadequate spectral resolution will yield a too small (lower bound) value of σ°_{ij}. An effective strategy is to pressure broaden the molecular line so that it becomes broader than the instrumental resolution (see Stark, *et al.*, 1992 and Murray, *et al.*, 1994).

Another quantity, the *line strength*, is defined assuming isotropic excitation and/or unpolarized radiation,

$$S_{i,v',J';j,v'',J''} \equiv \sum_{M',M''} \left|\langle i, v', J', M' \left|\mu\right| j, v'', J'', M''\rangle\right|^2 . \qquad (6.1.18)$$

The summation over all M', M'' sublevels takes into account the $(2J+1)$–fold degeneracy of a J–level in the absence of external fields. Because the line strength is defined as a sum over the degenerate components of *both* initial and final states

$$S_{i,v',J';j,v'',J''} = S_{j,v'',J'';i,v',J'}. \qquad (6.1.19)$$

The line strength is expressed as a product of three factors, one each for the vibrational, electronic, and rotational parts of the transition probability,

$$S_{i,v',J';j,v'',J''} = g_i q_{v'v''} \left|R_e^{i,j}\right|^2 \mathbf{S}_{J'J''}, \qquad (6.1.20)$$

where the initial state degeneracy factor is needed to make the linestrength factors for the $i \rightarrow j$ and $i \leftarrow j$ transitions equal, q is the *Franck-Condon factor*, $|\langle v'|v''\rangle|^2$, and $\mathbf{S}_{J'J''}$ is the *Hönl-London rotational line strength factor*, which satisfies the sum rules

$$\sum_{J'} \mathbf{S}_{J'J''} = (2 - \delta_{0,\Lambda'+\Lambda''})(2S''+1)(2J''+1) \qquad (6.1.21a)$$

$$\sum_{J''} \mathbf{S}_{J'J''} = (2 - \delta_{0,\Lambda'+\Lambda''})(2S'+1)(2J'+1) \qquad (6.1.21b)$$

(Whiting, *et al.*, 1980). Definitions and interrelationships between intensity factors are given by Hilborn (1982 and 2002), Tatum (1967), Whiting and Nicholls (1974), and Schadee (1978). The definitions of transition moments and intensity factors suggested by Whiting, *et al.*, (1980) are adopted in this chapter, except in Sections 6.3 and 6.4 where the factorization of Eq. (6.1.20) is not possible.

There are many frequently measured quantities that are related to the absolute strength of an absorption line or band (Pugh and Rao, 1976). The fundamental quantity to which the others are related is the integrated cross section, σ^0_{ij}, which has units of area times frequency (often $\text{cm}^2 \cdot \text{cm}^{-1}$ molecule^{-1} or, equivalently, cm/molecule). The integrated cross section is more fundamental than the peak cross section, $\sigma_{ij}(\nu_0)$, because it is insensitive to environmental factors (e.g., temperature, pressure, solvent or matrix effects), the effects of which are typically manifested in the width of the transition, $\Delta\nu_{ij}$.

One reason for the existence of many intensity-related quantities is the form of the Beer-Lambert law, which describes the attenuation of a beam of light as it traverses an absorbing medium,

$$I(\nu, x)/I_0(\nu) = e^{-\sigma(\nu)Nx}. \tag{6.1.22}$$

$I_0(\nu)$ is the incident light intensity at frequency ν, $I(\nu, x)$ is the intensity at frequency ν after the beam of light has traversed a distance x in the absorber, N is number of molecules per volume, and $\sigma(\nu)$ is the absorption cross-section at frequency ν. The quantity $\sigma(\nu)Nx$ is dimensionless. There are many conventions used to specify the number density of absorbers (e.g. molecules/cm^3, moles/liter, pressure in various units at $T = 273$ K or some other reference temperature), and each is associated with a different named quantity that expresses absorption strength.

The quantity

$$A = \log_{10} I/I_0 \tag{6.1.23}$$

is called the absorbance, but different families of absorption coefficients are defined for $\log_{10}$ vs. ln

$$\log I/I_0 = -\epsilon cl \tag{6.1.24}$$

$$\ln I/I_0 = -\kappa c'l' \tag{6.1.25}$$

where ϵ is the molar absorption coefficient (the concentration, c, is in moles/liter, and l is in cm, ϵ is in liter $\cdot$mol$^{-1}\cdot$cm^{-1}) and κ is the Napierian molar absorption coefficient (c' is in mol$\cdot$m^{-3}, l' is in m, κ is in m$^3\cdot$mol$^{-1}\cdot$ m^{-1}). Finally, when the integrated absorption strength is reported ($\int \sigma(\nu)d\nu$, $\int \epsilon(\nu)d\nu$, or $\int \kappa(\nu)d\nu$), there are three common choices of frequency units: cm^{-1}, Hz, and rad/s. Pugh and Rao (1976) give conversion factors among 14 of the most common integrated absorption strength quantities. Returning to the example of a transition at $\nu = 20000$ cm^{-1} with an oscillator strength of $f = 1$, if the FWHM of the absorption band is 5000 cm^{-1} (typical of large molecules in aqueous solution), the maximum molar absorption (or "extinction") coefficient is $\epsilon_{\max} = 4.4 \times 10^4$ liter $\cdot$ mol$^{-1}\cdot$ cm^{-1}.

The electronic transition moment is sometimes expressed as a matrix element of electron velocity $(\vec{\mathbf{p}}/m_e)$ rather than position, $\vec{\mathbf{r}}$. The relationship between the matrix elements of $\vec{\mathbf{r}}$ and $(\vec{\mathbf{p}}/m_e)$ is derived (see page 1310 of Cohen-Tannoudji, *et al.*, 1977) from the commutation rule between the electron coordinate and the electronic Hamiltonian

$$[\mathbf{r}_j, \mathbf{H}] = i\hbar \mathbf{p}_j/m_e \tag{6.1.26}$$

$$\langle \psi_f | (\mathbf{r}_j \mathbf{H} - \mathbf{H}\mathbf{r}_j) | \psi_i \rangle = i\hbar \langle \psi_f | \mathbf{p}_j/m_e | \psi_i \rangle \tag{6.1.27}$$

where ψ_i and ψ_f are eigenstates of $\mathbf{H}$ with eigen-energies E_i and E_f

$$\langle \psi_f | \mathbf{p}_j/m_e | \psi_i \rangle = (i/\hbar)(E_f - E_i) \langle \psi_f | \mathbf{r}_j | \psi_i \rangle \tag{6.1.28}$$

$$= i\omega_{fi} \langle \psi_f | \mathbf{r}_j | \psi_i \rangle . \tag{6.1.29}$$

There are also several perturbation-sensitive non-isolated molecule properties, such as collision-induced population anomalies (Radford and Broida, 1963) (Section 6.5.5) and differential pressure and power (Gottscho and Field, 1978) broadening efforts (Section 6.5.1). Since all of the isolated molecule properties are explicitly related to μ, the following discussion focuses on μ. Note, however, that the nature of a perturbation related intensity anomaly is profoundly dependent on whether $|\mu|^2, \tau$, or (A, B, f, σ) is being measured and on the state selectivity and spectral resolution of the specific experiment.

In this chapter, discussion will be restricted to intensity effects arising from bound∼bound interactions. Bound∼free predissociation and autoionization interactions are discussed in Chapters 7 and 8. Interactions between broadened states (or resonances) are dealt with by a complex-energy formalism in Section 9.3.

6.1.2 General Formulas for One-Photon and Multi-Photon Transition Strengths

For a one-color, N–photon transition in the weak-field limit, the cross section is

$$\sigma_{fi}^{(N)} \propto \left| \sum_{j,k\ldots l} \frac{\langle f|\vec{\epsilon}\cdot\vec{\mu}|j\rangle\langle j|\ldots|k\rangle\langle k|\ldots|l\rangle\langle l|\vec{\epsilon}\cdot\vec{\mu}|i\rangle}{[E_{ji}-(N-1)E]\ldots[E_{kl}-2E][E_{li}-E]} \right|^2 \tag{6.1.30}$$

E = photon energy, E_{fi} = transition energy between the i (initial) and f (final) rotation-vibration-electronic states, $\vec{\mu}$ is the electric dipole moment operator, $\vec{\epsilon}$ the polarization vector of the photon.

Lambropoulos (1976) has written a complete and definitive review of multiphoton processes in atoms. For a bound-bound N–photon transition, the differential transition rate (number of molecules excited per second at frequency ν), $W_{fi}^{(N)}(\nu)[\mathrm{s}^{-1}]$, is

$$W_{fi}^{(N)}(\nu) = I^{N-1}I(\nu)\sigma_{fi}^{(N)}(\nu)$$

where I is the number of photons per cm^2 per second ($I[\mathrm{cm}^{-2}\cdot\mathrm{s}^{-1}]$), $I(\nu)$ is the number of photons per cm^2 per second per unit bandwidth (Hz) ($I[\mathrm{cm}^{-2}\cdot\mathrm{s}^{-1}\cdot\mathrm{Hz}^{-1}]$), thus the units for $\sigma^{(N)}$ are $\mathrm{cm}^{2N}\mathrm{s}^{N-2}$. If $I(\nu)$ is instead expressed in photons per cm^2 per second per cm^{-1} bandwidth, then σ^N has units $\mathrm{cm}^{2N-1}\mathrm{s}^{N-1}$ and is smaller than $\sigma^{(N)}$ in $\mathrm{cm}^{2N}\mathrm{s}^{N-2}$ units by a factor of $1/c = 3.3356\times10^{-11}$ s/cm. For a free←bound N–photon transition, the differential transition rate is

$$W_{fi}^{(N)}(\nu) = \sigma_{fi}^{(N)}(\nu)I^N,$$

the units of $\sigma_{fi}^{(N)}(\nu)$ are $\mathrm{cm}^{2N}\mathrm{s}^{N-1}$, and $\sigma_{fi}^{(N)}(\nu)$ includes a factor that specifies the density of final states, $\rho(E_f/h)$.

i) Selection rules for electronic wave functions:

The symmetry of the dipole moment operator implies that for an N–odd-photon transition,

$\Sigma^{\pm} \leftrightarrow \Sigma^{\pm}$

$g \leftrightarrow u$

$\Delta\Lambda = 0, \pm 1, \cdots \pm N$

$\Delta S = 0$

since the electric dipole moment operator does not contain spin variables. This spin selection rule is satisfied if the wave functions are pure, i.e. in the absence of spin-orbit perturbations. Many nominally forbidden transitions become weakly allowed when states of different multiplicities are mixed by the spin-orbit operator (see Section 6.4).

ii) Selection rules for vibrational wave functions:

There is no strict selection rule for vibration in the case of electronic transitions (i.e. between two different electronic states). The Condon principle states that the R–variation of the $\langle \phi^f | \vec{\mu} | \phi^i \rangle$ transition moment electronic matrix elements can in general be neglected

$$\langle \phi^f \chi^f | \vec{\mu} | \phi^i \chi^i \rangle \approx \langle \phi^f | \vec{\mu} | \phi^i \rangle \langle \chi^f | \chi^i \rangle . \tag{6.1.31}$$

The square of the vibrational overlap $\left\langle \chi^f_{v_f} | \chi^i_{v_i} \right\rangle^2$ is the Franck-Condon factor. For transitions between Rydberg states converging to the same ion state, a very good selection rule is

$$\left\langle \chi^f_{v_f} | \chi^i_{v_i} \right\rangle = \delta_{v_f, v_i}. \tag{6.1.32}$$

iii) Selection rules for rotational wavefunctions:

$\Delta J = 0, \pm 1, \pm 2, \cdots \pm N$

The rotational branches are denoted $\ldots N, O, P, Q, R, S, T \ldots$ which correspond to $\Delta J = \cdots - 3, -2, -1, 0, 1, 2, 3 \ldots$. J is the rigorous, as opposed to *pattern forming* rotational quantum number. For transitions involving multiplet states ($S \neq 0$), it is often necessary to specify the change in pattern forming quantum number (which can be either integer or half-integer, depending on the Hund's cases of the initial and final states). [See Section 6.3.3].

The selection rules for parity are $\pm \leftrightarrow \pm$ for N even and $\pm \leftrightarrow \mp$ for N odd. Equation (6.1.30) will be explicitly discussed below for the examples of one-, two-, and three-photon transitions.

6.1.2.1 One-Photon Transitions

The one-photon integrated cross section is proportional to (see Eq. (6.1.11a))

$$\sigma^{\circ}_{fi} = \sigma^{(1)} \propto \langle f | \vec{\epsilon} \cdot \vec{\mu} | i \rangle^2 = \left\langle f | \mathbf{T}^{(1)}(\epsilon) \cdot \mathbf{T}^{(1)}(\mu) | i \right\rangle^2 . \tag{6.1.33}$$

This tensorial expression is very convenient for computer calculations (see Section 3.4.5). The tensor components $\boldsymbol{\epsilon}_{\mathbf{X}}, \boldsymbol{\epsilon}_{\mathbf{Y}}, \boldsymbol{\epsilon}_{\mathbf{Z}}$, are unit vectors defined relative to the laboratory frame.

The polarization tensor has the following components

$$\mathbf{T}_0^{(1)}(\epsilon) = \epsilon_{\mathrm{Z}} \tag{6.1.34}$$

for linear polarization,

$$\mathbf{T}_{\pm 1}^{(1)}(\epsilon) = \mp \frac{1}{\sqrt{2}}(\epsilon_{\mathrm{X}} + i\epsilon_{\mathrm{Y}}), \tag{6.1.35}$$

where the +1 component corresponds to right circular polarization and the −1 component corresponds to left circular polarization. $\mathbf{T}^{(1)}(\mu)$ is the electric dipole moment vector with components

$$\mathbf{T}_0^{(1)}(\mu) = \mu_{\mathrm{Z}} \tag{6.1.36a}$$

$$\mathbf{T}_{\pm 1}^{(1)}(\mu) = \mp \frac{1}{\sqrt{2}}(\mu_X + i\mu_Y). \tag{6.1.36b}$$

The scalar product of two tensors of k rank (see Eq. (3.4.40)) (for $k = 1$, here) is given by

$$\left\langle f|\mathbf{T}^{(1)}(\epsilon)\cdot\mathbf{T}^{(1)}(\mu)|i\right\rangle = \sum_p (-1)^p \left\langle f|\mathbf{T}_{-p}^{(1)}(\epsilon)\cdot\mathbf{T}_p^{(1)}(\mu)|i\right\rangle. \tag{6.1.37}$$

To evaluate the cross section defined in Eq. (6.1.33), it is necessary to re-express the matrix elements of $\vec{\mu}$ in the molecule-fixed coordinate system. The transformation between space-fixed and molecule-fixed coordinate systems is given by (see also Chapter 2, Section 2.3.3)

$$\mathbf{T}_p^{(k)} = \sum_q (-1)^{q-p}\mathbf{T}_q^{(k)}\mathbf{D}_{p,q}^{(k)}(\omega) \tag{6.1.38}$$

where $\mathbf{T}_p^{(k)}$ is the p–th component of $\mathbf{T}^{(k)}$ in the space-fixed system, $\mathbf{T}_q^{(k)}$ is the q–th component of $\mathbf{T}^{(k)}$ in the molecule-fixed system, and $\mathbf{D}_{p,q}^{(k)}$ are the matrix elements of the Wigner rotation matrix (i.e. the direction cosine matrix), and ω represents the Euler angles $(\phi, \theta, \pi/2)$ for a linear molecule (see Eq. (2.3.41)).

If the wavefunctions are written explicitly in terms of $|J, \Omega, M\rangle$ rotational basis-functions,

$$|J, \Omega, M\rangle = \left(\frac{2J+1}{8\pi^2}\right)^{1/2} \mathbf{D}_{\Omega,M}^{(J)}(\omega), \tag{6.1.39}$$

then

$$\left\langle f|\mathbf{T}^{(1)}(\epsilon).\mathbf{T}^{(1)}(\mu)|i\right\rangle =$$

$$\sum_p \sum_q (-1)^q \mathbf{T}_{-p}^{(1)}(\epsilon)\left\langle \phi^f\chi^f|\mathbf{T}_q^{(1)}(\mu)|\phi^i\chi^i\right\rangle \left\langle \Omega_f J_f M_f|\mathbf{D}_{pq}^{(1)}(\omega)|\Omega_i J_i M_i\right\rangle. \tag{6.1.40}$$

Using the well known relationships for rotation matrices (see Hougen, 1970; Edmonds, 1974; Zare, 1988)

$$\int\int\int \mathbf{D}^{(J_1)}_{M'_1M_1}(\omega)\mathbf{D}^{(J_2)}_{M'_2M_2}(\omega)\mathbf{D}^{(J_3)}_{M'_3M_3}(\omega)d(\omega)$$
$$= 8\pi^2 \begin{pmatrix} J_1 & J_2 & J_3 \\ M'_1 & M'_2 & M'_3 \end{pmatrix} \begin{pmatrix} J_1 & J_2 & J_3 \\ M_1 & M_2 & M_3 \end{pmatrix} \quad (6.1.41)$$

and

$$\mathbf{D}^{(J)}_{M'M}(\omega) = (-1)^{M'-M}\mathbf{D}^{(J)}_{-M'-M}(\omega), \quad (6.1.42)$$

one obtains,

$$\left\langle f|\mathbf{T}^{(1)}(\epsilon).\mathbf{T}^{(1)}(\mu)|i\right\rangle =$$
$$\sum_p\sum_q(-1)^q\mathbf{T}^{(1)}_{-p}(\epsilon)(2J_f+1)^{1/2}(2J_i+1)^{1/2}\begin{pmatrix} J_f & 1 & J_i \\ -M_f & p & M_i \end{pmatrix}\begin{pmatrix} J_f & 1 & J_i \\ -\Omega_f & q & \Omega_i \end{pmatrix}$$
$$\times \left\langle \phi^f\chi^f|\mathbf{T}^{(1)}_q(\mu)|\phi^i\chi^i\right\rangle \quad (6.1.43)$$

Equation (6.1.43) yields the well known selection rules for rotational transitions. For unpolarized light, the transition probability is calculated by taking the average of the transition probabilities for the three light polarization possibilities, $p = 0, +1$ and -1. To calculate a transition intensity, it is sufficient to evaluate the transition probability for one polarization component ($p = 0$ is usually most convenient) and one body-fixed μ–component ($q = 0$ for $\Delta\Omega = 0$ "parallel" transitions, $q = +1$ or -1 for $\Delta\Omega = +1$ "perpendicular" transitions). The total intensity is obtained by summing over the transition probabilities for all M values, i.e. over squares of 3-j coefficients. Due to the orthogonality relations among the 3-j coefficients,

$$\sum_{M_f}\sum_{M_i}\begin{pmatrix} J_f & 1 & J_i \\ -M_f & p & M_i \end{pmatrix}^2 = \frac{1}{3}. \quad (6.1.44)$$

For example, if $p = 0$, the selection rule for M is $M_f = M_i$, and, from the properties of the 3-j coefficients (see Section 3.4.5), the selection rule for J is

$$J_i - 1 \le J_f \le J_i + 1.$$

When $J_f = J_i + 1$, the transition is an R–line, for $J_f = J_i$, it is a Q–line, and for $J_f = J_i - 1$, it is a P–line. For the specific case of $p = 0$ and one value of q, the transition probability, P_{fi}, which is proportional to $\left|\langle f\,|\mathbf{T}^{(1)}(\epsilon).\mathbf{T}^{(1)}(\mu)|\,i\rangle\right|^2$, is given by

$$P_{fi} \propto \left|\mathbf{T}^{(1)}_0(\epsilon)\right|^2 \left|\left\langle \phi^f\chi^f|\mathbf{T}^{(1)}_q|\phi^i\chi^i\right\rangle\right|^2 \begin{pmatrix} J_f & 1 & J_i \\ -\Omega_f & q & \Omega_i \end{pmatrix}^2 \sum_{M_i}\begin{pmatrix} J_f & 1 & J_i \\ -M_f & 0 & M_i \end{pmatrix}^2$$
$$(6.1.45)$$

where the sum over M_i ($M_f = M_i$) reduces via

$$\begin{pmatrix} J & 1 & J \\ -M & 0 & M \end{pmatrix} = (-1)^{J-M} \frac{M}{[(2J+1)J(J+1)]^{1/2}} \quad (6.1.46)$$

Because

$$\sum_{M=-J}^{J} M^2 = \frac{(2J+1)(J+1)J}{3} \quad (6.1.47)$$

we have

$$\sum_M \begin{pmatrix} J & 1 & J \\ -M & 0 & M \end{pmatrix}^2 = \frac{1}{3} \quad (6.1.48)$$

The other squared 3-j coefficient, $\begin{pmatrix} J_f & 1 & J_i \\ -\Omega_f & q & \Omega_i \end{pmatrix}^2$ gives the Hönl-London rotational linestrength factors.

For a one-photon, electric dipole transition, the parity selection rule is always $+ \leftrightarrow -$. In terms of e/f symmetry, Fig. 6.1 illustrates a $^1\Pi - {}^1\Sigma^+$ transition with rotational lines Q_{fe}, P_{ee}, and R_{ee}.

6.1.2.2 Two-Photon Transitions

For two-photon $f \leftarrow\leftarrow i$ transitions excited by two identical (same frequency and polarization) photons, the transition probability is proportional to

$$P_{fi}^{(2)} \propto \left| \sum_j \frac{\langle f | \vec{\epsilon} \cdot \vec{\mu} | j \rangle \langle j | \vec{\epsilon} \cdot \vec{\mu} | i \rangle}{|E_{ji} - E|} \right|^2 \quad (6.1.49)$$

where the sum is over all "virtual" intermediate states, $|j\rangle$, $\vec{\mu}$ is an electric dipole transition operator expressed in a body-fixed coordinate system, $\vec{\epsilon}$ is the magnitude and direction of the linearly polarized laboratory-frame oscillating electric field of the photon, and E is the energy of the photon. Since $E_{fi} = 2E, |E_{fj} - E| = |E_{ji} - E|$, only one off-resonance energy denominator needs to be specified in Eq. (6.1.49). In this formalism, two-photon excitation is described by the squared sum of one-photon transition amplitudes via virtual intermediate states, j, where the virtual routes are weighted by the energy mismatch between the photon energy, E, and the transition energy, E_{ji} (Göppert-Mayer mechanism, 1931). Different virtual routes may interfere, depending on their one-photon transition matrix elements.

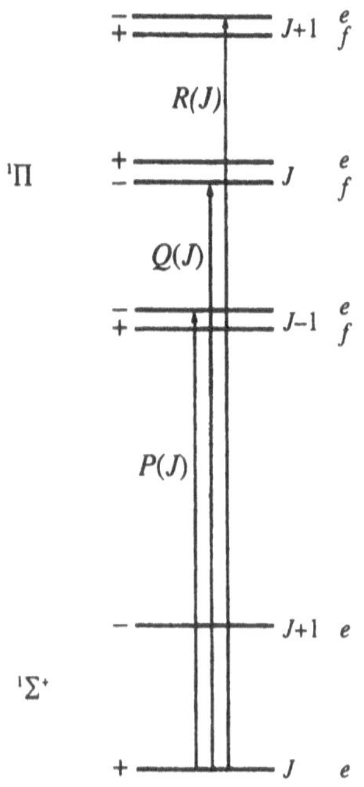

Figure 6.1: Parity and e/f labels for the rotational branches of a one-photon $^1\Pi - {}^1\Sigma^+$ transition.

A more general (and simplified) result than Eq. (6.1.49), expressed in terms of spherical tensors (Hippler, 1999) is

$$\sigma^{(2)}_{J_f,J_i} \propto (2J_f+1)(2J_i+1)\sum_k (2k+1)\left|\begin{pmatrix} 1 & 1 & k \\ -p_2 & -p_1 & p_1+p_2 \end{pmatrix}\right|^2 \left|\sum_{q_1+q_2} \mathbf{T}^{(k)}_{q_1+q_2}\right|^2, \tag{6.1.50}$$

where p_1 and p_2 specify the polarizations of the two photons ($p = 0$ linear polarization along the Z direction: $\Delta M_J = 0; p = \pm 1$ right or left circular polarization: $\Delta M_J = \pm 1$). q is the change of Ω and Λ during each virtual one-photon absorption step, $q_1 = \Omega_j - \Omega_i$ and $q_2 = \Omega_f - \Omega_j$. The $\mathbf{T}^k_{q_1+q_2}$ tensor components contain all information about virtual intermediate states (j), the rotationally averaged orientations of the body-frame axes relative to the laboratory-frame axes, and the body-frame eigenstate-to-eigenstate transition moments (see below).

For linearly polarized light ($p_1 = p_2 = 0$), the 3-j symbol in Eq. (6.1.50) is zero, except for a $k = 0$ and a $k = 2$ term, and for circularly polarized light ($p_1 = p_2 = \pm 1$), except for a $k = 2$ term, only:

$$\sigma_{J_f,J_i}^{(2),\text{linear}} \propto (2J_f+1)(2J_i+1)\left\{\frac{1}{3}\left|\sum_{q_1+q_2} \mathbf{T}_{q_1+q_2}^{(0)}\right|^2 + \frac{2}{3}\left|\sum_{q_1+q_2} \mathbf{T}_{q_1+q_2}^{(2)}\right|^2\right\} \tag{6.1.50a}$$

$$\sigma_{J_f,J_i}^{(2),\text{circular}} \propto (2J_f+1)(2J_i+1)\left|\sum_{q_1+q_2} \mathbf{T}_{q_1+q_2}^{(2)}\right|^2 \tag{6.1.50b}$$

From **Eq. (6.1.50a,b)** the well-known result is obtained that the ratio of line strengths for transitions excited by linearly polarized light compared with those excited by circularly polarized light is 2/3, provided that there is no contribution of rank $k = 0$ (see also Section 6.1.2.4).

The $\left|\sum_{q_1+q_2} \mathbf{T}_{q_1+q_2}^{(k)}\right|^2$ squared sum contains a $\mathbf{T}_0^{(2)} \big/ \mathbf{T}_{-2}^{(2)}$ interference effect that exclusively affects the contributions of $^{2S+1}\Sigma$ virtual states to $^{2S+1}\Pi - {}^{2S+1}\Pi$ two-photon transitions. Since the electronic angular momentum is associated with a *signed* quantum number, $\Pi - \Pi$ transitions are not only characterized by a $\Delta\Omega = 0$ component, but in addition also by a $\Delta\Omega = \pm 2$ component. These two components give rise to an interference effect, which has been neglected in otherwise definitive treatments of two-photon transition probabilities (Bray and Hochstrasser, 1976; Freedman, 1977).

Hippler (1999) simplifies **Eq. (6.1.50a,b)** by combining symmetry-related pairs of terms that involve opposite signs of body-frame Ω, Λ, Σ case (a) basis state and body-frame transition moment projection quantum numbers, q. These terms include products of pairs of transition moments $\left(\mu_{q_2}^{fj}\mu_{q_1}^{ji}\right)$, direction cosine matrix elements, and mixing coefficients for Hund's case (a) $|\Omega JM\rangle|\Lambda S\Sigma\rangle$ basis states in the upper (f) and lower (i) eigenstates $\left(a_{\Omega_f}^f, a_{\Omega_i}^i\right)$. Once the pairs of symmetry-related terms are combined, the signs of all non-zero projection quantum numbers for distinguishable transition paths are uniquely specified and the quantum numbers and pre-summed $T_{q_1+q_2}^{(k)}$ are labeled with a ($'$),

$$\mathbf{T}'^{(k)}_{q_1'+q_2'} = \sum_i \begin{pmatrix} 1 & 1 & k \\ -q_2' & -q_1' & q_1'+q_2' \end{pmatrix} \frac{\mu_{q_2'}^{fj}\mu_{q_1'}^{ji}}{E_{ji}-E} \times \sum_{\Omega_f',\Omega_i'} \bar{a}_{\Omega_f'}^f \bar{a}_{\Omega_i'}^i (-1)^{\Omega_i'-S} \begin{pmatrix} J_f & k & J_i \\ \Omega_f' & -q_1'-q_2' & -\Omega_i' \end{pmatrix} \tag{6.1.51}$$

where S is the total electron spin quantum number. The second 3-j coefficient in Eq. (6.1.51) uniquely determines $q_1' + q_2'$

$$q_1' + q_2' = \Omega_f' - \Omega_i' = \Delta\Lambda'$$

except for $^{2S+1}\Pi - {}^{2S+1}\Pi$ transitions via a $^{2S+1}\Sigma^\pm$ virtual intermediate state. For this $\Pi - (\Sigma) - \Pi$ special case, although only one sign of Λ_i' remains after the pre-summation, two distinct transition paths exist and are specified by $\Lambda'_f = \Lambda'_i$ and $\Lambda'_f = -\Lambda'_i$, respectively $\Delta\Lambda = 0$ and $\Delta\Lambda = -2$.

Equation (6.1.51) contains a product of two 3-j coefficients,

$$\begin{pmatrix} 1 & 1 & k \\ -q'_2 & -q'_1 & q'_1 + q'_2 \end{pmatrix} \text{ and } \begin{pmatrix} J_f & k & J_i \\ \Omega'_f & -q'_1 - q'_2 & -\Omega'_i \end{pmatrix},$$

which enforces a collapse of Eq. (6.1.51) for linearly polarized light to:

(i) for $\Delta\Lambda \neq 0$,

$$\sigma^{(2)}_{J_f,J_i} \propto (2J_f+1)(2J_i+1)\frac{2}{3}\left[Q^2_{\Delta\Lambda}P^2_{\Delta\Lambda}\right]^2, \tag{6.1.52}$$

(ii) for $\Delta\Lambda = 0$ but not $\Pi - \Pi$,

$$\sigma^{(2)}_{J_f,J_i} \propto (2J_f+1)(2J_i+1)\left[\frac{1}{3}\left(Q^0_0P^0_0\right)^2 + \frac{2}{3}\left(Q^2_0P^2_0\right)^2\right] \tag{6.1.53}$$

(iii) and for $\Pi - \Pi$

$$\sigma^{(2)}_{J_f,J_i} \propto (2J_f+1)(2J_i+1)\left[\frac{1}{3}\left(Q^0_0P^0_0\right)^2 + \frac{2}{3}\left(Q^2_0P^2_0 + Q^2_{-2}P^2_{-2}\right)^2\right], \tag{6.1.54}$$

where the Q^k terms contain all of the intrinsic molecular properties (a product of two body-frame transition moments and the $(E_{ji} - E)^{-1}$ energy denominator) and the P^k terms contain all of the basis function mixing coefficients and rotational properties that describe the rotationally averaged laboratory frame orientation of the molecule. For circularly polarized light, the terms of rank 2 have to be multiplied by 1 instead of 2/3, and the $\left(Q^0_0P^0_0\right)^2$ terms vanish. When $k = 0$, the two 3-j coefficients are nonzero only when $J' = J''$ (Q–branch) and $\Omega' = \Omega''$ ($\Delta\Lambda = 0$). When $k = 2$, the selection rules are $\Delta J = \pm 2, \pm 1, 0$ (O, P, Q, R, and S branches) and $|\Delta\Omega| = |\Delta\Lambda| = 2, 1$, and 0. Thus the transition probabilities for all rotational branches in a $\Delta\Omega \neq 0$ two-photon band are determined by the same squared sum over virtual states $\left(Q^2_{\Delta\Lambda}P^2_{\Delta\Lambda}\right)^2$ probability terms. For $\Delta\Lambda = 0$ transitions, there is an additional contribution exclusively to Q branches from $\left(Q^0_0P^0_0\right)^2$ probability terms. This contribution is responsible for the failure of Q branches in $\Delta\Lambda = 0$ transitions to follow the otherwise universal 2:3 linear:circular polarization line-by-line intensity ratio. However, uniquely for $\Pi - \Pi$ transitions that proceed via a Σ virtual state, there is an interference between two-photon transition amplitudes, $\left(Q^2_0P^2_0 + Q^2_{-2}P^2_{-2}\right)^2$, that alters the pattern of rotational branch relative intensities, particularly for pairs of branches that differ only in the e/f symmetry of the initial and final states (e.g., $O_{ee}(J)$ vs. $O_{ff}(J)$, $P_{ef}(J)$ vs. $P_{fe}(J)$, etc.). If only one virtual Σ state is relevant (single intermediate state approximation), then the sense of the interference effect is determined entirely by whether the virtual state is Σ^+ or Σ^-, not by the relative signs of the μ_{fj} and μ_{ji} transition moments. Hippler (1999) relates, for this case, all of the nonzero Q terms to a common factor, Q,

$$Q_0^0 = 3^{-1/2}Q \tag{6.1.55}$$

$$Q_0^2 = 30^{-1/2}Q \tag{6.1.56}$$

$$Q_{-2}^2 = (-1)^{s_i}5^{-1/2}Q \quad (s_i = 1 \text{ for } \Sigma^-,\ s_i = 0 \text{ for } \Sigma^+) \tag{6.1.57}$$

$$Q = \frac{\mu_1^{fj}\mu_{-1}^{ji}}{E_{ji} - E}. \tag{6.1.58}$$

The $\mathbf{T}_0^{(2)} \Big/ \mathbf{T}_{-2}^{(2)}$ interference effect is most clearly expressed in the O and S extreme ΔJ rotational branches because these two-photon transitions proceed through the virtual state by a single $P + P$ or $R + R$ transition path. When multiple paths (e.g., $P = P + Q$ and $P = Q + P$) contribute to a two-photon transition, the $\mathbf{T}_0^{(2)} \Big/ \mathbf{T}_{-2}^{(2)}$ interference effect can be obscured or partially cancelled.

The interference effect described here has been observed and confirmed in experimental spectra (Hippler, 1999): in the $C^2\Pi(v' = 0) \leftarrow X^2\Pi(v'' = 0)$ two-photon band of NO, some expected rotational branches are clearly absent (Gauyacq, *et al.*, 1991; Hippler and Pfab, 1998). The parity component O_{1f2f} is absent in the rotational branch O_{12} even though the splitting into two parity components should be large enough to be resolvable in the experimental spectrum, whereas in the Q branches, for example, the splitting into two parity components is evident.[†] It was proposed (Hippler and Pfab, 1998) that destructive interference of absorption tensor components might produce the quenching of Λ–doublet resolved rotational branches. This two-photon transition in NO is an ideal test case for the theory outlined by Hippler, 1999. The $A^2\Sigma^+$ electronic state is the dominant single intermediate state in the two-photon absorption, since the one-photon A–X absorption strength from the ground state, X, and the C–A absorption strength to the final state, C, are both rather strong, and the one-photon energy mismatch is not too severe. It is thus possible to simulate the line strengths of the two-photon spectrum without adjustment of any fit parameter and to compare the simulation with experiment. In the $\Pi \leftarrow \Pi$ transition, the rank 0 tensor contributes to the Q branch and the rank 2 tensors with the interfering components 0 and -2 to all rotational branches. Experiment and simulation are in very satisfactory agreement. The relative intensities of rotational branches are well reproduced (Hippler, 1999). As noted before (Gauyacq,

---

[†] The right subscripts 1, 2, 3 on the rotational branch designators specify the $F_i(J)$ rotational fine structure components (see Fig. 3.2) for the upper and lower electronic states according to the convention $i = 1 \leftrightarrow N = J - S, \ldots, i = 2S + 1 \leftrightarrow N = J + S$, or, when N is ill-defined, $i = 1$ and $i = 2S + 1$ correspond respectively to the lowest and highest energy eigenstates with the specified J–value (when $\Lambda \neq 0$ the $2S + 1$ i–values specify an e, f pair of eigenstates, but when $\Lambda = 0$ the $2S + 1$ i–values specify the energy-order of eigenstates with e, f symmetries in alternating energy order: $F_1 \leftrightarrow e$ for $^2\Sigma^+$, $^3\Sigma^-$, $^4\Sigma^+$, etc. and $F_{2S+1} \leftrightarrow f$ for all $^{2S+1}\Sigma^+$ states).

et al., 1991; Hippler and Pfab, 1998), the absence of branches can also be explained intuitively by parity selection of virtual one-photon absorption steps via the $A^2\Sigma^+$ state as the dominant intermediate state (see Fig. 6.2). An $O_{1f,1f}$ transition, for example, may only be accomplished by a virtual one-photon transition $P_{(2),1f}$ from the X to the intermediate A state and a $P_{1f,(2)}$ transition from the A to the C state, and similarly, an $O_{1e,1e}$ transition by $P_{(1),1e}$ and $P_{1e,(1)}$ (intermediate levels are shown in parentheses for clarity). Since the C state is in Hund's case (b), satellite rotational branches, such as $P_{1f,(2)}$, are very weak, and therefore the parity component $O_{1f,1f}$ has negligible transition strength relative to $O_{1e,1e}$. The O_{12} branch also exhibits parity selectivity. As before, an $O_{1e,2e}$ transition is only accomplished by virtual $P_{(1),2e}$ and $P_{1e,(1)}$ transitions, and $O_{1f,2f}$ by $P_{(2),2f}$ and $P_{1f,(2)}$. Since the satellite rotational branch $P_{1f,(2)}$ transition to the C state is very weak, $O_{1f,2f}$ parity components are missing in the spectrum. The X state, however, is in an intermediate coupling case, and hence the A–X satellite branch $P_{(1),2e}$ does carry intensity. The parity component $O_{1e,2e}$ thus appears in the two-photon spectrum. Similar parity selectivity arises for S branches. Other branches are less affected, since several one-photon virtual pathways contribute to the two-photon transition amplitude. These results are contained in the theoretical description based on the interference of tensor components discussed above.

There is an important difference between the mechanisms of the $T_0^{(2)}/T_{-2}^{(2)}$ interference effect for ${}^{2S+1}\Pi \leftarrow ({}^{2S+1}\Sigma^{\pm}) \leftarrow {}^{2S+1}\Pi$ transitions with integer vs. half-integer S. This difference is best illustrated by comparing the $\Delta\Lambda = -2$ and $\Delta\Lambda = 0$ transition paths in the case (a) limit:

$$\Delta\Lambda = -2:$$
$$|\Lambda_f = -1, \Sigma; \Omega_f = \Sigma - 1\rangle \leftarrow |\Lambda_j = 0, \Sigma; \Omega_j = \Sigma\rangle \leftarrow |\Lambda_i = 1, \Sigma; \Omega_i = \Sigma + 1\rangle$$
$$\Delta\Lambda = 0:$$
$$|\Lambda_f = 1, \Sigma; \Omega_f = \Sigma + 1\rangle \leftarrow |\Lambda_j = 0, \Sigma; \Omega_j = \Sigma\rangle \leftarrow |\Lambda_i = 1, \Sigma; \Omega_i = \Sigma + 1\rangle.$$

These two transition paths must terminate on substates with different values of $|\Omega_f|$, except when $\Sigma = 0$. However, $\Sigma = 0$ is possible only for integer–S (even number of electrons) molecules. For integer–S molecules, the interference effect demands a $\Sigma = 0$ virtual state (0^+ or e–symmetry for ${}^1\Sigma^+, {}^3\Sigma^-, {}^5\Sigma^+$; 0^- or f–symmetry for ${}^1\Sigma^-, {}^3\Sigma^+, {}^5\Sigma^-$). In the case (b) limit, there is no $\Sigma = 0$ character in the F_2 spin-components of ${}^3\Sigma$ states and in the F_2 and F_4 components of ${}^5\Sigma$ states. For half-integer-S molecules, the interference effect is most visible when the initial and final ${}^{2S+1}\Pi$ states are near opposite extremes of Hund's case (a) vs. (b) coupling.

For integer–S $\Pi - \Pi$ two–photon transitions, the $\mathbf{T}_0^{(2)}/\mathbf{T}_{-2}^{(2)}$ interference effect contributes to the intensity of $|\Delta J| = 2$ O and S branches as follows:

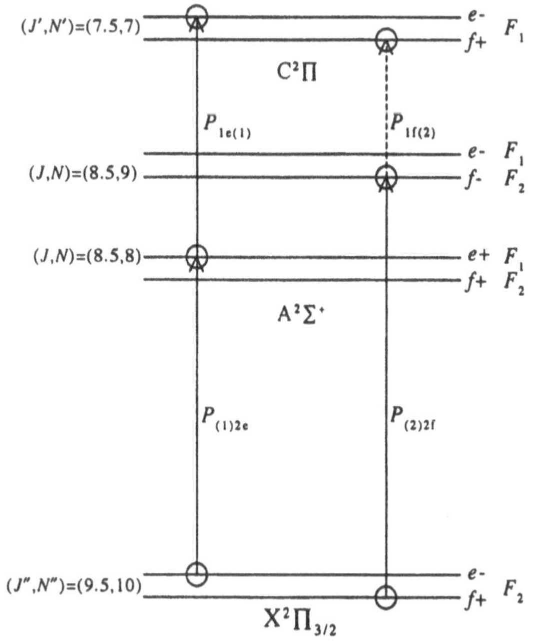

Figure 6.2: Allowed two-photon pathways via the NO $A^2\Sigma^+$ virtual state for the two-photon O_{1e2e} and O_{1f2f} rotational branches. Solid lines indicate a strongly allowed pathway and dashed lines indicate a weakly allowed pathway. The F_1, F_2 character of the virtual intermediate state is specified in parentheses in the one-photon branch labels. The A–X $P_{(1)2e}$ nominally satellite transition is strong because the A–X transition is case(b)-case(a) and the C–A $P_{1f(2)}$ satellite transition is weak because the C–A transition is case(b)-case(b).

$^1\Pi$ O_{ee} and S_{ee} via a $^1\Sigma^+$ virtual state
O_{ff} and S_{ff} via a $^1\Sigma^-$ virtual state

$^3\Pi$ (case (a) - case (a)) $^3\Pi_1 - {}^3\Pi_1$ sub-band, exclusively:
O_{ee} and S_{ee} via a $^3\Sigma^-$
O_{ff} and S_{ff} via a $^3\Sigma^+$

$^3\Pi$ (case (b) - case (b)):
O_{1e1e}, O_{3e3e}, S_{1e1e}, and S_{3e3e}, via $^3\Sigma^-$
O_{1f1f}, O_{3f3f}, S_{1f1f}, and S_{3f3f}, via $^3\Sigma^+$.

For half-integer S $\Pi - \Pi$ case (b) - case (a) two-photon transitions (see Fig. 6.2), the O and S branches that are most strongly affected by the $\mathbf{T}_0^{(2)} \Big/ \mathbf{T}_{-2}^{(2)}$ interference effect are, for $^2\Pi(\text{b}) - ({}^2\Sigma^+) - {}^2\Pi(\text{a})$:

O and S $1e2e, 1e1e, 2f1f$, and $2f2f$ enhanced
O and S $1f2f, 1f1f, 2e1e$, and $2e2e$ suppressed.

A $^2\Sigma^-$ virtual state would reverse the sense of the interference effect.

Two-photon transitions between $^2\Pi$ states in Hund's case (a), case (b) and in intermediate coupling cases are discussed in more detail in Hippler, (1999), where also a more detailed account of the theory can be found.

A special situation applies to two-photon transitions from Σ initial states: since the electronic selection rule is $\Delta\Lambda = 0, \pm1, \pm2$, transitions from a $^1\Sigma_g^+$ initial state can occur to $^1\Sigma_g^+$, $^1\Pi_g$ and $^1\Delta_g$ final states. For one-color, two-photon transitions, the $\mathbf{T}_0^{(0)}$ and $\mathbf{T}_0^{(2)}$ components belong to Σ^+ symmetry, the $\mathbf{T}_{\pm1}^{(2)}$ components to Π symmetry, and the $\mathbf{T}_{\pm2}^{(2)}$ component to Δ symmetry (Ashfold, 1986, Table 1).

Some general features of the results are (see Bray and Hochstrasser, 1976; McClain and Harris, 1977; and Metz, *et al.*, 1978):

i.) in $^1\Sigma^\pm - {}^1\Sigma^\pm$ transitions, P and R branches do not occur.

ii.) $\Sigma^+ - \Sigma^-$ transitions cannot occur in two-photon absorption spectra if the two photons have the same frequency and polarization. When the two photons have different frequencies, it is possible to observe $\Sigma^+ - \Sigma^-$ transitions in two photon absorption. (See Garetz and Kittrell (1984) for a discussion of the CO $\mathrm{I}^1\Sigma^- - \mathrm{X}^1\Sigma^+$ transition, which is forbidden in the limit of two photons of identical polarization and frequency.)

Unlike single photon absorption processes, the cross sections for multiphoton transitions are polarization dependent.

6.1.2.3 Three-Photon Transitions

Since the electronic selection rule is $\Delta\Lambda = 0, \pm1, \pm2, \pm3$, three-photon excitations from, for example, a $^1\Sigma_g^+$ state, can terminate in $^1\Delta_u$ and $^1\Phi_u$ states as well as the $^1\Sigma_u^+$ and $^1\Pi_u$ states accessible via one-photon transitions.

It can be shown that for one-color, three-photon transitions, (Dixon, *et al.*, 1984), the electronic transition operator has the general form

$$\mathbf{T}(\boldsymbol{\mu}) = \mathbf{T}^{(1)}(\boldsymbol{\mu}) + \mathbf{T}^{(3)}(\boldsymbol{\mu}), \qquad (6.1.59)$$

and similarly for the photon polarization operator, $\mathbf{T}(\boldsymbol{\epsilon})$.

The three-photon transition probability is proportional to $\sigma_{fi}^{(3)}$, as defined by Eq. (6.1.30), which includes terms involving the square modulus of scalar products of molecular transition moments and photon polarization tensors. The complicated looking expression is reduced as follows.

After summation over the $M_{J'}$ and $M_{J''}$ components of each J' and J'' level, and using the orthogonality relations for summations over products of 3-j coefficients, all cross terms involving $k \neq k'$, $p \neq p'$, and $q \neq q'$ may be eliminated (see Dixon, *et al.*, 1984),

$$\sum_{M'}\sum_{M''}\begin{pmatrix} J' & k' & J'' \\ M' & -p' & -M'' \end{pmatrix}\begin{pmatrix} J' & k & J'' \\ M' & -p & -M'' \end{pmatrix} = \frac{1}{(2k+1)}\delta_{kk'}.\delta_{pp'}. \qquad (6.1.60)$$

Thus,

$$\sigma^{(3)}_{\Omega',J',\Omega'',J''}$$
$$=\sum_p \frac{|\mathbf{T}^{(1)}_{-p}(\epsilon)|^2}{3}\sum_q \left|\left\langle \Omega'\left|\mathbf{T}^{(1)}_q(\mu)\right|\Omega''\right\rangle\right|^2 (2J'+1)(2J''+1)\begin{pmatrix} J' & 1 & J'' \\ \Omega' & -q & -\Omega'' \end{pmatrix}^2$$
$$+\sum_p \frac{|\mathbf{T}^{(3)}_{-p}(\epsilon)|^2}{7}\sum_q \left|\left\langle \Omega'\left|\mathbf{T}^{(3)}_q(\mu)\right|\Omega''\right\rangle\right|^2 (2J'+1)(2J''+1)\begin{pmatrix} J' & 3 & J'' \\ \Omega' & -q & -\Omega'' \end{pmatrix}^2, \tag{6.1.61}$$

where the factors $\begin{pmatrix} J' & k & J'' \\ \Omega' & -q & -\Omega'' \end{pmatrix}^2$ are generalized Hönl-London factors, from which the selection rules $\Delta J = 0, \pm 1, \ldots, \pm k, \Delta\Omega = \pm q$ are immediately evident. The $\mathbf{T}^{(1)}_0$ and $\mathbf{T}^{(3)}_0$ components belong to Σ^+ symmetry, the $\mathbf{T}^{(1)}_{\pm 1}$ and $\mathbf{T}^{(3)}_{\pm 1}$ components belong to Π symmetry, the $\mathbf{T}^{(3)}_{\pm 2}$ components belong to Δ symmetry, and the $\mathbf{T}^{(3)}_{\pm 3}$ components belong to Φ symmetry.

The relationship between the different components of the two tensor products in Eq. (6.1.61) and the contributions of the terms of the general expression Eq. (6.1.30) is explained in Table 3 of Dixon, *et al.*, (1984). For example, $\mathbf{T}^{(3)}_{\pm 3}(\mu)$ specifies the sequence of one-photon transitions, $\Sigma \to \Pi \to \Delta \to \Phi$ and $\mathbf{T}^{(3)}_{\pm 2}$ represents the three transition sequences, $\Sigma \to \Pi \to \Delta \to \Delta$ and $\Sigma \to \Sigma \to \Pi \to \Delta$, and $\Sigma \to \Pi \to \Pi \to \Delta$.

Equations (6.1.30) and (6.1.61) are valid when the initial and final states are in Hund's case (a). In the case of orbitally degenerate $S \neq 0$ states, if either or both of the initial and final states are intermediate between Hund's cases (a) and (b), interference effects can occur, as illustrated by Gauyacq, *et al.*, (1987) for the NO $C^2\Pi - X^2\Pi$ three-photon transition.

6.1.2.4 Comparisons between Excitation Schemes Involving Different Numbers of Photons

Comparisons between one-photon and multiphoton spectra are often useful for establishing the symmetry of an electronic state because the relative intensities of different rotational branches exhibit diagnostic polarization dependences which depend on photon number (see Table 2 of Ashfold, 1986). For example, (Maïnos, *et al.*, 1985), the ratio of cross sections for excitation using two circularly polarized photons relative to that for excitation using two linearly polarized photons of the same intensity is 3/2 for the $\mathbf{T}^{(2)}$ 2-photon transition tensor: for a transition with $\Delta\Omega = \pm 2$ or $\Delta\Omega = \pm 1$, this ratio 3/2 is for the branches O, P, Q, R, S, for $\Delta\Omega = 0$, this ratio 3/2 is for O, P, R, S only. This has been verified experimentally for the ${}^2\Sigma^+ \leftarrow\leftarrow {}^2\Pi$ transition of NO (Bray, *et al.*, 1975). This ratio is 5/2 for the $\mathbf{T}^{(3)}$ 3-photon transition tensor: for $\Delta\Omega = \pm 2, \pm 3$ transitions for all ΔJ branches, while for $\Delta\Omega = 0, \pm 1$, the ratio applies only for the $\Delta J = \pm 2, \pm 3$ branches (verified experimentally for the

$^1\Delta_u \leftarrow\leftarrow\leftarrow {}^1\Sigma_g^+$ transition in CO_2, Wu and Johnson, 1989).

Another striking example of the utility of a comparison of rotational branch intensity patterns in one- vs. multiphoton spectra is in the spectrum of O_2, where a mixed Rydberg-valence upper state exhibits different rotational branch interference effects which depend on its 1 vs. 3 photon excitation mode. When the O_2 $F^3\Pi_u$ state, (which is the result of strong interaction between the diabatic $4p\sigma_u$ Rydberg and repulsive $3\sigma_u$ valence states (Rydbergization of the $3\sigma_u$ orbital)) (see Fig. 5.12), is excited from the $X^3\Sigma_g^-$ state, the $(1,0)$ band is the strongest $(v',0)$ band in the one-photon spectrum. However, this $(1,0)$ band disappears in the three-photon $F^3\Pi_u \leftarrow X^3\Sigma_g^-$ spectrum, which instead is dominated by the $(0,0)$ band (Lewis, *et al.*, 1999). The character of the interference effects in transitions which terminate in such a Rydberg∼valence mixed state is essentially determined by the ratio of the valence and Rydberg transition moments (see Section 6.3.1). This transition moment ratio can differ in the one- vs. three-photon $F^3\Pi_u \leftarrow X^3\Sigma_g^-$ excitation schemes because, for the one-photon transition the interference depends on the ratio of valence and Rydberg transition amplitudes from the X-state, whereas, for the three-photon case, the interference depends on the valence and Rydberg transition amplitudes from the intermediate (virtual) state.

6.2 Intensity Borrowing

6.2.1 Perturbations by States with "Infinite" Radiative Lifetime; Simple Intensity Borrowing

Let the vibronic basis functions 1 and 2 correspond to a mutually interacting pair of electronic levels,

$$\begin{aligned} |1\rangle \quad \text{with} \quad E_1^0 &= \langle 1|\mathbf{H}^0|1\rangle \\ |2\rangle \quad \text{with} \quad E_2^0 &= \langle 2|\mathbf{H}^0|2\rangle \\ H_{12} &= \langle 1|\mathbf{H}|2\rangle \,. \end{aligned}$$

The simplest and most common situation is that basis functions 1 and 2 have, respectively, appreciable and negligible transition probability into a lower energy basis function 0,

$$I_{10}^0 \gg I_{20}^0.$$

I_{20}^0 will be negligible whenever the lifetime of basis function 2 is infinite,

$$\tau_2^0 = \left(\sum_i A_{2i}^0\right)^{-1} \rightarrow \infty$$

or, less stringently, whenever

$$\left(A_{20}^0\right)^{-1} \rightarrow \infty$$

(i.e., τ_2^0 is not infinite). The former case corresponds to Eq. (6.2.11). The latter case, often misleadingly referred to as "perturbation by a level without oscillator strength" (meaning $f_{20}^0 = 0$ but not that $f_{2i}^0 = 0$ for all i), corresponds to (Eq. (6.2.11)′).

The result of the H_{12} interaction is two mixed eigenstates, the higher-energy eigenstate, $|+\rangle$, corresponding to the eigenenergy

$$E_+ = \overline{E^0} + \left[\left(\Delta E^0/2\right)^2 + H_{12}^2\right]^{1/2}, \tag{6.2.1a}$$

and the lower energy eigenstate, $|-\rangle$, corresponding to

$$E_- = \overline{E^0} - \left[\left(\Delta E^0/2\right)^2 + H_{12}^2\right]^{1/2} \tag{6.2.1b}$$

$$\overline{E^0} = \left(E_1^0 + E_2^0\right)/2 \tag{6.2.2}$$

$$\Delta E^0 = E_1^0 - E_2^0 \tag{6.2.3}$$

and, if $E_1^0 > E_2^0$, the energy shift is

$$\delta E = E_+ - E_1^0 = -\left(E_- - E_2^0\right) \tag{6.2.4}$$

$$\delta E = \left[\left(\Delta E^0/2\right)^2 + H_{12}^2\right]^{1/2} - \Delta E^0/2. \tag{6.2.5}$$

The eigenfunctions are

$$|+\rangle = C_{1+}|1\rangle + C_{2+}|2\rangle \tag{6.2.6a}$$

where, by the normalization requirement,

$$C_{2+} = \pm\left[1 - C_{1+}^2\right]^{1/2},^\dagger \tag{6.2.6b}$$

and, by the orthogonality requirement,

$$|-\rangle = \mp\left[1 - C_{1+}^2\right]^{1/2}|1\rangle + C_{1+}|2\rangle^\dagger \tag{6.2.6c}$$

$$C_{1+} = 2^{-1/2}\left[1 + \left(\Delta E^0/2\right)\left[\left(\Delta E^0/2\right)^2 + H_{12}^2\right]^{-1/2}\right]^{1/2} \geq 0. \tag{6.2.7}$$

Note that, whenever $\Delta E^0 > 0$, regardless of the magnitude or sign of H_{12}, $C_{1+} \geq 2^{-1/2}$ and thus $|+\rangle$ and $|-\rangle$ are respectively the "nominal" perturbed $|'1'\rangle$ and $|'2'\rangle$ functions.

Both $|+\rangle$ and $|-\rangle$ have nonzero transition probability into level 0,

$$I_{+0} \propto |\langle +|\mu|0\rangle|^2$$

$$I_{+0} = C_{1+}^2 I_{10}^0 \tag{6.2.8a}$$

$$I_{-0} = \left(1 - C_{1+}^2\right) I_{10}^0. \tag{6.2.8b}$$

---

† The correct sign choice for the $\left[1 - C_{1+}^2\right]^{1/2}$ term (C_{2+}) in Eq. (6.2.6b) and Eq. (6.2.6c) is the top sign for $H_{12} > 0$. The opposite choice would cause $|+\rangle$ to belong to the eigenenergy E_-.

In the absence of the $1 \sim 2$ interaction ($H_{12} = 0$), only *one* line, $1 \leftrightarrow 0$, of intensity I_{10}^0 is observable. When the interaction is turned on ($H_{12} \neq 0$), *two* transitions become observable. These are called *main* and *extra* lines. For the specific case of a two-level interaction for which $I_{10}^0 \gg I_{20}^0$ (and $E_1^0 > E_2^0$), the level (E_+) giving rise to the main line is, by definition, the level with larger basis function 1 character. Thus the main line is necessarily more intense than the extra line. In addition, the main and extra lines lie above and below the calculated position of the (nonexistent) deperturbed $E_1^0 \leftrightarrow E_0$ transition, with the main line lying closer than the extra line to $E_1^0 - E_0$.

The occurrence of two perturbed lines where there would be only one unperturbed line is where the idea of *extra lines* originates. The metaphor, *intensity borrowing*, arises from the conservation of total intensity,

$$I_{+0} + I_{-0} = I_{10}^0, \tag{6.2.9}$$

which is a necessary consequence of a simple two-level interaction (provided $I_{10}^0 \gg I_{20}^0$). Basis function 1 character is conserved, as are the properties of basis function 1.

This is where important differences in the effect of perturbations on measured intensities and lifetimes become evident. Let $\tau_2^0 = \infty$ (the related case of $A_{20}^0 = 0$ and τ_2^0 finite will be discussed below).

In a high resolution experiment (transitions into levels E_+ and E_- are resolved), one finds

$$I_{+0} = C_{1+}^2 I_{10}^0 \tag{6.2.8a}$$

$$\tau_+ = \tau_1^0 C_{1+}^{-2} \tag{6.2.10a}$$

$$\tau_- = \tau_1^0 \left(1 - C_{1+}^2\right)^{-1} \tag{6.2.10b}$$

$$1/\tau_+ + 1/\tau_- = 1/\tau_1^0. \tag{6.2.11}$$

Note that Eq. (6.2.9) requires that intensity be conserved whereas Eq. (6.2.11) requires that $1/\tau$, not τ, be conserved. In a low-resolution experiment (levels E_+ and E_- not resolved), measurement of integrated intensity would show no perturbation effect, whereas a lifetime measurement (assuming broad bandwidth, nonsaturating excitation) would show biexponential decay,

$$I(t) = I_{+0}(t) + I_{-0}(t) \tag{6.2.12}$$

$$I_{+0}(t) = I_{10}^0 C_{1+}^2 \exp\left[-t\left(C_{1+}^2/\tau_1^0\right)\right] \tag{6.2.13a}$$

$$I_{-0}(t) = I_{10}^0 \left(1 - C_{1+}^2\right) \exp\left[-t\left(1 - C_{1+}^2\right)/\tau_1^0\right]. \tag{6.2.13b}$$

Note that the intensity decay from either mixed level separately is single-exponential [Eqs. (6.2.13a) and (6.2.13b)] but the unresolved fluorescence decay is biexponential [Eq. (6.2.12)]. Failure to recognize this biexponential character can yield an effective τ–value that depends artificially on the relative weighting of early versus late fluorescence, both in the actual experiment and in the fitting algorithm. When fluorescence from the extensively perturbed CO $A^1\Pi$ $v = 0$, 1, and 6 levels is excited by electron bombardment (Imhof and Read, 1971) or by 15 Å bandwidth synchrotron radiation (Field, *et al.*, 1983), highly nonexponential decay is observed. Also, if collisions transfer population from a single, selectively populated level into nearby levels (a process facilitated by perturbation, see Section 6.5.5), multiexponential decay is observed (Banic, *et al.*, 1981).

The main line can have intensity and lifetime in the ranges

$$I_{10}^0 \geq I_{+0} \geq 0.5 I_{10}^0 \tag{6.2.14}$$

$$\tau_1^0 \leq \tau_+ \leq 2\tau_1^0. \tag{6.2.15}$$

However, the effect of the perturbation on the extra line appears much more dramatic than a mere factor of 2 when one recalls that $I_{20}^0 = 0$ and $\tau_2^0 = \infty$. The perturbation can increase the intensity and radiative decay rate of an extra line by many orders of magnitude, whereas, for the main line, the perturbation decreases the intensity and decay rate by at most a factor of 2.

When $I_{20}^0 = 0$ but $\tau_2^0 \simeq \tau_1^0$ (e.g., suppose levels 2 and 1 are, respectively, upper levels of allowed triplet-triplet and singlet-singlet systems), all of the perturbation effects on intensities [Eqs. (6.2.8a), (6.2.8b), (6.2.9), and (6.2.14)] are unchanged from the $\tau_2^0 = \infty$ case, but the perturbation effects on lifetime are slightly different:

$$\tau_+ = \left[\left(1 - C_{1+}^2\right)/\tau_2^0 + C_{1+}^2/\tau_1^0\right]^{-1} \tag{6.2.10a$'$}$$

$$\tau_- = \left[\left(1 - C_{1+}^2\right)/\tau_1^0 + C_{1+}^2/\tau_2^0\right]^{-1} \tag{6.2.10b$'$}$$

$$1/\tau_+ + 1/\tau_- = 1/\tau_1^0 + 1/\tau_2^0 \tag{6.2.11$'$}$$

$$I_{+0}(t) = I_{10}^0 C_{1+}^2 \exp\left(-t/\tau_+\right) \tag{6.2.13a$'$}$$

$$I_{-0}(t) = I_{10}^0 \left(1 - C_{1+}^2\right) \exp\left(-t/\tau_-\right). \tag{6.2.13b$'$}$$

In contrast to the simple two-level problem, when level 1 is perturbed by many levels that do not have oscillator strength to level 0, the intensity and lifetime of nominal level 1 can be altered by far more than a factor of 2. Such a multistate perturbation is unusual for a diatomic molecule, but for polyatomic molecules with their huge density of vibronic perturbing levels, lifetime lengthening by factors much larger than 2 can occur. Douglas (1966) explained the paradoxical observation of lifetimes considerably longer than those predicted on the basis of integrated absorption strength, which is insensitive to perturbation effects. Recall that, in a low-resolution absorption experiment (i.e., absorption integrated over an entire band system), perturbation effects on intensities vanish whereas radiative decay becomes slower than τ_1^0 and nonexponential.

There are several reasons why a perturbing level might have negligible oscillator strength $\left(A_{20}^0 \ll A_{10}^0\right)$ to the level 0 of the monitoring transition. For example, the $A^1\Pi$ state of CO is perturbed by $a'^3\Sigma^+$, $e^3\Sigma^-$, $d^3\Delta$, $I^1\Sigma^-$, and $D^1\Delta$ states (see Fig. 3.19). Transitions from the a′, e and d states into $X^1\Sigma^+$ are spin-forbidden ($\Delta S \neq 0$), whereas the D–X and I–X transitions, although spin-allowed, are forbidden by $\Delta\Lambda = 0, \pm 1$ and $\Sigma^+ \leftarrow / \rightarrow \Sigma^-$ electric dipole selection rules, respectively. Although the a′, e and d states (unlike the D and I states) all have fully allowed transitions to the $a^3\Pi$ state, because of the ν^3 factor in the Einstein A–coefficient their radiative lifetimes are at least 10^2 times longer than $\tau(A^1\Pi) \simeq 10$ns, so the perturbations of the CO $A^1\Pi$ state may be treated as if $\tau_2^0 = \infty$.

In the spectrum of the CN molecule, two electric dipole allowed band systems, $B^2\Sigma^+ - X^2\Sigma^+$ and $A^2\Pi - X^2\Sigma^+$, are observed in two quite separate wavelength regions. B–X bands do not appear in the red region where A–X is found because Franck-Condon factors restrict strong B–X bands to $\Delta v = 0, \pm 1$ transitions. This leads to an unusual situation for the CN $A^2\Pi(v_A = 10) \sim B^2\Sigma^+(v_B = 0)$ perturbation, where, when monitoring the A–X 10-4 band, the perturbing level has a shorter intrinsic lifetime, $\tau_B^0 \ll \tau_A^0$, than the perturbed level. Even so, because the B–X 0-4 band has such a small Franck-Condon factor, $q_{0,4}^{BX} = 2.1 \times 10^{-5}$, the $B^2\Sigma^+$ $(v_B = 0)$ "extra" level must borrow transition strength from $A^2\Pi$ $(v_A = 10), q_{10,4}^{AX} = 3.5 \times 10^{-2}$.

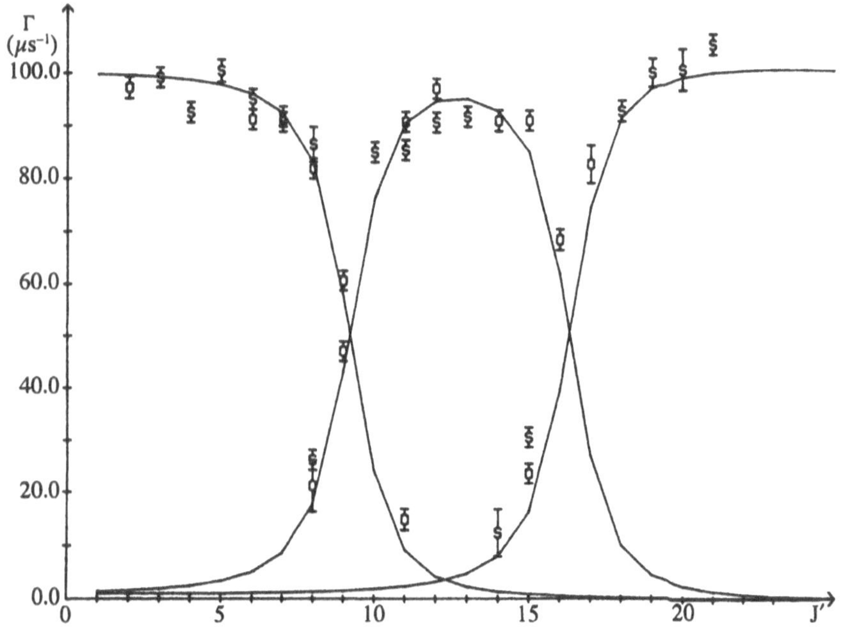

Figure 6.3: Radiative decay rates ($\Gamma = 1/\tau$) for single e–parity rotational levels of CO $A^1\Pi$ $(v = 0)$. The effects of perturbations by $e^3\Sigma^-(v = 1)$ are evident near $J = 9(F_1)$ and $J = 16(F_3)$. The points (O and S refer to branches in the two-photon spectrum) are measured and the solid curves depict values calculated for the nominal, $A^1\Pi$, $e^3\Sigma^-(F_1)$, and $e^3\Sigma^-(F_3)$ levels from the deperturbed $\tau_{v=0}^0 = 9.9$ ns of Field *et al.*, (1983) and mixing coefficients from the deperturbation analysis (Field, 1971). [From Girard, *et al.*, (1982).]

The N_2^+ ion, which is isoelectronic with CN, has similar $B^2\Sigma^+ \sim A^2\Pi$ perturbations. Dufayard, *et al.*, (1974) have measured radiative lifetimes for individual rotational lines of the $B^2\Sigma_u^+(v_B = 1) \sim A^2\Pi_u(v_A = 11)$ perturbation complex. The lifetimes of the perturbation-free $B^2\Sigma_u^+$ levels ($\tau_B^0 \simeq 60$ ns) are much shorter than those of $A^2\Pi_u$ ($\tau_A^0 \simeq 7\mu s$), and perturbed main lines of the B–X system are observed to have lifetimes as long as 95 ns. There is good qualitative agreement between measured lifetimes and those calculated using Eq. (6.2.10a) and B$\sim$A mixing coefficients derived from the molecular constants of Gottscho *et al.*, (1979). Provorov, *et al.*, (1977) and Girard, *et al.*, (1982) have measured single-rotational-level lifetimes for the CO $e^3\Sigma^-(v_e = 1) \sim A^1\Pi(v_A = 0) \sim d^3\Delta_1(v_d = 4)$ perturbation complex. Again, the agreement shown in Fig. 6.3 between measured (Maeda and Stoicheff, 1984; Girard, *et al.*, 1982) and calculated (Field, 1971; Field, *et al.*, 1983) perturbed lifetimes is satisfactory.

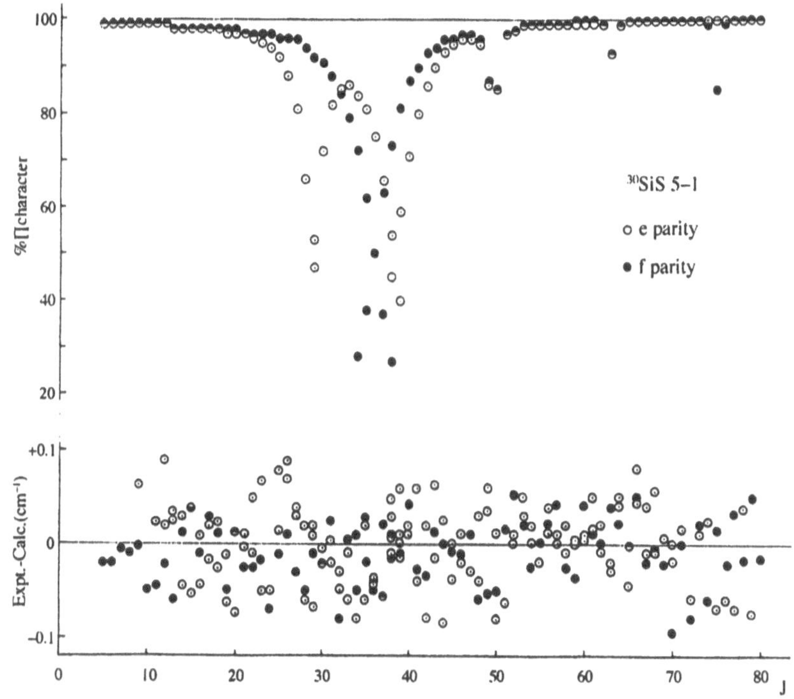

Figure 6.4: Mixing fractions in $^{30}Si^{32}S$ $A^1\Pi$ ($v = 5$). The deperturbation model fits all observed lines, main and extra, perturbed and unperturbed, in the spectrum without systematic residuals. Mixing coefficients computed from such a deperturbation model should be more accurate than from measurements of radiative lifetimes [From Harris, *et al.*. (1982).]

Kittrell, *et al.*, (1993), Kittrell and Garetz (1989), and Garetz, *et al.*, (1991) examined the CO $A^1\Pi(v = 6) \sim [D^1\Delta(v = 8) \sim d^3\Delta(v = 12), a'^3\Sigma^+(v = 17)]$ perturbation complex by VUV fluorescence-detected two-photon spectroscopy from the $X^1\Sigma^+$ state. Although both the A–X and D–X transitions are two-photon allowed, the only mechanisms by which VUV fluorescence can occur are via the fractional $A^1\Pi(v = 6)$ character in the two-photon prepared upper eigenstate or collision induced transfer from the two-photon prepared eigenstate into another eigenstate that has significantly larger fractional $A^1\Pi$ character. By recording the spectra at ~3 Torr (mean time between rotation-changing collisions ~ 30 ns) and simultaneously recording prompt (20 ns window, no delay after the excitation pulse) and delayed (100 ns window, 100 ns delay) VUV fluorescence, the spectrum could be easily sorted into pure $A^1\Pi$ states, states with significant $A^1\Pi$ character due to a perturbation, and unperturbed levels that are strongly collisionally coupled to locally perturbed levels.

It is almost invariably true that mixing coefficients, derived from measured frequencies of *correctly assigned main and extra lines*, provide at least as reliable a measure of the lifetime and relative oscillator strength of all perturbed main and extra lines as direct radiative decay measurements. This statement is supported indirectly by Fig. 6.4, which illustrates several perturbations of the ^{30}SiS $A^1\Pi$ ($v = 5$) level (Harris, *et al.*, 1982). Even though the percent $^1\Pi$ character drops to a minimum of less than 30%, no systematic deviations between the wave numbers of the observed and calculated levels are evident. From Eqs. (6.2.4), (6.2.5), and (6.2.7), one obtains

$$C_{1+}^2 = 1 - \frac{\delta E}{E_+ - E_-},$$

which implies that a 0.05 cm^{-1} error in a measured or computed transition frequency leads to an error of

$$\delta C_{1+}^2 \frac{0.05}{E_{\text{main}} - E_{\text{extra}}} < \frac{0.05}{\Delta E^0}$$

in the fractional character. If $E_+ - E_-$ were as small as 1 cm^{-1}, the errors in C_{1+}^2 and τ_+ would be only 5% and 0.05 τ_1^0, respectively.

The names "main" and "extra" should be applied only to transitions, not to levels. For example, the nominal $|'A^2\Pi', v_A = 10\rangle$ levels associated with main lines (i.e., stronger oscillator strength than extra lines) in the CN ($A, v_A = 10 \sim B, v_B = 0) \rightarrow X, v_X = 4$ band give rise to extra lines in the ($B, v_B = 0 \sim A, v_A = 10) \rightarrow X, v_X = 0$ band. Similarly, for a CO $A^1\Pi \sim d^3\Delta_1$ perturbation, the nominal $^1\Pi$ levels correspond to main lines in the $A^1\Pi - X^1\Sigma^+$ system and extra lines in the $d^3\Delta - a^3\Pi$ system.

It is not only the names "main" and "extra" or "perturbed" and "perturbing" that depend on which of two possible electronic transitions has oscillator strength in the spectral region being monitored; *the vibrational (Franck-Condon) and rotational (Hönl-London) linestrength factors that are applicable for both main and extra lines are those associated with the electronic-vibronic band with nonzero oscillator strength* . Complications occur when both bands of a perturbation complex have oscillator strength (as could happen for the CN $B \sim A \rightarrow X$ $0 \sim 10 \rightarrow 3$ band, but not for the CO singlet~triplet perturbations in the A–X or d–a bands). These effects are discussed in Section 6.3. For the general $1 \sim 2 \rightarrow 0$ problem where the electronic $1 \rightarrow 0$ system is allowed and $2 \rightarrow 0$ is forbidden, the linestrength factors defined by Eqs. (6.1.18) and (6.1.20) are

$$S^{\text{main}}_{1,v_1,J';0,v_0,J''} = \left[C_{1,\text{main}}(J')\right]^2 q_{v_1 v_0} \left|R_e^{1,0}\right|^2 \mathbf{S}_{J'J''} \tag{6.2.16a}$$

$$S^{\text{extra}}_{1,v_1,J';0,v_0,J''} = \left[1 - C_{1,\text{main}}(J')^2\right] q_{v_1 v_0} \left|R_e^{1,0}\right|^2 \mathbf{S}_{J'J''} \tag{6.2.16b}$$

$$S^{\text{extra}}_{1J';0J''} / S^{\text{main}}_{1J';0J''} = \left[C_{1,\text{main}}(J')\right]^{-2} - 1, \tag{6.2.17}$$

where q_{v_1,v_0}, $R_e^{1,0}$, and $\mathbf{S}_{J'J''}$ are the Franck-Condon, electronic transition moment, and Hönl-London factors for the unperturbed $1 \leftrightarrow 0$ electronic system.

Even if the mixed level has only 1% basis function 1 character, if $I^0_{10} \gg 100 I^0_{20}$, then the relevant Hönl-London, Franck-Condon, and transition moment factors for the nominal $2 \leftrightarrow 0$ transition are those of level 1, not those of level 2. The electronic and vibrational wavefunctions of basis function 2 are *completely irrelevant* to linestrengths in the $2 \rightarrow 0$ band system except insofar as they determine the magnitudes of H_{12} matrix elements.

If the intensity of the forbidden transition, into the v_{E} vibrational level is borrowed from several vibrational levels, v_{M}, of the main state, then

$$S^{extra}_{J'J''}(v_{\text{E}}) = \left[\sum_{v_{\text{M}}} C_{v_{\text{M}},v_{\text{E}}}(J') \langle v_{\text{M}}|v_{\text{X}}\rangle R_e^{\text{M},0}\right]^2 S_{J'J''}. \tag{6.2.18}$$

The mixing coefficient is given by first-order perturbation theory as

$$C_{v_{\text{M}},v_{\text{E}}}(J') = \frac{H_{\text{ME}} \langle v_{\text{E}}|v_{\text{M}}\rangle}{E^{\circ}_{v_{\text{E}}} - E^{\circ}_{v_{\text{M}}}} \tag{6.2.19}$$

thus

$$S^{extra}_{J'J''}(v_{\text{E}}) = \left[\sum_{v_{\text{M}}} \frac{H_{\text{ME}} \langle v_{\text{E}}|v_{\text{M}}\rangle \langle v_{\text{M}}|v_{\text{X}}\rangle}{E^{\circ}_{v_{\text{E}}} - E^{\circ}_{v_{\text{M}}}} R_e^{\text{M},0}\right]^2 S_{J'J''}. \tag{6.2.20}$$

An important point, noted by Rostas, *et al.*, 2000, is that the two vibrational overlap integrals must be calculated using the same phase convention. The signs of overlap integrals often go unnoticed, since these integrals are generally squared in expressions of transition intensities. However, the signs of products of overlap integrals become relevant when several transition amplitude contributions must be added before being squared to obtain a transition probability.

6.2.2 Multistate Deperturbation; The NO $^2\Pi$ States

The examples already cited from the spectra of CO, CN, and N_2^+ involve rather weak perturbations with highly J–dependent level shifts and mixing coefficients. The $^{14}N^{16}O$ $C^2\Pi$, $v_{\text{C}} = 3 \sim B^2\Pi, v_{\text{B}} = 15$ Rydberg$\sim$non-Rydberg (R$\sim$NR) interaction is so strong (interaction strength $H_{\text{C3;B15}} = 256$ cm^{-1}; deperturbed energy, spin-orbit, and rotational constants $E^0_{\text{B15}} = 59,478$ cm^{-1},

$A^0_{B15} = 53\ cm^{-1}$, $B^0_{B15} = 0.912\ cm^{-1}$, $E^0_{C3} = 59,373\ cm^{-1}$, $A^0_{C3} = 8\ cm^{-1}$, $B^0_{C3} = 1.903\ cm^{-1}$) that the result is two vibrational levels with rotational structure that, at first sight, appears to be perturbation-free (observed constants: $E^{obs}_{B15} = 59,663\ cm^{-1}$, $A^{obs}_{B15} = 38\ cm^{-1}$, $B^{obs}_{B15} = 1.280\ cm^{-1}$, $E^{obs}_{C3} = 59,224\ cm^{-1}$, $A^{obs}_{C3} = 33\ cm^{-1}$, $B^{obs}_{C3} = .360\ cm^{-1}$) (Gallusser and Dressler, 1982; Lagerqvist and Miescher, 1958). Every rotational level of this B~C complex is significantly mixed. Treated (inappropriately) as a two-level B ($v = 15$) ~ C ($v = 3$) interaction, the fractional admixed perturber character increases smoothly from ~ 40% at $J = \frac{1}{2}$ of $^2\Pi_{1/2}$ to 50% near $J = 10$ where the deperturbed B and C levels cross and then decreases slowly to ~ 20% by $J = 20.5$. However, the NO $B^2\Pi \sim C^2\Pi$ interaction is so strong that it is insufficient to think of pairwise $v_B \sim v_C$ vibrational level interactions. Gallusser and Dressler (1982) have simultaneously deperturbed the NO $B^2\Pi$, $C^2\Pi$, $K^2\Pi$, $L^2\Pi$, and $Q^2\Pi$ states (see Fig. 6.5) by diagonalizing two diabatic perturbation matrices (one for $^2\Pi_{1/2}$, the other for $^2\Pi_{3/2}$) of dimension 69 that include the $v_B = 0-37$, $v_C = 0-9$, $v_K = 0-4$, $v_L = 0-11$, and $v_Q = 0-3$ levels.

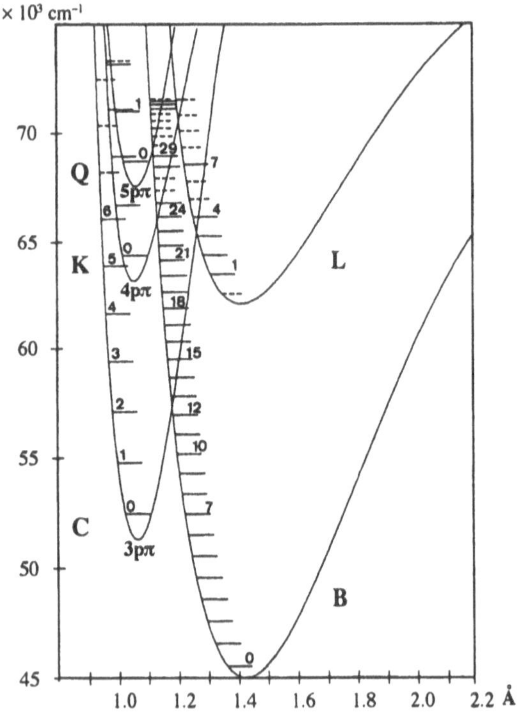

Figure 6.5: Deperturbed potential curves for the mutually interacting $^2\Pi$ valence and Rydberg states of NO. [From Gallusser and Dressler (1982).]

One product of this deperturbation is a set of calculated nonrotating-molecule absorption oscillator strengths (from $X^2\Pi$, $v_X = 0$) for the 69 × 2 vibronic eigenstates of the perturbation matrices corresponding to each of the three isotopic molecules $^{14}N^{16}O$, $^{15}N^{16}O$, and $^{14}N^{18}O$ (Gallusser and Dressler, 1982). Although computationally complex, it is conceptually simple, explaining all observables in terms of five diabatic Born-Oppenheimer potential energy curves, deperturbed electronic and vibrational properties of these five electronic states, and six R-independent R~NR electronic interaction parameters.

Gallusser and Dressler (1982) calculate that the deperturbed, nonrotating molecule oscillator strengths for several $B^2\Pi \sim C^2\Pi$ mixed pairs of vibrational levels differ by more than a factor of 40. Thus the $I_{10}^0 \gg I_{20}^0$ approximation is valid and the NO $C^2\Pi \sim B^2\Pi \rightarrow X^2\Pi$ $v_C = 2 \sim v_B = 12 \rightarrow v_X = 0$ and $v_C = 3 \sim v_B = 15 \rightarrow v_X = 0$ perturbation complexes may be understood in terms of simple intensity borrowing.

The $^{14}N^{16}O$ C~B←X (3~15,0) and B~C←X (15~3,0) bands appear with nearly equal intensity in the absorption spectrum of Lagerqvist and Miescher (1958). Gallusser and Dressler (1982) find, for the $^2\Pi_{1/2}$ levels of the nonrotating molecule:

Band	Deperturbed	Perturbed (observed)
C-X (3,0)	$f = 2.62 \times 10^{-3}$	$f = 0.98 \times 10^{-3}$
B-X (15,0)	$f = 0.06 \times 10^{-3}$	$f = 0.71 \times 10^{-3}$

The nearly identical perturbed f-values are consistent with the observation of similar absorption intensities. However, the fact that the sum of the deperturbed f-values, 2.68×10^{-3}, is different from the sum of the perturbed f-values, 1.69×10^{-3}, indicates that $C^2\Pi, v_C = 3$ character is distributed over several nominal $B^2\Pi$ vibrational levels, not simply the closest lying level, $v_B = 15$. Assuming that, at $J = \frac{1}{2}$, the nominal $v_C = 3$ and $v_B = 15$ levels obtain all of their oscillator strength from the $C^2\Pi_{1/2}$ ($v_C = 3$) basis function, the percent $v_C = 3$ character may be deduced to be 37 and 27% in the C~B and B~C levels, respectively. This means that the nominal $C^2\Pi_{1/2}, v_C = 3, J = \frac{1}{2}$ level is composed 63% of basis functions other than $v_C = 3$ (27% is $v_B = 15$, leaving 36% to be distributed among other basis functions, of which $v_B = 14$ and 16 are probably most important).

The $^{14}N^{16}O$ C~B←X (2~12,0) and B~C←X (12~2,0) bands illustrate one of the difficulties in comparing measured (Bethke, 1959) and calculated (Gallusser and Dressler, 1982) perturbed oscillator strengths.

Band	Deperturbed	Calculated $J = \frac{1}{2}$	Observed $\Omega = \frac{1}{2}$ and $\frac{3}{2}$, room temperature
$B^2\Pi_{1/2} - X^2\Pi_{1/2}(12,0)$	$f = 0.05 \times 10^{-3}$	2.45×10^{-3}	2.31×10^{-3}
$C^2\Pi_{1/2} - X^2\Pi_{1/2}(2,0)$	$f = 5.29 \times 10^{-3}$	2.59×10^{-3}	2.74×10^{-3}
sum	$f = 5.34 \times 10^{-3}$	5.04×10^{-3}	5.05×10^{-3}

Because of the large matrix element ($H_{B12,C2} = 161\ cm^{-1}$) and close energy proximity of the interacting basis functions ($\Delta E^0 = 117\ cm^{-1}$), the B~C mixing should be 74%/26% using Eq. (6.2.7) and assuming a perhaps too simple two-level interaction. The near equality of the calculated oscillator strengths is an excellent reminder that the $v_{\mathrm{B}} = 12 \sim v_{\mathrm{C}} = 2$ interaction is not a simple two-level problem, even though the deperturbed and calculated oscillator strength sums are nearly equal. Evidently, the nominal $v_B = 12$ level borrows some of its oscillator strength from the remote $v_{\mathrm{C}} = 1$ and 3 basis functions. The excellent agreement between the calculated and observed oscillator strength sums is a simple consequence of $\left|R_e^{\mathrm{BX}}\right|^2 \ll \left|R_e^{\mathrm{CX}}\right|^2$ and the use by Gallusser and Dressler (1982) of Bethke's (1959) measured *absolute* oscillator strengths to normalize the calculated *relative* oscillator strengths. The important point is that $f_{\mathrm{calc}} - f_{\mathrm{obs}}$ is $+0.14 \times 10^{-3}$ and -0.15×10^{-3} for the nominal B–X and C–X bands, respectively. This reflects that Bethke's measurements correspond to an average over a room temperature distribution of $\Omega = \frac{1}{2}, \frac{3}{2}$ and J–levels and that the calculated oscillator strengths are for $\Omega = \frac{1}{2}$ and $J = \frac{1}{2}$. However, the simple two-level $v_{\mathrm{B}} = 12 \sim v_{\mathrm{C}} = 2$ mixing should decrease rapidly as J increases (the mixing with remote vibrational levels will be approximately J–independent), going from 74%/26% at $J = \frac{1}{2}$ to 84%/16% at $J = 10.5$. This means that, as observed, the measured oscillator strength for a thermal J–distribution should be slightly smaller than the $J = \frac{1}{2}$ calculated value for the nominal $\mathrm{v_B} = 12$ level and larger by the same amount for the nominal $\mathrm{v_C} = 2$ level.

This NO B∼C example demonstrates that measured oscillator strengths for a perturbed band can be temperature-dependent, even though the f–value of a single resolved line and the oscillator strength sum over a group of interacting levels (bands) must be temperature independent. Other quantities, such as radiative lifetimes, will also exhibit temperature dependences when the measurement involves an average over a thermal distribution of levels. Although it is common practice to speak of the oscillator strength and radiative lifetime of an *entire vibrational level*, perturbations can introduce unexpected obstacles to the measurement of such averaged (and therefore over-simplified) intensity factors.

6.3 Interference Effects

The discussion in Section 6.2.1 was based on the assumption of either $\tau_2^0 = \infty$ or $I_{20}^0 = 0$. When two interacting basis functions have comparable transition probabilities to a common level, $I_{10}^0 \approx I_{20}^0$, then the perturbed intensities, I_{+0} and I_{-0}, exhibit a form of anomalous behavior that is explicable as a *quantum-mechanical interference effect.*

The transition probability is proportional to $|\mu_{ij}|^2$. By analogy with the discussion of the $1 \sim 2 \to 0$ problem discussed in Section 6.2.1, but with $\mu_{20}^0 \neq 0$,

$$\begin{aligned} \mu_{+0} &= C_{1+}^* \mu_{10}^0 + C_{2+}^* \mu_{20}^0 \\ \mu_{+0} &= C_{1+}^* \mu_{10}^0 \pm \left(1 - |C_{1+}|^2\right)^{1/2} \mu_{20}^{0\,\dagger} \end{aligned} \tag{6.3.1}$$

Equation (6.3.1) is written as if μ_{10}^0, μ_{20}^0, and C_{1+} were complex quantities. It is always possible to choose a self-consistent set of phases such that all off-diagonal matrix elements of **H**, transition moments (μ_{ij}), and mixing coefficients are real numbers (Whiting and Nicholls, 1974; Hougen, 1970). Therefore complex conjugation (*) and absolute value signs will be suppressed in the following equations. The μ^2 values for $+ \leftrightarrow 0$ and $- \leftrightarrow 0$ transitions are

$$\mu_{+0}^2 = C_{1+}^2 \left(\mu_{10}^0\right)^2 + \left(1 - C_{1+}^2\right) \left(\mu_{20}^0\right)^2 \pm 2C_{1+} \left(1 - C_{1+}^2\right)^{1/2} \mu_{10}^0 \mu_{20}^{0\,\dagger} \tag{6.3.2a}$$

$$\mu_{-0}^2 = \left(1 - C_{1+}^2\right) \left(\mu_{10}^0\right)^2 + C_{1+}^2 \left(\mu_{20}^0\right)^2 \mp 2C_{1+} \left(1 - C_{1+}^2\right) \mu_{10}^0 \mu_{20}^{0\,\dagger} \tag{6.3.2b}$$

The first two terms in Eqs. (6.3.2a) and (6.3.2b) are both positive and are simply the deperturbed transition moments weighted by the fractional basis function character. These are the simple intensity borrowing terms discussed in Section 6.2.1. The third term, the *interference term, may be positive or negative*. Note that

$$\left(\mu_{+0}\right)^2 + \left(\mu_{-0}\right)^2 = \left(\mu_{10}^0\right)^2 + \left(\mu_{20}^0\right)^2,$$

which implies that transition probability is conserved and that the interference term,

$$\Theta_{+0} \equiv 2C_{1+} \left(1 - C_{1+}^2\right)^{1/2} \mu_{10}^0 \mu_{20}^0, \tag{6.3.3}$$

for the $+ \leftrightarrow 0$ transition is equal in magnitude and opposite in sign to that for $- \leftrightarrow 0$. The effect of the interference term is to transfer probability from one transition to another in a manner that depends on the *relative signs* of off-diagonal matrix elements. Over and above the usual intensity borrowing terms, probability is transferred from the main to the extra line, or *vice versa*, such that

$$\Theta_{+0} = -\Theta_{-0}. \tag{6.3.4}$$

The magnitude of the intensity transfer is dependent on $|H_{12}|$, and the maximum value is reached when

$$C_{1+} = 2^{-1/2}$$

and is

$$|\Theta_{+0}| = |\Theta_{-0}| \leq \left|\mu_{10}^0 \mu_{20}^0\right|. \tag{6.3.5}$$

For the special case,

$$\mu_{10}^0 = +\mu_{20}^0, \qquad C_{1+} = +2^{-1/2}, \qquad \text{and} \quad E_1^0 > E_2^0,$$

$$\left(\mu_{+0}\right)^2 = 2\left(\mu_{10}^0\right)^2 \qquad \text{and} \qquad \left(\mu_{-0}\right)^2 = 0.$$

† The ± signs in Eqs. (6.3.1), (6.3.2a), and (6.3.2b) depend on whether H_{12} is positive (top sign) or negative (bottom sign). $C_{1+} \geq 0$ is assumed.

The direction of the intensity transfer depends on the sign of $H_{12}\mu_{10}^0\mu_{20}^0$. Since, for an explicitly defined phase choice, it is often possible to deduce the sign of the electronic factor of H_{12} from knowledge of the electronic configurations of the interacting states, the sign of an intensity interference effect can be useful for determining the relative signs of two transition moments. It is essential to verify that the phase used for the electronic wavefunctions is the same as that used in the calculation of transition moments.

6.3.1 Perturbations between States of the Same Symmetry; Vibrational-Band Intensity Anomalies

When two states of the same symmetry interact, interference affects the $\Delta J = 0, \pm 1$ and $\Delta\Omega = 0, \pm 1$ rotational transitions out of a common mixed eigenstate in exactly the same way. This is in contrast to the behavior described in Sections 6.3.2 and 6.3.3 in which intensity is transferred between rotational branches or Ω–subbands. For transitions between unperturbed levels, the linestrength factor [Eq. (6.1.20)] is expressed as a product of three terms, each of which is the absolute magnitude squared of an off-diagonal matrix element. This factorization of $\mathrm{S}_{i,v_i,J';0,v_0,J''}$ is no longer possible when the ith level is a $1\sim 2$ mixed eigenstate. However, when the interacting 1 and 2 states have identical symmetry ($\Delta\Omega = \Delta\Sigma = \Delta\Lambda = \Delta S = 0, \Sigma^{\pm} \leftrightarrow \Sigma^{\pm}$), the Hönl-London rotational factor can be factored out of $S_{\pm,J';0,v_0,J''}$,

$$S_{+,J';0,v_0,J''} = \mathbf{S}_{J'J''}\left\{C_{1+}^2\left(\mu_{10}^0\right)^2 + \left(1 - C_{1+}^2\right)\left(\mu_{20}^0\right)^2 \pm 2C_{1+}\left(1 - C_{1+}^2\right)^{1/2}\mu_{10}^0\mu_{20}^0\right\} \tag{6.3.6}$$

The implicit phase choice is always $C_{1+} \geq 0$ and the sign choice for the interference term is top sign for $H_{12} \geq 0$. The expression for $S_{-,J';0,v_0,J''}$ is identical to Eq. (6.3.6) except the locations of C_{1+}^2 and $\left(1 - C_{1+}^2\right)$ are interchanged and the interference term appears with $\mp$ signs.

The sign of the interference term may be determined experimentally from the ratio of the difference and sum of the intensities of main and extra lines,

$$\frac{S_{+,J';0,v_0,J''} - S_{-,J';0,v_0,J''}}{S_{+,J';0,v_0,J''} + S_{-,J';0,v_0,J''}} \equiv \frac{\Delta S}{\Sigma S} = \frac{\left(2C_{1+}^2 - 1\right)\left[\left(\mu_{10}^0\right)^2 - \left(\mu_{20}^0\right)^2\right] \pm 4C_{1+}\left(1 - C_{1+}^2\right)^{1/2}\mu_{10}^0\mu_{20}^0}{\left(\mu_{10}^0\right)^2 + \left(\mu_{20}^0\right)^2}. \tag{6.3.7}$$

If the measured $\Delta S/\Sigma S$ intensity ratio is plotted versus the quantity $(2C_{1+}^2 - 1)$ obtained from the deperturbation calculation, the $\Delta S/\Sigma S$ value may be interpolated to the position of the level crossing $(E_1^0 = E_2^0)$, at which point

$$2C_{1+}^2 - 1 = 0$$

$$\left(\frac{\Delta S}{\Sigma S}\right)_{E_1^0 = E_2^0} = \{\mathrm{sign}\,(H_{12})\}\frac{2\mu_{10}^0\mu_{20}^0}{\left(\mu_{10}^0\right)^2 + \left(\mu_{20}^0\right)^2} \tag{6.3.8}$$

where

$$\mu_{10}^{0} = \langle v_1|v_0\rangle\, R_e^{10}$$
$$\mu_{20}^{0} = \langle v_2|v_0\rangle\, R_e^{20}.$$

None of the five signed quantities ($H_{12}, \langle v_1|v_0\rangle, \langle v_2|v_0\rangle, R_e^{10}$, and R_e^{20}) that determine the sign of the intensity interference term are directly derivable from experiment. The two vibrational overlap factors may be calculated provided that the potential energy curves for the three electronic states have been adequately determined from independent experimental measurements. The H_{12} perturbation matrix element is actually the product of a calculable $\langle v_1|v_2\rangle$ vibrational factor and an electronic term. This means that each vibrational wavefunction appears exactly twice in the intensity interference term. Although the absolute phases of the three vibrational and three electronic wavefunctions are indeterminate, the sign of the interference term is a measurable quantity and must therefore be calculable *ab initio* . It is a straightforward matter to calculate the sign and magnitude of $H_{ij}^{e}(R)$ and $R_e^{i,j}(R)$ electronic quantities, but it is essential that the same phase choice be made for each computation. It is conceivable that knowledge of correct *ab initio* signs of $R_e^{i,j}$ and H_{jk}^{e} electronic quantities combined with observed intensity interference effects could distinguish between a correct and an incorrect deperturbation model.

One of the most dramatic manifestations of an interference effect is the vanishing of a line or of an entire band that, on the basis of known Franck-Condon factors and inappropriately simple intensity borrowing ideas, should be quite intense (see Fig. 6.6). This effect can easily be mistaken as an accidental predissociation (Section 7.13). Yoshino, *et al.*, (1979) have studied the valence~Rydberg N_2 $b'^1\Sigma_u^+ \sim c'_4{}^1\Sigma_u^+$ perturbations. Abrupt decreases in emission intensity for $c'_4 - X^1\Sigma_g^+(v' = 1$ and $4)$ and $b' - X$ $(v' = 4)$ bands had been attributed to weak predissociation rather than perturbation effects (Gaydon, 1944; Lofthus, 1957; Tilford and Wilkinson, 1964; Wilkinson and Houk, 1956). The b' $(v = 4) \sim c'_4$ $(v = 1)$ and b' $(v = 13) \sim c'_4$ $(v = 4)$ deperturbation models of Yoshino *et al.*, (1979) provide a predissociation-free unified account of both level shift and intensity effects. Weak predissociation effects cannot be ruled out, but are not needed to account for the present experimental observations.

It will be useful to derive the condition for the vanishing of transition probability for the general case

$$\mu_{10}^{0} \neq \mu_{20}^{0}$$

where

$$\mu_{i0}^{0} \equiv R_e^{i,0}\, \langle v_i|v_0\rangle\, .$$

From Eq. (6.3.1), the μ_{+0} transition moment will vanish if

$$\frac{C_{1+}}{C_{2+}} = -\frac{\mu_{20}^{0}}{\mu_{10}^{0}}, \tag{6.3.9}$$

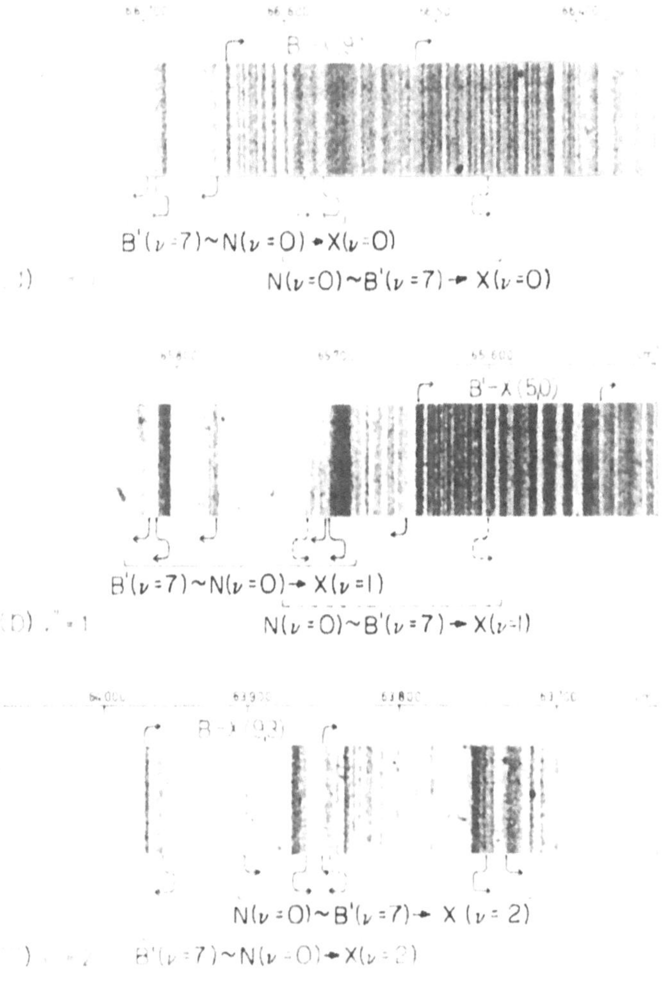

Figure 6.6: The vanishing of an entire band as a result of a homogeneous perturbation. The emission spectra from the valence~Rydberg $^{14}N^{16}O$ $B'^2\Delta$ $(v' = 7) \sim N^2\Delta$ $(v' = 0)$ perturbation complex into $X^2\Pi$ $(v'' = 0, 1$, and $2)$ (segments a, b, and c, respectively) illustrate the effect of a sign reversal on a near-perfect cancellation of B′−X and N−X transition amplitudes. The N $(v = 0)$ level crosses B′ $(v = 7)$ from below at $J \simeq 8.4$. The interaction matrix element, $H_{B'7;N0} = 45$ cm^{-1}, splits the B′ ~ N mixed levels into two series of rotational levels in which same-J levels are separated by at least 90 cm^{-1}. The higher energy series of levels, nominally $B'^2\Delta$ for $J' < 8.5$ and nominally $N^2\Delta$ for $J' > 8.5$, appears with enhanced intensity (constructive interference, highlighted by solid arrows below the spectra marking head-like features) in emission into $v'' = 0$ and $v'' = 1$ [labeled B′ $(v = 7) \sim$ N $(v = 0) \rightarrow X(v = v'')$ on the figure], but with diminished intensity (destructive interference – the weakened features are marked by dotted arrows below the spectra) in emission into $v'' = 2$. The intensity interference affects the lower energy series of levels [labeled N $(v = 0) \sim$ B′ $(v = 7) \rightarrow$ X $(v = v'')$ on the figure] in the opposite sense. [From Jungen (1966).]

where $C_{1+} \geq 0$ by definition and C_{2+} is understood to have the same sign as H_{12}. Thus Eq. (6.3.9) can be satisfied only if $H_{12}\mu_{10}^0\mu_{20}^0 < 0$.

Now, to obtain a relationship between C_{1+} and the observable separation between main and extra lines, $E_+ - E_-$, recall that if

$$|+\rangle = C_{1+}|1\rangle + C_{2+}|2\rangle \tag{6.3.10}$$

is the eigenfunction belonging to the E_+ eigenenergy (which is true as long as $C_{1+}C_{2+}H_{12} > 0$), then

$$|-\rangle = -C_{2+}|1\rangle + C_{1+}|2\rangle \tag{6.3.11}$$

is the eigenfunction belonging to E_-. Left-multiplying Eq. (6.3.10) by $\langle 1|\mathbf{H}$, one obtains

$$\begin{aligned} E_+C_{1+} &= C_{1+}E_1^0 + C_{2+}H_{12} \\ E_+ &= E_1^0 + \frac{C_{2+}}{C_{1+}}H_{12}, \end{aligned} \tag{6.3.12}$$

and similarly left-multiplying Eq. (6.3.11) by $\langle 1|\mathbf{H}$, one obtains

$$E_- = E_1^0 - \frac{C_{1+}}{C_{2+}}H_{12} \tag{6.3.13}$$

and thus

$$E_+ - E_- = \left[\frac{C_{2+}}{C_{1+}} + \frac{C_{1+}}{C_{2+}}\right]H_{12}. \tag{6.3.14a}$$

Alternatively, a similar expression,

$$E_1^0 - E_2^0 = \left[\frac{C_{1+}}{C_{2+}} - \frac{C_{2+}}{C_{1+}}\right]H_{12}, \tag{6.3.14b}$$

is obtained from Eq. (6.3.12) and by left-multiplying Eq. (6.3.10) by $\langle 2|\mathbf{H}$.

Inserting Eq. (6.3.9) into Eq. (6.3.14a) gives the general condition for the vanishing of the transition between level 0 and the *higher-energy* member of the pair of 1~2 interacting levels,

$$[E_+ - E_-]_{\mu_{+0}=0} = -\left(\frac{\mu_{10}^0}{\mu_{20}^0} + \frac{\mu_{20}^0}{\mu_{10}^0}\right)H_{12}. \tag{6.3.15}$$

Inserting Eq. (6.3.9) into Eq. (6.3.14b) gives

$$\left[E_1^0 - E_2^0\right]_{\mu_{+0}=0} = \left(\frac{\mu_{10}^0}{\mu_{20}^0} - \frac{\mu_{20}^0}{\mu_{10}^0}\right)H_{12}, \tag{6.3.16}$$

which determines whether the zero in μ_{+0} occurs before $(E_1^0 > E_2^0)$ or after $(E_1^0 < E_2^0)$ the level crossing $(E_1^0 = E_2^0)$. Equation (6.3.15) can be satisfied whenever

$$\mu_{10}^0\mu_{20}^0H_{12} < 0$$

and Eq. (6.3.16) can be satisfied for $E_1^0 > E_2^0$ whenever

$$|\mu_{20}^0| > |\mu_{10}^0|.$$

The four possible locations of the intensity null occur as follows:

$$\begin{array}{llll} E_+ & E_1^0 > E_2^0 & \mu_{10}^0\mu_{20}^0 H_{12} < 0 & |\mu_{20}^0| > |\mu_{10}^0| \\ E_+ & E_1^0 < E_2^0 & \mu_{10}^0\mu_{20}^0 H_{12} < 0 & |\mu_{20}^0| < |\mu_{10}^0| \\ E_- & E_1^0 > E_2^0 & \mu_{10}^0\mu_{20}^0 H_{12} > 0 & |\mu_{20}^0| < |\mu_{10}^0| \\ E_- & E_1^0 < E_2^0 & \mu_{10}^0\mu_{20}^0 H_{12} > 0 & |\mu_{20}^0| > |\mu_{10}^0| \end{array} \tag{6.3.17}$$

Equation (6.3.16) was first derived by Dressler (1970). Whenever it is possible to derive intensity parameters from measurements of frequency intervals, far greater accuracy is obtainable than from direct intensity measurements. Equation (6.3.15) allows extremely precise measurement of μ_{10}^0/μ_{20}^0 from the observed $E_+ - E_-$ difference at the J-value of the exact intensity null.

Figure 6.7 illustrates the vanishing of intensity at a homogeneous perturbation. Three cases are shown:

1. *Simple intensity borrowing*: $\mu_{10}^0 = 0$. Intensity vanishes when $E_+ - E_- \rightarrow +\infty$; in the higher-energy eigenvalue when $E_1^0 \gg E_2^0$ and in the lower-energy eigenvalue when $E_1^0 \ll E_2^0$.

2. *Interference at the crossing point* $(E_1^0 = E_2^0)$:

$$|\mu_{10}^0| = |\mu_{20}^0|.$$

If $\mu_{10}^0\mu_{20}^0 H_{12} > 0$, intensity vanishes for the lower energy eigenvalue at $(E_+ - E_-) = 2|H_{12}|$.

3. *General case*: $|\mu_{10}^0| \neq |\mu_{20}^0|$. Example shows $\mu_{10}^0\mu_{20}^0 H_{12} > 0$ and $|\mu_{20}^0| > |\mu_{10}^0|$.

In Fig. 6.7, the basis function energies E_1^0 and E_2^0 are shown as straight lines, plotted as

$$E_i^0 - \frac{E_1^0 + E_2^0}{2} \qquad \text{versus} \quad E_2^0 - E_1^0.$$

The eigenenergies and transition strengths are shown as double solid lines, plotted as $E_\pm - \overline{E}^0$ versus ΔE^0, where the width of the shaded region depicts the intensity of the $E_\pm \leftrightarrow E_0$ transition.

Interference effects are not limited to simple two-level interactions. Stahel, *et al.*, (1983) have performed a multilevel deperturbation ($^1\Sigma$–matrix: 41 levels; $^1\Pi$–matrix: 30 levels) of the N_2 $^1\Sigma_u^+$ and $^1\Pi_u$ states. Prior to the work of Dressler (1969), the numerous known N_2 vibronic levels of $^1\Sigma_u^+$ and $^1\Pi_u$ symmetry between 100,000 and 120,000 cm^{-1} had not been convincingly organized into separate electronic states. Dressler (1969) showed that the chaotic level and intensity patterns resulted from homogeneous perturbations of the $b^1\Pi_u$

and $b'^1\Sigma_u^+$ valence states by the $c^1\Pi_u$ (N_2^+ $X^2\Sigma_g^+$, 3pπ), $o^1\Pi_u$ (N_2^+ $A^2\Pi_u$, 3sσ), $e^1\Pi_u$ (X, 4$p\pi$), $c'^1\Sigma_u^+$ (X, 3$p\sigma$), and $e'^1\Sigma_u^+$ (X, 4$p\sigma$) Rydberg states.

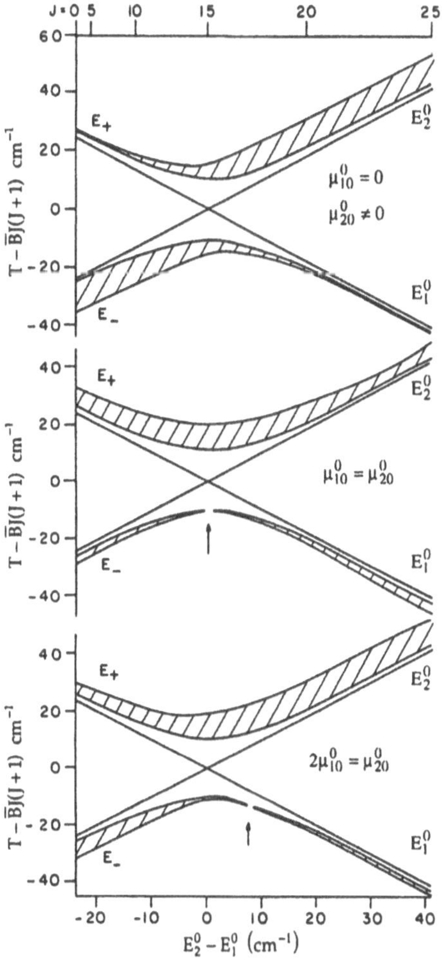

Figure 6.7: Schematic illustration of intensity interference effects. The straight lines are the basis function (i.e., deperturbed) reduced term values, $E_1^0 - \overline{B}J(J+1)$, where $\overline{B} = (B_1^0 + B_2^0)/2$. The level crossing occurs at $J = 15$ ($B_1^0 = 0.9$ cm^{-1}, $B_2^0 = 1.1$ cm^{-1}, $H_{12} = 10$ cm^{-1}). The inner edges and widths of the solid curves depict, respectively, the energies of the perturbed levels, E_+ and E_- [reduced by $\overline{B}J(J+1)$], and the intensities of the transitions into E_+ and E_-. The top frame illustrates simple intensity borrowing, $\mu_{10}^0 = 0$. The middle frame shows the vanishing of intensity at the crossing point ($J = 15$), $E_1^0 = E_2^0$, when $\mu_{10}^0 = \mu_{20}^0$. The bottom frame exemplifies the general case, $\mu_{10}^0 \neq \mu_{20}^0 \neq 0$ (here $2\mu_{10}^0 = \mu_{20}^0$), where the location of the intensity vanishing point ($J \simeq 17.3$) is given by Eq. (6.3.15) or Eq. (6.3.16). See also Fig. 5.8 for other properties of the same homogeneous perturbation illustrated here. [From Dressler (1970).]

Figure 6.8 illustrates a pronounced intensity interference effect for the two $^1\Pi_u$ levels at 104,700 and 105,346 cm^{-1}. These levels are nominally $\text{b}^1\Pi_u$ ($v = 5$ and 6); however both levels have significant $\text{c}^1\Pi_u$ ($v = 0$) and $\text{o}^1\Pi_u$ ($v = 0$) character. The dotted tie-lines on the figure depict the deperturbed intensity distributions for transitions from $\text{X}^1\Sigma_g^+$ ($v = 0$) (Franck-Condon factor times R-independent electronic transition moment squared); the heavy vertical lines are the observed (perturbed) relative intensities. The small intensities of the nominal b–X (5,0) and (6,0) bands, despite the large μ^0_{bX} and μ^0_{cX} transition moments for the $\text{b}^1\Pi_u$ and $\text{c}^1\Pi_u$ basis components of the two observed levels, are a result of destructive interference between b–X and c–X transition amplitudes. The nominal $\text{c}^1\Pi_u$ ($v = 0$) level lies below the b ($v = 5$ and 6) levels, and the interference effect is such that the intensity null occurs in the upper (E_+) component. This implies that

$$\mu^0_{\text{bX}}\mu^0_{\text{cX}}H_{\text{bc}} < 0$$

for the $v_{\text{b}} = 5, 6, v_{\text{c}} = 0$, and $v_{\text{X}} = 0$ basis levels. Intensity is transferred from the nominal b–X (5,0) and (6,0) bands into the nominal c–X (0,0) band. A similar explanation accounts for the excess intensity in the b–X (3,0) and (4,0) bands. The c–X (0,0) transition amplitude interferes with the b–X amplitude, again in the sense

$$\mu^0_{\text{bX}}\mu^0_{\text{cX}}H_{\text{bc}} < 0,$$

with the result that intensity is transferred into the lower (E_-) component. For a more complete discussion of intensity interference effects of the type shown in Fig. 6.8 as well as of other types, see Stahel, *et al.*, (1983).

The intensity interference pattern illustrated by Fig. 6.8 for bound-bound transitions is very similar to the Beutler-Fano lineshape for bound-free transitions discussed in Sections 7.9 and 8.9. The 101,000-106,000 cm^{-1} region of Fig. 6.8 is a band-by-band rather than a continuous representation of a Fano profile with $q < 0$ [see Fig. 7.26 and compare Eq. (7.9.6) to Eq. (6.3.17)].

6.3.2 $\Delta\Lambda = \pm 1$ Perturbations; Rotational-Branch Intensity Anomalies

There are two classes of transitions between pure case (a) basis functions. *Parallel* and *perpendicular* transitions are readily distinguished by their characteristic rotational branch intensity patterns. $\Delta\Lambda = 0$ transitions are called parallel bands. They have R and P branches of comparable intensity and a weak or absent Q branch. $\Delta\Lambda = \pm 1$ transitions are perpendicular bands and have a strong Q branch and R and P branches of approximately half the intensity of the Q branch.

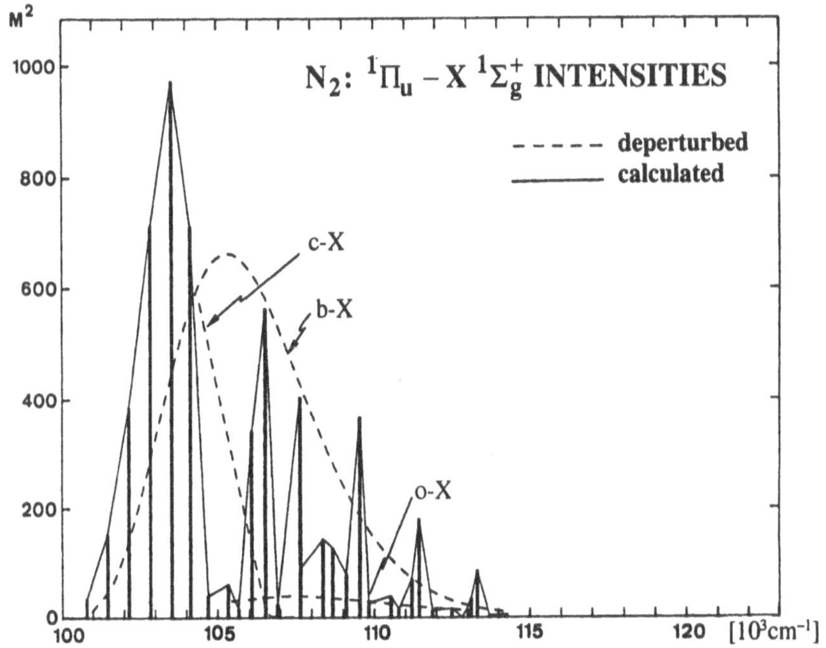

Figure 6.8: Calculated perturbed and deperturbed intensities for N_2 $^1\Pi_u \leftarrow X^1\Sigma_g$ $(v = 0)$ transitions. The valence $b^1\Pi_u$ and Rydberg $c^1\Pi_u(3p\pi)$ and $o^1\Pi_u$ $\left(N_2^+\ A^2\Pi_u,\ 3s\sigma\right)$ states interact strongly, giving rise to a chaotic appearing spectrum. Stahel, *et al.*, (1983) have deperturbed the N_2 $^1\Sigma_u^+$ and $^1\Pi_u$ states. The dotted lines show the Franck-Condon envelopes [from $X^1\Sigma_g^+$ $(v'' = 0)$] for the deperturbed states ($X^2\Sigma_g^+$–core Rydberg sharply peaked at $v' = 0$, valence and $A^2\Pi_u$–core Rydberg broadly peaked at high v'). The solid lines are the calculated intensities for transitions into strongly mixed eigenstates. Note the near perfect intensity cancellation for the three transitions into b∼c mixed levels near 105,000 cm^{-1} . [From Stahel, *et al.*(1983).]

When a $\Delta\Lambda = \pm 1$ perturbation occurs, the PQR intensity pattern for transitions terminating on the mixed levels is not necessarily intermediate between the parallel and perpendicular patterns. For example, consider a perturbation in the upper electronic state which affects the J' level. The intensity ratio for the $P(J'+1)$ and $R(J'-1)$ lines can depart from the approximate $1:1$ ratio typical of both parallel and perpendicular unperturbed transitions. In fact, one or the other line can vanish. Furthermore, in a fluorescence progression from a mixed upper level, $v'J'$, the $P(J'+1)$–line can be anomalously weak relative to $R(J'-1)$ in some (v', v'') bands and anomalously strong in others. The sign and magnitude of the effect can vary with v'' even though the perturbation is in the v' level!

The interference effects discussed in Section 6.3.1 could be explained without detailed examination of rotational line strength factors. Equation (6.3.6) shows that, for perturbations between states of the same symmetry, transition intensities can be expressed as a product of the usual Hönl-London rotational factor times a sum of terms involving products of electronic and vibrational matrix elements. Those interference effects were of electronic-vibrational origin, and it was unnecessary to consider the signs of matrix elements of the rotational part of the wavefunction, $|\Omega JM\rangle$. The interference effects resulting from $\Delta\Lambda = \pm 1$ perturbations always affect the intensities of $\Delta J = +1$ and -1 transitions out of each mixed level in equal and opposite amounts, because the transition amplitude phases for P and R lines of parallel transitions are identical while those for P and R lines of perpendicular transitions are opposite. If $\| \sim\perp$ interference is constructive for a P line, it will inevitably be destructive for the R line connected to the same mixed J–level (Klynning, 1974).

The operators responsible for electric dipole transitions are molecule-fixed components of $\boldsymbol{\mu}$ [Eqs. (6.1.2) and (6.1.38)]. $\boldsymbol{\mu}$ operates exclusively on the spatial coordinates of the electrons, and hence has selection rules $\Delta S = \Delta\Sigma = 0$ and, because it is a vector operator with respect to the spatial coordinates (see Section 3.4.5), $\Delta\Lambda = \Delta\Omega = 0, \pm 1$. $\Delta\Lambda = 0$ transitions arise from the $\boldsymbol{\mu}_z$ operator component. The relationship between $\Lambda \to \Lambda$ and $-\Lambda \to -\Lambda$ transition amplitudes may be derived from

$$\sigma_v(xz)\,|n\Lambda^s\rangle = (-1)^{\Lambda+s}\,|n,-\Lambda^s\rangle \quad \sigma_v(xz)\boldsymbol{\mu}_z = \boldsymbol{\mu}_z\sigma_v(xz)$$

[see Eqs. (3.2.79), (3.2.82), and (3.2.89) and also Hougen (1970)] by applying σ_v to the two wavefunctions and to the operator in $\left\langle n'\Lambda^{s'}\,|\boldsymbol{\mu}_z|\,n''\Lambda^{s''}\right\rangle$ to give

$$\left\langle n'\Lambda^{s'}\,|\boldsymbol{\mu}_z|\,n''\Lambda^{s''}\right\rangle = (-1)^{s'+s''}\left\langle n',-\Lambda^{s'}\,|\boldsymbol{\mu}_z|\,n'',-\Lambda^{s''}\right\rangle \equiv \mu_{\|}\delta_{s's''}.^{\dagger} \quad (6.3.18)$$

The $\Sigma^+ \leftarrow / \rightarrow \Sigma^-$ transition selection rule is proved by Eq. (6.3.18), since

$$\langle\Sigma^+\,|\boldsymbol{\mu}_z|\,\Sigma^-\rangle = -\langle\Sigma^+\,|\boldsymbol{\mu}_z|\,\Sigma^-\rangle = 0.$$

$\Delta\Lambda = \pm 1$ transitions arise from

$$\begin{aligned} \boldsymbol{\mu}^+ &= -(2)^{-1/2}(\boldsymbol{\mu}_x + i\boldsymbol{\mu}_y) \\ \boldsymbol{\mu}^- &= +(2)^{-1/2}(\boldsymbol{\mu}_x - i\boldsymbol{\mu}_y) \end{aligned}$$

[c.f. Eqs. (3.4.41), $\boldsymbol{\mu}^\pm = \mp T^{(1)}_{\pm 1}(\vec{\mathbf{r}})$] and

$$\left\langle n',(\Lambda\pm 1)^{s'}|\boldsymbol{\mu}^\pm|n'',\Lambda^s\right\rangle = -(-1)^{s+s'}\left\langle n',(-\Lambda\mp 1)^{s'}|\boldsymbol{\mu}^\mp|n'',-\Lambda^s\right\rangle \equiv 2^{1/2}\mu_\perp \quad (6.3.19)$$

where the relationship between the $+\Lambda$ and $-\Lambda$ matrix elements is given by Eq. (3.2.89) and

$$\sigma_v\boldsymbol{\mu}^\pm = \boldsymbol{\mu}^\mp\sigma_v.$$

† n is a shorthand for the electronic name and the vibrational quantum number of the state; s is the Σ^+/Σ^- symmetry, and $s = 0$ for all states except Σ^- states, for which $s = 1$.

Hougen (1970) has shown that it is always possible to define phases so that $\mu_{\parallel}$, $\mu_{\perp}$, and all perturbation mixing coefficients are real quantities. This fact has not always been recognized in previous work (Dieke, 1941) and has given rise to some errors in the literature.

Since transitions are observed in a space-fixed rather than molecule-fixed reference system, it is necessary to compute matrix elements of space-fixed components of $\boldsymbol{\mu}$, for example (for $\Delta M_J = 0$ transitions),

$$\mu_Z = \frac{1}{2}\alpha_Z^{+}\mu^{-} + \frac{1}{2}\alpha_Z^{-}\mu^{+} + \alpha_Z^{z}\mu_z, \quad (6.3.20)$$

where the α_I^j are direction cosine operator components (see Section 2.3.1) and

$$\alpha_Z^{\pm} = \alpha_Z^{x} \pm i\alpha_Z^{y}.$$

Matrix elements of μ_Z may be factored into empirically definable, *electronic transition-specific*, molecule-fixed, nonrotating molecule ($|n\Lambda^s S\Sigma\rangle$) matrix elements ($\mu_{\perp}$ or $\mu_{\parallel}$ parameters) and *universal*, tabulated (Hougen, 1970, p. 31, and Table 2.1), rotating molecule ($|\Omega JM\rangle$) matrix elements of $\boldsymbol{\alpha}$. Although all matrix elements of $\boldsymbol{\alpha}$ are tabulated, it is convenient to derive the relationship between (Ω', Ω'') and $(-\Omega', -\Omega'')$ matrix elements. Following Hougen (1970),

$$\sigma_v\alpha_Z^{z} = -\alpha_Z^{z}\sigma_v \qquad \sigma_v\alpha_Z^{\pm} = -\alpha_Z^{\mp}\sigma_v$$

$$\sigma_v|\Omega JM\rangle = (-1)^{J-\Omega}|-\Omega JM\rangle \qquad \text{(see Eq. (3.2.87))}$$

therefore, for $\Delta\Omega = 0$ transitions (parallel since $\Delta\Omega = \Delta\Lambda$)

$$\langle\Omega J'M|\alpha_Z^{z}|\Omega J''M\rangle = (-1)^{J'+J''-2\Omega+1}\langle-\Omega J'M|\alpha_Z^{z}|-\Omega J''M\rangle \quad (6.3.21a)$$

$(-1)^{J'+J''-2\Omega+1}$ is respectively $+1, -1, +1$ for P, Q, R transitions. For $\Delta\Omega = \pm1$ transitions (perpendicular),

$$\langle\Omega \mp 1, J'M|\alpha_Z^{\pm}|\Omega J''M\rangle = (-1)^{J'+J''-2\Omega\pm1+1}\langle-\Omega \pm 1, J'M|\alpha_Z^{\mp}|\Omega J''M\rangle \quad (6.3.21b)$$

$(-1)^{J'+J''-2\Omega\pm1+1}$ is respectively $-1, +1, -1$ for P, Q, R transitions. Combining Eqs. (6.3.19) and (6.3.21b), one obtains, for $\Delta\Omega = 0$ transitions,

$$\begin{aligned}
&\langle n'\Lambda S\Sigma\,|\langle\Omega J'M|\mu_Z|\Omega J''M\rangle|\,n''\Lambda S\Sigma\rangle \\
&\quad = \mu_{\parallel}(n'\Lambda, n''\Lambda)\,\langle\Omega J'M\,|\alpha_Z^{z}|\,\Omega J''M\rangle \\
&\quad = (-1)^{J'+J''-2\Omega+1}\,\langle n", -\Lambda, S, -\Sigma| \\
&\qquad \times \langle-\Omega J'M|\mu_Z|-\Omega J''M\rangle\,|n'', -\Lambda, S, -\Sigma\rangle
\end{aligned} \quad (6.3.22a)$$

and for $\Delta\Omega = \pm1$ transitions,

$$\begin{aligned}
&\left\langle n'(\Lambda \pm 1)^{s'}S\Sigma\,|\langle(\Omega \pm 1), J'M|\mu_Z|\Omega J''M\rangle|\,n''\Lambda^{s''}S\Sigma\right\rangle \\
&\quad = 2^{-1/2}\mu_{\perp}[(n'(\Lambda \pm 1)^{s'}, n''\Lambda^{s''}]\,\langle\Omega \pm 1, J'M\,|\alpha_Z^{\mp}|\,\Omega J''M\rangle \\
&\quad = (-1)^{J'+J''-2\Omega\pm1+2+s'+s''}\,\langle n', -\Lambda \mp 1, S, -\Sigma| \\
&\qquad \times \langle-\Omega \mp 1, J'M|\mu_Z|-\Omega J''M\rangle\,|n'', -\Lambda, S, -\Sigma\rangle.
\end{aligned} \quad (6.3.22b)$$

When one forms e/f–symmetrized basis functions [Eqs. (3.2.92b), (3.2.92c), (3.2.94), and (3.2.95),] by taking linear combinations of the form

$$2^{-1/2}\left[|\Omega JM\rangle\,|n\Lambda S\Sigma\rangle \pm |-\Omega JM\rangle\,|n,-\Lambda,S,-\Sigma\rangle\right],$$

one finds that, for both parallel and perpendicular transitions, R and P transitions are $e \leftrightarrow e$ or $f \leftrightarrow f$ and Q transitions are $e \leftrightarrow f$ or $f \leftrightarrow e$.

For parallel transitions, matrix elements of $\mu_Z \alpha_Z^z$ evaluated in the symmetrized $(e/f, |\Omega|, |\Lambda|)$ and unsymmetrized (signed Λ, Ω) basis sets are identical and equal to $\mu_{\|}\alpha_Z^z(\Omega J'M; \Omega JM)$. For perpendicular transitions between two non–Σ states, matrix elements of $\frac{1}{2}\left(\mu^+\alpha_Z^- + \mu^-\alpha_Z^+\right)$ evaluated in the symmetrized and unsymmetrized basis sets are identical and equal to $[2^{-1/2}\mu_\perp\alpha_Z^-(\Omega + 1, J'M, \Omega JM)]$ where, from the Eq. (6.3.19) definition,

$$\mu_\perp \equiv +2^{-1/2}\left\langle n', \Lambda+1\left|\mu^+\right|n\lambda\right\rangle \quad \text{for} \quad \Lambda > 0.$$

For perpendicular transitions involving one $\Sigma_0^\pm$ state, matrix elements of μ_Z evaluated in the unsymmetrized basis are smaller, by a factor of $2^{-1/2}$, than the corresponding symmetrized matrix element. In the unsymmetrized basis, matrix elements of μ_Z for $\Pi_{\Omega'} - \Sigma_\Omega^\pm$ transitions are equal to $[2^{-1/2}\mu_\perp\alpha_Z^-(\Omega + 1, J', M; \Omega JM)]$ times a phase factor as in the following table.

Transition	$\Omega' = 1 + \lvert\Sigma\rvert$	$\Omega' = 1 - \lvert\Sigma\rvert$
Even multiplicity		
R_{ff}^{ee} and P_{ff}^{ee}	+1	$\pm(-1)^{-S+s+1/2}$
Q_{fe}^{ef}	+1	$\mp(-1)^{-S+s+1/2}$
Odd multiplicity		
R_{ff}^{ee} and P_{ff}^{ee}	+1	$\pm(-1)^{-S+s}$
Q_{fe}^{ef}	+1	$\mp(-1)^{-S+s}$

The difference in phase factor for $\Omega' = 1 - |\Sigma|$, $\Pi_{\Omega'} - \Sigma_\Omega^+$ versus $\Pi_{\Omega'} - \Sigma_\Omega^-$ transitions is the source of the $\mu_\perp \sim \mu_\perp$ interference effect discussed in Section 6.3.4.

Equations (6.3.22a - 6.3.22b) can be simplified, for the case of isotropically oriented molecules, unpolarized radiation, and zero external magnetic or electric fields, by "summing" over M (see Hougen, 1970, p. 39).[†] The resultant M–independent $\langle\Omega' J'|\alpha|\Omega J\rangle$ direction cosine matrix elements are listed in Table 6.1. Note that the $\alpha^\pm$ $\Delta\Omega = \mp 1$ matrix elements have opposite signs for P versus R transitions, whereas the α^z $\Delta\Omega = 0$ matrix elements have the same signs for P and R transitions.

[†] In the absence of an external field, M_J is a good quantum number. Since **H** is a scalar operator, the **H**-matrix factors into $2J+1$ identical M–blocks for each J. In addition, the M–dependence of μ_Z matrix elements is contained in a (J', J'', M)–dependent factor that is identical for all initial and final electronic basis functions. The average over the M–dependence of *transition probabilities* is accomplished by squaring the M–dependent *transition amplitude* factor, summing over M, and taking the square root.

The rotational linestrength (Hönl-London) factors can also be expressed in terms of 3-j coefficients (see, for one-photon transitions, Eq. (6.1.45)).

Table 6.1: M–Independent $\langle \Omega' J' | \alpha | \Omega J \rangle$ Direction Cosine Matrix Elements

	$\alpha^z(\Omega' - \Omega)$	$\alpha^{\pm}(\Omega' = \Omega \mp 1)$
$R(J' = J+1)$	$\left[\frac{(J+\Omega+1)(J-\Omega+1)}{3(J+1)}\right]^{1/2}$	$\pm\left[\frac{(J\mp\Omega+1)(J\mp\Omega+2)}{3(J+1)}\right]^{1/2}$
$Q(J' = J)$	$\left[\frac{2J+1}{3J(J+1)}\right]^{1/2}$	$\left[\frac{(2J+1)(J\pm\Omega)(J\mp\Omega+1)}{3J(J+1)}\right]^{1/2}$
$P(J' = J-1)$	$\left[\frac{(J+\Omega)(J-\Omega)}{3J}\right]^{1/2}$	$\mp\left[\frac{(J\pm\Omega)(J\pm\Omega-1)}{3J}\right]^{1/2}$

It is possible to express the intensities of all rotational branches of any unperturbed electric dipole allowed $^{2S+1}\Lambda' - {}^{2S+1}\Lambda''$ transition in terms of a single $\mu_\perp$ or $\mu_\parallel$ parameter. This is true even for transitions in which one or both of the electronic states is in an intermediate case (a)-(b) Hund's coupling and the transition intensities cannot be reduced to simple closed form expressions. For example, the individual line intensities in all 54 rotational branches of a $^3\Delta - {}^3\Pi$ transition are proportional to a single $\mu_\perp$ parameter. The only nonzero case (a) *basis state transition moments* are for $\Delta\Omega = \Delta\Lambda = +1$, and their values are $2^{-1/2}\mu_\perp$ times the factor listed in the table below, which we will refer to as Eq. (6.3.23).

Transition	$R_{ee}(J)$ or $R_{ff}(J)$	$Q_{ef}(J)$ or $Q_{fe}(J)$	$P_{ee}(J)$ or $P_{ff}(J)$
$^3\Delta_3 - {}^3\Pi_2$	$-\left[\frac{(J+3)(J+4)}{(J+1)}\right]^{1/2}$	$+\left[\frac{(2J+1)(J-2)(J+3)}{J(J+1)}\right]^{1/2}$	$+\left[\frac{(J-2)(J-3)}{J}\right]^{1/2}$
$^3\Delta_2 - {}^3\Pi_1$	$-\left[\frac{(J+2)(J+3)}{(J+1)}\right]^{1/2}$	$+\left[\frac{(2J+1)(J-1)(J+2)}{J(J+1)}\right]^{1/2}$	$+\left[\frac{(J-1)(J-2)}{J}\right]^{1/2}$
$^3\Delta_1 - {}^3\Pi_0$	$-\left[\frac{(J+1)(J+2)}{(J+1)}\right]^{1/2}$	$+\left[\frac{(2J+1)J(J+1)}{J(J+1)}\right]^{1/2}$	$+\left[\frac{J(J-1)}{J}\right]^{1/2}$

(6.3.23)

The *eigenstate transition moments* may be obtained from Eq. (6.3.23) by finding the eigenfunctions of the $^3\Delta$ and $^3\Pi$ effective Hamiltonian matrices for each value of the good quantum numbers J and e/f. For example, the nominal $^3\Delta_{3e}$ and $^3\Pi_{0e}$ eigenfunctions are

$$\left|{}'^3\Delta'_{3e}, J'\right\rangle = C^{\Delta}_{3e,3e}(J')\,|^3\Delta_{3e}, J'\rangle + C^{\Delta}_{3e,2e}(J')\,|^3\Delta_{2e}, J'\rangle + C^{\Delta}_{3e,1e}(J')\,|^3\Delta_{1e}, J'\rangle \tag{6.3.24a}$$

$$\left|{}'^3\Pi'_{0e}, J''\right\rangle = C^{\Pi}_{0e,2e}(J'')\,|^3\Pi_{2e}, J''\rangle + C^{\Pi}_{0e,1e}(J'')\,|^3\Pi_{1e}, J''\rangle \; C^{\Pi}_{0e,0e}(J'')\,|^3\Pi_{0e}, J''\rangle \tag{6.3.24b}$$

where the mixing coefficients $C^{\Lambda}_{\Omega,\Omega'}(J)$ give the $|\Lambda_\Omega, J\rangle$ basis function character present in the nominal $|'\Lambda'_\Omega, J\rangle$ eigenstate. Thus, the nominally forbidden $^3\Delta_3 - {}^3\Pi_0$ $P_{ee}(J)$ transition probability is proportional to

$$\left|\left\langle {}'^3\Delta'_{3e}, J-1|\mu|\ {}'^3\Pi'_{0e}, J\right\rangle\right|^2$$
$$= \mu_\perp^2/2\left\{+C^{\Delta}_{3e,3e}(J-1)C^{\Pi}_{0e,2e}(J)\left[\frac{(J-2)(J-3)}{3J}\right]^{1/2}\right.$$
$$+\,C^{\Delta}_{3e,2e}(J-1)C^{\Pi}_{0e,1e}(J)\left[\frac{(J-1)(J-2)}{3J}\right]^{1/2}$$
$$\left.+C^{\Delta}_{3e,1e}(J-1)C^{\Pi}_{0e,0e}(J)\left[\frac{J(J-1)}{3J}\right]^{1/2}\right\}^2. \quad (6.3.25)$$

The signs and magnitudes of the $C^{\Delta}_{\Omega,\Omega'}$ and $C^{\Pi}_{\Omega,\Omega'}$ factors depend in a complicated way on the molecular constants that define the effective Hamiltonians, but most importantly on the sign and magnitude of A_Λ/B_Λ.

The transition amplitudes in Eq. (6.3.25) interfere with each other systematically, in a way that implicitly reflects the fine structures of the two multiplet states. At very low J, when both states are at the case (a) limit, there will be 18 strong branches (exclusively $\Delta\Omega = \Delta\Lambda = +1$); at very high J, when both states are at the case (b) limit, there will again be 18 strong branches (exclusively $\Delta N = \Delta J = 0, \pm 1$), few of which are continuations of the strong low-J branches. The branch intensity patterns display the same pattern-forming rotational quantum number [J, N, J, N^+ in cases (a), (b), (c), (d)], as does the energy-level structure (see Fig. 3.1a). This sort of systematic interference effect is important for transitions into Rydberg complexes. The NO $nf \leftarrow X^2\Pi$ system (Jungen and Miescher, 1969) is a beautiful example of this effect.

The relative intensities of the lines in the 147 $\Delta J = 0, \pm 1$ rotational branches (plus several $\Delta J = \pm 2$, $e \leftrightarrow f$ hyperfine induced branches) of the MnH $A^7\Pi - X^7\Sigma^+(0,0)$ band are described by a single $\mu_\perp$ transition moment, in combination with the eigenvectors of the $^7\Pi$ and $^7\Sigma^+$ **H** matrices. However, the predicted and observed low–J line spacings and intensities exhibit such erratic behavior (see Figs. 4 and 5 of Varberg, *et al.*, 1992) that they misleadingly invite the label "perturbed" (Nevin, 1942 and 1945). Members of 156 rotational branches (see Table VI of Varberg, *et al.*, 1992) are definitively assigned and the rotational levels in each fine structure component are followed to the lowest–J level (Varberg, *et al.*, 1991 and 1992). The most striking irregularities in the A–X (0,0) band are due to the rapid evolution of the $^7\Pi$ state ($A = 40.5$ cm^{-1}, $B = 6.346$ cm^{-1}) from the case (a) limit (J is the pattern forming rotational quantum number) to case (b) (N is the pattern forming rotational quantum number) and the irregular energy order of the diagonal matrix elements for the Ω–components of the $^7\Pi$ state:

$$\begin{aligned} E^{\circ}_{\Omega}(J) &= \langle \Lambda S \Sigma J \Omega \,| \mathbf{H}^{\mathrm{ROT}} + \mathbf{H}^{\mathrm{SO}} + \mathbf{H}^{\mathrm{SS}} |\, \Lambda S \Sigma J \Omega \rangle \\ &= B[J(J+1) - \Omega^2 + S(S+1) - \Sigma^2] + A\Lambda\Sigma + \frac{2}{3}\lambda[3\Sigma^2 - S(S+1)] \end{aligned} \quad (6.3.26)$$

$$\Omega = 4 \qquad E^{\circ}_{4}(J) = BJ(J+1) + 42.1\mathrm{cm}^{-1} \quad (6.3.27)$$

$$\Omega = 3 \qquad E^{\circ}_{3}(J) = BJ(J+1) + 74.7\mathrm{cm}^{-1} \quad (6.3.28)$$

$$\Omega = 2 \qquad E^{\circ}_{2}(J) = BJ(J+1) + 83.1\mathrm{cm}^{-1} \quad (6.3.29)$$

$$\Omega = 1 \qquad E^{\circ}_{1}(J) = BJ(J+1) + 67.4\mathrm{cm}^{-1} \quad (6.3.30)$$

$$\Omega = 0 \qquad E^{\circ}_{0}(J) = BJ(J+1) + 27.5\mathrm{cm}^{-1} \quad (6.3.31)$$

$$\Omega = -1 \qquad E^{\circ}_{-1}(J) = BJ(J+1) - 36.6\mathrm{cm}^{-1} \quad (6.3.32)$$

$$\Omega = -2 \qquad E^{\circ}_{-2}(J) = BJ(J+1) - 124.8\mathrm{cm}^{-1} \quad (6.3.33)$$

Note that although $A > 0$, the order of Ω–components is neither regular nor inverted. Often, a branch which is strong in the case (a) limit becomes weak in the case (b) limit and *vice versa*. This rapid turning off or on of the intensity in such a branch is due to intensity interference effects that are well understood by the molecule but demand considerable detective work by the spectroscopist.

Before discussing the intensity interference effects resulting from the multiple $\Delta\Lambda = \pm 1$ perturbations in an nf–complex ($f\phi \sim f\delta \sim f\pi \sim f\sigma$), it is useful to analyze several examples of isolated $\Delta\Lambda = \pm 1$ perturbations:

1. a $^1\Pi \sim \Sigma^+$ perturbation;

2. a case (c) $\Omega = \frac{1}{2}$ state;

3. a forbidden $^3\Sigma^- - {}^1\Sigma^+$ transition; and

4. an np Rydberg complex.

Consider transitions between a pair of mutually interacting $^1\Sigma^+$ and $^1\Pi$ upper states and an unperturbed $^1\Sigma^+$ lower state. The $^1\Sigma^+ - {}^1\Sigma^+$ system has only one P and one R branch, and the transition amplitudes are

$$R(J) \quad +\mu_{\|}(J+1)^{1/2} \qquad P(J) \quad +\mu_{\|}(J)^{1/2}. \quad (6.3.34)$$

The $^1\Pi - {}^1\Sigma^+$ transition has only three branches, a Q branch (from $^1\Pi_f$ levels) and R and P branches (from $^1\Pi_e$ levels). The $^1\Pi$ e/f basis functions are [Eq. (3.2.95)]

$$|^1\Pi(^e_f), J'\rangle = 2^{-1/2}\left[|J', \Lambda = \Omega = 1\rangle \pm |J', \Lambda = \Omega = -1\rangle\right],$$

and the $^1\Pi - {}^1\Sigma^+$ transition amplitudes are

$$\begin{aligned} &R(J) && -\mu_\perp(J+2)^{1/2} \\ &Q(J) && +\mu_\perp(2J+1)^{1/2} \\ &P(J) && +\mu_\perp(J-1)^{1/2}. \end{aligned} \tag{6.3.35}$$

Only the $^1\Pi_e$ levels can mix with $^1\Sigma^+$. The eigenfunctions are

$$\begin{aligned} |\text{`}^1\Pi_e\text{'}, J\rangle &= C_{\Pi\Pi}(J)\,|^1\Pi_e, J\rangle + C_{\Pi\Sigma}(J)\,|^1\Sigma^+, J\rangle \\ |\text{`}^1\Sigma^+\text{'}, J\rangle &= C_{\Sigma\Pi}(J)\,|^1\Pi_e, J\rangle + C_{\Sigma\Sigma}(J)\,|^1\Sigma^+, J\rangle \\ |\text{`}^1\Pi_f\text{'}, J\rangle &= |^1\Pi_f, J\rangle \end{aligned} \tag{6.3.36}$$

where

$$\begin{aligned} C_{\Sigma\Sigma}(J) &= \quad C_{\Pi\Pi}(J) \\ C_{\Sigma\Pi}(J) &= -C_{\Pi\Sigma}(J) = \mp\left[1 - C_{\Pi\Pi}(J)^2\right]^{1/2}. \end{aligned}$$

The perturbed transition amplitudes are given below, referred to as Eq. (6.3.37).

	$'^1\Pi' \rightarrow {}^1\Sigma^+$	
$R(J)$	$-C_{\Pi\Pi}(J+1)\mu_\perp(J+2)^{1/2} + C_{\Pi\Sigma}(J+1)\mu_\parallel(J+1)^{1/2}$	
$Q(J)$	$\mu_\perp(2J+1)^{1/2}$	(6.3.37a)
$P(J)$	$+C_{\Pi\Pi}(J-1)\mu_\perp(J-1)^{1/2} + C_{\Pi\Sigma}(J-1)\mu_\parallel(J)^{1/2}$	

	$'^1\Sigma' \rightarrow {}^1\Sigma^+$	
$R(J)$	$+C_{\Pi\Sigma}(J+1)\mu_\perp(J+2)^{1/2} + C_{\Pi\Pi}(J+1)\mu_\parallel(J+1)^{1/2}$	
$Q(J)$	0	(6.3.37b)
$P(J)$	$-C_{\Pi\Sigma}(J-1)\mu_\perp(J-1)^{1/2} + C_{\Pi\Pi}(J-1)\mu_\parallel(J)^{1/2}$.	

If $\mu_\perp = 0$, then the spectrum contains four lines for each J'–value, a pair of main $^1\Sigma^+ - {}^1\Sigma^+$ R and P lines and a pair of extra R and P lines. The ratio of transition probabilities for the two main lines terminating in a common J' level is

$$\frac{I[R(J-1)]}{I[P(J+1)]} = \frac{J}{J+1}, \tag{6.3.38}$$

which is identical to the R/P ratio for the extra lines and to the ratio expected for an unperturbed $^1\Sigma^+ - {}^1\Sigma^+$ transition. If $\mu_\parallel = 0$, then the spectrum contains five lines for each J', an unperturbed Q line, a pair of main $'^1\Pi' - {}^1\Sigma^+$ R and P lines, and a pair of extra R and P lines. The ratio of transition probabilities for the main lines is

$$\frac{I[R(J-1)]}{I[P(J+1)]} = \frac{J+1}{J}, \tag{6.3.39}$$

which is identical to the R/P ratio for the extra lines and to the ratio expected for an unperturbed $^1\Pi - {}^1\Sigma^+$ transition. In addition, there is a sum rule (for $\mu_{\|} = 0$),

$$I[Q(J)] = I[R_{\text{main}}(J-1)] + I[R_{\text{extra}}(J-1)] + I[P_{\text{main}}(J+1)] + I[P_{\text{extra}}(J+1)]. \tag{6.3.40}$$

If

$$C_{\Pi\Pi}\mu_{\perp} \simeq C_{\Pi\Sigma}\mu_{\|},$$

then the $'{}^1\Pi' - {}^1\Sigma^+$ R/P ratio is approximately zero. Similarly, if

$$C_{\Pi\Pi}\mu_{\|} \simeq C_{\Pi\Sigma}\mu_{\perp},$$

then the '$^1\Sigma^+$'$-{}^1\Sigma^+$ R/P ratio is very large. If $\mu_{\|} = \mu_{\perp}$, then the higher-energy P line and the lower-energy R line (or *vice versa* depending on the sign of the interaction matrix element, $H_{\Pi\Sigma}$, between $^1\Pi$ and $^1\Sigma^+$) vanish at the $^1\Pi - {}^1\Sigma^+$ level crossing J_0. Regardless of the signs and relative magnitudes of $\mu_{\|}$ and $\mu_{\perp}$, the sense of the R/P intensity anomaly in the nominal $'{}^1\Pi' - {}^1\Sigma^+$ transition will always be opposite (but not necessarily equal in magnitude) to that for the same J'–value in $'{}^1\Sigma' - {}^1\Sigma^+$. The sense of this anomaly will reverse, within a given *nominal* band, for the J–values below and above J_0.

Gottscho, *et al.*, (1978) observed R/P intensity anomalies in the BaO $C^1\Sigma^+ \rightarrow X^1\Sigma^+$ fluorescence spectrum. Along a (v', v'') progression, the magnitude and sense of the anomalies exhibited a strong dependence on the $X^1\Sigma^+$ state v'' level, even though the $^1\Pi \sim {}^1\Sigma^+$ mixing was in the $C^1\Sigma^+$ state. For several v'' members of the C–X (v', v'') progression, either the $P(J'+1)$ or the $R(J'-1)$ line was undetectable, even though the other remaining line was quite strong. Since, for a given J$'$, the fractional R/P intensity anomaly,

$$\frac{I[R(J-1)] - I[P(J+1)]}{I[R(J-1)] + I[P(J+1)]} \simeq \pm \frac{2C_{\Sigma\Sigma}(J)\left[1 - C_{\Sigma\Sigma}^2(J)\right]^{1/2}\mu_{\|}\mu_{\perp}}{C_{\Sigma\Sigma}^2(J)\mu_{\|}^2 + \left[1 - C_{\Sigma\Sigma}^2(J)\right]\mu_{\perp}^2}, \tag{6.3.41}$$

depends on v'' (the only v''–dependent terms are $\mu_{\|}$ and $\mu_{\perp}$), the intensity anomalies could be used to characterize the otherwise unobserved $^1\Pi$–perturber. A similar R/P intensity anomaly has been observed in fluorescence from Se_2 (Gouedard and Lehmann, 1976), and in N_2 (Walter, *et al.*, 2000).

In addition to direct detection in photon-detected absorption or emission spectra, interference effects may be detected indirectly, for example by monitoring photofragment atoms or ions. Walter, *et al.*, (2000) recorded spectra of the N_2 $(b'^1\Sigma_u^+ \sim c^1\Pi_u \sim c'^1\Sigma_u^+) \leftarrow a''^1\Sigma_g^+(v''=0)$ transition. A molecular beam of the metastable N_2 $a''^1\Sigma_g^+(v''=0)$ state is excited by a tunable laser to the $b' \sim c \sim c'$ perturbation complex and transitions are detected by a decrease in the $a''^1\Sigma_g^+(v''=0)$ beam flux and by the *coincidence* detection of two N atoms on a time and position sensitive detector. The coincidence detection scheme permits measurements of the relative probabilities of dissociation into the energetically accessible $N(^4S) + N(^2D)$ and $N(^4S) + N(^2P)$ photofragment channels. The probabilities of each $b' \sim c \sim c'$ (J', energy rank, e/f) $\leftarrow a''(J'', e)$ eigenstate transition (there are three e–symmetry and one f–symmetry eigenstate of the $b' \sim c \sim c'$ perturbation complex) and predissociation rate for each $b' \sim c \sim c'$ (J', energy rank, e/f) eigenstate are derived from diabatic (basis state) transition moments, predissociation rates, and basis state mixing coefficients from the multistate deperturbation model of Stahel, *et al.*, (1983), as extended to include rotation by Helm, *et al.*, (1993) and Edwards, *et al.*, (1995). Numerous $R(J-1) : P(J+1)$ $\mu_{\parallel}, \mu_{\perp}$ transition amplitude interference effects are observed. The deperturbation model correctly predicts the signs of most of these interference effects. Predissociation does not differentially affect the $R(J-1) : P(J+1)$ interference effects, even when the transition probabilities are monitored in the two-N-atom coincidence scheme (unless the continuum state contributes significantly to the parent molecule excitation transition amplitude). However, if two admixed basis states contribute amplitude for predissociation into the single channel associated with the same pair of photofragment atoms, then (J, energy rank, e/f)-*dependent* but excitation R, P branch *independent* interference effects will be observed in the branching ratios between the different photofragment channels.

Case (c) $\frac{1}{2} - \frac{1}{2}$ transitions have been discussed in detail by Kopp and Hougen (1967) [where the definition of $\mu_{\perp}$ differs by a factor of $2^{1/2}$ from that used here and by Hougen (1970)]. Case (c) corresponds to the strong spin-orbit limit where $\Delta\Omega = 0$ spin-orbit perturbations have destroyed both S and Λ as useful quantum numbers. For simplicity, the $\frac{1}{2} - \frac{1}{2}$ example to be discussed here will be a J–independent spin-orbit mixed $^2\Sigma^+ \sim {}^2\Pi_{1/2}$ $\Omega' = \frac{1}{2}$ upper level and a pure $^2\Sigma^+$ $\Omega'' = \frac{1}{2}$ lower level. $\frac{1}{2} - \frac{1}{2}$ transitions consist, in general, of six rotational branches: R_{ee}, R_{ff}, Q_{ef}, Q_{fe}, P_{ee}, and P_{ff}. Although a general case (c) $\Omega = \frac{1}{2}$ state can be composed of an unspecified mixture of $^2\Sigma^+$, $^2\Sigma^-$, $^2\Pi_{1/2}$, $^4\Sigma^+$, $^4\Sigma^-$, $^4\Pi$, $^4\Delta$, etc. case (a) basis states, the simplified example treated here contains all of the features of rotational linestrength anomalies in a $\frac{1}{2} - \frac{1}{2}$ case (c) transition.

Just as for the $^1\Pi \sim {}^1\Sigma^+ \rightarrow {}^1\Sigma^+$ example, two transition moments are needed to account for the rotational linestrengths in the $^2\Pi_{1/2} \sim {}^2\Sigma^+ \rightarrow {}^2\Sigma^+$ transition. Since case (c) implies that

$$\left| \frac{\langle {}^2\Pi_{1/2} | \mathbf{H}^{\mathrm{SO}} | {}^2\Sigma^+ \rangle}{E^0_{^2\Pi_{1/2}} - E^0_{^2\Sigma^+}} \right| \gg 1, \tag{6.3.42}$$

the $^2\Pi \sim {}^2\Sigma$ mixing can be considered to be J–independent and the mixing coefficients, $C_{\Pi\Pi}(J)$ and $C_{\Pi\Sigma}(J)$, can be absorbed into the transition moment factors, $\mu_{\parallel}$ and $\mu_{\perp}$. However, unlike the $^1\Pi \sim {}^1\Sigma^+$ example, the mixed state is composed entirely of basis functions belong to the same Ω–value; and, since the $\boldsymbol{\alpha}$ rotational matrix elements depend exclusively on J and Ω, the rotational linestrength factors can be reduced to simple closed form expressions (Table 6.2).

Table 6.2: Rotational Linestrengths for Case (c) $\frac{1}{2}-\frac{1}{2}$ Transitions

Transition	$\left|\langle n'\tfrac{1}{2}J'|\mu|n''\tfrac{1}{2}J\rangle\right|^2$
$P_{ee}(J)$	$\frac{(J+\frac{1}{2})(J-\frac{1}{2})}{J}\left[\mu_{\parallel}-2^{-1/2}\mu_{\perp}\right]^2$
$P_{ff}(J)$	$\frac{(J+\frac{1}{2})(J-\frac{1}{2})}{J}\left[\mu_{\parallel}+2^{-1/2}\mu_{\perp}\right]^2$
$Q_{ef}(J)$	$\frac{(J+\frac{1}{2})}{2J(J+1)}\left[\mu_{\parallel}+(2J+1)2^{-1/2}\mu_{\perp}\right]^2$
$Q_{fe}(J)$	$\frac{(J+\frac{1}{2})}{2J(J+1)}\left[\mu_{\parallel}-(2J+1)2^{-1/2}\mu_{\perp}\right]^2$
$R_{ee}(J)$	$\frac{(J+\frac{1}{2})(J+\frac{3}{2})}{(J+1)}\left[\mu_{\parallel}+2^{-1/2}\mu_{\perp}\right]^2$
$R_{ff}(J)$	$\frac{(J+\frac{1}{2})(J+\frac{3}{2})}{(J+1)}\left[\mu_{\parallel}-2^{-1/2}\mu_{\perp}\right]^2$

If $\mu_{\perp}=0$, the usual $^2\Sigma^+-{}^2\Sigma^+$ intensity pattern is obtained: two P and two R branches of almost equal intensities at high J and two very weak Q branches. If $\mu_{\parallel}=0$, the usual $^2\Pi-{}^2\Sigma^+$ intensity pattern emerges: two P and two R branches of comparable intensities and two Q branches that are stronger than each of the other four branches by a factor of approximately two. If $2^{-1/2}\mu_{\perp}=\mu_{\parallel}$, then the P_{ee} and R_{ff} branches vanish and Q_{fe} is weaker than Q_{ef} by a factor $[J/(J+1)]^2$, which is significantly different from unity only at low J. If $2^{-1/2}\mu_{\perp}=-\mu_{\parallel}$, the pattern reverses ($e\leftrightarrow f$). This is illustrated by the NO $H^2\Sigma^+$, $H'^2\Pi\leftarrow D^2\Sigma^+$ transitions. There is strong mixing between the H and H′ states (Ogi, *et al.*, 2000). Departures from the intensity pattern predicted by Table 6.2 could result from J-dependent $-B(R)\mathbf{J}\cdot(\mathbf{L}+\mathbf{S})$ mixing with $\Omega=\frac{3}{2}$ states or J-dependent $\left[E^0_{^2\Pi_{1/2}}(J)-E^0_{^2\Sigma^+}(J)\right]$ energy denominators [Eq. (6.3.42)].

$^3\Sigma^--{}^1\Sigma^+$ transitions are "doubly forbidden" in the sense that they violate two selection rules, $\Delta S=0$ and $\Sigma^+\leftarrow/\rightarrow\Sigma^-$. However, this terminology is misleading because only one perturbing state is required to lend intensity to this nominally forbidden transition (Watson, 1968; Hougen, 1970). There are four relevant perturbation mechanisms, each involving a $\Delta\Omega=0$ spin-orbit interaction:

$$\left.\begin{array}{l}^3\Sigma^-\sim{}^1\Pi\\ ^3\Sigma^-\sim{}^1\Sigma^+\end{array}\right\}\text{upper-state perturbations,}$$

$$\left.\begin{array}{l}^1\Sigma^+\sim{}^3\Pi\\ ^1\Sigma^+\sim{}^3\Sigma^-\end{array}\right\}\text{lower-state perturbations.}$$

The case (b) limit ($J \gg |\lambda/B|$) eigenstates are

$$\left|{}'^3\Sigma^{-\prime}_e F_1\right\rangle = \left(\frac{J+1}{2J+1}\right)^{1/2} [|^3\Sigma^-_{1e}\rangle + \alpha\,|^1\Pi_e\rangle]$$
$$+ \left(\frac{J}{2J+1}\right)^{1/2} [|^3\Sigma^-_0\rangle + \beta\,|^1\Sigma^+\rangle] \quad (6.3.43a)$$
$$\left|{}'^3\Sigma^{-\prime}_f F_2\right\rangle = \left[\left|^3\Sigma^-_{1f}\right\rangle + \alpha\,|^1\Pi_f\rangle\right] \quad (6.3.43b)$$
$$\left|{}'^3\Sigma^{-\prime}_e F_3\right\rangle = -\left(\frac{J}{2J+1}\right)^{1/2} [|^3\Sigma^-_{1e}\rangle + \alpha\,|^1\Pi_e\rangle]$$
$$+ \left(\frac{J+1}{2J+1}\right)^{1/2} [|^3\Sigma^-_0\rangle + \beta\,|^1\Sigma^+\rangle] \quad (6.3.43c)$$
$$\left|{}'^1\Sigma^{+\prime}_e\right\rangle = |^1\Sigma^+\rangle + \gamma\,|^3\Sigma^-_0\rangle + \delta\,|^3\Pi_{0e}\rangle . \quad (6.3.44)$$

(Watson(1968), using a spherical tensor phase convention [see Eqs. (3.4.41b) and (3.4.41c)] for $\mu_\perp$ of opposite sign to that used here and by Hougen (1970), has treated the general intermediate case (a)-(b) $^3\Sigma^- - {}^1\Sigma^+$ problem.)

There are four transition moments,

$$\mu_{3\perp} \equiv \langle ^3\Sigma^-|\mu|^3\Pi\rangle \quad (6.3.45a)$$
$$\mu_{3\parallel} \equiv \langle ^3\Sigma^-|\mu|^3\Sigma^-\rangle^{\dagger} \quad (6.3.45b)$$
$$\mu_{1\perp} \equiv \langle ^1\Pi|\mu|^1\Sigma^+\rangle \quad (6.3.45c)$$
$$\mu_{1\parallel} \equiv \langle ^1\Sigma^+|\mu|^1\Sigma^+\rangle ,^{\dagger} \quad (6.3.45d)$$

and five possible rotational branches with linestrength factors given in Table 6.3 where

$$\mu_\parallel \equiv \beta\mu_{1\parallel} + \gamma\mu_{3\parallel} \quad (6.3.46a)$$
$$\mu_\perp \equiv \alpha\mu_{1\perp} + \delta\mu_{3\perp} . \quad (6.3.46b)$$

The pattern of interference effects is identical to the $^1\Pi \sim {}^1\Sigma^+$ example discussed earlier.

It has been pointed out by Johns (1974) and Steimle, *et al.*, (1982) that the energy-level structure of a $^1\Pi \sim {}^1\Sigma^+$ perturbation complex is identical to that

† In the case where the $^3\Sigma^-$ upper state is the same state mixed into $'^1\Sigma^{+\prime}$, $\mu_{3\parallel}$ is the electric dipole moment of the $^3\Sigma^-$ state; similarly, $\mu_{1\parallel}$ can be the dipole moment of $^1\Sigma^+$ (see Section 6.4).

Table 6.3: Rotational Linestrengths for $^3\Sigma^- - {}^1\Sigma^+$ Transitions

$^OP(J)$ (F_1)	$\left\{\mu_\parallel\left[\frac{J(J-1)}{(2J-1)}\right]^{1/2} + \mu_\perp\left[\frac{J(J-1)}{(2J-1)}\right]^{1/2}\right\}^2$
$^QP(J)$ (F_3)	$\left\{\mu_\parallel\left[\frac{J^2}{(2J-1)}\right]^{1/2} - \mu_\perp\left[\frac{(J-1)^2}{(2J-1)}\right]^{1/2}\right\}^2$
$^QQ(J)$ (F_2)	$\left[\mu_\perp(2J+1)^{1/2}\right]^2$
$^QR(J)$ (F_1)	$\left\{\mu_\parallel\left[\frac{(J+1)^2}{(2J+3)}\right]^{1/2} - \mu_\perp\left[\frac{(J+2)^2}{(2J+3)}\right]^{1/2}\right\}^2$
$^SR(J)$ (F_3)	$\left\{\mu_\parallel\left[\frac{(J+1)(J+2)}{(2J+3)}\right]^{1/2} + \mu_\perp\left[\frac{(J+1)(J+2)}{(2J+3)}\right]^{1/2}\right\}^2$

of a $^3\Sigma^-$ state provided that

$$B\left({}^1\Pi\right) = B\left({}^1\Sigma^+\right) = B\left({}^3\Sigma^-\right) \equiv B \tag{6.3.47a}$$

$$E^0\left({}^1\Pi\right) - E^0\left({}^1\Sigma^+\right) = -B + 2\lambda\left({}^3\Sigma^-\right) - 3\gamma\left({}^3\Sigma^-\right) \tag{6.3.47b}$$

$$E \equiv E^0\left({}^1\Pi\right) = E^0\left({}^3\Sigma^-\right) + B + \frac{2}{3}\lambda - \frac{7}{3}\gamma \tag{6.3.47c}$$

$$\left\langle {}^1\Pi_e \left| BL^+ \right| {}^1\Sigma^+ \right\rangle = B \tag{6.3.47d}$$

$$q\left({}^1\Pi\right) = \gamma\left({}^3\Sigma^-\right)/2. \tag{6.3.47e}$$

These requirements are automatically satisfied for a Rydberg p–complex because

$$\left\langle \pi \left| l^+ \right| \sigma \right\rangle = 2^{1/2},$$

and, since Rydberg electrons do not contribute significantly to bonding, the potential curves for the $^1\Sigma$ and $^1\Pi$ components of a p–complex should be identical so that

$$B\left({}^1\Pi\right) = B\left({}^1\Sigma\right) = \langle v_\Pi | B(R) | v_\Sigma \rangle\, \delta_{v_\Pi v_\Sigma}.$$

The $^3\Sigma^-$ and p–complex structures resemble each other because both consist of one unit of spin or electronic angular momentum (**S** or **L**) coupled to the nuclear rotation (**R**). However, since $\boldsymbol{\mu}$ operates exclusively on electron spatial coordinates, any resemblance between the rotational-branch intensity patterns for $^3\Sigma^- - {}^1\Sigma^+$ and p–complex $-{}^1\Sigma^+$ transitions would seem to be coincidental. A $^3\Sigma^- - {}^1\Sigma^+$ transition will look exactly like a p–complex $-{}^1\Sigma^+$ transition if, in addition to satisfying Eqs. (6.3.47), the σ–orbital of the $^1\Sigma^+$ state is predominantly of $s\sigma$ united atom character. Then the transition moment ratio will be

$$\frac{\mu_\parallel}{\mu_\perp} \simeq \frac{\langle p\sigma \left| \mu_z \right| s\sigma \rangle}{2^{-1/2}\langle p\pi \left| \mu^+ \right| s\sigma \rangle} = \frac{\begin{pmatrix} 1 & 0 & 1 \\ 0 & 0 & 0 \end{pmatrix}}{\begin{pmatrix} 1 & 0 & 1 \\ -1 & 0 & 1 \end{pmatrix}} = -1 \tag{6.3.48a}$$

which is the atomic transition moment ratio (see Section 3.4.5). Similarly, if the σ orbital is approximately a $d\sigma$ orbital, then

$$\frac{\mu_{\parallel}}{\mu_{\perp}} \simeq \frac{\langle p\sigma\, |\mu_z|\, d\sigma\rangle}{2^{-1/2}\langle p\pi\, |\mu^+|\, d\sigma\rangle} = +2, \tag{6.3.48b}$$

and the transition will look like an $np \leftarrow nd\sigma$ transition.

Johns (1974) and Johns and Lepard (1975) have examined the intensity pattern of several bands in the Worley-Jenkins $np \leftarrow \mathrm{X}^1\Sigma_g^+$ Rydberg series of N_2. The $n = 6$ and 8 members of this series are clear examples of well-behaved p-complexes with $E(^1\Sigma) - E(^1\Pi)$ separations of 92 cm^{-1} (48B) and 40 cm^{-1} (21B), respectively. L-uncoupling effects are evident in both the level structure and intensity pattern, each suggestive of an approach to pure atomic electronic behavior. The Rydberg electron is beginning to forget that it is attached to a molecule. The energy levels follow the case (d) $BN^+(N^+ + 1)$ pattern rather than the case (a) $BJ(J+1)$ pattern. As J increases, the five rotational branches look increasingly like O, Q, Q, Q, and S form $(N^{+'} - J'')$ rather than $P_{\Pi\Sigma}$, $Q_{\Pi\Sigma}$, $R_{\Pi\Sigma}$, $P_{\Sigma\Sigma}$, and $R_{\Sigma\Sigma}$ $(J' - J'')$ branches. As N^+ (the rotational quantum number of the ion core) becomes the pattern-forming rotational quantum number, the three branches in the center of the complex become very compact, as if the molecular rotation were disappearing from the spectrum. This effect is heightened by the low intensity of the O and S form branches, which results from destructive interference between $\mu_{\parallel}$ and $\mu_{\perp}$ because Eq. (6.3.48a) is satisfied. The uncoupling of the Rydberg electron from the core causes the molecular spectrum to look like an atomic spectrum. All of the photon angular momentum goes into the Rydberg electron, hence the $\Delta N^+ = 0$ rotational branch propensity rule.

The failure of the N_2 $np \leftarrow \mathrm{X}^1\Sigma_g^+$ spectrum to go entirely over, at high J-values, into an atomic spectrum is primarily a consequence of the nonsphericity of the $\sigma_g 2p$ valence orbital (composed primarily from $2p$ orbitals on each N atom). $\sigma_g 2p$ is a bonding orbital, thus $B'' > B'$ and the Q-form branches do not collapse into a perfectly line-like feature. More importantly, the $\sigma_g 2p(3\sigma_g)$ orbital has a mixture of united atom characters, of which only $s\sigma_g$ and $d\sigma_g$ have the correct symmetry to combine with $np\lambda_u$ orbitals. Johns and Lepard (1975) conclude that the N_2 $np \leftarrow \mathrm{X}$ intensity pattern is consistent with significant united atom $s\sigma$ character in the $\sigma_g 2p$ orbital. This experimentally identified $s\sigma$ character is confirmed by calculations (Raoult, *et al.*, 1983), which show 54% s character and 42% d character for this $3\sigma_g$ orbital.

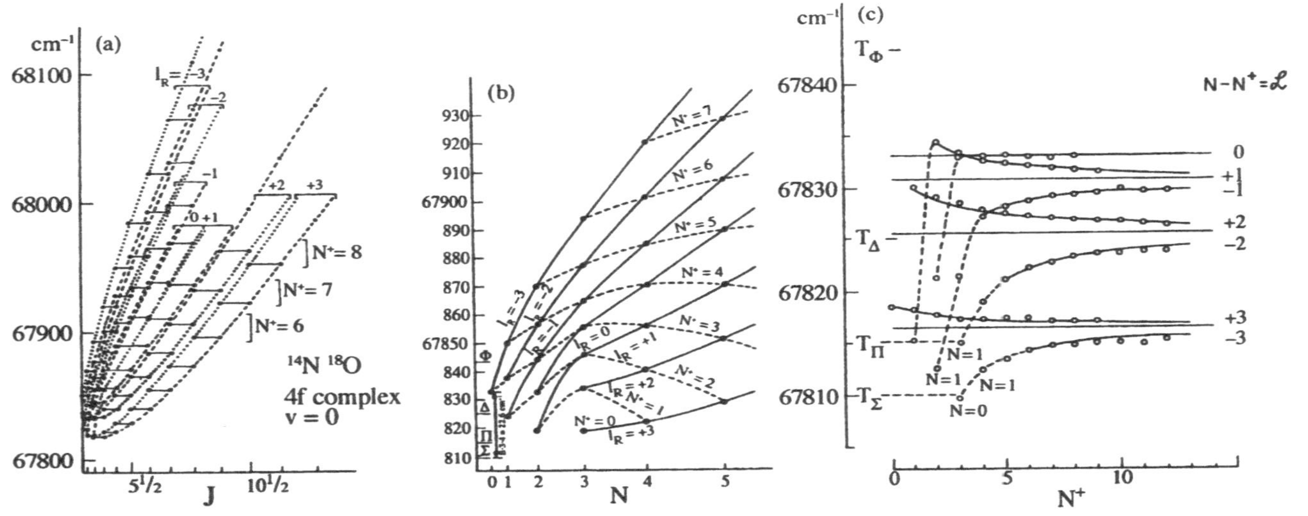

Figure 6.9: Natural rotational quantum numbers for the NO $4f$ Rydberg complex. (a) Energy of all 14 components of the $^{14}N^{18}O$ $4f$ complex plotted versus $J(J+1)$. The components are labeled by $l_R \equiv N - N^+$ and $S_R \equiv J - N$. The $e-$ and $f-$levels of a given l_R, S_R series are connected by dashed and dotted lines, respectively. The horizontal tie lines connect the near-degenerate same-N doublet components belonging to $S_R = \pm\frac{1}{2}$. The levels belonging to the same N^+ value are grouped by a vertical bracket. (b) Energies are plotted versus $N(N+1)$. Solid curves connect the N-levels belonging to the same l_R-value. The S_R-doublet structure has vanished owing to the presentation versus $N(N+1)$. Levels belonging to the same value of N^+ are connected by dashed lines. (c) A reduced term value plot, $T - BN^+(N^+ + 1)$ versus N^+, displays the approach to the case (d) pattern of levels varying as $N^+(N^+ + 1)$ and split into l_R components by the weak electric field of the nonspherical core. (From Jungen and Miescher, 1969, Figs. 3-5.)

Rydberg-Rydberg transitions often appear with no more structure than a few compact Q–form branches. Dressler, *et al.*, (1981) have observed the NO $5g \rightarrow 4f$ bands and suggested not only that rotation-vibration structure is essentially absent, but also that similar-appearing $5g \rightarrow 4f$ bands of many molecules will occur at the same wavelength. Herzberg and Jungen (1982) have rotationally analyzed the corresponding system in H_2 and D_2. For both NO and H_2/D_2, the rotational structure collapses at high rotational quantum numbers into atom-like features that correspond to transitions in which the core angular momentum quantum number ($J^+ = R^+ = N^+$ for NO^+ $X^1\Sigma^+, N^+$ for H_2^+ $X^2\Sigma_g^+$) does not change ($\Delta N^+ = 0$).

Jungen and Miescher (1969) were the first to analyze the rotational structure of an f–complex. The case (d) picture of an nf Rydberg electron interacting with an NO^+ $X^1\Sigma^+$ ion core is far more insightful than a computationally equivalent $^2\Phi \sim {}^2\Delta \sim {}^2\Pi \sim {}^2\Sigma^+$ multistate perturbation approach. This is illustrated by Fig. 6.9, which shows how the energy-level pattern is simplified when it is plotted versus $N^+(N^++1)$ rather than $J(J+1)$. At $N^+ \geq 10$, all 14 components of the $4f$ complex have rotational energies that go as $B^+N^+(N^++1)$. The limiting structure is well developed, showing seven pairs ($J - N = S_R = \pm\frac{1}{2}$) of doublet components. These, in turn, are arranged in four near-degenerate $l_R \equiv N - N^+ = \pm 3, \pm 2, \pm 1, 0$ groups corresponding to the $nf\lambda$ components split by the molecular quadrupole moment of the NO^+ ion core and mixed by ion-core rotation [$-B\mathbf{N}\cdot \mathbf{l}$ term, which causes the transition from case (b) to (d)]. For high N, the quantity $l_R = N - N^+$ can be interpreted as the projection of $\mathbf{l}$ onto the axis of rotation of the ion core (i.e. along $\vec{\mathbf{N}}^+$). l_R has been called l_J by Nikitin and Zare (1994), and $\mathcal{L}$ by Weizel, (1931) and Jungen and Miescher, (1969). The intensity pattern in this $nf \leftarrow X^2\Pi$ system is consistent with atomic $nf\lambda_u \leftarrow 3d\pi_g$ transition moments (no adjustable intensity parameters) and suggests that the singly occupied antibonding $\pi_g 2p$ orbital in the NO $X^2\Pi$ has the predominant $3d\pi_g$ united atom character predicted by Mulliken (1964). The transition from case (b) to case (d) rotational structure occurs at lower and lower N as n increases. At low–n, the rotational levels of the $f\sigma$, $f\pi$, $f\delta$, and $f\phi$ components, which belong to series converging to the same value of N^+, are so well separated in energy that the $B\mathbf{N}\cdot\mathbf{l}$ l–uncoupling term has a minor effect on the energy level and intensity patterns. As n becomes large, the $f\lambda/f\lambda+1$ splittings scale as n^{-3}, and l–uncoupling becomes complete at lower and lower values of N. The result is that N^+ becomes the pattern-forming rotational quantum number, and the rotational levels fall into clumps which follow $B^+N^+(N^++1)$.

$4f$–complexes have also been studied in ArH and KrH (Jungen and Roche, 1999).

The change in pattern-forming rotational quantum number is often the most easily recognized feature of a series of obviously related, but (temporarily) incompletely assigned rotational lines. Various rotational branch notations have been devised to highlight this change in pattern-forming quantum number. Jungen and Miescher (1969) used ${}^{(N^{+\prime}-J'')}(J'-J'')_{l'_R}(J'')$ to describe the rotational structure of the case(d) - case(a) NO $4f$–$X^2\Pi$ (0,0) band. For example, ${}^{-5/2}R_{+3}(9.5)$ implies $N^{+'} - J'' = -5/2, l'_R = +3$ (thus $N' = N^{+'} + 3$), $J'' = 9.5$ (thus $N^{+'} = 7$). Huber, *et al.*, (1987) used ${}^{N^{+'}-N''}(N' - N'')_{l'_R,\Lambda^{\pm''}}(N'')$ for a NO $4f \rightarrow 3d\delta$ band. For example, ${}^{-1}R_{+2,2^-}(10)$ implies $N'' = 10, N' = 11, N^{+'} = 9, l'_R = +2$, and $\Lambda'' = 2$ (where 2^- specifies the parity and e/f symmetry of the negative Kronig symmetry half of a case (b) $^2\Delta$ state that mimics a $^2\Sigma^-$ state, i.e. $(-1)^{N+1}$ parity, $F_1 \rightarrow f$–symmetry, $F_2 \rightarrow e$–symmetry). Arrowsmith, *et al.*, (1981) used ${}^{[N'-N''-l'_R]}\Delta N_{l'_R}(N'')$, which is the same as ${}^{[N^{+'}-N'']}\Delta N_{l'_R}(N'')$, to describe an NO $4f - F^2\Delta$ band.

6.3.3 Assignments Based on Pattern-Forming Rotational Quantum Numbers

It is often difficult to make secure rotational assignments of electronic transitions in which one of the electronic states is a core-nonpenetrating ($l \geq 3$) Rydberg state. Interference effects tend to suppress all but one rotational transition from a given initial rotational state, thus rotational combination differences cannot be relied upon for secure rotational assignments (Johns and Lepard, 1975; Herzberg and Jungen, 1982; Jungen, *et al.*, 1989). (The physical basis for the interference effects is the inability of the photon to transmit its 1 unit of angular momentum to the molecular ion-core by applying torque on the active electron.)

When the electronic transition is between a known rotational level (N'', e/f determined in a double resonance scheme) of a case (b) initial state and an unknown rotational level of a case (d) final state, the existence of pattern-forming rotational quantum numbers [[$B''N''(N''+1)$ for case (b) and $B^+N^+(N^++1)$ for case (d)] provides a basis for $N^+ - N''$ rotational assignments.

If $S = 0$ (or if $S \neq 0$ but $\vec{\mathbf{S}}$ is fully decoupled from the internuclear axis and the $\Delta N = \Delta J$ propensity rule is rigorously obeyed), then for a given value of l in the nonpenetrating state, there are $2l + 3$ possible values of $N^+ - N''$, from $N^+ = N'' - l - 1$ to $N^+ = N'' + l + 1$. These transitions are called "X–form" where X is the letter of the alphabet that corresponds to $N^+ - N''$. A "stacked plot" of spectra recorded from consecutive values of N'' immediately reveals the value of $N^+ - N''$, based on the standard expression for rotational branches

$$X(N'') = B^+N^+(N^+ + 1) - B''N''(N'' + 1) \tag{6.3.49}$$

$$m = N^+ - N'' \tag{6.3.50}$$

$$X(N'') = (B^+ - B'')N''^2 + [(2m+1)B^+ - B'']N'' + B^+(m^2 + m) \tag{6.3.51}$$

At low-N'', the spacing of successive lines in a branch, dX/dN'', is dominated by the term in $X(N'')$ linear in N''. The stacked plot, in which the spectra for excitation of $N'', N''+1, N''+2, \ldots$ are plotted, without any shift of wavenumber axis, each one above the previous one, reveals the "form" of the rotational branch, $m = N^+ - N''$. For example, if $B^+ \approx B''$, an N–form branch ($m = -3$) appears on the stacked plot as a series of lines of slope $-4B$.

Once a value for N^+ is established from the stacked plot, the three possibilities for N', e/f-symmetry, and l_R ($l_R \equiv N' - N^+$ is the projection of the Rydberg orbital angular momentum, $\vec{\mathbf{l}}$, on the axis of nuclear rotation, $\vec{\mathbf{N}}^+$), $N' = N'' \pm 1, e/f \leftarrow N'', e/f$ and $N' = N'', f/e \leftarrow N'', e/f$, can be examined by careful search for an upper state term value match of the N^+ level observed from the $(N'', e/f)$ lower level (lower state combination difference) in spectra recorded from the $(N''-2, e/f), (N''-1, f/e), (N''+1, f/e)$, and $(N''+2, e/f)$ lower levels. It is best to look at the lowest N^+ levels, because the transition to case (d) will be least complete at low-N^+ and the normally weak rotational branch needed for an N'–determining combination difference may have sufficient intensity to be observable. Once the values of N', N^+, l_R, and e/f symmetry are established, the $(-1)^{l-l_R}$ reflection symmetry of the $|ll_R\rangle$ Rydberg orbital determines whether l is even or odd and the requirement that $l \geq |l_R|$ puts a lower bound on l.

6.3.4 $\Sigma^+ \sim \Sigma^-$ Perturbations; Subband Intensity Anomalies in $\Sigma \leftrightarrow \Pi$ Transitions

A special class of intensity anomaly arises from $\Sigma^+ \sim \Sigma^-$ perturbations. Renhorn (1980) has shown that interference between two $\mu_\perp$ transition moments could account for the anomalous $\left({}^2\Sigma - {}^2\Pi_{3/2}\right) / \left({}^2\Sigma - {}^2\Pi_{1/2}\right)$ subband intensity ratios observed by Appleblad, *et al.*, (1981) in a CuO ${}^2\Sigma - \mathrm{X}^2\Pi$ transition at 767 nm. An unperturbed ${}^2\Sigma \rightarrow {}^2\Pi$ transition should have approximately equal intensity $\Sigma - \Pi_{3/2}$ and $\Sigma - \Pi_{1/2}$ subbands. The $\mu_\perp \sim \mu_\parallel$ interference effects discussed in Section 6.3.2 only affect R/P intensity ratios within a subband without changing the total subband intensity.

The effect identified by Renhorn can occur for $\Pi - \Sigma$ and $\Sigma - \Pi$ transitions with $S \geq \frac{1}{2}$ and $\Sigma \neq 0$ and has the *appearance* of an intensity transfer from all rotational branches of the $\Sigma - \Pi_{\Omega=1+|\Sigma|}$ subband into the corresponding branches of the $\Sigma - \Pi_{\Omega=1-|\Sigma|}$ subband. In fact, the intensity transfer is between corresponding subbands of $\Sigma^+ - \Pi$ and $\Sigma^- - \Pi$ transitions.

The subband intensity anomaly arises from $\Delta S = 0, \Delta\Omega = 0, \Omega \neq 0, \Sigma^+ \sim \Sigma^-$ spin-orbit perturbations combined with the opposite behavior of the phase factors for $\Sigma^+ - \Pi_{\Omega=1-|\Sigma|}$ versus $\Sigma^- - \Pi_{\Omega=1-|\Sigma|}$ transitions (Section 6.3.2). The nominal Σ^+ and Σ^- eigenstates are

$$
\begin{aligned}
|`\Sigma^{+}\text{'}, F_i, J\rangle &= \sum_{\Omega=-S}^{S} C_{i\Omega}^{+} \left[\left(1-\beta^2\right)^{1/2} |\Sigma_\Omega^+, J\rangle + \beta |\Sigma_\Omega^-, J\rangle\right] \\
|`\Sigma^{-}\text{'}, F_i, J\rangle &= \sum_{\Omega=-S}^{S} C_{i\Omega}^{-} \left[-\beta |\Sigma_\Omega^+, J\rangle + \left(1-\beta^2\right)^{1/2} |\Sigma_\Omega^-, J\rangle\right]
\end{aligned}
\tag{6.3.52}
$$

where the $C_{i\Omega}^{\pm}$ are mixing coefficients for $\Sigma^\pm$ states expressed in the case (a) basis and β is the Ω-independent $\Sigma^+ \sim \Sigma^-$ mixing coefficient,

$$\beta \equiv \frac{\langle \Sigma_\Omega^+ | \mathbf{H}^{\mathrm{SO}} | \Sigma_\Omega^- \rangle}{E^0(\Sigma^+) - E^0(\Sigma^-)} \left[1 - \delta_{\Omega,0}\right]. \tag{6.3.53}$$

The $1 - \delta_{\Omega,0}$ factor reflects the parity forbiddenness of $\Sigma_0^+ \sim \Sigma_0^-$ mixing. For a transition into a case (a) Π_Ω state, only the $|\Sigma| = \Omega + 1$ and $|\Sigma| = \Omega - 1$ character of the Σ-state contributes to the transition probability, thus:

$$\begin{aligned}\mu\left('\Sigma^+(F_i)' - \Pi_{\Omega=1+|\Sigma|}\right) \\ = 2^{-1/2}\langle \Omega = 1 + |\Sigma|, J_\Pi |\, \alpha^- \,|\, |\Sigma|, J_\Sigma\rangle C_{i|\Sigma|}^+(J_\Sigma)\left[(1-\beta^2)^{1/2}\mu_{\perp+} + \beta\mu_{\perp-}\right]\end{aligned} \tag{6.3.54a}$$

$$\begin{aligned}\mu\left('\Sigma^-(F_i)' - \Pi_{1+|\Sigma|}\right) \\ = 2^{-1/2}\langle 1 + |\Sigma|, J_\Pi |\, \alpha^- \,|\, \Sigma|, J_\Sigma\rangle C_{i|\Sigma|}^-(J_\Sigma)\left[-\beta\mu_{\perp+} + (1-\beta^2)^{1/2}\mu_{\perp-}\right]\end{aligned} \tag{6.3.54b}$$

$$\begin{aligned}\mu\left('\Sigma^+(F_i)' - \Pi_{1-|\Sigma|}\right) \\ = 2^{-1/2}\langle 1 - |\Sigma|, J_\Pi |\, \alpha^- \,| -|\Sigma|, J_\Sigma\rangle C_{i,-|\Sigma|}^+(J_\Sigma)\left[(1-\beta^2)^{1/2}\mu_{\perp+} - \beta\mu_{\perp-}\right]\end{aligned} \tag{6.3.54c}$$

$$\begin{aligned}\mu\left('\Sigma^-(F_i)' - \Pi_{1-|\Sigma|}\right) \\ = 2^{-1/2}\langle 1 - |\Sigma|, J_\Pi |\, \alpha^- \,| -|\Sigma|, J_\Sigma\rangle C_{i,-|\Sigma|}^-(J_\Sigma)\left[-\beta\mu_{\perp+} - (1-\beta^2)^{1/2}\mu_{\perp-}\right]\end{aligned} \tag{6.3.54d}$$

where

$$\mu_{\perp\pm} = 2^{-1/2}\langle \Pi | \mu^+ | \Sigma^\pm \rangle.$$

Equation (6.3.53) implies that transitions terminating in $\Pi_{\Omega=1}$ levels will not exhibit the $\mu_{\perp+} \sim \mu_{\perp-}$ interference effect. The interference term for the $'\Sigma^+(F_i)' - \Pi_{1+|\Sigma|}$ subbands,

$$\left[\langle 1 + |\Sigma|, J_\Pi |\, \alpha^- \,|\, |\Sigma|, J_\Sigma\rangle\right]^2 \left[C_{i|\Sigma|}^+(J_\Sigma)\right]^2 (1-\beta^2)^{1/2}\beta\mu_{\perp+}\mu_{\perp-},$$

is of comparable magnitude but opposite sign to that for the $'\Sigma^-(F_i)' - \Pi_{1+|\Sigma|}$ and $'\Sigma^+(F_i)' - \Pi_{1-|\Sigma|}$ subbands. If $|\beta| \ll 2^{-1/2}$, the interference effect will be most noticeable in the nominal $\Sigma^\pm - \Pi$ system associated with the smaller transition moment.

6.3.5 F_1 vs. F_2 Intensity Anomalies in $^2\Sigma$ States

When a $^2\Sigma^+$ state is predissociated (see Sections 7.8.2 and 7.8.3) by a $^2\Pi$ state, interference between the spin-orbit ($\Delta\Omega = 0$) and L-uncoupling ($\Delta\Omega = \pm 1$, but in the e, f basis set, the $^2\Sigma^+_{|1/2|} \sim {}^2\Pi_{1/2}$ interaction is effectively $\Delta\Omega = 0$) interaction pathways give rise to a diagnostically unique difference in radiative lifetimes (or fluorescence quantum yields) for the F_1 vs. F_2 components of the same N–value (Field, *et al.*, 2002). Among all of the electronic states capable of perturbing a $^2\Sigma^+$ state ($^2\Sigma^+, {}^2\Sigma^-, {}^2\Pi, {}^4\Pi$), a difference in lifetimes between same–N F_1 and F_2 components can only be caused by a $^2\Pi$ state.

6.4 Forbidden Transitions; Intensity Borrowing by Mixing with a Remote Perturber

It is always possible to express the eigenfunction corresponding to a nominal i, v_i, J' electronic-vibration-rotation level as a sum of rovibronic basis functions,

$$|'i, v_i, J'\,'\rangle = \sum_{k, v_k, \Omega_k} C_{i;k,v_k,\Omega_k} \, |v_k\rangle \, |k\Lambda_k S_k \Sigma_k\rangle \, |\Omega_k J' M\rangle \,, \tag{6.4.1}$$

and to express the transition moment matrix element between $|'i, v_i, J'\,'\rangle$ and an unperturbed basis function $|0, v_0, J\rangle$ as [see Eq. (6.1.20)]

$$\langle 'i, v_i, J'\,'|\mu|0, v_0, J\rangle = \sum_{k, v_k, \Omega_k} C_{i;k,v_k,\Omega_k} \, \langle v_k|v_0\rangle \, R_e^{k0}\left(R_{v_k v_0}\right) \langle \Omega_k J'|\alpha|\Omega_0 J\rangle \,. \tag{6.4.2}$$

This equation can be generalized to include a mixed $|'\,0, v_0, J\,'\rangle$ eigenfunction. Whenever more than one term in the Eq. (6.4.2) summation is significant, the simple factorization of the transition probability into electronic $\left(R_e^2\right)$, Franck-Condon $\left(q_{v'v''} = \langle v'|v''\rangle^2\right)$, and Hönl-London rotational factors $(\mathbf{S}_{J'\Omega', J''\Omega''})$ breaks down. In Section 6.2.1 the case was discussed in which only the $k = i$ term in the summation was nonzero. Interference effects arise when two or more terms in Eq. (6.4.2) are significant and have comparable magnitudes. This section deals with the situation where the $k = i$ term is zero and hence the transition is nominally forbidden. One example of this type, the forbidden $^3\Sigma^- - {}^1\Sigma^+$ transition, was discussed in Section 6.3.2.

There has been a great deal of confusion about which Franck-Condon factors should be used to model the vibrational intensities in a v''–progression in fluorescence originating from a perturbed, nominally nonfluorescing, upper level. Three cases must be considered: (1) the $|'i, v'\rangle$ vibronic level borrows all of its $i \rightarrow 0$ oscillator strength from a single, remote, electronic-vibrational level, $|j, v_j\rangle$; (2) intensity is borrowed from all vibrational levels of a single remote electronic state; (3) more than one remote electronic state contributes to the $i \rightarrow 0$ transition probability. The Franck-Condon factors appropriate to case (1) are those for the remote perturbing $|j, v_j\rangle$ basis state $\left(q^{j,0}_{v_j,v''}\right)$ whereas for case (2) the $q^{i,0}_{v_i,v''}$ are appropriate even though $R_e^{i,0} = 0$. In case (3), interference effects will make calculation of vibrational and rotational linestrengths difficult, but not impossible.

The phrase "remote perturber" is intended to imply that

$$\left|E_i^0 - E_j^0\right| \gg |H_{ij}| \tag{6.4.3a}$$

$$\left|E_i^0 - E_j^0\right| \gg \left|B_i^0 - B_j^0\right| J(J+1) \tag{6.4.3b}$$

so that the C_{ij} mixing coefficients are not J–dependent. The limits specified by Eqs. (6.4.3a) and (6.4.3b) are frequently applicable for homogeneous interactions ($\Delta\Omega = 0$) because H_{ij} is J–independent. However, this limit can never be achieved for heterogeneous ($\Delta\Omega = \pm1$) perturbations because H_{ij} is proportional to J. However, $\Delta\Omega = \pm1$ mixing is often a less important intensity borrowing mechanism than $\Delta\Omega = 0$ mixing (up to some threshold J–value) and can never contribute directly to the oscillator strength of a nominally spin-forbidden transition.

Case (1). *Perturbation by a single remote* $|j, v_j\rangle$ *vibronic state with nonzero oscillator strength to* $|0, v_0\rangle$. The only nonzero term in Eq. (6.4.2) is the j, v_j term. There may be many terms with $C_{i;j,v_j,\Omega_j} \neq 0$ but only one term where both $R_e^{j,0}\langle v_j|v_0\rangle$ and C are nonzero. Eq. (6.4.2) becomes

$$\langle' i, v_i, J'\,'|\boldsymbol{\mu}|0, v_0, J\rangle = C_{i;j,v_j,\Omega_j}\langle v_j|v_0\rangle R_e^{j,0}\left(R_{v_j,v_0}\right)\langle\Omega_j J'|\boldsymbol{\alpha}|\Omega_0 J\rangle. \tag{6.4.4}$$

and the transition probabilities can be expressed in terms of

$$\left[\mu_{'i,v_i,J'\,';0,v_0,J}\right]^2 = \mu^2 q_{v_j,v_0}^{j,0}\mathbf{S}_{\Omega_j J';\Omega_0 J}, \tag{6.4.5}$$

where

$$\mu^2 \equiv \left[C_{i;j,v_j,\Omega_j}\right]^2\left[R_e^{j,0}\left(R_{v_j,v_0}\right)\right]^2. \tag{6.4.6}$$

Equation (6.4.5) looks like the usual product of electronic, vibrational, and rotational factors except that μ^2 is an effective parameter, which may be strongly dependent on $v_i, \Omega_i, e/f$, and J (rather than a function of the R_{v_j,v_0} R–centroid), and q and $\mathbf{S}$ are the Franck-Condon and Hönl-London factors appropriate for the $j, v_j, \Omega_j \rightarrow 0, v_0, \Omega_0$ transition rather than the $i, v_i, \Omega_i \rightarrow 0, v_0, \Omega_0$ transition suggested by the label of the $'i'$ state. This result is counterintuitive and has been forgotten many times in published papers.

Case (2). *Perturbation by a single remote electronic state with nonzero oscillator strength to* $|0, v_j\rangle$. The only nonzero terms in Eq. (6.4.2) are those for *all* v_j of the jth electronic state. If Eqs. (6.4.3a) - (6.4.3b) are satisfied, the sum over remote states can be simplified by expressing the mixing coefficients as

$$C_{i;j,v_j,\Omega_j} = \frac{\langle i, \Omega_i|\mathbf{H}|j, \Omega_j\rangle\langle v_i|v_j\rangle}{E_{i,v_i}^0 - E_{j,v_j}^0}, \tag{6.4.7}$$

assuming that $\langle i, \Omega_i|\mathbf{H}|j, \Omega_j\rangle / \left(E_{i,v_i}^0 - E_{j,v_j}^0\right)$ is independent of v_j and equal to $H_{ij}/\left(E_{i,0}^0 - E_{j,0}^0\right)$, assuming that $R_e^{j,0}$ is independent of R–centroid, and recognizing that

$$\sum_{v_j} \langle v_i|v_j\rangle \langle v_j|v_0\rangle = \langle v_i|v_0\rangle \tag{6.4.8}$$

by the completeness of the $|v_j\rangle$ vibrational basis set. The transition probabilities can then be expressed as

$$|\langle 'i, v_i, J' \,'|\boldsymbol{\mu}|0, v_0, J\rangle|^2 = \mu^2 q^{i,0}_{v_i,v_0} \mathbf{S}_{\Omega_j J';\Omega_0 J} \tag{6.4.9}$$

where

$$\mu^2 \equiv \left(H_{ij}/\Delta E^0_{ij}\right)^2 \left(R^{j,0}_e\right)^2 . \tag{6.4.10}$$

The assumptions required to obtain Eqs. (6.4.9) and (6.4.10) are more drastic than those for Eqs. (6.4.5) and (6.4.6), but their failure can often be absorbed into a strongly R–centroid dependent effective μ^2 parameter. Note that Eq. (6.4.9) contains the "intuitive" Franck-Condon factors but the Hönl-London factors are those appropriate to the Ω–value of the remote perturber.

Case (3). *Several remote electronic states contribute to the 'i' → 0 oscillator strength.* A frequently encountered example is the nominally forbidden $^3\Pi$– $X^1\Sigma^+$ transition. The $^3\Pi$ state can become mixed with singlet states,

$$\left|{}'^3\Pi'_1\right\rangle = |^3\Pi_1\rangle + \frac{\langle ^3\Pi_1|\mathbf{H}^{SO}|^1\Pi_1\rangle}{E^0(^3\Pi_1) - E^0(^1\Pi_1)} |^1\Pi_1\rangle \tag{6.4.11}$$

$$\left|{}'^3\Pi'_{0e}\right\rangle = |^3\Pi_{0e}\rangle + \frac{\langle ^3\Pi_{0e}|\mathbf{H}^{SO}|^1\Sigma^+_0\rangle}{E^0(^3\Pi_{0e}) - E^0(^1\Sigma^+_0)} |^1\Sigma^+_0\rangle , \tag{6.4.12}$$

and the $X^1\Sigma^+$ state can be mixed with triplet states (the possibility of mixing $|^3\Sigma^-_0\rangle$ and states other than $|^3\Pi_{0e}\rangle$ into $X^1\Sigma^+$ will be ignored here),

$$|'X^1\Sigma^+_0\,'\rangle = |X^1\Sigma^+_0\rangle - \frac{\langle X^1\Sigma^+|\mathbf{H}^{SO}|^3\Pi_{0e}\rangle}{E^0(^3\Pi_{0e}) - E^0(X^1\Sigma^+)} |^3\Pi_{0e}\rangle . \tag{6.4.13}$$

The $^1\Pi$ state most likely to mix with $^3\Pi_1$ is the isoconfigurational state (see Eq. (3.4.14)) and

$$\langle ^3\Pi_1|\mathbf{H}^{SO}|^1\Pi\rangle = \pm A(^3\Pi).$$

Although $^3\Pi_{0e}$ can mix with many excited $^1\Sigma^+$ states, mixing with the $X^1\Sigma^+$ state introduces a novel feature, namely, the appearance of permanent electric dipole moments as well as transition moments in the intensity borrowing model. For $^3\Pi$ in the case (a) limit ($A \gg 2^{1/2}BJ$),

$$\begin{aligned}\mu_\perp &\equiv 2^{-1/2}\left\langle {}'^3\Pi'_1|\boldsymbol{\mu}^+|'X^1\Sigma^{+\prime}\right\rangle \\ &= 2^{-1/2}\frac{\langle ^3\Pi_1|\mathbf{H}^{SO}|^1\Pi_1\rangle}{E^0\left(^3\Pi_1\right) - E^0\left(^1\Pi\right)} \langle ^1\Pi|\boldsymbol{\mu}^+|X^1\Sigma^+\rangle ,\end{aligned} \tag{6.4.14}$$

$$\mu_{\|} \equiv \left\langle {}'^3\Pi_{0e}{}' | \mu_z | {}'X^1\Sigma^{+\prime} \right\rangle$$
$$= \frac{\langle X^1\Sigma_0^+ | \mathbf{H}^{SO} | ^3\Pi_{0e} \rangle}{E^0\left(^3\Pi_{0e}\right) - E^0(X^1\Sigma^+)} \left[\langle X^1\Sigma^+ | \mu_z | X^1\Sigma^+ \rangle - \langle ^3\Pi_{0e} | \mu_z | ^3\Pi_{0e} \rangle \right]. \quad (6.4.15)$$

The term in square brackets is the difference in permanent electric dipole moments of the $X^1\Sigma^+$ and $^3\Pi$ states. In the case of CO, $\mu(X^1\Sigma^+) \simeq 0$, $\mu(a^3\Pi) \simeq 1.4$ D, and $\langle X^1\Sigma^+ | \mathbf{H}^{SO} | a^3\Pi_{0e} \rangle \simeq 60$ cm^{-1} (Schamps, 1973). The $\mu_\perp$ term expresses the transfer of intensity from $^1\Pi - X^1\Sigma^+$ to $^3\Pi_1 - X^1\Sigma^+$ and makes the $^3\Pi_1 - X$ subband look exactly like a $^1\Pi - ^1\Sigma^+$ transition. The $\mu_\|$ term expresses the intensity transferred from $^3\Pi_{0e} - {}^3\Pi_{0e}$ and $^1\Sigma^+ - {}^1\Sigma^+$ transitions (in this case, pure rotational and rotation-vibration transitions within the $X^1\Sigma^+$ and $a^3\Pi$ states) into the $^3\Pi_{0e} - X^1\Sigma^+$ subband. The $^3\Pi_{0f} - X^1\Sigma^+$ and $^3\Pi_2 - X^1\Sigma^+$ transitions remain forbidden in the case (a) limit. The case (a) $^3\Pi_{0e} - X^1\Sigma^+$ transition looks like an $\Omega = 0 \rightarrow \Omega = 0$ transition in which the Q–branch is rigorously forbidden. For the analogous $a^3\Pi_{0e} - X^1\Sigma^+$ transition in CS, the Q branch appears only for $J' > 10$ and is a consequence of spin-uncoupling in the $a^3\Pi$ state (Cossart, *et al.*, 1977).

When $J \neq 0$, the **S**-uncoupling operator $(-B\mathbf{J}^\pm\mathbf{S}^\mp)$ causes the Ω–components of the $^3\Pi$ state to mix [see Eqs. (6.3.23) and (6.3.24a)-(6.3.24b) for the effect of spin-uncoupling on an unperturbed $^3\Delta - {}^3\Pi$ transition],

$$\left| {}'^3\Pi'_{\Omega e/f} J \right\rangle = \sum_{\Omega'=0}^{2} C_{\Omega\Omega'}(J, e/f) \left| ^3\Pi_{\Omega' e/f} J \right\rangle. \quad (6.4.16)$$

The general spin-uncoupling expression for $\mu({}'^3\Pi' - {}^1\Sigma^+)$ is then

$$\left\langle {}'^3\Pi'_{\Omega e/f} J | \mu | X^1\Sigma^+ \right\rangle = C_{\Omega 0}(J, e)\mu_\| + C_{\Omega 1}(J, e/f) 2^{1/2} \mu_\perp. \quad (6.4.17)$$

When $\mu_\perp \gg \mu_\|$, one needs only keep track of the $\Omega = 1$ character in each of the $'\Omega' e/f$ substates. When $2^{1/2}BJ < A$, approximate expressions for C_{21} and C_{01} can be obtained from first-order perturbation theory:

$$C_{21}(J) \simeq -2^{1/2}BJ/A \simeq -C_{01}(J). \quad (6.4.18)$$

The intensity of $^3\Pi_{0e/f} - X^1\Sigma^+$ and $^3\Pi_{2e/f} - X^1\Sigma^+$ subbands will increase as J^2, and the $R : Q : P$ branch intensities will be $1 : 2 : 1$. When $\mu_\| \gg \mu_\perp$, the $\Omega = 0$ character determines the subband intensity. The $^3\Pi_{1e} - X^1\Sigma^+$ subband will consist exclusively of equal-intensity R and P branches with intensity proportional to J^2. The $^3\Pi_{2e} - X^1\Sigma^+$ subband will only appear at very high J (intensity proportional to J^4). The general case of $\Omega, J, e/f$–dependent $^3\Pi - {}^1\Sigma^+$ transition probabilities and radiative lifetimes has been treated theoretically by James (1971a) and experimentally for CO by James (1971b), Slanger and Black (1971), and Jongma, *et al.*, (1997). The radiative lifetime of each $\Omega, J, e/f$ level is different,

$$\tau_{v\Omega e/f}(J) \propto \left[\left(\sum_{v_X} q_{v_a v_X} \nu^3_{v_a v_X} \right) \left[C_{\Omega 0}(J,e)\mu_{\parallel} + C_{\Omega 1}(J,e/f) 2^{1/2} \mu_{\perp} \right]^2 \right]^{-1} \tag{6.4.19}$$

and it is misleading to speak of "the radiative lifetime of the vth vibrational level." The predicted J, Ω–dependence of the radiative lifetime in the CO $a^3\Pi$ state (setting $\mu_{\parallel} = 0$) is illustrated by Fig. 6.10.

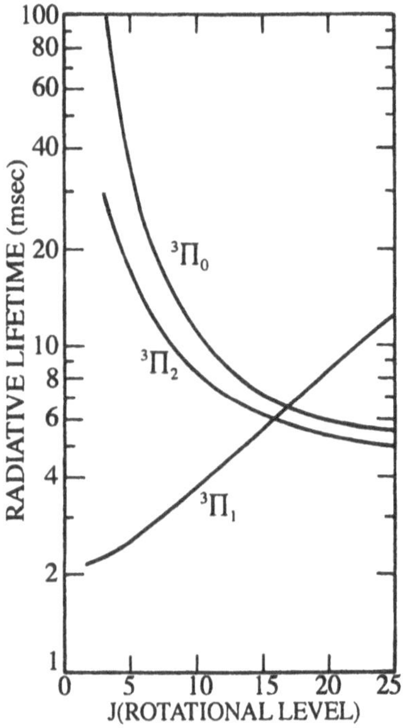

Figure 6.10: Variation of the CO $a^3\Pi$ radiative lifetime versus J and Ω. The CO $a^3\Pi$ state borrows its oscillator strength from $A^1\Pi$ and, to a much lesser extent, from $X^1\Sigma^+$. The effective lifetime of an entire vibrational level will depend in a complex way on pressure, temperature and the mode of excitation. From data of Fairbairn (1970) corrected by Slanger and Black (1971). Jongma, *et al.*, (1997) measure $\tau = 3.8ms$ for $v = 0, \Omega = 1, J = 2e$ and $2f$ levels.

Measurements of τ that are partially averaged over $J, \Omega, e/f$ will not agree with each other and will exhibit unexpected temperature and pressure dependences. However, since the $^3\Pi \sim {}^1\Pi$ and $^3\Pi \sim {}^1\Sigma^+$ perturbation matrix elements are not J–dependent, the fractional $^1\Pi$ and $^1\Sigma^+$ characters, summed over the six $\Omega, e/f$ $^3\Pi$ components of a given J–value, are J–independent. This provides an unambiguous definition of an average, J–independent, vibrational-level radiative lifetime:

$$6/\overline{\tau}_{v'} \equiv \sum_{\Omega,e/f} 1/\tau_{v'\Omega e/f}(J). \tag{6.4.20}$$

It is very difficult to measure rotation-vibration state-specific radiative lifetimes or single rovibronic transition oscillator strengths for states as long-lived as the CO $a^3\Pi$ state. In addition, owing to the strong dependence of lifetimes and oscillator strengths on $J, e/f$, and Ω, caution must be used in exploiting some of the usual relationships between radiative properties. Jongma, *et al.*, (1997) measured the radiative lifetimes of the CO $a^3\Pi$ $v = 0, \Omega = 1, J = 1e$ and $1f$ levels by two different ways and, from these measurements, inferred the nominally J-independent oscillation strength sum over the $a^3\Pi - X^1\Sigma^+(0,0)$ band transitions into the three fine structure components of specified J-value and parity. Jongma, *et al.*, (1997) solved the difficult problem of measuring a peak absorption cross section as small as 4×10^{-19} cm^2 by using an ingenious fluorescence detection scheme rather than a direct measurement of absorption. A 0.005 cm^{-1} FWHM beam of tunable 206 nm radiation traverses a 1.76 m long, room temperature CO absorption cell (0.113 cm^{-1} Doppler width). Absorption in the cell is monitored using fluorescence from a supersonic jet CO molecular beam (< 0.008 cm^{-1} FWHM Doppler width) as a narrow bandwidth detection system. By monitoring the intensity of CO $a^3\Pi \to X^1\Sigma^+$ fluorescence in the molecular beam as the pressure in the absorption cell is varied (0, 5, 10 Torr), the absorption cross sections (oscillator strength, Einstein A-coefficient) for selected rotational transitions ($R_2(1), Q_2(1), Q_2(2)$) in the $a^3\Pi_1 - X^1\Sigma^+(0,0)$ band are measured. To convert a measured A-coefficient to a single-rovibronic level radiative lifetime, it is necessary to use measured $(0, v'')$ fluorescence branching ratios to compute the A-coefficient sum

$$1/\tau_{v'=0} = \sum_{v''} A_{v'=0,v''}. \tag{6.4.21}$$

The lifetimes derived from absorption measurements for the $a^3\Pi, v = 0, \Omega = 1, J = 2e$ and $2f$ levels, are respectively $\tau = 3.78 \pm 0.20$ ms (via $R_2(1)$) and $\tau = 3.79 \pm 0.20$ ms (via $Q_2(2)$). These indirectly measured lifetimes agree well with directly measured values for $J = 1e$ (via $R_2(0)$) and $J = 1f$ (via $Q_2(1)$), respectively 3.4 ± 0.4 ms and 3.8 ± 0.5 ms. The lifetimes were measured directly by monitoring fluorescence in the molecular beam at two locations downstream from the excitation region, and taking the ratio of fluorescence intensities at these two locations as the velocity of the molecular beam is systematically charged by using He, Ne, Ar, Kr, and Xe as carrier gas. In order to avoid collision induced rotational equilibration (transfer from short-lived low-J, $\Omega = 1$ levels to longer lived high-J and $\Omega = 0, 2$ levels) the background gas pressure in the molecular beam was maintained at $< 10^{-9}$ Torr. (To ensure that rotation changing self-collisions have a $< 1\%$ probability at $\tau \approx 10$ ms and a typical self-collision rate of "100MHz/Torr", the background gas pressure must be $< 10^{-8}$ Torr.)

$\mu_{\parallel} \sim \mu_{\perp}$ interference effects in a $^3\Pi \sim {}^1\Sigma$ transition appear as anomalies in the ratios of R and P branch intensities, primarily in the $^3\Pi_1 - {}^1\Sigma^+$ and $^3\Pi_0 - {}^1\Sigma^+$ subbands. The Q lines can never be affected because the $\mu_{\parallel}$ mechanism cannot contribute to their intensity. If $|\mu_{\parallel}| > |\mu_{\perp}|$, the R/P anomalies will be most pronounced in the weaker $\left({}^3\Pi_1 - {}^1\Sigma^+\right)$ subband as is observed by Lee, *et al.*, (2000) for the Rb_2 $2^3\Pi_u - X^1\Sigma_g^+$ transition. If $|\mu_{\parallel}| < |\mu_{\perp}|$, the anomalies will appear in the weak $\left({}^3\Pi_0 - {}^1\Sigma^+\right)$ subband. At some J' value, either the $P(J+1)$ or the $R(J-1)$ line in the weaker subband will vanish, having transferred all of its transition probability to its same-subband partner.

Cossart, *et al.*, (1994) observed two kinds of interference effects in the CS $a^3\Pi - X^1\Sigma^+(0,0)$ band: a $\mu_{\perp}, \mu_{\parallel}$ interference, which is manifest as $R(J-1)$: $P(J+1)$ intensity anomalies in the $^3\Pi_0 - {}^1\Sigma^+$ subband, and a $\mu_{\parallel}, \mu_{\parallel}$ interference, which causes the parallel transition to appear with much less intensity than would naively be expected. The $a^3\Pi \sim A^1\Pi$ spin–orbit interaction accounts for the dominant contribution to the $\mu_{\perp}$ transition moment, while $\mu_{\parallel}$ arises from both the $a^3\Pi \sim X^1\Sigma^+$ and $a^3\Pi \sim A'^1\Sigma^+$ spin-orbit interactions. The observation that the interference effect is evident primarily in the $^3\Pi_0 - {}^1\Sigma^+$ subband implies that $|\mu_{\perp}| >> |\mu_{\parallel}|$, and the observation that the interference is destructive for the P branch (see Fig. 6.11) implies that $\mu_{\perp}$ and $\mu_{\parallel}$ have opposite signs. A fit yields $\mu_{\perp}/\mu_{\parallel} = -0.14$.[†] The value of $|\mu_{\parallel}|$ derived from the fitted value of $\mu_{\perp}/\mu_{\parallel}$ and an estimated value of $\mu_{\perp}$ (obtained from the known $\langle a^3\Pi|\mathbf{H}^{SO}|A^1\Pi\rangle$ spin-orbit matrix element, the μ_{A-X} electronic transition moment deduced from the $A^1\Pi$ state radiative lifetime, and the experimental $\Delta E(a, A)$ energy difference) is much smaller than that expected (based on semi-empirically predicted values of the $a^3\Pi \sim X^1\Sigma^+$ and $a^3\Pi \sim A'^1\Sigma^+$ spin-orbit matrix elements together with *ab initio* and experimental values, respectively, for the permanent electric dipole moments of the $a^3\Pi$ and $X^1\Sigma^+$ states). This implies that the $a^3\Pi \sim X^1\Sigma^+$ and $a^3\Pi \sim A'^1\Sigma^+$ contributions to $\mu_{\parallel}$ are of almost equal magnitude but opposite signs. This is an especially clear example of a $\mu_{\parallel}, \mu_{\parallel}$ interference effect. If the sense of the $\mu_{\parallel}, \mu_{\parallel}$ interference had instead been constructive, the $R(J-1)$:$P(J+1)$ rotational branch interference effect would have been most evident in the $^3\Pi_1 - {}^1\Sigma^+$ subband, because in that case $|\mu_{\parallel}| >> |\mu_{\perp}|$.

The forbidden $^3\Sigma^- - {}^1\Sigma^+$ transition has been discussed earlier [Section 6.3.2, Eqs. (6.3.43a) - (6.3.46b), Table 6.3]. It is useful to return to a specific example here. $^3\Sigma^- - {}^1\Sigma^+$ transitions have been observed between two states belonging to the same π^2 configuration (molecules with 6 valence electrons, NH, PH, etc. and heteronuclear molecules with 12 valence electrons, SO, NF, NCl, PF). Wayne and Colbourn (1977) have discussed the spin-orbit interaction between this pair of isoconfigurational states. Inserting the first-order perturbation theory definitions of the β and γ mixing coefficients of Eq. (6.3.46a), an expression analogous to Eq. (6.4.15) is obtained,

---

[†] The sign of $\mu_{\perp}/\mu_{\parallel}$ was incorrectly printed in the paper by Cossart, *et al.*, (1994).

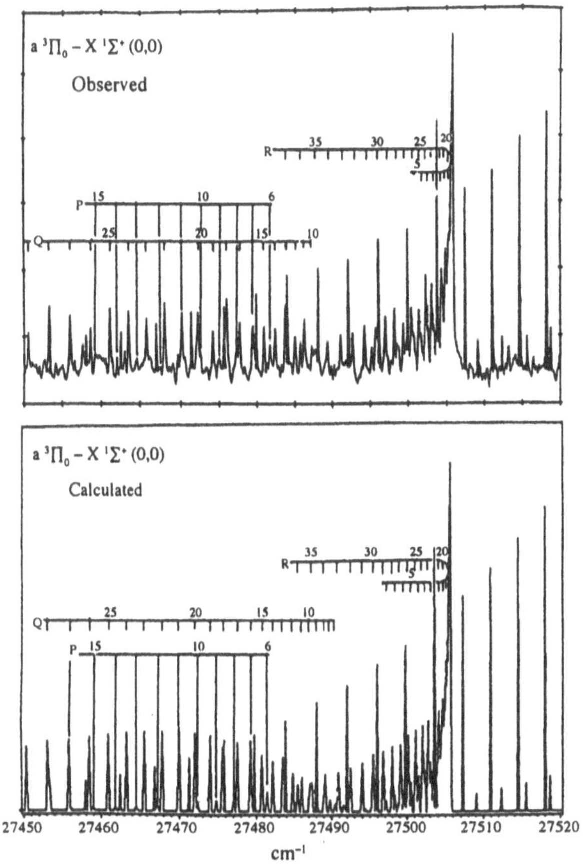

Figure 6.11: The $a^3\Pi_0 - X^1\Sigma^+(0,0)$ sub-band of CS. A $\mu_\perp/\mu_\parallel$ interference effect results in unusual weakness of the P–branch peaks, which are labeled by longer tie lines. Peaks without tie line assignments belong to the $a^3\Pi_1 - X^1\Sigma^+(0,0)$ sub-band [from Cossart, *et al.*, (1994)].

$$\mu_\parallel = \frac{\langle {}^3\Sigma_0^- | \mathbf{H}^{\mathrm{SO}} | {}^1\Sigma^+ \rangle}{E^0({}^3\Sigma^-) - E^0({}^1\Sigma^+)} \left[\langle {}^1\Sigma^+ | \mu_z | {}^1\Sigma^+ \rangle - \langle {}^3\Sigma_0^- | \mu_z | {}^3\Sigma_0^- \rangle \right]. \tag{6.4.22}$$

Similarly, Eq. (6.3.46b) can be rewritten as

$$\mu_\perp = \left[\frac{\langle {}^3\Sigma_1^- | \mathbf{H}^{\mathrm{SO}} | {}^1\Pi_1 \rangle}{E^0({}^3\Sigma^-) - E^0({}^1\Pi)} \mu_\perp({}^1\Pi - {}^1\Sigma) + \frac{\langle {}^1\Sigma_0^+ | \mathbf{H}^{\mathrm{SO}} | {}^3\Pi_{0e} \rangle}{E^0({}^1\Sigma^+) - E^0({}^3\Pi)} \mu_\perp({}^3\Sigma^- - {}^3\Pi) \right]. \tag{6.4.23}$$

Using the branch intensity formulas in Table 6.3 and observed intensity ratios, a value of $\mu_\perp/\mu_\parallel$ can be obtained and compared with *ab initio* predictions. For this specific π^2 example, $\mu_\parallel$ is the difference of permanent electric dipole moments for two states belonging to the same configuration. In the same-orbital single-configuration approximation, $\mu\left(\pi^2\ {}^1\Sigma^+\right) = \mu\left(\pi^2\ {}^3\Sigma^-\right)$, and thus $\mu_\parallel$ should be very small. However, a more accurate calculation (Havriliak and Yarkony, 1985) shows that $\mu({}^3\Sigma^-) < \mu({}^1\Sigma^+)$, thus $\mu_\parallel$ is not as small as predicted on the basis of the single configuration approximation. Wayne and Colbourn (1977) have shown that $\mu_\perp$ [Eq. (6.4.23)] involves a sum of two terms with the same sign if ${}^1\Sigma^+$ and ${}^3\Sigma^-$ belong to π^2 and ${}^1\Pi$ and ${}^3\Pi$ both belong to another configuration $\left(\pi^3\sigma \text{ or } \pi\sigma\right)$.[†] For NF, $\mu_\parallel/\mu_\perp = -3.8$ (Vervloet and Watson, 1986), which is reproduced by calculations (Yarkony, 1986). Mixing with other remote states cannot be ignored, especially when the radiative lifetime is very long $\left(\tau \gtrsim 10^{-4}\ s\right)$.

Forbidden transitions borrow their oscillator strength, in principle, from an infinite number of allowed transitions. However, when one or two terms dominate Eq. (6.4.2), isoconfigurational estimates of perturbation parameters and dipole moments will be capable *a priori* of identifying those terms. Semiempirical parameter estimates can also provide a warning that no specific terms will be sufficiently large to dominate the $\mu_\perp$ and $\mu_\parallel$ transition moments. Consider a typical value of the interelectronic energy separation, a spin-orbit matrix element, and an allowed radiative lifetime of respectively 10,000 cm^{-1}, 100 cm^{-1}, and 10^{-8} s. The borrowed radiative rate would then be approximately 10^4 s^{-1}, corresponding to a lifetime of 10^{-4} s. Therefore, in such a molecule, any electronic state with a lifetime longer than 10^{-4} s is likely to have borrowed its ability to radiate from many remote states.

When one remote perturber, $|j\rangle$, dominates the radiative lifetime of state $|i\rangle$, where the $j \to X$ transition is allowed, the lifetime of the $|i\rangle$ state, τ_i, can be easily estimated using experimental data, by the formula:

$$\tau_i = \left[\frac{E_j - E_i}{\langle j|\mathbf{H}^{\mathrm{SO}}|i\rangle}\right]^2 \frac{\nu^3(i-\mathrm{X})}{\nu^3(j-\mathrm{X})}\tau_j \tag{6.4.24}$$

Molecule	Forbidden transition	Allowed Transition	τ_j	τ_i(est.)	τ_i(obs.)
CH$^+$	$a^3\Pi - X^1\Sigma^+$	$A^1\Pi - X^1\Sigma^+$	850ns	3.3s[a]	7.0±1s[a]
NO$^+$	$a^3\Sigma^+ - X^1\Sigma^+$	$A^1\Pi - X^1\Sigma^+$	55ns	226ms[b]	530ms[c] , 720ms[d]

[a] Amitay, *et al.*(1996)
[b] using $\langle A^1\Pi|\mathbf{H}^{\mathrm{SO}}|a^3\Sigma^+\rangle = \frac{1}{4}A(b^3\Pi) = 17.5\mathrm{cm}^{-1}$
[c] Kuo, *et al.*(1990)
[a] Calamai and Yoshino (1994)

A dark vibrational-electronic state, $|\mathrm{D}\rangle|v_\mathrm{D}\rangle$, can borrow transition amplitude

[†] This is true provided that both Π states lie above or below both Σ states.

from several vibrational levels of the same electronic bright state, $|\mathrm{B}\rangle|v_\mathrm{B}\rangle$,

$$\begin{aligned}\langle \mathrm{D}|\langle v_\mathrm{D}|\mu|v_\mathrm{X}\rangle\,|\mathrm{X}\rangle &= \\ \sum_{v_\mathrm{B}} \frac{\langle \mathrm{D}|\mathbf{H}'|\mathrm{B}\rangle\,\langle v_\mathrm{D}|v_\mathrm{B}|\rangle\,\langle \mathrm{B}|\mu|\mathrm{X}\rangle\,\langle v_\mathrm{B}|v_\mathrm{X}\rangle}{E^{(0)}_{\mathrm{D},v_\mathrm{D}} - E^{(0)}_{\mathrm{B},v_\mathrm{B}}} &= \\ \sum_{v_\mathrm{B}} H'_{\mathrm{DB}}\mu_{\mathrm{BX}} \frac{\langle v_\mathrm{D}|v_\mathrm{B}\rangle\,\langle v_\mathrm{B}|v_\mathrm{X}\rangle}{\Delta E^{(0)}_{v_\mathrm{D},v_\mathrm{B}}}.\end{aligned} \tag{6.4.25}$$

Usually, the sum over v_B states is dominated by a single term that has the smallest energy denominator. However, a small $\langle v_\mathrm{D}|v_\mathrm{B}\rangle$ overlap factor for the term with the smallest $\Delta E^{(0)}_{v_\mathrm{D},v_\mathrm{B}}$ will sometimes prevent the collapse of the sum to a single dominant term. When two or more terms are of comparable magnitude, interference effects will result and must be taken into account. Rostas, *et al.*, (2000) have successfully modeled the CO $(\mathrm{a}'^3\Sigma^+, \mathrm{e}^3\Sigma^-, \mathrm{d}^3\Delta) \leftarrow \mathrm{X}^1\Sigma^+$ intersystem transition vibrational band absorption oscillator strengths by taking into account the (J-dependent) *transition amplitude* contributions of several vibrational levels of the $\mathrm{A}^1\Pi$ state to each $^3\Lambda, v' \leftarrow \mathrm{X}^1\Sigma^+, v'' = 0$ vibrational band *transition probability*. The calculations are based on the $\mathrm{A}^1\Pi - \mathrm{X}^1\Sigma^+$ transition moment function, the vibrational overlaps $\langle v_\mathrm{A}|v_\mathrm{d}\rangle$, $\langle v_\mathrm{A}|v_\mathrm{e}\rangle$, and $\langle v_\mathrm{A}|v_{\mathrm{a}'}\rangle$ calculated from deperturbed RKR potential curves, and the electronic factors of the A∼d, A∼e, and A∼ a′ spin-orbit matrix elements. The agreement between the observed and calculated transition probabilities for the nominally forbidden $(\mathrm{d}^3\Delta, \mathrm{e}^3\Sigma^-, \mathrm{a}'^3\Sigma^+) \leftarrow \mathrm{X}^1\Sigma^+$ bands is sufficient that it is likely that the absolute intensity of any $(v_\mathrm{d}, v_\mathrm{e}, \text{ or } v_{\mathrm{a}'}) \leftarrow v_\mathrm{X}$ transition can be calculated more accurately than it can be measured. Owing to the extremely large oscillator strength for the CO $\mathrm{A}^1\Pi - \mathrm{X}^1\Sigma^+$ transition, it is often necessary, for the measurement of interstellar CO column densities, to have access to transitions, with accurately known absorption strength, that are as much as a factor of 10^6 times weaker than nearby A–X bands (Eidelsberg and Rostas, 2003).

6.5 Special Effects

A variety of unusual experimental schemes have been employed to detect perturbations and to characterize the perturbing state. The methods described in this section involve subjecting the molecule to external perturbations such as an intense monochromatic radiation field (Section 6.5.1), a static magnetic or electric field (Sections 6.5.2, 6.5.3 and 6.5.4), multiple static, oscillatory, or pulsed electromagnetic fields (Sections 6.5.2 and 6.5.3), weak bimolecular collisions (Section 6.5.5), or confining the molecule in a high pressure collisional (Section 6.5.6) or matrix (Section 6.5.7) cage. External perturbations can make observable extremely weak or exotic internal perturbations or can create intramolecular interactions that do not exist in the isolated molecule.

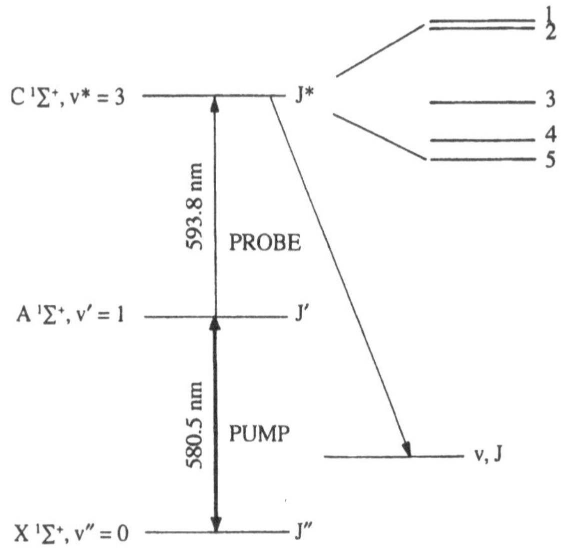

Figure 6.12: Level scheme for optical-optical double resonance on BaO $C^1\Sigma^+ \leftarrow A^1\Sigma^+ \leftarrow X^1\Sigma^+$. The level structure for $C^1\Sigma^+(v^* = 3, J^* = 50)$ is expanded and drawn to scale illustrating the five observed $J^* = 50$ levels. Undispersed $v^*, J^* \rightarrow v, J$ ultraviolet fluorescence is detected as the probe dye laser is scanned [from Gottscho and Field (1978).]

6.5.1 Differential Power Broadening

Perturbations often provide information about previously unknown perturbing states. Extra lines, arising from levels with predominant perturber character, are perhaps the most information-rich features of a perturbed band spectrum. When the perturbation interaction is very weak, one is lucky to locate an extra line and even luckier to be able to prove that this line is *extra* rather than *extraneous* (rare isotope or impurity).

An optical-optical double resonance (OODR) scheme exists, utilizing two continuous-wave (cw), monochromatic, tunable lasers, whereby the rotational quantum numbers of all observed lines may be established without ambiguity, prior knowledge of B–values, trial-and-error searches for consistent combination differences, or redundant confirmation; lines that are weak because of level population may be distinguished from those that are weak because of intrinsic linestrength; and forbidden transitions may be made to appear with comparable peak intensities but considerably narrower widths than allowed transitions.

Figure 6.12 illustrates the level scheme for OODR. The PUMP laser prepares a single v', J' level of the intermediate electronic state. Since the PUMP laser spectral width (~1 MHz) is smaller than the Doppler width (~1 GHz) of the A,$v', J' \leftarrow$ X,v'', J'' transition, the prepared v', J' molecules are partially velocity-selected. Only the component of velocity along the PUMP laser propagation direction is selected. The v', J' molecules are prepared with a longitudi-

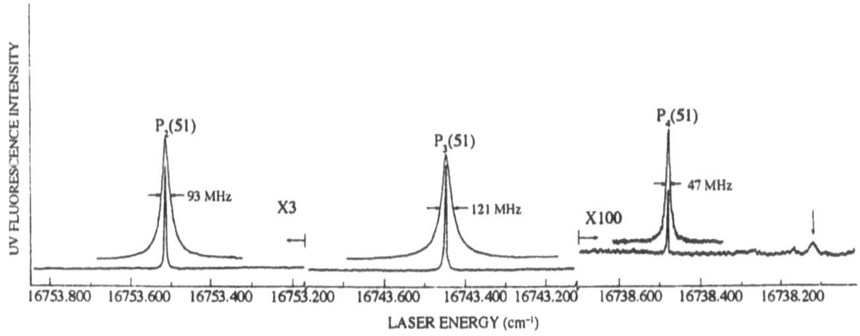

Figure 6.13: OODR differential power broadening in BaO. The PUMP laser is tuned to the $A^1\Sigma^+ \leftarrow X^1\Sigma^+$ (1,0) $R(50)$ line; the PROBE is scanned in the region near $P(51)$ of $C^1\Sigma^+ \leftarrow A^1\Sigma^+(3,1)$. The main line ($P_3$) and two of four extra lines (P_2 and P_4) are shown. The unassigned collisional satellite line marked with an arrow has area equal to that of the P_4 extra line [from Gottscho and Field (1978).]

nal velocity distribution much narrower than the full thermal distribution. Thus the linewidth of the C,$v^*, J^* \leftarrow$ A,v', J' transition, recorded by scanning the co- or counterpropagating PROBE laser and recording undispersed fluorescence at shorter wavelength than λ_{PUMP} and λ_{PROBE}, is significantly sub-Doppler. This means that differences in OODR lineshapes reflect differences in the radiative and collisional properties of the various v_i^*, J_i^* levels and, most importantly, the relative ease of power-broadening the $v_i^*, J^* \leftarrow v', J'$ transitions. Extra lines (partially forbidden transitions) power-broaden more slowly than main lines (allowed transitions) as the intensity of the PROBE laser is increased.

Figure 6.13 shows the main line and two of four extra lines observed upon excitation from BaO $A^1\Sigma^+(v' = 1, J' = 51)$ by Gottscho and Field (1978) as the PROBE laser was scanned in the region of the $C^1\Sigma^+ - A^1\Sigma^+$ (3,1) $P(51)$ line. A similar spectrum was obtained from $A^1\Sigma^+(v = 1, J = 49)$ near the C–A $R(49)$ line to prove that all observed lines belong to $J^* = 50$ levels. The main line, $P_3(51)$, has an FWHM of 121 MHz, considerably larger than the FWHM of 47 MHz of the weakest extra line, $P_4(51)$.

At high PROBE power, where the limit of $n_{v^*,J^*} \simeq n_{v',J'}$ population saturation is approached, the peak intensities of main and extra lines would be comparable but their integrated line intensities should continue to reflect the large difference in intrinsic linestrengths. In an experiment where the resolution limit is set by the molecular linewidth rather than the measuring device, it is easier to detect a tall, narrow line than a broad line of equal integrated intensity. In fact, it is possible to imagine a situation where the PROBE laser is so intense that main lines become so broad as to become unrecognizable, leaving only sharp features associated with extra and forbidden transitions.

6.5.2 Effects of Magnetic and Electric Fields on Perturbations

Molecular rotational levels can be split into M–components and shifted in static electric and magnetic fields. Stark and Zeeman effects have been very useful in revealing otherwise unobservable perturbations and in diagnosing the electronic symmetry of previously unknown perturbers. Long before the phenomena of perturbations and the Zeeman effect were understood, there were several studies of the effect of a magnetic field on perturbed lines (Fortrat, 1913; Bachem, 1920).

Between 1929 and 1935, several perturbations in the CO $A^1\Pi$ and N_2^+ $B^2\Sigma^+$ states were studied in magnetic fields up to 36 kG (Crawford, 1929, 1934; Watson, 1932; Parker, 1933; Schmid and Gerö, 1935). Lines were observed to split, to broaden symmetrically and asymmetrically, to gain or lose intensity; each line was a special case. Spectral resolution and sensitivity were seldom adequate to resolve individual M–components. Although many qualitative features were satisfactorily explained, several perturbing states were conclusively but incorrectly assigned, and no quantitative theory of the effect of an external field on perturbed line positions, shapes, and intensities emerged.

Wood and Hackett (1909) observed anomalies in the magnetic rotation spectrum of the Na_2 $A^1\Sigma_u^+-X^1\Sigma_g^+$ system. [A magnetic rotation (Faraday effect) spectrum differs from an ordinary absorption spectrum in that the sample is placed between crossed polarizers and in a magnetic field directed parallel to the propagation direction of the incident radiation.] Fredrickson and Stannard (1933) suggested that the isolated rotational transitions that displayed significant magnetic field induced polarization rotations were perturbed $A^1\Sigma_u^+ \sim b^3\Pi_{0u}-X^1\Sigma_g^+$ lines. Carroll (1937) developed the general theory of perturbed and unperturbed magnetic rotation spectra of diatomic molecules.

Radford (1961, 1962) and Radford and Broida (1962) presented a complete theory of the Zeeman effect for diatomic molecules that included perturbation effects. This led to a series of detailed investigations of the CN $B^2\Sigma^+$ ($v = 0$) $\sim$ $A^2\Pi$ ($v = 10$) perturbation in which many of the techniques of modern high-resolution molecular spectroscopy and analysis were first demonstrated: anticrossing spectroscopy (Radford and Broida, 1962, 1963), microwave optical double resonance (Evenson, *et al.*, 1964), excited-state hyperfine structure with perturbations (Radford, 1964), effect of perturbations on radiative lifetimes and on inter-electronic-state collisional energy transfer (Radford and Broida, 1963). A similarly complete treatment of the effect of a magnetic field on the CO $a'^3\Sigma^+ \sim A^1\Pi$ perturbation complex is reported by Sykora and Vidal (1998). The $\Delta S = 0$ selection rule for the Zeeman Hamiltonian leads to important differences between the CN $B^2\Sigma^+ \sim A^2\Pi$ and CO $a'^3\Sigma^+ \sim A^1\Pi$ perturbation plus Zeeman examples, primarily in the absence in the latter case of interference effects between the Zeeman and intramolecular perturbation terms.

Moehlmann, *et al.*, (1972) observed both magnetic and electric field induced perturbations in HCP and discussed the theory of such perturbations.

A crucial feature of the CN system is that, when CN molecules are formed in a flame of < 1 Torr active nitrogen plus CH_2Cl_2, the $A^2\Pi$ state is populated preferentially to $B^2\Sigma^+$. The result is a considerably larger population flow rate into the red-emitting, longer-lived $A^2\Pi$ ($v = 10$) level than the violet-emitting, shorter-lived $B^2\Sigma$ ($v = 0$) level. Figure 6.14 illustrates the complex magnetic field and pressure dependence of a portion of the $B^2\Sigma^+-X^2\Sigma^+(0,0)$ band. The collisional aspects of this figure are discussed in Section 6.5.5. The perturbed lines, marked on the figure by M (main) and E (extra), vary rapidly in relative intensity, lineshape, and frequency as the magnetic field strength changes. Most of this variation is caused by Zeeman tuning of the M_J components (each at a different rate) through various $\Delta M_J = 0$ $E_{\mathrm{A},J_\mathrm{A},M} = E_{\mathrm{B},J_\mathrm{B},M}$ degeneracies. As this occurs, a nominal A level (extra line) acquires appreciable B state character and becomes able to radiate in the violet. Its violet/red fluorescence branching ratio increases drastically. At the same time, a nominal B level acquires more A character, thus enabling the reaction to populate it more rapidly. However, because the Einstein A-coefficient for B$-$X (0,0) violet fluorescence is much larger than for A$-$X $(10, v'')$ red fluorescence, the fractional change in violet photon yield from a nominal B-level is smaller than the increase in its chemical formation rate and the net effect is an increase in both main and extra line violet fluorescence intensity.

Figure 6.14 conceals a tremendous amount of information about the many individual M_J crossings. Figure 6.15, which shows the energy versus magnetic field behavior for each M_J level at the four B$\sim$ A crossings ($J = 3.5, 7.5, 10.5$ and 15.5), reveals the true complexity typical of all perturbation plus static field problems. Figure 6.15 also suggests a variety of ways of recording spectra of field induced level crossings: scan wavelength at fixed field (high-resolution absorption or resolved fluorescence spectroscopy), scan field at fixed excitation and/or detection wavelength (anticrossing spectroscopy, laser magnetic resonance, laser Stark spectroscopy), fixed field and fixed wavelength with scanning radiofrequency (rf) or microwave radiation (rf or microwave-optical double resonance), short-pulse preparation of $\Delta M = 0$ or $\Delta M = 2$ coherent superposition state with time-resolved fluorescence decay (quantum-beat spectroscopy). The observability of double-resonance transitions or quantum-beat signals between nominal A and B levels can depend on setting the magnetic field strength to cause appreciable A$\sim$B mixing in at least one of the two levels involved. Levy (1972) and Cook and Levy (1973a) combined a magnetic field with a perpendicular electric field in order to observe CN B$\sim$ A $\Delta M = \pm 1$ and $+ \leftrightarrow -$ level crossings which are unobservable in a magnetic field alone.

The remainder of this section is devoted to a simplified two-level treatment of the Zeeman and Stark effects in the presence of zero-field *Stark effect* and field-dependent interactions between basis functions $|1M\rangle$ and $|2M\rangle$. In the presence of a static field directed along the space Z-axis, M_J remains a good quantum number. The Zeeman and Stark Hamiltonians involve the interaction between a magnetic field or electric dipole, $\boldsymbol{\mu}$, in the molecule-fixed axis system and the space-fixed magnetic or electric field, $\vec{\mathrm{F}}$, parallel to the laboratory direction $\hat{\mathrm{K}}$. The interaction can be expressed in terms of direction cosines

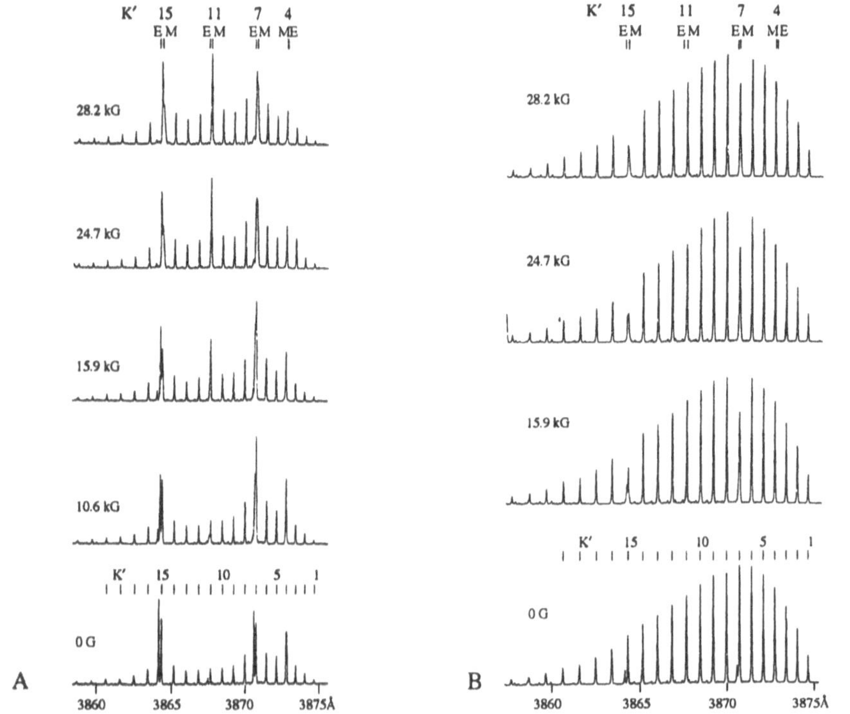

Figure 6.14: Effect of magnetic field and pressure on the CN $B^2\Sigma^+ - X^2\Sigma^+(0,0)$ R branch lines. (a) Spectra recorded at 0.6 Torr in an active nitrogen plus CH_2Cl_2 flame. (b) Spectra recorded at 30 Torr. Main and extra lines are marked by M and E. [From Radford and Broida (1963), in which Figs. 1 and 2 were inadvertently interchanged.]

$$\begin{aligned}\mathbf{H}^{\text{field}} &= -\vec{\mu}\cdot\vec{\mathrm{F}} = -(F\hat{\mathrm{K}})\cdot\left(\mu_x\hat{\imath} + \mu_y\hat{\jmath} + \mu_z\hat{k}\right) \\ &= -F\left(\alpha_Z^x\mu_x + \alpha_Z^y\mu_y + \alpha_Z^z\mu_z\right). \end{aligned} \tag{6.5.1}$$

The $\Delta J = 0$ matrix elements of $\boldsymbol{\alpha}$ are listed in Table 2.1. All $\Delta J = 0, \Delta M = 0$ $\boldsymbol{\alpha}$ matrix elements are proportional to M. The following discussion deals exclusively with $\Delta J = 0$ matrix elements. Note that nonzero $\Delta J = \pm 1$ matrix elements of $\boldsymbol{\alpha}_Z^i$ exist and that these all include the factor

$$\langle J+1M|\alpha_Z^i|JM\rangle \propto \left[(J+1)^2 - M^2\right]^{1/2}. \tag{6.5.2}$$

Thus $\Delta J = \pm 1$ Stark and Zeeman matrix elements have a qualitatively different M–dependence from the $\Delta J = 0$ examples discussed below. This difference is useful for distinguishing $\Delta J = 0$ from $\Delta J = \pm 1$ field-induced interactions.

Three cases must be considered when a magnetic or electric field of magnitude F is applied and the energies of basis functions $|1M\rangle$ and $|2M\rangle$ are tuned through exact degeneracy,

$$E^0_{1M}(F) = E^0_{2M}(F),$$

where

$$\begin{aligned} E^0_{iM}(F) &= \langle iM | \mathbf{H}^{zf} + \mathbf{H}^{\text{field}} | iM \rangle \\ &= \langle iM | \mathbf{H}^{zf} | iM \rangle + \langle iM | \mathbf{H}^{\text{field}} | iM \rangle \\ &= E^0_i + MF\mu_i.^{\dagger} \end{aligned} \tag{6.5.3}$$

$\mathbf{H}^{zf}$ and $\mathbf{H}^{\text{field}}$ are respectively the zero-field and field-dependent parts of $\mathbf{H}$ and $M\mu_i$ is the tuning rate for the energy of basis function $|iM\rangle$ in the field F. The Hamiltonian matrix for this $\Delta J = 0$ two-level problem is

$$\mathbf{H} = \begin{pmatrix} E^0_1 + MF\mu_1 & H_{12} + MF\mu_{12} \\ H_{12} + MF\mu_{12} & E^0_2 + MF\mu_2 \end{pmatrix} \tag{6.5.4}$$

where

$$\langle 1M | \mathbf{H}^{zf} | 2M \rangle \equiv H_{12}, \tag{6.5.5}$$

and, for $\Delta J = 0$ matrix elements only,

$$\langle 1M | \mathbf{H}^{\text{field}} | 2M \rangle \equiv MF\mu_{12},^{\dagger} \tag{6.5.6}$$

The three cases are:

(1) $H_{12} \neq 0, \mu_{12} = 0$;

(2) $H_{12} = 0, \mu_{12} \neq 0$; and

(3) $H_{12} \neq 0, \mu_{12} \neq 0$.

Cases (1) and (2) correspond to simple anticrossings, whereas interference effects can occur for case (3).

The above two-level treatment conceals a simplifying approximation for the electric field case. All diagonal matrix elements of $\mathbf{H}^{\text{Stark}}$ of parity basis functions are rigorously zero because $\mathbf{H}^{\text{Stark}}$ has odd parity (see Section 3.2.2). However, if two zero-field basis functions of opposite parity are degenerate (or if $[E^0(+) - E^0(-)] \ll \mathcal{E}\langle +|\mu|-\rangle \equiv \mathcal{E}\mu_{+-}$), a nonzero electric field ($\mathcal{E}$) will destroy parity and cause the energies of the resultant parity-mixed functions to tune as $\pm|M|\mathcal{E}\mu_{+-}$. The degeneracy between the energies of the $|+, J, M\rangle$ and $|+, J, -M\rangle$ basis functions is not lifted but, for every degenerate $+M, -M$ pair of levels tuning to higher energy as $\mathcal{E}$ increases, there will be another degenerate pair with identical $|\mu_{+-}|$ tuning to lower energy. Aside from the present caveat about $|M|$ versus M and some peculiarities associated with incomplete parity mixing, the two-level treatment provides a framework for understanding Stark as well as Zeeman anticrossings.

† All J–dependence is implicitly included in the μ_i and μ_{12} tuning-rate factors.

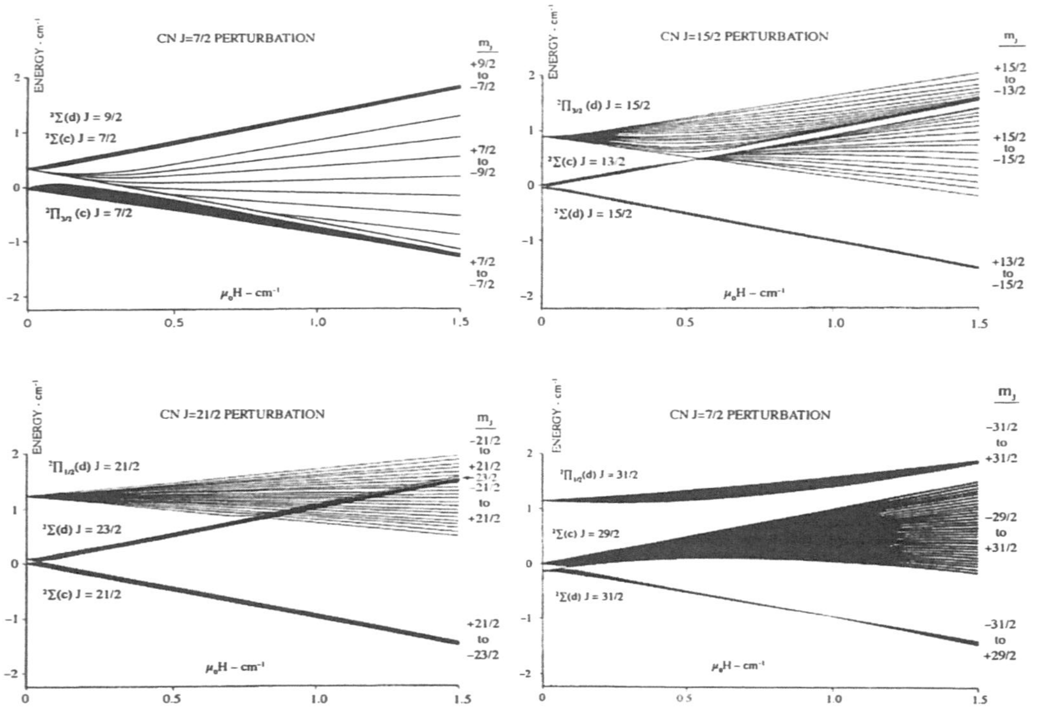

Figure 6.15: Calculated Zeeman patterns for the perturbed rotational levels of CN $A^2\Pi$ $(v = 10) \sim B^2\Sigma^+$ $(v = 0)$. Only the $^2\Pi$ Λ–doublet and Ω–components involved in the perturbation are shown. The horizontal axis is labeled in units of $\mu_0(4.6688 \times 10^{-5}\ \mathrm{cm}^{-1}/G)$ multiplied by the magnetic field (H) to give reciprocal centimeters rather than Gauss. [From Radford and Broida (1962).]

Case (1). $H_{12} \neq 0, \mu_{12} = 0$. As F is increased from zero there will be a series of $\Delta J = 0, \Delta M = 0$ anticrossings, one for each value of $|M| \neq 0$, which occur at field strengths of

$$F_{|M|} = \frac{E_1^0 - E_2^0}{M(\mu_2 - \mu_1)}. \tag{6.5.7}$$

The lowest field anticrossing will involve $|M| = J$, and the $|M|$ and $|M-1|$ anticrossings will occur at field strengths differing by

$$F_{|M-1|} - F_{|M|} = \frac{E_1^0 - E_2^0}{\mu_2 - \mu_1} \frac{1}{M(M-1)}. \tag{6.5.8}$$

The lineshape of an isolated anticrossing is related to the variation with F of $|\langle 2M|1M, F\rangle|^2$, the fractional $|2M\rangle$ basis function character admixed into the nominal $|1M, F\rangle$ eigenfunction. The shape of an anticrossing signal depends on the specific methods of excitation and detection (Miller, 1973), but $|\langle 2M|1M, F\rangle|^2$ is a crucial factor. For case (1) crossings, each isolated anticrossing signal will have a symmetric lineshape,

$$\begin{aligned} \sigma_M(F) &\simeq 2|\langle 2M|1M, F\rangle|^2 \\ &= 1 - \left\{1 + \left[\frac{2H_{12}}{M(\mu_1 - \mu_2)(F - F_M)}\right]^2\right\}^{-1/2}, \end{aligned} \tag{6.5.9}$$

and the linewidth (FWHM), Γ_M, in units of field strength rather than energy, is

$$\Gamma_M = \left|\frac{(16/3)^{1/2} H_{12}}{M(\mu_1 - \mu_2)}\right|. \tag{6.5.10}$$

Note that the width is proportional to H_{12} and inversely proportional to $|M|$ and to the differential tuning rate, $|\mu_1 - \mu_2|$. The sharpest anticrossings are those with small H_{12} and $|M| = J$. See Fig. 6.18 for an example of ultra-narrow, forbidden (small H_{12}), singlet$\sim$triplet anticrossings in H_2. These anticrossings are observed by monitoring the intensity of the specified spectrally-selected emission line as the magnetic field strength is varied. See Section 6.5.3 for further discussion of anticrossing spectroscopy.

Case (2). $H_{12} = 0, \mu_{12} \neq 0$. The only difference from case (1) is that each isolated anticrossing signal does not have a symmetric lineshape. Since the coupling matrix element increases in proportion to F, these anticrossings exhibit level-shift and mixing-coefficient (lineshape) behavior versus F similar to that versus J for heterogeneous perturbations where the coupling matrix element is proportional to J. However, for $\mu_{12} \neq 0$ anticrossings the differential tuning rate is proportional to $(\mu_1 - \mu_2)\ MF$ versus F rather than ΔBJ^2 versus J as for heterogeneous perturbations. The lineshape is

$$\sigma_M(F) \simeq 1 - \left\{1 + \left[\frac{2\mu_{12}}{(\mu_1 - \mu_2)(1 - F_M/F)}\right]^2\right\}^{-1/2}. \tag{6.5.11}$$

and the low-field FWHM will be narrower than the high-field FWHM. Note that the $|1M\rangle \sim |2M\rangle$ mixing does not vanish as $F \to \infty$,

$$\sigma_M(\infty) = 1 - \left[1 + \left(\frac{2\mu_{12}}{\mu_1 - \mu_2}\right)^2\right]^{-1/2}. \tag{6.5.12}$$

The linewidth

$$\Gamma_M = (16/3)^{1/2}\mu_{12}(\mu_1 - \mu_2)F_M[(\mu_1 - \mu_2)^2 - 4\mu_{12}^2/3]^{-1} \tag{6.5.13}$$

is inversely proportional to $|M|$, as for case (1), because the only M–dependent factor in Γ_M, F_M, is proportional to $|M|^{-1}$ [Eq. (6.5.7)].

Case (3). $\mu_{12} \neq 0, H_{12} \neq 0$. The interference effect is most evident for $E_1^0 = E_2^0 = E^0$. For that situation, all M–levels are anticrossed at $F = 0$. The energies and eigenfunctions are

$$\begin{aligned} &E_{+,M}, \quad \text{which belongs to} \quad |+, M, F = 0\rangle = 2^{-1/2}[|1M\rangle + |2M|], \\ &E_{-,M}, \quad \text{which belongs to} \quad |-, M, F = 0\rangle = 2^{-1/2}[|1M\rangle - |2M|], \end{aligned}$$

and

$$E_{+,M} - E_{-,M} = 2H_{12}.$$

As soon as $F \neq 0$, an asymmetry develops in the energies

$$|E_{\pm,M} - E^0| \neq |E_{\pm,-M} - E^0|$$

and in the mixing fractions

$$|\langle 1M|\pm, M, F\rangle|^2 \neq |\langle 1, -M|\pm, -M, F\rangle|^2$$

because the total off diagonal matrix element has a different absolute magnitude for $M > 0$ and $M < 0$. If

$$F = \frac{H_{12}}{M\mu_{12}}$$

then the off-diagonal matrix element vanishes for M but has the magnitude $2H_{12}$ for $-M$. This has the effect of transferring all of the interaction strength, which would normally be shared equally, from M to $-M$.

Another consequence of the interference between$\mathbf{H}^{z\!f}$ and $\mathbf{H}^{\text{field}}$ is manifest as an asymmetry in the behavior of rotational levels above and below a zero-field level crossing. Let $E_1^0 - E_2^0 = \Delta > 0$; then the crossings occur at

$$F_M = \frac{\Delta}{M(\mu_2 - \mu_1)}, \tag{6.5.14}$$

and, if F and $(\mu_2 - \mu_1)$ are positive, the crossings will involve the $M > 0$ levels. The width of each crossing will be approximately proportional to the total coupling matrix element at F_M,

$$\Gamma_M \approx \frac{(16/3)^{1/2}[H_{12} + \Delta\mu_{12}/(\mu_2 - \mu_1)]}{M(\mu_2 - \mu_1)}. \tag{6.5.15}$$

Now let $E_1^0 - E_2^0 = -\Delta < 0$. The crossings at positive F will involve the $M < 0$ levels, and their widths will be

$$\Gamma_{-M} \approx \frac{(16/3)^{1/2}[H_{12} - \Delta\mu_{12}/(\mu_2 - \mu_1)]}{|M|(\mu_2 - \mu_1)}. \tag{6.5.16}$$

The anticrossings will be much narrower in rotational levels either above $\left(E_1^0 > E_2^0\right)$ or below $\left(E_1^0 < E_2^0\right)$ the zero-field level crossing, depending on the relative signs of H_{12} and μ_{12}. The sign of $H_{12}\mu_{12}$ is thus an experimental observable and, provided that consistent wavefunction phases are used in computing H_{12} and μ_{12}, should be *a priori* predictable.

Rather than measuring the widths and locations of anticrossings, another kind of useful information is obtained by direct measurement of the tuning rate of each eigenstate in the applied field,

$$\frac{\partial}{\partial F}\langle 1M, F|\mathbf{H}|1M, F\rangle \equiv \partial E_{1M,F}/\partial F, \tag{6.5.17}$$

where

$$|1M, F\rangle = (1 - C_{21}^2)^{1/2}|1M\rangle + C_{21}|2M\rangle. \tag{6.5.18}$$

It is convenient to make use of the Hellmann-Feynman theorem, which states that

$$\frac{\partial E_i}{\partial \lambda} = \left\langle i \left| \frac{\partial \mathbf{H}}{\partial \lambda} \right| i \right\rangle \tag{6.5.19}$$

if $|i\rangle$ and E_i are, respectively, an eigenstate and eigenvalue of $\mathbf{H}$. For

$$\begin{aligned} \mathbf{H} &= \mathbf{H}^{zf} + \mathbf{H}^{\text{field}}, \\ \frac{\partial \mathbf{H}}{\partial F} &= \mathbf{H}^{\text{field}}/F \equiv \mathbf{H}^f/F.^\dagger \end{aligned} \tag{6.5.20}$$

Thus

$$\begin{aligned} \frac{\partial E_{1M,F}}{\partial F} &= \langle 1M, F|\mathbf{H}^f/F|1M, F\rangle \\ &= (1 - C_{21}^2)\langle 1M|\mathbf{H}^f/F|1M\rangle + C_{21}^2\langle 2M|\mathbf{H}^f/F|2M\rangle \\ &\quad + (1 - C_{21}^2)^{1/2}C_{21}\langle 1M|\mathbf{H}^f/F|2M\rangle \\ &= M\left[\mu_1 + C_{21}^2(\mu_2 - \mu_1) + C_{21}(1 - C_{21}^2)^{1/2}\mu_{12}\right] \\ &\equiv Mg_1. \end{aligned} \tag{6.5.21}$$

The important point to notice is that, since the sign of the mixing coefficient C_{21} is determined by the signs of $E_1^0 - E_2^0$ and H_{12}, the field tuning rate, g_J,

† This requires that all matrix elements of $\mathbf{H}^{\text{field}}$ be proportional to F.

for an M, J level above the zero-field crossing point will be different from that for an M, J' level an equal distance below the crossing.

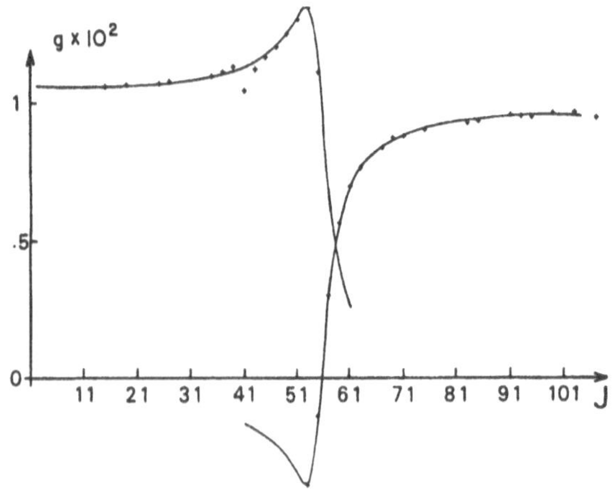

Figure 6.16: g_J-values for $^{80}Se_2$ $B0_u^+$ ($v = 4$). Points are observed by $\Delta M = 2$ Zeeman quantum-beat spectroscopy. The solid curves display the g_J-values calculated from the results of a fit to the observed g values (including both main and extra lines at $J = 53$ and 55), from which g_J values for the $B0_u^+$ state and the perturbing 1_u state as well as the off-diagonal g-value (g_{12}) and $B\mathbf{J} \cdot \mathbf{J}_a$ matrix element were determined. For each J there are two calculated g_J values, the larger one corresponding to the main level. The failure of the high-J g_J values to return to the low-J value is proof that the perturbation is heterogeneous. The asymmetry (including a sign change) in the main-line g_J curve about $J = 55$ is proof that $g_{12} \neq 0$. [From Gouedard and Lehmann, (1981).]

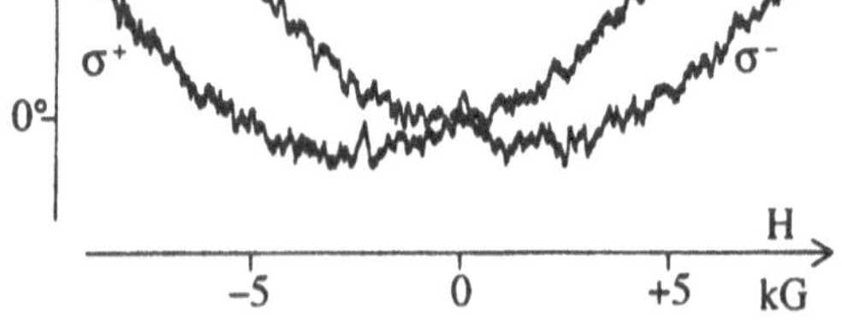

Figure 6.17: Difference in lifetime for σ^+ versus σ^- polarized fluorescence from I_2 $B0_u^+$ ($v = 40, J = 77$). Lifetimes are measured by a phase-shift technique (the excitation radiation is amplitude modulated at 100 kHz) and the arrow points in the direction of increasing lifetime. Note that the σ^+/σ^- difference vanishes at $H = 0$ and reverses between $H > 0$ and $H < 0$. [From Vigué, *et al.*(1975).]

This asymmetric variation of g_J near the J–value of the zero-field crossing was observed by Gouedard and Lehmann (1979, 1981) in the B 0_u^+ state of $^{80}Se_2$ and is illustrated by Fig. 6.16. They measured Zeeman-tuned quantum beats between the J, M and $J, M+2$ nominal B 0_u^+ levels. If the μ_{12} term had been zero, their g_J–values would have been symmetrically affected on either side of each J–crossing point. The sharp onset and slow disappearance of the deviation in $v' = 1$ at $J' = 69$ and in $v' = 3$ at $J' = 107$ (see Figs. 1-7 of Gouedard and Lehmann, 1979) and the sign change in $v' = 4$ at $J' = 55$ prove that the direct magnetic coupling term is nonzero. Measurement of g–values can be a very sensitive probe for perturbations, especially if the perturbation-free g_J–value is very small, as it is for the Se_2 B 0_u^+ state.

Another rather surprising effect arises from an interference between $\mathbf{H}^{\text{Zeeman}}$ and $\mathbf{H}^{\text{ROT}}$. Although this example involves predissociation, which is not discussed until Chapter 7, a brief description here is appropriate. Vigué, *et al.*, (1974) observed that when the I_2 B $0_u^+ - X^1\Sigma_g^+$ (40,0) $R(76)$ line was excited using linearly polarized radiation with the $\mathcal{E}$–vector parallel to an applied magnetic field, H, the resultant fluorescence propagating along the field direction was found to be circularly polarized whenever $H \neq 0$. The degree of circular polarization depended on the magnitude of H, and the dominant polarization, σ^+ versus σ^-, depended on the sign of H. This could be explained if the $+M$ and $-M$ levels had different lifetimes. Figure 6.17 illustrates that for $H > 0$, the σ^+ polarized fluorescence ($M' - M'' = +1$, thus sampling more $M' > 0$ than $M' < 0$) has a shorter lifetime than σ^-.

This $+M, -M$ lifetime asymmetry is explained by noting that the B 0_u^+ state could interact with a 1_u state via both $\mathbf{H}^{\text{ROT}}$ (the $-B\mathbf{J}\cdot\mathbf{L}$ term) and $\mathbf{H}^{\text{Zeeman}}$. The coupling matrix element is a sum of two terms, one of which depends on the sign of M and the sign of the magnetic field. For $+M$, the two terms add, whereas for $-M$, they partially cancel. The fact that the perturber is a repulsive state simply means that the unbound character mixed into the nominal B 0_u^+ ($v = 40, J = 77$) level is larger for $+M$ than $-M$. The lifetime difference is responsible for the more rapid nonradiative decay (predissociation) of $+M$ than $-M$.

6.5.3 Anticrossing, Quantum-Beat, and Double-Resonance Experiments

When the perturbation matrix element, H_{12}, is small compared to $\left|E_1^0 - E_2^0\right|$, perturbation effects may be undetectably weak or too small to provide useful information about the perturbing state. By applying a static magnetic or electric field, it may be possible to Zeeman- or Stark-tune the $|1M\rangle$ and $|2M\rangle$ *basis* functions into the crossing region where

$$\left|E_{1M}^0 - E_{2M}^0\right| < |H_{12} + MF\mu_{12}|.$$

If this condition is met, *regardless of the size of* $H_{12} + MF\mu_{12}$, the radiative properties of both $|+, M\rangle$ and $|-, M\rangle$ *eigenfunctions* become profoundly different from those of the *basis* functions $|1M\rangle$ and $|2M\rangle$. [The Stark and Zeeman tuning coefficients are also affected; see Eq. (6.5.21).] For example, if an M_J component of a CN $A^2\Pi$ ($v = 10$) rotational level is Zeeman-tuned into resonance with a suitable ($\Delta M = 0, \Delta J = \pm 1, 0, + \nleftrightarrow -$) component of $B^2\Sigma^+(v = 0)$, the radiative lifetime will decrease from $\tau_A^0 \approx 600$ ns to approximately $2\tau_B^0 \approx 130$ ns and, more importantly, the fluorescence will switch abruptly from the red to the violet spectral region.

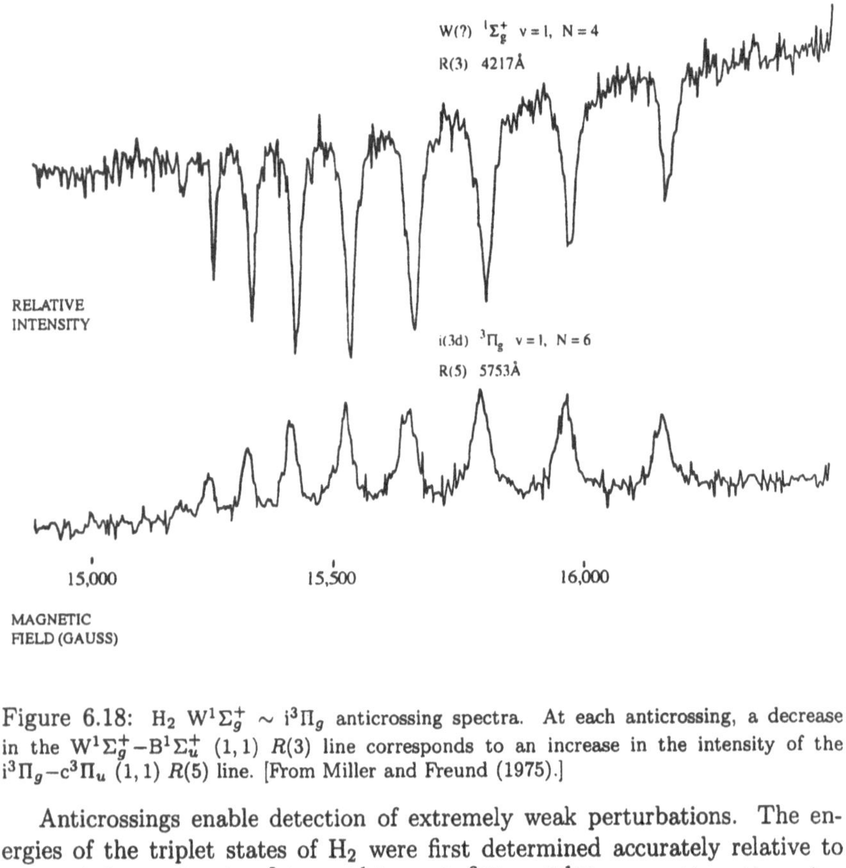

Figure 6.18: H_2 $W^1\Sigma_g^+ \sim i^3\Pi_g$ anticrossing spectra. At each anticrossing, a decrease in the $W^1\Sigma_g^+ - B^1\Sigma_u^+$ (1,1) $R(3)$ line corresponds to an increase in the intensity of the $i^3\Pi_g - c^3\Pi_u$ (1,1) $R(5)$ line. [From Miller and Freund (1975).]

Anticrossings enable detection of extremely weak perturbations. The energies of the triplet states of H_2 were first determined accurately relative to those of the singlets by $j^3\Delta_g \sim J^1\Delta_g$ and $r^3\Pi_g \sim R^1\Pi_g$ magnetic anticrossing spectroscopy (Miller and Freund, 1974; Jost and Lombardi, 1974).

These anticrossings were observable despite the extremely small spin-orbit interaction associated with $3d\delta$ or $4d\pi$ Rydberg orbitals of the H_2 molecule. In fact, it is advantageous for H_{12} to be small because this results in extremely narrow and fully resolved anticrossings [See Eqs. (6.5.10), (6.5.13), and (6.5.16)]. Figure 6.18 shows spectra of the $\Delta M_N = 0$ anticrossings between the $i^3\Pi_g(v = 1, N = 6)$ and $W^1\Sigma_g^+(v = 1, N = 4)$ levels. This is a forbidden anticrossing because all $\Delta N = 2$ matrix elements vanish between $^3\Pi$ and $^1\Sigma$ case (b) *basis* functions. The selection rule for $\Delta S = 1$ spin-orbit interactions between case (b) basis functions is $\Delta N = 0, \pm 1$. (The hyperfine Hamiltonian gives no $\Delta N = 2$ matrix elements for para-H_2 because the two $I_H = \frac{1}{2}$ nuclear spins are paired to form a nuclear spin singlet, $I = 0$, which has zero magnetic dipole-dipole interactions with all other angular momenta.) The anticrossing becomes observable because the magnetic field mixes some $N = 5$ character into either the nominal $^3\Pi$ $N = 6$ or $^1\Sigma$ $N = 4$ level, thereby turning on a spin-orbit interaction between these levels. The coupling matrix element between the two anticrossing levels is ~ 1 MHz.

Figure 6.18 shows two complementary anticrossing spectra. Intensity is transferred from the $\mathrm{W}^1\Sigma_g^+ - \mathrm{B}^1\Sigma_u^+$ (1, 1) $R(3)$ line to the $\mathrm{i}^3\Pi_g - \mathrm{c}^3\Pi_u$ (1, 1) $R(5)$ line. The H_2 molecules are excited by electron bombardment, resulting in slightly greater population of $\mathrm{W}^1\Sigma_g^+$ than $\mathrm{i}^3\Pi_g$. Each intensity decrease in the W–B line corresponds to an increase in the i–c line.

Note that the horizontal axis of Fig. 6.18 is labeled in Gauss, which is *not* a unit of energy. It is a nontrivial matter to go from the anticrossing signal to a zero-field energy-level diagram, a set of Zeeman tuning coefficients, and H_{12}. The analysis of Zeeman (Stark) anticrossing spectra is always more complicated than analysis of a zero-field perturbation, in part because each J level splits into $2J+1$ ($J+1$ or $J+\frac{1}{2}$ for Stark because of the $+M, -M$ degeneracy) M_J-components, but primarily because at least two additional parameters must be determined, namely, the rates at which both interacting levels tune in the applied field. For most molecules, the magnetic and electric dipole moments determine the Zeeman and Stark tuning rates.

A fundamental difference exists with respect to *a priori* knowledge of electric versus magnetic dipole moments. Electric moments depend on the detailed shape of electronic wavefunctions and must always be measured experimentally or computed from accurate *ab initio* wavefunctions. In contrast, magnetic moments are directly proportional to angular momenta and may be computed from the eigenvectors of an effective Hamiltonian without knowledge of the electronic wavefunctions.

In the case of electronic angular momenta, the proportionality constants g_L and g_S in

$$\boldsymbol{\mu}_L \equiv \mu_B g_L \mathbf{L} \tag{6.5.22a}$$

$$\boldsymbol{\mu}_S \equiv \mu_B g_S \mathbf{S} \tag{6.5.22b}$$

are known exactly,

$$g_L = 1 \tag{6.5.23}$$

$$g_S = 2.00232, \tag{6.5.24}$$

and

$$\mu_B \equiv 1.39967 \text{ MHz/G}$$

is the Bohr magneton.

For nuclear angular momenta, **I** and **R**, the g–values are typically much smaller (by the ratio of the proton to electron masses, $1836.11 = \mu_B/\mu_n$, where μ_n is the nuclear magneton) and are not fundamental physical quantities. However, the g_I values for all stable $I \neq 0$ nuclei have been accurately measured,

$$\boldsymbol{\mu}_I \equiv \mu_n g_I \mathbf{I} \tag{6.5.22c}$$

where the quantity ($g_I = \mu_I/\mu_n I$) is tabulated (Townes and Schawlow, 1955, p. 644).

The only magnetic dipolar quantities that cannot be derived from *a priori* known g–values are g_R,

$$\boldsymbol{\mu}_R \equiv \mu_n g_R \mathbf{R}, \tag{6.5.22d}$$

and matrix elements of $\boldsymbol{\mu}$ between two electronic states.[†] See Townes and Schawlow (1955, p. 292) for an approximate calculation of g_R. There is a direct proportionality between $B\mathbf{L}^+$ and $B\mathbf{S}^+$ perturbation parameters and vibronically off-diagonal matrix elements of $\boldsymbol{\mu}$, for example,

$$\begin{gathered}\frac{\langle v_{\Lambda+1}\Lambda+1|B\mathbf{L}^+|v_\Lambda\Lambda\rangle}{\langle v_{\Lambda+1}|\hbar/2\mu R^2|v_\Lambda\rangle} = \frac{\langle v_{\Lambda+1}\Lambda+1|\boldsymbol{\mu}|v_\Lambda\rangle}{\mu_B g_L \langle v_{\Lambda+1}|v_\Lambda\rangle} \\ \langle\Lambda+1|\mathbf{L}^+|\Lambda\rangle = \mu_{\Lambda+1,\Lambda}/(\mu_B g_L)\end{gathered} \tag{6.5.25}$$

(Radford, 1962; Radford and Broida, 1962; Cook and Levy, 1973b.) This means that, for a rotation-electronic perturbation, knowledge of the zero-field perturbation parameter, H_{12}, implies knowledge of the off-diagonal Zeeman perturbation parameter, μ_{12}, and of the sign of μ_{12} relative to H_{12}. Any discrepancy would imply inadequacy of the deperturbation model resulting from the involvement of an additional perturbing state or the nonnegligibility of a neglected perturbation mechanism ($\mathbf{H}^{\mathrm{SO}}$, $\mathbf{H}^{\mathrm{SS}}$, $\mathbf{H}^{\mathrm{hyperfine}}$).

Anticrossing spectroscopy (see review by Lombardi, 1988) is complementary to quantum beat (QB) and radio frequency-optical and microwave-optical double resonance (MODR) techniques. When an anticrossing is broad (because of large H_{12} or small $|\mu_1 - \mu_2|$), the useful information is concealed under a complicated lineshape, which is a composite of anticrossings between many pairs of M–levels. A broad anticrossing does not necessarily imply broad M–levels. The splittings between M levels can be measured directly in megaHertz (not in Gauss or Volts/centimeter, which is quite an advantage) by QB or MODR spectroscopy.

Quantum-beat spectroscopy (see Sections 9.2.1, 9.2.2 and 9.3.2 and reviews by Lombardi, 1988, and by Hack and Huber, 1991) requires preparation of a coherent superposition state, $\Psi_{+,M;-,M'}$, composed of two eigenstates, $|+,M\rangle$ and $|-,M'\rangle$, by pulsed excitation from a single initial eigenstate $|i,M''\rangle$,

$$\begin{aligned}\Psi_{+,M;-,M'}(t) = {} & \mu_{+i}|+,M\rangle \exp[t(iE_{+,M}/\hbar - 1/2\tau_+)] \\ & + \mu_{-i}|-,M'\rangle \exp[t(iE_{-,M'}/\hbar - 1/2\tau_-)],^{\ddagger}\end{aligned} \tag{6.5.26}$$

[†] $\boldsymbol{\mu}$ is the total of $\boldsymbol{\mu}_L, \boldsymbol{\mu}_S, \boldsymbol{\mu}_R, \boldsymbol{\mu}_I$.

where τ_+ and τ_- are the spontaneous fluorescence lifetimes of $|+, M\rangle$ and $|-, M'\rangle$, and μ_{+i} and μ_{-i} are transition moments [Eq. (6.3.1)] between initial ($|i, M''\rangle$) and final eigenstates. (See Sections 9.3.1 and 9.3.2 for a discussion of quantum beats based on the complex **H** formalism. The example discussed here corresponds to the "strong coupling limit" of Section 9.3.1.) This superposition state decays by spontaneous fluorescence into $|0, M'''\rangle$ at a rate proportional to

$$\begin{aligned} P(t) &= |\langle \Psi_{+,M;-,M'}(t)|\mu|\rangle|^2/|\langle \Psi_{+,M;-,M'}(t=0)|\Psi_{+,M;-,M'}(t=0)\rangle|^2 \\ &= f_+|\mu_{+0}|^2 \exp(-t/\tau_+) + f_-|\mu_{-0}|^2 \exp(-t/\tau_-) \\ &\quad + 2(f_+f_-)^{1/2}\mu_{+0}\mu_{-0}\cos[(E_+ - E_-)t/\hbar]\exp[-t(1/2\tau_+ + 1/2\tau_-)], \end{aligned} \tag{6.5.27}$$

where the fractional $|+, M\rangle$ and $|-, M'\rangle$ eigenstate characters, f_+ and f_-, are

$$f_\pm = \frac{|\mu_{\pm i}|^2}{|\mu_{+i}|^2 + |\mu_{-i}|^2}. \tag{6.5.28}$$

The fluorescence rate is modulated at the frequency splitting of eigenstates $|+, M\rangle$ and $|-, M'\rangle$. This modulation is the quantum beat. The quantum beat decays at a rate

$$1/\tau_{QB} = \frac{1}{2}(1/\tau_+ + 1/\tau_-), \tag{6.5.29}$$

which is the average of the two eigenstate decay rates. The depth of modulation is

$$\Phi_{QB} = \frac{2(f_+f_-)^{1/2}\mu_{+0}\mu_{-0}}{f_+|\mu_{+0}|^2 + f_-|\mu_{-0}|^2}. \tag{6.5.30}$$

The modulation depth is the ratio of the oscillatory component of $P(t)$ to the sum of the two exponentially decaying eigenstate components: 100% modulation corresponds to the case where (if $\tau_+ = \tau_-$) $P(t) = 0$ once every beat period, $\Delta t = 2\pi\hbar/(E_+ - E_-)$. In the optimal case where $f_+\mu_{+0}^2 = f_-\mu_{-0}^2$, the modulation depth will be 100%.

In order to *prepare* $\Psi_{+,M;-,M'}$, the exciting radiation must be capable of simultaneously populating both $|+, M\rangle$ and $|-, M\rangle$ from the common initial level $|i, M''\rangle$. Furthermore, in order to *detect* the quantum beat, the experimental apparatus must be arranged so that fluorescence from both $|+, M\rangle$ and $|-, M'\rangle$ into a common final level $|0, M'''\rangle$ is detectable. These requirements lead to quantum number, temporal, and geometric (i.e., polarization) constraints.

‡ $\Psi_{+,M;-,M'}(t)$ in Eq. (6.5.26) is not normalized. The $t = 0$ normalization factor, which appears in Eq. (6.5.27), is

$$|\langle \Psi_{+,M;-,M'}(t=0)|\Psi_{+,M;-,M'}(t=0)\rangle|^{-1/2} = [|\mu_{+i}|^2 + |\mu_{-i}|^2]^{-1/2}.$$

Typical N_2 and Nd:YAG pumped dye lasers have a pulse duration of ~ 6 ns FWHM. This means that quantum beats between levels split by more than $(6\times 10^{-9}s)^{-1} \simeq 170$ MHz will be unobservable with a preparation pulse duration longer than 6 ns. Recall that, if the two beating levels correspond to main and extra lines, the $\lesssim 170$ MHz limit restricts the perturbation interaction to $H_{12} \lesssim 85$ MHz.

The quantum number constraint leads to the two quite distinct polarization classes of QB spectra: $\Delta M = 0$ beats (polarization of excitation and detection parallel to static field, $M = M' = M'' = M'''$) and $\Delta M = 2$ beats (perpendicular polarization, $M = M' + 2 = M'' + 1 = M''' + 1$). The former type is particularly useful for examining an anticrossing between two perturbing states (provided H_{12} is small enough). The latter displays both intrastate splittings (useful for measurement of g–values and dipole moments) and interstate splittings.

In order to observe quantum beats, all four transition moments $\mu_{+i}, \mu_{+0}, \mu_{-i}$, and μ_{-0} must be nonzero. If $|+, M\rangle$ and $|-, M'\rangle$ are simply field-split M–components of the same electronic state, there is no problem. However, if either $|+, M\rangle$ or $|-, M'\rangle$ is a mixed eigenstate composed of basis functions $|1, M\rangle$ and $|2, M\rangle$ belonging to two electronic states, for example

$$|+, M\rangle = \left(1 - C_{2+}^2\right)^{1/2} |1, M\rangle + C_{2+}|2, M\rangle$$

(see Eqs. (6.2.6a) - (6.2.6c)) and

$$|-, M'\rangle = |2, M'\rangle$$

where $|1, M\rangle \leftarrow |i, M''\rangle$ and $|1, M\rangle \rightarrow |0, M'''\rangle$ transitions are forbidden, then

$$\Phi_{QB} = \frac{2C_{2+}^2\mu_{2M,i}\mu_{2M,0}\mu_{2M',i}\mu_{2M',0}}{2C_{2+}^2\mu_{2M,i}^2\mu_{2M,0}^2 + \mu_{2M',i}^2\mu_{2M',0}^2} \tag{6.5.31}$$

thus the QB modulation depth is approximately $2C_{2+}^2/(1 + C_{2+}^2)$. The beats vanish if $C_{2+} = 0$. Near-degeneracy between the $|1, M\rangle$ and $|2, M'\rangle$ basis functions is not sufficient for observation of a quantum beat. There must be a $|1, M\rangle \sim |2, M\rangle$ perturbation to allow detection of forbidden inter-electronic-state quantum beats. A great deal of information about both electronic states and their interaction (zero-field and field-induced) is contained in the variation of QB frequency, modulation depth, and decay rate versus the applied magnetic or electric field.

Zeeman quantum beat spectroscopy was used by Gouedard and Lehmann (1979, 1981) to measure the effect of various 1_u perturbing states on the g_J–values [Eq. (6.5.21)] of more than 150 rotational levels of the Se_2 B 0_u^+ state (see Section 6.5.2 and Fig. 6.16). In that experiment, the excitation polarization was perpendicular to the applied magnetic field so that quantum beats were observed between nominal B-state components differing in M by 2. The frequencies of these beats increase linearly from 0 MHz at 0 G until the $\Delta M = 2$ splitting falls outside the coherence width of the excitation source and the beats disappear. For each J' level of the B-state, all pairs of $M, M+2$ levels are split by the same amount[†] (except for J' levels near the perturbations where the Zeeman pattern will become slightly asymmetric owing to M–dependent $\mathbf{H}^{\mathrm{zf}} \times \mathbf{H}^{\mathrm{Zeeman}}$ interference and energy denominator effects). This near-perfect superposition of all $\Delta M = 2$ beats from a given J' level vastly enhances the sensitivity of QB spectroscopy for high J–levels. Because the beats are between levels differing in M by 2, the beats are at twice the Larmor frequency (the $\Delta M = 1$ level spacing). At high magnetic field, a mixed $|0_u, M\rangle \sim |1_u, M\rangle$ level could be tuned through a nominal $|\mathrm{B}\ 0_u, M \pm 2\rangle$ level and a $\Delta M = 2$ interstate QB would result. This effect would be difficult to detect because only one of the $2J+1$ M–levels would be affected and a lengthy search for the proper magnetic field would be required.

Ring, *et al.*, (1998 and 1999) have used a time-dependent magnetic field and the combination of a static magnetic field in a direction perpendicular to that of a time-dependent field to create and manipulate novel coherences and to monitor the quantum beats associated with specifiable details of the time evolution of these coherences. The frequencies and decay rates of different classes of coherence ($\Delta M_J = \pm 2$ and ± 1 polarization beats, $\Delta M_J = 0$ singlet$\sim$triplet population beats) may be sampled and modified selectively.

Stark quantum-beat (SQB) spectroscopy was used by Brieger, *et al.*, (1980) for LiH $\mathrm{A}^1\Sigma^+$ and by Schweda, *et al.*, (1980, 1985) for BaO $\mathrm{A}^1\Sigma^+$ to measure electric dipole moments of several rotation-vibration levels. In a $\Lambda = 0$ state, the electric-field-induced shift of a J, M–level is proportional to $\mathcal{E}^2/B$ because $\mathbf{H}^{\mathrm{Stark}}$ has only $\Delta J = \pm 1, \Delta M = 0$ matrix elements:

$$\begin{aligned} E_{J=1,M=0} &= \sum_{J'} \frac{|\langle J,0|\mu\mathcal{E}|J',0\rangle|^2}{E_J^0 - E_{J'}^0} \\ &= \mathcal{E}^2\mu_\Sigma^2 \left[\frac{|\langle 1,0|\alpha_Z^z|2,0\rangle|^2}{2B_v - 6B_v} + \frac{|\langle 1,0|\alpha_Z^z|0,0\rangle|^2}{2B_v - 0B_v}\right] \\ &= \frac{\mathcal{E}^2\mu_\Sigma^2}{B_v}\left[-\frac{1}{15} + \frac{1}{6}\right] = +\frac{\mathcal{E}^2\mu_\Sigma^2}{10B_v} \end{aligned} \tag{6.5.32}$$

$$E_{J=1,M=1} = -\frac{\mathcal{E}^2\mu_\Sigma^2}{20B_v} \tag{6.5.33}$$

so, for an unperturbed $^1\Sigma$ state,

$$(E_{J=1,M=0} - E_{J=1,M=1})/\hbar = \frac{3}{20}\frac{\mathcal{E}^2\mu_\Sigma^2}{\hbar B_v} = 7.9692\frac{\mathcal{E}^2\mu_\Sigma^2}{B_v}\mathrm{Hz} \tag{6.5.34}$$

for $\mathcal{E}$ (V/cm), μ_Σ (D), and B_v (cm^{-1}). Figure 6.19 shows the SQB superimposed on the exponential decay of the fluorescence from BaO $\mathrm{A}^1\Sigma^+(v = 2, J = 1)$ at 500 V/cm.

[†] Neglecting $\Delta J = \pm 1$ matrix elements of $\mathbf{H}^{\mathrm{Zeeman}}$.

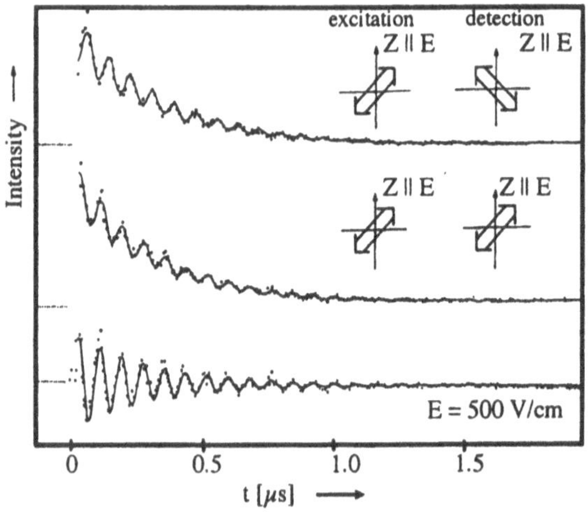

Figure 6.19: Stark quantum beats in BaO $A^1\Sigma^+(v = 2, J = 1)$. The $J = 1$ level is excited via the $R(0)$ line by radiation from an N_2–laser-pumped dye laser. The pump radiation is linearly polarized at 45° to the $\mathcal{E}$–field direction in order to produce a coherent superposition of $M = 0$ with $|M| = 1$ components. The top trace shows the signal resulting when the polarization of the detected fluorescence is selected to be at 45° to $\mathcal{E}$ and at 90° to the excitation polarization. The middle trace is for parallel excitation and detection polarizations. The bottom trace is the difference between the two detection geometries. [From Schweda, *et al.*(1985).]

The BaO $A^1\Sigma^+$ state is perturbed by the $b^3\Pi$ and $A'^1\Pi$ states. The effect of perturbations on the Stark tuning coefficients of a $^1\Sigma^+$ state is more complicated than the situation described by Eq. (6.5.21) (because parity remains an almost good quantum number in $\Lambda = 0$ states at $\mathcal{E}$–fields that are sufficiently large to cause complete parity mixing in $|\Lambda| > 0$ states). There will be Stark tuning terms proportional to (suppressing all J, M–dependent factors)

$$(1 - C_{\Sigma\Pi}^2)^2\mathcal{E}^2\ \mu_\Sigma^2/B_\Sigma \quad \text{direct } ^1\Sigma \text{ term} \tag{6.5.35a}$$

$$C_{\Sigma\Pi}(1 - C_{\Sigma\Pi}^2)^{3/2}\mathcal{E}^2\ \mu_\Sigma\mu_{\Sigma\Pi}/B_\Sigma \quad \text{cross term} \tag{6.5.35b}$$

$$C_{\Sigma\Pi}^2(1 - C_{\Sigma\Pi}^2)\mathcal{E}^2\ \mu_{\Sigma\Pi}^2/\Delta E_{\Sigma\Pi} \quad \text{transition moment term} \tag{6.5.35c}$$

$$C_{\Sigma\Pi}^2(1 - C_{\Sigma\Pi}^2)\mathcal{E}^2\ \mu_\Sigma\mu_\Pi/B_\Sigma \quad \text{borrowed } ^1\Pi \text{ term,} \tag{6.5.35d}$$

where $C_{\Sigma\Pi}$ is the zero-field mixing coefficient, $\mu_{\Sigma\Pi}$ is the electric dipole transition moment between Σ and Π levels, and

$$\Delta E_{\Sigma\Pi} = E_{\Sigma J}^0 - E_{\Pi J'}^0.$$

Since $^1\Sigma^+ - {}^3\Pi$ transitions are dipole-forbidden, the three terms involving $\mu_{\Sigma\Pi}$ are zero for the $b^3\Pi$ perturbers. However, the $\mu_{\Sigma\Pi}$ terms will not be zero for the $A^1\Sigma^+ – A'^1\Pi$ transition. The $C_{\Sigma\Pi}(1 - C_{\Sigma\Pi}^2)^{3/2}\mathcal{E}^2\mu_\Sigma\mu_{\Sigma\Pi}/B_\Sigma$ term is the most important perturbation-related term because it is linear rather than quadratic in $C_{\Sigma\Pi}$.

MODR spectroscopy is not limited to levels separated by $\lesssim$ 170 MHz (the coherence width of a 6ns excitation pulse). What is needed is a radiation field of sufficient intensity that

$$\mu\mathcal{E}/\hbar > 1/\tau \quad \text{or} \quad \mu_0 g_J H/\hbar > 1/\tau,$$

so that the microwave-induced transition rate is competitive with radiative decay. MODR can be detected in a variety of ways.

MODR experiments differ in the way the initial population inequality between the initial and final states of the microwave transition is created and in the way the resonant interaction of the microwave radiation within the initially prepared molecular system is detected. In the vast majority of MODR experiments, the initial population inequality is created when a tunable laser populates or depopulates a selected level (a difference between diagonal elements of the density matrix, $\rho_{ii} - \rho_{ff}$, see Section 9.1.5) and the resonant interaction between the molecular sample and the microwave field is detected as a change in the intensity, polarization, or spectral content of the molecular fluorescence. This is often called a trigger scheme because the absorption of a puny microwave photon, $h\nu << kT$, is signalled by a gigantic optical photon, $h\nu >> kT$. If one believes exclusively in the photons as bullets, molecules as targets picture (see Sections 1.2.6 and 9.1.6), then it might seem that optically detected MODR would be much more sensitive than any microwave detected MODR scheme. But this is not true. Lehmann and Coy (1984 and 1985) exploit a microwave detected MODR scheme to rotationally assign vibrational overtone transitions in NH_3 and electronic-vibrational transitions in NO_2.

The microwave detected MODR scheme closely resembles pulsed nuclear magnetic resonance (Hahn, 1950), optical coherent transients by Stark switching (Brewer and Shoemaker, 1971) and laser frequency switching (Brewer and Genack, 1976). The on-resonance microwave radiation field, $\omega_0 = (E_2 - E_1)/\hbar$, creates an oscillating bulk electric dipole polarization (off-diagonal element of the density matrix, $\rho_{12}(t)$). The oscillation is at $\omega_0 \pm \omega_R$, where ω_R is the (M_J-dependent) Rabi frequency,

$$\Omega_R = \mu\epsilon_0/\hbar, \tag{6.5.36}$$

μ is the transition moment, $\epsilon_0 e^{i\omega_0 t}$ is the oscillating microwave electric field, and the amplitude of this oscillating polarization is proportional to the Boltzmann equilibrium population difference between levels E_2 and E_1. A pulsed laser abruptly depopulates either level 1 or level 2, resulting in a sudden change in the amplitude of the polarization oscillating at $\omega_0 \pm \omega_R$. The sign of this change depends on whether the upper or lower level is depopulated. This "transient nutation" effect is detected on a square law detector when the radiation at $\omega_0 \pm \omega_R$ generated by the bulk polarization is mixed with the applied microwave radiation at ω_0. The detected power is proportional to $|\epsilon(t)|^2$ where ϵ is the electric field at the detector

$$P(t) \propto \left| \epsilon_0 e^{i\omega_0 t} + \epsilon_+ e^{i(\omega_0+\omega_R)t} + \epsilon_- e^{i(\omega_0-\omega_R)t} \right|^2$$
$$= [\epsilon_0^2 + \epsilon_+^2 + \epsilon_-^2] + 2(\epsilon_0\epsilon_+ + \epsilon_0\epsilon_-)\cos\omega_R t$$
$$+ 2\epsilon_+\epsilon_- \cos 2\omega_0 t. \quad (6.5.37)$$

The time-independent terms in [ ] and the high frequency term at $2\omega_0$ are outside the detection bandwidth, thus only the $\cos\omega_R t$ term is detected. Since each of the degenerate M_J components of the $2 \leftrightarrow 1$ transition has a different Rabi frequency, the nutation signal dephases rapidly and appears as a coherent transient rather than a cw oscillation at ω_R. (The coherent transient signal is maximized by increasing the pressure until the collision-induced dephasing rate is larger than the M_J-dependent Rabi-dephasing rate but not larger than the maximum value of ω_R, which is the rise-time of the coherent transient.) This homodyne detection scheme has the advantage that the signal is proportional to $\epsilon_0(\epsilon_+ + \epsilon_-)$ rather than to $(\epsilon_+^2 + \epsilon_-^2)$, because $|\epsilon_0| >> |\epsilon_+ + \epsilon_-|$. By mixing the molecular field at $\omega_0 \pm \omega_R$ with the local oscillator field at ω_0, the detected signal is amplified by a factor of $|\epsilon_0/\epsilon_+| >> 1$. In many cases, detection of a signal as a coherent transient rather than as a population transient can be both more sensitive and provide more information about the molecule$\leftrightarrow$radiation interaction.

Evenson, *et al.*, (1964) took advantage of the preferential chemical population of the CN $A^2\Pi$ state. Although the $A^2\Pi \to B^2\Sigma^+(10,0)$ band, which occurs in the microwave region, is both electric and magnetic dipole-allowed, its small Franck-Condon factor of 0.01 makes microwave A$\to$B transitions between unperturbed levels much weaker than pure rotational transitions within either state ($\mu_A \approx 0.6$ Debye, $\mu_B \approx 1.2$ Debye) and also weaker than interelectronic transitions involving a mixed level. When the microwave frequency is tuned to resonance with a nominal A$\to$B transition (or a magnetic field is used to tune a transition between M–components into resonance with a fixed microwave field), fluorescence intensity is transferred from the red (A$-$X) to the violet (B$-$X) region, exactly as in an anticrossing experiment. Since the microwave transition affects a small fraction of the chemically populated levels of the $A^2\Pi$ state, a satisfactory signal/noise is obtained by monitoring a specific B$-$X (0,0) rotational line, selected using a monochromator with ~ 0.4Å bandpass.

Miller, *et al.*, (1974) used MOMRIE (Microwave Optical Magnetic Resonance Induced by Electrons) complementarily with anticrossing spectroscopy in experiments on He, H_2, and other molecules. Small differences in electron bombardment excitation cross-sections combined with different radiative lifetimes lead to steady-state population differences between initial and final levels of the microwave transition. By monitoring fluorescence from one of the two involved electronic states, a change in fluorescence intensity signals a microwave resonance. Electron bombardment excitation tends to produce smaller population differences than the chemical pumping scheme exploited in the CN experiments, but the MOMRIE technique is more generally applicable.

Atomic and molecular resonance lamps were used by Silvers, *et al.*, (1970) and Field and Bergeman (1971) to optically pump the CS $A^1\Pi(v = 0)$ level, which is perturbed by $a'^3\Sigma^+(v = 10)$. Electric-dipole-allowed Λ–doubling transitions ($\Delta J = 0, \Delta M = 0, |M| = J, e \leftrightarrow f$) in $A^1\Pi$ were observed in electric fields up to 8 kV/cm. The MODR signal was detected as a change in the polarization of undispersed side fluorescence. Stark tuning coefficients precise to $\sim$ 1 part in 10^3 were obtained for $J = 3 - 9$. The J–dependence of the Stark coefficients, combined with $A^1\Pi \sim a'^3\Sigma^+$ mixing coefficients determined from a deperturbation analysis of the MODR Λ–doubling and the optical data of Barrow, *et al.*(1960), permitted determination of the dipole moment of the perturbing $a'^3\Sigma^+$ state as well as that of the $A^1\Pi$ state.

Tamanis, *et al.*, (1997) measured the NaK $D^1\Pi$ $v = 7$, $J = 23$ and $v = 12, J = 7$ $(J, e \rightarrow J, f)$ Λ–doublet transitions by radiofrequency optical double resonance (RFODR). The radiofrequency radiation transfers population between the J, e level, which is initially populated by an $R(J-1)$ transition, and the J, f level, which fluoresces via a $Q(J)$ transition. In the absence of the resonant rf radiation, the dispersed fluorescence spectrum from the $R(J-1)$ excited J, e rotational level contains only the $R(J-1)$ and $P(J+1)$ lines. The $Q(J)$ fluorescence line may alternatively be turned on by application of a DC electric field ($\Delta J = 0, \Delta M = 0, e \leftrightarrow f$ Stark effect selection rules), which mixes some J, M, f basis state character into the optically bright (via the $R(J-1)$ excitation transition) J, M, e basis states (Stark Anticrossing Spectroscopy, see Silvers, *et al.*, 1970, and Moore, *et al.*, 1984). However, the dependence of the $Q(J)$ line intensity (all $2J+1$ unresolved M_J–components) on the electric field strength is described by a complicated sum over M_J sublevels. The complexity arises from the M_J–dependences of both the Stark matrix elements and the individual M_J–sublevel populations. Such Stark-mixing experiments primarily provide information about the ratio of dipole moment to Λ–doubling. Thus, by combining a zero-field RFODR measurement of the Λ–doublet frequency with results from a Stark-mixing experiment, the electric dipole moment may be determined. In the case of the NaK $D^1\Pi$ state, strong, v–dependent $D^1\Pi \sim d^3\Pi$ spin-orbit interaction affects both the Λ–doubling (indirectly, by diluting the $^1\Pi$ character hence reducing the L–uncoupling $D^1\Pi$ interaction with remote $^1\Sigma^+$ states) and the dipole moment (by mixing in some character of the $d^3\Pi$ state, which has a dipole moment opposite in sign to that of the $D^1\Pi$ state).

When the Λ–doublet splitting and dipole moment are both known, as they are in the BCl $A^1\Pi$ state, the Stark-mixing experiment [excite $R(J-1)$, detect Stark-induced $Q(J)$ fluorescence] provides a non-invasive method for measuring spatially and temporally varying electric fields. Since at low–J $\Delta J = 0, e \leftrightarrow f$ Stark matrix elements are invariably large and Λ–doublet splittings are invariably small, low–J levels provide sensitivity to small electric fields ($\sim$ 50 V/cm) [Moore, *et al.*, 1984, Mandich, *et al.*, 1985].

German and Zare (1969) populated the OH $A^2\Sigma^+$ $(v = 0, N = 2, J = \frac{3}{2})$ level using an atomic lamp. This level is doubled by the $I = \frac{1}{2}$ proton nuclear spin. Radiofrequency magnetic dipole transitions within the Zeeman-split M_F

components of the $F = 1$ and 2 levels were observed as changes in undispersed fluorescence polarization. German and Zare measured magnetic g–values that were found to be in satisfactory agreement with g_F values predicted for isolated $N = 2, J = \frac{3}{2}, F = 1$ and 2 levels of a case (b) $^2\Sigma^+$ state.

Field, *et al.*, (1972, 1973) used Ar^+ and cw dye lasers to observe $\Delta J = \pm 1$ MODR pure rotational transitions in the BaO $A^1\Sigma^+$ state. The MODR effect is observable either as a change in undispersed fluorescence polarization or intensity. The MODR B_v–values for $A^1\Sigma^+(v = 0 - 5, 7)$ show large deviations from a smooth B_v versus v variation owing to perturbations by $b^3\Pi$ and $A'^1\Pi$ levels (Field, *et al.*, 1973). No interelectronic MODR transitions were found in BaO, but several "mystery" MODR lines were reported in electronically excited NO_2 by Tanaka, *et al.*, (1974) and by Solarz and Levy (1973). These incompletely assigned lines are certainly perturbation-facilitated transitions between two vibronic levels. The radiative lifetimes and pressure-dependent photon yields for the initial and final levels of the MODR transition are quite dissimilar, giving rise to a complex pressure-dependence of the sign and magnitude of the MODR signal.

An extremely sensitive MODR scheme, microwave optical polarization spectroscopy (MOPS), was introduced by Ernst and Törring (1982). The most important features of MOPS are that it requires respectively 100 and 10 times lower laser and microwave intensities than MODR and results in 10 times narrower lines. This means that it will be possible to take full advantage of differential power broadening effects (Section 6.5.1) and to utilize low-power, frequency-doubled dye lasers and low-power, broadly tunable microwave sources (backward wave oscillators) in order to gain access to and systematically study perturbations.

A perturbation acts as a window through which normally unobservable states may be viewed. Anticrossing, quantum-beat, and MODR spectroscopies are capable of examining the states in the immediate vicinity of the perturbed level. Optical-optical double resonance (OODR) allows systematic exploration of states far above and below the perturbed level. Gottscho, (1979) utilized an OODR scheme to observe the BaO $a^3\Sigma^+$ state. He used two single-mode cw dye lasers to excite a singlet~triplet mixed level of the $C^1\Sigma^+$ state *via* the $A^1\Sigma^+$ state. Figure 6.20 shows the resolved fluorescence spectrum originating from the $C^1\Sigma^+$ $(v = 2, J = 16)$ level. The fluorescence pattern immediately identifies the lower state as a $^3\Sigma^+$ state and provides approximate values for its rotational and spin-spin constants. This OODR-resolved fluorescence scheme is a rather general method for observing low-lying excited states, because, at high energies where the vibronic density of states is large, perturbations are much more common than in the lowest excited states. Thus one is likely to find the necessary type of perturbation among the high-energy states in order to gain access via fluorescence to virtually any suspected low-lying state.

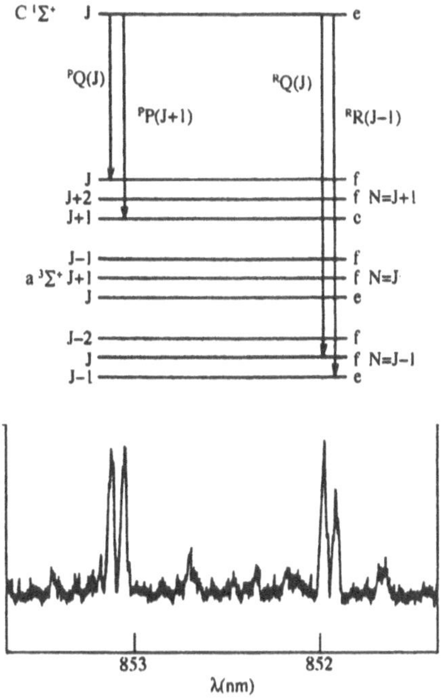

Figure 6.20: BaO $C^1\Sigma^+ - a^3\Sigma^+$ (2,12) fluorescence excited by OODR. The triplet contaminated $J = 16, v = 2$ level of $C^1\Sigma^+$ is excited using two single-mode cw dye lasers. [From Gottscho, (1979).]

In a complementary experiment, Li and Field (1983) have utilized Na_2 $A^1\Sigma_u^+ \sim b^3\Pi_u$ perturbations to survey, via OODR excitation spectroscopy, excited triplet *gerade* levels with T_v–values in the 33,000 cm^{-1} region. By tuning the pump laser to a transition into a singlet∼triplet mixed level, the probe laser is equally capable of exciting transitions into pure triplets or pure singlets. These two spin classes of excited states are experimentally separable by restricting the wavelength region of detected fluorescence and by comparing the OODR intensities obtained with the pump laser tuned to main (singlet strong, triplet weak) versus extra (singlet weak, triplet strong) lines.

6.5.4 Rydberg States and the Zeeman Effect

The diagonal and off-diagonal matrix elements associated with the Zeeman effect ($\sim$ 1cm^{-1} at 1 Tesla) are so small that it is tempting to think of an external magnetic field as only generating small but diagnostically useful level splittings, small modifications of perturbations caused by other more robust operators, and other mostly esoteric effects. However, high–n Rydberg states provide an opportunity for the Zeeman effect to profoundly affect the spectrum and the dynamics encoded in the spectrum. Two things conspire to lend unexpected importance to the Zeeman effect: the electronic density of states increases so that level splittings associated with the non-spherical nature of the molecular-ion-core become comparable to the ordinary linear Zeeman effect interaction energies and, at high–n, the quadratic (diamagnetic) Zeeman effect (which scales as $n^4\ \mathrm{B}_Z^2$) can no longer be ignored. The Zeeman Hamiltonian for a one-electron molecule is

$$\mathbf{H}^{\mathrm{Zeeman}} = \mu_B(\mathbf{L}_Z + 2\mathbf{S}_Z)B_Z + \frac{e^2}{8m_e}B_Z^2 r^2 \sin^2\theta \tag{6.5.38}$$

where μ_B is the Bohr magneton, $\mathbf{L}_Z$ and $\mathbf{S}_Z$ are the orbital and spin angular momentum components along the laboratory Z axis, B_Z is the magnetic field, e and m_e are the charge and mass of the electron, and r and θ are the polar coordinates of the electron in the laboratory frame. The linear Zeeman term destroys the N quantum number [by introducing off-diagonal matrix elements between the near degenerate (quadrupole-split) N–components of an N^+ cluster]. The quadratic Zeeman term destroys both n (Δn unrestricted) and l ($\Delta l = 0, \pm 2$). Gauyacq, *et al.*, (1999) illustrate the effects of a magnetic field on the Rydberg states of NO.

There is a close analogy between the Rydberg series of alkali atoms, A, converging to the $\mathrm{A}^+\ {}^1\mathrm{S}$ ion-core state and Rydberg series of NO converging to the $\mathrm{NO}^+\ \mathrm{X}^1\Sigma^+$ state. Both are, to a very good approximation, "one-electron" systems. The electronic properties of all Rydberg states of such one-electron systems depend on the atom-like properties of the $nl\lambda$ (body-frame) or nlm_l (laboratory-frame) Rydberg orbitals. The atom-like $nl\lambda$ or nlm_l characters of the Rydberg orbital may be perturbed by the axially symmetric electric field of the ion-core (intramolecular field), the rotation of the ion-core (l-uncoupling), and an external magnetic or electric field (spatial anisotropy). In this section three limiting cases are discussed: case (b), where the three competing effects have relative importance: intramolecular field > rotation > linear Zeeman effect; case (d), where: rotation > intramolecular field > Zeeman; and *uncoupled* case (d), where: rotation > Zeeman > intramolecular field (see Fig. 6.21). At high n and l, the diamagnetic Zeeman effect ($E \propto n^4 B_Z^2$) can overpower both the linear Zeeman effect and the Coulomb structure ($\Delta E \approx 2\Re/n^3$) of the Rydberg series, with the result that n– and l–mixing effects significantly increase the range of possibilities for the structure and dynamics of Rydberg states. Owing to the *a priori* known n, l, N^+, and B_Z–scaling properties of each of the competing interaction terms, one can design experiments to measure the parameter that controls the strength of each interaction term under conditions maximally free of complexity due to the other terms. More importantly, it is possible to select or manipulate the hierarchy of interaction strengths merely be selecting the n, l, N^+ region of the level diagram and the strengths and directions of external fields.

The examples used here to illustrate the effects of a moderately strong magnetic field ($B_Z \approx 1$ Tesla) are from an experimental study of the NO $nf \leftarrow \mathrm{A}^2\Sigma^+$ transition probed by double resonance excitation via a selected rotational-Zeeman sublevel of the $\mathrm{A}^2\Sigma^+(v = 1)$ state. Individual $nf(N^+)$

rotation-Zeeman clusters of transitions ($\sim$ 1 cm^{-1} wide) can be resolved up to $n \approx 50$ [$E(n+1/2) - E(n-1/2) \approx 2\Re/n^3 = 1.7$ cm^{-1} at $n = 50$] (see Fig. 6.22.

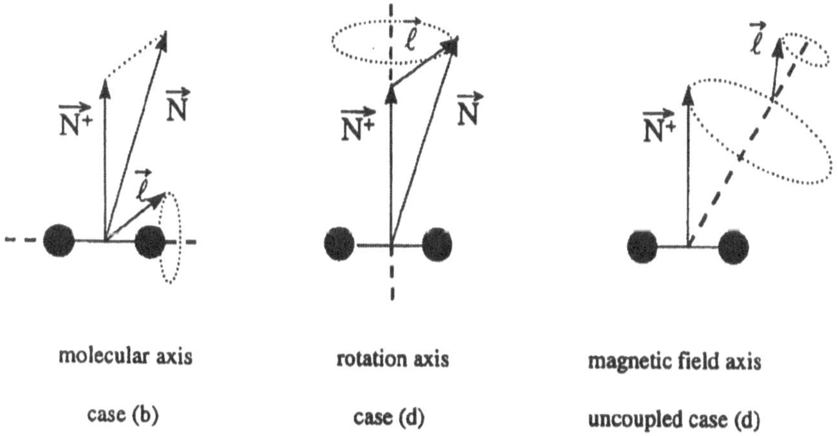

Figure 6.21: Angular Momentum Coupling Schemes for Rydberg States of a Diatomic Molecule with a closed-shell Molecular-Ion-Core. In low–n, low–l states, the orbital angular momentum of the Rydberg electron, $\vec{l}$, is quantized according to its projection, λ, along the internuclear axis, owing to the effect of the axially symmetric intramolecular field. This *strong intramolecular field* limit corresponds to Hund's case (b), $|nl\lambda NN^+M_N\rangle$. At high–$n$ and/or high l, in the absence of an external field, $\vec{l}$ is quantized according to its projection, l_R, along the $\vec{\mathbf{N}}^+$ rotational axis of the ion-core. This *weak intramolecular field* limit corresponds to Hund's case (d), $|nll_RNN^+M_N\rangle$. At high–n and/or high–l in the presence of an external magnetic field, both $\vec{\mathbf{N}}$ and $\vec{l}$ precess independently of each other around the laboratory field axis (Z), as described by the M_{N^+} and m_l laboratory frame projection quantum numbers. This *strong external field* limit corresponds to *uncoupled* Hund's case (d), $|nlm_lN^+M_{N^+}\rangle$. The quantization axes in the three coupling cases appear in the figures as dashed lines (from Gauyacq, *et al.*, 1999).

In order to understand the effect of a magnetic field on Rydberg-Rydberg transitions, it is necessary first to examine the zero-field situation. The upper levels are nf Rydberg states near the Hund's case (d) limit in which the rotational level structure, $B^+N^+(N^++1)$, is described by the pattern-forming N^+ quantum number, which is the ion-core total angular momentum exclusive of electron spin. Each value of N^+ is split into $2l+1$ N–values (if $N \geq l$). This electronic fine structure results from the coupling between $\vec{\mathbf{N}}^+$ and $\vec{\mathbf{l}}$, $\vec{\mathbf{N}} = \vec{\mathbf{N}}^+ + \vec{\mathbf{l}}$. The fine structure exhibited by transitions into a case (d) l, N^+ state-cluster is due to $\Delta N = \pm 1, 0$ R, Q, P transitions from a selected N–level in the lower electronic state, which is near the case (b) limit. The familiar R, Q, P rotational branch structure is replaced by an electronic fine structure produced by the axially symmetric intramolecular electric field (molecular ion-core anisotropy). The effect of the core anisotropy on the electronic fine structure in a Rydberg

nl complex depends on the penetrating character of the Rydberg orbital and decreases roughly as $n^{-3}l^{-3}$. The width of the N^+ cluster, fractionated into $2l+1$ N–components, collapses as n and l increase.

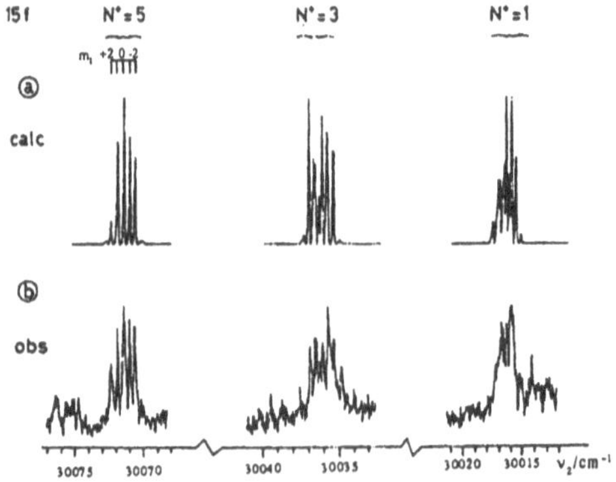

Figure 6.22: Resolved Rotational (N^+) and Zeeman (m_l) Structures in the NO $15f - A^2\Sigma^+$ Electronic Transition that originates from an $N = 3$, $M_S = -1/2$ level (with unresolved λ, M_N Zeeman splittings) of the $A^2\Sigma^+$ state. Parts a and b are, respectively, the observed and calculated spectra (from Guizard, *et al.*, 1991).

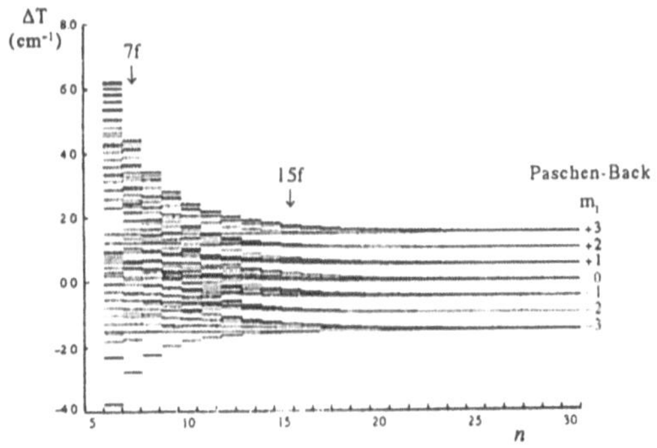

Figure 6.23: The Zeeman Patterns of $n = 6-30$ $nf(N^+ = 3)$ complexes of NO in a 1 Tesla Magnetic Field (from Guizard, *et al.*, 1991). ΔT is due to the combined effects of the axially symmetric intramolecular electrnic field (core anisotropy) and the linear Zeeman effect. At low–n, most of the $(2l+1)(2N^+ + 1) = 49$ m_l, M_{N^+} Zeeman components are resolved. As n increases, the Zeeman pattern collapses to the Paschen-Back pattern of $2l+1$ equally spaced m_l–clusters, each of which is $2N^+ + 1$–fold degenerate. The reason for this collapse is that the m_l–splittings are proportional to the Bohr magneton and the M_{N^+} splittings reflect much smaller nuclear magnetic moments. In the absence of perturbations by interloper states, there are two identical manifolds (separated by $2\mu_B B_Z$) of $|nlm_lN^+M_{N^+}\rangle$ states, one each for $M_S = +1/2$ and $-1/2$ (from Gauyacq, *et al.*, 1999).

An external magnetic field creates a spatial anisotropy in the laboratory frame, and the $\Delta N = \pm 1$, $\Delta M_N = 0$ off-diagonal matrix elements of the linear term in the Zeeman effect Hamiltonian (Eq. (6.5.38)) destroy N (but not M_N) as a good quantum number. The external magnetic field causes a transition from the usual Hund's case (d) $|nll_R N^+ N\rangle$[†] scheme to the *uncoupled* case (d) $|nlm_l N^+ M_{N^+}\rangle$ scheme. The uncoupled case (d) limit was first introduced by Monteiro and Taylor (1990) and is depicted in Fig. 6.21. This uncoupling of angular momenta from each other ($\vec{\mathbf{N}}$ and $\vec{\mathbf{l}}$ precess independently of each other about the direction of the external magnetic field) is analogous to the Paschen-Back effect in atoms. The Rydberg electron loses its molecular character. Its angular momentum, $\vec{\mathbf{l}}$, is completely decoupled from the molecular frame and precesses almost freely about the magnetic field axis. Both m_l and M_{N^+} become good quantum numbers (at the expense of N). Since N is destroyed as a quantum number, the $\Delta N = 0, \pm 1$ selection rule is no longer applicable and, as shown in Figs. 6.22 and 6.23, many more fine-structure components appear in an uncoupled case (d) - case (b) transition than in an ordinary case (d) – case (b) transition.

Figure 6.23 shows, as n increases, the evolution of the fine structure within the $N^+ = 3$ rotational level in an nf complex in a 1 Tesla magnetic field. The width of the splitting due to the intramolecular field decreases as n^{-3} from ~ 8 cm^{-1} at $n = 6$ to < 0.1 cm^{-1} at $n = 30$. The n–independent Zeeman splitting between $m_l, m_l - 1$ components, $\mu_0 B_Z \approx 0.5$ cm^{-1}, appears clearly at high–n. Each N^+ level is split into $(2l+1)(2N^+ + 1)$ Zeeman m_l, M_{N^+} (or, at low–n, l_R, M_N) components. Because N is destroyed by the magnetic field, most of the $(2l+1)(2N^+ + 1)$ fine structure components of an N^+ level in the $nf \leftarrow \mathrm{A}^2\Sigma^+$ transition have appreciable intensity. This is in marked contrast to the $\Delta N = \Delta l$ propensity rule in a case (d) – case (b) transition.

Figure 6.22 displays the rotational plus electronic fine structure of the NO $15f \leftarrow \mathrm{A}^2\Sigma^+(v=1)$ transition from the $N = 3$, $M_S = -1/2$ Zeeman component of the intermediate level (Guizard, *et al.*, 1991). The parity selection rule permits transitions from the – parity $N = 3$ initial level of the $\mathrm{A}^2\Sigma^+$ state to the three + parity $N^+ = 1, 3, 5$ rotational clusters (separated by $10B^+$ and $18B^+$) of an nf complex. The structure of an $nf \leftarrow \mathrm{A}^2\Sigma^+ N = 3, M_S = +1/2$ transition (not shown) is identical to that originating from the $M_S = -1/2$ Zeeman component. The electric dipole transition operator operates exclusively on the spatial coordinates of the electron, thus $\Delta M_S = 0$ is a rigorous selection rule. Since the d–character of the $\mathrm{A}^2\Sigma^+$ state is exclusively responsible for making the $nf \leftarrow \mathrm{A}^2\Sigma^+$ transition allowed, one expects Z–polarized transitions that terminate on $m_l = -2, -1, 0, +1, +2$ Zeeman components in each N^+ cluster. The observed intensity patterns in Fig. 6.22b are in excellent agreement with those calculated for an uncoupled case (d) $\leftarrow$ case (b) pure $f \leftarrow d$ transition (Guizard, *et al.*, 1991).

The quadratic (diamagnetic) term in the Zeeman Hamiltonian, Eq. (6.5.38),

---

[†] The l_R (or $\mathcal{L}$) quantum number is defined at the end of Section 6.3.2 and in the caption of Fig. 6.21

makes a negligible contribution to the Rydberg spectrum at low n, l values. However, at 1 Tesla, the $n^4 B_Z^2$ scaling of the quadratic Zeeman term results in interaction energies of $\sim 0.5\ \text{cm}^{-1}$ at $n \approx 30$, comparable to those of the n–independent linear Zeeman term. Above $n \approx 30$, the diamagnetic Zeeman effect becomes dominant over first the linear Zeeman effect, then the intramolecular field, next the rotational (l–uncoupling term), and finally the monopole Coulomb ($\Delta E_n \approx 2\Re/n^3$) interaction. As for Rydberg atoms, a "strong field" situation occurs when the magnitude of the diamagnetic perturbation is on the order of the Coulomb structure (Kleppner, *et al.*, 1983). This occurs when

$$\frac{\langle H^Z_{\text{diam}} \rangle}{\Delta E_n} \propto n^7 B_Z^2 \approx 1$$

or, for B_Z =1 Tesla, at $n \approx 40$. The selection rules for off-diagonal matrix elements of the diamagnetic perturbation term are

$$\begin{aligned} &\Delta n \quad \text{unrestricted} \\ &\Delta l = 0, \pm 2 \\ &+ \not\leftrightarrow -. \end{aligned}$$

The $\Delta m_l = 0$ selection rule, which is rigorous for Rydberg atoms, is not valid for Rydberg molecules. The intramolecular electric field, which induces both l–mixing and m_l–mixing for penetrating Rydberg states, is in competition with the m_l–preserving Zeeman terms. However, for non-penetrating ($l \gtrsim 2$) states, the $\Delta m_l = 0$ selection rule is approximately valid.

The diamagnetic Zeeman interaction is responsible for l–mixing at moderate–n and for both l– and n–mixing at higher–n. For NO Rydberg states, the diamagnetic term causes both $p \sim f$ and $f \sim h$ mixing. The high–l nonpenetrating states have such small quantum defects that they are all nearly perfectly degenerate (except for the effects of the $\approx n^{-3} l^{-3} l_R^2$–scaling diagonal matrix elements of the interaction term between the Rydberg electron and the ion-core quadrupole moment. This core-quadrupole effect lifts all of the otherwise perfect Δl = even, $\Delta l_R = 0$ degeneracies). Thus, once the diamagnetic Zeeman interaction overpowers the quadrupole-core splittings, the diamagnetic perturbation is profoundly effective in destroying l among the high–n nonpenetrating Rydberg states. Gauyacq, *et al.*, (1999) have incorporated both linear and diamagnetic Zeeman effects in an MQDT treatment of Rydberg states of NO, similar to the model first used by Monteiro and Taylor (1990) for H_2 in a strong magnetic field.

An external magnetic field can be a sensitive probe or a profound modifier of the multiplicity of possible couplings between the Rydberg electron and the ion-core. The range of possible hierarchies of coupling is vastly larger for molecules than atoms. At high–n the diamagnetic Zeeman term becomes dominant over the intramolecular field and causes profound l– and n–mixings within each ion-core rotational–N^+ channel. However, the $B^+ N^+(N^+ + 1)$ rotational structure of the ion-core is negligibly affected by the magnetic perturbation. Profound modification of the rotational structure by an external magnetic field would require strong channel mixing due to the core-anisotropy. N^+–channel mixing is due to the axially symmetric intramolecular electric field (and is most pronounced at low–l). N–mixing is due to the axially symmetric external magnetic field. The two kinds of anisotropy, ion-core and external field, are responsible for complementary channel mixing effects.

6.5.5 Nonthermal Population Distributions; Chemical and Collisional Effects

Perturbations affect the rate of absorption and emission of radiation in a fully understood and exactly calculable manner. They also affect the rates of chemical and collisional population/depopulation processes, but in a less easily estimated way. Perturbation effects on steady-state populations can be very large and level-specific. Although collision-induced transitions and chemical reactions are not governed by rigorous selection rules as are electric dipole transitions and perturbation interactions, some useful propensity rules have been suggested theoretically and confirmed experimentally. Gelbart and Freed (1973) suggested that the cross sections for collision-induced transitions between two different electronic states, E and E′, are

$$\sigma_{\mathrm{E},J;\mathrm{E}',J'} \simeq \sigma_{\mathrm{E},J;\mathrm{E},J'} C_{\mathrm{E},\mathrm{E}'}(J')^2 + \sigma_{\mathrm{E}',J;\mathrm{E}',J'} C_{\mathrm{E},\mathrm{E}'}(J)^2 \qquad (6.5.39)$$

where $\sigma_{\mathrm{E},J;\mathrm{E}',J'}$ and $\sigma_{\mathrm{E}',J;\mathrm{E}',J'}$ are the $J \rightarrow J'$ purely rotation-changing cross-sections within the E and E′ electronic states, respectively, and $C_{\mathrm{EE}'}(J)$ is the isolated-molecule E, $J \sim$ E′, J mixing coefficient. In this model, collision-induced inter-electronic-state transfer is thus viewed as a rotational relaxation process (typical values for $\sigma_{\mathrm{E},J;\mathrm{E},J\pm 1}$ and $\Sigma_{J'}\ \sigma_{\mathrm{E},J;\mathrm{E},J'}$ are 10 and 100 Å^2, respectively) reduced by the fractional mixing in either the initial or final J–level. When only one rotation-vibration level is appreciably mixed, this level should act as a doorway through which all population flows on its way from one electronic state to the other. The terms "doorway" and "gateway" are both widely used in the literature to describe the same phenomenon. As discussed in a review of state-resolved collision-induced electronic transitions by Dagdigian (1997) and illustrated in particular for CN $\mathrm{A}^2\Pi \rightarrow \mathrm{X}^2\Sigma^+$ and N_2^+ $\mathrm{A}^2\Pi_u \rightarrow \mathrm{X}^2\Sigma_g^+$ transitions, collision-induced electronic transitions can also be enabled by mixing of the electronic states induced by the approach of the collision partner. For CN and N_2^+, the collision-induced mixing can be strong because of favorable Franck-Condon factors, as the outer limbs of the A and X state potential energy curves cross. In fact, collisional quenching rate constants which are a significant fraction of the gas-kinetic value have been measured for the excited electronic states of many diatomic molecules despite the absence of obvious spectroscopic perturbations.

The Gelbart-Freed model is based on the assumption that this direct, electronically inelastic process is significantly less probable than the perturbation-facilitated one. This is a reasonable assumption in the case of electronic transitions involving a change of electron spin multiplicity ($\Delta S \neq 0$) or a small vibrational overlap ($\langle v|v'\rangle < 10^{-3}$). The Gelbart-Freed model should thus be particularly applicable when the E and E′ vibronic levels are connected by a spin-forbidden transition and the $C_{\mathrm{EE}'}$ isolated-molecule mixing coefficients are nonzero. The validity of the model rests on two additional assumptions: the collision must be "sudden" and "weak." A collision is sudden (Mukamel 1979) when

$$\tau_{\mathrm{coll}} H_{\mathrm{E},v,J;\mathrm{E}',v',J'} < \hbar \tag{6.5.40}$$

where τ_{coll} is the collision duration [typically 10 Å/(5 $\times 10^4$ cm/s) = 2 ps, which puts an upper limit, $H_{\mathrm{E},v,J;\mathrm{E}'v,J'} << 3\ \mathrm{cm}^{-1}$, on the isolated-molecule perturbation matrix element]. If Eq. (6.5.40) is not satisfied, additional terms describing the evolution of the system during the lifetime of the complex must be taken into account. A collision is weak (Freed and Tric, 1978) when the relative shifts of the basis function energies, $E^0_{\mathrm{E},J} - E^0_{\mathrm{E}',J'}$ during the collision can be neglected. Such level shifts, resulting from differences in the strength of the interaction of the E and E′ electronic states with the collision partner, would cause the mixing coefficients to evolve over a collision trajectory because of the dependence of the interaction energies on the instantaneous distance between and relative orientations of the collision partners. After averaging over all collision trajectories, the effective mixing coefficients $\langle C_{\mathrm{EE}'}(J)\rangle$ could become quite different from the isolated-molecule coefficients $C_{\mathrm{EE}'}(J)$.

We can distinguish between spin-conserving and nonconserving electronically inelastic collision processes. This is due to the much stronger effect of a collision partner on the spatial (orbital) rather than the spin part of the electronic wavefunction. Below, we consider in detail the collision-induced mixing of $^2\Pi$ and $^2\Sigma^+$ electronic states. In the absence of isolated-molecule $\Delta S \neq 0$ mixing via $\mathbf{H}^{\mathrm{SO}}$, spin-changing ($\Delta S \neq 0$) collisions are not very probable except in the case of a heavy collision partner ("external heavy atom effect"), where $\mathbf{H}^{\mathrm{SO}}$ can cause significant spin-mixing in the collision complex. Since the collision-induced mixing of electronic states is usually small for $\Delta S \neq 0$ collision-induced electronic transitions, such transitions often show strong doorway effects, where the doorway is centered on an isolated-molecule perturbation.

It has been difficult to verify the predictions of the Gelbart-Freed (1973) model, or more generally to determine the importance of isolated-molecule spectroscopic perturbations in enabling collision-induced electronic transitions, at the state-to-state level. Measurement of the electronically inelastic E,$J \rightarrow$ E′, J' process can be frustrated by spectroscopic difficulties in preparing the initial level and probing the populations of final levels. Moreover, the rate of the electronically inelastic process must be measured in the presence of the more rapid rotationally inelastic E,$J \rightarrow$ E, J' and E′, $J \rightarrow$E′, J' processes. This is especially difficult when the radiative lifetime of one of the two electronic states is a factor of more than 10^2 longer than the other. Collisions equilibrate the rotational populations in the longer-lived state so rapidly that the specific $\sigma_{\mathrm{E},J;\mathrm{E}',J'}$ cross-section is not measurable by the usual selective, pulsed-excitation, time-gated dispersed fluorescence detection technique. Problems associated with rotational equilibration within the longer lived electronic state may be circumvented by the use of a pulsed OODR pump-probe scheme. By varying the delay between the pump and probe pulses, the evolution of rotational level populations may be sampled on a time scale far shorter than the radiative lifetime.

Dagdigian (1997) reviewed state-resolved studies of collision-induced electronic transitions carried out up to 1997, with a particular emphasis on understanding the role of spectroscopic perturbations in enabling this process. Three different types of collision experiments have been carried out: (1) those in which the initial vibronic level is selectively populated, (2) those in which the final vibronic level is selectively detected, and (3) those in which both the initial and final levels are resolved. The third type of experiment is the most informative in revealing the importance of spectroscopic doorways, but is the most demanding. Such experiments usually employ the OODR technique, and appropriate optical transitions must be available to prepare (populate or label) and detect the initial and final levels. The role of spectroscopic perturbations can be revealed in experiments which define only the initial or final level if there is a strong doorway effect in the system.

Early optically detected microwave transition linewidth studies (Pratt and Broida, 1969) suggested that CN $A^2\Pi(v = 10) \rightarrow B^2\Sigma^+(v = 0)$ collision-induced transitions through perturbed levels occur at gas-kinetic collision rates, faster than pure rotational relaxation. The pressure dependence of chemiluminescence in the Ba + N_2O system also implied BaO $b^3\Pi(v = 10) \rightarrow A^1\Sigma^+(v = 1)$ collisional transfer through spectroscopic doorways (Field, *et al.*, 1974).

Grimbert, *et al.*, (1978) found a monotonic but nonlinear relationship between the observed $\sigma_{A^1\Pi,J;T}$ (T = $a'^3\Sigma^-$, or $e^3\Sigma^-$, or $d^3\Delta$) cross-sections and the rotationally averaged mixing fractions for the $v = 0 - 7$ vibrational levels of CO $A^1\Pi$, $\overline{C^2}_{A^1\Pi,J:T,J}$ [calculated from the mixing fractions derived from the deperturbation eigenvectors of Field (1971)]. In this case, the isolated molecule inter-state coupling is so strong that Eq. (6.5.40) is not satisfied. A model proposed by Grimbert, *et al.*, (1978) explained the apparent saturation of $\sigma_{A^1\Pi,J:T}$ by taking into account the finite duration of the collision.

Ottinger and co-workers recorded dispersed fluorescence spectra when molecular beams of metastable electronic states impinged upon a scattering gas. In those experiments, the final levels are well defined, but a broad range of initial levels were populated. They observed collision-induced transitions within the triplet states of CO, in particular $a^3\Pi \rightarrow a'^3\Sigma^+, d^3\Delta$ (Ottinger, *et al.*, 1995a). For most collision partners and initial and final levels studied, the transitions did not proceed through spectroscopic doorways. However, formation of the $d^3\Delta(v = 9)$ level involved such a doorway, and the doorway mechanism was found to be more important for He than Xe collision partners. Perturbation-irrelevant N_2 $A^3\Sigma_u^+$, $W^3\Delta_u \rightarrow B^3\Pi_g$ collision-induced transitions were observed with this molecular beam technique (Bachmann, *et al.*, 1992, Ottinger, *et al.*, 1995b). Contributions from the two initial electronic states were separated by a time-of-flight method which is based on the substantial difference between the radiative lifetimes of the A and W states. Transitions out of specific vibrational levels in the $A^3\Sigma_u^+$ state were subsequently resolved by optical pumping state selection (Bachmann, *et al.*, 1993). In contrast to transitions among these triplet states, the N_2 $A'^5\Sigma_u^+(v=0) \rightarrow B^3\Pi_g(v=12)$ transition was found to proceed through spectroscopically perturbed levels (Ottinger, *et al.*, 1994). The Ottinger group also observed perturbation-facilitated $\Delta S \neq 0$ transitions in NO $[a^4\Pi(v=12) \rightarrow B^2\Pi(v=3)]$ (Ottinger and Vilesov, 1994) and NH $[b^1\Sigma^+(v=5) \rightarrow A^3\Pi(v=2)]$ (Mo, *et al.*, 1999). Perturbation-enhanced collisional electronic energy transfer in $[a^4\Pi(v=12) \rightarrow B^2\Pi(v=3)]$ transitions was also observed by Copeland, *et al.*, (1997) in experiments in which the $a^4\Pi(v=12)$ level was prepared by direct laser excitation from the $X^2\Pi$ ground state and the final vibrational levels were resolved in the dispersed fluorescence spectrum.

Collision-induced $\Delta S \neq 0$ transitions do not necessarily involve perturbation-facilitated transitions. A good example of a perturbation-irrelevant transition is NH $a^1\Delta \rightarrow X^3\Sigma^-$. The room-temperature NH(a) collisional quenching rate constants for Xe and CO are large (Adams and Pasternack, 1991, Hack and Rathman, 1992) and the quenching proceeds through this collision-induced electronic transition, despite the absence of identified a$\sim$ X spectroscopic perturbations. (In this case of CO on NH, physical quenching competes with the chemical reaction which forms H + NCO.) Patel-Misra and Dagdigian (1992) detected the collisionally populated final NH $X^3\Sigma^-(v=0-4)$ rotational levels in a molecular beam experiment. The rotational distributions were cold, reflecting a $\Delta J \approx 0$ propensity, with a near statistical distribution in the fine-structure levels. This contrasts sharply with the essentially complete absence of NH $X^3\Sigma^- F_2(J=N)$ levels in the spin-forbidden decomposition of $HN_3(\tilde{X})^1A'$ studied by Stephenson, *et al.*,(1988). The exclusive formation of only the symmetric F_1 and F_3 components was ascribed (Alexander, *et al.*, 1988) to a symmetry restriction in the unimolecular decomposition process mediated by $\mathbf{H}^{SO}$ (Yarkony, 1990).

Electronic $A^2\Pi_u(v=3,4) \rightarrow X^2\Sigma_g^+(v'=v+4)$ transitions in N_2^+ due to collisions with He have been observed, with initial and final level specificity, by the OODR technique by Katayama and co-workers (Katayama, 1984, 1985, Katayama and Dentamaro, 1989, Dentamaro and Katayama, 1991). The doorway model does not apply to this system since isolated-molecule A $\sim$ X perturbations are forbidden by the rigorous $g \not\leftrightarrow u$ selection rule, and collision-induced electronic transitions are enabled by electronic-state mixing induced by the collision partner. While ratios of initial and final state populations cannot be directly compared because of spectroscopic selection rules, the deduced rate constants were found to be large. A propensity toward small ΔJ changes was also observed. The s, a nuclear permutation symmetry is preserved in the collision, so that only even- or odd-N final rotational levels are populated, depending upon the symmetry of the initial state. As discussed below, Berning and Werner (1994) have carried out quantum scattering calculations with *ab initio* potential energy surfaces for N_2^+A $\rightarrow$X transitions. They find good agreement with the measured final rotational level distributions but disagree on the magnitude of the cross sections.

Dentamaro and Katayama (1989) have also employed the OODR method to study $A^2\Pi(v = 0) \rightarrow X^2\Sigma^+(v = 10)$ transitions in the isoelectronic CO^+ ion induced by collisions with He. In contrast to the He collision-induced transitions in N_2^+ and CN, these transitions are much less efficient because of a poor Franck-Condon factor. Significant collisional transitions are observed only for perturbed F_2f initial levels.

Dagdigian and co-workers (Furio, *et al.*, 1986; Jihua, *et al.*, 1986; Ali, *et al.*, 1987; Dagdigian, *et al.*, 1993; Nizamov, *et al.*, 2002) have conducted an extensive series of OODR studies on CN $A^2\Pi(v = 3, 7, 8) \rightarrow X^2\Sigma^+(v' = v + 4)$ transitions involving both low-J and high-J initial levels, induced by collisions with He and Ar. Perturbation-irrelevant collisional transfer was observed for low-J levels of all initial vibrational levels investigated. However, it should be noted that doorway effects involving the most perturbed $v = 7$ low-J initial levels could not be directly probed on strongly perturbed levels involving the $A^2\Pi(v = 7) \rightarrow X^2\Sigma^+(v = 11)$ transition because of spectral overlaps (Furio, *et al.*, 1986). These results on CN are discussed further below in conjunction with the comparison with quantum scattering calculations of the cross sections.

Electronic transitions have also been observed in "half-collision" experiments involving the predissociation of electronically excited CN $A^2\Pi(v = 3)$–Ne van der Waals complexes (Fei and Heaven, 1993; Lawrence, *et al.*, 1997). The upper $^2\Pi_{1/2}(v = 3)$ spin-orbit component decays by formation of free CN $A^2\Pi_{3/2}$, while the lower $^2\Pi_{3/2}$ component yields CN $X^2\Sigma^+(v = 7)$ fragments. In analogy with the bimolecular collision-induced transitions discussed in more detail below, the rotational level distributions displayed an even-odd alternation of the populations *vs.* N. One difference between the bimolecular *vs.* half-collision dynamics is that electronic relaxation is observed to be much less efficient in the latter process than spin-orbit relaxation, while these two processes occur at comparable rates in bimolecular collisions. Yang and Alexander (1997) found that the perturber-induced coupling of the $A^2\Pi$ and $X^2\Sigma^+$ states is small at the approximately T-shaped geometry of the van der Waals complex.

The OODR method has also been used to study collision-induced electronic transitions between singlet electronic states of the neutral nitrogen molecule, in particular N_2 $a^1\Pi_g \rightarrow a'^1\Sigma_u^-$ (Katayama and Dentamaro, 1992, Katayama, *et al.*, 1996). These transitions occur by mixing induced by the He or Ar collision partner, as isolated-molecule perturbations are prohibited by the $g \not\leftrightarrow u$ selection rule. The rate constants were found to be smaller than for N_2^+–He collisions, reflecting the weaker interactions in neutral systems.

Using 2-photon excitation and 1+1 resonance-enhanced multiphoton ionization detection, Sha and co-workers (Sha, *et al.*, 1995, Chen, *et al.*, 1996) have measured cross-sections for CO $A^1\Pi(v = 0) \rightarrow e^3\Sigma^-(v = 1)$ transitions. The transition rates were found to be enhanced at spectroscopic perturbations between these vibronic levels and, moreover, displayed interference effects similar to those displayed in calculations on CaO collisions, described below.

Li, *et al.*, (1992) used OODR to study collisional energy transfer from Na_2 $A^1\Sigma_u^+(v = 26)$. A dramatic doorway effect was observed through a spectroscopic perturbation with the $b^3\Pi_u(v = 28)$ level. Li, *et al.*, (1996) found that singlet

→ triplet transitions from Li_2 $F^1\Sigma_g^+$ and $G^1\Pi_g$ required singlet-triplet mixed doorway levels.

SiF $C^2\Delta$ →$B^2\Sigma^+$ and SiCl $B'^2\Delta$ →$B^2\Sigma^+$ transitions have been studied by McKendrick and co-workers (Singleton and McKendrick, 1993, Watson and McKendrick, 1994) by selective laser excitation and dispersed fluorescence detection of the final levels. The final rotational level distributions are broad and smooth, reflecting the fact that these transitions are induced by collision-induced mixing of the electronic states.

Collision-induced electronic transitions between ion-pair states of the iodine molecule have been investigated in several laboratories (Ubachs, *et al.*, 1993, Akopyan, *et al.*, 2001, Fecko, *et al.*, 2002). Initial levels in the $E(0_g^+)$ state were prepared by two-step laser excitation through the $B^3\Pi_{0e}$ state, and final levels were monitored by dispersed fluorescence. Both near resonant and broad vibrational level distributions were observed for transfer to the $D(0_u^+)$ state, depending upon the initial vibrational level and collision partner.

Bondybey and Miller (1978) and Katayama, *et al.*, (1979) proposed that the rates of electronically inelastic processes in the gas phase should follow a Franck-Condon rate law,

$$\sigma_{1,v_1;2,v_2} \approx \sigma_{12}^{\mathrm{el}} \langle v_1|v_2\rangle^2 e^{-\Delta E_{1,v_1;2,v_2}/kT} \tag{6.5.41}$$

similar to that proposed by Bondybey (1977) (see Section 6.5.7). The σ_{12}^{el} electronic factor is precisely the term assumed to be negligible by Gelbart and Freed (1973). Katayama's (1984, 1985) observations for N_2^+ A ↔ X transfer are consistent with neither a perturbation doorway model Eq. (6.5.39) nor an energy gap law Eq. (6.5.41).

Detailed theoretical treatments of perturbation-facilitated collision-induced electronic transitions have been carried out for only a few systems. Alexander (1982a) has proposed a dipolar model for electronically inelastic processes that exhibits limiting behavior consistent with both Eqs. (6.5.39) and (6.5.41). This model is based on a time-dependent Born approximation treatment of the leading term in the multipolar expansion of the collisional interaction potential. In the case of CaO ($A^1\Sigma^+$, $A'^1\Pi$, $a^3\Pi$)+ N_2O discussed by Alexander (1982a), this leading term is the interaction between the permanent electric dipole moment of N_2O and the electric dipole transition moment for transitions between isolated CaO molecular eigenstates. There are several contributions to the CaO transition moments: the permanent electric dipole moments of the $A^1\Sigma^+$, $A'^1\Pi$, and $a^3\Pi$ electronic states (μ_A, $\mu_{A'}$, and μ_a) and the A ↔ A′ dipole allowed transition moment $\mu_{AA'}$. A unique feature of Alexander's (1982a) model is that interference between transition amplitudes leads to an even stronger J–dependence of the transition probabilities than predicted by the Gelbart-Freed model (1973) [(Eq. (6.5.39))]. The solid lines in Fig. 6.24 depict the calculated rate constants for the process

$$\mathrm{CaO}(\text{`}a^3\Pi_{0e}, v\text{'}, J) + \mathrm{N_2O} \rightarrow \mathrm{CaO}(\text{`}A^1\Sigma^+, v = 0\text{'}, J+1) + \mathrm{N_2O},$$

where the '' denote the dominant character of a mixed state. The sharp upward spikes occur at the J-values where the $A^1\Sigma^+(v = 0)$ level crosses successive $a^3\Pi_{0e}$ vibrational levels. The sharp dips near each spike arise from destructive interference between the μ_A and μ_a contributions to the $J \to J+1$ transition amplitudes,

$$\sigma_{a,v_a,J;A,v_A=0,J+1} \propto \left|\langle \text{`}a^3\Pi_{0e}\text{'}, v_a, J|\mu|\text{`}A^1\Sigma^+\text{'}, 0, J+1\rangle\right|^2$$

$$\begin{aligned}\propto \Big| &\left[1 - C_{aA}(J+1)^2\right]^{1/2} C_{aA}(J)\mu_A \\ &- \left[1 - C_{aA}(J)^2\right]^{1/2} C_{aA}(J+1)\mu_a \\ &+ \left\{\left[1 - C_{aA}(J+1)^2\right]^{1/2}\left[1 - C_{aA}(J)^2\right]^{1/2} \right. \\ &\left. - C_{aA}(J)C_{aA}(J+1)\right\}\mu_{aA}\Big|^2. \end{aligned} \tag{6.5.42a}$$

where $C_{aA}(J)$ is the mixing coefficient for the $|A^1\Sigma^+, v = 0, J\rangle$ basis function in the nominal $|\text{`}a^3\Pi_{0e}, v\text{'}, J\rangle$ eigenfunction. Setting the dipole transition moment $\mu_{aA} = 0$ for a $\Delta S \neq 0$ transition, one obtains

$$\begin{aligned}\sigma_{a,v_a J;A,0,J+1} \propto\ & [1 - C_{aA}(J)^2]C_{aA}(J+1)^2\mu_a^2 + [1 - C_{aA}(J+1)^2]C_{aA}(J)^2\mu_A^2 \\ & - 2[1 - C_{aA}(J)^2]^{1/2}[1 - C_{aA}(J+1)^2]^{1/2}C_{aA}(J)C_{aA}(J+1)\mu_a\mu_A. \end{aligned} \tag{6.5.42b}$$

The mixing coefficient $C_{aA}(J)$ has opposite signs above and below the level crossing. This means that the $\mu_a\mu_A$ term will interfere constructively or destructively for all J-values *except* when J is below and $J+1$ is above the crossing. At this point, where the coefficients of the μ_a^2, μ_A^2, and $\mu_a\mu_A$ terms approach their maximum values of 1/4, 1/4, and 1/2, respectively, the sign of the interference effect will be opposite to that for all other values of J. Even more complicated interference patterns could occur at $A^1\Sigma^+ \sim A'^1\Pi$ perturbations because of the nonzero (but in the case of CaO, negligibly small) $\mu_{AA'}$ transition moment term. Note that, in the case when there is no state mixing ($C_{12} = 0$) and $\mu_{12} \neq 0$, as for example $N_2^+ A^2\Pi_u \to X^2\Sigma_g^+$, an equation of the form of Eq. (6.5.42a) reduces to $\sigma_{1,J;2,J+1} \propto \mu_{12}^2$.

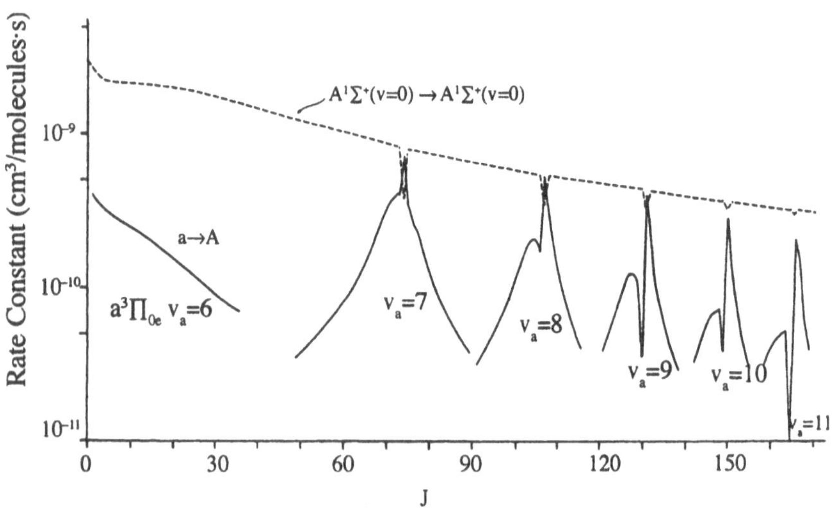

Figure 6.24: Calculated $J \to J+1$ rate constants for CaO $+N_2O$ collisions. The dotted curve describes transitions within the CaO nominal $A^1\Sigma^+(v=0)$ level. The rates for transitions between nominal $a^3\Pi_{0e}$ $(v = 6-11)$ and $A^1\Sigma^+(v=0)$ levels are presented as solid curves. The $A^1\Sigma^+$ state has a larger B-value than $a^3\Pi$, hence the $A^1\Sigma^+$ $(v=0)$ level overtakes $a^3\Pi_{0e}$ $(v=7)$ at $J \simeq 73$ and $(v=8)$ at $J \simeq 107$, etc. The downward (upward) spikes in the dotted (solid) curves occur at the level crossings. [From Alexander (1982a).]

Alexander's (1982a) transition dipole model would reduce to the Gelbart-Freed (1973) model if the $\mu_a\mu_A$ term in Eq. (6.5.42b) could be neglected ($\mu_a^2 >> \mu_A^2$, or $\mu_a^2 << \mu_A^2$) and if the transition moment term ($\mu_{AA'}$ or μ_{Aa}) were to vanish. Alternatively, one would obtain a model similar to Eq. (6.5.41) in the limit that both permanent moments vanish. The CN, CO^+, and N_2^+ $A^2\Pi \to X^2\Sigma^+$ transfer rates seem to follow Eq. (6.5.41) because μ_{AX} is larger than both μ_A and μ_X. Collisional transfer in a number of other systems, including CO $A^1\Pi \to T$ ($T = a'^3\Sigma^+, e^3\Sigma^-, d^3\Delta$), follow a doorway model because $\mu_{AT} = 0$ and $\mu_T > \mu_A$.

Quantum scattering calculations with *ab initio* computed potential energy surfaces have been carried out by Alexander, Werner, and co-workers (Werner, *et al.*, 1989; Dagdigian, *et al.*, 1993; Berning and Werner, 1994; Yang and Alexander, 1997; Alexander, *et al.*, 2000; Yang, *et al.*, 2000; Nizamov, *et al.*, 2001, 2002) for collision-induced electronic transitions for several exemplary systems, involving collisions of CN or N_2^+ with rare gases. Unlike the dipolar model (Alexander, 1982a) presented above, a number of terms in the multipolar expansion of the full potential energy surfaces (PESs) can contribute to inelastic transitions. Alexander and Corey (1986) have presented the formalism for the quantum treatment of collision-induced $^2\Pi \leftrightarrow {}^2\Sigma^+$ electronic transitions. The approach of a spherical collision partner (e.g. rare gas atom) lifts the electronic degeneracy of the Π state, and the interaction of this state can be described by two PESs, of A′ and A″ symmetry in C_s geometry. Likewise, the interaction of the Σ state can be described by an A′ PES. It is convenient for quantum scattering calculations to employ an electronically diabatic electronic basis to facilitate matching of the scattering and asymptotic atom-molecule wavefunctions. In this diabatic description of the electronic states, electrostatic coupling induced by the collision partner will occur between the two PESs of A′ symmetry. These couplings will occur independent of whether there are any perturbation-facilitated collisional pathways coupling the two electronic states, as for example in $N_2^+ A^2\Pi_u \leftrightarrow X^2\Sigma_g^+$ transitions (Katayama, 1984, 1985, Berning and Werner, 1994), for which the isolated-molecule perturbation matrix elements are zero. This collision-induced coupling of electronic states can also occur for $|\Delta\Lambda| > 1$ transitions, e.g. CH $A^2\Delta \leftrightarrow B^2\Sigma^-$ (Randall, *et al.*, 2000).

Alexander and co-workers (Dagdigian, *et al.*, 1993, Nizamov, *et al.*, 2002) have carried out quantum scattering calculations with the inclusion of the isolated-molecule perturbation matrix elements (Kotlar, *et al.*, 1980) for CN $A^2\Pi$ ($v = 3$ and 7) $\rightarrow X^2\Sigma^+(v' = v + 4)$ transitions in collisions with He and Ar. In this treatment, it was assumed that the $C_{AX}(J)$ mixing coefficients were not modified by the approach of the rare gas atom. While significant cross sections were found for transitions between unperturbed initial and final levels, the computed cross sections did exhibit enhancements for transitions involving strongly perturbed initial or final levels.

In the OODR experiments of Dagdigian and co-workers (Furio, *et al.*, 1986, Jihua, *et al.*, 1986) on CN $A^2\Pi$ ($v = 3$) $\rightarrow X^2\Sigma^+(v = 7)$ and $A^2\Pi$ ($v = 7$) $\rightarrow X^2\Sigma^+(v = 11)$ transitions from low-J initial rotational levels, the final rotational level populations displayed an even-odd alternation vs N, with the phase of the oscillations dependent upon the e/f parity of the initial level. The enhanced levels are those that would be expected for a strictly homonuclear diatom-atom interaction, as for the isoelectronic N_2^+ ion, and reflects the near homonuclear nature of the CN–rare gas PESs. More recently, the Dagdigian group used OODR to study collisions of high-N levels ($N \approx 60$) in CN $A^2\Pi(v = 3)$ to probe the effect of a spectroscopic perturbation at $J = 62.5$ (Nizamov, *et al.*, 2002). The perturbation-facilitated pathway was found to yield a significant enhancement in the total removal bimolecular rate constant in the most perturbed level in $A^2\Pi(v = 3)$. The most populated final rotational/fine-structure level was not the most strongly perturbed level in $X^2\Sigma^+(v = 7)$, in contrast to the predictions of the Gelbart-Freed model.

The rates of perturbation-facilitated collision-induced transitions between electronic states should exhibit strong dependences on initial and final states whenever $\mu_{12}^2 << \mu_1^2$ or μ_2^2 (within the dipolar model discussed above) and the perturbation matrix element is small so that only a few $C_{12}(J)$ are nonnegligible. This can give rise to highly non-Boltzmann steady-state population distributions, as has been observed in CN flames (Evenson and Broida, 1966, Savadatti and Broida, 1967). Schemes for utilizing or manipulating these unusual distributions include electronic transition chemical lasers, magnetic or electric field switching of perturbations via anticrossings, electronic-state-specific photochemistry, and isotope separation. A chemical laser, for example, could be designed around storage of the energy released in the initially formed, energetic, nonradiating levels (typically high vibrational levels of the electronic ground state) followed by collisional transfer through a "perturbation funnel" into a perturbed level from which emission is electric-dipole-allowed. Although no such laser has been developed, several pure rotational transitions in the CH_2 $\tilde{a}^1A_1$ state as well as a few nominal $\tilde{a}^1A_1 - \tilde{X}^3B_1$ transitions have been observed as stimulated emission in the laser magnetic spectrum (McKellar, *et al.*, 1983). The population inversions result from perturbation-facilitated rapid depopulation of the lower energy level.

The propensity rules for collision-induced transitions between electronic states and among the fine-structure components of non-$^1\Sigma^+$ states depend on the identity of the leading term in the multipole expansion of the molecule/collision-partner interaction potential. Alexander (1982a) has considered the dipole-dipole term, which included both permanent and transition dipole contributions. In the limit that first-order perturbation theory applies (not the usual circumstance for thermal molecular collisions), the following *collisional propensity rules for the permanent dipole term* can be enumerated from the selection rules for both perturbations and pure rotational transitions ($\mu_{\|}$):

$$\begin{array}{ll}
s \not\leftrightarrow a & \\
\Delta J = 0, \pm 1 & (\Delta J = 0 \text{ occurs only at low-}J \text{ in } \Omega \neq 0) \\
e \not\leftrightarrow f & (\text{except for } \Delta J = 0 \text{ at low-}J) \\
g \not\leftrightarrow u & \\
\Delta S = 0, \pm 1 & \\
\Delta\Omega = \Delta\Sigma = 0 & [\text{homogeneous perturbation, case (a) limit}] \\
\Delta\Omega = \Delta\Sigma = \pm 1 & [\text{heterogeneous perturbation, case (a) limit}].
\end{array}$$

The *collisional propensity rules for the transition dipole term* follow from the selection rules for $\mu_{\|}$ or $\mu_{\perp}$ electronic transitions:

$$s \not\leftrightarrow a$$
$$\Delta J = 0, \pm 1$$

(for $\mu_{\perp}$, $\Delta J = 0$ is stronger than $\Delta J = \pm 1$, and for $\mu_{\|}$, $\Delta J \neq 0$ except at low–J in $\Omega \neq 0$)

$$\begin{array}{ll}
+ \leftrightarrow - & \\
\Sigma^+ \not\leftrightarrow \Sigma^- & \\
\Delta S = 0 & \\
\Delta\Omega = \Delta\Lambda = \pm 1, 0 & [\text{case (a) limit}] \\
\Delta J = \Delta N & [\text{case (b) limit}] \\
g \leftrightarrow u. &
\end{array}$$

The most important difference between the permanent and transition moment terms are $e \not\leftrightarrow f$ versus $+ \leftrightarrow -$ and $g \not\leftrightarrow u$ versus $g \leftrightarrow u$.

Propensity rules for collision-induced transitions with non–$^1\Sigma^+$ states have been discussed by Gottscho (1981), Nedelec and Dufayard (1984), Alexander

(1982b) for $^2\Sigma$ and Alexander (1982c) for $^2\Pi$ states, Alexander and Dagdigian (1983 for $^3\Sigma$ states and 1984 for Λ–doublet transitions in Π states), and Alexander and Pouilly (1983) for $^3\Pi$ states. Dagdigian, *et al.*, (1989) considered differences in Λ–doublet propensities for $^2\Pi$ states in the Hund's case (b) limit for electronic states of π vs. π^3 orbital occupancy.

Experimental studies with resolution of the initial and final levels of rotationally inelastic collisions have been carried out for a number of molecules in non–$^1\Sigma^+$ states, and these studies have been reviewed by Dagdigian (1995). For some systems, the experimentally derived state-to-state cross sections have been compared with results from quantum scattering calculations on the basis of *ab initio* potential energy surfaces. Many of the propensity rules deduced from a formal analysis of the interaction with a collision partner are displayed. For collisions of high–J rotational levels of molecules in $^2\Pi$ states, rather strong Λ–doublet propensities can be seen because of interference between the scattering flux on the A′ and A″ potential energy surfaces (Dagdigian, *et al.*, 1989), which describe the interaction of the orbitally degenerate $^2\Pi$ state with the collision partner. Moreover, for such high–J levels, the preferred final Λ–doublet level is often better predicted by consideration of the crossings of adiabatic bender effective potential energy curves emanating from the asymptotic rotational levels, rather than examination of the coupling between the initial and final levels due to the interaction with the collision partner (Alexander and Dagdigian, 1994, Yang, *et al.*, 2000).

6.5.6 "Deperturbation" at High Pressure and in Matrices

The purpose of a deperturbation calculation is to define a computational model that faithfully reproduces every detail of observable spectra in terms of deperturbed potential energy curves and the interaction terms that are responsible for the perturbation. Deperturbation calculations can be extremely complicated. When faced with such a calculation, one is tempted to doubt whether a specific set of deperturbed potentials has any physical significance.

Perhaps one of the most convincing demonstrations that molecules know about deperturbed potential curves and that these curves are not arbitrary constructs that result from an abstract and prejudiced computation is the phenomenon of pressure-induced configuration demixing (Miladi, *et al.*, 1975, 1978). The electronic spectrum of the NO molecule is especially well suited to illustrate this demixing phenomenon because of the enormous structural difference between valence states (typically ω_e = 1000 cm^{-1}, B$_e$ = 1.1 cm^{-1}) and Rydberg states (typically ω_e = 2400 cm^{-1}, B$_e$ = 2.0 cm^{-1}) and the numerous valence~Rydberg perturbations. These perturbations among the NO $^2\Pi$ states (B and L valence states, C, K, and Q Rydberg states, see Section 6.2.2) have been treated by Gallusser and Dressler (1982).

Because of the considerably larger size [Eq. (5.2.1)] of Rydberg ($\langle r\rangle_{3p}$ = 3.7 Å) than valence ($\langle r\rangle_{2p}$ = 2.6 Å) orbitals, Rydberg states have much larger pressure broadening cross-sections than valence states, as shown in the following data at 5 bar neon (Miladi, *et al.*, 1978):

Rydberg	$D^2\Sigma^+ - X^2\Pi$	(0,0)	FWHM=1.95 cm^{-1}
Valence	$B^2\Pi - X^2\Pi$	(9,0)	FWHM=0.80 cm^{-1}
50/50 Mixed	$B^2\Pi(v=15) \sim C^2\Pi(v=3) - X^2\Pi(v=0)$		FWHM=1.30 cm^{-1}
	$C^2\Pi(v=3) \sim B^2\Pi(v=15) - X^2\Pi(v=0)$		FWHM=1.30 cm^{-1}

Miladi, *et al.*, (1975, 1978) have shown that, as a sample of NO is pressurized by an admixture of an inert gas, the Rydberg transitions are progressively broadened and shifted until, by ~ 1000 bar, the Rydberg states have all but vanished from the spectrum. This pressure-induced deperturbation process can be carried essentially to completion in inert-gas matrices (Boursey and Roncin, 1975; Boursey, 1976), where the frequencies and intensities of the observed bands correspond almost perfectly (except for a constant energy shift) with the deperturbed (Gallusser and Dressler, 1982; Jungen, 1966) quantities for valence ← $X^2\Pi$ absorption bands. Table 6.4 and Fig. 6.25 show that the $B^2\Pi - X^2\Pi$ bands from 5 K Ne and Ar matrices display regular vibrational intervals and smoothly varying intensities, in marked contrast to the behavior of the observed gas-phase B–X bands. Buxton and Duley (1970) have observed, in the reflectance spectrum from solid N_2, a change from the gas-phase pattern of perturbations of the N_2 $b^1\Pi_u$ valence state by the $c^1\Pi_u$ and $o^1\Pi_u$ Rydberg states (Stahel *et al.*, 1983). Despite the large deperturbed oscillator strength of the $c^1\Pi_u - X^1\Sigma_g^+$ 0 – 0 band (Stahel, *et al.*, 1983), there appears to be no trace of this band nor any other transitions into Rydberg levels in the solid N_2 spectrum. However, unlike the situation for the NO $B^2\Pi - X^2\Pi$ system, the deperturbation is not complete.

Table 6.4: The NO $B^2\Pi_{1/2} - X^2\Pi_{1/2}$ Transition in the Gas Phase and in Inert Gas Matrices

	Observed[a] gas phase		Deperturbed[b]		Observed[c] Ne matrix, 5 K		Observed[c] Ar matrix, 5 K	
v_B	E_{gas}	$f \times 10^{4d}$	E^b_{dep}	$I^{c,e}$	$E_{obs} - E_{dep}$	I^e	$E_{obs} - E_{dep}$	I
5	50,452[f]	0.2	50,479[f]	–	-101[f]	–	-200[f]	–
6	51,405	0.4	51,437	2.7	-108	0.9	-221	–
7	52,347	3.6	52,381	3.5	-102	1.5	-200	3.1
8	53,274	1.2	53,312	5.3	-95	2.0	-186	3.8
9	54,181	3.4	54,231	6.9	-104	5.5	-203	5.4
10	55,091	0.4	55,136	8.2	-103	6.5	-203	7.0
11	55,961	3.5	56,028	9.3	-93	7.8	-207	7.7
12	56,737	24.5	56,905	9.8	-89	8.4	-196	9.2
13	57,746	0.1	57,769	10.0	-82	9.1	-202	9.5
14	58,539	1.5	58,618	9.8	-91	10.0	-200	10.0
15	59,644	7.1	59,451	9.5	-90	9.6	-193	4.7
16	60,313	0.6	60,268	8.9	-81	8.1	-163	3.1
17	61,075	0.0	61,068	8.2	-85	7.8	-189	7.2
18	61,862	0.8	61,850	7.5	-95	5.6	(-216)[g]	1.3
19	62,592	0.0	62,613	6.7	-105	2.6	–	0.8

[a]Tabulated as E_0 and F in Table II of Gallusser and Dressler (1982).
[b]Tabulated as H in Table II of Gallusser and Dressler (1982).
[c]From Chergui and Schwentner (1992)
[d] f is the oscillator strength.
[e] I is a relative intensity normalized to $I = 10$ for the most intense band.
[f]Energy units are reciprocal centimeters.
[g]Uncertain value.

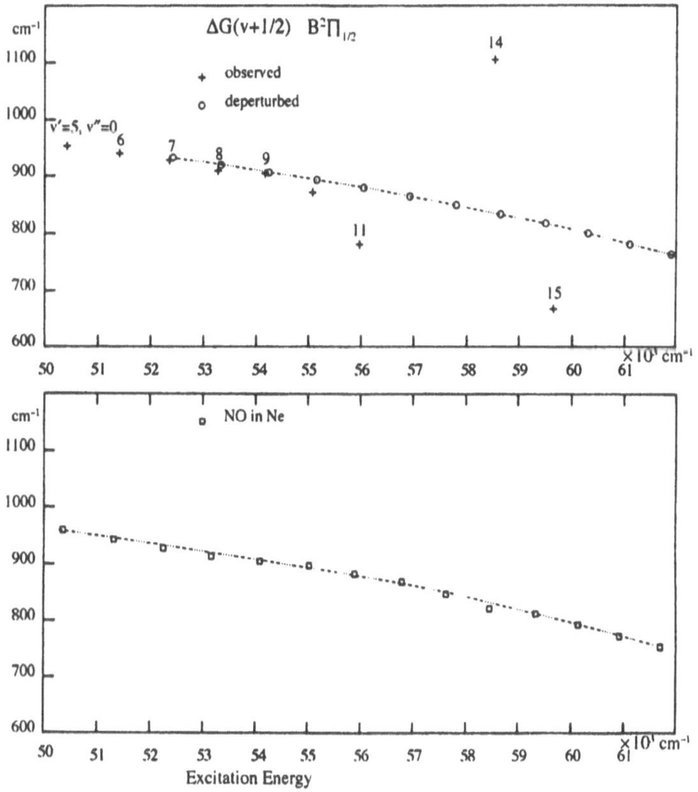

Figure 6.25: Comparison of observed (+) and deperturbed (○) ΔG values for the NO $B^2\Pi_{1/2}$ state with ΔG values observed in a 5 K Ne matrix (□). [From Boursey and Roncin (1975).]

The pressure-induced and inert-gas matrix demixing phenomenon is not general, but it illustrates that the computational deperturbation process recovers physically significant potential curves. Demixing depends on the existence of two classes of interacting basis states characterized by orbitals of drastically different spatial extent. Two valence states will not be systematically demixed in a matrix, but differential matrix shifts can alter perturbation patterns (level shifts and intensity borrowing). For valence~Rydberg perturbations, the ease of demixing depends on the strength of the H_{12} perturbation interaction in the free molecule. The Rydberg state must either be broadened (Γ_{matrix}) or differentially shifted (δE_{matrix}) so that $\Gamma_{\text{matrix}} \gg H_{12}$ or $\delta E_{\text{matrix}} \gg H_{12}$. In no sense do the Rydberg states cease to exist; either they become so broad as to disappear from the high-resolution spectrum, or they are shifted to a higher (or lower) energy region.

Chergui, *et al.*, (1985) have used synchrotron radiation to observe the NO $C^2\Pi$ state in an argon matrix and found the $C^2\Pi(v = 1)$ level near 61,500 cm^{-1}

above $v = 17$ of the $B^2\Pi$ state, as compared to the near degeneracy of $C^2\Pi(v = 1)$ with $v = 7$ of $B^2\Pi$ state in the gas phase. This huge blue shift (~ 6700 cm^{-1}) can be explained by repulsive overlap between the spatially extended Rydberg orbital and the closed shell rare gas atoms (Gross and Spiegelmann, 1998). Above $v = 17$ of the $B^2\Pi$ state, new perturbations show up in the matrix spectrum due to accidental degeneracies with vibrational levels of the blue-shifted $C^2\Pi$ state.

The $E^2\Sigma^+(4s\sigma)$ Rydberg state, which lies above $D^2\Sigma^+(3p\sigma)$, is shifted to the red in a xenon matrix. The higher Rydberg states ($n \geq 5$), which are red-shifted, are strongly compressed in an energy region just below the ionization limit and are therefore not clearly resolved (Vigliotti and Chergui, 2000).

The repulsive interaction between a Rydberg orbital and closed-shell rare gas atoms leads, in case of the $A^2\Sigma^+(3s\sigma)$ state of NO, which has largely spherical character, to the formation of a "microcavity" or "bubble" (Goodman and Brus, 1978b). Owing to the finite time required to form such a microcavity, delayed fluorescence from the $A^2\Sigma^+$ state occurs, which shows a lifetime lengthening with respect to the gas phase (Vigliotti, *et al.*, 1998).

6.5.7 Matrix Effects

When a diatomic molecule (guest) is present as a dilute impurity in an inert matrix (host), the selection rules for perturbations and electric dipole allowed transitions can be altered by guest-host interactions. For example, the inevitable absence of cylindrical symmetry ($C_{\infty v}$ or $D_{\infty h}$) at a matrix site destroys the distinction between π and δ orbitals; thus $\Delta\Lambda = 2$ transitions [S_2 and SO $A'^3\Delta - X^3\Sigma^-$ and $c^1\Sigma^- - a^1\Delta$ (Lee and Pimentel, 1978, 1979); NO $^2\Phi - X^2\Pi$ (Chergui, *et al.*, 1988); N_2 $w^1\Delta_u - X^1\Sigma_g^+$ (Kunsch and Boursey, 1979)] and perturbations (Goodman and Brus, 1977) are quite common.

The precise nature of a matrix site may be inferred from the occurrence of processes that are forbidden in the gas phase, the removal of a gas-phase degeneracy (e.g., lifting of e/f–orbital degeneracy in $\Lambda > 0$ states), or observation of polarization behavior (in amorphous matrices, the degree of fluorescence polarization resulting from linearly polarized excitation or, in single-crystal hosts, the dependence of the absorption cross-section for linearly polarized radiation on the orientation of the crystal axes).

The relaxation of Σ^+/Σ^- symmetry restrictions implies a site in which there are no reflection planes containing the molecular axis. Relaxation of $\Delta\Lambda$ selection rules need not imply destruction of Σ^+/Σ^- restrictions. For example, although the O_2 $C^3\Delta_u - X^3\Sigma_g^-$ and $c^1\Sigma_u^- - a^1\Delta_g$ transition strengths are considerably enhanced in an inert matrix [for $c^1\Sigma_u^-$, Rossetti and Brus (1979) report $\tau_{\text{radiative}} \approx 2 \times 10^{-4}s$ versus 25 s gas phase], the Σ^+/Σ^- forbidden $A^3\Sigma_u^+ - X^3\Sigma_g^-$ transition does not appear as a sharp zero phonon line except when the $A^3\Sigma_u^+$ ($v = 4$) level is perturbed by $C^3\Delta_u$ ($v = 6$) [in an N_2 matrix (Goodman and Brus, 1977)].

Removal of the distinction between $\Lambda = 0$ and $\Lambda > 0$ states (and the lifting of e/f degeneracy in $\Lambda > 0$ states) implies a site with less than threefold rotation symmetry about the molecular axis (except D_{2d}). Violations of g, u symmetry restrictions are unusual in inert gas atomic matrices, but should be commonplace in polar molecular lattices. Spin selection rules are broken by the "external heavy atom effect," whereby xenon is much more effective than argon in making spin-forbidden transitions observable.

A rich variety of perturbation-mediated energy transfer phenomena are observed in inert matrices. Bondybey (1977) has shown that $B^2\Sigma^+ \leftrightarrow A^2\Pi \leftrightarrow X^2\Sigma^+$ internal conversion processes in CN are governed by an exponential energy gap law and can be much more rapid than vibrational relaxation within a single electronic state (Fig. 6.26). B~X and A~X perturbation strengths and vibrational overlap factors do not appear to be as important as the amount of energy that must be converted into lattice phonons. Conversely, Goodman and Brus (1978a) and Rossetti and Brus (1979) have shown that the energy gap law is violated for the anomalously slow spin-forbidden processes, of Fig. 6.27,

$$\begin{array}{lll} {}^{15}N^{16}O & B^2\Pi_{1/2}(v=0) \rightarrow a^4\Pi_{5/2}\ (v=8) & \sim +20\ \text{cm}^{-1} \\ {}^{18}O_2 & C^3\Delta_u(v=0) \rightarrow c^1\Sigma^-\ (v=2) & \sim +60\ \text{cm}^{-1} \end{array}$$

An unusually good energy match is required before these intersystem relaxation paths compete with more rapid processes such as spin-relaxation within a multiplet state, vibrational relaxation, or spontaneous fluorescence (NO B→X). It is interesting that the $B^2\Pi_{1/2} \rightarrow a^4\Pi_{5/2}$ $\Delta\Omega = 2$ process seems to require a better energy match than $\Delta\Omega = \pm 1$ or 0 processes even though the quantum numbers Ω, Λ, and Σ are destroyed by the noncylindrical matrix site.

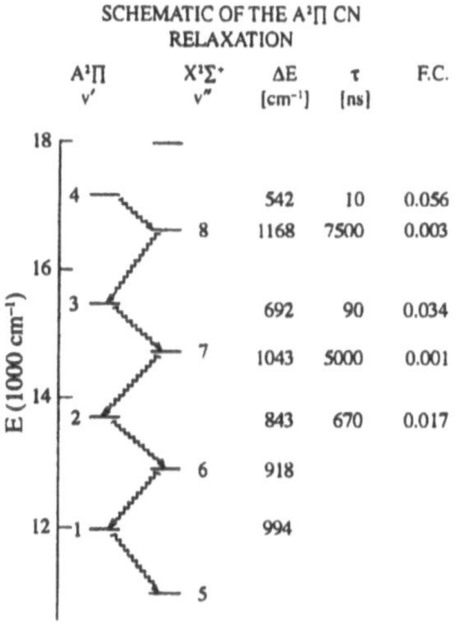

Figure 6.26: Schematic of CN $A^2\Pi \leftrightarrow X^2\Sigma^+$ relaxation in a neon matrix. Radiationless relaxations follow the A↔X path shown. The energy gap (ΔE), observed lifetime (τ), and Franck-Condon factors (F.-C.) are also displayed. [From Bondybey (1977).]

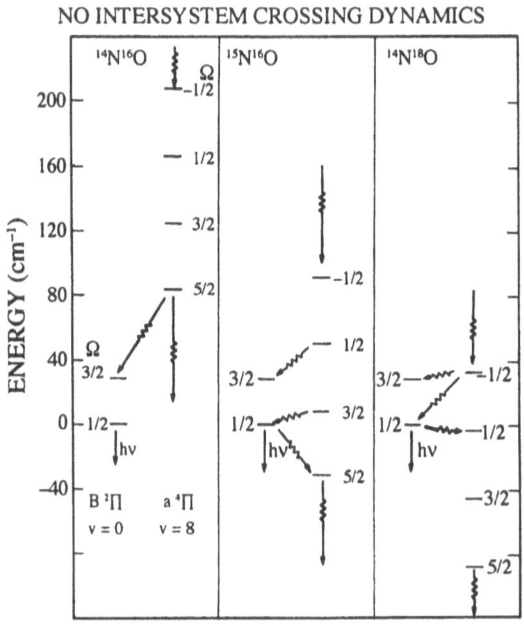

Figure 6.27: NO $a^4\Pi \rightarrow B^2\Pi$ intersystem crossing mechanism in Ar at 5 K. Thick (thin) arrows represent high (low) quantum yield processes. Fluorescence from $^{15}N^{18}O$ $B^2\Pi_{1/2}$ is strong because of the inefficiency of $B^2\Pi_{1/2}$–$a^4\Pi_{5/2}$ transfer. Fluorescence form $^{14}N^{18}O$ $B^2\Pi$ is anomalously weak because it receives population *via* the slow $a^4\Pi_{-1/2} \rightarrow B^2\Pi_{3/2}$ process and is efficiently depopulated by transfer from $B^2\Pi_{1/2}$ into the slightly lower-lying $a^4\Pi_{1/2}$ level. The $a^4\Pi$ levels shown are calculated from the constants of Vichon, *et al.*, (1978) ($T_e = 38,698$ cm^{-1}, $\omega_e = 1000$ cm^{-1}, $\omega_e x_e = 10.5$ cm^{-1}), A= 41.5 cm^{-1}, and a matrix shift of +135 cm^{-1} for $a^4\Pi$ relative to $B^2\Pi$. [Revised from Goodman and Brus (1978a).]

6.6 References

Adams, J. S., and Pasternack, L. (1991), *J. Phys. Chem.* **95**, 2975.

Akopyan, M. E., Bibnov, N. K., Kokh, D. B., Pravilov, A. M., Sharova, O. L., and Stepanov, M. B. (2001), *Chem. Phys.* **263**, 459.

Alexander, M. H. (1982a), *J. Chem. Phys.* **76**, 429.

Alexander, M. H. (1982b), *J. Chem. Phys.* **76**, 3637.

Alexander, M. H. (1982c), *J. Chem. Phys* **76**, 5974.

Alexander, M. H. and Dagdigian, P. J. (1983), *J. Chem. Phys.* **79**, 302.

Alexander, M. H. and Pouilly, B. (1983), *J. Chem. Phys.* **79**, 1545.

Alexander, M. H. and Dagdigian, P. J. (1984), *J. Chem. Phys.* **80**, 4325.

Alexander, M. H., and Corey, G. C. (1986), *J. Chem. Phys.* **84**, 100.

Alexander, M. H., Werner, H.-J., and Dagdigian, P. J. (1988), *J. Chem. Phys.* **89**, 1388.

Alexander, M. H., and Dagdigian, P. J. (1994), *J. Chem. Phys.* **101**, 7468.

Alexander, M. H., Yang, M., Dagdigian, P. J., Berning, A., and Werner, H.-J. (2000), *J. Chem. Phys.* **112**, 781.

Ali, A., Jihua, G., and Dagdigian, P. J. (1987), *J. Chem. Phys.* **87**, 2045.
Amitay, Z., Zajfman, D., Forck, P., Hechtfischer, U., Seidel, B., Grieser, M., Habs, D., Repnow, R., Schwalm, D., and Wolf, A. (1996), *Phys. Rev. A* **54**, 4032.
Appleblad, O., Lagerqvist, A., Renhorn, I., and Field, R. W. (1981), *Phys. Scr.* **22**, 603.
Arrowsmith, P., Jones, W. J., and Tuckett, R. P. (1981), *J. Mol. Spectrosc.* **86**, 216.
Ashfold, M. N. R. (1986), *Mol. Phys.* **58**, 1.
Bachem, A. (1920), *Z. Phys.* **3**, 372.
Bachman, R., Li, X., Ottinger, C., and Vilesov, A. F. (1992), *J. Chem. Phys.* **96**, 5151.
Bachman, R., Li, X., Ottinger, C., Vilesov, A. F., and Wulfmeyer, V. (1993), *J. Chem. Phys.* **98**, 8606.
Banic, J. R., Lipson, R. H., Efthimiopoulos, T., and Stoicheff, B. P. (1981), *Opt. Lett.* **6**, 461.
Barrow, R. F., Dixon, R., Lagerqvist, A., and Wright, C. (1960), *Ark. Fys.* **18**, 543.
Berning, A., and Werner, H.-J. (1994), *J. Chem. Phys.* **100**, 1953
Bethke, G. W. (1959), *J. Chem. Phys.* **31**, 662.
Bondybey, V. E. (1977), *J. Chem. Phys.* **66**, 995.
Bondybey, V. E., and Miller, T. A. (1978), *J. Chem. Phys.* **69**, 3597.
Boursey, E., and Roncin, J.-Y. (1975), *J. Mol. Spectrosc.* **55**, 31.
Boursey, E. (1976), *J. Mol. Spectrosc.* **61**, 11.
Bray, R. G., Hochstrasser, R. M., and Sung, H. N. (1975), *Chem. Phys. Lett.* **33**, 1.
Bray, R. G. and Hochstrasser, R. M. (1976), *Mol. Phys.*, **31**, 1199.
Brewer, R. G., and Shoemaker, R. L. (1971), *Phys. Rev. Lett.* **27**, 631.
Brewer, R. G., and Genack, A. Z. (1976), *Phys. Rev. Lett.* **16**, 959.
Brieger, M., Hese, A., Renn, A., and Sodeik, A. (1980), *Chem. Phys. Lett.* **76**, 465.
Buxton, R. A. H., and Duley, W. W. (1970), *Phys. Rev. Lett.* **25**, 801.
Calamai, A. G., and Yoshino, K. (1994), *J. Chem. Phys.* **101**, 9480.
Carroll, T. (1937), *Phys. Rev.* **52**, 822.
Chen, X., Sha, G., Jiang, B., He, J., and Zhang, C. (1996), *J. Chem. Phys.* **105**, 8661.
Chergui, M., Schwentner, N., Böhmer, W., and Haensel, R. (1985), *Phys. Rev. A* **31**, 527.
Chergui, M., Schwentner, N., and Chandrasekharan, V. (1988), *J. Chem. Phys.* **89**, 7094.
Chergui, M., and Schwentner, N. (1992), *J. Chem. Phys.* **97**, 2881.
Cohen-Tannoudji, C., Diu, B., and Laloë, F. (1977), "Quantum Mechanics", John Wiley and Sons, New York.
Cook, T. J., and Levy, D. H. (1973a), *J. Chem. Phys.* **59**, 2387.
Cook, T. J., and Levy, D. H. (1973b), *J. Chem. Phys.* **58**, 3547.
Copeland, R. A., Dyer, M. J., Bloemink, H. I., and Slanger, T. G. (1997), *J. Chem. Phys.* **107**, 2257.
Cossart, D., Horani, M., and Rostas, J. (1977), *J. Mol. Spectrosc.* **67**, 283.
Cossart, D., Cossart-Magos, C., Vervloet, M., and Bergeman, T., (1994), *J. Mol. Spectrosc.* **163**, 587.
Crawford, F. H. (1929), *Phys. Rev.* **33**, 341.
Crawford, F. H. (1934), *Rev. Mod. Phys.* **6**, 90.
Dagdigian, P. J., Alexander, M. H., and Liu, K. (1989), *J. Chem. Phys.* **91**, 839.
Dagdigian, P. J., Patel-Misra, D., Berning, A., Werner, H.-J., and Alexander, M. H. (1993), *J. Chem. Phys.* **98**, 8580.
Dagdigian, P. J. (1995), p. 315 in "The Chemical Dynamics and Kinetics of Small Radicals," Part I (Eds. K. Liu and A. Wagner), World Scientific Press.
Dagdigian, P. J. (1997), *Ann. Rev. Phys. Chem.* **48**, 95.

Dentamaro, A. V., and Katayama, D. H. (1989), *J. Chem. Phys.* **90**, 91.
Dentamaro, A. V., and Katayama, D. H. (1991), *Phys. Rev. A* **43**, 1306.
Dixon, R. N., Bayley, J. M., and Ashfold, M. N. R. (1984), *Chem. Phys.*, **84**, 21.
Douglas, A. E. (1966), *J. Chem. Phys.* **45**, 1007.
Dieke, G. H. (1941), *Phys. Rev.* **60**, 523.
Dressler, K. (1969), *Can. J. Phys.* **47**, 547.
Dressler, K. (1970), Journées d'études sur les intensités dans les spectres électroniques, Paris (unpublished).
Dressler, K., Jungen, C., and Miescher, E. (1981), *J. Phys. B* **14**, L701.
Dufayard, J., Negre, J. M., and Nedelec, O. (1974), *J. Chem. Phys.* **61**, 3614.
Edmonds, A. R. (1974), "Angular Momentum in Quantum Mechanics", Princeton University Press, Princeton, New Jersey.
Edwards, S. A., Tchang-Brillet, W.-Ü. L., Roncin, J.-Y., Launay, F., and Rostas, F. (1995), *Planet. Sci.* **43**, 67.
Eidelsberg, M., and Rostas, F. (2003), *Ap. J. Supplement* **145**, 89.
Ernst, W. E., and Törring, T. (1982), *Phys. Rev. A* **25**, 1236.
Evenson, K. M., Dunn, J. L, and Broida, H. P. (1964), *Phys. Rev. A* **136**, 1566.
Evenson, K. M., and Broida, H. P. (1966), *J. Chem. Phys.* **44**, 1637.
Fairbairn, A. F. (1970), *J. Quant. Spectrosc. Radiat. Transfer* **10**, 1321.
Fecko, C. J., Freedman, M. A., and Stephenson, J. C. (2002), *J. Chem. Phys.* **116**, 1361.
Fei, S., and Heaven, M. C. (1993), *J. Chem. Phys.* **98**, 753.
Field, R. W. (1971), Ph.D. thesis, Harvard University, Cambridge, Massachusetts.
Field, R. W., and Bergeman, T. H. (1971), *J. Chem. Phys.* **54**, 2936.
Field, R. W., Bradford, R. S., Broida, H. P., and Harris, D. O. (1972), *J. Chem. Phys.* **57**, 2209.
Field, R. W., English, A. D., Tanaka, T., Harris, D. O., and Jennings, D. A. (1973), *J. Chem. Phys.* **59**, 2191.
Field, R. W., Jones, C. R., and Broida, H. P., (1974), *J. Chem. Phys.* **60**, 4377.
Field, R. W., Benoist d'Azy, O., Lavollée, M., Lopez-Delgado, R., Tramer, A. (1983), *J. Chem. Phys.* **78**, 2838.
Field, R. W., Clevenger, J. O., and Dodd, J. A. (2003), *J. Chem. Phys.* **XXX**, xxxx.
Fortrat, R. (1913), *Compt. Rend.* **156**, 1452.
Fredrickson, W. R., and Stannard, C. R. (1933), *Phys. Rev.* **44**, 632.
Freed, K. R., and Tric, C., (1978), *Chem. Phys.* **33**, 249.
Freedman, P. A. (1977), *Canad. J. Phys.* **55**, 1387.
Furio, N., Ali, A., and Dagdigian, P. J. (1986), *J. Chem. Phys.* **85**, 3560.
Gallusser, R., and Dressler, K. (1982), *J. Chem. Phys.* **76**, 4311.
Garetz, B. A. and Kittrell, C. (1984), *Phys. Rev. Lett.* **53**, 156.
Garetz, B. A., Kittrell, C., and Le Floch, A. C. (1991) *J. Chem. Phys.* **94**, 843.
Gauyacq, D., Fredin, S., and Jungen, C. (1987), *Chem. Phys.* **117**, 457.
Gauyacq, D., Raoult, M., and Shafizadeh, N. (1999), p. 393 in "The Role of Rydberg States in Spectroscopy and Photochemistry" (Ed. C. Sándorfy), Kluwer Academic, Dordrecht.
Gaydon, A. G. (1944), *Proc. Roy. Soc. London* **182**, 286.
Gelbart, W. M., and Freed, K. F. (1973), *Chem. Phys. Lett.* **18**, 470.
German, K., and Zare, R. (1969), *Phys. Rev. Lett.* **23**, 1207.
Girard, B., Billy, N., Vigué, J., and Lehmann, J. C. (1982), *Chem. Phys. Lett.* **92**, 615.
Goodman, J., and Brus, L. E. (1977), *J. Chem. Phys.* **67**, 1482.
Goodman, J., and Brus, L. E. (1978a), *J. Chem. Phys.* **69**, 1853.
Goodman, J., and Brus, L. E. (1978b), *J. Chem. Phys.* **69**, 4083.

Göppert-Mayer, M. (1931), *Annalen Phys.* **9**, 273.
Gottscho, R. A., and Field, R. W. (1978), *Chem. Phys. Lett.* **60**, 65.
Gottscho, R. A., Koffend, J. B., Field, R. W., and Lombardi, J. R. (1978), *J. Chem. Phys.* **68**, 4110.
Gottscho, R. A. (1979), *J. Chem. Phys.* **70**, 3554.
Gottscho, R. A. (1981), *Chem. Phys. Lett.* **81**, 66.
Gottscho, R. A., Field, R. W., Dick, K. A., and Benesch, W. (1979), *J. Mol. Spectrosc.* **74**, 435.
Gouedard, G., and Lehmann, J. C. (1976), *J. Phys. B* **9**, 2113.
Gouedard, G., and Lehmann, J. C. (1979), *J. Phys. (Orsay, Fr.) Lett.* **40**, L119.
Gouedard, G., and Lehmann, J. C. (1981), *Faraday Discuss. Roy. Soc. Chem.* **71**, 11.
Grimbert, D., Lavollée M., Nitzan, A., and Tramer, A. (1978), *Chem. Phys. Lett.* **57**, 45.
Gross, M. and Spiegelmann, F. (1998), *Euro. Phys. J. D.* **4**, 219.
Guizard, S., Shafizadeh, N., Chapoulard, D., Horani, M., and Gauyacq, D. (1991), *Chem. Phys.* **156**, 509.
Hack, E., and Huber, J. R. (1991), *Int. Revs. Phys. Chem.* **10**, 287.
Hack, W., and Rathman, K. (1992), *J. Phys. Chem.* **96**, 47.
Hahn, E. L. (1950), *Phys. Rev.* **80**, 580.
Harris, S. M., Gottscho, R. A., Field, R. W., and Barrow, R. F. (1982), *J. Mol. Spectrosc.* **91**, 35.
Havriliak, S. J., and Yarkony, D. R. (1985), *J. Chem. Phys.* **83**, 1168.
Heidner, R. F., Sutton, D. G., and Suchard, S. N. (1976), *Chem. Phys. Lett.* **37**, 243.
Helm, H., Hazell, I., and Bjerre, N. (1993), *Phys. Rev. A* **48**, 2762.
Herzberg, G., and Jungen, C. (1982), *J. Chem. Phys.* **77**, 5876.
Hilborn, R. C. (1982), *Am. J. Phys.* **50**, 982.
Hilborn, R. C. (2002), http://xxx.lanl.gov/pdf/physics/0202029.
Hippler, M., and Pfab, J. (1998), *Mol. Phys.* **94**, 313.
Hippler, M. (1999), *Mol. Phys.* **97**, 105.
Hougen, J. T., (1970), "The Calculation of Rotational Energy Levels and Rotational Line Intensities in Diatomic Molecules," Nat. Bur. Stand. (U.S.), Monograph 115.
Huber, K. P., Vervloet, M., Jungen, C., and Roche, A. L. (1987), *Mol. Phys.* **61**, 501.
Imhof, R. E., and Read, F. H. (1971), *Chem. Phys. Lett.* **11**, 326.
James, T. C. (1971a), *J. Chem. Phys.* **55**, 4118.
James, T. C. (1971b), *J. Mol. Spectrosc.* **40**, 545.
Jihua, G., Ali, A., and Dagdigian, P. J. (1986), *J. Chem. Phys.* **85**, 7098.
Johns, J. W. C. (1974), in "Molecular Spectroscopy, Specialist Periodical Report", (R. F. Barrow, D. A. Long and D. J. Millen, eds.) Vol. 2, p. 513, The Chemical Society, London.
Johns, J. W. C., and Lepard, D. W. (1975), *J. Mol. Spectrosc.* **55**, 374.
Jongma, R. T., Berden, G., and Meijer, G. (1997), *J. Chem. Phys.* **107**, 7034.
Jost, R., and Lombardi, M. (1974), *Phys. Rev. Lett.* **33**, 53.
Jungen, C. (1966), *Can. J. Phys.* **44**, 3197.
Jungen, C., and Miescher, E. (1969), *Can. J. Phys.* **47**, 1769.
Jungen, C., Dabrowski, I., Herzberg, G., and Kendall, D. J. W. (1989), *J. Chem. Phys.* **91**, 3926.
Jungen, C. and Roche, A. L. (1999), *J. Chem. Phys.* **110**, 10784.
Katayama, D. H., Miller, T. A., and Bondybey, V. E. (1979), *J. Chem. Phys.* **71**, 1662.
Katayama, D. H., Miller, T. A., and Bondybey, V. E. (1980), *J. Chem. Phys.* **72**, 5469.
Katayama, D. H., (1984), *J. Chem. Phys.* **81**, 3495.

Katayama, D. H., (1985), *Phys. Rev. Lett.* **54**, 657.
Katayama, D. H., and Dentamaro, A. V. (1989), *J. Chem. Phys.* **91**, 4571.
Katayama, D. H., and Dentamaro, A. V. (1992), *J. Chem. Phys.* **97**, 2820.
Katayama, D. H., Dentamaro, A. V., and Welsh, J. A. (1996), *J. Phys. Chem.* 100, 7854.
Kittrell, C., and Garetz, B. (1989), *Spectrochim. Acta* **45A**, 31.
Kittrell, C., LeFloch, A. C., and Garetz, B. A. (1993), *J. Phys. Chem.* **97**, 2221.
Kleppner, D., Littman, M. G., and Zimmerman, M. L. (1983), in "Rydberg States of Atoms and Molecules," (R. F. Stebbings and F. B. Dunning, eds.) p. 73, Cambridge University Press.
Klynning, L. (1974), in "Atoms, Molecules and Lasers", p. 464, International Atomic Energy Agency, Vienna.
Kotlar, A. J., Field, R. W., Steinfeld, J. I., and Coxon, J. A. (1980), *J. Mol. Spectrosc.* **80**, 86.
Kopp, I., and Hougen, J. T. (1967), *Can. J. Phys.* **45**, 2581.
Kunsch, P. L., and Boursey, E. (1979), *J. Chem. Phys.* **70**, 731.
Kuo, C. H., Wyttenbach, T., Beggs, C. G., Kemper, P. R., and Bowers, M. T. (1990), *J. Chem. Phys.* **92**, 4849.
Lagerqvist, A., and Miescher, E. (1958), *Helv. Phys. Acta* **31**, 221.
Lambropoulos, P. (1976), "Topics on Multiphoton Processes in Atoms," *Adv. At. Mol. Phys.* **12**, 87.
Larsson, M. (1983), *Astron. Astrophys.* **128**, 291.
Lawrence, W. G., Chen, Y., and Heaven, M. C. (1997), *J. Chem. Phys.* **107**, 7163.
Lee, Y., Yoon, Y., Baek, S. J., Joo, D.-L., Ryu, J.-S., and Kim, B. (2000), *J. Chem. Phys.* **113**, 2116.
Lee, Y.-P., and Pimentel, G. C. (1978), *J. Chem. Phys.* **69**, 3063.
Lee, Y.-P., and Pimentel, G. C. (1979), *J. Chem. Phys.* **70**, 692.
Lehmann, K. K., and Coy, S. L. (1984), *J. Chem. Phys.* **81**, 3744.
Lehmann, K. K., and Coy, S. L. (1985), *J. Chem. Phys.* **83**, 3290.
Levy, D. H. (1972), *J. Chem. Phys.* **56**, 5493.
Lewis, B. R., Gibson, S. T., Copeland, R. A., and Bressler, C. G. (1999), *Phys. Rev. Lett.* **82**, 4212.
Li Li and Field, R. W. (1983), *J. Phys. Chem.* **87**, 3020.
Li Li, Zhu, Q., Lyyra, A. M., Whang, T.-J., Stwalley, W. C., Field, R. W., and Alexander, M. H. (1992), *J. Chem. Phys.* **97**, 8835.
Li Li, Antonova, S., Yiannopolou, A., Urbanski, K., and Lyyra, A. M. (1996), *J. Chem. Phys.* **105**, 9859.
Lofthus, A. (1957), *Can. J. Phys.* **35**, 216.
Lombardi, M. (1988), p. 163 in "Excited States: Rotational Effects on the Behaviour of Excited Molecules, Vol. 7," Academic Press, San Diego.
Maeda, M., and Stoicheff, B. P. (1984), "Proc. Laser Techniques in the Extreme Ultraviolet" (S. E. Harris and T. B. Lucatorto, eds.), American Institute of Physics, New York.
Maïnos, C., Le Duff, Y., and Boursey, E. (1985), *Mol. Phys.* **56**, 1165.
Mandich, M. L., Gaebe, C. E., and Gottscho, R. A. (1985), *J. Chem. Phys.* **83**, 3349.
McClain, W. M., and Harris, R. A. (1977), p. 1 in "Excited States," Vol. 3 (Lim, E. C., editor), Academic Press, New York.
McKellar, A. R. W., Bunker, P. R., Sears, T. J., Evenson, K. M., Saykally, R. J., and Langhoff, S. R. (1983), *J. Chem. Phys.* **79**, 5251.
Metz, F., Howard, W. E., Wunsch, L., Neusser, H. J., and Schlag, E. W. (1978), *Proc.*

Roy. Soc. Lond. A **363**, 381.
Miladi, M., le Falher, J.-P., Roncin, J.-Y., and Damany, H. (1975), *J. Mol. Spectrosc.* **55**, 81.
Miladi, M., Roncin, J.-Y., and Damany, H. (1978), *J. Mol. Spectrosc.* **69**, 260.
Miller, T. A. (1973), *J. Chem. Phys.* **58**, 2358.
Miller, T. A., and Freund, R. S. (1974), *J. Chem. Phys.* **61**, 2160.
Miller, T. A., and Freund, R. S. (1975), *J. Chem. Phys.* **63**, 256.
Mo, Y., Ottinger, Ch., and Shen, G. (1999), *J. Chem. Phys.* **111**, 4598.
Moehlmann, J. G., Hartford, Jr., A., and Lombardi, J. R. (1972), *J. Chem. Phys.*.**57**, 4764.
Monteiro, T. S., and Taylor, K. T. (1990), *J. Phys. B* **23**, 427.
Moore, C. A., Davis, G. P., and Gottscho, R. A. (1984), *Phys. Rev. Lett.* **52**, 538.
Mukamel, S. (1979), *Chem. Phys. Lett.* **60**, 310.
Mulliken, R. S. (1964), *J. Am. Chem. Soc.* **86**, 3784.
Murray, J. E., Yoshino, K., Esmond, J. R., Parkinson, W. H., Sun, Y., Dalgarno, A., Thorne, A. P., and Cox, G. (1994), *J. Chem. Phys.* **101**, 62.
Nedelec, O., and Dufayard, J. (1984), *Chem. Phys.* **84**, 167.
Nevin, T. E. (1942), *Proc. Roy. Irish Acad.* **48A**, 1.
Nevin, T. E. (1945), *Proc. Roy. Irish Acad.* **50A**, 123.
Nikitin, E. E. and Zare, R. N. (1994), *Mol. Phys.* **82**, 85.
Nizamov, B., Dagdigian, P. J., and Alexander, M. H. (2001), *J. Chem. Phys.* **115**, 8393.
Nizamov, B., Yang, X., Dagdigian, P. J., and Alexander, M. H. (2002), *J. Phys. Chem. A* **106**, 8345.
Noda, C., and Zare, R. N. (1982), *J. Mol. Spectrosc.* **95**, 254.
Ogi, Y., Takahashi, M., Tsukiyama, K., and Bersohn, R. (2000), *Chem. Phys.* **255**, 379.
Ottinger, Ch., Smirnova, L. G., and Vilesov, A. F. (1994), *J. Chem. Phys.* **100**, 4848.
Ottinger, Ch., and Vilesov, A. F. (1994), *J. Chem. Phys.* **100**, 1805.
Ottinger, Ch., Vilesov, A. F., and Xu, D. D. (1995a), *J. Chem. Phys.* **96**, 1673.
Ottinger, Ch., Vilesov, A. F., and Xu, D. D. (1995b), *J. Phys. Chem.* **99**, 15642.
Parker, A. E. (1933), *Phys. Rev.* **44**, 84.
Patel-Misra, D., and Dagdigian, P. J. (1992), *J. Chem. Phys.* **97**, 4871.
Pratt, D. W., and Broida, H. P. (1969), *J. Chem. Phys.* **50**, 2181.
Provorov, A. C., Stoicheff, B. P., and Wallace, S. (1977), *J. Chem. Phys.* **67**, 5393.
Pugh, L. A., and Rao, K. N. (1976) p. 165 in "Molecular Spectroscopy: Modern Research." Vol. II (Rao, K. N., editor), Academic Press, New York.
Radford, H. E. (1961), *Phys. Rev.* **122**, 114.
Radford, H. E. (1962), *Phys. Rev.* **126**, 1035.
Radford, H. E. (1964), *Phys. Rev. A* **136**, 1571.
Radford, H. E., and Broida, H. P. (1962), *Phys. Rev.* **128**, 231.
Radford, H. E., and Broida, H. P. (1963), *J. Chem. Phys.* **38**, 644.
Randall, C. J., Murray, C., and McKendrick, K. G. (2000), *Phys. Chem. Chem. Phys.* **2**, 461.
Raoult, M., Le Rouzo, H., Raşeev, G., and Lefebvre-Brion, H., (1983), *J. Phys. B* **16**, 4601.
Renhorn, I. (1980), *Mol. Phys.* **41**, 469.
Ring, H., Carter, R. T., and Huber, J. R. (1998), *Eur. Phys. J. D* **4**, 73.
Ring, H., Carter, R. T., and Huber, J. R. (1999), *Laser Physics* **9**, 253.
Rossetti, R., and Brus, L. E. (1979), *J. Chem. Phys.* **71**, 3963.
Rostas, F., Eidelsberg, M., Jolly, A., Lemaire, J. L., Le Floch, A., and Rostas, J. (2000), *J. Chem. Phys.* **112**, 4591.

Savadatti, M. I., and Broida, H. P. (1967), *J. Chem. Phys.* **45**, 2390
Schadee, A. (1978), *J. Quant. Spectrosc. Radiat. Transfer* **19**, 451.
Schamps, J. (1973), Thesis, Université des Sciences et Techniques de Lille, Lille, France.
Schmid, R., and Gerö, L. (1935), *Z. Phys.* **94**, 386.
Schweda, H. S., Renn, A., and Hese, A. (1980), *Symp. Mol. Spectrosc. 35th, Columbus, Ohio.* Talk MF8.
Schweda, H. S., Renn, A., Büssener, H. and Hese, A. (1985), *Chem. Phys.* **98**, 157.
Sha, G., He, J., Jiang, B., and Zhang, C. (1995), *J. Chem. Phys.* **102**, 2772.
Silvers, S., Bergeman, T., and Klemperer, W. (1970), *J. Chem. Phys.* **52**, 4385.
Singleton, S., and McKendrick, K. G. (1993), *J. Phys. Chem.* **97**, 1389.
Slanger, T. G., and Black, G. J. (1971), *J. Chem. Phys.* **55**, 2164.
Solarz, R., and Levy, D. H. (1973), *J. Chem. Phys.* **58**, 4026.
Stahel, D., Leoni, M., and Dressler, K., (1983), *J. Chem. Phys.* **79**, 2541.
Stark, G., Smith, P. L., Huber, K. P., Yoshino, K., Stevens, M. H., and Ito, K. (1992), *J. Chem. Phys.* **97**, 4809.
Steimle, T. C., Brazier, C. R., and Brown, J. M. (1982), *J. Mol. Spectrosc.* **91**, 137.
Stephenson, J. C., Casassa, M. P., and King, D. S. (1988), *J. Chem. Phys.* **89**, 1378.
Sykora, T., and Vidal, C. R. (1998), *J. Chem. Phys.* **108**, 6320.
Tamanis, M., Auzinsh, M., Klincare, I., Nikolayeva, O., Stolyarov, A. V., and Ferber, R. (1997), *J. Chem. Phys.* **106**, 2195.
Tanaka, T., Field, R. W., and Harris, D. O. (1974), *J. Chem. Phys.* **61**, 3401.
Tatum, J. B. (1967), *Astrophys, J. Suppl.* **14**, 21.
Tilford, S. G., and Wilkinson, P. G. (1964), *J. Mol. Spectrosc.* **12**, 231.
Tellinghuisen, J. (1984), *J. Mol. Spectrosc.***103**, 455.
Townes, C. H., and Schawlow, A. L. (1955), "Microwave Spectroscopy", McGraw-Hill, New York.
Ubachs, W., Aben, I., Milan, J. B., Somsen, G. J., Stuiver, A. G., and Hogervorst, W. (1993), *Chem. Phys.* **174**, 285.
Varberg, T. D., Field, R. W., and Merer, A. J. (1991), *J. Chem. Phys.* **95**, 1563.
Varberg, T. D., Gray, J. A., Field, R. W., and Merer, A. J. (1992), *J. Mol. Spectrosc.* **156**, 296.
Vervloet, M. and Watson, J. K. G. (1986), *Can. J. Phys.* **64**, 1529.
Vichon, D., Hall, R. I., Gresteau, F., and Mazeau, J. (1978), *J. Mol. Spectrosc.* **69**, 341.
Vient, A., Shafizadeh, N., Fillion, J. H., Gauyacq, D., Horani, M., and Lemaire, J. L. (1995), *J. Phys. Chem.* **99**, 1666.
Vigliotti, F., Zerga, G., Chergui, M., and Rubayo-Soneira, J. (1998), *J. Chem. Phys.* **109**, 3508.
Vigliotti, F. and Chergui, M. (2000), *Euro. Phys. J. D.* **10**, 379.
Vigué, J., Broyer, M., and Lehmann, J. C. (1974), *J. Phys. B* **7**, L158.
Vigué, J., Broyer, M., and Lehmann, J. C. (1975), *J. Chem. Phys.* **62**, 4941.
Walter, C. W., Cosby, P. C., and Helm, H. (2000), *J. Chem. Phys.* **112**, 4621.
Watson, C. W., and McKendrick, K. G. (1994), *Chem. Phys.* **187**, 79.
Watson, J. K. G. (1968), *Can. J. Phys.* **46**, 1637.
Watson, W. W. (1932), *Phys. Rev.* **41**, 378; **42**, 509.
Wayne, F. D., and Colbourn, E. A. (1977), *Mol. Phys.* **34**, 1141; erratum in (1984) **51**, 531.
Werner, H.-J., Follmeg, B., and Alexander, M. H. (1989), *J. Chem. Phys.* **91**, 5425.
Whiting, E. E., and Nicholls, R. W. (1974), *Astrophys. J. Suppl.* **27**, 1.
Whiting, E. E., Schadee, A., Tatum, J. B., Hougen, J. T., and Nicholls, R. W. (1980), *J. Mol. Spectrosc.* **80**, 249.

Wilkinson, P. G., and Houk, N. B. (1956), *J. Chem. Phys.* **24**, 528.
Wood, R. W., and Hackett, F. E. (1909), *Astrophys. J.* **30**, 339.
Wu, M., and Johnson, P. M. (1989), *J. Chem. Phys.* **91**, 7401.
Yang, M., and Alexander, M. H. (1997), *J. Chem. Phys.* **107**, 7148.
Yang, X., Dagdigian, P. J., and Alexander, M. H. (2000), *J. Chem. Phys.* **112**, 4474.
Yarkony, D. R. (1986), *J. Chem. Phys.* **85**, 7261.
Yarkony, D. R. (1990), *J. Chem. Phys.* **92**, 320.
Yoshino, K., Freeman, D. E., and Tanaka, Y. (1979), *J. Mol. Spectrosc.* **76**, 153.
Zare, R. N., (1988), "Angular Momentum, Understanding Spatial Aspects in Chemistry and Physics", (Wiley Interscience Inc., New York).

Chapter 7

Photodissociation Dynamics

Contents

7.1 Photofragmentation 470

7.2 Direct Dissociation 471

7.2.1 Photodissociation: Wigner-Witmer Rules 471

7.2.2 Photodissociation Cross Sections 476

7.2.3 Photofragment Branching Ratios for Photodissociation 485

7.2.4 Photofragment Angular Distribution 486

7.2.5 Alignment of the Photofragment 491

7.3 Introduction to Predissociation 493

7.4 Experimental Aspects of Predissociation 495

7.4.1 Measurement of Lifetimes 496

7.4.2 Measurement of Linewidths 498

7.4.3 Energy Shifts 503

7.4.4 Detection of Fragments 505

7.5 Theoretical Expressions for Widths and Level Shifts 505

7.6 The Vibrational Factor 510

7.7 Mulliken's Classification of Predissociations . . . 514

7.8 The Electronic Interaction Strength 518

7.8.1 Electrostatic Predissociation 519

7.8.2 Spin-Orbit Predissociation 520

7.8.3 Rotational or Gyroscopic Predissociation 521

7.8.4 Hyperfine Predissociation 521

7.9 Fano Lineshape 522

7.10 Isotope Effects 526

7.11 Examples of Predissociation 528

7.11.1 Examples of Spin-Orbit Predissociation 528

7.11.2 Examples of Nonadiabatic Predissociation 531

7.12 Case of Intermediate Coupling Strength 535
7.13 Indirect (Accidental) Predissociation and Interference Effects 538
7.14 Some Recipes for Interpretation 543
7.15 References 545

7.1 Photofragmentation

Photodissociation and photoionization are two aspects of a more general process: photofragmentation (Greene and Zare (1982)). Photodissociation of a diatomic molecule

$$\mathrm{AB} + h\nu \rightarrow \mathrm{AB}^* \rightarrow \mathrm{A} + \mathrm{B} \tag{7.1.1}$$

results in the formation of two atoms. The rate, the partitioning of excess energy between translation and internal excitation of either or both atoms, the expression in product atom anisotropies and vector correlations of the spatial anisotropies of the optical excitation and the interaction between separating atoms, and the mechanism of the photodissociation process will be the subject of present chapter. Photoionization

$$\mathrm{AB} + h\nu \rightarrow \mathrm{AB}^{**} \rightarrow \mathrm{AB}^+ + e^- \tag{7.1.2}$$

results in the formation of an electron and a molecular ion and will be the subject of Chapter 8. Dynamical parameters such as branching ratios, photofragment angular distributions, polarizations, and vector correlations are defined identically for photodissociation and photoionization. Dissociative photoionization

$$\mathrm{AB} + h\nu \rightarrow \mathrm{AB}^{**} \rightarrow \mathrm{A}^+ + \mathrm{B}^* + e^- \tag{7.1.3}$$

is a process where the photofragments are an atom (possibly electronically excited), an atomic-ion, and an electron. Three particle vector correlation experiments are beginning, for example, the reaction

$$\mathrm{NO} + h\nu \rightarrow \mathrm{NO}^+(c^3\Pi) + e^- \rightarrow \mathrm{N}^+(^3P) + \mathrm{O}(^3P) + e^- \tag{7.1.4}$$

(Lafosse, *et al.*, 2000), (see also, for H_2, Sanchéz and Martín, 1999, for NaI, Charron and Suzor-Weiner, 1998, and for O_2, Lafosse, *et al.*, 2001).

The properties measured in photofragmentation experiments (photodissociation or photoionization) may be divided into *scalar* quantities, such as photofragmentation cross sections and branching ratios, and *vectorial* quantities, such as angular distributions of the photofragments and their alignment or orientation. Moreover, these properties may be measured independently or "in coincidence" where two or more properties associated with either or both photofragmentation products are measured simultaneously, event-by-event.

7.2 Direct Dissociation

The distinction between direct dissociation processes discussed in the present section and indirect dissociation or predissociation processes discussed in Section 7.3 to Section 7.14 is that in a direct process photoexcitation occurs from a bound state (typically $v = 0$ of the electronic ground state) directly to a repulsive state (or to an energy region above the dissociation asymptote of a bound state) whereas in an indirect process the photoexcitation is to a nominally bound vibration-rotation level of one electronically excited state which in turn is "predissociated" by perturbative interaction with the continuum of another electronic state. Direct dissociation, often termed a "half collision" is much faster and dynamically simpler than indirect dissociation. In a direct dissociation process the distance between atoms increases monotonically and the time required for the two atoms to separate is shorter than a typical vibrational or rotational period (Beswick and Jortner, 1990).

The discussion of direct dissociation begins (Section 7.2.1) with a framework for deciding which dissociation channels are energetically open, what is the single dissociation channel that would be followed if the atoms were to separate on the single initially-photoselected adiabatic potential curve, and what are the expected product branching ratios that would result if the dissociation were so fast (diabatic) that changes in the electronic wavefunction of the parent molecule are minimized. In other words, the adiabatic and diabatic correlations provide two extreme models for what might be expected, thereby providing a framework for recognizing behavior that is surprising and therefore deserving of further study. In Section 7.2.2 the electronic selection rules and free-bound Franck-Condon distributions are discussed with the goal of showing how the intensity pattern in a spectrum provides information about the repulsive potential energy curves. The branching ratios into atomic L-S-J states are shown, in Section 7.2.3, to provide information about whether the dissociation process follows the diabatic (crossing) vs. adiabatic (non-crossing) potential energy curves. The branching ratios are *scalar quantities*, averaged over all spatial directions for the departing atoms (relative to both the initial orientation of the parent molecule and the polarization of the exciting radiation) and over the M_J levels of both atoms. In Section 7.2.4 the information provided about the dissociation mechanism by *vector quantities* and *multi-property correlations*, such as the photofragment angular distribution and photofragment imaging experiments is reviewed.

The photodissociation cross sections and angular distributions can be obtained experimentally by different methods: for example, time of flight (TOF) photofragment translational spectroscopy (Busch, *et al.*, 1969), Doppler analysis (Xu, *et al.*, 1986), and ion imaging.

7.2.1 Photodissociation: Wigner-Witmer Rules

The non-crossing rule requires that adiabatic potential energy curves which belong to electronic states of the same symmetry species (same Λ, S for nonrelativistic curves, same Ω for relativistic curves) do not cross. The Wigner-Witmer rules (see Herzberg, 1950 p. 315) specify what molecular symmetry species result when two atoms in definite L, M_L, S, M_S or L, S, J, M_J states are brought together from $R = \infty$. Since there are a large number of possible states of two combining atoms, the Wigner-Witmer rules and the insights they provide into photodissociation will be illustrated by example.

When atom-1 approaches atom-2, the axial electric field created by atom-1 lifts the $2L+1$ M_L–degeneracy of atom-2 and vice-versa. Here M_L is the projection of **L** into the internuclear axis and is related to the molecular Λ quantum number as

$$\Lambda = |M_{L_1} + M_{L_2}| . \tag{7.2.1}$$

The spin is not affected by the axial electric field, thus the atomic spins, S_1 and S_2, couple to form the molecular spin quantum number as

$$\vec{\mathbf{S}} = \vec{\mathbf{S}}_1 + \vec{\mathbf{S}}_2 \tag{7.2.2a}$$

$$S = S_1 + S_2, S_1 + S_2 - 1, \ldots |S_1 - S_2| . \tag{7.2.2b}$$

An example of this *weak spin-orbit* limit nonrelativistic adiabatic correlation is given by the HBr^+ system. The lowest energy dissociation asymptote is $Br^+(^3P) + H(^2S)$, thus $S_1 = 1, S_2 = 1/2$ and S can be 3/2 (quartet state) or 1/2 (doublet state); $M_{L_1} = 0, \pm 1, M_{L_2} = 0$ and $\Lambda = 0$ or 1. The $\Lambda = 0$ possibility gives a $^4\Sigma$ and a $^2\Sigma$ state, but Σ–states must be classified according to Σ^+/Σ^- symmetry (see Section 3.2.2, Eq. (3.2.89)). Each separated atom asymptote will give rise to an odd number of molecular Σ–states of each S value (spin multiplicity); all but one of these Σ–states occur in non-degenerate Σ^+, Σ^- pairs, and the Σ^+/Σ^- symmetry of the left over Σ–state is determined by the even/oddness of $L_1 + \sum_i l_{i1} + L_2 + \sum_i l_{i2}$. Since the electronic configuration of Br^+ ^3P is $(4p)^4$, the unique Σ–states must be of Σ^- symmetry. The $\Lambda = 1$ possibility gives a $^4\Pi$ and a $^2\Pi$ state. Thus the states of HBr^+ which correlate adiabatically to the lowest energy dissociation limit, $H(^2S)+Br^+(^3P)$, are $X^2\Pi$, $^4\Pi, ^4\Sigma^-$, and $^2\Sigma^-$. Note that it is the lowest energy state of each of these four symmetries that correlates adiabatically to the lowest energy dissociation limit.

A good way to check that the adiabatic correlation has been done correctly is to check whether the total degeneracy of the molecular states is equal to that of the atomic states. In the case of $HBr^+ \to H(^2S) + Br^+(^3P)$, the molecular degeneracy is $4+8+4+2 = 18$ and the atomic degeneracy is $2 \times 3 \times 3 = 18$.

The $Br^+(4p)^4$ configuration also gives rise to ^1D and ^1S states. The H+Br^+ (^1D) limit gives $^2\Delta, ^2\Pi$, and $^2\Sigma^+$ (total degeneracy $4+4+2 = 2 \times 5$) and the H + $Br^+(4p^4\ ^1S)$ limit gives only $^2\Sigma^+$. However, the $H(^2S)+Br^+(4p^4\ ^1S)$ limit lies at higher energy than Br $(^2P) + H^+(^1S)$, which gives a $^2\Pi$ and a $^2\Sigma^+$ state (total degeneracy $4+2 = 2 \times 3 \times 1$). The first and third HBr^+ $^2\Sigma^+$ states correlate with H + Br^+ in the $(4p)^4$ configuration, but the second $^2\Sigma^+$ state correlates with H^++$Br(4p^5\ ^2P)$.

Because of the strength of the spin-orbit coupling in Br^+, it is more appropriate to consider the *relativistic adiabatic correlations* in the *intermediate spin-orbit* limit. When the spin-orbit splitting(s) of the separated atom states is comparable to the energy separations between the different $(L_1, S_1) + (L_2, S_2)$ states, it is necessary to perform the adiabatic correlations with respect to each of the separated atom J states. For HBr^+, the spin-orbit splittings of Br^+ ^3P are large $\left(^3P_2 - {}^3P_1 = 3139\ cm^{-1}, {}^3P_1 - {}^3P_0 = 3840\ cm^{-1}\right)$ and it is necessary to use the (J_1, J_2) correlation scheme (Herzberg 1950, p.319) (see Table 5.17) to specify the molecular states that arise from each (J_1, J_2) pair of separated atom states. The good quantum number is Ω (instead of Λ and S),

$$|\Omega| = |M_{J_1} + M_{J_2}| \tag{7.2.3}$$

where the $2J_1 + 1$ M_{J_1} components $-J_1 \leq M_{J_1} \leq J_1$ of atom-1 are split by the axial electric field of atom-2 and *vice versa*. Every state with $|\Omega| \neq 0$ appears twice and is doubly degenerate. For $HBr^+ \leftrightarrow H(^2S) + Br^+(^3P_J)$, the lowest asymptote is $^2S_{1/2} + {}^3P_2$, thus $J_1 = 1/2, J_2 = 2$, and the result is $\Omega = 5/2, 3/2, 3/2, 1/2, 1/2$ (5 doubly degenerate molecular components). When all three J components of ^3P_J are considered, the result is $5/2, 3/2(3), 1/2(5)$ $(2 + 2\times 3 + 2\times 5 = 2\times 3\times 3)$. The $\Omega = 5/2$ substate is pure $^4\Pi_{5/2}$, the three $\Omega = 3/2$ substates are explicit mixtures of $^4\Pi_{3/2}, {}^2\Pi_{3/2}$, and $^4\Sigma^-_{3/2}$, and the five $\Omega = 1/2$ substates are mixtures of $^4\Pi_{1/2}, {}^4\Pi_{-1/2}, {}^2\Pi_{1/2}, {}^4\Sigma^-_{1/2}$, and $^2\Sigma^-_{1/2}$ (see Fig. 153 of Herzberg, 1950).

Another example of adiabatic correlation is the lowest dissociation limit of HCl: $HCl \rightarrow H(^2S) + Cl(^2P)$. Here $S_1 = 1/2, S_2 = 1/2$, thus $S = 0$ or 1 and $\Lambda = 0$ or 1 as for HBr^+. This gives $^1\Pi, {}^3\Pi, {}^1\Sigma^+, {}^3\Sigma^+$ molecular states and the Σ–states are Σ^+ because the Cl ^2P electronic configuration is $(3p)^5$. Since the Cl spin-orbit splitting is relatively large $\left(^2P_{1/2} - {}^2P_{3/2} = 881\ cm^{-1}\right)$, an intermediate spin-orbit relativistic adiabatic correlation is appropriate. The resultant Ω–values are, for the $^2P_{3/2}$ limit, $\Omega = 2, 1, 1, 0^+$, and 0^-, and, for the $^2P_{1/2}$ limit, $\Omega = 1, 0^+$, and 0^-. The specific correlations with molecular $^{2S+1}\Lambda_\Omega$ states depends on the energy order of the molecular states (see Fig. 7.1). For HCl $X^1\Sigma^+ < {}^3\Pi < {}^1\Pi < {}^3\Sigma^+$ and the specific adiabatic correlations are:

$$\begin{aligned} H(^2S) + Cl(^2P_{3/2}) \rightarrow\ & \Omega = 2(^3\Pi_2), \\ & \Omega = 1(^3\Pi_1, {}^1\Pi_1), \\ & \Omega = 0^+(X^1\Sigma^+), \\ & \Omega = 0^-(^3\Pi_{0^-}) \end{aligned} \tag{7.2.4a}$$

and

$$\begin{aligned} H(^2S) + Cl\,(^2P_{1/2}) \rightarrow\ & \Omega = 1\,(^3\Sigma^+_1)\,, \\ & \Omega = 0^-(^3\Sigma^+_{0^-}), \\ & \Omega = 0^+(^3\Pi_{0^+}). \end{aligned} \tag{7.2.4b}$$

(The $0^+, 0^-$ symmetry of each $\Omega = 0$ state is decided as follows: when the number of $\Omega = 0$ states is even, half are 0^+ and half 0^-; if J_1 and J_2 are both integer, then the number of $\Omega = 0$ states can be odd, and the $0^+/0^-$ symmetry of the unique state is determined by $\sum_i l_{i_1} + J_1 + \sum_i l_{i_2} + J_2$.)

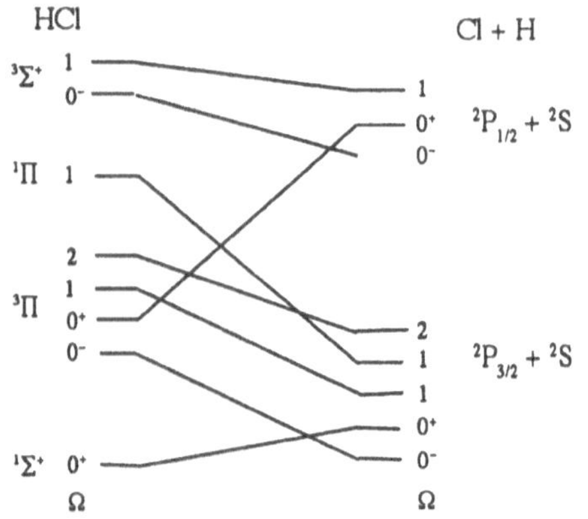

Figure 7.1: Transition from (Λ, S) to (J_1,J_2) coupling for the ^{2}P + ^{2}S separated atom states which correlate adiabatically with the molecular states of HCl. The specific adiabatic correlations shown are determined by the energy order of the molecular states and the noncrossing rule (from Matsumi, *et al.*, 1990).

These adiabatic correlations imply that the HCl $X^1\Sigma^+$ ground state belongs to the $\sigma^2\pi^4$ configuration at R_e but at $R = \infty$ belongs to the $\sigma\pi^4\sigma^*$ configuration. The $^{1,3}\Pi$ states belong to $\sigma^2\pi^3\sigma^*$ at all R and the $^3\Sigma^+$ state belongs to $\sigma\pi^4\sigma^*$ at all R. The σ "bonding" and σ^* "antibonding" orbitals become respectively Cl 3pσ and H 1sσ as $R \to \infty$.

Table 7.1 summarizes the correlations between separated ($R \to \infty$) atomic L-S-J-Ω and molecular $^{2S+1}\Lambda_\Omega$ states for the ^{2}P + ^{2}S separated atom example. It also gives the explicit linear combinations of molecular $^{2S+1}\Lambda_\Omega$ states that correlate to each separated atom L-S-J-Ω combination and *vice versa.* These molecular → separated atom linear combinations describe the expected fine structure branching ratios that would result without any electronic orbital rearrangements, i.e., the diabatic limit.

The adiabatic correlations for homonuclear molecule photodissociation are more complicated because of the presence of g, u symmetry. For example, in O_2 the O(^{3}P) + O(^{3}P) ground state atoms (total degeneracy 81) give rise to 8 ungerade Λ–S states ($A'^3\Delta_u$, $A^3\Sigma_u^+$, $c^1\Sigma_u^-$, $^3\Pi_u$, $^1\Pi_u$, $^5\Sigma_u^-$, $^5\Pi_u$, $^3\Sigma_u^+$; total degeneracy of 36) and 10 gerade Λ–S states ($X^3\Sigma_g^-$, $b^1\Sigma_g^+$, $^3\Pi_g$, $a^1\Delta_g$, $^5\Pi_g$, $^1\Pi_g$, $^5\Delta_g$, (2)$^5\Sigma_g^+$, $^1\Sigma_g^+$; total degeneracy of 45) (Leahy, *et al.*, 1995). (Note that in Table II of Leahy, *et al.*, (1995) the $a^5\Delta_g$ state must be replaced by $a^1\Delta_g$.)

Often, in the discussion of photodissociation phenomena, a "Hund's case (e)" basis set is used to describe the dissociation products. However that basis set is in fact an *atomic* basis set (Band, *et al.*, (1987)) and is distinct from the *molecular* case (e) basis set described in Section 3.2.1.

Table 7.1: Correlation between molecular and atomic wave functions for HX at R=∞ (X=Halogen)[c]

$(\Omega,\ J)^a$	Atomic $L-S-J-\Omega$ functions	Molecular $^{2S+1}\Lambda_\Omega$ states
	$^2P_{\frac{3}{2}} + {}^2S_{\frac{1}{2}}$	
$(2, \frac{3}{2})$	$\lvert s\alpha p_+\alpha p_+\beta p_0\alpha p_0\beta p_-\alpha\rvert$	$^3\Pi_2$
$(1, \frac{3}{2})^b$	$\frac{1}{\sqrt{3}}\{\lvert s\alpha p_+\alpha p_+\beta p_0\alpha p_0\beta p_-\beta\rvert - \sqrt{2}\lvert s\alpha p_+\alpha p_+\beta p_0\alpha p_-\alpha p_-\beta\rvert\}$	$\frac{1}{\sqrt{6}}({}^3\Pi_1 - {}^1\Pi_1) - \sqrt{\frac{2}{3}}\ {}^3\Sigma_1^+$
$(1', \frac{3}{2})^b$	$\lvert s\beta p_+\alpha p_+\beta p_0\alpha p_0\beta p_-\alpha\rvert$	$\frac{1}{\sqrt{2}}({}^3\Pi_1 + {}^1\Pi_1)$
$(0^+, \frac{3}{2})$	$\frac{1}{\sqrt{6}}\{\lvert s\alpha p_+\alpha p_0\alpha p_0\beta p_-\alpha p_-\beta\rvert - \sqrt{2}\lvert s\alpha p_+\alpha p_+\beta p_0\beta p_-\alpha p_-\beta\rvert$	
	$-\lvert s\beta p_+\alpha p_+\beta p_0\alpha p_0\beta p_-\beta\rvert + \sqrt{2}\lvert s\beta p_+\alpha p_+\beta p_0\alpha p_-\alpha p_-\beta\rvert\}$	$-\sqrt{\frac{2}{3}}\,{}^1\Sigma_{0+}^+ + \frac{1}{\sqrt{3}}\ {}^3\Pi_{0+}$
$(0^-, \frac{3}{2})$	$\frac{1}{\sqrt{6}}\{\lvert s\alpha p_+\alpha p_0\alpha p_0\beta p_-\alpha p_-\beta\rvert - \sqrt{2}\lvert s\alpha p_+\alpha p_+\beta p_0\beta p_-\alpha p_-\beta\rvert$	
	$+\lvert s\beta p_+\alpha p_+\beta p_0\alpha p_0\beta p_-\beta\rvert - \sqrt{2}\lvert s\beta p_+\alpha p_+\beta p_0\alpha p_-\alpha p_-\beta\rvert\}$	$-\sqrt{\frac{2}{3}}\,{}^3\Sigma_{0-}^+ + \frac{1}{\sqrt{3}}\ {}^3\Pi_{0-}$
	$^2P_{\frac{1}{2}} + {}^2S_{\frac{1}{2}}$	
$(1, \frac{1}{2})$	$\frac{1}{\sqrt{3}}\{\lvert s\alpha p_+\alpha p_+\beta p_0\alpha p_-\alpha p_-\beta\rvert + \sqrt{2}\lvert s\alpha p_+\alpha p_+\beta p_0\alpha p_0\beta p_-\beta\rvert\}$	$\frac{1}{\sqrt{3}}\,{}^3\Sigma_1^+ + \sqrt{\frac{1}{3}}({}^3\Pi_1 - {}^1\Pi_1)$
$(0^+, \frac{1}{2})$	$\frac{1}{\sqrt{6}}\{\lvert s\alpha p_+\alpha p_+\beta p_0\beta p_-\alpha p_-\beta\rvert + \sqrt{2}\lvert s\alpha p_+\alpha p_0\alpha p_0\beta p_-\alpha p_-\beta\rvert$	
	$-\lvert s\beta p_+\alpha p_+\beta p_0\alpha p_-\alpha p_-\beta\rvert + \sqrt{2}\lvert s\beta p_+\alpha p_+\beta p_0\alpha p_0\beta p_-\beta\rvert\}$	$\frac{1}{\sqrt{3}}\,{}^1\Sigma_{0+}^+ + \sqrt{\frac{2}{3}}\ {}^3\Pi_{0+})$
$(0^-, \frac{1}{2})$	$\frac{1}{\sqrt{6}}\{\lvert s\alpha p_+\alpha p_+\beta p_0\beta p_-\alpha p_-\beta\rvert + \sqrt{2}\lvert s\alpha p_+\alpha p_0\alpha p_0\beta p_-\alpha p_-\beta\rvert$	
	$+\lvert s\beta p_+\alpha p_+\beta p_0\alpha p_-\alpha p_-\beta\rvert + \sqrt{2}\lvert s\beta p_+\alpha p_+\beta p_0\alpha p_0\beta p_-\beta\rvert\}$	$\frac{1}{\sqrt{3}}\,{}^3\Sigma_{0-}^+ + \sqrt{\frac{2}{3}}\ {}^3\Pi_{0-})$
	Molecular functions	Atomic $(\Omega,\ J)$ states
$^3\Pi_2$	$\lvert s\sigma\alpha p\pi^+\alpha p\pi^+\beta p\sigma\alpha p\sigma\beta p\pi^-\alpha\rvert$	$(2, \frac{3}{2})$
$^3\Pi_1$	$\frac{1}{\sqrt{2}}\{\lvert s\sigma\beta p\pi^+\alpha p\pi^+\beta p\sigma\alpha p\sigma\beta p\pi^-\alpha\rvert + \lvert s\sigma\alpha p\pi^+\alpha p\pi^+\beta p\sigma\alpha p\sigma\beta p\pi^-\beta\rvert\}$	$\frac{1}{\sqrt{6}}(1, \frac{3}{2}) + \frac{1}{\sqrt{2}}(1', \frac{3}{2}) + \frac{1}{\sqrt{3}}(1, \frac{1}{2})$
$^1\Pi_1$	$\frac{1}{\sqrt{2}}\{\lvert s\sigma\beta p\pi^+\alpha p\pi^+\beta p\sigma\alpha p\sigma\beta p\pi^-\alpha\rvert - \lvert s\sigma\alpha p\pi^+\alpha p\pi^+\beta p\sigma\alpha p\sigma\beta p\pi^-\beta\rvert\}$	$-\frac{1}{\sqrt{6}}(1, \frac{3}{2}) + \frac{1}{\sqrt{2}}(1, \frac{3}{2}) - \frac{1}{\sqrt{3}}(1, \frac{1}{2})$
$^3\Pi_{0+}$	$\frac{1}{\sqrt{2}}\{\lvert s\sigma\alpha p\pi^-\alpha p\pi^-\beta p\sigma\alpha p\sigma\beta p\pi^+\alpha\rvert - \lvert s\sigma\beta p\pi^+\alpha p\pi^+\beta p\sigma\alpha p\sigma\beta p\pi^-\beta\rvert\}$	$\frac{1}{\sqrt{3}}(0^+, \frac{3}{2}) + \sqrt{\frac{2}{3}}(0^+, \frac{1}{2})$
$^3\Pi_{0-}$	$\frac{1}{\sqrt{2}}\{\lvert s\sigma\alpha p\pi^-\alpha p\pi^-\beta p\sigma\alpha p\sigma\beta p\pi^+\alpha\rvert + \lvert s\sigma\beta p\pi^+\alpha p\pi^+\beta p\sigma\alpha p\sigma\beta p\pi^-\beta\rvert\}$	$\frac{1}{\sqrt{3}}(0^-, \frac{3}{2}) + \sqrt{\frac{2}{3}}(0^-, \frac{1}{2})$
$^3\Sigma_1^+$	$\lvert s\sigma\alpha p\pi^+\alpha p\pi^+\beta p\pi^-\alpha p\pi^-\beta p\sigma\alpha\rvert$	$-\sqrt{\frac{2}{3}}(1, \frac{3}{2}) + \frac{1}{\sqrt{3}}(1, \frac{1}{2})$
$^3\Sigma_{0-}^+$	$\frac{1}{\sqrt{2}}\{\lvert s\sigma\alpha p\pi^+\alpha p\pi^+\beta p\pi^-\alpha p\pi^-\beta p\sigma\beta\rvert + \lvert s\sigma\beta p\pi^+\alpha p\pi^+\beta p\pi^-\alpha p\pi^-\beta p\sigma\alpha\rvert\}$	$-\sqrt{\frac{2}{3}}(0^-, \frac{3}{2}) + \frac{1}{\sqrt{3}}(0^-, \frac{1}{2})$
$^1\Sigma_{0+}^+$	$\frac{1}{\sqrt{2}}\{\lvert s\sigma\alpha p\pi^+\alpha p\pi^+\beta p\pi^-\alpha p\pi^-\beta p\sigma\beta\rvert - \lvert s\sigma\beta p\pi^+\alpha p\pi^+\beta p\pi^-\alpha p\pi^-\beta p\sigma\alpha\rvert\}$	$-\sqrt{\frac{2}{3}}(0^+, \frac{3}{2}) + \frac{1}{\sqrt{3}}(0^+, \frac{1}{2})$

[a] Ω is defined as the sum of the projections (M_{J_1} of the ^{2}P atom, M_{J_2} of the hydrogen atom) of the two atomic J's on the molecular axis. J is the total angular momentum associated with the ^{2}P atom.

[b] There are two $\Omega = 1$ states that arise from the $^2P_{3/2} + {}^2S_{1/2}$ combination. One $\Omega = 1$ state, called $(1, 3/2)$ corresponds to $M_{J_1} = 1/2$, $M_{J_2} = 1/2$.; the other $\Omega = 1$ state, called $(1', 3/2)$ corresponds to $M_{J_1} = 3/2$, $M_{J_2} = -1/2$.

[c] The top and middle sections of the table give the explicit combinations of separated atom Slater determinants and molecular $\Lambda - S - \Omega$ states that correspond to each separated atom (Ω, J) state. The bottom section of the table gives the explicit combinations of molecular spin-orbital Slater determinants and separated atom (Ω, J) states that correspond to each molecular $\Lambda - S - \Omega$ state.

7.2.2 Photodissociation Cross Sections

Once the adiabatic correlations are established between molecular electronic states and the separated atom asymptotes, it is possible to ask what states contribute to the photodissociation cross section and how do these cross sections vary with excitation energy, initial state, and excitation and detection scheme. For free←bound transitions into a single continuum, the energy dependent photoexcitation cross sections depend on the product of R–dependent electronic and vibrational factors. Often these R–dependencies have a simple physical explanation and may be used to identify the dissociative state, determine whether the photodissociation proceeds via single or multiple electronic states, or to determine the shape of a repulsive potential energy curve. Important information about which dissociative states are accessed in the photodissociative process is often obtained by examining the same energy region via one- and two-photon transitions, by comparing spectra originating from $v'' = 0$ and $v'' = 1$ of the electronic ground state, or by recording the spectrum with selective detection of a specific atomic photoproduct $L - S - J$ state.

A complete picture of the R–dependence of the electronic character of the repulsive state provides a basis for planning and interpreting time domain wavepacket experiments. A feature of femtosecond spectroscopy that is particularly useful for the systematic study of photodissociation processes is the ability to prepare, at $t = 0$, an initially highly localized time-evolving state, $\Psi'(t = 0)$,

$$\Psi'(t) = \psi'_d(r; R) \int \mathrm{a}(E)\chi_{d,E}(R)e^{-iEt/\hbar}dE \tag{7.2.5}$$

where ψ'_d is the wavefunction of the dissociative electronic state, $\mathrm{a}(E)$ is the E–dependent (complex) amplitude of the $\chi_{d,E}(R)$ continuum wavefunction prepared at $t = 0$, and $e^{-iEt/\hbar}$ is the standard time-dependent factor that describes the time-evolution of an energy eigenstate with energy E. Simple ideas, based on the stationary phase approximation, permit the nature of this initial localization (encoded in $\mathrm{a}(E)$) to be varied over a useful range of excitation energy

$$\langle\Psi'(t)|\mathbf{H}|\Psi'(t)\rangle = \langle E\rangle \tag{7.2.6}$$

and initial internuclear distance

$$\langle\Psi'(t = 0)|\mathbf{R}|\Psi'(t = 0)\rangle = \langle R\rangle\,(t = 0) \tag{7.2.7}$$

merely by exciting to the repulsive potential, $V'_d(R)$, from different initial vibrational levels, v'', of a bound potential, $V''(R)$, and by choosing the center-energy of the femtosecond pulse, $\overline{E}_{\mathrm{pulse}}$, in order to select whether the excitation occurs localized near the inner, $R_-(E_{v''})$, or outer, $R_+(E_{v''})$, turning point

$$E_{v''} = V''(R_\pm(E_{v''})) \tag{7.2.8}$$

of the $V''(R)$ potential. The stationary phase approximation specifies that transition amplitude accumulates near the value of R at which the difference potential

$$\Delta V(R) = V_d'(R) - V''(R) \tag{7.2.9}$$

is equal to the transition frequency. Since the values of the difference potential at $R_-(E_{v''})$ and $R_+(E_{v''})$ are usually quite different

$$\Delta V(R_\pm) = V_d'(R_\pm) - V''(R_\pm), \tag{7.2.10}$$

either

$$\langle R\rangle\,(t=0) \approx R_+$$

or

$$\langle R\rangle\,(t=0) \approx R_-$$

may be selected by adjusting $\overline{E}_{\text{pulse}}$ to be near $\Delta V(R_+)$ or $\Delta V(R_-)$. The time-independent energy of the time-evolving excited state is then

$$\langle E\rangle_\pm = E_{v''} + \left[V_d'(R_\pm) - V''(R_\pm)\right]. \tag{7.2.11}$$

The values of $\langle E\rangle_\pm$ are dependent on *a priori* known values of $E_{v''}$ and $R_\pm(E_{v''})$, and only very weakly dependent on $\overline{E}_{\text{pulse}}$ (except for the ability to use $\overline{E}_{\text{pulse}}$ to select either $\langle E\rangle_+$ or $\langle E\rangle_-$) and the pulse shape (duration, chirp, etc.).

Here, in order to illustrate the ingredients of a zero-order model for a photodissociation mechanism, are some simple predictions about the not yet observed, longest wavelength photodissociation transitions from the HBr^+ $X^2\Pi_i$ electronic ground state. Figure 7.2 summarizes all of the relativistic adiabatic correlations of molecular states with the three $Br^+(^3P_J) + H(^2S)$ dissociation asymptotes. Note that both components of the $X^2\Pi$ ground state, separated at R_e by $E\left(^2\Pi_{1/2}\right) - E\left(^2\Pi_{3/2}\right) = 2651\ \text{cm}^{-1}$, correlate with the lowest atomic fine structure asymptote: $Br^+\left(^3P_2\right)$ +H. The noncrossing rule does not prevent these two Ω–components from going to the same dissociation limit because they belong to different Ω–values, and only Ω is a good quantum number all the way from $R = R_e$ to ∞. The R–dependent electronic characters of each electronic state are obtained by diagonalizing the electronic plus spin-orbit Hamiltonians at each R for each Ω–block (Banichevitch, *et al.*, (1992)). Although each potential curve shown in Fig. 7.2 is labeled with a single dominant Λ–S–Ω character, all states are of mixed Λ–S (same Ω) character, and this mixture varies with R in a manner obtainable from the eigenvectors of $\mathbf{H}^{\text{el}} + \mathbf{H}^{\text{SO}}$.

The crucial features of a simple model for the R–dependent characters of the Br^+ $(4p^4\ ^3P)$+ H molecular states are obtained from four diabatic potential curves, $^{2,4}\Pi_i$ and $^{2,4}\Sigma^-$, which correspond respectively to the four $4p$ electrons on $Br^+(^3P)$ arranged $\pi^3(^2\Pi)\sigma$ (the $^2\Pi_i$ state is the bonding $X^2\Pi_i$ state, because the two singlet-coupled electrons in $\sigma(4p)$ and $\sigma(\text{H}1s)$ combine to make a doubly occupied bonding σ orbital; the $^4\Pi_i$ state is repulsive because the two electrons in σ orbitals must remain triplet coupled) and $\pi^2(^3\Sigma^-)\sigma^2$ (both $^4\Sigma^-$ and $^2\Sigma^-$ are weakly bonding because the $\sigma^2\sigma^*$ sigma orbitals yield $\sim 1/2$ a bond). These four diabatic states are split and mixed by an R–independent spin-orbit coupling parameter, $\zeta_{\mathrm{Br}^+}(4p) = 2560\ \mathrm{cm}^{-1}$.

Absorption cross sections can describe two qualitatively different situations (Merzbacher, p. 495, 1998):

1. The initial and final states are bound (or quasibound) and the absorption spectrum consists of sharp (or slightly broadened, but in principle resolvable) lines, as described in Section 6.1.1.

2. The initial state is discrete, but the final state is in the continuum and transitions are described by a spectral density, σ(E), which is a slowly varying function of the energy or frequency.

The expressions for differential cross sections (units of area), σ(E), for these two cases differ in how the density of states is treated. For sharp line transitions, the transition is betweeen single eigenstates (or quasi-eigenstates) and the spectral density corresponds to one eigenstate-to-eigenstate transition per spectral linewidth (either homogeneous or inhomogeneous) and is expressed by a lineshape function (Eqs. (6.1.9a) - (6.1.9d), Eqs. (6.1.10a) - (6.1.10d) and Eqs. (6.1.12a) - (6.1.12b)). For continuum←bound transitions, the spectral density reflects the density of states in the continuum, as expressed by appropriately normalized continuum←bound matrix elements, vibrational overlap density $\langle v_{E_{\mathrm{final}}}|v_{\mathrm{initial}}\rangle$ or electronic transition moment density $\langle E_{\mathrm{final}}|\mu|n\Lambda S\Sigma_{\mathrm{initial}}\rangle$.

The photodissociation cross section for a transition between the v_i, J_i level of the bound initial state, g, to the v_E vibrational level of the continuum state, e, at energy E is given by

$$\sigma(E) = \frac{4\pi^2 E}{\hbar c}\left|\langle e, v_E, \Omega\,|\mathbf{T}(\epsilon)\cdot\mathbf{T}(\mu)|\,g, v_i, J_i\rangle\right|^2 \tag{7.2.12a}$$

$\sigma(E)$ is of dimensionality area: the quantity $\langle e, v_E, \Omega\,|\mathbf{T}(\epsilon)\cdot\mathbf{T}(\mu)|\,g, v_i, J_i\rangle$, where $\mathbf{T}(\epsilon)$ is the first rank tensor (defined relative to the laboratory coordinate system) that describes the polarization of the photoexcitation radiation and $\mathbf{T}(\mu)$ is the first rank tensor (defined relative to the molecule fixed coordinate system) that describes the electronic transition moment, has been discussed in Sections 6.1.1 and 6.1.2.1 for the case of one-photon transitions between bound states. Here the final state is a dissociative state and an incoherent sum over all allowed $J, J\pm 1 \leftarrow J_i$ rotational transitions is implied. The photodissociation cross section may be expressed in terms of quantities with typically used units,

$$\sigma(E)[\mathrm{cm}^2] = 2.869\times 10^{-18} E[\mathrm{cm}^{-1}]\left|\langle v_E|M_{eg}(R)[\mathrm{a.u.}]|v_i\rangle\right|^2, \tag{7.2.12b}$$

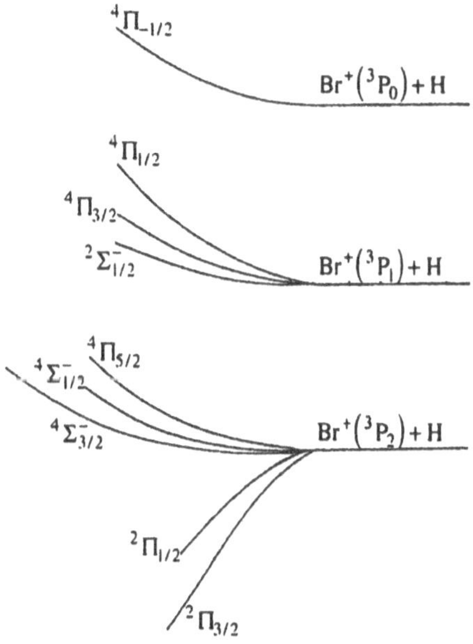

Figure 7.2: Schematic description of the dissociation channels $Br^+(^3P_J)$ + $H(^2S)$ for the molecular states $X^2\Pi_i$ (configuration $\sigma^2\pi^3$ at R_e), $^4\Sigma^-$ and $^2\Sigma^-(\sigma^2\pi^2\sigma^*)$, and $^4\Pi(\sigma\pi^3\sigma^*)$ (from Banichevitch, *et al.*, (1992))

M_{eg} (in atomic units, 1Debye = 0.3935 a.u.) is the electronic transition matrix element between the e and g electronic states, assuming the dipole length approximation, $\langle v_E|$ is the energy normalized nuclear continuum wavefunction, and $|v_i\rangle$ is the initial state bound vibrational wavefunction. The overlap integral $\langle v_E|v_i\rangle$ has units of $cm^{1/2}$ (see Section 7.5). Note that 10^{-18} cm^2=1Mb (Megabarn).

The theoretical cross section, σ(E), must be convoluted with the experimental resolution before being compared to the experimental cross section.

Above the ionization threshold (see Chapter 8), the absolute photodissociation cross section can be obtained as the difference between the absolute total absorption cross section and the absolute photoionization cross section (see for example, for N_2, Fig. 6 of Shaw, *et al.*, 1992). Another experimental quantity is the ionization efficiency defined as the total photoionization cross section divided by the total absorption cross section.

In order to compute or interpret photodissociation $\sigma(E)$ functions, it is necessary to know: (i) the potential energy curve of the initial state; (ii) the potential energy curves of all energetically accessible excited electronic states (both purely repulsive states and the above dissociation repulsive region of bound states); (iii) the electric dipole transition moment function between the initial and final state; (iv) the (R–dependent) interactions among the set of electronic basis functions in terms of which each relativistic adiabatic state is expressed. The simplest approach is to assume that (at least in the energy and/or internuclear distance region that it is sampled) the electronic transition moment is independent of R and to replace the repulsive vibrational eigenfunction by a delta-function located at the turning point (the "reflection" of $|\chi_{v''}(R)|^2$ is indicated by the broken-line curve on Fig. 7.3). This is known as the " Condon *reflection* approximation". The reason for the name "reflection approximation" is that the E dependence of $\sigma(E)$ mimics the R-dependence of the initial, bound state, vibrational wavefunction, $|\chi_{v''}(R)|^2$ (see Fig. 7.3.). If the initial state is $v'' = 0$, then $|\chi_{v''}(R)|^2$ is a Gaussian and $\sigma(E)$ will resemble a Gaussian. If the initial state is $v'' = 1$, $\sigma(E)$ will have two lobes separated by a node.

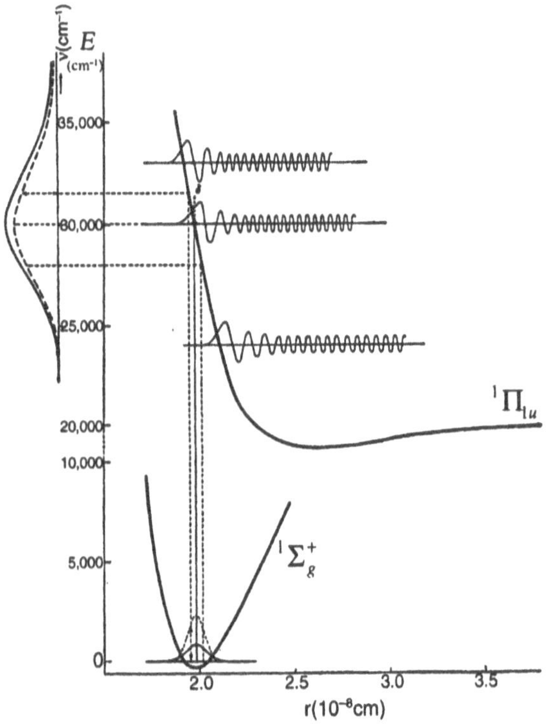

Figure 7.3: Reflection approximation: Potential energy curves, repulsive wavefunctions, and intensity distribution for Cl_2 (from Herzberg, 1950, p.392). The wavefunctions are shown as thin full lines and $|\chi_{v''=0}|^2$ is shown as a dashed curve. The lower state $|\chi|^2$ is reflected on the upper state potential curve to give the spectral intensity distribution. The reflection approximation amounts to replacing the upper state wavefunction by a δ-function (shown as a narrow dashed curve labeled δ) at the turning point.

Figure 7.3 illustrates the reflection approximation. A detailed discussion of the stationary phase approximation and how that approximation permits the derivation of analytical and intuitive relationships between the Franck-Condon overlap integrals and the shape of an unknown potential energy curve is given in Section 5.1.1. To a good approximation, (because the innermost lobe of the continuum vibrational wavefunction $\chi_E(R)$ occurs near the turning point), for a free←bound transition, the E dependence of $\sigma(E)$ is given by drawing a vertical line of length E from the energy of the initial vibrational level. This vertical line is shifted left or right in R until it terminates on the repulsive potential energy curve $V'(R)$, at $R(E)$. The relative intensity at E is given by $|\chi_{v''}(R)|^2$ at $R(E)$. Figure 7.3 shows this vertical line as being reflected by 90° off the intersection point with the $V'(R)$ curve and the length of the reflected line is given by $|\chi_{v''}(R(E))|^2$. The maximum of $\sigma(E)$ for a transition originating from a $v'' = 0$ level occurs at the energy of a vertical transition from R''_e. The width of $\sigma(E)$ is determined by the slope of $V'(R)$ at $R = R''_e$. When the intersection is on the rapidly rising, strongly repulsive, inner wall of $V'(R)$ (large negative slope) the $\sigma(E)$ curve is very wide and, if $v'' = 0$, nearly perfectly Gaussian. As the slope of V' approaches zero, $\sigma(E)$ gets narrower and more distorted from a Gaussian shape. The reflection approximation is illustrated on Fig. 7.4 for free←bound transitions into three repulsive states of OH: $1^2\Sigma^-, 1^2\Delta$, and $B^2\Sigma^+$. The $\sigma(E)$ curves exhibit the expected increase in width (and decrease in maximum intensity) as the negative slope of $V'(R)$ at R''_e becomes larger.

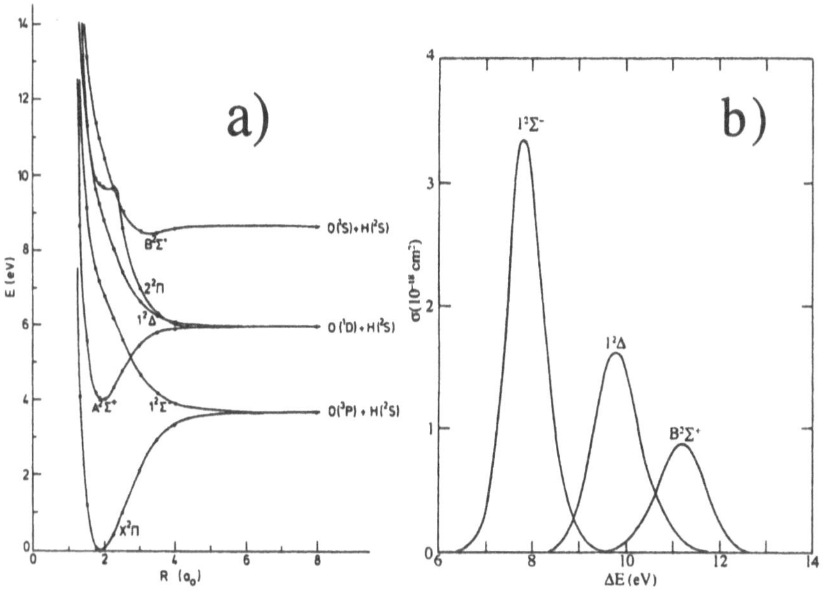

Figure 7.4: a) Potential energy curves of the bound $X^2\Pi$ and $A^2\Sigma^+$ and dissociative $1^2\Sigma^-$, $1^2\Delta$, and $B^2\Sigma^+$ states of OH. b) Calculated photodissociation cross sections, $\sigma(E)$, from $X^2\Pi$ v"=0 (from Van Dishoek and Dalgarno, 1983)

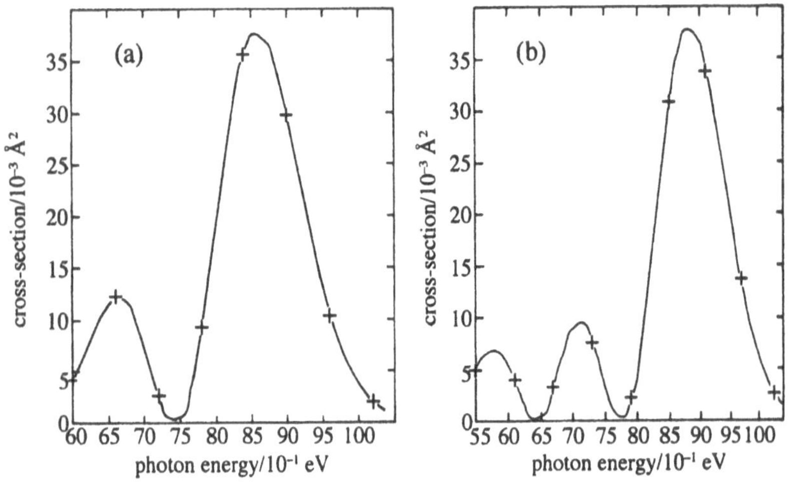

Figure 7.5: Total photodissociation cross sections, $\sigma(E)$, for HCl starting from the $v = 1$ and 2 vibrationally excited levels of HCl in its ground electronic state: (a) $v = 1$;(b) $v = 2$ (from Givertz and Balint-Kurti, 1986).

If the photodissociation starts from a $v'' \neq 0$ level, the $\sigma(E)$ will be an oscillatory curve with $v'' + 1$ lobes (v'' nodes) (see Fig. 7.5). The possibility of destructive interference arising from multiple stationary phase points (e.g. R_1, R_2) (Tellinghuisen, 1984) is negligible for direct dissociation because the difference potential $V'(R) - V''(R) \equiv \Delta V(R)$ will be single valued (i.e. $\Delta V(R_1) \neq \Delta V(R_2)$) over the Franck-Condon region, $R''_{\text{MIN}} \leq R \leq R''_{\text{MAX}}$.

The slope of the repulsive potential at R''_e (or at the R'' values of the two maxima in the $v'' = 1$ probability distribution) may be determined from the width of $\sigma(E)$. The vertical excitation energy of the repulsive state at R''_e is determined by the E at which $\sigma(E)$ reaches its maximum value. In this semiclassical approximation, the repulsive potential curve can be determined from $\sigma(E)$ provided that $|\mu(R)|$ varies no more rapidly than linearly in R (Child, *et al.*, 1983). When a sufficient quantity of $\sigma(E)$ data is obtained from free-bound absorption or emission transitions originating from several bound vibrational levels, it is then also possible to determine the shape of the bound potential (Le Roy, *et al.*, 1988). The R-dependence of $|\mu(R)|^2$ can arise from two sources: (i) the R-dependence of the fractional contributions of several different Λ–S basis states to a single relativistic adiabatic Ω-state (ii) R-variation of the transition moment between Λ–S basis states arising from the molecule to separated atom evolution of the LCAO characters of the occupied orbitals; (iii) R-variation of the configurational character (Configuration Interaction) of either electronic state.

Let us consider how R–dependent $|\mu(R)|^2$ effects contribute to the $\sigma(E)$ for the hydrogen halides, HX (X = I, Br, and Cl). The curves shown on Fig. 7.6 correspond to relativistic adiabatic potential energy curves (respectively: 0^- dotted, 0^+ dashed, 1 and 2 solid) for HI obtained after diagonalization of the electronic plus spin-orbit Hamiltonians (see Section 3.1.2.2). The strong R–dependence of the electronic transition moment reflects the R–dependence of the relative contributions of the case(a) Λ–S–Ω basis states to each relativistic adiabatic Ω–state. The R–dependent experimental photodissociation cross sections are plotted as solid curves in Fig. 7.7 for HI and HBr. Note that, in addition to the R–dependent variations in the $\Lambda - S$ characters of each Ω–state caused by $\Delta\Omega = 0$ spin-orbit interactions, all transitions from the $X^1\Sigma^+$ state to states that dissociate to the $X(^2P) + H(^2S)$ limit are forbidden in the separated atom limit because they are at best $\left(^2P_{1/2} \leftarrow {}^2P_{3/2}\right)$ parity forbidden electric dipole transitions on the X atom. In the case of the continuum region of an attractive potential, the energy dependence of the dissociation cross section exhibits continuity in the Franck-Condon factor density (see Fig. 7.18; Allison and Dalgarno, 1971; Smith, 1971; Allison and Stwalley, 1973).

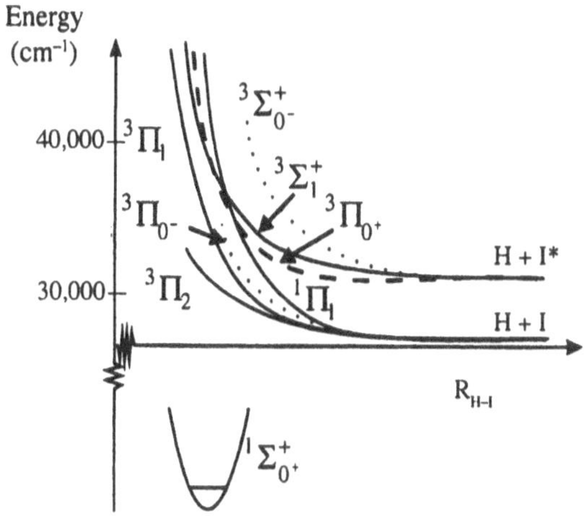

Figure 7.6: Schematic relativistic adiabatic potential curves of HI (from Langford, *et al.*, 1998). The ground state dissociates to the H + I asymptote.

In direct absorption experiments, it is difficult to sample a very wide range of internuclear separations, because the Franck-Condon region of R defined by the $v'' = 0$ or 1 vibrational levels of the electronic ground state is very narrow. In double resonance experiments, where $e, v_E \leftarrow e', v' \leftarrow X, v''$ free←bound←bound spectra can be recorded via many selectively populated e', v' intermediate levels, a wide range of internuclear separations can be systematically sampled. It is primarily through such two-color double resonance experiments that meaningful experimental information can be obtained about the R–dependence of Λ–S basis state mixing coefficients in a relativistic adiabatic dissociative state or the R–dependence of transition moments arising from R–dependent changes in LCAO-MO character or from configuration interaction mixing. There is a price to pay, however; it is difficult to make reliable absolute or even relative intensity measurements in two-color, pulsed laser double resonance experiments.

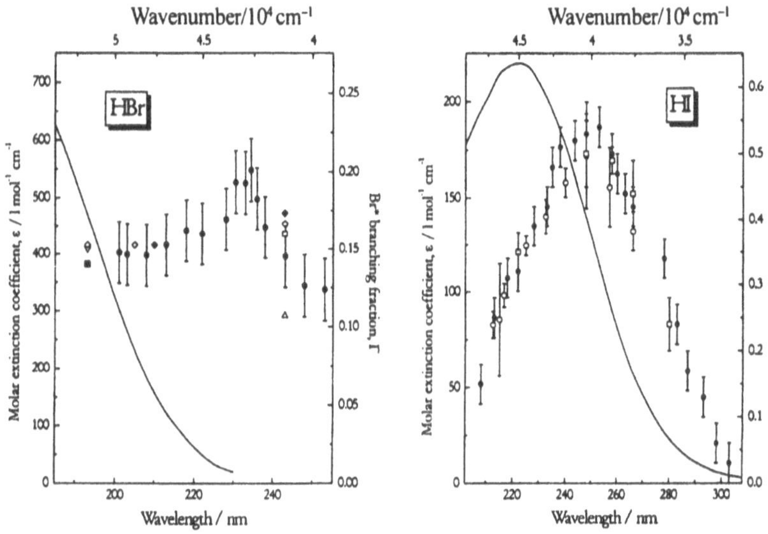

Figure 7.7: $X*/(X*+X)$ photofragment branching ratios (right hand scale) that result from photodissociation of HBr (left panel) and HI (right panel) molecules. For HBr: open down triangle (Magnotta, *et al.*, 1981), filled square (Xu, *et al.*, 1988), open up triangle (Kinugawa and Arikawa, 1992), open diamond (Baumfalk, *et al.*, 1999), open square (Fernandez-Alonso, *et al.*, 1999), filled diamond (Scheill and Hepburn, 1999), filled circle (Ashfold, *et al.*, 1998). For HI: filled circle (corrected results from Ashfold, *et al.*, 1998), open square (Regan, *et al.*, 1999), open circle (Gendron and Hepburn, 1998). The branching ratios and the respective parent absorption spectra (solid curves, left hand scale) are plotted as function of the excitation wavelength. (Courtesy P. Regan).

Additional information about the dynamics of the dissociation process can often be obtained by monitoring the internal $L-S-J$ state, velocity, and/or angular distribution (both vector velocity and vector J) of the photoproducts rather than $\sigma(E)$ integrated over all product states. Methods for measurements of these quantities fall into at least four categories (see Section 1.2.2.4): two-color double resonance excitation followed by REMPI probing of atomic fragment(s), time-of-flight methods (TOF) in which the internal states of both atom photoproducts are inferred from the excess energy available as kinetic energy, Doppler lineshape measurements, and photofragment imaging (for a review, see Houston (1996)).

7.2.3 Photofragment Branching Ratios for Photodissociation

The ability to record $\sigma(E)$ while detecting a single $L-S-J$ state of one of the photofragments or to measure the $L-S-J$ internal state distribution vs. excitation energy can be used to separate two distinct but spectrally overlapping free←bound transitions or to gain insight into the R–dependent evolution of the $\Lambda-S-\Omega$ diabatic (crossing) characters in a single-Ω relativistic adiabatic (non-crossing) state.

To illustrate the significance of measurements of internal state branching ratios, we will turn once again to the example of the photodissociation of the hydrogen halides, HX. The fine structure ratio is the branching ratio of X^*/X populations: $^2P_{1/2}/^2P_{3/2}$. In the *non-relativistic* adiabatic representation, this branching ratio would be predicted to be zero because the only case (a) basis state which has a non-zero transition moment from the $X^1\Sigma^+$ state is the $^1\Pi_1$ state which correlates adiabatically with the $X(^2P_{3/2})$ +$H(^2S)$ separated atom limit. However, in the more realistic relativistic adiabatic representation, $\Delta\Omega = 0$ $^1\Pi \sim {}^3\Sigma_1^+$, $^1\Pi_1 \sim {}^3\Pi_1$, and $^3\Pi_1 \sim {}^3\Sigma_1^+$ spin-orbit matrix elements cause the branching ratio to be different from zero. For HCl, for which the atomic spin-orbit coupling constant $\left(\zeta_{Cl}(3p) = 587.3\ \text{cm}^{-1}\right)$ is relatively small $\left[E\left(^2P_{1/2}\right) - E\left(^2P_{3/2}\right) = 881\ \text{cm}^{-1}\right]$, the excitation is mainly to the nominal '$^1\Pi_1$' and '$^3\Pi_1$' states (via the $^1\Pi_1$ basis state character) which dissociate adiabatically to the Cl $^2P_{3/2}$ ground state (see Fig. 7.1), but there is also some excitation into the nominal '$^3\Sigma_1^+$' state (which has some $^1\Pi_1$ character obtained directly via $^3\Sigma_1^+ \sim {}^1\Pi$ spin-orbit interaction and indirectly via $^3\Sigma_1^+ \sim {}^3\Pi_1 \sim {}^1\Pi$ interaction) which dissociates adiabatically to the Cl* $^2P_{1/2}$ excited atom limit. But this mixing of $^1\Pi$ into $^3\Sigma_1^+$ is relatively small and the $^2P_{1/2}/^2P_{3/2}$ branching ratio is non-zero (~0.5). This value corresponds to the diabatic limit or to the statistical limit based on the 2 : 4 M_J–degeneracy ratio for $J = 1/2 : J = 3/2$. Adiabatic limit behavior (branching ratio equal to zero) is observed at very low energies (Lambert, *et al.*, 1998). At higher energies and over a large range of energies the branching ratio becomes greater than the statistical limit of 0.5 (Liyanage, *et al.*, 1995).

If we consider the heavier HX molecules, for which the spin-orbit interaction is considerably stronger $\left[E\left(^2P_{1/2}\right) - E\left(^2P_{3/2}\right) = 3748\ \text{cm}^{-1}(\text{HBr}) \text{ or } 7603\ \text{cm}^{-1} (\text{HI})\right]$, the mixing induced by the spin-orbit interaction among all of the $\Omega = 1$ states is considerably more important relative to HCl and the stronger R–dependence of the $^1\Pi_1$, $^3\Pi_1$, $^3\Sigma_1^+$ mixings affects the energy dependence of the X^*/X branching ratio. Figure 7.7 left and right sides show, respectively for HBr and HI, the experimental branching ratios measured by photofragment translational spectroscopy (Ashfold, *et al.*, 1998).

The measured branching ratio for HBr is not very different from that for HCl. However, for HI the I^*/I branching ratio is strongly energy dependent. The maximum near 40,000 cm^{-1} in the branching ratio for HI (Fig. 7.7, right side) is explained (Levy and Shapiro, 1988) by the shapes of the repulsive potential energy curves; at 40,000 cm^{-1} excitation energy the nominal '$^3\Pi_{0^+}$' and

'$^3\Sigma_1^+$' states, which correlate with the I* $(^2P_{1/2})$ limit, have better vibrational overlap with $X^1\Sigma^+$ $v''=0$ than does the nominal '$^1\Pi$' state (see Fig. 7.6). Consequently, at 250 nm (40,000 cm^{-1}), the partial photoabsorption to the '$^3\Pi_{0^+}$' and '$^3\Sigma_1^+$' states exceeds that to the '$^1\Pi$' state. At higher excitation energy, the contribution of the '$^1\Pi$' state to $\sigma(E)$ becomes dominant and is responsible for the maximum near 220 nm in the absorption spectrum (see Fig. 7.7). This borrowing by the '$^3\Pi_{0^+}$' and '$^3\Sigma_1^+$' states of photoabsorption intensity from the energetically remote $^1\Pi$ state (for $^3\Pi_{0^+}$, the intensity borrowing is indirect through the $^3\Sigma_1^+ \sim {}^3\Pi_{0^+}$ L–uncoupling interaction), yet the persistence of direct dissociation into I* via the dominant (nominal) character of each state is a nice illustration of intensity borrowing by mixing with a remote perturber (see Section 6.4 case (2)). When the intensity-providing perturber is remote, the Franck-Condon factors appropriate are those calculated using the nominal state's own potential curve.

Here the X*/X branching ratio provides information about the relative contributions of the '$^3\Sigma_1^+$', '$^3\Pi_{0^+}$', and '$^1\Pi_1$' states to the photodissociation $\sigma(E)$. Although all of the absorption is due to the $^1\Pi_1$ basis state character in each continuum electronic state, the X*/X branching ratio is primarily determined by the dominant Λ–S–Ω basis state character in each relativistic adiabatic Ω–state. Once again, a simple model explaining all observations can be built at large–R by assuming two diabatic potential curves, one (π–hole) with the five np electrons on X arranged $\pi^3\sigma^2$ ($^{1,3}\Pi$: strongly repulsive) and the other (σ–hole) $\pi^4\sigma$ ($^3\Sigma^+$: less strongly repulsive) which interact via a single, R–independent spin-orbit coupling constant $\zeta_X(np)$.

Le Roy, *et al.*, (2002) have reviewed all of the different types of experimental observations and theoretical calculations for HI. By an empirical analysis, they have shown that, because in HI the spin-orbit interaction is especially important, the adiabatic *relativistic* potential curves can explain all of the experimental data without introducing residual nonadiabatic coupling. For the lighter halogen hydrides, the $J = 1/2 : J = 3/2$ branching ratio can be obtained from the solution of inhomogeneous coupled equations with a "source" term representing the initial vibrational wavefunction multiplied by the electronic transition moment (Band, *et al.*, 1981). These calculations are based on adiabatic electronic (or diabatic relativistic) potential curves (see, for example, for HCl, Alexander, *et al.*, 1993 and for HBr, Péoux, *et al.*, 1997).

7.2.4 Photofragment Angular Distribution

Until now we have been discussing experiments in which the *total* photodissociation cross sections were measured for all photoproducts ejected in all directions and integrated over all relative velocities, spatial directions, and fragment polarizations (M_J components). If the *differential* cross section is measured as a function of direction (relative to either the laboratory fixed polarization axis or the propagation direction of the photodissociating radiation), supplementary information is provided about the dissociation mechanism. Indeed, the incident light, either through its polarization or its direction of propagation, introduces a privileged axis in space. This spatial anisotropy is transmitted to the angular momentum distributions of the fragments. This anisotropy manifests itself as:

(i) anisotropic angular distributions of the photofragments;

(ii) alignment (or orientation) of the photofragment angular momenta.

There are two kinds of photofragment angular distribution experiments, one in which the angular distribution is measured relative to a coordinate system in which the Z axis is the propagation direction of the light that excites the photodissociative transition (specified by the unprimed θ, ϕ coordinates defined in Fig. 7.8), and another in which the angular distribution is measured relative to a coordinate system in which the Z axis is the linear polarization direction of the photolysis light beam (specified by primed θ', ϕ' coordinates). In both cases an angular distribution of photofragments is measured in which the anisotropy of the distribution is described by a single parameter, β, and the same value of β will be shown to result from either experimental scheme.

Let us begin with the scheme for measuring differential photodissociation cross sections using a beam of unpolarized radiation propagating in the +Z direction as defined by Fig. 7.8. Photofragments are detected along $\hat{\mathbf{k}}$ which has spherical polar coordinates θ, ϕ (θ is the inclination relative to Z, ϕ is the angle relative to X made by the projection of $\hat{\mathbf{k}}$ into the XOY plane). The differential photodissociation cross section $d\sigma/d\Omega$ can be written (Beswick, 1993)

$$\frac{d\sigma}{d\Omega} = \frac{\sigma}{4\pi}[1 - (\beta/2)P_2(\cos\theta)] \tag{7.2.13}$$

in the case of one-photon excitation by unpolarized light. σ is the total photodissociation cross section, $P_2(\cos\theta)$ is the second Legendre polynomial

$$P_2(\cos\theta) = \frac{3\cos^2\theta - 1}{2} \tag{7.2.14}$$

and β is the *asymmetry parameter*, which contains the crucial information about the dissociation mechanism. One can verify that

$$\int \frac{d\sigma}{d\Omega} d\Omega = \int_0^{\pi} \frac{d\sigma}{d\Omega} d\theta \sin\theta \int_0^{2\pi} d\phi = \frac{\sigma}{4\pi} 4\pi = \sigma \tag{7.2.15}$$

because $P_2(\cos\theta)$ averaged over all space is zero. Eq. (7.2.13) is derived by replacing the continuum wavefunction in Eq. (7.2.12a), $|v_E, \Omega\rangle$, by

$$|v_E, \Omega\rangle = (1/4\pi)\left| v_E \sum_{J,M} (2J+1)^{1/2} \mathcal{D}^J_{\Omega,M}(\theta,\phi) \right\rangle \tag{7.2.16}$$

and using the properties of integration over a product of three rotation matrices (see Edmonds, Sect. 4.6, 1974). When $\theta = 0$, $\hat{\mathbf{k}}$ (the location of the particle detector) points in the forward direction and $P_2(\cos\theta) = 1$. For $\theta = \pi$, $\hat{\mathbf{k}}$

points in the backward or recoil direction and again $P_2(\cos\theta) = 1$. $\theta = \pi/2$ corresponds to sideways ejection and $P_2(\cos\theta) = -1/2$. $P_2(\cos\theta)$ can have any value between $+1$ and $-1/2$. Because $\frac{d\sigma}{d\Omega}$ must be non-negative for all θ, β in Eq. (7.2.13) is limited to the range $-1 \le \beta \le +2$.

Another equation, analogous to Eq. (7.2.13), may be derived for experiments in which a beam of linearly polarized light (propagating along X′, polarized along Z′) is used to photodissociate a diatomic molecule. For this scheme, the photofragment ejection direction θ', ϕ' is defined relative to the linear polarization direction, Z′, and the resultant photofragment distribution is described by

$$\frac{d\sigma}{d\Omega'} = (\sigma/4\pi)\,[1 + \beta P_2(\cos\theta')]\,. \tag{7.2.17}$$

Note that, although Eqs. (7.2.13) and (7.2.17) are expressed in terms of the same parameters, the coefficients of β in two equations are not identical. The form of Eq. (7.2.17) also implies that $-1 \le \beta \le +2$. Since σ and β appear in both Eqs. (7.2.13) and (7.2.17), the same information is obtained from both experimental schemes for measuring differential photodissociation cross sections $d\sigma/d\Omega$ (relative to the propagation direction) and $d\sigma/d\Omega'$ (relative to the polarization direction). However, it is usually more convenient to scan θ', by varying the angle of the linear polarization axis (keeping the position of the excitation beam and detector fixed), than to scan the detector position or laser propagation direction, θ, so most differential photodissociation cross sections are measured by varying the polarization direction.

Figure 7.9 illustrates the θ' dependences of $d\sigma/d\Omega'$ for three characteristic values of the asymmetry parameter. For $\beta = 2$, Eqs. (7.2.13) and (7.2.17) respectively reduce to

$$\frac{d\sigma}{d\Omega} = (\sigma/4\pi)\left(\frac{3}{2}\right)\sin^2\theta \tag{7.2.13a}$$

$$\frac{d\sigma}{d\Omega'} = (\sigma/4\pi)3\cos^2\theta' \tag{7.2.17a}$$

and for $\beta = -1$,

$$\frac{d\sigma}{d\Omega} = (\sigma/4\pi)\left(\frac{3}{4}\right)[1 + \cos^2\theta] \tag{7.2.13b}$$

$$\frac{d\sigma}{d\Omega'} = (\sigma/4\pi)\left(\frac{3}{2}\right)\sin^2\theta' \tag{7.2.17b}$$

and for $\beta = 0$, $d\sigma/d\Omega = d\sigma/d\Omega' = (\sigma/4\pi)$. $\beta = 2$ corresponds to photofragments preferentially ejected perpendicular to the light propagation direction or parallel to the light polarization axis (i.e. $\theta = \pi/2$ and any ϕ or $\theta' = 0$ and any ϕ'). $\beta = -1$ corresponds to photofragments preferentially ejected parallel to the light propagation direction or perpendicular to the polarization direction (i.e. $\theta = 0$, any ϕ or $\theta' = \pi/2$, any ϕ'). When $\beta = 0$, the photofragment distribution is isotropic. If either θ or θ' is selected to be the "magic angle" of 54.7°, then $\cos\theta = 0.578$, $P_2(\cos\theta) = 0$, and either $\frac{d\sigma}{d\Omega}$ or $\frac{d\sigma}{d\Omega'}$ is independent of β. This can be useful because all β–dependence of the photofragment distribution is eliminated. In practice, photodissociation product branching ratios are measured at the magic angle because any bias in the branching ratio due to different angular dependences of the detected photofragment J–state is eliminated.

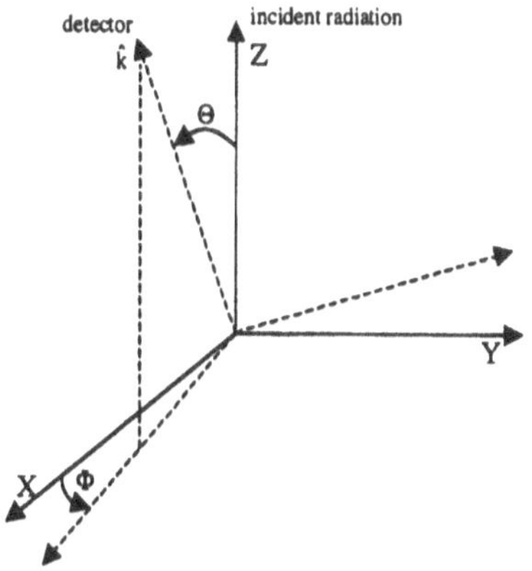

Figure 7.8: Coordinate system defining the direction of the incident radiation ($+Z$) and location of the photofragment detector (along the $\hat{\mathbf{k}}$ axis). The $\hat{\mathbf{k}}$ axis points along the Θ, Φ direction.

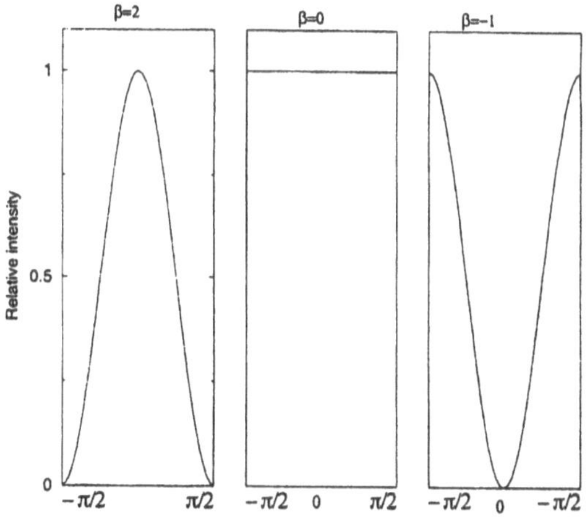

Figure 7.9: Intensity distributions for three important values of β: 2, 0, and -1.

As will be seen in Chapter 8, equations identical to Eq. (7.2.13) and Eq. (7.2.17) describe the photoelectron angular distributions observed in photoionization processes. However, owing to the difference in photofragment masses (atom vs. electron), the photoelectron angular distributions sample the photofragmentation dynamics very differently from photodissociation angular distributions.

For photodissociation, when it is assumed that the dissociation is rapid relative to the rotational period of the molecule (axial recoil approximation) (Zare and Herschbach, 1963) a limiting value of $\beta = +2$ is obtained when dissociation is induced by a $\|$ transition ($\Delta\Omega = 0$, μ parallel to the internuclear axis) and $\beta = -1$ when the dissociative transition is $\perp$ ($\Delta\Omega = \pm 1, \mu$ perpendicular to the internuclear axis).

How do the angular distribution experiments provide insight into the HX photodissociation experiments? We expect that the nominal '$^1\Pi_1$' and '$^3\Pi_1$' states will dissociate adiabatically to $X(^2P_{3/2})$ and will be excited from $X^1\Sigma^+$ via a $\perp$ transition, hence $\beta = -1$. The other nominal states ('$^3\Pi_{0^-}$', '$^3\Pi_2$') which correlate adiabatically to $X(^2P_{3/2})$ should not play a major role in the photodissociation process because they have negligible transition strength from the $X^1\Sigma^+$ state. The nominal '$^3\Pi_{0^+}$', '$^3\Sigma_1^+$', and '$^3\Sigma_{0^-}^+$' state correlate adiabatically to X^* $(^2P_{1/2})$. The '$^3\Sigma_1^+$' state is accessed via a $\perp$ transition (oscillator strength borrowed from $^1\Pi$ via $^1\Pi \sim {}^3\Sigma_1^+$ spin-orbit interaction) with $\beta = -1$. The '$^3\Pi_{0^+}^+$' state is excited via a parallel transition, by means of the diagonal dipole moment of the $X^1\Sigma^+$ state (with which it is mixed by spin-orbit interaction (Péoux, *et al.*, 1997)), with $\beta = +2$. The '$^3\Sigma_{0^-}^+$' state does not contribute significantly to $\sigma(E)$ because it cannot borrow oscillator strength directly from the $^1\Pi$ state. However, the '$^3\Sigma_{0^-}^+$' state does borrow some $^1\Pi$ character indirectly (at $J \geq 1$) from $^3\Sigma_1^+$ via a spin-uncoupling interaction.

Figure 7.10a (HBr) and Fig. 7.10b (HI) illustrate the wavelength dependences of the measured β parameter for HBr and HI. For both HBr and HI we find the expected value of $\beta = -1$ for the $^2P_{3/2}$ photofragment, but for the excited $^2P_{1/2}$ photofragment from HBr β decreases from near +2 at 250nm toward -1 as energy increases because the $\perp$ transition to the '$^3\Sigma_1^+$' state makes an increasingly important contribution at short wavelength. The reason for this dominance of the HBr$\rightarrow$ H + Br* photodissociation via the '$^3\Sigma_{0^+}^+$' and '$^3\Pi_{0^+}^+$' states at longer λ is the mixing with the $X^1\Sigma^+$ ground state. For HI, the $^2P_{1/2}$ production is dominated at all λ by the '$^3\Pi_{0^+}$'$\leftarrow X^1\Sigma^+$ $\|$ transition with $\beta \approx 2$.

Photofragment imaging experiments combine measurements of photofragment kinetic energy (via time-of-flight), β-dependent anisotropy in photofragment angular distribution, and (sometimes) coincidence measurements in which the time and position at which each photofragment, arising from a single photodissociation event, strikes the position-sensitive particle detector (see, for example, Buijsse, *et al.*, 1998 for O_2 and Samartzis, *et al.*, 1997 for Cl_2), and also photofragment alignment and orientation (Section 7.2.5) (see a review paper by Wouters, *et al.*, (2000)). Similar experiments have permitted the observation of a breakdown of the axial recoil approximation in the case of a slow dissociation (e.g., IBr and Br_2). When the dissociation is slow, the molecule can rotate before it dissociates (Wrede, *et al.*, 2002).

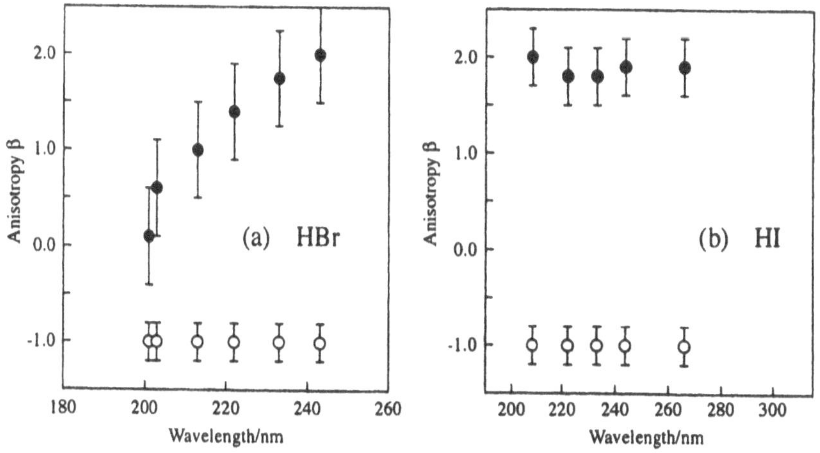

Figure 7.10: Asymmetry parameters, β, determined from the measured ground state H + X (o) and spin-orbit excited H + X^*(•) product angular distributions resulting from the photolysis of (a) HBr and (b)HI, plotted as a function of excitation wavelength (adapted by P. Regan from Ashfold, *et al.*, 1998).

Similar information can also be provided by "fast beam photofragment translational spectroscopy" where the two photofragments from the same fragmentation event are detected in coincidence (Leahy, *et al.*, 1995).

7.2.5 Alignment of the Photofragment

An atom (or a molecule) is said to be polarized with respect to a particular coordinate system (e.g., laboratory or parent molecule body fixed system) if any of its internal angular momenta are either aligned or oriented. The limit of a non-polarized fragment corresponds to equal populations in all the M_J sublevels of a given J. The system exhibits alignment when the populations are unequal in the M_J levels for a given J but cylindrical symmetry is preserved by equal populations in $+M_J$ and $-M_J$ levels. Orientation corresponds to the case where the $+M_J$ and $-M_J$ populations are not equal for any one of the $|M_J|$ values (see Fig. 7.11).

The polarization can be expressed in tensorial form (see Zare, 1988, p. 226-242). Since for any value of J, there are $2J+1$ M_J components, a complete description of the M_J populations and phases within a given J requires information about $(2J+1) \times (2J+1)$ components of the density matrix, ρ_J (see Section 9.1.3), or independent observables. It is convenient to express these quantitites in terms of tensorial components (Section 3.4.5). For $J=0$, there is only 1 component of the ρ_J, and it is a scalar, the population $T^{(0)}(\rho)$. For $J = 1/2$, there are 4 components, a scalar (population), and a vector (orientation) $T^{(1)}(\rho)$. For $J = 1$, there are 9 components, scalar, 1 component (population), vector, 3 components (orientation), second-rank tensor, 5 components, (alignment) $T^{(2)}(\rho)$. As J gets larger, more and more tensors are needed to describe the ρ_J. But if the molecule is probed by absorption of one photon, all of the obtainable information can, in principle, be written in terms of tensors of ranks 0 (population), 1 (orientation), and 2 (alignment).

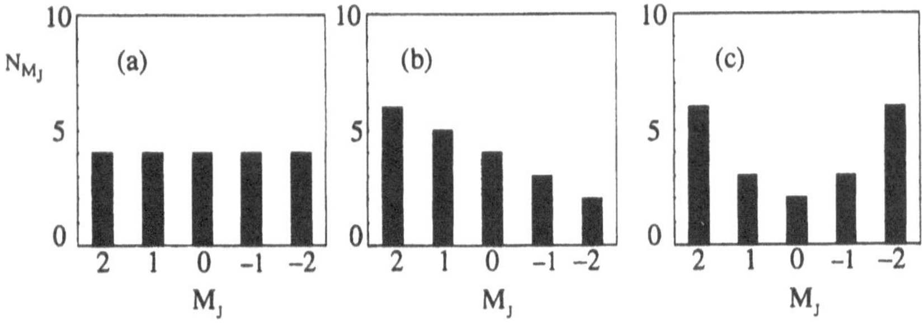

Figure 7.11: M_J–sublevel populations for $J = 2$. (a) no polarization (isotropic); (b) orientation; (c) alignment but no orientation. (from Greene and Zare, 1982).

Aligned photofragments can be produced either by linearly or circularly polarized light. If one uses an absorption-based detection scheme, the orientation of photofragments can be measured only using circularly polarized probe radiation (Siebbeles, *et al.*, 1994). Note that alignment can be created by n–photon excitation. This alignment can be sampled, if a photofragment is produced in an excited state, by the polarization of its fluorescence. This polarization of the emission from aligned photofragments has been predicted by Van Brunt and Zare (1968) and observed in several molecular dissociation processes (see for example the dissociation of Ca_2: Vigué, *et al.*, 1983). This polarization has the form

$$P = \frac{I_\parallel - I_\perp}{I_\parallel + I_\perp}, \qquad (7.2.18)$$

where $I_\parallel$ and $I_\perp$ are respectively the fluorescence intensity polarized parallel and perpendicular to the polarization of the incident photon.

Measurement of the degree of polarization of the fluorescence from an excited atomic fragment can give more information than β about the relative absorption amplitudes for excitation to different dissociative states. The alignment of photofragment emission can give information above the relative phases of the transition moments for photoexcitation of the parent molecule (β samples only the squared transition moments), due to interference effects (Vigué, *et al.*, 1983; Siebbeles, *et al.*, 1994) and about the nonadiabatic interactions in photodissociation (see, for example, in Cl_2, Alexander, *et al.*, 2000).

7.3 Introduction to Predissociation

A predissociation is an indirect dissociation. It manifests itself by decomposition of the molecule when it is excited into a state that is quasi-bound with respect to the dissociation continuum of the separated atoms:

$$AB + hv \rightarrow AB^* \rightarrow A + B. \tag{7.3.1}$$

See Katô and Baba (1995) for a review of predissociation which includes nearly 400 references. This quasi-bound state, AB^*, is called a resonance or a resonant state.

Two cases of predissociation may be distinguished in diatomic molecules.

1. *Predissociation by rotation.* The non-rotating molecule can dissociate only if it is excited above the highest bound vibrational level. This is normal dissociation. If the molecule rotates, a centrifugal potential is added to the electronic potential. This can result in a centrifugal barrier to dissociation whereby the molecule is quasi-bound in rotation-vibration levels at energies above the rotationless dissociation limit but below the top of this centrifugal barrier. Such quasi-bound levels can dissociate through the centrifugal rotational barrier by a tunnelling effect. This is called *predissociation by rotation* (see Fig. 7.12) or tunnelling-predissociation, and is described in detail in Herzberg's book (1950, p. 425).† Predissociation by rotation involves a single potential energy curve and is an example of a "*shape resonance*". The theory of this phenomenon will not be discussed here.

2. *Electronic predissociation.* This type of predissociation is a form of perturbation. A molecule in a vibrational level v_1 of a bound state, which lies at lower energy than the dissociation limit of this state, can dissociate into atoms by mixing with the continuum "level" v_E that belongs to another potential curve. These two potential curves may or may not cross. For the crossing curve case, see Fig. 7.13. This two-electronic-state process is the type of predissociation discussed in this chapter. Electronic predissociation involves two potential energy curves and is an example of a "*Feschbach resonance.*" Different curve-crossing cases are distinguished in Section 7.7. The experimental characteristics of predissociation are presented first. This is followed by a discussion of the sorts of information that can be deduced from experimental data, pertaining to both the nature of the repulsive state and the mechanism of the predissociation.

† When there is a barrier in the rotationless potential, there will also be quasi-bound rotation-vibration levels that dissociate by tunnelling through the combined intrinsic and centrifugal barrier (Herzberg, 1950, p. 429).

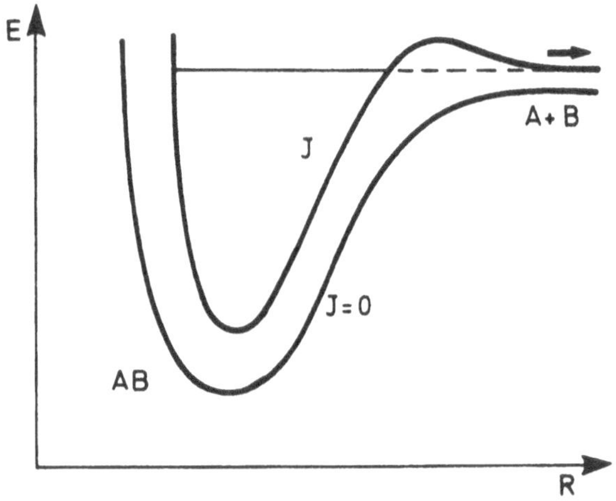

Figure 7.12: Predissociation by rotation. The $J \neq 0$ potential is obtained by adding the centrifugal energy, $\hbar^2 J(J+1)/2\mu R^2$, to the $J=0$ potential.

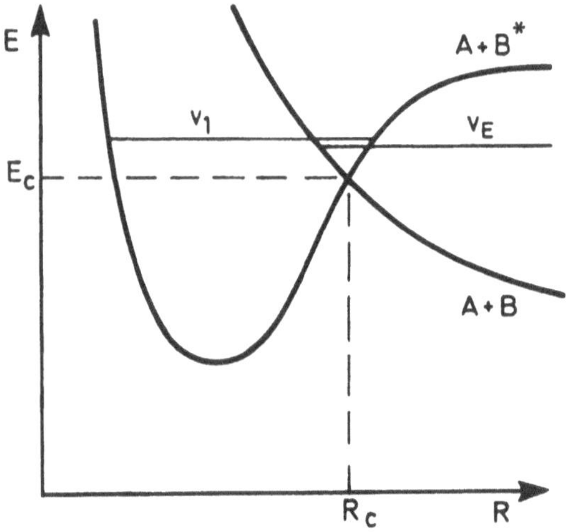

Figure 7.13: Predissociation. The bound v_1 vibrational level of an electronic state dissociating into excited atoms (A+B*) is predissociated by the continuum (v_E) of another state dissociating to ground state atoms. The two potential energy curves intersect at E_C, R_C.

7.4 Experimental Aspects of Predissociation

There are two complementary ways of looking at the phenomenon of predissociation: *time-dependent* and *time-independent* pictures.

1. *Time-dependent picture.* Experimental observations can be discussed in terms of a competition between two processes: the radiative process (absorption or emission), characterized by a radiative rate τ_r^{-1}, and a nonradiative process, predissociation, characterized by a nonradiative rate τ_{nr}^{-1}. There are two possible decay pathways for the excited state,

$$\mathrm{AB}^* \rightarrow \mathrm{AB} + h\upsilon \qquad \text{radiative} \tag{7.4.1}$$

or

$$\mathrm{AB}^* \rightarrow \mathrm{A} + \mathrm{B} \qquad \text{nonradiative.} \tag{7.4.2}$$

Predissociation is a nonradiative process because some of the internal energy of the predissociated state is transformed into kinetic energy of the fragment atoms rather than into radiation.

Only the total lifetime τ of a level can be measured. τ is related to the rate of decrease of the number of molecules initially in a given level via both radiative and nonradiative routes. Let k_r be the radiative rate constant (the probability per unit time that a molecule will leave the level as a result of emission of a quantum of light) and k_{nr} the predissociation rate (the dissociation probability per unit time). Recall that the pressure is assumed to be low enough that the rates are not affected by collisions. The number of molecules leaving the initial state during the time interval dt is given by

$$dN = -k_r N dt - k_{nr} N dt, \tag{7.4.3}$$

with the solution

$$N(t) = N_0 e^{-(k_r + k_{nr})t}, \tag{7.4.4}$$

N_0 being the initial population, and the emitted radiation has intensity proportional to $N(t)$. The lifetime is defined by

$$N(t) = N_0 e^{-t/\tau}, \tag{7.4.5}$$

and thus

$$1/\tau = k_r + k_{nr} = 1/\tau_r + 1/\tau_{nr}. \tag{7.4.6}$$

Assume that the continuum state involved in the predissociation is not radiatively coupled to the initial state of the quasi-bound ← bound transition (a restriction that is relaxed in Section 7.9). This is a very common situation that often occurs because the continuum-state turning point is located outside of the Franck-Condon range of allowed vertical transitions out of the initial bound state. The continuum $\leftarrow$ bound optical transition may also be forbidden by electric dipole selection rules. If the molecule is excited by a pulse of light of shorter duration than τ_{nr}, it can be shown that the system is prepared in the bound *basis* function. Owing to the perturbation interaction between bound and continuum basis states, the prepared nonstationary state decays nonradiatively into the continuum. The rate of this bound $\rightarrow$ continuum radiationless transition, $1/\tau_{nr}$, may be measured in a time-resolved experiment.

A relative magnitude for the predissociation rate can be obtained by measuring the fluorescence quantum yield (see Eq. (1.2.3)),

$$\phi_f = k_r/(k_r + k_{nr}) = \tau/\tau_r. \tag{7.4.7}$$

The dissociative quantum yield is

$$\phi_d = 1 - \phi_f = \tau/\tau_{nr}. \tag{7.4.8}$$

Since experiments generally determine k_{nr} relative to k_r, it is natural to assess the strength of a predissociation relative to τ_r.

2. *Time-independent picture.* The opposite extreme from short-pulse excitation involves the use of nearly monochromatic radiation. Practically, this means that the interaction between molecule and radiation field is of longer duration than τ_{nr}. In this limit, the quantity measured is the absorption lineshape. It will be shown below that the linewidth observed in an energy-resolved experiment is related in a very simple way to the predissociation lifetime in the time-resolved experiment.

The effect of predissociation on spectral features depends on whether it is the initial or final state of the transition that is predissociated. Predissociation can be detected either by direct measurements of lifetimes (τ), linewidths (Γ), or level shifts (δE), or indirectly by observation of fragments. Table 7.2 surveys the range of predissociation rates sampled by different methods. Erman (1979) has reviewed the experimental methods for characterizing predissociation phenomena.

7.4.1 Measurement of Lifetimes

Consider first predissociation of the initial state of a transition (i.e., the upper state in emission or the lower state in absorption).

If there is only a barely observable weakening of the lines in emission, the nonradiative lifetime has the same order of magnitude as the radiative lifetime. For an allowed transition, the usual order of magnitude for τ_r is 10^{-8} s; thus, a weakening of the emission can be seen if predissociation corresponds to τ_{nr} around 10^{-8} s, as for example in the predissociation of the N_2^+ $C^2\Sigma_u^+$ state (see Table 7.3). If the transition is weak or nominally forbidden, much weaker predissociations can be detected. In the case of the I_2 $B^3\Pi_{0_u^+}-X^1\Sigma_g^+$ transition ($\Delta S = 1$), with a lifetime of 10^{-6}s, weak predissociations ($\tau_{nr} = 10^{-6} - 10^{-7}$s) have been observed. This also depends on the sensitivity of the technique. Brzozowski, *et al.*, (1976) have detected a very small decrease in the measured lifetimes of CH $A^2\Delta$ for rotational levels above the dissociation limit. This decrease corresponds to a predissociation rate of $8 \times 10^5\ s^{-1}$ (Fig. 7.14). Measurements of the lifetime of a level are useful for detecting weak predissociation, but only if a reliable model exists for the v, J–dependence of the radiative lifetime, τ_r.

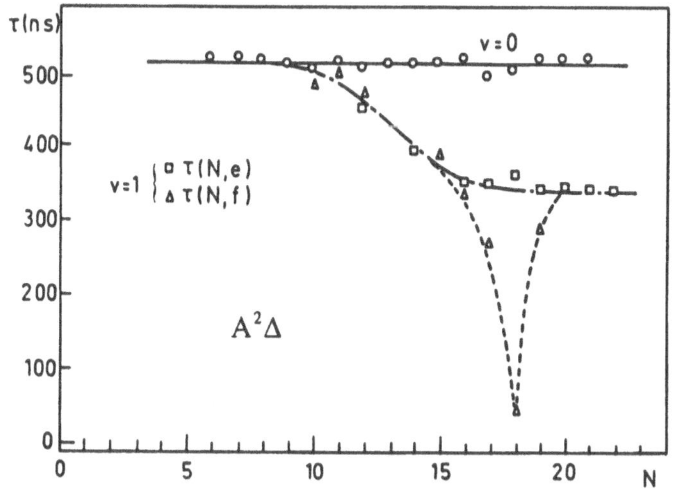

Figure 7.14: Measured lifetimes of the CH $A^2\Delta$ state. The decay of the $v = 0$ level is purely radiative. The lifetime of the $v = 1$ level indicates a weak predissociation which has been attributed to the continuum of the $X^2\Pi$ state. The shorter lifetimes for the f–levels of the F_1 component probably reflect an X∼ B∼ A accidental predissociation (see Section 7.13) through the $B^2\Sigma^-$ state. [From Elander, *et al.*(1979) based on data of Brzozowski, *et al.*(1976).]

The mechanism of a predissociation may be characterized by measurements of the lifetime, τ_v, of each vibrational level or, even better, τ_{vJ}, of every rotational level, carefully extrapolated to zero pressure. When the two unknown rates that appear in Eq. (7.4.6) have similar magnitudes, it is necessary to partition the observed total decay rate into τ_r and τ_{nr}. If the radiative lifetime is known for a non-predissociated level of the same electronic state, τ_r can be calculated for predissociated levels assuming an R–independent value for the electronic transition moment. The nonradiative lifetime is then deduced by subtraction of $1/\tau_r$ from the experimental $1/\tau$ value as follows:

$$\frac{1}{\tau(\text{expt})} - \frac{1}{\tau_r(\text{calc})} = \frac{1}{\tau_{nr}}. \qquad (7.4.9)$$

To obtain the exact variation with v and J of τ_{nr}, the dependence of the electronic transition moment on R must be taken into account in the calculation of τ_r. Note that the lifetime of a state generally varies very little with v.

A rapid decrease of the lifetime with v reveals either an unusually large variation of the electronic transition moment with R or, more likely, the presence of predissociation (or of perturbation, see Section 6.3.1).

A direct measure of the ratio of the radiative $(1/\tau_r)$ and dissociative $(1/\tau_{nr})$ rates is possible when the nonradiative channel exclusively forms an electronically excited atom (see Section 7.4.4).

If the nonradiative lifetime becomes much shorter than the radiative lifetime, lines that are quite sharp and strong in an absorption spectrum can be entirely absent from an emission spectrum (Fig. 2.3a and b). The loss of emission intensity can occur quite abruptly, and the rotational lines of a band or a vibrational progression appear to be cut off. The molecule follows a dissociative route rather than a radiative one. The weakness or the absence of such strongly predissociated emission lines prevents direct measurement of their lifetimes. Only an upper limit for the nonradiative lifetime can then be obtained.

When the predissociation rate is so much larger than the radiative decay rate that the fluorescence quantum yield is too low to measure a radiative decay rate directly, it is possible to infer the decay rate of the parent molecule from the effect of a static magnetic field on the polarization of a photofragment (Buijsse and van der Zande, 1997).

Recall that $10^{-9}s$ = 1 nanosecond = $1ns$; $10^{-12}s$ = 1 picosecond = $1ps$; $10^{-15}s$ = 1 femtosecond = $1fs$; $10^{-18}s$ = 1 attosecond = $1as$.

7.4.2 Measurement of Linewidths

Consider now predissociation of the final state of a transition (upper state in absorption or lower state in emission). It is often the upper state of an absorption transition that is predissociated, but the effect of the predissociation on spectral linewidths is identical to the case when the lower state of an emission transition is predissociated (for example, the $B^2\Sigma^+ \rightarrow A^2\Sigma^+$ system of OH where the A state is strongly predissociated). In either case, the spectrum consists of diffuse lines (see Fig. 7.15 for predissociation of the Se_2 $B0_u^+$ state). The predissociated state obtains its diffuse character from the continuum state.

To treat predissociation quantitatively, precise measurements of the width (full width at half maximum, FWHM, or Γ, often misleadingly referred to as the half-width) of each rotational level are necessary. In the absence of Doppler broadening, the lineshape $\sigma_a(E)$ is usually Lorentzian,

$$\sigma_a(E) = \frac{\sigma_{\max}(\Gamma/2)^2}{(E - E_r)^2 + (\Gamma/2)^2}, \tag{7.4.10}$$

where Γ is related to the lifetime of the state (see Section 7.5) by

$$\Gamma(\text{cm}^{-1}) = (2\pi c\tau)^{-1} = 5.3 \times 10^{-12}/\tau(s). \tag{7.4.11}$$

Table 7.2: Order of Magnitude of Predissociation Rates

$\tau_{nr}(s)$	$\tau_r(s)$	$\Gamma_{nr}(\text{cm}^{-1})$	Molecule	State	Ref.[a]	v'	J' or N'	Detection Technique
				Measurements of τ				
10^{-5}	5×10^{-8}	5×10^{-7}	K_2	$C^1\Pi_u$	(1)	2 and 10		Selectively detected laser-induced fluorescence (K atom $4^2P_{3/2} \rightarrow 4^2S$)
1.2×10^{-6}	5×10^{-6}	4×10^{-6}	CH	$A^2\Delta$	(2)	1	16	High-frequency deflection
10^{-9}	10^{-8}	5×10^{-3}	N_2^+	$C^2\Sigma_u^+$	(3)	6		Fluorescence quantum yield
				Measurements of Γ				
0.5×10^{-9}		10^{-2}	O_2^+	$b^4\Sigma_g^-$	(4)	5	9	Laser photofragment spectroscopy
$2 \times 10^{-10} - 4\times 10^{-11}$	10^{-6}	0.025-0.125	ICl	$B^3\Pi_{0^+}$	(5)			Fabry-Perot interferometric spectroscopy
1.8×10^{-11}		0.3	H_2	$B''^1\Sigma_u^+$	(6)	2	2	V.U.V. laser spectroscopy
5×10^{-12}		1.0	O_2	$B^3\Sigma_u^-$	(7)	4		Photoelectrically recorded absorption spectroscopy
2×10^{-14}		300	N_2	$C^3\Pi_u$	(8)	6		Electron energy loss spectroscopy

[a] References: (1) Meiwes and Engelke (1982); (2) Brzozowski, *et al.*(1976); (3) Govers, *et al.*(1975); (4) Carré, *et al.*(1980) and references therein; (5) Olson and Innes (1976); (6) Rothschild, *et al.*(1981); (7) Lewis, *et al.*(1980); (8) Mazeau, private communication.

Table 7.3: Well-Characterized Weak Predissociations

Molecule	Ref.[a]	Predissociated State	Predissociating State	$\tau(s)$	Electronic Matrix Element (calc.)	Origin of the Interaction
OH	(1)	$A^2\Sigma^+$	$^4\Sigma^-$	2×10^{-10} $(v = 3, N = 0)$	17 cm^{-1}	Spin-orbit
N_2^+	(2)	$C^2\Sigma_u^+$	$B^2\Sigma_u^+$	10^{-9} $(v = 6)$	3.6 $Å^{-1}$	Nuclear kinetic energy operator
I_2	(3)	$B^3\Pi_{0_u^+}$	$^1\Pi_{1_u}$	10^{-6} $(v = 9, J = 10)$	3×10^{-2} cm^{-1}	Hyperfine
				7×10^{-7} $(v = 9, J = 20)$	3×10^{-4} dimensionless	Gyroscopic (second-order effects)
O_2^+	(4)	$b^4\Sigma_g^-$	$d^4\Sigma_g^+$	10^{-10} $(v = 5, N = 9)$	11 cm^{-1}	Spin-orbit (configuration-interaction)
NH	(5)	$A^3\Pi$	$^5\Sigma^-$	7×10^{-7} $(v = 2, N = 2)$	34 cm^{-1}	Spin-orbit
	(6)	$C^1\Pi$	$^5\Sigma^-$	7×10^{-8} $(v = 1)$	0.88 cm^{-1}	Spin-spin and indirect predissociation
Li_2	(7)	$A^1\Sigma_u^+$	$b^3\Pi_u$	1.8×10^{-8} $(v = 3, J = 48)$	[b]	Indirect Predissociation
H_2	(8)	$D'\ ^1\Pi_u$	$B'\ ^1\Sigma_u^+$	10^{-9}	[b]	Indirect Predissociation

[a] References: (1) Yarkony (1992); (2) Roche and Tellinghuisen (1979) (factorization of the vibrational part not possible due to strong R–variation of H^e); (3) Broyer, *et al.*(1976); (4) Carré, *et al.*(1980), (5) Patel-Misra, *et al.*(1991); (6) Bohn, *et al.*(1992); (7) Schmidt, *et al.*(1988); (8) Gao (1997).

[b] Two matrix elements are needed to describe an indirect predissociation.

Table 7.4: Well-Characterized Strong Predissociations

Molecule	Ref.[a]	Predissociated State	Predissociating State	$\Gamma_{max}(cm^{-1})$ Experimental (except Se_2,BeH)[b]	$\langle\chi_v\|\chi_E\rangle^2$ Differential Franck-Condon Factor (cm)	H^e Electronic Matrix Element (calc.)	Origin of the Interaction
H_2	(1)	$D^1\Pi_u$	$B'^1\Sigma_u^+$	4 ($v = 4, J = 1$)		$(2)^{1/2}$ (dimensionless)	Gyroscopic
OH	(2)	$A^2\Sigma^+$	$^4\Pi$	1.47 ($v = 6$)	7.5×10^{-5}	57 cm^{-1}	Spin-orbit
O_2	(3)	$B^3\Sigma_u^-$	$^5\Pi_u$	3 ($v = 4$)	1.1×10^{-4}	65 cm^{-1}	Spin-orbit
Se_2	(4)	$B0_u^+$	0_u^+	180 ($v = 24$)	2×10^{-4}	373 cm^{-1}	Spin-orbit
NO	(5)	$I^2\Sigma^+$	$A'^2\Sigma^+$	9 ($v = 6$)	3.6×10^{-5}	200 cm^{-1}	Electrostatic
NO	(6)	$Q^2\Pi$ (Rydberg)	$B^2\Pi$ (valence)	8 ($v = 2$)	3.5×10^{-6}	660 cm^{-1}	Electrostatic
BeH	(7)	$^2\Sigma^+$ (Rydberg)	$^2\Sigma^+$ (valence)	250 ($v = 0$)	1.1×10^{-4}	600 cm^{-1}	Electrostatic
O_2	(8)	$F^3\Pi_u$ (Rydberg)	$^3\Pi_u$ (valence)	6.4 ($v = 2$)	[c]	7100 cm^{-1}	Electrostatic
	(9)	$C^3\Pi_g$ (Rydberg)	$^3\Pi_g$ (valence)	90 ($v = 3$)	[c]	750 cm^{-1}	Electrostatic
	(9)	$d^1\Pi_g$ (Rydberg)	$^1\Pi_g$ (valence)	70	[c]	860 cm^{-1}	Electrostatic
HBr^+	(10)	$A^2\Sigma^+$	$^4\Pi, {}^4\Sigma^-, {}^2\Sigma^-$	800 ($v = 4$)	[d]	790 cm^{-1}	Spin-orbit
CO	(11)	$B^1\Sigma^+$ (Rydberg)	$D'\ {}^1\Sigma^+$ (valence)	1 ($v = 2$)	[c]	2900 cm^{-1}	Electrostatic

[a] References: (1) Herzberg (1971); (2) Czarny, *et al.*(1971); (3) Julienne and Krauss (1975); (4) Atabek and Lefebvre (1972); (5) Gallusser (1976); (6) Miescher (1976); (7) Lefebvre-Brion and Colin (1977); (8) Li, *et al.*(1998); (9) Li, *et al.*(1997); (10) Penno, *et al.*(1998); (11) Tchang-Brillet, *et al.*(1992).

[b] Theoretical calculations for Se_2 from ref. (4); BeH from ref. (7).

[c] The Golden Rule is not valid.

[d] There are several states which contribute to the predissociation.

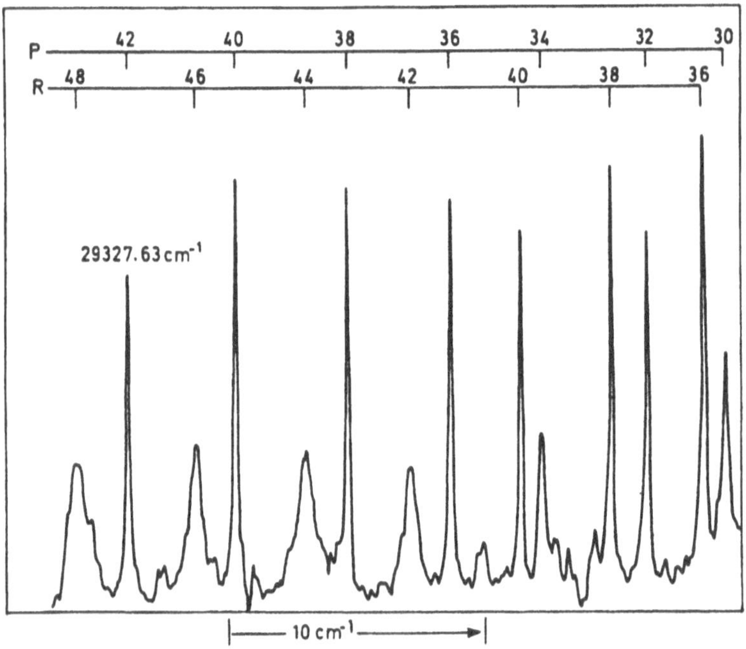

Figure 7.15: The diffuse 15-0 Band of the B 0_u^+ –X 0_g^+ system of $^{78}Se_2$ observed in absorption. The linewidths increase as J' increases. [Courtesy R.F. Barrow from Barrow, *et al.*(1966).]

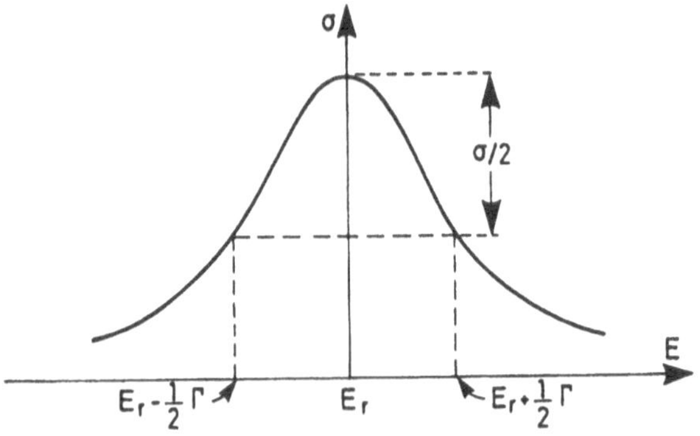

Figure 7.16: A Lorentzian lineshape.

For $E = E_r$ (the energy including the predissociation-induced energy shift, called the resonance energy), the absorption has its maximum value, $\sigma_a(E_r) =$

σ_{max}. For $E = E_r \pm \Gamma/2$, the absorption is $\sigma_{max}/2$ (see Fig. 7.16). Asymmetric line profiles will be discussed in Section 7.9.

In typical experiments, only strong predissociations that result in a nonradiative lifetime much shorter than the radiative lifetime can be detected by line broadening, since the usual radiative lifetime of 10^{-8}s corresponds to a width of only 5×10^{-4} cm^{-1}. Doppler-free spectroscopic techniques have made it possible to measure extremely small predissociation linewidths (Carré, *et al.*, 1980; Carrington, *et al.*, 1978).

For a nonradiative lifetime of 10^{-13}s, the linewidth will be 50 cm^{-1}, causing the rotational structure to disappear and the band to become diffuse. For a shorter nonradiative lifetime, 10^{-15}s, the vibrational structure disappears and the spectrum becomes similar to a continuous spectrum. As spectral lines become broader and broader, their intensity is spread out and becomes lost in the background. However, low-resolution techniques, such as photoelectron spectroscopy or electron impact spectroscopy, can enable detection of strongly predissociated bands (see Table 7.2).

Doppler broadening has a Gaussian lineshape, and its convolution with the Lorentzian natural lineshape yields a Voigt profile. In typical experiments, this effect can be neglected since the Doppler width is usually much smaller than the resolution of the apparatus. Collisional line broadening is also Lorentzian, and the Lorentzian component of measured lines must be carefully extrapolated to zero pressure.

Measurements by photographic photometry require careful calibration due to the nonlinear response of photographic plates: saturation effects can lead to erroneous values. Line profiles can be recorded photoelectrically, if the stability of the source intensity and the wavelength scanning mechanism are adequate. Often individual rotational lines are composed of incompletely resolved spin or hyperfine multiplet components. The contribution to the linewidth from such unresolved components can vary with J (or N). In order to obtain the FWHM of an individual component, it is necessary to construct a model for the observed lineshape that takes into account calculated level splitttings and transition intensities. An average of the widths for two lines corresponding to predissociated levels of the same parity and J-value (for example the P and R lines of a $^1\Pi - {}^1\Sigma^+$ transition) can minimize experimental uncertainties. A theoretical Lorentzian shape is assumed here for simplicity, but in some cases, as explained in Section 7.9, interference effects with the continuum can result in asymmetric Fano-type lineshapes.

7.4.3 Energy Shifts

In addition to line broadening, the predissociation process can cause line shifts. Each discrete or diffuse level can be shifted by its interaction with the entire continuum of the predissociating state, but this effect is considerably smaller than level shifts caused by interactions between discrete levels. The orders of magnitude of predissociation-induced level shifts and linewidths are comparable.

Thus, in classical spectroscopy, only predissociations observable in absorption, having a width of at least 10^{-1} cm^{-1}, can be accompanied by detectable level shifts. Despite the common practice of ignoring predissociative level shifts, the shifts can occasionally be larger than the corresponding level widths. However, determination of level shifts requires a good model for the unshifted line positions. This effect is best detected by studying, not the level shift of the band origin, but the second vibrational difference $\Delta^2 G_v$, where the presence of irregularities may be due to this effect. An example is given in Fig. 7.17 from the very careful study of the predissociation of the O_2 $B^3\Sigma_u^-$ state by Julienne and Krauss (1975).

In principle, rotational constants and fine structure parameters of the predissociated state may also be affected by the predissociating state, as in the case of perturbations. Levels below the dissociation limit, which therefore cannot be predissociated, will also be shifted by their interaction with the continuum. In the case of strong predissociation, only this level shift affecting bound (sharp) levels will be seen, because the levels above the crossing point become too diffuse to be observed. This is the situation for the homogeneously predissociated Se_2 $B0_u^+$ (Atabek and Lefebvre, 1972) and NO $G^2\Sigma^-$ states (Ben-Aryeh, 1973).

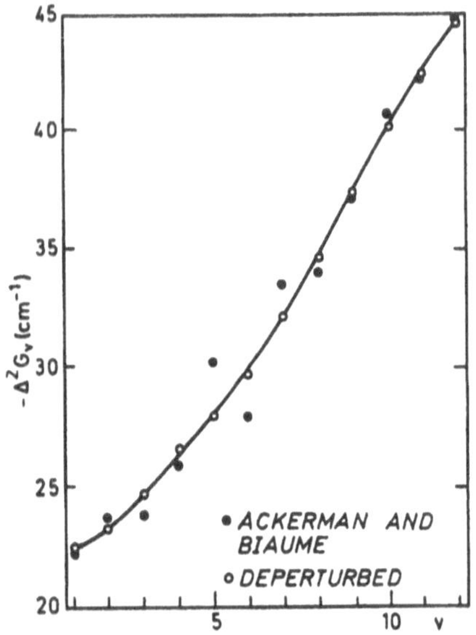

Figure 7.17: $\Delta^2 G(v)$ differences in the O_2 $B^3\Sigma_u^-$ state. The filled circles show the measured values of Ackerman and Biaume (1970). The open circles show the deperturbed values obtained by subtracting the calculated level shifts from the observed G_v values. [From Julienne and Krauss (1975).]

7.4.4 Detection of Fragments

Laser photofragment spectroscopy is a sensitive indirect method for characterizing certain predissociation processes. A beam of molecular ions is excited by a laser while the intensity variation of a mass-selected photofragment ion is monitored versus laser frequency. By measuring the kinetic energy (i.e., time-of-flight) of each photofragment, the electronic states of the dissociation products may be identified. Extremely small predissociation rates may be detected, since only dissociation can result in a signal at the selected fragment-ion mass. Structure in the photodissociation cross-section versus laser frequency reflects the vibrational (and possibly rotational) structure of the predissociated state. An example is given by the predissociation of the $b^4\Sigma_g^-$ state of O_2^+ (Carrington, *et al.*, 1978; Moseley, *et al.*, 1979; Carré, *et al.*, 1980).

The photofragment angular distribution can indicate the electronic symmetry of the predissociating state (Zare, 1972).

Another possibility involves monitoring the light emitted by a specific electronically excited atomic dissociation product versus laser excitation frequency (Borrell, *et al.*, 1977; Meiwes and Engelke, 1982).

7.5 Theoretical Expressions for Widths and Level Shifts

Mixing of the bound $\Psi_{1,v,J}$ and continuum $\Psi_{2,E,J}$ states is governed by an interaction matrix element,

$$H_{v,J;E,J} = \langle \Psi_{1,v,J} | \mathbf{H} | \Psi_{2,E,J} \rangle = \langle \phi_1(r,R)\chi_{v,J}(R) | \mathbf{H} | \phi_2(r,R)\chi_{E,J}(R) \rangle \,. \quad (7.5.1)$$

The vibrational energy of the continuum state is not quantized; consequently, the vibrational wave functions for energies above the dissociation limit are labeled by the good quantum number, E. These continuum functions are "energy-normalized," rather than "space-normalized" as are the bound vibrational wavefunctions.

One way to understand the necessity for energy normalization and the relationship between energy- and space-normalized functions is to convert the dissociation continuum artificially into a discrete (quasi-continuous) spectrum by adding a vertical and infinite barrier to dissociation at very large internuclear distance. If the location of this outer wall is moved to larger and larger internuclear distances, the amplitude of each discrete, space-normalized wavefunction becomes smaller and smaller as the density of discrete levels, ρ (inverse of the energy interval between two successive levels), becomes larger and larger. The amplitude decrease occurs because the wavefunction occupies an increasingly large region of space and is normalized so that there is unit probability of finding the system in this spatial region. As the barrier is moved to infinity, the amplitude becomes zero and the density of states becomes infinite (the spectrum becomes continuous). Fortunately, the product of the amplitude times the density of states goes to a constant, finite value as the barrier goes to infinite internuclear distance.

The idea of energy normalization can be extended to levels in the discrete region of the spectrum as follows:

$$\chi_E(R) = [\rho(E)]^{1/2}\chi_v(R) \tag{7.5.2}$$

or

$$R^{-1}\xi_E(R) = [\rho(E)]^{1/2}R^{-1}\xi_v(R) \tag{7.5.3}$$

where the density, ρ, is the inverse of the separation between two successive vibrational levels or, more rigorously, dv/dE. The space-normalized (with respect to dR) function, $\xi_v(R)$, has the dimensionality of $R^{-1/2}$, and ρ has that of E^{-1}; thus the energy-normalized function, ξ_E of Eq. (7.5.3) must have the same dimensionality as the energy-normalized continuum function, namely, $E^{-1/2}R^{-1/2}$. Matrix elements of this type of function can only be compared to the corresponding matrix elements of the continuum function. For example, it is well known that the differential oscillator strength, $df/d\nu$, varies continuously through a dissociation limit (Allison and Dalgarno, 1971; Smith, 1971). Figure 7.18 shows the continuous variation of differential oscillator strength for the O_2 Schumann-Runge system. Below the dissociation limit, the differential oscillator strength for absorption from initial v'' to final discrete v',

$$\rho(E_{v'})f_{v'v''} = (dv'/dE)f_{v'v''}, \tag{7.5.4}$$

is plotted versus E; above the limit, the corresponding continuum quantity,

$$df_{E,v''}/dE,$$

is plotted.

Fano's theory (1961) relates the width of a predissociated level to the interaction matrix element of Eq. (7.5.1). It can be shown that the discrete state amplitude, $a(E)$, in the continuum eigenfunction,

$$\Psi_{E,J} = a(E)\Psi_{1,v,J} + \int dE' b_{E'}(E)\Psi_{2,E',J} \tag{7.5.5}$$

is given by

$$a(E) = \frac{H_{v,J;E,J}}{E - E_r + i\pi\left|H_{v,J;E,J}\right|^2} \tag{7.5.6}$$

where $H_{v,J;E,J}$ is the matrix element between discrete and *energy*-normalized continuum wavefunctions. When $H_{v,J;E,J}$ varies slowly with energy, $|a(E)|^2$ is a Lorentzian function [cf. Eq. (7.5.6)] with FWHM

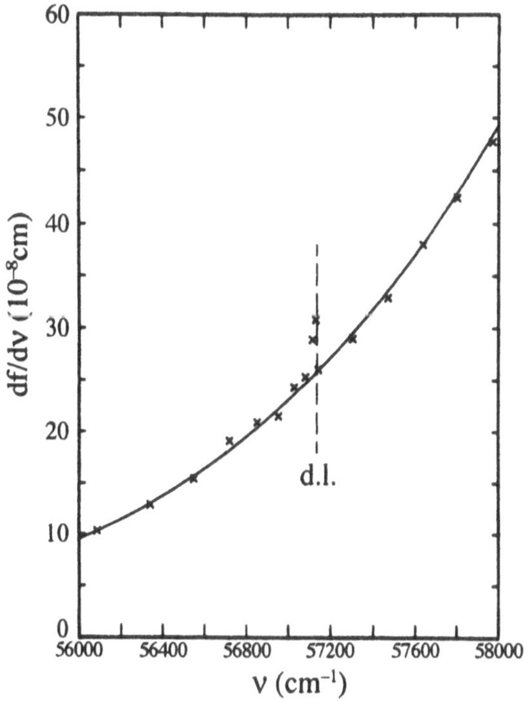

Figure 7.18: Differential oscillator strengths obtained from band and continuum measurements of the $v'' = 0-1$ progressions of the $B^3\Sigma_u^- - X^3\Sigma_g^-$ system of O_2 near the dissociaton limit (d.l.). Continuity is observed across the limit (from Lewis, *et al.*, 1986).

$$\Gamma_{v,J} = 2\pi \left|H_{v,J;E,J}\right|^2 \tag{7.5.7a}$$

with the prescription that the coupling is calculated with E equal to the energy of the discrete state. Another equivalent expression for Γ explicitly includes $\rho(E)$,

$$\Gamma_{v,J} = 2\pi\rho(E) \left|H_{v,J;E,J}\right|^2, \tag{7.5.7b}$$

but this expression implies a matrix element of $\mathbf{H}$ evaluated between space-normalized wavefunctions. This expression for $\Gamma_{v,J}$ is the well-known Fermi-Wentzel Golden Rule.

Similarly, Fano's theory requires that the continuum levels in the interval E' to $E' + dE'$, where $E' \neq E^{\circ}_{v,J}$, contribute an energy shift

$$dF(E') = \frac{\left|H_{v,J;E,J}\right|^2}{E^0_{v,J} - E'} dE'. \tag{7.5.8}$$

This formula is analogous to second-order perturbation theory for the case of bound states, but the total level shift, $\delta E_{v,J}$, is obtained by replacing the summation over discrete perturbers by an integration.

$$E_{v,J} = E^0_{v,J} + \delta E_{v,J} = E^0_{v,J} + \mathcal{P} \int dF(E'). \tag{7.5.9}$$

Care must be taken to avoid the singularity in the integration at $E = E'$, for example, by analytic integration in the region of the singularity (Atabek and Lefebvre, 1972), where $\mathcal{P}$ means "principal part."

If one views predissociation as a time-dependent process, then it is possible to derive the relationship between the FWHM of the predissociated line (assuming no oscillator strength from the continuum basis functions) and the predissociation rate. At time $t = 0$, let the system be prepared with unit amplitude in the discrete state,

$$A(0) = \langle \Psi_{1,vJ} | \Psi(t = 0) \rangle = 1.$$

Then, at time t, the amplitude of the discrete state component is

$$A(t) = \langle \Psi_{1,vJ} | \Psi(t) \rangle = \int_{-\infty}^{\infty} dE \exp[-(i/\hbar)Et] |a(E)|^2, \tag{7.5.10}$$

where $a(E)$ has been given in Eq. (7.5.6). By the definition of $A(t)$, $|a(E)|^2$ and $A(t)$ are related by a Fourier transform and (Fontana, 1982, p. 30)

$$A(t) \propto \exp[-\Gamma t/2\hbar] \exp[-(i/\hbar)E_r t]. \tag{7.5.11}$$

The time evolution of the discrete state *probability* is given by

$$|A(t)|^2 \propto \exp(-\Gamma t/\hbar), \tag{7.5.12}$$

and, by comparison with Eq. (7.4.5),

$$1/\tau = \Gamma/\hbar \quad \text{or} \quad \Gamma\tau = \hbar, \tag{7.5.13}$$

which is the minimum linewidth compatible with the Heisenberg uncertainty principle. Achievement of this minimum Γ (or maximum τ) is a consequence of assuming either a perfect Lorentzian lineshape or perfect single-exponential decay and level shift of zero. This ideal level shift of zero would occur if H were independent of E and the continuum extended to infinity above and below E.

If the electronic matrix element in Eq. (7.5.1) is assumed to be independent of R, or to vary linearly with R, the matrix element $H_{v,J;E,J}$ can be factored into two parts:

$$\begin{aligned} H_{v,J;E,J} &= \langle \phi_1(r,R) | \mathbf{H} | \phi_2(r,R) \rangle \, \langle \chi_{v,J}(R) | \chi_{E,J}(R) \rangle \\ &= H^e \, \langle \chi_{v,J}(R) | \chi_{E,J}(R) \rangle \end{aligned} \tag{7.5.14}$$

where H^e is either the constant electronic interaction or its value taken at the R–centroid defined by

$$\overline{R}_{v,E} = \frac{\langle \chi_{v,J}(R)|R|\chi_{E,J}(R)\rangle}{\langle \chi_{v,J}(R)|\chi_{E,J}(R)\rangle}. \tag{7.5.15}$$

When two potential curves cross, the value of the R–centroid between isoenergetic levels is, as for bound-state perturbations, the internuclear distance at the curve crossing point, R_C. This approximation predicts a FWHM of

$$\Gamma_{v,J} = 2\pi |H^e|^2 \langle \chi_{v,J}|\chi_{E,J}\rangle^2 . \tag{7.5.16}$$

$\langle \chi_{v,J}|\chi_{E,J}\rangle^2$ is a differential Franck-Condon factor because the continuum function $\chi_{E,J}$ is energy-normalized. It has the dimensionality of E^{-1}, whereas the usual Franck-Condon factor is dimensionless. Contrary to the slow variation of H^e with energy, the differential Franck-Condon factor often varies rapidly and in an oscillatory manner with energy.

In the case of J–dependent $\Delta\Omega \neq 0$ interaction matrix elements, the same approximation can be made, giving the following expression for the width:

$$\Gamma_{v,J} = 2\pi (H^e)^2 \langle \chi_{v,J}|\hbar^2/2\mu R^2|\chi_{E,J}\rangle^2 [J(J+1) - \Omega(\Omega \pm 1)] \tag{7.5.17}$$

and

$$\langle \chi_{v,J}|\hbar^2/2\mu R^2|\chi_{E,J}\rangle \simeq (\hbar^2/2\mu R_C^2) \langle \chi_{v,J}|\chi_{E,J}\rangle = B_C \langle \chi_{v,J}|\chi_{E,J}\rangle . \tag{7.5.18}$$

Predissociation effects may be treated in either the diabatic (crossing potential curves) or adiabatic (noncrossing curves) representation. Criteria for choosing the more convenient representation are discussed in Section 3.3.4. When the adiabatic representation is appropriate, predissociative interactions between states of the same symmetry result from the nuclear kinetic energy operator. The matrix elements of this operator cannot be factored in the same way as Eq. (7.5.14) because the electronic element

$$W_{12}^e(R) = \langle \Phi_1|\partial/\partial R|\Phi_2\rangle \tag{7.5.19}$$

can vary rapidly with R. In fact, for a simple two-state avoided crossing, $W_{12}^e(R)$ can be shown to be a Lorentzian function of R [Section 3.3.3, Eq. (3.3.14)]. The $W_{12}^e(R)$ function cannot be factored outside of the vibrational integral

$$H_{v_1,J:E_2,J} = \left\langle \chi_{v_1,J} \left| W_{12}^e(R)\frac{d}{dR} \right| \chi_{E_2,J} \right\rangle, \tag{7.5.20}$$

and the magnitude of the adiabatic interaction matrix element is very sensitive to the nodal structure of $\chi_{v_1,J}$ and $\chi_{E_2,J}$ in the R region near the maximum of $W_{12}^e(R)$.

An example, the N_2^+ $B^2\Sigma_u^+ \sim C^2\Sigma_u^+$ interaction, is discussed using the adiabatic picture in Section 7.11.2. An example where the adiabatic picture is applicable has been given for the $A^1\Sigma^+$ state of LiH embedded in the continuum of the $X^1\Sigma^+$ ground state. The nonradiative decay rate is identical to that obtained from a diabatic coupled equation calculation (Gadéa, *et al.*, 1997). In the case of spin-orbit predissociation, the choice is between the relativistic diabatic and relativistic adiabatic limits. These limits have been discussed for the example of the predissociation of the O_2 Schumann-Runge bands (Leahy, *et al.*, 1995).

Calculations of predissociation level shifts in the diabatic representation are simplified by the Eq. (7.5.14) factorization,

$$\delta E_{v,J} = E_{v,J} - E^0_{v,J} = (H^e)^2 \mathcal{P} \int \frac{|\langle \chi_{v,J} | \chi_{E',J} \rangle|^2}{E^\circ_{v,J} - E'} dE'. \quad (7.5.21)$$

One need only consider the energy dependence of the differential Franck-Condon factor. For small H^e, this factorization may be justified in the same way as in the case of second-order energy shifts of bound states. The variation of Γ and δE with v and J provides information about the initially unknown shape of the repulsive potential curve.

In general the bound (predissociat*ed*) potential curve is much better characterized experimentally than the repulsive (predissociat*ing*) curve. Predissociation linewidths and shifts are usually the best available experimental information about the repulsive state. Indeed, as for bound~bound interactions, the vibrational variation of the overlap factor is related to the relative locations of the nodes of the bound and continuum vibrational wavefunctions near R_C (the point of stationary phase of the product $\chi^*_{v,J}\chi_{E',J}$, which is where $\chi^*_{v,J}\chi_{E',J}$ oscillates most slowly). The $\chi_{v,J}$ and $\chi_{E,J}$ functions are solutions of the following nuclear Schrödinger equations expressed in atomic units:

$$\left\{ -\frac{1}{2\mu}\frac{d^2}{dR^2} + V_1(R) + \frac{1}{2\mu R^2}[J(J+1)] - E \right\} \chi_{v,J}(R) = 0$$
$$\left\{ -\frac{1}{2\mu}\frac{d^2}{dR^2} + V_2(R) + \frac{1}{2\mu R^2}[J(J+1)] - E \right\} \chi_{E,J}(R) = 0. \quad (7.5.22)$$

In these equations, $V_1(R)$ is the potential energy curve of the predissociated state and $V_2(R)$ is the potential curve of the predissociating state. Note that these equations determine χ_v and χ_E independently from each other. These equations are uncoupled equations. For a coupled equation approach, see Section 7.12.

7.6 The Vibrational Factor

Spectroscopically determined potential energy curves for bound electronic states may be extrapolated into the repulsive region by an expression $V_2(R) = aR^{-n} + b$ where $n = 12$ is typically chosen. (A repulsive wall with $n > 12$ would cause the overlap factor to be even smaller than for $n = 12$ because the amplitude of the continuum function near the turning point would decrease.) One of two analytical forms is usually chosen to model repulsive potentials:

$$V_2(R) = D_e + Ae^{-bR} \tag{7.6.1}$$

$$V_2(R) = D_e + A'R^{-n'}. \tag{7.6.2}$$

As long as the two parameters A and b or A' and n' are chosen so that the repulsive curve crosses the bound curve at the same point, R_C, and with the same slope, overlap factors obtained from either form of the potential are nearly identical (Julienne and Krauss, 1975). In practice, the two parameters of Eqs. (7.6.1) or (7.6.2) are varied until optimal agreement with the experimental vibration-rotation dependence of Γ (or τ) is obtained.

Analytic formulas can be very useful if linewidths have been measured for many vibrational levels (for example, Child, 1974). It is convenient to represent $\chi_{v,J}$ and $\chi_{E,J}$ in a uniform semiclassical approximation (Section 5.1.1). As for the bound~bound case, the overlap integral between bound and continuum wavefunctions can be expressed as an Airy function (taking into account the proper normalization factor). The linewidth is then

$$\Gamma_{v,J} = \frac{4\pi\hbar\overline{\omega}_{v,J}|H^e|^2}{\hbar\underset{\sim}{v}\Delta F}\xi_{v,J}^{1/2}Ai(-\xi_{v,J})^2 = \pi\Gamma^0_{v,J}\xi_{v,J}^{1/2}Ai(-\xi_{v,J})^2,^{\dagger} \tag{7.6.3}$$

where ΔF is the difference between the slopes of the two potentials at R_C,

$$\Delta F = \left(\frac{dV_1}{dR}\right)_{R=R_C} - \left(\frac{dV_2}{dR}\right)_{R=R_C}; \tag{7.6.4}$$

$\overline{\omega}_{v,J}$ is the local vibrational spacing,

$$\hbar\overline{\omega}_{v,J} \simeq \frac{dE_{v,J}}{dv} = \frac{E_{(v+1),J} - E_{(v-1),J}}{2}; \tag{7.6.5}$$

$\underset{\sim}{v}$ is the classical velocity at R_C,

$$\underset{\sim}{v} = \frac{\hbar k(R_C)}{\mu}, \tag{7.6.6}$$

where $k(R_C)$ is the common value of the wavenumber for the two wavefunctions at the point of stationary phase, R_C,

$$k_1(E_{v,J}, R_C) = k_2(E_{v,J}, R_C) = \frac{(2\mu)^{1/2}}{\hbar}[E_{v,J} - V_J(R_C)]^{1/2}, \tag{7.6.7}$$

† Eq. (7.6.3) may be rewritten in the usual spectroscopic units as

$$\Gamma_{v,J}(\text{cm}^{-1}) = \frac{1.5307|H^e(\text{cm}^{-1})|^2[\mu(\text{Aston units})]^{1/2}\overline{\omega}_{v,J}(\text{cm}^{-1})\xi_{v,J}^{1/2}Ai\,(-\xi_{v,J})^2}{[E(\text{cm}^{-1}) - V_J(R_C)]^{1/2}\Delta F}$$

where, in the calculation of ΔF, V and R are expressed in reciprocal centimeters and angströms, respectively.

with

$$V_{J1}(R_C) = V_{J2}(R_C) = V_J(R_C); \tag{7.6.8}$$

and

$$\xi_{v,J} = \left[\frac{3}{2}\phi(E_{v,J})\right]^{2/3}. \tag{7.6.9}$$

$\phi(E_{v,J})$ is a phase difference, which, for the case of an outer wall crossing (Fig. 7.19a), is

$$\phi(E_{v,J}) = \int_{a_2}^{R_C} dR\ k_2(E_{v,J}, R) + \int_{R_C}^{b_1} dR\ k_1(E_{v,J}, R) \tag{7.6.10}$$

where a and b are inner and outer turning points and the subscripts 1 and 2 refer to the bound and repulsive potentials, respectively. ϕ is one-half the difference in phase between two classical paths for the nuclear motion at energy E_v. Path I is from a_1 to R_C on V_1 and from R_C to ∞ on V_2; path II is from a_1 to b_1 on V_1, b_1 to R_C on V_1, R_C to a_2 on V_2, and a_2 to ∞ on V_2 (Schaefer and Miller, 1971). When $\xi_v > 1.5$, the Airy function of Eq. (7.6.3) approaches a sine function and

$$\Gamma_{v,J} = \Gamma^0_{v,J} \sin^2\left[\phi(E_{v,J}) + \pi/4\right], \tag{7.6.11}$$

which is the result obtained using the JWKB approximation (Child, 1980b, Appendix).

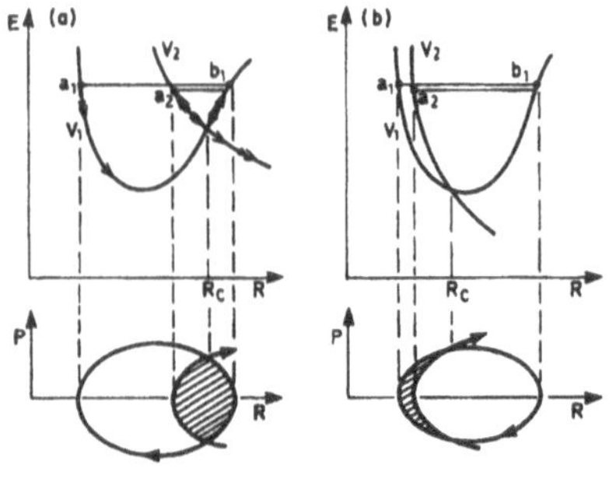

Figure 7.19: Pictorial descriptions of the phase difference between bound and continuum vibrational wavefunctions. The top part of the figure shows the crossing bound and repulsive potential curves and the two paths between which the phase shift is to be determined. The lower part of the figure represents the classical phase-space trajectories for motion on V_1 (ellipse) and V_2 (parabola). The shaded area is the phase difference between the two paths. (a) Outer wall crossing. Path I (single arrows) is the most direct dissociation path: a_1 to R_C on V_1, R_C to ∞ on V_2. Path II (double arrows) is the shortest indirect path: a_1 to b_1 to R_C on V_1, R_C to a_2 to ∞ on V_2. (b) Inner wall crossing. The phase difference is between the shortest (b_1 to R_C on V_1, R_C to a_2 to ∞ on V_2) and next longer (b_1 to a_1 to R_C on V_1, R_C to ∞ on V_2) path.

Figure 7.19a is a pictorial description of $\phi(E_v)$ for an outer wall curve crossing. In phase space, bound motion in the vth vibrational level appears as an ellipse in the harmonic approximation. Motion on a linear unbound potential is represented as a parabola. The shaded area is $2\phi(E_v)$. As v increases, the area of the ellipse increases [the unbound motion parabola shifts to the left so that the minimum value of R at $P = 0$ occurs at $V_2(R_{\text{min}}) = E_v$]; consequently, ϕ increases with v for $E > E_C$. Whenever the value of $\phi(E_v) = (2v+1)\pi/4$, the predissociation linewidth has its maximum value [except for the first maximum, $v = 0$, at which Eq. (7.6.11) is invalid] (Child, 1980b).

If V_1 and the form of $\phi(E_v)$ are known, the repulsive curve V_2 can be determined by an RKR-like method that computes individual turning points (Child, 1973, 1974). This method is useful for obtaining an initial approximation for the repulsive potential. However, if only a few experimental Γ_v–values are known, it is difficult to identify unambiguously the oscillatory frequency of Γ versus v. For example, in Fig. 7.20 the number of vibrational levels sampled is insufficient to determine the actual shape of Γ_v.

The predissociation level shift, $\delta E_{v,J}$, for $E_{v,J} > E_C$ has also been treated semiclassically (Bandrauk and Child, 1970):

$$\delta E_{v,J} = (1/2)\Gamma^0_{v,J} \sin\left[\phi(E_{v,J}) + \frac{\pi}{4}\right] \cos\left[\phi(E_{v,J}) + \frac{\pi}{4}\right] + \Delta_{v,J} \qquad (7.6.12)$$

where $\Delta_{v,J}$ is a correction term arising from the nonlinearity of the potentials and is nearly independent of v and J. Note that a relative maximum of $|\delta E_{v,J}|$ will occur between each maximum and minimum value of $\Gamma_{v,J}$. $\delta E_{v,J}$ oscillates at the same frequency as $\Gamma_{v,J}$ but leading in phase by $\pi/2$ (Fig. 7.20). For $E < E_C$, δE is always negative for an outer crossing and positive for an inner crossing because the continuum $\chi_{E,J}$ functions having maximum overlap with $\chi_{v,J}$ lie above and below $E_{v,J}$, respectively, for outer and inner crossings. Level shifts are even observable below the dissociation limit (i.e., for $E < D^0 < E_C$), where it is impossible for line broadening to occur (because $E < D^0$).

Alternatively, the level shift for $E < E_C$ can be understood in terms of a noncrossing curve (adiabatic) representation. The outer- (inner-) crossing case corresponds to an adiabatic curve that is wider (narrower) below E_C than the diabatic curve. Broadening (narrowing) the potential curve has the effect of shifting each v, J-level to lower (higher) energy.

The difference between the behavior of Γ at inner versus outer wall crossings can be understood by examining Eq. (7.6.11). For an outer crossing, the phase difference, $\phi(E)$, varies rapidly with v and the $\sin^2\phi(E)$ function will oscillate rapidly. For an inner crossing, the phase difference does not change very much because the potential curves are nearly parallel. This is evident from the definition of $\phi(E)$,

$$\phi(E) = \int_{a_1}^{R_C} k_1 dR - \int_{a_2}^{R_C} k_2 dR, \qquad (7.6.13)$$

and Fig. 7.19b. Thus $\phi(E)$ and $\sin^2\phi(E)$ are nearly independent of energy. The level shift varies slowly; consequently, it will be difficult to detect inner-crossing level shifts experimentally from the $\Delta^2 G(v)$ variation. For the third case, noncrossing or parallel potential curves, the variation of the phase difference is even slower and the function $\sin^2 \phi(E)$ becomes very flat. (See also Section 7.14.)

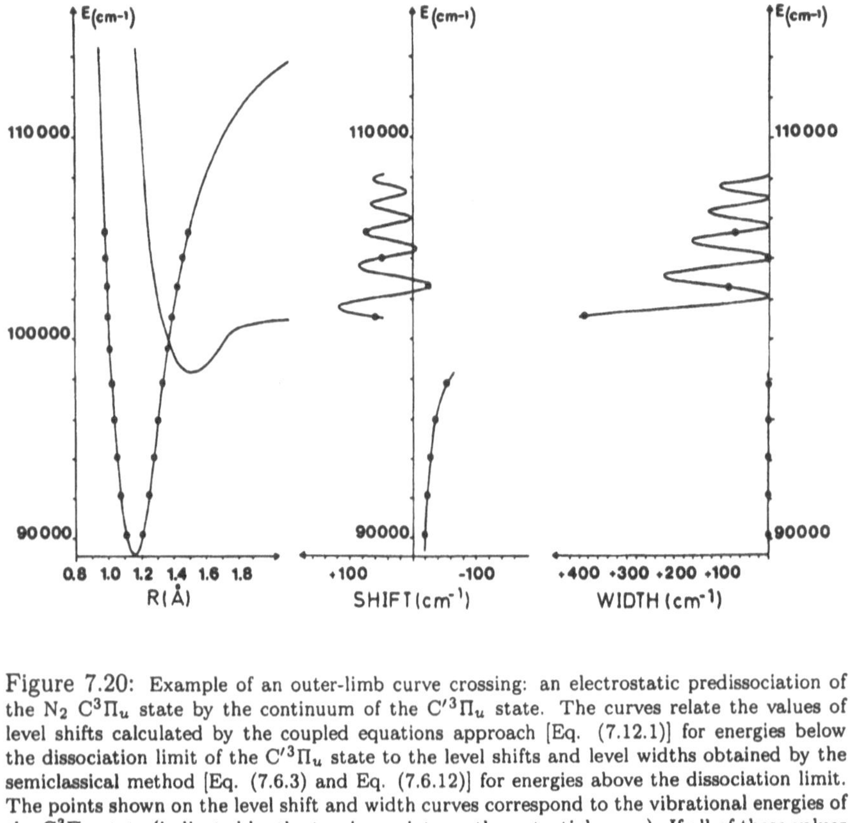

Figure 7.20: Example of an outer-limb curve crossing: an electrostatic predissociation of the N_2 $C^3\Pi_u$ state by the continuum of the $C'^3\Pi_u$ state. The curves relate the values of level shifts calculated by the coupled equations approach [Eq. (7.12.1)] for energies below the dissociation limit of the $C'^3\Pi_u$ state to the level shifts and level widths obtained by the semiclassical method [Eq. (7.6.3) and Eq. (7.6.12)] for energies above the dissociation limit. The points shown on the level shift and width curves correspond to the vibrational energies of the $C^3\Pi_u$ state (indicated by the turning points on the potential curve). If all of these values had been observed, they would have been insufficient to suggest the actual shape of the δE and Γ curves. [Courtesy of J.M. Robbe from data of Robbe (1978).]

7.7 Mulliken's Classification of Predissociations

Mulliken (1960) has distinguished several classes of predissociation based on the location of the curve crossing (E_C, R_C) relative to the dissociation energy (D^0) and the R_e of the predissociated state. Cases b, a, and c correspond to $E_C < D^0, E_C = D^0$, and $E_C > D^0$, respectively. The superscripts $-, i, +$, and 0 specify the cases $R_C < R_e, R_C = R_e, R_C > R_e$, and noncrossing potentials, respectively. For the present discussion, it is useful to retain only Mulliken's $-, +$, and 0 classifications. It is immaterial whether the predissociating curve is partly attractive or completely repulsive. These three classes of predissociation manifest profound differences in the variation of linewidth versus the vibrational quantum number of the predissociated level.

1. If the potential curves cross,

the crossing can be on the right side of the bound potential (attractive branch). This is an *outer* crossing or, in Mulliken's notation, a case a^+, c^+, or b^+ predissociation. In this case, the vibrational dependence of the linewidth and level shift exhibits large fluctuations. An example is given in Fig. 7.20, which shows the expected rapidly oscillatory behavior. One even encounters significant variation with J within a given v. Figure 7.21 shows this J–variation for the OD $A^2\Sigma^+$ state. Figure 7.22 is a pictorial representation of similar variations. The energies of rotational levels are plotted versus $J(J+1)$ as solid lines with slope B_v. Dashed and dotted lines, respectively, connect linewidth minima (Δ) and maxima ($\bigcirc$). The slopes of these lines are found to be $\hbar^2/2\mu R_C^2$, where R_C is the location of the curve crossing.

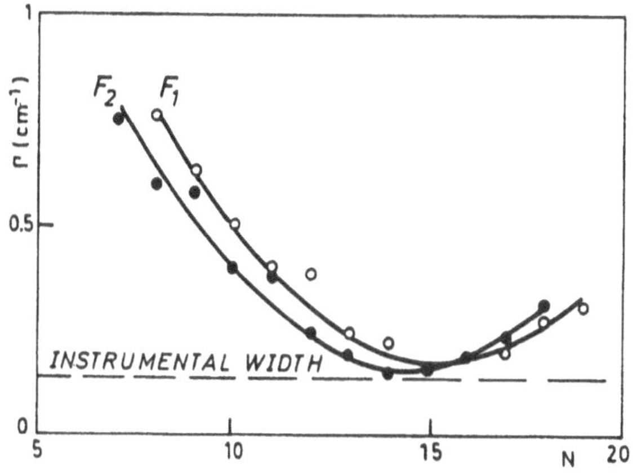

Figure 7.21: Variation of linewidth versus N in the OD $A^2\Sigma^+$ state. The resolved spin components of $v = 11$ are shown. [Data from Czarny, *et al.*, (1971).]

Since, at any R–value, the centrifugal energy is identical for all potential curves, R_C does not change with J. Consequently, the J–dependent energy of the curve crossing is given by

$$\begin{aligned} E_C(J) &= E_C(J=0) + \frac{\hbar^2}{2\mu R_C^2} J(J+1) \\ &= E_C(0) + B_C J(J+1). \end{aligned} \tag{7.7.1}$$

At the right of Fig. 7.22 is a plot of the calculated variation of Γ versus E at a constant J–value, designated as J_{MAX}. The oscillations in Γ versus J within a given v–level are rapid for hydrides because of their large rotational constants. For heavier molecules such as O_2, the linewidths show only a slight J–dependence (Lewis, *et al.*, 1980). The linewidth maximum occurs at an energy just above E_C. The curve of the energy of the first maximum in Γ versus $J(J+1)$ (the lowest energy dotted line on Fig. 7.22) is similar to the "limiting curve of dissociation" from which the dissociation energy may be inferred in the case of dissociation by rotation (cf. Herzberg, 1950, p. 428).

To a good approximation, the vibrational overlap integral is determined in the region near R_C, the point of stationary phase. This means that $\langle \chi_{v,J} | \chi_{E,J} \rangle$ will be a function of the energy difference, $E_{v,J} - E_C(J)$, and will be largely insensitive to the values of v and J. However,

$$E_{v,J} - E_C(J) = E_{v,0} - E_C(0) + (B_v - B_C)J(J+1), \qquad (7.7.2)$$

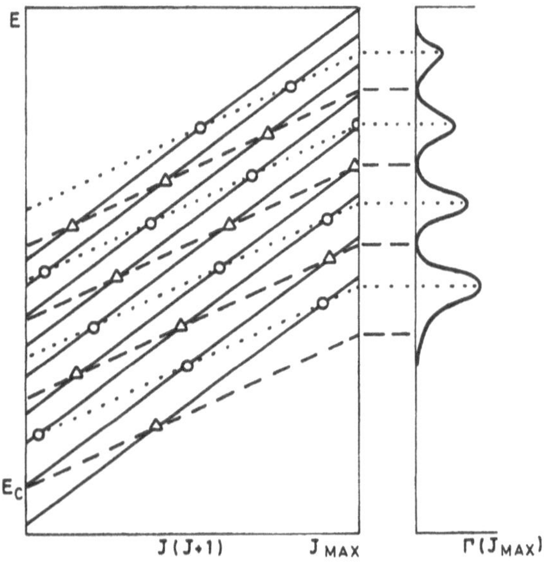

Figure 7.22: Schematic illustration of the variation of Γ versus v and J. The dashed and dotted lines correspond, respectively, to the minima and maxima of the level widths, except for the first dashed line, which corresponds to the inflection points for the first maximum of the linewidth variation. The Γ curve depicts the variation of the linewidth at J_{MAX} versus energy. The solid lines correspond to rotational levels of successive v levels of the bound state. The points, △ and ○, are the observed locations of linewidth minima and maxima. [Courtesy of M.S. Child.]

which implies that if there is a relative maximum in $\langle \chi_{v,J} | \chi_{E,J} \rangle$ for the v, J–level, a series of corresponding maxima will occur at the v, J–values shown on Fig. 7.22 connected by each line with slope B_C. At the energy of an outer-wall curve crossing, $E_{v,J} = E_C(J)$, the first maximum in $|\chi_{v,J}(R)|$ occurs at $R_{\text{MAX}} < R_C$. As $E_{v,J} = E_C(J)$ increases, R_{MAX} is swept through R_C, causing $\langle \chi_{v,J} | \chi_{E,J} \rangle$ to pass through its absolute maximum value. As $E_{v,J} - E_C(J)$ continues to increase, the vibrational overlap will continue to oscillate. One might naively expect each successive relative maximum to be smaller than the previous one because of the decrease in the amplitudes of both $\chi_{v,J}(R_C)$ and $\chi_{E,J}(R_C)$ as the kinetic energy at R_C increases, but this effect is reduced by the increase in continuum density of states. The net result is the production of secondary maxima.

The rates of oscillation of the linewidth and level shift versus $E_{v,J} - E_C(J)$ depend on the difference in the slopes of the two potentials at R_C [Eq. (7.6.3)]. Murrell and Taylor (1969) describe many examples of the variation of Γ with v. Several exceptions to the expected strong fluctuation in Γ of outer limb crossing will be discussed in Section 7.14.

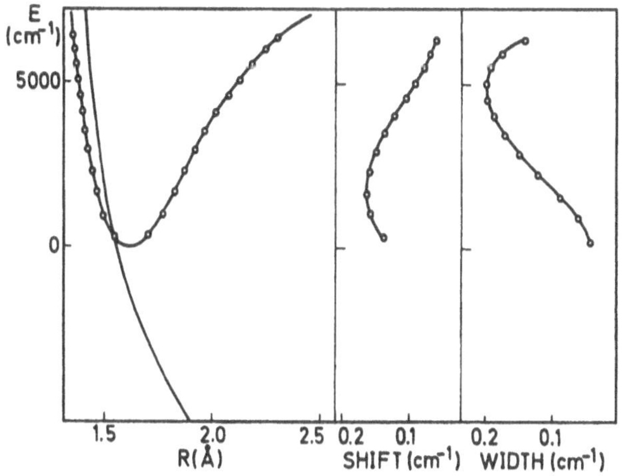

Figure 7.23: Example of an inner-limb curve crossing: a spin-orbit predissociation of the O_2 $B^3\Sigma_u^-$ state by the continuum of the $^3\Pi_u$ state. [From data of Julienne and Krauss (1975).]

2. For a crossing on the left side (repulsive branch) of the bound potential, an *inner* crossing or one of Mulliken's a^-, c^-, or b^- predissociation cases, a single oscillation of Γ_v will extend over many vibrational levels so that, in practice, only a small variation with v (and J) can be sampled (see Fig. 7.23). The reason for this is that the slopes of the bound and repulsive curves at R_C are much larger and more nearly equal for an inner rather than an outer wall crossing. Consequently, as $E_{v,J} - E_C$ increases, the innermost maximum of $\chi_{v,J}(R)$ moves to smaller R much more slowly than the outermost maximum sweeps to larger R.

3. For the case of *noncrossing* or parallel curves (cases a^0, b^0, or c^0 of Mulliken), the oscillatory character of Γ is difficult to detect (Durmaz and Murrell, 1971). This is actually a limiting case of an inner crossing. This situation results typically when the predissociating states are the continua (inner walls) of the lowest-energy bound states, for example, the ground state. The differential Franck-Condon factor between the $\chi_{v,J}$ and $\chi_{E,J}$ vibrational functions is typically about 10^{-2} to 10^{-6} times smaller than in the case of crossing curves. For example, the predissociation of the CH $A^2\Delta$ states probably results from

coupling with the dissociative continuum of the $X^2\Pi$ ground state (see Fig. 7.14 and Fig. 7.24).

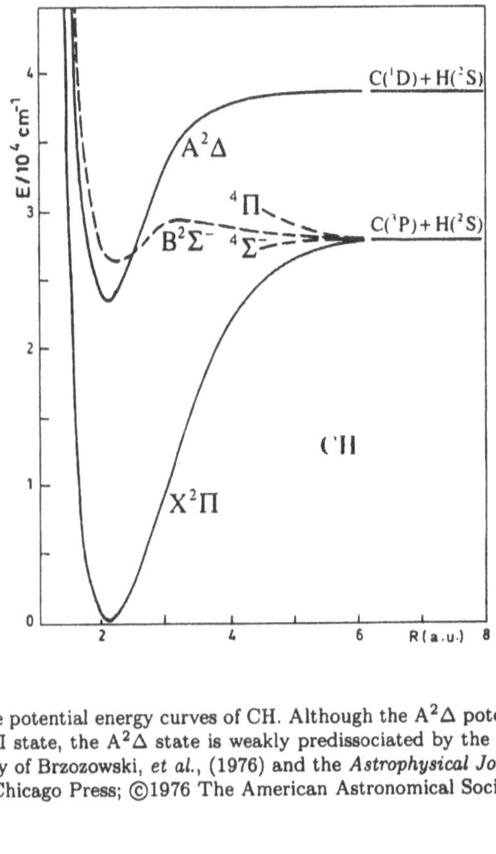

Figure 7.24: The potential energy curves of CH. Although the $A^2\Delta$ potential is not crossed by that of the $X^2\Pi$ state, the $A^2\Delta$ state is weakly predissociated by the continuum of $X^2\Pi$. [Reprinted courtesy of Brzozowski, *et al.*, (1976) and the *Astrophysical Journal*, published by the University of Chicago Press; ©1976 The American Astronomical Society.]

7.8 The Electronic Interaction Strength

It is insufficient, having obtained an approximate shape for the dissociative curve from the study of the *relative* variation of Γ (or τ), for the interpretation of the predissociation to be complete. An equally important but often quite difficult step consists of accounting for the *absolute* value of Γ. The mere presence of a crossing by a repulsive curve is not sufficient to cause predissociation of bound levels. A value for the electronic interaction H^e may be deduced from an experimental value for $\Gamma_{v,J}$ if the calculated value of $\langle\chi_{v,J}|\chi_{E,J}\rangle^2$ is introduced into Eq. (7.5.16). This semiexperimental value for H^e may be compared to a calculated or estimated value. The few cases that have been completely interpreted are listed in Tables 7.3 and 7.4.

The coupling operator, $\mathbf{H}$, for predissociation in Eq. (7.5.1) has exactly the same origin as for perturbations (see Table 3.2). It can be an electronic, spin-orbit, or rotational operator. It is easiest to calculate the electronic part of the predissociation matrix element in the Hund's case (a) basis, for which Ω is a good quantum number. The case (b) matrix elements can be obtained from the case (b) $\rightarrow$ case (a) transformations [for example, Eqs. (3.2.61a), (3.2.61b), (3.2.68a), and (3.2.68b)]. Examples of case (b) predissociation matrix elements are given by Julienne and Krauss (1975) for triplet states and by Carrington, *et al.*, (1978) for quartet states.

The selection rules and v, J–dependence of predissociation effects depend on the identity of the operator responsible for the predissociation. From knowledge of the selection rules, qualitative information can immediately be obtained from the variation of the total interaction with v or J. For example, if lines from low–J levels are missing in emission, the predissociation is certainly not due to a gyroscopic ($\Delta\Omega \neq 0$) interaction, which would be zero for $J = 0$, but must arise from a homogeneous ($\Delta\Omega = 0$) interaction.

When a predissociation is weak, its interpretation is often difficult: small first-order effects can be masked by second-order effects. If only a few lines are missing or weakened, it is necessary to consider the possibility of an accidental predissociation, or, in other words, a three-state interaction involving a local perturbation by a weakly predissociated level (See Section 7.13). Predissociation of normally long-lived (metastable) states detected in emission may originate from very small interactions such as spin-spin or hyperfine interaction, as is the case for the I_2 $B^3\Pi_{0_u^+}$ state (Broyer, *et al.*, 1976).

Some general conclusions may be drawn from Tables 7.3 and 7.4.

7.8.1 Electrostatic Predissociation

There are two ways of interpreting the interactions responsible for homogeneous predissociations between states of the same symmetry and multiplicity. The N_2^+ $B^2\Sigma_u^+ \sim C^2\Sigma_u^+$ interaction must be interpreted using an adiabatic model (noncrossing curves); thus the coupling is attributed to the d/dR operator (see Section 7.11.2) and the predissociation will be called nonadiabatic. Rydberg$\sim$non-Rydberg predissociations in NO are represented by a diabatic model (crossing curves), and consequently the coupling is due to the $1/r_{12}$ electrostatic interaction operator. The NO $^2\Pi$ Rydberg states are strongly predissociated (Miescher, 1976). These predissociations are J–independent and clearly due to the $B^2\Pi$ valence state, which is also responsible for perturbations of the lowest Rydberg states below the dissociation limit (Giusti-Suzor and Jungen, 1984) (see Fig. 7.25). For light molecules, the most important predissociation mechanism is via electrostatic interaction. The affected levels often completely disappear in the absorption spectrum. For example, the N_2 $C^3\Pi_u(v = 6)$ level (cf. Fig. 7.20) is observed only by electron impact spectroscopy (Mazeau, private communication). Its width is found to be $\Gamma = 300\ \text{cm}^{-1}$.

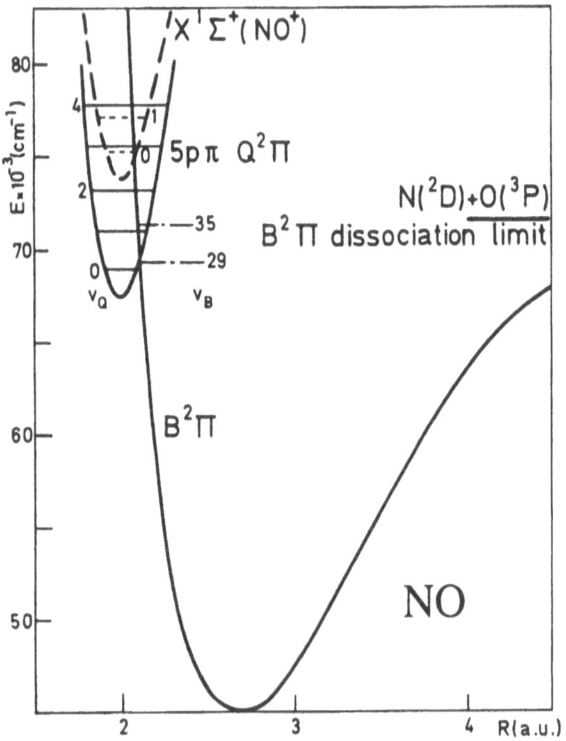

Figure 7.25: Example of $^2\Pi \sim {}^2\Pi$ Rydberg$\sim$valence electrostatic interaction in NO. The $v = 0$ and 1 levels of the $5p\pi$ $Q^2\Pi$ Rydberg state are perturbed by bound levels ($v = 29$ and 35) of the $B^2\Pi$ valence state, but the $v \geq 2$ levels of $Q^2\Pi$ are predissociated by the continuum of the same $B^2\Pi$ state. (Adapted from Gallusser and Dressler (1982).]

7.8.2 Spin-Orbit Predissociation

The spin-orbit interaction, especially between states of different spin, is the *most frequent* cause of predissociation. In Tables 7.3 and 7.4, seven out of twenty-one cases result from this interaction. This means that the $\Delta S = 0$ selection rule, so often invoked for predissociation, is *totally invalid.* The direct spin-orbit interaction causes predissociations, the strengths of which increase with the molecular weight. Compare predissociations of corresponding states for the isovalent molecules OH and HBr^+ (Γ_{max} = 1.5 cm^{-1}, and 800 cm^{-1}, Table 7.4) or those of O_2 and Se_2 (Γ_{max} = 3 cm^{-1} and 180 cm^{-1}). The difference in the widths is mainly a manifestation of the difference between the spin-orbit constant of a light atom (ζ_O = 150 cm^{-1}) and that of a heavy atom (ζ_{Br^+} = 2500 cm^{-1} or ζ_{Se} = 1600 cm^{-1}). A spin-orbit interaction due to a slight departure from the single-configuration limit is exemplified by the predissociation of the

O_2^+ $b^4\Sigma_g^-$ state by a $^4\Sigma_g^+$ state (Carré, *et al.*, 1980) (see Section 7.11.1) and that of the first vibrational levels of the OH $A^2\Sigma^+$ state by the $^4\Sigma^-$ state (Bergeman, *et al.*, 1981).

7.8.3 Rotational or Gyroscopic Predissociation

Predissociations induced by terms from $\mathbf{H}^{\text{ROT}}$ $[\mathbf{B}(\mathbf{J}^+\mathbf{L}^- + \mathbf{J}^-\mathbf{L}^+)]$ are designated as gyroscopic predissociations, to avoid confusion with predissociation by rotation (Fig. 7.12). Experimentally, it is difficult to distinguish between a gyroscopic interaction and predissociation by rotation, because both exhibit linewidths which increase monotonically with J. For example, tunneling through a barrier in the *rotationless* potential [an intrinsic rather than centrifugal barrier as in the Na_2 $B^1\Pi_u$ state; see Vedder, *et al.*, (1984)] can be confused with a predissocation that involves gyroscopic effects; in both cases, Γ varies with J (see Fig. 5 of Helm, *et al.*, 1980). Only careful study of predissociation effects in several vibrational levels can eliminate this ambiguity.

Gyroscopic coupling only connects states of the same multiplicity and can be important, especially at high J in light molecules such as H_2, but its importance for heavy molecules decreases in proportion to the rotational constant. Note, however, that thermal access to extremely high J–levels for heavy molecules can overcome the effect of small $\langle v, J|\mathbf{B}(R)|E, J\rangle$ matrix elements.

Predissociation rates proportional to $J(J+1)$ (or J) can also result from an indirect predissociation mechanism (see Section 7.13). A state, which is predissociated via an l–uncoupling interaction with a repulsive state, may perturb and thereby indirectly predissociate another state. This indirect mechanism is exemplified in the NO $H'^2\Pi$ ($v=2$) state, which interacts with the $H^2\Sigma^+$ state via l–uncoupling, and the $H^2\Sigma^+$ state is indirectly homogeneously predissociated by the A' $^2\Sigma^+$ continuum via the strongly predissociated $I^2\Sigma^+$ state (Ishii, *et al.*, 1995). This predissociation with a linewidth on the order of 0.2 cm^{-1} for $N=3$, exhibits the expected rotational dependence for indirect predissociation (see Section 7.13). A similar indirect predissociation mechanism has been observed in Cs_2. In that case, the $D^1\Sigma_u^+$ state borrows its rate of predissociation, which is proportional to $J(J+1)$, by spin-orbit mixing with the $(2)^3\Pi_{0u}$ state, which in turn is predissociated by l–uncoupling interaction with the continuum of the $c^3\Sigma_u^+$ state (Katô, *et al.*, 1991). It is the l–uncoupling $(2)^3\Pi_{0u} \sim c^3\Sigma_u^+$ interaction that is responsible for the $J(J+1)$ dependence of the predissociation rate. The linewidth in this case is of the order of 4-6 MHz. Recall that 1 $cm^{-1} = 30{,}000$ MHz.

7.8.4 Hyperfine Predissociation

The first example of hyperfine predissociation appears simultaneously with gyroscopic predissociation in the I_2 $B^3\Pi_{0_u^+}$ state (Broyer, *et al.*, 1976). The predissociation due to gyroscopic coupling is very small in this particular case. Taking its effect into account, a residual effective radiative lifetime (nonzero for $J=0$) has been found that shows a strong variation with v. This is actually a predominantly nonradiative lifetime, weakly dependent on J, due to hyperfine interaction between the $^3\Pi_{0_u^+}$ and $^1\Pi_u$ states. This hyperfine predissociation rate depends on the quantum number **F**=**I**+**J** (**I** is the nuclear spin angular momentum), but its interpretation goes beyond the scope of this book. The determination of the predissociation rate due to the hyperfine interaction has been made possible by separate observations of the lifetime of individual hyperfine components (Vigué, *et al.*, 1977, 1981; Pique, *et al.*, 1983; Martinez, *et al.*, 1988). A similar hyperfine predissociation has been observed in Br_2 (Koffend, *et al.*, 1983; Siese, *et al.*, 1985). Another type of hyperfine predissociation mechanism, due to the electric quadrupole interaction, has been observed for Br_2 (Booth, *et al.*, 1994).

7.9 Fano Lineshape

Constructive and destructive interference of bound-bound vs. free-bound transition amplitudes on opposite sides of linecenter of a predissociatively broadened transition result in an asymmetric Fano type line shape, characterized by Fano's index q_v. The absorption cross-section in the case of dissociation has the form

$$\sigma_a(\epsilon) = \sigma_i \frac{(q+\epsilon)^2}{1+\epsilon^2}, \tag{7.9.1}$$

where

$$\epsilon(E) = (E - E_r)/(\Gamma/2) \tag{7.9.2}$$

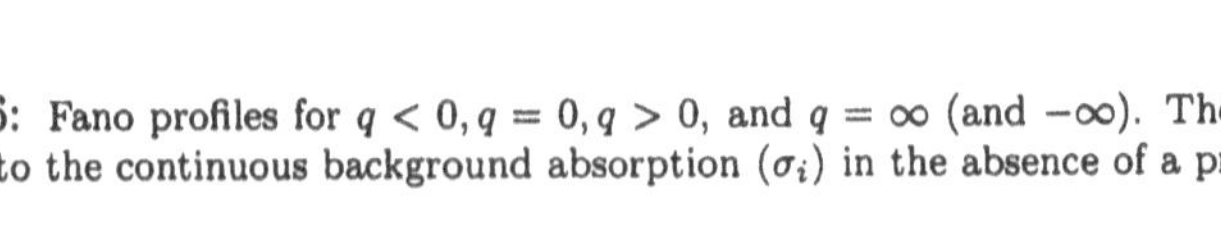

Figure 7.26: Fano profiles for $q < 0, q = 0, q > 0$, and $q = \infty$ (and $-\infty$). The dotted line corresponds to the continuous background absorption (σ_i) in the absence of a predissociated level.

is the dimensionless energy-offset from line-center and σ_i corresponds to the dissociative continuum cross-section. A negative q–value results in a "window" on the high-frequency side of an absorption line; $q > 0$ is associated with a window on the low-frequency side of the line; and $q = 0$ corresponds to an absorption feature that looks as if it were an emission line but is actually an absorption window superimposed on a continuous absorption background. Figure 7.26 displays the different types of Fano line profiles. $q \to \infty$ is the most frequently encountered situation, and occurs when the continuum state has zero oscillator strength from the initial state (level 0). Since $q^2\sigma_i \propto \sigma_{10}$ [q is defined in Eq. (7.9.5) and σ_{10} is the absorption cross-section for the bound ← bound 1 ← 0 transition], Eq. (7.9.1) takes the form of Eq. (7.4.10) (with $\sigma_{\max} = \sigma_{10}$) as $q \to \infty$ and one finds a Lorentzian line of FWHM= Γ.

Let the bound and continuum Born-Oppenheimer upper-state basis functions be $\phi_1\chi_{v,J}$ and $\phi_2\chi_{E,J}$ and the bound lower state be $\phi_0\chi_0$. The mixed upper-state wavefunction is then

$$\Psi_{E,J} = a(E)\phi_1\chi_{v,J} + \int dE' b_{E'}(E)\phi_2\chi_{E',J}, \tag{7.9.3}$$

and the $\Psi_E \leftarrow \phi_0\chi_0$ transition moment is

$$\begin{aligned}&\langle\Psi_{E,J}|\boldsymbol{\mu}|\phi_0\chi_0\rangle \\ &= a^*(E)\langle\phi_1|\boldsymbol{\mu}|\phi_0\rangle\langle\chi_{v,J}|\chi_0\rangle + \langle\phi_2|\boldsymbol{\mu}|\phi_0\rangle\int dE' b^*_{E'}(E)\langle\chi_{E',J}|\chi_0\rangle.\end{aligned} \tag{7.9.4}$$

Fano (1961) showed that

$$|\langle\Psi_{E,J}|\boldsymbol{\mu}|\phi_0\chi_0\rangle|^2$$

has the form of Eq. (7.9.1) if q is defined as

$$q = \frac{\langle\phi_1\chi_{v,J}|\boldsymbol{\mu}|\phi_0\chi_0\rangle + \mathcal{P}\int \frac{\langle\phi_1\chi_{v,J}|\mathbf{H}|\phi_2\chi_{E',J}\rangle\langle\phi_2\chi_{E',J}|\boldsymbol{\mu}|\phi_0\chi_0\rangle}{E-E'}dE'}{\pi\langle\phi_1\chi_{v,J}|\mathbf{H}|\phi_2\chi_{E,J}\rangle\langle\phi_2\chi_{E,J}|\boldsymbol{\mu}|\phi_0\chi_0\rangle} \tag{7.9.5}$$

The principal part in the expression of q can play an important role in particular when the transition moment to the bound state is zero (see Kim, *et al.*, 1994).

If this principal part is neglected and if $\langle\phi_1\chi_{v,J}|\mathbf{H}|\phi_2\chi_{E,J}\rangle = H_{vJ;E,J}$ and the electronic matrix element of $\boldsymbol{\mu}$ are R–independent, then Eq. (7.9.5) can be factored,

$$q_{v,J} = \frac{1}{\pi}\frac{\langle\phi_1|\boldsymbol{\mu}|\phi_0\rangle\langle\chi_{v,J}|\chi_0\rangle}{\langle\phi_2|\boldsymbol{\mu}|\phi_0\rangle\langle\chi_{E,J}|\chi_0\rangle}\frac{1}{\langle\phi_1|\mathbf{H}|\phi_2\rangle\langle\chi_{v,J}|\chi_{E,J}\rangle}. \tag{7.9.6}$$

Note that q is dimensionless because $\langle\chi_{E,J}|\chi_0\rangle$ and $\langle\chi_{v,J}|\chi_{E,J}\rangle$ have units of (energy)$^{-1/2}$ and $\langle\phi_1|\mathbf{H}|\phi_2\rangle$ has units of (energy)$^{+1}$. Again, as for bound∼bound perturbations, the sign of the interference index, q, does not depend on arbitrary phase choices for ϕ and χ functions. The definition of q contains each of the three relevant electronic and vibrational wavefunctions exactly twice.

The first observations of Fano lineshapes in predissociation have been in the spectrum of the H_2 molecule (Fig. 7.27 and Fig. 7.28). The $v = 3, J = 1$ level of H_2 $D^1\Pi_u$ is predissociated by the $B'^1\Sigma_u^+$ state (Herzberg, 1971). Also Fano profiles are found in $v = 1$ of $B''^1\Sigma_u^+$ which also mixes with the continuum of $B'^1\Sigma_u^+$. Since Franck-Condon factors for absorption from $X^1\Sigma_g^+(v = 0)$ into the continuum of $B'^1\Sigma_u^+$ are favorable, the overlap term ($\langle\chi_{E,J}|\chi_0\rangle$) in the denominator of Eq. (7.9.6) is nonzero and q has a finite value. q is positive for $D^1\Pi_u$ and negative for $B''^1\Sigma_g^+$.

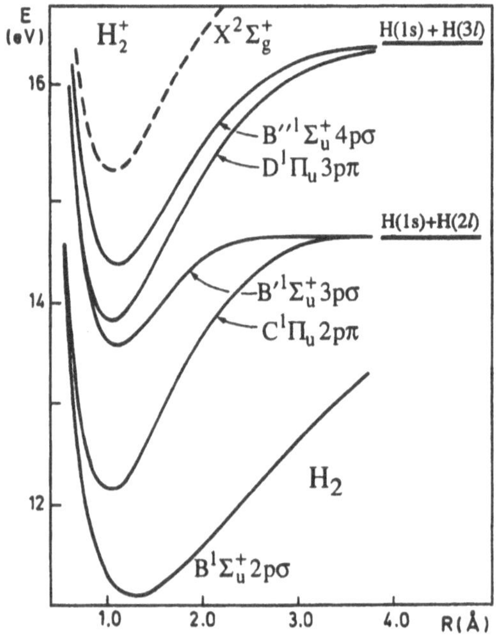

Figure 7.27: The potential energy curves for some excited states of H_2. The zero of energy is the $v = 0, J = 0$ level of the $X^1\Sigma_g^+$ ground state. [From Sharp (1970).]

Owing to photographic plate saturation effects, precise q–values cannot be measured from Herzberg's spectra. Figure 7.28 shows a saturation-free Fano profile for H_2 $D^1\Pi_u(v = 5)$ obtained using synchrotron radiation by simultaneously recording the absorption spectrum and the H–atom Lyman-α ($^2P \rightarrow {}^1S$) fluorescence excitation spectrum resulting from the $H_2^* \rightarrow H(^1S) + H(^2P)$ dissociation process (Glass-Maujean, *et al.*, 1979).

Other predissociation profiles which exhibit the Fano lineshape have been observed in H_2 ($j^3\Delta_g \leftarrow c^3\Pi_u$, Lembo, *et al.*, 1988) and in O_2 ($E^3\Sigma_u^- \leftarrow X^3\Sigma_g^-$, Lewis, *et al.*, 1988a,b; $D^3\Sigma_u^+ \leftarrow X^3\Sigma_g^-$, Lewis and Gibson, 1990; $f^1\Sigma_u^+ \leftarrow X^3\Sigma_g^-$, Lewis, *et al.*, 1995a; $f'\,{}^1\Sigma_u^+ \leftarrow X^3\Sigma_g^-$, Lewis, *et al.*, 1995b), including the first example of a window resonance (Lewis, *et al.*, 1995b). The transitions to the $E^2\Sigma^+$ Rydberg state of NO also include examples of asymmetric lineshapes (Ashfold, *et al.*, 1986). The $E^2\Sigma^+$ state is indirectly predissociated by interaction with the $I^2\Sigma^+$ state, which itself is strongly predissociated via the repulsive $A'^2\Sigma^+$ state. Because both the bound-bound $E^2\Sigma^+ \leftarrow X^2\Pi$ and free-bound $I^2\Sigma^+ \leftarrow X^2\Pi$ transitions are allowed, q has a finite value. The same situation occurs for the $8p$ and $9p$ Rydberg states of NO ($v = 1$), which are predissociated by the $B^2\Pi$

valence state (McCormack, *et al.*, 1998).

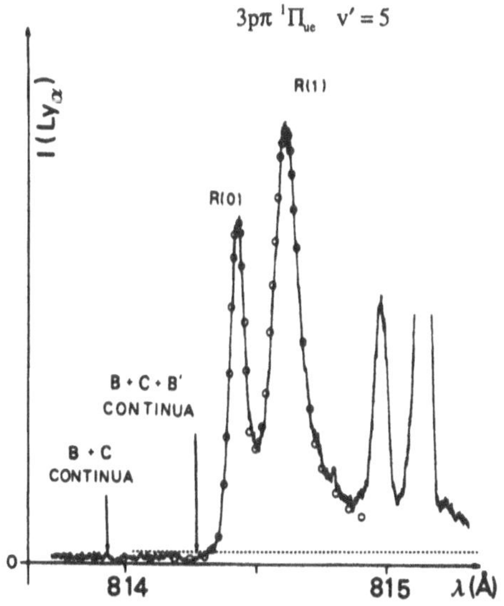

Figure 7.28: Fano lineshape in H_2. The predissociation of the N=2 [R(1) line] and N=1 [R(0) line] levels of the $D^1\Pi_{ue}(v = 5)$ state by the continuum of $B'^1\Sigma_u^+$ is detected by monitoring the Lyman-α emission from one of the fragment atoms. The dots represent the lineshape calculated from the Fano formula [Eq. (7.9.1)] with parameter values $\Gamma(N = 2) = 14.5\ \text{cm}^{-1}, q(N = 2) = -9$; $\Gamma(N = 1) = 4.8\ \text{cm}^{-1}, q(N = 1) = -18$. These lineshapes should be compared to the symmetric profile of Fig. 7.16 ($q = \infty$). The horizontal dotted line separates the interacting continuum σ_i from the noninteracting continua [σ_d of Eq. (8.9.1)]. [From Glass-Maujean, *et al.*(1987).]

In order to find a value for q that is different from ∞, the transition into the continuum must have its own absorption strength. Usually, the transition into the continuum state, ϕ_2, has no absorption because ϕ_2 has a spin multiplicity different from that of the initial state, i.e., $\langle \phi_2 |\mu| \phi_0 \rangle = 0$, therefore $q \to \infty$, and the continuum is observable because ϕ_2 interacts with the bound state by spin-orbit coupling. In the case of H_2, however, q has a finite value because the continuum state is $^1\Sigma_u^+$. In the case of O_2, the Rydberg state, $E^3\Sigma_u^-$, is predissociated by the valence state, $B^3\Sigma_u^-$, of the same symmetry. In the case of Cs_2, the continuum state is $^3\Sigma_u^+$, but due to the relatively weak spin-orbit coupling (primarily with a $^1\Pi_u$ state) (30 cm^{-1}, Kim, *et al.*, 1994) it borrows some absorption strength.

Changes of sign of the Fano asymmetry parameter, q, with v and isotope, have been found in the predissociation of O_2 (Lewis, *et al.*, 1988a, b, 1995b) and with v for Cs_2 (Kim and Yoshihara, 1993).

This phenomenon is called "q reversal" and has only quite recently been observed in predissociation. It has been shown that q reversal can occur even in the absence of interference effects, e.g., with only one closed and one open channel (Kim, *et al.*, 1994).

q reversal occurs when, within the Golden-Rule framework,

$$\Gamma_{v,J} = 2\pi(H_{v,E,J})^2 = 2\pi H_e^2 \langle \chi_{v,J}|\chi_{E,J}\rangle^2 \tag{7.9.7}$$

the width becomes zero, i.e., when the overlap $\langle \chi_{v,J}|\chi_{E,J}\rangle$ between the bound, $\chi_{v,J}$, and the continuum, $\chi_{E,J}$, vibrational wave functions becomes zero for a particular value of v, v_0, of the bound state vibrational quantum number. When this occurs, q becomes very large or infinite. The overlap integral changes sign as v passes through v_0. It is well known that, in the case of an outer wall crossing of bound and continuum curves, $\langle \chi_v|\chi_E\rangle$ is a rapidly oscillatory function of the energy (see, e.g., Fig. 5.5). It is also well known that, for an outer crossing, the transition widths oscillate with v and therefore the overlap factor will pass through zero (see e.g., Fig. 7.20).

Much work has been done to predict the locations of the zeroes of the line width. Child (1976) has demonstrated that, in the case of intermediate coupling strength, a zero occurs in the linewidth whenever the energies of the diabatic and adiabatic levels coincide (and this conclusion is also valid in the case of weak coupling). However, until the work of Kim, *et al.*, (1994) and recently of Cornett, *et al.*, (1999), it seems that the connection between zeroes of predissociation linewidth and q reversal was not appreciated.

Assuming that the two other overlap integrals in the Eq. (7.9.6) definition of q, $\langle \chi_{E,J}|\chi_0\rangle$ and $\langle \chi_{v,J}|\chi_0\rangle$, do not oscillate rapidly in sign with v and E and are non-zero, it is clear that if $\langle \chi_{v,J}|\chi_{E,J}\rangle$ changes sign with v, q also changes sign and produces the q reversal (see example of Cs_2 in Kimura, *et al.*, 2000).

7.10 Isotope Effects

Even more so than for perturbations, the isotopic dependence of predissociations is useful for identifying the electronic symmetry of the unbound state. The naive assertion that the lightest isotopic molecule is predissociated most rapidly must be examined with caution. It is valid only in the cases of gyroscopic predissociation and predissociation by rotation through a centrifugal barrier. In Eq. (7.5.17), the matrix element

$$\langle \chi_{v,J}|\hbar^2/2\mu R^2|\chi_{E,J}\rangle$$

is approximately equal to $B_C \langle \chi_{v,J}|\chi_{E,J}\rangle$, where B_C is the value of the rotational constant at the crossing point R_C. The ratio of the linewidths for two isotopic species is thus proportional to the square of the ratio of their rotational constants or inversely proportional to their μ^2 ratio. In the case of hydrides, for example,

$$\frac{\Gamma(\mathrm{XD})}{\Gamma(\mathrm{XH})} = \frac{\tau(\mathrm{XH})}{\tau(\mathrm{XD})} \simeq \left[\frac{\mu(\mathrm{XH})}{\mu(\mathrm{XD})}\right]^2 \left(\frac{\langle \chi_v|\chi_E\rangle_{\mathrm{XD}}}{\langle \chi_v|\chi_E\rangle_{\mathrm{XH}}}\right)^2 \approx \frac{1}{4}. \tag{7.10.1}$$

If isotopic effects on the differential Franck-Condon factors are neglected, it is indeed found that the hydride is four times more strongly predissociated than the deuteride, but only when the predissociation is due to a gyroscopic interaction.

The isotope effect on differential Franck-Condon factors has been investigated by Child (1974) (see Fig. 7.29), and it can be predicted that the absolute maximum value of Γ varies roughly as $\mu^{1/6}$. This $\mu^{1/6}$ dependence for the magnitude of Γ is weaker and in the opposite sense to the μ^{-2} dependence for gyroscopic predissociations. The oscillation frequency of $\Gamma(E)$ versus E is also sensitive to the reduced mass. Since the phase difference, $\phi(E_{v,J})$ [Eq. (7.6.10)] increases approximately in proportion to the area,

$$\Delta_v = \int \hbar k(R)dR = \left(v + \frac{1}{2}\right) h/2, \tag{7.10.2}$$

of the bound-state circle in Fig. 7.19b,

$$\frac{d\phi(E_{v,J})}{dE} \simeq \frac{d\Delta_v}{dv}\frac{dv}{dE} = \left(\frac{1}{\overline{\omega}_{v,J}}\right) \propto \mu^{1/2}, \tag{7.10.3}$$

one expects that successive maxima in $\Gamma(E)$ will be closer together in energy for the heavier isotopic molecule. The $\Gamma(E)$ functions four light and heavy isotope molecules start out in phase at E_C. Γ_{light} will lag farther and farther behind Γ_{heavy} as $E - E_C$ increases. Eventually, far above E_C, one isotopic species will exhibit a minimum in Γ at the same energy that the other has a linewidth maximum (Fig. 7.29).

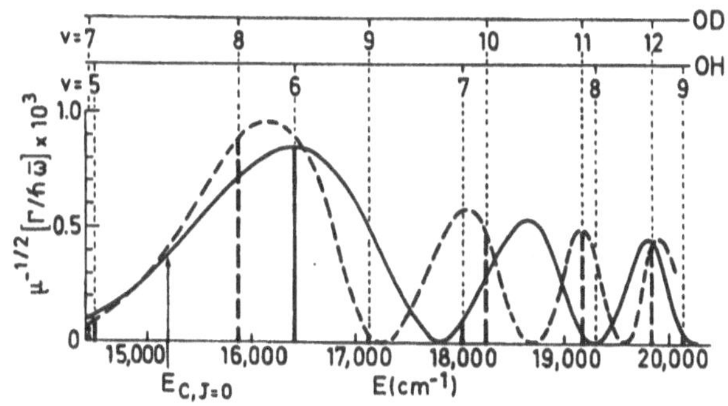

Figure 7.29: Reduced mass dependence of the reduced (dimensionless) linewidth. The solid and dashed vertical lines depict the reduced linewidths for the $A^2\Sigma^+$ state of OH and OD, respectively, as calculated by Czarny, *et al.*, (1971). The curves represent the values calculated using Eqs. (7.6.10) and (7.6.11) [from Child (1974)].

Anomalous isotope effects occur at accidental or indirect predissociations, which are discussed in Section 7.13. The accidentally predissociated v, J–level is perturbed by a v, J–level that is directly predissociated by a third (unbound) state. The accidentally predissociated level, having acquired an admixture of the perturber's wavefunction, borrows part of the characteristics of the perturber, in particular the linewidth. Line broadening at an accidental predissociation depends, in a complex way, on both the magnitudes of coupling matrix elements and accidental near-degeneracies. Consequently, the observed linewidths will be strongly influenced by isotopic substitution (Lorquet and Lorquet, 1974; Lefebvre-Brion and Colin, 1977).

Another type of anomalous isotope effect, which is observed in the N_2^+ $C^2\Sigma_u^+$ state, is discussed in the following section.

A spectacular isotope shift effect in photodissociation spectra is shown in Fig. 7.30. The Σ Rydberg levels of CO built on the $X^2\Sigma^+$ core $[(X^+(v^+)nl\sigma)$ in the notation used on the figure] are predissociated by the repulsive part of the $D'^1\Sigma^+$ valence state (Tchang-Brillet, *et al.*, 1982). The vibrational levels of $^{13}C^{16}O$ are isotopically shifted to lower energy relative to those of $^{12}C^{16}O$. Because $\mu(^{12}C^{18}O) \approx \mu(^{13}C^{16}O)$, the $^{12}C^{18}O$ and $^{13}C^{16}O$ spectra are similar. For $^{13}C^{18}O$, the isotope shift is largest and the peaks with $n = 4, 4p\sigma, v^+ = 3$ and $4s\sigma, v^+ = 4$ overlap the $p\pi$ and $p\sigma$ peaks $n = 6, v^+ = 0$.

7.11 Examples of Predissociation

7.11.1 Examples of Spin-Orbit Predissociation

If a predissociation is strong, the observed widths can often be explained in terms of a single-configuration picture for the two states. Consider, for example, the OD $A^2\Sigma^+$ state (see Fig. 7.31). The $v > 7$ vibrational levels of the $A^2\Sigma^+$ state of OD are strongly predissociated. The rotational lines of the $B^2\Sigma^+ - A^2\Sigma^+$ emission system are broadened by predissociation of $A^2\Sigma^+$, the lower state of the transition, and the linewidth is found to vary rapidly with v and N. Figure 7.21 shows this variation for the $v = 11$ level of OD $A^2\Sigma^+$.

This rapid variation of Γ means that the outer wall of the $A^2\Sigma^+$ potential is crossed by a repulsive state. By adjusting the location of the crossing point, R_C, and the slope of the repulsive curve at the crossing, so that the computed and observed linewidth variations match, a repulsive curve has been shown to cross the $A^2\Sigma^+$ potential between $v = 7$ and 8. The experimental value for Γ_{max} is 1.65 cm^{-1} (at $v = 8, N = 5$). The calculated value of $\langle \chi_{v,N} | \chi_{E,N} \rangle^2$ is 1.20×10^{-4} cm. From Eq. (7.5.16), an electronic matrix element of 47 cm^{-1} is obtained (Czarny, *et al.*, 1971). The Wigner-Witmer rules (see Section 7.2.1) require that four electronic states originate from the $O(^3P)+H(^2S)$ dissociation limit: $X^2\Pi, {}^4\Sigma^-, {}^2\Sigma^-$, and $^4\Pi$. The $X^2\Pi$ potential does not cross $A^2\Sigma^+$ and can be disregarded. The $^4\Sigma^-$ and $^2\Sigma^-$ states arise mainly from the $\sigma^2\pi^2\sigma^*$ configuration, which differs by two spin-orbitals from the predominant $\sigma\pi^4$ configuration of the $A^2\Sigma^+$ state. However, the $^4\Pi$ state arises from the $\sigma\pi^3\sigma^*$ configuration, which differs by only one spin-orbital from the configuration of the $A^2\Sigma^+$ state. Consequently, with the wave functions

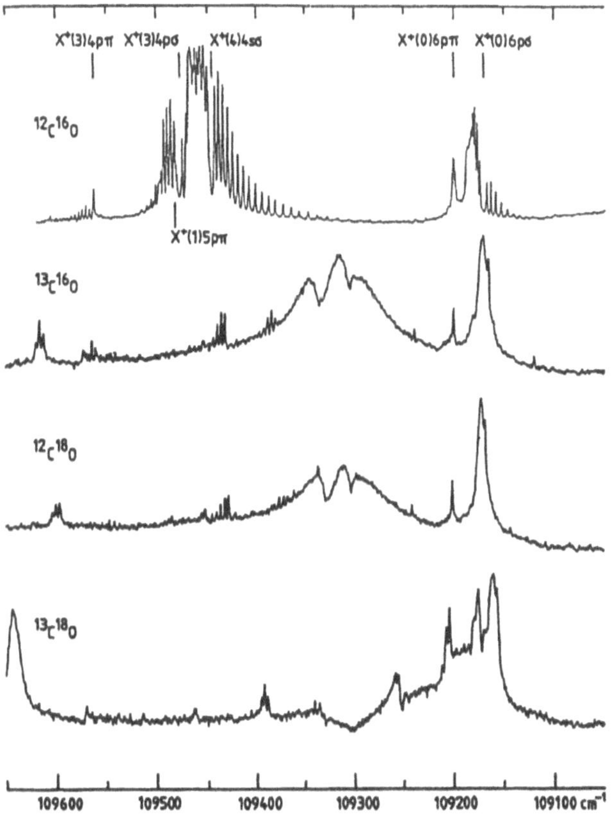

Figure 7.30: Jet absorption spectra of CO isotopomers in the region from 109100 to 109600 cm^{-1} (10 meter spectrograph of the Meudon Observatory), (Courtesy K. P. Huber). Rydberg states built on the $CO^+X^2\Sigma^+$ core, are predissociated by the continuum of the $D'^1\Sigma^+$ state. Notation: $X^+(v^+)nl\lambda$.

$$\Psi\left({}^4\Pi_{1/2}\right) = 3^{-1/2}[|\sigma\beta\sigma^*\beta\pi^-\alpha\pi^+\alpha\pi^+\beta| + |\sigma\beta\sigma^*\alpha\pi^-\beta\pi^+\alpha\pi^+\beta| + |\sigma\alpha\sigma^*\beta\pi^-\beta\pi^+\alpha\pi^+\beta|] \quad (7.11.1)$$

$$\Psi\left({}^2\Sigma^+_{1/2}\right) = |\sigma\alpha\pi^-\alpha\pi^-\beta\pi^+\alpha\pi^+\beta|, \quad (7.11.2)$$

the spin-orbit interaction is (Section 3.4.2)

$$\left\langle A^2\Sigma^+_{1/2}|\mathbf{H}^{SO}|{}^4\Pi_{1/2}\right\rangle = 3^{-1/2}[0 + 0 + \frac{1}{2}a_+] \quad (7.11.3)$$

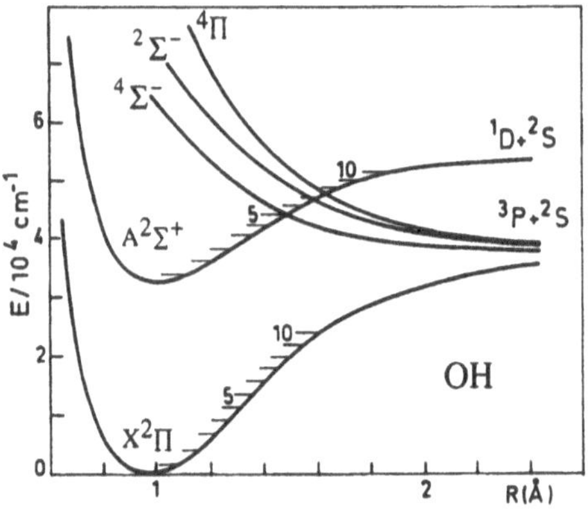

Figure 7.31: The potential energy curves of OH. [From Bergeman, *et al.*, (1981).]

where

$$\begin{aligned} a_+ &= \langle \pi | \hat{a}1^+ | \sigma^* \rangle \simeq \langle 2p\pi_O | \hat{a}1^+ | 2p\sigma_O \rangle = -2^{1/2} A(X^2\Pi) \\ &= 2^{1/2}(140 \text{ cm}^{-1}) = 197 \text{ cm}^{-1}. \end{aligned} \tag{7.11.4}$$

The hypothesis that the σ^* and π orbitals are in a pure precession (Section 5.5) relationship to each other is supported by *ab initio* calculations (Czarny, *et al.*, 1971), which find dominant $2p_O$ character for these orbitals in the $A^2\Sigma^+$ and $^4\Pi$ states at the internuclear distance, R_C, of the curve crossing. The pure precession value of the $^4\Pi_{1/2} \sim A^2\Sigma^+$ interaction is $3^{-1/2}a_+/2 = 57$ cm^{-1}, in satisfactory agreement with the value of 47 cm^{-1} obtained from the observed linewidths.

For SH, with a calculated spin-orbit interaction of 162 cm^{-1} between the $A^2\Sigma^+$ and $^4\Pi$ states (Wheeler, *et al.*, 1997), the ratio between the interaction strengths for SH and OH is in good agreement with the ratio of the atomic spin-orbit constants of sulfur and oxygen, namely about 2.5.

A similar mchanism explains the predissociation of the $A^2\Sigma^+$ state of the hydrogen halide cations by a $^4\Pi$ state. For HBr^+, where the spin-orbit interaction strength is very large, the width of the $v = 4$ level is equal to 800 cm^{-1} (exp: Penno, *et al.*, 1998, calc.: Banichevich, *et al.*, 1992). For HI, the width is so large that the spectrum becomes continuous (Lefebvre-Brion, 1991).

Sometimes a predissociation cannot be understood in such a simple single-configuration picture. An example is the predissociation of the O_2^+ $b^4\Sigma_g^-$ state (see Fig. 7.32). This predissociation is very weak but has been detected as a line broadening by a Doppler-free laser technique. The width of the upper state of the $b^4\Sigma_g^-$-$a^4\Pi_u$ transition in absorption is 0.05 cm^{-1} for $v = 5, N = 9$, a factor of 25 smaller than the width in OD. As the oxygen atom spin-orbit constant is responsible for both of these predissociations, a direct mechanism for O_2^+ seems improbable. The $b^4\Sigma_g^-$ state is most likely predissociated by a $^4\Sigma_g^+$ state (Carré, *et al.*, 1980). The dominant configuration of the $^4\Sigma_g^+$ state, $3\sigma_g^2 1\pi_u^3 1\pi_g 3\sigma_u$, differs by two spin-orbitals from the main configuration of the $\mathrm{b}^4\Sigma_g^-$, which is $3\sigma_g 1\pi_u^4 1\pi_g^2$. The interaction between these two states is explained by the fact that the electronic wave function of $\mathrm{b}^4\Sigma_g^-$, in the region of the curve crossing, contains a small fractional admixture (6×10^{-3}) of the configuration $3\sigma_g^2 1\pi_u^3 1\pi_g 3\sigma_u$, the dominant configuration of the $^4\Sigma_g^+$ state. The electronic matrix element is

$$\begin{aligned}\langle {}^4\Sigma_g^+ |\mathbf{H}^{\mathrm{SO}}| \mathrm{b}^4\Sigma_g^- \rangle &= 0.99\,\langle \sigma_g^2\pi_u^3\pi_g\sigma_u |\mathbf{H}^{\mathrm{SO}}| \sigma_g\pi_u^4\pi_g^2 \rangle \\ &\quad + 0.08\,\langle \sigma_g^2\pi_u^3\pi_g\sigma_u |\hat{a}\mathbf{l}_z\mathbf{s}_z| \sigma_g^2\pi_u^3\pi_g\sigma_u \rangle \qquad (7.11.5)\end{aligned}$$

The first matrix element is zero and the second is calculated to be 140 cm^{-1}. Thus the overall electronic matrix element has a value of about 11cm^{-1}, which is sufficient to partly explain the observed width.

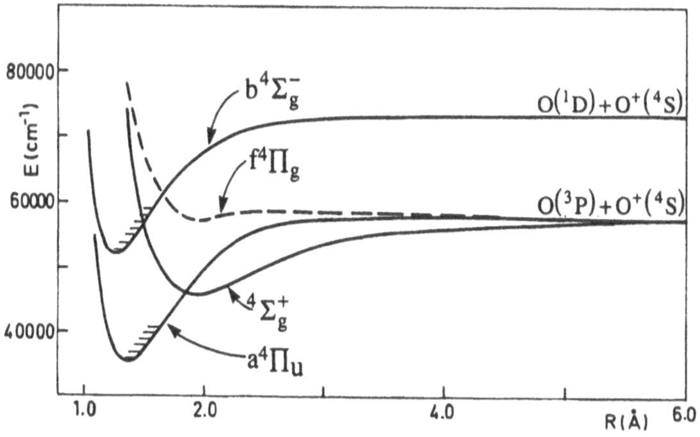

Figure 7.32: The potential energy curves of O_2^+. [Calculations by A.L. Roche from Carré, *et al.*, (1980).]

7.11.2 Examples of Nonadiabatic Predissociation

The N_2^+ $\mathrm{C}^2\Sigma_u^+(v \geq 3)$ levels are subject to very weak predissociation. The predissociation rate is comparable to the purely radiative decay rate. Figure 7.33 illustrates the strong isotopic dependence of the observed decay rates. Roche and Tellinghuisen (1979) have shown that this effect results mainly from an interaction with the continuum of the $\mathrm{B}^2\Sigma_u^+$ state. The *ab initio* potential energy curves for the $\mathrm{B}^2\Sigma_u^+$ and $\mathrm{C}^2\Sigma_u^+$ states (Fig. 7.34) reflect the very strong, two-state avoided curve crossing predicted many years ago by Douglas (1952). Consequently, the adiabatic picture is the ideal starting point for analyzing the B$\sim$C predissociation.

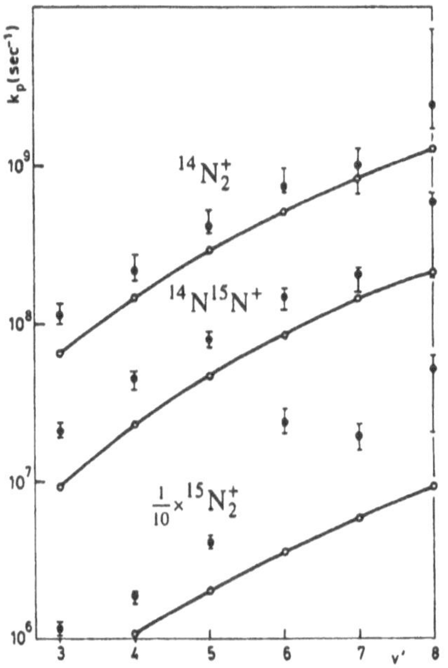

Figure 7.33: Vibration dependence of the observed (solid circles with error bars) and calculated (open circles) predissociation rates for the $C^2\Sigma_u^+$ state of $^{14}N_2^+$, $(^{14}N^{15}N)^+$, and $^{15}N_2^+$. [From the data given in Table 2 of Roche and Tellinghuisen (1979).]

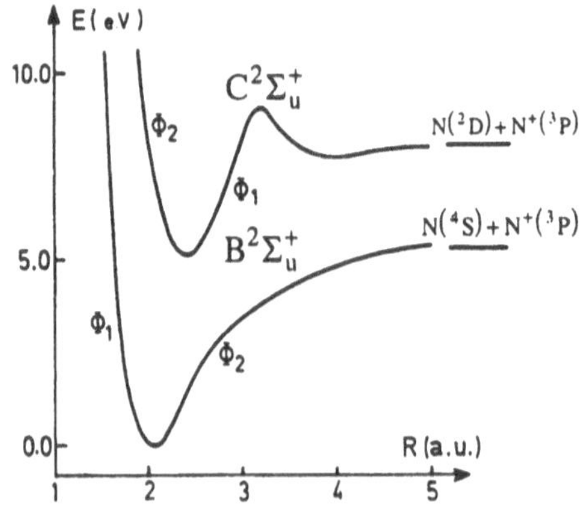

Figure 7.34: Calculated adiabatic potential energy curves for the $B^2\Sigma_u^+$ and $C^2\Sigma_u^+$ states of N_2^+. The curves are labeled by the dominant character, ϕ_1 versus ϕ_2, of the wavefunctions [from Roche and Tellinghuisen (1979)].

In the adiabatic picture, the electronic coupling matrix element involves the nuclear kinetic energy operator. The wavefunctions for the two adiabatic $^2\Sigma_u^+$ states are mixtures of two dominant configurations,

$$\begin{aligned}\phi_1 &= 2\sigma_u 3\sigma_g^2 1\pi_u^4 \\ \phi_2 &= 2\sigma_u^2 3\sigma_g 1\pi_u^3 \pi_g\end{aligned} \tag{7.11.6}$$

which differ by two orbitals. (ϕ_2 actually represents two isoconfigurational $^2\Sigma_u^+$ states derived from the $\pi_u^3\pi_g$ $^1\Sigma_u^+$ and $^3\Sigma_u^+$ parents.) The B and C state wavefunctions are

$$\begin{aligned}\Psi_{\mathrm{B}} &= C_{1\mathrm{B}}(R)\phi_1 + C_{2\mathrm{B}}(R)\phi_2 \\ \Psi_{\mathrm{C}} &= C_{1\mathrm{C}}(R)\phi_1 + C_{2\mathrm{C}}(R)\phi_2,\end{aligned} \tag{7.11.7}$$

where the C–coefficients vary slowly with R. At small R, Ψ_{B} is dominated by ϕ_1, whereas at large R it is dominated by ϕ_2 (*vice versa* for Ψ_{C}).† The electronic matrix element (Section 3.3.3) is

$$\left\langle \Psi_{\mathrm{B}} \left| \frac{d}{dR} \right| \Psi_{\mathrm{C}} \right\rangle_r = W_{\mathrm{BC}}^e(R) \tag{7.11.8}$$

$$W_{\mathrm{BC}}^e(R) = C_{1\mathrm{B}}\frac{d}{dR}C_{1\mathrm{C}} + C_{2\mathrm{B}}\frac{d}{dR}C_{2\mathrm{C}}. \tag{7.11.9}$$

There are no $C_1(d/dR)C_2$ cross terms because ϕ_1 and ϕ_2 are mutually orthogonal. The matrix elements of the form

$$\langle \phi_1 | d/dR | \phi_2 \rangle$$

must vanish since ϕ_1 and ϕ_2 differ by two orbitals (Section 3.3.2).

The $W^e(R)$ function, resulting from a simple two-state interaction, has a Lorentzian form [Eq. (3.3.14)]. For the N_2^+ B$\sim$ C interaction, $W^e(R)$ has its maximum in an R–region of numerous constructive and destructive interferences between the vibrational wavefunctions. Thus, slight changes in the vibrational functions resulting from isotopic substitution can drastically alter the vibronic matrix element,

$$H_{\mathrm{B},E;\mathrm{C},v} = \langle \chi_E^{\mathrm{B}}(R) | W_{\mathrm{BC}}^e(R) d/dR | \chi_v^{\mathrm{C}}(R) \rangle . \tag{7.11.10}$$

Figure 7.35a shows the electronic factor $W_{\mathrm{BC}}^e(R)$ (light dashed line) superimposed on the $^{14}N_2^+$ vibrational wavefunctions χ_E^{B} (heavy dashed line) and $\chi_{v=4}^{\mathrm{C}}$ (solid lines).

---

†The R–value where ϕ_1 and ϕ_2 are equally mixed is where the diabatic potential curves cross. Note that in the diabatic picture the dominant configurational character of a given state is identical on both sides of R_C, whereas in the adiabatic picture the configurational parentage reverses.

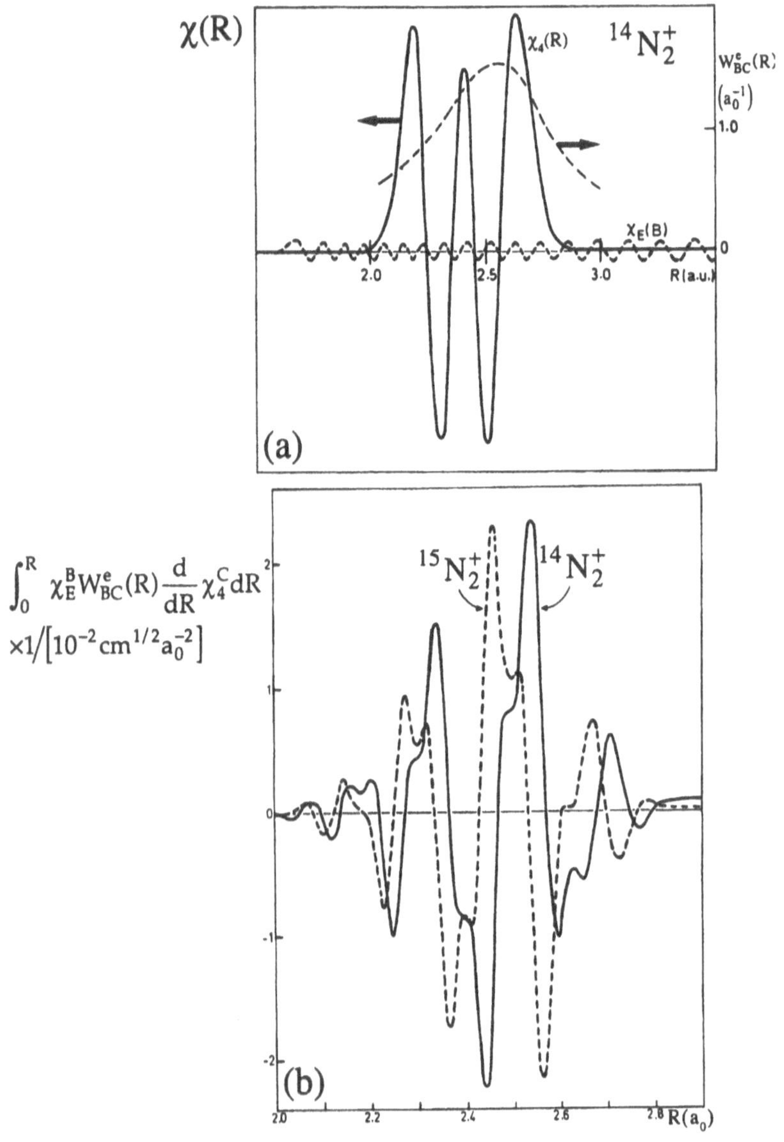

Figure 7.35: (a) Vibrational wavefunctions χ_4^C and χ_E^B for the N_2^+ $C^2\Sigma_u^+ v = 4$ level and the isoenergetic continuum of $B^2\Sigma_u^+$. The electronic matrix element, $W_{BC}^e(R)$, is drawn as a dashed line.(b) Isotopic dependence in the accumulation of $\langle \chi_E^B | W_{BC}^e(R) d/dR | \chi_4^C \rangle$ vibronic integral [Eq. (7.11.10)] [from Roche and Tellinghuisen (1979).]

Figure 7.35b shows how the value of the vibronic integral [Eq. (7.11.10)] accumulates as the integration proceeds from $R = 0$ to larger R. Results for the $^{14}N_2^+$ and $^{15}N_2^+$ isotopic molecules are shown. After rapid, large-amplitude oscillation, the integral eventually stabilizes to a very small value at an internuclear distance where $\chi_{v=4}^{C}$ is zero. The calculated predissociation rates are displayed on Fig. 7.33 and agree well with the experimental isotopic dependence.

Another example is in Li_2 (Antonova, *et al.*, 2000) where the vibrational, rotational, and isotope dependence of the linewidths is studied. However, the coupling term is taken as constant rather than a Lorentzian form, giving a not so good fit.

7.12 Case of Intermediate Coupling Strength

The Golden Rule formula Eq. (7.5.16) for the FWHM and Eq. (7.5.9) for the level shift are expressed in terms of the unperturbed vibrational wavefunctions. For strong predissociations, this approximation becomes untenable. Exact methods exist that can determine both the linewidth and the level shift. One method consists of numerically solving the following coupled equations (Lefebvre-Brion and Colin, 1977; Child and Lefebvre, 1978):

$$\left[-(1/2\mu)\frac{d^2}{dR^2} + V_1(R) + (1/2\mu R^2)J(J+1) - E\right]\chi'_{1,J}(R) = H^e(R)\chi'_{2,J}(R)$$
$$\left[-(1/2\mu)\frac{d^2}{dR^2} + V_2(R) + (1/2\mu R^2)J(J+1) - E\right]\chi'_{2,J}(R) = H^e(R)\chi'_{1,J}(R) \tag{7.12.1}$$

where the χ' vibrational functions are unknown exact solutions. Note that the Golden Rule formula uses vibrational wavefunctions that are solutions of the uncoupled Eq. (7.5.22) where the terms on the right-hand side of Eq. (7.12.1) have been disregarded.

Very strong predissociations can result from homogeneous interactions: electrostatic interactions for light molecules ($I^2\Sigma^+$ state of NO) or spin-orbit interactions for heavy molecules ($B^3\Pi_{0^+}$ state of IBr). Above the energy of the curve crossing, the expectation that levels should belong to one or the other potential curve seems completely unsound. One finds only numerous very broad levels, arranged without obvious rotation-vibration structure, interspersed with a smaller number of sharp lines. These sharp lines are the key to an understanding of such a case, as shown by Child (1976), who has developed a method for treating strong predissociations by a semiclassical approach.

Even for this case of strong coupling, the adiabatic picture of two potential curves that avoid crossing is inappropriate. Child has introduced an intermediate coupling picture that takes advantage of both diabatic and adiabatic characteristics. The diabatic curve of the predissociated state is displayed (solid lines) in Fig. 7.36a The corresponding diabatic vibrational levels, E_d, are plotted versus $J(J+1)$ (solid lines) in Fig. 7.36b.

The adiabatic picture provides an alternative point of view. In Fig. 7.36a the adiabatic curves are plotted as dashed lines. The upper adiabatic potential supports a set of adiabatic levels, $E_{\rm ad}$. As the rotational constants for the levels of the adiabatic curve are smaller than those for the corresponding levels of the diabatic curve, the adiabatic levels (dashed lines on Fig. 7.36) will cross the diabatic levels. Note that there is no relationship between the numbering of v_d and v_{ad}.

By Child's semiclassical approach, it has been shown (Child, 1976) that the actual levels E (dotted-dashed lines in Fig. 7.36b) in the region of the curve crossing lie intermediate in energy between the two sets of approximate levels

$$E = (E_d + xE_{ad})/(1+x)^{\dagger} \tag{7.12.2}$$

where x is a dimensionless coupling-strength parameter,

$$x = (\overline{\omega}_d/\overline{\omega}_{ad})\left(\lambda^{-2} - 1\right) \tag{7.12.3}$$

($\overline{\omega}_d$ and $\overline{\omega}_{ad}$ are the local vibrational spacing for the diabatic and adiabatic potentials, respectively, and $(\lambda^{-2} - 1)$ is the Landau-Zener coupling strength parameter (see for example, Preston, *et al.*, 1974) with

$$\lambda^2 = \exp[-2\pi(H^e)^2/\hbar \underset{\sim}{v} \Delta F] \tag{7.12.4}$$

and $\underset{\sim}{v}$ and ΔF are defined in Eqs. (7.6.6) and (7.6.4), respectively. Similarly, the values of the observed rotational constants are

$$B = (B_d + xB_{ad})/(1+x). \tag{7.12.5}$$

Figure 7.36b shows that certain J–values exist for which the energies of the three levels – diabatic, adiabatic, and exact – are identical. For this J–value, the linewidth is predicted and found to be zero. For the neighboring levels, the linewidths are shown to vary as

$$\Gamma = 2\pi x(a+x)(E_d - E_{ad})^2/\hbar\overline{\omega}_{ad}(1+x)^3, \tag{7.12.6}$$

where $a = \overline{\omega}_d/\overline{\omega}_{ad}$, and $\overline{\omega}_d$ and $\overline{\omega}_{ad}$ are the local vibrational spacing for diabatic and adiabatic potentials, respectively. This semiclassical approach has been found to be in excellent agreement with exact numerical solution of the coupled Eq. (7.12.1) (Child and Lefebvre, 1978).

---

†A small correction must be added to the diabatic and adiabatic levels to satisfy this equation. This correction for the diabatic levels is the term $\Delta_{v,J}$ of Eq. (7.6.12). For adiabatic levels this shift is, at most, one quarter of the adiabatic vibrational interval, $\Delta G_{\rm ad}(v)/4$.

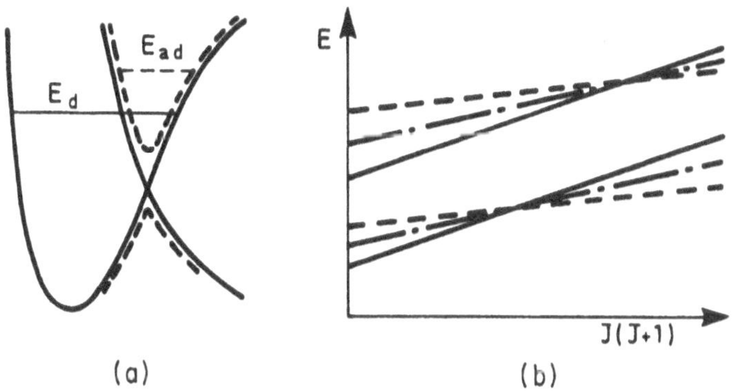

Figure 7.36: Predissociation for a pair of states intermediate between adiabatic and diabatic coupling limits. [From Child (1980b).] (a) Diabatic (solid line) and adiabatic (dashed line) potential curves and the corresponding vibrational levels. (b) Term values plotted versus $J(J+1)$: diabatic (solid line), adiabatic (dashed line), and actual term values (dotted-and-dashed line).

In order to make practical use of Child's semiclassical method, it is necessary to have absolute rotational assignments for the sharp lines. This is a nontrivial problem since the usual rotation-vibration branch and band structures are shattered beyond recognition. However, the sharpest lines will have appreciable fluorescence quantum yields, thus enabling rotational assignments to be made on the basis of resolved $R(J-1)-P(J+1)=B''(4J'+2)$ splittings in laser-excited fluorescence spectra. It is fortunate that the theoretical analysis is based on the sort of assignment information that is most accessible in the simplest tunable laser experiment.

Once rotational constants are obtained from the sharpest lines, trial diabatic and adiabatic potentials are refined until the rotational constants, B_d and B_{ad}, calculated from them satisfy Eq. (7.12.5). Figure 7.37 shows that the exact solutions of the coupled equations give a width that is not proportional to $(H^e)^2$ when the coupling becomes very strong. One can understand this by varying the electronic matrix element. As H^e increases, the adiabatic level shifts. When it has shifted into coincidence with the diabatic level, the width is zero according to Eq. (7.12.6).

For very large widths, slight deviations from a Lorentzian lineshape are predicted (Child and Lefebvre, 1978).

It becomes necessary to solve coupled equations in more complicated cases. Here are two examples.

1. *Several states cross the dissociated state (multichannel problem).* In general, the total width is assumed to be the sum of the partial widths arising separately from interactions with individual states, but this is valid only if the continua of the dissociative states do not interact with each other. If the continua are coupled to each other by an interaction, I (I is a dimensionless

parameter), the width is

$$\Gamma = \frac{\Gamma_1 + \Gamma_2}{1 + \pi^2 I^2}, \qquad (7.12.7)$$

where Γ_1 and Γ_2 are the partial widths 'from each separate dissociative state (Beswick and Lefebvre, 1973). I^2 will be, in general, small. The interaction strength between a discrete state and a continuum can be extrapolated to the case of a continuum$\sim$continuum interaction by dividing it by ΔG, the separation between two successive vibrational levels of the discrete state [see Eq. (7.5.3)]. Then

$$\pi^2 I^2 = \frac{\pi\Gamma}{2} \frac{1}{\Delta G}; \qquad (7.12.8)$$

assuming $\Gamma = 10$ cm^{-1} and $\Delta G = 1500$ cm^{-1}, $\pi^2 I^2 \simeq 0.01$.

2. *Indirect or accidental predissociation*, which is treated in the following section.

7.13 Indirect (Accidental) Predissociation and Interference Effects

In Section 7.8 the possibility of predissociation of isolated lines was mentioned. This is usually called accidental predissociation and can be interpreted as perturbation of a nominally bound rotational level by a predissociated level that lies nearby in energy for this value of J. This type of predissociation should more generally be called indirect predissociation, since the predissociation takes place through an intermediate state (or "doorway" state, see Section 9.2).

Let H_{12} be the coupling between level 1 and level 2, where level 2 is predissociated by the continuum of a third state and has a linewidth Γ_2. Direct predissociation of level 1 by state 3 is assumed to be negligible.

Kovács and Budó (1947) have argued that one should first compute the mixing of the two discrete states. Then, the width of level v_1 of state 1, Γ_{1,v_1}, is borrowed from state 2 in proportion to the mixing coefficient $a_{1,v_1;2,v_2}$, which expresses the level 2 character mixed into the nominal level 1 eigenstate, $'\Psi'_{1,v_1}$:

$$'\Psi'_{1,v_1} = a_{1,v_1;2,v_2}\Psi_{1,v_1} + (1 - a^2)^{1/2}\Psi_{2,v_2}. \qquad (7.13.1)$$

The width is then

$$\Gamma_{1,v_1} = a^2_{1,v_1;2,v_2}\Gamma_{2,v_2}. \qquad (7.13.2)$$

If the energy separation between the two discrete states, ΔE_{12}, is large compared to the interaction matrix element, H_{12}, this mixing coefficient is given by the perturbation formula,

$$a_{1,v_1;2,v_2} = \frac{H_{12}}{\Delta E_{12}}. \qquad (7.13.3)$$

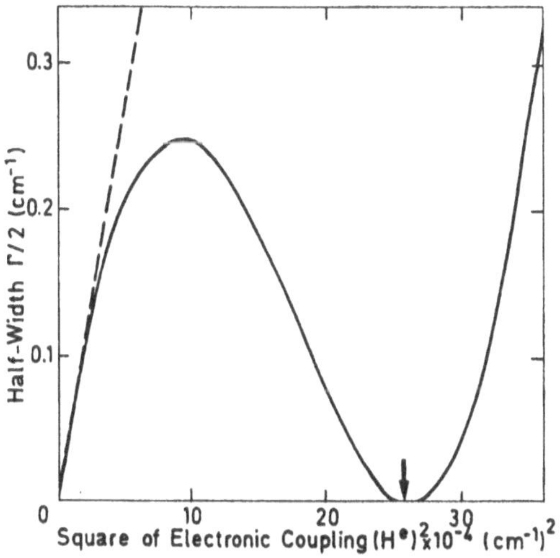

Figure 7.37: Variation of Γ versus $(H^e)^2$. The full line corresponds to the $v = 21$ diabatic level of a model molecule. The arrow indicates the value of H^e for which the adiabatic $v = 0$ level coincides with the diabatic $v = 21$ level. The dotted line gives the Γ-values expected from the Golden Rule formula. [Adapted from data of the model of Child and Lefebvre (1978) and unpublished results.]

However, if state 1 borrows its width from *several* vibrational levels of state 2, Eq. (7.13.2) must be replaced by

$$\Gamma_{1,v_1} = 2\pi \left[\sum_{v_j} a_{1,v_1;2,v_j} H_{2,v_j;3} \right]^2 , \tag{7.13.4}$$

where $H_{2,v_j;3}$ is related to the width of level v_j by

$$\Gamma_{2,v_j} = 2\pi H^2_{2,v_j;3}. \tag{7.13.5}$$

This formula shows how interference between terms in the sum over predissociated perturbers can increase or decrease the accidental width (Lefebvre-Brion and Colin, 1977).

If levels 1 and 2 are degenerate before interaction, the linewidth Γ might be expected to be divided equally between them, and

$$\Gamma_{1,v_1} = \Gamma_{2,v_2} = \Gamma. \tag{7.13.6}$$

Atabek, *et al.*, (1980) have pointed out that this conclusion is incorrect when $H_{12} \ll \Gamma_{2,v_2}$. Then it would not be proper to prediagonalize the interaction between the two discrete states. Indeed level 2, after its interaction with the continuum of state 3, must be considered as a pseudo-continuum with respect to state 1. In the weak coupling ($H_{12} \ll \Gamma_{2,v_2}$) limit, the width of the accidental predissociation is then given by

$$\Gamma_{1,v_1} = \frac{H_{12}^2 \Gamma_{2,v_2}}{\Delta E_{12}^2 + (\Gamma_{2,v_2}/2)^2} \tag{7.13.7}$$

If Γ_{2,v_2} is small compared to ΔE_{12}, this formula is equivalent to Eq. (7.13.2) when levels 1 and 2 are far from coincidence. However, if $\Delta E = 0$, then

$$[\Gamma_{1,v_1} = \frac{4H_{12}^2}{\Gamma_{2,v_2}} \quad \text{for} \quad \Gamma_{2,v_2} \gg H_{12}. \tag{7.13.8}$$

Paradoxically, the accidental width is found to decrease as Γ_{2,v_2} increases. This resonance narrowing phenomenon corresponds to an accidental decoupling of level 1 from the continuum of state 3, when $\Delta E_{12} \ll \Gamma_{2,v_2}/2$.

An accidental predissociation occurs in the N_2 $b^1\Pi_u$ state (Robbe, 1978, Ubachs, *et al.*, 1989). A possible explanation for the strong predissociation of the $b^1\Pi_u(v = 3)$ level is shown in Fig. 7.38 Its coupling with the continuum of the $C'^3\Pi_u$ state is negligible, but it is near degenerate with $C^3\Pi_u(v = 8)$,[†] which is strongly predissociated by the continuum of the $C'^3\Pi_u$ state. The situation is represented schematically in Fig. 7.39. Fig. 7.40 shows the difference between the value of $\Gamma_{1,v}$ calculated using the Kovács-Budó model and the exact value, obtained by solution of coupled equations. If the coupling between the C and C′ states is increased, the $\Gamma_C \gg H_{bC}$ limiting case is reached and the Γ_{b,v_b} value is given by Eq. (7.13.5).

If the mutually perturbing levels 1 and 2 are *both* predissociated by the *same* continuum (of the $|\phi_0\chi_E\rangle$ state), then the direction in which the width is transferred between levels 1 and 2 depends on the relative signs of three matrix elements: the $1 \sim 2$ matrix element and the $1 \sim |\phi_0\chi_E\rangle$ and $2 \sim |\phi_0\chi_E\rangle$ matrix elements.

The width of one of the mixed levels is given by

$$\Gamma_v = 2\pi \left| \left\langle \left(a\phi_1\chi_{v_1} + \left(1 - a^2\right)^{1/2} \phi_2\chi_{v_2} \right) |\mathbf{H}| \phi_0\chi_E \right\rangle \right|^2. \tag{7.13.9a}$$

The mixing coefficient, a, is given by nondegenerate perturbation theory as

$$a = \frac{\langle \phi_1\chi_{v_1} |\mathbf{H}| \phi_2\chi_{v_2} \rangle}{\Delta E_{12}}. \tag{7.13.9b}$$

The sign of the interference term depends on the product:

$$a \langle \chi_{v_1} | \chi_{v_2} \rangle \langle \chi_E | \chi_{v_1} \rangle \langle \chi_{v_2} | \chi_E \rangle. \tag{7.13.9c}$$

[†] Recent calculations show that this near degeneracy is with the $v = 9$ level of the $C^3\Pi_u$ state (Lewis, *et al.*, 2001).

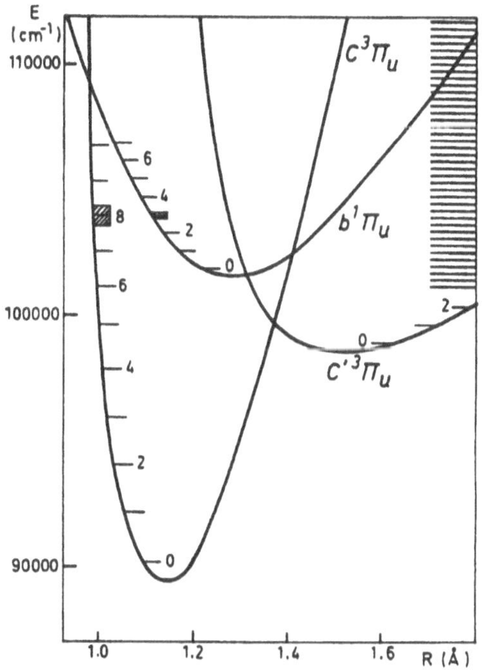

Figure 7.38: The potential energy curves of some states of N_2. The indirect predissociation of $b^1\Pi_u(v=3)$ by the $C'^3\Pi_u$ continuum via the $C^3\Pi_u(v=8)$ level is shown. The widths of the directly and indirectly predissociated levels are indicated.

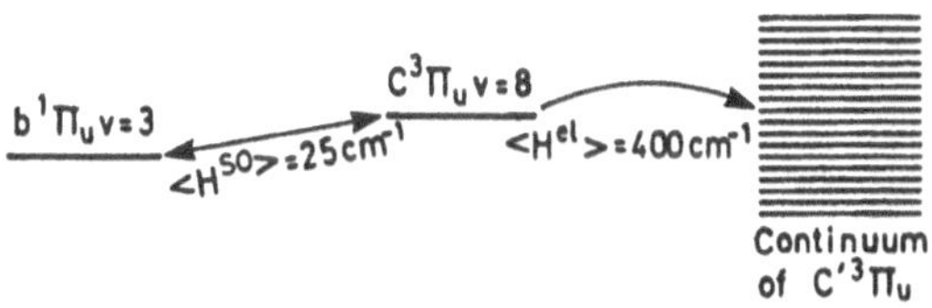

Figure 7.39: Schematic mechanism of the indirect or accidental predissociation described in Fig. 7.38.

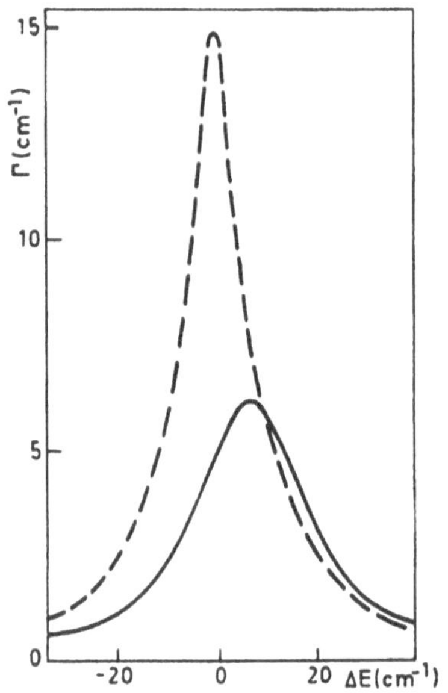

Figure 7.40: Comparison between the Budó-Kovács (dashed line) and coupled equations (solid line) models for indirect predissociation. The width of the indirectly predissociated N_2 $b^1\Pi_u(v=3)$ level is plotted versus the energy separation from the $C^3\Pi_u(v=8)$ intermediate predissociated perturbing level. The electronic coupling between the C and C′ states is taken as 400 cm^{-1}. [Courtesy J.M. Robbe from data of Robbe (1978).]

Note that, in this product, each vibrational wave function appears twice, therefore the sign of the interference term is independent of phase convention for the vibrational wavefunctions. Consequently, the sense of the interference effect will reverse when the energy denominator which appears in a switches sign. When levels E_1 and E_2 cross, $\Delta E_{12} = E_1 - E_2$ switches sign below and above the crossing point and consequently, depending on the sign of the matrix element product term, there is an increase (or decrease) of the width as the crossing point is approached from below and the opposite situation from above the crossing point. A nice example of this type of interference in predissociation is given by the predissociation of the $B^1\Pi$ state of NaK in the vicinity of a perturbation by the $b^3\Pi$ state (Kasahara, *et al.*, 1997).

A simple accidental predissociation has been investigated in detail for some rotational levels of the $v = 0$ and 1 vibrational levels of the CO $E^1\Pi$ Rydberg state. These levels are near degenerate with levels of the predissociated $k^3\Pi$ valence state. There is a very weak interaction between these two states, with an interaction matrix element of about 2 cm^{-1}. The dominant configurations of these two states differ by two orbitals, but the $E^1\Pi$ Rydberg state interacts by spin-orbit with the isoconfigurational $c^3\Pi$ state, which in turn interacts by electrostatic interaction with the $k^3\Pi$ valence state (Cacciani, *et al.*, 1995, Ubachs, *et al.*, 2000, where this accidental predissociation has been examined for six CO isotopomers).

7.14 Some Recipes for Interpretation

This chapter concludes with a brief outline of the information crucial to selection and testing of a mechanistic model for an observed predissociation.

It is frequently possible to eliminate all but a few plausible candidates for the electronic identity of the predissociating state simply by considering the symmetries (Wigner-Witmer rules) of the electronic states associated with the lowest energy dissociation limit. For predissociation of very high-lying electronic states, it is often necessary to consider all of the atomic dissociation asymptotes that lie at lower energy than the predissociated level. Knowledge of the configurational character of the predissociated and plausible predissociating states can be valuable because, in addition to the usual electronic selection rules for perturbations, orbital selection rules and estimates of relevant orbital matrix elements can frequently eliminate all but one candidate.

Once a hypothesis about the symmetry and configurational parentage of a predissociating state is made, a variety of qualitative and quantitative tests must be made. The approximate size of the electronic interaction factor may be derived from an experimental Γ (or τ) value and an estimate of the vibrational factor. Some differential Franck-Condon factors for crossing potential curves are listed in Table 7.4, column 6. Recall that the isotopic dependence of the differential Franck-Condon factor is $\propto \mu^{1/6}$, and it is unusual to find a value larger than $\mu^{1/6} \times 10^{-4}$cm. If the electronic interaction is of spin-orbit origin, it is a simple matter to estimate its magnitude from the spin-orbit parameters of the constituent atoms (Section 5.3). If the predissociation is gyroscopic, then the electronic factor will be approximately $B_C[l(l+1)]^{1/2}[J(J+1)]^{1/2}$. (Strictly speaking, the electronic factor is only $[l(l+1)]^{1/2}$, as listed in Table 7.3.)

If experimental data concerning the variation of Γ with v, J, and μ are available, then a detailed interpretation of the predissociation may be attempted making use of computer programs for calculating vibrational factors. These calculations may involve either semiclassical formulas [Eq. (7.6.3)] or numerical integration of vibrational wavefunctions.

Greater caution is required when dealing with weak predissociations. In general, the weaker the effect, the larger the number of plausible mechanisms. As for perturbations, second-order effects (especially spin-orbit) can give rise to weak interactions between states differing by $\Delta\Lambda = \pm 2$ (and other spin-spin selection rules). Second-order predissociations will have very small electronic matrix elements. On the other hand, very small vibrational factors occur for interactions between noncrossing potential curves. Table 7.3 summarizes several examples of weak predissociations.

There are several pitfalls to be avoided in formulating a predissociation mechanism. A linear variation of the linewidth with $J(J+1)$ can be caused by either gyroscopic interaction or penetration through a centrifugal barrier. Another ambiguous indication is given by the J–dependence of the vibrational factor for the case of homogeneous, outer-crossing predissociations. This is illustrated (see Fig. 7.41) by the OH $A^2\Sigma^+(v = 8)$ level, which, when predissociated by the $^4\Pi$ state (Section 7.11.1), shows a J–variation of the linewidth similar to that expected for a gyroscopic interaction.

When the linewidth exhibits no oscillations, this suggests the occurrence of an inner crossing, but two cases exist where an outer crossing is shown to display no linewidth oscillation. The first example concerns the OD molecule. Below the energy of the curve crossing, the bound~free vibrational overlap comes only from the tail of the discrete wavefunction (tunnelling). The nonradiative decay rate is very slow, but it increases smoothly with J [predissociation of the OD $A^2\Sigma^+(v = 0 - 2)$ levels by the $^4\Sigma^-$ state (Bergeman, *et al.*, 1981)].

A second example is given by the heterogeneous predissociation of the Br_2 $B^3\Pi_{0_u^+}$ state by the $^1\Pi_{1_u}$ state (Child 1980a). Oscillations in the linewidth disappear for high vibrational levels because, in this energy region, the repulsive curve is nearly parallel to the bound potential curve and the outer crossing case behaves as a noncrossing case.

These ambiguities can be eliminated if levels belonging to successive values of v or J are examined. Isotope effects are also useful for confirming the nature of a predissociation. In the absence of detailed information, one must be cautious about inferring the origin of observed predissociation effects.

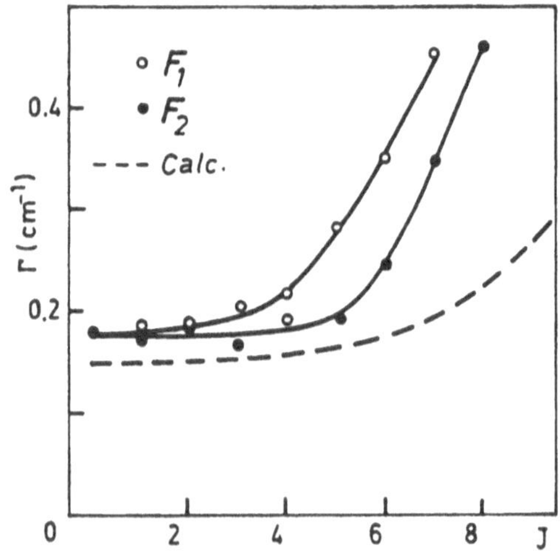

Figure 7.41: Linewidth variation in the OH $A^2\Sigma^+(v = 8)$ level for the resolved F_1 and F_2 spin components. The solid lines represent the experimental data of Czarny and Felenbok (1968). The dotted line shows the calculated linewidths from Czarny, *et al.*(1971).

7.15 References

Ackerman, M., and Biaume, F. (1970), *J. Mol. Spectrosc.* **35**, 73.

Allison, A. C., and Dalgarno, A. (1971), *J. Chem. Phys.* **55**, 4342.

Allison, A. C., and Stwalley, W. C. (1973), *J. Chem. Phys.* **58**, 5187.

Alexander, M. H., Pouilly, B., and Duhoo, T. (1993), *J. Chem. Phys.* **99**, 1752.

Alexander, A. J., Kim, Z. H., Kandel, S. A., Zare, R. N., Rakitzis, T. P., Asano, Y., and Yabushita, S. (2000), *J. Chem. Phys.* **113**, 9022.

Antonova, S., Lazarov, G., Urbanski, K., Lyyra, A. M., Li Li, Jeung, G.-H., and Stwalley, W. C. (2000), *J. Chem. Phys.* **112**, 7080.

Ashfold, M. N. R., Dixon, R. N., Prince, J. D., Tutcher, B., and Western, C. M. (1986), *J. Chem. Soc. Faraday Trans. 2* **82**, 1257.

Ashfold, M. N. R., Cook, P. A., Langford, S. R., Orr-Ewing, A. J., and Regan, P. M. (1998), *Laser Techniques for State-Selected and State-to-State Chemistry IV Proceedings of SPIE p. 84.*

Atabek, O., and Lefebvre, R. (1972), *Chem. Phys. Lett.* **17**, 167.

Atabek, O., Lefebvre, R., and Requena, A. (1980), *J. Mol. Spectrosc.* **82**, 364.

Band, Y. B., Freed, K. F., and Kouri, D. J. (1981), *J. Chem. Phys.* **74**, 4380.

Band, Y. B., Freed, K. F., Singer, S. J., and William, C. J. (1987), *J. Phys. Chem.* **91**, 5402.

Bandrauk, A. D., and Child, M. S. (1970), *Mol. Phys.* **19**, 95.

Banichevich, A., Klotz, R., and Peyerimhoff, S. (1992), *Mol. Phys.* **75**, 173.

Barrow, R. F., Chandler, G. G., and Meyer, C. B. (1966), *Philos. Trans. R. Soc. London, Ser. A* **260**, 395.

Baumfalk, R., Buck, U., Frischkorn, C., Nahler, N. H., and Huwel, L. (1999), *J. Chem. Phys.* **111**, 2595.

Ben-Aryeh, Y. (1973), *J. Quant. Spectrosc. Radiat. Transfer* **13**, 1441.

Bergeman, T., Erman, P., Haratym, Z., and Larsson, M. (1981), *Phys. Scr.* **23**, 45.

Beswick, J. A., and Lefebvre, R. (1973), *Mol. Phys.* **29**, 1611.

Beswick, J. A., and Jortner, J. (1990), *Chem. Phys. Lett.* **168**, 246.

Beswick, J. A., (1993) in *Dynamical Processes in Molecular Physics*, Ed. Delgado-Barrio, IOP (Bristol) p. 183.

Bohn, B., Stuhl, F., Parlant, G., Dagdigian, P. J., and Yarkony, D. R. (1992), *J. Chem. Phys.* **96**, 5059.

Booth, J. L., Ozier, I., and Dalby, F.W. (1994), *Phys. Rev. Lett.* **72**, 2371.

Borrell, P., Guyon, P. M., and Glass-Maujean, M. (1977), *J. Chem. Phys.* **66**, 818.

Broyer, M., Vigué, J., and Lehmann, J. C. (1976), *J. Chem. Phys.* **64**, 4793.

Brzozowski, J., Bunker, P., Elander, N., and Erman, P. (1976), *Astrophys. J.* **207**, 414.

Buijsse, B. and van der Zande, W. J. (1997), *Phys. Rev. Lett.* **79**, 4558.

Buijsse, B., van der Zande, W.J., Eppink, A. T. J. B., Parker, D. H., Lewis, B. R., and Gibson, S. T. (1998), *J. Chem. Phys.* **108**, 7229.

Busch, G. E., Mahoney, R. T., Morse, R. I., and Wilson, K. R. (1969), *J. Chem. Phys.* **51**, 449.

Cacciani, P., Hogervost, W., and Ubachs, W. (1995), *J. Chem. Phys.* **102**, 8308.

Carré, M., Druetta, M., Gaillard, M. L., Bukow, H. H., Horani, M., Roche, A. L., and Velghe, M. (1980), *Mol. Phys.* **40**, 1453.

Carrington, A., Milverton, D. R. J., and Sarre, P. (1978), *Mol. Phys.* **35**, 1505.

Charron, E., and Suzor-Weiner, A. (1998), *J. Chem. Phys.* **108**, 3922.

Child, M. S. (1973), *J. Mol. Spectrosc.* **45**, 293.

Child, M. S. (1974), in "Molecular Spectroscopy", Vol. 2 (R.F. Barrow, D.A. Long, and

D.J. Millen, eds.), p. 466. Chemical Soc., London, Specialist Periodical Report.
Child, M. S. (1976), *Mol. Phys.* **32**, 1495.
Child, M. S. (1980a), *J. Phys. B.* **13**, 2557.
Child, M. S. (1980b), *in* "Semi-Classical Methods in Molecular Scattering and Spectroscopy" (M.S. Child, ed.) p. 127, Reidel Publ. Dordrecht, Holland.
Child, M. S., and Lefebvre, R. (1978), *Chem. Phys. Lett.* **55**, 213.
Child, M. S., Essén, H., and Le Roy, R. J. (1983), *J. Chem. Phys.* **78**, 6732.
Cornett, S. T., Sadeghpour, H. R. and Cavagnero, M. J. (1999), *Phys. Rev. Lett.* **82**, 2488.
Czarny, J., and Felenbok, P. (1968), *Ann. Astrophys.* **31**, 141.
Czarny, J., Felenbok, P., and Lefebvre-Brion, H. (1971), *J. Phys. B* **4**, 124.
Douglas, A. E. (1952), *Can. J. Phys.* **30**, 302.
Durmaz, S., and Murrell, J. N. (1971), *Trans. Faraday Soc.* **67**, 3395.
Edmonds, A. R. (1974), "Angular Momentum in Quantum Mechanics," Princeton University Press, Princeton, New Jersey.
Elander, N., Hehenberger, M., and Bunker, P. R. (1979), *Phys. Scr.* **20**, 631.
Erman, P. (1979), *in* "Molecular Spectroscopy", Vol. 6 (R.F. Barrow, D.A. Long, and J. Sherican, eds.), p. 174. Chemical Soc., London, Specialist Periodical Report.
Fano, U. (1961), *Phys. Rev.* **124**, 1866.
Fernandez-Alonso, F., Bean, B. D., and Zare, R. N. (1999), *J. Chem. Phys.* **111**, 1022.
Fontana, P. R. (1982), "Atomic Radiative Processes", Academic Press, New York.
Gadéa, F. X., Berriche, H., Roncero, O., Villarreal, P., and Delgado Barrio, G. (1997), *J. Chem. Phys.* **107**, 10515
Gallusser, R. (1976), Thesis, Physical Chemistry Laboratory, ETH Zurich, Switzerland.
Gallusser, R., and Dressler, K. (1982), *J. Chem. Phys.* **76**, 4311.
Gao, H. (1997), *J. Chem. Phys.* **107**, 7278.
Gendron, D. J., and Hepburn, J. W. (1998), *J. Chem. Phys.* **109**, 7205.
Giusti-Suzor, A., and Jungen, C. (1984), *J. Chem. Phys.* **80**, 986.
Givertz, S. C., and Balint-Kurti, G. G. (1986), *J. Chem. Soc. Far. Trans. 2* **82**, 1231.
Glass-Maujean, M., Breton, J., and Guyon, P. M. (1979), *Chem. Phys. Lett.* **63**, 591.
Glass-Maujean, M., Guyon, P. M., and Breton, J. (1987), *Zeits. für Phys* **D5**, 189.
Govers, T. R., van de Runstraat, C. A., and de Heer, F. J. (1975), *Chem. Phys.* **9**, 285.
Greene, C. H., and Zare, R. (1982), *Ann. Rev. Phys. Chem.* **33**, 119.
Herzberg, G. (1950), "Molecular Spectra and Molecular Structure. I. Spectra of Diatomic Molecules," Van Nostrand-Reindhold, Princeton, New Jersey, reprinted 1988 by Krieger, Malabar.
Herzberg, G. (1971), *in* "Topics in Modern Physics," Colorado Assoc. Univ. Press, p. 191, Boulder, Colo.
Helm, H., Cosby, P. C., and Huestis, D. L. (1980), *J. Chem. Phys.* **73**, 2629.
Houston, P. L. (1996), *J. Phys. Chem.* **100**, 12757.
Hudson, R. D., Carter, V. L., and Stein, J. A. (1966), *J. Geophys. Res.* **7**, 2295.
Ishii, J., Uehara, K. and Tsukiyama, K. (1995), *J. Chem. Phys.* **102**, 9174.
Julienne, P. S., and Krauss, M. (1975), *J. Mol. Spectrosc.* **56**, 270.
Kasahara, S., Shibata, M., Baba, M., and Katô, H. (1997), *J. Phys. Chem. A*, **101**, 422.
Katô, H., Kobayashi, T., Chosa, M., Nakahori, T., Iida, T., Kasahara, S. and Baba, M. (1991), *J. Chem. Phys.* **94**, 2600.
Katô, H. and Baba, M. (1995), *Chem. Rev.* **95**, 2311.
Kim, B., and Yoshihara, K. (1993), *J. Chem. Phys.* **99**, 1433.
Kim, B., Yoshihara, K., and Lee, S. (1994), *Phys. Rev. Lett.* **73**, 424.
Kimura, Y., Lefebvre-Brion, H., Kashahara, S., Katô, H., Baba, M., and Lefebvre, R.

(2000), *J. Chem. Phys.* **113**, 8637.
Kinugawa, T., and Arikawa, T. (1992), *J. Chem. Phys.* **96**, 4801.
Koffend, J. B., Bacis, R., Churassy, S., Gaillard, M. L., Pique, J. P., and Hartmann, F. (1983), *Laser Chem.* **1**, 185.
Kovács, I., and Budó, A. (1947), *J. Chem. Phys.* **15**, 166.
Lafosse, A., Lebech, M., Brenot, J. C., Guyon, P. M., Jagutzki, O., Spielberger, L., Vervloet, M., Houver, J. C., and Dowek, D. (2000), *Phys. Rev. Lett.* **84**, 5987.
Lafosse, A., Brenot, J. C., Golovin, A. V., Guyon, P. M., Hoejrup, K., Houver, J. C., Lebech, M., and Dowek, D. (2001), *J. Chem. Phys.* **114**, 6605.
Lambert, H. M., Dagdigian, P. J., and Alexander, M. H. (1998), *J. Chem. Phys.* **108**, 4460.
Langford, S. R., Regan, P.M., Orr-Ewing, A. J., and Ashfold, M. N. R. (1998), *Chem. Phys.* **231**, 245.
Leahy, D. J., Osborn, D. L., Cyr, D. R., and Neumark, D. H. (1995), *J. Chem. Phys.* **103**, 2495.
Lefebvre-Brion, H., and Colin, R. (1977), *J. Mol. Spectrosc.* **65**, 33.
Lefebvre-Brion, H., American Institute of Physics Conference Proceedings 225, eds. M. Garcia-Sucre, G. Raşeev and S.C. Ross (American Institute of Physics, New York, 1991), p. 375.
Lembo, L. J., Huestis, D. L., Keiding, S. R., Bjerre, N., and Helm, M. (1988), *Phys. Rev. A* **38**, 3447.
Le Roy, R. J., Keogh, W. J., and Child, M. S. (1988), *J. Chem. Phys.* **89**, 4564.
Le Roy, R. J., Kraemer, G. T., and Manzhos, S. (2002), *J. Chem. Phys.* **117**, 9353.
Levy, I., and Shapiro, M. (1988) *J. Chem. Phys.* **89**, 2900.
Lewis, B. R., Carver, J. H., Hobbs, T. I., McCoy, D. G., and Gies, H. P. F. (1980), *J. Quant. Spectrosc. Radiat. Transfer* **24**, 365.
Lewis, B. R., Gibson, S. T., Emami, M., and Carver, J. H. (1988a), *J. Quant. Spectrosc. Radiat. Transfer* **40**, 1.
Lewis, B. R., Gibson, S. T., Emami, M., and Carver, J. H. (1988b), *J. Quant. Spectrosc. Radiat. Transfer* **40**, 469.
Lewis, B. R., and Gibson, S. T. (1990), *Can. J. Phys.* **68**, 231.
Lewis, B. R., Banergee, S. S., and Gibson, S. T. (1995a), *J. Chem. Phys.* **102**, 6631.
Lewis, B. R., England, J. P., Winkel, R. J., Banerjee, S. S., Dooley, P. M., Gibson, S.T., and Baldwin, K.G. H. (1995b), *Phys. Rev. A* **52**, 2717.
Lewis, B. R., Gibson, S. T., Zhang, W., Robbe, J. M., and Lefebvre-Brion, H. (2003), *J. Chem. Phys.* **000**, 0000.
Li, Y., Petsalakis, D., Liebermann, H. P., Hirsch, G., and Buenker, R. J. (1997), *J. Chem. Phys.* **106**, 1123.
Li, Y., Hirsch, G. and Buenker, R. J. (1998), *J. Chem. Phys.* **108**, 8123.
Liyanage, R., Yang, Y., Hashimoto, S., Gordon, R. J., and Field, R. W. (1995), *J. Chem. Phys.* **103**, 6811.
Lorquet, A. J., and Lorquet, J. C. (1974), *Chem. Phys. Lett.* **26**, 138.
Magnotta, F., Nesbitt, D. J., and Leone, S. R. (1981), *Chem. Phys. Lett.* **80**, 21.
Martinez, E., Martinez, M. T., and Castano, F. J. (1988), *J. Mol. Spectrosc.* **128**, 554.
Matsumi, Y., Tonokura, K., Kawasaki, M., and Ibuki, T. (1990), *J. Chem. Phys.* **93**, 7981.
McCormack, E. F., Di Teodoro, F., Grochochinski, J. M., and Pratt, S. T. (1998), *J. Chem. Phys.* **109**, 63.
Meiwes, K. H., and Engelke, F. (1982), *Chem. Phys. Lett.* **85**, 409.
Merzbacher, E. (1998), "Quantum Mechanics", third edition, John Wiley, New York.

Miescher, E. (1976), *Can. J. Phys.* **54**, 2074.
Mulliken, R. S. (1960), *J. Chem. Phys.* **33**, 247.
Moseley, J. T., Cosby, P. C., Ozenne, J. B., and Durup, J. (1979), *J. Chem. Phys.* **70**, 1974.
Murrell, J. N., and Taylor, J. M. (1969), *Mol. Phys.* **16**, 609.
Olson, C. D., and Innes, K. K. (1976), *J. Chem. Phys.* **64**, 2405.
Patel-Misra, D., Parlant, G., Sauder, D. G., Yarkony, D. R., and Dagdigian, P. J. (1991), *J. Chem. Phys.* **94**, 1913.
Penno, M., Holzwarth, A., and Weitzel, K.-M. (1998), *J. Phys. Chem. A* **102**, 1927.
Péoux, G., Monnerville, M., Duhoo, T., and Pouilly, B. (1997), *J. Chem. Phys.* **107**, 70.
Pique, J. P., Bacis, R., Hartmann, F., Sadeghi, N., and Churassy, S. (1983), *J. Phys. (Orsay, Fr.)* **44**, 347.
Preston, R. K., Sloane, C., and Miller, W.H. (1974), *J. Chem. Phys.* **60**, 4961.
Regan, P. M., Ascenzi, D., Clementi, C., Ashfold, M. N. R., and Orr-Ewing, A. J. (1999), *Chem. Phys. Lett.* **315**, 187.
Robbe, J. M. (1978), Ph.D. thesis, Université des Sciences et Techniques de Lille, Lille, France, unpublished.
Roche, A. L., and Tellinghuisen, J. (1979), *Mol. Phys.* **38**, 129.
Rothschild, M., Egger, H., Hawkins, R. T., Bokor, J., Pummer, H., and Rhodes, C. K. (1981), *Phys. Rev. A* **23**, 206.
Samartzis, P. C., Sakellariou, I., Gougousi, T., and Kitsopoulos, T. (1997), *J. Chem. Phys.* **107**, 43.
Sánchez, I., and Martín, F. (1999), *Phys. Rev. A* **60**, 2200.
Schaefer, H. F., III, and Miller, W. H. (1971), *J. Chem. Phys.* **55**, 4107.
Schmidt, I., Meyer, W., Krüger, B., and Engelke, F. (1988), *Chem. Phys. Lett.* **143**, 353.
Sharp, T. E. (1970), *At. Data* **2**, 119.
Shaw, D. A., Holland, D. M. P., MacDonald, M. A., Hopkirk, A., Hayes, M. A., and McSweeny, S. M. (1992), *Chem. Phys.* **166**, 379, erratum (1993), **172**, 187.
Sheill, R. C., and Hepburn, J. W. (1999), unpublished results cited in Baumfalk, *et al.*, 1999.
Siebbeles, L. D. A., Glass-Maujean, M., Vasyutinskii, O. S., Beswick, J. A., and Roncero, O. (1994), *J. Chem. Phys.* **100**, 3610.
Siese, M., Tiemann, E., and Wulf, U. (1985), *Chem. Phys. Lett.* **117**, 208.
Smith, A. L. (1971), *J. Chem. Phys.* **55**, 4344.
Tchang-Brillet, W.-Ü. L., Julienne, P. S., Robbe, J.-M., Letzelter, C., and Rostas, F. (1992), *J. Chem. Phys.* **96**, 6735.
Tellinghuisen, J. (1984), *J. Mol. Spectrosc.* **103**, 455.
Ubachs, W., Tashiro, L., and Zare, R. N. (1989), *Chem. Phys.* **130**, 1.
Ubachs, W., Velchev, I., and Cacciani, P. (2000), *J. Chem. Phys.* **113**, 547.
Van Brunt, R. J., and Zare, R. N. (1968), *J. Chem. Phys.* **48**, 4304.
Van Dishoek, E. F., and Dalgarno, A. (1983), *J. Chem. Phys.* **79**, 873.
Vedder, H. J., Chawla, G. K., and Field, R. W. (1984), *Chem. Phys. Lett.* **111**, 303.
Vigué, J., Broyer, M., and Lehmann, J. C. (1977), *J. Phys. B.* **10**, L379.
Vigué, J., Broyer, M., and Lehmann, J. C. (1981), *J. Physique (Orsay, Fr.)* **42**, 937.
Vigué, J., Beswick, J. A., and Broyer, M. (1983), *J. Physique (Orsay, Fr.)* **44**, 1225.
Wheeler, M. D., Orr-Ewing, A. J., and Ashfold, M. N. R. (1997), *J. Chem. Phys.* **107**, 7591.
Wouters, E. R., Ahmed, M., Peterka, D. S., Bracker, A. S., Suits, A. G., and Vasyutinskii, O. S. (2000), in "Imaging in Chemical Physics" (Suits, A. G. and Continetti, R.

E., eds.), p. 238 ACS Symposium Series 770, Washington.
Wrede, E., Wouters, E. R., Beckert, M., Dixon, R. N., and Ashfold, M. N. R. (2002), *J. Chem. Phys.* **116**, 6064.
Xu, Z., Koplitz, B., Buelow, S., Baugh, D., and Wittig, C. (1986), *Chem. Phys. Lett.* **127**, 534.
Xu, Z., Koplitz, B., and Wittig, C. (1988), *J. Phys. Chem.* **92**, 5518.
Yarkony, D. R. (1992), *J. Chem. Phys.* **97**, 1838.
Zare, R. N., and Herschbach, D. R. (1963), *Proc. IEEE* **51**, 173.
Zare, R. N. (1972), *Mol. Photochem.* **4**, 1.
Zare, R. N. (1988), "Angular Momentum. Understanding Spatial Aspects in Chemistry and Physics," Wiley Interscience, New York.

Chapter 8

Photoionization Dynamics

Contents

8.1 Direct Ionization 552

8.1.1 Photoelectron Spectroscopy 552

8.1.2 ZEKE Spectroscopy 557

8.1.3 Shape resonances 559

8.1.4 Cooper minima 561

8.2 Experimental Aspects of Autoionization 564

8.3 The Nature of Autoionized States 568

8.4 Autoionization Widths 569

8.5 Rotational Autoionization 572

8.6 Vibrational Autoionization 576

8.7 Spin-Orbit Autoionization 581

8.8 Electronic (or Electrostatic) Autoionization . . . 586

8.9 Validity of the Approximations 588

8.10 Influence of Autoionization on ZEKE Peak Intensities 591

8.11 Photoelectron Angular Distribution, Photoion Alignment, and Spin Polarization 595

8.11.1 Photoelectron Angular Distribution 595

8.11.2 Photoion Alignment 600

8.11.3 Spin-Polarization 602

8.12 Competition between Autoionization and Predissociation 604

8.12.1 Superexcited State Decay Pathways 604

8.12.2 Theoretical Treatment 608

8.13 Coherent Control of Photofragmentation Product Branching Ratios 609

8.14 References 615

8.1 Direct Ionization

If the photon (or multiphoton) energy is sufficiently large to excite the molecule to an energy above the first ionization limit of the neutral molecule, the molecule can ionize. Similar to photodissociation processes, there are two possibilities:

direct ionization:

$$AB + h\nu \rightarrow AB^+ + e^-(KE) \tag{8.1.1}$$

or indirect ionization or autoionization:

$$AB + h\nu \rightarrow AB^{**} \rightarrow AB^+ + e^-(KE) \tag{8.1.2}$$

where $e^-(KE)$ is the kinetic energy of the electron. The "superexcited" state of the neutral molecule, denoted AB^{**}, is situated above the first ionization limit. In general, AB^{**} is a Rydberg state, which autoionizes by mixing with the electronic continuum of the ion-core plus free electron system.

8.1.1 Photoelectron Spectroscopy

As specified by Eq. (8.1.1), as the photon energy $h\nu$ is varied and either the ion or the electron is detected, a photoionization spectrum is obtained. If the monitored rovibronic state of the AB^+ ion-core is known and kept fixed, as $h\nu$ is scanned, the obtained spectrum is called a "*constant ionic state*" (CIS) spectrum. Alternatively, if $h\nu$ is held fixed, and the $e^-(KE)$ is scanned, the peaks in the $e^-(KE)$ spectrum correspond to successive energy levels of the ion. This is known as *photoelectron spectroscopy* (PES).

$$E^+_{v^+,J^+} = h\nu + E''_{v'',J''} - e^-(KE). \tag{8.1.3}$$

Since the development of photoelectron spectroscopy by Turner and Al-Jabory, (1962), continual efforts have been invested to improve the spectral resolution. In the beginning, the resolution was sufficient to resolve vibrational structure and, by 1970, resolution of 80 cm^{-1} has been obtained by Edqvist *et al.*, (1970). The first photoelectron spectrum in which partial rotational structure was revealed was reported by Asbrink for H_2 (1970). But 80 cm^{-1} resolution only permits observation of the rotational structure of H_2 (Pollard *et al.*, 1982) and high rotational levels of heavier molecules, such as NO, because the rotational spacing increasing as $2B^+J^+$, where B^+ is the rotational constant of the ion (Wilson *et al.*, 1984).

More recently, a resolution of $5meV$ = 40 cm^{-1} has been obtained by Edvardsson *et al.*, (1995) and the PES showing the rotational structure of the $A^2\Sigma^+$ state of HCl^+ is given in Fig. 8.1.

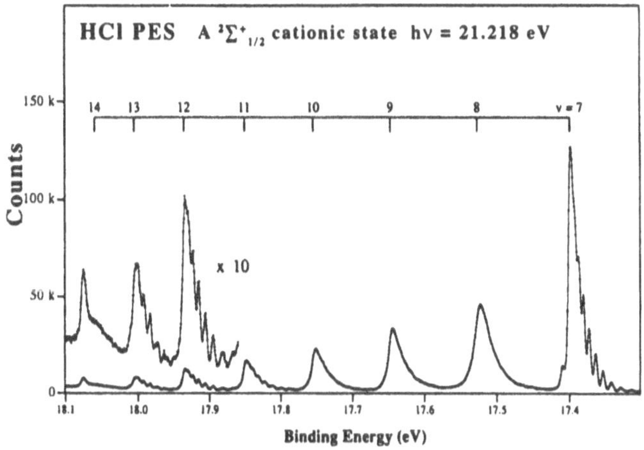

Figure 8.1: A Photoelectron Spectrum (PES) of HCl, excited by the He I resonance line. This PES displays rotational substructure in $A^2\Sigma^+$ state vibrational transitions. The intensity of the spectrum in dashed lines is multiplied by 10. The sharp peak at 18.07 eV is due to CO_2 (from Edvardsson, *et al.*, 1995).

Some of the vibrational peaks and high resolution traces of the HCl^+ $A^2\Sigma^+$ state are broadened because the $A^2\Sigma^+$ state is predissociated by the $a^4\Pi$ state by a spin-orbit mechanism described in Section 7.11.1.

The presently achievable PES resolution, $2meV = 16$ cm^{-1}, has resolved the rotational structure of heavier molecules than hydrides, such as N_2 (Öhrwall *et al.*, 1998).

Conventional photoelectron spectroscopy uses a rare-gas discharge lamp to produce radiation at the wavelength of the He $2p \leftarrow 1s$ atomic transition ($h\nu = 21.218$ eV). Synchrotron radiation is now widely used for PES because its photon energy is widely tunable yet monochromatic. The initial state, in the first PES experiments, has been the molecular ground state but now, by exploiting Resonance Enhanced Multi-Photon Ionization (REMPI) excitation/detection schemes (see Section 1.2.2.3), any excited state of the molecule can be used as the initial state for PES (for a review, see Pratt, 1995).

Molecular anions, AB^-, have also been studied for many years by various forms of photoelectron and photodetachment spectroscopy,

$$AB^- + h\nu \rightarrow AB + e^-(KE) \tag{8.1.4}$$

The $e^-(KE)$ peaks in the PES (fixed $h\nu$) provide information about the relative energies of electronic states of the AB neutral molecule (for example AsO^-, Lippa *et al.*, 1998) that, owing to spin or orbital selection rules, are not easily observed by direct study of the neutral molecule spectrum. An anion photodetachment spectrum is recorded by detecting the low-energy electrons ejected as the photon energy is scanned through the successive photodetachment thresholds associated with the energies of each neutral molecule rovibronic state.

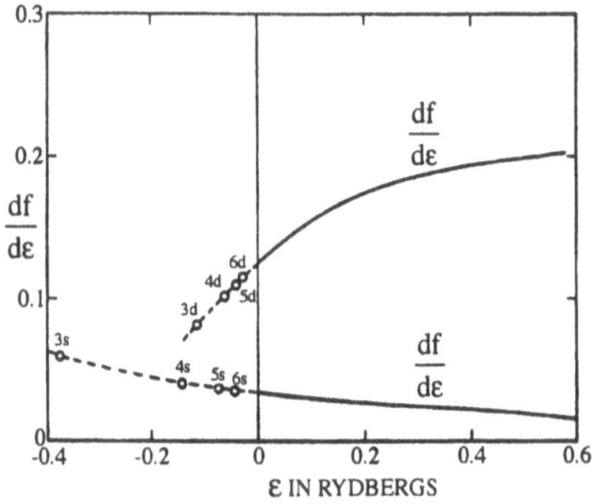

Figure 8.2: Differential cross sections for p→d and p→s transitions in both discrete and continuous spectra (from Cooper, 1962).

The electronic part of the photoionization cross section (in Megabarns, 1Mb= $10^{-18}cm^2$) for photoionization is given by:

$$\sigma(E)[\text{Mb}] = 2.689\Delta E \sum_{l,\lambda} |M_{l,\lambda}|^2 \tag{8.1.5}$$

where ΔE and $M_{l,\lambda}$ are given in atomic units. $M_{l,\lambda}$ is the electronic transition moment matrix element between the molecular orbital of the initial state, λ, and the ejected electron wavefunction, l. ΔE is the energy difference between the initial and final states of the system. A complete expression for the differential cross section has been given by Buckingham, Orr, and Sichel (1970) (denoted BOS in many papers). In the hydrogenic approximation, the electronic transition moment, $M_{l,\lambda}$, decreases monotonically with increasing electron kinetic energy. Consequently, the photoionization cross section decreases with energy, as can be seen on Fig. 8.2 for $2p \to ns$ transitions in Ne. However, in the central field (nonhydrogenic) approximation, the cross section can increase monotonically above the ionization threshold as it does for $2p \to nd$ transitions in Ne (Fig. 8.2). The monotonic trends in the variation of σ with $h\nu$ are disrupted by *shape resonances* (see Section 8.1.3), *autoionization* (see Section 8.2), or *Cooper minima* (see Section 8.1.4) which can produce strong local variations of the cross section. [See experimentally measured *absolute* cross sections for N_2 (Shaw, *et al.*, 1992, 1993) and for O_2 (Holland, *et al.*, 1993), and the review of Gallagher, *et al.*, (1988).]

In the absence of shape resonances and autoionization effects, the Franck-Condon factors, $\left\langle v_f^+ | v_i \right\rangle^2$, between the initial state of the neutral molecule, $|v_i\rangle$, and the final state of the ion, $\left\langle v_f^+ \right|$, determine the vibrational state population distribution of the molecular ion.

If the initial state is a Rydberg state, a one-photon ionizing transition can only alter the electronic configuration by a single orbital promotion, resulting in a final state in which the Rydberg electron has been excited into the electronic continuum. The final ion state must have the same ion-core electronic state as the Rydberg state. Since Rydberg electrons are essentially nonbonding, the vibrational selection rule is $v_f^+ = v_i$. Deviations from this restrictive rule imply that the Rydberg state is perturbed by a valence state. Examples of such perturbations will be given in Section 8.1.2.

If the photon energy, $h\nu$, coincides with a superexcited state, AB^{**}, autoionization contributes extra intensity to the direct ionization signal. As a result, vibrational intensity distributions will not be described by Franck-Condon factors (see, for example, Caprace *et al.*, 1976). These vibrational intensity anomalies may be explained similarly to the effect of perturbations on intensity borrowing (Section 6.2).

Starting from a specific initial rotational state of the neutral molecule, the rotational distribution for the final ionic state can be predicted. The rotational propensity rules depend not only on the symmetries of the two states involved in the photoionization process but also on the orbital angular momentum of the ejected electron, l. Complete selection rules, including the parity (e or f) of the rotational levels of the initial neutral and final ionic states, are given by Xie and Zare, (1990), Raşeev and Cherepkov, (1990), and Dixit and McKoy, (1986). Note that the apparent difference between the sets of selection rules given by different authors (see for example Eq. (8.1.8a) is a factor $(-1)^x$ where x is an even integer, consequently the selection rules are equivalent.

The first selection rule is based on conservation of the total angular momentum

$$\vec{N}'' + \vec{1} = \vec{N}^+ + \vec{l} \tag{8.1.6}$$

where N'', N^+ and 1 are the angular momenta of the initial state, ionic state and of the photon respectively, and $\vec{l}$ is the angular momentum of the ejected electron. This vector equation implies the following constraint on the angular momentum magnitudes:

$$|l-1| \leq N^+ - N'' \leq l+1 \tag{8.1.7}$$

In the case of a one-photon excitation of H_2 in its rovibronic ground state, $N'' = 0$ and $l = 1$ (because the H_2 $X^1\Sigma_g^+$ state resembles a He atom in the $1s^2$ configuration) thus

$$\Delta N = N^+ - N'' = 0, \pm 1, \pm 2.$$

However, ± 1 is excluded for homonuclear molecules when the initial and final states have the same g, u symmetry. Consequently for H_2, $\Delta N = 0, \pm 2$. This same rotational selection rule holds for NO, which is nearly homonuclear (see Fig. 8.3).

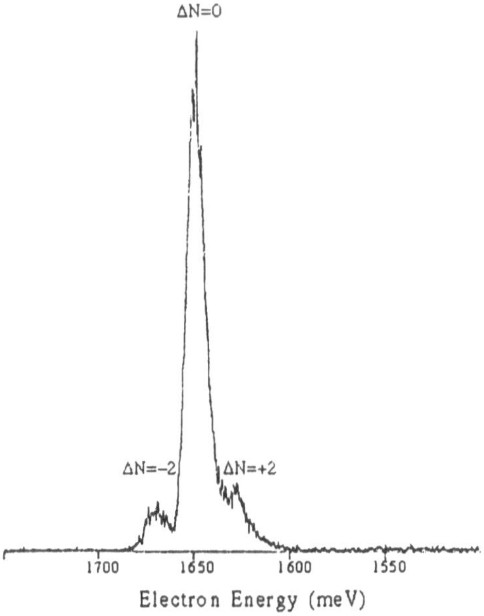

Figure 8.3: Photoelectron spectrum of NO obtained by resonant one-photon excitation of $N = 21$ of the $A^2\Sigma^+(v = 0)$ state (from Wilson *et al.*, 1984).

The general selection rule is, when both the initial molecular and final ion states are in case (a):

$$(J^+ - J'') + (S^+ - S'') + (p^+ - p'') + l = even \tag{8.1.8a}$$

or

$$\Delta J + \Delta S + \Delta p + l = even \tag{8.1.8b}$$

$p = 0$ for e–symmetry,
$p = 1$ for f–symmetry.

Additional selection rules for homonuclear molecules are specified in Xie and Zare (1990). The selection rule for parity can be understood as follows:
The total parity of the final state is equal to the parity of the ionic level times the parity of the electron partial wave (which is even for even l and odd for odd l). For example, for a transition from a $^1\Sigma^+$ molecular state to a $^2\Pi$ ion state, starting from $J'' = 4$ (e–level, + parity), the $J^+ = 7/2$ rotational level of the ion has two components: one e–level (− parity), one f–level (+ parity) (see Fig. 8.16). The selection rule for allowed one-photon transitions is $+ \rightarrow -$. Consequently, for the transition into the ion e–level, the partial wave of the ejected electron is $l = 0$ (s); for the transition into the ionic f–level, the partial wave of the ejected electron is $l = 1$ (p). Equation (8.1.8a), with $S^+ = 1/2$, is satisfied for these l values.

The usual rotational branch notation in PES is that the branches corresponding to $N^+ - N''$ or $J^+ - J''$ transitions are named in the same way as the usual rotational branches: $\Delta N = 0$ Q–branch; $\Delta N = 1$, R–branch; and $\Delta N = -1, P$–branch.

However, when the PES transition is between molecular and ion states which belong to different Hund's cases, then the rotational branches are often labeled according to the numerical value of the change in pattern-forming rotational quantum number (respectively J, N, J, N^+, and J^+ for Hund's cases (a), (b), (c), (d), and (e)), which can have either integer or half-integer value.

Another approach to obtaining high resolution photoelectron spectra has been to use time-of-flight photoelectron detectors to restrict the detected electrons to those near "threshold", i.e., photoelectrons with nearly zero kinetic energy (Villajero *et al.*, 1967), hence the name "*Threshold Photoelectron Spectroscopy*" or TPES, (for a review see Baer and Guyon, 1995). In TPES the electron kinetic energy $e^-(KE)$ is held fixed while $h\nu$ is varied. $h\nu$ frequently coincides with a transition to an autoionization resonance, for example a Rydberg level converging to a vibrationally excited state of the ion (see Section 8.6). This results in large deviations from Franck-Condon intensities.

8.1.2 ZEKE Spectroscopy

A more sensitive yet experimentally simpler way of selecting photoelectrons of low or zero kinetic energy (ZEKE) was developed by Müller-Dethlefs *et al.*, (1984). The key idea consists of extracting the photoelectrons only at a specified time delay (0.5 - 5 μs) after the excitation pulse. A pulsed electric field is then applied in the ionization region, which field-ionizes all Rydberg states with n–quantum number above a selected threshold value. During the delay time, energetic electrons escape from the ionization region and are not detected, thus only ZEKE electrons are detected, (Fig. 8.4) (for a review see Merkt and Softley, 1993).

The ZEKE detection scheme is equivalent to ionization of high-n Rydberg states by Pulsed Field Ionization (PFI). If one assumes that the pulsed field ionization of the Rydberg electron follows a diabatic process (Chupka, 1993), then the ionization threshold is lowered by

$$\Delta IP\ (\text{cm}^{-1}) = 3.1[F\ (V/cm)]^{1/2} = \frac{\Re}{n^2} \tag{8.1.9}$$

where F is the pulsed extraction electric field in V/cm (Merkt, 1997). Equation (8.1.9) provides an estimate of the minimum n–value of the ionized high Rydberg states. The commonly obtained resolution of about 2 cm^{-1}, for example in the ZEKE spectrum of O_2 (Braunstein *et al.*, 1992), can be improved to 0.2 cm^{-1} and has permitted resolution of the spin splittings of the $X^3\Sigma_g^-$ state of O_2 (Palm and Merkt, 1998).

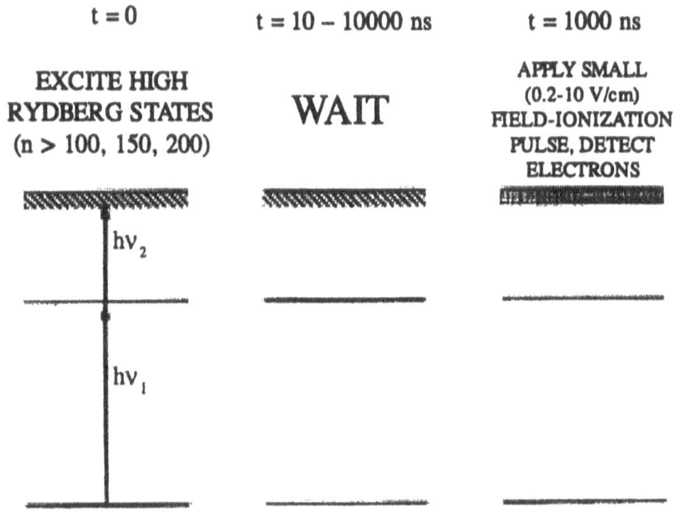

Figure 8.4: Mechanism and timing of a ZEKE spectroscopy experiment, (Courtesy of W.A. Chupka).

Numerous intensity anomalies occur in ZEKE spectra. Most of them are due to autoionization effects and will be discussed later (Section 8.10). We will speak here only about perturbations due to Rydberg$\sim$valence mixing (Section 5.2.2). One example is given by the ZEKE-PFI spectrum recorded from the $v = 4$ level of the $B^2\Pi$ valence state of NO. The ionization transition to the ion ground state is a forbidden two-electron excitation: $(\pi)^4\epsilon l \leftarrow (\pi)^3(\pi^*)^2$. Nevertheless, rotational branch intensity distributions in the $B^2\Pi$ state ZEKE spectrum (Mackenzie *et al.*,, 1996) are similar to those observed in the ZEKE spectrum recorded from the $C^2\Pi$ $(\pi)^4(3p\pi)$ Rydberg state, because ionization from the Rydberg states (or via the Rydberg character in a mixed state) is a one-electron process. This can be explained by Rydberg$\sim$valence $B^2\Pi \sim C^2\Pi$ mixing (see Table 5.5). A complementary example is given by White, *et al.*, (1982), in which NO $C^2\Pi(v = 0)$ is two-photon ionized to $NO^+X^1\Sigma^+$. Photoelectron spectra show, in addition to the expected $v^+ = 0$ peak, some intensity in $v^+ = 3, 4$, and 5 due to the Rydberg$\sim$valence mixing of $C^2\Pi$ $v = 0$ with $B^2\Pi$ $v = 7$.

Deviations from predicted rotational intensity distributions are very common in ZEKE spectra. This is due to random near coincidence between extremely numerous rapidly- and slowly-autoionizing resonances (Rydberg series converging to excited rovibronic states of the ion). Since the waiting time between excitation and pulsed field ionization is long, and the very weak DC and stray electric fields present during the ZEKE waiting period can induce weak interactions

($+ \sim -$ parity, $\Delta l \neq 0$) between these near degenerate autoionizing resonances, population leaks out (by field assisted autoionization) of some levels and not others, resulting in some anomalously weak rotational lines in ZEKE spectra. l–mixing due to weak electric fields (Chupka, 1993) must be considered in order to explain the relative intensities observed in ZEKE spectra (Softley, *et al.*, 1997). These rotational intensity anomalies will be illustrated in Section 8.10 where the PES and ZEKE spectra of HCl are compared.

8.1.3 Shape resonances

"Shape resonances" appear as broad maxima in the photoionization cross sections. Their width varies from about 4 eV (for O_2) to 20 eV (for N_2), but a shape resonance is often hidden beneath multiple, much narrower, and more intense autoionization resonances.

When there is a local maximum (barrier) on the effective potential energy curve, $V_l(r)$, which describes the interaction between an electron, with orbital angular momentum l, and an atomic- or molecular-ion (A^+ or AB^+, r is the distance between the electron and the center of mass of the ion), the result is one or more quasibound states (resonances) embedded in the electronic continuum. These short-lived states, stabilized behind the barrier on $V_l(r)$, are called "shape resonances" because their existence (and properties) reflect the shape of the barrier on $V_l(r)$. Such barriers often occur when the repulsive centrifugal $l(l+1)/(2r^2)$ term (Dehmer, *et al.*, 1985) overpowers the shielded-nuclei attractive $-Z(r)/r$ term. (Z(r) is the r–dependent effective nuclear charge, and at large–r, the leading term in $Z(r)/r \sim r^{-n}$ must have $n \geq 2$ in order for a centrifugal barrier to occur.)

This description of shape resonances is valid for atoms and, in some cases, for molecules, despite the lack of spherical symmetry in the electron-ion potential. Shape resonances are observed in the inner shell photoionization of hydrogen halides where a $d \rightarrow \epsilon f$ transition exhibits an "atom-like" shape resonance (Keller and Lefebvre-Brion, 1986). "Molecular" shape resonances are known for O_2, where the partial photoionization cross section exhibits a shape resonance that corresponds to $3\sigma_g(2p) \rightarrow \epsilon\sigma_u$ ionization. In this case, the l–values of the nominally optically accessible orbitals are too small to generate a centrifugal barrier. The l–mixing due to the nonspherical molecular potential is essential, both to increase the strength of the repulsive term (by mixing in high–l character), and to maintain the electric dipole allowedness of the $3\sigma_g \rightarrow \epsilon\sigma_u$ transition (Raşeev, *et al.*, 1981). The transition moment is significantly increased by the mixing of partial waves with different values of l, especially the p– and f–waves. This l–mixing is strongly R–dependent and is essentially complete (50/50) at the R–value of the maximum cross section for the transition to the shape resonance.

As noted in Section 7.3, shape resonances also occur in the spectrum of vibrational transitions. They correspond to predissociation by rotation (resonances stabilized behind the nuclear motion centrifugal barrier, $V_J(R)$). Here the name "shape resonance" comes from the shape of the barrier on $V_J(R)$.

Both electronic and vibrational shape resonances arise from a direct process and can be explained by a single potential (McKoy, *et al.*, 1984). Shape resonances (single $V_l(r)$ or $V_J(R)$) differ from autoionization resonances and predissociation (with the exception of predissociation by rotation), which involve two potentials or two states with different quantum numbers.

An important characteristic of shape resonances is that they cause non-Franck-Condon effects in vibrationally resolved photoionization spectra. These effects are a consequence of the strong R–variation of the transition moment caused by the R–dependence of the form of the continuum molecular orbital.

Shape resonances in the photoionization spectrum can be pictorially related to profound R–dependent changes in the nature of a singly occupied molecular orbital. For example, an orbital which is of σ^* (or σ_u) antibonding character at large R becomes a $p\sigma$ Rydberg orbital at shorter R (Rydbergization) and, at the same time, an orbital which is of $p\sigma$ Rydberg character at intermediate R becomes antibonding at short–R (valenation). (The same sort of interaction can affect a $\pi^*, d\pi$ orbital pair.) The result is two electronic states, built on the same ion-core electronic state, which have electronic configurations differing by a single orbital. The complementary and strongly R–dependent evolution of the $\sigma^* \rightarrow p\sigma$ and $p\sigma \rightarrow \sigma^*$ orbitals is due to the very strong interaction between diabatic basis states $(\text{AB}^+ \ {}^{2S^++1}\Lambda^+)\sigma^* \sim (\text{AB}^+ \ {}^{2S^++1}\Lambda^+)p\sigma$ which differ by a single orbital. The diabatic basis states have the "expected" shape potential curves, are located at the expected excitation energies, and participate in transitions with the expected patterns of intensities. These diabatic valence and Rydberg states have been denoted Rydberg/valence shell conjugates by Robin (1985). However, because of the strong interaction between the pair of complementary diabatic states, two complementary adiabatic states are formed, which have abnormal potentials, excitation energies, and intensity patterns.

Since, for most molecules the ionization energy is larger than the lowest dissociation energy, the diabatic repulsive state will start out at large–R at lower energy than its complementary Rydberg state. As R decreases, the potential curve of the diabatic repulsive state will cross through from below the potential of the complementary Rydberg state. The strong interaction between these two diabatic states produces a peculiar shaped but repulsive lower state (the "Rydbergized" state), and a peculiar shaped but bound upper state (the "Complementary" state). It is this upper "Complementary" state that gives rise to the shape resonances that appear in spectra. Since its mixing coefficients for valence and Rydberg characters are strongly R–dependent, the transition moment function is strongly R–dependent.

Depending on the energy of the intersection between the diabatic states, the complementary state can be located below the ionization threshold, as it is for O_2 in the $\left(O_2^+ \ X^2\Pi_g\right) 3p\sigma_u$ F-state (see Fig. 5.12) and for F_2 (Orel *et al.*, 1980) where the complementary state is a real bound state. However, when the complementary state is embedded in the electronic continuum, it gives rise to a shape resonance, and cannot be represented by a bound state. This complementary state corresponds to a very short lived resonance (the width corresponds to a lifetime of 10^{-16} to $10^{-17}s$). Consequently, the molecule lives for too short a time to vibrate. Calculations of electronic continuum spectra and their associated shape resonances require very situation-specific techniques (Raşeev, *et al.*, 1978; Watson, *et al.*, 1980).

For the NO $X^2\Pi$ ground state ($\sigma^2\pi^*$ configuration), the $\sigma^2\sigma^*$ configuration is that of the $D^2\Sigma^+$ $3p\sigma$ Rydberg state, which becomes the repulsive $A'^2\Sigma^+$ valence state at large-R. The higher lying complementary state is responsible for a σ shape resonance which has been observed by Southworth, *et al.*, (1988). On the other hand, ionization by removal of an electron from the $\sigma 2p$ molecular orbital gives rise to two states, $b^3\Pi$ and $A^1\Pi$ ($\sigma\pi^*$ configuration), in the continua of which the effects of shape resonances are well characterized (Stratmann, *et al.*, 1996). For CO, the effects of shape resonances are not very clear (Smith, *et al.*, 1986). For the O_2 molecule, the first $^3\Sigma_u^-$ Rydberg states derived from a $3p\sigma_u$ Rydberg orbital and the O_2^+ ion-core $b^4\Sigma_g^-$ or $B^2\Sigma_g^-$ states are repulsive states, but their complementary states are responsible for the observed and well characterized shape resonances (Raşeev *et al.*, 1981, Morin *et al.*, 1982). Figure 8.5 shows, in the continuum of O_2^+ $B^2\Sigma_g^-$, both the Rydbergized and the complementary state.

For light hydrides (e.g. OH), the situation is different: the first excited state is a Rydbergized state (see Section 5.2.2), the external orbital is $3s\sigma$ at small R and becomes σ^* at large R, giving rise to a repulsive state, which has not been observed, but the complementary state is embedded among the other Rydberg states. The Rydberg orbital of the second Rydberg state is $3p\sigma$ at small internuclear distances and becomes $3s\sigma$ at large R.

We have considered here only the case of a σ^* orbital because, in general for diatomic molecules, excitation to a nominal π^* orbital lies in the discrete part of the spectrum. Examples of π^* shape resonances have been observed in polyatomic molecules as, for example, the cyanogen molecule, NCCN (see Lynch *et al.*, 1986).

One consequence of the R and electron energy dependence of the transition moment is that the R-averaged calculated cross sections for resolved final vibrational states do not follow Franck-Condon predictions. In Fig. 8.6, the vibrational state branching ratio, $\sigma(v_f = 1)/\sigma(v_f = 0)$, for the $3\sigma_g$ photoionization cross section in N_2 is shown (Raşeev *et al.*, 1980). The departure of this ratio, as a function of the photon energy, from the constant Franck-Condon factor compares well with the experimental results obtained by West *et al.*, (1980). Even better agreement was obtained later by Lucchese and McKoy (1981).

Shape resonances are also responsible for intensity anomalies in the PES and ZEKE spectra of O_2 (for example, Braunstein, *et al.*, 1992).

8.1.4 Cooper minima

Cooper minima are also responsible for non Franck-Condon vibrational intensity distributions and intensity anomalies observed in both PES and ZEKE spectra. As for atoms, when the photoionization transition corresponds to excitation from a Rydberg orbital having at least one radial node in its wavefunction, the ionization cross section can pass through a minimum value at an energy known as a "Cooper minimum" (Cooper, 1962).

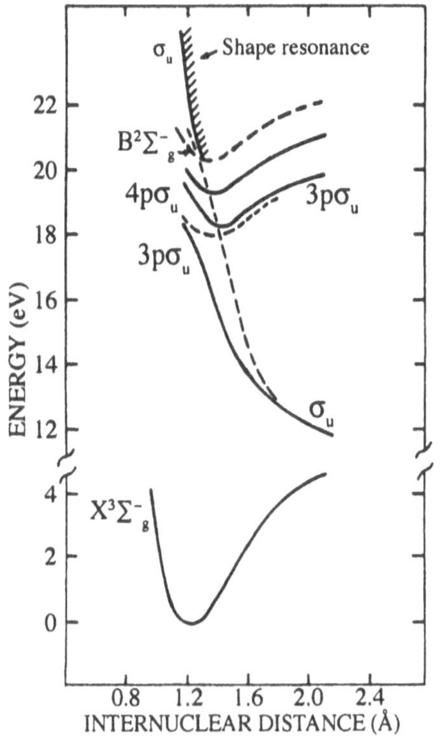

Figure 8.5: A pair of Rydbergized and complementary states in O_2. The diabatic states, $\left(O_2^+\ B^2\Sigma_g^-\right)\sigma_u$ and $\left(O_2^+\ B^2\Sigma_g^-\right)3p\sigma_u$, are shown in dashed lines (from Lefebvre-Brion, 1988) and the resultant pair of adiabatic states (the "$4p\sigma_u \rightarrow 3p\sigma_u$" and the shape resonance) are shown as solid lines. The ion-core $B^2\Sigma_g^-$ state is shown as a heavy dashed line.

A Cooper minimum arises from a zero in an electronic radial overlap integral, in close analogy to zeroes in vibrational overlap integrals observed as one follows a vibrational progression. Figure 8.7 shows a nodeless $3d$ radial wavefunction, a slowly oscillating, near threshold $\epsilon f(\epsilon \approx 0\ a.u.)$ continuum wavefunction, and a more rapidly oscillating $\epsilon f(\epsilon \approx 3\ a.u.)$ continuum wavefunction. The $3d$ radial wavefunction overlaps primarily with the positive region of the $\epsilon f(\epsilon \approx 3\ a.u.)$ continuum wavefunction, but there is a small region of negative overlap beyond $r \approx 2a_0$. As ϵ increases, the negative overlap increases until the sign of the transition moment matrix element changes sign. The continuum energy, ϵ_{Cooper}, at which the transition moment passes through zero is the Cooper minimum. Several Cooper minima can occur and can be located in both bound ($n < \infty$) and ionization continuum ($\epsilon \geq 0\ a.u.$) regions of transitions into a Rydberg channel.

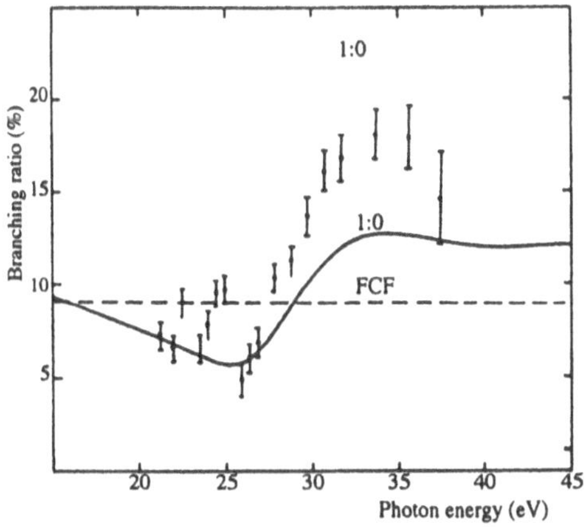

Figure 8.6: Vibrational state branching ratio for the photoionization of N_2: full line, calculations of Raşeev, *et al.*, (1980) compared to the experimental results of West *et al.*, (1980). The dashed FCF line is the ratio of Franck-Condon factors for the excitation transitions into the $v = 1$ and $v = 0$ vibrational levels.

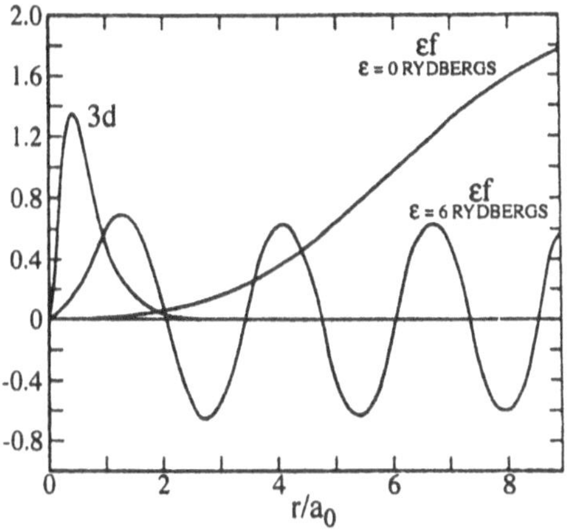

Figure 8.7: The 3d bound orbital and two examples of ϵf continuum wavefunctions (at $\epsilon \approx 0$ *a.u.* and $\epsilon \approx 3$ *a.u.* (from Manson and Cooper, 1968).

For molecules, the problem is more complex than for atoms because the continuum must usually be expressed as a sum of many partial waves. It is highly unlikely that the Cooper minima for all of the important l–partial wave components will occur near the same value of ϵ. An example of a Cooper minimum in a molecular photoionization cross section is observed for HI, corresponding to ionization from the $5p\pi$ orbital (Carlson *et al.*, 1984). Other examples of Cooper minima have been observed in REMPI-PES excited out of the OH $D^2\Sigma^-(3p\sigma)$ state (de Beer, *et al.*, 1991) and the $3p\pi$ and $3p\sigma$ Rydberg states of NH (de Lange, 1995).

In Section 8.11, the influences of shape resonances and Cooper minima on photoelectron angular distributions will be discussed.

8.2 Experimental Aspects of Autoionization

Above ionization limits, lines in the absorption spectrum are often broad or diffuse. This diffuse character results from an interaction of a very highly excited neutral molecule state, AB^{**}, with the continuum of an ionized molecule, AB^+, plus an electron. This continuum reflects the fact that the electron can be ejected from the molecule over a continuous range of kinetic energy:

$$AB + h\nu \longrightarrow AB^{**} \longrightarrow AB^+ + e^-(KE) \tag{8.2.1}$$

The state AB^{**}, often called the "superexcited" state, is an autoionized or resonance state. Autoionization is called preionization by Herzberg (1950). This can be justified by the analogy between preionization and predissociation. In predissociation, the interaction of a discrete state with the vibrational continuum of the nuclei allows this discrete state a finite probability of dissociation. In preionization, it is the mixing of a discrete state with the electronic continuum that provides a finite ionization probability.

Equation (8.2.1) is not the only mechanism governing the decay of AB^{**}. Indeed, three main mechanisms compete:

$$AB^{**} \begin{array}{lll} \nearrow & AB^+ + e^- & \text{autoionization} \\ \rightarrow & A + B^* & \text{dissociation} \\ \searrow & AB + h\nu & \text{radiative decay} \end{array}$$

We consider in the following the case of a single isolated "resonance," far removed from other resonances. If an absorption line has a Lorentzian profile, the linewidth,[†] Γ, is related to the total lifetime, τ, of the state by the usual relation: $\Gamma \propto \tau^{-1}$ [cf. Eq. (7.4.11)]. This width includes contributions from each of these three different types of decay. If interactions between continua of the electron, nuclei, and radiation field are neglected, one can write

$$\Gamma \simeq \Gamma_a + \Gamma_d + \Gamma_r \tag{8.2.2}$$

or

$$\tau^{-1} \simeq \tau_{nr}^{-1} + \tau_r^{-1}, \tag{8.2.3}$$

[†]Width here means the homogeneous part of the width, in the absence of any pressure effect and corresponds to the full width at half maximum (FWHM).

where

$$\tau_{nr}^{-1} = \tau_a^{-1} + \tau_d^{-1}. \tag{8.2.4}$$

These partial widths, Γ_a, Γ_d, and Γ_r, associated with autoionizing, dissociative, and radiative decay, respectively, and the corresponding partial lifetimes τ_a, τ_d, and τ_r, cannot be directly observed. Only the total width Γ can be observed experimentally. This width is identical whether the line is observed by absorption, photoionization, or fluorescence spectroscopy, but the associated cross-sections can be different. Here we address only the problem of autoionization in absorption spectroscopy.

The orders of magnitude of these partial lifetimes can be very different. The radiative lifetime is longer than 10^{-8}s for most diatomic molecule excited states. Generally, the rates of nonradiative processes above an ionization limit are much faster than radiative decay, and consequently no emission lines are easily detected above such limits. Two exceptions have been found in the spectrum of H_2, where Q–branch lines of the first members of the $^1\Pi_u$ Rydberg series are seen in emission (Sroka, 1969, Larzillière, *et al.*, 1980) above the ionization limit. The emitting levels of these Q–lines, which correspond to f–parity levels of $^1\Pi_u$ states, are very weakly predissociated. In contrast, the e–levels (upper levels of the P and R branches) are strongly predissociated by $^1\Sigma_u^+$ Rydberg states ($^1\Sigma^+$ states are exclusively of e–parity with perturbation selection rule $f \nleftrightarrow e$). These f–levels, in addition to being weakly predissociated, are also very weakly autoionized (see also Fig. 8.23). Another example appears in the spectrum of Li_2 (Chu and Wu, 1988). In many other cases, decay by predissociation can be very fast, particularly for electrostatic predissociations (see Tables 7.3 and 7.4). For example, in the spectrum of the NO molecule, most $^2\Pi$ Rydberg states are predissociated by the vibrational continuum of the $B^2\Pi$ valence state so rapidly that autoionization cannot compete (Giusti-Suzor and Jungen, 1984).

In general, complex features may arise in the absorption spectrum due to competition between the two processes, autoionization and predissociation (see Section 8.12). For example, as will be shown in Section 8.8, the partial width due to electronic autoionization is independent of v for a given $nl\lambda$ Rydberg state, but the partial width due to predissociation can vary considerably from one vibrational level to another because of oscillatory variations of the vibrational part of the interaction with the predissociating state.

Another problem that affects any interpretation of a diffuse peak is instrumental resolution. Recall that the resolution is given by

$$\Delta\lambda \quad (\text{Å}) = 10^{-8}\lambda^2\Delta\nu \quad (\text{cm}^{-1}).$$

For example, at 800 Å, a very good resolution of $\Delta\lambda = 0.02$Å corresponds to $\Delta\nu = 3$ cm^{-1}. Such resolution is sufficient to resolve individual rotational lines of only the H_2 molecule.

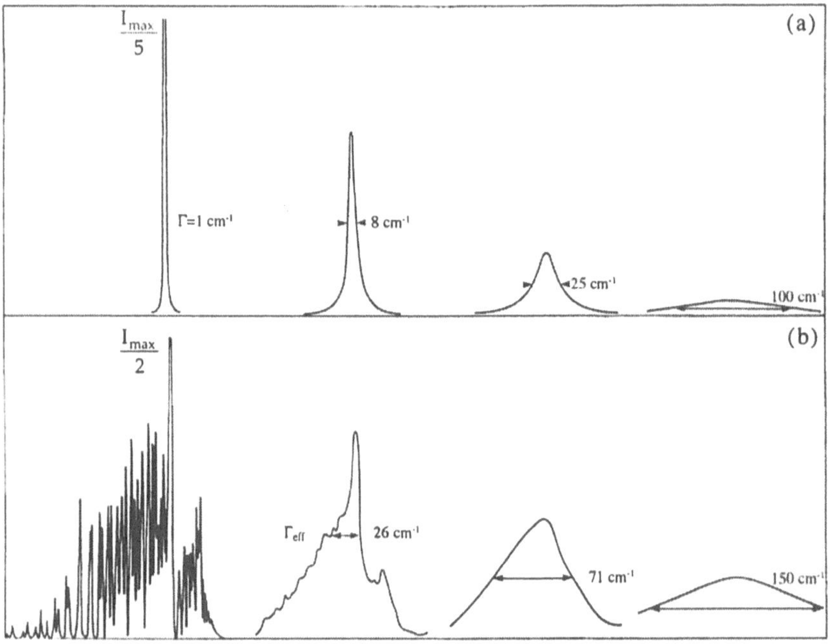

Figure 8.8: Relationship between full width at half maximum (FWHM) of individual lines and band contours. (a) Single lines with FWHM consistent with band contour shown below. (b) Band contours are constructed by convolution of individual lines (each with the lineshape shown above) with an instrumental resolution of 0.01 cm^{-1}. The spectrum is calculated for a $^2\Pi$ [case (b)] $\leftarrow X^2\Pi_{1/2}$ [case (a)] transition of the NO molecule observed in absorption at 78 K. [From Giusti-Suzor and Jungen (1984).]

When autoionized rotational lines are broad, it is difficult to extract from the observed band contour the fundamental quantity of interest, the width of a single rotational line. The width of the rotational contour is, for example, about 40 cm^{-1}, for the $n = 13$ member of the Worley-Jenkins Rydberg series of N_2 (Carroll, 1973) at liquid nitrogen temperature (77 K). A different example is given in Fig. 8.8, which shows a band contour in NO obtained from individual lines of a given width. When the width of a single line (25 cm^{-1}) is of the order of magnitude of the rotational *envelope*, the observed band contour (71 cm^{-1}) masks the true linewidth. Recent experiments using supersonic molecular beams, where the rotational temperature is lowered to a few Kelvins, have been able to reduce this effect of rotational congestion.

The band contour depends not only on the temperature but also on Hund's coupling cases. Jungen (1980) has pointed out that the rotational structure becomes narrower as n increases due to a change from Hund's case (b) to case (d). This envelope-narrowing effect must not be construed as a variation of autoionization widths with effective quantum number n^*. Fortunately, double-resonance techniques in the vacuum ultraviolet region are now able to isolate individual rotational line profiles.

Sometimes, the autoionized lines for very large values of n near the ionization threshold present a regular pattern characterized by the appearance of "fringes". This phenomenon has been attributed to a "stroboscopic" effect between the periodic orbital motion of the Rydberg electron and that of the ion rotation (in Na_2, Labastie, *et al.*, 1984, in Li_2, Schwarz, *et al.*, 1988) and an interference effect resulting from the relative phases of the l–mixing matrix elements and transition moments (in NO, Fredin, *et al.*, 1987), or an effect of finite resolution (in HCl, Drescher, *et al.*, 1993).

Another pernicious effect of limited resolution is a reduction of the peak height. Theoretically (i.e., at infinite instrumental resolution), the maximum peak height is obtained at $E = E_r$ and its value is proportional to μ^2/Γ, where μ is the transition moment. Since both μ^2 and Γ vary as $(n^*)^{-3}$, the maximum peak height is independent of n^* along an autoionized Rydberg series. However, because the linewidth, Γ, decreases as $(n^*)^{-3}$, the apparent peak height begins to be affected by limited resolution as n^* increases. The narrower the peak, the more its apparent peak height is reduced. Consequently, the experimental peak height appears to decrease as $(n^*)^{-3}$† This resolution effect is clearly evident in Fig. 8.9.

Spectra with higher resolution are now available. The development of new experimental techniques such as tunable, vacuum ultraviolet radiation, generated by gas-phase four-wave-mixing processes, permits the resolution of rotational structure.

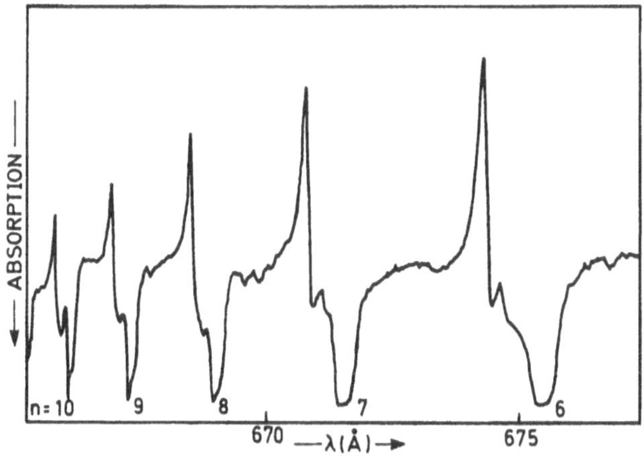

Figure 8.9: Absorption spectrum showing the autoionization of the N_2 Hopfield Rydberg series. (Courtesy K. Yoshino.)

† Note that the integrated intensity of a line $[\sim (\mu^2/\Gamma)\Gamma]$ decreases as $(n^*)^{-3}$, regardless of instrumental resolution.

8.3 The Nature of Autoionized States

Autoionized states are Rydberg states converging to an ionization limit that lies above the lowest ionization threshold. The general formula for the energy of Rydberg states is

$$E_n = E_{ion} - \frac{\Re}{(n^*)^2} = E_{ion} - \frac{\Re}{(n-a)^2} \tag{8.3.1}$$

where n and n^* are the principal and effective principal quantum numbers, and a is the quantum defect, which is characteristic of a particular Rydberg series. E_{ion} must be carefully defined, especially if the ion electronic state has a spin-orbit splitting (see Section 8.7). The Rydberg constant for a molecule of finite mass is defined as

$$\Re = \Re_\infty \left(1 + \frac{m_e}{M}\right)^{-1} = \Re_\infty \left(\frac{M}{m_e + M}\right) \tag{8.3.2}$$

where $\Re_\infty = 109{,}737.318\ \text{cm}^{-1} = 0.5$ a.u. m_e is the electron mass and M the mass of the molecule. For example, for HCl with a mass of 36 a.m.u. $(m_e/m_p = 1/1836), \Re^{\text{HCl}} = 109{,}735.64\ \text{cm}^{-1}$. This correction can become important for states with large n.

The quantum defect is about 1.0 for $ns\sigma$ orbitals of molecules composed of atoms between Li and N, since these orbitals "penetrate" into the molecular core (cf. Mulliken, 1964) and consequently are strongly modified by the presence of inner electrons of the same $s\sigma$ symmetry. The quantum defect is of the order 0.7 for $np\sigma$ and $np\pi$ orbitals, which are less modified by the core, and nearly zero for $nd\sigma$, $nd\pi$, and $nd\delta$ orbitals, which have no "precursors" in the core. For a given molecule this quantum defect is roughly independent of the ion state to which the Rydberg series converges (see, for example, for O_2, Table 1 of Wu, 1987).

The quantum defect, a, is defined modulo 1. It is tempting, especially when one wants to interpret the magnitude of a, to use the observed magnitude of a to decide whether a particular Rydberg series is primarily derived from s, p, or d orbitals, or to deduce how many members of a Rydberg series lie at lower energy than the lowest observed member. Such temptations are fraught with danger for molecular Rydberg states primarily because, in the absence of a well defined l quantum number, it is impossible to specify an unambiguous number of radial nodes in the electronic wavefunction (for an atom, the number of radial nodes is $n - l - 1$) and from that node count deduce a dominant n or l value.

Rydberg series can converge to excited rotational or vibrational levels of the ground state of the ion or to excited electronic states of the ion. Each Rydberg state can be described as one external electron in the field of the ion core state that forms the limit of the corresponding $nl\lambda$ Rydberg series. Autoionization of a Rydberg state involves formation of a state of the ion that lies lower than the Rydberg series limit of the autoionized state, plus an ejected electron, which carries away the excess energy.

Since autoionization involves a nonradiative change in energy of the core ion, the ejected electron obtains its kinetic energy from the superexcited (autoionized) state by conversion of

1. rotational energy,
2. vibrational energy,
3. spin-orbit energy,
4. electronic energy,

depending on which processes are energetically possible. The first three types of core-to-ejected-electron energy transfer are observed for autoionization of discrete states by the continuum that belongs to the same electronic state but to a different vibrational, rotational, or spin-orbit state of the core ion. Approximate selection rules exist for vibrational transfer. The last type of core-to-electron energy transfer occurs between a discrete state built on an electronically excited core and a continuum associated with a lower-energy electronic state of the ion.

The first two types of interaction are peculiar to molecules. The last two also appear in atoms, but nuclear motion introduces a level of complexity beyond that of the atomic case.

Some general rules for the autoionization process are derived next. These rules are based on the following approximations. The spectral feature in which autoionization effects are manifested must be a single, well-resolved line associated with a well-characterized rovibronic level, for which the autoionization process can be described to a good approximation by a single predominant mechanism. The consequences of the breakdown of these approximations are discussed in Section 8.9.

8.4 Autoionization Widths

The observed autoionization width is the result of decay into several different continua. Therefore, it is useful for calculations to define a quantity, the partial autoionization width, in which the initial and final states of the autoionization process are defined perfectly. One can then describe the total width, the only experimentally observable autoionization width, by summing over all possible final states. The initial state is specified by 1, n, and v, the final state by 2, v_+ of the resultant ion, and ϵ of the ejected electron. The partial autoionization width can be expressed by the Golden Rule formula:

$$\Gamma_{1,n,v;2,\epsilon,v_+} = 2\pi \left| H_{1,n,v;2,\epsilon,v_+} \right|^2 . \tag{8.4.1}$$

The interaction H_{12} can often be factored into electronic and vibrational parts, as in the case of predissociation, but here it is the electronic part that has the continuous character. In some cases, the vibrational part is simply an overlap factor and

$$H_{1,n,v;2,\epsilon,v_+} = \left\langle \psi_{1,n,v} | \mathbf{H} | \psi_{2,\epsilon,v_+} \right\rangle = \langle \phi_{1,n} | \mathbf{H} | \phi_{2,\epsilon} \rangle \langle v | v_+ \rangle , \tag{8.4.2}$$

where the function $\phi_{2,\epsilon}$ for the continuum electron is energy-normalized, but the energy-normalization factor is different for nuclear vs. electronic continua due to the difference in masses.

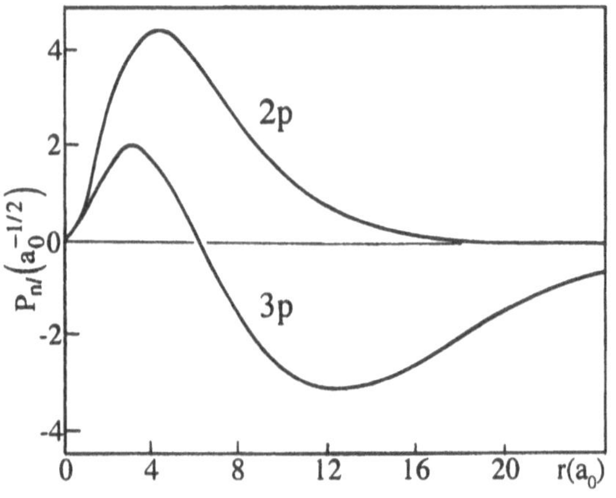

Figure 8.10: Hydrogenic radial wavefunctions for successive $2p$ and $3p$ Rydberg orbitals; r is the distance of the Rydberg electron from the molecular center of mass.

The interactions responsible for autoionization have the same origins as those for perturbations between discrete states. A useful relationship between perturbation and autoionization matrix elements can be obtained as follows: the autoionized state is the nth member of a given Rydberg series converging to a limit higher in energy than the lowest ionization limit. If the first members of this Rydberg series lie below the lowest ionization limit, they can perturb a higher n' member of the Rydberg series that converges to the lowest ionization limit. See Fig. 8.13, where the $n = 3$ member of a Rydberg series that converges to the vibrationally excited H_2^{*+} state of the ion perturbs the $n' = 10$ member of a Rydberg series that converges to the ground state of the ion.

The interaction between two Rydberg states belonging to different series comes from the part of the Rydberg electron wave functions near the ion core and, for light molecules, is restricted to a distance within about 2 Å from the center of mass. In this region, the first lobes (i.e., the region within the first nodal surface) of successive Rydberg orbitals of a given series are very similar in form (because they must be orthogonal to the lowest-energy, nodeless orbital), but differ only by an amplitude scale factor (see Fig. 8.10 and Slater, 1960, p. 223, Fig. 9.2). Indeed, as n increases, the amplitude of this inner part of the Rydberg electron wavefunction decreases proportionally with the normalization factor, $(n^*)^{-3/2}$, of the wavefunction.

The interaction between two Rydberg states of the same n, which belong to two different series, $\phi_{1,n}$ and $\phi_{2,n}$, will be derived using a convenient Hund's case basis set and by expressing the interaction as part of a residual *electronic* interaction of the phenomenological form

$$H^e_{1,n;2,n} = \langle \phi_{1,n} | \mathbf{H} | \phi_{2,n} \rangle = \frac{2\Re}{(n^*)^3} I, \tag{8.4.3}$$

where $(n^*)^3$ plays the role of a scale factor and I is a dimensionless parameter that, to a good approximation, is independent of n and is characteristic of the type of interaction. This interaction will be generalized to states of different n by the formula

$$H^e_{1,n;2,n'} = \langle \phi_{1,n} | \mathbf{H} | \phi_{2,n'} \rangle = \frac{2\Re}{(n^*)^{3/2}(n'^*)^{3/2}} I, \tag{8.4.4}$$

where I is the same dimensionless parameter and is, to a good approximation, independent of n and n'. This formula is valid even if state 2 is a valence state. In Section 5.2, Table 5.5 shows how the "experimental" value of the $(n^*)^{3/2} H^e_{1,n;2}$ electronic parameter is found to be nearly constant for different Rydberg states, $1, n$, of series 1 interacting with successive vibrational levels of the same valence electronic state, 2.

The electronic part, $\langle \phi_{1,n} | \mathbf{H} | \phi_{2,n'} \rangle$, of the perturbation interaction between two discrete states cannot be compared directly with the electronic part $\langle \phi_{1,n} | \mathbf{H} | \phi_{2,\epsilon} \rangle$ of the autoionization matrix element. Only the energy-normalized perturbation interaction (see Section 7.5), $\rho_n^{1/2} \langle \phi_{1,n} | \mathbf{H} | \phi_{2,n'} \rangle$, which has the same dimensionality as $\langle \phi_{1,n} | \mathbf{H} | \phi_{2,\epsilon} \rangle$, can be compared to this last quantity and should have the same magnitude.

For Rydberg states, the density of states, ρ_n, is given by the inverse separation between two successive states:

$$\frac{1}{\rho_{n'}} = \Delta E_{n'} = \frac{\Re}{(n'^*)^2} - \frac{\Re}{(n'^* + 1)^2} \simeq \frac{2\Re}{(n'^*)^3}. \tag{8.4.5}$$

For a Rydberg state of a given n, the quantity I from Eq. (8.4.4), assumed previously to be constant for all n' members of a second Rydberg series, will now be assumed constant in the limit $n' \to \infty$, which corresponds to a continuum electron with zero kinetic energy. Thus, by combining Eq. (8.4.4) and Eq. (8.4.5),

$$\langle \phi_{1,n} | \mathbf{H} | \phi_{2,\epsilon=0} \rangle = \rho_{n'}^{1/2} \langle \phi_{1,n} | \mathbf{H} | \phi_{2,n'} \rangle = \frac{(2\Re)^{1/2}}{(n^*)^{3/2}} I, \tag{8.4.6}$$

the partial width becomes

$$\Gamma_{n,v;\epsilon=0,v_+} = 2\pi \frac{(2\Re)}{(n^*)^3} I^2 \langle v | v_+ \rangle^2. \tag{8.4.7}$$

With the assumption that I is independent of n and therefore also independent of the energy of the ejected electron, Eq. (8.4.7) is valid for any energy ϵ. If I varies smoothly with the energy of the ejected electron, it must be replaced by $I(\epsilon)$ in Eq. (8.4.7).

Two conclusions may be drawn from this general formula.

1. For successive autoionized states of the same series, the widths decrease as $(n^*)^{-3}$. This law is general regardless of the type of electronic interaction and assumes that I is independent of (or weakly dependent on) the energy of the continuum electron. Expressions for I, valid for different types of autoionization, are discussed in the following sections.

2. The magnitude of the width depends on the interval between ionization thresholds. Indeed, autoionization into a given continuum is possible only if the superexcited state lies above the ionization threshold of that continuum,

$$E_n = E(\mathrm{AB}^{*+}) - \frac{\Re}{(n^*)^2} \geq E(\mathrm{AB}^+). \tag{8.4.8}$$

Equation (8.4.8) sets a lower bound on the value of n^* required for autoionization:

$$n^* \geq \left(\frac{\Re}{\Delta E}\right)^{1/2}, \tag{8.4.9}$$

where $\Delta E = E(\mathrm{AB}^{*+}) - E(\mathrm{AB}^+)$. This in turn, when inserted into Eq. (8.4.7), sets an upper limit for the autoionization width. Table 8.1 summarizes several such bounds on n associated with different types of autoionization.

8.5 Rotational Autoionization

In rotational autoionization, the coupling matrix element is J–dependent and has the same origin as the coupling that occurs for heterogeneous rotational interactions where the selection rule is $\Delta\Omega = \pm 1$ (cf. Section 3.5.3). Rotational autoionization has been studied, especially in the H_2 molecule (Herzberg and Jungen, 1972; Jungen and Dill, 1980).

Let us first consider the rotational interaction between discrete Rydberg states. For example, the $^1\Pi_u$ and $^1\Sigma_u^+$ levels of an np–complex of H_2 are mixed by the l-uncoupling term in $\mathbf{H}^{\mathrm{ROT}}$. Since $\Delta\Sigma = 0$, the selection rule $\Delta\Omega = \pm 1$ implies that $\Delta\lambda = \pm 1$. The 2×2 $^1\Pi \sim {}^1\Sigma^+$ perturbation secular determinant for the Jth rotational levels, expressed in the case (a) basis, is, for $l = 1$,

$$\begin{vmatrix} E_{n,\Sigma} + B[J(J+1)+2] - E & -2B[J(J+1)]^{1/2} \\ -2B[J(J+1)]^{1/2} & E_{n,\Pi} + B[J(J+1)] - E \end{vmatrix} = 0. \tag{8.5.1}$$

The $\Delta J = 0$ energy separation between the Π and Σ states of the $l = 1$ complex is proportional to $(n^*)^{-3}$; thus, at high values of n, the Π and Σ levels of the same rotational quantum number become nearly completely mixed. At the limit $(n = \infty)$, the energy separation between the Π and Σ levels is $2B$, because

$$E_{(n=\infty),\Sigma} = E_{(n=\infty),\Pi} = E_\infty.$$

Table 8.1: Minimum n–values Required for Different Types of Autoionized States[a]

Interaction	Molecular Ion	Ionization thresholds	Interval ΔE between the thresholds (cm^{-1})	n_{min}
Electronic	N_2^+	$X^2\Sigma_g^+$ $A^2\Pi_u$	8950	4
	NO^+	$X^1\Sigma^+$ $b^3\Pi$	58,793	2
Spin-Orbit	N_2^+	$A^2\Pi_{3/2}$ $A^2\Pi_{1/2}$	80	37
	HI^+	$^2\Pi_{3/2}$ $^2\Pi_{1/2}$	5352	5
Vibrational	H_2^+	$X^2\Sigma_g^+(v=0)$ $X^2\Sigma_g^+(v=1)$	2300	7
Rotational	$ortho-H_2^+$	$X^2\Sigma_g^+(v=0,N=0)$ $X^2\Sigma_g^+(v=0,N=2)$	180	25
	$ortho-N_2^+$	$X^2\Sigma_g^+(v=0,N=0)$ $X^2\Sigma_g^+(v=0,N=2)$	12	96

[a]From an original idea of Jungen (private communication)

In this case, the eigenvalues of Eq. (8.5.1) are

$$E_1 = [B + B(2J+1)] + E_\infty + BJ(J+1)$$
$$E_2 = [B - B(2J+1)] + E_\infty + BJ(J+1), \tag{8.5.2}$$

which correspond to the two ionization limits associated with the $N = 2$ and 0 ion core rotational levels. The $\Delta J = 0$ energy separation between these two series of $N = J+1$ (E_1) and $N = J-1$ (E_2) rotational levels is $2B(2J+1)$. The corresponding eigenfunctions in this complete $\Pi \sim \Sigma$ mixing limit,

$$\Psi_1 = \left(\frac{J+1}{2J+1}\right)^{1/2} \psi(^1\Sigma) \quad - \quad \left(\frac{J}{2J+1}\right)^{1/2} \psi(^1\Pi)$$
$$\Psi_2 = \left(\frac{J}{2J+1}\right)^{1/2} \psi(^1\Sigma) \quad + \quad \left(\frac{J+1}{2J+1}\right)^{1/2} \psi(^1\Pi), \tag{8.5.3}$$

are the case (d) basis functions. The mixing coefficients in Eq. (8.5.3) form the transformation matrix, which is convenient for describing high values of n [which are intermediate between cases (d) and (a)]. Using this transformation matrix, the residual $\Delta J = 0$ interaction between the different levels of identical n becomes, for large n, a consequence of the small electronic energy separation between Σ and Π basis components of the $l = 1$ complex, which vanishes at the case (d) limit. The off-diagonal matrix element, evaluated in the case (d) limit, is

$$\langle \Psi_{1,n}|\mathbf{H}|\Psi_{2,n}\rangle = \frac{[J(J+1)]^{1/2}}{2J+1}\left(\langle n^1\Sigma|\mathbf{H}|n^1\Sigma\rangle - \langle n^1\Pi|\mathbf{H}|n^1\Pi\rangle\right) + \frac{(J+1)-J}{J+1}\langle n^1\Sigma|\mathbf{H}|n^1\Pi\rangle \tag{8.5.4}$$

or, since $\langle n^1\Sigma|\mathbf{H}|n^1\Pi\rangle = -2B[J(J+1)]^{1/2}$,

$$\begin{aligned}\langle \Psi_{1,n}|\mathbf{H}|\Psi_{2,n}\rangle &= \frac{[J(J+1)]^{1/2}}{2J+1}\left(E_{n\Sigma} + 2B - E_{n\Pi} - 2B\right)\\ &= \frac{[J(J+1)]^{1/2}}{2J+1}\left(E_{n\Sigma} - E_{n\Pi}\right).\end{aligned} \tag{8.5.5a}$$

Similarly,

$$\langle \Psi_{1,n}|\mathbf{H}|\Psi_{1,n}\rangle = \frac{E_{n\Sigma}+E_{n\Pi}}{2} - \frac{E_{n\Pi}-E_{n\Sigma}}{2(2J+1)} + \frac{4BJ(J+1)}{2J+1} \tag{8.5.5b}$$

$$\langle \Psi_{2,n}|\mathbf{H}|\Psi_{2,n}\rangle = \frac{E_{n\Sigma}+E_{n\Pi}}{2} + \frac{E_{n\Pi}-E_{n\Sigma}}{2(2J+1)} - \frac{4BJ(J+1)}{2J+1}. \tag{8.5.5c}$$

Now, in order to re-express Eq. (8.5.5a) in terms of quantum defects, let

$$E_{n\Pi} = E_\infty - \frac{\Re}{(n-a_\pi)^2} \quad \text{and} \quad E_{n\Sigma} = E_\infty - \frac{\Re}{(n-a_\sigma)^2}, \tag{8.5.6}$$

where a_π and a_σ are the quantum defects of the unperturbed Π and Σ series, respectively. Expressed in terms of the very small difference of quantum defects, $a' = \frac{1}{2}(a_\pi - a_\sigma)$, Eq. (8.5.6) becomes

$$E_{n\Pi} = E_\infty - \frac{\Re}{(n^*-a')^2} \quad \text{and} \quad E_{n\Sigma} = E_\infty - \frac{\Re}{(n^*+a')^2} \tag{8.5.7}$$

where n^* is defined as a mean effective principal quantum number for the two series:

$$n^* \equiv n - \frac{1}{2}(a_\pi + a_\sigma).$$

Using a truncated Taylor expansion,

$$\frac{\Re}{(n^* \pm a')^2} \simeq \frac{\Re}{(n^*)^2} \mp \frac{2\Re}{(n^*)^3}a', \tag{8.5.8}$$

Eq. (8.5.5a) becomes

$$\langle \Psi_{1,n}|\mathbf{H}|\Psi_{2,n}\rangle = \frac{[J(J+1)]^{1/2}}{2J+1}\frac{2\Re}{(n^*)^3}2a' = \frac{[J(J+1)]^{1/2}}{2J+1}\frac{2\Re}{(n^*)^3}(a_\pi - a_\sigma). \tag{8.5.9}$$

Further, by analogy with Eq. (8.4.4), the interaction ($\Delta J = 0, \Delta N = \pm 2$) between two states, one belonging to an np complex, the other to an $n'p$ complex, is

$$\langle \Psi_{1,n,p} | \mathbf{H} | \Psi_{2,n',p} \rangle = \frac{2\Re}{(n^*)^{3/2}(n^{*\prime})^{3/2}} I \tag{8.5.10}$$

with

$$I = \frac{[J(J+1)]^{1/2}}{2J+1}(a_\pi - a_\sigma). \tag{8.5.11}$$

As this interaction occurs between states that converge to the same electronic state of the ion (but to different N levels), their vibrational functions are nearly identical. Thus, the vibrational factor is almost unity for $\Delta v = 0$ and zero for $\Delta v \neq 0$.

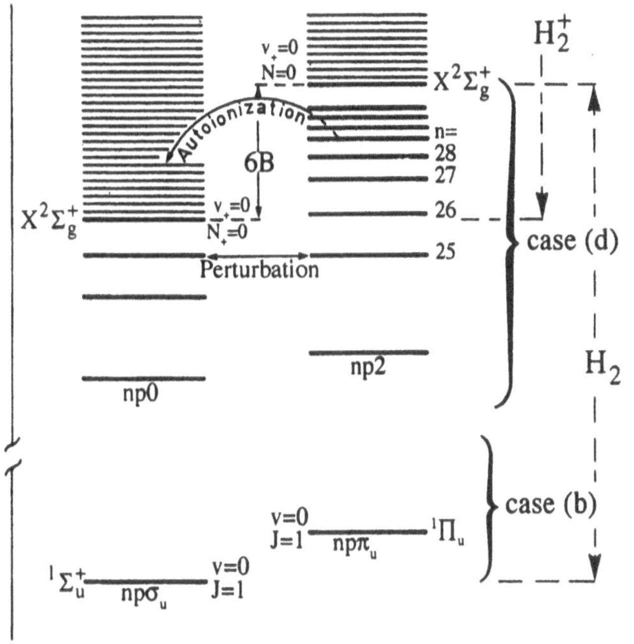

Figure 8.11: Schematic illustration of rotational autoionization in *para*–H_2 showing the transition from case (b) to case (d). The electronic continua associated with H_2^+ $X^2\Sigma_g^+$ ($v_+ = 0$, $N^+ = 0$ and 2) are depicted by light horizontal lines. The stable and autoionizing levels of the H_2 np Rydberg series with $v = 0$ and $J = 1$ are shown as heavy horizontal lines.

For autoionization of the Jth level, which belongs to the series converging to the second ($N^+ = 2$) ionization limit and autoionized by the continuum of $H_2^+(N^+ = 0)$, Eq. (8.4.7) gives (see Herzberg and Jungen, 1972)

$$\Gamma_{n,J} = 2\pi \frac{2\Re}{(n^*)^3} \frac{J(J+1)}{(2J+1)^2}(a_\pi - a_\sigma)^2, \tag{8.5.12}$$

which, for $J = 1$, becomes

$$\Gamma_{n,J} = 2\pi \frac{2\Re}{(n^*)^3} \left(\frac{2}{9}\right) (a_\pi - a_\sigma)^2. \qquad (8.5.13)$$

Thus, it is possible to view rotational autoionization as arising from the departure of the molecular-ion core from spherical symmetry ($a_\pi \neq a_\sigma$), which causes the np complex to split into Π and Σ levels. Figure 8.11 illustrates rotational autoionization in H_2. The right side of Fig. 8.12 displays lines of the $v = 0, J = 0$ series converging to the $N^+ = 2$ limit of the ion (called $np2$) that are autoionized by the continuum of H_2^+ $X^2\Sigma_g^+(v_+ = 0, N^+ = 0)$ resulting in the appearance of "emission windows" (see Section 8.9).

The measured difference between the quantum defects for Σ and Π Rydberg series is, for $H_2(v_+ = 1) : a_\sigma - a_\pi = 0.230 - (-0.085) = 0.315$. From this and Eq. (8.4.4), $I = 0.148$ for $J = 1$ (see Table 8.2). For *para*–H_2, the interval between the $N^+ = 2$ and $N^+ = 0$ ionization limits is equal to $6B = 108$ cm^{-1} for $J = 1$, where B is the rotational constant of H_2^+. Thus, the smallest possible value of n (for $J = 1, \Delta N = 2$ rotational autoionization) is 25 (see Table 8.1). For the rotational line $26p2$, the width is 2.3 cm^{-1} (Table 8.2). The f–parity levels cannot be autoionized by this mechanism because there is only one such level in an np complex for each value of J.

The interval $2B(2J+1)$ between the two ionization limits becomes small for heavy molecules and low values of J.

8.6 Vibrational Autoionization

Autoionizing interactions between high Rydberg levels converging to one vibrational level of the ion and the continuum associated with a lower vibrational level of the same ion core electronic state are more easily observable than rotational autoionization. This process is illustrated schematically in Fig. 8.13 for $v_+ = 0$ and $v_+ = 1$. At sufficiently high n, the $v = 1$ Rydberg levels converging to the ion core $v_+ = 1$ level of the electronic ground state are embedded in the electronic continuum of the $v_+ = 0$ level of this ground state. Just as for rotational autoionization, this vibrational interaction can be related to homogeneous perturbations between discrete Rydberg levels.

Because these states have the same symmetry and identical orbital angular momentum projection quantum number (Λ), spin, and core electronic wavefunctions, but different vibrational wavefunctions, the nuclear kinetic energy operator, which depends on R, is the only operator that can couple these two vibronic states (see Section 3.3). This interaction can be understood using a very simple model (Herzberg and Jungen, 1972). The potential energy curve, $U_{n\Lambda}(R)$, of a Rydberg state is not strictly parallel to that of the ion, $U^+(R)$, and this deviation, which reflects the minuscule bonding/antibonding effect of the Rydberg electron, can be expressed by an R–dependent quantum defect as follows (cf. Mulliken, 1969, Fig. 1):

$$U_{n\Lambda}(R) = \langle \phi_{n,\Lambda}|\mathbf{H}|\phi_{n,\Lambda}\rangle = U^+(R) - \Re/[n - a_\lambda(R)]^2 \tag{8.6.1}$$

(with the definition $\Lambda = \Lambda^+ + \lambda$, where Λ^+ and λ correspond respectively to the orbital angular momentum projection quantum number of the ion-core and the Rydberg electron) where $a_\lambda(R)$ varies slowly with R and can be represented by a truncated Taylor expansion about the equilibrium internuclear distance, R_e^+, of the ion:

$$a_\lambda(R) = a_{\lambda,(R=R_e^+)} + \left(\frac{da_\lambda}{dR}\right)_{R=R_e^+} (R - R_e^+) + \frac{1}{2}\left(\frac{d^2a_\lambda}{dR^2}\right)_{R=R_e^+} (R - R_e^+)^2 + \ldots \tag{8.6.2}$$

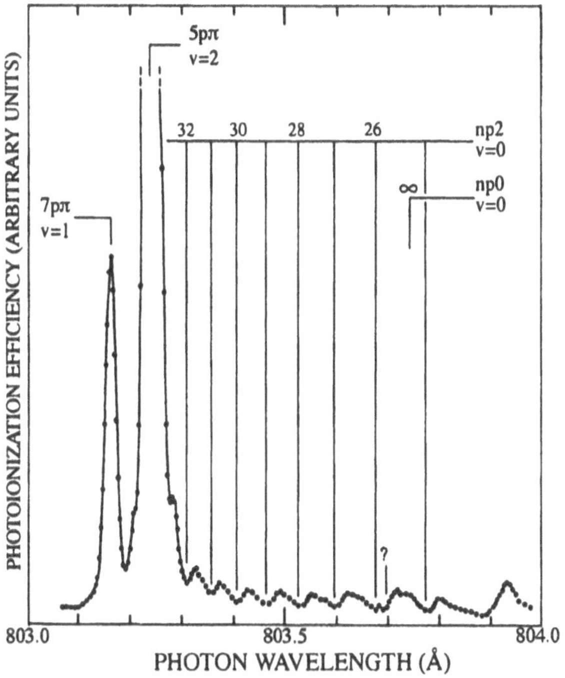

Figure 8.12: Relative photoionization cross section of *para*–H_2, at 78 K, in the region of the ionization threshold, recorded at a wavelength resolution of 0.016 Å. On the right-hand side of the figure, the rotationally autoionized lines of the $np2$ series appear as emission windows. The large peaks on the left are the result of vibrational autoionization (see Section 8.6) [from Dehmer and Chupka (1976)].

Table 8.2: Typical Examples of Autoionization Linewidths

Type of autoionization	Molecule	State	$\Gamma(\text{cm}^{-1})$ (obs)	I(calc)	Vibrational selection rules
Rotational	H_2	$26p2, v=1, J=1$	2.3	0.148	$\Delta v = 0$
Vibrational	H_2	$8p\sigma, v=2, J=1$	9.7	0.05	$\Delta v \simeq 1$
Spin-orbit	HI	$n=6$	370	$(0.24)^a$	$\Delta v = 0$
Electronic	N_2	$^1\Sigma_u^+ \ (n=5)$	$(2)^b$	0.05	any Δv (Franck-Condon factors)
Electronic	NO	$(b^3\Pi)3p\pi$	500	$(0.05)^a$	

[a]Deduced from experimental value of Γ and Eq. (8.7.7), Eq. (8.7.8), and Eq. (8.8.1).
[b]Calculated by Duzy and Berry (1976).

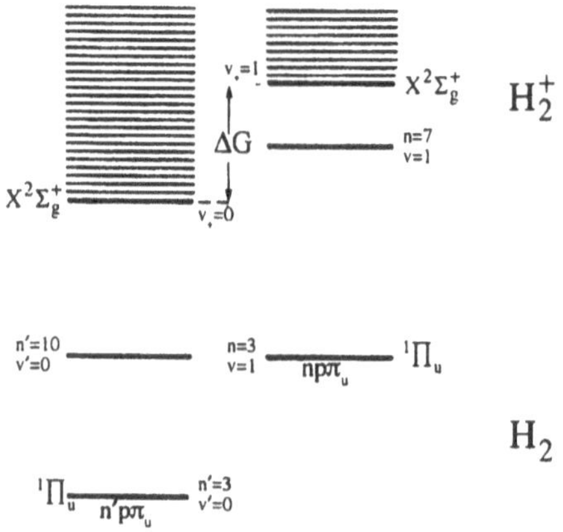

Figure 8.13: Schematic illustration of vibrational autoionization of the H_2 $np\pi_u$ $^1\Pi_u$ $(v=1)$ Rydberg series by the continuum of H_2^+ $X^2\Sigma_g^+$ $(v_+=0)$.

In this expression, da_λ/dR reflects the difference in the R_e values of the ion and Rydberg state potential curves and d^2a_λ/dR^2 reflects the difference in the harmonic constants (ω_e) or in the dissociation energies of these two curves. If one defines $n^* = n - a_{\lambda,(R=R_e^+)}$, using a truncated Taylor expansion as in Eq. (8.5.8) and keeping only the linear term of Eq. (8.6.2), one obtains

$$U_{n\Lambda}(R) = U^+(R) - \frac{\Re}{(n^*)^2} - \frac{2\Re}{(n^*)^3}\frac{da_\lambda}{dR}(R - R_e^+). \tag{8.6.3}$$

Converting to the usual off-diagonal form [Eq. (8.4.4)] of Rydberg~Rydberg matrix elements between members of the same $l\Lambda$ series but with different vibrational factors, the matrix element can be written as

$$\langle\Psi_{n,\Lambda,v}|\mathbf{H}|\Psi_{n',\Lambda,v'}\rangle = \frac{2\Re}{(n^*)^{3/2}(n^{*\prime})^{3/2}}\frac{da_\lambda}{dR}\langle v|R - R_e^+|v'\rangle. \tag{8.6.4}$$

In the harmonic approximation, the only nonzero matrix elements of $(R - R_e^+)$

are $\Delta v = v' - v = \pm 1$ (see, for example, Wilson, *et al.*, 1980, Appendix III),

$$\langle \Psi_{n,\Lambda,v}|\mathbf{H}|\Psi_{n',\Lambda,v-1}\rangle = \frac{2\Re}{(n^*)^{3/2}(n^{*\prime})^{3/2}}\frac{da_\lambda}{dR}\left[\frac{h}{8\pi^2\mu\omega c}\right]^{1/2}(v)^{1/2} \qquad (8.6.5)$$

where μ is the reduced mass and ω the harmonic vibrational frequency. The linewidth for the n^*, v–level autoionized by the continuum of the $v-1$ level is obtained from Eq. (8.6.5) and the Golden Rule [Eq. (8.4.1)], eliminating the $n^{*\prime}$ dependence by inserting the density of states factor,

$$\left(\frac{dE}{dn}\right)^{-1} = \left[\frac{2\Re}{(n^{*\prime})^3}\right]^{-1}, \qquad (8.6.6)$$

to obtain (Herzberg and Jungen, 1972),

$$\Gamma_{n_{v,v-1}}(\text{cm}^{-1}) = 2\pi\frac{2\Re}{(n^*)^3}\left[\frac{da_\lambda}{dR}(\text{Å}^{-1})\right]^2 \frac{16.8576}{\mu(\text{amu})\ \ \omega(\text{cm}^{-1})}v, \qquad (8.6.7)$$

where v and $v-1$ are, respectively, the vibrational quantum numbers of the nth autoionized Rydberg state and the product ion.

Vibrational autoionization corresponds to an exchange between the kinetic energy of the ejected electron and the vibrational energy of the core ion. This process occurs because the potential curves of the ion and Rydberg states are slightly different. Eq. (8.6.7) can be derived from a more sophisticated model using multichannel quantum defect theory (Raoult and Jungen, 1981).

Equation (8.6.7) shows that the widths decrease as $(n^*)^{-3}$ and are proportional to v if only the first derivative of the quantum defect is retained. The approximate selection rule, $\Delta v = -1$, known as the propensity rule of Berry (1966), is well verified experimentally, at least for H_2. The width, for $\Delta v = -2$ in H_2, is typically two orders of magnitude smaller than for $\Delta v = -1$. The effect of the second derivative is neglected for H_2, but second-order effects proportional to $\langle v-1|v\rangle\,\langle v|v+1\rangle$ (and higher order effects) can produce small widths for any Δv value (Dehmer and Chupka, 1976).

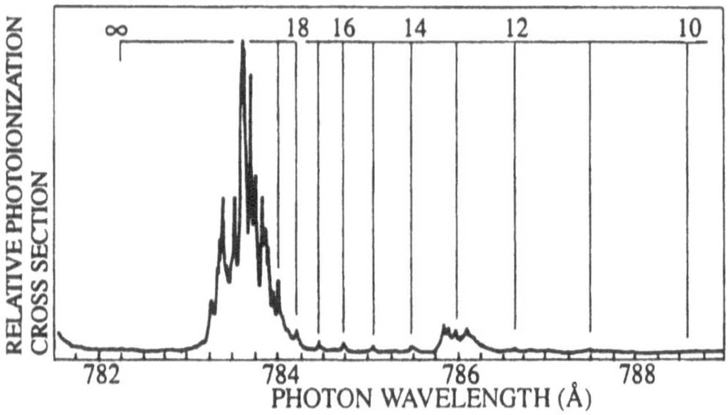

Figure 8.14: Photoionization cross-section of N_2 taken at 77 K with a resolution of 0.016 Å (from Dehmer, *et al.*, 1984). The first members of the $v = 1$ series are vibrationally autoionized by the $v_+ = 0$ continuum of N_2^+ $X^2\Sigma_g^+$. For $n \geq 14$, anomalies result from indirect electrostatic autoionization (from Giusti-Suzor and Lefebvre-Brion, 1984).

For H_2, the width is larger for Σ states than for Π states because the quantum defect a_σ varies more rapidly with internuclear distance than a_π (see Fig. 2 of Jungen and Atabek, 1977). This, in turn, can be understood by the fact that, at $R = \infty$ the $3p\sigma$ state of H_2 dissociates into H(2p)+ H(1s) atoms. The value of n is therefore "promoted" by one unit on going from the separated atoms to the united atom limit. Consequently, a_σ must vary rapidly with R. This does not occur for Π Rydberg states. A typical autoionization width is given in Table 8.2 for the level $8p\sigma$ $v = 2$ autoionized by the $v_+ = 1$ continuum of H_2^+ $X^2\Sigma_g^+$. In Fig. 8.12, one can see a peak that corresponds to R(0) $7p$ $\pi(v = 1)$, for which the calculated width (without convolution with the experimental resolution) is only 0.15 cm^{-1}. The peak corresponding to the level $5p\pi$ ($v = 2$) autoionized by the $v_+ = 0$ continuum of H_2^+, illustrates a deviation from Berry's propensity rule. This deviation is due to interference between rotational and vibrational autoionization (Jungen and Dill, 1980). The proportionality of the widths to v is often well verified in pure vibrational autoionization cases (see, for example, Dehmer and Chupka, 1976).

In HF (Guyon, *et al.*, 1976) vibrational autoionization peaks are clearly seen in the threshold photoelectron spectrum. For heavy molecules, the da_λ/dR factors appear to be small, as suggested by the calculations for N_2 by Duzy and Berry (1976) and Lefebvre-Brion and Giusti-Suzor (1983) and verified in the photoionization spectrum of N_2 (Fig. 8.14), where the members of the series converging to the $v_+ = 1$ level of the $X^2\Sigma_g^+$ state are very weakly autoionized by the continuum of the X state ($v_+ = 0$) level (see, for example, $n = 11$ or 12). For the NO molecule (Ng, *et al.*, 1976), fine structure is also visible in the photoionization curve. An apparent deviation from the $\Delta v = -1$ propensity rule is observed that cannot be explained simply by a non-negligible second derivative term. This anomalous line broadening has been shown to be induced by predissociation (Giusti-Suzor and Jungen, 1984). For Li_2 (Roche and Jungen,

1993) the 24d ($v = 1, J = 8$) state has a width of only 1.7×10^{-2} cm^{-1}.

8.7 Spin-Orbit Autoionization

This type of autoionization can affect Rydberg levels that lie between the ionization limits associated with different multiplet components of a given ion core electronic state. A well known example concerns the Rydberg states of the Xenon atom (Connerade, 1983). This type of autoionization has been recently studied in molecules. Consider, for example, the Rydberg states converging to the O_2^+ $X^2\Pi_{3/2}$ substate, which can be autoionized by the continuum of the $X^2\Pi_{1/2}$ substate. The $X^2\Pi$ state is regular, which means that $E\left(^2\Pi_{3/2}\right) > E\left(^2\Pi_{1/2}\right)$. The spin-orbit interaction will appear between discrete Rydberg states of different series having the same Ω value (see Section 3.4.2).

Consider the $(p\pi)(ns\sigma)$ $^3\Pi_1$ and $^1\Pi_1$ Rydberg series of O_2 converging to the $X^2\Pi$ state of O_2^+, as described by the Rydberg formulas

$$\begin{aligned} E[(ns\sigma)\,^1\Pi] &= E(X^2\Pi) - \frac{\Re}{(n-a_1)^2} = E(X^2\Pi) - \frac{\Re}{(n^*+a')^2} \\ E[(ns\sigma)\,^3\Pi] &= E(X^2\Pi) - \frac{\Re}{(n-a_3)^2} = E(X^2\Pi) - \frac{\Re}{(n^*-a')^2}. \end{aligned} \quad (8.7.1)$$

$E(X^2\Pi)$ is the electronic energy in the absence of spin-orbit splitting:

$$E(X^2\Pi) = \frac{1}{2}[E(X^2\Pi_{1/2}) + E(X^2\Pi_{3/2})]$$

$$a' = \frac{1}{2}(a_3 - a_1) \qquad \text{and} \qquad n^* = n - \frac{1}{2}(a_3 + a_1).$$

The spin-orbit interaction associated with the unpaired $p\pi$ core electron mixes the $^3\Pi_1$ and $^1\Pi_1$ components of the same n value ($\Delta\Omega = 0$). In the case (a) basis, the secular equation is [see Eq. (3.4.14)]

$$\begin{vmatrix} E\left[(ns\sigma)\,^3\Pi_1\right] - E & A(^3\Pi) \\ A(^3\Pi) & E\left[(ns\sigma)\,^1\Pi_1\right] - E \end{vmatrix} = 0. \quad (8.7.2)$$

The difference between the triplet and singlet energies is approximately $2K_n$, where K_n is an exchange integral proportional to $(n^*)^{-3}$. As discussed in Section 3.4.2.1.2, the mixing between $^3\Pi_1$ and $^1\Pi_1$ increases with n, and when n becomes infinite the states are completely mixed and are described by a case (c) basis that corresponds to the limiting solutions,

$$\begin{aligned} E_1 &= E(X^2\Pi) + A(^3\Pi) \qquad & \Psi_1 &= 2^{-1/2}(^1\Pi_1 + {}^3\Pi_1) \\ E_2 &= E(X^2\Pi) - A(^3\Pi) \qquad & \Psi_2 &= 2^{-1/2}(^1\Pi_1 - {}^3\Pi_1) \end{aligned} \quad (8.7.3)$$

See also Fig. 5.1 of Lefebvre-Brion (1995). These case (a)→(c) transformation coefficients will now be used to describe Rydberg states for high values of n. Two new series are obtained. One of the two $\Omega = 1$ series converges to the

$X^2\Pi_{1/2}$ component of the ion and the other to the $X^2\Pi_{3/2}$ component, with energies and new quantum defects

$$E_{1n} = E(X^2\Pi_{1/2}) - \frac{\Re}{(n - a_{1/2})^2}$$
$$E_{2n} = E(X^2\Pi_{3/2}) - \frac{\Re}{(n - a_{3/2})^2}. \tag{8.7.4}$$

From Eq. (8.7.3) and Eq. (8.5.8) and noting that $E(X^2\Pi_{3/2}) - E(X^2\Pi_{1/2}) = 2A(^3\Pi)$ it is easy to show that

$$a_{1/2} = a_{3/2} = \frac{1}{2}(a_1 + a_3).$$

The residual interaction between these same-n Rydberg levels, expressed in the case (c) basis, but intermediate between cases (c) and (a), can now be viewed as a consequence of the electronic part of the Hamiltonian, since the case (c) basis functions are exact eigenfunctions only when the energy difference between isoconfigurational triplet and singlet states is zero. The off-diagonal matrix element in the case (c) basis is

$$\left\langle 2^{-1/2}\ [(ns\sigma)^1\Pi_1 + (ns\sigma)^3\Pi_1]|\mathbf{H}|2^{-1/2}[(ns\sigma)^1\Pi_1 - (ns\sigma)^3\Pi_1]\right\rangle$$
$$= \frac{1}{2}\left\{E[(ns\sigma)^1\Pi_1] - E[(ns\sigma)^3\Pi_1]\right\}$$
$$= K_n \simeq \frac{\Re}{(n^*)^3}2a' = \frac{\Re}{(n^*)^3}(a_3 - a_1). \tag{8.7.5}$$

The approximate equality is obtained from a Taylor expansion of Eq. (8.4.4) analogous to that of Eq. (8.5.8). The interaction between the nth term of the series converging to the $X^2\Pi_{1/2}$ limit and the (n')th term of the $X^2\Pi_{3/2}$ series can be written by analogy with Eq. (8.4.4), as

$$\langle\Psi_{1,n}|\mathbf{H}|\Psi_{2,n'}\rangle = \frac{\Re}{(n^*)^{3/2}(n'^*)^{3/2}}(a_3 - a_1). \tag{8.7.6}$$

From the general result, Eq. (8.4.4), the I parameter may be identified here as

$$I = \frac{a_3 - a_1}{2}. \tag{8.7.7}$$

To a good approximation, the vibrational wavefunctions of the two series, which converge to the same vibrational level of the $^2\Pi_{1/2}$ and $^2\Pi_{3/2}$ limits, are identical, leading to a $\Delta v = 0$ propensity rule and a vibrational factor equal to unity. Finally, the autoionization width given by Eq. (8.4.7) is

$$\Gamma_n = 2\pi\frac{2\Re}{(n^*)^3}\left(\frac{a_3 - a_1}{2}\right)^2. \tag{8.7.8}$$

For O_2, two Rydberg series corresponding to $\left(O_2^+\ X^2\Pi_g\right) ns\sigma$ have been observed by Sur, *et al.*, (1985, 1986). The difference between the two quantum defects for $v = 0$ ($a_1 \simeq 1.12$ and $a_3 \simeq 1.14$) gives $I \simeq 0.01$. The interval between the $O_2^+\ X^2\Pi_{3/2}$ and $X^2\Pi_{1/2}$ limits is small (200 cm^{-1}). Consequently, the minimum value of n for autoionization obtained from Eq. (8.4.9) is very large, $n_{\text{min}} \geq 24$; thus the maximum value of Γ will be about 0.01 cm^{-1}. Actually, electronic (or, more precisely, electrostatic) autoionization is considerably stronger and conceals this spin-orbit autoionization mechanism.

For molecules containing an atom heavier than O, the spin-orbit splitting of the ion will be larger and n_{min} can become small. Consequently, the spin-orbit interaction is then the most important cause of autoionization, as it is already well known to be for predissociation. The HI photoionization spectrum (Eland and Berkowitz, 1977) is a good example. Figure 8.15 shows the Rydberg levels as they converge to the $^2\Pi_{1/2}$ limit. (For HI the $^2\Pi$ state of the ion is inverted.) The width of the $n = 6$ member of the series is estimated from the spectrum to be 370 cm^{-1}. As this series is certainly very perturbed even at low n by spin-orbit interaction, it is difficult to extract deperturbed values for a_1 and a_3 and, from them, I. However, using the experimental value for Γ, Eq. (8.7.8) gives $I = 0.24$. This value is similar to the value that can be deduced for the Xe atom from the widths of autoionized levels between the $^2P_{1/2}$ and $^2P_{3/2}$ thresholds.

The magnitude of the spin-orbit splitting of the ion-core state affects the autoionization width most importantly through the minimum value of n, not through the size of I. This result is not surprising, since the present treatment was based on case (c) basis functions, which are eigenfunctions of the spin-orbit Hamiltonian (i.e., which diagonalize the spin-orbit part of the total Hamiltonian). (See Lefebvre-Brion, *et al.*, (1985) for a theoretical treatment of the HI spin-orbit autoionization.)

Experimental progress has made it possible to obtain rotationally resolved photoionization spectra (see Mank, *et al.*, 1991). Thus it becomes necessary to define Rydberg series which converge to a given rotational level, J^+, of the ion-core. Following the selection rules for ionization given in Section 8.1.1, Rydberg series with different l values converge to rotational levels of different parities. Considering the levels of an inverted $^2\Pi$ ion state with Λ–doubling described by $p(J+1/2)$ (see Fig. 8.16), the series belonging to even l values (s or d) converge to rotational levels of negative parity and the series of odd l (p or f) converge to levels of positive parity. The different Rydberg series must now be described in Hund's case (e) (see Section 3.2.1).

Because the number of states for a configuration such as $(\pi)^3 nl\lambda$ is large, the expressions of the eigenfunctions in Hund's case (e) as linear combinations of eigenfunctions in Hund's case (a) becomes very complicated, even for $l = 0$. For $l = 0$, $J' = 1$, there are three states in Hund's case (a): $^1\Pi_1$, $^3\Pi_1$, $^3\Pi_0$. Among the corresponding three states in Hund's case (e), two of these states give rise to two series converging to $^2\Pi_{1/2}$ with $J^+ = 1/2$ and $3/2$. The wavefunctions are (Lefebvre-Brion, 1990):

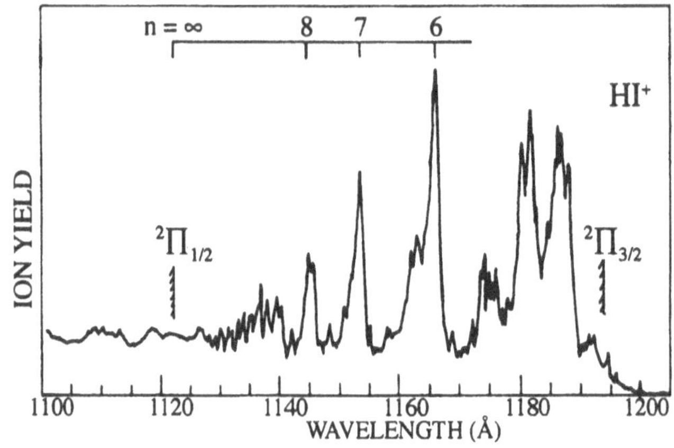

Figure 8.15: Photoionization of HI at 0.14 Å resolution showing the spin-orbit autoionization between the two $\Omega = \frac{3}{2}$ and $\Omega = \frac{1}{2}$ thresholds of the ground state of the HI^+ ion [from Eland (1980)].

$$3^{-1/2}\left[\left({}^1\Pi_1 + {}^3\Pi_1\right) + {}^3\Pi_0\right] \tag{8.7.9}$$

which converges to $J^+ = 1/2$, and

$$6^{-1/2}\left[\left({}^1\Pi + {}^3\Pi_1\right) - 2\,{}^3\Pi_0\right] \tag{8.7.10}$$

which converges to $J^+ = 3/2$ of ${}^2\Pi_{1/2}$. The third series is

$$2^{-1/2}\left({}^1\Pi_1 - {}^3\Pi_1\right) \tag{8.7.11}$$

which converges to ${}^2\Pi_{3/2}$ with $J^+ = 3/2$. The two series that converge to ${}^2\Pi_{1/2}$ are autoionized in the continuum of the ${}^2\Pi_{3/2}$ state. Their intensities and widths become equal as n increases and their energy separation for a given value of n goes towards ($3B^+ - 3/2\ p$) (see Fig. 3 of Mank, *et al.*, 1991). For $J' = 2$, $l = 0$, there are 4 wavefunctions.

For $l = 1$ and $l = 2$, the wavefunctions are given respectively in Lefebvre-Brion (1990 and 1995). For example, for $l = 1, J = 0$ one series converges to ${}^2\Pi_{3/2}$ with $J^+ = 3/2, j = l + s = 3/2$

$$2^{-1/2}\left({}^1\Sigma_0^+ - {}^3\Sigma_0^-\right) p\pi \tag{8.7.12}$$

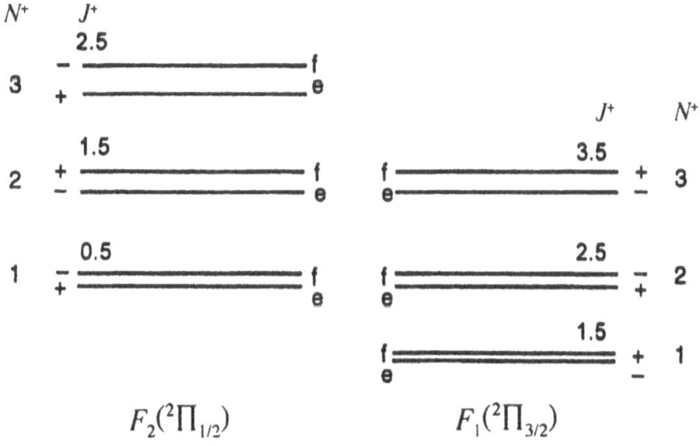

Figure 8.16: Energy level diagram for a $^2\Pi_i$ state: $F_1, N^+ = J^+ - 1/2$; $F_2, N^+ = J^+ + 1/2$.

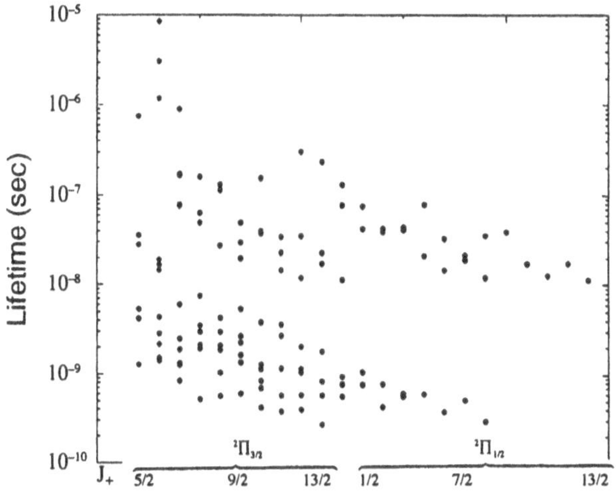

Figure 8.17: Calculated lifetimes of rotational levels of HCl $n = 93$ for $J \leq 3$ and $l \leq 3$ (from Lefebvre-Brion, 1999).

and two series converge to $^2\Pi_{1/2}$: one with $J^+ = 1/2, j = 1/2$

$$3^{-1/2}\left(\left({}^1\Sigma_0^+ + {}^3\Sigma_0^-\right)p\pi - {}^3\Pi_0\ p\sigma\right) \tag{8.7.13}$$

the other with $J^+ = 3/2, j = 3/2$

$$6^{-1/2}\left({}^1\Sigma_0^+ + {}^3\Sigma_0^-\right)p\pi + 2^{1/2}3^{-1/2}\ {}^3\Pi_0\ p\sigma \tag{8.7.14}$$

The general formula for the width of a rotational level is given in Eq. 17 of Lefebvre-Brion (1995).

MQDT calculations see Section 8.9) have been able to reproduce the observed widths of the rotational levels which lie between the $^2\Pi_{3/2}$ and $^2\Pi_{1/2}$ ionization limits, in particular for the HBr absorption spectrum observed from the intermediate $F^1\Delta$ state (Wales, *et al.*, 1996). Following the usual Rydberg scaling rule, the lifetimes increase along a series as n^3 but are very different for different series, depending not only on the value of l but also on the value of the J^+ rotational level of the ion to which the series converges. In Fig. 8.17 the calculated lifetimes are given for $n = 93$ of HCl for $l \leq 3$ and $J \leq 3$. It can be seen that the rotational levels converging to the $J^+ = 3/2$ level of the $^2\Pi_{3/2}$ lowest limit cannot be autoionized because they lie below the lowest energy ionization channels. The levels converging to the higher rotational levels of the $^2\Pi_{3/2}$ state are weakly autoionized, but their lifetimes decrease as the number of open channels increases, i.e. as J^+ increases. The rotational levels of the $n = 93$ Rydberg complex that converge to the rotational levels of $^2\Pi_{1/2}$ have shorter lifetimes (than those converging on $^2\Pi_{3/2}$) which further decrease as the value of J^+ increases. Spin-orbit autoionization can, in principle, also be observed in Rydberg states converging to different components of a $^3\Pi$ ion state, as in NO^+, or a $^4\Pi$ ion state, as in O_2^+.

8.8 Electronic (or Electrostatic) Autoionization

Electronic autoionization occurs between the nth Rydberg state converging to an excited electronic state of the ion in its v vibrational level and the continuum of a lower electronic state of the ion in its v_+ vibrational level. In Eq. (8.4.7), the general expression for the width,

$$\Gamma_{n,v;\epsilon,v_+} = 2\pi \frac{2\Re}{(n^*)^3} I^2 \langle v|v_+\rangle^2 \tag{8.8.1}$$

the vibrational functions, χ_v and χ_{v_+}, correspond to different potential curves of the ion. There is thus no selection rule for the vibrational factor, and the overlap-squared factor is analogous to a Franck-Condon factor. In the electronic part, the parameter I corresponds to an electrostatic interaction between states of identical symmetry (same values of Λ, S, and Σ). In a simple picture, these two states must be derived from electronic configurations which differ by at most two orbitals.

The unknown parameter, I, can be related to perturbation interactions between two Rydberg states converging to different states of the ion, provided that homogeneous perturbations are experimentally observed in the discrete spectrum below the ionization limit. Figure 8.18 illustrates such a situation for N_2. The first two members of two different $^1\Pi_u$ Rydberg series, the $\left[(X^2\Sigma_g^+)\, 3p\pi_u\right] c^1\Pi_u$ and the $\left[(A^2\Pi_u)\, 3s\sigma_g\right] o^1\Pi_u$ states, which converge, respectively, to the N_2^+ $X^2\Sigma_g^+$ $(v_+ = 0)$ and $A^2\Pi_u$ $(v_+ = 0)$, states are found at similar energies and perturb each other. These homogeneous perturbations have been studied by Stahel, *et al.*, (1983). They have deduced a value of about 120 cm^{-1} for the effective electronic parameter, from which Eq. (8.4.4) gives

$$I = (n^*)^{3/2}(n'^*)^{3/2} \times 120/2\Re \simeq 0.01.$$

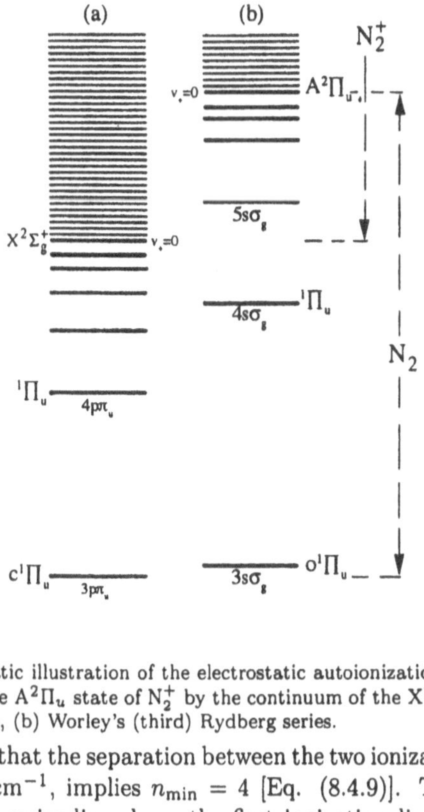

Figure 8.18: Schematic illustration of the electrostatic autoionization of the $^1\Pi_u$ Rydberg states converging to the $A^2\Pi_u$ state of N_2^+ by the continuum of the $X^2\Sigma_g^+$ state. (a) Worley-Jenkins Rydberg series, (b) Worley's (third) Rydberg series.

Table 8.1 shows that the separation between the two ionization limits, $E(A^2\Pi_u) - E(X^2\Sigma_g^+) = 9000$ cm^{-1}, implies $n_{min} = 4$ [Eq. (8.4.9)]. The $n = 5$ member of the $(A^2\Pi_u)ns\sigma_g$ series lies above the first ionization limit $(X^2\Sigma_g^+, v^+ = 0)$ and appears strongly in the ionization spectrum (Fig. 8.14). Unfortunately, the peak is much broader than expected from $I \approx 0.01$ ($\Gamma_{obs} \approx 80$ cm^{-1}, $\Gamma_{calc} \approx 0.6$ cm^{-1}). This results from an enhancement by indirect electronic autoionization (analogous to accidental predissociation) of the higher members of the Rydberg series converging to the $v^+ = 1$ level of the $X^2\Sigma_g^+$ state (Giusti-Suzor and Lefebvre-Brion, 1984).

For NO, the separation between the first two excited states of NO^+ is considerably larger than for N_2^+, giving $n_{min} \geq 2$. A linewidth of 500 cm^{-1} has been observed in the ultraviolet spectrum of NO (Takezawa, 1977), for a $3p\pi$ Rydberg state converging to the NO^+ $b^3\Pi$ excited state. If that width is caused exclusively by electronic autoionization and not by predissociation, it corresponds to a value of about 0.05 for the I parameter.

In conclusion, Table 8.2 shows that the strongest autoionization process is electronic autoionization for light molecules and spin-orbit autoionization for heavy molecules. Nothing has been said yet about level shifts in autoionization. As the coupling responsible for autoionization is even less energy-dependent than that for predissociation, the level shift should be less important. Autoionization level shifts have not yet been experimentally detected for molecules, but the possible occurrence of such shifts should not be forgotten.

8.9 Validity of the Approximations

In the preceding sections, we have assumed that an absorption line has a Lorentzian shape. If this is not true, then the linewidth cannot be defined as the full width at half maximum intensity. Transitions from the ground state of a neutral molecule to an ionization continuum often have appreciable oscillator strength, in marked contrast to the situation for ground state to dissociative continuum transitions. The absorption cross-section near the peak of an autoionized line can be significantly affected by interference between two processes: (1) direct ionization or dissociation, and (2) indirect ionization (autoionization) or indirect dissociation (predissociation). The line profile must be described by the Beutler-Fano formula (Fano, 1961):

$$\sigma_a = \sigma_d + \sigma_i \frac{(q+\epsilon)^2}{1+\epsilon^2}, \tag{8.9.1}$$

where σ_d is the cross-section for the portion of the ionization or dissociative continuum that, by symmetry or selection rules, does not interact with the resonance state, and σ_i corresponds to the part of the ionization or dissociative continuum that interacts with this state. The lineshape is characterized by two parameters, ϵ and q:

1. $\epsilon = (E - E_r)/(\Gamma/2)$, where E_r and Γ are the energy and full width of the resonance state (and ϵ must not be confused with the kinetic energy of the electron).

2. q is defined similarly to Eq. (7.9.5):

$$q = \frac{1}{\pi} \frac{\langle \phi_r|\boldsymbol{\mu}|\phi_0\rangle \langle v|v_0\rangle}{\langle \phi_i|\boldsymbol{\mu}|\phi_0\rangle \langle v_+|v_0\rangle \langle \phi_r|\mathbf{H}|\phi_i\rangle \langle v|v_+\rangle}, \tag{8.9.2}$$

   where $\langle \phi_r|\boldsymbol{\mu}|\phi_0\rangle$ and $\langle \phi_i|\boldsymbol{\mu}|\phi_0\rangle$ are the electronic transition moments for transitions from the ground state ϕ_0 in its vibrational level v_0 into the resonance and continuum states, respectively.[†] We neglect here the contribution to q of the principal part because, in the case of autoionization, the matrix elements generally vary slowly with E.

Figure 8.9 reproduces Hopfield's Rydberg series in the N_2 absorption spectrum, consisting (in absorption) of a series with a very large q and an apparent "emission" series with $q \simeq 0$. This means that the bound Rydberg states of the

[†] Since ϕ_i is an energy-normalized continuum function, q is dimensionless, as is ϵ.

latter series have near-zero transition moments from the ground state of the N_2 molecule. This series appears because its intensity is borrowed from the continuum [see Raoult, *et al.*, (1983) for a theoretical interpretation, which has been revised by Ueda, *et al.*, (1993)]. Such transitions into states with zero nominal transition moments will appear less strongly below the ionization limit, since they can only borrow intensity from nearby discrete states (see Chapter 6).

In the preceding sections, the case of one resonance isolated from the others and interacting with a single continuum of a specific nature has been discussed. A brief treatment of more general situations follows.

If the resonance state interacts with several continua, Eq. (8.9.1) remains valid provided that one defines q by summing over each of these continua in Eq. (8.9.2). In the case where multiple continua are involved – for example, those corresponding to all vibrational levels of a specific electronic state of the ion – it can be shown (Smith, 1970) that the q and Γ parameters become independent of the vibrational quantum number of the Rydberg state. Note that the lineshape asymmetry parameter, q, does not depend on the quantum number n; in contrast, the width Γ depends on $(n^*)^{-3}$.

This generalized form of Eq. (8.9.1) can be applied only when a *single discrete state* interacts with one or several continua *that do not interact with each other*. This situation is typical for a predissociated state. Unfortunately, it does not often occur for an autoionized Rydberg state. When there are several resonances that have large widths relative to their separations, Fano's formula must be replaced by the formula given by Mies (1968):

$$\sigma = \sigma_d + \sigma_i \frac{\left[1 + \sum_n (q_n/\epsilon_n)\right]^2}{1 + \left(\sum_n 1/\epsilon_n\right)^2}, \tag{8.9.3}$$

where q_n and ϵ_n are defined as previously for a single resonance, n. This formula takes into account interference between several near-degenerate resonances.

Typically, each resonance is coupled to several continua, and these continua interact with each other. In the case where two continua are coupled by an interaction I [I is a dimensionless parameter analogous to that in Eq. (8.4.4)], the width of the resultant Lorentzian peak is given by the same formula as in Eq. (7.12.7) (Beswick and Lefebvre, 1973):

$$\Gamma = \frac{\Gamma_1 + \Gamma_2}{1 + \pi^2 I^2}, \tag{8.9.4}$$

where Γ_1 and Γ_2 represent the widths arising from each independent continuum. This formula is valid regardless of whether the interacting continua belong to different electronic states of the ion or correspond to ionization and dissociative continua.

In reality, one encounters overlapping multiple resonances interacting with multiple continua. The analytical expressions given here represent an idealized situation. Several examples involving interferences between different autoionization processes [for example, in H_2, interference between rotational and vibrational autoionization (Fig. 8.12); in N_2, interference between vibrational and electronic autoionization (Fig. 8.14)] have been discussed. Multichannel quantum defect theory (MQDT) is capable of treating such problems without introducing the concept of a width (Dill and Jungen, 1980).

The basic concepts of MQDT are unfamiliar to most molecular spectroscopists. The crucial ideas of *frame transformation* and *phase shift* are briefly explained in the next two paragraphs.

In this chapter on autoionization processes several examples have been given where the members of a Rydberg series follow one Hund's case when the principal quantum number, n, is small and follow another Hund's case when n is large. For spin-orbit autoionization, the small- and large-n Rydberg states are well represented by Hund's cases (a) and (c), respectively. For rotational autoionization (of H_2, for example), the small- and large-n Rydberg states are well represented by Hund's cases (b) and (d), respectively. One of the basic concepts of MQDT is that the space in which the outer (Rydberg) electron moves may be divided into two parts: the *inner region* near the nuclei, where the maximum in the radial probability distribution for a low–n Rydberg orbital occurs, and the *outer region*, far from the core, where large -n Rydberg orbitals have their maximum radial probability. In the inner region the Rydberg electron is strongly affected by the field of the molecule; exchange interactions between the Rydberg and core electrons are particularly important. In the outer region, exchange effects become negligible and the Rydberg electron becomes weakly coupled to the core electrons; the molecular ion-core appears to the outer electron as a definite $|\Lambda^+, S^+, J^+, \Omega^+\rangle$ [case (a) inner region $\rightarrow$ case (c) outer region] or $|\Lambda^+, S^+, N^+, J^+\rangle$ [case (b) inner region $\rightarrow$ case (d) outer region] state. Thus, the appropriate Hund's coupling case depends on the instantaneous coordinate of the electron in the Rydberg orbital. When the electron moves from the inner to the outer region, a *frame transformation* is performed to adapt the total electronic wavefunction (core plus Rydberg electron) to the appropriate Hund's case.

A second crucial idea of MQDT is the relationship between the *quantum defect, a,* of a Rydberg series and the *phase shift* of the radial part of the outer electron wavefunction. At large r, the radial wavefunction of the external electron would be a pure Coulomb function if the potential were of the hydrogenic atom form

$$V(r) = -eZ/r.$$

The asymptotic form of a Coulomb function is a sine function. For a nonhydrogenic atom or any molecule, the external electron wavefunction is more complicated than a Coulomb function. However, at large r this wavefunction can be well approximated by a Coulomb radial function with an asymptotic form that is shifted in phase by ϕ with respect to the hydrogenic atom wavefunction. Figure 8.19 illustrates the effect of this phase shift on the Coulomb wavefunction of a nonhydrogenic atom. The phase shift summarizes the net effect of all interactions experienced by the external electron when it enters the inner region. These interactions are strong relative to the binding energy of the external electron. Therefore the phase shift, ϕ, varies slowly as the energy of the outer electron is increased, even when an ionization threshold is crossed and the electron energy changes from negative (bound electron) to positive (continuum electron). For a bound electron, the quantum defect, a, can be related to the phase shift, ϕ, of the radial wavefunction by

$$\phi = \pi a,$$

and ϕ is, in general, nearly independent of n. For a molecule, ϕ can depend on internuclear distance, and it is the R–dependence of $a(R)$ that is the critical factor controlling vibrational autoionization (see Section 8.6).

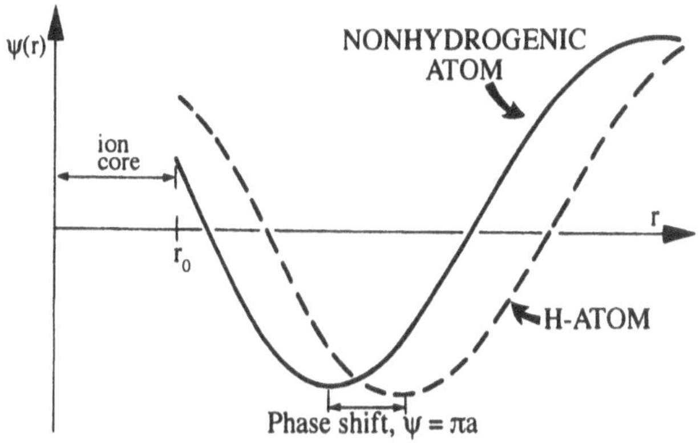

Figure 8.19: Diagram representing the phase shift induced by the electric field of the core ion in a nonhydrogenic atom; $r > r_0$ denotes the asymptotic region. [From Fano (1975).]

MQDT uniformly treats all members of a Rydberg series with given $l^{\dagger}$, λ, and v and the adjoining continuum to which this series converges as a single entity, a "channel." Based on relationships such as Eq. (8.4.4) and Eq. (8.4.6), MQDT replaces separate treatments of the many interactions between individual states by a global treatment of interchannel interactions. The underlying unity of the phenomena of perturbations, predissociation, and autoionization, which has been the central theme of this book, appears to be optimally expressed by MQDT (Fano, 1970; Jungen and Atabek, 1977; Jungen, 1982; Giusti-Suzor and Jungen, 1984; Green and Jungen, 1985; Jungen, 1996).

8.10 Influence of Autoionization on ZEKE Peak Intensities

As noted in Section 8.1.2, numerous intensity anomalies are observed in ZEKE spectra. Most of them are due to autoionization. For the effects of rotational, vibrational, and electronic autoionizations, we refer to the work of Merkt and

† l can never be a rigorously good quantum number in a molecule.

Softley (1993). Here we will analyze the effect of spin-orbit autoionization on ZEKE line intensities, for the example of HCl, where the ion-core ground state is a $^2\Pi_i$ state.

In photoionization, the final states correspond to the ionization continua of the ion state, i.e. states consisting of an ion plus an electron. In the case of a doublet ion state, when an electron of defined $\epsilon l\lambda$ enters the core, it gives rise to singlet and triplet states. Since these continuum singlet and triplet states are degenerate in energy, the mixing of the singlet and triplet states is complete (see Fig. 3.13 and Lefebvre-Brion, *et al.*, 1985). For example, if the continuum electron is $\epsilon s\sigma$, the continuum associated with the substate $X^2\Pi_{1/2}$ is $2^{-1/2}\left({}^1\Pi_1 + {}^3\Pi_1\right)$, and that associated with $X^2\Pi_{3/2}$ is $2^{-1/2}\left({}^1\Pi_1 - {}^3\Pi_1\right)$.

If the initial state in a photoionization transition is the $X^1\Sigma^+$ ground state, the transition moments into the continuum states are:

$$\left\langle X^1\Sigma^+|\mu|\left(X^2\Pi_{1/2}+\epsilon s\sigma\right)\right\rangle = 2^{-1/2}\left\langle X^1\Sigma^+|\mu|{}^1\Pi_1\right\rangle \quad (8.10.1)$$

$$\left\langle X^1\Sigma^+|\mu|\left(X^2\Pi_{3/2}+\epsilon s\sigma\right)\right\rangle = 2^{-1/2}\left\langle X^1\Sigma^+|\mu|{}^1\Pi_1\right\rangle . \quad (8.10.2)$$

The transition intensities at the $X^2\Pi_{1/2}$ and $X^2\Pi_{3/2}$ thresholds are equal, if the ion-core is at the Hund's case (a) limit. Experimentally, in conventional photoelectron spectra from the $X^1\Sigma^+$ ground state of HCl, the intensity ratio ${}^2\Pi_{1/2}/{}^2\Pi_{3/2}$ is equal to 1.06 ± 0.05 (Yencha, *et al.*, 1989). This slight deviation from 1 can be explained because the $^2\Pi$ state is intermediate between Hund's cases (a) and (b) (Wang and McKoy, 1991).

If the initial state is a Rydberg state, it is well known that such a state, built on a $^2\Pi$ ion-core, is neither a pure singlet nor triplet state as is the ground state. For example, the Rydberg configuration $(\pi)^3 p\pi$ gives rise to a $^1\Delta$ and a $^3\Delta$ state. Due to the strong spin-orbit interaction of the core $\left[A(\mathrm{HCl},{}^2\Pi) = -650\mathrm{cm}^{-1}\right]$, the observed states are a mixture of singlet and triplet states: the nominal singlet state, denoted by $`1'\Lambda_\Omega$, is

$$\left\langle F`1'\Delta_2\right\rangle = \alpha\,{}^1\Delta_2 + \beta\,{}^3\Delta_2 \quad (8.10.3)$$

and the nominal triplet, denoted by $`3'\Lambda_\Omega$ is

$$\left\langle f`3'\Delta_2\right\rangle = \beta\,{}^1\Delta_2 - \alpha\,{}^3\Delta_2 \quad (8.10.4)$$

where α and β are mixing coefficients. The photoionization cross section may be expressed as

$$\left\langle F`1'\Delta_2|\mu|(X^2\Pi_{1/2}+\epsilon s\sigma)\right\rangle = \frac{\alpha}{\sqrt{2}}\left\langle {}^1\Delta_2|\mu|{}^1\Pi_1\right\rangle + \frac{\beta}{\sqrt{2}}\left\langle {}^3\Delta_2|\mu|{}^3\Pi_1\right\rangle \quad (8.10.5)$$

$$\left\langle f`3'\Delta_2|\mu|(X^2\Pi_{1/2}+\epsilon s\sigma)\right\rangle = \frac{\beta}{\sqrt{2}}\left\langle {}^1\Delta_2|\mu|{}^1\Pi_1\right\rangle - \frac{\alpha}{\sqrt{2}}\left\langle {}^3\Delta_2|\mu|{}^3\Pi_1\right\rangle . \quad (8.10.6)$$

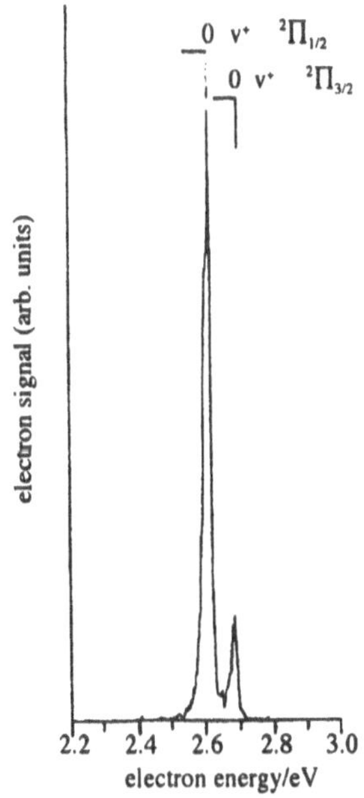

Figure 8.20: PES of HCl arising from the two-photon excitation $^1\Delta_2(v' = 0, J' = 2) \leftarrow X^1\Sigma^+(v'' = 0, J'' = 2)$, followed by one-photon ionization to the $^2\Pi$ ionic ground state (from de Beer, *et al.*, 1990).

Because the transition moments between triplet states can be, especially for Rydberg states, assumed to be nearly equal to those between the corresponding singlet states, and if $\beta = \alpha$, transitions originating from the $F^{'1'}\Delta_2$ state can access only the $^2\Pi_{1/2}$ component of the cation and those from the $f^{'3'}\Delta_2$ state can access only the $^2\Pi_{3/2}$ component. In fact, in HCl, the singlet-triplet spin-orbit mixing within the $4p\pi$ configuration is not complete. Starting from the $F^{'1'}\Delta_2$ state, the conventional photoelectron spectra of de Beer, *et al.*, 1993 show a $^2\Pi_{1/2}/{}^2\Pi_{3/2}$ ratio of about 6.5, which corresponds to a singlet-triplet mixing of 22% as estimated by Xie, *et al.*, 1991 (see Fig. 8.20).

The ZEKE-PFI spectrum, recorded from the HCl $F^1\Delta$ intermediate state, shows a completely different $^2\Pi_{1/2}/{}^2\Pi_{3/2}$ subband branching ratio, equal to 0.12 (de Beer, *et al.*, 1990 and Fig. 8.21, upper panel). A calculation, using the BOS formalism (see Section 8.1.1) with *ab initio* transition moments and a singlet-triplet mixing of 22% reproduces the ratio observed by conventional spectroscopy but not that in the ZEKE spectrum (Fig. 8.21, middle panel).

This opposite branching behavior observed in the ZEKE spectrum can be explained if one considers the effect of spin-orbit autoionization. The contribution of spin-orbit autoionization to the decay by autoionization (hence loss of those molecules to the ZEKE signal) of the high-n Rydberg states, during the delay time, differently affects the intensity of the two spin-orbit components.

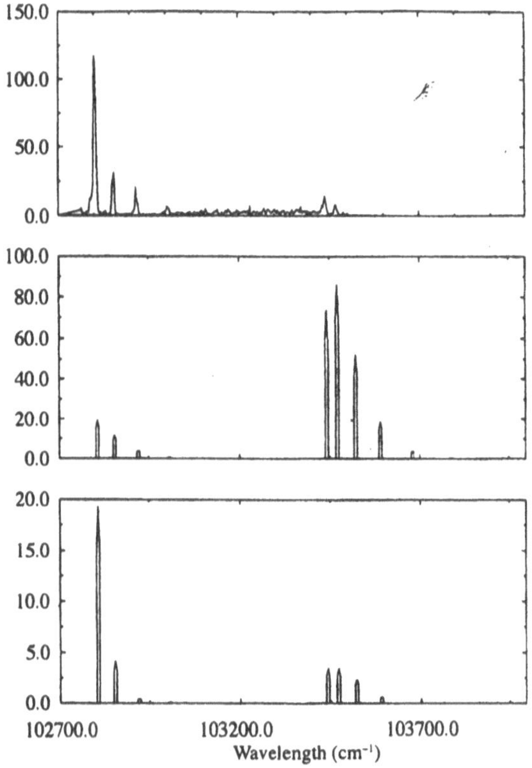

Figure 8.21: ZEKE-PFI spectrum resulting from ionization via the $P(3)$ rotational transitions associated with the $F^1\Delta_2(v'' = 0, J'' = 2e)$ state of HCl. Upper panel: observed (de Beer, *et al.*, 1993); middle panel: calculated for direct ionization; lower panel: calculated taking into account the population decay by spin-orbit and rotational autoionization after a delay of 200 ns (from Lefebvre-Brion, 1996)

Those Rydberg states converging to the $J^+ = 3/2$ first rotational level of $^2\Pi_{3/2}$ cannot decay because all autoionization channels are closed. Those states converging to the other rotational levels of $^2\Pi_{3/2}$ can decay in the lower J^+ levels of $^2\Pi_{3/2}$ only by what has improperly been called rotational autoionization (improper because it is due in part to spin-orbit interaction), with lifetimes for the d–complex ($n = 600$) between 100 ns and 10^{-6} s. However, the high–n Rydberg levels converging to $^2\Pi_{1/2}$ rotational levels can decay into the rotational levels of both $^2\Pi_{3/2}$ and $^2\Pi_{1/2}$ and their lifetimes are shorter than 10^{-6} s.

Consequently, after a delay time of 200ns, the decay for the $^2\Pi_{1/2}$ component is so strong that the $^2\Pi_{3/2}$ subband becomes more intense than the $^2\Pi_{1/2}$ subband, in agreement with experiment (Fig. 8.21, lower panel). Note that the small DC electric field present in this ZEKE experiment, also causes a lifetime lengthening by l–mixing. In the present calculations, this l–mixing effect is taken into account by using an effective value of n, the n–value of the Rydberg states ionized by the pulsed electric extraction field (Lefebvre-Brion, 1996). A more rigorous calculation has been completed in which the l–mixing due to the DC field was introduced (Lefebvre-Brion, 1997). Cockett (1995) has proposed a similar explanation of the intensity anomalies in the ZEKE spectrum of I_2.

Many experimental and theoretical studies have been published on this subject, for example the decay kinctics of the ZEKE intensity due to rotational autoionization in NO (Remacle and Vrakking, 1998).

8.11 Photoelectron Angular Distribution, Photoion Alignment, and Spin Polarization

A simple measurement of thc total photoionization cross section (isotropic sample, cross section measured at a specified photon energy, integrated over all photoejection angles, without specification of the internal state of the photoion, without determination of m_s of the ejected electron) contains no information about the distribution of l, m_l–values of the ejected electron. However, measurable properties of the photoionization event can provide information about the mechanism of the photoionization process. The frequently measured quantities included β, $\mathcal{A}_0^{(2)}$, and **A**. The β quantity describes the angular distribution of the photoelectrons and is defined analogously to the β for photodissociation (see Section 7.2.4), $\mathcal{A}_0^{(2)}$ is the alignment ($|M_N|$ distribution) of the photoion. **A** (not to be confused with the spin-orbit constant) or alternatively $\overline{P}$, is the spin-polarization of the ejected electron, which is relevant when the photoion has nonzero spin.

8.11.1 Photoelectron Angular Distribution

The general expressions given by Yang (1948), Eqs. (7.2.13) and (7.2.17), provide a valid description for all photofragmentation products, atoms in the case of photodissociation, electrons in the case of photoionization. In photoionization, the masses of the photofragments are in the ratio m/M.

For photoionization by linearly polarized radiation,

$$\frac{d\sigma}{d\Omega} = \frac{\sigma}{4\pi}[1 + \beta P_2(\cos\Theta)] \tag{8.11.1}$$

where Θ is the photoejection angle of the departing electron relative to the linear polarization direction of the photoionizing radiation and β is the asymmetry parameter.

Recall, as for photodissociation, the requirement that the cross section be positive defines the limiting values of β:

(i) $\beta = 2$ means the electron is ejected preferentially along the polarization direction of the photoionizing radiation,

(ii) $\beta = -1$ means that the electron is ejected preferentially in a plane perpendicular to the polarization direction,

Table 8.3: The asymmetry parameter, β, for each angular momentum transfer, j_t.

j_t	l(odd)	l(even)	$\beta(j_t)$	parity
0	1		2	PF
1	1		-1	UF
1		0	0	PF
1		2	1	PF
2	1		1/5	PF
2	3		4/5	PF
2		2	-1	UF

(iii) $\beta = 0$ means that the photoelectron is ejected isotropically.

The total angular momentum (both magnitude and projection) of the reaction

$$AB(\Lambda_0) + h\nu \rightarrow AB^+(\Lambda_c) + e^-(l) \tag{8.11.2}$$

is conserved. Since Λ_0 and Λ_c are (usually) well defined components of a (usually) ill defined total electron orbital angular momentum, angular momentum conservation does not significantly constrain the l–values of the photoelectron. However, there are propensity rules and limiting cases, as follows.

Several methods have been proposed to predict β when only a few values of l are assumed to contribute to the value of β.

(i) *The method of "transferred" angular momentum.* Dill and Fano, (1972) have formulated a simple model, which is well illustrated by the example of H_2. They have shown that the value of β depends only on j_t, the angular momentum transferred between the photon (which has 1 unit of angular momentum) and the molecule

$$j_t = |\Lambda_c - \Lambda_0| = |\vec{l} - \vec{1}|. \tag{8.11.3}$$

Each value of j_t corresponds to an *a priori* known value of β, $\beta(j_t)$ (see Table 8.3) (Dill and Fano, 1972). The measured value of β is the cross section-weighted average of $\beta(j_t)$ values,

$$\beta = \frac{\sum_{j_t} \beta(j_t)\sigma(j_t)}{\sum_{j_t} \sigma(j_t)}. \tag{8.11.4}$$

Of equal importance to j_t, the transferred angular momentum, is the concept of the parity, π_t, of the photoionization transition (Dill and Fano, 1972, Dill, 1973). π_t is defined by

$$\pi_t = (-1)^{j_t+l+1}$$

where $\pi_t = +1$ and -1 correspond respectively to *parity favored* and *parity unfavored* transitions. Since the selection rule for single-orbital electric dipole transitions in atoms is $\Delta l = \pm 1$, molecular transitions in which j_t and l are both even or odd are parity unfavored whereas those where j_t, l are even,odd or odd, even are parity favored. Vector addition in Eq. (8.11.3) restricts j_t to $j_t = l$ (parity unfavored, UF) and $j_t = l \pm 1$ (parity favored, PF). When there are two parity favored values of l for the same value of j_t (i.e., $j_t \neq 0$), then there will be interference between the two values of $l = j_t \pm 1$,

$$\beta^{PF}(j_t) = \frac{1}{\sigma(j_t)} \{(j_t - 1)|R_{j_t-1}|^2 + (j_t + 2)|R_{j_t+1}|^2 - 6\,[j_t(j_t+1)]^{1/2}\mathrm{Re}[R_{j_t-1}R_{j_t+1}]\} \quad (8.11.5)$$

$$\sigma(j_t) = (2j_t + 1)[|R_{j_t-1}|^2 + |R_{j_t+1}|^2] \quad (8.11.6)$$

where $R_{j_t \pm 1}$ is the radial transition moment for the photoionization transition that produces $l = j_t \pm 1$.

According to Chang (1978), a photoionization transition that originates from a Σ^+–state and terminates in a Σ^+ ion–state is parity favored. In contrast, a photoionization transition from a Σ–state to a Π ion-state comprises a mixture of parity favored and parity unfavored transitions. Consequently, the β–values for photoionization to a Π ion-state are intermediate between the extreme values of -1 and 2. See Table 1 of Hancock and Samson (1976) for experimental illustration of predictions for β based on j_t and π_t.

It has been observed that β values vary for photoionization transitions that originate from various vibrational-rotational levels of the parent molecule. For example, the initial (N_0) and final (N^+) rotational quantum numbers determine j_t

$$j_t = |N^+ - N_0| \quad (8.11.7)$$

For homonuclear molecules, the initial and final rotational states must belong to the same ortho/para nuclear permutation symmetry. For example, in the $H_2(X^1\Sigma_g^+) \rightarrow H_2^+(X^2\Sigma_g^+) + e^-$ photoionization transition, the even N_0 and N^+ levels are *para* ($I = 0$ nuclear spin singlets, statistical weight 1) whereas the odd N_0 and N^+ levels are *ortho* ($I = 1$ nuclear spin triplets, statistical weight 3). The allowed transitions are $N^+ - N_0 =$ even:

$$N^+ - N_0 = 0 \text{ (Q branch)}, j_t = 0, l = 1, \beta(j_t = 0) = 2$$
$$N^+ - N_0 = 2 \text{ (S branch)}, j_t = 2, l = 1, \beta(j_t = 2) = 1/5$$

where the $\beta(j_t)$ values are from Table 8.3. For a rotationally unresolved vibrational $v^+ - v$ band of rotationally cold para–H_2, Eq. (8.11.4) gives

$$\beta_{v^+} = \frac{2\sigma(v^+, N^+ = 0) + 0.2\sigma(v^+, N^+ = 2)}{\sigma(v^+, N^+ = 0) + \sigma(v^+, N^+ = 2)} \approx 2 \quad (8.11.8)$$

This $\beta_{v^+} \approx 2$ prediction is verified experimentally (Dehmer, *et al.*, 1992).

Measured values of the asymmetry parameter, β, for autoionization resonances exhibit considerable variation with the frequency of the photoionizing radiation. However, the shape of a $\beta(\nu)$ curve can be quite different from that of the photoionization cross section, $\sigma(\nu)$. The extremum in $\beta(\nu)$ is often displaced from that in $\sigma(\nu)$. As for the Fano index, q (see Section 8.9), the extremum value of β is the same for all autoionization peaks within the same Rydberg series (except when an interaction with an interloping series disrupts the nominally n–independent pattern).

Approximate values of β for autoionization resonances have been predicted, using the method of transferred angular momentum, for H_2 (Dill, 1972) and O_2 (Parr, *et al.*, 1998).

(ii) *The partitioning method.*

Another approach consists of expressing the β parameter as a sum over the partial waves of the ejected electron, of products of an energy-independent geometrical factor and an energy-dependent dynamical factor. This is the partitioning scheme of Thiel (1982 and 1983). This scheme is based on the expectation that, in the case of a resonance (shape or autoionization), only one contribution is dominant. In such a case, β reduces to a purely geometrical factor. The geometrical factors, G_l, are listed in Table 8.4 for the case where the ionic state is Σ and the partial wave is $l\sigma$.

As an example, for N_2^+ $X^2\Sigma_g^+ + e^- \leftarrow N_2$ $X^1\Sigma_g^+$ ionization, in the energy region of the well known shape resonances (see Section 8.1.3), the photoelectron partial wave is mainly $f\sigma_u(l = 3)$, consequently β is expected to be close to 0.53. A Cooper minimum has a strong influence on β: as analyzed by Thiel (1984), the β parameter exhibits a minimum not only because some transition moments vanish (due to the Cooper minimum) but also because the $s\sigma(l = 0)$ partial wave dominates in the region of the Cooper minimum.

However, it is often necessary to consider off-diagonal interactions between different partial waves. Recall (see Section 8.1.1) that, to first approximation, the vibrational photoionization intensities for *direct photoionization* are proportional to the Franck-Condon factors (e.g. $\langle v_X|v_X^+\rangle^2$) between the initial state vibrational level (e.g. CO $X^1\Sigma^+ v_X$) and the $v^+ = 0, 1, 2, \ldots$ vibrational levels of the photo-ion electronic state (e.g., CO^+ $X^2\Sigma^+$ v_X^+). For *electronic autoionization*, the final state vibrational branching fractions are proportional to the Franck-Condon factors (e.g. $\langle v_X^+|v_A^{+2}\rangle$) between the resonant autoionized vibrational states (e.g. CO^+, $A^2\Pi$, v_A^+) and the final photo-ion state (e.g., CO^+,$X^2\Sigma^+$ v_X^+) (Smith, 1970).

An example, from the photoionization spectrum of the CO $X^1\Sigma^+$ $v = 0$ state, illustrates how photo-ion vibrational branching ratios and β values sample different aspects of the photoionization dynamics. Figure 8.22 displays how interactions between autoionized high–n members of Rydberg series converging to the CO^+ $A^2\Pi$ state and the $3p\pi$ $v' = 0$ Rydberg state that belongs to a series converging to the $B^2\Sigma^+$ state affect the $X^2\Sigma^+$ $v^+ = 0, 1$ and 2 vibrational branching fractions and the resonance structure of the β parameter. The agreement between experiment (heavy line) and theory (light line) is satisfactory for the $v^+ = 0, 1$ and 2 vibrational branching fractions but unsatisfactory for β.

Table 8.4: The geometrical factors for the β parameter for a Σ ion-core (from Table 1 of Thiel, 1983).

l	0	1	2	3
G_l	0	0.8	0.57	0.53

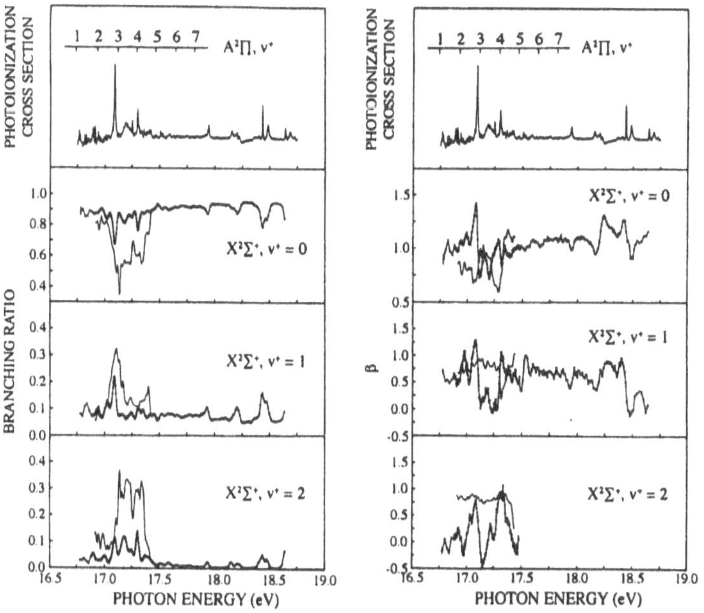

Figure 8.22: Photoion vibrational branching fractions (left side) and photoelectron asymmetry parameters (right side) for CO^+ $X^2\Sigma^+$ $v^+ = 0-2$ produced by photoionization of CO $X^1\Sigma^+$ $v = 0$ (Hardis, *et al.*, 1988). The β parameters are obtained from the photoelectron spectra measured simultaneously at two observation angles. The heavy lines are the experimental values and the light lines are the MQDT results from Leyh and Raşeev (1988), convoluted to the experimental energy resolution. The peak at 17.02 eV corresponds to the $3p\pi$ $v^+ = 0$ Rydberg state converging to the CO^+ $B^2\Sigma^+$ state.

The partitioning scheme of Thiel (1982) has been extended to molecules with a $^2\Pi$ ion-core (Raseev, *et al.*, 1987). In this case, the electrons can be ejected from either the $^2\Pi_{3/2}$ or the $^2\Pi_{1/2}$ sublevel. For the same electron kinetic energy:

$$\beta\left({}^2\Pi_{3/2}\right) = \beta\left({}^2\Pi_{1/2}\right). \tag{8.11.9}$$

For shape resonances (see Fig. 8.6), as for the partial vibrational cross section, the photoelectron β parameter is very sensitive to the photo-ion vibrational state (Dehmer, *et al.*, 1979).

In a resonance enhanced multiphoton ionization (REMPI) spectrum the differential cross section may be expanded in Legendre polynomials, but this expansion does not collapse to a single term. For $(n+m)$ REMPI:

$$\frac{d\sigma}{d\Omega} = \frac{\sigma}{4\pi} \sum_{L=0}^{2(n+m)} \beta_L P_L(\cos\Theta), \tag{8.11.10}$$

where $P_L(\cos\Theta)$ is the L–th Legendre polynomial. For $(1+1')$ photoionization, the angular distribution must thus be expressed in terms of two parameters: β_2 and β_4.

Molecular alignment (see Section 8.11.2) can be probed by measurement of the circular dichroism in the photoelectron angular distribution (CDAD). CDAD spectra are obtained by taking the difference between angle-resolved photoelectron spectra for left and right circularly polarized light (Dubs, *et al.*, 1986).

The study of the photoelectrons produced in the $(1+1')$ REMPI spectrum of NO via several rotational levels of the $A^2\Sigma^+$ state has been called a "complete experiment" because the CDAD and the angular distribution measurements were used to extract twelve parameters, corresponding to the magnitudes and phase shifts of the transition moments to all of the important $l\lambda$ partial waves. The measured parameters were compared to corresponding theoretical values (Reid and Leahy, 1995, and Allendorf, *et al.*, 1989).

8.11.2 Photoion Alignment

In photoionization, analogous to photodissociation, the photofragment, here the photoion, can be aligned. The alignment of the photoion, $\mathcal{A}_0^{(2)}$, (Greene and Zare, 1982) may be measured if the molecular photoion is produced in a fluorescing excited electronic state. The polarization of the fluorescence determines the alignment of the nascent photoion. The alignment of the J^+ photoion rotational level is defined as

$$\mathcal{A}_0^{(2)}(J^+) = \frac{\sum_{M_J^+}[3(M_{J^+})^2 - J^+(J^+ + 1)]\sigma(J^+, M_{J^+})}{J^+(J^+ + 1)\sum_{M_{J^+}} \sigma(J^+, M_{J^+})} \tag{8.11.11}$$

where $\sigma(J^+, M_{J^+})$ is the total cross section for production of the M_{J^+} sublevel of the J^+ photoion rotational level.

Kinematic constraints for photoionization of a linear molecule limit the alignment parameter to the range

$$-1 \leq \mathcal{A}_0^{(2)} \leq 4/5, \tag{8.11.12a}$$

which is reduced to

$$-2/5 \leq \mathcal{A}_0^{(2)} \leq 4/5, \tag{8.11.12b}$$

in the limit of large J^+. In the special case where $\sigma(J^+, M_{J^+})$ is independent of M_{J^+} (all $2J^+ + 1$ M_{J^+}–components formed with equal probability), then

$$\sum_{M_{J^+}} 3(M_{J^+})^2 = \frac{3(2J^+ + 1)J^+(J^+ + 1)}{3} \quad (8.11.13)$$

$$\sum_{M_{J^+}} 1 = 2J^+ + 1 \quad (8.11.14)$$

thus

$$\mathcal{A}_0^{(2)}(J^+) = \frac{J^+(J^+ + 1)[2J^+ + 1 - (2J^+ + 1)]\sigma(J^+)}{J^+(J^+ + 1)\sigma(J^+)} = 0 \quad (8.11.15)$$

In the high–J^+ limit, the value of $\mathcal{A}_0^{(2)}$ is directly determined from the fluorescence polarization, P,

$$\mathcal{A}_0^{(2)}(J^+) = -8P/(3 - P) \quad (8.11.16)$$

where

$$P = \frac{1 - R}{7 + 13R} \quad (8.11.17)$$

and

$$R = \frac{\sigma_\Pi}{\sigma_\Sigma}$$

is the cross section ratio for the separate Π and Σ contributions to the total cross section. If there is a contribution only from σ_Σ, then the J^+–averaged photoion alignment is

$$\mathcal{A}_0^{(2)} = -2/5\ [P = 1/7]. \quad (8.11.18a)$$

If there is a contribution only from σ_Π, then

$$\mathcal{A}_0^{(2)} = 1/5\ [P = -1/13]. \quad (8.11.18b)$$

For example, for photoionization of the N_2 $X^1\Sigma_g^+$ state to yield N_2^+ $B^2\Sigma_u^+$, the $B^2\Sigma_u^+ \rightarrow X^2\Sigma_g^+$ fluorescence polarization, at the threshold for production of the $B^2\Sigma_u^+$ state, is observed to be

$$P = 0.079,$$

from which

$$R = 0.22$$

is determined (Guest, *et al.*, 1983). This implies that the $\epsilon s\sigma$ and $\epsilon d\sigma$ photoelectron channels $[N_2^+\ B^2\Sigma_u^+\epsilon l\sigma\ ^1\Sigma_u^+]$ are dominant over the $\epsilon d\pi$ channel $[N_2^+\ B^2\Sigma_u^+\epsilon l\pi\ ^1\Pi_u]$. Raoult, *et al.*, (1983) showed that it is necessary to introduce continuum$\sim$continuum interaction between all of the $N_2^+\epsilon l\lambda$ channels of the same symmetry to obtain agreement between MQDT calculations and experiment. Also, the influence of autoionization on the polarization of fluorescence is evident in O_2^+ A$\rightarrow$X fluorescence (Lefebvre-Brion, 1988).

8.11.3 Spin-Polarization

The polarization of the photoelectron spin in atomic and molecular photoionization processes encodes supplementary information about the relative magnitudes of electron transition moments. In atoms, the spherical symmetry permits complete characterization of the transition moments that contribute to the total excitation cross section for each resonance (Heinzmann, 1980). However, in molecules, the l–mixing caused by the absence of spherical symmetry makes it impossible to completely characterize a resonance excitation mechanism from partial cross sections, angular distribution, and spin-polarization measurements.

The information obtainable from photoelectron polarization measurements is reviewed, for both atoms and molecules, by Heinzmann and Cherepkov (1996). Even at non-relativistic excitation energy, photoelectrons can be spin-polarized (Fano, 1969). For $l \neq 0$ atoms, due to the spin-orbit splitting of the initial atomic and/or the final ionic state, photoelectrons are in most cases highly spin-polarized (up to 100%) when photoexcited with circularly polarized light. Analogous effects occur in molecular photoionization, but systematic studies have only been made for hydrogen halide molecules, HX. The electronic ground state of HX^+ is $X^2\Pi$.

The first prerequisite for measurement of photoelectron spin-polarization is the ability to separately detect the photoelectrons ejected from the different fine-structure levels (e.g., $^2\Pi_{3/2}$ and $^2\Pi_{1/2}$ for HX^+ $X^2\Pi$). When the molecule contains a heavy atom (e.g., large spin-orbit splitting), it becomes easier to use the electron kinetic energy to distinguish the photoelectrons ejected from the different fine structure channels. For spin-polarization analysis, the accelerated electron beam (20-120 keV) can be scattered by a thin gold foil in a Mott-detector. The spin-polarization is determined from the left-right (or up-down) asymmetry in the intensities of the scattered electrons (Heinzmann, 1978). Spin polarization experiments, however, are difficult because the differential spin-up/spin-down flux of photoelectrons is typically one thousandth that obtained when recording a total photoionization spectrum.

Similarly to the angular asymmetry parameter, β, the spin-polarization parameters have signed values and their expressions, as function of contributing transition moments, in principle permit determination of the relative signs of these transition moments. However, in contrast to the situation for atoms, the number of relevant unknown parameters exceeds the number of experimentally measurable quantities. Nevertheless, measurement of both resonance widths and spin-polarization parameters can considerably narrow the assignment possibilities. This is particularly true in the region between the two ion-core fine-structure thresholds, for example, $^2\Pi_{3/2}$ and $^2\Pi_{1/2}$ of an AB^+ $^2\Pi$ state.

The spin-polarization parameters have the property of having opposite signs for different fine-structure components if the photoelectrons have the same kinetic energy, for example:

$$\mathbf{A}(^2\Pi_{3/2}) = -\mathbf{A}(^2\Pi_{1/2}) \qquad (8.11.19)$$

where $\mathbf{A}$ specifies the degree of polarization of the total photoelectron current ejected by circularly polarized light,

$$\mathbf{A} = \frac{N\uparrow - N\downarrow}{N\uparrow + N\downarrow} \tag{8.11.20}$$

where $N\uparrow$ and $N\downarrow$ correspond respectively to electrons with $M_S = 1/2$ (α spin) or $M_S = -1/2$ (β spin) in a coordinate system where the $+Z$ axis is parallel to the light propagation direction if the light is right circularly polarized (or antiparallel if the light is left circularly polarized). Equation (8.11.19) explains why the spin-polarization is zero when the two fine-structure components of the ion are not resolved.

Recall Eq. (8.11.9), which specifies that the angular asymmetry parameter, β, does not exhibit this opposite sign behavior,

$$\beta(^2\Pi_{3/2}) = \beta(^2\Pi_{1/2}). \tag{8.11.21}$$

When the rotational structure in the photoionization spectrum is not resolved and the molecular ion-core state can be treated as belonging to Hund's case (a), the spin polarization parameter is

$$\mathbf{A}(^2\Pi_{3/2}) = \frac{(\sigma_\sigma - \sigma_\delta)}{2\sigma_\Pi} \tag{8.11.22}$$

where

$$\sigma_\Pi = \sigma_\sigma + \sigma_\pi + \sigma_\delta \tag{8.11.23}$$

is the total photoionization cross section into the $^2\Pi_{3/2}$ substate, and σ_λ is the contribution of the λ partial wave to the total cross section σ_Π.

For Rydberg series converging to an inverted $^2\Pi$ state, $^2\Pi_{3/2}$ below $^2\Pi_{1/2}$, which is the case for the hydrogen halides (the only example studied), an autoionized resonance belongs to a Rydberg series

$$\mathrm{AB}^+(^2\Pi_{1/2})ns\sigma \text{ or } nd\sigma$$

which is autoionized by the $^2\Pi_{3/2}$ ionization continuum. For this case the **A** parameter is negative relative to that of the $^2\Pi_{3/2}$ background. If the resonance belongs instead to $\mathrm{AB}^+(^2\Pi_{1/2})nd\delta$, **A** is expected to be positive.

If the rotational structure is resolved between the two $^2\Pi_{3/2}$ and $^2\Pi_{1/2}$ ionization thresholds, it is possible to assign definite ion-core rotational J^+–values to each autoionized resonance, and the resonances are then described by Hund's case (e). Their wavefunctions are explicitly known as linear combinations of the case (a) wave functions, due to the mixing by the rotational operator (the j–uncoupling operator, see Section 8.7, Eqs. (8.7.9) - (8.7.14)). Consequently, the resonances no longer have a well-defined λ–value [for example $(\mathrm{AB}^{+\,2}\Pi)d\delta\ ^1\Pi_1$ can be mixed with $(\mathrm{AB}^{+\,2}\Pi)d\pi\ ^1\Sigma_0^+)$] and the value of **A** cannot be predicted without calculations. Such a study has been performed for HBr (Irrgang, *et al.*, 1998).

8.12 Competition between Autoionization and Predissociation

8.12.1 Superexcited State Decay Pathways

As discussed in Section 8.2, superexcited states, AB^{**}, can decay by both autoionization and dissociation (more specifically, by predissociation). Decay by spontaneous fluorescence can be neglected for superexcited states because, generally, the predissociation or autoionization rates ($1/\tau_{nr} \sim 10^{12}$ to $10^{14} s^{-1}$) are much faster than the fluorescence rate ($1/\tau_r \lesssim 10^8 s^{-1}$). Only two examples of detected spontaneous fluorescence from superexcited states have been reported (for H_2, Glass-Maujean, *et al.*, 1987, for Li_2, Chu and Wu, 1988). The H_2 $D^1\Pi_u$ e–symmetry component is predissociated by an L–uncoupling interaction with the B' $^1\Sigma_u^+$ state (see Section 7.9 and Fig. 7.27). Since a $^1\Sigma_u^+$ state has no f–symmetry levels, the f–components of the $D^1\Pi$ Λ–doublets cannot interact with the $B^1\Sigma_u^+$ state and are not predissociated. The $v = 8$ level of the $D^1\Pi_u$ state, which lies just above the H_2^+ $X^2\Sigma_g^+$ $v^+ = 0$ ionization threshold, could in principle be autoionized (both e and f components) by the $X^2\Sigma_g^+$ $v^+ = 0$ $\epsilon\pi$ continuum. However, the $\Delta v = 1$ propensity rule for vibrational autoionization implies that the $v = 8$ level will be only weakly autoionized. Consequently, the nonradiative decay rate, $1/\tau_{nr}$, is slow only for the f–symmetry component of the $D^1\Pi_u$ $v = 8$ state. Thus, in the LIF spectrum of the $D^1\Pi_u - X^1\Sigma_g^+$ $8-0$ band, Q lines ($f \leftarrow e$ transitions) are present while R and P lines ($e \leftarrow e$ transitions) are absent in the LIF spectrum. Figure 8.23 shows how this D–X 8-0 band appears for various detection schemes [detection of H atom photofragments via Lyman–α excitation, LIF via VUV fluorescence, absorption, and H_2^+ ion-detection]. The H atom Lyman–α channel (see also Fig. 7.28) samples the predissociation rate, which is much faster for R and P than Q lines.

Several examples of competition between autoionization and predissociation, from the spectra of HCl and H_2, are discussed below.

Superexcited Rydberg states of HCl converging to the excited HCl^+ $A^2\Sigma^+$ state may be autoionized by the continuum of the $X^2\Pi$ state, predissociated by repulsive Rydberg states built on the dissociative HCl^+ $a^4\Pi$ ion-core state (see Section 7.11.1), and predissociated by the $V^1\Sigma^+$ ion-pair state that dissociates to H^+ and Cl^-:

$$(A^2\Sigma^+)nl\sigma \rightarrow (X^2\Pi)\epsilon l\pi \rightarrow (X^2\Pi) + e^- \tag{8.12.1a}$$

autoionization,

$$\rightarrow (a^4\Pi)nl\lambda \rightarrow H + Cl^*(3p^4(^3P)4s) \tag{8.12.1b}$$

predissociation into neutral atoms (Cl excited),

$$\rightarrow H^* + Cl \tag{8.12.1c}$$

predissociation into neutral atoms (H excited), and

$$\rightarrow V^1\Sigma^+ \rightarrow H^+ + Cl^- \qquad (8.12.1d)$$

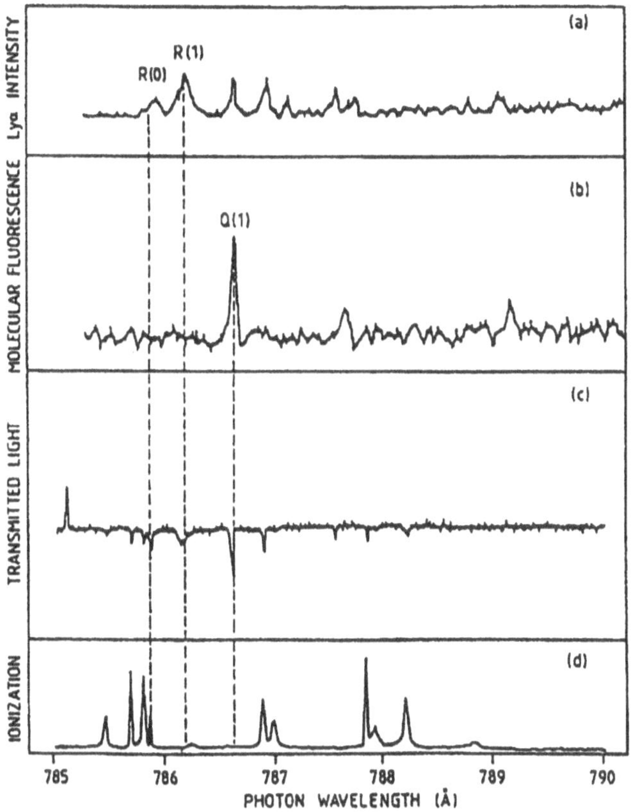

Figure 8.23: Spectra illustrating various decay pathways for the H_2 $D^1\Pi_u$ $v = 8$ level: (a) photodissociation monitored by Lyman-α excitation of H atom photofragment, (b) VUV fluorescence detected LIF spectrum, (c) direct absorption, and (d) direct ionization to H_2^+ $X^2\Sigma_g^+$ $v^+ = 0$. [(a) and (b) from Glass-Maujean, *et al.*, 1987; (c) and (d) from Dehmer and Chupka, 1976; composite figure from Lefebvre-Brion, 1991.]

predissociation into atomic ions (see Fig. 8.24). These different decay pathways may be selectively detected by fluorescence from Cl* ($3p^4(^3P)4s$ ^2P) (White, *et al.*, 1987), fluorescence from $H^*(n = 2)$ (Frohlich and Glass-Maujean, 1990), detection of Cl^- (Yencha, *et al.*, 1993), and detection of HCl^+ $X^2\Pi$ (Frohlich and Glass-Maujean, 1990). Spectra which display these four distinct photofragmentation channels are shown, on a common wavelength (excitation energy) scale, in Fig. 8.25. The peaks in the dissociation cross sections (spectra b, c, and d) do not correspond to any features in the photoionization spectrum (spectrum a) (Lefebvre-Brion and Keller, 1989). The peaks in the photoionization spectrum (Fig. 8.25 (a)) correspond to $(A^2\Sigma^+)nd\pi\ ^1\Pi$ Rydberg states, which are electronically autoionized in the $(X^2\Pi)^1\Pi$ continuum. The peaks detected in H* or Cl* fluorescence (Fig. 8.25 (b) and (c)) correspond to $(A^2\Sigma^+)^1\Sigma^+$ Rydberg states, which are predissociated by spin-orbit interaction with the repulsive $(a^4\Pi)4s\sigma\ ^3\Pi_0$ state. However, these $(A^2\Sigma^+)^1\Sigma^+$ Rydberg states are only weakly autoionized by the $(X^2\Pi)^1\Sigma^+$ continuum and are therefore not directly detectable in the autoionization spectrum.

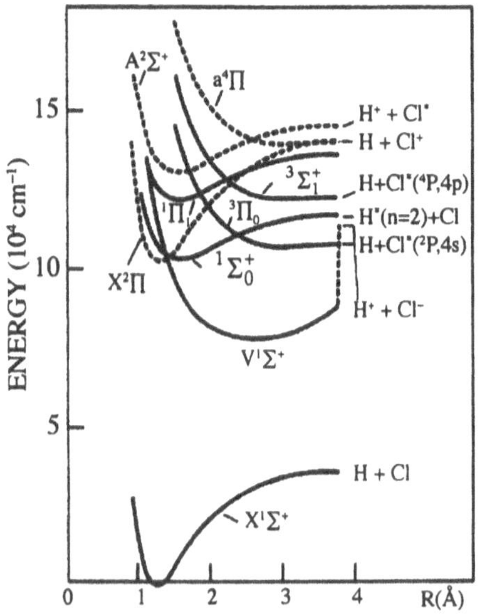

Figure 8.24: Potential energy curves of HCl (solid curves) and HCl^+ (dashed curves). It is clear that the potential curves for Rydberg states converging to HCl^+ $A^2\Sigma^+$ are intersected by potential curves dissociating to H^++ Cl^-, H+Cl*$(3p^44s\ ^2P)$, H*$(n=2)$+ Cl, and H+Cl*$(3p^44p\ ^4P)$. The potential curve with non-labeled asymptote is the $(A^2\Sigma^+)3d\pi\ ^1\Pi$ state (from Lefebvre-Brion and Suzor-Weiner, 1994).

The total width, Γ, is the sum of partial widths, which can be calculated but not observed separately. Only the total width can be observed experimentally. This width does not depend on whether the line is observed in an absorption, photoionization, photodissociation, or emission spectrum because the width (or the lifetime) is characteristic of a given state (or resonance). In contrast, the peak profile can have different line shapes in different channels: the line profile, q, is dependent on the excitation and decay mode (see Sections 7.9 and 8.9). For predissociation into H+Cl*, the transition moment from the $X^1\Sigma^+$ state to the $^3\Pi$ (or $^3\Sigma^+$) predissociating state is zero, consequently $q = \infty$ and the lineshape is Lorentzian. In contrast, the ratio of the two transition moments for transitions to the $^1\Sigma^+$ continuum of the $X^2\Pi$ state and to the $(A^2\Sigma^+)^1\Sigma^+$ Rydberg states leads to $q \approx 0$ for the autoionized peaks (see Fig. 8.26) (Lefebvre-Brion and

Keller, 1989).

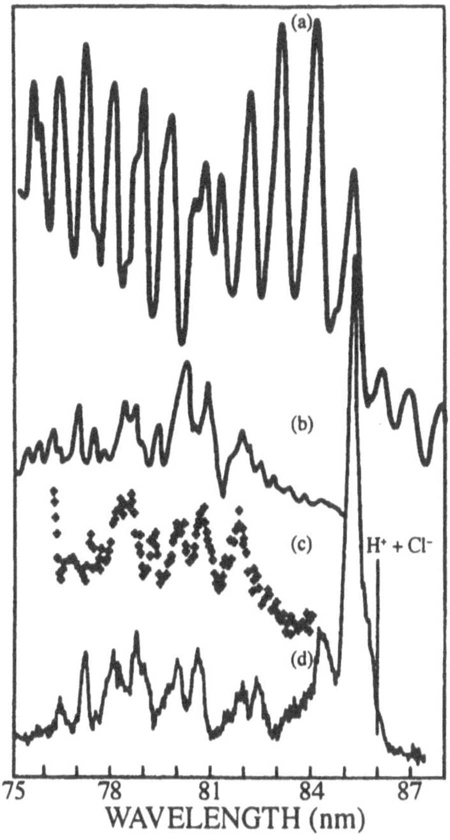

Figure 8.25: Four spectra which sample different decay pathways for superexcited states of HCl: (a) autoionization of HCl^+ $(A^2\Sigma^+)nl\lambda$ (Frohlich and Glass-Maujean, 1990), (b) dissociation into $H^*(n = 2)$+Cl (Frohlich and Glass-Maujean, 1990), (c) dissociation into H+$Cl^*(3p^4(^3P)4s)$ (White, *et al.*, 1987), and (d) dissociation into H^++ Cl^- (Yencha, *et al.*, 1993) [from Lefebvre-Brion and Suzor-Weiner, 1994].

The cross section for excitation into the H^++ Cl^- ion pair state is relevant to the line profile for excitation into the $^1\Sigma^+$ Rydberg states. The $V^1\Sigma^+$ state (configuration $\pi^4\sigma\sigma^*$) is the only state that dissociates into H^++ Cl^-. This state is involved in a $\sigma^* \sim 4s\sigma$ electrostatic interaction with the $A^2\Sigma^+$ $(\pi^4\sigma)4s\sigma$ $^1\Sigma^+$ Rydberg state (See Fig. 8.24).

The nonradiative decay pathways for the $(X^2\Sigma_g^+)np$ Rydberg states of H_2 have been examined by McCormack, *et al.*, (1993). Selected vibrational-rotational levels of the E, $F^1\Sigma_g^+$ state were excited by two-photon transitions from the $X^1\Sigma_g^+$ state and the ungerade Rydberg states were examined by a double resonance excitation scheme via the selected levels of the E, F state.

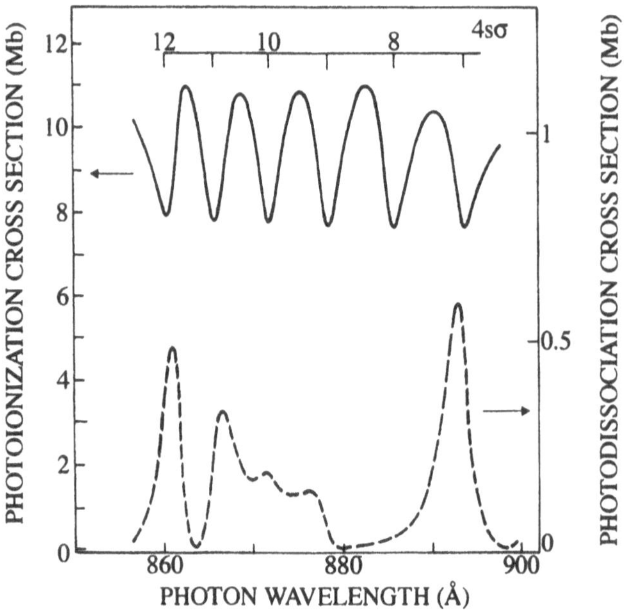

Figure 8.26: Calculated photoionization of HCl into the HCl^+ $X^2\Pi$ state (solid curve) and total photodissociation (dashed curve) cross sections for transitions into the $v = 7 - 12$ vibrational levels of the $A^2\Sigma^+ 4s\sigma\ ^1\Sigma^+$ Rydberg state. The transitions into $v = 7$ and $v = 12$ levels, which are free from overlapping features in the two spectra, exhibit profoundly different profiles (from Lefebvre-Brion and Keller, 1989).

The nonradiative decay pathways sampled include

$$(\mathrm{X}^2\Sigma_g^+)np, v^+ = 9, N^+ = 0 \rightarrow \mathrm{H}_2^+(\mathrm{X}^2\Sigma_g^+)v^+ = 8 + e^- \quad (8.12.2a)$$

$$\rightarrow \mathrm{H}(1s) + \mathrm{H}(3l) \quad (8.12.2b)$$

$$\rightarrow \mathrm{H}(1s) + \mathrm{H}(4l) \quad (8.12.2c)$$

$$\rightarrow \mathrm{H}^+ + \mathrm{H}^- \quad (8.12.2d)$$

These processes have not yet been theoretically discussed.

8.12.2 Theoretical Treatment

When a mechanism for each of the relevant decay channels is assumed and a set of specific interaction matrix elements is calculated by *ab initio* or semi-empirical methods, multichannel quantum defect theory (MQDT) (see Section 8.9) provides a basis for a unified and global treatment of the competing nonradiative decay processes and how these competing processes are sampled by specific experiments. Giusti-Suzor and Jungen (1984) were the first to apply MQDT to competing nonradiative processes. In Fig. 7.25 it was shown that the NO $Q^2\Pi$ $5p\pi$ Rydberg state is predissociated by electrostatic interaction with the vibrational continuum of the $B^2\Pi$ valence state. Above the ionization threshold to NO^+ $X^1\Sigma^+$ $v^+ = 0$, the $v \geq 3$ levels of the $Q^2\Pi$ state undergo vibrational autoionization by the $X^1\Sigma^+$ $v^+ \geq 0$ continua. Giusti-Suzor and Jungen (1984) showed that the electronic continuum of the $X^1\Sigma^+$ state and the vibrational continuum of the $B^2\Pi$ valence state interact. Recall that the electronic part of the continuum$\sim$continuum interaction matrix element is dimensionless (see Eq. (8.5.4)). In the limit where the predissociation rate via the $B^2\Pi$ vibrational continuum is faster than the vibrational autoionization via the $X^1\Sigma^+$ state, the autoionization channel is reduced further by the interaction between the ionization and dissociation channels ("final state interaction"). Conversely, even if the direct $Q^2\Pi \sim B^2\Pi$ vibrational interaction vanishes, an indirect inter-continuum interaction can contribute to $\Delta v > 1$ vibrational autoionization.

Theoretical calculations show that there is no coupling between dissociation and ionization continua in HCl (see Fig. 5 of Lefebvre-Brion and Keller, 1989).

8.13 Coherent Control of Photofragmentation Product Branching Ratios

The coherence properties of laser radiation provide an opportunity to exert some external control over intramolecular dynamics. Control over photofragmentation product branching ratios has been achieved in both time and frequency domain experiments.

Many time domain control experiments are based on a two-color pump/probe scheme proposed by Tannor and Rice (1988) (see review by Brixner, *et al.*, 2001). A pump pulse creates a vibrational wavepacket that evolves on an electronically excited bound state potential surface. The probe pulse is capable of inducing transitions to either of two potential surfaces that yield detectably distinct photofragments (e.g. neutral fragments, molecular cation plus electron, fragment cation plus neutral fragment plus electron, or ion pair). Control over the branching ratio between the two (or more) photofragment channels is achieved by varying the time delay between pump and probe pulses.

An experiment on the Na_2 molecule by Baumert and Gerber (1994) illustrates the Tannor-Rice scheme. A pair of ultrashort (70-110 fs) laser pulses is used to excite selectively either the

$$Na_2\,(X^1\Sigma_g^+) + h\nu(\text{pump}) \rightarrow Na_2\,(2^1\Sigma_u^+) + h\nu(\text{probe}) \rightarrow Na_2^+\,(X^2\Sigma_g^+) + e^- \tag{8.13.1}$$

or the

$$Na_2\,(X^1\Sigma_g^+) + h\nu(\text{pump}) \rightarrow Na_2\,(2^1\Sigma_u^+) + h\nu(\text{probe}) \rightarrow Na\,(3s) + Na^+ + e^- \tag{8.13.2}$$

photofragmentation channel. The 340 nm pump pulse excites from the $v = 0$ level of the Na_2 $X^1\Sigma_g^+$ state and creates a vibrational wavepacket initially located at the inner turning point of the $2^1\Sigma_u^+$ double minimum state. A 540 nm probe pulse is capable of producing either Na_2^+ $(X^2\Sigma_g^+) + e^-$ by excitation near the inner turning point or Na $(3s)$ + $Na^+ + e^-$ by excitation near the outer turning point. At $t = 0$ (and at integer multiples of the vibrational period, $T_{\text{vib}} = 1/\overline{\omega}_v c = 1$ ps), the center of the wavepacket is located at the inner turning point, and excitation by the probe pulse in the molecular-ion channel is favored. (For a non-harmonic potential, the vibrational frequency is v–dependent, $\overline{\omega}_v \approx [\Delta G(v + 1/2) + \Delta G(v - 1/2)]/2$.) At $t = T_{\text{vib}}/2$ (and at odd integer multiples of $T_{\text{vib}}/2$) the center of the wavepacket is located at the outer turning point and excitation by the probe pulse in the atomic-ion channel is favored. The usual "dynamics in real time" explanation of the molecular-ion vs. atomic-ion selectivity is illustrated by Fig. 8.27. However, reality, expressed in terms of the frequency domain concepts and transition propensity rules developed in this book, is more complicated than this simple picture would suggest. The excitation by the probe pulse at the inner turning point (dashed vertical arrow) of the $2^1\Sigma_u^+$ state has poor Franck-Condon overlap with the high vibrational levels of the Na_2^+ $X^2\Sigma_g^+$ state suggested by the position of the probe pulse arrowhead.

Instead, the vibrational wavepacket created at zero pump/probe delay on the Na_2^+ $X^2\Sigma_g^+$ state is built from low–v^+ levels centered near the intersection of the probe pulse dotted arrow with the $X^2\Sigma_g^+$–state potential. The excess energy from the probe pulse must go into the kinetic energy of the ejected electron (in even–l partial waves). The excitation by the probe pulse at the outer turning point (solid vertical arrow) has good Franck-Condon overlap with the $^2\Sigma_u^+$–state repulsive potential. The $g \leftrightarrow u$ optical selection rule requires that the low kinetic energy ejected electron depart in odd–l (i.e., u symmetry) partial waves. Both electronic and vibrational transition probability factors favor excitation at the outer turning point.

Figure 8.28 shows the variation of the atomic-ion/molecular-ion branching ratio vs. the delay between the pump and probe pulses. At $t = 0$, the Na^+/Na_2^+ branching ratio should be at its minimum value. The branching ratio oscillates with a period determined by the effective harmonic frequency of the center of the wavepacket, $\overline{\omega}$. The center of the wavepacket is determined by the maximum of Franck-Condon overlap for $2^1\Sigma_u^+ \leftarrow X^1\Sigma_g^+(v'' = 0)$ transitions, not the center frequency of the pump pulse. The oscillatory temporal structure of the branching ratio exhibits a dephasing (decaying amplitude, broadening, and chirp) related to the non-harmonic nature of the $2^1\Sigma_u^+$ potential. More advanced pump/probe schemes (exploiting multi-element phase and amplitude controlling masks, see Section 9.1.10) exist, where the phases and amplitudes of individual frequency components of the pump pulse are crafted to cause the wavepacket evolving on the highly dispersive $2^1\Sigma_u^+$ potential to refocus at either the inner or outer turning point. Considerably enhanced control over branching ratio can be achieved with such crafted pulses.

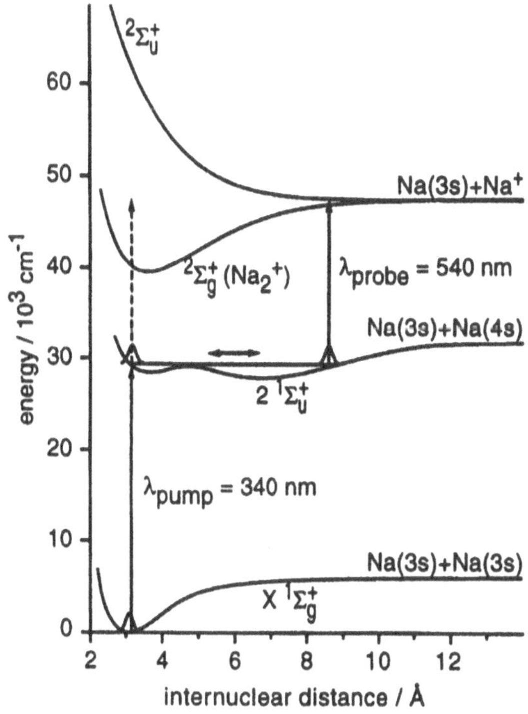

Figure 8.27: Potential energy diagram of Na_2 displaying the control mechanism for direct ionization and dissociative ionization of Na_2 (from Brixner, *et al.*, 2001)

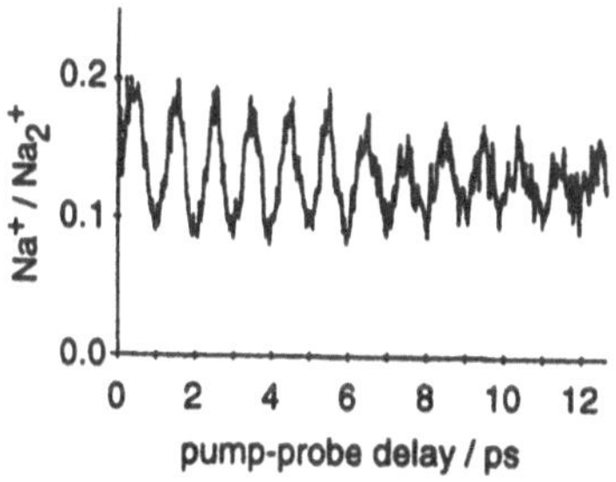

Figure 8.28: Ratio of atomic to molecular products (Na^+ / Na_2^+) as a function of the delay between the 340-nm pump and 540-nm probe pulses (from Brixner, *et al.*, 2001).

Brumer and Shapiro (1989) proposed a frequency domain control scheme based on quantum mechanical interference between two transition paths. The two paths involve 1–photon and 3–photon transitions where the radiation that excites the 1–photon transition, ω_1, is generated by frequency tripling the laser radiation, ω_L, that excites the 3–photon transition,

$$\omega_1 = 3\omega_L. \tag{8.13.3}$$

In order for the 1–photon and 3–photon transition amplitudes to interfere, both transition paths must begin at the same initial rovibronic state and terminate on final states of the same energy and rigorous symmetry (total parity, J, M_J). In order for the interference to be experimentally controllable, the amplitudes and the relative phase of the ω_1 and ω_L radiation fields must be systematically adjustable. Since ω_1 is generated by tripling ω_L, a definite phase relationship exists between the ω_1 and ω_L photon fields. The relative phase is adjusted by exploiting the difference between the index of refraction at ω_L and at $\omega_1 = 3\omega_L$ for a non-absorbing gas, such as H_2 or Ar. As the pressure×pathlength of the phase-shifting gas is varied, the relative phase of the ω_1 and ω_L beams may be adjusted continuously through several cycles of 2π.

When a superexcited molecular state of energy E is created by either 1–photon absorption of ω_1 or 3–photon absorption of ω_L, molecular photofragmentation can occur into either of two qualitatively distinct channels, photodissociation (D) and photoionization (I). The excitation probabilities, as a function of E for the D and I channels, are

$$P^D(E) = P_1^D(E) + P_3^D(E) + 2\left|P_{13}^D(E)\right|\cos\left(\phi + \delta_{13}^D(E)\right) \tag{8.13.4a}$$

$$P^I(E) = P_1^I(E) + P_3^I(E) + 2\left|P_{13}^I(E)\right|\cos\left(\phi + \delta_{13}^I(E)\right) \tag{8.13.4b}$$

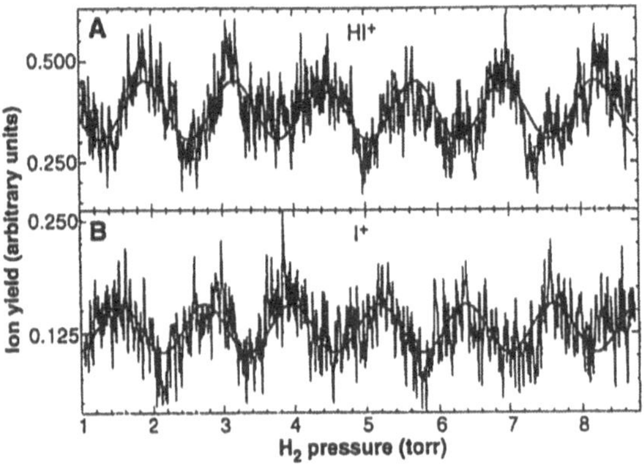

Figure 8.29: Detection of ionization (HI^+ upper panel) and dissociation (I^+, lower panel) for excitation frequency $\omega_L = 353.69$ nm (from Zhu, *et al.*, 1995).

where ϕ is the relative phase of the two laser fields and

$$\Delta\delta(E) = \delta_{13}^I(E) - \delta_{13}^D(E) \tag{8.13.5}$$

is the phase difference (phase lag) between the molecular phase due to channel I (δ_{13}^I) and that due to channel D (δ_{13}^D). The numbers of photons and photofragmentation channels are specified, respectively, by superscripts 1 and 3 and subscripts D and I.

Figure 8.29 shows the variation of ionization yield (upper panel) and dissociation yield (lower panel) observed for HI at $\omega_L = 353.69$ nm as the pressure of the H_2 phase-shifting gas is varied. These two photofragment yield curves appear to be approximately π out of phase, but as shown on Fig. 8.31, this molecular contribution to the phase lag, $\Delta\delta(E)$, is strongly dependent on ω_L.

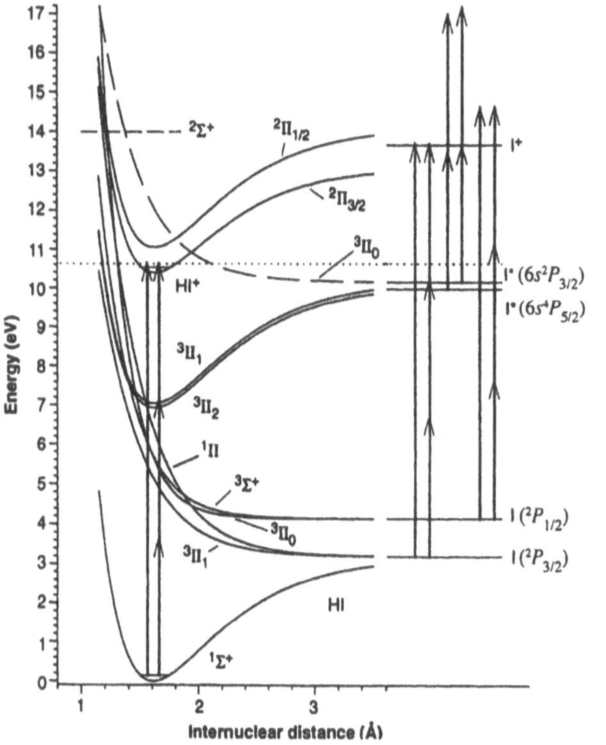

Figure 8.30: Potential energy diagram of HI showing the control mechanism for ionization vs. dissociation in the energy region between the two ionization thresholds (from Zhu, *et al.*, 1995).

Zhu, *et al.*, (1995) were the first to demonstrate experimentally the existence of the molecular phase lag between photoionization and photodissociation. A molecular beam of HI was excited to the superexcited $5d\delta$ Rydberg state (converging to the HI^+ $^2\Pi_{1/2}$ substate) by superimposed $\omega_1 = 3\omega_L$ and ω_L beams of radiation. The ω_1 and ω_L beams traversed a phase shifting cell filled with H_2 before crossing the HI molecular beam. The $5d\delta(X^2\Pi_{1/2})$ state is autoionized by the continuum of the $X^2\Pi_{3/2}$ substate (spin-orbit autoionization; see Section 8.6) and is predissociated by a repulsive state going to the $I^*(6s\,^2P_{3/2}) + H$ separated atom limit. The I atom dissociation product is detected by

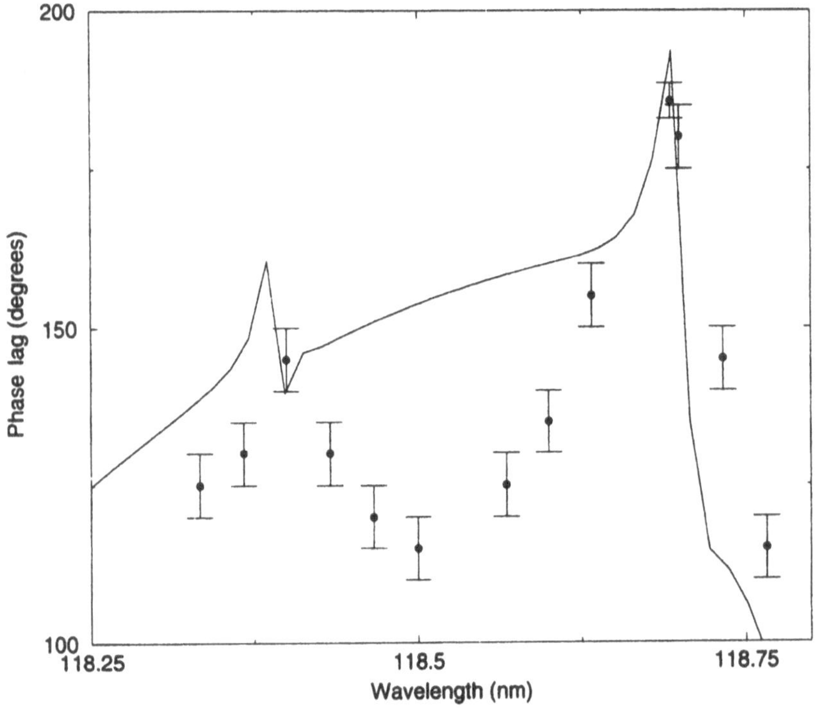

Figure 8.31: Calculated phase lag in the region of the 5d δ resonance of HI: solid line (Lefebvre-Brion, 2003), experimental measurements: circles (Fiss, *et al.*, 1999).

$$I^* + 2\omega_L \rightarrow I^+ + e^-. \tag{8.13.6}$$

The control mechanism and detection scheme are summarized on Fig. 8.30.

The existence of the molecular phase lag, $\Delta\delta(E)$, was initially controversial, but now, its physical origin has been well understood (Seideman, 1999; Gordon, *et al.*, 2001). The HI results are explained using unified Multichannel Quantum Defect Theory (MQDT; see Section 8.9) in which the autoionization and predissociation processes are treated simultaneously (Giusti, 1980). The calculated phase lag (Lefebvre-Brion, *et al.*, 2001) agrees well with experimental observations, as shown in Fig. 8.31. Recent unified MQDT calculations, including rotation, have reproduced the experimental phase lag in the region of the 5sσ resonance (Lefebvre-Brion, 2002). It has been shown that, in this particular case of spin-orbit autoionization, the phase of a channel A, δ_{13}^A, is non-zero only if the two fragmentation continua, A and B, are coupled, directly or indirectly.

Other control schemes, based on modifications of the spectral content, spectral phases, and intensity of a laser pulse, have been implemented (see reviews by Gordon and Rice, 1997, Rice and Zhao, 2000, and Shapiro and Brumer, 2003.)

8.14 References

Allendorf, S. W., Leahy, D. J., Jacobs, D. C., and Zare, R. N. (1989), *J. Chem. Phys.* **91**, 2216.
Åsbrink, L. (1970), *Chem. Phys. Letters* **7**, 549.
Baer, T., and Guyon, P.-M., (1995) in *High Resolution Laser Photoionization and Photoelectron Studies*, page 1, edited by I. Powis, T. Baer, and C.Y. Ng, (John Wiley & Sons).
Baumert, T., and Gerber, G. (1994), *Isr. J. Chem.* **34**, 103.
de Beer, E., Koenders, B. G., Koopmans, M. P., and de Lange, C. A. (1990), *J. Chem. Soc. Farad. Trans.* **86**, 2035.
de Beer, E., de Lange, C. A., Stephens, J. A., Wang, K., and McKoy, V. (1991), *J. Chem. Phys.* **95**, 714.
de Beer, E., Buma, W. J., and de Lange, C. A. (1993), *J. Chem. Phys.* **99**, 3252.
Berry, R. S. (1966), *J. Chem. Phys.* **45**, 1228.
Beswick, J. A., and Lefebvre, R. (1973), *Mol. Phys.* **29**, 1611.
Braunstein, M., McKoy, V., and Dixit, S. N. (1992), *J. Chem. Phys.* **96**, 5726.
Brixner, T., Damrauer, N., and Gerber, G. (2001), *Adv. At. Mol. Opt. Phys.* **46**, 1.
Brumer, P., and Shapiro, M. (1989), *Accts. Chem. Res.* **22**, 407.
Buckingham, A. D., Orr, B. J., and Sichel, J. M., (1970), *Phil. Trans. Roy. Soc. (London)* **A 268**, 147.
Caprace, G., Delwiche, J., Natalis, P., and Collin, J. E. (1976) *Chem. Phys.* **13**, 43.
Carlson, T. A., Fahlman, A., Krause, M. O., Keller, P. R., Taylor, J. W., Whitley, T. A., and Grimm, F. A. (1984), *J. Chem. Phys.* **80**, 3521.
Carroll, P. M. (1973), *J. Chem. Phys.* **58**, 3597.
Chang, E. S. (1978) in *Photoionization of atoms and molecules*, p. 30, ed. B. Buckley (Daresbury Laboratory, report DL/SCI/R11, Atomic and Molecular).
Chu, T. C., and Wu, C. W. R. (1988), *J. Chem. Phys.* **88**, 5237.
Chupka, W. A., (1993), *J. Chem. Phys.* **98**, 4520.
Cockett, M. C. R. (1995), *J. Phys. Chem. A* **99**, 16,228.
Connerade, J. P. (1983), *J. Phys. B* **16**, L329.
Cooper, J. W., (1962), *Phys. Rev.* **128**, 681.
Dehmer, P. M., and Chupka, W. A. (1976), *J. Chem. Phys.* **65**, 2243.
Dehmer, J. L., Dill, D., and Wallace, S. (1979), *Phys. Rev. Lett.* **43**, 1005.
Dehmer, P. M., Miller, P. J., and Chupka, W. A. (1984), *J. Chem. Phys.* **80**, 1030.
Dehmer, J. L., Dill, D., and Parr, A. C., (1985) in *Photophysics and Photochemistry in the Vacuum Ultraviolet*, p. 341, eds. S. McGlynn, G. Findley and R. Huebner (D. Reidel Publishing Company, Dordrecht, Holland).
Dehmer, J. L., Dehmer, P. M., West, J. B., Hayes, M. A., Siggel, M. R. F., and Parr, A. C. (1992), *J. Chem. Phys.* **97**, 7911.
Dill, D., and Fano, U. (1972), *Phys. Rev. Lett.* **29**, 1203.
Dill, D. (1972), *Phys. Rev. A* **6**, 160.
Dill, D. (1973), *Phys. Rev. A* **7**, 1976.
Dill, D., and Jungen, C. (1980), *J. Phys. Chem.* **84**, 2116.
Dixit, S. N., and McKoy, V., (1986), *Chem. Phys. Letters* **128**, 49.
Drescher, M., Brockhinke, A., Böwering, N., Heinzmann, U., and Lefebvre-Brion, H.

(1993), *J. Chem. Phys.* **99**, 2300.
Dubs, R. L., Dixit, S. N., and McKoy, V. (1986), *J. Chem. Phys.* **85**, 656.
Duzy, C., and Berry, R. S. (1976), *J. Chem. Phys.* **64**, 2431.
Edqvist, O., Lindholm, E., Selin, L. E., and Åsbrink, L., (1970) *Physica Scripta* **1**, 25.
Edvardsson, D., Baltzer, P., Karlsson, L., Lundqvist, M., and Wannberg, B., (1995) *J. Electron. Spectrosc.* **73**, 105.
Eland, J. H. D. (1980), *J. Chim. Phys. Phys. Chim. Biol.* **77**, 613.
Eland, J. H. D., and Berkowitz, J. (1977), *J. Chem. Phys.* **67**, 5034.
Fano, U. (1961), *Phys. Rev.* **124**, 1866.
Fano, U. (1969), *Phys. Rev.* **178**, 131.
Fano, U. (1970), *Phys. Rev. A.* **2**, 353.
Fano, U. (1975), *J. Opt. Soc. Am.* **65**, 979.
Fiss, J. A., Zhu., L., Gordon, R. J., and Seideman, T. (1999), *Phys. Rev. Lett.* **82**, 65.
Fredin, S., Gauyacq, D., Horani, M., Jungen, C., Lefèvre, G., and Masnou-Seeuws, F. (1987) *Mol. Phys.* **60**, 825.
Frohlich, H., and Glass-Maujean, M. (1990), *Phys. Rev. A* **42**, 1396.
Gallagher, J. W., Brion, C. E., Samson, J. A. R., and Langhoff, P. W. (1988), *J. Phys. Chem. Ref. Data* **17**, 9.
Giusti, A. (1980), *J. Phys. B*, **13**, 3867.
Giusti-Suzor, A., and Jungen, C. (1984), *J. Chem. Phys.* **80**, 986.
Giusti-Suzor, A., and Lefebvre-Brion, H. (1984), *Phys. Rev. A* **30**, 3057.
Glass-Maujean, M., Breton, J., and Guyon, P. M. (1987), *Z. Phys. D* **5**, 189.
Gordon, R. J., and Rice, S. A. (1997), *Annu. Rev. Phys. Chem.* **48**, 601.
Gordon, R. J., Zhu, L., and Seideman, T. (2001), *J. Phys. Chem. A* **105**, 4387.
Greene, C. H., and Zare, R. N. (1982), *Annu. Rev. Phys. Chem.* **33**, 119.
Greene, C. H., and Jungen, C. (1985), *Adv. At. Mol. Phys.* **21**, 51.
Guest, J. A., Jackson, K. H., and Zare, R. N. (1983), *Phys. Rev. A* **28**, 2217.
Guyon, P. M., Spohr, R., Chupka, W. A., and Berkowitz, J. (1976), *J. Chem. Phys.* **65**, 1650.
Hancock, W. H., and Samson, J. A. R. (1976), *J. Electron Spectrosc.* **9**, 211.
Hardis, J. E., Ferrett, T. A., Southworth, S. H., Parr, A. C., Roy, P., Dehmer, J. L., Dehmer, P. M., and Chupka, W. A. (1988), *J. Chem. Phys.* **89**, 812.
Heinzmann, U. (1978), *J. Phys. B.* **11**, 399.
Heinzmann, U. (1980), *J. Phys. B* **13**, 4353, 4367.
Heinzmann, U., and Cherepkov, N. A. (1996) in "VUV and Soft X-Ray Photoionization" (Becker, U., and Shirley, D. A., ed.) Plenum, New York, p. 521.
Herzberg, G. (1950), "Spectra of Diatomic Molecules," Van Nostrand-Reinhold, Princeton, New Jersey.
Herzberg, G., and Jungen, C. (1972), *J. Mol. Spectrosc.* **41**, 425.
Holland, D. M. P., Shaw, D. A., McSweeney, S. M., MacDonald, M. A., Hopkirk, A., and Hayes, M. A. (1993), *Chem. Phys.* **173**, 315.
Irrgang, R., Drescher, M., Spieweck, M., Heinzmann, U., Cherepkov, N.A., and Lefebvre-Brion, H. (1998) *J. Chem. Phys.* **108**, 10070.
Jungen, C., and Atabek, O. (1977), *J. Chem. Phys.* **66**, 5584.
Jungen, C. (1980), *J. Chim. Phys. Phys. Chim. Biol.* **77**, 27.
Jungen, C., and Dill, D. (1980), *J. Chem. Phys.* **73**, 3338.
Jungen, C. (1982), in "Physics of Electronic and Atomic Collisions" (S. Datz, ed.), p. 455. North-Holland-Publ., Amsterdam.
Jungen, C. (ed.) (1996), "Molecular Applications of Quantum Defect Theory," Institute of Physics Publishing, Bristol.

Keller, F., and Lefebvre-Brion, H. (1986), *Z. Phys. D* **4**, 15.
Labastie, P., Bordas, M. C., Tribollet, B., and Broyer, M. (1984) *Phys. Rev. Lett.* **52**, 1681.
de Lange, C. A. (1995) in "High Resolution Laser Photoionization and Photoelectron Studies", p. 195, edited by Powis, I., Baer, T., and Ng, C. Y., John Wiley and Sons.
Larzillière, M., Launay, F., and Roncin, J. Y. (1980), *J. Phys. (Orsay, Fr.)* **41**, 1431.
Lefebvre-Brion, H., and Giusti-Suzor, A., (1983), in "Electron-Atom and Electron-Molecule Collisions," (J. Hinze, ed.), p. 215, Plenum.
Lefebvre-Brion, H., Giusti-Suzor, A., and Raşeev, G. (1985), *J. Chem. Phys.* **83**, 1557.
Lefebvre-Brion, H., (1988), in "Molecules in Physics, Chemistry and Biology", Vol. 11, p. 257 (Kluwer Academic Publishers, Dordrecht).
Lefebvre-Brion, H., (1988), *J. Chem. Phys.* **89**, 2691.
Lefebvre-Brion, H., and Keller, F. (1989) *J. Chem. Phys.* **90**, 7176.
Lefebvre-Brion, H. (1990), *J. Chem. Phys.* **93**, 5898.
Lefebvre-Brion, H. (1991), AIP Conference Porceedings **225**, editors Garcia-Sucre, M., Raşeev, G., and Ross, S. C. (American Institute of Physics, New York) p. 275.
Lefebvre-Brion, H., and Suzor-Weiner, A. (1994), *Comments At. Mol. Phys.* **29**, 305.
Lefebvre-Brion, H. (1995), in "High Resolution Laser Photoionization and Photoelectron Studies", edited by I. Powis, T. Baer, and C.Y. Ng (John Wiley Sons, New York) p. 171.
Lefebvre-Brion, H. (1996), *Chem. Phys. Letters* **253**, 43.
Lefebvre-Brion, H. (1997), *Chem. Phys. Letters* **268**, 107.
Lefebvre-Brion, H. (1999), in "The Role of Rydberg States in Spectroscopy and Photochemistry", (Ed. C. Sándorfy), p. 267, Kluwer Academic Publishers, Dordrecht.
Lefebvre-Brion, H., Seideman, T., and Gordon, R. J. (2001), *J. Chem. Phys.* **114**, 9402.
Lefebvre-Brion, H., (2002), *J. Chem. Phys.* **117**, 4751.
Lefebvre-Brion, H., (2003), "Quantum Control: Mathematical and Numerical Challenges," CRM Proceedings and Lectures Notes Series, American Mathematical Society, Publications, Providence, RI.
Leyh, B., and Raşeev, G. (1988), *J. Chem. Phys.* **89**, 820.
Lippa, T. P., Xu, S.-J., Lyapustina, S. A., and Bowen, K. H., (1998) *J. Chem. Phys.* **109**, 9263.
Lucchese, R. R., and McKoy, V. (1981), *J. Phys. B* **14**, L629.
Lynch, D. L., Dixit, S. N., and McKoy, V., (1986) *J. Chem. Phys.* **84**, 5504.
Mackenzie, S. R., Halse, E. J., Gordon, E., Rolland, D., and Softley, T. P., (1996), *Chem. Phys.* **209**, 127.
Mank, A., Drescher, M., Hüth-Fehre, T., Böwering, N., Heinzmann, U., and Lefebvre-Brion, H. (1991), *J. Chem. Phys.* **95**, 1676.
Manson, S. T., and Cooper, J. W., (1968) *Phys. Rev.* **165**, 126.
McCormack, E. F., Pratt, S. T., Dehmer, P. M., and Dehmer, J. L. (1993), *J. Chem. Phys.* **98**, 8370.
McKoy, V., Carlson, T. A., and Lucchese, R. R. (1984), *J. Phys. Chem.* **88**, 3188.
Merkt, F., and Softley, T. P., (1993), *Int. Rev. in Phys. Chem.* **12**, 205.
Merkt, F. (1997), *Ann. Rev. Phys. Chem.* **48**, 675.
Mies, F. H. (1968), *Phys. Rev.* **175**, 164.
Morin, P., Nenner, I., Adam, M. Y., Hubin-Franskin, M. J., Delwiche, J., Lefebvre-Brion, H., and Giusti-Suzor, A., (1982), *Chem. Phys. Lett.* **92**, 609.
Müller-Dethlefs, K., Sander, M., and Schlag, E. W., (1984), *Z. Naturforsch.* **39**, 1089; *Chem. Phys. Lett.* **112**, 291.

Mulliken, R. S. (1964), *J. Am. Chem. Soc.* **86**, 3183.
Mulliken, R. S. (1969), *J. Am. Chem. Soc.* **91**, 4615.
Ng, C. Y., Mahan, B. H., and Lee, Y. T. (1976), *J. Chem. Phys.* **65**, 1956.
Ogawa, M., and Tanaka, Y. (1962), *Can. J. Phys.* **40**, 1593.
Öhrwall, G., Baltzer, P., and Bozek, J., (1998), *Phys. Rev. Lett.* **81**, 546.
Orel, A. E., Rescigno, T. N., McKoy, V., and Langhoff, P. W. (1980), *J. Chem. Phys.* **72**, 1265.
Palm, H., and Merkt, F. (1998), *Phys. Rev. Lett.* **81**, 1385.
Parr, A. C., West, J. B., Marr, G. V., and Dehmer, J. L. (1998), *J. Phys. B* **31**, 5161.
Pollard, J.E., Trevor, D.J., Reutt, J.E., Lee, Y.T., and Shirley, D.A., (1982), *J. Chem. Phys.* **77**, 34.
Pratt, S. T., (1995), *Rep. Prog. Phys.* **58**, 821.
Raşeev, G., Giusti-Suzor, A., and Lefebvre-Brion, H. (1978), *J. Phys. B* **11**, 2735.
Raşeev, G., Le Rouzo, H., and Lefebvre-Brion, H., (1980), *J. Chem. Phys.* **72**, 5701.
Raşeev, G., Lefebvre-Brion, H., Le Rouzo, H., and Roche, A. L. (1981), *J. Chem. Phys.* **74**, 6686.
Raşeev, G., Keller, F., and Lefebvre-Brion, H. (1987), *Phys. Rev.* **36**, 4759.
Raşeev, G., and Cherepkov, N. A. (1990), *Phys. Rev. A* **42**, 3948.
Raoult, M., and Jungen, C. (1981), *J. Chem. Phys.* **74**, 3388.
Raoult, M., LeRouzo, H. Raşeev, G., and Lefebvre-Brion, H. (1983), *J. Phys. B* **16**, 4601.
Reid, K. L., and Leahy, D. J. (1995) in "High Resolution Laser Photoionization and Photoelectron Studies", p. 247, edited by Powis, I., Baer, T., and Ng, C. Y. (John Wiley & Sons).
Remacle, F., and Vrakking, M. J. J. (1998), *J. Phys. Chem. A* **102**, 9507.
Rice, S. A., and Zhao, M. (2000), "Optical Control of Molecular Dynamics", Wiley, New York.
Robin, M. B. (1985) in "Higher Excited States of Polyatomic Molecules", Vol. III, p. 30 (Academic Press, Orlando, Florida).
Roche, A. L., and Jungen, C. (1993), *J. Chem. Phys.* **98**, 3637.
Schwarz, M., Duchowicz, R., Demtröder, W., and Jungen, C. (1988), *J. Chem. Phys.* **89**, 5460.
Seideman, T. (1999), *J. Chem. Phys.* **111**, 9168.
Shapiro, M., and Brumer, P. (2003), "Principles of the Quantum Control of Molecular Processes," Wiley, New York.
Shaw, D. A., Holland, D. M. P., McDonald, M. A., Hopkirk, A., Hayes, M. A., and McSweeney, S. M. (1992), *Chem. Phys.* **166**, 379; erratum (1993) *ibid.* **172**, 187.
Slater, J. C. (1960), "Quantum Theory of Atomic Structure," McGraw-Hill, New York.
Smith, A. L. (1970), *Philos. Trans. R. Soc. London. A* **268**, 169.
Smith, M. E., Lynch, D. L., and McKoy, V. (1986), *J. Chem. Phys.* **85**, 6455.
Softley, T. P., Hudson, A. J., and Watson, R. (1997), *J. Chem. Phys.* **106**, 1041.
Southworth, S. H., Parr, A. C., Hardis, J. E., and Dehmer, J. L. (1988), *J. Chem. Phys.* **87**, 5125.
Sroka, W. (1969), *Phys. Lett.* **22A**, 784.
Stahel, D., Leoni, M., and Dressler, K. (1983), *J. Chem. Phys.* **79**, 2541.
Stratmann, R. E., Zurales, R. W., and Lucchese, R. R. (1996), *J. Chem. Phys.* **104**, 8989.
Sur, A., Ramana, C. V., and Colson, S. D. (1985), *J. Chem. Phys.* **83**, 904.
Sur, A., Ramana, C. V., Chupka, W. A., and Colson, S. D. (1986), *J. Chem. Phys.* **84**,

69.
Takezawa, S. (1977), *J. Mol. Spectrosc.* **66**, 121.
Tannor, D. J., and Rice, S. A. (1988), *Adv. Chem. Phys.* **70**, 441.
Thiel, W. (1982), *Chem. Phys. Lett.* **87**, 249.
Thiel, W. (1983), *Chem. Phys.* **77**, 103.
Thiel, W. (1984), *J. Electron Spectrosc.* **34**, 399.
Turner, D. W., and Al-Jabory, M. I. (1962), *J. Chem. Phys.* **37**, 3007.
Ueda, K., West, J. B., Hayes, M. A., Siggel, M. R. F., Parr, A. C., and Dehmer, J. L. (1993), *J. Phys. B* **26**, L601.
Villajero, D., Herm, R. R., and Inghram, M. G., (1967), *J. Chem. Phys.* **46**, 4995.
Wales, N. P. L., Buma, W. J., de Lange, C. A., and Lefebvre-Brion, H. (1996), *J. Chem. Phys.* **105**, 2978.
Wang, K. and McKoy, V. (1991), *J. Chem. Phys.* **95**, 4977.
Watson, D. K., Lucchese, R. R., McKoy, V., and Rescigno, T. N. (1980), *Phys. Rev. A* **21**, 738.
West, J. B., Parr, A.C., Cole, B.E., Ederer, D.L., Stockbauer, R., and Dehmer, J.L., (1980), *J. Phys. B* **13**, L105.
White, M. G., Seaver, M., Chupka, W., and Colson, S. D. (1982), *Phys. Rev. Lett.* **49**, 28.
White, M. G., Leroi, G. E., Ho, M. H., and Poliakoff, E. D. (1987), *J. Chem. Phys.* **87**, 6553.
Wilson, E. B., Decius, J. C., and Cross, P. C. (1980), "Molecular Vibrations," Dover, New York.
Wilson, W. G., Viswanathan, K. S., Sekreta, E., and Reilly, J. P. (1984), *J. Phys. Chem* **88**, 672.
Wu, C.Y. (1987), *J. Quant. Spect. Rad. Transfer* **37**, 1.
Xie, J., and Zare, R. J. (1990), *J. Chem. Phys.* **93**, 3033 (see comment by Watson, J. K. G. (1998) *J. Chem. Phys.* **108**, 820).
Xie, Y., Reilly, P. T. A., Chilikiri, S., and Gordon, R. J. (1991), *J. Chem. Phys.* **95**, 854.
Yang, C. N. (1948), *Phys. Rev.* **74**, 764.
Yencha, A. J., Ganz, J., Ruf, M.-W., and Hotop, H. (1989), *Z. Phys. D.* **14**, 57.
Yencha, A. J., Kaur, D., Donovan, R. J., Kvaran, A., Hopkirk, A., Lefebvre-Brion, H., and Keller, F. (1993), *J. Chem. Phys.* **99**, 4986.
Zhu, L., Kleiman, V., Li, X., Lu, S. and Gordon, R. J. (1995), *Science* **270**, 77.

Chapter 9

Dynamics

Contents

9.1 Dynamical Concepts, Tools, and Terminology . . 622

9.1.1 The Time-Dependent Picture: Terminology 623

9.1.2 Solution of the Time-Dependent Schrödinger Equation 624

9.1.3 Frequency Domain Spectra Treated as the Fourier Transform of the Autocorrelation Function 626

9.1.4 Dynamical Quantities 635

9.1.5 General Density Matrix Formulation of a Dynamical Experiment: Excitation, Evolution, and Detection Matrices 639

9.1.6 Particle (Photon) vs. Wave Pictures of Spectroscopy 643

9.1.7 Motion of the Center of the Wavepacket 644

9.1.8 Equations of Motion for Resonance Operators 646

9.1.9 The One-Color Pump-Probe Experiment 649

9.1.10 Crafted Pulses for Detailed Manipulation of Molecular Dynamics 655

9.2 From Quantum Beats to Wavepackets 656

9.2.1 Polarization Quantum Beats 657

9.2.2 Population Quantum Beats 658

9.2.3 Nuclear Wavepackets 659

9.2.3.1 Vibrational Wavepackets 661

9.2.3.2 Rotational Wavepackets 667

9.2.4 Rydberg Wavepackets: Kepler and Precessional Periods 668

9.3 Relaxation into a Quasi-Continuum: A Tool for Dimensionality Reduction 671

9.3.1 The Complex-Energy Effective Hamiltonian 672

9.3.1.1 Treatment of Two-State Interaction by Nondegenerate Perturbation Theory 674

9.3.1.2 Treatment by Quasidegenerate Perturbation Theory: $\mathbf{2} \times \mathbf{2}$ Diagonalization 675
9.3.2 Quantum Beats Between Two Decaying Quasi-Eigenstates679
9.3.3 The Use of the Complex $\mathbf{H}^{\mathrm{eff}}$ in Reduced-Dimension Models 681
9.4 *Beyond* the Spectra and Dynamics of Diatomic Molecules683
9.4.1 Basis States 683
9.4.2 What is Deperturbation Anyway? 684
9.4.3 Visualization of Dynamics 685
9.4.4 Beyond Diatomic Molecules: Polyatomic Molecule Vibrational Dynamics 687
9.4.5 Polyads 689
9.4.6 Creation and Annihilation Operators 690
9.4.7 Dynamics in State Space 692
9.4.8 Number Operator 693
9.4.9 Resonance Energy and Energy Transfer Rate Operators 694
9.4.10 The Use of Expectation Values of Resonance Operators to Visualize Dynamic Processes 697
9.4.11 $\mathbf{Q}$ and $\mathbf{P}$ 700
9.4.12 Transformation Between Local and Normal Mode Limits 702
9.4.12.1 Classical Mechanical Treatment 703
9.4.12.2 The Morse Oscillator 705
9.4.12.3 Quantum Mechanical Minimal Model for Two Anharmonically Coupled Local Stretch Morse Oscillators 706
9.4.12.4 Transformation between 4-Parameter Forms of the Normal and Local Mode Basis Sets . 710
9.4.12.5 Transformation between 6-Parameter Forms of the Normal Mode and Local Mode $\mathbf{H}^{\mathrm{eff}}$ 714
9.4.13 From Quantum Mechanical $\mathbf{H}^{\mathrm{eff}}$ to Classical Mechanical $\mathcal{H}(\mathbf{Q}, \mathbf{P})$ 717
9.4.14 Polyatomic Molecule Dynamics 733
9.4.15 Inter-System Crossing, Internal Conversion, and Intramolecular Vibrational Distribution 733
9.5 References 736

9.1 Dynamical Concepts, Tools, and Terminology

Since "Perturbations in The Spectra of Diatomic Molecules" was published in 1986, new concepts and experimental capabilities have altered profoundly the way we think about both the structure and the dynamics of molecules (Heller,

1981; Zewail, 1996 and 2000). Because it better engages our experience of the macroscopic world, the *time-evolving* localized state or "wavepacket" concept has begun to displace the more abstract concepts of basis states and *stationary* eigenstates. Since insights generated by the older time-independent and the newer time-dependent pictures are complementary, it would be foolish to permit the enthusiasm for things new to sweep things old into oblivion. This chapter is intended to aid in the process of building bridges between the stationary state and wavepacket communities. It also takes a tentative first step into the area explicitly prohibited by the title of this book, the dynamics of polyatomic molecules. A central justification for a book dedicated to the spectra and dynamics of diatomic molecules is that all of the concepts, models, and analysis tools developed for diatomic molecules are applicable to polyatomic molecules.

9.1.1 The Time-Dependent Picture: Terminology

Just as a satisfactory approximation to the exact solution of the time-independent Schrödinger equation for a many-electron molecule must be expressed in a form that belies the apparent simplicity of the spectrum, so too is the exact solution of the time-dependent Schrödinger equation often opaquely complex. For the time-independent problem, insight comes from defining a suitable *zero-order model* $\left(\mathbf{H}^{(0)}, \{\psi_i^{(0)}\}, \{E_i^{(0)}\}\right)$, specifying the essential *coupling terms* $(\mathbf{H}^{(1)})$ and *coupled quasidegenerate states*, and then *decoupling* the states of interest in the specific experiment from the infinite number of remote-perturber states by an implicit or explicit *Van Vleck transformation* (see Section 4.2). The result is a *spectroscopic effective Hamiltonian model*, $\mathbf{H}^{\text{eff}}$, which is of a profoundly different nature than the exact Hamiltonian, $\mathbf{H}$, even though the notation and terminology used to describe both $\mathbf{H}^{\text{eff}}$ and $\mathbf{H}$ are misleadingly similar (see Section 4.4.2). $\mathbf{H}^{\text{eff}}$ is a vastly simplified model in which microscopic forces and the response of the system to those forces are identified, the essential coupling terms are arranged in a hierarchy of importance, and the dimensionality of the spectroscopically relevant part of state space is reduced from that of the full state space so that intuition can complement and focus computation.

When the state of a molecule evolves under the influence of a time-independent $\mathbf{H}$, structure and dynamics are two sides of the same coin. Since the same $\mathbf{H}$ appears in the time-independent and time-dependent Schrödinger equations, an $\mathbf{H}^{\text{eff}}$-based intuitive picture of the molecular structure generates an intuitive, causal, *mechanistic* picture of molecular dynamics.

A dynamical mechanism consists of the following elements: (*i*) creation of a *nonstationary state*; (*ii*) departure of the nonstationary state from its initial location in coordinate or state space; (*iii*) long time evolution, including partial rephasings and decay.

(*i*) A *nonstationary state* is created at $t = 0$. This initial state, which is localized in coordinate or state space, is specified perfectly in terms of a linear combination of a small number (often only one) of zero-order states,

$$\Psi(t=0) = \sum_i a_i \psi_i^{(0)}. \tag{9.1.1}$$

Typically, $\Psi(0)$ is called the "*bright state.*" This name is chosen precisely because this $\Psi(0)$ is a member of the special class of localized states that can be created at $t = 0$ by exciting from a purposefully selected initial eigenstate with a short pulse ($\sim 100fs$) of electromagnetic radiation. The excitation pulse that creates the bright state is called the "*pluck.*" [Heller and Gelbart, 1980] (*ii*) The initially prepared localized state, $\Psi(t)$, moves away from its $t = 0$ starting location, either in state-space (dephasing: $a_i(t) = \left\langle \psi_i^{(0)} | \Psi(t) \right\rangle \neq a_i(0)$) or coordinate-space ($\langle \mathbf{x}(t) \rangle \equiv \langle \mathbf{x} \rangle_t = \langle \Psi(t) | \mathbf{x} | \Psi(t) \rangle \neq \langle \mathbf{x} \rangle_0$) or both. This moving away from itself is conveniently expressed as an *autocorrelation function,* $\langle \Psi(t) | \Psi(0) \rangle$ or as a *survival probability,* $P(t) \equiv |\langle \Psi(t) | \Psi(0) \rangle|^2$. The reason for the initial decay of the autocorrelation function is either a force (the negative of the gradient of the potential energy surface) acting on the center of a spatially localized $\Psi(0)$ or a coupling term, $\mathbf{H}^{(1)}$, which causes population to "flow" (in state space) from the bright state into a manifold of "*dark states*" (see Fig. 9.1). It is usually possible (and instructive) to identify the specific causes for all of the qualitative features (initial decay rate, times at which of partial recurrences occur, and the amplitudes of these recurrences) in the early-time behavior of $\langle \Psi(t) | \Psi(0) \rangle$. (*iii*) The evolution of $\Psi(t)$ can embody many classes of dynamical behavior, including recurrent oscillation between the initial bright state and a small number of dark states (quantum beats), localized evolution among a small group of states that exhibit complementary brightness/darkness for different detector set-ups, irreversible dephasing (exponential or multi-exponential decay) into a (quasi)-continuum, and localized evolution into a "doorway state" followed by irreversible decay.

A dynamical "mechanism" is specified by answering the classic questions about the time-evolution of an isolated system: *where does it start, how fast does it leave, what causes it to leave, where does it go next, and why does it go there?* Answers to all of these questions are obtainable by constructing an appropriate $\mathbf{H}^{\text{eff}}$ model following the techniques presented in Chapters 1-8. The parameters that define the $\mathbf{H}^{\text{eff}}$ may equally well be defined by fitting the information contained in a frequency domain (transition frequencies, intensities, and linewidths) or a time domain (arrival times of the wavepacket at a specified location in coordinate or state space, decay rates, transfer rates, and recurrence times and amplitudes) experiment.

9.1.2 Solution of the Time-Dependent Schrödinger Equation

The time-dependent Schrödinger equation is

$$i\hbar \frac{\partial \Psi}{\partial t} = \mathbf{H}\,\Psi. \tag{9.1.2}$$

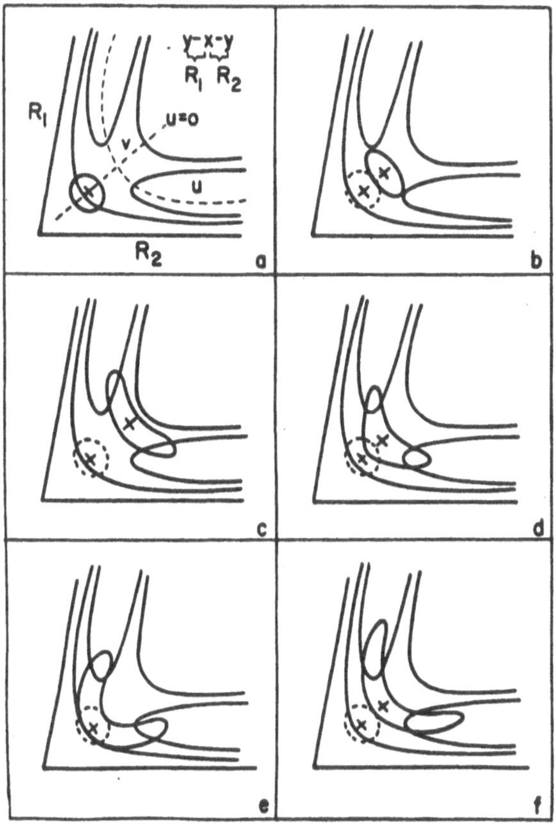

Figure 9.1: Motion of a wavepacket in coordinate space. Panels (a) - (f) depict six times in the evolution of an initially Gaussian wavepacket. The wavepacket is launched at the Franck-Condon point of a repulsive potential surface of a Y-X-Y triatomic molecule. The v coordinate (symmetric stretch) is bound and the u coordinate (antisymmetric stretch) is unbound. The wavepacket oscillates along v (the first partial recurrence is in frame e) and spreads along u (from Heller, 1978).

If **H** is time-independent, then $\Psi(t)$ may be expressed as

$$\Psi(t) = \sum_j a_j \psi_j e^{-iE_j t/\hbar}, \tag{9.1.3}$$

where the wavefunction, ψ_j, and the energy, E_j, are an eigenstate and an eigenenergy, respectively, of the Hamiltonian operator, **H**, that appears in the time-independent Schrödinger equation,

$$\mathbf{H}\,\psi = E\psi. \tag{9.1.4}$$

By completeness of the set of eigenstates of **H**, $\{\psi_i\}$, it always is possible to express any initial state (at $t = 0$) as a superposition of eigenstates,

$$\Psi(0) = \sum_j a_j \psi_j, \tag{9.1.5}$$

where

$$a_j = \langle \psi_j | \Psi(0) \rangle . \tag{9.1.6}$$

But the unadorned Eqs. (9.1.5) and (9.1.6) do not make sufficiently clear the physical significance of the nonstationary,initially-localized, $\Psi(0)$ states that are most easily created by a short, Fourier transform limited pulse of electromagnetic radiation. "Short" is not an absolute quality. For a sufficiently short pulse, the nature of the initial localization prepared and the specific dynamical processes sampled depend primarily on the duration of the preparation pulse and secondarily on its spectral content (Heller, *et al.*, 1982; Johnson, *et al.*, 1996).

The electronic excitation that results from the absorption of a photon corresponds to transfer of the energy and the one unit of angular momentum of the photon into the spatial part of the spin-orbital occupied by one electron. This short-pulse *photon-pluck* creates a $\Psi(0)$ that is described well by one of the zero-order basis states with which we have become familiar in Chapters 1-8. These basis states are not merely members of some arbitrarily chosen complete basis set. The "bright states" are members of the same family of basis states used to describe the pre-excitation initial state, ψ_{initial}, which is an eigenstate of $\mathbf{H}$. The selection rules, which we know are approximate for eigenstate→eigenstate transitions, become exact for short-pulse $\psi_{\text{initial}} \rightarrow \Psi(0)$ plucks: the electronic transition moment operator $\boldsymbol{\mu}$ *is* a one-electron operator (see Section 6.1.1), electron spin magnitude (S) and projections ($\Sigma, \alpha/\beta$) *are* conserved, Hund's coupling case rotational wavefunctions *are* conserved (see Section 3.2.1), and the pre-excitation vibrational wavefunction (see Sections 5.1.1 and 5.1.3) is transferred, in the "vertical" manner ($\Delta x \approx 0, \Delta p = 0$) prescribed by the Franck-Condon principle, onto the final electronic state potential energy surface (see Sections 5.1, 6.1.1, 7.5 and 7.6). (If the electronic transition moment does not depend on internuclear distance, then $\Delta x = 0$.)

A bright state is prepared at $t = 0$. Lee and Heller (1979) derive a rigorous description of the wavepacket that is *actually* prepared when the excitation pulse is of finite duration. It is not a single eigenstate. It evolves in time subject to the spectroscopic time-independent $\mathbf{H}^{\text{eff}}$, as specified by the time-dependent Schrödinger equation Eq. (9.1.2).

9.1.3 Frequency Domain Spectra Treated as the Fourier Transform of the Autocorrelation Function

In the frequency domain, it is natural to think of the spectrum that originates from eigenstate ψ_k,

$$I_k(\omega) = \sum_j c_{kj}\delta(\omega - \omega_{jk}) = \sum_j c_{kj}\delta\left(\frac{1}{\hbar}(E - (E_j - E_k))\right), \tag{9.1.7}$$

as a collection of eigenstate to eigenstate transitions with frequencies,

$$\omega_{jk} = (E_j - E_k)/\hbar, \tag{9.1.8}$$

and intensities, c_{kj}. $\delta(a-b) = \delta(a,b)$ is a Dirac delta function which is zero except when $a = b$. In Eq. (9.1.7) the spectrum is represented as a sum of Dirac delta functions. The rules for calculating the $\{c_{kj}\}$ intensity factors and the models that generate the energy level patterns in $\{E_j\}$ are second nature to frequency domain spectroscopists. But eigenstates, taken individually, are stationary – they seem unrelated to the ball-and-spring motions with which every curious child is familiar. An alternative formulation of the spectrum, $I_k(\omega)$, entails the time-to-frequency Fourier transform of the time-evolving autocorrelation function of the initial state produced by the photon pluck, $\langle\Psi_i(t)|\Psi_i(0)\rangle$. In this form the eigenstates and transition intensities are replaced by a particle-like object, the evolution of which may be described by ball-and-spring pictures (see Section 9.1.7). (Heller, 1975; Heller, *et al.*, 1982; Imre, *et al.*, 1982 and 1984)

Although the eigenstate-to-eigenstate and autocorrelation function formulations of the spectrum, $I_k(\omega)$, are mathematically equivalent, they focus attention on complementary features. The most readily interpretable features in the autocorrelation function picture are *early-time* features (initial decay rate, the times at which partial recurrences occur, the magnitudes of the earliest and largest partial recurrences) which primarily sample the potential surface at the highly localized and *a priori* known initial position of the wavepacket, $\Psi(0)$, before it has had time to explore the entire dynamically accessible region of the potential surface. This early time information is encoded in the broad envelope (low resolution) of the $I_k(\omega)$ spectrum (see Fig. 9.2).

In contrast, the eigenstate-to-eigenstate transition picture focuses on eigenstates. Eigenstates sample a vastly larger region of the potential surface than the highly localized region sampled by the early time behavior of the autocorrelation function. But the global sampling of the potential surface by eigenstates obscures insights into causality or mechanism: where in phase space is the initially localized wavepacket created, what forces cause the wavepacket coordinate and momentum expectation values, $\langle\mathbf{R}\rangle$ and $\langle\mathbf{P}\rangle$, to change from their $t = 0$ values, in what regions of R and P does wavefunction amplitude leak from one potential surface to another? These are inherently local effects that are best explained by locally sampled features of the potential (see Sections 5.1.1 and 7.6).

The autocorrelation function picture of a frequency domain spectrum (Heller, 1981) is derived as follows. Consider the $\{e, v'_e\} \leftarrow g, v''_g$ electronic *absorption* spectrum. In the Franck-Condon limit (constant transition moment) and neglecting rotation, the eigenstate spectrum observed from the v''_g vibrational level

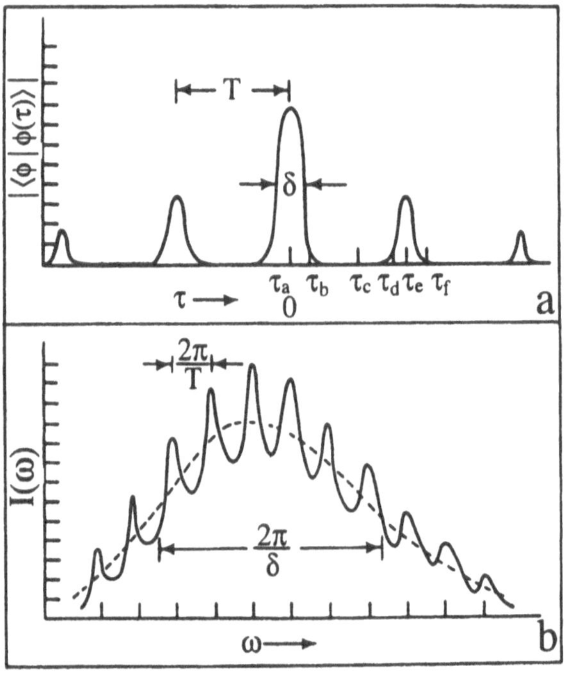

Figure 9.2: Relationships between features in the autocorrelation function, $\langle\phi|\phi(\tau)\rangle$, and lineshapes in the frequency domain spectrum, $I(\omega)$. The times indicated as $\tau_a, \tau_b, \ldots \tau_f$ in the top frame correspond to the wavepacket depicted at six times in Fig. 9.1. The width of the initial decay of the autocorrelation function, δ, is related to the overall width of $I(\omega)$. The time delay between the initial launch of the wavepacket and the first partial recurrence of the autocorrelation function, T, is encoded as oscillations at $\omega = 2\pi/T$ in $I(\omega)$. (from Heller, 1978)

of the electronic ground state, g, is given by

$$I_{v''_g}(\omega) = |\mu_{eg}|^2 \sum_{v'_e} q_{v'_e v''_g} \delta(\omega - \omega_{v'_e, v''_g}), \tag{9.1.9}$$

in which the homogeneously broadened lineshapes of individual transitions are replaced by δ–functions, each centered at the frequency of an electronic-vibrational transition, $\omega_{v'_e, v''_g}$. The pertinent frequencies are: the frequency of the electromagnetic radiation,

$$\omega = E/\hbar, \tag{9.1.10}$$

and the transition frequency,

$$\omega_{v'_e, v''_g} = (E_{e, v'_e} - E_{g, v''_g})/\hbar. \tag{9.1.11}$$

The Franck-Condon factors are

$$q_{v'_e v''_g} = \left|\langle v'_e | v''_g \rangle\right|^2. \tag{9.1.12}$$

The $\delta(\omega - \omega_{v'_e, v''_g})$ δ–function is replaced by its integral definition

$$\delta(\omega - \omega_{v'_e, v''_g}) = \frac{1}{2\pi} \int_{-\infty}^{\infty} dt e^{-i(\omega - \omega_{v'_e, v''_g})t} \tag{9.1.13}$$

and, using the completeness of the vibrational basis set for the e–electronic state, the g–state (initial electronic state) vibrational eigenfunction, $\chi_{v''_g}(R)$, is replaced by a superposition of e–state (final electronic state) vibrational eigenfunctions, $\chi_{v'_e}(R)$.

$$\chi_{v''_g}(R) = \langle R | v''_g \rangle \tag{9.1.14}$$

$$= \sum_{v'_e} \langle R | v'_e \rangle \langle v'_e | v''_g \rangle$$

$$= \sum_{v'_e} \chi_{v'_e} \langle v'_e | v''_g \rangle . \tag{9.1.15}$$

Now, all of the factors in Eq. (9.1.9) may be rearranged into the form

$$I_{v''_g}(\omega) = \frac{1}{2\pi} \int_{-\infty}^{\infty} dt e^{-i(\omega + E_{g, v''_g}/\hbar)t} \langle \Psi_i(t) | \Psi_i(0) \rangle \tag{9.1.16}$$

where

$$|\Psi_i(0)\rangle = |e\rangle \langle e|\mu|g\rangle \sum_{v'_e} |v'_e\rangle \langle v'_e | v''_g \rangle \tag{9.1.17}$$

and

$$\langle\Psi_i(t)| = \langle g|\boldsymbol{\mu}|e\rangle\,\langle e|\sum_{v'_e}\langle v''_g|v'_e\rangle\,\langle v'_e|\,e^{iE_{e,v'_e}t/\hbar}$$
$$= \langle g|\boldsymbol{\mu}|e\rangle\,\langle e|\sum_{v'_e}\langle v''_g|v'_e\rangle\,\langle v'_e|e^{i\mathbf{H}_e t/\hbar}. \qquad (9.1.18)$$

The quantity $\langle\Psi_i(t)|\Psi_i(0)\rangle$ is the autocorrelation function of the electronic-vibrational state function created at time $t = 0$ with a replica of itself evolved from time 0 to time t under the influence of the excited state vibrational Hamiltonian. Thus, the standard Eq. (9.1.9) expression, in terms of individual transition frequencies and transition relative intensities, for a complete $e, \{v'_e\} \leftarrow g, v''_g$ absorption spectrum, may alternatively be viewed as the Fourier transform of the autocorrelation function, Eq. (9.1.16).

The Eq. (9.1.16) alternative form for $I_{v''_g}(\omega)$ is mathematically equivalent to the standard Eq. (9.1.9) form, but in many cases Eq. (9.1.16) is both more convenient to use and provides deeper insights into how local features of the e–state potential energy surface affect the dynamical processes that are encoded in the absorption spectrum (see Figs. 9.1 and 9.2). The keys to these insights are: (i) that $|\Psi_i(0)\rangle$ is the initially localized state-function that *would be produced* by a sufficiently short excitation pulse and (ii) that the early time evolution of $|\Psi_i\rangle$ generates the most prominent features of the absorption spectrum, $I_{v''_g}(\omega)$.

The state-function, $|\Psi_i(0)\rangle$, that appears in Eq. (9.1.16), is the v''_g vibrational wavefunction of the electronic ground state, $\chi_{v''_g}(R) = \langle R|v''_g\rangle$, translated vertically (with respect to both coordinates and momenta) by the excitation pulse at $t = 0$ onto the excited state potential surface, where $\chi_{v''_g}(R)$ is no longer an eigenstate. The state-function, $\langle\Psi_i(t)|$ describes the time-evolved form of the vibrational wavepacket, $\langle R\,|\Psi_i^*(0)\Psi_i(0)|\,R\rangle$, created at $t = 0$, subject to the e–state Hamiltonian.

This may be demonstrated as follows. There are three factors in $\langle\Psi_i(t)|$; an electronic factor

$$\langle g|\boldsymbol{\mu}|e\rangle\,\langle e|\,,$$

which is the electric dipole transition amplitude from $\langle g|$ to $|e\rangle$ multiplying the e–state electronic eigen-bra $\langle e|$; a vibrational factor

$$\sum_{v'_e}\langle v''_g|v'_e\rangle\,\langle v'_e|\,,$$

which is the expansion of $\langle v''_g|$ in the e–state vibrational eigen-basis; and a time-dependent factor,

$$e^{+i\omega_{v'_e,v''_g}t}$$

that expresses the energy difference, $E_{e,v'_e;g,v''_g} = \hbar\omega_{v'_e,v''_g}$, between the e, v'_e and g, v''_g eigenstates. Equation (9.1.16) may alternatively be derived directly by time-dependent perturbation theory in the rotating-wave approximation (provided that the excitation pulse is of sufficiently short duration) (Fayer, 2001).

When the absorption transition originates from the g–state $v''_g = 0$ level, the wavepacket created at $t = 0$ on the e–state potential surface is a Gaussian (Heller, *et al.*, 1982),

$$\Phi_i(R, t = 0) = \langle R | v''_g = 0 \rangle = \left(\frac{2\mu\omega_g}{h}\right)^{1/4} \exp\left[\left(\frac{-\mu\omega_g}{2\hbar}\right)(R - R_{e_g})^2\right] \tag{9.1.19}$$

where Φ_i is normalized in the sense

$$\int dR' \, |\Phi_i(R')|^2 = 1$$

and the FWHM of Φ_i is

$$\mathrm{FWHM}(\Phi_{v''_g=0}) = \left[\frac{\hbar 8 ln 2}{\mu\omega_g}\right]^{1/2} \tag{9.1.20}$$

and ω_g is the (assumed to be harmonic) vibrational frequency of the g–state,

$$\omega_g = [k_g/\mu]^{1/2} \tag{9.1.21}$$

$$k_g = \left.\frac{d^2V_g}{dR^2}\right|_{R=R_{e_g}}, \tag{9.1.22}$$

where R_{e_g} is the g–state equilibrium internuclear distance.

Heller, *et al.*, (1982) give the early-time form (neglecting terms of order t^2 and higher, $O(t^2)$) of the equation of motion for a Gaussian nuclear motion wavepacket on a smooth potential surface,

$$\Phi_i(R, t) = \left(\frac{2\mu\omega_g}{\hbar}\right)^{1/4}$$
$$\exp\left[\left(-\frac{\mu\omega_g(t)}{2\hbar}\right)(R - R(t))^2 + \left(\frac{i}{\hbar}\right)P(t)(R - R(t)) + \left(\frac{i}{\hbar}\right)\gamma(t)\right] \tag{9.1.23}$$

where

$$\omega_g(t) = \omega_g - i\left(\omega_g^2 - \frac{1}{\mu}\left.\frac{d^2V_e}{dR^2}\right|_{R=R(t)}\right)t + O(t^2) \tag{9.1.24}$$

$$R(t) = R_{e_g} + O(t^2) \tag{9.1.25}$$

$$P(t) = -\left.\frac{dV_e}{dR}\right|_{R=R(t)} t + O(t^2) \tag{9.1.26}$$

(note that for excitation from a $v''_g = 0$ initial state, $\langle P(0) \rangle = 0$), and

$$\gamma(t) = \gamma_0 - [\Delta E_{eg} + V_e(R_{e_g})]t + O(t^2), \tag{9.1.27}$$

where ΔE_{eg} is the energy difference between the minima of the e–state and g–state surfaces, and the zero of energy is at the minimum of the e–state potential, $V_e(R_{e_e})$. The $\omega_g(t)$ term shows that the spatial width (square root of the variance in R), $\Delta R(t)$, of the wavepacket does not change at early time, but that the momentum width, $\Delta P(t)$, initially increases at a rate proportional to the difference between the curvatures, at R_{e_g}, in the e–state and g–state potentials. The $R(t)$ and $P(t)$ terms imply that the dominant early-time change in the phase space location of the wavepacket is in the momentum. The $\gamma(t)$ term expresses the overall phase of Φ_i and the $e^{-iEt/\hbar}$ oscillatory term appropriate for the center energy (relative to $E = 0$ at $V_g(R_{e_g})$) of the wavepacket. Rearranging terms, one obtains

$$\begin{aligned}\Phi_i(R,t) &= \left(\frac{2\mu\omega_g}{\hbar}\right)^{1/4}\left(\exp\left[-\left(\frac{\mu\omega_g}{2\hbar}\right)(R-R_{e_g})^2\right]\right)\\ &\times\left(\exp\left\{\left(\frac{it}{\hbar}\right)\left[\left(-\left.\frac{dV_e}{dR}\right|_{R=R_{e_g}}\right)(R-R_{e_g})+\Delta E_{eg}+V_e(R_{e_g})\right]\right\}\right)\\ &\times\left(\exp\left[\left(\frac{it\mu}{\hbar}\right)(R-R_{e_g})^2\left(\omega_g^2-\frac{1}{\mu}\left.\frac{d^2V_2}{dR^2}\right|_{R=R_{e_g}}\right]\right)e^{i\gamma_0/\hbar}.\end{aligned} \tag{9.1.28}$$

The first term is real and encodes the early time-dynamics of the spatial part of the wavepacket. The second and third terms have imaginary exponentials, hence encode momentum. The second and third terms describe the acceleration, $\left\langle \dot{\mathbf{P}} \right\rangle(t)$, and spreading, $\Delta P(t)$, of the momentum part of the wavepacket.

Similar formulations of spontaneous fluorescence, stimulated emission pumping, and Raman spectra as the (full- or half-) Fourier transform of an autocorrelation exist (Tannor, 2003) and provide unique insights into the forces responsible for the early-time dances of atoms subsequent to an artfully designed and implemented initially localized excitation.

The key mathematical point in this autocorrelation function formulation of the frequency domain spectrum is the separation of time scales for the electronic excitation (or de-excitation) process (on the order of the time it takes a large and fuzzy photon to move past a compact molecule, $\delta t = \lambda/c = 1/\nu$, which, for $\lambda =$ 300 nm, gives $\delta t \approx 1 \times 10^{-15} s$) and nuclear motion (for a vibrational frequency of 1000 cm^{-1}, the vibrational period is $3 \times 10^{-14} s$). The key physical insight is that a perfectly known initial-state nuclear vibrational-rotational eigenstate is shifted in energy by the photon (photon pluck) and transfered onto an electronic potential surface where it is no longer an eigenstate, hence no longer stationary.

Because the initial vibrational state for absorption spectra often is $v''_g = 0$, the vibrational nonstationary state typically produced initially is an only slightly distorted Gaussian wavepacket centered at R''_{e_g}. Conservation of momentum requires that this approximately minimum-uncertainty wavepacket be launched at the turning point on the upper surface, $R'_{\pm_e} = R''_{e_g}$, which lies vertically above R''_{e_g}. [This is a consequence of the stationary phase condition, see Sections 5.1.1 and 7.6 and Tellinghuisen's (1984) discussion of the classical Franck-Condon principle.] The initial decay of the autocorrelation function, which is sampled experimentally by the inverse Fourier transform of the spectrum, provides a measure of the gradient of the upper state potential at $R = R''_{e_g}$, $\left.\frac{dV_e}{dR}\right|_{R=R_{e_g}}$ (see Eqs. (5.1.12), (5.1.16), and (5.1.17)).

For a polyatomic molecule, the intensity distribution along the vibrational progressions in each of the totally symmetric, Franck-Condon-active normal modessamples the projections of the gradient of the upper state potential $\left.\frac{\partial V'}{\partial \mathbf{Q}''_i}\right|_{\mathbf{Q}''_e}$ along each of the lower-state normal coordinates, $\mathbf{Q}''_i$, at $\mathbf{Q} = \mathbf{Q}''_e$. Here, $\mathbf{Q}$ specifies the locations of all atoms of an N–atom polyatomic molecule, $\mathbf{Q}'_e$ and $\mathbf{Q}''_e$ are respectively the equilibrium positions of all atoms on the upper and lower electronic state potentials, $V'(\mathbf{Q})$ and $V''(\mathbf{Q})$, and $\mathbf{Q}''_i - \mathbf{Q}''_e$ specifies the displacements from equilibrium of all N atoms along the i–th normal mode of the lower state. (For a nontotally symmetric normal coordinate displacement, $\mathbf{Q}_n$, the first derivative of $V'(\mathbf{Q})$ along the $\mathbf{Q}''_n$ direction is zero by symmetry at $\mathbf{Q} = \mathbf{Q}''_e$, thus the ratio of intensities for absorption transitions terminating in $v'_n = 2$ and 0 samples $\partial^2 V'(\mathbf{Q}''_e)/\partial \mathbf{Q}''^2_n$.) [Imre, *et al.*, 1982; Heller 1975; Lee and Heller 1979; Heller, *et al.*, 1982.]

The earliest partial recurrences of $\langle\Psi(t)|\Psi(0)\rangle$ occur at the times $T_n (n = 1, 2, \ldots)$, when the center of the wavepacket (or a piece of the wavepacket) returns to $\mathbf{Q} = \mathbf{Q}''_e$ (and momentum $\mathbf{P} = 0$, as is required for a turning point). This provides a measure of the vibrational frequency of one of the normal modes on $V'(\mathbf{Q})$ at the vibrational excitation energy, E_{VIB}, specified by

$$E_{\text{VIB}} = V'(\mathbf{Q}''_e) - V'(\mathbf{Q}'_e) \tag{9.1.29}$$

$$\omega(E_{\text{VIB}}) = 2\pi/T_1, \tag{9.1.30}$$

where T_1 is the time of the first partial recurrence of $\langle\Psi(t)|\Psi(0)\rangle$. Given a model (e.g., harmonic or Morse oscillator) for the i–th normal mode on the excited potential, $V'(\mathbf{Q}_i)$, measurements of both $\frac{dV'}{d\mathbf{Q}_i}(\mathbf{Q}''_e)$ and $\omega(E_{\text{VIB}})$, combined with prior knowledge of $\mathbf{Q}''_e$, approximately determines the width of $V'(\mathbf{Q}_i)$ along the i–th normal coordinate at E_{VIB},

$$\Delta\mathbf{Q}'_i(E_{\text{VIB}}) = \mathbf{Q}'_{i+}(E_{\text{VIB}}) - \mathbf{Q}'_{i-}(E_{\text{VIB}}), \tag{9.1.31}$$

where $\mathbf{Q}'_{i\pm}(E_{\text{VIB}})$ are the turning points for the i–th mode on V' at E_{VIB}. For a polyatomic molecule, the initial wavepacket samples the frequencies of several of the Franck-Condon active normal modes on $V'(\mathbf{Q})$ via the earliest time partial recurrences.

All of this information about vibrational frequencies and transition intensities is observable directly in the frequency domain absorption spectrum, $I_{v''=0}(\omega)$. The autocorrelation function picture is an alternative way of deriving a ball-and-spring physical picture (or dynamical mechanism) from the raw experimental data. Although there is a simple Fourier transform relationship between $I_{v''}(\omega)$ and $\langle\Psi(t)|\Psi(0)\rangle$, profoundly different intuitive pictures are used to make sense of experimental results and to guide the design of new experiments.

Figure 9.3 illustrates the information obtainable, based on interpretation via the Fourier transform of an autocorrelation function, about motions of the atoms of a molecule caught in the act of direct photodissociation (see Section 7.2). Photoexcitation transfers the $v_g'' = 0$ vibrational wavefunction of the bound electronic ground state vertically (at $\mathbf{Q}' = \mathbf{Q}_e''$ and $\mathbf{P}' = 0$) to a turning point on the multidimensional repulsive potential surface. The force that acts on the initially created wavepacket has magnitude and direction determined by $-\left.\frac{\partial V'}{\partial \mathbf{Q}}\right|_{\mathbf{Q}=\mathbf{Q}_e''}$, the gradient of the upper state potential surface (a vector quantity) at $\mathbf{Q} = \mathbf{Q}_e'$, the equilibrium geometry of the electronic ground state potential. In response to the force, the wavepacket is accelerated at different rates along each of the normal mode displacement coordinates of the ground state surface. The wavepacket also spreads at rates determined by the differences between the normal mode curvatures of the excited state and ground state surfaces. As the wavepacket moves and spreads, it sequentially overlaps the turning points of various vibrational levels of the electronic ground state. As a consequence of the mode-specific wavepacket accelerations and spreading rates, the various emission transitions from the initially prepared dissociative state to many excited vibrational levels of the ground state surface acquire relative intensities from which local information about both surfaces may be derived.

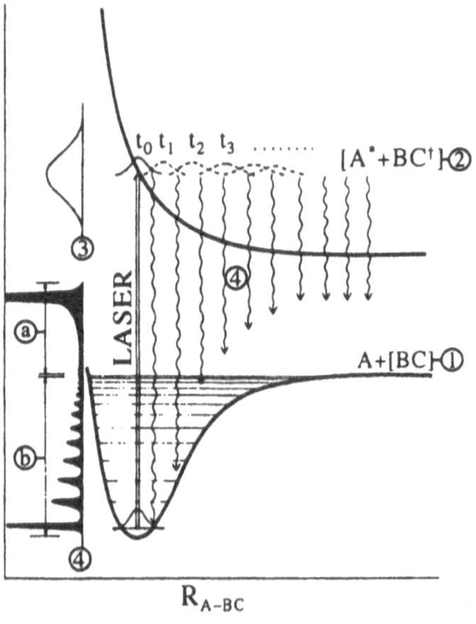

Figure 9.3: Photodissociation Spectra of ABC. A laser pulse transfers the ground state vibrational wavefunction to the repulsive electronically excited potential energy surface. The wavepacket moves and spreads (t_0, t_1, t_2, t_3 snapshots) on the excited surface, being accelerated in the steepest descent direction (force is negative gradient of the potential) toward $A^* + BC^\dagger$ products (where $*$ and $\dagger$ refer respectively to electronic and vibrational excitation). Information about the photodissociation mechanism may be obtained from the (structureless)

absorption spectrum ③, the internal state (v, J, velocity $\vec{v}, \vec{v} : \vec{J}$ correlation) distribution of the photofragments ②, and the dispersed fluorescence (DF) spectrum ④ which consists of discrete ⓑ and continuum red wing ⓐ emission. The relative intensities in the DF spectrum determine the gradient of the potential in the vertical excitation (Franck-Condon) region (from Imre, *et al.*, 1984).

9.1.4 Dynamical Quantities

Eigenstates of a time-independent **H** are stationary. When a molecular system is prepared in a *single eigenstate*, there is *no motion*. The probability amplitude distribution of electron and nuclear positions is frozen in time. The classical ball-and-spring motions with which we are all familiar are encoded in the frequency domain spectrum of transitions between eigenstates in ways that are not obvious. However, by using a short pulse, δt, of electromagnetic radiation, a molecule may be excited from any one of several thermally (incoherently) populated eigenstates to a coherent superposition, $\Psi_I(0)$, of eigenstates that belong to different eigen-energies, provided that these eigen-energies span an energy region smaller than $1/(2\pi\delta t)$,

$$\Psi_I(0) = \sum_i a_i \psi_i \tag{9.1.32a}$$

$$\Psi_I(t) = \sum_i a_i \psi_i e^{-iE_i t/\hbar} \tag{9.1.32b}$$

where $\{\psi_i, E_i\}$ are eigenstates and eigenvalues of **H**. The resultant coherent superposition state moves and this motion is reflected in several pictorially and diagnostically useful quantities:

(*i*) The *probability density*

$$\Psi_I^*(t)\Psi_I(t) = \sum_i \left[|a_i|^2|\psi_i|^2 + \sum_{j>i} \left(a_i a_j^* \psi_i \psi_j^* e^{i(E_j - E_i)t/\hbar} + \text{c.c.} \right) \right]$$
$$= \sum_i \left\{ |a_i|^2|\psi_i|^2 + \sum_{j>i} \left[2\text{Re}(a_i a_j^* \psi_i \psi_j^*) \cos\omega_{ji} t \right.\right.$$
$$\left.\left. -2\text{Im}\left(a_i a_j^* \psi_i \psi_j^*\right) \sin\omega_{ji} t \right] \right\}; \tag{9.1.33}$$

(*ii*) The *density matrix*

$$\rho_I(t) \equiv |\Psi_I(t)\rangle\langle\Psi_I(t)| = \sum_i \left\{ |a_i|^2 |\psi_i\rangle\langle\psi_i| + \sum_{j>i} \left[a_i a_j^* e^{-i\omega_{ij}t} |\psi_i\rangle \right.\right.$$
$$\left.\left. \times \langle\psi_j| + a_i^* a_j e^{i\omega_{ij}t} |\psi_j\rangle\langle\psi_i| \right] \right\};$$
$$(9.1.34)$$

(*iii*) The *autocorrelation function*

$$\langle \Psi_I(t)|\Psi_I(0)\rangle = \sum_i |a_i|^2 e^{iE_i t/\hbar}; \tag{9.1.35}$$

(*iv*) The *survival probability*

$$P_I(t) = |\langle \Psi_I(t)|\Psi_I(0)\rangle|^2 = \sum_i \left[|a_i|^4 + \sum_{j>i} 2|a_i|^2|a_j|^2 \cos\omega_{ij}t \right]; \tag{9.1.36}$$

(*v*) The *probability of transfer* from initial state I to final state F, or more precisely, the probability of arriving at the final (non-evolving) state, $\Psi_F(0)$, after preparing the initial, nonstationary state, $\Psi_I(t)$, at $t = 0$,

$$\begin{aligned}
\Psi_I(0) &= \sum_i a_i\psi_i \\
\Psi_F(0) &= \sum_i b_i\psi_i \\
P_{I\to F}(t) &= |\langle \Psi_I(t)|\Psi_F(0)\rangle|^2 \\
&= \sum_i \Big\{ |a_i|^2|b_i|^2 + \sum_{j>i} \big[2\mathrm{Re}(a_i^* a_j b_i b_j^*)\cos\omega_{ij}t \\
&\qquad - 2\mathrm{Im}(a_i^* a_j b_i b_j^*)\sin\omega_{ij}t\big]\Big\};
\end{aligned} \tag{9.1.37}$$

(*vi*) *expectation values* of electronic (**r**,**p**), vibrational (**R**,**P**), and rotational *coordinates* $(\boldsymbol{\theta}, \boldsymbol{\phi}, \boldsymbol{\chi})$ *and conjugate momenta*

$$\langle \vec{\mathbf{r}}\rangle_{It} = \langle \Psi_I(t)|\vec{\mathbf{r}}|\Psi_I(t)\rangle \quad \text{and} \quad \langle \vec{\mathbf{p}}\rangle_{It} = \frac{\hbar}{i}\left\langle \Psi_I(t)|\vec{\boldsymbol{\Delta}}_{\mathbf{r}}|\Psi_I(t)\right\rangle \tag{9.1.38}$$

$$\langle R\rangle_{It} = \langle \Psi_I(t)|\mathbf{R}|\Psi_I(t)\rangle \text{ and } \langle \mathbf{P}\rangle_{It} = \frac{\hbar}{i}\left\langle \Psi_I(t)|\vec{\boldsymbol{\Delta}}_{\mathbf{R}}|\Psi_I(t)\right\rangle \tag{9.1.39}$$

$$\langle \theta, \phi, \chi\rangle_{It} = \langle \Psi_I(t)|\boldsymbol{\alpha}|\Psi_I(t)\rangle \tag{9.1.40a}$$

$$\langle \mathbf{J} - \mathbf{L} - \mathbf{S}\rangle_{It} = \langle \Psi_I(t)|\mathbf{J} - \mathbf{L} - \mathbf{S}|\Psi_I(t)\rangle; \tag{9.1.40b}$$

where $\boldsymbol{\alpha}$ is the direction cosine matrix which describes the orientation of the molecule fixed coordinate system relative to the space fixed coordinate system and $\mathbf{R} = \mathbf{J} - \mathbf{L} - \mathbf{S}$ is the rotational angular momentum vector of the molecule relative to a molecule fixed axis system. In addition to the above dynamical quantities, there are several others which will be defined in Sections 9.4.6-9.4.11

(especially expectation values of "number operators", $\mathbf{a}_i^\dagger \mathbf{a}_i$ and perturbation or "resonance" operators, $\mathbf{H}^{(1)}$, e.g., $\mathbf{a}_j^\dagger \mathbf{a}_j^\dagger \mathbf{a}_i \mathbf{a}_i + c.c.$).

Each of the above dynamical quantities generates useful insights into the molecular dynamics, especially those quantities in which detailed r– and R–dependent spatial information about eigenfunctions, $\psi_i(r, R)$, has been removed by integration over all internal coordinates. These dynamical quantities, rather than being functions of many internal coordinates, are simple functions of a single variable, t. Such reduced dynamical quantities may be divided into two classes, those describing dynamics in *state space* (*iii, iv,* and *v*) and those describing dynamics in coordinate and/or momentum space (*vi*).

Some insights into the information content of the quantities that describe dynamics in state space are obtained by choosing the simplest possible pair of superposition states

$$\Psi_I(0) = 2^{-1/2}[\psi_1 + \psi_2] \tag{9.1.41a}$$

$$\Psi_F(0) = 2^{-1/2}[\psi_1 - \psi_2], \tag{9.1.41b}$$

for which

$$\langle \Psi_I(t) | \Psi_I(0) \rangle = \frac{1}{2}\left[e^{iE_1 t/\hbar} + e^{iE_2 t/\hbar}\right] = \frac{1}{2}\left(1 + e^{i\omega_{12} t}\right) e^{iE_2 t/\hbar} \tag{9.1.42}$$

$$P_I(t) = \frac{1}{2}(1 + \cos\omega_{12} t) = \cos^2(\omega_{12} t/2) \tag{9.1.43}$$

$$P_{I\to F}(t) = \frac{1}{2}(1 - \cos\omega_{12} t) = \sin^2(\omega_{12} t/2). \tag{9.1.44}$$

The autocorrelation function and survival probability are both 1 at $t = 0$ and undergo 100% oscillations between 1 and 0 with an oscillation period $T = \frac{2\pi}{\omega_{12}}$. $P_{I\to F}(t)$ also undergoes 100% oscillations, but it oscillates π out-of-phase with $P_I(t)$, and, since probability is conserved,

$$P_I(t) + P_{I\to F}(t) = 1. \tag{9.1.45}$$

If more general forms for the two orthogonal superposition states, $\Psi_I(0)$ and $\Psi_F(0)$, composed of two eigenstates are considered,

$$\Psi_I(0) = \cos\theta\ \psi_1 + \sin\theta\ \psi_2 \tag{9.1.46}$$

$$\Psi_F(0) = -\sin\theta\ \psi_1 + \cos\theta\ \psi_2, \tag{9.1.47}$$

then

$$\langle \Psi_I(t) | \Psi_I(0) \rangle = (\sin^2\theta + \cos^2\theta e^{i\omega_{12} t}) e^{iE_2 t/\hbar} \tag{9.1.48}$$

$$P_I(t) = 1 - \frac{1}{2}\sin^2 2\theta(1 - \cos\omega_{12} t) = 1 - \sin^2 2\theta \sin^2(\omega_{12} t/2) \tag{9.1.49}$$

$$P_{I\to F}(t) = \frac{1}{2}\sin^2 2\theta(1 - \cos\omega_{12} t) = \sin^2 2\theta \sin^2(\omega_{12} t/2). \tag{9.1.50}$$

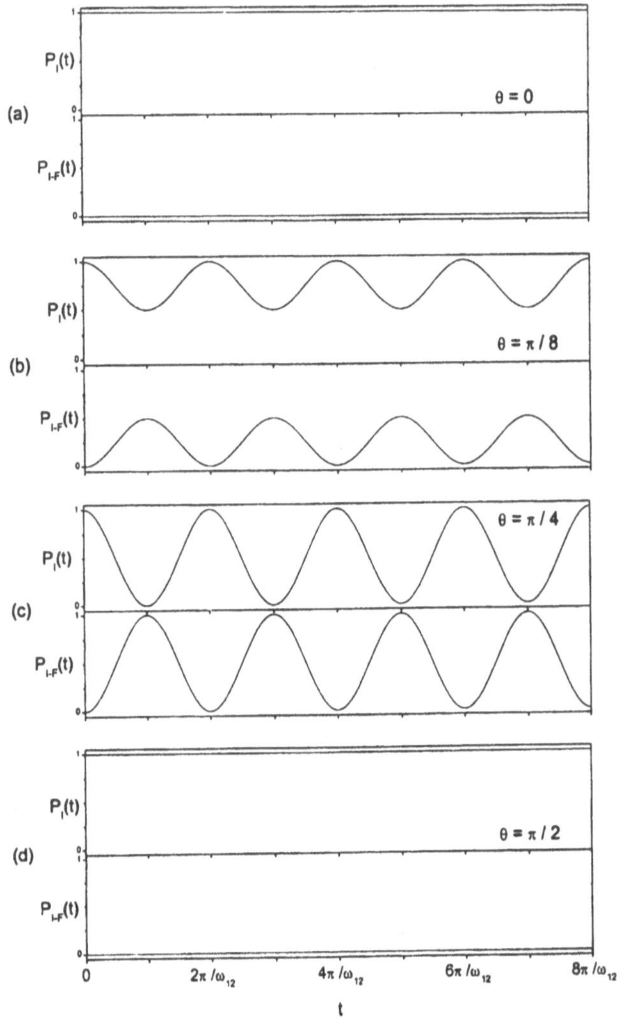

Figure 9.4: Dependence of the amplitude and phase of the survival and transfer probability on mixing angle in $\Psi_I(0)$ (see Eqs. (9.1.46), (9.1.47), (9.1.49), and (9.1.50)). $P_I(t)$ and $P_{I\to F}(t)$ are shown for mixing angles $\theta = 0, \pi/8, \pi/4$ (maximum amplitude), and $\pi/2$ (figure prepared by Kyle Bittinger).

Thus the oscillations in P_I and $P_{I\to F}$ shown in Fig. 9.4 have maximum amplitude when $\theta = \pi/4$ ($\Psi_I(0)$ is a 50/50 mixed state), zero amplitude when $\theta = 0$ or π (Ψ_I is pure ψ_1) or $\pi/2$ (Ψ_I is pure ψ_2), and initial values $P_I(0) = 1, P_{I\to F}(0) = 0$ for all choices of θ.

When the state space spanned by $\Psi_I(0)$ includes more than 2 eigenstates, the dynamics in state space can begin to look very complicated. However, the concepts of "bright state," "dark state," "state-selective detection," and suitable dynamical measures, provide a useful and intuitively simple basis for visualization and interpretation.

9.1.5 General Density Matrix Formulation of a Dynamical Experiment: Excitation, Evolution, and Detection Matrices

Most dynamical experiments are compactly and insightfully describable by applying three transformations to the density matrix of the system, $\rho(t)$. These transformations correspond to *excitation* at $t = 0$, $\mathbf{E}$, evolution from $t = 0$ to $t = \tau$, $e^{-i\mathbf{H}\tau/\hbar}$, and detection at $t = \tau$, $\mathbf{D}$. The detected signal is given by

$$I(\tau) = \mathbf{Tr}\left[\mathbf{D}^{\dagger}e^{i\mathbf{H}^{\dagger}\tau/\hbar}\mathbf{E}^{\dagger}\rho(0)\mathbf{E}e^{-i\mathbf{H}\tau/\hbar}\mathbf{D}\right]. \tag{9.1.51}$$

$\mathbf{H}$ is usually Hermitian,

$$\mathbf{H}^{\dagger} = \mathbf{H} \tag{9.1.52}$$

except when complex diagonal elements are included in $\mathbf{H}$ to account for decay (see Section 9.3). For example, when $\mathbf{E}$, $\mathbf{H}$, and $\mathbf{D}$ are diagonal but $\mathbf{H}$ includes imaginary $-i\Gamma_n/2$ terms on the diagonal, then

$$I(\tau) = \sum_n \rho_{nn}(0)e^{-\Gamma_n\tau/\hbar}, \tag{9.1.53}$$

which is the expected result (consistent with Eq. (9.3.4)). When $\mathbf{H}$ is time-independent, it is convenient to express $\mathbf{E}$, $\mathbf{H}$, and $\mathbf{D}$ in the eigenbasis of $\mathbf{H}$.

Two-state quantum beats (discussed in Section 6.5.3 in a simpler but less powerful notation, and in Sections 9.2.1, 9.2.2 and 9.3.2), provide the simplest illustration of the density matrix formalism. The key ideas are the creation by $\mathbf{E}$ of a coherent superposition state, which produces off-diagonal elements of $\rho(t)$ (known as "coherences"), and the simultaneous selection of the desired coherence and destruction of unwanted coherences by $\mathbf{D}$. It is instructive to consider the essential features of $\mathbf{E}$ and $\mathbf{D}$ by which specific coherences are selectively created and detected.

When an electronic transition of a thermally equilibrated molecule is excited by a short pulse of electromagnetic radiation at $t = 0$, amplitude is removed from rotational-vibrational states of the electronic ground state, g, and amplitude is created in rotational-vibrational states of an electronically excited state, e. At $t < 0$, the molecular density matrix is entirely diagonal and time-independent,

$$\rho(t) = \sum_j |j\rangle\langle j|\rho(t)|j\rangle\langle j| = \sum_j \rho_{jj}(t)|j\rangle\langle j| \tag{9.1.54}$$

$$Q = \sum_j g(E_{gj})e^{-E_{gj}/kT}\rho_{jj}, \tag{9.1.55}$$

where Q is the canonical partition function (Hill, 1986) for the molecule, $g(E_{gj})$ is the degeneracy of the j–th rotational-vibrational level of the g–th electronic state, with eigen-energy E_{gj}, k is the Boltzmann constant, and T is the temperature. The nonzero elements of $\rho(t)$ are located exclusively along the diagonal and are called *populations.* There are no nonzero off-diagonal elements of $\rho(t)$, because, for a thermally-equilibrated system, there are no phase relationships between different energy eigenstates of the same molecule. The phase factors for the eigenstates of different molecules are unselected, hence random.

At $t = 0$, new nonzero diagonal and off-diagonal elements of the density matrix are created by the excitation pulse, but the contributions to $\rho(t)$ that originate from different initially-populated E_{gj} eigenstates have uncorrelated phases. Thus $\rho(t)$ can be partitioned into separate, independent, additive pieces, each piece, $\rho^{(gj)}(t)$, originating from one E_{gj} eigenstate. Thus there is no loss in generality (as far as dynamics on the e–state potential surface is concerned) if all contributions to $\rho(t)$, except those that originate from one E_{gj} eigenstate, are neglected. (When transitions from several thermally populated g, j eigenstates to a single e, k eigenstate are excited, g–state "hole" coherences are produced, but will not be considered here.)

Consider a four-state basis set that consists of the initial state, $|gi\rangle$, which is an eigenstate belonging to the electronic ground state, two intermediate states, $|e1\rangle$ and $|e2\rangle$, which are eigenstates belonging to an electronically excited state, and a final state, $|gf\rangle$, which is an eigenstate distinct from $|gi\rangle$ that belongs to the electronic ground state. In this basis set, $\rho(t < 0), \mathbf{E}, \mathbf{D}$, and $e^{-i\mathbf{H}t/\hbar}$ are each 4×4 matrices,

$$\rho(t < 0) = |gi\rangle\langle gi|, \tag{9.1.56}$$

$$\mathbf{E} = \alpha_{e1gi}\mu_{e1gi}|e1\rangle\langle gi| + \alpha_{e2gi}\mu_{e2gi}|e2\rangle\langle gi| + c.c., \tag{9.1.57}$$

where α_{e1gi} and α_{e2gi} are complex amplitudes related to the amplitude and phase of the radiation field (at ω_{e1gi} and ω_{e2gi}, respectively) at $t = 0$. The transition moments are,

$$\mu_{e1gi} = \langle e1|\mu|gi\rangle \tag{9.1.58}$$

$$\mu_{e2gi} = \langle e2|\mu|gi\rangle\,, \tag{9.1.59}$$

and the time evolution matrix is

$$e^{-i\mathbf{H}t/\hbar} = e^{-iE_{gi}t/\hbar}|gi\rangle\langle gi| + e^{-iE_{gf}t/\hbar}|gf\rangle\langle gf| + e^{-iE_{e1}t/\hbar}|e1\rangle\langle e1| + e^{-iE_{e2}t/\hbar}|e2\rangle\langle e2|. \tag{9.1.60}$$

Three $\mathbf{D}$ matrices are examined here,

$$\mathbf{D}^{e1} = \mu_{gfe1}|gf\rangle\langle e1| + c.c. \tag{9.1.61}$$

and

$$\mathbf{D}^{\pm} = 2^{-1/2}\left[\mu_{gfe1}|gf\rangle\langle e1| \pm \mu_{gfe2}|gf\rangle\langle e2| + c.c.\right]. \tag{9.1.62}$$

The $I(\tau)$ that results from $\mathbf{D}^{e1}$ (detection at $t = \tau$) is shown here to be time independent because a $\mathbf{D}$ matrix with a single complex conjugate pair of off-diagonal elements destroys all coherences,

$$\begin{aligned}\mathbf{E}e^{-i\mathbf{H}\tau/\hbar}\mathbf{D}^{e1} &= \mathbf{E}\left(\mu_{gfe1}e^{-iE_{e1}\tau/\hbar}|gf\rangle\left\langle e1\right| + \mu^*_{gfe1}e^{-iE_{gf}\tau/\hbar}\left|e1\right\rangle\langle gf|\right)\\ &= \left[\alpha^*_{e1gi}\mu^*_{e1gi}|gi\rangle\langle e1|\right]\left[\mu^*_{gfe1}e^{-iE_{gf}\tau/\hbar}|e1\rangle\langle gf|\right]\\ &= \alpha^*_{e1gi}\mu^*_{e1gi}\mu^*_{gfe1}e^{-iE_{gf}\tau/\hbar}|gi\rangle\langle gf|. \end{aligned} \tag{9.1.63}$$

Multiplying Eq. (9.1.63) on the left by

$$\begin{aligned}\mathbf{D}^{e1\dagger}e^{i\mathbf{H}\tau/\hbar}\mathbf{E}^\dagger\rho(0) &= \alpha_{e1gi}\mu_{e1gi}\mu_{gfe1}e^{iE_{gf}\tau/\hbar}|gf\rangle\langle e1|e1\rangle\langle gi|gi\rangle\langle gi|\\ &= \alpha_{e1gi}\mu_{e1gi}\mu_{gfe1}e^{-iE_{gf}\tau/\hbar}|gf\rangle\langle gi| \end{aligned} \tag{9.1.64}$$

to get $I(\tau)$, one obtains

$$\begin{aligned}I(\tau) &= \mathbf{Tr}\left[|\alpha_{e1gi}|^2|\mu_{e1gi}|^2|\mu_{gfe1}|^2e^{i0\tau/\hbar}|gf\rangle\langle gi|gi\rangle\langle gf|\right]\\ &= |\alpha_{e1gi}|^2|\mu_{e1gi}|^2|\mu_{gfe1}|^2. \end{aligned} \tag{9.1.65}$$

The $\mathbf{D}^{e1}$ matrix destroys the coherence created by $\mathbf{E}$ and the resultant $I(\tau)$ is τ–independent. In contrast, as shown below, the $\mathbf{D}^+$ and $\mathbf{D}^-$ matrices preserve the coherences created by $\mathbf{E}$, and the resultant $I(\tau)$ signals are τ–dependent. $\mathbf{D}^+$, the structure of which resembles that of $\mathbf{E}$, produces a "phased up" signal that has maximum intensity at $\tau = 0$. In contrast, $\mathbf{D}^-$ produces a complementary, "phased out" minimum intensity signal at $\tau = 0$. For $\mathbf{D}^\pm$,

$$\begin{aligned}\mathbf{E}e^{-i\mathbf{H}\tau/\hbar}\mathbf{D}^\pm &= \mathbf{E}2^{-1/2}\left[e^{-iE_{gf}\tau/\hbar}\left(\mu_{gfe1}|gf\rangle\langle e1| \pm \mu_{gfe2}|gf\rangle\langle e2|\right)\right.\\ &\quad\left. + e^{-iE_{e1}\tau/\hbar}\mu^*_{gfe1}|e1\rangle\left\langle gf\right| \pm e^{-iE_{e2}\tau/\hbar}\mu^*_{gfe2}\left|e2\right\rangle\langle gf|\right]\\ &= 2^{-1/2}\left[e^{-iE_{e1}\tau/\hbar}\mu^*_{gfe1}\alpha^*_{e1gi}\mu^*_{e1gi}|gi\rangle\langle gf|\right.\\ &\quad\left.\pm e^{-iE_{e2}\tau/\hbar}\mu^*_{gfe2}\alpha^*_{e2gi}\mu^*_{e2gi}|gi\rangle\langle gf|\right]. \end{aligned} \tag{9.1.66}$$

Thus,

$$\begin{aligned}I(\tau) &= \mathbf{Tr}\left[\mathbf{D}^{\pm\dagger}e^{i\mathbf{H}\tau/\hbar}\mathbf{E}^\dagger\rho(0)\mathbf{E}e^{-i\mathbf{H}\tau/\hbar}\mathbf{D}^\pm\right]\\ &= \frac{1}{2}\mathbf{Tr}\left[|\mu_{gfe1}|^2|\alpha_{e1gi}|^2|\mu_{e1gi}|^2 + |\mu_{gfe2}|^2|\alpha_{e2gi}|^2|\mu_{e2gi}|^2\right.\\ &\quad\left.\pm\left\{e^{i\omega_{e2e1}\tau}\mu_{gfe2}\mu^*_{gfe1}\alpha_{e2gi}\alpha^*_{e1gi}\mu_{e2gi}\mu^*_{e1gi}|gf\rangle\langle gf| + \text{c.c.}\right\}\right], \end{aligned} \tag{9.1.67}$$

or

$$\begin{aligned}I(\tau) &= \frac{1}{2}\left[|\mu_{gfe1}|^2|\alpha_{e1gi}|^2|\mu_{e1gi}|^2 + |\mu_{gfe2}|^2|\alpha_{e2gi}|^2|\mu_{e2gi}|^2\right]\\ &\quad\pm\left[(\text{Re}A)\cos\omega_{e2e1}\tau + (\text{Im}A)\sin\omega_{e2e1}\tau\right];\\ A &= \mu_{gfe2}\mu^*_{gfe1}\alpha_{e2gi}\alpha^*_{e1gi}\mu_{e2gi}\mu^*_{e1gi}. \end{aligned} \tag{9.1.68}$$

If all of the μ and α factors are real and positive and, for the special cases,

$$\mu_{gfe2} = \mu_{gfe1} = \mu_f \tag{9.1.69}$$

$$\mu_{e2gi} = \mu_{e1gi} = \mu_i \tag{9.1.70}$$

$$\alpha_{e2gi} = \alpha_{e1gi} = \alpha, \tag{9.1.71}$$

then

$$I(\tau) = \alpha^2 \mu_f^2 \mu_i^2 \left[1 \pm \cos \omega_{e2e1} \tau\right]. \tag{9.1.72}$$

The relative phases and amplitudes of the (μ_{gfe2}, μ_{gfe1}) and (μ_{e2gi}, μ_{e1gi}) transition moments are controlled by vibrational overlap and/or direction cosine matrix elements. It is possible to excite phased up quantum beats by selecting (using a filter or monochromator) one gf final state to monitor $e1, e2 \rightarrow gf$ transitions and phased out quantum beats by selecting a different rotational-vibrational level in the electronic ground state (see Fig. 9.5). If $e1$ and $e2$ are $J' = J'' + 1$ and $J' = J'' - 1$ rotational levels excited via $R(J'')$ and $P(J'')$ transitions of a perpendicular band, then both (μ_{gfe2}, μ_{gfe1}) and (μ_{e2gi}, μ_{e1gi}) transition moment pairs will have opposite signs, and the overall sign of A in Eq. (9.1.68) will be positive.

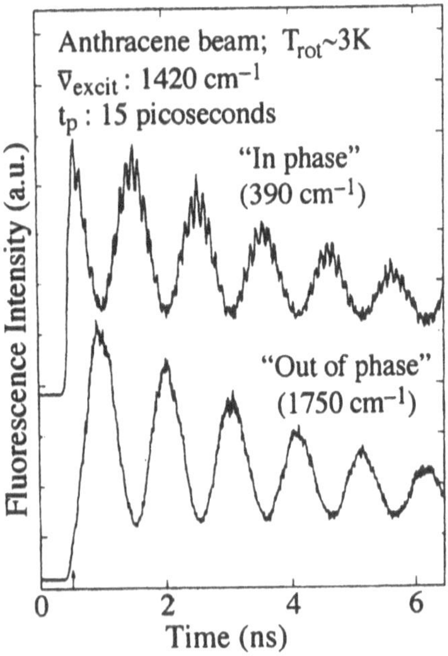

Figure 9.5: The phase of a vibrational quantum beat depends on whether the bright state for the excitation step is bright or dark in the fluorescence detection step. A molecular beam of anthracene, rotationally cooled to 3K, is excited by a 15 picosecond pulse at 1420 cm^{-1} above the $S_1 \leftarrow S_0$ 0_0^0 origin band. Fluorescence is detected in a selected wavelength region through

a monochromator. When the monochromator is tuned to the wavelength of a strong Franck-Condon transition from the bright state (red shifted by 390 $cm^{-1}$from the excitation band) an in-phase quantum beat (I_{MAX} at $t = 0$) is observed. When fluorescence (at a red shift of 1750 $cm^{-1}$from the excitation band) from the single dark state that is anharmonically coupled to the bright state is selected, an out-of-phase quantum beat ($I = 0$ at $t = 0$) is observed. (from Felker and Zewail, 1984)

The utility of the $\mathbf{E}$, ρ, $\mathbf{D}$, $e^{-\mathbf{H}t/\hbar}$ formalism is that excitation, evolution, and detection matrices can be rationally designed to detect selectively any desired coherence, even for level schemes which involve many more than four levels and for systems where all of the levels have different relaxation rates.

9.1.6 Particle (Photon) vs. Wave Pictures of Spectroscopy

Embodied in the $\mathbf{D}$ operator is a profound difference between two forms of molecular spectroscopy. In one form, electromagnetic radiation is thought of as "photons", the photons are treated as independent particles, and there is no need to think about any phase relationships between photons. Molecules are treated as independent targets, and any coherences created by radiation field↔molecule interactions are of a single molecule rather than a bulk sample nature. The $\mathbf{D}$ operator in this *photons as bullets, molecules as targets* form of spectroscopy is mostly based on sampling the diagonal elements (populations) of individual-molecule density matrices. Although the absorption and emission of electromagnetic radiation is always mediated by off-diagonal elements of the density matrix that oscillate at the frequency of the absorbed/emitted electromagnetic radiation, most small molecule, gas phase, frequency domain spectroscopists build their mental picture of how their experiments work almost entirely from populations. Quantum beats are an exception, but the coherence term responsible for quantum beats is a single-molecule coherence that oscillates at a frequency much lower than the optical frequency that creates those beats.

In the other form of spectroscopy, the concepts of independent photons and independent molecules have at best secondary interpretive value (see Section 3.3.5). In nuclear magnetic resonance, the wavelength of the electromagnetic radiation is large compared to the spatial extent of the entire irradiated molecular sample. The energy of a radio frequency photon is much smaller than kT, which makes incoherent detection of absorption or emission of radiofrequency radiation problematic at best. All molecules in the sample see radiation of the same phase, the radiation field induces bulk material polarizations, and the entire, coherently driven sample communicates with the outside world through off-diagonal elements of the many-molecule density matrix. The $\mathbf{D}$ matrix is designed to select or interrogate particular off-diagonal elements of the bulk sample density matrix.

Even when the wavelength of the electromagnetic radiation is small relative to the length of the bulk molecular sample, it is often necessary to think of the matter↔radiation interaction in terms of a bulk polarization induced by the radiation and detected via one or more coherence terms of the sample polarization (Mukamel, 1995). When the wavelength is small relative to sample size, these coherent responses are detected in special phase-matching directions. For example, absorption is detected exclusively in the forward direction because the radiation that generates the bulk polarization and the phase-retarded oscillating field system response travel through the sample at the same velocity. Detected off-axis, radiation from the oscillating bulk polarization undergoes nearly perfect destructive interference. What is left is Rayleigh scattering. There are numerous four-wave mixing schemes (see Section 1.2.6) where three radiation fields induce a bulk polarization that radiates in a unique ($\vec{k}_1 + \vec{k}_2 + \vec{k}_3 = -\vec{k}_4$) wave-vector conserving direction which is *not* coincident with one of the inducing beams (Levenson and Kano, 1988).

The independent-photon and bulk-polarization spectroscopic communities use different aspects of the matter↔radiation interaction to form the cartoons that help them to understand their experiments. The concepts and language are so different that communication between members of the different communities is often difficult.

9.1.7 Motion of the Center of the Wavepacket

Our experience with dynamics in the world of macroscopic objects is based on the observation of motions of spatially well-defined (i.e., localized) objects. We have little direct experience with the wave character of particles, as is manifested in interference and spreading. Introductory quantum mechanics courses often tell us to forget about the localized particle motions (planetary orbits, harmonic oscillators, rotating tops) that are an essential part of our ability to understand, interrogate, and manipulate the macroscopic world. This is in exchange for the promise that all insight and analysis instead is to be based on an inherently-unobservable construct, the set of eigenstates of **H**, $\{\psi_i\}$. Everything we know and understand is mysteriously encoded into Schrödinger eigenstates of **H** or the arrays of numbers with which the Heisenberg **H** matrices are filled. But localized states do exist in quantum mechanics and much of our experience of the macroscopic world does translate directly to interrogation and manipulation of the microscopic world.

Newton's equations

$$\frac{d\vec{r}}{dt} = \frac{1}{m}\vec{p} \tag{9.1.73}$$

$$\frac{d\vec{p}}{dt} = -\nabla V(\vec{r}) \tag{9.1.74}$$

are at the core of our understanding of the macroscopic world. Ehrenfest's theorem

$$\frac{d}{dt}\langle\vec{\mathbf{r}}\rangle = \frac{1}{m}\langle\vec{\mathbf{p}}\rangle \tag{9.1.75}$$

and

$$\frac{d}{dt}\langle\vec{\mathbf{p}}\rangle = -\langle\boldsymbol{\nabla} V(\vec{\mathbf{r}})\rangle \tag{9.1.76}$$

tells us that Newton's equations apply, with some cautions, to non-stationary quantum mechanical states (superposition of eigenstates belonging to several different energy eigenvalues). The more localized the wavepacket, the less stringent are the cautions. Ehrenfest's theorem applies to expectation values of $\vec{\mathbf{r}}, \vec{\mathbf{p}}, \boldsymbol{\nabla}\mathbf{V}(\vec{\mathbf{r}})$, which often are colloquialized to "*the center of the wavepacket.*" If the wavepacket were infinitely sharp, then there would be no difference between expectation value and center in the sense that $\langle\boldsymbol{\nabla}\mathbf{V}(\vec{\mathbf{r}})\rangle$ would be the same as $\boldsymbol{\nabla} V(\langle\vec{\mathbf{r}}\rangle)$. That is,

$$V(q) = V(\mathbf{q}_0) + \frac{dV}{d\mathbf{q}}(\mathbf{q}-\mathbf{q}_0) + \frac{1}{2}\frac{d^2V}{d\mathbf{q}^2}(\mathbf{q}-\mathbf{q}_0)^2 + \ldots \tag{9.1.77}$$

where

$$\langle\mathbf{q}\rangle = \mathbf{q}_0. \tag{9.1.78a}$$

Thus the coefficient of $\left.\frac{dV}{d\mathbf{q}}\right|_{\mathbf{q}=\mathbf{q}_0}$ is zero and

$$\langle V\rangle = V(\mathbf{q}_0) + \frac{1}{2}\left.\frac{d^2V}{d\mathbf{q}^2}\right|_{\mathbf{q}=\mathbf{q}_0}\left(\langle\mathbf{q}^2\rangle - \mathbf{q}_0^2\right) + \ldots \tag{9.1.78b}$$

Similarly for the Taylor series expansion of $\frac{dV}{d\mathbf{q}}$ in the expression for the expectation value $\left\langle\frac{dV}{d\mathbf{q}}\right\rangle$, the coefficient of the $\left.\frac{d^2V}{d\mathbf{q}^2}\right|_{\mathbf{q}=\mathbf{q}_0}$ term vanishes and

$$\left\langle\frac{dV}{d\mathbf{q}}\right\rangle = \left.\frac{dV}{d\mathbf{q}}\right|_{\mathbf{q}=\mathbf{q}_0} + \frac{1}{2}\left.\frac{d^3V}{d\mathbf{q}^3}\right|_{\mathbf{q}=\mathbf{q}_0}\left(\langle\mathbf{q}^2\rangle - \mathbf{q}_0^2\right) + \ldots \tag{9.1.78c}$$

thus

$$\left.\begin{array}{l}\langle V\rangle \neq V(\mathbf{q}_0) \\ \left\langle\frac{dV}{d\mathbf{q}}\right\rangle \neq \left.\frac{dV}{d\mathbf{q}}\right|_{\mathbf{q}=\mathbf{q}_0}\end{array}\right\} \text{ the error for both is } \propto \frac{1}{2}\left(\langle\mathbf{q}^2\rangle - \mathbf{q}_0^2\right) = \frac{1}{2}\sigma_{\mathbf{q}}, \tag{9.1.79}$$

where $\sigma_{\mathbf{q}}$ is the variance (colloquially squared "width") of the wavepacket. But the wavepacket cannot be infinitely sharp in both $\mathbf{q}$ and $\mathbf{p}$, so we are stuck with the cautions. In quantum mechanics the force that acts on a particle is a function of both $\mathbf{q}_0$ and $\sigma_{\mathbf{q}}$; the force on a wavepacket centered at $\mathbf{q}$ is the expectation value of $-\boldsymbol{\nabla} V(\mathbf{q})$, not the negative gradient of the potential evaluated at $\mathbf{q}_0$.

In addition to localization in coordinate ($\mathbf{q}$) or phase ($\mathbf{q}, \mathbf{p}$) space, wavepackets can be localized in abstract basis-state space. Though this kind of localization is less clearly related to the macroscopic world than $\mathbf{q}, \mathbf{p}$ localized wavepackets, state-space localization does have a macroscopic correspondence. When the behavior of a "real life" macroscopic system is observed to deviate slightly from the behavior of a well understood but idealized model, the rate at which the deviation grows may be used to identify the specific cause of the deviation. Weak couplings between parts of a macroscopic system often are understood by building a "toy" model that ignores the couplings and then comparing the evolution of the simplified model to that of the real system. The rate at which energy flows from one part of a system to another reveals the intra-system coupling mechanism and the coupling strength. Systematic experiments in which all of the initial excitation is localized sequentially in each of the different subsystems can be used to characterize the coupling mechanisms that are most important dynamically. A perturbative picture, in which the dynamics is controlled by a well-understood $\mathbf{H}^{(0)}$ and an incompletely-understood perturbation term $\mathbf{H}^{(1)}$, is equally valuable for characterizing macroscopic and microscopic worlds. There is no fundamental difference between wavepacket dynamics in state space and in $\mathbf{q}, \mathbf{p}$ space. The only difference is in the nature of the mental picture one uses to describe and attempt to understand the dynamics. Typically, mechanistic or causal models are built and refined by using well-understood, early-time, localized dynamics in an idealized model to focus one's attention on the most important but initially uncharacterized aspects of the real behavior.

There are many useful tools to help focus attention on specifiable parts of the behavior of a many-body system, and many of these were listed in Section 9.1.4. The concepts of "bright state," "doorway state," and "dark state," the experimental tools of state-selective excitation and detection, and the analysis tools of wavepackets and survival and transfer probabilities, can provide insights into the causes, preferred pathways, and possibilities for control of intramolecular dynamics.

9.1.8 Equations of Motion for Resonance Operators

An enormous amount of dynamical information is encoded in the $\Psi(r, R, t)$ that arises from any non-eigenstate preparation of a molecular system at $t = 0$. One way of displaying intelligible portions of this information is to plot, versus time, the expectation values of various coordinates and momenta, as discussed in Section 9.1.7. There are, however, other kinds of *equations of motion.*

There are several standard results from time-dependent quantum mechanics that will prove useful. The equation of motion of the expectation value of any operator, $\mathbf{O}$,

$$\langle \mathbf{O} \rangle_t = \langle \Psi(r, R, t) | \mathbf{O} | \Psi(r, R, t) \rangle \tag{9.1.80}$$

is obtainable from the commutator of that operator with the exact Hamiltonian,

$$[\mathbf{O}, \mathbf{H}] = \mathbf{OH} - \mathbf{HO} \tag{9.1.81}$$

$$i\hbar \frac{d}{dt} \langle \mathbf{O} \rangle_t = \langle [\mathbf{O}, \mathbf{H}] \rangle_t . \tag{9.1.82}$$

This equation is especially useful when **H** can be written as a sum of simpler operators

$$\mathbf{H} = \sum_i \mathbf{h}_i + \sum_{i>j} \mathbf{h}_{ij} \tag{9.1.83}$$

where, for example, the $\mathbf{h}_{ij}$ coupling term is expressed as a product of operators $\mathbf{a}_i$ and $\mathbf{b}_j$

$$\mathbf{h}_{ij} = c_{ij}\mathbf{a}_i\mathbf{b}_j \tag{9.1.84}$$

that are associated with the i and j subsystems of the molecule. Often, when **H** can be expressed in this way, the $[\mathbf{O}, \mathbf{H}]$ commutator yields a very simple operator, which can be thought of as the "*cause of the dynamics*" of **O**, and for which the expectation value is evaluated easily. In this way the cause of the dynamics may be extracted, purely by operator algebra, without reference to the nature of the specific state prepared at $t = 0, \Psi(r, R, 0)$.

The expectation value of any operator, **O**, can be evaluated by

$$\langle \mathbf{O} \rangle_t = \mathbf{Tr}[\rho(t)\mathbf{O}] \tag{9.1.85}$$

$$\rho(t) = |\Psi(t)\rangle \langle \Psi(t)| = \sum_{i,j} a_i a_j^* |\psi_i\rangle \langle \psi_j| \tag{9.1.86}$$

where Ψ, ρ, and **O** are expressed in the $\{\psi_i\}$ basis set (which can be either eigenstates or basis states) and the expansion coefficients $\{a_i\}$ are time-dependent. The trace of the matrix product $\rho\mathbf{O}$ picks out *only those nonzero products of two expansion coefficients,* $\rho_{ij} = a_i a_j^*$, *to which the operator,* **O**, *is sensitive.* **O** is sensitive to ρ_{ij} only if $\mathbf{O}_{ji} \neq 0$. Once again, operator algebra yields an *a priori* known, simplified sample of the dynamics, independent of the quantitative nature of the specific state prepared at $t = 0$.

The density matrix, $\rho(t)$, is an $N \times N$ array of matrix elements, each of which is time-dependent (but $\rho_{ij} = \rho_{ji}^*$). The $N(N+1)/2$ independent equations of motion, one for each of the $i \leq j \leq N$ elements of ρ, are

$$i\hbar \frac{d}{dt}\rho = [\mathbf{H}, \rho] \tag{9.1.87}$$

$$i\hbar\dot{\rho}_{ij} = \sum_k (H_{ik}\rho_{kj} - \rho_{ik}H_{kj}), \tag{9.1.88}$$

which typically leads to a system of coupled differential equations. (Note that the sign of the commutator in Eq. (9.1.87) is opposite that in Eq. (9.1.82).) Once again, the presence of specific non-zero off-diagonal elements in **H** and ρ focuses our attention on the dynamics of specific *populations* (ρ_{ii}) or *coherences* (ρ_{ij}).

The system of coupled differential equations embodied in Eq. (9.1.87) describes the dynamics of the system in a numerically explicit but mechanistically opaque manner. A mechanism is a short list of the dynamically most important coupling terms, a description of how, where, and why these terms act, and the temporal signature (dominant Fourier components) of each term. Jacobson and Field (2000a) propose a "resonance operator" and "rate operator" formalism to tease out the most important cause-and-effect mechanisms from the exact numerical dynamics.

Any effective Hamiltonian can be partitioned into two parts,

$$\mathbf{H}^{\text{eff}} = \mathbf{H}_{\text{diag}} + \mathbf{H}_{\text{res}}, \tag{9.1.89}$$

a diagonal part and a nondiagonal or "resonance" part. The form (and mechanistic transparency) of this partitioning is dependent on the choice of basis set. The expectation value of the energy of any $\Psi(t)$, subject to a time-independent $\mathbf{H}^{\text{eff}}$, is constant

$$E = \langle \mathbf{H}_{\text{diag}} \rangle_t + \langle \mathbf{H}_{\text{res}} \rangle_t = E_{\text{diag}}(t) + E_{\text{res}}(t). \tag{9.1.90}$$

Energy may be viewed as "flowing" between E_{diag} and E_{res}. The cause of this energy flow may be traced to specific additive terms in $\mathbf{H}_{\text{res}}$,

$$\mathbf{H}_{\text{res}} = \sum_k \mathbf{O}_k, \tag{9.1.91}$$

where each $\mathbf{O}_k$ is constructed to be Hermitian. It is useful to partition each $\mathbf{O}_k$ into two non-Hermitian operators

$$\mathbf{O}_k = \boldsymbol{\Omega}_k + \boldsymbol{\Omega}_k^\dagger \tag{9.1.92}$$

where $\boldsymbol{\Omega}_k$ and $\boldsymbol{\Omega}_k^\dagger$ are matrices composed of all matrix elements $(\mathbf{O}_k)_{ij}$ where $i < j$ and $i > j$, respectively. All other elements of $\boldsymbol{\Omega}_k$ and $\boldsymbol{\Omega}_k^\dagger$ are zero. Since $\boldsymbol{\Omega}_k$ and $\boldsymbol{\Omega}_k^\dagger$ are not Hermitian, their expectation values can be complex numbers, but because $\mathbf{O}_k$ is Hermitian, $\langle i|\mathbf{O}_k|j\rangle = \langle j|\mathbf{O}_k|i\rangle^*$,

$$\langle \boldsymbol{\Omega}_k \rangle_t = \left\langle \boldsymbol{\Omega}_k^\dagger \right\rangle_t^*. \tag{9.1.93}$$

Thus

$$\langle \mathbf{O}_k \rangle_t = \left\langle \boldsymbol{\Omega}_k + \boldsymbol{\Omega}_k^\dagger \right\rangle = \text{real} \tag{9.1.94}$$

and

$$\left\langle \boldsymbol{\Omega}_k - \boldsymbol{\Omega}_k^\dagger \right\rangle = \text{imaginary}. \tag{9.1.95}$$

The expectation values of a specific resonance operator, $\mathbf{O}_k$, and the associated rate operator, $\boldsymbol{\Omega}_k - \boldsymbol{\Omega}_k^\dagger$, can provide insights into the causes of features in the survival and transfer probabilities, $P_I(t)$ and $P_{I\to F}(t)$. The importance of a specific resonance operator, $\mathbf{O}_k$, to the dynamics is given by the time average of the expectation value

$$\lim_{T\to\infty}\frac{1}{T}\int_0^T dt\,\langle\mathbf{O}_k\rangle \equiv \overline{E}_{res,k}. \tag{9.1.96}$$

The *fractional dynamical importance* of the k–th resonance operator relative to all of the coupling terms in $\mathbf{H}_{\mathrm{res}}$ is

$$f_k = \left|\frac{\overline{E}_{\mathrm{res},k}}{\overline{E}_{\mathrm{res}}}\right|. \tag{9.1.97}$$

The mechanistic origin for each Fourier component of the total resonance energy, $E_{\mathrm{res}}(t)$, may be inferred from the existence of Fourier components in the contribution of the k-th coupling term to the resonance energy, $E_{\mathrm{res},k}(t)$, of the same frequency, phase, and similar amplitude. The expectation value of $\boldsymbol{\Omega}_k-\boldsymbol{\Omega}_k^\dagger$ provides information about the *rates* of specific dynamical processes. The same Fourier components will appear in $E_{\mathrm{res},k}(t)$ and $\left\langle\boldsymbol{\Omega}_k-\boldsymbol{\Omega}_k^\dagger\right\rangle_t$, but the phases of the Fourier components of $E_{\mathrm{res},k}(t)$ will lag those of $\left\langle\boldsymbol{\Omega}_k-\boldsymbol{\Omega}_k^\dagger\right\rangle_t$, often by $\pi/2$.

Both $\mathbf{O}_k$ and $\boldsymbol{\Omega}_k-\boldsymbol{\Omega}_k^\dagger$ provide useful insights into the causes and rates of specific dynamical processes. Sections 9.4.9 and 9.4.10 provide analyses of the dynamics of the $\mathbf{S}$–uncoupling operator in a $^{2S+1}\Lambda$ state and the $1:2$ anharmonic coupling operator that contributes to Intramolecular Vibrational Redistribution (IVR) in a polyatomic molecule and illustrate the diagnostic power of the $\boldsymbol{\Omega}_k+\boldsymbol{\Omega}_k^\dagger$ and $\boldsymbol{\Omega}_k-\boldsymbol{\Omega}_k^\dagger$ resonance and rate operators.

9.1.9 The One-Color Pump-Probe Experiment

One of the most important yet simple ideas that ignited excitement about femtochemistry is wavepacket interferometry (Salour and Cohen-Tannoudji, 1977; Scherer, *et al.*, 1990, 1991, 1992; Jonas and Fleming, 1995; Weinacht, *et al.*, 1999), an optical form of Ramsey-fringe spectroscopy (Ramsey, 1990). A molecular system is subjected to two identical optical pulses created by splitting one pulse at a beam splitter. The two pulses are called the "pump" and the "probe". The time delay between pump and probe pulses is scanned systematically using an optical delay line. The optical arrangement is very similar to that of a Fourier Transform Spectrometer (Heller, 1990). The difference in the paths traveled by the pump and probe pulses, Δd, before the two pulses are recombined at a second beam splitter corresponds to a time delay, $\Delta t = \Delta d/c$, where c is the speed of light.

The fundamental idea is that the pump and probe pulses create wavepackets, which evolve on the excited state potential surface. Interference between the excited state wavefunction amplitudes created by the two pulses affects the population transferred to the excited state. The population that is measureable in a typical incoherent experiment (spontaneous fluorescence, field ionization, excitation to a different excited state by a nanosecond pulsed laser) is proportional to the total probability of population transfer, $P(\Delta t)$,

$$P(\Delta t) = |[\alpha^*_{\text{pump}} \langle \Psi(t+\Delta t)| + \alpha^*_{\text{probe}} \langle \Psi(t)|] \times [\alpha_{\text{pump}} |\Psi(t+\Delta t)\rangle + \alpha_{\text{probe}} |\Psi(t)\rangle]|^2 \tag{9.1.98}$$

where α_{pump} and α_{probe} are the transition amplitudes induced by the pump and probe pulses. If the delay is chosen so that $\Psi(t+\Delta t) = \Psi(t)$ (the wavepacket has returned, in-phase, to its $t = 0$ position) and $\alpha_{\text{pump}} = \alpha_{\text{probe}}$, then

$$P(\Delta t = 0) = \left|2|\alpha|^2 \langle \Psi(t)|\Psi(t)\rangle\right|^2 = 4|\alpha|^4. \tag{9.1.99}$$

If the spatial overlap between the two wavepackets is poor

$$|\langle \Psi(t)|\Psi(t+\Delta t)\rangle| \ll 1 \tag{9.1.100}$$

$$\text{P} = 2|\alpha|^4 \,|[1 + \langle \Psi(t+\Delta t)|\Psi(t)\rangle]|^2 \approx 2|\alpha|^4. \tag{9.1.101}$$

The strong dependence of the excited state probability on the choice of Δt gives one the sense that the motion of the wavepacket in real time is actually visible. This is a far more engaging and appealing picture than that of non-moving eigenstates.

Things are not quite as simple as they seem. In order for the constructive interference, which is at the core of wavepacket interferometry, to occur, not only must $\Psi(t+\Delta t) = \Psi(t)$, but also the phases of α_{pump} and α_{probe}, which depend on the *optical* phase of the femtosecond laser rather than the *molecular* phase, must match. A rigorous treatment of the *phase coherent* pump/probe scheme using optically phase-locked pulse pairs is presented by Scherer, *et al.*, [1990, 1991, 1992] and refined by Albrecht, *et al.*, (1999), who discuss the distinction between and consequences of pulse envelope delays vs. carrier wave phase shifts (see Fig. 9.6). A simplified treatment, valid only for weak optical pulses is presented here.

Consider a three level system, the ground state, ψ_g, and two closely spaced excited states, ψ_{e1} and ψ_{e2}. At $t < 0$ the system is in the ground state

$$\Psi(t < 0) = \psi_g e^{-iE_g t/\hbar}. \tag{9.1.102}$$

The pump pulse arrives at $t = 0$,

$$\Psi(0) = (1 - |\alpha|^2)^{1/2}\psi_g + \alpha(\psi_{e1} + \psi_{e2}) \tag{9.1.103}$$

$$\alpha = |\alpha| e^{i\omega_L t}, \tag{9.1.104}$$

where it is assumed that the transition moments, $\mu_{e1g} = \mu_{e2g}$, are equal and that $|\alpha|^2 \ll 1$ and $e^{i\omega_L t}$ is the optical phase, which is a function of ω_L within the bandwidth of the laser pulse. This choice of the optical phase as $e^{i\omega_L 0} = 1$ at $t = 0$ for all ω_L frequencies within the pump pulse is an arbitrary phase choice that is convenient for a simplified view of the effects of a short, smooth

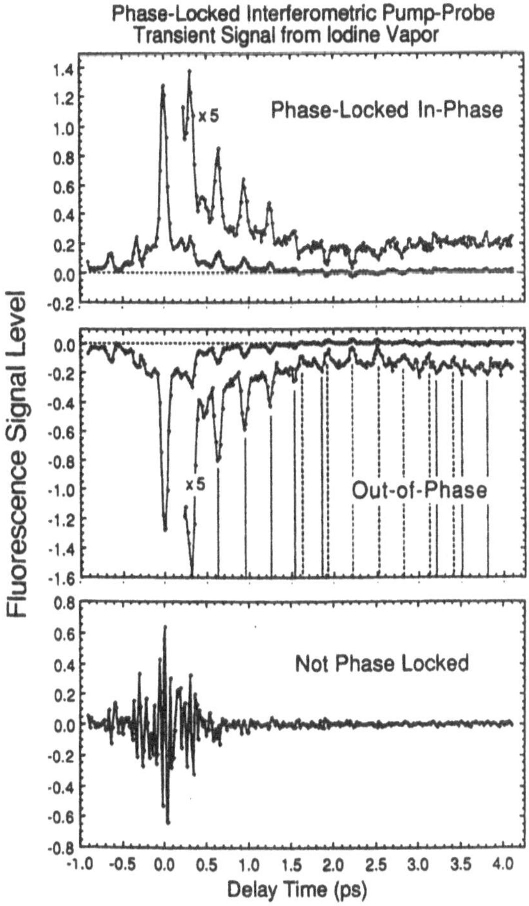

Figure 9.6: Wavepacket Interferometry in I_2. A fluorescence detected one-color pump/probe experiment (65 fs pulses) on the I_2 B–X system illustrates the difference between phase-locked (pump/probe delay stepped in integer or integer +1/2 multiples of the carrier wavelength, 611 nm) and non-phase locked signals. The optical phase (carrier as opposed to envelope phase) is strobed out in the top two phase-locked panels but appears as large amplitude oscillations in the non-phase-locked bottom panel. In the two phase-locked signals, the series of molecular partial rephasings upon return to the Franck-Condon (vertical excitation) region exhibit a phase reversal after ~5 vibrational periods. This periodic reversal of molecular phase is due to anharmonicity. Although the center of the wavepacket oscillates at a constant period, the unequal energy spacings of its constituent eigenstates result in a slow modulation of the wavepacket phase when the wavepacket is sampled at the Franck-Condon point (from Scherer, *et al.*, 1990).

pulse. What really matters for linear spectroscopy [Scherer, *et al.*, 1990, 1991, and 1992] is the phase that accumulates at each ω_L within the pulse between the arrival of the pump pulse at $t = 0$ and the arrival of the probe pulse at $t = \Delta t$. The time-evolving (normalized to 1) wavefunction is

$$\Psi(t) = (1 - 2|\alpha|^2)^{1/2}\psi_g e^{-iE_g t/\hbar} + \alpha\left[\psi_{e1}e^{-iE_{e1}t/\hbar} + \psi_{e2}e^{-iE_{e2}t/\hbar}\right]. \quad (9.1.105)$$

The probe pulse transfers additional amplitude into the excited states. The amplitude transferred from the ground state into each excited state is $\alpha(1 - 2|\alpha|^2)^{1/2} \approx \alpha$, which is much larger than the amplitude transferred from the excited states back to the ground state, $2|\alpha|^2$,

$$|\alpha| \gg 2|\alpha|^2 \qquad \text{because} \qquad |\alpha|^2 \ll 1, \quad (9.1.106)$$

hence transfer back to the ground state by the probe pulse will be ignored, except for the interference effect discussed below.

The probe pulse arrives at $t = \Delta t$, hence the *molecular phase* of the new piece of Ψ created in the excited state will *lag* that of the original piece. The *optical phase* of the ω_L frequency component of the probe pulse will differ from that of the pump pulse by $e^{i\omega_L\Delta t}$, and hence, for $t > \Delta t$ and ignoring the amplitude transferred out of ψ_g,

$$\begin{aligned}\Psi(t) &= \psi_g e^{-iE_g t/\hbar} + \alpha e^{i\omega_L\Delta t}\left[\psi_{e1}e^{-iE_{e1}(t-\Delta t)/\hbar} + \psi_{e2}e^{-iE_{e2}(t-\Delta t)/\hbar}\right] \\ &\quad + \alpha\left[\psi_{e1}e^{-iE_{e1}t/\hbar} + \psi_{e2}e^{-iE_{e2}t/\hbar}\right] \quad (9.1.107)\\ &= \psi_g e^{-iE_g t/\hbar} + \alpha\left[\psi_{e1}\left(e^{i\omega_L\Delta t}e^{-iE_{e1}(t-\Delta t)/\hbar} + e^{-iE_{e1}t/\hbar}\right)\right. \\ &\quad \left. + \psi_{e2}\left(e^{i\omega_L\Delta t}e^{-iE_{e2}(t-\Delta t)/\hbar} + e^{-iE_{e2}t/\hbar}\right)\right]. \quad (9.1.108)\end{aligned}$$

After the probe pulse has passed through the sample ($t > \Delta t$), a total population (sum of the populations in levels E_{e1} and E_{e2}) has been transferred to the excited state. This total population does *not* change for $t > \Delta t$ (ignoring radiative decay and collisions) until the "measurement" is performed. The signal that is measured is proportional to the *probability of population transfer* into both states, E_{e1} and E_{e2},

$$P(t, \Delta t) = P(\Delta t) = |\langle\psi_{e1}|\Psi(t)\rangle|^2 + |\langle\psi_{e2}|\Psi(t)\rangle|^2. \quad (9.1.109)$$

We calculate the sum of probabilities (rather than the squared sum of the amplitudes) because the detection operation destroys the coherence between the $e1$ and $e2$ states (see Section 9.1.5) . It would, in principle, be possible to measure the two level populations separately by dispersing the fluorescence from the $e1$ and $e2$ levels. If that were done, all information about the molecular phase would be destroyed (Jones, *et al.*, 1995).

The population in level E_{e1} is

$$\begin{aligned}|\langle\psi_{e1}|\Psi(t)\rangle|^2 &= |\alpha|^2\left|e^{i\omega_L\Delta t}e^{-iE_{e1}(t-\Delta t)/\hbar} + e^{-iE_{e1}t/\hbar}\right|^2 \quad (9.1.110)\\ &= |\alpha|^2\left[2 + 2\cos(\omega_L\Delta t + E_{e1}\Delta t/\hbar)\right] \quad (9.1.111)\end{aligned}$$

Similarly,

$$|\langle\psi_{e2}|\Psi(t)\rangle|^2 = |\alpha|^2\,[2 + 2\cos(\omega_L \Delta t + E_{e2}\Delta t/\hbar)]\,. \tag{9.1.112}$$

and

$$\begin{aligned} P(\Delta t) &= |\alpha|^2\,[4 + 2\cos(\omega_L\Delta t + E_{e1}\Delta t/\hbar) + 2\cos(\omega_L\Delta t + E_{e2}\Delta t/\hbar)] \\ &= 4|\alpha|^2\left[1 + \cos\left(\frac{E_{e1} - E_{E2}}{2\hbar}\Delta t\right)\cos\left(\omega_L\Delta t + \frac{E_{e1} + E_{E2}}{2\hbar}\Delta t\right)\right]. \end{aligned} \tag{9.1.113}$$

$P(\Delta t)$ is a rapidly oscillating, 100% amplitude modulated function of Δt. It is the product of a slowly oscillating term that contains the molecular information, $\omega_{e1e2} = (E_{e1} - E_{e2})/\hbar$, and a rapidly oscillating term. The usual rotating-wave approach would be to throw away or average over the rapidly oscillating term. However, such an approximation is inappropriate here because $P(\Delta t)$ is sampled discretely rather than continuously. (High frequency terms are ignorable only when the measured signal has been explicitly or implicitly time averaged. The time resolution of a femtosecond pump/probe experiment is set by the pulse durations, not by the temporal response of the detection electronics.) Each laser pulse pair probes a specific Δt. $P(\Delta t)$ contains 100% amplitude oscillations that are extremely sensitive to Δt. To avoid sampling these optical-phase-dependent oscillations of $P(\Delta t)$, Δt is varied in steps which are integer multiples of the oscillation period that corresponds to a selected value of ω_L within the bandwidth of the pulse,

$$\Delta t_n = n\left(\frac{2\pi}{\omega_L}\right). \tag{9.1.114}$$

For 900 nm radiation, the path difference between pump and probe is stepped in 0.9 μm increments (or integer multiples), which corresponds to $\Delta t_1 = 3$ fs. This crucial modification is called *phase coherent* wavepacket interferometry (Scherer, *et al.*, 1990, 1991, 1992) and it results in a profound simplification of $P(\Delta t)$,

$$P(\Delta t) = 4|\alpha|^2\left\{1 + \cos\left(\frac{n\pi}{\omega_L}\omega_{e1e2}\right)\cos\left[\frac{n\pi}{\hbar\omega_L}(E_{e1} + E_{e2})\right]\right\}. \tag{9.1.115}$$

Let δ be the detuning of ω_L from the midpoint between E_{e1} and E_{e2},

$$\delta = \omega_L - \frac{E_{e1} + E_{e2}}{2\hbar} \tag{9.1.116}$$

(note that one chooses δ by selecting the phase coherent time-step-size), then

$$\begin{aligned} P(\Delta t) &= 4|\alpha|^2\,\{1 + \cos(n\pi\omega_{e_1e_2}/\omega_L)\cos[2n\pi - 2n\pi(\delta/\omega_L)]\} \\ P(\Delta t) &= 4|\alpha|^2\left[1 + \cos\left(n\pi\frac{\omega_{e_1e_2}}{\omega_L}\right)\cos 2n\pi(\delta/\omega_L)\right]. \end{aligned} \tag{9.1.117}$$

Reinserting the definition of $n = \Delta t_n \left(\frac{\omega_L}{2\pi}\right)$, the excitation probability simplifies to

$$P(\Delta t) = 4|\alpha|^2 \left[1 + \cos(\omega_{e_1 e_2}\Delta t_n/2)\cos\delta\Delta t_n\right], \tag{9.1.118}$$

where the $\cos(\omega_{e_1 e_2}\Delta t_n/2)$ term is analogous to the ordinary two-level quantum beat and the $\cos\delta\Delta t_n$ is an envelope term that permits deconvolution of the $P(\Delta t)$ signals arising from multiple pairs of levels at different average detunings from ω_{L} [e.g. (E_{e1}, E_{e2} at δ_1), (E_{e3}, E_{e4} at δ_2), etc.] into signals associated with the average energy of each level pair.

If the Δt time step in the pump/probe experiment is only approximately constant, imperfect specification of the optical phase in Eq. (9.1.117) results in 100% pulse-to-pulse amplitude noise. The ω_{e1e2} molecular frequency is still present in $P(\Delta t)$ vs. Δt, but the signal/noise ratio is reduced by the necessity to average over the uncontrolled optical phase (phase averaging). In addition, it is no longer possible to associate each ω_{ij} beat frequency with a level pair at an average energy $\frac{E_i+E_j}{2}$.

If, instead of there being only two near-degenerate levels coherently excited by the pump and probe pulses, there are N levels, it is straightforward to show that $P(\Delta t = 0)$ will be a factor of $4N$ larger than the signal that results from pump-only excitation of a single level and a factor of 2 larger than the signal obtained from time-averaged (via continuous variation of Δt) or incoherent excitation of N levels. If *all* of the N levels are separated by *integer multiples of a common frequency*, as, for example, the vibrational levels of a harmonic oscillator ($\Delta E = n\hbar\omega$) or the quadrupole-split rotational levels of a core-nonpenetrating ($l \geq 3$) Rydberg complex ($\Delta E =$ constant$[l_R^2 - l'^2_R]$ where l_R is the integer-valued projection of the orbital angular momentum of the Rydberg electron onto the ion-core rotation axis), then, when $\Delta t = n\left(\frac{2\pi\hbar}{\Delta E}\right)$, a *grand rephasing* will occur and $P(\Delta t) = P(0)$. The entire wavepacket will rephase and $P(\Delta t)$ will be once again twice as large as the signal that would have resulted from two-pulse incoherent excitation. If a single bright state is fractionated into N eigenstates, each with a transition moment $N^{-1/2}$ as large as that of the unfractionated bright state, then $P(\Delta t_n)$ will decay to a time average value of $1/N$ of its value at $\Delta t = 0$. But if a grand rephasing can occur, $P(\Delta t_n)$ will oscillate between $P(0)$ and 0, a profound contrast with the dephased value, $P(0)/N$.

The phase-coherent pump/probe scheme is equally well suited for viewing the dynamics of a system where the wavepacket moves in Q, P coordinate, momentum space or in $\mathbf{a}_i^\dagger\mathbf{a}_i$ quantum number or state space. One advantage of a time-domain approach to a molecular system in which the individual eigenstates are, in principle, resolvable in the frequency domain spectrum accrues when the time-domain experiment samples the relatively fast dynamics that is encoded as a pattern of relative intensities and frequency splittings that span a wide frequency region. Such large scale patterns often are difficult to uncover in a high resolution frequency domain spectrum. It is a question of being unable to see the whole forest rather than the individual trees. However, fast dynamics reflects the short-time behavior of a system. Often molecules are well described for a short time by an *approximately good quantum number.* The dynamics are more regular and better understandable when the system is confined to a lower dimensionality of state space (e.g. a polyad, see Section 9.4.5) by an approximately good quantum number (the polyad quantum number(s)). It is essential to be able to understand the mechanisms for the destruction of approximately conserved constants of motion. As a system evolves, it samples increasingly larger regions of $V(\mathbf{Q})$ in coordinate space (see Figs. 9.1 and 9.3) and in $\mathbf{a}_i^\dagger \mathbf{a}_i$ state space (see Sections 9.4.6-9.4.10 and Fig. 9.11). When the system initially is localized, it is much easier to relate the dynamical observable in a femtosecond pump/probe experiment to a specific property of a restricted region of the potential surface, $V(\mathbf{Q})$, or to a specific coupling term in $\mathbf{H}$. *Mechanism* consists of a few early time events for which the causes can be determined. These early time events are followed by essentially statistical behavior at later times. It is the early time when the system is still behaving simply that provides the opportunity for insight, visualization, and rationally-based external control. As experimental methods become increasingly sophisticated, it may be possible to push the limits of understanding and control to longer and longer times, but the essential beauty of molecular spectra and dynamics is elegant simplicity, not indescribable complexity.

9.1.10 Crafted Pulses for Detailed Manipulation of Molecular Dynamics

It is possible to encode an enormous amount of amplitude and phase information into a short pulse of electromagnetic radiation. The pulse can be designed to direct the intra-molecular dynamics toward an *a priori* specified target state, the pulse can be modified empirically based on some sort of learning algorithm to maximize a desired dynamical outcome (e.g., a specified photofragment) (Judson and Rabitz, 1992; Bardeen, *et al.*, 1997; Feurer, *et al.*, 2001; Rice and Zhao, 2000; Levis, *et al.*, 2001; Weinacht, *et al.*, 2001; Levis and Rabitz, 2002), or the pulse can be tailored to create a $\Psi(t)$ in which the amplitudes and phases of individual electronic or rotation-vibration eigenstates could be used as a multi-bit memory element in a quantum computer (Ahn, *et al.*, 2000; Ballard, *et al.*, 2002).

A short, smooth, Fourier-transform-limited pulse of electromagnetic radiation may be thought of as being composed of a sum of many longer pulses, each with its own center frequency, amplitude, and phase. By controlling the amplitudes and phases of each of these component pulses, the original short, smooth pulse can be converted into a series of temporally displaced, frequency-chirped sub-pulses (Kawashima, *et al.*, 1995; Cao and Wilson, 1997). Such crafted pulses have been used to accomplish a variety of control and information storage objectives.

When a short ($\Delta t \approx 30\ fs, \Delta L = c\Delta t \approx 10\ \mu m$) laser pulse is expanded by a telescope and input on a grating, its frequency components are separated in the transverse direction and the pulse is stretched in duration and spatial length (30 $ps \sim 1\ cm$). The specific spatio-temporal effects of the grating on the laser pulse depend on the grating characteristics (grooves/mm, tilt angle) and the size of the expanded beam. If the stretched beam is input on a computer controlled, 2–D liquid crystal array device, the amplitude and/or phase of the radiation transmitted through each pixel may be modified. The modified pulse is then recompressed by a second grating plus telescope combination. The result is a pulse capable of exerting exquisite control over molecular dynamics and the post-pulse state of the molecular system.

Optimal chirped-pulse schemes for achieving population inversion ("molecular π pulses") and to explain the chirp-dependence of multiphoton absorption yields have been described by Cao (Cao and Wilson, 1997; Cao, *et al.*, 1998; Cao, *et al.*, 2000). The learning algorithm approach has been reviewed by Levis, *et al.*, (2001) and Rabitz, *et al.*, (2000). The use of masks, arrays, and computer controlled liquid crystal devices for phase and amplitude control has been described by Kawashima, *et al.*, (1995); Weiner, (1995); Krause, *et al.*, (1997); and Tull, *et al.*, (1997). Schemes for storing information in the rotation-vibration levels of diatomic molecules have been implemented by Ballard, *et al.*, (2002) and Stauffer, *et al.*, (2002).

9.2 From Quantum Beats to Wavepackets

The strong- and weak-coupling limits of the two-state quantum beat problem (see Section 6.5.3) are discussed in Section 9.3.2. A short excitation pulse creates a coherent superposition of several nondegenerate eigenstates or quasi-eigenstates (i.e. decaying states described in the complex energy formalism of Section 9.3.1). Since the eigenstates coherently excited by an excitation pulse at $t = 0$ do not have the same energy, the wavefunction $\Psi(t)$ for each molecule evolves after the excitation pulse in a manner determined by the pulsed molecule↔radiation field interaction. The spectral distribution, total intensity, spatial distribution, and polarization of the molecular spontaneous fluorescence can exhibit quantum beats at some or all of the energy separations of the eigenstates included in $\Psi(t)$ (Aleksandrov, 1964; Dodd, *et al.*, 1964; Haroche, *et al.*, 1973; Haroche, 1976; Wallenstein, *et al.*, 1974; Hack and Huber, 1991). Alternatively, the ability of the coherently-prepared molecule to absorb or emit electromagnetic radiation under stimulation by a probe pulse also can exhibit temporal oscillations in spectral content and spatial anisotropy.

Although a simple Fourier transform relationship can exist between a high spectral resolution frequency domain experiment and a time-domain quantum beat experiment, whenever the excitation and detection steps involve electromagnetic radiation of different spectral, spatial, or temporal characteristics, the intrinsic information content of time and frequency domain experiments needs not be identical. The format in which the information is presented may be more transparently interpretable in either the time or frequency domain.

A two-level quantum beating system exhibits one beat frequency. An N-level beating system can exhibit as many as $N(N-1)/2$ beat frequencies.

This suggests that a quantum beat spectrum of an N-level system will, because it contains redundant information, be more complicated than the corresponding frequency domain spectrum. However, when the level spacings are approximately integer multiples of a common factor, such as $2B'$ for upper-state $\Delta_2F(J) = B'(4J+2)$ rotational combination differences, then *each* upper state $(J'+1, J'-1)$ pair of rotational levels coherently excited from *all* thermally populated lower-state J'' levels contributes to a "grand rephasing" at $t_n = n\left[\frac{\pi}{Bc}\right] \quad (n = 1, 2, \ldots)$. This is Rotational Coherence Spectroscopy (RCS) (Felker and Zewail, 1987 and 1995; Felker, 1992). It provides upper state rotational constants without the need for a rotational analysis.

The earliest pulsed laser quantum beat experiments were performed with nanosecond pulses (Haroche, *et al.*, 1973; Wallenstein, *et al.*, 1974; see review by Hack and Huber, 1991). Since the coherence width of a temporally smooth Gaussian 5 ns pulse is only $\sim 0.003\ \text{cm}^{-1}$, ($121 fs \leftrightarrow 121\ \text{cm}^{-1}$ for a Gaussian pulse) nanosecond quantum beat experiments could only be used to measure very small level splittings [e.g. Stark (Vaccaro, *et al.*, 1989) and Zeeman effects (Dupré, *et al.*, 1991), hyperfine, and extremely weak perturbations between accidentally near degenerate levels (Abramson, *et al.*, 1982; Wallenstein, *et al.*, 1974)]. The advent of sub-picosecond lasers has expanded profoundly the scope of quantum beat spectroscopy. In fact, most pump/probe wavepacket dynamics experiments are actually quantum beat experiments cloaked in a different, more pictorial, interpretive framework.

9.2.1 Polarization Quantum Beats

Stark and Zeeman polarization quantum beats are discussed in Section 6.5.3. An external electric or magnetic field destroys the isotropy of space. As a result, the amplitudes for two transition sequences $J'', M'' \to J', M' = M'' \pm 1 \to J''', M''$ interfere, and the intensity of X or Y (but not Z) polarized fluorescence is modulated at $(E_{J'M'=M''+1} - E_{J'M'=M''-1})/\hbar$. However, it is not necessary to destroy the isotropy of space in order to observe polarization quantum beats.

An isotropic sample of gas phase molecules is excited by a pulse from a linearly polarized beam of light. It is convenient to choose the direction of the linear polarization as the quantization (Z) axis ($\Delta M_J = 0$ selection rule). If the duration of the light pulse is sufficiently long that at most one J', $M'_J = M''_J$ excited eigenstate can be excited from each J'', M''_J thermally populated eigenstate, then an anisotropic but *stationary* distribution of excited molecules is prepared. One molecule in the J'', M_J initial state is excited by a Z–polarized photon to the J', M_J state. The $\vec{J}$ vector of this excited molecule lies on a cone which makes an angle $\theta = \arccos(M/[J(J+1)]^{1/2})$ relative to the $+Z$ axis in the laboratory. The azimuthal angle of $\vec{\mathbf{J}}$ on this cone is stationary, but the spatial isotropy of the sample of many molecules results in all values of this azimuthal angle being equally populated.

The transition moment is the antenna by which radiation enters and leaves a molecule. It may also be viewed as the mechanical handle upon which the radiation field applies the torque that transfers the one unit of photon angular momentum into the molecule. The initial state $|\Omega'' J'' M_J\rangle$ rotational wavefunction specifies the distribution of orientations of this $\Omega' J' M \leftarrow \Omega'' J'' M$ molecule frame transition moment relative to the laboratory frame. Since $|\Omega'' J'' M_J\rangle$ is a single eigenstate, the laboratory frame transition moment distribution for a single molecule in the $|\Omega'' J'' M_J\rangle$ state is stationary (but anisotropic), and there is no $t = 0$ laboratory frame localization of transition amplitude beyond that specified by the $|\Omega J M_J\rangle$ rotational wavefunction.

If the photon pulse is sufficiently short and transitions from one initial J'', M_J state to two or more excited eigenstates belonging to different J' (same-M_J) quantum numbers lie within the photon spectral width, then a nonstationary state is prepared at $t = 0$. If the coherently excited $J' = J'', J'' \pm 1, M_J$ eigenstates are not degenerate, the transition amplitudes for the transitions into common J'', M''_J eigenstates will interfere in a time-dependent manner. In particular, the laboratory frame orientation of the transition moment for spontaneous fluorescence evolves in time. The intensities of $z-$ and $(x, y)-$polarized fluorescence are modulated $\pi/2$ out of phase, but the intensity of the total $x + y + z$ polarized fluorescence is not modulated. This is the physical basis for polarization quantum beats (Aleksandrov, 1964; Dodd, *et al.*, 1964) and Rotational Coherence Spectroscopy (Felker and Zewail, 1995).

The temporally oscillatory signal associated with the laboratory frame motion of the fluorescence transition moment may be detected by monitoring fluorescence of selected polarization state. It may also be detected by using a second polarized photon field to induce transitions to another state, from which fluorescence (Ducas, *et al.*, 1975; Lange and Mlynek, 1978; Coté, *et al.*, 1989) or an ion signal is detected (Zygan-Maus and Wolter, 1978; Georges and Lambropoulos, 1978).

Spectroscopic perturbations follow $\Delta J = 0, \Delta M_J = 0$ selection rules. Therefore perturbations are *not* a necessary feature of polarization quantum beats. However, such perturbations *are* essential to a major class of population quantum beats.

9.2.2 Population Quantum Beats

The language of *bright state* and *dark state* is central to population quantum beats and also to the related polyatomic molecule radiationless decay processes (Bixon and Jortner, 1968; Rhodes, 1983), Intramolecular Vibrational Redistribution (IVR) (Parmenter, 1983; Nesbitt and Field, 1996; Wong and Gruebele, 1999; Keske and Pate, 2000), Inter-System Crossing (ISC), and Internal Conversion (IC), discussed in Section 9.4.15.

The reason for the name "population" quantum beats is that the signal intensity (fluorescence, REMPI), *integrated over all solid angles and polarization states*, oscillates in time after the preparation pulse. It appears as if the population prepared in the excited state at $t = 0$ vanishes and returns periodically (see Fig. 9.5) (Felker and Zewail, 1984). In fact, the population does not oscillate, but the radiative capability of the time-evolving state prepared at $t = 0$, $\Psi(t)$, does oscillate.

At $t = 0$, a photon pulse excites a molecule to a basis state via an allowed transition. But the basis state is not a single eigenstate of the photon-free Hamiltonian. The bright basis state is constructed at $t = 0$ as a coherent superpositon of all of the eigenstates, accessible within the bandwidth of the excitation pulse, that contain appreciable character of that basis-state. The quantum beating intensities follow the survival probability of the bright state

$$I_{\rm QB}(t) \propto P_{\rm BRIGHT}(t) = \left|\langle\Psi(t)|\Psi(0)\rangle\right|^2 . \tag{9.2.1}$$

In the frequency domain, the absorption spectrum in the region of the transition into the bright state exhibits *fractionation* into a cluster of eigenstate components. The intensity weighted average energy and integrated absorption strength are, respectively, the zero-order (deperturbed) energy and transition strength of the bright basis state.

As the bright state mixes with more and more dark states, the resultant lineshape evolves toward a smooth Lorentzian shape (Bixon and Jortner, 1968). The width of this composite line approaches that predicted by Fermi's Golden rule formula

$$\Gamma = 2\pi \left\langle \mathbf{H}^2_{\rm BD}\right\rangle \rho_{\rm D} \tag{9.2.2}$$

where $\left\langle \mathbf{H}^2_{\rm BD}\right\rangle$ is the average of the squared bright$\sim$dark state matrix element and $\rho_{\rm D}$ is the density of dark states. As $\rho_{\rm D}$ increases, the time-domain signal changes from a few-level quantum beat to a many-level quantum beat and finally to an exponential decay with lifetime $\tau_{\rm nr} = \hbar/\Gamma$, where $1/\tau_{\rm nr}$ is a dephasing rate, not a spontaneous fluorescence rate. The radiative lifetimes of the individual eigenstate components of an N–component fractionated line are $\sim N\tau_{\rm r}$, where $1/\tau_{\rm r}$ is the radiative decay rate ($\sum_f A_{if}$, A is the Einstein coefficient and the sum is over all lower states, see Section 6.1.1) of the bright state. More precisely, the radiative decay rate of the bright state, summed over the decay rates of the N eigenstates that comprise the fractionated line, is $1/\tau_{\rm r}$.

A crucial difference between diatomic molecules and polyatomic molecules is that the density of nonfragmenting (bound) dark states can become so large that individual eigenstates can no longer be resolved. As a result, fast radiationless processes that do not involve photofragmentation (see Section 9.4.14) are the rule rather than the exception. However, most of the interpretive concepts, perturbation mechanisms, and dynamical models are well exemplified in the spectra and dynamics of diatomic molecules.

9.2.3 Nuclear Wavepackets

The two essential components of a nuclear wavepacket experiment in a diatomic molecule are an *excitation pulse* that is of sufficiently short duration that it creates, from a single v'', J'' initial eigenstate, a state $\Psi(t)$ that contains amplitudes in *at least two* $J' = J'' + 1$ and $J' = J'' - 1$ rotational eigenstates (a rotational wavepacket with motion in the rotational coordinates, θ, ϕ, which specify the orientation of the molecule fixed coordinate system relative to the laboratory fixed coordinate system) or *at least two* $v' = v, J'$ and $v' = v + n, J'$ vibrational eigenstates (a vibrational wavepacket with motion in the vibrational coordinate, R), e.g.,

$$\Psi(\theta, \phi, R; t) = \sum_{\substack{J'=J''+1, J''-1 \\ v'=v, v+n}} a_{v'J'} \langle \theta\phi | \Omega' J' M' \rangle \langle R | v' \rangle e^{-iE_{v'J'}t/\hbar}, \tag{9.2.3}$$

and a *detection scheme* that is capable of detecting the arrival of the center of the wavepacket at specifiable values of θ and ϕ or R

$$\begin{aligned} \theta_{\text{detect}} &= \langle \theta \rangle_t \\ \phi_{\text{detect}} &= \langle \phi \rangle_t \\ R_{\text{detect}} &= \langle R \rangle_t . \end{aligned}$$

For the four-eigenstate superposition

$$[(v' = v, J' = J'' + 1), (v, J'' - 1), (v + n, J'' + 1), (v + n, J'' - 1)],$$

$$\begin{aligned} \langle \theta \rangle_t = & \sum_{J'=J''-1}^{J''+1} \left(|a_{vJ'}|^2 + |a_{v+nJ'}|^2 \right) \langle \Omega' J' M' | \theta | \Omega' J' M' \rangle + \\ & \sum_{v'=v}^{v+n} \Big[a^*_{v'J''+1} a_{v'J''-1} \langle \Omega' J'' + 1 M' | \theta | \Omega' J'' - 1 M' \rangle e^{i(E_{v'J''+1} - E_{v'J''-1})t/\hbar} \\ & + \text{c.c.} \Big], \end{aligned} \tag{9.2.4}$$

a similar expression for $\langle \phi \rangle_t$, and

$$\begin{aligned} \langle R \rangle_t = & \sum_{v'=v}^{v+n} \left(|a_{v'J''+1}|^2 + |a_{v'J''-1}|^2 \right) \langle v' | R | v' \rangle + \\ & \sum_{J'=J''-1}^{J''+1} \left[a^*_{vJ'} a_{v+nJ'} \langle v | R | v + n \rangle e^{i(E_{vJ'} - E_{v+nJ'})t/\hbar} + \text{c.c.} \right]. \end{aligned} \tag{9.2.5}$$

To a good approximation, there are no $v, v + n$ cross terms in $\langle \theta \rangle_t$ and $\langle \phi \rangle_t$ and no $J'' - 1, J'' + 1$ cross terms in $\langle R \rangle_t$. As a result of the dependence of the upper state rotational constant on v', the $(E_{v'J''+1} - E_{v'J''-1})$ energy difference in the time dependent terms for $\langle \theta \rangle_t$ and $\langle \phi \rangle_t$ is weakly v'–dependent and the $(E_{vJ'} - E_{v+nJ'})$ energy difference in the time dependent terms for $\langle R \rangle_t$ is weakly J'–dependent. The result will be a secondary, long period dephasing and rephasing of the two, time-dependent contributions to the primary oscillation periods of $\langle \theta \rangle_t$, $\langle \phi \rangle_t$, and $\langle R \rangle_t$. The primary oscillation periods are

$$T_\theta = T_\phi \approx \frac{1}{c\left(B'_v + B'_{v+n}\right)(1/2)(4J'' + 2)} \tag{9.2.6}$$

$$T_R \approx \frac{1}{c(n\omega'_e)}. \tag{9.2.7}$$

9.2.3.1 Vibrational Wavepackets

The following discussion of vibrational wavepackets is unconventional in its emphasis on vibrational eigenstates and on the simple ideas used by frequency domain spectroscopists to understand their spectra. It is written with the hope of beginning to de-mystify the apparently disparate concepts and tools used by the frequency and time domain communities.

In most vibrational wavepacket experiments, the time-evolving $\Psi(t)$ generally contains more than two vibrational eigenstates. The vibrational quantum numbers and $t = t_0$ amplitudes of these vibrational eigenstates in $\Psi(t)$ are determined by the nature of the pluck that creates $\Psi(t)$ at t_0. If the excitation-pulse is a simple, smooth, and short-time Gaussian, the result is a *Franck-Condon pluck*

$$|\Psi(t_0)\rangle \propto \sum_{v'} \langle v'|v''\rangle\, |v'\rangle\, \mathcal{E}\left(\omega_{v'v''}, t_0\right) \tag{9.2.8}$$

where $\langle v'|v''\rangle$ is a vibrational overlap integral and $\mathcal{E}(\omega_{v'v''}, t_0)$ is the complex amplitude of the excitation-pulse at frequency $\omega_{v'v''}$ and time t_0. If the excitation pulse is sufficiently short and smooth (the amplitude and phase of $\mathcal{E}(\omega_{v'v''}, t_0)$ are independent of v'; also, the spectral content of the pulse does not change during the pulse, i.e. no "chirp") then the coherent superposition state created at t_0 is a near perfect replica of $\psi_{v''}(R) = \langle R|v''\rangle$. Since $\psi_{v''=0}$ has a Gaussian shape, the vibrational wavepacket created on the electronically excited potential curve from a $v'' = 0$ initial state also has a Gaussian shape at $t = t_0$. As the wavepacket evolves on a non-harmonic potential curve, it loses its Gaussian shape. The successive $v', v' + 1$ eigenstate pairs within $\Psi(t)$ have slightly different oscillation periods

$$\begin{aligned} \Delta T = T_{v',v'+1} - T_{v'+1,v'+2} &= \frac{1}{c[\omega'_e - 2\omega_e x'_e(v'+1)]} \\ - \frac{1}{c[\omega'_e - 2\omega_e x'_e(v'+2)]} &\approx -\frac{\omega_e x'_e}{c\omega'^2_e} \end{aligned} \tag{9.2.9}$$

thus the fractional change in period is

$$\Delta T/T \approx -\, \omega_e x'_e / \,\omega'_e \tag{9.2.10}$$

and the phase dispersion per round trip is $2\pi\omega_e x'_e/\omega'_e$. If the excitaiton pulse is negatively chirped (the center frequency decreases during the pulse), the slower moving, high–v' part of the vibrational wavepacket is created first and gets a head start relative to the faster moving, low–v' part of the wavepacket (Cao and Wilson, 1997). It is possible to design a chirped pulse to create a vibrational wavepacket that reaches a state of maximal compression at a time, t_{focus}, and position $\langle R\rangle_{t_{\text{focus}}}$ chosen by the experimentalist. The vibrational wavepacket achieves this state of maximal compression only once.

When the vibrational wavepacket is launched at t_0 from a $v'' = 0$ initial eigenstate on the electronic ground state potential surface,

$$\chi_{v''=0}(R) = \langle R|v'' = 0\rangle\,,$$

its t_0 shape on the electronically excited potential surface is approximately that of a single Gaussian

$$\Psi'(t_0) \approx \langle R|v'' = 0\rangle \tag{9.2.11a}$$

located at

$$\langle R\rangle_{t_0} \approx R''_e \tag{9.2.11b}$$
$$\langle P\rangle_{t_0} \approx 0 \tag{9.2.11c}$$

and (t–independent) total energy

$$\langle E\rangle \approx E''_{v''=0} + [V'(R''_e) - V''(R''_e)]\,. \tag{9.2.11d}$$

Provided that the detuning, δ, of the center of the pump-pulse, $\langle\omega_L\rangle$, from the vertical transition energy

$$\delta = \langle\omega_L\rangle - [V'(R''_e) - V''(R''_e)]/\hbar \tag{9.2.12}$$

is less than the $1/\Delta t$ Fourier bandwidth of the laser pulse (Δt is the temporal FWHM of the pulse), then $\langle R\rangle_{t_0}$ and $\langle E\rangle$ are insensitive to δ.

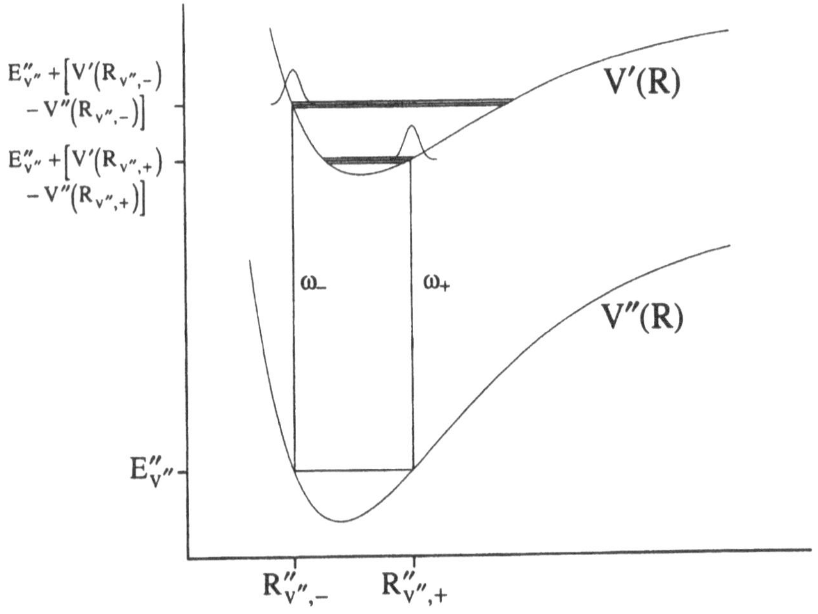

Figure 9.7: Vibrational wavepackets launched from a $v'' > 0$ initial state. The vertical excitations from the inner ($R_{v'',-}$) and outer ($R_{v'',+}$) turning points of the $v'' > 0$ initial state on the $V''(R)$ potential curve, centered respectively at ω_- and ω_+, create two spatially and energetically distinct wavepackets on the excited $V'(R)$ potential curve.

The situation is much more complicated for a vibrational wavepacket launched from a $v'' \neq 0$ initial eigenstate (see Fig. 9.7). However, a simple picture based on the classical Franck-Condon (stationary phase) principle (see Section 5.1.1) captures the essential details of the wavepacket produced at t_0 on the electronically excited potential surface. First, there is the limiting case of an excitation pulse sufficiently short that an exact replica of the electronic ground state vibrational eigenstate, $\langle R|v'' \neq 0\rangle$, is created at t_0 on the excited potential surface,

$$\Psi'(t_0) = \langle R|v'' \neq 0\rangle \,. \tag{9.2.13a}$$

If the V' and V'' potential surfaces have nearly identical shapes, then the Franck-Condon factors impose a very restrictive $\Delta v = 0$ propensity rule, with the result that the $\Psi'(t_0)$ specified by Eq. (9.2.13a) is approximately an eigenstate on the $V'(R)$ potential surface and there is no motion!

If the V' and V'' potential surfaces have very different shapes, then the vertical excitation frequencies, ω_+ and ω_-, from the outer and inner turning points of the $\langle R|v'' \neq 0\rangle$ launch eigenstate, $R_{v'',+}$ and $R_{v'',-}$

$$\omega_\pm = \left[V'(R_{v'',\pm}) - V''(R_{v'',\pm})\right]/\hbar \tag{9.2.14}$$

are very different. In such a case, experimental limits on pulse duration frequently make it impossible to transfer a replica of the $v'' \neq 0$ eigenstate onto the V' potential surface. The reason for this is that the Fourier bandwidth, $1/\Delta t$, of the excitation pulse is insufficient to cover the two dominant lobes of the distribution of Franck-Condon factors from the $v'' \neq 0$ initial eigenstate,

$$1/\Delta t < |\omega_+ - \omega_-|. \tag{9.2.15}$$

An intermediate duration excitation pulse, which satisfies the lower and upper bounds on the bandwidth

$$2\omega'_{v'_\pm} < 1/\Delta t < |\omega_+ - \omega_-|, \tag{9.2.16}$$

where $\omega'_{v'_\pm}$ is the average vibrational frequency

$$\omega'_{v'_\pm} = \left[E_{v'_\pm+1} - E_{v'_\pm-1}\right]\Big/ 2\hbar \tag{9.2.17}$$

in the neighborhood of the $v'_\pm$ upper state vibrational eigenstate (eigenenergy $E'_{v'_\pm}$), may be used to create selectively a wavepacket on the excited potential surface. This wavepacket, at t_0, resembles either the outermost or innermost lobe of the $v'' \neq 0$ vibrational eigenstate. To a good approximation, the t_0 wavepacket that is generated when $\langle\omega_L\rangle \approx \omega_\pm$, has the properties

$$\langle R\rangle_{t_0} \approx R''_{v'',\pm} \tag{9.2.13b}$$

$$\langle P\rangle_{t_0} \approx 0 \tag{9.2.13c}$$

$$\langle E\rangle_{t_0} \approx E''_{v''} + \left[V'(R''_{v'',\pm}) - V''(R''_{v'',\pm})\right]. \tag{9.2.13d}$$

This approximate picture is based on the assumptions that either the innermost or outermost lobe of $\langle R|v'' \neq 0\rangle$ is transferred to the excited potential surface and the features in $\Psi'(t_0)$ that originate from internal lobes of the $\langle R|v'' = 0\rangle$ vibrational wavefunction have negligible amplitudes. It becomes easier to satisfy both of these assumptions as the launch state v'' quantum number increases (provided that the V' and V'' potentials are not too similar).

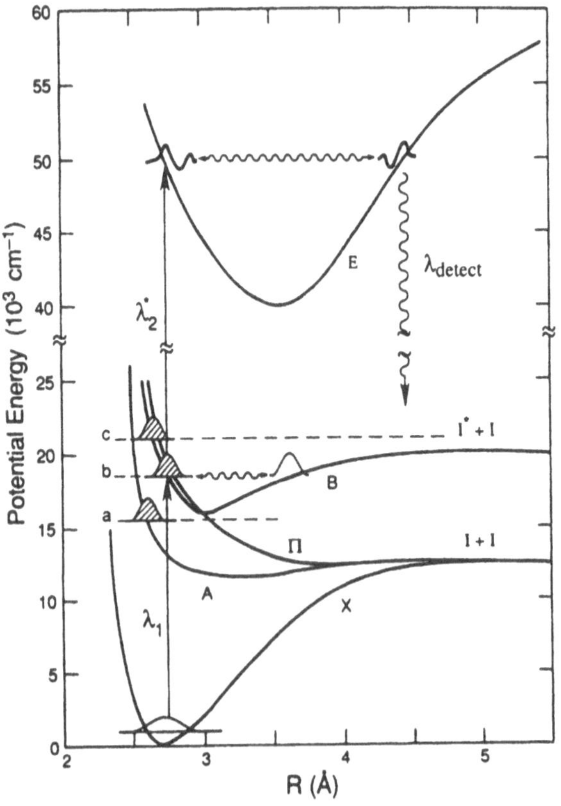

Figure 9.8: Femtosecond Transition State (FTS) spectroscopy of I_2. By adjusting the center wavelength of the pump pulse (λ_1) a wavepacket is launched on the repulsive wall of the weakly bound $A^3\Pi_{1u}$ state ($\lambda_1 \approx 700$ nm, wavepacket labelled **a**), the Franck-Condon vertical excitation region of the $B^3\Pi_{0_u^+}$ state ($\lambda_1 \approx 620$ nm, wavepacket labelled **b**), or on the repulsive wall of the $B^3\Pi_{0_u^+}$ state (and the repulsive $^1\Pi_u$ state) near the $I(^2P_{3/2}) + I^*(^2P_{1/2})$ dissociation limit ($\lambda_1 \approx 505$ nm, wavepacket labelled **c**). The internuclear distance at which the evolving wavepacket is sampled is selected by adjusting the wavelength of the probe pulse, λ_2^*. The λ_2^* pulse excites population to the bound ion-pair E–state from which fluorescence (λ_{detect}) is monitored. The fluorescence signal is monitored as a function of the time delay, τ, between the λ_1 and λ_2^* pulses. The choice of λ_2^* has no effect on the oscillation periods in the $\lambda_1\tau\lambda_2^*$ (λ_{detect}) intensity vs. delay data records, but it does affect the amplitude and time delay to the first coherent transient. The choice of λ_{detect} has no effect on the temporal structure of the observed FTS signal (from Bowman, *et al.*, 1989).

The usual cartoon descriptions of vibrational wavepacket experiments depict the $t = t_0$ creation, by a pump pulse, of a Gaussian wavepacket on an excited electronic state potential energy curve and the subsequent detection, by the probe pulse, of this still Gaussian wavepacket when it arrives at an internuclear separation, usually the inner or outer turning point, specified by the probe pulse (see Fig. 9.8). Such a picture is over-simple for several reasons. Even if the molecule is initially in the $v'' = 0$ eigenstate of the electronic ground state, the $t = t_0$ shape of the vibrational wavepacket on the excited potential curve is dependent on the duration of the pump pulse and is Gaussian only in the limit of a very short pump pulse. Even when the initially created wavepacket resembles a Gaussian, it rapidly develops a complicated, non-Gaussian structure. If the probe pulse samples the evolving wavepacket at a turning point, it "sees" the wavepacket once per vibrational period. However, a probe pulse with the wrong center frequency and insufficient bandwidth (too long duration) to permit probing of the wavepacket by a vertical, turning-point-to-turning-point transition, samples the wavepacket twice per period as it traverses the internal sampling point. If many vibrational levels of the electronic ground state are thermally populated, then the pump pulse creates a wavepacket on the excited potential and a hole on the ground state. The wavepacket and hole evolve independently and both can contribute to signals produced by the probe pulse.

There are many schemes for detecting the arrival of the vibrational wavepacket at a specifiable position on the excited state potential energy curve. These include one-color and two-color schemes, with detection by absorption, fluorescence, ionization, four-wave mixing, and many other methods. Two simple schemes are discussed here.

In the *phase-coherent, one-color pump/probe scheme* (see Section 9.1.9) the wavepacket is detected when the center of the wavepacket returns to its t_0 position, $\langle x \rangle_{t_0+nT} = \langle x \rangle_{t_0}$, after an integer number of vibrational periods. The pump pulse creates the wavepacket. The probe pulse creates another identical wavepacket, which may add constructively or destructively to all or part of the original pump-produced wavepacket. If the envelope delay and optical phase of the probe pulse (Albrecht, *et al.*, 1999) are both chosen correctly, near perfect constructive or destructive interference occurs and the total spontaneous fluorescence intensity (detected after the pump and probe pulses have traversed the sample) is either *quadrupled* (relative to that produced by the pump pulse alone) or *nulled.* As discussed in Section 9.1.9, the probe pulse is delayed, relative to the pump pulse, in discrete steps of $\Delta t = 2\pi/\omega_L$. ω_L is selected by the experimentalist from within the range $\langle \omega_L \rangle \pm 1/\Delta t$ (Δt is the temporal FWHM of the pulse) to define the optical phase of the probe pulse relative to that of the pump pulse and the average excitation frequency. However, $[\langle E \rangle - E_{v''}]/\hbar$ is selected by the molecule in accord with the classical Franck-Condon principle (Tellinghuisen, 1984), also within the $\langle \omega_L \rangle \pm 1/\Delta t$ range. When the envelope delay is chosen so that the probe pulse arrives simultaneously with the return of the center of the vibrational wavepacket to its position at t_0, a relative maximum (optical phase at ω_L delayed by $2n\pi$) or minimum (optical phase at ω_L delayed by $(2n + 1)\pi$) in the fluorescence intensity is observed.

Using the *two-color pump/probe scheme*, similar to the one depicted in Fig. 9.8, it is possible to measure the time required for the *center of the vibrational wavepacket created at* t_0 *on the* B*–state to arrive* at either the inner R_{B-} or outer R_{B+} turning point of the bound, electronically excited B–state. Owing to their different center frequencies, $\langle\omega_L\rangle_{\text{pump}} \neq \langle\omega_L\rangle_{\text{probe}}$, and their non-interchangeable roles in the

$$\mathrm{B} \xleftarrow{\omega_{\text{pump}}} \mathrm{X}$$
$$\mathrm{E} \xleftarrow{\omega_{\text{probe}}} \mathrm{B}$$

sequential excitation scheme, it is unnecessary to attempt to control the optical phase of the probe pulse relative to that of the pump pulse. The pump pulse creates a vibrational wavepacket on the B–state at t_0. The properties of this wavepacket are described by Eqs. (9.2.11a–d). After a selectable delay, the probe pulse excites the vibrational wavepacket from the B–state bound electronic potential curve to a highly excited E–state curve. Because the B–state and E–state potential curves have very different shapes, the Franck-Condon regions accessed from each of the v_B eigenstates that are components of the B–state vibrational wavepacket are narrow (in energy and internuclear separation) but mutually overlapping. As a result, if the probe pulse arrives when the bound state wavepacket is at its inner turning point, a compact, near-Gaussian wavepacket is created on the E–state potential. However, if the center of the bound state wavepacket is not at its inner turning point when the probe pulse arrives, destructive interferences among the E–state nuclear motion $\langle R|v_B\rangle$ amplitudes that originate from each of the B–state $|v_B\rangle$ eigenstate components in $\Psi(t)$ reduce the probability of creating a wavepacket on the E–state. Instead, absorption and emission processes, stimulated by different frequency components of the probe pulse, transfer amplitude among the various $|v_B\rangle$ eigenstate components of the B–state vibrational wavepacket. (This modification of $\Psi(t)$ by the probe pulse remains undetected unless a second identical, phase coherent probe pulse interrogates $\Psi(t)$.) The detailed shape and excitation energy of the E–state potential curve are of negligible consequence to the two-color pump/probe experiment. All that is required is that the frequency of a vertical E←B electronic transition that originates from the outer turning point on the B–state potential lies within the spectral width of the probe pulse,

$$\langle\omega_L\rangle_{\text{probe}} - 1/\Delta t \leq [V_E(R_{B+}) - V_B(R_{B+})] \leq \langle\omega_L\rangle_{\text{probe}} + 1/\Delta t \qquad (9.2.18)$$

To measure the oscillation period of a vibrational wavepacket located at a different average energy on the B–state, it is necessary to prepare it from an X–state vibrational level other than $v'' = 0$ [populated, for example, by a nanosecond Stimulated Emission pumping scheme (Kittrell, *et al.*, 1981)] or from an electronic state other than the X–state. It is not sufficient merely to change the center frequency of the pump or probe pulse. (However, if the duration of the pump pulse is too long to create a replica of the $v'' = 0$ wavefunction at the stationary phase point on the excited potential energy surface, then the center energy, $\langle E\rangle$, of the $t = 0$ wavepacket can be pulled to energies slightly higher or lower than the stationary phase energy, at a cost of non-Gaussian shape and imperfectly specified $\langle E\rangle$ and $\langle R\rangle_0$.) The classical Franck-Condon principle (no change in nuclear positions or momenta, $\Delta Q = 0$, $\Delta P = 0$) is equally valid in the frequency and time domains. The validity of the Franck-Condon principle rests on the stationary phase approximation (see Sections 5.1.1 and 7.6).

9.2.3.2 Rotational Wavepackets

The rotational selection rule for electronic transitions between states at the same Hund's case (see Section 3.2.1) limit is $\Delta J = \Delta N = \pm 1, 0$. If either electronic state is not at a Hund's limiting case or if the limiting cases are not identical in the upper and lower electronic states, less restrictive rotational selection rules apply. However, the number of rotational eigenstate J' or N' components present in a $\Psi(t_0)$ created by a typical spectroscopically realizable pluck (i.e., a single, non-chirped, not-saturating excitation pulse) is small, typically 2 or 3. The possibility that each of these rotational components is split by spin fine structure (see Section 3.4) is neglected in the following discussion.

The $\Psi(t_0)$ prepared by a Z-polarized pump pulse from each thermally-populated J'' (or N'') initial state has the form

$$\Psi_{J''}(t_0) = \sum_{J'=J''-1}^{J''+1} a_{J'} \left|\Omega' J' M' = M''\right\rangle \qquad (9.2.19)$$

where the $a_{J'}$ are proportional to direction cosine matrix elements (see Section 2.3.3),

$$a_{J'} \propto \alpha_{J'\Omega' M'', J''\Omega'' M''}.$$

The density matrix associated with this $\Psi_{J''}(t_0)$ is

$$\begin{aligned}\rho_{J''}(t) = &\sum_{J'=J''-1}^{J''+1} |a_{J'}|^2 \left|\Omega' J' M''\right\rangle \left\langle \Omega' J' M''\right| \\ &+ \sum_{J_i'=J''-1}^{J''+1} \sum_{\substack{J_f'=J''-1 \\ J_f'\neq J_i'}}^{J''+1} \Big[a_{J_i'} a_{J_f'}^* \left|\Omega J_i' M''\right\rangle \\ &\times \left\langle \Omega' J_f' M''\right| e^{-i2\pi c B_{v'}[J_i'(J_i'+1) - J_f'(J_f'+1)](t-t_0)} + c.c.\Big]\end{aligned} \qquad (9.2.20)$$

There are off-diagonal elements in $\rho_{J''}(t)$ which oscillate as $e^{\pm icB_v[n](t-t_0)}$ where $n = 4J''+2, 2J''+2$, and $2J''$. But, since many J'' levels are populated thermally, the total $\rho(t)$ is a sum of many independent $\rho_{J''}$ parts,

$$\rho(t) = \sum_{J''} n_{J''} \rho_{J''}(t) \qquad (9.2.21)$$

where $n_{J''}$ is the population density (molecules/volume) in the J'' level,

$$n_{J''} = N(2J''+1)e^{-hcB''J''(J''+1)/kT} \Big/ [kT/hcB''], \qquad (9.2.22)$$

N is the total population density, and $[kT/hcB'']$ is the rotational partition function. This complete $\rho(t)$ has terms that oscillate as $e^{\pm i2\pi cB_v[2m](t-t_0)}$ where $m = 1, 2, \ldots$. Because the frequencies of all of the oscillating terms in $\rho(t)$ have a common multiplicative factor of $2cB_v$, there will be a series of grand recurrence times, $t_{\text{rot},n} = t_0 + n/(2cB_v)$ at which $\rho(t_{\text{rot}}) \approx \rho(t_0)$. Centrifugal distortion effects will result in a very slow dephasing of successive rotational recurrences. The rate of this dephasing is determined by D_v/B_v.

Rotational recurrences may be detected in polarization selected spontaneous fluorescence (provided the photodetector has a sufficiently fast response) or by a variety of sub-nanosecond pump/probe schemes (Felker and Zewail, 1987; Felker, 1992; Hartland, *et al.*. 1992; Joireman, *et al.*. 1992; Smith, *et al.*, 2003a,b).

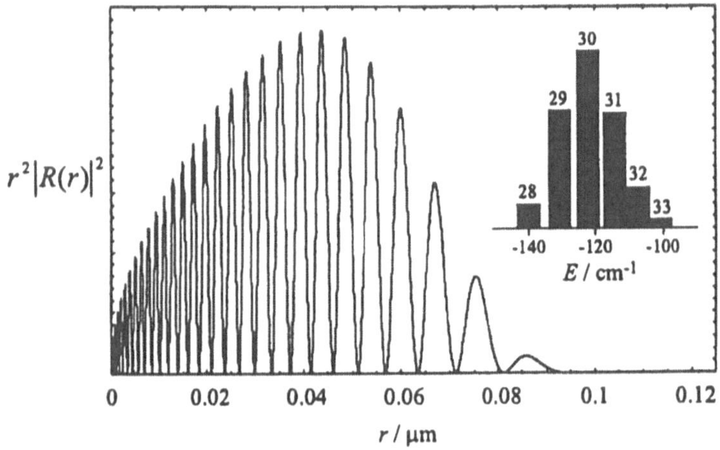

Figure 9.9: The radial probability distribution of an $n \approx 30$ wavepacket in a hydrogen atom. The populations and binding energy of the eigenstates in the wavepacket are shown in the inset. Note that the figure shows the t_0 probability distribution. A plot of the t_0 wavepacket amplitude would show a single dominant feature in the region of the innermost lobe of the initial state (1*s*), with small wiggles at larger r (from Smith, *et al.*, 2003b).

9.2.4 Rydberg Wavepackets: Kepler and Precessional Periods

The energy spacing between consecutive $(n^* - 1/2, n^* + 1/2)$ members of a Rydberg series is $2hc\Re/n^{*3}$, where $\Re$ is the Rydberg constant (109737 cm^{-1}) and n^* is the (non-integer) effective principal quantum number. The intensities of transitions between valence and Rydberg states are controlled by a one-electron $\langle \text{valence}|r|\text{Rydberg}\rangle$ transition moment orbital integral. This integral accumulates near a stationary phase point located well inside the ion-core. Owing to the $n^{*-3/2}$ amplitude scaling of the otherwise n^*–invariant intra-core region of Rydberg series orbitals, the transition probabilities for one-electron Rydberg←valence transitions scale as n^{*-3}. This means that a sufficiently short excitation pulse will create a $\Psi(t_0)$ which contains many n^*–eigenstates belonging to at least one Rydberg series,

$$\Psi(t_0) \propto \sum_{m=0}^{M_{\max}} (n^* + m)^{-3/2} \mathcal{E}\left(\omega_{n^*+m,\text{valence}}, t_0\right) \tag{9.2.23}$$

where m is an integer and $(n^* + m)^{-3/2}$ is the transition moment scaling factor and $\mathcal{E}\left(\omega_{n^*+m,\text{valence}}, t_0\right)$ is the complex amplitude of the excitation pulse at frequency $\omega_{n^*+n,\text{valence}}$ and time t_0 (see Fig. 9.9). If the excitation pulse is sufficiently short and smooth, then the coherent superposition state, $\Psi(t_0)$ amplitude (not $|\Psi(t_0)|^2$ probability), created at t_0 is a near perfect replica of the portion of $\psi_{\text{valence}}(r)$ near the stationary phase point. The stationary phase point for a low-n to high-n transition is always inside the ion-core. One important difference between a Rydberg pluck and a Franck-Condon pluck is that the average excitation energy of the Rydberg wavepacket is controlled simply by the center frequency, ω_{center}, of the light pulse, whereas a Franck-Condon pluck creates a vibrational wavepacket with $\langle E\rangle$ and $\langle R\rangle_{t_0}$ consistent with the classical Franck-Condon principle, regardless of ω_{center}.

The Rydberg pluck creates several Rydberg wavepackets (one for each Rydberg series that is connected to the initial electronic state by an allowed transition), each localized at $t = t_0$ with $\langle r\rangle_{t_0}$ near the appropriate radial stationary phase point and with kinetic energy determined by ω_{center}. This wavepacket initially moves away from the ion-core and then returns to the ion-core in highly eccentric Kepler orbits. If the wavepacket has high angular momentum, l, the $l(l+1)/2\mu r^2$ centrifugal barrier prevents the wavepacket from penetrating inside the ion-core on any of its closest approaches to the core region. (Such core-nonpenetrating high–l wavepackets cannot be excited from a low–n^* valence orbital.) A low–l wavepacket will penetrate inside the highly-anisotropic ion-core on each of its returns to the core region. Each impulsive collision of the low–l, "core-penetrating" Rydberg wavepacket with the ion-core results in the exchange of energy and angular momentum between the Rydberg electron and the ion-core. This energy and angular momentum exchange is encoded in an R–dependent, non-diagonal quantum defect matrix, $a_{l\lambda,l'\lambda'}(R)$ (see Sections 3.2.1.3.2 and 8.6). It is also encoded in the nodal structure of the intra-core part of the Rydberg orbital, which is a compact region of space ideally suited for accurate electronic structure computations.

The center of the Rydberg wavepacket, $\langle r\rangle_t$, undergoes periodic radial oscillations, analogous to the planetary motions (Kepler orbits). The Kepler period of $\langle r\rangle_t$ for the Rydberg wavepacket with center–n^*, n^*_{center}, is

$$T_{n^*_{\text{center}}} = n^{*3}_{\text{center}} / (2\Re c). \tag{9.2.24}$$

Unlike vibrational and rotational wavepackets, the Rydberg wavepacket does not undergo a near perfect grand rephasing because the energy spacings of the component eigenstates are not *integer multiples of a common factor.* The energy width, $\Delta E = \left[\langle E^2\rangle - \langle E\rangle^2\right]^{1/2}$, of the Rydberg wavepacket is set by the duration of the excitation pulse and is independent of time thereafter. The $t = t_0$ radial width, $\Delta r(t) = \left[\langle r^2\rangle_t - \langle r\rangle_t^2\right]^{1/2}$, of the wavepacket is comparable to the diameter of the atomic cation obtained by removing all of the valence electrons from the more electropositive atom. This radial spread of the wavepacket oscillates, reaching local maximum and minimum values at half-integer and integer multiples of $T_{n^*_{\text{center}}}$, respectively.

The Rydberg electron is moving in the nonspherical field of a molecular ion. In the limit where the ion-core rotational period, $T^J_{\text{rot}} = [cB_v(2J+1)]^{-1}$, is long relative to $T_{n^*_{\text{center}}}$, when the Rydberg wavepacket is outside the ion-core it moves in a spherically symmetric Coulomb potential, with origin at the ion-core center of charge, plus an axially symmetric (prolate) quadrupole potential, with symmetry axis along the molecular axis. This quadrupole field splits each l–state into $l+1$ $|\lambda|$–components, with energy splittings

$$E_{n^*l\lambda+1} - E_{n^*l\lambda} \propto Qn^{*-3}l^{-3}\left[(|\lambda|+1)^2 - (\lambda)^2\right] \qquad (9.2.25)$$

where Q is the quadrupole moment of the molecular-ion. Note that these energy splittings are integer multiples of a common factor. Thus, if a Rydberg wavepacket containing several $|\lambda|$–components is excited at $t = t_0$, it will undergo precessional motion about the internuclear axis and will exhibit grand rephasings at $T_{\text{precession}} \propto n^{*3}l^3/Q$. In the opposite limit, where $T_{\text{rot}} < T_{n^*}$, then, when the Rydberg wavepacket is outside the ion-core, it moves in a spherically-symmetric Coulomb field with origin at the molecular center of mass, plus an axially symmetric (oblate) quadrupole potential, with symmetry axis along the ion-core nuclear rotation ($\vec{N}^+$) axis. This quadrupole field splits each l–state into $l+1$ $|l_R|$–components, with energy splittings

$$E_{n^*ll_R+1} - E_{n^*ll_R} \propto \frac{-Q}{2}n^{*-3}l^{-3}\left[(|l_R+1|)^2 - l_R^2\right]. \qquad (9.2.26)$$

Once again, the quadrupole splittings are integer multiples of a common factor, and if several $|l_R|$–components are excited at $t = t_0$, the Rydberg wavepacket will precess about $\vec{N}^+$ and exhibit grand rephasings.

If the Kepler period of a Rydberg wavepacket is an integer multiple of the rotational wavepacket period, then a stroboscopic effect (Labastie, *et al.*, 1984; Greene and Jungen, 1985; Lombardi, *et al.*, 1988; Altunata, *et al.*, 2002; Smith, *et al.*, 2003a) occurs. At $t = t_0$, both Rydberg and rotational wavepackets are created. Their positions at $t = t_0$ are uniquely determined by details of the excitation pulse. At $t = T_{n^*_{\text{center}}} + t_0$, both wavepackets have returned to their $t = t_0$ positions. The time-dependent interactions between the electronic and nuclear wavepackets during the $[t_0, T_{n^*}+t_0]$ and $[T_{n^*}+t_0, 2T_{n^*}+t_0]$ time intervals are identical. As a result, the temporal behavior of the Rydberg wavepacket (and the spectrum in which this behavior is encoded) is considerably simplified relative to the non-stroboscopic n^*, N^+ combinations. This stroboscopic effect is most perfect when $T_{n^*_{\text{center}}}$ is an integer multiple of the rotational grand rephasing time, $t_{\text{rot}} = (2cB_v)^{-1}$, because then all details of the evolution of the rotational wavepacket are exactly reproduced from one Rydberg wavepacket period to the next. (Labastie, *et al.*, 1984; Altunata, *et al.*, 2002)

All of the vector precession pictures associated with the various Hund's coupling cases correspond to experimentally realizable wavepacket motions, provided that molecules with energy level patterns well described by one of the Hund's cases (see Section 3.2.1) can be suitably plucked. (Texier and Jungen, 1998 and 1999; Smith, *et al.*, 2003a).

9.3 Relaxation into a Quasi-Continuum: A Tool for Dimensionality Reduction

A wide variety of photofragmentation processes were discussed in Chapters 7 and 8. The initially prepared *nonstationary* state, $\Psi(r, R, t)$, of the exact $\mathbf{H}$, evolves in such a way that its survival probability, $P(t) = |\langle\Psi(t)|\Psi(0)\rangle|^2$, decays irreversibly. Radiative and nonradiative (photofragmentation) processes contribute to the decay of $P(t)$. For polyatomic molecules, processes which involve neither radiation nor fragmentation can contribute to the decay of $P(t)$. For polyatomic molecules at internal energies well below the lowest dissociation or ionization threshold, the density of vibrational states can be so large that the decay of a bright state into the *quasi-continuum* of dark (but non-fragmenting) states can mimic the decay of a bright state into a true continuum. Many of the conceptual models and computational schemes developed in Chapters 7 and 8 to describe diatomic molecule photofragmentation processes can provide valuable insights into the mechanisms of polyatomic molecule nonradiative decay processes.

Two limiting cases are commonly invoked to describe the coupling between a bright state and a quasicontinuum of dark states: *direct coupling*, where the bright state couples indiscriminately (statistically) to all of the dark states that comprise the quasicontinuum, and *doorway mediated coupling*, where the bright state interacts exclusively with one special dark state, the doorway state, which in turn is coupled statistically to all of the other dark states (Dupré, *et al.*, 1991; Altunata and Field, 2000 and 2001). This artifice reduces the enormous number of interacting zero-order states to two (the "sharp" bright state and one "broad" continuum state) or three (the sharp bright and doorway states and one broad continuum state). Such a reduction of the number of interacting states can provide great interpretive and predictive insights into the mechanism of the non-radiative process. Direct coupling is analogous to simple predissociation (Section 7.3) or autoionization (Section 8.2). Doorway mediated coupling is analogous to *accidental* predissociation (Section 7.13) or autoionization.

Between the direct and doorway-mediated limiting cases one finds *sequential relaxation* or *tier* models (Sibert, *et al.*, 1982a and b). The bright state is strongly coupled (for mechanistic reasons or due to an accidental degeneracy) to a small number of "first tier" dark states. These first tier states are each, in turn, strongly coupled (again for mechanistic or near degeneracy reasons) to a small number of "second tier" states. The sequence of tiers continues until the full density of dark states participates in the dynamics. Typically, the coupling matrix elements and the density of participating states increase rapidly from one tier to the next. As the relaxation proceeds deeper into the manifold of tiers, it gets faster, less capable of being described by a small number of deterministic mechanisms, thus more nearly statistical or "quantum ergodic." The *early time dynamics* as sampled in an ultrafast experiment or in a low-resolution, frequency domain experiment, often conveys simple mechanistic information. In order to have any hope of understanding successive dynamical tiers, it is essential to develop a mechanistic picture for the preceeding tier. (Sibert, *et al.*, 1982a and b; Stuchebrukhov and Marcus, 1993; Stuchebrukhov, *et al.*, 1993; Stuchebrukhov and Marcus, 1994; Mehta, *et al.*, 1995; Lehmann, *et al.*, 1994; Beil, *et al.*, 1997; Quack and Stohner, 1993; Quack, 1990).

9.3.1 The Complex-Energy Effective Hamiltonian

The usual *Hermitian* effective Hamiltonian matrix may be generalized to deal with bound-free interactions by permitting the diagonal elements to be complex (Child, 1991; Dupré, *et al.*, 1991; Tric, 1973; Moiseyev, 1998). When this is done, the generalized $\mathbf{H}^{\mathrm{eff}}$ is no longer Hermitian, its eigenvalues are complex, its eigenvectors are not orthogonal, and special care must be taken in finding the eigenvalues and eigenvectors. However, the interpretive benefits of being able to use the ideas and computational methods of ordinary non-degenerate and quasidegenerate perturbation theory are worth the price of admission.

For a time-independent $\mathbf{H}$, the eigenstates, ψ_j, with energy, E_j, that satisfy the time-dependent Schrödinger equation are

$$\Psi_j(t) = \psi_j e^{-iE_j t/\hbar}. \tag{9.3.1}$$

If E_j is allowed to be complex,

$$E_j = \epsilon_j - i\Gamma_j/2, \tag{9.3.2}$$

where E_j and Γ_j are expressed in the same units (consistent with Chapters 7 and 8)

$$\Psi_j(t) = \psi_j e^{-i\epsilon_j t/\hbar} e^{-t\Gamma/2\hbar} \tag{9.3.3}$$

and the survival probability decays as

$$P_j(t) = |\langle\widetilde{\Psi}_j(0)|\Psi(t)\rangle|^2 = e^{-t\Gamma_j/\hbar}, \tag{9.3.4}$$

where the meaning of and necessity for the tilde on $\left\langle\widetilde{\Psi}_j(0)\right|$ is discussed below, and the lifetime, τ_j, is the $1/e$ decay time for the population in the j–th state,

$$P_j(t) = e^{-t/\tau_j}, \tag{9.3.5a}$$

where

$$\tau_j = \frac{\hbar}{\Gamma_j}. \tag{9.3.5b}$$

So the first thing we get by allowing **H** to have complex diagonal elements is that state populations (diagonal elements of the density matrix) can decay exponentially with the expected decay rate, $(1/\tau_j = \Gamma_j/\hbar)$, and this exponential decay corresponds to a Lorentzian broadened level (as opposed to a δ–function) centered at ϵ_j with the expected full width at half maximum

$$\text{FWHM} = \Gamma_j = \hbar/\tau_j = h/(2\pi\tau_j). \tag{9.3.5c}$$

The convenience of being able to generalize $\mathbf{H}^{\text{eff}}$ to include complex diagonal matrix elements means that the usual orthonormality and completeness relationships for the eigenvectors of the non-Hermitian $\mathbf{H}^{\text{eff}}$ are invalid. Morse and Feshbach (1953) (pages 884-886) show that "biorthogonal" basis vectors must be used to describe the eigenstates and dynamics of a non-Hermitian operator (see also Lefebvre, 1973, pages 125-135, and Moiseyev, 1998). For an $\mathbf{H}^{\text{eff}}$ in which all but the diagonal elements are real numbers, it is necessary to consider two complete sets of eigenvectors, one for **H** and one for $\mathbf{H}^*$

$$\mathbf{H}|i\rangle = \mathbf{H}\begin{pmatrix} a_1^i \\ a_2^i \\ \vdots \\ a_N^i \end{pmatrix} = E_i \begin{pmatrix} a_1^i \\ a_2^i \\ \vdots \\ a_N^i \end{pmatrix} \tag{9.3.6a}$$

$$\mathbf{H}^*|\tilde{i}\rangle = \mathbf{H}^*\begin{pmatrix} a_1^{i*} \\ a_2^{i*} \\ \vdots \\ a_N^{i*} \end{pmatrix} = E_i^* \begin{pmatrix} a_1^{i*} \\ a_2^{i*} \\ \vdots \\ a_N^{i*} \end{pmatrix} \tag{9.3.6b}$$

The usual orthonormality and completeness relationships must be replaced by

$$\langle \tilde{i}|i\rangle = \sum_j (a_j^i)^2, \tag{9.3.7}$$

which can be complex,

$$\frac{\langle \tilde{j}|i\rangle}{\langle \tilde{j}|j\rangle} = \delta_{ji} \tag{9.3.8}$$

$$\mathbb{I} = \sum_j \frac{|j\rangle\langle \tilde{j}|}{\langle \tilde{j}|j\rangle} \tag{9.3.9}$$

where the $\langle \tilde{j}|j\rangle$ in the denominator ensures that time-dependent (exponentially decreasing or increasing) state vectors remain normalized. The replacement of the usual orthonormalization and completeness relationships by biorthogonal relationships does not alter the formulas for $E_i^{(0)}, E_i^{(1)}, E_i^{(2)}, \psi_i^{(0)}$ and $\psi_i^{(1)}$ from ordinary nondegenerate perturbation theory.

Nondegenerate perturbation theory enables algebraically transparent analysis of the interactions between any number of decaying quasi-eigenstates provided that $\left|E_1^{(0)} - E_2^{(0)}\right| >> H_{12}$. However, analysis of the two-level problem (related to accidental predissociation, discussed in Section 7.13, autoionization, discussed in Section 8.4, and Intramolecular Vibrational Redistribution, discussed in Section 9.4.14) provides insights into the unique effects that derive from widths and decay rates of the basis states.

Consider the two-level problem

$$\mathbf{H} = \begin{pmatrix} E_1 & V \\ V & E_2 \end{pmatrix} = \begin{pmatrix} \bar{\epsilon} - i\bar{\Gamma}/2 & 0 \\ 0 & \bar{\epsilon} - i\bar{\Gamma}/2 \end{pmatrix} + \begin{pmatrix} \delta\epsilon - i\delta\Gamma/2 & V \\ V & -\delta\epsilon + i\delta\Gamma/2 \end{pmatrix} \tag{9.3.10}$$

where the off-diagonal matrix element, V, is real and the quantities in Eq. (9.3.10) are defined below:

$$E_1 = \epsilon_1 - i\Gamma_1/2 \tag{9.3.11a}$$
$$E_2 = \epsilon_2 - i\Gamma_2/2 \tag{9.3.11b}$$
$$\bar{\epsilon} = (\epsilon_1 + \epsilon_2)/2 \tag{9.3.11c}$$
$$\bar{\Gamma} = (\Gamma_1 + \Gamma_2)/2 \tag{9.3.11d}$$
$$\delta\epsilon = (\epsilon_1 - \epsilon_2)/2 \tag{9.3.11e}$$
$$\delta\Gamma = (\Gamma_1 - \Gamma_2)/2. \tag{9.3.11f}$$

9.3.1.1 Treatment of Two-State Interaction by Nondegenerate Perturbation Theory

Nondegenerate perturbation theory gives

$$E_1 = E_1^{(0)} + \frac{|H_{12}|^2}{E_1^{(0)} - E_2^{(0)}} = \left[\epsilon_1 + \frac{V^2}{2(\delta\epsilon + \delta\Gamma^2/4\delta\epsilon)}\right] - i\left[\Gamma_1/2 - \frac{V^2}{2(\delta\Gamma/2 + 2\delta\epsilon^2/\delta\Gamma)}\right] \tag{9.3.12a}$$

where the $\delta\Gamma^2/4\delta\epsilon$ and $2\delta\epsilon^2/\delta\Gamma$ terms in the denominators arise when the definitions of $\mathrm{E}_1^{(0)}$ and $\mathrm{E}_2^{(0)}$ are inserted in the $1\Big/\left(\mathrm{E}_1^{(0)} - \mathrm{E}_2^{(0)}\right)$ term and the denominator is rationalized,

$$\psi_1 = \psi_1^{(0)} + \frac{H_{12}}{E_1^{(0)} - E_2^{(0)}}\psi_2^{(0)}$$

$$= \psi_1^{(0)} + \psi_2^{(0)}\left[\frac{V}{2(\delta\epsilon + \delta\Gamma^2/4\delta\epsilon)} + \frac{iV}{2(\delta\Gamma/2 + 2\delta\epsilon^2/\delta\Gamma)}\right] \tag{9.3.13a}$$

$$E_2 = \left[\epsilon_2 - \frac{V^2}{2(\delta\epsilon + \delta\Gamma^2/4\delta\epsilon)}\right] - i\left[\Gamma_2/2 + \frac{V^2}{2(\delta\Gamma/2 + 2\delta\epsilon^2/\delta\Gamma)}\right] \tag{9.3.12b}$$

$$\psi_2 = \psi_2^{(0)} - \psi_1^{(0)}\left[\frac{V}{2(\delta\epsilon + \delta\Gamma^2/4\delta\epsilon)} + \frac{iV}{2(\delta\Gamma/2 + 2\delta\epsilon^2/\delta\Gamma)}\right]. \tag{9.3.13b}$$

Note that, if $\delta\Gamma = 0$ (interaction of two levels of the same width, $\Gamma_1 = \Gamma_2 = \Gamma$),

$$E_1 = \epsilon_1 + \frac{V^2}{2\delta\epsilon} - i\Gamma/2 \tag{9.3.14a}$$

$$E_2 = \epsilon_2 - \frac{V^2}{2\delta\epsilon} - i\Gamma/2 \tag{9.3.14b}$$

and, if instead $\delta\epsilon = 0$ ($\epsilon_1 = \epsilon_2 = \epsilon$), and $|\delta\Gamma| >> V$ (to permit the use of nondegenerate perturbation theory)

$$E_1 = \epsilon - i\left[\Gamma_1/2 - \frac{V^2}{\delta\Gamma}\right] \tag{9.3.15a}$$

$$E_2 = \epsilon - i\left[\Gamma_2/2 + \frac{V^2}{\delta\Gamma}\right]. \tag{9.3.15b}$$

These results are very similar to the normal two-level problem, except: (*i*) the real parts of the energies exhibit *level repulsion*, but this repulsion is reduced (but never reversed) by the $\delta\Gamma^2/2\delta\epsilon$ term in the denominator of Eqs. (9.3.12a) and (9.3.12b); (*ii*) the imaginary parts of the energies exhibit *level attraction* (the interaction causes the widths of the mixed states to become more similar), and this attraction is reduced (but never reversed) by the $4\delta\epsilon^2/\delta\Gamma$ term in the denominator of Eqs. (9.3.12a) and (9.3.12b); (*iii*) when $\delta\Gamma = 0$, the level widths are unaffected by the interaction; (*iv*) when $\delta\epsilon = 0$ but $|\delta\Gamma| >> |V|$, the level positions are unaffected by the interaction (because the narrower level is symmetrically surrounded by the much broader level and the usual level shift by $\pm V$ is suppressed). This amounts to an intuitively-sensible extension of the normal (real E, Hermitian $\mathbf{H}$) two-level problem.

9.3.1.2 Treatment by Quasidegenerate Perturbation Theory: 2 × 2 Diagonalization

To eliminate the restrictions on $\delta\epsilon$ and $\delta\Gamma$ demanded by non-degenerate perturbation theory, it is necessary to find the eigenvalues and eigenvectors of the complex 2×2 energy matrix, which gives

$$E_\pm = \bar{\epsilon} - i\bar{\Gamma}/2 \pm \mathcal{E} \tag{9.3.16}$$

$$\mathcal{E} = [\delta\epsilon^2 - \delta\Gamma^2/4 + V^2 - i\delta\epsilon\delta\Gamma]^{1/2} \tag{9.3.17}$$

$$\psi_\pm = \alpha_\pm\psi_1 + \beta_\pm\psi_2 \tag{9.3.18}$$

$$\alpha_+ = \beta_- = \left(\frac{\mathcal{E} + \delta\epsilon - i\delta\Gamma/2}{2\mathcal{E}}\right)^{1/2} \tag{9.3.19a}$$

$$\beta_+ = -\alpha_- = \left(\frac{\mathcal{E} - \delta\epsilon + i\delta\Gamma/2}{2\mathcal{E}}\right)^{1/2} \tag{9.3.19b}$$

Note that the $\psi_\pm$ functions are neither normalized nor orthogonal in the usual sense

$$\langle +|+\rangle = |\alpha_+|^2 + |\beta_+|^2 \neq 1 \tag{9.3.20a}$$

$$\begin{aligned}\langle -|+\rangle &= \alpha_-^*\alpha_+ + \beta_-^*\beta_+ \\ &= (-\beta_+^*\alpha_+) + \alpha_+^*\beta_+ \neq 0,\end{aligned} \tag{9.3.20b}$$

but, in the biorthogonal sense,

$$\langle\tilde{\pm}|\pm\rangle = (\alpha_\pm)^2 + (\beta_\pm)^2 = 1 \tag{9.3.20c}$$

and

$$\langle\tilde{\mp}|\pm\rangle = -\beta_\pm\alpha_\pm + \alpha_\pm\beta_\pm = 0. \tag{9.3.20d}$$

These expressions are valid for all combinations of $\delta\epsilon$, $\delta\Gamma$, and V, and are consistent with the four special cases discussed above for $|\delta\epsilon|$ or $|\delta\Gamma|$ very large relative to V. However, this expression shows that there is a *strong coupling* regime, where $V^2 >> \delta\Gamma^2/4 + \delta\epsilon^2$, and a *weak coupling* regime, where $V^2 << |\delta\Gamma^2/4 - \delta\epsilon^2|$.

In order to separate the complex eigenvalues of the complex 2×2 **H** matrix into their real and imaginary parts, let

$$\mathcal{E} = [\delta\epsilon^2 - \delta\Gamma^2/4 + V^2 - i\delta\epsilon\delta\Gamma]^{1/2} \tag{9.3.21}$$

$$= K + i\kappa \tag{9.3.22}$$

and

$$\mathcal{E}^2 = K^2 - \kappa^2 + 2iK\kappa \tag{9.3.23}$$

The equation for the real part of $\mathcal{E}^2$ is

$$K^2 - \kappa^2 = \delta\epsilon^2 - \delta\Gamma^2/4 + V^2, \tag{9.3.24}$$

and for the imaginary part,

$$2iK\kappa = -i\delta\epsilon\delta\Gamma. \tag{9.3.25}$$

Therefore,

$$\kappa = -\delta\epsilon\delta\Gamma/2K. \tag{9.3.26}$$

Inserting this equation for κ into the real part equation, Eq. (9.3.24),

$$\begin{aligned}
0 &= K^2 - \delta\epsilon^2 + \delta\Gamma^2/4 - V^2 - (\delta\epsilon\delta\Gamma/2K)^2 \\
0 &= K^4 + K^2[-\delta\epsilon^2 + \delta\Gamma^2/4 - V^2] - \delta\epsilon^2\delta\Gamma^2/4 \\
K^2 &= \frac{\delta\epsilon^2 - \delta\Gamma^2/4 + V^2 \pm [(-\delta\epsilon^2 + \delta\Gamma^2/4 - V^2)^2 + \delta\epsilon^2\delta\Gamma^2]^{1/2}}{2}.
\end{aligned} \tag{9.3.27}$$

Since K is defined to be real, the numerator of Eq. (9.3.27) must be positive, making the only two physically acceptable values for K those with the positive sign in front of the internal square root term,

$$K_\pm = \pm\left\{\frac{\delta\epsilon^2 - \delta\Gamma^2/4 + V^2 + [(-\delta\epsilon^2 + \delta\Gamma^2/4 - V^2)^2 + \delta\epsilon^2\delta\Gamma^2]^{1/2}}{2}\right\}^{1/2} = \pm\chi \tag{9.3.28}$$

where, by definition, $\chi \geq 0$, thus

$$\kappa_\pm = \mp\delta\epsilon\delta\Gamma/2\chi \tag{9.3.29}$$

$$E_\pm = (\bar{\epsilon} \pm \chi) - i(\bar{\Gamma}/2 \pm \delta\epsilon\delta\Gamma/2\chi). \tag{9.3.30}$$

These are algebraically complicated equations. It is instructive to simplify them by considering two limiting cases: *strong coupling*,

$$V^2 >> |\delta\epsilon^2 + \delta\Gamma^2/4| > |\delta\epsilon^2 - \delta\Gamma^2/4| \tag{9.3.31}$$

and *weak coupling*,

$$V^2 << |\delta\epsilon^2 - \delta\Gamma^2/4|. \tag{9.3.32}$$

For the strong coupling limit, begin by rearranging χ^2 as

$$\chi^2 = \frac{\delta\epsilon^2 - \delta\Gamma^2/4 + V^2 + V^2\left[\left(\frac{\delta\Gamma^2/4 - \delta\epsilon^2}{V^2} - 1\right)^2 + \frac{\delta\epsilon^2\delta\Gamma^2}{V^4}\right]^{1/2}}{2}. \tag{9.3.33}$$

In the strong coupling limit the $\delta\epsilon^2\delta\Gamma^2/V^4$ term is negligible and the two values of χ are

$$\chi = [\delta\epsilon^2 - \delta\Gamma^2/4 + V^2]^{1/2} \approx |V|\left[1 + \frac{\delta\epsilon^2 - \delta\Gamma^2/4}{2V^2}\right] \tag{9.3.34a}$$

and

$$\chi \approx 0, \tag{9.3.34b}$$

but $\chi = 0$ is not physically reasonable (because the imaginary part of $E_{\pm}$ would be infinite), and

$$\frac{1}{\chi} \approx \frac{1}{|V|}\left[1 - \frac{\delta\epsilon^2 - \delta\Gamma^2/4}{2V^2}\right]. \tag{9.3.35}$$

Thus

$$E_{\pm} = \left[\bar{\epsilon} \pm |V|\left(1 + \frac{\delta\epsilon^2 - \delta\Gamma^2/4}{2V^2}\right)\right] - i\left[\bar{\Gamma}/2 \pm \left(\frac{\delta\epsilon\delta\Gamma}{2|V|}\right)\left(1 - \frac{\delta\epsilon^2 - \delta\Gamma^2/4}{2V^2}\right)\right]. \tag{9.3.36}$$

The real part of $\Delta E = E_+ - E_-$ differs from the familiar two-level perturbation result in the $\delta\Gamma = 0$ case, which is,

$$\Delta E = 2[V^2 + \delta\epsilon^2]^{1/2} \approx 2V + \frac{\delta\epsilon^2}{|V|}, \tag{9.3.37}$$

by a small level attraction due to the $\delta\Gamma^2/4$ term. The imaginary part of ΔE shows that the interaction *modifies* the difference in the widths of the zero-order levels by a multiplicative factor smaller or larger (depending on the sign of $\delta\epsilon^2 - \delta\Gamma^2/4$) than $\left|\frac{\delta\epsilon}{V}\right|$, but if *either* $\delta\epsilon$ or $\delta\Gamma$ is zero (while remaining in the $V^2 >> \delta\epsilon^2 + \delta\Gamma^2/4$ strong coupling limit), the two mixed states have the same width (50-50 mixing of the imaginary parts).

In the weak coupling limit, χ^2 may be rearranged as

$$\chi^2 = \frac{\delta\epsilon^2 - \delta\Gamma^2/4 + V^2 + [(\delta\Gamma^2/4 + \delta\epsilon^2)^2 + V^4 + 2V^2(\delta\epsilon^2 - \delta\Gamma^2/4)]^{1/2}}{2} \tag{9.3.38}$$

$$\chi^2 = \frac{\delta\epsilon^2 - \delta\Gamma^2/4 + V^2 + (\delta\epsilon^2 + \delta\Gamma^2/4)\left[1 + \frac{V^4}{(\delta\epsilon^2+\delta\Gamma^2/4)^2} + \frac{2V^2(\delta\epsilon^2-\delta\Gamma^2/4)}{(\delta\epsilon^2+\delta\Gamma^2/4)^2}\right]^{1/2}}{2}$$

$$\chi^2 \approx \frac{\delta\epsilon^2 - \delta\Gamma^2/4 + V^2 + (\delta\epsilon^2 + \delta\Gamma^2/4)\left[1 + \frac{V^2(\delta\epsilon^2-\delta\Gamma^2/4)}{(\delta\epsilon^2+\delta\Gamma^2/4)^2}\right]}{2} \tag{9.3.39}$$

by neglecting the $\frac{V^4}{(\delta\epsilon^2+\delta\Gamma^2/4)^2}$ term and retaining only the first two terms in a Taylor series expansion of the square root term. Thus

$$\chi^2 = \delta\epsilon^2 + V^2/2 + \frac{V^2(\delta\epsilon^2 - \delta\Gamma^2/4)}{2(\delta\epsilon^2 + \delta\Gamma^2/4)} \tag{9.3.40}$$

$$\chi^2 = \delta\epsilon^2 + \frac{V^2\delta\epsilon^2}{\delta\epsilon^2 + \delta\Gamma^2/4}. \tag{9.3.41}$$

Note that $\chi = [\delta\epsilon^2 + V^2]^{1/2}$ when $\delta\Gamma = 0$ and $\chi = 0$ when $\delta\epsilon = 0$. Now, to get the imaginary part of the complex energy, we need $\delta\epsilon\delta\Gamma/2\chi$,

$$\frac{\delta\epsilon\delta\Gamma/2}{2\chi} = \frac{\delta\epsilon\delta\Gamma/2}{\left[\delta\epsilon^2 + \frac{V^2\delta\epsilon^2}{\delta\epsilon^2+\delta\Gamma^2/4}\right]^{1/2}} = \frac{\delta\Gamma/2}{\left[1 + \frac{V^2}{\delta\epsilon^2+\delta\Gamma^2/4}\right]^{1/2}} \tag{9.3.42}$$

$$\approx \frac{\delta\Gamma}{2}\left[1 - \frac{V^2}{2(\delta\epsilon^2 + \delta\Gamma^2/4)}\right], \tag{9.3.43}$$

which is smaller than $\frac{\delta\Gamma}{2}$ at all values of $\delta\epsilon$ and goes to zero when $\delta\Gamma = 0$ (provided that $\delta\epsilon^2 >> V^2$). Thus, in the weak coupling limit

$$E_\pm = \left[\bar{\epsilon} \pm \delta\epsilon\left(1 + \frac{V^2/2}{\delta\epsilon^2 + \delta\Gamma^2/4}\right)\right] - i\left[\bar{\Gamma}/2 \pm \frac{\delta\Gamma}{2}\left(1 - \frac{V^2/2}{\delta\epsilon^2 + \delta\Gamma^2/4}\right)\right]. \tag{9.3.44}$$

In the weak coupling limit, the real part of ΔE differs from the familiar $\delta\Gamma = 0$ two-level perturbation result by a small reduction in level repulsion due to the $\delta\Gamma^2$ term and the imaginary part shows that, as intuitively expected, the difference in zero-order level widths is reduced by the interaction, except that, when $\delta\Gamma^2/4 >> V^2$, a narrrow level can tune through resonance ($\delta\epsilon = 0$) with a broad level *without any significant change in width*!

9.3.2 Quantum Beats Between Two Decaying Quasi-Eigenstates

One dynamical example, two-level quantum beats *including decay* (but not including collision induced depopulation and dephasing), illustrates the power of this complex **H** formalism. (See Section 6.5.3 for a detailed discussion of quantum beats in the strong coupling limit.) Consider two zero-order states, $E_1 = \epsilon_1 - i\Gamma_1/2$ and $E_2 = \epsilon_1 - i\Gamma_2/2$, where state 1 is bright and narrow and state 2 is dark and broad ($\Gamma_2 >> \Gamma_1$).

A short pulse excites a coherent superposition of the eigenstates ψ_+ and ψ_- which result from the interaction between the zero-order states ψ_1 and ψ_2. This coherent superposition exhibits quantum beats at frequency and decay rate, respectively, $\omega_{QB} = (\epsilon_+ - \epsilon_-)/\hbar$ and $\Gamma_{QB}/\hbar = \frac{1}{2}(\Gamma_+ + \Gamma_-)/\hbar$,

$$I_{QB}(t) = I_{QB}e^{-\Gamma_{QB}t/\hbar}\cos(\omega_{QB}t + \phi_{QB}), \tag{9.3.45}$$

where I_{QB}, Γ_{QB}, ω_{QB}, and ϕ_{QB} are discussed below. In addition to the quantum beating fluorescence, there will be unmodulated biexponentially decaying fluorescence

$$I(t) = I_+e^{-t(\Gamma_+)/\hbar} + I_-e^{-t(\Gamma_-)/\hbar}, \tag{9.3.46}$$

where the magnitudes, I_+ and I_-, of the Γ_+ and Γ_- decaying components are discussed below.

The complex **H** formalism provides a compact way of deriving the analytic $\delta\Gamma$ and V dependences of ω_{QB}, Γ_{QB}, Γ_+, and Γ_- as the narrow state is tuned through resonance (i.e., the sequence of cases, $\delta\epsilon > V, \delta\epsilon \approx 0, \delta\epsilon < -V$) with the broad state. At $t = 0$

$$\Psi(0) = \psi_1, \tag{9.3.47}$$

which is not an eigenstate of **H**. In order to express the time-evolution of such a state, this state must be expressed (using Eq. (9.3.9)) as a linear combination of eigenstates,

$$|\psi_1\rangle = |\psi_+\rangle\langle\tilde{\psi}_+|\psi_1\rangle + |\psi_-\rangle\langle\tilde{\psi}_-|\psi_1\rangle = \alpha_+|\psi_+\rangle + \alpha_-|\psi_-\rangle \tag{9.3.48}$$

where α_+ and α_- are the mixing coefficients (at $t = 0$) of the two eigenstates, ψ_+ and ψ_-, in the zero-order state, ψ_1. These mixing coefficients are, in the weak coupling limit,

$$\alpha_+ = \left\langle\tilde{\psi}_+|\psi_1\right\rangle = 1 - \frac{V^2}{8(\delta\epsilon - i\delta\Gamma/2)^2} \tag{9.3.49a}$$

$$\alpha_- = \left\langle\tilde{\psi}_-|\psi_1\right\rangle = -\frac{V^2}{4(\delta\epsilon - i\delta\Gamma/2)^2} \tag{9.3.49b}$$

or, in the strong coupling limit,

$$\alpha_+ = 2^{-1/2}\left(1 + \frac{\delta\epsilon - i\delta\Gamma/2}{2|V|}\right) \tag{9.3.50a}$$

$$\alpha_- = -2^{-1/2}\left(1 - \frac{\delta\epsilon - i\delta\Gamma/2}{2|V|}\right) \tag{9.3.50b}$$

The time-evolution of the coherent (and decaying) superposition state is given by

$$\begin{aligned}\Psi_1(t) &= \alpha_+\psi_+e^{-iE_+t/\hbar} + \alpha_-\psi_-e^{-iE_-t/\hbar} \\ &= \alpha_+\psi_+e^{-i\epsilon_+t/\hbar}e^{-\Gamma_+t/2\hbar} + \alpha_-\psi_-e^{-i\epsilon_-t/\hbar}e^{-\Gamma_-t/2\hbar}.\end{aligned} \tag{9.3.51}$$

The fluorescence intensity will be proportional to the survival probability,

$$P_1(t) = \left|\left\langle\tilde{\Psi}_1(0)|\Psi_1(t)\right\rangle\right|^2 \tag{9.3.52}$$

because $\Psi_1(0)$ is the pure bright state and $\left\langle\tilde{\Psi}_1(0)|\Psi_1(t)\right\rangle$ is the amplitude of the bright state in the evolving superposition state.

$$\begin{aligned}
P_1(t) &= \left(\alpha_+^2 e^{-iE_+t/\hbar} + \alpha_-^2 e^{-iE_-t/\hbar}\right)\left(\alpha_+^{*2} e^{+iE_+^*t/\hbar} + \alpha_-^{*2} e^{+iE_-^*t/\hbar}\right) \\
&= |\alpha_+|^4 e^{-\Gamma_+t/\hbar} + |\alpha_-|^4 e^{-\Gamma_-t/\hbar} \\
&+ \alpha_+^2\alpha_-^{*2} e^{-i(E_+-E_-^*)t/\hbar} + \alpha_-^2\alpha_+^{*2} e^{-i(E_--E_+^*)t/\hbar} \\
&= I_+ e^{-\Gamma_+t/\hbar} + I_- e^{-\Gamma_-t/\hbar} \\
&+ e^{-\Gamma_{QB}t/\hbar}\left[2\mathrm{Re}(\alpha_+^2\alpha_-^{*2})\cos\omega_{QB}t - 2\mathrm{Im}(\alpha_+^2\alpha_-^{*2})\sin\omega_{QB}t\right] \qquad (9.3.53a) \\
&= e^{-\Gamma_{QB}t/\hbar}\left[I_{QB}\cos(\omega_{QB}t + \phi_{QB})\right] \qquad (9.3.53b) \\
I_{QB} &= 2\left|\alpha_+^2\alpha_-^{*2}\right| \qquad (9.3.53c) \\
\phi_{QB} &= \tan^{-1}\frac{\mathrm{Im}(\alpha_+^2\alpha_-^{*2})}{\mathrm{Re}(\alpha_+^2\alpha_-^{*2})} \qquad (9.3.53d)
\end{aligned}$$

$$\omega_{QB} = \epsilon_+ - \epsilon_- = 2\chi \qquad (9.3.54)$$

$$\Gamma_{QB} = \frac{1}{2}(\Gamma_+ + \Gamma_-) = \overline{\Gamma}. \qquad (9.3.55)$$

Fig. 9.10 illustrates the dependences of $\omega_{QB}, \Gamma_+ - \Gamma_-$, and the $t = 0$ modulation depth, I_{QB}, of a quantum beating signal on $\delta\Gamma, \delta\epsilon$, and V. The modulation depth is the fraction of the total decaying fluorescence signal that is modulated at ω_{QB}.

This two-state quantum beat example is identical to the doorway mediated non-radiative decay problem frequently encountered in polyatomic molecule Intramolecular Vibrational Redistribution (IVR), Inter-System Crossing (ISC), Internal Conversion (IC), and compound anticrossings. There is a single, narrow bright state. It couples to a single, broad, and dark doorway state. The width of the doorway state is determined by the rate of its Fermi Golden Rule decay into a quasi-continuum of dark states.

9.3.3 The Use of the Complex H^{eff} in Reduced-Dimension Models

It is often computationally efficient and interpretively effective to divide a complicated dynamical problem into a part that must be solved exactly and understood mechanistically and another part that need only be treated in a perfunctory, phenomenological, or statistical way. One divides configuration space or state-space into a small, low-dimensional active region surrounded by an absorbing boundary and a large, high-dimensional inactive or bath region (Miller and George, 1972; Stine and Marcus, 1972; Child, 1991; Schinke, 1993; Moiseyev, 1998). Some of the probability flux that approaches near the boundary of the active region is irreversibly absorbed. The irreversibility of this absorption reflects the physically reasonable assumption that the enormously larger density of states in the inactive than the active region ensures statistical behavior within the inactive region and negligible return of probability flux to the active region.

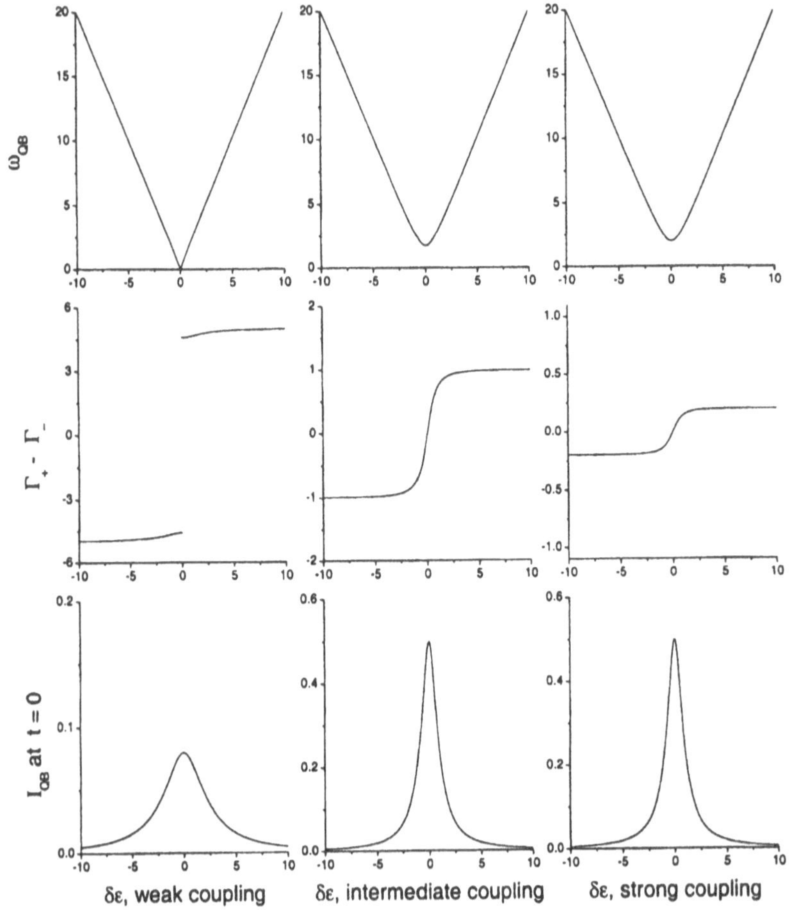

Figure 9.10: Level Anticrossings between states with complex energies in the strong and weak coupling limits. The quantum beat frequency, ω_{QB}, the difference between the widths of the levels at higher and lower real energy, $\Gamma_+ - \Gamma_-$, and the intensity of the quantum beat, I_{QB}, are plotted vs. the difference between the real part of the zero-order energy, $\delta\epsilon$, at constant coupling strength (V), for three values of the difference between the imaginary parts of the zero-order energy, $\delta\Gamma$: strong coupling ($V = 5\delta\Gamma$), intermediate coupling ($V = \delta\Gamma$), and weak coupling ($V = \delta\Gamma/5$). Note the use of different vertical scales for the three $\Gamma_+ - \Gamma_-$ and I_{QB} plots. The ω_{QB} plots illustrate the reduction in level repulsion from the strong coupling value of minimum(ω_{QB}) $= 2V$ that occurs in the weak coupling limit. That the sharp level tunes, without level repulsion, through the broad level in the weak coupling limit, is illustrated by the linearity of the vee-shaped ω_{QB} curve and the discontinuity in the $\Gamma_+ - \Gamma_-$ curve. The I_{QB} curves illustrate that, in the weak coupling limit, the product of mixing fractions is significantly reduced at the level crossing (dynamical decoupling) but the FWHM of the $I_{QB} = 2\left|\alpha_+^2 \alpha_-^{*2}\right|$ mutually-mixed region is increased (figure prepared by Kyle Bittinger).

Whether the inactive region is a true continuum (e.g., photofragmentation) or a quasi-continuum comprised of an enormous density of rigorously bound eigenstates (polyatomic molecule dynamics, Section 9.4.14) is often of no detectable consequence. The dynamical quantities discussed in Section 9.1.4 (probability density, density matrix, autocorrelation function, survival probability, transfer probability, expectation values of coordinates and conjugate momenta) describe the active space dynamics without any reference to the detailed nature of the inactive space.

A complex $\mathbf{H}^{\text{eff}}$ model is constructed by associating an amplitude decay rate, $\Gamma_j/2$, with the zero-order energy, $\epsilon_j - i\Gamma_j/2$, of each active-space basis state. The Γ_j values may be derived from a state-space Fermi Golden Rule treatment of the average squared interaction strength of the j th active-space basis state with the approximately isoenergetic basis states in the inactive space (I)

$$\Gamma_j = 2\pi\rho_I \left\langle H_{jI}^2 \right\rangle. \tag{9.3.56}$$

Each of the anharmonic coupling terms in the $V(\mathbf{Q})$, e.g., $k_{jKK}Q_jQ_K^2$ (where Q_K is a coordinate in the inactive region of space), generates an explicitly calculable contribution to ρ_I and $\left\langle H_{jI}^2 \right\rangle$. The $\{\Gamma_j\}$ values may be computed in a coordinate space picture by expressing the potential energy surface as a complex function of real coordinates. The negative imaginary contribution at $\mathbf{Q}$ is large in regions of the active space from which amplitude leaks out rapidly. For example, if molecules are confined to the active region of space by a potential energy barrier, the negative imaginary part of $V(\mathbf{Q})$ would be largest in the region of $\mathbf{Q}$ near the minimum energy path over the barrier [Germann and Miller, 1998]. The magnitudes of the various Q-dependent imaginary terms in $V(\mathbf{Q})$ would be computed by evaluating one-dimensional WKB barrier penetration integrals.

9.4 *Beyond* the Spectra and Dynamics of Diatomic Molecules

9.4.1 Basis States

A complete set of basis states is much more than a mathematical tool that permits an "exact" description of reality. Although the choice of basis set is arbitrary, a physically appropriate basis set is of enormous practical, fundamental, and metaphoric value. A physically appropriate basis set expresses the diagnostically significant patterns that facilitate assignment of a spectrum, specifies the control parameters that are important in both preserving and destroying the characteristic patterns, and provides the language and pictures by which an apparently indescribably complex many-body reality may be understood.

Basis states are equally important in understanding frequency- and time-domain experiments. In the frequency domain, the pattern is comprised of the frequencies and relative intensities of the vast majority of observed spectral features. This majority status enables the pattern to be recognized and draws attention to the specific subset of pattern breaking transitions. The experimentalist is challenged to identify the outside-the-model states ("local perturbers") responsible for disrupting the pattern. In the time-domain, one typically finds a regular, interpretable pattern immediately following the initiation of the dynamics by a suitably localized pluck. This regular, early-time pattern either decays or evolves into a system response of rapidly (but not necessarily monotonically) increasing complexity. Once again, the early-time regularity defines the interpretive model and demands characterization of the specific pattern-destroying mechanism.

It appears that there is a fundamental difference between pattern breaking in the frequency and time domains. In the frequency domain, the pattern breaking is usually subtle and associated with small terms in the $\mathbf{H}^{\text{eff}}$. In the time-domain, the pattern breaking is typically global, caused by large terms in the $\mathbf{H}^{\text{eff}}$. But this appearance is misleading. Frequency domain spectra need not be resolved to the eigenstate level. By recording a frequency domain spectrum at lower resolution ($\Delta t \Delta \nu \sim 1$), most of the subtle pattern-disruptions vanish and the spectrum samples the same mechanisms responsible for the global pattern-disruptions in the time-domain. It also is true that long-time regular dynamics can coexist with and even out-live the rapidly evolving early-time pattern.

There is no "best way" to observe and understand molecular dynamics. Sometimes a mechanistic feature is sampled more transparently in the time or the frequency domain. The design of an experiment (quantum number pre- or post-selection, an ingeniously designed pluck) is far more important than whether the experiment is performed in the time- or frequency-domain.

9.4.2 What is Deperturbation Anyway?

An effective Hamiltonian is profoundly different from an exact Hamiltonian. This is a reason for imperfect communication between experimentalists and *ab initio* theorists. The two communities use the same symbols and language to refer to often quite different molecular properties. The main difference between effective and exact Hamiltonians is that the molecule gives experimentalists an empirical basis set that has been prediagonalized implicitly to account for the infinite number of "remote perturbers". This is the Van Vleck or "contact" transformation, but it is performed by the molecule, not by a graduate student. The basis set is truncated and the dynamics occurs in a reduced-dimension state space.

The molecule also chooses the physically appropriate basis set. Sometimes the molecule suggests to the experimentalist one basis set to describe an experimentally-realizable pluck and a different basis set to describe the post-pluck dynamics. No matter what basis set the molecule chooses, there always will be some pairs of basis states connected by an off-diagonal matrix element of $\mathbf{H}^{(1)}$ larger than their difference in zero-order energies. Under such circumstances, nondegenerate perturbation theory fails and a quasi-degenerate block of the $\mathbf{H}^{\text{eff}}$

must be diagonalized. This deperturbation procedure creates a choice between two artificially simplified worlds. One is the world where the problematic off-diagonal matrix elements of $\mathbf{H}^{(1)}$ have been turned off. This is useful if the perturbation is deemed to be accidental, irrelevant to the more global dynamics being modeled (e.g. a perturbation of a Rydberg series by an "interloper" state that belongs to a different ion-core vibrational, v^+, or rotational, N^+, level). The other is the very low-dimensional world of uniquely fast dynamics among the small subset of strongly coupled states. The latter choice is useful, for example, if one wants to exploit the rapidly evolving, recurrent dynamics subsequent to a special class of pluck in order to probe the more complex but more slowly evolving dynamics within a broad region of state space.

A Hamiltonian can be "dressed" (see Section 3.3.5) to simplify the interpretation of the time or frequency domain spectrum. The Van Vleck transformation removes from consideration the coupling to energetically remote perturbers. But this coupling is real. In the time domain the neglected remote perturbers contribute to $\Psi(t)$ a huge number of component states with very fast, very small amplitude oscillations. Once a dynamically-interesting region of state space is excised from the complete state space by a Van Vleck transformation, many off-diagonal coupling terms often remain in $\mathbf{H}^{\text{eff}}$. The choice of the physically-appropriate basis set involves dressing by prediagonalizing a mutually commuting subset of these coupling operators. However, and this is especially true for polyatomic molecules, several incompatible (non-commuting) coupling terms often are of comparable magnitude. Therefore one has the choice of dressing the $\mathbf{H}^{\text{eff}}$ by prediagonalizing alternative sets of coupling operators. One choice might provide insight into the early time dynamics in a restricted state space by "removing" all but a small number of interaction mechanisms, either by causing interacting pairs of basis states to be detuned from near degeneracy or by causing coupling matrix elements to be turned off by destructive interference between coupling pathways. Another choice could provide insight into the longer time dynamics by prediagonalizing all of the strong interactions that affect the restricted region of state space sampled by the selected pluck. In Section 3.3.5 the example of potential energy curves dressed by a strong electromagnetic field is discussed. The transformation into the dressed state representation displays all of the dynamically significant interactions between the radiation field and the molecule by selectively tuning the interacting molecule-field states into near degeneracy.

9.4.3 Visualization of Dynamics

The information contained in a diatomic molecule rotation-vibration-electronic wavefunction is enormous. But this is dwarfed by the information content of a time-evolving wavefunction that originates from a non-eigenstate pluck. A simplified, reduced-dimension representation, rather than an exact numerical description, is prerequisite to visualization and understanding. The concepts and techniques presented in this book, developed explicitly for diatomic molecule spectra and dynamics, are applicable to larger molecules. Indeed, any attempt to understand the dynamics of complex systems should begin with mastery of simpler systems, because most interpretive schemes strive to reduce the dimensionality of the problem, at least for a chemically relevant short time. A critical difference between simple and complex systems is that, although the dynamics of both simple and complex systems can be described for a short time by reduced-dimension cause-and-effect mechanisms, it is necessary to resort to phenomenological or statistical models for the longer-time dynamics of complex systems.

Eigenstates are stationary and wavefunctions have numerous and odd-shaped nodal surfaces. Our experience of the world is of predictable (as opposed to chaotic) motion and continuously varying probability distributions. Time-domain experiments make direct contact with experience-based instincts. A central goal of this book has been to enable the incorporation of frequency domain $\mathbf{H}^{\text{eff}}$ models, scaling rules, and methods for dimensionality reduction into the toolkit and worldview of time domain spectroscopists.

In designing an experiment to determine the mechanism of a dynamical process, the experimentalist must consider the excitation (the "pluck"), evolution (the "resonance"), and detection (the "probe") components of an experiment. The excitation involves selection of an *a priori* perfectly specified non-eigenstate "bright state". In order for the evolution to be neither boringly simple nor indescribably complex, the pluck should occur in a region of state space selected so that the bright state is in near resonance with a specified class of dark state, $\left|H_{12}/\left(E_1^{(0)} - E_2^{(0)}\right)\right| > 1$. The important quantity is the coupling matrix element *divided by* the energy difference, *not* the matrix element itself. As a result, the early time dynamics is dominated by a single, easily visualized process, the flow of probability between the bright state and the selected resonant class of dark states.

A low resolution spectrum is a spectrum that consists exclusively of transitions into non-eigenstate bright states. If the frequency resolution, $\Delta\nu$, is low enough (see Fig. 9.2) each bright state is captured before probability has had time to leak from the initially localized excitation into other classes or combinations of internal motions. The dynamical processes (IVR, ISC, IC) are allowed insufficient time, $t < (2\pi\Delta\nu)^{-1}$ to occur. As the resolution is increased, the early time dynamics of the selected bright state is revealed. The broad spectral envelope of the bright state splits into finer "feature state" structures. When the relative frequencies and intensities of these (non-eigenstate) feature states are used to least squares adjust the parameters that define an $\mathbf{H}^{\text{eff}}$ model, one obtains a description of the early-time dynamics. Low frequency resolution is equivalent to high time resolution. A low resolution frequency domain experiment captures what would be seen in the simplest possible time domain experiment in which the state formed at $t = 0$, $\Psi(0) = \psi_{\text{bright}}$, is the bright state.

The best characterization of the selected dynamical process is obtained by devising selective detection schemes capable of monitoring the flow of probability out of the bright state and into several candidate dark states. Bright and dark are not absolute quantities; rather, as illustrated by Fig. 9.5, the excitation (pluck) and detection (probe) schemes devised by the experimentalist determine which classes of states are endowed with brightness and detectability. Major insights into intramolecular dynamics accrue each time a new class of pluck/resonance/probe scheme is devised. Examples include high overtone spectroscopy, stimulated emission pumping, and vibrationally mediated photodissociation spectroscopy.

While it is possible to extract, by a carefully designed series of experiments, a detailed description of how and why energy flows out of a specified bright state, the existence of such an early-time description does not rule out long-time statistical or "ergodic" behavior. Mechanism and ergodicity are not incompatible. The early-time evolution of a localized pluck occurs in a reduced dimension coordinate or state space. Often, the causes and nature of the early-time evolution are easily visualized and described. This is "mechanism." At long-time the initially localized excitation fractionates into an enormous number of eigenstates. When every (symmetry accessible) eigenstate under the broad spectral envelope of the bright state has an equal share, in a statistical sense, of the bright state, that is "ergodicity."

9.4.4 Beyond Diatomic Molecules: Polyatomic Molecule Vibrational Dynamics

Most of the concepts, models, and tools described in this book are applicable to polyatomic molecules. Perhaps the most significant difference between diatomic and polyatomic molecules is the number of vibrational degrees of freedom: one for diatomic molecules vs. $3N-6$ for N–atom polyatomic molecules. This difference in the number of vibrational modes has several profound consequences for polyatomic molecules: (*i*) there exists no semiclassical, RKR-like inversion from the rotation-vibration energy levels to a potential energy surface; (*ii*) owing to the existence of $3N-6$ vibrational modes, the density of vibrational energy levels increases very rapidly with E_{VIB} ($\rho_{\mathrm{VIB}} \approx E_{\mathrm{VIB}}^{3N-7}$); (*iii*) there is the possibility of systematic near-degeneracies between integer-multiple combinations of normal mode frequencies (e.g., a $2:1$ resonance $\omega_1 \approx 2\omega_2$ or a $1:(1\pm1\pm1)$ resonance, such as $\omega_1 = \omega_2+\omega_3+\omega_4$ or $\omega_2 = \omega_3-\omega_4-\omega_5$).

The unavailability of an RKR-like inversion (hence the impossiblity of obtaining the potential energy surface, $V(\mathbf{Q})$, and exact vibrational eigenfunctions directly from experimental data) makes it convenient to use products of simple harmonic or Morse-oscillator basis functions as vibrational basis states

$$\psi_{v_1 v_2 \ldots v_{3N-6}} = \psi_{\mathbf{v}} = \prod_{i=1}^{3N-6} \phi_{v_i}, \tag{9.4.1}$$

where the set of $3N-6$ individual mode vibrational quantum numbers, $\{v_i\}$, is denoted by $\mathbf{v}$, and to use simple products of vibrational normal mode displacement operators to model the anharmonic interactions between vibrational basis states,

$$V(Q_1, Q_2, \ldots Q_{3N-6}) = V(\mathbf{Q}) = \sum_{i=1}^{3N-6} (k_{ii}/2)Q_i^2$$
$$+ \sum_{i,j,k} k_{ijk} Q_i Q_j Q_k + [\text{quartic}] + [\text{quintic}] + \ldots \qquad (9.4.2)$$

so that all anharmonic coupling matrix elements may be evaluated trivially based on harmonic oscillator matrix element selection and scaling rules

$$\langle v_i + n | Q_i^a | v_i \rangle = \left[\frac{\hbar}{\mu_i \omega_i} \right]^{a/2} \left\{ v_i^{a/2} \right\}$$
$$n = a, a-2, \ldots, -a \qquad (9.4.3)$$

where $\{v^{a/2}\}$ denotes a sum of terms in powers of v where the highest power of v is $v^{a/2}$ and μ_i is a mass and geometry dependent factor with units of mass that comes from an **F**,**G** matrix normal mode analysis (Wilson, *et al.*, 1955). Child, *et al.*, (2001) discuss modified matrix element scaling laws for extremely anharmonic systems. The rapid increase with E_{VIB} of the vibrational density of states results in a dense manifold of vibrational eigenstates, which is often called a quasicontinuum. Although quasicontinua are not true continua, all of the methods developed in this book for bound↔continuum interactions are applicable to bright state↔quasicontinuum energy flow processes, e.g., vibration↔vibration (Intramolecular Vibrational Redistribution, IVR) and vibration↔electronic (Inter-System Crossing, ISC, and Internal Conversion, IC).

The collision-free nonradiative (and non-photofragmenting) decay of a polyatomic molecule is described by the time evolution of the bound, bright, basis-state (not an eigenstate) prepared at $t = 0$ (Heller and Gelbart, 1980; Avouris, *et al.*, 1977). This bright state is actually a coherent superposition of many degenerate and nondegenerate eigenstates.

The nonradiative decay consists of the irreversible dephasing of the bright state character in the time-evolving superposition state (Bixon and Jortner, 1968). Total energy is conserved, but the energy initially localized in one special basis state becomes distributed over many non-special basis states. Even when the density of dark vibrational states is too low to generate a quasicontinuum, and even when the spectral resolution is sufficient to resolve individual eigenstates, vibrational dynamics is encoded (via level splittings and relative intensities) as "fractionation" of the zero-order bright state character into a cluster of eigenstates. It turns out that the Fourier transform of this piece of fractionated eigenstate spectrum is the survival probability of the bright state that was excited at $t = 0$.

9.4.5 Polyads

The existence of systematic near degeneracies between groups of anharmonically-coupled vibrational basis states results in a convenient separation of the intramolecular dynamics into fast, deterministic, near-resonant energy flow within a select group of vibrational basis states, called a "polyad", and the much slower, statistical energy flow between polyad basis states and "bath" basis states. Intrapolyad energy flow is an example of dynamics in state space that provides an excellent opportunity to demonstrate the insights obtainable from several of the concepts and tools described in this chapter.

A polyad is a group of near-degenerate vibrational states coupled by one or more low-order anharmonic terms. For example, the very common 2 : 1 "Fermi resonance", which is due to an anharmonic coupling term like $k_{122}Q_1Q_2^2$, where $k_{122} = \frac{1}{2}\left(\frac{\partial^3 V}{\partial^2 Q_2 \partial Q_1}\right)_{\mathbf{Q}_e}$, becomes dynamically important when $\omega_1 \approx 2\omega_2$. The number of systematically near-degenerate coupled states grows with excitation energy. Let $N = 2v_1 + v_2$ be the "polyad number", ($E_{\mathrm{VIB}} \approx \omega_1 N/2 \approx \omega_2 N$), which is an approximately conserved quantity, the conservation of which is not guaranteed by any fundamental symmetry. Denoting basis-states by (v_1, v_2), we have, for $N = 0$ only $(0,0)$, for $N = 1$ only $(0,1)$, for $N = 2$, $(1,0)$ and $(0,2)$, ... and, for e.g., $N = 10$ the polyad includes 6 basis-states $(5,0), (4,2), (3,4), (2,6), (1,8)$, and $(0,10)$. Not only does the number of basis states grow with polyad number, but so too does the size of the off-diagonal coupling matrix elements. The largest matrix element in the N= 2, 4, and 10 polyads is equal to a common multiplicative constant times the following factor obtained from harmonic oscillator matrix element scaling rules:

$$N = 2 \qquad \langle 1,0|\mathbf{H}|0,2\rangle \propto [1]^{1/2}[2\cdot 1]^{1/2} \tag{9.4.4a}$$

$$N = 4 \qquad \langle 1,2|\mathbf{H}|0,4\rangle \propto [1]^{1/2}[4\cdot 3]^{1/2} \tag{9.4.4b}$$

$$N = 10 \qquad \langle 2,6|\mathbf{H}|1,8\rangle \propto [2]^{1/2}[8\cdot 7]^{1/2}. \tag{9.4.4c}$$

When more than one anharmonic interaction term couples near-degenerate, zero-order levels, a simple vector orthogonalization technique can be used to generate a complete set of the dynamically important (i.e., approximately conserved) polyad quantum numbers (Fried and Ezra, 1987; Kellman, 1990). For example, in acetylene, HC $\equiv$ CH, where the ratios of normal mode frequencies $\omega_1 : \omega_2 : \omega_3 : \omega_4 : \omega_5$ are approximately $5 : 3 : 5 : 1 : 1$, modes 1, 2, and 3 are stretching modes (respectively symmetric CH stretch, and CC stretch, and antisymmetric CH stretch), modes 4 and 5 are bending modes (*trans*-bend and *cis*-bend), each polyad is labeled by 3 polyad quantum numbers,

$$N_{\mathrm{resonance}} = 5v_1 + 3v_2 + 5v_3 + v_4 + v_5 \tag{9.4.5a}$$

$$N_{\mathrm{stretch}} = v_1 + v_2 + v_3 \tag{9.4.5b}$$

$$l_{\mathrm{total}} = l_4 + l_5, \tag{9.4.5c}$$

where l_4 and l_5 are the vibrational angular momenta associated with the two doubly-degenerate bending modes.

9.4.6 Creation and Annihilation Operators

At this point it is useful to define creation ($\mathbf{a}^\dagger$) and annihilation ($\mathbf{a}$) operators, which are analogous to angular momentum raising and lowering operators. These $\mathbf{a}^\dagger, \mathbf{a}$ operators profoundly simplify the algebra needed to set up the polyad $\mathbf{H}^{\text{eff}}$ matrices, to apply some of the dynamics diagnostics discussed in Sections 9.1.4 and 9.1.7, and to transform between basis sets (e.g., between normal and local modes). They also provide a link between the quantum mechanical $\mathbf{H}^{\text{eff}}$ model, which is expressed in terms of $\mathbf{a}^\dagger, \mathbf{a}$ operators and adjustable molecular constants (evaluated by least squares fits of spectra), and a reduced-dimension classical mechanical $\mathcal{H}^{\text{eff}}$ model.

The harmonic oscillator matrix elements for a normal coordinate displacement, $\mathbf{Q}$, and the conjugate momentum, $\mathbf{P}$, operators are converted into dimensionless quantities, $\widehat{\mathbf{Q}}$ and $\widehat{\mathbf{P}}$,

$$\widehat{\mathbf{Q}} = \left[\frac{2\pi c\mu\omega}{\hbar}\right]^{1/2} \mathbf{Q} \tag{9.4.6}$$

$$\widehat{\mathbf{P}} = [\hbar 2\pi c\mu\omega]^{-1/2}\mathbf{P} \tag{9.4.7}$$

$$\omega = \frac{1}{2\pi c}[k/\mu]^{1/2}. \tag{9.4.8}$$

[The generalization of a force constant, k, and reduced mass, μ, from a one dimensional harmonic oscillator to the normal mode oscillators of a polyatomic molecule is accomplished by the $\mathbf{F}$, $\mathbf{G}$ matrix methods of Wilson, *et al.*, (1955). For the present discussion it is sufficient to know that a force constant and a reduced mass may be uniquely defined for each of $3N - 6$ linearly independent sets of internal coordinate displacements in a polyatomic molecule (see also Section 9.4.12).] The harmonic oscillator Hamiltonian

$$\mathbf{H} = \frac{1}{2}k\mathbf{Q}^2 + \mathbf{P}^2/2\mu \tag{9.4.9}$$

may be re-expressed in dimensionless form,

$$\widehat{\mathbf{H}} = \frac{1}{\hbar(2\pi c\omega)}\mathbf{H} = \frac{1}{2}[\widehat{\mathbf{Q}}^2 + \widehat{\mathbf{P}}^2]. \tag{9.4.10}$$

The nonzero matrix elements of $\widehat{\mathbf{Q}}, \widehat{\mathbf{P}}$ and $\widehat{\mathbf{H}}$ are, in the Harmonic Oscillator basis set $\{|v\rangle\}$,

$$\left\langle v+1|\widehat{\mathbf{Q}}|v\right\rangle = \left\langle v|\widehat{\mathbf{Q}}|v+1\right\rangle = 2^{-1/2}[v+1]^{1/2} \tag{9.4.11}$$

$$\left\langle v+1|\widehat{\mathbf{P}}|v\right\rangle = -\left\langle v|\widehat{\mathbf{P}}|v+1\right\rangle = i2^{-1/2}[v+1]^{1/2} \tag{9.4.12}$$

$$\left\langle v|\widehat{\mathbf{H}}|v\right\rangle = (v+1/2). \tag{9.4.13}$$

Note that $\widehat{\mathbf{Q}}$ and $\widehat{\mathbf{P}}$ have selection rules $\Delta v = +1$ *and* -1, which leads to minor inconveniences in deriving matrix elements of integer powers of $\widehat{\mathbf{Q}}$ and $\widehat{\mathbf{P}}$ by matrix multiplication. The $\mathbf{a}^\dagger, \mathbf{a}$ operators

$$\mathbf{a}^\dagger = 2^{-1/2}[\widehat{\mathbf{Q}} - i\widehat{\mathbf{P}}] \quad (9.4.14)$$

$$\mathbf{a} = 2^{-1/2}[\widehat{\mathbf{Q}} + i\widehat{\mathbf{P}}] \quad (9.4.15)$$

$$\widehat{\mathbf{Q}} = 2^{-1/2}(\mathbf{a}^\dagger + \mathbf{a}) \quad (9.4.16)$$

$$\widehat{\mathbf{P}} = 2^{-1/2}i(\mathbf{a}^\dagger - \mathbf{a}) \quad (9.4.17)$$

have the convenient property that their selection rules are $\Delta v = +1$ *or* -1, respectively, and that their only nonzero matrix elements are

$$\langle v+1|\mathbf{a}^\dagger|v\rangle = [v+1]^{1/2} \quad (9.4.18)$$

$$\langle v|\mathbf{a}|v+1\rangle = [v+1]^{1/2} \quad (9.4.19)$$

$$\langle v|\mathbf{a}^\dagger\mathbf{a}|v\rangle = v. \quad (9.4.20)$$

Thus $\mathbf{a}^\dagger$ *creates* one quantum of vibration, $\mathbf{a}$ *destroys* one quantum of vibration, and $\mathbf{a}^\dagger\mathbf{a}$, called the "number operator" ($\widehat{\mathbf{N}}$), is a measure of the *number of quanta* of vibration.

For a molecule with $3N-6$ vibrational modes, an $\mathbf{a}_i^\dagger, \mathbf{a}_i$ pair is defined for each mode, and

$$\mathbf{H}^\circ = \sum_{i=1}^{3N-6} \hbar 2\pi c\omega_i(\mathbf{a}_i^\dagger\mathbf{a}_i + 1/2) \quad (9.4.21)$$

$$\langle \mathbf{v}|\mathbf{H}^\circ|\mathbf{v}\rangle = \sum_{i=1}^{3N-6} \hbar 2\pi c\omega_i(v_i + 1/2). \quad (9.4.22)$$

Anharmonic terms, both diagonal and off-diagonal, are expressed compactly in terms of $\mathbf{a}_i^\dagger\mathbf{a}_i$ and combinations of products of $\mathbf{a}_i^\dagger, \mathbf{a}_i$ and $\mathbf{a}_j^\dagger, \mathbf{a}_j$ operators. For example, the usual Dunham-like expansion of the diagonal elements of $\mathbf{H}$ may be re-expressed as

$$\begin{aligned} E_\mathbf{v}/hc &= \sum_i \omega_i(v_i + 1/2) + \sum_{i\geq j} x_{ij}(v_i + 1/2)(v_j + 1/2) + \ldots \\ &= \sum_i \omega_i[\mathbf{a}_i^\dagger\mathbf{a}_i + 1/2] + \sum_{i\geq j} x_{ij}[\mathbf{a}_i^\dagger\mathbf{a}_i + 1/2][\mathbf{a}_j^\dagger\mathbf{a}_j + 1/2], \end{aligned} \quad (9.4.23)$$

and the off-diagonal elements of the form $\widehat{\mathbf{Q}}_i^n\widehat{\mathbf{Q}}_j^m$ are expressed as simple products

$$\widehat{\mathbf{Q}}_i^n\widehat{\mathbf{Q}}_j^m = (2^{-1/2})^{n+m}[\mathbf{a}_i^\dagger + \mathbf{a}_i]^n[\mathbf{a}_j^\dagger + \mathbf{a}_j]^m, \quad (9.4.24)$$

which can be manipulated into simple sums of operator products by exploiting the commutation rules

$$[\mathbf{a}_i, \mathbf{a}_j^\dagger] = \delta_{ij} \tag{9.4.25}$$

$$[\mathbf{a}_i, \mathbf{a}_j] = [\mathbf{a}_i^\dagger, \mathbf{a}_j^\dagger] = 0. \tag{9.4.26}$$

For example, $\widehat{\mathbf{Q}}_1\widehat{\mathbf{Q}}_2^2$, the dimensionless coupling term for the ubiquitous 2 : 1 anharmonic resonance (that arises from the $\frac{k_{122}}{2} = \mathbf{Q}_1\mathbf{Q}_2^2$ term in the Taylor series expansion of $V(\mathbf{Q}_1, \mathbf{Q}_2)$), becomes

$$\widehat{\mathbf{Q}}_1\widehat{\mathbf{Q}}_2^2 = 2^{-3/2}[\mathbf{a}_1^\dagger + \mathbf{a}_1][\mathbf{a}_2^{\dagger 2} + \mathbf{a}_2^2 + 2\mathbf{a}_2^\dagger\mathbf{a}_2 + 1] \tag{9.4.27}$$

and the two terms that couple near-degenerate levels are

$$H_{v_1,v_2;v_1-1,v_2+2}/hc = k'_{122}\left\langle v_1 v_2 \left|\mathbf{a}_1^\dagger\mathbf{a}_2^2\right| v_1 - 1 v_2 + 2\right\rangle$$

$$= k'_{122}[v_1(v_2+2)(v_2+1)]^{1/2} \tag{9.4.28a}$$

$$H_{v_1,v_2;v_1+1,v_2-2}/hc = k'_{122}\left\langle v_1 v_2 \left|\mathbf{a}_1\mathbf{a}_2^{\dagger 2}\right| v_1 + 1 v_2 - 2\right\rangle$$

$$= k'_{122}[(v_1+1)v_2(v_2-1)]^{1/2} \tag{9.4.28b}$$

$$k'_{122} = \frac{1}{2}\left(\frac{\hbar}{\mu_1 2\pi c\omega_1}\right)^{1/2}\left(\frac{\hbar}{\mu_2 2\pi c\omega_2}\right)(2)^{-3/2}k_{122}/hc, \tag{9.4.28c}$$

where the values of the only nonzero matrix elements of successive **a**'s and $\mathbf{a}^\dagger$'s (working from right to left) may be written by inspection as the square root of the larger of the two quantum numbers. k'_{122} is in units of cm^{-1}.

9.4.7 Dynamics in State Space

The anharmonic coupling term, $k'_{122}\widehat{\mathbf{Q}}_1\widehat{\mathbf{Q}}_2^2$, may be manipulated trivially so that the two dynamically important terms appear in the form discussed in Section 9.1.8,

$$\mathbf{O} = \boldsymbol{\Omega} + \boldsymbol{\Omega}^\dagger, \tag{9.4.29}$$

where $\boldsymbol{\Omega} = \mathbf{a}_1^\dagger\mathbf{a}_2^2$ and $\boldsymbol{\Omega}^\dagger = \mathbf{a}_1\mathbf{a}_2^{\dagger 2}$. These two terms are dynamically important because they conserve the polyad quantum number, $N = 2v_1 + v_2$, hence act only within the quasidegenerate state space of a single polyad. The other terms in $\widehat{\mathbf{Q}}_1\widehat{\mathbf{Q}}_2^2$ are of minimal dynamical consequence because all intrapolyad (i.e., $\Delta N = 0$) matrix elements of each term in

$$\mathbf{H}^{(1)} = k'_{122}[\mathbf{a}_1^\dagger\mathbf{a}_2^{\dagger 2} + \mathbf{a}_1\mathbf{a}_2^2 + \mathbf{a}_1^\dagger(2\mathbf{a}_2^\dagger\mathbf{a}_2 + 1) + \mathbf{a}_1(2\mathbf{a}_2^\dagger\mathbf{a}_2 + 1)] \tag{9.4.30}$$

are zero. The ΔN selection rules for the four terms in Eq. (9.4.30) are, respectively, +4, −4, +2, and −2.

The $\mathbf{a}, \mathbf{a}^\dagger$ algebra makes it possible to simplify the operator forms of the various dynamics diagnostics discussed in Sections 9.1.4 and 9.1.8: the expectation value of a number operator (e.g., $\langle \mathbf{a}_i^\dagger \mathbf{a}_i \rangle$, which tells us how many quanta or how much energy is in mode i as a function of time), the resonance energy term, $\mathbf{O}$, the transfer rate term, $\boldsymbol{\Omega} - \boldsymbol{\Omega}^\dagger$, and $\mathbf{Q}_i$ and $\mathbf{P}_i$.

9.4.8 Number Operator

The equation of motion of the expectation value of any operator is given by the expectation value of the commutator of that operator with $\mathbf{H}$, e.g.,

$$i\hbar \frac{d}{dt} \left\langle \mathbf{a}_i^\dagger \mathbf{a}_i \right\rangle = \left\langle \left[\mathbf{a}_i^\dagger \mathbf{a}_i, \mathbf{H} \right] \right\rangle. \tag{9.4.31}$$

Exploiting

$$[\mathbf{a}_i^\dagger \mathbf{a}_i, (\mathbf{a}_i)^n] = [\mathbf{a}_i \mathbf{a}_i^\dagger, (\mathbf{a}_i)^n] = -n(\mathbf{a}_i)^n \tag{9.4.32}$$

$$[\mathbf{a}_i^\dagger \mathbf{a}_i, (\mathbf{a}_i^\dagger)^n] = [\mathbf{a}_i \mathbf{a}_i^\dagger, (\mathbf{a}_i^\dagger)^n] = n(\mathbf{a}_i^\dagger)^n \tag{9.4.33}$$

and writing $\mathbf{O}$ in its general form

$$\mathbf{O} = \prod_j (\mathbf{a}_j)^{n_j} \prod_k (\mathbf{a}_k^\dagger)^{n'_k} + \text{h.c.} = \boldsymbol{\Omega} + \boldsymbol{\Omega}^\dagger, \tag{9.4.34}$$

(h.c. means Hermitian conjugate) it is possible to show that

$$[\mathbf{a}_j^\dagger \mathbf{a}_j, \mathbf{O}] = \Delta n_j (\boldsymbol{\Omega} - \boldsymbol{\Omega}^\dagger), \tag{9.4.35}$$

where the change in number of quanta of excitation in mode j is

$$\Delta n_j = (n_j - n'_j). \tag{9.4.36}$$

For our 2 : 1 resonance ($\omega_1 \approx 2\omega_2$) example

$$\Delta n_1 = 1 \tag{9.4.37a}$$

$$\Delta n_2 = -2 \tag{9.4.37b}$$

$$[\mathbf{a}_1^\dagger \mathbf{a}_1, \mathbf{O}] = 1(\boldsymbol{\Omega} - \boldsymbol{\Omega}^\dagger) = \mathbf{a}_1^\dagger \mathbf{a}_2^2 - \mathbf{a}_1 \mathbf{a}_2^{\dagger 2} \tag{9.4.38}$$

$$[\mathbf{a}_2^\dagger \mathbf{a}_2, \mathbf{O}] = -2(\boldsymbol{\Omega} - \boldsymbol{\Omega}^\dagger), \tag{9.4.39}$$

thus

$$i\hbar \frac{d}{dt} \left\langle \mathbf{a}_1^\dagger \mathbf{a}_1 \right\rangle = \left\langle \mathbf{a}_1^\dagger \mathbf{a}_2^2 - \mathbf{a}_1 \mathbf{a}_2^{\dagger 2} \right\rangle. \tag{9.4.40}$$

According to Eq. (9.1.95), $\left\langle \boldsymbol{\Omega} - \boldsymbol{\Omega}^\dagger \right\rangle$ is pure imaginary, and thus

$$\frac{d}{dt} \left\langle \mathbf{a}_1^\dagger \mathbf{a}_1 \right\rangle = \dot{v}_1(t) = \frac{-i}{\hbar} \left\langle \mathbf{a}_1^\dagger \mathbf{a}_2^2 - \mathbf{a}_1 \mathbf{a}_2^{\dagger 2} \right\rangle \tag{9.4.41}$$

is real. The rate of change of the number of quanta in mode 1, $\dot{v}_1(t)$, is given by the expectation value of $\boldsymbol{\Omega} - \boldsymbol{\Omega}^\dagger$. Conservation of polyad quantum number implies $\dot{v}_2 = -2\dot{v}_1$.

For the 2 : 1 resonance,

$$\begin{aligned}\mathbf{H}/hc = {}& \omega_1(\mathbf{a}_1^\dagger\mathbf{a}_1 + 1/2) + \omega_2(\mathbf{a}_2^\dagger\mathbf{a}_2 + 1/2) + x_{11}(\mathbf{a}_1^\dagger\mathbf{a}_1 + 1/2)^2 \\ & + x_{22}(\mathbf{a}_2^\dagger\mathbf{a}_2 + 1/2)^2 + x_{12}(\mathbf{a}_1^\dagger\mathbf{a}_1 + 1/2)(\mathbf{a}_2^\dagger\mathbf{a}_2 + 1/2) \\ & + k'_{122}[\mathbf{a}_1^\dagger\mathbf{a}_2^2 + \mathbf{a}_1\mathbf{a}_2^{\dagger 2} + \mathbf{a}_1^\dagger\mathbf{a}_2^{\dagger 2} + \mathbf{a}_1\mathbf{a}_2^2 + (\mathbf{a}_1^\dagger + \mathbf{a}_1)(2\mathbf{a}_2^\dagger\mathbf{a}_2 + 1)] \end{aligned} \quad (9.4.42)$$

where the x_{11}, x_{22}, and x_{12} Dunham constants are diagonal anharmonicity constants. The commutator $[\mathbf{a}_1^\dagger\mathbf{a}_1, \mathbf{H}]$ is evaluated easily by noting that number operators commute with themselves and each other. Thus no terms in $\mathbf{H}$, except for the $\mathbf{a}_1^\dagger\mathbf{a}_2^2$ and $\mathbf{a}_1\mathbf{a}_2^{\dagger 2}$ terms, yield terms from the commutator that act within a polyad (because, by inspection, all but the $\mathbf{a}_1^\dagger\mathbf{a}_2^2$ and $\mathbf{a}_1\mathbf{a}_2^{\dagger 2}$ terms have selection rules $\Delta N = \Delta(2v_1 + v_2) \neq 0$). Thus

$$[\mathbf{a}_1^\dagger\mathbf{a}_1, \mathbf{H}] = k'_{122}\left([\mathbf{a}_1^\dagger\mathbf{a}_1, \mathbf{a}_1^\dagger\mathbf{a}_2^2] + [\mathbf{a}_1^\dagger\mathbf{a}_1, \mathbf{a}_1\mathbf{a}_2^{\dagger 2}]\right) = k'_{122}[\hat{v}_1, \mathbf{O}], \quad (9.4.43)$$

which identifies $\mathbf{O}$ as the *cause* of the intrapolyad dynamics of v_1. This is not a surprising result. For this single resonance case, all of the intrapolyad motion of $\langle v_1\rangle(t)$ is caused by the two parts of the resonance operator, $\boldsymbol{\Omega}$ and $\boldsymbol{\Omega}^\dagger$. Note that conservation of the polyad quantum number imposes the constraint

$$N = 2\left\langle \mathbf{a}_1^\dagger\mathbf{a}_1\right\rangle + \left\langle \mathbf{a}_2^\dagger\mathbf{a}_2\right\rangle. \quad (9.4.44)$$

Hence

$$\left\langle \mathbf{a}_2^\dagger\mathbf{a}_2\right\rangle_t = \langle v_2\rangle_t = N - 2\langle v_1\rangle_t. \quad (9.4.45)$$

Note also that the expectation value of the energy in mode 1 is given by

$$\langle E_1\rangle_t = hc\omega_1\left[\left\langle \mathbf{a}_1^\dagger\mathbf{a}_1\right\rangle_t + 1/2\right], \quad (9.4.46)$$

neglecting the x_{ij} terms, or by

$$\begin{aligned}\langle E_1\rangle_t = {}& hc\left[\omega(\langle v_1\rangle_t + 1/2)\right. \\ & \left. + x_{11}(\langle v_1\rangle_t + 1/2)^2 + \frac{1}{2}x_{12}(\langle v_1\rangle_t + 1/2)(\langle v_2\rangle_t + 1/2)\right] \end{aligned} \quad (9.4.47)$$

where the contribution from the x_{12} term is divided equally between modes 1 and 2.

9.4.9 Resonance Energy and Energy Transfer Rate Operators

Once a basis set is specified, the $\mathbf{H}^{\text{eff}}$ may be partitioned into a diagonal term, $\mathbf{H}_{\text{diag}}$, and exclusively non-diagonal terms, $\mathbf{H}_{\text{res}}$,

$$\mathbf{H}^{\text{eff}} = \mathbf{H}_{\text{diag}} + \mathbf{H}_{\text{res}}, \quad (9.4.48)$$

where

$$\mathbf{H}_{\text{res}} = \sum_k \mathbf{O}_k = \sum_k \boldsymbol{\Omega}_k + \boldsymbol{\Omega}^\dagger{}_k. \tag{9.4.49}$$

$\boldsymbol{\Omega}_k$ and $\boldsymbol{\Omega}^\dagger{}_k$ are non-Hermitian matrices into which any exclusively non-diagonal $\mathbf{O}_k$ may be partitioned. $\mathbf{O}_k$ and $\boldsymbol{\Omega}_k - \boldsymbol{\Omega}^\dagger{}_k$ are respectively the resonance energy and energy transfer rate operators associated with a particular non-diagonal term in $\mathbf{H}^{\text{eff}}$ (see Section 9.1.8).

The relationship between the time evolution of $\langle \mathbf{O}_k \rangle_t$ and $\left\langle \boldsymbol{\Omega}_k - \boldsymbol{\Omega}^\dagger{}_k \right\rangle_t$ is discussed here for two examples: *(i)* the $^{2S+1}\Lambda$ multiplet state $\mathbf{H}^{\text{eff}}$; *(ii)* a polyad $\mathbf{H}^{\text{eff}}$ for 1 : 2 anharmonically coupled $[(K_{sbb}/2)\mathbf{Q}_s\mathbf{Q}_b^2$, where K_{sbb} is the cubic anharmonic coupling term] stretch and bend normal modes where $\omega_s \approx 2\omega_b$.

Since $\mathbf{O}_k$ is a resonance energy operator and $\boldsymbol{\Omega}_k - \boldsymbol{\Omega}^\dagger{}_k$ is an energy transfer rate operator, one expects to find a relationship between their time dependent expectation values of the general form

$$\frac{d}{dt} \langle \mathbf{O}_k \rangle_t = \left\langle \boldsymbol{\alpha}(\boldsymbol{\Omega}_k - \boldsymbol{\Omega}^\dagger{}_k) \right\rangle_t \tag{9.4.50}$$

where $\boldsymbol{\alpha}$ is an operator defined by the specific problem. The time dependence of $\langle \mathbf{O}_k \rangle$ is given by

$$i\hbar \frac{d}{dt} \langle \mathbf{O}_k \rangle_t = \langle [\mathbf{O}_k, \mathbf{H}^{\text{eff}}] \rangle . \tag{9.4.51}$$

For the situation of a case (a)↔(b) $^{2S+1}\Lambda$ multiplet state

$$\mathbf{O}_k = -B[\mathbf{J}^+\mathbf{S}^- + \mathbf{J}^-\mathbf{S}^+] \tag{9.4.52}$$

$$\boldsymbol{\Omega}_k = -B\mathbf{J}^+\mathbf{S}^- \tag{9.4.53}$$

$$\boldsymbol{\Omega}^\dagger{}_k = -B\mathbf{J}^-\mathbf{S}^+ \tag{9.4.54}$$

and

$$\mathbf{H}_{\text{diag}} = B\left[(\mathbf{J}^2 - \mathbf{J}_z^2) + (\mathbf{S}^2 - \mathbf{S}_z^2) + \mathbf{L}_\perp^2\right] + A\mathbf{L}_z\mathbf{S}_z. \tag{9.4.55}$$

The $[\mathbf{O}_k, \mathbf{H}^{\text{eff}}]$ commutator is reduced in two steps,

$$\begin{aligned}
[\mathbf{O}_k, \mathbf{H}^{\text{eff}}] &= [\mathbf{O}_k, \mathbf{H}_{\text{diag}}] = -B[\mathbf{O}_k, (\mathbf{J}_z^2 + \mathbf{S}_z^2)] + A\mathbf{L}_z[\mathbf{O}_k, \mathbf{S}_z] \\
&= B^2[(\mathbf{J}^+\mathbf{S}^- + \mathbf{J}^-\mathbf{S}^+), (\mathbf{J}_z^2 + \mathbf{S}_z^2)] \\
&\quad - AB\mathbf{L}_z[(\mathbf{J}^+\mathbf{S}^- + \mathbf{J}^-\mathbf{S}^+), \mathbf{S}_z] \\
&= i\hbar[2B^2\mathbf{L}_z(\mathbf{J}^-\mathbf{S}^+ - \mathbf{J}^+\mathbf{S}^-) - AB\mathbf{L}_Z(\mathbf{J}^-\mathbf{S}^+ - \mathbf{J}^+\mathbf{S}^-)] \\
&= i\hbar 2B(B - A)\mathbf{L}_z(\mathbf{J}^-\mathbf{S}^+ - \mathbf{J}^+\mathbf{S}^-).
\end{aligned} \tag{9.4.56}$$

Thus, for $\mathbf{O}_k = -B[\mathbf{J}^+\mathbf{S}^- + \mathbf{J}^-\mathbf{S}^+]$,

$$\frac{d}{dt} \langle \mathbf{O}_k \rangle_t = 2B(B - A) \langle \mathbf{L}_z(\mathbf{J}^-\mathbf{S}^+ - \mathbf{J}^+\mathbf{S}^-) \rangle_t \tag{9.4.57a}$$

$$\frac{d}{dt} \langle \mathbf{O}_k \rangle_t = 2B(B - A) \left\langle \boldsymbol{\alpha}(\boldsymbol{\Omega} - \boldsymbol{\Omega}^\dagger) \right\rangle_t \tag{9.4.57b}$$

where

$$\alpha = \mathbf{L}_z$$

In this example, α has a special form that results in a simplification of $\frac{d}{dt}\langle\mathbf{\Omega}_k\rangle$. Every eigenstate out of which $\Psi(0)$ must be composed is a linear combination of basis states with definite parity (see Section 3.2.2). All eigenstates may be expressed as a linear combination of Hund's case (a) basis states (see Section 3.2.1), each of the symmetrized form

$$2^{1/2}[|\Lambda S\Sigma J\Omega\rangle \pm |-\Lambda S-\Sigma J-\Omega\rangle].$$

Equation (9.4.57a) may therefore be simplified to

$$\frac{d}{dt}\langle\mathbf{O}_k\rangle_t = 2B(B-A)\left\langle|\mathbf{L}_z|(\mathbf{J}^-\mathbf{S}^+ + \mathbf{J}^+\mathbf{S}^-)\right\rangle_t. \tag{9.4.58a}$$

For the time evolution of $\langle\mathbf{O}_k\rangle$ exclusively within a $^{2S+1}\Lambda$ multiplet state,

$$\frac{d}{dt}\left\langle -B(\mathbf{J}^+\mathbf{S}^- + \mathbf{J}^-\mathbf{S}^+)\right\rangle_t = -2|\Lambda|(B-A)\left\langle -B(\mathbf{J}^+\mathbf{S}^- + \mathbf{J}^-\mathbf{S}^+)\right\rangle_t \tag{9.4.58b}$$

therefore

$$\left\langle -B(\mathbf{J}^+\mathbf{S}^- + \mathbf{J}^-\mathbf{S}^+)\right\rangle_t = \left\langle -B(\mathbf{J}^+\mathbf{S}^- + \mathbf{J}^-\mathbf{S}^+)\right\rangle_0 \cos[2|\Lambda|(B-A)t]. \tag{9.4.58c}$$

The expectation value of the resonance operator that drives a system from case (a) toward case (b) is a simple cosinusoid, with period

$$T = \frac{\pi}{|\Lambda|(B-A)} \tag{9.4.58d}$$

and 100% modulation depth. The initial preparation of the system determines the amplitude (but not the phase) of the oscillation. If $\Lambda = 0$ or $B = A$ there is no time dependence of $\langle\mathbf{O}_k\rangle_t$.

For the $1:2$ $\omega_s \approx 2\omega_b$ stretch-bend polyad,

$$\begin{aligned}\mathbf{H}_{\text{diag}}/hc &= \omega_s(v_s+1/2) + x_{ss}(v_s+1/2)^2 + \omega_b(v_b+1/2)\\ &\quad + x_{bb}(v_b+1/2)^2 + x_{sb}(v_s+1/2)(v_b+1/2)\end{aligned} \tag{9.4.59}$$

$$\mathbf{H}_{\text{res}}/hc = \mathbf{O}_k = (K_{sbb}/4\cdot 2^{1/2}hc)[\mathbf{a}_s^\dagger\mathbf{a}_b^2 + \mathbf{a}_b^{\dagger 2}\mathbf{a}_s] \tag{9.4.60}$$

$$v_s \to \mathbf{a}_s^\dagger\mathbf{a}_s$$

$$v_b \to \mathbf{a}_b^\dagger\mathbf{a}_b$$

and

$$[\mathbf{O}_k, \mathbf{H}^{\text{eff}}] = [\mathbf{O}_k, \mathbf{H}_{\text{diag}}] \tag{9.4.61}$$

$$\begin{aligned}&= \omega_s[\widehat{\mathbf{O}}_k, \mathbf{a}_s^\dagger\mathbf{a}_s] + \omega_b[\widehat{\mathbf{O}}_k, \mathbf{a}_b^\dagger\mathbf{a}_b]\\ &+ x_{ss}[\widehat{\mathbf{O}}_k, (\mathbf{a}_s^\dagger\mathbf{a}_s)^2 + \mathbf{a}_s^\dagger\mathbf{a}_s]\\ &+ x_{bb}[\widehat{\mathbf{O}}_k, (\mathbf{a}_b^\dagger\mathbf{a}_b)^2 + \mathbf{a}_b^\dagger\mathbf{a}_b]\\ &+ x_{sb}[\widehat{\mathbf{O}}_k, \mathbf{a}_b^\dagger\mathbf{a}_b\mathbf{a}_s^\dagger\mathbf{a}_s] + \frac{x_{sb}}{2}[\widehat{\mathbf{O}}_k, \mathbf{a}_b^\dagger\mathbf{a}_b + \mathbf{a}_s^\dagger\mathbf{a}_s].\end{aligned} \tag{9.4.62}$$

To obtain $\left\langle \mathbf{a}_b^\dagger \mathbf{a}_b \right\rangle_t$ and $\langle \mathbf{a}_s^\dagger \mathbf{a}_s \rangle_t$, it is necessary to evaluate the $[\mathbf{a}^\dagger \mathbf{a}, \mathbf{H}]$ commutators that appear in the equations of motion for $\mathbf{a}_s^\dagger \mathbf{a}_s$ and $\mathbf{a}_b^\dagger \mathbf{a}_b$,

$$[\mathbf{a}_s^\dagger \mathbf{a}_s, \mathbf{H}^{\text{eff}}] = [\mathbf{a}_s^\dagger \mathbf{a}_s, \mathbf{O}_k] = (K_{sbb}/4 \cdot 2^{1/2}) \left\langle \mathbf{a}_s^\dagger \mathbf{a}_b^2 - \mathbf{a}_b^{\dagger 2} \mathbf{a}_s \right\rangle \tag{9.4.63}$$

(see Eq. 18 of Jacobson and Field, 2000a)

$$[\mathbf{a}_b^\dagger \mathbf{a}_b, \mathbf{H}^{\text{eff}}] = [\mathbf{a}_b^\dagger \mathbf{a}_b, \mathbf{O}_k] = -2(K_{sbb}/4 \cdot 2^{1/2}) \left\langle \mathbf{a}_s^\dagger \mathbf{a}_b^2 - \mathbf{a}_b^{\dagger 2} \mathbf{a}_s \right\rangle. \tag{9.4.64}$$

Thus

$$\begin{aligned} \frac{d}{dt} \langle \mathbf{O}_k \rangle_t = \frac{1}{i\hbar} \Big\langle \Big[&(\omega_s - 2\omega_b) + x_{ss}(2\mathbf{a}_s^\dagger \mathbf{a}_s + 1) \\ &- x_{bb}(2\mathbf{a}_b^\dagger \mathbf{a}_b + 1) \\ &- x_{sb}(2\mathbf{a}_s^\dagger \mathbf{a}_s - \mathbf{a}_b^\dagger \mathbf{a}_b + \frac{1}{2}) \Big] \\ &(\mathbf{a}_s^\dagger \mathbf{a}_b^2 - \mathbf{a}_b^{\dagger 2} \mathbf{a}_s) \Big\rangle_t . \end{aligned} \tag{9.4.65}$$

But the polyad number is conserved, $N = 2v_s + v_b$, therefore

$$\left\langle \mathbf{a}_b^\dagger \mathbf{a}_b \right\rangle_t = N - 2 \langle \mathbf{a}_s^\dagger \mathbf{a}_s \rangle_t \tag{9.4.66}$$

$$\frac{d}{dt} \langle \mathbf{O}_k \rangle_t = \frac{1}{i\hbar} \left\langle (C + D\mathbf{a}_s^\dagger \mathbf{a}_s)(\mathbf{\Omega}_k - \mathbf{\Omega}^\dagger{}_k) \right\rangle_t \tag{9.4.67}$$

$$C = \omega_s - 2\omega_b + x_{ss} - x_{bb} - \frac{1}{2} x_{sb} + N(x_{sb} - 2x_{bb}) \tag{9.4.68}$$

$$D = 2x_{ss} + 4x_{bb} - 4x_{sb} \tag{9.4.69}$$

$$\mathbf{\Omega}_k = (K_{ssb}/4 \cdot 2^{1/2} hc) \mathbf{a}_s^\dagger \mathbf{a}_b^2 \tag{9.4.70}$$

The right side of Eq. (9.4.67) is real (see Eq. (9.1.93)),

$$\frac{d}{dt} \langle \mathbf{O}_k \rangle_t = \frac{2C}{\hbar} \text{Im} \langle \mathbf{\Omega}_k \rangle_t + \frac{2D}{\hbar} \text{Im} \langle \mathbf{a}_s^\dagger \mathbf{a}_s \mathbf{\Omega}_k \rangle_t .$$

Each Fourier component of $\langle \mathbf{O}_k \rangle$ is causally related to those of $\text{Im} \langle \mathbf{\Omega}_k \rangle_t$ and $\text{Im} \langle \mathbf{a}_s^\dagger \mathbf{a}_s \mathbf{\Omega}_k \rangle_t$. The causal structure of the dynamics of the anharmonically coupled 2 : 1 resonant system is specified independent of the initial preparation of the system, $\Psi(t = 0)$. The initial preparation of the system may be designed to accomplish a desired dynamical objective.

9.4.10 The Use of Expectation Values of Resonance Operators to Visualize Dynamic Processes

The 5 vibrational normal modes of the HCCH $\tilde{X}^1\Sigma_g^+$ state (two of which are doubly degenerate) have frequencies in the approximate ratios 5 : 3 : 5 : 1 : 1.

As a result, the early time dynamics of acetylene could be affected by numerous near resonant anharmonic interactions. Nine such anharmonic resonances are well characterized and known to be dynamically relevant. One needs a framework to relate the Fourier amplitudes and phases of the time-dependence of any observable quantity to the expectation values of each of the resonance operators in order to establish the relative importance of each of the resonances.

The 9 important anharmonic resonances destroy all of the 7 normal mode quantum numbers (v_1, v_2, v_3, v_4, l_4, v_5, l_5), but three approximately conserved polyad quantum numbers remain,

$$N_{\text{stretch}} = v_1 + v_2 + v_3 \tag{9.4.71}$$

$$N_{\text{resonance}} = 5v_1 + 3v_2 + 5v_3 + v_4 + v_5 \tag{9.4.72}$$

$$l_{\text{total}} = l_4 + l_5 \tag{9.4.73}$$

The three polyad quantum numbers serve to block diagonalize the total vibrational $\mathbf{H}$ into individual, scaling-related polyads, $\mathbf{H}^{\text{eff}}\left([N_{\text{stretch}}, N_{\text{resonance}}]^{l_{\text{total}}}\right)$. The coordinate space or state space dimensionality is reduced from 7 to $(7-3) = 4$.

Figure 9.11 displays the time-dependence of several dynamical quantities computed for the initially prepared $\left(v_1, v_2, v_3, v_4^{l_4}, v_5^{l_5}\right)^{l_{\text{total}}}$ state, $(0, 1, 0, 10^0, 0^0)^{0_g^+}$, which is a member of the $\left([N_{\text{stretch}}, N_{\text{resonance}}]^{l_{\text{total}}}\right)$ polyad, $[1, 13]^{0_g^+}$. 27 basis states are members of this polyad, hence $\mathbf{H}^{\text{eff}}$ is of dimension 27×27. The top panel displays the survival probability, $|\langle\Psi(t)|\Psi(0)\rangle|^2$, of

$$|\Psi(0)\rangle = |0, 1, 0, 10^0, 0^0\rangle^{0_g^+}. \tag{9.4.74}$$

The resonance energy and energy transfer rate operators are diagnostics that serve to apportion the features in the survival probability among the various resonances, thereby determining the mechanism by which a particular bright state explores the accessible state space. It also permits design of optimal control schemes by revealing the critical times when the freely evolving system most closely approaches a target state.

Three of the 9 possible resonance terms are explicitly represented in frames 2-4 of Fig. 9.11:

$$DD1 = \text{Darling-Dennison-1 } (\Delta v_4 = -\Delta v_5 = \pm 2, \Delta l_4 = \Delta l_5 = 0)$$

$$\left\langle v_4^{l_4} v_5^{l_5} |\mathbf{H}^{\text{eff}}| (v_4-2)^{l_4} (v_5+2)^{l_5} \right\rangle = \frac{s_{45}}{4} \left[(v_4^2 - l_4^2)(v_5 + l_5 + 2)(v_5 - l_5 + 2)\right]^{1/2} \tag{9.4.75}$$

$$3,245 = \text{Anharmonic } (-\Delta v_3 = \Delta v_2 = \Delta v_4 = \Delta v_5 = \pm 1, \Delta l_4 = \Delta l_5 = 0)$$

$$\left\langle v_2 v_3 - 1 v_4^{l_4} v_5^{l_5} |\mathbf{H}^{\text{eff}}| (v_2-1) v_3 (v_4-1)^{l_4} (v_5-1)^{l_5} \right\rangle \frac{K_{3,245}}{4} [v_2 v_3 v_4 v_5]^{1/2} \tag{9.4.76}$$

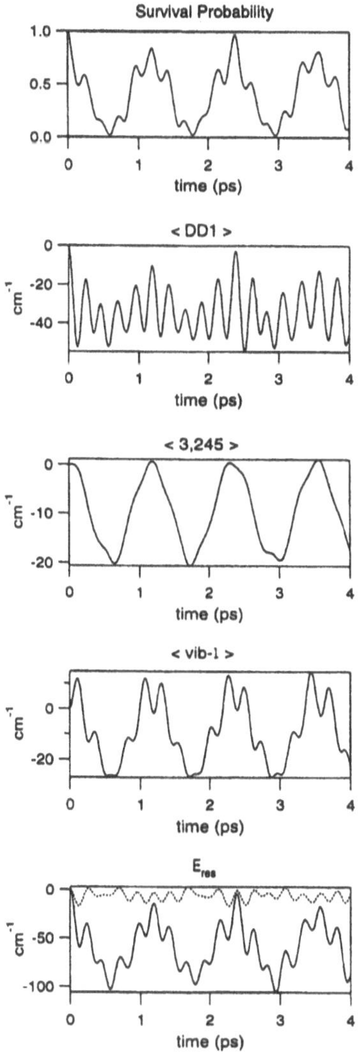

Figure 9.11: Comparison of survival probability to the resonance energy for various non-diagonal terms in the $\mathbf{H}^{\rm eff}$ for the HCCH $[N_{\rm stretch} = 1, N_{\rm resonance} = 13]^{l_{\rm total}=0_g^+}$ polyad. $|\Psi(0)\rangle = |0,1,0,10^0,0^0\rangle$ and the survival probability is $|\langle\Psi(t)|\Psi(0)\rangle|^2$. The top panel is the survival probability. Panels 2-4 are the time-dependence of the expectation value of a resonance energy operator $\frac{1}{hc}\langle\mathbf{O}_k\rangle_t$, where $\mathbf{O}_k = \boldsymbol{\Omega}_k + \boldsymbol{\Omega}^\dagger{}_k$. Three of the nine resonance terms make the dominant contribution to the time dependence of the survival probability. The bottom panel shows the sum of panels 2-4 (solid line) and the sum of the contributions from the remaining six resonance terms (dotted line) (from Jacobson and Field, 2000a).

$$\text{vib}-l = \text{Vibrational } l\text{-Resonance } (\Delta v_4 = \Delta v_5 = 0)$$

$$\left\langle v_4^{l_4} v_5^{l_5} |\mathbf{H}^{\text{eff}}| v_4^{l_4 \pm 2} v_5^{l_5 \mp 2} \right\rangle = \frac{r_{45}}{4} \left[(v_4 \mp l_4)(v_4 \pm l_4 + 2)(v_5 \pm l_5)(v_5 \mp l_5 + 2)\right]^{1/2}. \tag{9.4.77}$$

It is clear that $\langle DD1 \rangle$, $\langle 3, 245 \rangle$, and $\langle \text{vib}-l \rangle$ account for all of the important features of the survival probability. This is illustrated by the bottom frame in which the solid curve is the sum of the expectation values for the three resonances and the dotted curve is the computed sum of the expectation values of all of the other 6 resonance energy terms.

Figure 9.12 displays the contribution of the $DD1$ and 3, 245 terms in $\mathbf{H}^{\text{eff}}$ to the transfer rate and expectation value for the number of quanta in mode 4

$$\langle v_4 \rangle_t = \left\langle \mathbf{a}_{4d}^{\dagger} \mathbf{a}_{4d} + \mathbf{a}_{4g}^{\dagger} \mathbf{a}_{4g} \right\rangle \tag{9.4.78}$$

(see Section 9.4.12, Eq. (9.4.189)). The left column is $\frac{d}{dt} \langle v_4 \rangle$ and the right column is

$$\Delta \langle v_4 \rangle_t = \langle v_4 \rangle_t - \langle v_4 \rangle_0 = \int_0^t [d \langle v_4 \rangle / dt] dt. \tag{9.4.79}$$

It is clear that DD1 makes the dominant contribution to both $\frac{d}{dt} \langle v_4 \rangle_t$ and $\Delta \langle v_4 \rangle_t$.

9.4.11 Q and P

The equations of motion of $\mathbf{Q}_i$ and $\mathbf{P}_i$ for the i-th vibrational mode are given by

$$\widehat{\mathbf{Q}}_i = 2^{-1/2}(\mathbf{a}_i^{\dagger} + \mathbf{a}_i) \tag{9.4.80}$$

$$\widehat{\mathbf{P}}_i = 2^{-1/2} i (\mathbf{a}_i^{\dagger} - \mathbf{a}_i) \tag{9.4.81}$$

for which the commutators

$$[\mathbf{Q}_i, \mathbf{H}] \text{ and } [\mathbf{P}_i, \mathbf{H}] \tag{9.4.82}$$

are needed. However, the intra-polyad matrix elements of all of the terms generated by these commutators are zero! This is a remarkable result. It means that if $\Psi(0)$ is composed exclusively of basis states that belong to one polyad, there will be no motion in coordinate or momentum space. This is in marked contrast to the extremely complicated motions of $\langle v_1 \rangle(t), \langle v_2 \rangle(t), \langle \mathbf{O} \rangle(t)$, and $\left\langle \boldsymbol{\Omega} - \boldsymbol{\Omega}^{\dagger} \right\rangle(t)$ that occur in basis-state space. The pictures of $\langle v_1 \rangle(t)$ and $\langle v_2 \rangle(t)$ are extremely informative, much more so than the survival probability, $|\langle \Psi(t) | \Psi(0) \rangle|^2$.

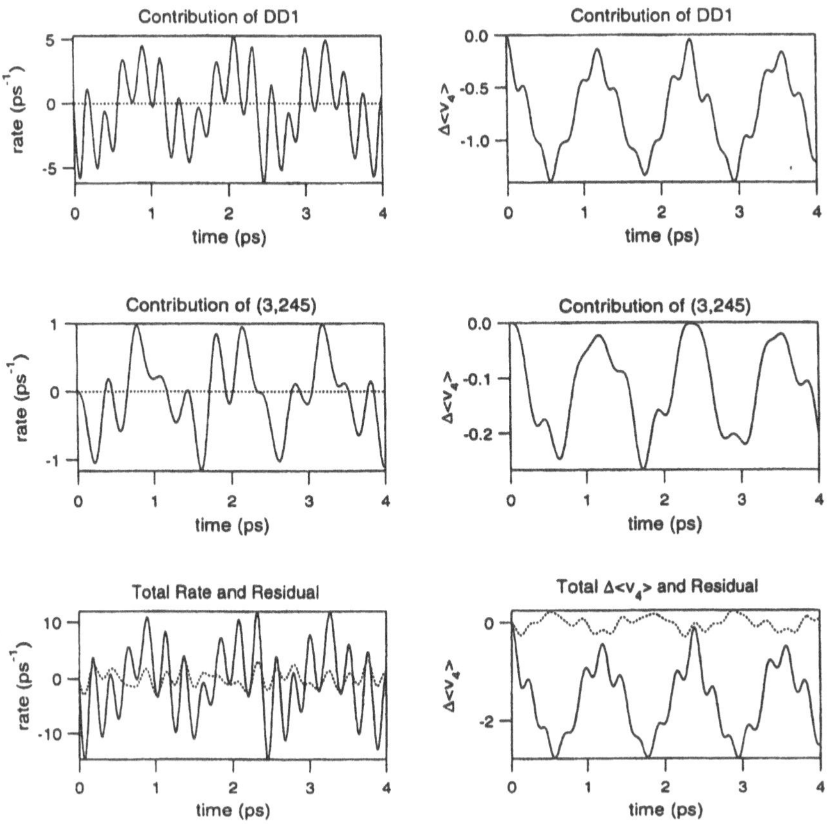

Figure 9.12: Contributions of DD1 and 3,245 to the transfer rate and accumulated population in mode 4. Left column is $d\langle v_4\rangle/dt$ and right column is $\langle v\rangle_{4t} - \langle v_4\rangle_0 = \int_0^t dt(d\langle v_4\rangle/dt)$. The bottom panels show the sum of contributions from DD1 and 3,245 (solid curve) and the sum of the contributions from the other seven resonance terms (dotted curve) (from Jacobson and Field, 2000a).

Of course it is possible to excite coherently several bright states that belong to different polyads. To obtain motion of $\mathbf{Q}_1$ and $\mathbf{P}_1$ associated with mode 1, one needs to excite members of polyads belonging to polyad numbers, $N = 2v_1 + v_2$, that differ in steps of 2 in N, whereas to get motion of $\mathbf{Q}_2$ and $\mathbf{P}_2$ one needs to excite polyads that differ in steps of 1 in N. This offers an interesting possibility for achieving motion of $\mathbf{Q}_1$ and $\mathbf{P}_1$ but not $\mathbf{Q}_2$ and $\mathbf{P}_2$ or *vice versa.*

9.4.12 Transformation Between Local and Normal Mode Limits

When two intramolecular motions, for example vibrations, have equal or nearly equal zero-order frequencies, the coupling between these two near degenerate internal degrees of freedom is uniquely important in both the spectrum and early-time dynamics. The couplings between nondegenerate states may be treated by second-order perturbation theory, but the systematically near-degenerate states arising from similar frequency intramolecular motions demand special treatment (see Eq. (3.2.2)). An effective Hamiltonian, which describes the interactions between systematically near degenerate basis states, must be set up and diagonalized. Such a group of near degenerate vibrational states is called a polyad (see Section 9.4.5), and the effective Hamiltonian that describes these states is called a polyad $\mathbf{H}^{\text{eff}}$. If v_i and v_j are the quantum numbers for the two nearly resonant $\omega_i \approx \omega_j$ vibrations, then all of the near degenerate basis states are specified by a common polyad quantum number, $N = v_i + v_j$, and there are $N + 1$ (v_i, v_j) near degenerate basis states in the polyad. As the vibrational energy increases, so does N, thus the dimension of the polyad $\mathbf{H}^{\text{eff}}$ increases. However, harmonic oscillator matrix element scaling rules permit the $N + 1$st polyad $\mathbf{H}^{\text{eff}}$ to be derived from the N'th polyad $\mathbf{H}^{\text{eff}}$, without the necessity for additional fit parameters or spectroscopic data. This explicit N–scaling of the vibrational polyad $\mathbf{H}^{\text{eff}}$ is analogous to the explicit J–scaling of $\mathbf{H}^{\text{ROT}}$ for diatomic molecules (see Sections 3.1, 3.2 and 3.5).

An excellent example of non-accidental intramolecular resonance is provided by a polyatomic molecule with two chemically identical bonds. The spectrum and dynamics of such a molecule may equally well be described by an effective Hamiltonian expressed in basis sets corresponding to either of two opposite limiting cases: *normal mode* ($\mathbf{H}^{\text{eff}}_{\text{NORMAL}}$) and *local mode* ($\mathbf{H}^{\text{eff}}_{\text{LOCAL}}$).

As in all situations described by opposite limiting cases, a special pair of case-preserving and case-destroying parameters exists (see Section 3.2.1.6). The case-preserving parameter in one limiting case plays a case-destroying role in the other limiting case. Insight into the physical basis of what might otherwise appear to be only phenomenological fit-parameters (see Sections 4.2 and 4.4.2) is obtained by comparing maximally simple $\mathbf{H}^{\text{eff}}$ models for the two limiting cases, thus extracting the relationships between the fit parameters for the two models and, in addition, unexpected inter-relationships between nominally unrelated fit parameters of one or the other primitive model (Lehmann, 1983; Mills and Robiette, 1985; Mills, 1972).

A change of basis set, for both $\mathbf{H}^{\text{eff}}$ and $|\psi^\circ\rangle$, can provide valuable insights into dynamics. What might appear to be indescribably complex dynamics in one basis set could become very simple and understandable dynamics in the other basis set. Plucking the system in a basis set that is nearly the eigenbasis results in relatively simple dynamics, whereas a wrong-basis pluck will result in extremely complex dynamics. This is the time-domain manifestation of the well known frequency domain fact that transitions between electronic states that are near the same Hund's coupling limit have extremely simple rotational branch structure (strong $\Delta J = \Delta N$ main lines; weak $\Delta J \neq \Delta N$ satellite lines) whereas transitions between electronic states that are near different Hund's coupling limits have complex rotational branch structure (strong satellite lines) (Herzberg, 1950). It is valuable to know how to construct a pluck of the system that will give the desired dynamics and to know which basis set will give the most intelligible view of that dynamics.

9.4.12.1 Classical Mechanical Treatment

The classical mechanical problem of two coupled identical harmonic oscillators is described by the classical mechanical Hamiltonian

$$\mathcal{H} = \mathcal{T}(P_R, P_L) + \mathcal{V}(Q_R, Q_L) \tag{9.4.83}$$

where the subscripts R and L refer to the harmonic oscillator on the right and left,

$$\begin{aligned}\mathcal{T} &= \frac{1}{2}(P_R\ P_L)\mathbf{G}\begin{pmatrix}P_R\\P_L\end{pmatrix}\\ &= \frac{1}{2}\left[G_{rr}(P_R^2 + P_L^2) + 2G_{rr'}P_RP_L\right]\end{aligned} \tag{9.4.84}$$

$$\begin{aligned}\mathcal{V} &= \frac{1}{2}(Q_R\ Q_L)\mathbf{F}\begin{pmatrix}Q_R\\Q_L\end{pmatrix}\\ &= \frac{1}{2}\left[F_{rr}(Q_R^2 + Q_L^2) + 2F_{rr'}Q_RQ_L\right],\end{aligned} \tag{9.4.85}$$

where the subscripts r and r' refer to simple bond stretching vibrations, thus

$$\begin{aligned}\mathcal{H} = &\left[\frac{1}{2}G_{rr}P_R^2 + \frac{1}{2}F_{rr}Q_R^2\right] + \left[\frac{1}{2}G_{rr}P_L^2 + \frac{1}{2}F_{rr}Q_L^2\right]\\ &+ G_{rr'}P_RP_L + F_{rr'}Q_RQ_L.\end{aligned} \tag{9.4.86}$$

The **G** (mass and geometry dependent) and **F** (mass and geometry independent) matrices are discussed by Wilson *et al.*, (1955). For two bond stretches coupled by a shared atom, the relevant **F** and **G** matrix elements are defined by

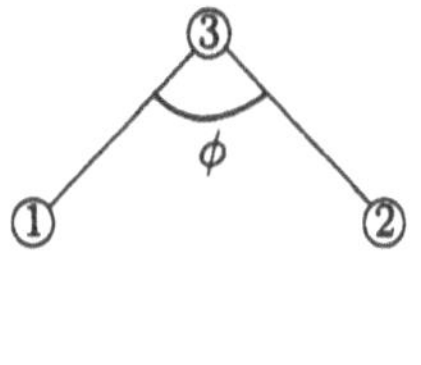

$$F_{rr} = k \tag{9.4.87}$$

$$F_{rr'} = k_{RL} \tag{9.4.88}$$

$$G_{rr} = \frac{1}{m_1} + \frac{1}{m_3} = \frac{m_1 + m_3}{m_1 m_3} \equiv \frac{1}{\mu} \tag{9.4.89}$$

$$G_{rr'} = \frac{1}{m_3} \cos\phi \tag{9.4.90}$$

(note that, for linear geometry, $\phi = \pi$ and $G_{rr'} = -\frac{1}{m_3}$). If the two identical bonds do not share a common atom, the coupling between them is indirect and is not considered here. The Eq. (9.4.86) classical mechanical Hamiltonian corresponds to two identical classical local harmonic oscillators with frequency (in cm^{-1})

$$\omega_0 = \frac{1}{2\pi c}[F_{rr}G_{rr}]^{1/2} = \frac{1}{2\pi c}\left[\frac{k}{\mu}\right]^{1/2} \tag{9.4.91}$$

coupled by 1 : 1 kinetic energy, $G_{rr'}P_R P_L$, and potential energy, $k_{RL}Q_R Q_L$, terms. The two identical harmonic oscillators may be uncoupled to form the "normal mode oscillators" by transforming to symmetric

$$Q_s = 2^{-1/2}[Q_R + Q_L] \tag{9.4.92a}$$

$$P_s = 2^{-1/2}[P_R + P_L] \tag{9.4.93a}$$

and antisymmetric

$$Q_a = 2^{-1/2}[Q_R - Q_L] \tag{9.4.92b}$$

$$P_a = 2^{-1/2}[P_R - P_L] \tag{9.4.93b}$$

coordinates and conjugate momenta, thus

$$\begin{aligned} \mathcal{H} = &\left[\frac{1}{2}\left(\frac{1}{\mu} + G_{rr'}\right)P_s^2 + \frac{1}{2}(k + k_{RL})Q_s^2\right] \\ &+ \left[\frac{1}{2}\left(\frac{1}{\mu} - G_{rr'}\right)P_a^2 + \frac{1}{2}(k - k_{RL})Q_a^2\right], \end{aligned} \tag{9.4.94}$$

and the normal mode frequencies are obtained as

$$\omega_s = \frac{1}{2\pi c}\left[\left(\frac{1}{\mu} + G_{rr'}\right)(k + k_{RL})\right]^{1/2} \tag{9.4.95}$$

$$\omega_a = \frac{1}{2\pi c}\left[\left(\frac{1}{\mu} - G_{rr'}\right)(k - k_{RL})\right]^{1/2}, \tag{9.4.96}$$

which may be simplified to

$$\omega_s = \omega_0 + \beta + \lambda \tag{9.4.97}$$

$$\omega_a = \omega_0 + \beta - \lambda \tag{9.4.98}$$

$$\beta = \frac{k_{RL} G_{rr'}}{2(2\pi c)^2 \omega_0} \tag{9.4.99}$$

$$\lambda = \frac{\omega_0}{2}\left(1 - \frac{\beta}{\omega_0}\right)\left[k_{RL}/k + \mu G_{rr'}\right]. \tag{9.4.100}$$

The normal mode vibrational frequencies for the symmetric and antisymmetric vibrations are denoted by ω_s and ω_a. Notc that $|\beta| << 1$, but, owing to its dependence on $\cos\theta$, β is usually negative. Whether λ is positive or negative depends on whether the 1 : 1 potential or kinetic energy coupling term is larger. λ is usually negative (dominated by kinetic energy coupling).

9.4.12.2 The Morse Oscillator

It is useful and instructive to treat the local bond stretch oscillators (but not the normal mode oscillators)[†] as Morse oscillators (Jaffé and Brumer, 1980; Sibert, *et al.*, 1982a and b; Child, 1991; Mills and Robiette, 1985),

$$V_{\text{MORSE}}(r) = D_e[1 - e^{-ar}]^2 \tag{9.4.101}$$

$$r = R - R_e \tag{9.4.102}$$

$$V_{\text{MORSE}}(r) \approx \frac{1}{2} f_{rr} r^2 + \frac{1}{6} f_{rrr} r^3 + \frac{1}{24} f_{rrrr} r^4 \tag{9.4.103}$$

$$f_{rr} = \left(\frac{d^2 V_{\text{MORSE}}}{dr^2}\right)_{r=0} = 2a^2 D_e \tag{9.4.104}$$

$$f_{rrr} = \left(\frac{d^3 V_{\text{MORSE}}}{dr^3}\right)_{r=0} = -6a^3 D_e \tag{9.4.105}$$

$$f_{rrrr} = \left(\frac{d^4 V_{\text{MORSE}}}{dr^4}\right)_{r=0} = 14a^4 D_e. \tag{9.4.106}$$

The Morse oscillator has the convenient quantum mechanical property that the vibrational energy levels of a Morse oscillator have the simple form of

$$E_{\text{MORSE}}(v)/hc = E_0^{\text{MORSE}}/hc + \omega_m(v + 1/2) + x_m(v + 1/2)^2. \tag{9.4.107}$$

[†] Although the symmetric stretch can dissociate (by simultaneously breaking both bonds) similarly to a single bond local stretch, the antisymmetric stretch cannot dissociate to a knowable dissociation asymptote, thus there is no physical basis for treating an antisymmetric stretch (or any bend) as a Morse oscillator.

An exact solution of the Morse oscillator Schrödinger equation yields

$$E_0^{\text{MORSE}}/hc = 0 \tag{9.4.108}$$

$$\omega_m = \frac{1}{2\pi c}\left(\frac{2a^2 D_e}{\mu}\right)^{1/2} \tag{9.4.109}$$

$$x_m = \frac{-a^2\hbar}{4\pi c\mu}. \tag{9.4.110}$$

Exactly the same relationship between (ω_m, x_m) and (a, D_e) is derived by perturbation theory,

$$\mathbf{H}^{(0)} = \frac{1}{2} f_{rr}\mathbf{r}^2 + \frac{1}{2\mu}\mathbf{p}^2 \tag{9.4.111}$$

$$\mathbf{E}_v^{(1)} = \frac{1}{24} f_{rrrr}\langle v|r^4|v\rangle \tag{9.4.112}$$

$$\mathbf{E}_v^{(2)} = \left(\frac{f_{rrr}}{6}\right)^2 \frac{1}{\omega_m}\left[\frac{\langle v-1|\mathbf{r}^3|v\rangle - \langle v+1|\mathbf{r}^3|v\rangle}{1} + \frac{\langle v-3|\mathbf{r}^3|v\rangle - \langle v+3|\mathbf{r}^3|v\rangle}{3}\right] \tag{9.4.113}$$

provided that the Taylor series expansion is terminated after the $\mathbf{r}^4$ term and that the $\mathbf{r}^4$ term is treated exclusively by first-order perturbation theory (the effects of $\Delta v = \pm 2, \pm 4$ matrix elements are neglected).

The Morse potential provides a useful model for the anharmonic stretching vibrations of a polyatomic molecule. It is superior to a harmonic oscillator perturbed by cubic and quartic anharmonicity terms in terms of both convergence (a $V(\mathbf{r})$ that includes an $\mathbf{r}^3$ term cannot be bound, thus cannot have any rigorously bound vibrational levels) and the need for a smaller number of adjustable parameters to describe both the potential energy curve (a and D_e for Morse; f_{rr}, f_{rrr}, and f_{rrrr} for the cubic plus quartic perturbed harmonic oscillator) and the energy levels.

9.4.12.3 Quantum Mechanical Minimal Model for Two Anharmonically Coupled Local Stretch Morse Oscillators

The quantum mechanical $\mathbf{H}^{\text{eff}}_{\text{LOCAL}}$ is derived from the classical mechanical $\mathcal{H}_{\text{LOCAL}}$ by converting the coordinate displacements, $\mathbf{Q}_i$, and conjugate momenta, $\mathbf{P}_i$, into dimensionless quantities,

$$\mathbf{Q}_i = \alpha_i^{-1/2}\widehat{\mathbf{Q}}_i \tag{9.4.114}$$

$$\mathbf{P}_i = \hbar\alpha_i^{1/2}\widehat{\mathbf{P}}_i \tag{9.4.115}$$

$$\alpha_i = \frac{2\pi c\omega_i\mu_i}{\hbar} \tag{9.4.116}$$

$$\omega_i = \frac{1}{2\pi c}\left[k_i/\mu_i\right]^{1/2}. \tag{9.4.117}$$

Thus

$$\mathbf{H}_{\text{LOCAL}} = \left[\frac{1}{2\mu}\mathbf{P}_R^2 + \frac{1}{2}k\mathbf{Q}_R^2 + V^{anh}(\mathbf{Q}_R)\right] + \left[\frac{1}{2\mu}\mathbf{P}_L^2 + \frac{1}{2}k\mathbf{Q}_L^2 + V^{anh}(\mathbf{Q}_L)\right] + G_{rr'}\mathbf{P}_R\mathbf{P}_L + k_{RL}Q_RQ_L \tag{9.4.118}$$

where

$$V^{anh}(\mathbf{Q}) = V^{\text{MORSE}}(\mathbf{Q}) - \frac{1}{2}k\mathbf{Q}^2. \tag{9.4.119}$$

Convert to dimensionless quantities,

$$\mathbf{H}_{\text{LOCAL}} = \hbar(2\pi c\omega_m)\left[\frac{1}{2}\left(\widehat{\mathbf{P}}_R^2 + \widehat{\mathbf{Q}}_R^2\right) + \left(\widehat{\mathbf{P}}_L^2 + \widehat{\mathbf{Q}}_L^2\right)\right] - a^3D_e\alpha^{-3/2}\left(\widehat{\mathbf{Q}}_R^3 + \widehat{\mathbf{Q}}_L^3\right) + \frac{7}{12}a^4D_e\alpha^{-2}\left(\widehat{\mathbf{Q}}_R^4 + \widehat{\mathbf{Q}}_L^4\right) + G_{rr'}\hbar^2\alpha\widehat{\mathbf{P}}_R\widehat{\mathbf{P}}_L + k_{RL}\alpha^{-1}\widehat{\mathbf{Q}}_R\widehat{\mathbf{Q}}_L \tag{9.4.120}$$

$$\widehat{\mathbf{H}}_{\text{LOCAL}} = \frac{\mathbf{H}_{\text{LOCAL}}}{\hbar(2\pi c\omega_m)} \tag{9.4.121a}$$

$$= \left[\frac{1}{2}(\widehat{\mathbf{P}}_R^2 + \widehat{\mathbf{Q}}_R^2) + \frac{1}{2}(\widehat{\mathbf{P}}_L^2 + \widehat{\mathbf{Q}}_L^2)\right] + A(\widehat{\mathbf{Q}}_R^3 + \widehat{\mathbf{Q}}_L^3) + B(\widehat{\mathbf{Q}}_R^4 + \widehat{\mathbf{Q}}_L^4) + C\widehat{\mathbf{P}}_R\widehat{\mathbf{P}}_L + D\widehat{\mathbf{Q}}_R\widehat{\mathbf{Q}}_L, \tag{9.4.121b}$$

where the dimensionless quantities A, B, C, and D are

$$A = -\frac{a^3D_e\alpha^{-3/2}}{\hbar 2\pi c\omega_m} = -(2^{-5/4})(\hbar a)^{1/2}(\mu D_e)^{-1/4} \tag{9.4.122a}$$

$$B = \frac{7a^4D_e\alpha^{-2}}{12\hbar(2\pi c\omega_m)} = -\frac{7}{24}2^{-1/2}(\hbar a)(\mu D_e)^{-1/2} \tag{9.4.122b}$$

$$C = \frac{G_{rr'}\hbar^2\alpha}{\hbar(2\pi c\omega_m)} = G_{rr'}\mu \tag{9.4.122c}$$

$$D = \frac{k_{RL}}{\alpha\hbar(2\pi c\omega_m)} = \frac{k_{RL}}{2D_ea^2} = \frac{k_{RL}}{k_m}. \tag{9.4.122d}$$

The dimensionless $\widehat{\mathbf{H}}_{\text{LOCAL}}$ may be expressed in terms of $\mathbf{a}^\dagger, \mathbf{a}$ creation, annihilation operators

$$\widehat{\mathbf{Q}}_R = 2^{-1/2}\left(\mathbf{a}_R^\dagger + \mathbf{a}_R\right) \tag{9.4.123a}$$

$$\widehat{\mathbf{P}}_R = 2^{-1/2} i\left(\mathbf{a}_R^\dagger - \mathbf{a}_R\right) \tag{9.4.123b}$$

$$\widehat{\mathbf{Q}}_L = 2^{-1/2}\left(\mathbf{a}_L^\dagger + \mathbf{a}_L\right) \tag{9.4.123c}$$

$$\widehat{\mathbf{P}}_L = 2^{-1/2} i\left(\mathbf{a}_L^\dagger - \mathbf{a}_L\right) \tag{9.4.123d}$$

$$\begin{aligned}
\widehat{\mathbf{H}}_{\text{LOCAL}} = {} & \left[\left(\mathbf{a}_R^\dagger \mathbf{a}_R + \frac{1}{2}\right) + \left(\mathbf{a}_L^\dagger \mathbf{a}_L + \frac{1}{2}\right)\right] \\
& + 2^{-3/2} A\left[\left(\mathbf{a}_R^{\dagger 3} + \mathbf{a}_R^3 + 3\mathbf{a}_R^\dagger \mathbf{a}_R^2 + 3\mathbf{a}_R^{\dagger 2}\mathbf{a}_R\right)\right. \\
& \left. + \left(\mathbf{a}_L^{\dagger 3} + \mathbf{a}_L^3 + 3\mathbf{a}_L^\dagger \mathbf{a}_L^2 + 3\mathbf{a}_L^{\dagger 2}\mathbf{a}_L\right)\right] \\
& + \frac{1}{4}B\left[\left(\mathbf{a}_R^{\dagger 4} + \mathbf{a}_R^4 + 2\mathbf{a}_R^{\dagger 2} + 2\mathbf{a}_R^2\right.\right. \\
& \left. + (6\mathbf{a}_R^{\dagger 2} - 2\mathbf{a}_R^2)\mathbf{a}_R^\dagger \mathbf{a}_R + 6\mathbf{a}_R^\dagger \mathbf{a}_R \mathbf{a}_R^\dagger \mathbf{a}_R + 6\mathbf{a}_R^\dagger \mathbf{a}_R + 3\right) \\
& \quad \left(\mathbf{a}_L^{\dagger 4} + \mathbf{a}_L^4 + 2\mathbf{a}_L^{\dagger 2} + 2\mathbf{a}_L^{\dagger 2}\right. \\
& \left.\left. + (6\mathbf{a}_L^{\dagger 2} - 2\mathbf{a}_L^2)\mathbf{a}_L^\dagger \mathbf{a}_L + 6\mathbf{a}_L^\dagger \mathbf{a}_L \mathbf{a}_L^\dagger \mathbf{a}_L + 6\mathbf{a}_L^\dagger \mathbf{a}_L + 3\right)\right] \\
& - (C/2)\left[\mathbf{a}_R^\dagger \mathbf{a}_L^\dagger + \mathbf{a}_R \mathbf{a}_L - \mathbf{a}_R^\dagger \mathbf{a}_L - \mathbf{a}_L^\dagger \mathbf{a}_R\right] \\
& - (D/2)\left[\mathbf{a}_R^\dagger \mathbf{a}_L^\dagger + \mathbf{a}_R \mathbf{a}_L + \mathbf{a}_R^\dagger \mathbf{a}_L + \mathbf{a}_L^\dagger \mathbf{a}_R\right].
\end{aligned} \tag{9.4.124}$$

Thus, exploiting the built-in v–dependence of the Morse oscillator eigen-energies (Eq. (9.4.107)),

$$\begin{aligned}
\widehat{\mathbf{H}}_{\text{LOCAL}} = {} & |v_R v_L\rangle \\
& \times \langle v_R v_L|\left[(v_R + 1/2) + (v_L + 1/2) + F(v_R + 1/2)^2 + F(v_L + 1/2)^2\right] \\
& + |v_R \pm 1 v_L \mp 1\rangle\langle v_R v_L|(\tfrac{D+C}{2})\left[(v_R + 1/2 \pm 1/2)(v_L + 1/2 \mp 1/2)\right]^{1/2} \\
& + |v_R \pm 1 v_L \pm 1\rangle\langle v_R v_L|(\tfrac{D-C}{2})\left[(v_R + 1/2 \pm 1/2)(v_L + 1/2 \pm 1/2)\right]^{1/2},
\end{aligned} \tag{9.4.125}$$

where the dimensionless parameter, F, is

$$F = -\frac{a^2\hbar}{4\pi c\mu}\frac{1}{\hbar(2\pi c\omega_m)} = -\frac{2^{-1/2}(\hbar a)}{4\pi(\mu D_e)^{1/2}}. \tag{9.4.126}$$

All zero-order states with $N = v_R + v_L$ are near degenerate and form an $N{+}1$ member polyad (see Section 9.4.5). The polyad $\mathbf{H}^{\text{eff}}_{\text{LOCAL}}$ must be corrected by second-order perturbation theory [Van Vleck (see Section 4.2) or "contact" transformation (Nielsen, 1951; Amat, *et al.*, 1971; Sibert, 1988 and 1990; Fried and Ezra, 1987)] for the effects of the out-of-polyad $|v_R \pm 1 v_L \pm 1\rangle\langle v_R v_L|$ matrix elements,

$$\begin{aligned}\widehat{\mathbf{H}}^{(2)}_{\mathrm{LOCAL}} &= |v_R v_L\rangle\langle v_R v_L| \frac{1}{2}\left(\frac{D-C}{2}\right)^2 [v_R v_L - (v_R+1)(v_L+1)] \\ &= |v_R v_L\rangle\langle v_R v_L| - \frac{(D-C)^2}{8}(v_R + v_L + 1), \end{aligned} \qquad (9.4.127)$$

thus

$$\begin{aligned}\widehat{\mathbf{H}}^{\mathrm{eff}}_{\mathrm{LOCAL}} &= |v_R v_L\rangle\langle v_R v_L| \left\{ (v_R + v_L + 1)\left[1 - \frac{(D-C)^2}{8}\right] \right. \\ &\quad \left. + \frac{F}{2}\left[(v_R + v_L + 1)^2 + (v_R - v_L)^2\right] \right\} \\ &\quad + |v_R \pm 1 v_L \mp 1\rangle\langle v_R v_L| \left(\frac{D+C}{2}\right) \\ &\quad \times \left[(v_R + 1/2 \pm 1/2)(v_L + 1/2 \mp 1/2)\right]^{1/2}. \end{aligned} \qquad (9.4.128)$$

The term $\frac{F}{2}(v_R - v_L)^2$ lifts the degeneracy of the members of the polyad, hence it provides the driving force toward the local mode limit. The overall zero-order energy spread of the Nth polyad is $\frac{F}{2}N^2$. The zero-order states (v_R, v_L) and $(v_R + 1,\, v_L - 1)$ are most nearly degenerate near the center of the polyad, where $v_R \approx N/2$ and the adjacent level spacing is $2F$.

The local mode preserving effects of the diagonal $(F/2)\ (v_R - v_L)^2$ term are opposed by the off-diagonal term, $\left(\frac{D+C}{2}\right)\left[(v_R + 1/2 \pm 1/2)(v_L + 1/2 \mp 1/2)\right]^{1/2}$, which provides the driving force toward the normal mode limit. Near the center of the polyad, where $v_R \approx v_L \approx N/2$, the off-diagonal term reaches its maximum value, $\sim \left(\frac{D+C}{2}\right)\left(\frac{N}{2}\right)$.

Two high-N limiting behaviors emerge. At high N, the off-diagonal term becomes small relative to the overall zero-order energy width of the polyad. This implies that all coupled-identical-oscillator systems approach local mode limiting behavior at high N. Departures from local mode limiting behavior are largest near the center of each polyad, where the N-dependent off-diagonal $\left(\frac{D+C}{2}\right)\left(\frac{N}{2}\right)$ term can overpower the N-independent minimum level spacing, $2F$.

The Eq. (9.4.128) $\widehat{\mathbf{H}}^{\mathrm{eff}}_{\mathrm{LOCAL}}$ corresponds to the simplest possible local-mode fit model

$$\begin{aligned}\mathbf{H}^{\mathrm{eff}}_{\mathrm{LOCAL}}/hc &= |v_R v_L\rangle\langle v_R v_L| \left[\omega_R(v_R + 1/2) + \omega_L(v_L + 1/2)\right. \\ &\quad + x_R(v_R + 1/2)^2 + x_L(v_L + 1/2)^2 \\ &\quad \left. + x_{RL}(v_R + 1/2)(v_L + 1/2)\right] \\ &\quad + |v_R \pm 1 v_L \mp 1\rangle\langle v_R v_L|(H_{RL}/hc) \\ &\quad \times \left[(v_R + 1/2 \pm 1/2)(v_L + 1/2 \mp 1/2)\right]^{1/2} \end{aligned} \qquad (9.4.129)$$

where

$$\omega_R = \omega_L = \omega_m \left[1 - \frac{(D-C)^2}{8}\right] = \omega' \qquad (9.4.130)$$

$$x_R = x_L = x_m = \omega_m F = -\frac{a^2\hbar}{4\pi c\mu} \qquad (9.4.131)$$

$$x_{RL} = 0 \qquad (9.4.132)$$

$$H_{RL}/hc = \omega_m \left[\frac{D+C}{2}\right] \qquad (9.4.133)$$

and the H_{RL} coupling term (and the correction to ω) contains contributions from both (mass and geometry dependent) kinetic energy and (mass and geometry independent) potential energy terms. Note that the "harmonic" term for both local oscillators, $\omega_R = \omega_L = \omega'$, differs from the isolated Morse oscillator harmonic term, ω_M, due to kinetic and potential energy coupling effects similar to the β correction term in Eqs. (9.4.97) - (9.4.99) for the classical mechanical normal mode frequencies.

9.4.12.4 Transformation between 4-Parameter Forms of the Normal and Local Mode Basis Sets

The local mode basis functions and $\mathbf{H}^{\text{eff}}_{\text{LOCAL}}$ may be transformed to the normal mode basis set by exploiting the properties of the creation and annihilation operators, $\mathbf{a}_i^\dagger$ and $\mathbf{a}_i$.

All vibrational basis states are generated by multiple applications of $\mathbf{a}_i^\dagger$ on $|v_i\rangle$. For an N-atom molecule in the normal mode product basis state,

$$|v_1, v_2, \ldots v_{3N-6}\rangle \equiv |v_1\rangle|v_2\rangle \ldots |v_{3N-6}\rangle,$$

$$|v_1 v_2 \ldots v_{3N-6}\rangle = \left[\prod_{i=1}^{3N-6} (v_i!)\right]^{-1/2} \left(\mathbf{a}_1^\dagger\right)^{v_1} \left(\mathbf{a}_2^\dagger\right)^{v_2} \ldots \left(\mathbf{a}_{3N-6}^\dagger\right)^{v_{3N-6}} |00\ldots 0\rangle, \qquad (9.4.134a)$$

exploiting the equivalence of the zero point states in both basis sets

$$|00\ldots 0\rangle_N = |00\ldots 0\rangle_L. \qquad (9.4.134b)$$

The transformation from the normal mode $|v_s v_a\rangle_N$ to the local mode $|v_R v_L\rangle_L$ basis set is accomplished by replacing the normal mode $\mathbf{a}^\dagger, \mathbf{a}$ operators by the corresponding local mode operators

$$\mathbf{a}_s^\dagger = 2^{-1/2}\left(\mathbf{a}_R^\dagger + \mathbf{a}_L^\dagger\right) \qquad (9.4.135a)$$

$$\mathbf{a}_a^\dagger = 2^{-1/2}\left(\mathbf{a}_R^\dagger - \mathbf{a}_L^\dagger\right) \qquad (9.4.135b)$$

$$\mathbf{a}_R^\dagger = 2^{-1/2}\left(\mathbf{a}_s^\dagger + \mathbf{a}_a^\dagger\right) \qquad (9.4.135c)$$

$$\mathbf{a}_L^\dagger = 2^{-1/2}\left(\mathbf{a}_s^\dagger - \mathbf{a}_a^\dagger\right) \qquad (9.4.135d)$$

and similar relationships for the **a** operators.

It is straightforward to transform the normal mode basis states, $|v_s, v_a\rangle_N$ into local mode basis states, $|v_R, v_L\rangle_L$ (or *vice versa*). For example, the $|0,4\rangle_N$ basis state, which corresponds to $v_s = 0, v_a = 4$, is expressed in terms of local mode basis states,

$$|0,4\rangle_N = [4!]^{-1/2}[2^{-1/2}(\mathbf{a}_R^\dagger - \mathbf{a}_L^\dagger)]^4|0,0\rangle_L \tag{9.4.136}$$

$$= [24]^{-1/2}\frac{1}{4}\left[\mathbf{a}_R^{\dagger 4} - 4\mathbf{a}_R^{\dagger 3}\mathbf{a}_L^\dagger + 6\mathbf{a}_R^{\dagger 2}\mathbf{a}_L^{\dagger 2} - 4\mathbf{a}_R^\dagger\mathbf{a}_L^{\dagger 3} + \mathbf{a}_L^{\dagger 4}\right]|0,0\rangle_L$$

$$= \frac{1}{4}|4,0\rangle_L - \frac{1}{2}|3,1\rangle_L + (3/8)^{1/2}|2,2\rangle_L - \frac{1}{2}|1,3\rangle_L + \frac{1}{4}|0,4\rangle_L. \tag{9.4.137}$$

Since the transformation to local mode $\mathbf{a}_R^\dagger$ and $\mathbf{a}_L^\dagger$ from normal mode $\mathbf{a}_s^\dagger$ and $\mathbf{a}_a^\dagger$ in Eqs. (9.4.135c) - (9.4.135d) may be obtained by interchaning $\mathbf{a}_s^\dagger \leftrightarrow \mathbf{a}_R$ and $\mathbf{a}_a^\dagger \leftrightarrow \mathbf{a}_L$ in Eqs. (9.4.135a) - (9.4.135b), the inverse transformation is

$$|0,4\rangle_L = \frac{1}{4}|4,0\rangle_N - \frac{1}{2}|3,1\rangle_N + (3/8)^{1/2}|2,2\rangle_N - \frac{1}{2}|1,3\rangle_N + \frac{1}{4}|0,4\rangle_N. \tag{9.4.138}$$

Terms in the $\mathbf{H}^{\text{eff}}$ are transformed by replacing each vibrational quantum number, v_i, wherever it appears along the diagonal of $\mathbf{H}^{\text{eff}}$ by the number operator, $\mathbf{a}_i^\dagger\mathbf{a}_i$ and by expressing the off-diagonal elements explicitly in $\mathbf{a}^\dagger, \mathbf{a}$ form. The $\mathbf{H}^{\text{eff}}_{\text{LOCAL}}$ of Eq. (9.4.129) becomes

$$\begin{aligned}\mathbf{H}^{\text{eff}}_{\text{LOCAL}}/hc = {}& |v_R v_L\rangle\langle v_R v_L|\left\{\omega'(\mathbf{a}_R^\dagger\mathbf{a}_R + \mathbf{a}_L^\dagger\mathbf{a}_L + 1)\right.\\ &\left. + x_M\left[(\mathbf{a}_R^\dagger\mathbf{a}_R + 1/2)^2 + (\mathbf{a}_L^\dagger\mathbf{a}_L + 1/2)^2\right]\right\}\\ & + |v_R + 1 v_L - 1\rangle\langle v_R v_L|(H_{RL}/hc)\mathbf{a}_R^\dagger\mathbf{a}_L\\ & + |v_R - 1 v_L + 1\rangle\langle v_R v_L|(H_{RL}/hc)\mathbf{a}_R\mathbf{a}_L^\dagger.\end{aligned} \tag{9.4.139}$$

Exploiting the Eqs. (9.4.135a) - (9.4.135d) relationships between normal and local mode $\mathbf{a}^\dagger, \mathbf{a}$ operators, one obtains

$$[\mathbf{a}_R^\dagger\mathbf{a}_R + \mathbf{a}_L^\dagger\mathbf{a}_L + 1] = [\mathbf{a}_s^\dagger\mathbf{a}_s + \mathbf{a}_a^\dagger\mathbf{a}_a + 1] \tag{9.4.140}$$

$$[\mathbf{a}_R^\dagger\mathbf{a}_L + \mathbf{a}_L^\dagger\mathbf{a}_R] = [\mathbf{a}_s^\dagger\mathbf{a}_s - \mathbf{a}_a^\dagger\mathbf{a}_a] \tag{9.4.141}$$

$$\begin{aligned}[(\mathbf{a}_R^\dagger\mathbf{a}_R + 1/2)^2 + (\mathbf{a}_L^\dagger\mathbf{a}_L + 1/2)^2] = {}& \left[\frac{3}{4}(\mathbf{a}_s^\dagger\mathbf{a}_s + \mathbf{a}_a^\dagger\mathbf{a}_a + 1)^2\right.\\ & - \frac{1}{4}(\mathbf{a}_s^\dagger\mathbf{a}_s - \mathbf{a}_a^\dagger\mathbf{a}_a)^2 - \frac{1}{4}\\ & \left. - \frac{1}{2}(\mathbf{a}_s^{\dagger 2}\mathbf{a}_a^2 + \mathbf{a}_s^2\mathbf{a}_a^{\dagger 2})\right]\end{aligned} \tag{9.4.142}$$

as well as several other potentially useful relationships,

$$[(\mathbf{a}_R^\dagger \mathbf{a}_R + 1/2)^2 - (\mathbf{a}_L^\dagger \mathbf{a}_L + 1/2)^2] = [(\mathbf{a}_L^\dagger \mathbf{a}_R + \mathbf{a}_R^\dagger \mathbf{a}_L)(\mathbf{a}_R^\dagger \mathbf{a}_R + \mathbf{a}_L^\dagger \mathbf{a}_L + 1)] \tag{9.4.143}$$

$$\begin{aligned}[(\mathbf{a}_R^\dagger \mathbf{a}_R + 1/2)(\mathbf{a}_L^\dagger \mathbf{a}_L + 1/2)] = &\left[\frac{1}{8}(\mathbf{a}_s^\dagger \mathbf{a}_s + \mathbf{a}_a^\dagger \mathbf{a}_a + 1)^2\right. \\ &+ \frac{1}{8}(\mathbf{a}_s^\dagger \mathbf{a}_s - \mathbf{a}_a^\dagger \mathbf{a}_a)^2 + \frac{1}{8} \\ &\left. - \frac{1}{4}(\mathbf{a}_s^{\dagger 2} \mathbf{a}_a^2 + \mathbf{a}_s^2 \mathbf{a}_a^{\dagger 2})\right]\end{aligned} \tag{9.4.144}$$

$$\begin{aligned}[(\mathbf{a}_R^\dagger + \mathbf{a}_R)^2 (\mathbf{a}_L^\dagger + \mathbf{a}_L)^2] = &\left\{\frac{1}{4}[\mathbf{a}_s^{\dagger 4} + \mathbf{a}_s^4 + \mathbf{a}_a^{\dagger 4} + \mathbf{a}_a^4]\right. \\ &- \frac{1}{2}[\mathbf{a}_s^{\dagger 2}\mathbf{a}_a^{\dagger 2} + \mathbf{a}_s^2 \mathbf{a}_a^2] \\ &+ \left[\mathbf{a}_s^{\dagger 2} + \mathbf{a}_s^2 + \mathbf{a}_a^{\dagger 2} + \mathbf{a}_a^2\right] \\ &+ \left[\mathbf{a}_s^{\dagger 2} + \mathbf{a}_s^2 - \mathbf{a}_a^{\dagger 2} - \mathbf{a}_a^2\right](\mathbf{a}_s^\dagger \mathbf{a}_s - \mathbf{a}_a^\dagger \mathbf{a}_a) \\ &+ \left[\frac{1}{4}\left(\mathbf{a}_s^\dagger \mathbf{a}_s + \mathbf{a}_a^\dagger \mathbf{a}_a + 1\right)^2 + \right. \\ &\qquad \left. \frac{5}{4}\left(\mathbf{a}_s^\dagger \mathbf{a}_s - \mathbf{a}_a^\dagger \mathbf{a}_a\right)^2 + \frac{3}{4}\right] \\ &\left. - \frac{1}{2}\left[\mathbf{a}_s^{\dagger 2}\mathbf{a}_a^2 + \mathbf{a}_s^2 \mathbf{a}_a^{\dagger 2}\right]\right\}.\end{aligned} \tag{9.4.145}$$

These relationships between normal and local mode expressions are inverted by replacing $(\mathbf{a}_R^\dagger, \mathbf{a}_R)$ by $(\mathbf{a}_s^\dagger, \mathbf{a}_s)$ and $(\mathbf{a}_L^\dagger, \mathbf{a}_L)$ by $(\mathbf{a}_a^\dagger, \mathbf{a}_a)$.

The normal mode $\mathbf{H}^{\text{eff}}_{\text{NORMAL}}$, derived from the local mode $\mathbf{H}^{\text{eff}}_{\text{LOCAL}}$ for two 1 : 1 coupled Morse oscillators, is

$$\begin{aligned}\mathbf{H}^{\text{eff}}_{\text{NORMAL}}/hc = &|v_s v_a\rangle\langle v_s v_a| \{\omega'(v_s + v_a + 1) \\ &+ x_m \left[\frac{3}{4}(v_s + v_a + 1)^2 - \frac{1}{4}(v_s - v_a)^2 - \frac{1}{4}\right] \\ (H_{RL}/hc)[v_s - v_a]\} & \\ &+ |v_s \pm 2v_a \mp 2\rangle\langle v_s v_a| \frac{1}{2} x_m \\ &\times [(v_s + 1/2 \pm 1/2)(v_s + 1/2 \pm 3/2) \\ &\times (v_a + 1/2 \mp 1/2)(v_a + 1/2 \mp 3/2)]^{1/2}\end{aligned} \tag{9.4.146}$$

which may be arranged to more traditional form

$$\begin{aligned}
\mathbf{H}^{\text{eff}}_{\text{NORMAL}}/hc = |v_s v_a\rangle\langle v_s v_a| \Bigg\{ & \left(\omega' + \frac{H_{RL}}{hc}\right)\left(v_s + \frac{1}{2}\right) \\
& + \left(\omega' - \frac{H_{LR}}{hc}\right)\left(v_a + \frac{1}{2}\right) \\
& + x_m \left[\frac{1}{2}(v_s + 1/2)^2 + \frac{1}{2}(v_a + 1/2)^2 + 2(v_s + 1/2)(v_a + 1/2)\right]\Bigg\} \\
& + |v_s \pm 2v_a \mp 2\rangle\langle v_s v_a| \frac{x_m}{2} \\
& \times [(v_s + 1/2 \pm 1/2)(v_s + 1/2 \pm 3/2) \\
& \times (v_a + 1/2 \mp 1/2)(v_a + 1/2 \mp 3/2)]^{1/2}.
\end{aligned} \tag{9.4.147}$$

Thus

$$\omega_s = \omega' + \frac{H_{RL}}{hc} = \omega' + \lambda \tag{9.4.148}$$

$$\omega_a = \omega' - \frac{H_{RL}}{hc} = \omega' - \lambda \tag{9.4.149}$$

$$x_{ss} = x_{aa} = x_m/2 \tag{9.4.150}$$

$$x_{sa} = 2x_m \tag{9.4.151}$$

$$\frac{K_{ssaa}}{16hc} = \frac{x_m}{2}, \tag{9.4.152}$$

where K_{ssaa} is the coupling constant $\left(\frac{\partial^4 V}{\partial \mathbf{Q}_s^2 \partial \mathbf{Q}_a^2}\right)$ for the Darling-Dennison (Darling and Dennison, 1940) 2 : 2 anharmonic coupling term in $V(\mathbf{Q}_s, \mathbf{Q}_a)/hc$,

$$\frac{K_{ssaa}}{4hc}\mathbf{Q}_s^2\mathbf{Q}_a^2 = \frac{K_{ssaa}}{16hc}\left[\left(\mathbf{a}_s^{\dagger 2} + \mathbf{a}_s^2 + 2\mathbf{a}_s^\dagger \mathbf{a}_s + 1\right)\left(\mathbf{a}_a^{\dagger 2} + \mathbf{a}_a^2 + 2\mathbf{a}_a^\dagger \mathbf{a}_L + 1\right)\right]. \tag{9.4.153}$$

Starting from the local mode picture, it is evident that K_{ssaa} originates from the diagonal anharmonicity of the individual local mode Morse oscillators. However, the

$$\frac{K_{ssaa}}{16hc}\left[\left(\mathbf{a}_s^{\dagger 2}\mathbf{a}_a^2 + \mathbf{a}_s^2\mathbf{a}_a^{\dagger 2}\right)\right] \tag{9.4.154}$$

term is only a part of the full

$$\frac{K_{ssaa}}{4hc}\mathbf{Q}_s^2\mathbf{Q}_a^2 \tag{9.4.155}$$

anharmonic coupling term. The neglected parts of this $\mathbf{Q}_s^2\mathbf{Q}_a^2$ term, an out-of-polyad term

$$\frac{K_{ssaa}}{16hc}\left[\mathbf{a}_s^{\dagger 2}\mathbf{a}_a^{\dagger 2} + \mathbf{a}_s^2\mathbf{a}_a^2 + 2(v_s + 1/2)(\mathbf{a}_a^{\dagger 2} + \mathbf{a}_a^2) + 2(v_a + 1/2)(\mathbf{a}_s^{\dagger 2} + \mathbf{a}_s^2)\right], \tag{9.4.156}$$

and a diagonal term,

$$\frac{K_{ssaa}}{4hc}(v_s + 1/2)(v_a + 1/2), \quad (9.4.157)$$

would lead to small corrections in the definitions of the ω', λ', x_{ss}, x_{aa}, and x_{sa} fit parameters in terms of coupled local mode Morse oscillator parameters if one were to insist on the microscopic origin of the K_{ssaa} term in the 2 : 2 $\mathbf{Q}_s^2\mathbf{Q}_a^2$ anharmonic coupling between the symmetric and antisymmetric normal modes.

It is clear from Eq. (9.4.146) that the $H_{RL}/hc = \lambda$ term provides the driving force toward preservation of the normal mode limit by lifting the degeneracy of the members of the $N = v_s + v_a = v_R + v_L$ polyad. The $(x_m/4)(v_s - v_a)^2$ term provides an additional push toward the normal mode limit, especially at the highest and lowest energy regions of the polyad. However, the primary driving force toward the local mode limit also comes from the x_m constant in its role as the coefficient of the off-diagonal coupling term

$$(x_m/2)(\mathbf{a}_s^{\dagger 2}\mathbf{a}_a + \mathbf{a}_s^2\mathbf{a}_a^{\dagger 2}).$$

The off-diagonal matrix elements are largest near the center of the polyad, $\sim \frac{x_m}{8}N^2$, and the overall energy width of the polyad is $2N\lambda$. Thus, at high N, the normal-mode-destroying $\frac{x_m}{8}N^2$ off-diagonal term overpowers the normal-mode-preserving $2N\lambda$ term (which is consistent with the high-N behavior of $\mathbf{H}^{\text{eff}}_{\text{LOCAL}}$). Similarly to the situation for the $\mathbf{H}^{\text{eff}}_{\text{LOCAL}}$, the most problematic off-diagonal matrix elements in $\mathbf{H}^{\text{eff}}_{\text{NORMAL}}$ are in the center of the polyad. Whatever representation one chooses, basis state mixing is always most severe in the center of a polyad (and this is where classical chaos first appears) and the least mixed basis states have zero-order energies that lie at the top and bottom of the polyad.

9.4.12.5 Transformation between 6-Parameter Forms of the Normal Mode and Local Mode $\mathbf{H}^{\text{eff}}$

In Section 9.4.12.4 the simplest possible local mode $\mathbf{H}^{\text{eff}}_{\text{LOCAL}}$, expressed in terms of four independently adjustable parameters (the Morse D_e and a parameters, and two 1 : 1 kinetic and potential energy coupling parameters, $G_{rr'}$ and k_{RL}), is transformed to the simplest possible normal mode $\mathbf{H}^{\text{eff}}_{\text{NORMAL}}$, which is also expressed in terms of four independent parameters. However, the interrelationships between parameters, based on the 1 : 1 coupled local Morse oscillator model, result in only 3 independent fit parameters. This paradox is resolved when one realizes that the 4 parameter local-Morse model generates the Darling-Dennison 2 : 2 coupling term in the normal mode model. However, the full effects of this $(K_{ssaa}/16hc)[(\mathbf{a}_s^\dagger + \mathbf{a}_s)^2(\mathbf{a}_a^\dagger + \mathbf{a}_a)^2]$ coupling term are not taken into account in the local mode model.

The transformation between the standard spectroscopic normal mode model, expressed in terms of 6 independently adjustable fit parameters, to an equivalent 6 parameter local mode model, is considered here. The standard $\mathbf{H}^{\text{eff}}_{\text{NORMAL}}$ model is

$$\begin{aligned}\mathbf{H}^{\text{eff}}_{\text{NORMAL}}/hc = &|v_s v_a\rangle\langle v_s v_a| \{\omega_s(v_s+1/2)+\omega_a(v_a+1/2) \\ &+ x_{ss}(v_s+1/2)^2 + x_{aa}(v_a+1/2)^2 + x_{sa}(v_s+1/2)(v_a+1/2)\} \\ &+ |v_s \pm 2 v_a \mp 2\rangle\langle v_s v_a| \left[(K_{ssaa}/16hc)(\mathbf{a}_s^{\dagger 2}\mathbf{a}_a^2 + \mathbf{a}_a^{\dagger 2}\mathbf{a}_s^2)\right] \quad (9.4.158)\end{aligned}$$

and the 6 fit parameters are ($\omega_s, \omega_a, x_{ss}, x_{aa}, x_{sa}$ and K_{ssaa}). The out-of-polyad contributions from $(K_{ssaa}/4hc)Q_s^2 Q_a^2$ are ignored because the $|v_s \pm 2 v_a \pm 2\rangle\langle v_s v_a|$, $|v_s \pm 2 v_a\rangle\langle v_s v_a|$, and $|v_s v_a \pm 2\rangle\langle v_s v_a|$, matrix elements are small relative to the zero-order energy differences between the coupled basis states (except possibly at very high $N = v_s + v_a$). An alternative (but equivalent form) of $\mathbf{H}^{\text{eff}}_{\text{NORMAL}}$, which is especially convenient for analysis of the intrapolyad dynamics is

$$\begin{aligned}\mathbf{H}^{\text{eff}}_{\text{NORMAL}}/hc = |v_s v_a\rangle\langle v_s v_a| \Bigg\{ &\left(\frac{\omega_s+\omega_a}{2}\right)(v_s+v_a+1) \\ &+ \left[\frac{\omega_s-\omega_a}{2} + \frac{x_{ss}-x_{aa}}{2}(v_s+v_a+1)\right](v_s-v_a) \\ &+ \left(\frac{x_{ss}+x_{aa}+x_{sa}}{4}\right)[(v_s+v_a)(v_s+v_a+1)+1] \\ &+ \left(\frac{x_{ss}+x_{aa}-x_{sa}}{4}\right)(v_s-v_a)^2 \Bigg\} \\ &+ (K_{ssaa}/16hc)\left[\mathbf{a}_s^{\dagger 2}\mathbf{a}_a^2 + \mathbf{a}_a^{\dagger 2}\mathbf{a}_s^2\right]. \quad (9.4.159)\end{aligned}$$

The equivalent standard spectroscopic $\mathbf{H}^{\text{eff}}_{\text{LOCAL}}$ is

$$\begin{aligned}\mathbf{H}^{\text{eff}}_{\text{LOCAL}}/hc = |v_s v_a\rangle\langle v_s v_a| \Bigg\{ &\left(\frac{\omega_s+\omega_a}{2}\right)(v_R+v_L+1) \\ &+ \left(\frac{x_{ss}+x_{aa}+x_{sa}+K_{ssaa}/8hc}{4}\right)[(v_R+1/2)^2+(v_L+1/2)^2] \\ &+ \left(\frac{5x_{ss}+5x_{aa}-x_{sa}-K_{ssaa}/2hc}{4}\right)[(v_R+1/2)(v_L+1/2)] \Bigg\} \\ &+ |v_R \pm 1 v_L \mp 1\rangle\langle v_R v_L| \left[\left(\frac{\omega_s-\omega_a}{2}\right) + \left(\frac{x_{ss}-x_{aa}}{2}\right)(n_R+n_L+1)\right] \\ &\times \left[\left(v_R+\frac{1}{2}\pm\frac{1}{2}\right)\left(v_L+\frac{1}{2}\mp\frac{1}{2}\right)\right]^{1/2} \\ &+ |v_R \pm 2 v_L \mp 2\rangle\langle v_R v_L| \left[\left(\frac{x_{ss}+x_{aa}-x_{sa}+K_{ssaa}/32hc}{4}\right)\right] \\ &\times \left[\left(v_R+\frac{1}{2}\pm\frac{1}{2}\right)\left(v_R+\frac{1}{2}\pm\frac{3}{2}\right)\left(v_L+\frac{1}{2}\mp\frac{1}{2}\right)\left(v_L+\frac{1}{2}\mp\frac{3}{2}\right)\right]^{1/2}. \quad (9.4.160)\end{aligned}$$

An alternative form of $\mathbf{H}^{\text{eff}}_{\text{LOCAL}}$, which is convenient for analysis of intrapolyad dynamics, is

$$\begin{aligned}
\mathbf{H}^{\text{eff}}_{\text{LOCAL}}/hc &= |v_R v_L\rangle\langle v_R v_L| \left\{ \left(\frac{\omega_s + \omega_a}{2}\right)(v_R + v_L + 1) \right. \\
&+ \left(\frac{3K_{ssaa}/16hc - x_{ss} - x_{aa} + x_{sa}}{4}\right)(n_R - n_L)^2 \\
&+ \left. \left(\frac{3x_{ss} + 3x_{aa} + x_{sa} - K_{ssaa}/8hc}{2}\right)\right\} \\
&+ |v_R \pm 1 v_L \mp 1\rangle\langle v_R v_L| \left[\left(\frac{\omega_s - \omega_a}{2}\right) + \left(\frac{x_{ss} - x_{aa}}{2}\right)(v_R + v_L + 1)\right] \\
&\times \left[\left(v_R + \frac{1}{2} \pm \frac{1}{2}\right)\left(v_L + \frac{1}{2} \mp \frac{1}{2}\right)\right]^{1/2} \\
&+ |v_R \pm 2 v_L \mp 2\rangle\langle v_R v_L| \left[\frac{K_{ssaa}/32hc + x_{ss} + x_{aa} - x_{sa}}{4}\right] \\
&\times \left[\left(v_R + \frac{1}{2} \pm \frac{1}{2}\right)\left(v_R + \frac{1}{2} \pm \frac{3}{2}\right)\left(v_L + \frac{1}{2} \mp \frac{1}{2}\right)\left(v_L + \frac{1}{2} \mp \frac{3}{2}\right)\right]^{1/2}.
\end{aligned} \tag{9.4.161}$$

The parameters that define either the $\mathbf{H}^{\text{eff}}_{\text{NORMAL}}$ or the $\mathbf{H}^{\text{eff}}_{\text{LOCAL}}$ are determined by fitting spectroscopic observations of energy levels and relative transition intensities within one or several scaling-related polyads. There are a variety of tools capable of providing insights into the quantum mechanical intrapolyad dynamics (see Sections 9.1.4, 9.1.7 and 9.1.8 and Sections 9.4.3-9.4.10). Complementary insights may be obtained by converting via the Heisenberg form of the Correspondence Principle, the quantum mechanical spectroscopic $\mathbf{H}^{\text{eff}}$ into a classical mechanical $\mathcal{H}^{\text{eff}}$ (Heisenberg, 1925; Kellman and Xiao, 1989; Child, 1991).

The phase space of a coupled, two-identical-anharmonic oscillator system is four-dimensional. Conservation of energy and polyad number reduces the number of independent variables from four to two. At specified values of E and $N = v_R + v_L = v_s + v_a$ (in classical mechanics, N need no longer be restricted to integer values nor E to eigenenergies), accessible phase space divides into several distinct regions of regular, qualitatively describable motions and (for more general dynamical systems) chaotic, indescribable motions. Systematic variation of E and N reveals bifurcations in the number of forms of these describable motions. Examination of the classical mechanical form of the polyad $\mathbf{H}^{\text{eff}}$ often reveals the locations and causes of such bifurcations.

9.4.13 From Quantum Mechanical $\mathbf{H}^{\text{eff}}$ to Classical Mechanical $\mathcal{H}(\mathbf{Q},\mathbf{P})$

The reason for including this section about polyatomic molecule vibrations in a book about diatomic molecules is to show that it is always possible to convert a quantum mechanical $\mathbf{H}^{\text{eff}}$ fit model, which accounts for every detail in a frequency domain spectrum or in any conceivable time-domain experiment, to a classical mechanical $\mathcal{H}^{\text{eff}}$. And, more importantly, information about the qualitative structures in the exploration of accessible phase space and the mechanistic reasons for changes in these structures may be extracted from the $\mathcal{H}^{\text{eff}}$ using the techniques of non-linear classical dynamics. Although "quantum chaos" cannot exist, the classical $\mathcal{H}^{\text{eff}}$derived from the quantum mechanical $\mathbf{H}^{\text{eff}}$ generates large scale chaos in measurable fractions of systematically map-able phase space.

There is a profound but subtle difference between an exact vibrational $\mathbf{H}$ and a polyad $\mathbf{H}^{\text{eff}}$ fit model. Many of the off-diagonal matrix elements in the exact $\mathbf{H}$ couple states that satisfy the convergence criterion for nondegenerate perturbation theory

$$\left|\frac{\mathbf{H}_{ij}}{E_i^{(0)} - E_j^{(0)}}\right| << 1.$$

These nonresonant couplings may be dealt with by a Van Vleck (see Section 4.2) or contact transformation (Nielsen, 1951), which folds their effects into systematically quasi-degenerate groups of states called polyads (see Section 9.4.5). The polyad $\mathbf{H}^{\text{eff}}$ fit model accounts accurately for the observed energy levels (and many other properties) of an entire family of scaling-related polyads. In effect, the dimensionality of the exact $\mathbf{H}$ is drastically reduced in the polyad $\mathbf{H}^{\text{eff}}$. This reduction is due to the existence of several approximate constants of motion which permit $\mathbf{H}$ to be block diagonalized into families of dynamically decoupled polyad $\mathbf{H}^{\text{eff}}$ matrices.

It is possible to derive a classical mechanical $\mathcal{H}$ from the quantum mechanical $\mathbf{H}^{\text{eff}}$ spectroscopic fit model. The relationship between the exact, full dimensional quantum mechanical $\mathbf{H}$ and the approximate, reduced dimensional $\mathbf{H}^{\text{eff}}$ fit model is, at best, numerically complicated. The $\mathcal{H}$ derived from $\mathbf{H}^{\text{eff}}$ is based on an accurate numerical representation of a spectrum. It is vastly simpler than an exact classical mechanical $\mathcal{H}$ derived from the exact quantum mechanical $\mathbf{H}$.

A classical mechanical $\mathcal{H}$ provides insights into intramolecular dynamics, guided by the powerful diagnostic tools of nonlinear classical dynamics. The "action–angle" representation of a classical mechanical problem is easily derived from the spectroscopic $\mathbf{H}^{\text{eff}}$. In fact, the action-angle picture lies at the core of Heisenberg's version of the Correspondence Principle (Heisenberg, 1925). The prescription is that

$$\mathbf{a}_i^\dagger \rightarrow (v_i + 1/2)^{1/2} e^{i\phi_i} = I_i^{1/2} e^{i\phi_i} \tag{9.4.162a}$$

$$\mathbf{a}_i \rightarrow (v_i + 1/2)^{1/2} e^{-i\phi_i} = I_i^{1/2} e^{-i\phi_i} \tag{9.4.162b}$$

$$(v_i + 1/2) = \frac{1}{2}\left[\mathbf{a}_i^\dagger \mathbf{a}_i + \mathbf{a}_i \mathbf{a}_i^\dagger\right] \rightarrow I_i \tag{9.4.162c}$$

where I_i and ϕ_i are the action and angle associated with the i–th vibrational mode. The action, I_i, and the angle, ϕ_i, associated with it are conjugate variables, analogous to position and momentum in the sense that the time derivative of one variable is related to the partial derivative of the $\mathcal{H}$ with respect to the other variable, and *vice versa*,

$$\dot{Q}_i = \frac{\partial \mathcal{H}}{\partial P_i} \quad \text{and} \quad \dot{P}_i = -\frac{\partial \mathcal{H}}{\partial Q_i} \tag{9.4.163}$$

$$\dot{\phi}_i = \frac{\partial \mathcal{H}}{\partial I_i} \quad \text{and} \quad \dot{I}_i = -\frac{\partial \mathcal{H}}{\partial \phi_i}. \tag{9.4.164}$$

The similarity of these classical mechanical Hamilton's equations of motion to their quantum mechanical Ehrenfest's Theorem counterpart, (see Section 9.1.7), is an expression of the Correspondence Principle equivalence of a quantum mechanical commutator, $[\mathbf{A}, \mathbf{B}]$, to a classical mechanical Poisson bracket,

$$\{\mathcal{A}, \mathcal{B}\}_{q,p} = \sum_k \left(\frac{\partial \mathcal{A}}{\partial q_k} \frac{\partial \mathcal{B}}{\partial p_k} - \frac{\partial \mathcal{A}}{\partial p_k} \frac{\partial \mathcal{B}}{\partial q_k} \right) \tag{9.4.165}$$

$$\{q_i, p_j\}_{q,p} = \delta_{ij} \tag{9.4.166}$$

$$\{p_i, \mathcal{H}\}_{q,p} = -\frac{\partial \mathcal{H}}{\partial q_i} = \dot{p}_i \tag{9.4.167}$$

$$\{q_i, \mathcal{H}\}_{q,p} = \frac{\partial \mathcal{H}}{\partial p_i} = p_i/m = \dot{q}_i \tag{9.4.168}$$

where Eqs. (9.4.167) and (9.4.168) are Newton's equations of motion.

In order to illustrate the relationship between the quantum mechanical $\mathbf{H}^{\text{eff}}$ and the classical mechanical $\mathcal{H}$ derived from it by applying Heisenberg's (1925) version of the Correspondence Priniciple, we return to the problem of two coupled identical anharmonic oscillators (see Section 9.4.12 and Xiao and Kellman, 1989). The quantum mechanical $\mathbf{H}^{\text{eff}}_{\text{LOCAL}}$ is,

$$\begin{aligned}\mathbf{H}^{\text{eff}}_{\text{LOCAL}}/hc = {} & \omega_0(v_R + v_L + 1) + \frac{1}{2}\left[\beta + (\epsilon/2)(v_R + v_L + 1)\right]\left[\mathbf{a}_R^\dagger \mathbf{a}_L + \mathbf{a}_L^\dagger \mathbf{a}_R\right] \\ & + (\alpha_1/4)(v_R + v_L + 1)^2 + (\alpha_2/4)[v_R - v_L]^2 \\ & + \delta'\left[\mathbf{a}_R^{\dagger 2}\mathbf{a}_L^2 + \mathbf{a}_L^{\dagger 2}\mathbf{a}_R^2\right],\end{aligned} \tag{9.4.169}$$

where the six independently adjustable parameters (ω_0, β, ϵ, α_1, α_2, and δ') are defined in terms of traditional normal mode spectroscopic constants (ω_s, ω_a,

x_{ss}, x_{aa}, x_{sa}, and K_{ssaa}) in Eq. (9.4.161). This quantum mechanical $\mathbf{H}^{\text{eff}}_{\text{LOCAL}}$ is converted into the equivalent classical mechanical $\mathcal{H}_{\text{LOCAL}}$ by making the following replacements

$$v_R + v_L + 1 = 2I \tag{9.4.170a}$$

$$v_R - v_L = 2I_z \tag{9.4.170b}$$

$$\mathbf{a}_R^\dagger = [v_R + 1/2]^{1/2} e^{i\phi_R} \tag{9.4.170c}$$

$$\mathbf{a}_R = [v_R + 1/2]^{1/2} e^{-i\phi_R} \tag{9.4.170d}$$

$$\mathbf{a}_L^\dagger = [v_L + 1/2]^{1/2} e^{i\phi_L} \tag{9.4.170e}$$

$$\mathbf{a}_L = [v_L + 1/2]^{1/2} e^{-i\phi_L} \tag{9.4.170f}$$

$$(v_R + 1/2)(v_L + 1/2) = I^2 - I_z^2 \tag{9.4.170g}$$

$$\phi_R - \phi_L = \psi, \tag{9.4.170h}$$

thus

$$\mathcal{H}_{\text{LOCAL}} = 2\omega_0 I + [\beta + \epsilon I][I^2 - I_z^2]^{1/2} \cos\psi + \alpha_1 I^2 + \alpha_2 I_z^2 + 2\delta'[I^2 - I_z^2] \cos 2\psi. \tag{9.4.171}$$

$\mathcal{H}_{\text{LOCAL}}$ is expressed in terms of two actions, I and I_z, and one angle, ψ, which is associated with the I_z action. The angle, Θ, associated with the action, I, does not appear in $\mathcal{H}_{\text{LOCAL}}$ because I is a conserved quantity

$$\frac{\partial \mathcal{H}_{\text{LOCAL}}}{\partial \Theta} = \dot{I} = 0. \tag{9.4.172a}$$

However, the conservation of I implies

$$\frac{\partial \mathcal{H}_{\text{LOCAL}}}{\partial I} = \dot{\Theta}, \tag{9.4.172b}$$

thus $\dot{\Theta}$ is time-independent and

$$\Theta(t) = \Theta_0 + \frac{\partial \mathcal{H}_{\text{LOCAL}}}{\partial I} t \tag{9.4.172c}$$

where, in the weak coupling limit: $|\beta| << \omega_0$, $|\delta'| << \omega_0$, $|\epsilon| << \alpha_1$,

$$\frac{\partial \mathcal{H}_{\text{LOCAL}}}{\partial I} = 2\omega_0 + 2\alpha_1 I. \tag{9.4.172d}$$

Since both energy and I are conserved, setting $\mathcal{H}_{\text{LOCAL}} = E$ and specifying I gives an equation that uniquely relates each value of the nonconserved action, I_z, to a value of the associated angle, $\psi(I_z)$ at that value of I_z. This $1:1$ relationship between $I_z(E, I, \psi)$ and $\psi(E, I, I_z)$ is a trajectory in a two-dimensional I_z, ψ action-angle space. The time dependence for motion along this trajectory is obtained by integrating either

$$\frac{\partial \mathcal{H}_{\text{LOCAL}}}{\partial \psi} = -\dot{I}_z \tag{9.4.173a}$$

or

$$\frac{\partial \mathcal{H}_{\text{LOCAL}}}{\partial I_z} = \dot{\psi}. \tag{9.4.173b}$$

Similarly the classical mechanical $\mathcal{H}_{\text{NORMAL}}$ is obtained from

$$\begin{aligned} \mathbf{H}^{\text{eff}}_{\text{NORMAL}}/hc = {} & \omega_0(v_s + v_a + 1) + \frac{1}{2}[\beta + (\epsilon/2)(v_s + v_a + 1)](v_s - v_a) \\ & + (\gamma_1/4)(v_s + v_a + 1)^2 + (\gamma_2/4)(v_s - v_a)^2 \\ & + \delta(\mathbf{a}_s^{\dagger 2}\mathbf{a}_a^2 + \mathbf{a}_a^{\dagger 2}\mathbf{a}_s^2), \end{aligned} \tag{9.4.174}$$

where the six independently adjustable parameters (ω_0, β, ϵ, γ_1, γ_2, and δ) are defined in terms of traditional normal mode spectroscopic constants (ω_s, ω_a, x_{ss}, x_{aa}, x_{sa}, and K_{ssaa}) in Eq. (9.4.159). This quantum mechanical $\mathbf{H}^{\text{eff}}_{\text{NORMAL}}$ is converted into the equivalent classical mechanical $\mathcal{H}_{\text{NORMAL}}$ by making the replacements

$$v_s + v_a + 1 = 2\overline{I} \quad (\overline{I} = I) \tag{9.4.175a}$$

$$v_s - v_a = 2\overline{I}_z \quad (\overline{I}_z = I_x) \tag{9.4.175b}$$

$$\mathbf{a}_s^\dagger = [v_s + 1/2]^{1/2} e^{i\phi_s} \tag{9.4.175c}$$

$$\mathbf{a}_s = [v_s + 1/2]^{1/2} e^{-i\phi_s}, \tag{9.4.175d}$$

similarly for $\mathbf{a}_a^\dagger$ and $\mathbf{a}_a$, and

$$(v_s + 1/2)(v_a + 1/2) = \overline{I}^2 - \overline{I}_z^2 \tag{9.4.175e}$$

$$\overline{\psi} = \phi_s - \phi_a. \tag{9.4.175f}$$

(The overbar specifies actions and angles in the normal mode representation, which differs from the local mode representation only by a rotation of $\pi/2$ about the y axis). Thus,

$$\begin{aligned} \mathcal{H}_{\text{NORMAL}}/hc = {} & 2\omega_0\overline{I} + (\beta + \epsilon\overline{I})\overline{I}_z + \gamma_1\overline{I}^2 \\ & + \gamma_2\overline{I}_z^2 + 2\delta(\overline{I}^2 - \overline{I}_z^2)\cos 2\overline{\psi}. \end{aligned} \tag{9.4.176}$$

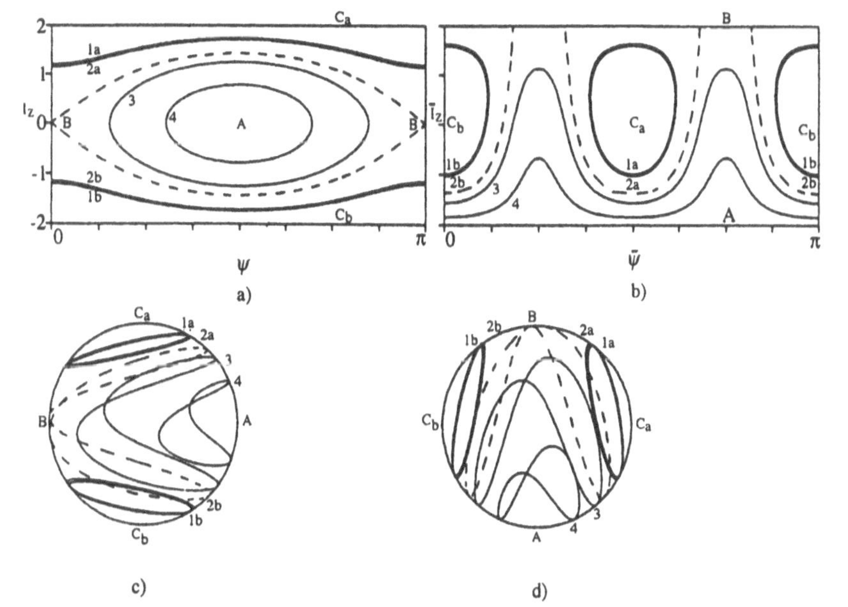

Figure 9.13: Phase space trajectories and polyad phase spheres for polyad $N = 3$ of H_2O. H_2O is an example of a molecule with two identical coupled anharmonic bond stretch oscillators. The local mode (I_z, ψ) phase space trajectories in part (a) contain the same information as the normal mode $(\overline{I}_z, \overline{\psi})$ trajectories in part (b). The corresponding polyad phase spheres in parts (c) and (d) are identical except for a rotation by $\pi/2$ about the y axis (from Xiao and Kellman, 1989).

Figure 9.13 displays the phase space structure of $\mathcal{H}_{\text{LOCAL}}$ and $\mathcal{H}_{\text{NORMAL}}$ for the $N = v_s + v_a = v_R + v_L = 3$ $[I = (N+1)/2 = 2]^{\dagger}$ polyad of H_2O. Just as $\mathbf{H}^{\text{eff}}_{\text{LOCAL}}$ and $\mathbf{H}^{\text{eff}}_{\text{NORMAL}}$ provide identical quality representations of the observed spectrum, so too do $\mathcal{H}_{\text{LOCAL}}$ and $\mathcal{H}_{\text{NORMAL}}$. The phase space structures displayed in parts (a), LOCAL, and (b), NORMAL, of Fig. 9.13 are equivalent. The appearance of qualitatively different structures in the LOCAL and NORMAL representations is largely due to the mapping of the information onto an I_z, ψ (or $\overline{I}_z, \overline{\psi}$) planar rectangle rather than a "polyad phase sphere." As shown in parts (c) and (d) of Fig. 9.13, the structures from parts (a) and (b) differ only by a rotation of the phase sphere by $\pi/2$ about the y axis.

The phase space structures displayed in Fig. 9.13(a) and (b) provide insights into the classical dynamics encoded in the quantum spectrum. The solid curves labeled 1a, 1b, 2a, 2b, 3, and 4 are the phase space I_z, ψ trajectories associated with the four eigenstates that belong to the $N = 3$ polyad (obtained equivalently from an $\mathbf{H}^{\text{eff}}$ expressed in terms of the four $(v_s, v_a) = (3,0), (2,1), (1,2), (0,3)$ normal or $(v_R, v_L) = (3,0), (2,1), (1,2), (0,3)$ local mode basis states).

---

†The $+1$ in the $(N+1)$ term includes the $1/2$ quantum of zero point excitation in each mode.

Although in classical mechanics E may be varied continuously, the trajectories are displayed only for the values of E that are quantum eigen-energies. The dotted $I_z(\psi)$ line is a *separatrix*, which is a dividing surface that no trajectory can cross and which divides the accessible phase space into qualitatively distinct regions (exhibiting normal vs. local mode behavior) filled with qualitatively different trajectories.

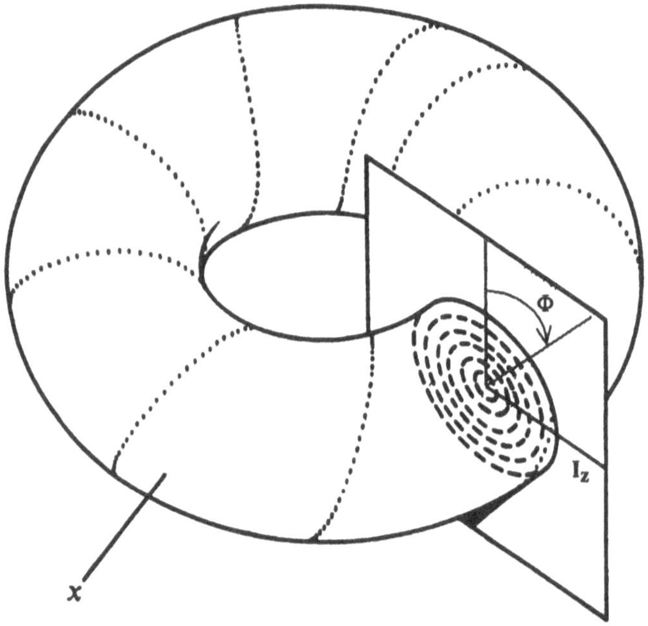

Figure 9.14: Nested invariant tori and a surface of section. Each dotted circle corresponds to a quasiperiodic trajectory at different E. When the radius of the circle collapses to zero, the trajectory is periodic. The dotted trajectory shown looping about the outermost invariant torus, will eventually fill the surface of that torus and generate a solid circle on the surface of section (from Child, 1991).

Trajectories 1a, 1b and 2a, 2b are characteristic of local mode behavior because they explore the full range of ψ in Fig. 9.13(a) and trajectories 3 and 4 are characteristic of normal mode behavior because they explore the full range of $\overline{\psi}$ in Fig. 9.13(b).

Although the phase space trajectories appear as simple curves on the two-dimensional I_z, ψ phase space diagram (the Θ coordinate is suppressed) most trajectories are actually quasiperiodic. The actual trajectories lie on the 2-dimensional surface of a 3-dimensional invariant torus in 4-dimensional phase space. Fig. 9.14 shows such a torus. Any point on the surface of the torus is specified by two angles, Θ and Φ. The Θ and Φ circuits about the torus are shown, respectively, as large and small diameter circles. The diameter of the Θ

circuit is determined by I. The "diameter" of the Φ circuit is determined by both I and E (a different invariant torus for every value of E). A trajectory on the I_z, ψ phase space diagram is a projection of the full trajectory onto a plane that intersects the invariant torus at a specified value of Θ. A circle on the I_z, ψ phase space diagram or polyad phase sphere is a quasiperiodic trajectory on the invariant torus. Each such circle corresponds to a trajectory at a different E. As E is varied systematically over the range compatible with $I, E_{\text{min}}(I) \leq E \leq E_{\text{max}}(I)$, a family of circles may be observed to shrink into a single point and then vanish. The point at the center of a family of circles is a stable fixed point and corresponds to a periodic orbit (constant $I_z, \psi, 0 \leq \Theta \leq 2\pi$). Another family of circles is observed to expand as E is varied until the outermost one contacts a point on the separatrix and then vanishes, possibly reappearing as a member of a qualitatively different class of family of circles on the other side of the separatrix. This vanishing point is an unstable fixed point, which also corresponds to a periodic oribt.

The fixed points on the phase space diagrams or phase spheres in Fig. 9.13 are labeled A, B, C_a, and C_b. Each corresponds to a periodic orbit that is said to organize the surrounding region of phase space that is filled with topologically similar quasiperiodic trajectories.

The volume of the surrounding organized region is related to the stability of the fixed point periodic orbit trajectory. Stability is measured by the rate of divergence between two trajectories that originate from two nearby points in phase space.

The local mode trajectories on Fig. 9.13 come in pairs because there are two identical local oscillators. However, the near perfect superposition of the $(1a, 1b)$ pair of trajectories with the $(2a, 2b)$ pair of trajectories illustrates a characteristic feature of local mode limiting behavior. This corresponds to the quantum mechanical signature of the local mode limit, the near perfect degeneracy of pairs of local mode eigenstates.

The central region of the local mode representation of the phase space trajectories is called the "resonance region." In this resonance region of phase space the trajectories (#3 and #4) are not free to explore the full $0 \leq \psi \leq \pi$ range and are threfore classified as normal mode trajectories. Points A and B are fixed points which lie at the maximum and minimum E extremes, $E_A(I)$ and $E_B(I)$, of the resonance region for a particular value of I. Point A at $I_z = 0$ $(v_R = v_L)$ and $\psi = \pi/2$ (out-of-phase motion of the R and L oscillators) is stable and corresponds to a pure antisymmetric stretch. Point B at $I_z = 0$ $(v_R = v_L)$ and $\psi = 0$ and π (in-phase motion) is unstable (because it lies on the separatrix) and corresponds to a pure symmetric stretch. Quasiperiodic trajectories that circulate about a stable fixed point resemble the fixed point periodic trajectory. At $E > E_A(I)$ no trajectories of any type can exist. At $E < E_B(I)$ the B–like trajectories vanish and are replaced by trajectories that circulate about the C_a, C_b fixed points and are therefore C–like. The C_a and C_b lines ($I_z = \pm I = \pm 2$, $0 \leq \psi < \pi$) are actually the north and south poles on the local mode polyad phase sphere (Fig. 9.13(c)). The stable fixed points lie near the poles and trajectories 1a and 2a circulate about the fixed point near C_a and trajectories 1b and 2b circulate about the fixed point near C_b.

The normal mode representation of the phase space trajectories contains the same information as the local mode representation. However, the resonance region on the normal mode phase space map contains the local mode trajectories (1a, 1b; 2a, 2b) and the stable fixed points C_a and C_b. The trajectories contained within the resonance zone are not free to explore the entire $0 \leq \overline{\psi} \leq \pi$ range whereas the trajectories outside the resonance zone do explore the $0 \leq \overline{\psi} \leq \pi$ range and are therefore classified as normal mode trajectories. The fixed point B ($\overline{I}_z = I = +2$) is unstable, because it lies on a separatrix, and is located at the north pole of the normal mode polyad phase sphere. The stable fixed point A ($\overline{I}_z = -I = -2$) is located at the south pole.

Figure 9.15 shows the evolution of the local mode representation of the polyad phase sphere for $I = 2$ ($N = v_s + v_a = v_R + v_L = 3$) as the strength of the 1 : 1 coupling term ($\beta + \epsilon I$ in Eq. (9.4.171)) is increased from 0 (part (a)) to ∞ (part (f)). (The 1 : 1 coupling term is antagonistic to local mode behavior whereas the 2 : 2 coupling term is antagonistic to normal mode behavior.) In part (b), an unstable fixed point (at $I_z = 0$, $\psi = 0, \pi$) and a separatrix appear, but no eigenstate trajectory lies within the resonance region bounded by the separatrix. As the 1 : 1 coupling strength increases, first the level at highest E, level 4 (part (c)), then level 3 (part (d)), then all of the levels (part (e)) pass through fixed point B and fall into the resonance region. Finally (part (f)) the resonance region fills the entire phase sphere and all the trajectories are in the normal mode limit.

Figure 9.16 shows the evolution of the local mode polyad phase spheres for H_2O as I increases from 1 ($N = v_s + v_a = 1$) to 3 ($N = v_s + v_a = 5$). As I increases the lowest energy levels sequentially pass through the unstable fixed point B, depart from the normal mode resonance region, and become local mode states. At $I = 1$ (part (a)) there are only 2 levels and both are on the normal mode side of the separatrix. At $I = 3$ (part (e)) there are 6 levels and the lowest 4 of these have departed the resonance region and are local mode states.

Trajectories in action/angle polyad phase space convey all of the most important qualitative relationships between a quantum spectrum and classical intramolecular dynamics. However, coordinate space trajectories are both more easily visualized and more directly comparable to quantum probability densities, $|\psi(Q_1, Q_2, \ldots Q_{3N-6})|^2$. Xiao and Kellman (1989) describe how the action/angle phase space trajectories for each eigenstate may be converted into a coordinate space trajectory. The key to this is the exact relationship between Morse oscillator displacement coordinates, r_i, and the action, angle variables, I_i, ϕ_i (Rankin and Miller, 1971). Figure 9.17 shows, for the 6 eigenstates in the $I = 3$ ($N = v_s + v_a = 5$) polyad of H_2O, the correspondences between the phase space trajectories, the coordinate space trajectories, and the probability densities. The resemblance between the classical coordinate space trajectories and the quantum probability densities is striking!

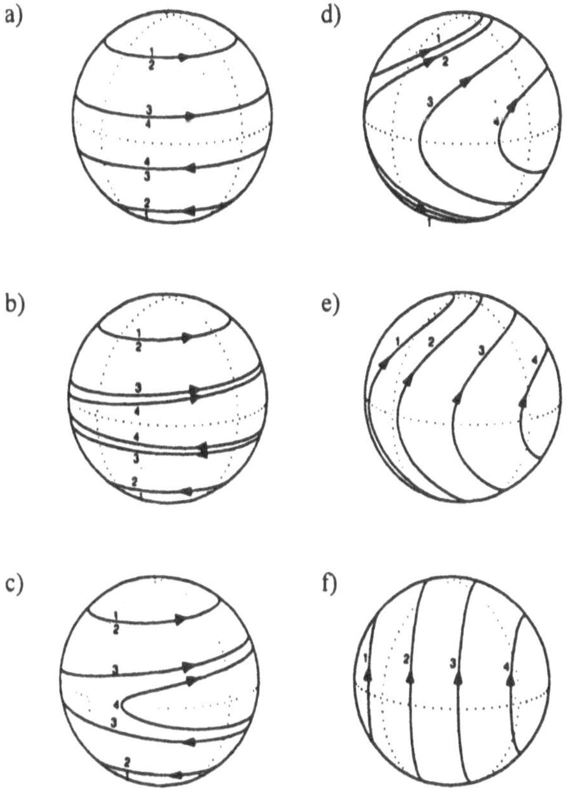

Figure 9.15: Evolution of the polyad phase sphere from the local mode to the normal mode limit as the strength of the 1 : 1 coupling term (antithetical to the local mode limit) is increased from 0 (part a) to ∞ (part f). As the coupling term, δ in Eq. (9.4.174) increases from 0, first in part (c) one trajectory (level 4, at highest E) falls through the unstable fixed point into the normal mode region (antisymmetric stretch); eventually, in part (e), the resonance zone fills the entire phase sphere; finally, in part (f), the normal mode limit is reached (from Xiao and Kellman, 1989).

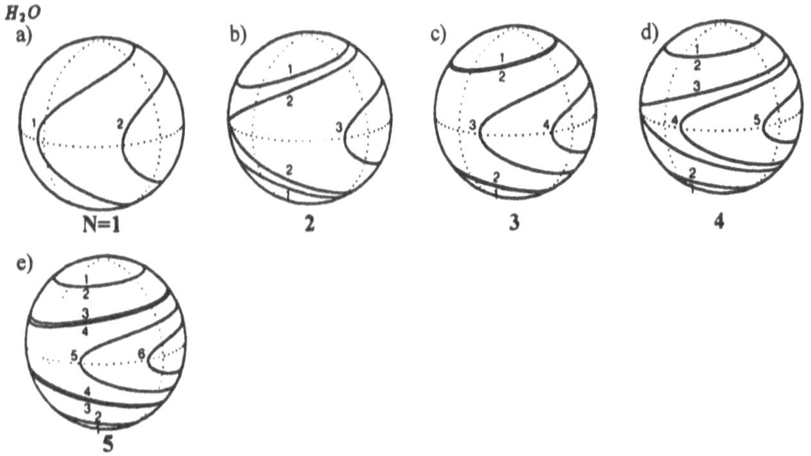

Figure 9.16: Polyads $N = 1$ through $N = 5$ for H_2O. As N increases, the new states that appear at the bottom of the polyad fall on the local mode side of the separatrix (from Xiao and Kellman, 1989).

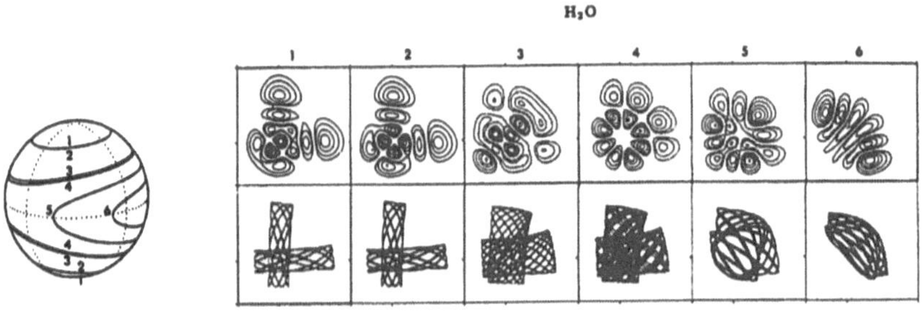

Figure 9.17: The probability densities, $|\psi|^2$, in the top row are found to resemble the configuration space trajectories in the bottom row. The states are from the $N = 5$ polyad of H_2O and are numbered in increasing energy order from 1 to 6 in correspondence with the numbered trajectories on the polyad phase sphere (from Xiao and Kellman, 1989).

The phase space structures for two identical coupled anharmonic oscillators are relatively simple because the trajectories lie on the surface of a 2–dimensional manifold in a 4–dimensional phase space. The phase space of two identical 2–dimensional isotropic benders is 8–dimensional, the qualitative forms of the classifying trajectories are far more complicated, and there is a much wider range of possibilities for qualitative changes in the intramolecular dynamics. The classical mechanical polyad $\mathcal{H}^{\text{eff}}$ conveys unique insights into the dynamics encoded in the spectrum as represented by the $\mathbf{H}^{\text{eff}}$ fit model.

The bending vibration of a linear ABC molecule may be treated as a two-dimensional isotropic harmonic oscillator. The $|v, l\rangle$ basis set is particularly convenient, where v and l are respectively the vibrational and vibrational angular momentum quantum numbers. The allowed values of l are $-v$, $-v+2$, $\ldots$ v. Since there are two quantum numbers, two pairs of creation, annihilation operators are needed to generate all basis states from the $|v=0,\ l=0\rangle$ zero-point state. These are, following the notation of Cohen-Tannoudji, *et al.*, (1977), $\mathbf{a}_d^\dagger, \mathbf{a}_d$ and $\mathbf{a}_g^\dagger, \mathbf{a}_g$, where

$$N_g = (v+l)/2 \tag{9.4.177}$$

$$N_d = (v-l)/2 \tag{9.4.178}$$

$$\mathbf{a}_d = |v,l\rangle = \mathbf{a}_d|N_d\rangle = [N_d]^{1/2}|N_d - 1\rangle = \left(\frac{v+l}{2}\right)^{1/2}|v-1, l-1\rangle \tag{9.4.179}$$

$$\mathbf{a}_g = |v,l\rangle = \mathbf{a}_g|N_g\rangle = [N_g]^{1/2}|N_g - 1\rangle = \left(\frac{v-l}{2}\right)^{1/2}|v-1, l+1\rangle \tag{9.4.180}$$

and the number operators are

$$\mathbf{a}_d^\dagger\mathbf{a}_d|v,l\rangle = N_d|v,l\rangle = \left(\frac{v+l}{2}\right)|v,l\rangle \tag{9.4.181}$$

$$\mathbf{a}_g^\dagger\mathbf{a}_g|v,l\rangle = N_g|v,l\rangle = \left(\frac{v-l}{2}\right)|v,l\rangle. \tag{9.4.182}$$

Thus

$$\left(\mathbf{a}_d^\dagger\mathbf{a}_d + \mathbf{a}_g^\dagger\mathbf{a}_g\right)|v,l\rangle = v|v,l\rangle \tag{9.4.183}$$

$$\left(\mathbf{a}_d^\dagger\mathbf{a}_d - \mathbf{a}_g^\dagger\mathbf{a}_g\right)|v,l\rangle = l|v,l\rangle. \tag{9.4.184}$$

The Heisenberg (1925) quantum→classical correspondence specifies

$$\mathbf{a}_d \to [N_u + 1/2]^{1/2}e^{-i\phi_u} = \left[\frac{v+l+1}{2}\right]^{1/2} e^{-i\phi_d}, \tag{9.4.185}$$

$$\mathbf{a}_g \to [N_g + 1/2]^{1/2}e^{-i\phi_g} = \left[\frac{v-l+1}{2}\right]^{1/2} e^{-i\phi_g}, \tag{9.4.186}$$

and similarly for $\mathbf{a}_g^\dagger$ and $\mathbf{a}_d^\dagger$, except with an $e^{+i\phi}$ factor.

Acetylene, H–C≡C–H, has two identical HCC two-dimensional isotropic local-bender harmonic oscillators. In the normal mode basis set, the bending basis states are specified by four quantum numbers, $|v_4 l_4 v_5 l_5\rangle$, where mode 4 is the π_g–symmetry *trans*-bend and mode 5 is the π_u-symmetry *cis*-bend. (The normal mode basis states do not have identical harmonic frequencies, $\omega_5 - \omega_4 = 121\ \text{cm}^{-1}$.) There are four pairs of normal mode creation, annihilation operators, two for mode 4, $\mathbf{a}_{4d}^\dagger$, $\mathbf{a}_{4d}$, and $\mathbf{a}_{4g}^\dagger$, $\mathbf{a}_{4g}$, and two for mode 5, $\mathbf{a}_{5d}^\dagger$, $\mathbf{a}_{5d}$, and $\mathbf{a}_{5g}^{\dagger}$, $\mathbf{a}_{5g}$. The matrix elements and number operators for modes 4 and 5 are defined exactly as in Eqs. (9.4.177) - (9.4.184).

The $|v_4 l_4 v_5 l_5\rangle$ normal mode basis states form polyads (see Section 9.4.5), which are labeled by the approximately good polyad quantum numbers, N and l,

$$N = v_4 + v_5 \tag{9.4.187}$$

$$l = l_4 + l_5. \tag{9.4.188}$$

These polyad quantum quantum numbers are associated with number operators

$$[(\mathbf{a}_{4d}^{\dagger}\mathbf{a}_{4d} + \mathbf{a}_{4g}^{\dagger}\mathbf{a}_{4g}) + (\mathbf{a}_{5d}^{\dagger}\mathbf{a}_{5d} + \mathbf{a}_{5g}^{\dagger}\mathbf{a}_{5g})]|v_4 l_4 v_5 l_5\rangle$$

$$= (v_4 + v_5)|v_4 l_4 v_5 l_5\rangle \tag{9.4.189}$$

$$[(\mathbf{a}_{4d}^{\dagger}\mathbf{a}_{4d} - \mathbf{a}_{4g}^{\dagger}\mathbf{a}_{4g}) + (\mathbf{a}_{5d}^{\dagger}\mathbf{a}_{5d} - \mathbf{a}_{5g}^{\dagger}\mathbf{a}_{5g})]|v_4 l_4 v_5 l_5\rangle$$

$$= (l_4 + l_5)|v_4 l_4 v_5 l_5\rangle. \tag{9.4.190}$$

N and l correspond to the conserved classical actions,

$$K_a = [N_{4d} + N_{4g} + N_{5d} + N_{5g} + 2]^{\dagger}/4 \tag{9.4.191}$$

$$K_b = [N_{4d} - N_{4g} + N_{5d} - N_{5g}]/4, \tag{9.4.192}$$

and angles,

$$\phi_a = \phi_{4d} + \phi_{4g} + \phi_{5d} + \phi_{5g} \tag{9.4.193}$$

$$\phi_b = \phi_{4d} - \phi_{4g} + \phi_{5d} - \phi_{5g}, \tag{9.4.194}$$

but the conserved angles, ϕ_a and ϕ_b, do not appear in the classical mechanical polyad $\mathcal{H}^{\text{eff}}$. There remain two non-conserved actions, J_a and J_b,

$$J_a = [(N_{4d} + N_{4g}) - (N_{5d} + N_{5g})]/4 \tag{9.4.195}$$

$$J_b = (N_{4d} - N_{4g}) - (N_{5d} - N_{5g})]/4, \tag{9.4.196}$$

which correspond, respectively, to $(v_4 - v_5)/4$ and $(l_4 - l_5)/4$, and their conjugate angles, ψ_a and ψ_b,

$$\psi_a = \phi_{4d} + \phi_{4g} - \phi_{5d} - \phi_{5g} \tag{9.4.197}$$

$$\psi_b = \phi_{4g} - \phi_{4d} + \phi_{5d} - \phi_{5g}. \tag{9.4.198}$$

Thus, the quantum mechanical polyad $\mathbf{H}^{\text{eff}}$, defined by 16 spectroscopic parameters (Table II of Jacobson, *et al.*, 1998) obtained from a fit to 83 observed vibrational levels, that appears as Eq. 17 in Jacobson and Field (2000b), may be converted to the classical mechanical polyad $\mathcal{H}^{\text{eff}}$ model, that appears as Eqs. (12-15) of Jacobson, *et al.*, (1999).

†The additional +2 term in K_a comes from the 1/2 quantum of zero-point excitation in each oscillator.

Analysis of this $\mathcal{H}^{\text{eff}}$ using the techniques of nonlinear classical dynamics reveals the structure of phase space (mapped as a continuous function of the conserved quantities E, K_a, and K_b) and the qualitative nature of the classical trajectory that corresponds to every eigenstate in every polyad. This analysis reveals qualitative changes, or *bifurcations*, in the dynamics, the onset of classical chaos, and the fraction of phase space associated with each qualitatively distinct class of regular (quasiperiodic) and chaotic trajectories.

This analysis is complicated, owing to the relatively high dimensionality of the phase space of two coupled 2-dimensional isotropic anharmonic bending oscillators. But the analysis is rich in insights, some of which may guide the formulation of novel schemes for external control over intramolecular dynamics.

The 8-dimensional phase space of two 2-dimensional oscillators is reduced by the existence of two conserved actions, K_a and K_b, and by the absence of the conjugate angles, ϕ_a and ϕ_b, from the classical mechanical polyad $\mathcal{H}^{\text{eff}}$. The conserved actions appear parametrically in $\mathcal{H}^{\text{eff}}$, thus the phase space accessible at specified values of K_a and K_b is four dimensional. Since energy is conserved, in addition to K_a and K_b, all trajectories lie on the surface of a 3-dimensional energy shell.

A trick, the "surface of section," is needed to visualize the essential features of these trajectories. The surface of section is constructed by exploiting a planar cut through the 3-dimensional energy shell. One is free to choose the location and orientation of this plane. The choice of a J_b, ψ_b plane located at $\psi_a = 0$ is convenient.

The structure of phase space accessible at specified values of K_a, K_b, and E is revealed on the surface of section by launching a series of trajectories and, for each trajectory, plotting a point on the (J_b, ψ_b) plane at each location where the trajectory crosses the plane in the $\frac{d\psi_a}{dt} > 0$ direction. If the trajectory is quasiperiodic, the set of all of its intersections with the (J_b, ψ_b) plane forms a closed curve on the surface of section.

Each trajectory is launched at chosen initial values of J_b and ψ_b and at $\Psi_a = 0$. Since any point on the 3-dimensional energy shell may be specified by three linearly independent coordinates, selection of initial values J_b°, ψ_b°, and ψ_a°, implies a definite value of J_a°. Thus trajectories are launched at various $[J_b^\circ,\ \psi_b^\circ,\ \psi_a^\circ = 0,\ J_a^\circ(J_b^\circ,\ \psi_b^\circ,\ \psi_{a}^\circ,)\ ]$ initial values until all of the qualitatively distinct regions of phase space are represented on the surface of section by either a family of closed curves (quasiperiodic trajectories) that surround a fixed point (a periodic trajectory that defines the qualitative topological nature of the neighboring quasiperiodic trajectories) or an apparently random group of points (chaos). Often, color is used to distinguish points on the surface of section that belong to different trajectories.

Figure 9.18 shows surfaces of section for the $[N, l] = [4, 0]$ and $[8, 0]$ polyads. The $[4, 0]$ polyad is represented by $K_a = (4+2)/4 = 1.5$, $K_b = 0$, and $E = 2461$ cm^{-1}, which is near the middle of the polyad. The surface of section reveals only two families of trajectories, one surrounding a periodic orbit at $J_b = 0$, $\psi_b = \pm\pi$ (*cis*-bend) and the other at $J_b = 0$, $\psi_b = 0$ (*trans*-bend) with a dividing surface (separatrix) at $\psi_b = \pm\pi/2$. This surface of section is typical of all surfaces of section at $K_a \leq 2$ ($N_b \leq 6$) over the entire range of E compatible with each value of K_a. At low E and K_a, all of phase space is divided between *trans*-bending and *cis*-bending trajectories.

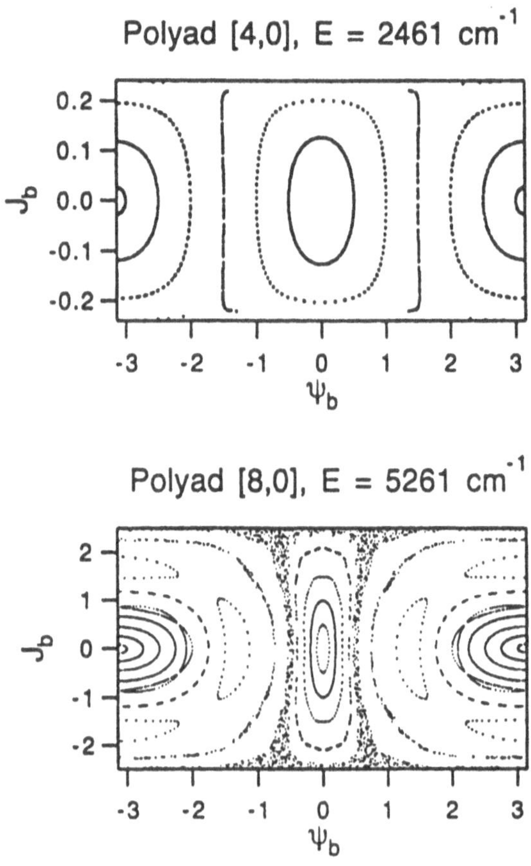

Figure 9.18: Surfaces of section for the HCCH $[N_b, l] = [4,0]$ and $[8,0]$ polyads. The surface section for the $[4,0]$ polyad shows that all of phase space is divided between *cis*-bend and *trans*-bend normal modes. The phase space structure for the $[8,0]$ polyad contains large scale chaos as well as at least two new qualitative behaviors (from Jacobson, *et al.*, 1999).

The surface of section for the $[8,0]$ ($K_a = 2.5$) polyad contains several structures that are not present for the $[4,0]$ polyad. Most importantly, a significant fraction of the surface of section is filled with the apparently random dots symptomatic of chaos. The fraction of phase space organized by the periodic *trans*- and *cis*-bending vibrations is considerably reduced relative to that for the $[4,0]$ polyad. New closed loops have appeared (centered at $J_b \approx \pm 1.7$, $\psi_b = \pm\pi$ and $J_b = 0$, $\psi_b \approx \pm\pi/2$), and the qualitatively new trajectories represent the first appearance of localized motions in which the two ends of the molecule are involved in different kinds of motions.

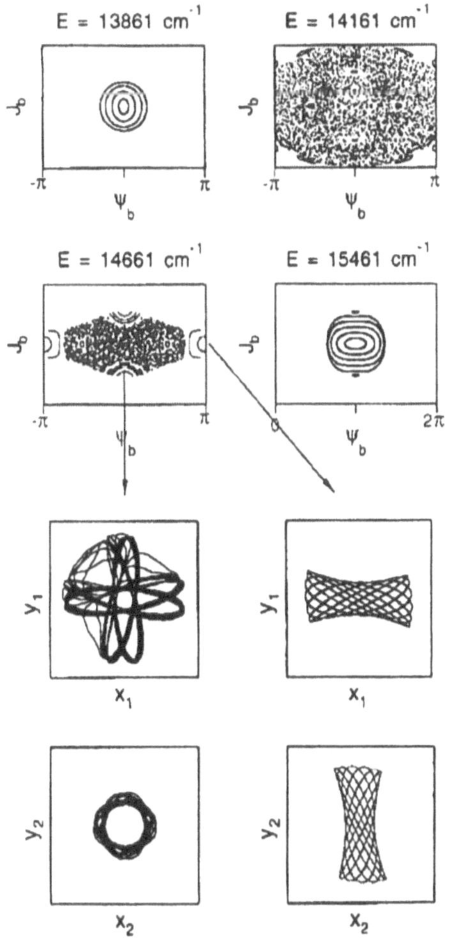

Figure 9.19: Overview of the phase space and configuration space dynamics associated with the HCCH $[N_b = 22, l = 0]$ polyad. The top four plots are surfaces of section for four energies within the polyad. Only simple structures are found near the bottom (local bender) and top (counter-rotators) of the polyad. Chaos dominates at energies near the middle of the polyad, but two classes of stable trajectories coexist with chaos. The bottom four plots show the coordinate space motions of the left- and right-hand H-atoms that correspond to the two different periodic orbits (from Jacobson and Field, 2000b).

Figure 9.19 shows four surfaces of section for four different energies within the $N_b = 22$ $(K_a = 6)$, $l = 0$ polyad and the individual atom trajectories associated with the two qualitatively distinct regular features present at an energy of 14,661 cm^{-1} near the center of the polyad. The surfaces of section at the bottom and top of the polyad are organized respectively around periodic trajectories that turn out to involve local-bender and counter-rotator motions.

In contrast, phase space is essentially entirely chaotic at ~300 cm^{-1} above the bottom of the polyad. At ~800 cm^{-1} above the bottom of the polyad, two regular regions have emerged. The one centered at $\psi_b = \pm\pi$ corresponds to local bender like motions where the left and right benders oscillate in two perpendicular planes. The regular region centered at $\psi_b = 0$ corresponds to one bender engaged in a purely circular "rotator" motion and the other attempting to bend in a plane, but forced to rotate in order to maintain $K_b = 0$ $(l = 0)$.

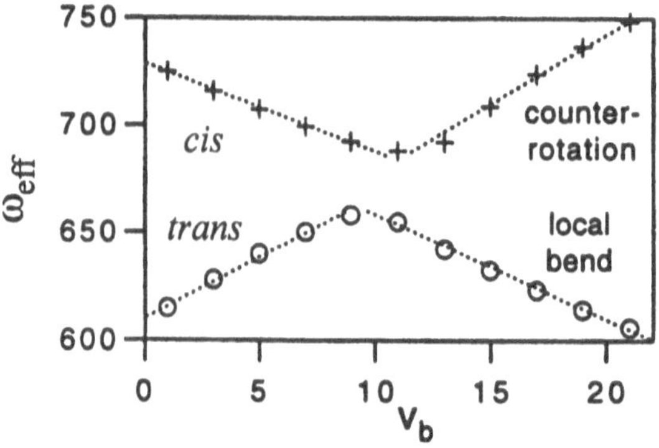

Figure 9.20: The *cis*-bend and *trans*-bend normal mode frequencies tune into resonance at $N_b \approx 10$. Similar to an inhomogeneous L-uncoupling perturbation, de-mixing after the level crossing cannot occur, and the qualitative character of the vibrational modes changes irreversibly from normal to local (from Jacobson and Field, 2000b).

Figure 9.20 reveals the mechanistic reason for the profound changes in the intramolecular dynamics in acetylene that occur near $N_b = 10$ $(K_a = 3)$. The opposite sign anharmonicities of the *cis*- and *trans*-bending vibrations cause the effective vibrational frequencies for the *cis* and *trans* normal mode benders, $\omega + 2vx$, to become equal near $v_{\text{bend}} = 10$,

$$\omega_4 + 20x_{44} \approx \omega_5 + 20x_{55}$$

[$\omega_4 = 609$ cm^{-1}, $x_{44} = 3.5$ cm^{-1}, $\omega_5 = 729$ cm^{-1}, $x_{55} = -2.4$ cm^{-1}]. This tuning of two normal modes into resonance causes a profound change in the spectrum and dynamics. It is analogous to an inhomogenous **L**–uncoupling perturbation (see Sections 3.5.3 and 5.1.5). The coupling matrix element is approximately proportional to v_4v_5. This v–dependence of the coupling matrix element, analogous to the J–dependence of an **L**–uncoupling matrix element, prevents the two basis states from demixing by detuning after the level crossing. As a result, the *trans*-bend, initially at the bottom of the polyad, becomes the counter-rotator at the top of the polyad and the *cis*-bend, initially at the top of the polyad, becomes the local bender at the bottom of the polyad.

9.4.14 Polyatomic Molecule Dynamics

When a bright basis-state is embedded in a dense manifold (quasi-continuum) of dark basis-states, a variety of dynamical processes ensue (Bixon and Jortner, 1968; Rhodes, 1983). These include Intramolecular Vibrational Redistribution (IVR), Inter-System Crossing (ISC), and Internal Conversion (IC). At $t = 0$, the bright basis state, which is not an eigenstate of $\mathbf{H}$, is prepared, $\Psi(0) = \psi^{(0)}_{\text{BRIGHT}}$.

The time-evolution of the prepared state is computed by expanding $\psi^{(0)}_{\text{BRIGHT}}$ in a complete basis set of eigenstates,

$$\Psi(t) = \sum_j a_j \psi_j e^{-iE_jt/\hbar}$$

where

$$a_j = \left\langle \psi_j | \psi^{(0)}_{\text{BRIGHT}} \right\rangle .$$

The early time evolution of $\Psi(t)$ often is understandable in terms of a causal mechanism: where does the system start out, where does it go next, how fast, and why? However, the long-time behavior of $\Psi(t)$ typically is understandable only in a statistical sense: irreversible decay into a dense manifold of essentially indistinguishable states, describable only in a Fermi Golden Rule sense in terms of $\langle \text{H}^2_{\text{BD}} \rangle$ and ρ_D where $\langle \text{H}^2_{\text{BD}} \rangle$ is the average squared coupling matrix element between the bright state and the isoenergetic dark states and ρ_D is the density of dark states.

It is the early time dynamics, while the system remains highly localized in state space, the specific coupled states are known *a priori*, and the coupling matrix elements are experimentally measurable or theoretically predictable, for which our study of diatomic molecule spectra and dynamics prepares us. At early time, the system evolves in a reduced dimensionality. The excitation initially is localized spatially on a part of the molecule (a chromophore) and restricted by approximate constants of motion to a tiny fraction of energetically-accessible state space.

Typically, although strong couplings (large off-diagonal matrix elements) may exist between many sub-systems within a molecule, the couplings that are important to the early time dynamics are those that are *near resonant*. The *resonant* coupling matrix elements are those that are of magnitude comparable to the zero-order energy separations between coupled basis states. This is the criterion for the inapplicability of ordinary nondegenerate perturbation theory. Energy resonance (accidental or systematic) picks out those coupling terms that are important in the early time dynamics, with the result that sometimes small coupling terms are dynamically more important than much larger ones.

9.4.15 Inter-System Crossing, Internal Conversion, and Intramolecular Vibrational Distribution

Inter-System Crossing (ISC) is a dynamical process that arises from spin-orbit coupling between electronic states belonging to different electron spin multiplicities. Typically, a vibrational level of an excited singlet state is spin-orbit coupled to the dense manifold of vibrational levels of a lower-lying triplet state, or a triplet vibrational level is coupled to the dense manifold of vibrational levels of a lower-lying singlet state (usually the electronic ground state).

Spin-orbit perturbations are well understood in diatomic molecules. A singlet$\sim$triplet mixed eigenstate can act as a spectroscopic doorway between the singlet and triplet manifolds. Perturbation Facilitated Optical-Optical Double Resonance Spectroscopy (Li Li and Field, 1983) exploits such spin-mixed doorway levels to gain systematic spectroscopic access to a nominally spin-forbidden manifold of states. Such a spin-mixed state can act to transfer spontaneous fluorescence intensity from a short wavelength $S_1 \rightarrow S_0$ transition to the longer wavelength of an $S_1 \sim T_n \rightarrow T_1$ transition. The strength of an $S \sim T$ spin-orbit matrix element is determined by the product of a usually estimable electronic factor (using the methods of Section 5.3 and the atomic coupling strengths in Table 5.6) and an accurately calculable vibrational overlap factor.

For polyatomic molecules, the vibrational overlap integrals are not so readily calculable, but the Franck-Condon (and stationary phase) principle provides a basis for systematic predictions. For example, the "energy gap" rule applies to two potential surfaces of similar shape. A large energy gap leads to vanishingly small Franck-Condon overlap factors and negligible (i.e., very slow) ISC. However, when the S and T potential surfaces have quite different shapes, certain normal modes will be "Franck-Condon active" in the sense that a few special dark states with vibrational quantum numbers in the active modes, those very different from those of the bright state, will have uniquely large $\langle \mathbf{v}_S | \mathbf{v}_T \rangle$ vibrational overlap integrals. These Franck-Condon active modes promote ISC and generate a sparse manifold of "doorway states" that are uniquely effective in promoting ISC. The rates of Doorway-Mediated ISC are highly state- and mode-specific. This is an example of a "lumpy quasicontinuum." Each lump in the quasicontinuum is a clump of eigenstates among which the character of an assignable bright state is distributed. The effects of perturbations between two bright states on the energy centroid, width, and integrated intensity of the two associated lumps may be treated using the complex Hamiltonian formalism presented in Section 9.3. The study of spin-orbit perturbations in diatomic molecules is excellent preparation for understanding ISC in polyatomic molecules.

Internal Conversion (IC) is a spin-conserving dynamical process. Although for diatomic molecules, spin-orbit, **L**–uncoupling, and **S**–uncoupling interactions can have nonzero $\Delta S = 0$ electronically off-diagonal matrix elements, it is the electrostatic (diabatic picture) or nuclear kinetic energy (adiabatic picture) interaction mechanism that is most germane for building an understanding of IC in polyatomic molecules.

The electronic selection rule for nonzero electrostatic or nuclear kinetic energy coupling matrix elements is quite simple for diatomic molecules: only states with identical electronic quantum numbers can perturb each other. For polyatomic molecules, the situation is not always so simple. It is possible that two electronic states will have different electronic quantum numbers in a high-symmetry nuclear geometry, but identical electronic quantum numbers in regions of lower geometric symmetry. The result is a *conical intersection* between potential surfaces. The potential surfaces cross without interacting in the high symmetry region, but in the lower-symmetry regions that surround the surface intersection, strong couplings between the electronic states can occur. As a result, the vibrational levels of the two electronic states often are strongly mixed near the conical intersection. The most strongly mixed vibrational states are those with significant amplitude near the nuclear geometry of the conical intersection.

As is ISC, IC is very slow for electronic states with similarly shaped potential energy surfaces. When the potential surfaces have very different shapes, there will be a small number of vibrational doorway states that are especially effective in coupling to the bright state. Conical intersections are a special class of potential surfaces of very different shapes. But even when potential surfaces have very different shapes, many normal coordinate displacements and the associated vibrational normal modes will have nearly identical forms on both surfaces. These normal modes are Franck-Condon inactive and do not contribute to IC. The normal coordinate displacements that express the differences in shapes of the potential surfaces are embodied in vibrational normal modes that are Franck-Condon active. These modes are called promoting modes because, when such a mode on one potential surface is plucked from an eigenstate on the other surface, intramolecular dynamics is "promoted" or initiated.

A crucial issue that determines whether the intramolecular dynamics is direct (statistical distribution of coupling strengths to all of the dark states) or doorway mediated (cause-and-effect mechanisms) is whether the coupling of the doorway state to the dense manifold of dark states is sufficiently strong that the Fermi-Golden-Rule broadened doorway states overlap. When that occurs, the lumpy continuum becomes a true continuum. There is, however, another important factor. When a doorway state couples to a dense manifold of dark states, its (quasi)-homogeneous width gets large. Strong and weak coupling limits in the interaction between decaying states (represented by complex zero-order energies) are discussed in Section 9.3. When the difference in the widths of two interacting states is larger than the coupling matrix element, the weak-coupling limit applies. A bifurcation in the bright$\sim$dark coupling occurs. In the weak-coupling limit the sharp bright state becomes nearly decoupled from the quasicontinuum of dark states. Not only do the lumps in the quasi-continuum disappear, but the coupling of the sharp state with the continuum essentially turns off (Tric, 1973; Rhodes, 1983).

Intramolecular Vibrational Redistribution (IVR) is fundamentally different from ISC and IC because IVR rates are not governed by Franck-Condon vibrational overlap factors. Since all vibrational eigenstates of the same $V(\mathbf{Q})$ are by definition orthogonal, all off-diagonal vibrational overlap integrals are zero. The most important IVR processes are those that are near resonant (see Section 9.4.5). IVR is inherently nonstatistical. The bright state is strongly anharmonically coupled to a very sparse manifold of first-tier doorway states. These doorway states are anharmonically coupled to a denser manifold of second-tier doorway states. IVR often proceeds through a sequence of tiers (Sibert, *et al.*, 1982a and b). However, the transition between IVR proceeding via an identifiable sequence of causal mechanisms and a purely statistical process is based on the bifurcation that occurs when a doorway state interacts so strongly with an underlying quasi-continuum that the bright state∼doorway state coupling switches from the strong to the weak coupling limit (Section 9.3).

9.5 References

Abramson, E., Kittrell, C., Kinsey, J. L., and Field, R. W. (1982), *J. Chem. Phys.* **76**, 2293.

Ahn, J., Weinacht, T. C., and Bucksbaum, P. H. (2000), *Science* **287**, 463.

Albrecht, A. W., Hybl, J. D., Gallagher-Faeder, S. M., and Jonas, D. M. (1999), *J. Chem. Phys.* **111**, 10934.

Aleksandrov, E. B. (1964), *Opt. Spec.* **17**, 522.

Altunata, S., and Field, R. W. (2000), *J. Chem. Phys.* **113**, 6640.

Altunata, S., and Field, R. W. (2001), *J. Chem. Phys.* **114**, 6557.

Altunata, S. N., Cao, J., and Field, R. W. (2002), *Phys. Rev. A* **65**, 3415.

Amat, G., Nielsen, H. H., and Tarago, G. (1971), "Rotation-Vibration Spectra of Molecules" (Marcel Dekker, New York).

Avouris, Ph., Gelbart, W. M., and El-Sayed, M. A. (1977), *Chem. Revs.* **77**, 793.

Ballard, J. B., Stauffer, H. U., Amitay, Z., and Leone, S. R. (2002), *J. Chem. Phys.* **116**, 1350.

Bardeen, C. J., Yakovlev, V. V., Wilson, K. R., Carpenter, S. D., Weber, P. M., and Warren, W. S. (1997), *Chem. Phys. Lett.* **280**, 151.

Beil, A., Luckhaus, D., and Quack, M. (1997), *Ber. der Bunsenges. Phys. Chem.* **101**, 311.

Bixon, M., and Jortner, J. (1968), *J. Chem. Phys.* **48**, 715.

Bowman, R. M., Dantus, M., and Zewail, A. H. (1989), *Chem. Phys. Lett.* **161**, 297.

Cao, J., and Wilson, K. R. (1997), *J. Chem. Phys.* **107**, 1441.

Cao, J., Che, J., and Wilson, K. R. (1998), *J. Phys. Chem.* **102**, 4248.

Cao, J., Bardeen, C. J., and Wilson, K. R. (2000), *J. Chem. Phys.* **113**, 1898.

Child, M. S. (1991), "Semiclassical Mechanics with Molecular Applications," Chapter 7, Clarendon Press, Oxford.

Child, M. S., Jacobson, M. P., and Cooper, C. D. (2001), *J. Phys. Chem. A* **105**, 10,791.

Clark, A. P., Dickinson, A. S., and Richards, D. (1977), *Adv. Chem. Phys.* **36**, 63.

Cohen-Tannoudji, C., Diu, B., and Laloë, F. (1977), "Quantum Mechanics", John Wiley and Sons, New York.

Coté, M. J., Kauffman, J. F., Smith, P. G., and McDonald, J. D. (1989), *J. Chem. Phys.* **90**, 2865.

Darling, B. T., and Dennison, D. M. (1940), *Phys. Rev.* **57**, 128.

Dodd, J. N., Kaul, R. D., and Warrington, D. M. (1964), *Proc. Phys. Soc.* **84**, 176.

Ducas, T. W., Littman, M. G., and Zimmerman, M. L. (1975), *Phys. Rev. Lett.* **35**, 1752.

Dupré, P., Jost, R., Lombardi, M., Green, P. G., Abramson, E., and Field, R. W. (1991), *Chem. Phys.* **152**, 293.

Fayer, M. D. (2001), "Elements of Quantum Mechanics," Oxford University Press, New York.

Felker, P. M., and Zewail, A. H. (1983), *Chem. Phys. Lett.* **102**, 113.
Felker, P. M., and Zewail, A. H. (1984), *Phys. Rev. Lett.* **53**, 501.
Felker, P. M., and Zewail, A. H. (1987), *J. Chem. Phys.* **86**, 2460.
Felker, P. M., and Zewail, A. H. (1988), *Adv. Chem. Phys.* **70**, 265.
Felker, P. M. (1992), *J. Phys. Chem.* **96**, 7844.
Felker, P. M., and Zewail, A. H. (1995), p. 193 in "Femtosecond Chemistry" [eds. J. Manz and L. Wöste), VCH, Weinheim.
Feurer, T., Glass, A., Rozgonyi, T., Sauerbrey, R., and Szabo, G. (2001), *Chem. Phys.* **267**, 223.
Fried, L. E., and Ezra, G. S. (1987), *J. Chem. Phys.* **86**, 6270.
Georges, A. T., and Lambropoulos, P. (1978), *Phys. Rev. A*, **18**, 1072.
Germann, T., and Miller, W. H. (1998), *J. Chem. Phys.* **109**, 94.
Greene, C. H., and Jungen, C. (1985), *Adv. At. Mol. Phys.* **21**, 51.
Hack, E., and Huber, J. R. (1991), *Int. Rev. Phys. Chem.* **10**, 287.
Haroche, S., Paisner, J. A., and Schawlow, A. L. (1973), *Phys. Rev. Lett.* **30**, 948.
Haroche, S., (1976), Chapter 7, in "Topics in Applied Physics, Vol. 13, High Resolution Spectroscopy" (ed. K. Shimoda), Springer, Berlin.
Hartland, G. V., Connell, L. L., and Felker, P. M. (1991), *J. Chem. Phys.* **94**, 7649.
Heisenberg, W. (1925), *Z. Phys.* **33**, 879 (translated in Souces of Quantum Mechanics, ed. R. L. van der Waerden, Dover, New York, 1967).
Heller, E. J. (1975), *J. Chem. Phys.* **62**, 1544.
Heller, E. J. (1978), *J. Chem. Phys.* **68**, 3891.
Heller, E. J., and Gelbart, W. M. (1980), *J. Chem. Phys.* **73**, 626.
Heller, E. J. (1981), *Acc. Chem. Res.* **14**, 368.
Heller, E. J., Sundberg, R. L., and Tannor, D. (1982), *J. Phys. Chem.* **86**, 1822.
Heller, E. J. (1990), in "Ultrafast Phenomena VII" (eds. C. B. Harris, E. P. Ippen, G. A. Mourou, and A. H. Zewail), Springer-Verlag, Berlin.
Herzberg, G. (1950), "Spectra of Diatomic Molecules," van Nostrand-Reinhold, Princeton, New Jersey, reprinted in 1989 by Krieger, Malabar.
Hill, T. L. (1986), "An Introduction to Statistical Thermodynamics," Dover, New York.
Imre, D. G., Kinsey, J. L., Field, R. W., and Katayama, D. H. (1982), *J. Phys. Chem.* **86**, 2564.
Imre, D., Kinsey, J. L., Sinha, A., and Krenos, J. (1984), *J. Phys. Chem.* **88**, 3956.
Jacobson, M. P., O'Brien, J. P., Silbey, R. J., and Field, R. W. (1998), *J. Chem. Phys.* **109**, 121.
Jacobson, M. P., Jung, C., Taylor, H. S., and Field, R. W. (1999), *J. Chem. Phys.* **111**, 600.
Jacobson, M. P., and Field, R. W. (2000a), *Chem. Phys. Lett.* **320**, 553.
Jacobson, M. P., and Field, R. W. (2000b), *J. Phys. Chem.* **104**, 3073.
Jaffé, C., and Brumer, P. (1980), *J. Chem. Phys.* **73**, 5646.
Johnson, B. R., Kittrell, C., Kelley, P. B., and Kinsey, J. L. (1996), *J. Phys. Chem.* **100**, 7743.
Joireman, P. W., Connell, L. L., Ohline, S. M., and Felker, P. M. (1992), *J. Chem. Phys.* **96**, 4118.
Jonas, D. M., and Fleming, G. R., (1995), in "Ultrafast Processes in Chemistry and Photobiology" (eds. M. A. El-Sayed, I. Tanaka, and Y. Molin), Blackwell, Oxford.
Jones, R. R., Schumacher, D. W., Gallagher, T. F., and Bucksbaum, P. H. (1995), *J. Phys. B: At. Mol. Opt. Phys.* **28**, L405.
Judson, R. S., and Rabitz, H. (1992), *Phys. Rev. Lett.* **68**, 1500.
Kawashima, H., Wefers, M. M., Nelson, K. A. (1995), *Ann. Rev. Phys. Chem.* **46**, 627.

Kellman, M. E. (1985), *J. Chem. Phys.* **83**, 3843.
Kellman, M. E., and Xiao, L. (1989), *J. Chem. Phys.* **90**, 6086.
Kellman, M. E. (1990), *J. Chem. Phys.* **93**, 6630.
Keske, J. C., and Pate, B. H. (2000), *Ann. Rev. Phys. Chem.* **51**, 323.
Kittrell, C., Abramson, E., Kinsey, J. L., McDonald, S., Reisner, D. E., Katayama, D., and Field, R. W. (1981), *J. Chem. Phys.* **75**, 2056.
Krause, J. L., Schafer, K. J., Ben-Nun, M., and Wilson, K. R. (1997), *Phys. Rev. Lett.* **79**, 4978.
Labastie, P., Bordas, C., Tribollet, B., and Broyer, M. (1984), *Phys. Rev. Lett.* **52**, 1681.
Lange, W., and Mlynek, J. (1978), *Phys. Rev. Lett.* **40**, 1373.
Lee, S.-Y., and Heller, E. J. (1979), *J. Chem. Phys.* **71**, 4777.
Lefebvre, R. (1973), "Selected Topics in Physics, Astrophysics, and Biophysics," Proc. XIVth Latin American School of Physics, (E. Abecassis de Laredo and N. K. Jurisc, editors), p. 87, "Radiative and Non-radiative Processes in Molecules," D. Reidel, Dordrecht.
Lehmann, K. K. (1983), *J. Chem. Phys.* **79**, 1098.
Lehmann, K. K., Pate, B. H., and Scoles, G. (1994), *Ann. Rev. Phys. Chem.* **45**, 241.
Levenson, M. D., and Kano, S. S. (1988), "Introduction to Nonlinear Laser Spectroscopy," Academic Press, San Diego.
Levis, R. J., Menkir, G. M., and Rabitz, H. (2001), *Science* **292**, 709.
Levis, R. J., and Rabitz, H. A. (2002), *J. Phys. Chem.* **106**, 6427.
Li Li, and Field, R. W. (1983), *J. Phys. Chem.* **87**, 3020.
Lombardi, M., Labastie, P., Bordas, M. C., and Broyer, M. (1988), *J. Chem. Phys.* **89**, 3479.
Mehta, A., Stuchebrukhov, A. A., and Marcus, R. A. (1995), *J. Phys. Chem.* **99**, 2677.
Miller, W. H., and George, T. F. (1972), *J. Chem. Phys.* **56**, 5637 and 5668 and **57**, 5019.
Mills, I. M. (1972), p. 115, "Molecular Spectroscopy: Modern Research," Vol. 1 (eds. K. N. Rao and C. W. Mathews), Academic Press, San Diego.
Mills, I. M., and Robiette, A. G. (1985), *Mol. Phys.* **56**, 743.
Moiseyev, N. (1998), *Phys. Reports* **302**, 211.
Morse, P. M., and Feshbach, H. (1953), "Methods of Theoretical Physics," McGraw-Hill, New York.
Mukamel, S. (1995), "Principles in Nonlinear Optical Spectroscopy," Oxford University Press, New York.
Nesbitt, D. J., and Field, R. W. (1996), *J. Phys. Chem.* **100**, 12735.
Nielsen, H. H. (1951), *Rev. Mod. Phys.* **23**, 90.
Parmenter, C. S. (1983), *Faraday Disc. Chem. Soc.* **75**, 7.
Quack, M. (1990), *Ann. Rev. Phys. Chem.* **41**, 839.
Quack, M., and Stohner, J. (1993), *J. Phys. Chem.* **97**, 12,574.
Rabitz, H., de Vivie-Riedle, R., Motzkus, M., and Kompa, K. (2000), *Science* **288**, 824.
Ramsey, N. F. (1990), *Rev. Mod. Phys.* **62**, 541.
Rankin, C. C., and Miller, W. H. (1971), *J. Chem. Phys.* **35**, 3150.
Rice, S. A., and Zhao, M. (2000), "Optical Control of Molecular Dynamics", John Wiley and Sons, New York.
Rhodes, W. (1983), *J. Phys. Chem.* **87**, 30.
Salour, M. M., and Cohen-Tannoudji, C. (1977), *Phys. Rev. Lett.* **38**, 757.
Scherer, N. F., Ruggiero, A. J., Du, M., and Fleming, G. R. (1990), *J. Chem. Phys.* **93**, 856.
Scherer, N. F., Carlson, R. J., Matro, A., Du, M., Ruggiero, A. J., Romero-Rochin, V.,

Cina, J. A., Fleming, G. R., and Rice, S. A. (1991), *J. Chem. Phys.* **95**, 1487.
Scherer, N. F., Matro, A., Ziegler, L. D., Du, M., Carlson, R. J., Cina, J. A., and Fleming, G. R. (1992), *J. Chem. Phys.* **96**, 4180.
Schinke, R. (1993), "Photodissociation Dynamics," Cambridge University Press, Cambridge.
Sibert, E. L., III, Reinhardt, W. P., and Hynes, J. T. (1982a), *Chem. Phys. Lett.* **92**, 455.
Sibert, E. L., III, Hynes, J. T., and Reinhardt, W. P. (1982b), *J. Chem. Phys.* **77**, 3595.
Sibert, E. L. (1988), *J. Chem. Phys.* **88**, 4378.
Sibert, E. L. (1990), *Int. Rev. Phys. Chem.* **9**, 1.
Smith, R. A. L., Verlet, J. R. R., and Fielding, H. H. (2003b), *Phys. Chem. Chem. Phys.* **5**, 3567.
Smith, R. A. L., Stavros, V. G., Verlet, J. R. R., Fielding, H. H., Townsend, D., and Softley, T. P. (2003a), *J. Chem. Phys.* **119**, 3085.
Stauffer, H. U., Ballard, J. B., Amitay, Z., and Leone, S. R. (2002), *J. Chem. Phys.* **116**, 946.
Stine, J. R., and Marcus, R. A. (1972), *Chem. Phys. Lett.* **15**, 536.
Stuchebrukhov, A. A., and Marcus, R. A. (1993), *J. Chem. Phys.* **98**, 6044.
Stuchebrukhov, A. A., Mehta, A., and Marcus, R. A. (1993), *J. Phys. Chem.* **97**, 12,491.
Stuchebrukhov, A. A., and Marcus, R. A. (1994), *J. Phys. Chem.* **98**, 3285.
Tannor, D. J. (2003), "Introduction to Quantum Mechanics: A Time Dependent Perspective," University Science Books, Sausalito.
Tellinghuisen, J. T. (1984), *J. Mol. Spectrosc.* **103**, 455.
Texier, F., and Jungen, C. (1998), *Phys. Rev. Lett.* **81**, 4329.
Texier, F., and Jungen, C. (1999), *Phys. Rev. A* **59**, 412.
Tric, C. (1973), *Chem. Phys. Lett.* **21**, 83.
Tull, J. X., Dugan, M. A., and Warren, W. S. (1997), *Adv. Magn. Opt. Reson.* **20**, 1.
Vaccaro, P. H., Zabludoff, A., Carrera-Patiño, M. E., Kinsey, J. L., and Field, R. W. (1989), *J. Chem. Phys.* **90**, 4150.
Wallenstein, R., Paisner, J. A., and Schawlow, A. L. (1974), *Phys. Rev. Lett.* **32**, 1333.
Weinacht, T. C., Ahn, J., and Bucksbaum, P. H. (1999), *Nature* **397**, 233.
Weinacht, T. C., Bartels, R., Backus, S., Bucksbaum, P. H., Pearson, B., Geremia, J. M., Rabitz, H., Kapteyn, H. C., and Murnane, M. M. (2001), *Chem. Phys. Lett.* **344**, 333.
Weiner, A. M. (1995), *Prog. Quant. Electr.* **19**, 161.
Wilson, E. B., Jr., Decius, J. C., and Cross, P. C. (1955), "Molecular Vibrations. The Theory of Infrared and Raman Vibrational Spectra", McGraw-Hill, New York.
Wong, V., and Gruebele, M. (1999), *J. Phys. Chem.* **103**, 10083.
Xiao, L., and Kellman, M. E. (1989), *J. Chem. Phys.* **90**, 6086.
Zewail, A. H. (1996), Nobel Symposium on Femtochemistry, (ed. V. Sundström), Imperial College Press, London.
Zewail, A. H. (2000), *Angewandte Chemie, Internatl. Ed.* **39**, 2586.
Zygan-Maus, R., and Wolter, H. H. (1978), *Phys. Lett. A* **64**, 351.

Index

A_D or A_J, 337
a_+ and l-uncoupling matrix element, 267
a_+ spin-orbit matrix element, 267
a_λ
  quantum defect, 128
$\mathbf{a}^\dagger, \mathbf{a}$ creation, annihilation operators, 708
Above Threshold Dissociation (ATD), 180
absorption cross-section, 522
absorption spectrum, 27, 627, 635
accidental perturbation
  interloper in Rydberg series, 685
accidental predissociation, 671, 674
acetylene
  chaos, 731
  local↔normal benders, 727
  phase space structure, 730
    surface of section, 730, 731
  polyads, 730
  quasiperiodic trajectory, 731
  resonance between
    cis- and *trans*-bend, 732
adiabatic curves, 94, 168, 173
adiabaticity parameter, 177
  in H_2, 171
  in N_2, 171
  in N_2^+, 171
  in O_2, 171
  in SiO, 171
Airy function, 281
algebraic solutions
  secular equations, 243, 247
alignment of the photofragment, 491
alkaline earth monohalides, 332
angular momentum, 61, 72
  electron spin, $\vec{\mathbf{S}}$, 72
  electronic orbital, $\vec{\mathbf{L}}$, 72
  ion-core, $\vec{\mathbf{N}}^+$, 72
  matrix elements, 73
  nuclear framework, $\vec{\mathbf{R}}$, 72, 79
  operators, 72
  projection, Ω, 73
  total electronic, $\vec{\mathbf{J}}_a$, 72, 79
  total exclusive of spin, $\vec{\mathbf{N}}$, 72
  total, $\vec{\mathbf{J}}$, 70, 72, 96
  transfer, 596
anharmonic coupling
  Darling-Dennison, 698, 714
  Fermi resonance, 689
  harmonic oscillator scaling, 689
  matrix elements, 688, 691
anharmonic resonances, 698
anions, 39, 553
anticrossings, 681
ArH
  f-complex in, 402
AsN
  spin-orbit constant in, 321
Aston mass units, 285
asymmetry parameters, 487, 602
  HBr, 491
  HI, 491
atomic spin-spin parameter, 324
attosecond (as), 498
autocorrelation function, 624, 630, 633, 634, 636, 637
  envelope of $I(\omega)$, 628
  Fourier Transform of, 626, 627
  frequency domain spectrum, 626
  photon pluck, 627
  recurrences, 628
  spectrum, $I(\omega)$, 628

autoionization, 671, 674
cross-section, 588
electrostatic, 586
in H_2, 565, 572, 573, 578, 598
in HBr, 603
in HF, 580
in HI, 578, 584
level shifts, 588
in Li_2, 565, 580
in N_2, 567, 573, 578, 580, 587, 588
in NO, 578, 580
in NO^+, 573
in O_2^+, 581
in O_2, 314, 598
rotational, 572
spin-orbit, 581
vibrational, 576
autoionization and predissociation
competition between, 604
in H_2, 604, 607
in HCl, 604
in NO, 609
axial recoil approximation, 490
axis systems, 72, 83

b, orbital perturbation matrix element, 327
l-uncoupling matrix element, 267
BaF
Rydberg supercomplex in, 127
ball-and-spring motions, 627, 633
band intensity anomalies, 380
bandhead
BaO, 9
BaO
bandhead, 9
electronically inelastic collision in, 447
intensity anomalies in, 35, 395
MODR in, 438
Optical Optical Double Resonance (OODR) in, 416
perturbations in, 296
Stark quantum-beats (SQB) in, 433, 434
BaS
perturbation parameters, 305
perturbations in $A^1\Sigma^+$ state, 302
potential energy curve, 303
basis sets
choice of, 683
coupled, 105, 106
normal↔local mode, 710, 712
transformation
$\mathbf{a}^\dagger, \mathbf{a}$, 710
uncoupled, 105, 106
unitary transformation, 106
Beer-Lambert law, 354
BeF, 332
BeH
electrostatic parameter in, 313
predissociation in, 501
Berry's propensity rule, 580
Beutler-Fano formula, 588
Beutler-Fano lineshape, 386
bifurcations, 729
biorthogonal, 673, 674, 676
Bohr magneton, 429
bond-softening, 179
Born-Oppenheimer approximation, 66, 113
clamped nuclei, 128
potential curve, 113
principal quantum number, 128
terms neglected, 92
Born-Oppenheimer product function, 90
Br_2
predissociation in, 522, 544
branching diagram, 150
branching ratios, 485
bright state, 624, 626, 638, 643, 646, 658, 671, 681, 733, 734
bulk polarization, 643

C_2
perturbations between $X^1\Sigma_g^+$ and $b^3\Sigma_g^-$ states, 340
CaO
electronically inelastic collision in, 450

perturbations in, 296
cause of dynamics
intrapolyad, 694
Cavity Ringdown Spectroscopy (CRDS), 28
center of gravity rule, 246
center of the wavepacket, 645
centrifugal distortion terms, 242
centrifugal distortion, D_e
Kratzer, 286
CH
collision induced transition in, 453
Λ-doubling in, 227
potential energy curves, 518
predissociation in, 497, 499, 517
Rydbergization in, 311
CH^+
forbidden transition in, 414
CH_2
collision effect in, 454
chirped pulse, 661
negatively, 661
circular dichroism (CDAD), 600
Cl_2
potential energy curves, 480
classical chaos, 717
onset, 729
classical dynamics, 721
classical mechanics, 703
actions, conserved, 728
angles, conserved, 728
coupled identical oscillators, 703
Poisson bracket, 718
polyad $\mathcal{H}^{\text{eff}}$ model, 716, 728
Clebsch-Gordan coefficients
case (b)$\rightarrow$(d), 120
CN
anticrossing or level crossing, 418, 419, 428
collision induced transition in, 445, 449, 452
perturbed radiative lifetime in, 372
relaxation in matrix, 459
Stark quantum-beat (SQB) in, 436
Zeeman effect in, 422, 428
CO
accidental predissociation, 542
anticrossing in Zeeman effect in, 418
atomic populations in, 322
$B^1\Sigma^+-X^1\Sigma^+$, 64
Cameron system, 340
collision induced transition in, 449
electronically inelastic collision in, 447
electrostatic parameter in, 313
forbidden transition in, 409
isotope effect in, 528
nonexponential decay in, 370
perturbations in, 216, 296, 341
perturbed radiative lifetime in, 371, 373
photoionization spectrum of, 599
potential energy curves of, 217
predissociation in, 266, 501, 542
shape resonances in, 561
spin interactions in, 323
spin-orbit constant in, 321
spin-orbit perturbation parameter, 325, 327
transfer rate in, 452
CO^+
collision induced transitions in, 449
CO_2
Rydberg states in Hund's case (e), 125
three-photon transition, 368
coherence
creation, 639
destruction, 639, 641, 652
detection, 639
coherent control
in HI, 613
on Na_2, 609
photofragmentation product branching ratios, 609

coherent superposition state, 430, 635, 659, 679, 680
motion, 635
coincidence experiments, 39, 46
collision effect
doorway or gateway, 445
electronic transition, 447
in CH_2, 454
in CN, 445
in N_2^+, 445
lifetime, 445
collision induced transition
in CH, 453
in CN, 445, 449, 452
in CO, 449
in CO^+, 449
in I_2, 450
in Li_2, 450
in N_2^+, 450, 452
in N_2, 448, 449
in Na_2, 449
in NH, 448
in SiF, 450
propensity rule, 454
collisions between molecules
$CaO+N_2O$, 452
combination differences, 12, 25
complementary states, 312
complete experiment, 600
complex energy formalism, 656, 672
complex $\mathbf{H}^{\text{eff}}$ formalism, 639, 672, 680, 683
biorthogonal, 673, 674
bright state, 680, 681
completeness, 673, 674
Fermi Golden Rule, 683
orthonormal, 674
strong coupling limit, 680
weak coupling limit, 680
WKB barrier penetration, 683
Condon and Shortley phase convention, 74, 143
matrix elements, 142
$\boldsymbol{\sigma}_v$ on $|\lambda\rangle$, 142
Condon reflection approximation, 480
configuration
interaction, 340, 530
configuration (electronic), 149
π^2, 160, 188, 197
$\pi^3\pi'$, 201
$\sigma\pi^3$, 198
configuration space
trajectories
relationship to probability density, 726
conical intersection, 735
contact transformation, 709
continuum function, 505
energy-normalized, 506
continuum$\sim$continuum interaction, 538, 601
control
chirped pulse, 655, 656
crafted pulse, 655
learning algorithm, 656
target state, 698
Cooper minima, 561
in HI, 563
in NH, 564
in OH, 564
coordinate and momentum operators
dimensionless $\widehat{\mathbf{Q}}$ and $\widehat{\mathbf{P}}$, 690
equation of motion
of $\mathbf{Q}$ and $\mathbf{P}$, 700
coordinate system
electronic, 192
inversion, $\mathbf{I}$ or $\mathbf{E}^*$, 139
laboratory fixed, 139
nuclear, 83
Coriolis perturbations, 98
correlation diagram
between case (a) and case (c), 189
correlation effect of an electric field, 336
correlation matrix, 250
correlations between parameters, 255
Correspondence Principle, 717, 718
$\mathbf{H}^{\text{eff}} \leftrightarrow \mathcal{H}^{\text{eff}}$, 716
Heisenberg, 716
Coulomb Explosion, 41
Coulomb function, 590

Coulomb integrals, 155
coupled equations, 264, 535, 537
  in H_2^+, 266
  in N_2, 265
coupling
  (J_1, J_2), 473
  kinetic energy, 705, 714
  potential energy, 705, 714
creation, annihilation operators
  2-D isotropic oscillator, 727
  $\mathbf{a}^\dagger, \mathbf{a}$, 690, 708
  $\mathbf{a}_d^\dagger, \mathbf{a}_d, \mathbf{a}_g^\dagger, \mathbf{a}_g$, 727
  Heisenberg quantum→classical, 727
  number operators, 727
CRIES (Cross-Correlation Ionization Energy Spectroscopy), 19
CrN
  spin-orbit constant, 322
cross section
  in absorption, 351, 356
  in autoionization, 588
  in dissociation, 522
CS
  forbidden transitions in, 409, 412
  MODR in, 437
  MOMRIE in, 437
  spin-orbit constant in, 321
  spin-orbit perturbation matrix element in, 327
CS^+, 244
  Λ-doubling in $A^2\Pi_i$ state of, 215
  spin-orbit perturbation matrix element in, 327
Cs_2
  predissociation, 521
  q reversal in, 525
CS_2
  Rydberg states in Hund's case (e), 125
CSe
  spin-orbit constant in, 321
CTe
  spin-orbit constant in, 321
CuO
  $^2\Sigma - X^2\Pi$ transition in, 404
D_e and D_v, rotational constant, 242, 286
$^2\Delta$ states, 183
dark state, 624, 638, 643, 646, 658, 671, 681, 733
Debye, 350, 479
decay
  irreversible, 624
δ–function
  integral definition, 629
$^3\Delta$ states
  spin-orbit constant of $\pi^3\pi'$ configuration, 202, 321
density matrix, ρ, 635, 647, 667
  coherences, ρ_{ij}, 639, 647
  detection, **D**, 639
  equations of motion, 647
  evolution, **H**, 639
  excitation, **E**, 639
  populations, ρ_{ii}, 639, 647
deperturbation, 684
  local perturber, 684
  NO $^2\Pi$, 262
  remote perturber, 684
deperturbation at high pressure and in matrices, 455
dephasing, 624
detection operator, **D**, 643
diabatic curves, 93, 163
differential oscillator strength, 506
  in O_2, 507
dimensionality reduction, 686, 717
dimensionless $\widehat{\mathbf{Q}}$ and $\widehat{\mathbf{P}}$, 690
dipole moment, 214
Dirac delta function, 627
direct approach, 259
direct dissociation, 471
direction cosine
  matrix, 82
  matrix elements, 75, 391, 667
discrete variable representation (DVR), 291
  method, 266
Dispersed Fluorescence (DF), 25, 34, 635
  I_2, 47

dissociation
in HBr^+, 472
in HCl, 473
in HX, 475
limiting curve, 516
dissociation energy, D^e, 285
dissociative photoionization, 470
doorway state, 9, *see also* gateway state, 445, 624, 646, 671, 681, 734
mediated ISC, 734
nonradiative decay, 681
Doppler broadening, 352
Doppler width, 503
dressed $\mathbf{H}^{\text{eff}}$, 685
prediagonalization, 685
radiation field, 685
Van Vleck transformation, 685
Dunham expansion, 236
dynamical Hund's cases, 136
dynamical quantities
autocorrelation function, 635
coherent superposition state, 635
density matrix, 635
expectation values of, 635
coordinates and momenta, 635
number operator, 635
probability density, 635
resonance operator, 635
survival probability, 635
transfer probability, 635
dynamics
Bixon-Jortner, 733
cause of, 647, 697
in coordinate/momentum space, 637
early time, 646, 672
localized, 646
reduced-dimension, 685
resonance operator, 697
in state space, 637, 689
tier models, 672
visualization, 685, 697

early time dynamics
near resonant coupling, 733
early time information, 627
e/f symmetry, 139
case (a) basis states, 144
labels for $^3\Sigma^+$ and $^1\Pi$, 146
effective and true parameters, 261
effective molecular constants, 234
Ehrenfest's theorem, 644, 645, 718
eigenfunctions of $\mathbf{S}^2$, 151
Einstein coefficient, 659
stimulated emission/absorption, B_{ij}, 350
electric field in ZEKE experiment, 594
electronic coordinate system, 192
electronic wavefunctions, 148
matrix elements between, 156
electronically inelastic collision
in BaO, 447
in CaO, 450
in CO, 447
electrostatic interaction, *see* perturbation, electrostatic
electrostatic parameter
in BeH, 313
in CO, 313
in N_2, 313
in NO, 313, 314
in O_2, 313
in PO, 313
emission spectra, 33
emission windows, 576, 588
energy flow, 648
bright state↔quasicontinuum, 688
energy level diagram
$^2\Pi$, 585
energy normalization, 505
energy rank, 256
equation of motion, 631, 646, 693
of expectation values, 646
Hamilton's, 718
Newton's, 718
of $\mathbf{Q}$ and $\mathbf{P}$, 700
of resonance operators, 646
étalon, 43
Euler angles, 81
evolution operator, 643

exchange integrals, 155
excitation operator, 643
expectation value, 693
  of coordinates and momenta, 636
  equation of motion, 138, 647
  of operator, 647
    trace with ρ, 647
eXtended spectral Cross-Correlation (XCC), 19
extra lines, 246, 370, 374
eXtreme UltraViolet (XUV), 26

f-complex
  l_R, 402
  in ArH, 402
  in KrH, 402
  in MnH, 392
  in NO, 392
F,**G** matrix methods of Wilson, 688, 690, 703
F_i spin-component
  convention, 131, 140
Fano lineshape, 522, 588
  in H_2, 524, 525
  in NO, 524
  in O_2, 524
  parameter, q, 525
    in autoionization, 588
    in predissociation, 386, 522
  q reversal, 525, 526
Femtosecond Transition State (FTS) spectroscopy, 665
Fermi's Golden Rule formula, 535, 539, 569, 659, 681, 733, 735
Fermi-Wentzel Golden Rule, 507
Feschbach resonance, 493
figure of merit, 258
fine-structure constant
  α, 181
fit model, $\mathbf{H}^{\text{eff}}$, 717
  $\mathbf{H}^{\text{eff}} \rightarrow \mathcal{H}^{\text{eff}}$, 717
fluorescence decay, non-exponential, 370
fluorescence polarization, *see* polarization of fluorescence
  in N_2^+, 601
  in O_2^+, 601
fluorescence quantum yield, 28, 496
forbidden transition, 406
  $^1\Sigma^+ - {}^3\Sigma^-$
    in CH^+, 414
    in CO, 409
    in CS, 409, 412
    in NCl, 412
    in NF, 412
    in NH, 412
    in NO^+, 414
    in PF, 412
    in PH, 412
    in SO, 412
    intensity borrowing, 368, 414
Fortrat parabola, 6
four-wave-mixing, 51
Fourier transform, 44
  autocorrelation function, 632
Fourier Transform spectrum, 44
  I_2, 20, 21
  Spectrometer (FTS), 33, 649
fractional dynamical importance, 649
fractionation
  approach to Lorentzian, 659
  of bright state, 659
frame transformation, 590
Franck-Condon
  active normal modes, 633, 734
  bright state, 734
  energy gap rule, 734
  factor, 353, 629
    differential, 509
  pluck, 661
  point, 625, 651
  principle, 626
    classical, 633, 665, 667
  region, 651
  stationary phase, 663
frequency calibration, 43
frequency domain, 683
frequency generation, 50
Frequency Modulation Spectroscopy (FMS), 28
FWHM, full width at half maximum, 498

g_J–values, 426
gateway state, *see* doorway state
Gauss-Newton method, 254
Gaussian lineshape, 351
Gaussian wavepacket, 665
  distortion, 665
  equation of motion, 631
  vibrational, 631
GeO
  spin-orbit constant in, 321
GeS
  spin-orbit constant in, 321
global perturbation
  polyad, 685
  scaling, 685
golden rule formula, 535, 539, 569, 659, 681, 733, 735
good quantum numbers, 71, 112, 256
gradient of the potential, 635
  Fourier transform
    of spectrum, 633
grand rephasing, 654, 670
graphical methods for deperturbation, 243

$\mathcal{H}$
  action-angle representation, 717
H
  dimensionality reduction, 717
  exact→$\mathbf{H}^{\text{eff}}$ fit model, 717
  rotational part, 96
H vs. $\mathbf{H}^{\text{eff}}$
  Van Vleck transformation, 623
$\mathcal{H}_{\text{LOCAL}}$, 719
$\mathcal{H}_{\text{NORMAL}}$, 720
H_2^+
  coupled equations, 266
  potential energy curves of, 179, 180
H_2
  adiabatic parameter, 171
  anticrossing in, 428
  asymmetry parameter, 597
  autoionization in, 565, 573, 578, 598
  competition between autoionization and predissociation in, 604, 607
  Fano lineshape in, 524, 525
  MOMRIE in, 436
  nonadiabatic calculations, 236
  perturbations in ($^1\Pi_u$), 173
  photoelectron spectrum of, 555
  potential energy curves in, 173, 524
  predissociation in, 499–501, 523
  $^1\Sigma_g^+$ perturbations, 265, 266
  $X^1\Sigma_g^+$ state, 236
Hamiltonian
  $^3\Pi$ state, 242
  complex formalism, 734
  effective, 69, 84, 237, 255, 261
  exact, 70, 71
  microscopic, 69, 83, 84
  molecular, 66
  neglected terms, 67
  rotational operator, 79
  spin-orbit, 94, 181
  spin-spin, 94
Harmonic Generation, 49
harmonic oscillator
  two-dimensional isotropic, 727
HBr
  absorption spectrum, 586
  asymmetry parameter in, 491
  autoionization in, 603
  photofragment branching ratios in, 484
  Rydberg states in Hund's case (e), 125
  spin-polarization in, 603
HBr^+
  dissociation in, 472
  photodissociation in, 477
  predissociation in, 501, 520, 530
HCl
  competition between autoionization and predissociation, 604
  dissociation in, 473
  lifetime in, 585
  photoelectron spectrum of, 553

potential energy curves, 606
Rydberg states in Hund's case (e), 125
Rydbergization in, 312
Zero Electron Kinetic Energy (ZEKE) in, 593, 594
HCl^+, 332
HCP
anticrossing in, 418
$\mathbf{H}^{\text{eff}}$, 623, 648, 702
complex energy, 672
harmonic oscillator Hamiltonian, 690
$\mathbf{H}^{\text{eff}} \rightarrow \mathcal{H}^{\text{eff}}$, 717, 718
$\mathbf{H}^{\text{eff}}_{\text{LOCAL}}$, 706, 708
standard spectroscopic, 715
$\mathbf{H}^{\text{eff}}_{\text{NORMAL}}$
standard spectroscopic, 715
Heisenberg Correspondence Principle, 716
$HeKr^+$, 100, 136
$\mathbf{H}^{\text{el}}$
case (d)
non-spherical, 122
spherical (atom-like), 122
$\mathbf{H}^{\text{el}(0)}$ spherical term, 127
$\mathbf{H}^{\text{el}(1)}$ nonspherical term, 127, 129
case (b)→(d), 129
cases (d), (e), 127
exchange splitting, 129
non-diagonal in N^+, 127
nondiagonal in case (c), 129
opposed by l-uncoupling term, 129
transformation case (b)→(d), 121
Hellmann-Feynman theorem, 252, 254, 425
Hermitian operator, 648
HF
autoionization in, 580
HI
asymmetry parameter in, 491
autoionization in, 578, 584
coherent control in, 613
Cooper minima in, 563
photodissociation in, 483
photofragment branching ratios in, 484
potential energy curves, 483, 613
predissociation in, 530
Rydberg states in Hund's case (e), 125
$\mathbf{H}^{\text{ROT}}$
case (a), 107
$E^{\text{ROT}}_{(a)}(J)$, 107, 114
case (b), 108
$E^{\text{ROT}}_{(b)}(N)$, 108, 114
case (c), 108
$E^{\text{ROT}}_{(c)}(J)$, 108
case (d), 109, 121
$E^{\text{ROT}}_{(d)}(N)$, 109, 121
$E^{\text{ROT}}_{(d)}(N^+)$, 109
case (e), 109
$E^{\text{ROT}}_{(e)}(N^+)$, 110
pattern destroying term: $\mathbf{H}^{\text{el}}$, 121
$\mathbf{H}^{\text{SO}}$
in case (b), 117
in case (a), 116
selection rules for matrix elements, 182
$\mathbf{H}^{\text{SR}}$, 114
$\mathbf{H}^{\text{SS}}$
in case (b), 116
Hund's case, 667, 671
case (a) basis states, 696
rotational selection rules, 703
Hund's case (a), 103, 113
Hund's case (b), 103, 114, 441
Hund's case (c), 103, 123
$\mathbf{H}^{(0)}$, 123
$\mathbf{J}_a$-uncoupling, 124
Hund's case (d), 103, 104, 117, 126, 441
(a^+, d) and (b^+, d)
good quantum numbers, 119, 120
$\mathbf{H}^{\text{el}}$, 122
$\mathbf{H}^{\text{ROT}}$
high-N limit, 123

pattern destroying term, 123
pattern forming quantum number, 123
$\vec{\mathbf{R}}$-axis form, 123
z-axis form, 123
$\mathbf{H}^{\mathrm{SO}}$, 122, 123
$\mathbf{H}^{\mathrm{SR}}$, 122
N^+-multiplet, 122
Hund's case (e), 100, 103, 104, 125, 474, 583
case (a^+, e), 125
case (b^+, e), 125
case (c)↔case (e), 125
$\mathbf{H}^{(0)}$, 125
j–uncoupling, 125
$^2\Pi$ ion core, 125
Hund's cases, 101
(a), (b), (c) vs. (d), (e), 113
approximate constants of motion, 101
case (a), 112
$E^{\mathrm{ROT}}_{(a)}(J)$, 107, 114, 115
$\mathbf{H}^{\mathrm{SO}}$, 116
case (b), 112
$\mathbf{H}^{\mathrm{SO}}$, 117
$E^{\mathrm{ROT}}_{(b)}(N)$, 108, 114
$\mathbf{H}^{\mathrm{SR}}$, 114
$\mathbf{H}^{\mathrm{SS}}$, 116
case (b)→case (d) level pattern, 119
case (c), 112
case (d), 112, 119
good quantum numbers, 119
case (e), 112
choice of basis set, 101
dynamical, 136
ion, Rydberg e^-, 136
photofragment angular momenta, 136
separated atom, 136
good quantum number, 112
$\mathbf{H}^{\mathrm{el}}$: non-diagonal in N^+
in cases (d), (e), 127
$\mathbf{H}^{\mathrm{el}(0)}$ spherical term, 127
$\mathbf{H}^{\mathrm{el}(1)}$ non-spherical term, 127
hierarchy
of $\mathbf{H}^{\mathrm{el}}$, $\mathbf{H}^{\mathrm{SO}}$, $\mathbf{H}^{\mathrm{ROT}}$, 99, 137
intermediate, 126
between (a) and (d), 127
ion-core, Rydberg e^-, 104
l-uncoupling vs. l-mixing, 127
$\mathbf{H}^{\mathrm{eff}}$, 127
level pattern, 113
case (a), 113
mixed-l supercomplexes, 127
BaF s∼p∼d, 127
NO s∼d, 127
Nikitin and Zare, 99, 100
$\vec{\mathbf{J}}$-axis, 100
molecule frame projection quantum numbers, 100
z-axis, 100
pattern-destroying term
case (a), 115
case (b), 116
pattern-preserving term
case (a), 115
patterns in a spectrum, 101
traditional, 111, 112
case (a), 112
case (b), 112
case (c), 112
case (d), 112
case (e), 112
transformations between, 130
vector precession, 99, 101
Watson, 101
coupling sub-cases, 101
ion-core, Rydberg e^-, 101
$\vec{\mathbf{R}}$-axis, 101
z-axis, 101
zero-order pattern, 102
pattern-forming rotational quantum number, 115
hydrogenic wavefunction, 570
Hönl-London factor, 353

I_2
anticrossing in, 427
collision induced transition in, 450

Dispersed Fluorescence (DF), 47
Fourier Transform spectrum, 20, 21
lifetime in
Zeeman effect in, 426
predissociation in, 500, 519, 521
Zeeman effect
anticrossing in, 427
lifetime in, 426
Zero Electron Kinetic Energy (ZEKE) spectrum of, 595
ICl
predissociation in, 499
initial localization
on a chromophore, 733
intensity, 348, *see also* transition moment
anomalies, 380, 386, 404
in BaO, 35, 395
in N_2, 395
in Rb_2, 412
in Se_2, 395
borrowing, 370
rotational, 386
Inter-System Crossing (ISC), 658, 681, 688, 733
in NO $a^4\Pi \rightarrow B^2\Pi$, 460
interaction matrix
$^2\Pi(\pi) \sim {}^2\Sigma^+(\sigma)$, 219
$^2\Pi(\sigma^2\pi) \sim (\sigma\pi^2)$, 331
interactions
between g and u states, 99
interference effect, 363, 378
Internal Conversion (IC), 658, 681, 688, 733, 734
conical intersection, 735
electrostatic coupling, 734
nuclear kinetic energy coupling, 734
promoting mode, 735
intramolecular dynamics
control, 655
Intramolecular Vibrational Redistribution (IVR), 649, 658, 674, 681, 688, 733, 735
bifurcation, 736
bright state, 735
dark state, 735
doorway state, 735
strong and weak coupling limits, 735
tiers, 735
inverse perturbation approach (IPA), 289
ion
doubly charged, 34
ion spectroscopy, 28
ionization efficiency, 479
isotope effect
in CO, 528
in N_2^+, 528

JWKB approximation, 512
JWKB wavefunction, 279

K_2
predissociation in, 499
Kepler
orbits, 669
period, 669, 670
Ketene, 40
Kratzer relation, 66, 286
KrH
f–complex in, 402

l–complex, 128
l–mixing, 214, 400, 559, 594
by magnetic field, 444
l_R (or $\mathcal{L}$) quantum number, 105, 120
useful label vs. good quantun number, 105, 106
$\mathbf{L}^+$ operator, 325
Λ–doubling
in CH, 227
in CS^+ $A^2\Pi_i$ state, 215
in Δ state, 214
in OH, 227
parameter
o, 243, 328
p, 243, 328
q, 243, 328
in $^1\Pi_1$ state, 214

in $^2\Pi$ state, 214, 243, 328
in SH, 227
in SiH, 332
L-uncoupling operator, 116, 213
l-uncoupling matrix element
b, 267
Landau-Zener parameter, 536
Landé electronic factor, 181
Laser Induced Fluorescence (LIF), 27
least-squares fitting, 237, 248
procedure for coupled-equations, 265
Legendre polynomial, 487
level anticrossing
quantum beats, 682
strong vs. weak coupling, 682
level shift, 244
in autoionization, 588
in predissociation, 504, 510, 513
in perturbations, 296
Li_2
collision induced transition in, 450
linewidth in, 535
predissociation in, 500
lifetime, 495
in HCl, 585
radiative, 659
LiH
predissociation in, 510
Stark quantum-beat (SQB) in, 433
limiting case
destruction, 709
preservation, 709
line profile, q, 588, 606
line strength, 353
lineshape, *see* Fano lineshape
linewidth, 68, 503, 564
in Li_2, 535
in O_2, 515
in OD, 515
in OH, 527
relation with band contour, 566
variation, 516
localized state, 623, 624
bright state, 626
early time evolution, 630
photon-pluck, 626
short pulse, 626
$\Psi(0)$, 626
Loomis-Wood plot, 15, 44
Lorentzian function, 351

Magnetic Rotation Spectroscopy (MRS), 28
spectrum of Na_2, 418
Mass Analyzed Threshold Ionization (MATI), 31
matrix
of a $^2\Pi$ state, 252
matrix effects, 458
in O_2, 458
in N_2, 456
in NO, 455
relaxation, CN, 459
matrix element method, 293
measurement
destruction of coherence, 652
measurement of lifetimes, 496
measurement of linewidths, 498
mechanism, 623, 624, 633, 646–648, 655, 671, 681, 733, 735
cause of bifurcation, 732
Megabarn (Mb), 479, 554
Megahertz (MHz), 521
metastable states, 341
method of merging, 259
method of steepest descents, 254
MgO
perturbations in, 341
mixing coefficient, 235, 268, 271, 374
MnH, 392
MODR
in BaO, 438
in CS, 437
in NO_2, 438
Modulated Population Spectroscopy (MPS), 19, 35
molecular field anisotropy, 113

cases (a), (b), (c) vs. (d), (e), 113
molecular orbitals, 148
molecular spin-orbit constants, 318
molecular vs. optical phase, 652
molecule sources, 41
MOMRIE, 436
in CS, 437
in H_2, 436
MOPS, 438
Morse oscillator, 705, 706, 714
$E(v)$, 706
via perturbation theory, 706
two coupled
local stretch oscillator, 706
$V(r)$, 705
μ
reduced mass, 89
Multi-Photon Ionization (MPI), 29, 269
Multichannel Quantum Defect Theory (MQDT), 113, 236, 265, 267, 313, 444, 579, 590, 608
Born-Oppenheimer region, 113
calculations, 586
multiphoton transition, 355
multiplet state
S–uncoupling dynamics, 695
case (a)↔(b), 695
mutually commuting operators, 104, 111
restrictions on basis sets, 104, 105

N_2^+
adiabatic parameter, 171
anticrossing in, 418
collision effect in, 445
collision induced transition in, 450, 452
fluorescence polarization in, 601
isotope effect in, 528
perturbed intensities in, 372
perturbed radiative lifetimes in, 372
potential energy curves, 532
predissociation in, 499, 500, 531
spin-orbit constant in, 319
Zeeman effect
anticrossing in, 418
N_2
adiabatic parameter, 171
autoionization in, 573, 578, 587, 588
collision induced transitions in, 448, 449
coupled equations in, 265
electrostatic parameter in, 313
electrostatic perturbations
in matrix, 456
in $^1\Pi_u$ states, 314, 384
in $^1\Sigma_u^+$ states, 162, 166, 384
valence∼Rydberg, 162
intensity anomalies in, 395
p–complex in, 400
perturbation in, 384
perturbed intensities in, 381
potential energy curves, 541
predissociation in, 499, 514, 519
Rydberg emission series in, 588
Rydberg state, 188
shape resonance in, 561
spin-orbit constant in, 319
vibrational branching ratio in photoionization, 563
Na_2
coherent control in, 609
collision induced transition in, 449
Magnetic Rotation Spectrum (MRS) of, 418
perturbation facilitated OODR in, 439
predissociation tunneling barrier in, 521
NaK
predissociation in, 542
NBr
spin-spin constant in, 201
NCCN
shape resonances in, 561
NCl

forbidden $^1\Sigma^+ - ^3\Sigma^-$ transition in, 412

Newton's equations, 644, 645

NF

forbidden $^1\Sigma^+ - ^3\Sigma^-$ transition in, 412

NH

collision induced transition in, 448

Cooper minima in, 564

forbidden $^1\Sigma^+ - ^3\Sigma^-$ transition in, 412

predissociation in, 500

Rydbergization in, 311

NO

autoionization and predissociation, competition between, 609

autoionization in, 573, 578, 580

$^2\Delta$ homogeneous perturbation in, 167, 314, 382

electrostatic parameter in, 313, 314

f-complex in, 392

Fano lineshape in, 524

heterogeneous perturbation, 267

homogeneous perturbation, 236

in $^2\Pi$, 313

in $^2\Sigma^+$, 313

indirect mechanism in, 269

intensity anomalies in, 397

in matrix, 455

$\pi^3\pi^{*2}$ configuration, 187

$^2\Pi$ states in, 314

$^2\Pi$ states in, 375

$^2\Pi - ^4\Pi$ intersystem crossing mechanism in matrix, 460

$^2\Pi \sim ^2\Delta$ perturbation, 269

perturbations, doublet~quartet in, 341

photoelectron spectrum of, 556

$^2\Pi$ autoionization and predissociation, competition between, 609

$^2\Pi$ homogeneous perturbations, 262, 269, 376

potential energy curves in, 167, 376, 520

predissociation in, 501, 504, 519, 521

REMPI spectrum of, 600

Rydberg supercomplex in, 127

$^2\Sigma^+$ homogeneous perturbations in, 313

shape resonances in, 561

spin-orbit coupling constants in, 309

three-photon transitions in, 367

two-photon transitions in, 365

Zeeman effect in Rydberg states in, 440

Zero Electron Kinetic Energy (ZEKE) spectrum of, 558, 595

NO^+

forbidden transition in, 414

spin-orbit, 339

NO_2

MODR in, 438

nominal character, 67, 235, 257, 272

nonadiabatic calculations

for H_2^+ and H_2, 236

nonadiabatic coupling term, 94

nonadiabatic functions, 235

noncommuting terms in **H**, 137

expectation value, 138

equation of motion, 138

operator antagonism, 138

dynamics, 137, 138

pattern destroying, 137

pattern preserving, 137

nonradiative decay, 659, 671, 688

Bixon-Jortner, 733

dephasing of bright state, 688

direct vs. doorway mediated, 671

doorway mediated, 681

fractionation, 688

photofragmentation, 671

survival probability, 688

nonstationary state, 623

normal matrix, 250, 255, 257

normal mode limit
destruction, 714
driving force, 714
preservation, 714
normal mode of vibration, 633
antisymmetric, 704
Franck-Condon active, 633
frequencies, 705
nontotally symmetric, 633
normal↔local transformation, 702, 712, 714
symmetric, 704
normal↔local mode $\mathbf{H}^{\text{eff}}$
transformation, 702, 712, 714
normalization factor
of bound wavefunctions, 280
of continuum wavefunctions, 506
NS
doublet∼quartet perturbations in, 341
NSe
doublet∼quartet perturbations in, 341
perturbations in $X^2\Pi$ state, 213
number operator, 691, 693
$\mathbf{a}_i^\dagger \mathbf{a}_i$, 637
numerical methods
deperturbation, 375
numerical solutions, 264
Numerov-Cooley numerical integration, 291

o parameter, 243, 328
O_2^+
$^4\Pi$ state, spin-orbit, 186
fluorescence polarization, 601
potential energy curves, 531
predissociation in, 499, 500, 505, 521, 530
states of, 155
O_2
3-photon transition, 368
adiabatic parameter, 171
autoionization in, 581, 598
β in, 598
differential oscillator strength, 507
dissociation of, 474
electrostatic parameter in, 313
Fano lineshape, 524
linewidth in, 515
matrix effects in, 458
predissociation in, 499, 501, 504, 510, 517, 520, 524, 525
pure precession approximation in, 332
$^3\Pi_u$ Rydberg state of, 186
Rydberg states of, 156
Rydbergization in, 311
shape resonances in, 561
spin-orbit constant in, 186, 319
spin-rotation in, 193
spin-spin constant in, 326
Zero Electron Kinetic Energy (ZEKE) spectrum of, 557, 561
OD
linewidth, 515
predissociation in, 528, 544
off-diagonal anharmonic matrix elements, 691
OH
anticrossing in, 437
Cooper minima, 564
Λ–doubling in, 226, 227, 331
photodissociation in, 481
potential energy curves, 530
predissociation in, 266, 498, 500, 501, 520, 521, 544
Rydbergization in, 311
shape resonances in, 561
spin-orbit constant in, 335
Zeeman effect, anticrossing in, 437
one-photon transitions, 356
open-core atom, 104
Optical Optical Double Resonance (OODR), 25, 35, 416, 438
in BaO, 416
optical phase, 653, 654
optical vs. molecular phase, 650

Optically Pumped Laser (OPL), 53
orbital perturbation matrix element
  b, 216, 222

p–complex, 399
  N_2, 400
$^2\Pi$ states in NO, 314
$^1\Pi$
  perturbation in, 384
$^1\Pi$ states of SiO, 175
$^1\Pi - {}^1\Sigma^+$ transition, 394
$^2\Pi$ state, 184
  Λ–doubling, 328
  energy level diagram, 585
  π configuration, 184
  π^3 configuration, 184
$^3\Pi$ state, 185
  matrix elements, 242
  spin-orbit, 185
  spin-spin, 198
$^3\Pi$– $X^1\Sigma^+$ transition, 408
$^4\Pi$ state
  spin-orbit, 185
P_2^+
  spin-orbit constant in, 319
P_2
  second-order spin-orbit in, 202
  spin-orbit constant in, 319
  spin-spin constant in, 202
parallel and perpendicular transitions, 386
parameters correlated, 257
parity, 138
  case (a) basis states, 143
  case (b^+, d), 147
  determined by l, 147
  e/f basis states, 144
  favored and unfavored transition, 596
  of $|l, l_R\rangle$ states, 127
  labels for $^3\Sigma^+$ and $^1\Pi$, 146
  of levels, 70, 139
  not affected by S_R, l_R, s_R, j_R, 147
partition function
  canonical, 639, 640
partitioning method, 598
pattern destroying quantum number
  case (b), 120
pattern destroying term: $\mathbf{H}^{\mathrm{el}}$, 121
pattern forming quantum number
  case (a), 117, 118
  case (b), 117, 119, 120
  reduced term value plot, 117
patterns, 684
  frequency domain, 684
  pattern breaking, 684
  time-domain, 684
patterns in spectra, 137
  approximate quantum numbers, 137
  reduced term value plot, 137
Pekeris relation, 66, 286
perturbation, 62–66, 68, 83
  between g and u states, 99
  between Rydberg states, 175, 314
  electrostatic, 98, 161
    in N_2, 162, 166, 314, 384
    in NO, 314
  heterogeneous, 98
  homogeneous, 98
    in $^2\Delta$ NO, 167
  in CO, 216
  in H_2 ($^1\Pi_u$), 173
  inhomogeneous, 732
  $\mathbf{L}$–uncoupling, 732
  level shift, 296
  model, 67
  in N_2, 384
  on lifetimes, 370
  in $^1\Pi$, 384
  profound change in
    dynamics, 732
    spectrum, 732
  by a $^3\Sigma^-$ state, 299
  in $^1\Sigma_u^+$, 384
  in SiS, 374
  in the SiO $A^1\Pi$ state, 296
  spin-orbit, 734

Perturbation Facilitated Optical Optical Double Resonance (PFOODR), 26, 35, 734
in Na_2, 439
perturbation selection rule, 98, 145
perturbation theory
choice of basis set, 103
convergence criterion, 102
$\mathbf{H}=\mathbf{H}^{(0)}+\mathbf{H}^{(1)}$, 103
Hund's cases (a)-(e), 103
nondegenerate, 674, 717
quasidegenerate, 675
strong coupling limit, 677
weak coupling limit, 677, 678
perturbations
electrostatic
in N_2 $^1\Pi_u$, 384
perturbed f-values, 377
perturbed intensities
in N_2, 381
perturbed radiative lifetime
in CN, 372
in CO, 371, 373
in N_2^+, 372
PF
forbidden $^1\Sigma^+ - ^3\Sigma^-$ transition in, 412
PH
forbidden $^1\Sigma^+ - ^3\Sigma^-$ transition in, 412
phase convention, 143, 375
summary, 143
phase difference, 512
phase factor
in the rotational basis functions, 82
in vibrational wave functions, 291
phase lag, 612
phase matching directions, 644
phase shift, 590
phase space, 716
action/angle, 724
bifurcations, 716
chaotic, 716
diagram, 723
early time motion, 632
fixed point
resonance zone, 723, 724
stable, 723, 724
unstable, 723
invariant tori, 722
periodic orbit, 722, 723
polyad, 723, 724
quantum spectrum, 724
regular, 716
separatrix, 722
structure, 721, 724, 729
trajectories, 721, 722
quasiperiodic, 722, 723
Photo-Ion Photo-electron COincidence (PIPECO), 41
photodissociation
bond-softening, 179
cross section, 478
direct, 634
in HBr, 479
in HBr^+, 477
in HCl, 482
in HI, 483
ionization efficiency, 479
in O_2, 474
in OH, 481
population trapping, 179
snapshots, 635
photoelectron spectroscopy (PES), 552
angular distribution, 595
spectrum
of HCl, 553, 607
of H_2, 555
of NO, 556
photofragment angular distribution, 486
photofragment branching ratios
in HBr, 484
in HI, 484
PHOtofragment Fluorescence EXcitation (PHOFEX), 40
photofragment imaging experiments, 484, 490
photofragments, 635

photoion alignment, 600
photoionization
  of CO, 599
  cross section, 554
photon pluck, 626, 632, 661, 684
  Hund's basis state, 102
photon vs. wave pictures, 643
  bullets, 643
  molecules as targets, 643
pluck, 624, 684
  post-pluck dynamics, 684
  Rydberg vs. Franck-Condon, 669
  wrong basis, 703
PN
  spin-orbit constant in, 321
PO
  electrostatic parameter in, 313
  perturbation between $^2\Pi$ states, 277
  perturbation between $^2\Sigma^+$ states, 313
  perturbations, Rydberg~Rydberg, 263
  spin-orbit constant, 309, 335
polarization of fluorescence, 492, 601
polyad, 655, 689
  $\mathbf{H}^{\text{eff}}$
    local mode, 702
    normal mode, 702
    scaling rules, 702
    stretch-bend, 696
  number, 689, 697
  phase sphere, 721
    for H_2O, 724
  quantum numbers, 698, 728
  scaling, 717
polyatomic molecules
  anharmonic coupling matrix elements, 687
  Morse oscillator, 687
  normal mode, 687
  vibrational basis set, 687
populations, 640
potential energy curves, 65, 66, 90
  adiabatic, 164
  of BaS, 303
  case (d) l–complex, 128
  of CH, 518
  of Cl_2, 480
  of CO, 217
  diabatic, 164
  dressed, 177
  of H_2^+, 179, 180
  of H_2, 173, 524
  harmonic, 285
  of HCl, 606
  of HI, 483, 613
  Morse, 285
  of N_2^+, 532
  of N_2, 541
  of NO, 167, 376, 520
  of O_2^+, 531
  of OH, 530
power broadening, 416, 438
predissociation, 61, 64, 68, 83, 84, 493, 519, 671
  accidental, *see also* indirect, 527
    in CO, 542
    in N_2, 540
  in BeH, 501
  in Br_2, 522, 544
  in CH, 497, 499, 517
  in CO, 266, 501
  in Cs_2, 521
  electrostatic, 519
  energy shift, 507
  gyroscopic, 521
  in H_2, 499–501, 523
  in HBr^+, 501, 520, 530
  in HI, 530
  hyperfine, 521
  in I_2, 500, 519, 521
  in ICl, 499
  indirect (accidental), 538
  intermediate coupling strength, 535
  isotope effects, 526
  in K_2, 499
  in Li_2, 500
  in LiH, 510
  level shift, 504, 510, 513

Mulliken's classification, 514
in N_2^+, 499, 500, 531
in N_2, 499, 514, 519
in NaK, 542
in NH, 500
in NO, 501, 504, 519–521
nonadiabatic, 531
nonradiative rate, 495
in O_2^+, 499, 500, 505, 521, 530
in O_2, 499, 501, 504, 510, 517, 520, 525
in OD, 528, 544
in OH, 266, 498, 500, 501, 520, 521, 544
overlap factor, 510
by rotation, 493
radiative rate, 495
resonance energy, 502
resonant state, 493
rotational, 521
in Se_2, 501, 502, 504, 520
in SH, 530
spin-orbit, 520, 528
tunneling, 521
in Na_2, 521
principal part, 508, 523, 588
probability
density, 635
squared sum vs. sum of squares, 652
probe, 649
promoting mode, 735
Pulsed Field Ionization (PFI), 557
pump, 649
pump/probe
coordinate space, 654
one-color, 649, 651
phase coherent, 650, 654, 665
phase locked, 651
state space, 654
pure precession, 104
approximation, 191, 267, 327
hypothesis, 226

q reversal, 525, 526
quadrupole field split, 670
quantum beat, 639, 642, 656, 657, 679
decaying quasi-eigenstates, 679
exponential decay, 659
few level, 659
fluorescence intensity, 656
hyperfine, 657
in-phase, 643
many level, 659
modulation depth, 681
out-of-phase, 643
phase, 643
polarization, 656, 658
population, 658
pump/probe, 657
recurrences, 624
spatial angular distribution, 656
spectral distribution, 656
Stark, 657
survival probability, 659
wavepacket dynamics, 657
Zeeman, 657
quantum chaos, 717
quantum defect, 5, 136, 308, 574, 576, 590
a_λ, 128, 129
R–dependent, 128
quantum number
good, 70
in Hund's cases, 112
ion-core rotation, N^+, 103
ion-core total angular momentum, J^+, 103
photofragments, 136
$\vec{\mathbf{R}}$–axis projection, 104
S_R, l_R, s_R, j_R, 104
quasi-continuum, 681, 683

R–centroid approximation, 164, 349, 509
R–dependence
of rotational perturbation parameters, 325
of the spin interaction parameters, 333

R–dependent quantum defect, 128, 576
ρ–doubling, 226
Rabi frequency, 435
radial stationary phase point, 669
radiation sources, 47
radiative lifetime, 68, 349, 659
raising and lowering operators, 73
Raman spectra, 632
rate operator, 648
Rb_2
intensity anomalies in, 412
recurrence
first partial, 625
partial, 633
reduced mass
μ, 89
reduced term value plot, 117, 137, 244
reflection symmetry, σ_v, 70
relativistic potential curves, 94
remote perturber, 407
renormalized Numerov method, 265
rephasing
grand, 654, 657, 670
resolution, 15, 565
resonance, 564
resonance energy, 649, 693, 698, 699
operator, 637, 646, 648, 649, 694, 695
fractional dynamical importance, 649
Resonance Enhanced Multi-Photon Ionization (REMPI), 26, 29, 37, 45, 553
resonance narrowing, 540
reversed angular momentum, 74
RFODR, 437
RKR
not for polyatomic molecules, 687
Rotational Coherence Spectroscopy (RCS), 657, 658
rotational constant operator, 107
B_v, 98
rotational coordinates
conjugate momenta, 636
rotational Hamiltonian ($\mathbf{H}^{\mathrm{ROT}}$), 107
rotational intensity anomalies, 386
rotational perturbations, 92, 210, 325
rotational quantum number
natural
case (a), 117–119
case (b), 117–119
rotational wavefunction, 81, 142
RuC, 7, 10
spin-orbit constant, 322
Rydberg
vs. Franck-Condon wavepacket, 669
n^{*-3} scaling, 669
pluck, 669
wavepacket, 669, 670
core-nonpenetrating, 669
core-penetrating, 669
effect of ion-core quadrupole, 670
grand rephasing, 670
precessional period, 670
stroboscopic effect, 670, 671
Rydberg n^{-3}–scaling, 128
Rydberg constant, 5, 308, 568
Rydberg states, 61, 72, 308, 568
of N_2, 188
of O_2 $^3\Pi_u$, 186
Zeeman effect, 439
of NO, 440
Rydberg states in Hund's case (e)
CO_2, 125
CS_2, 125
Hbr, 125
HCl, 125
HI, 125
Rydberg$\sim$Rydberg interactions, 174
Rydberg-Klein-Rees (RKR) potential, 288
not for polyatomic molecules, 687
Rydberg-Rydberg transitions, 402
Rydbergization, 173, 560
in CH, 311
in HCl, 312

in NH, 311
in O_2, 311
in OH, 311

σ_v operator, 139, 159
A', A'', 140
e/f symmetry, 139, 140
effect on case (a-e) basis states, 145
on $|\lambda\rangle$, 141
Condon and Shortley phase convention, 141
phase choice, 141
Σ^+, Σ^- and $0^+, 0^-$, 139, 140
on total wavefunction, 143
total parity, 139, 140
laboratory inversion, **I** or **E***, 138
$^1\Sigma^+ - {}^1\Sigma^+$ system, 393
$^1\Sigma_g^+$ states of H_2, 236
$^1\Sigma_u^+$
perturbations in, 384
$^2\Sigma - X^2\Pi$ transition
in CuO, 404
$^2\Sigma \sim {}^2\Pi$ interactions, 328
$^3\Sigma^- - {}^1\Sigma^+$ transitions, 397, 399, 412
$^3\Sigma^-$ state, 399
S–uncoupling, 649
dynamics, 695
operator, 212
R–dependent, 212
S_2
spin-orbit perturbation in, 341
spin-spin constants, 326
Schrödinger equation
time-dependent, 177, 624, 672
time-independent, 625
Se_2
intensity anomalies in, 395
predissociation in, 501, 502, 504, 520
spin-spin constants, 326
Zeeman effect in, 432
Second Harmonic Generation (SHG), 49
second order effects, 327, 328, 543
spin-orbit effects, 183, 196
spin-spin parameters, 323
second order term (or correction), 240, 243, 262
secular equation, 235, 262
selection rules
for electric dipole transitions, 145
for perturbations, 145
for spin-orbit operator, 182
Self-Consistent-Field (SCF) molecular orbitals, 320
semiclassical
approach, 536
approximation, 511
in bound states, 279
sensitivity, 27
SeO
spin-spin constants, 326
SeS
spin-spin constants, 326
SH
Λ–doubling in, 227
predissociation in, 530
shape resonances, 312, 559, 599
in CO, 561
in N_2, 561
in NCCN, 561
in NO, 561
in O_2, 561
in OH, 561
short time dynamics
approximate quantum number, 655
SiF
collision induced transition in, 450
SiH
Λ–doubling in, 332
SiO
$A^1\Pi$ state, 244
adiabatic parameter, 171
$H^1\Sigma$ state, perturbations in, 63
perturbations in $^1\Pi$ states, 175
perturbations in $A^1\Pi$ state, 293, 298
spin-orbit constant in, 321

spin-orbit perturbation matrix elements, 327
SiS
perturbations in $A^1\Pi$ state, 301, 374
spin-orbit constant in, 321
spin-orbit perturbation matrix elements, 327
SiSe
spin-orbit constant in, 321
SiTe
spin-orbit constant in, 321
six (6-j) coefficients, 110, 111
intermediate angular momenta, 111
SO
forbidden $^1\Sigma^+ - {}^3\Sigma^-$ transition in, 412
spin-spin constants, 326
spectroscopic studies
of molecular ions, 28
spin eigenfunctions of $\mathbf{S}^2$, 151
spin interactions
in CO, 323
spin-component, F_i
convention, 140
spin-orbit
coupling constant A, 183
diagonal matrix elements, 184
effects, second-order, 199
isoconfigurational, 199
Inter-System Crossing (ISC), 658, 681, 688, 733
off-diagonal matrix elements, 187, 202
operator, 206
second-order
$\pi\pi'$ configuration, 201
$\pi^3\pi'$ configuration, 201
$\pi^3\pi'^3$ configuration, 201
in $^3\Delta$ state, 202
in P_2, 202
phase choice, 202
spin-orbit constant
in AsN, 321
in CO, 321
in CrN, 322
in CS, 321
in CSe, 321
in CTe, 321
for $\sigma\pi^*$ configuration, 320
in GeO, 321
in GeS, 321
in N_2^+, 319
in N_2, 319
in NO, 309
in NO^+, 339
in O_2, 319
of atoms and ions, 316
in $\pi^{*3}\pi$ $^3\Delta$ states, 321
in P_2^+, 319
in P_2, 319
in PN, 321
in PO, 309, 335
in RuC, 322
in SiO, 321
in SiS, 321
in SiSe, 321
in SiTe, 321
in TeH, 320
in VO, 185
spin-orbit coupling constants
in NO, 309
of Rydberg states, 309
spin-orbit perturbation matrix elements
in CO, 325, 327
in CS, 327
in CS^+, 327
in SiO, 327
in SiS, 327
spin-other-orbit interaction, 182
spin-polarization, 602
in HBr, 603
spin-rotation
$^3\Sigma^-$ of O_2, 193
constant, 194, 330
operator, 191
spin-rotation parameters
correlation between A_D and γ^π, 337
spin-spin

constant
α, 198
$^3\Delta$ configuration, 200
η, 324
λ, 197, 323
in NBr, 201
in P_2, 202
$\pi\pi^*$ configuration, 200
π^2 configuration, 200
$\pi^3\pi^*$ configuration, 200
$\pi^3\pi^{*2}$ configuration, 200
$\pi^3\pi^{*3}$ configuration, 200
$\sigma\pi$ configuration, 200
$\sigma\pi^2$ configuration, 200
$\sigma\pi^3$ configuration, 200
effects, second-order
λ^{SO}, 199
λ^{SS}, 199
λ_{eff}, 199
π^2 configuration, 201
operator, 196
second-order spin-orbit effects, 196
spin-spin constants
in O_2, 326
in S_2, 326
in Se_2, 326
in SeO, 326
in SeS, 326
in SO, 326
in Te_2, 326
in TeO, 326
in TeS, 326
in TeSe, 326
spontaneous fluorescence, 632
SrO
perturbations in, 296
standard deviation, 250
Stark effect, 419
Stark quantum-beat (SQB), 433
in BaO, 433, 434
in CN, 436
in LiH, 433
state space
dynamics, 692
resonance energy operator, 693
transfer rate operator, 693
state-selective
detection, 638, 646
excitation, 646
stationary phase, 511, 632, 667
approximation, 667
point, 669
steepest descents, 254
Stimulated Emission Pumping (SEP), 26, 36, 103, 632
stretch-bend polyad, 696
strong and weak coupling limits, 656, 735
bifurcation in dynamics, 735
subband intensity anomalies, 404
Sum Frequency Generation (SFG), 49
supercomplex, 214
Rydberg
in BaF , 127
in NO, 127
superexcited state, 552, 564
decay pathways, 604
superposition state
pair of, 637
supersonic jet, 42
Surface Electron Ejection by Laser Excited Metastable molecules (SEELEM), 45
surface of section, 722, 729, 730
survival probability, 103, 624, 636–638, 672, 680, 699, 700
transfer probability, 646
σ_v operator, 70
symmetry properties, 138
case (a) basis set, 138
parity, 138

Te_2
spin-spin constants in, 326
TeH
spin-orbit constant in, 320
tensorial operators, 203
TeO
spin-spin constants in, 326
term-value method, 259

TeS
spin-spin constants in, 326
TeSe
spin-spin constants in, 326
theoretical expressions for widths, 505
third-order effect, 269
three (3-j) coefficients, 110
coupled↔uncoupled, 111
selection rules, 204
use of, for transformations between Hund's cases, 130
three-photon transition, 366
in CO_2, 368
in NO, 367
in O_2, 368
threshold photoelectron spectroscopy (TPES), 557
Time-of-Flight methods (TOF), 22, 484
trajectories
phase space, 721
transfer probability, 636, 638, 646
transfer rate
in CO, 452
transfer rate operator, 694, 695, 698, 701
transformation
$(a^+, a) \leftrightarrow (a^+, e)$, 132, 135
of basis set, 132
of $\mathbf{H}$, 135
of $\mathbf{H}^{\text{eff}}$, 132
$v_i \leftrightarrow \mathbf{a}^\dagger \mathbf{a}_i$, 711
between Hund's cases, 130
by diagonalizing $\mathbf{H}^{\text{eff}}$, 130
case (a)↔case (b), 116
$^2\Pi$, 132
case (b)↔case (d), 121
of basis states, 121
of $\mathbf{H}^{(d)} \leftrightarrow \mathbf{H}^{(b)}$, 121
of $\mathbf{H}^{\text{el}}$, 121
case (d)↔case (e), 110
Clebsch-Gordan (CG) coefficients, 130
coupled↔uncoupled, 133
F_i spin-component convention, 131
normal↔local mode, 702
three (3-j) coefficients, 131, 133
(3-j)↔CG, 130
properties, 133
reversed angular momentum, 134
transition
in $^3\Delta - {}^3\Pi$, 391
matrix elements, 390
in $^1\Pi - {}^1\Sigma^+$, 393
in $^1\Sigma^+ - {}^1\Sigma^+$, 394
in $^3\Sigma^- - {}^1\Sigma^+$, 399, 412
transition moment, 348
case (b) $np - {}^1\Sigma^+$, 398
case (c) $\frac{1}{2} - \frac{1}{2}$, 396
transition probabilities
Einstein, 349, 659
truncation of basis set
contact transformation, 684
Van Vleck transformation, 684
tunneling, 521
two-level problem, 674, 675, 678
autocorrelation function, 637
complex energies, 674
level attraction, 675, 678
level repulsion, 674, 675
nondegenerate perturbation theory, 674
quasidegenerate perturbation theory, 675
strong coupling, 676, 677
survival probability, 637
transfer probability, 637
weak coupling limit, 676–679
two-photon
transitions, 359
in NO, 365

ultracold collisions, 43
ultrashort pulses, 54
unique perturber approximation, 225, 259, 261
unitary transformation
trace invariance, 111
transformation coefficients, 107
notation, 107

valence state, 262, 308, 315
valence~Rydberg mixing, 308, 310
valence~Rydberg perturbation, 262
Van Vleck transformation, 237, 247, 684, 709, 717
variance of the fit, 251, 256
variance-covariance matrix, 250
variational calculations, 234
vector coupling coefficient, 117
Velocity Imaging Photo-electron CO-incidence Spectroscopy (VIPCO), 41
velocity modulation spectroscopy, 29
vibrational eigenfunction, 91
vibrational overlap integrals, 282
vibrational quasi-continuum, 671
  relaxation into
    direct vs. doorway mediated, 671
    nonstationary state, 671
vibrational state branching ratio for the photoionization of N_2, 563
vibrational wavefunction
  phase, 292
vibrational wavepacket, 630, 633
VO, spin-orbit constant in, 185
Voigt profile, 503

wavefunction
  time-dependent, 625
wavenumber, 3
wavepacket, 476, 610, 623, 646, 656
  center, 645
  compression, 661
  ΔR vs. ΔP motion, 632
  Gaussian, 661
  interference, 649
  interferometry, 650, 651
    phase coherent, 653
    Ramsey-fringe spectroscopy, 649
  localized state, 644
  motion, 650
    of center, 644
    in coordinate space, 625
  nuclear, 659
    rotational, 659
    vibrational, 660
  radial, 668
    electronic, 668
    Rydberg, 668
  rephasing, 651
  rotational, 667
    recurrences, 668
  Rydberg
    effect of ion-core quadrupole, 670
    vs. Franck-Condon, 669
    Kepler period, 668
    precessional period, 668
    time evolution, 623
  variance, σ, 645
  vibrational, 661
  width, 645
weighting function, 258
Wigner rotation matrix element, 107
Wigner-Eckart theorem, 71, 204
Wigner-Witmer rules, 471, 528
Worley's third series, 314

Zeeman effect, 418
  anticrossing, 421, 427
    in CN, 418, 419, 428
    in CO, 418
    in H_2, 428
    in HCP, 418
    in I_2, 427
    in N_2^+, 418
    in NO, 418, 440
    in OH, 437
  g-value, 430, 438
  g_J-value in Se_2, 432
  lifetime in I_2, 426
  quantum beat, 430, 432
    in Se_2, 432
Zero Electron Kinetic Energy (ZEKE), 38, 45
  intensities, 591
  spectroscopy, 557
  spectrum, 558
    of HCl, 593, 594

of I_2, 595
of NO, 558, 595
of O_2, 557, 561

Made in United States
Troutdale, OR
10/03/2024

23394991R00448